5th International Workshop on Environment and Geoscience

IOP Conference Series: Earth and Environmental Science
Volume 1087

Online
16 – 18 July 2022

Editor:

Youmin Chen

ISBN: 978-1-7138-6434-9
ISSN: 1755-1307

Printed from e-media with permission by:

Curran Associates, Inc.
57 Morehouse Lane
Red Hook, NY 12571

CURRAN ASSOCIATES INC. proceedings.com

Some format issues inherent in the e-media version may also appear in this print version.

Institute of Physics
Dirac House, Temple Back
Bristol BS1 6BE UK

Phone: 44 1 17 929 7481
Fax: 44 1 17 920 0979

techtracking@iop.org

Additional copies of this publication are available from:

Curran Associates, Inc.
57 Morehouse Lane
Red Hook, NY 12571 USA
Phone: 845-758-0400
Fax: 845-758-2633
Email: curran@proceedings.com
Web: www.proceedings.com

TABLE OF CONTENTS

IOP Conf. Series: Earth and Environmental Science 1087 (2022) 011001 doi:10.1088/1755-1315/1087/1/011001

Preface

Youmin Chen

Editor, Proceedings of the Fifth International Workshop on Environment and Geoscience (IWEG2022)

Professor, Henan University, China

Email: youmin.chen@henu.edu.cn

The Fifth International Workshop on Environment and Geoscience (IWEG 2022) was held successfully on 16-17 July 2022. Under the effect of COVID-19, the conference was chosen to be online, which gave everyone access to attend. As an annual international conference, IWEG aims at being a fast and efficient platform for researchers and scholars worldwide to discuss recent developments in the area of Environment and Geoscience.

The program was divided into 3 sessions: keynote speeches, oral presentations and poster presentations, attracted over 50 participants from 13 different countries and regions, including China, Australia, Egypt, Morocco, Sri Lanka, Russia etc. Details about the presentations can be found in the Conference Overview part. We greatly appreciate Prof. Youmin Chen from Henan University, who presided over the conference, and all participants for attending and sharing!

This conference proceedings included 82 accepted articles selected from 211 submissions, all the papers have been through rigorous review and process to meet the requirements of International publication standard. We would like to express our gratitude to the reviewers of these manuscripts, who provided constructive criticism and stimulated comments and suggestions to the authors. We are extremely grateful as organizers, technical program committee and editors and extend our most sincere thanks to all the authors for their excellent contribution and work. Our sincere gratitude also goes to the IOP Publishing editors and managers for their helpful cooperation during the preparation of the proceedings.

On behalf of the Organizing Committees of IWEG2022.

IOP Conf. Series: Earth and Environmental Science 1087 (2022) 011001 doi:10.1088/1755-1315/1087/1/011001

Technical Program Committee

Conference Chair
Youmin Chen, Henan University, China

Conference Co-chair
R.A. Stephenson, King's College University of Aberdeen, United Kingdom
Tianpeng Gao, School of Biological & Environmental Engineering of Xi'an University, Eco Xi'an Academy, mining area pollution control and ecological restoration engineering research center in Gansu Province, Urban Development Academy of Gansu province, and Lanzhou Smart City Academy, P.R. China

TPC Chair
Xiaxin Tao, School of Civil Engineering, Harbin Institute of Technology, China

TPC member
Ramesh Agarwal, Washington University, USA
KEMAL ÖZGÜR HASTAOĞLU, Sivas Cumhuriyet University, The Republic of Turkey
Tao Zhu, China University of Mining and Technology (Beijing), China
Hongzhu Cai, China University of Geosciences (Wuhan), China
Mahmoud Reza Delavar, University of Tehran, Iran
Lei Zhao, Xi'an University of Architecture & Technology, China
Vallam Sundar, Ocean Department, IIT Madras, India
Chien-Yuan Chen, National Chiayi University
Waheed Ullah, Nanjing University of Information Science and Technology, China
Chunrong Jia, University of Memphis, USA
Jaroslaw Krzywanski, Jan Dlugosz University in Czestochowa, Poland
Jean-Pierre TOUMAZET, Environmental Geography Laboratory, University Clermont – Auvergne, France
Miklas Scholz, Faculty of Engineering, Lund University, Sweden
Wenjie Liao, Sichuan University, China
Zhaofei Fan, Forest Biometrics and Spatial Ecology Lab, School of Forestry and Wildlife Sciences, Auburn University

IOP Conf. Series: Earth and Environmental Science 1087 (2022) 011001 doi:10.1088/1755-1315/1087/1/011001

Conference Overview

IWEG is an annual academic event since 2018, aims to provide a high-level platform for experts and participants from academia and industry, through workshops, invited lectures, and oral and poster sessions, to present research findings in the most state-of-the-art areas of Environment and Geoscience.

Due to the continuing impact of pandemic, IWEG2022 was held online during July 16-17, 2022. The program was divided into 3 sessions: keynote speeches, oral presentations and poster presentations. The event attracted over 50 participants from 13 different countries and regions, including China, Australia, Egypt, Morocco, Sri Lanka, Russia, Brazil, USA, Portugal, Kuwait, Iraq, Vietnam and Paraguay.

Thank all the reporters, who have done excellent jobs and attendees for being with us. Except that, we'd like to extend a "thank you" for all reviewers for providing comments which are very useful for the authors and give special thanks to Prof. Youmin Chen from Henan University, who presided over the conference.

Conference Schedule

Date	Time	Program	Online Platform
July 16	10:00-17:00	Online Registration	VooV Meeting (腾讯会议) ID: 847 1860 3572
July 17	09:00-09:10	Opening Ceremony	
	09:10 12:10	Keynote Speech	
	14:00-18:00	Oral Presentation	
	17:40-18:10	Poster Presentation	

IOP Conf. Series: Earth and Environmental Science 1087 (2022) 011001 doi:10.1088/1755-1315/1087/1/011001

It was our great honor to have the following five impressive keynote speeches and two invited speakers.

Keynote Speeches

Prof. Youmin Chen, Henan University
Speech Title: On study of the atmospheric ozone with its distribution and variation

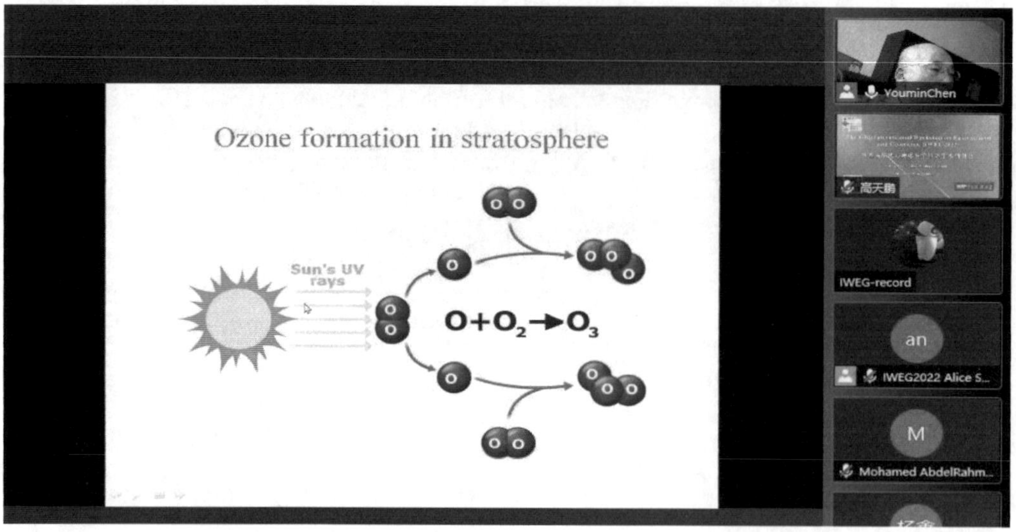

Prof. Tianpeng Gao, Dean of School of Biological & Environmental Engineering of Xi'an University, Director of Eco Xi'an Academy, the director of mining area pollution control and ecological restoration engineering research center in Gansu Province, A vice director of Urban Development Academy of Gansu province, and vice president of Lanzhou Smart City Academy, P.R. China
Speech Title: The Removal of Heavy Metals and Petroleum Pollution by Plant, Microorganisms and Ecological Remediation Technologies in Mining Areas

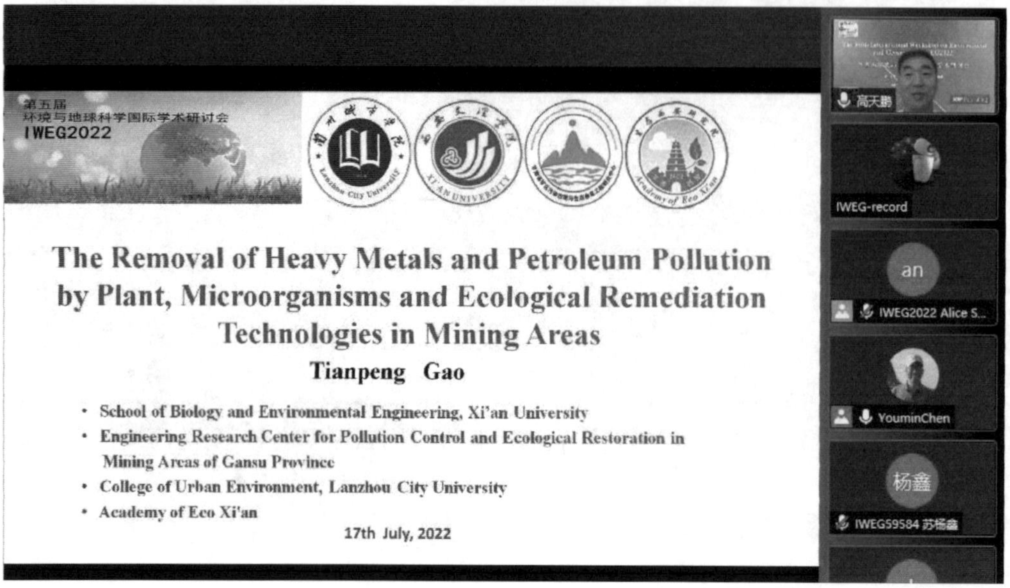

IOP Conf. Series: Earth and Environmental Science 1087 (2022) 011001 doi:10.1088/1755-1315/1087/1/011001

Dr. Monzur Imteaz, Associate Professor, Head, Urban Environmental and Transport Systems School of Engineering, Swinburne University of Technology, Melbourne, Australia
Speech Title: Urban stormwater quality improvement devices: are we over-spending?

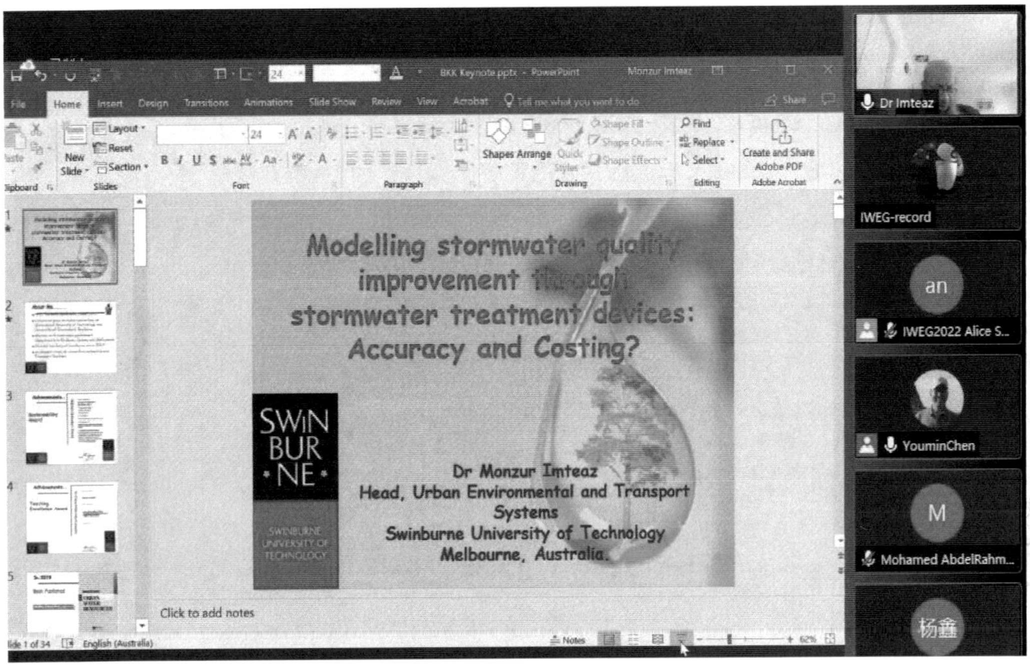

Dr. Mohamed A.E. AbdelRahman, Associate Professor, Division of Environmental Studies and Land Use, National Authority for Remote Sensing and Space Sciences (NARSS), Egypt
Speech Title: Potentiality of Geomatics for Assessment Climate Change and Anthropogenic Impacts on Soil in-North Delta, Egypt

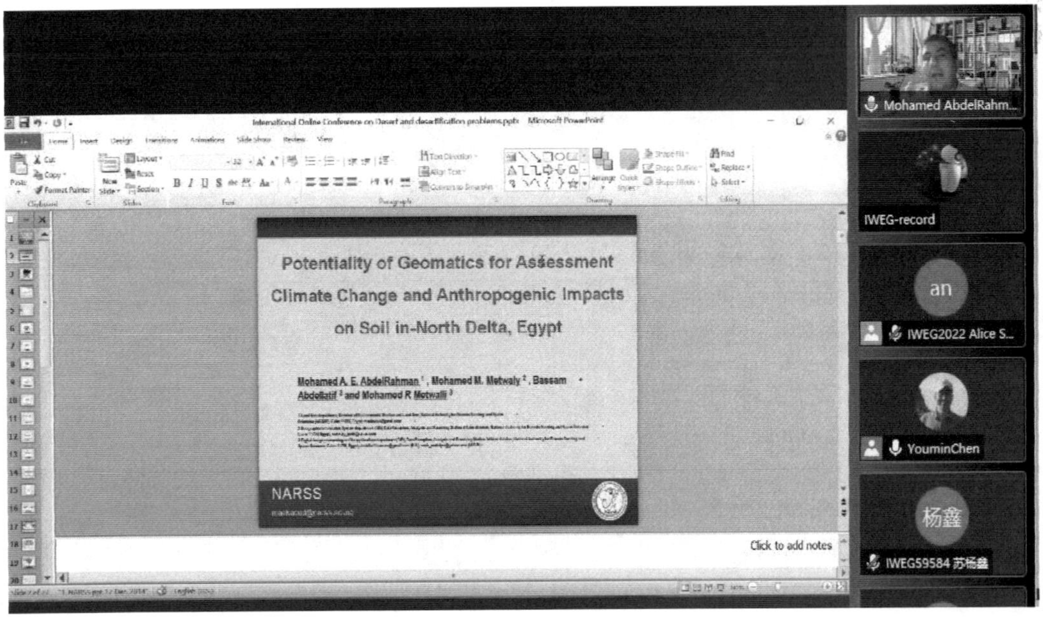

IOP Conf. Series: Earth and Environmental Science 1087 (2022) 011001 doi:10.1088/1755-1315/1087/1/011001

Dr. Muhammad Jawad Sajid, Associate Professor, Xuzhou University of Technology, Xuzhou, Jiangsu, China
Speech Title: The equitable distribution of the burden of carbon emissions mitigation among industries and final consumers

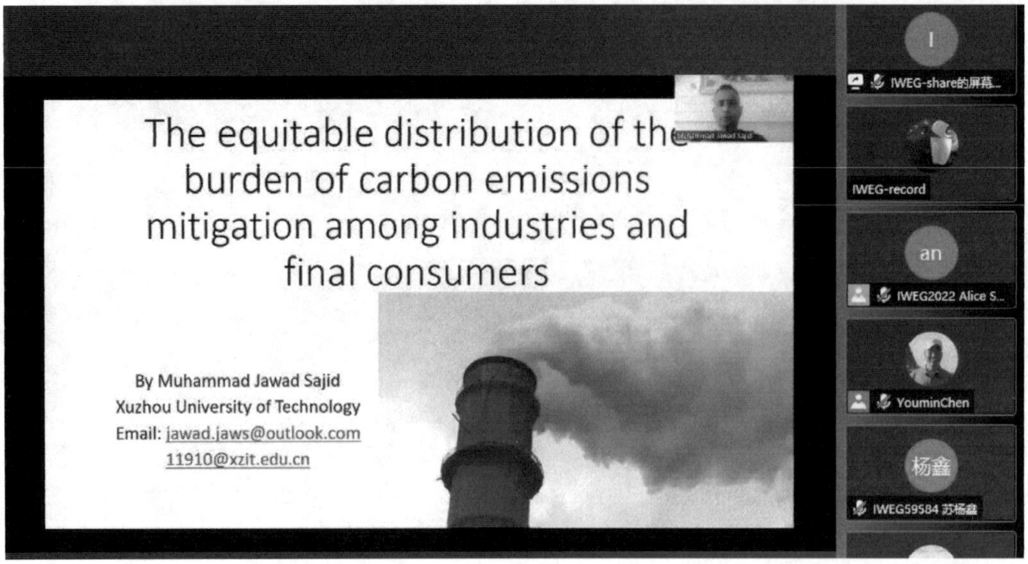

Invited Speakers:
Dr. Yongze Song, Curtin University, Australia
Speech Title: Advanced Spatial Models for Geographical Factor Exploration

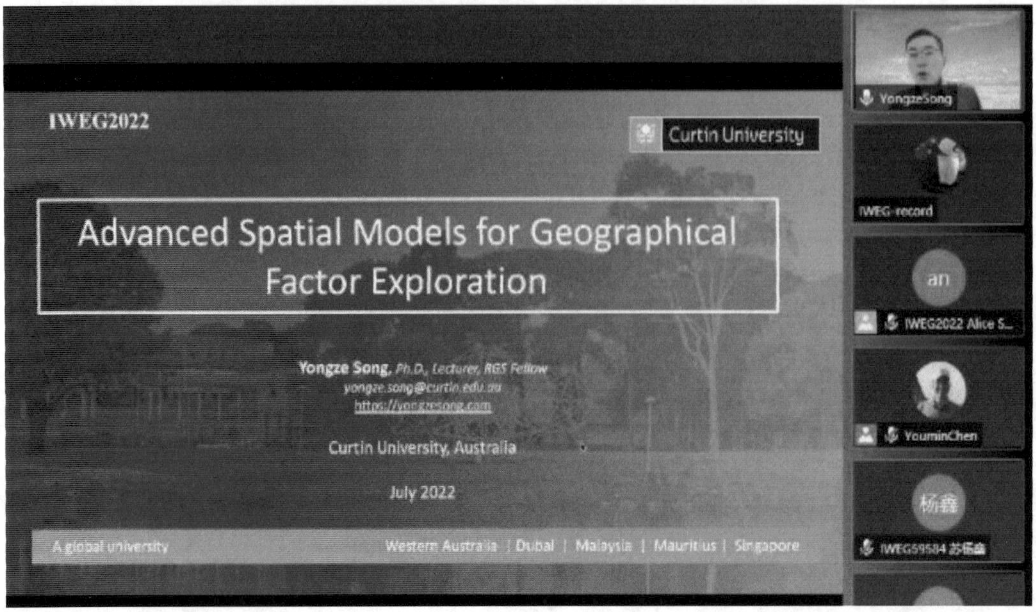

IOP Conf. Series: Earth and Environmental Science 1087 (2022) 011001 doi:10.1088/1755-1315/1087/1/011001

Dr. Soufiane Haddout, Ibn Tofail University, Kenitra, Morocco
Speech Title: Water issues in developing countries: Management and challenges for the future

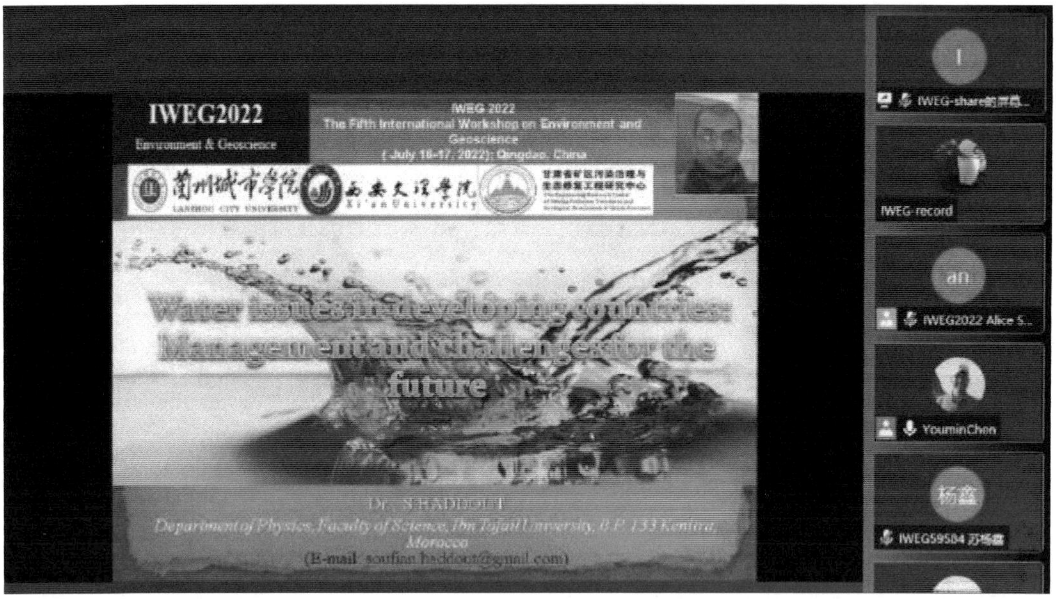

Oral Presentations

Following the keynote speakers, 21 scholars made the oral presentations respectively in the afternoon. Information of speakers and their speeches can be found as below.

Speech Title	Speaker	Affiliation
On the Legal Mechanism of Nuclear Safety in China under the Background of Environmental Protection	Jiu Liu	Harbin Engineering University
Effects of Lime Nitrogen on the Fungi Community of Tobacco Soil and its Correlation Analysis	Susu Zhang	Zhengzhou University
Kalman Filtering for SAR Satellite Yaw Steering	Tao Chen	Aviation Industry of China (AVIC)
Zhejiang University	Xia Cheng	Ultrasensitive detection of Cr (VI) using screen-printed carbon electrodes modified by carbon nanotubes and gold nanoparticles
Spatiotemporal Characteristics Analysis of Ground subsidence in Wuhan Based on SBAS-InSAR	Weiping Zhang	Guilin University of Technology

Review & Latest Progress in the Usage and protection of geological relics in China	Bo Zhang	Hebei GEO University
Metal Organic Framework MIL-101(Cr) Coupled with Solid Phase Extraction Technique for Separation and Detection of Phenoxycarboxylic acids (PCAs) in Environmental Water Samples	Zehui Deng	Nanjing University
An Efficient and stable nanoscale zero-valent iron coated by N-doped carbon for the reduction of bromate	Li Long	Nanjing University
High-Throughput Method for Natural Organic Matter Hydrophobicity Assessment Using an Ionic Liquid-Based Aqueous Two-Phase System	Jiaxue Wang	Nanjing University
Mechanism study of single atom decorated metal oxides for high-performance gas sensors	Lingyue Liu	Nanjing University
Genesis of the Heishan'gou BIFs in the Yudongzi complex of the Bikou terrane, central China	Xiaohui Sun	Chang'an University
Determination of U, Th and K for Quaternary Sediments by Inductively Coupled Plasma Atomic Emission Spectrometry-Mass Spectrometry	Jinhua Du	Chang'an University
Investigating the correlation of nitrates and phosphates to the effective removal of Fats, Oils, and Grease in Two Mangaung Municipal Wastewater Treatment Plants	Ms. Mpho Gladness Sello	Student at Central University of Technology
SEM and XRF Characterizations of Sri Lankan Dolomites for Some Advanced Chemical and Process Applications	Mr. Suresh Aluvihara	University of Peradeniya, Peradeniya, Sri Lanka
Geoinformation representation of maritime knowledge flows: new frontiers of coastalization	Mikhaylova A.	Immanuel Kant Baltic Federal University, Kaliningrad, Russia
Effects of some agroecological practices on bacterial abundance, soil nutrients, and faba bean yield under no chemical input conditions in Meknes region, Morocco	Dr. Abdellatif Boutagayout	Faculty of Sciences-Moulay Ismail University, Morocco
Paleoclimatology and paleoecology of eastern tropical South America during the last 50,000 years	Jaqueline Quirino Ferreira	Institute of Geosciences, University of São Paulo, São Paulo, Brazil

Shift of Paradigm in Urban Stormwater Management	Yuming (Jim) Su	State Key Laboratory of Silica Sand Resources Utilization
Improving sustainability on rural villages water supply systems: evolutionary game theory perspective	Christian E. Schaerer	Polytechnic School, National University of Asuncion, Paraguay
A comparative study of evaluating the ability on Eichhornia crassipes and Enydra fluctuans to remove heavy metals in water in Vietnam	Huynh The An	Da-Yeh University
Study of Unio shells in the Quaternary deposits, southern Iraq	Mohanad Hamid Al-Jaberi	University of Basrah-College of Science-Geology department

Poster Presentations

Fifteen poster presentations came after oral presentations. Below is a brief list of presenters and their poster titles.

Presenter	Affiliation	Presentation Title
HaiJing Wang	Wenzhou University	Big Data Analytics for Adaption to Climate Change
Alidu Rashid	Universiti Teknologi PETRONAS, 32610, Seri Iskandar, Perak Darul Ridzuan, Malaysia	Field Characteristics of Sedimentary Rocks in the Dohol Formation, Johor, Peninsular Malaysia
Chelnokov G. A.	Geological institute Russian Akademy of Sciences	Rare Earth Elements behaviour in different types of mineral waters within Baikal Lake
Xuanyu Chen	Natural Resources Information Center of Guangxi Zhuang Autonomous Region, Nanning 530028, China	3D model construction and accuracy analysis based on UAV tilt photogrammetry
XinfaWang	Henan Institute of science and technology	Research on Intelligent Building Greenhouse Plant Factory and "3-Positions and 1-Entity" Development Mode
Ferreira-Gomes L M	University of Beira Interior, 6200-001 Covilhã, Portugal	Hydrogeological studies in the São Torcato Zone -Tabuaço with aim of optimizing the location of a special groundwater abstraction to plan a new Medical Spa in Portugal
Wenhan Zhong	Zhejiang Light Industrial Products Inspection and Research Institute	Electrochemical determination of formamide for EVA splicing mats using activated CPE
Yu Liu	Life Sciences Institute Zhejiang University	Application of a modified graphite paste electrode to the detection of fluorene in black latices

Jingwei Yang	Shenzhen Water Planning & Design Institute Co., Ltd.	Developing comprehensive resilient water systems and their driving factors for the 21st Century Society in China
Maoqing Duan	Center of Eco-environmental Monitoring and Scientific Research, Administration of Ecology and Environment of Haihe River Basin and Beihai Sea Area, Ministry of Ecology and Environment of PRC, Tianjin 300170, China	Nitrogen and Phosphorus Concentration Zoning Control Method of Surface Water in the Haihe River Basin in China
Le Van Dang	College of Agriculture, Can Tho University, Can Tho City, Vietnam	Enhancing soil fertility and lowbush blueberry (Vaccinium angustifolium) growth using bio-organic fertilizer
Suresh Aluvihara	University of Peradeniya, Peradeniya, Sri Lanka	SEM and XRF Characterizations of Sri Lankan Dolomites for Some Advanced Chemical and Process Applications
Modi Ahmed	Kuwait Institute for Scientific Research	Ecological study on native plant diversity in Kuwait desert ecosystem
Tang Chi Fong	Macao Polytechnic University	A Study of the Urban Heat Island Effect in Guangzhou
M.S. Shipaeva	Institute of Geology and Petroleum Technologies, Kazan Federal University, Kazan, Russian Federation	Geochemical analysis of formation water as a tool for better understanding of water flooding

This conference has made a complete success because of their supports and contributions. We are expecting more experts and scholars from academia and industry to join IWEG2023 in Kunming, China July 15-17, 2023.

Peer Review Statement

All papers published in this volume have been reviewed through processes administered by the Editors. Reviews were conducted by expert referees to the professional and scientific standards expected of a proceedings journal published by IOP Publishing Publishing.

Type of peer review: Double anonymous
Conference submission management system: Morressier
Number of submissions received: 211
Number of submissions sent for review: 110
Number of submissions accepted: 82
Acceptance Rate (Submissions Accepted / Submissions Received × 100): 38.9 %
Average number of reviews per paper: 6.00
Total number of reviewers involved: 586
Contact person for queries:
Name: Alice Shan
Affiliation: Wuhan Heruirong Culture Media Co., Ltd.
Email: Secretary@iwegconf.org

Evaluation of ecological environment carrying capacity of mineral resources exploitation - Take Shenzhen as an example

Jianqiao Qin[1,2], Hao Liu[2], Huarong Zhao[3,6], Fuxing Gu[4], Donghang Jiang[2], Zhixi Zhu[5] and Jinxiang Hou[5]

[1] Guangdong Provincial Key Laboratory of Environmental Health and Land Resource, Zhaoqing University, Zhaoqing, 526061, China;
[2] College of Environmental and Chemical Engineering, Zhaoqing University, Zhaoqing 526061, China;
[3] School of Environmental Science and Engineering, Guilin University of Technology, Guilin 541004, China;
[4] Guangdong Zhonghao survey and Design Consulting Co., Ltd, Zhaoqing, 526061, China;
[5] Guangdong Academy of Environmental Sciences, Guangzhou 510045, China.

[6] Corresponding author's e-mail: zhaohuar@mail3.sysu.edu.cn

Abstract. The assessment of eco-environmental carrying capacity is the focus of planning environmental impact assessment and an important basis for determining whether the planning can be implemented. Based on the theory and method of ECO-environmental impact assessment, this paper applies the analytic hierarchy process (AHP) to analyze the ecological resilience, resource-environment carrying capacity and mining pressure of mineral resources. A comprehensive evaluation model of environmental impact of mineral resources exploitation is established. Taking Shenzhen City, Guangdong Province as an example, the difference of eco-environmental carrying capacity of the main planned mining areas is evaluated. The results show that the pressure bearing degree of most of the assessed blocks is less than 1, indicating that the pressure on the ecosystem is less than the carrying capacity. The ecosystem is not easy to change and can be restored to the original state in a short time after change, and the ecosystem can maintain normal development. This is of great guiding significance to the scientific formulation of mineral resources exploitation strategy.

1. Introduction

With the increasing demand for energy, mineral resources have been exploited and utilized on a large scale, and environmental pollution and damage caused by mineral resources exploitation have become increasingly serious. Prevention and control of environmental pollution and improvement of mining environment have received widespread attention [1,2]. Due to the process of mineral resources exploitation and the environmental pollution and destruction caused by the extremely complex, although many experts at home and abroad have done a lot of research about this, but not yet form a universal system of evaluation methods, especially for mineral resources exploitation of quantitative study of the difference of bearing capacity of the ecological environment has hardly involved [3, 4]. Ecological environment carrying capacity: it is to determine the maximum bearing capacity of an ecosystem to human activities. The so-called maximum bearing capacity of an ecosystem to human activities refers to the intensity of human activities that an ecosystem can bear without destroying ecosystem services [5, 6]. For this, the author based on the characteristics of mineral resources exploitation, ecological environment carrying capacity evaluation model is put forward, and in shenzhen city, guangdong province, for example, according to the shenzhen general mineral resources planning (2008-2015), the ecological carrying capacity of the main planned mining areas is evaluated to provide theoretical basis for the subsequent development of mineral resources development strategy.

2. Establishment of ecological environmental carrying capacity assessment model

2.1. Establishment of comprehensive evaluation index system

Ecological carrying capacity is the threshold of the ability of natural systems to maintain and regulate systems. If the threshold is exceeded, the natural system will lose its ability to maintain balance, be destroyed or destroyed, from a higher level of natural system to a lower level of natural system.

Based on the theory and method of eco-environmental impact assessment, the comprehensive evaluation index system of eco-environmental carrying capacity of mineral resources exploitation was established by AHP (Table 1).

The index system has three levels, namely target layer, criterion layer and index layer.Among them, the target layer is the evaluation object, including ecological elastic force, resource-environmental carrying capacity and mining pressure of mineral resources [7, 8]: (1) Ecological resilience, that is, the stability degree of the ecosystem under the current natural conditions, as the first meaning of environmental carrying capacity; (2) Resource-environment carrying capacity, resource carrying capacity reflects the richness and occurrence of resources, environmental carrying capacity is the generalization of pollutant carrying capacity and resilience capacity, because resources and environment are related to each other, constitute a resource-environment carrying capacity evaluation criteria layer; (3) Mining pressure of mineral resources, the development and utilization of bearing objects is shown as pressure, namely, mining pressure of mineral resources, which is regarded as the third meaning of environmental carrying capacity [9, 10].

2.2. Determination and assignment of index weight

The determination of index weight must follow the principle of reducing subjectivity and improving objectivity as well as giving consideration to flexibility and operability. Based on Analytic Hierarchy Process (AHP), experts' opinions are widely solicited and the weight of each evaluation index is determined in combination with the actual situation of the evaluation area [11, 12]. The weights of each indicator are shown in Table 1.

Index assignment, for specific data of the index, according to the relevant standards for classification; For indexes without standards, the ideal value or target expectation value is taken as reference standard, and the following formula is used to calculate and determine:

$$Ci=Fi / Fo \times 100 \quad (1)$$

Where, Ci is the score of factor I; Fi is the actual value; Fo is the standard or ideal value.

Indexes without specific data are divided according to index classification, and different scores are assigned to different types [13].

Table 1. Comprehensive evaluation index system of eco-environmental carrying capacity.

The target layer	Rule layer	The weight	Index layer	The weight
Ecological elastic force A1	Geological landforms B1	0.045	Geomorphic type C1	0.009
			Geological hazard susceptibility C2	0.036
	Climate B2	0.143	Annual rainfall C3	0.107
			Annual precipitation change C4	0.036
	Hydrological B3	0.241	Hydrogeological condition C5	0.241
	Vegetation B4	0.366	Forest coverage C6	0.244
			Vegetation types C7	0.122
	Land B5	0.128	Soil erosion intensity C8	0.064
			Land use type C9	0.064
	Human geography B6	0.076	The population density C10	0.019
			Location advantage C11	0.057
Resource ~ environment carrying capacity A2	The water resources B7	0.441	Degree of water resources assurance C12	0.353
			The water quality conditions C13	0.088
	Land resources B8	0.155	Proportion of cultivated area C14	0.129
			Types of productive potential of cultivated land C15	0.026
	Mineral resources B9	0.329	Mineral resources and reserves C16	0.208
			Engineering geological condition C17	0.035
			Tectonic condition of mining area C18	0.086
	The atmospheric environment B10	0.075	The TSP concentration C19	0.062
			Air quality condition C20	0.013
Mineral resources exploitation pressure A3	Land pressure B11	0.230	Ground collapse rate C21	0.192
			Pressure ratio of solid waste C22	0.038
	Water pressure B12	0.648	Water resource destruction rate C23	0.540
			Waste water discharge C24	0.108
	The atmospheric pressure B13	0.122	Emissions C25	0.098
			Dust emission C26	0.024

2.3. Classification standard and meaning

The higher the ecological elastic force score, the better the stability of the ecosystem. The higher the score of resource-environment carrying capacity is, the greater the actual carrying capacity is. The higher the mineral resource exploitation pressure score is, the more pressure the ecosystem is subjected to. The classification standards and meanings of each grade are shown in Table 2.

Table 2. Standard of ecological environmental bearing capacity hierarchy and meaning.

Indicators	Level	Parameter values	Meaning
The ecological elastic force A1	Weak stability	<20	The stability of the ecosystem is very poor, the ability to resist all kinds of pressure and disturbance is very weak, and the ability of self-recovery and regulation is very poor
	Low stability	21～40	The stability of the ecosystem is poor, the ability of resisting various pressures and disturbances is weak, and the ability of self-recovery and regulation is poor
	Moderate stability	41～60	The stability of the ecosystem is average, the ability of resisting all kinds of pressure and disturbance is average, and the ecosystem has certain self-recovery and adjustment ability
	Higher stability	61～80	The ecosystem has good stability, strong ability to resist all kinds of pressure and disturbance, and good ability of self-recovery and regulation
	High stability	>80	The ecosystem has good stability, strong resistance to all kinds of pressure and disturbance, and good self-recovery and adjustment ability
Resource ~ environment carrying capacity A2	Weak bearing capacity	<20	The overall carrying capacity level is extremely low, and generally resources are scarce and environmental support capacity is extremely poor
	The low bearing capacity	21～40	The overall carrying capacity level is low, generally speaking, the overall resource advantage is not obvious, and the environmental support capacity is poor
	Medium bearing capacity	41～60	The overall carrying capacity level is medium, generally speaking, it has a certain advantage in resource endowment, and the environmental support capacity is average
	Higher bearing capacity	61～80	The overall carrying capacity level is high, generally speaking, it has better resource endowment advantages and environmental support ability
	High bearing capacity	>80	The overall carrying capacity level is high. Generally speaking, it has a good ability of resource observation and environmental support
Mineral resources exploitation pressure A3	Weak pressure	<20	The ecosystem is under very low pressure, the development and utilization of regional resources is very low, and the impact on the environment is small
	The low pressure	21～40	The ecological system is under low pressure, and the regional resources are developed and utilized in a low degree or with advanced technology and reasonable utilization, with little impact on the environment
	Medium pressure	41～60	The pressure on the ecosystem is average, the development and utilization degree of regional resources is medium, or the development and utilization degree and development mode have caused certain impact on the environment
	The high pressure	61～80	The pressure on the ecosystem is high, the development and utilization of regional resources is high or the way

			of utilization is unreasonable, and the technology is backward. The environment is greatly affected
	High pressure	>80	The pressure on the state system is very high. Regional resources development and utilization degree is high. It has a big impact on the environment
Pressure on the degree of A4	overload	>1.1	Ecosystems are under more pressure than they can carry, and they are being destroyed
	Low load	<1	The pressure borne by the ecosystem is less than the carrying capacity, so the ecosystem can maintain normal development
	balance	1～1.1	The pressure borne by the ecosystem is equal to the carrying capacity. The ecosystem is in an excessive position, and any disturbance may cause carrying overload

3. Case Study of Shenzen

3.1. Existing Environmental Conditions

In Guangdong Province, Shenzhen water system is located in the lower reaches of the Dongjiang River system. Most of the water systems under its jurisdiction belong to the Dongjiang River system and the Pearl River Delta system, including 9 drainage basins, more than 310 rivers, 242 reservoirs, the Pearl River Estuary and 3 bays. Water resources can be divided into three water systems: Dongjiang river system, Pearl River Estuary system and eastern gulf system according to the three levels of water resources and their final direction.

The quality of drinking water source in Shenzhen has been rising steadily and getting better. In 2009, Shenzhen Reservoir, Xili reservoir, Meilin reservoir, Tiegang reservoir, Shiyan reservoir, Chiao reservoir, Songzikeng reservoir, Jingxin reservoir, Tongluojing reservoir, Fengmulang reservoir, Sanzhoutian reservoir, Luotian reservoir, Qinglinjing reservoir, 13 main reservoir drinking water quality standard rate of 100%, the water quality is good. Compared with last year, the water quality of Luotian reservoir changed from class iii to class ii, Songzikeng reservoir changed from class ii to Class i, and the water quality of other reservoirs remained basically stable. River, with the development of river regulation work, Shenzhen Longgang river, stream water quality improvement trends such as ping river, Shahe, MaoZhou river, Longgang river, ping river, Fengtang and Shawan river water quality basic stable, mainly in the middle and lower reaches of the river water is polluted, water quality inferior to surface water V class standard, the main pollutants are ammonia nitrogen, total phosphorus and biochemical oxygen demand.

3.2. Mineral mining activities

By the end of 2006, the city had completed the rectification of 169 gaps in exposed mountains, and the area of soil erosion decreased to 74.8km^2. The total amount of sediment erosion decreased from 4.1112 million tons to less than 1 million tons. Serious soil erosion was brought under fundamental control, and the urban ecological environment significantly improved.

In terms of geological disasters, slope geological disasters will gradually weaken as human engineering activities slow down and prevention and control work deepens. The intensity of karst collapse may be more intense in the karst development zone where human engineering activities are intense. In the future, the Shenzhen fault structure will not cause moderate or strong earthquake disasters [14,15].

3.3. Analysis of ecological environment carrying capacity of mineral resources development and utilization zoning planning

According to the development and utilization of mineral resources of Shenzhen layout, in order to optimize the building stone mining layout and effectively guide the mining right setting, on the basis of the existing mining information, according to the area can be used as building stone granite and gneiss distribution characteristics and environmental factors, the planning building stone blocks defined limit mining area is 9. Mineral resources such as mineral water and underground hot water are not divided into planned blocks [16,17].

By nine planning building stone blocks as the basic evaluation units, the evaluation of the final score according to the product of the indicators of weight and assignment, according to the ecological elastic force, resource environmental bearing capacity, mineral resource mining stress classification summary, and according to the standard Table 2 hierarchies and meaning, the classification of the mining results as shown in Table 3.

Table 3. Analysis results of evaluation blocks.

The serial number	Serial number	Name of planned block	Ecological elastic force	Resource - environmental carrying capacity	Mineral resources exploitation pressure	Pressure bearing degree
1	CQ001	Shipo mining planning block in Bao'an District	Moderate stability	Higher bearing capacity	Medium isobaric	0.85
2	CQ002	Shiyanyuanling mining planning block	Low stability	Low bearing capacity	Higher pressure	0.95
3	CQ003	Longgang District changchangkeng mining planning block	Moderate stability	Higher bearing capacity	Medium isobaric	0.80
4	CQ004	Central South mining planning block of Pingshan New Area	Low stability	Low bearing capacity	Higher pressure	0.95
5	CQ005	Longgang District Kuiyong mining planning block	Low stability	Low bearing capacity	Higher pressure	0.98
6	CQ006	Longgang Shibi mining planning block	Moderate stability	Higher bearing capacity	Medium isobaric	0.85
7	CQ007	Longgang Furong mining planning block	Moderate stability	Higher bearing capacity	Medium isobaric	0.80
8	CQ008	Longgang District honghualing mining planning block	Moderate stability	Higher bearing capacity	Medium isobaric	0.88
9	CQ009	Longgang District huangniuding mining planning block	Low stability	Medium bearing capacity	Higher pressure	1.00

(1) Ecological elasticity. Among the 9 planned mining blocks, CQ002 Shiyanyuanling mining planning block, CQ004 Pingshan New Area south Central mining planning block, CQ005 Longgang

District Guiyong mining planning block, CQ009 Longgang District Huangniuding mining planning block are low stability, and the other 5 evaluation blocks are medium stability. It can be seen that the stability of the ecosystem in most evaluation areas is average, with certain ability to resist pressure and disturbance, as well as self-recovery and adjustment after disturbance, but relatively weak. The ecosystem is easy to deviate from the original equilibrium state, and it takes a long time to recover to the original state after being changed.

(2) Resource - environmental carrying capacity. Among the 9 planned mining blocks, CQ002 Shiyanyuanling mining planning block, CQ004 Pingshan New Area south Central mining planning block, CQ005 Longgang District Kwai Yong mining planning block are low bearing capacity; CQ009 Huangniuding mining planning block in Longgang District is medium bearing capacity; The other five evaluation blocks all have high bearing capacity. The weight of mineral resources in the criterion layer is high, but the overall resource-environmental carrying capacity is medium level.

(3) Mineral resources exploitation pressure. Among the 9 planned mining blocks, CQ002 Shiyanyuanling mining planning block, CQ004 Pingshan New Area south central mining planning block, CQ005 Longgang District Guiyong mining planning block, CQ009 Longgang District Huangniuding mining planning block are relatively high pressure, and the other 5 evaluation blocks are medium pressure. Therefore, it is necessary to change the current way of resource exploitation and utilization, improve the level of science and technology, and reduce the adverse impact of mining on the environment.

3.4. *Overall ecological carrying capacity analysis*

The overall ecological carrying capacity is measured by pressure degree, which refers to the ratio of mining pressure of mineral resources to resource-environment carrying capacity, and is divided into three grades. The classification standards and meanings of each grade are shown in Table 4.

Table 4. Classification standard and meaning of overall ecological environment carrying capacity.

Index	Level	Index value	Meaning
Pressure bearing degree	Overload	>1.1	The pressure on the ecosystem is greater than the carrying capacity, and the ecosystem is being destroyed
	Low load	<1	The pressure on the ecosystem is less than the carrying capacity, and the ecosystem can maintain normal development
	balance	1～1.1	The pressure on the ecosystem is equal to the carrying capacity. The ecosystem is in an excessive position, and any disturbance may cause overload

Pressure according to the analysis results, calculation of mineral resources exploitation and resources - the ratio of environmental bearing capacity, the vast majority of the evaluation of blocks on degree of < 1, show that the pressure in ecosystem state of bearing capacity, the ecosystem is not easy to change and change can in a relatively short period of time after the return to original state, ecological system can maintain normal development. Therefore, reasonable arrangement of mining sequence and quantity of mineral resources, and coordination and unity of mining of mineral resources and environmental protection can promote sustainable development of local economy [17,18].

4. Conclusions

1. The exploitation of mineral resources affects the regional ecological environment and in turn is restricted by the environment. Different natural environment conditions in different regions lead to differences in supporting capacity of ecosystem. Similarly, if the mining methods are different, the level of mining technology is different, and the intensity of environmental input is different, the mining pressure of mineral resources is also different. Shenzhen ecological environmental bearing

capacity of mineral resources exploitation evaluation shows that the vast majority of the evaluation of blocks on degree of < 1, show that the pressure in ecosystem state of bearing capacity, the ecosystem is not easy to change and change can in a relatively short period of time after the return to original state, ecological system can maintain normal development. Therefore, it is necessary to rationally arrange the sequence and quantity of mineral resources exploitation to ensure the coordination and unity of mineral resources exploitation and environmental protection and promote the sustainable development of local economy.

2. New methods and technologies will be applied to the study of ecological carrying capacity. The complexity of ecosystem determines the complexity of research methods and means of carrying capacity. Mature models combined with remote sensing and geographic information system technology will provide more accurate and comprehensive research results for carrying capacity research.

Acknowledgments
Guangdong Provincial Key Laboratory of Environmental Health and Land Resource (Project No: 2020B121201014); Special Project on Key Areas of General Universities in Guangdong Province (Science and Technology Services for Rural Revitalization) "Research and Development of Key Technologies for Resourceful Utilization of Large-scale Livestock and Poultry Breeding Manure in Rural Areas of Western Guangdong" (No. 2021ZDZX4023); Notice of the List of Teaching Quality and Teaching Reform Project Construction Projects in Undergraduate Colleges and Universities in Guangdong Province in 2021 (YJGH [2021] No. 29) Construction Project of Social Practice Teaching Base for College Students "Zhaoqing College-Zhongrun Green Ecological Environment Group (Shenzhen) Collaborative Innovation Social Practice Teaching Base"; Zhaoqing University 2020 Annual Curriculum Ideological Construction Project (ZXY[2020] No.95), Environmental Engineering Professional Course Ideological Construction Project; Zhaoqing University 2020 Annual Curriculum Ideological Construction Project (ZXY[2020] No.95), "Physical Pollution Control Engineering" Course Ideological and Political Construction Projects.

References

[1] Huang Qing, Ren Zhiyuan 2004 On ecological carrying capacity and ecological security *Resources and environment in arid areas* **18(2)** 11-17
[2] Yan Xuqian, Xu Junyan 2005 Study on evaluation method of resource and environment carrying capacity in mining area *Metal mines* **(6)** 56-59
[3] Lu Lianhong, Zhang Zheng, lidaofeng 2005 Applying analytic hierarchy process to construct the evaluation index system of ecological environment quality of coal cities in China *Energy and environmental protection* **19(5)** 53-56
[4] Gao Jixi 2001 The theory of sustainable development; Theory, method and application of ecological carrying capacity *Beijing: China Environmental Science Press* 20-40
[5] Li Huan, Duan Peili, Shao Xigao, Yin Peng 2021 Analysis on the relationship between urban agglomeration development intensity and eco-environmental carrying capacity in Yangtze River Economic Belt *[J] Statistics and Decision* **37(10)** 70-74
[6] Wang Wenju, Tian Yongjie 2020 Coupling analysis of new urbanization quality and eco-environmental carrying capacity in henan province *[J] Chinese journal of agricultural resources and regional planning* **41(04)** 21-26
[7] Zhang Fanming, Ye Jinping, WanYan Xiaopan 2021 Coupling and coordination analysis of new urbanization quality and ecological environment carrying capacity - Taking the central region as an example *Ecological economy* **37(04)** 63-69
[8] Wangzhenguo, Xuliangliang, lijubao, jiangjiyi, zhouzhiming 2021 Evaluation of eco-environmental carrying capacity in the middle and upper reaches of Qujiang River Basin *Journal of College of disaster prevention and technology* **23(04)** 67-77
[9] Wang Jin, Haojinmin, Chen aiqi 2018 Evaluation of ecological environment carrying capacity

of Tianjin Based on fuzzy cluster analysis *Journal of China Agricultural University* **23(10)** 112-121
[10] WANG Jiaji, YAO Xiaohong, LI Jinrong, et al. 2000 Estimation of ecological carrying capacity in Heihe River Basin. Research of Environmental Sciences **13(2)** 44-48
[11] Yang Zhifeng, Sui Xin 2005 Assessment of ecological carrying capacity based on ecosystem health *Acta Scientiae Environmenta Sinica* **25(5)** 586-594
[12] Li Jinping, Wang Zhishi 2003 Ecological footprint analysis of Macao in 2001 *Journal of Natural Resources* **18(2)** 197-203
[13] LI Tianwei, Li Wei. Case Review of Strategic Environmental Impact Assessment. Beijing: China Environmental Science Press, 2006,100-125.
[14] Zhu Jiawei, Xie Xiaotong, Li Xinhui 2017 Evaluation of eco-environmental carrying capacity: a case study of henan province *Acta ecologica sinica* **37(21)** 7039-7047
[15] Wang Ning, Liu Ping, Huang Xihuan 2004 Research progress of ecological carrying capacity *Chinese Agricultural Science Bulletin* **20(6)** 278-281
[16] Zhaoxiangui, Xiaoling 2005 Dynamic study on Ecological Footprint and ecological carrying capacity of Shaanxi Province *China Agricultural Science* **38(4)** 746-753
[17] Wangkuifeng, Li Na, Yu Xuefeng 2014 Construction and application of evaluation index system for ecological environment carrying capacity of Shandong Peninsula *Chinese geology* **41(03)** 1018-1027
[18] Zhang Yin, Yang Qingyuan, Min Jie 2016 Coupling analysis of the quality of new urbanization and eco-environmental carrying capacity in chongqing *Acta geographica sinica* **71(05)** 817-828

IOP Conf. Series: Earth and Environmental Science 1087 (2022) 012002 doi:10.1088/1755-1315/1087/1/012002

Magnetic activated carbon optimization using response surface approach for Cr(VI)

Wenxuan Yao[1], Dezhi Tai[1], Honghu Zeng[1,2], Hua Zhang[1,2,3] and Haiyi Huang[1,2]

[1] College of Environmental Science and Engineering, Guilin University of Technology, Guilin, 541006, China
[2] Collaborative Innovation Center for Water Pollution Control and Water Safety in Karst Area, Guilin University of Technology, Guilin 541006, China

[3] Corresponding author's e-mail: 454267091@qq.com

Abstract. Magnetic activated carbon (Fe_3O_4-AC) is made using Fe_3O_4 and activated carbon as primary materials to adsorb Cr(VI) in simulated wastewater. The Response Surface Method (RSM) optimizes the experimental parameters (pH, temperature, initial solution concentration, and activated carbon dosage). The impacts of various elements are explored to determine the ideal reaction conditions for maximum elimination. According to the data, the interaction between pH and the other three parameters is exceptionally significant ($P<0.0001$) in the experimental response. Furthermore, a maximum removal rate of 83.88 percent may be attained at pH=2, temperature 62°C, starting Cr(VI) concentration 179mg/L, and activate carbon dosage 0.14g/50mL.

1. Introduction

In effluent from mining, electroplating, printing, dyeing, and tanning, chromium, and its derivatives are predominantly Cr(III) and Cr(VI) [1]. Cr(VI) is 100 times more poisonous than Cr(III) and is more mobile than Cr(III) [2]. It will ruin the transmission of blood and nerves after it has collected in the human body, and it has a terrible carcinogenic effect [3]. Controlling the amount of chromium in the environment is therefore critical. In recent years, photocatalysis, adsorption, and chemical reduction have been utilized to remove Cr(VI) from water [4]. Many academics have been drawn to the adsorption technique because of its ease of use, high removal rate, and low cost [5]. Fe_3O_4 in magnetic activated carbon (MBC) was discovered by D Zhong [6] et al. to successfully increase the adsorption sites of Cr (VI) and the pore volume of MBC, which is helpful to improve the removal rate. YAB Neolaka [7] et al. used magnetic graphene adsorption of Cr(VI) in water (Fe_3O_4-GO). However, several parameters influence Cr(VI) adsorption experiments using magnetic materials as adsorbents, and there are little data on how these factors interact and how much influence they have.

The response surface method (RSM) is a data-fitting methodology that establishes a model that can provide the optimal response effect for trials with numerous components [8]. Common techniques include CCD, BBD, and FFD [9]. MN Chollom [10] et al. compared response surface optimization methods and found the BBD model more reliable. Box-Behnken Design (BBD) is a three-level fractional factorial design with a center point and the midpoint of the cube edges inscribed on a sphere as an efficient RSM [11]. The feature is that the experiment can be optimized with the least effort, resulting in significant time and cost savings.

The interaction of many components was described in this study, which used magnetic activated carbon Fe_3O_4-AC to adsorb Cr(VI) in water. The ideal process parameters for Fe_3O_4-AC to adsorb Cr(VI) in water were determined using a BBD-based response surface methodology (RSM) optimization experiment.

2. Materials and methods

2.1. *Raw materials*

The analytical grades include potassium dichromate, acetone, diphenylcarbazide, hydrochloric acid, sodium hydroxide, sulfuric acid, phosphoric acid, etc. This research received activated carbon raw material from Shuye Activated Carbon Company, a commercial wood-activated carbon provider.

2.2. *Sorbent preparation*

First, strain the commercial activated carbon through a 100-mesh filter and soak it for 12h in a (1+9) HNO_3 solution. Boil for two hours after washing with ultrapure water. Then wash till neutral with ultrapure water and dry in an oven at 105℃ for later use. Then put $FeSO_4 \cdot 7H_2O$ and $FeCl_3 \cdot 6H_2O$ in a beaker with a molar mass of 1:2, add 150mL ultrapure water, and stir to dissolve. The pH was then corrected using 4mol/L NaOH solution after adding a particular amount of activated carbon to the beaker. The mixture was maintained in a water bath before being removed. After cooling, the top solution was filtered off, rinsed with ultrapure water until neutral, and then filtered. Finally, the resultant solid material was dried to constant weight in an electric heating blast oven at 110℃. Grind it through a 100-mesh sieve once it has cooled to obtain a magnetic ferric oxide/activated carbon adsorbent (Fe_3O_4-AC).

2.3. *Response surface test scheme and optimization*

The four input elements in this experiment were the solution pH, adsorption temperature, the beginning concentration of the solution, and Fe_3O_4-AC dosage, with the adsorption removal rate (%) of Cr (VI) serving as response value. Every characteristic was put through its paces on three levels: low, medium, and high (denoted as -1, 0, and 1) [12]. Table 1 shows the range of independent variables in the BBD four-factor three-level response surface optimization experiment.

Table 1. Independent variables and levels.

Variables	Levels		
	-1	0	+1
pH	2	4	6
Temperature（℃）	25	45	65
Cr (VI) concentration（mg/L）	150	250	350
Adsorbent dose（g/50mL）	0.05	0.10	0.15

3. Results and discussion

3.1. *Model building*

Table 2 shows the experimental and fitting data, with removal rate Y as the dependent variable, pH (A), adsorption temperature(B), Cr(VI) initial concentration(C), and Fe_3O_4-AC dosage(D) as independent variables. Regression analysis was performed and referred to equation 1:

$$\begin{aligned} Y &= 99.049 - 44.330A + 0.737B - 0.182C + 535.433D - 0.062AB + 0.028AC \\ &- 78.950AD - 0.0012BC + 0.475BD + 0.0175CD + 4.651A^2 - 0.0005B^2 \\ &+ 0.00015C^2 - 121.333D^2 \end{aligned} \tag{1}$$

Table 2. Design of RSM and its experimental and predicted values.

Run	pH	Temperature (℃)	Concentration (mg/L)	Dose (g/50mL)	Removal(%)	
					Actural	Predicted
1	1	-1	0	0	14.89	14.47
2	0	0	1	1	27.51	26.87
3	0	0	-1	-1	13.40	14.06
4	0	0	0	0	19.47	19.13
5	-1	1	0	0	65.14	65.57
6	-1	-1	0	0	52.76	53.04
7	0	0	0	0	18.86	19.13
8	0	0	-1	1	35.39	35.99
9	0	0	1	-1	5.17	4.58
10	1	1	0	0	17.33	17.06
11	0	1	-1	0	29.84	30.07
12	-1	0	0	1	77.32	76.92
13	0	-1	1	0	13.20	13.22
14	1	0	0	1	17.64	17.59
15	0	0	0	0	18.25	17.89
16	-1	0	0	-1	38.71	39.02
17	0	1	1	0	16.34	15.96
18	0	-1	-1	0	17.06	17.70
19	0	0	0	0	18.55	17.89
20	1	0	0	-1	10.61	11.27
21	-1	0	1	0	50.20	50.82
22	0	1	0	-1	11.22	10.83
23	0	1	0	1	33.52	33.90
24	-1	0	-1	0	72.44	71.20
25	0	-1	0	1	25.27	25.39
26	0	-1	0	-1	4.87	4.23
27	0	0	0	0	18.25	19.09
28	0	0	0	0	18.86	19.09
29	1	0	-1	0	17.47	16.58
30	1	0	1	0	17.39	18.36

The R_{adj}^2=0.9982 coefficient of determination of the regression model adjusted by this model indicates that at least 1% of the experimental outcomes are unaccounted for. The observed signal-to-noise ratio is 114.528, the standard deviation is 0.84, and the coefficient of variation is 3.17, indicating that the equation is correct and usable and can adequately explain the relationship between the response and the independent variable.

3.2. Analysis of variance (ANOVA)
The model was subjected to analysis of variance, and the findings are presented in Table 3. F=1053.64, P<0.0001, indicating that the model is highly significant. The P values for the single factor in the experiment were all less than 0.0001, indicating that multiple parameters had a substantial impact on the Cr(VI) removal rate. As demonstrated by the size of the F value, the pH effect is the most significant. F=8000.05. In terms of the degree of response, the following is the order of each component: pH > dosage > Cr(VI) starting concentration > temperature.

Furthermore, the P values of the interaction terms of pH and the other three parameters were all less than 0.0001, showing that they responded strongly to the rate of Cr(VI) removal. The interaction factors between dosage, temperature, and beginning concentration yielded P values greater than 0.05, indicating that they did not affect Cr(VI) clearance rate.

Table 3. ANOVA for response surface quadratic model.

Source	SS	DF	MS	F value	P value prob
Model	10486.33	14	749.02	1053.64	<0.0001
A	5687.19	1	5687.19	8000.05	<0.0001
B	171.31	1	1171.31	240.98	<0.0001
C	259.38	1	259.38	364.86	<0.0001
D	1466.78	1	1466.78	2063.28	<0.0001
AB	24.70	1	24.70	34.75	<0.0001
AC	122.77	1	122.77	172.69	<0.0001
AD	249.32	1	249.32	350.72	<0.0001
BC	23.23	1	23.23	32.68	<0.0001
BD	0.90	1	0.90	1.27	0.2802
CD	0.031	1	0.031	0.043	0.8388
A^2	2373.68	1	2373.68	3339.00	<0.0001
B^2	0.28	1	0.28	0.39	0.5411
C^2	16.35	1	16.35	23.00	0.0003
D^2	0.63	1	0.63	0.89	0.3633
Residual	9.24	13	0.71		
Lack of fit	8.82	10	0.88	6.35	0.0776
Pure error	0.42	3	0.14		
Cor total	10505.48	29			

3.3. Analysis and optimization of key response surfaces

3.3.1. The relationship between pH and temperature. Figure 1 depicts the interplay of pH and temperature. However, improving the Cr(VI) elimination rate is good by reducing the pH and boosting the temperature. On the other hand, the pH effect was more pronounced, and the temperature had no impact on the Cr (VI) elimination rate's changing trend. Could be due to the shift in hydroxyl groups in solid acid conditions, which makes the surface of Fe_3O_4-AC positively charged, allowing the negatively charged Cr(VI) ions in the solution to be readily adsorbed.

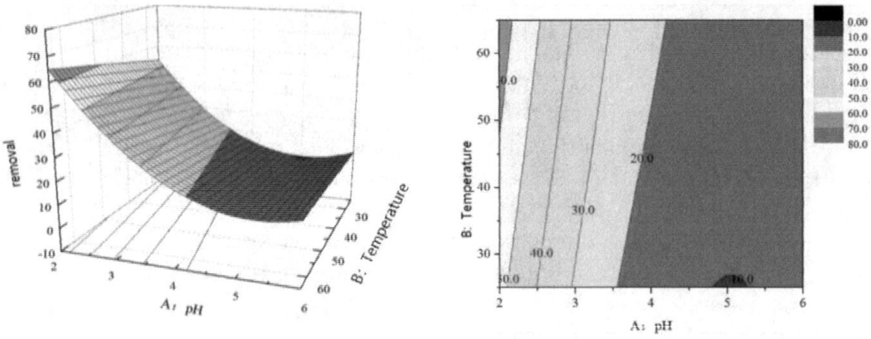

Figure 1. pH and temperature of Cr(VI) predicted removal rate.

3.3.2. The interaction between pH and dosage. Figure 2 depicts the interaction of pH and dose on Cr(VI) adsorption by Fe_3O_4-AC. With decreasing pH, the Cr(VI) removal rate increases dramatically. At the same time, the clearance rate increased consistently when the dosage was increased. Furthermore, the best clearance rate was achieved when the pH=2.09 dosage was 0.140g/50mL. Because the acidic environment promotes Cr(VI) ion adsorption in water; Increasing Fe_3O_4-AC dose will surely increase the specific surface area and give more adsorption/active sites, increasing Cr(VI) removal rate.

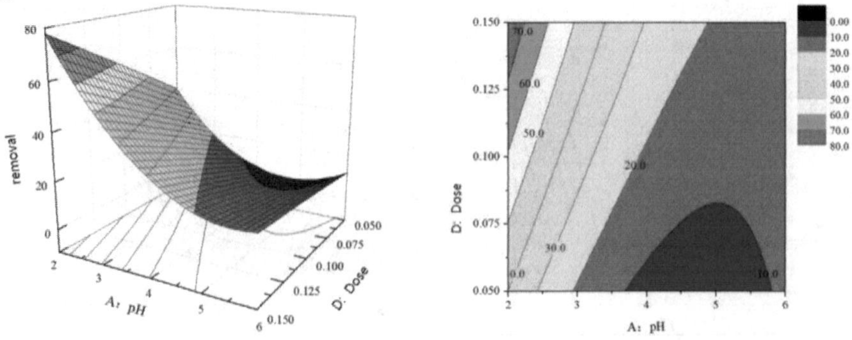

Figure 2. pH and dose of Cr(VI) predicted removal rate.

3.3.3. Temperature and initial concentration interaction. Figure 3 depicts the effect of temperature and starting concentration on Cr(VI) adsorption by Fe_3O_4-AC. In the single factor experiment, however, both temperature and beginning concentration have a massive impact on adsorption. However, when the two were combined, $P>0.05$ was observed, and there was no ellipse in the contour plot, indicating that the interaction between the two had no significant influence on the removal rate.

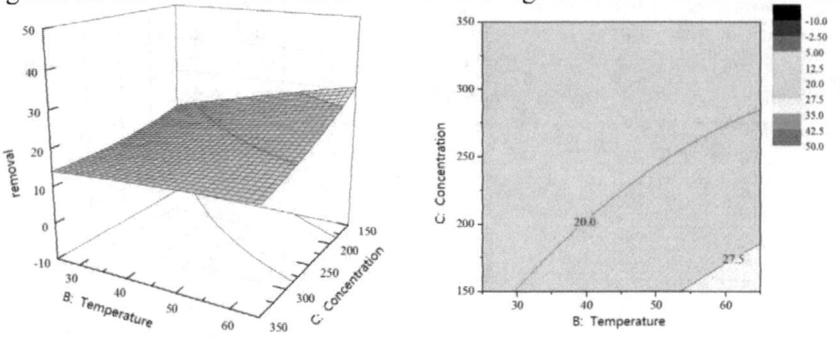

Figure 3. temperature and concentration of Cr(VI) predicted removal rate.

3.4. Validation and optimal conditions

The estimated Cr(VI) removal rate can reach 84.63% under the parameters of pH=2.09, temperature 62.05℃, initial Cr(VI) concentration of 178.90mg/L Fe_3O_4-AC dose of 0.14g/50mL, according to the established adsorption model.

Based on the actual operation, it was verified and tweaked according to optimal circumstances. The average removal rate of Cr(VI) was 83.88%, which was close to the expected value, demonstrating that RSM may be used to model the adsorption process. Its accuracy and effectiveness are great, and it can reflect the effects of many conditions on adsorption.

4. Conclusions

Four parameters having significant effects in the single factor test (pH, temperature, beginning concentration, and adsorbent dosage) were chosen as independent variables. Among single components, pH showed the most significant effect (F=8000.05), according to the study. The connection between pH and the other three variables was crucial ($P<0.0001$). R2=0.9991, F=1053.64, and $P<0.0001$ are the regression coefficients for the model, indicating that the data is intelligible. After response surface adjustment with pH=2.09, temperature 62.05°C, initial Cr(VI) concentration 178.90mg/L, and Fe_3O_4-AC dosage 0.14g/50mL, the Cr(VI) elimination rate can reach 84.63 percent. The error is less than 1% compared to the actual removal rate of 83.88 percent, suggesting acceptable repeatability.

References

[1] Buaisha, M., Balku, S., & Özalp-Yaman, S 2020 Heavy metal removal investigation in conventional activated sludge systems *Civil Engineering Journal* **6(3)** 470-477

[2] Zhu, L., Tong, L., Zhao, N., Li, J., & Lv, Y 2019 Coupling interaction between porous biochar and nano zero valent iron/nano α -hydroxyl iron oxide improves the remediation efficiency of cadmium in aqueous solution *Chemosphere* **219** 493-503

[3] Xu, H., Liu, Y., Liang, H., Gao, C., Qin, J., You, L., ... & Yang, S 2021 Adsorption of Cr (VI) from aqueous solutions using novel activated carbon spheres derived from glucose and sodium dodecylbenzene sulfonate *Science of The Total Environment* **759** 143-457

[4] Wang, Y., Chen, S. Y., Yang, X., Wu, Y., Huang, X. F., He, E. K., ... & Wang, S 2019 Enhanced removal of Cr(VI) in the Fe(III)/natural polyphenols system: role of the in situ generated Fe(II) *Journal of hazardous materials* **377** 321-329

[5] Zhang, W., Wang, H., Hu, X., Feng, H., Xiong, W., Guo, W., ... & Peng, Y 2019 Multicavity triethylenetetramine-chitosan/alginate composite beads for enhanced Cr(VI) removal *Journal of Cleaner Production* **231** 733-745

[6] Zhong, D., Zhang, Y., Wang, L., Chen, J., Jiang, Y., Tsang, D. C., ... & Crittenden, J. C 2018 Mechanistic insights into adsorption and reduction of hexavalent chromium from water using magnetic biochar composite: key roles of Fe3O4 and persistent free radicals *Environmental Pollution* **243** 1302-1309

[7] Neolaka, Y. A., Lawa, Y., Naat, J. N., Riwu, A. A., Iqbal, M., Darmokoesoemo, H., & Kusuma, H. S 2020 The adsorption of Cr(VI) from water samples using graphene oxide-magnetic (GO-Fe3O4) synthesized from natural cellulose-based graphite (kusambi wood or Schleichera oleosa): Study of kinetics, isotherms and thermodynamics *Journal of Materials Research and Technology* **9(3)** 6544-6556

[8] Li, Z., Lu, D., & Gao, X 2021 Optimization of mixture proportions by statistical experimental design using response surface method-A review *Journal of Building Engineering* **36** 102-101

[9] Prajapati, N., Kodgire, P., & Kachhwaha, S. S 2022 Comparison of RSM Based FFD and CCD Methods for Biodiesel Production Using Microwave Technique *Materials Today* **62(13)** 6985-6991 *Proceedings*

[10] Chollom, M. N., Rathilal, S., Swalaha, F. M., Bakare, B. F., & Tetteh, E. K 2020 Comparison of response surface methods for the optimization of an upflow anaerobic sludge blanket for the

treatment of slaughterhouse wastewater *Environmental Engineering Research* **25(1)** 114-122
[11] Fu, J. F., Zhao, Y. Q., Xue, X. D., Li, W. C., & Babatunde, A. O 2009 Multivariate-parameter optimization of acid blue-7 wastewater treatment by Ti/TiO2 photoelectrocatalysis via the Box - Behnken design *Desalination* **243(1-3)** 42-51
[12] Hu, Y. Y., Pan, C., Zheng, X., Hu, F., Xu, L., Xu, G., ... & Peng, X 2021 Prediction and optimization of adsorption properties for Cs+ on NiSiO@ NiAlFe LDHs hollow spheres from aqueous solution: kinetics, isotherms, and BBD model *Journal of Hazardous Materials* **401** 123374

Bioaccumulation of silver nanoparticles in marine copepod *Tigriopus japonicus* under waterborne and dietary exposure

Chenge Liu[1,2] and Wenxia Liu[1]

[1] College of Environmental Science and Engineering, Ocean University of China, Qingdao 266100, China.

[2] Corresponding author's e-mail: lcg1110@foxmail.com

Abstract. At present, the emission of silver nanoparticles (AgNPs) will affect a variety of aquatic organisms. AgNPs may be transferred to the food web through the bioaccumulation or biomagnification of the food chain, which is harmful to at higher trophic level organisms, and even threatens human health. Therefore, in this study, we tested the acute toxicity effect of AgNPs to marine copepod *Tigriopus japonicus*, and analyzed the Ag bioaccumulation in *T. japonicus* under short-term waterborne (exposure of AgNPs to *T. japonicus*) and dietary exposure (AgNP-accumulated algae cells of *Isochrysis galbana* are used to feed *T. japonicus*). The results showed that, after 48 h exposure, significant effects of AgNPs on the mortalities were observed (LC_{50}=6.330 mg·L^{-1}). The bioaccumulation of Ag by *T. japonicus* under different exposure modes was positively correlated with exposure concentration and time, and the bioaccumulation level under waterborne exposure was significantly higher. The biomagnification of *T. japonicus* was not observed under 24 h dietary exposure of different AgNP concentrations (BMFs<1), but the transfer efficiency was higher under low exposure concentration ($BMF_{low\ concentration}>BMF_{high\ concentration}$). This study is helpful to understand the toxicity mechanism and effect of AgNPs on aquatic organisms.

1. Introduction

Silver nanoparticles (AgNPs) are one of the most widely used and consumed nanomaterials. Due to its excellent physical and chemical properties, such as antibacterial and conductive, AgNPs are widely used in medical, food industry, biotechnology and environmental remediation. With the widespread production and application of AgNPs, more and more AgNPs are discharged into aquatic environment. They may eventually enter the ocean along the rivers and threaten the aquatic organisms. Studies have shown that instead of fresh water, AgNPs are extremely unstable in seawater [1]. Therefore, the potential impact of AgNPs on ecological environment has attracted increasing attention, especially the marine toxicology and ecological risk assessment of AgNPs have become one of the research hotspots.

In recent years, most studies on the ecological toxicity of NPs to aquatic organisms focus on the direct bioaccumulation and toxicity effect of single organism [2, 3]. The most obvious toxic effect of AgNPs on algae is to inhibit the growth, material metabolism and photosynthesis of algae cells [4, 5]. AgNPs also have toxic effects on aquatic vertebrates. AgNPs will enter the body of fish and accumulate in their livers, gills and other regions, which may lead to oxidative damage, apoptosis, respiratory damage, neurotoxicity, and even genotoxicity [6, 7]. In addition, AgNPs may also have toxic effects on aquatic invertebrates at different levels such as survival, feeding metabolism, reproductive development, antioxidant defense system, lipid peroxidation and DNA damage [8, 9].

However, various organisms in the water environment constitute the food chain/web due to predation. Thus, NPs also affect the interaction between them. They are transferred to the food web through bioaccumulation/biomagnification of the food chain, and poses a threat to higher trophic level organisms and even human [10]. NP exposure and their subsequent bioaccumulation in aquatic organisms are important mechanisms for their migration in aquatic ecosystems, which can be produced through different uptake modes, including water environmental media and diet [11]. Especially when consumer diet increase, in order to meet the additional energy requirements for detoxification, waterborne exposure does not fully explain the biological effects of pollutants. Therefore, dietary exposure is closer to natural conditions, and can more comprehensively reflect the potential hazards of AgNPs. The accumulation of NPs in organisms is still controversial mainly through waterborne or diet uptake [12]. Therefore, in order to evaluate the ecological effects of NPs, more attention should be paid to the transmission and magnification of NPs in aquatic food chains and ecosystems in the future. Bioaccumulation cannot directly reflect the toxicity of AgNPs, so the effectiveness of bioaccumulation should be considered more in environmental evaluation. The assessment of biomagnification can be used to quantify the risk of specific pollutants uptake and accumulation through waterborne/dietary exposure. Studies have shown that AgNPs also have different degrees of effective transfer through dietary exposure [13, 14], but no matter what exposure mode, it poses a threat to the entire species and even the food chain (web). In addition, studies on dietary exposure of NPs are mainly concentrated in freshwater [15, 16]. But the different indicators of seawater will affect AgNPs aggregation and sedimentation, making it easy to transfer in the benthic food chain [17]. AgNP aggregation may also indirectly affect their toxicity [18]. Therefore, it is necessary to further study the transfer of AgNPs in the marine benthic food chain.

As a primary producer and a typical ecological receptor of pollutants, *Isochrysis galbana* may be an important way for AgNPs to accumulate into the food chain [19]. *Tigriopus japonicus* is a kind of common intertidal benthic copepod, which is one of the standard organisms for ecological toxicity experiments and monitoring [20]. As a primary consumer, *T. japonicus* can feed on algae, and is an important part of food chain.

In this study, we studied the acute toxicity effect of AgNPs to *T. japonicus* and Ag bioaccumulation in *T. japonicus* under different exposure modes (waterborne and dietary exposure). This study can compare and analyze the effects of AgNPs under different exposure modes on *T. japonicus* from the perspectives of bioaccumulation. It is of great practical significance to understand the toxic mechanism and potential ecological effects of AgNPs.

2. Materials and Methods

2.1. Culture of I. galbana and T. japonicus

I. galbana and *T. japonicus* were provided by the Ecotoxicological Laboratory of College of Marine Life Sciences, Ocean University of China, and then expanded in the laboratory. *I. galbana* was cultured in f/2 medium (culture conditions: temperature=25±1 ℃; salinity=30±1; light intensity=5000 lx; light/dark ratio=12 h:12 h), and cultured to logarithmic growth phase for test. *T. japonicus* was cultured in the natural seawater after aerated, filtered and sterilized. The culture conditions were the same as *I. galbana*. *T. japonicus* was fed with *I. galbana* once every 2 d. Individuals with complete morphology and sensitive response were randomly selected from the adults of *T. japonicus* with the same age as tested organisms.

2.2. Preparation of AgNPs

AgNPs (particle size <100 nm) were purchased from Sigma-Aldrich. The surface was coated with PVP, and the purity was 99.5%. Stock solution of AgNPs (100 $mg \cdot L^{-1}$) was prepared with ultrapure water and ultrasonically treated.

2.3. Acute Toxicity Tests

10 individuals of *T. japonicus* in each well with 5 mL solution (sterilized seawater added with AgNP solution) were selected to 6-well plates under the above culture conditions. The treatment concentration gradients were set to 0, 2, 4, 6, 8, 10, 12 and 14 $mg \cdot L^{-1}$. Each group was set to 3 parallel groups. The tests lasted for 48 h without feeding. The solution was changed (about 90%), and the numbers of dead individuals are observed and recorded every 24 h. Approximately 90° bending of the head, chest and abdomen, or no moving during 10 s after touching with a dissecting needle is considered dead.

2.4. Bioaccumulation Tests

According to the results of acute toxicity tests, two treatment concentrations, including a low concentration (1/100 48 h $LC_{50} \approx 0.06$ $mg \cdot L^{-1}$) and a high concentration (1/10 48 h $LC_{50} \approx 0.6$ $mg \cdot L^{-1}$), were selected. The waterborne and dietary exposure groups were set up with 4 treatment groups and 2 control groups, with 3 parallels in each group. At 24 h before the beginning of the exposure tests, the individuals of *T. japonicus* were selected to empty the intestine in sterilized seawater. The specific methods are as follows.

2.4.1. Waterborne Exposure. *T. japonicus* was transferred to sterilized seawater with 60 individuals in each group, added with AgNP solution, and exposed under the above culture conditions. After exposure for 3, 6, 12 and 24 h, *T. japonicus* and 10 mL medium in each group were collected for test.

The collected *T. japonicus* was dried overnight. Then the samples of *T. japonicus* and medium were digested with concentrated HNO_3. After cooling and constant volume, the concentration of Ag was determined by inductively coupled plasma-mass spectrometry (ICP-MS). The bioconcentration factor (BCF) of AgNPs was calculated according to the following formula:

$$BCF=\frac{\text{Ag concentration in } T. japonicus \ (mg \cdot g^{-1})}{\text{Ag concentration in the culture medium } (mg \cdot L^{-1})} \tag{1}$$

2.4.2. Dietary Exposure. A certain volume of *I. galbana* was loaded, and the initial algae density was adjusted to 1.0×10^6 $cells \cdot mL^{-1}$, added with AgNP solution (*I. galbana* without AgNPs as the control groups) and pre-exposed under the above culture conditions. After pre-exposure for 48 h, *I. galbana* was centrifuged, washed with ultrapure water, centrifuged repeatedly to remove the loose attached AgNPs on the surface. Algae cells were collected and resuspended in the medium.

T. japonicus was transferred to sterilized seawater with 60 individuals in each group, fed with pre-exposed algae (feeding density was 1.0×10^6 $cells \cdot mL^{-1}$) under the above culture conditions. After exposure for 3, 6, 12 and 24 h, *T. japonicus* and 50 mL medium in each group was collected for test. At the same time, the concentration of algal cells in the medium was quantified.

The collected *T. japonicus* was dried overnight; the culture medium was centrifuged, washed with ultrapure water, and then centrifuged again. Algae cells in the medium were collected and dried. The samples of *T. japonicus* and *I. galbana* were digested with concentrated HNO_3. After cooling, the concentration of Ag was determined by ICP-MS. The biomagnification factor (BMF) of AgNPs was calculated according to the following formula:

$$BMF=\frac{\text{Ag concentration in } T. japonicus \ (mg \cdot g^{-1})}{\text{Ag concentration in } I. galbana \ (mg \cdot g^{-1})} \tag{2}$$

BMF>1 indicated that biomagnification occurred.

3. Results

3.1. Acute Toxicity

As shown in Figure 1, no dead individuals were observed within 48 h in the control group. The mortalities at 48 h gradually increased with the increase of AgNP concentration. When the concentration was 14 mg·L^{-1}, the mortality at 48 h reached 100%. The 48 h LC_{50} of *T. japonicus* exposed to AgNPs was 6.330 mg·L^{-1} (95% confidence interval was 5.582-7.155 mg·L^{-1}). The regression equation of AgNPs concentration and mortality of *T. japonicus* is $y=-0.1587x^2+9.9206x-4.1667$. The coefficient of determination (R^2) is 0.9876.

3.2. Bioaccumulation

The accumulation of AgNPs by *T. japonicus* in the first 12 h was low, and increased significantly ($P<0.05$) after 12 h (Figure 2). *T. japonicus* could accumulate Ag through waterborne exposure of AgNPs, and the accumulation of AgNPs was positively correlated with the exposure concentration. The 24 h BCF values (Table 1) showed that, *T. japonicus* had a strong ability to accumulate AgNPs. The BCF value. decreased with increasing exposure concentration.

With the increase of exposure time, the accumulation increased gradually regardless of exposure concentration (Figure 2). It indicated that, AgNPs could be transferred from *I. galbana* to *T. japonicus*, and the accumulation under high concentration exposure was always higher. Compared with the waterborne exposure, the Ag accumulation in *T. japonicus* under dietary exposure for 24 h decreased sharply (by 55% under 0.06 mg·L^{-1} exposure, and 51% under 0.6 mg·L^{-1} exposure). Therefore, the results of this study showed that under relatively short-term experimental conditions, waterborne exposure was still the main mode for the bioaccumulation of AgNPs by *T. japonicus*. The 24 h BMF values (Table 1) were both less than 1, indicating that there was no biomagnification, but the transfer efficiency was higher at low concentration.

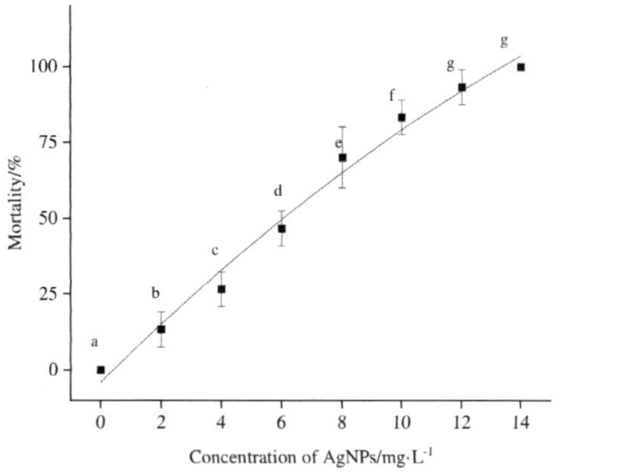

Figure 1. The mortality of *T. japonicus* exposed to different concentrations of AgNPs.

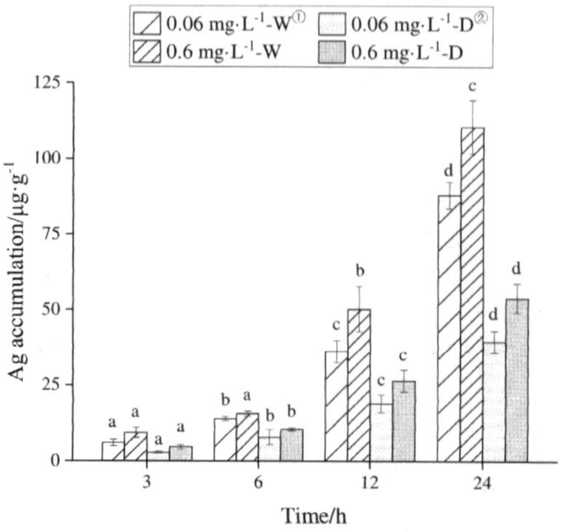

Figure 2. Ag accumulation of *T. japonicus* exposed to AgNPs for 24 h under different exposure modes (①Waterborne exposure; ② Dietary exposure).

Table 1. BCF and BMF values of different concentrations of AgNPs in *T. japonicus*.

Concentrations of AgNPs ($mg \cdot L^{-1}$)	BCF (L·g-1)	BMF
0.06	6.94	0.93
0.6	4.46	0.63

4. Discussions

T. japonicus is commonly used model organism in aquatic ecotoxicological studies. Toxic effects of other NPs on *T. japonicus* have been studied [21, 22], but the data on the acute toxicity of AgNPs to *T. japonicus* are still lacking. Therefore, in this study, we preliminarily discussed the short-term effects of AgNPs on *T. japonicus*. It can be found that *T. japonicus* has relatively higher tolerance to AgNPs compared with other copepods [23, 24].

So far, the bioaccumulation of NPs by *T. japonicus* has not been reported. Our results of the bioaccumulation and influence of exposure concentration on it are consistent with previous studies on other small invertebrates [11, 25], and the reason may be that the uptake of AgNPs at low concentrations is controlled by endocytosis. A number of previous studies [26, 27] indicated that, due to the different properties of AgNPs, exposure conditions and concentrations, the BCF data of AgNPs in aquatic organisms are quite different and lack of comparison. In addition, the bioaccumulation of AgNPs in aquatic organisms under dietary exposure has been studied [28]. Similarly, they found AgNPs can be effectively accumulated into aquatic organisms through dietary exposure, thus affecting higher trophic level organisms, and this process is affected by exposure concentration. According to Fabianne Ribeiro et al, the accumulation of Ag in *Daphnia magna* was 15.6 $\mu g \cdot g^{-1}$ after dietary exposure (*Raphidocelis subcapitata*) to AgNPs, and even reached 55.0 $\mu g \cdot g^{-1}$ after concomitantly waterborne and dietary exposure to AgNPs for 48 h [11].

By comparison in this study, it was found that the waterborne exposure was dominant when *T. japonicus* accumulated AgNPs. This conclusion is consistent with some previous research results. Fabianne Ribeiro et al found that, under combined exposure of waterborne and diet, the contribution rates of Ag under the two exposure modes to the accumulation of *D. magna* were 62.5% and 37.5%, respectively, so the waterborne exposure of AgNPs is the main reason for the increase in Ag accumulation in *D. magna* [11]. The reason for this phenomenon may be that different exposure modes have different accumulation mechanisms, and algae have limited absorption of Ag. However, the assimilation efficiency of aquatic organisms to Ag through diet may be relatively higher [29]. Due to the large specific surface area of NPs, it is most likely to adsorb inorganic/organic particles from the aquatic environment, which will increase the opportunity for NPs to transfer along the food chain. Therefore, in the long run, the role of diet as the carrier of AgNPs cannot be ignored.

Although bioaccumulation is worthy of attention, biomagnification has been attributed to the obvious adverse effects on organisms at higher trophic levels, which is generally considered to be a larger problem. Similar to this study, generally, AgNPs also do not exhibit biomagnification in other aquatic food chain [27, 30]. One reason may be that pollutants can only be biomagnified if they are soluble in fat or lipophilic substances [31], while AgNPs are not. For another thing, organisms may accumulate AgNPs from contaminated diet, but they can also discharge them by excretion, etc. At the same time, algal cells may induce the purification of AgNPs, thereby hindering the long-term bioaccumulation and causing limited biomagnification of AgNPs in organisms [32].

5. Conclusions

To our knowledge, our study investigated the bioaccumulation of AgNPs in the marine copepod *T. japonicus* for the first time. This study contributes to the debate about the potential effects of NPs in the marine environment. We found that, compared with dietary exposure, waterborne exposure seemed to dominate the bioaccumulation of AgNPs. However, this study cannot observe the specific transfer forms and behaviors of AgNPs in the food chain. Different bioaccumulation characteristics of AgNPs under different exposure modes may be due to these complex processes. Therefore, in order to detect the potential transfer and biomagnification of AgNPs in the aquatic food chain, further studies including different related species are needed.

References

[1] N Akaighe, S W Depner, S Banerjee and M Sohn 2013 *Chemosphere* **92(4)** 406-412
[2] L M Rossbach, D H Oughton, E Maremonti, D M Eide and D A Brede 2021 *Ecotoxicol Environ Saf* **216** 112178
[3] H H Mahboub, R R Beheiry, S E Shahin, A Behairy, M H E Khedr, S M Ibrahim, G E Elshopakey, W M Daoush, D E Altohamy, T A Ismail and W El-Houseiny 2021 *Aquat Toxicol* **235** 105828
[4] M Behzadi Tayemeh, M Esmailbeigi, I Shirdel, H S Joo, S A Johari, A Banan, H Nourani, H Mashhadi, M J Jami and M Tabarrok 2020 *Chemosphere* **238** 124576
[5] C J Dedman, G C Newson, G-L Davies and J Christie-Oleza 2020 *Science of The Total Environment* **747** 141229
[6] A Ale, A S Rossi, C Bacchetta, S Gervasio, F R de la Torre and J Cazenave 2018 *Ecological Indicators* **93** 1190-1198
[7] C Bacchetta, A Ale, M F Simoniello, S Gervasio, C Davico, A S Rossi, M F Desimone, G Poletta, G López, J M Monserrat and J Cazenave 2017 *Ecological Indicators* **76** 230-239
[8] R Cui, Y Chae and Y J An 2017 *Chemosphere* **185** 205-212
[9] F R Khan, K B Paul, A D Dybowska, E Valsami-Jones, J R Lead, V Stone and T F Fernandes 2015 *Environmental Science & Technology* **49(7)** 4389-4397
[10] D Batista, D P Giling, A Pradhan, C Pascoal, F Cassio and M O Gessner 2021 *Environ Pollut* **290** 118088
[11] F Ribeiro, C A M Van Gestel, M D Pavlaki, S Azevedo, A M V M Soares and S Loureiro 2017 *Science of The Total Environment* **574** 1633-1639
[12] A Bour, F Mouchet, J Silvestre, L Gauthier and E Pinelli 2015 *J Hazard Mater* **283** 764-777
[13] J McTeer, A P Dean, K N White and J K Pittman 2013 *Nanotoxicology* **8(3)** 305-316
[14] J M Lacave, A Fanjul, E Bilbao, N Gutierrez, I Barrio, I Arostegui, M P Cajaraville and A Orbea 2017 *Comp Biochem Physiol C Toxicol Pharmacol* **199** 69-80
[15] M Khoshnamvand, Z Hao, O O Fadare, P Hanachi, Y Chen and J Liu 2020 Chemosphere **258** 127346
[16] X Chen, Y Zhu, K Yang, L Zhu and D Lin 2019 *Environ Pollut* **247** 421-430
[17] J Zhao, L Xu, S Yin, Y Li, Z Wang, R Hou, H Yao, Y Dai and Z Liu 2018 *Chinese Science Bulletin* **63(9)** 790-800
[18] T Fang, L P Yu, W C Zhang and S P Bao 2015 *Ecotoxicology* **24(10)** 2054-2066
[19] M A Trenfield, J W van Dam, A J Harford, D Parry, C Streten, K Gibb and R A van Dam 2015 *Environ Toxicol Chem* **34(8)** 1833-1840
[20] S Raisuddin, K W Kwok, K M Leung, D Schlenk and J S Lee 2007 *Aquat Toxicol* **83(3)** 161-173
[21] X Yi, T Chi, B Liu, C Liu, G Feng, X Dai, K Zhang and H Zhou 2019 *Comp Biochem Physiol C Toxicol Pharmacol* **222** 118-124
[22] A Broccoli, S Anselmi, A Cavallo, V Ferrari, D Prevedelli, P Pastorino and M Renzi 2021 *Chemosphere* **279** 3-4
[23] M S Pedroso, G L Pinho, S C Rodrigues and A Bianchini 2007 *Aquat Toxicol* **82(3)** 173-180
[24] J Farkas, V Cappadona, A J Olsen, B H Hansen, W Posch, T M Ciesielski, R Goodhead, D Wilflingseder, M Blatzer, D Altin, J Moger, A M Booth and B M Jenssen 2020 *Aquat Toxicol* **227** 105582
[25] C M Zhao and W X Wang 2012 *Nanotoxicology* **6(4)** 361-370
[26] F Gagné, J Auclair, M Fortier, A Bruneau, M Fournier, P Turcotte, M Pilote and C Gagnon 2013 *Journal of Toxicology and Environmental Health, Part A* **76(13)** 767-777
[27] N Yan and W X Wang 2021 *Environ Sci Technol* **55(8)** 5143–5151

[28] C M Zhao and W X Wang 2010 *Environmental Science & Technology* **44(19)** 7699-7704
[29] Y T Ng, C Amiard-Triquet, P S Rainbow, J C Amiard and W X Wang 2005 *Marine Ecology Progress Series* **299** 179-191
[30] B Xiao, R Yang, P Chen, J Yang, B Sun, K Wang, T Zhang and L Zhu 2022 *Sci Total Environ* **804** 150228
[31] X Zhu, J Wang, X Zhang, Y Chang and Y Chen 2010 *Chemosphere* **79(9)** 928-933
[32] Y Y Liu, W B Guo, Y T Zhao, S Xu, L Y Yang and A J Miao 2019 *Environ Sci Technol* **53(13)** 7803-7811

Reaction of Japanese media to discharging Fukushima nuclear waste water into ocean and the analysis----taking relevant reports of Japanese media as examples

Guangzhen Li [1] and Zilong Li [2,3]

[1] Japanese Department, School of Foreign Languages, Shandong Normal University, Jinan, Shandong 250014, China;
[2] Jinan Pediatric Research Institute, Qilu Children's Hospital, Cheeloo College of Medicine, Shandong University, Jinan, Shandong 250022, China.

[3] Corresponding author's e-mail: shxlzl@163.com

Abstract. Japanese government published the decision that the stored nuclear waste water in Fukushima Daiichi Nuclear Power Plant would be discharged into ocean, which met with violent objection from surrounding countries. Japanese domestic media reported in succession, and expressed a strong disagreement. And Fukushima local media were full of worries. They believed that local life and fishery production would be influenced for long. Being made a general observation of Japanese media, million tons of treated nuclear waste water discharged into ocean would be quite harmful to the environment. Japanese domestic attitude presented a distinct objection to the decision about discharging nuclear waste water into ocean made by Japanese government.

1. Introduction

Fukushima Daiichi Nuclear Power Plant (FDNPP), the first nuclear power plant of Tokyo Electric Power Company (TEPCO), is located at Okumacho and Futaba-machi, Futaba county, Fukushima industrial district. This plant was proposed in 1960 with 6 boiling-water reactor units. After being constructed for years, the first unit, namely the first nuclear power plant at present, operated in 1971. On March 11, 2011, The 2011 Earthquake of the Pacific coast of Tōhoku, whose earthquake intensity was Mw 9.0, occurred in the eastern Pacific Ocean to the ease of Miyagi Prefecture. The earthquake triggered huge tsunami, which destroyed Miyagi Prefecture, Iwate Prefecture and Fukushima Prefecture in northeast coast severely, where the FDNPP was damaged seriously. According to the website of Fukushima Prefecture Government, primary damage included that 1-5 units lost power because of water flooding power supply and battery, radioactive substances entered into atmosphere because of the fuel meltdown in 1-3 units, hydrogen explosion of unit 1, 3 and 4 resulted in reactor building damage, as well as spent fuel pools of unit 1-6 lost cooling function [1]. Namely, leading to nuclear power plant leak, seismic sea wave ruined unit 1-6 permanently. Moreover, evacuation area around FDNPP reached to 371 km^2. With a mass of radioactive substances getting into atmosphere and ocean, the nuclear leakage accident influenced half people of Fukushima Prefecture, which was as many as 120 thousand people. International Atomic Energy Agency (IAEA) payed great attention to Fukushima nuclear leak in Japan, and definitely indicated that the situation of FDNPP was "very serious".TEPCO and Japanese government took many measures to rescue and improve FDNPP damaged in earthquake as far as possible, but it brought little effect as severe damage. Especially the water originally used to cool molten nuclear fuel in units, mixing rainwater and underground water flowed into damaged reactor, and formed a mass of waste water. The nuclear contaminated water was stored in storage tanks provisionally. Because of the limited capacity, storage tanks would be full soon. It became a hard nut to crack to treat the water.

On April 13, 2021, decided by Japanese Cabinet Meeting officially, nuclear waste water (NWW), treated water called by Japan, which was stored in FDNPP, TEPCO, was going to be discharged into ocean by constructing harbor tunnel. TEPCO planned to discharge treated NWW to the short sea 1 km away from coast from the spring of 2023, when million tons of treated NWW would flow into Pacific Ocean. The NWW has been treated subtly, but the discharge would influence ecological environment severely because of the special nature. As long as the news presented, international community payed close attention intensively, and bitter against came from surrounding countries including China and South Korea. Meanwhile, this issue resulted in a great disturbance in Japan domestically. As well as, Japanese media published a great deal of correlative reports, and followed persistently. It is highly conducive to figure out the Japanese domestic attitude towards to the issue that sorting out and analyzing the relevant reports from Japanese media. As the first one studying Japanese domestic reaction to the issue, we introduced and analyzed the Japanese domestic opposing attitude to the decision of government clearly.

2. Methods

Data in this paper was based on resources from Japanese media and government public websites: Asahi Shimbun, Yomiuri Shimbun, Mainichi Shimbun, Tokyo Shimbun, Fukushima Minyou, Fukushima min Pao, the website of Fukushima Prefecture Government, and Greenpeace Japan. The media and correlative websites were collected because of following reasons. Asahi Shimbun, Yomiuri Shimbun and Mainichi Shimbun are the three most general newspapers. Asahi Shimbun started publication in 1879, Yomiuri Shimbun started publication in 1874, and Mainichi Shimbun started publication in 1872. With more than 100 years history, these three newspapers have high reputation and influence in domestic and overseas. Three local newspapers were collected: Tokyo Shimbun, Fukushima Minyou and Fukushima min Pao. Tokyo Shimbun is a local newspaper of Tokyo, the capital city of Japan. While Fukushima Minyou and Fukushima min Pao are local newspapers of Fukushima, where the accident occurred. The three newspapers are local, but they represent local attitudes to the issue. Additionally, reports involved in the issue from Fukushima Prefecture Government and Greenpeace Japan also were collected to be data of this study.

Data in this study were collected from relevant reports mainly from April 13, 2021, when Japanese government published the decision that discharging FDNPP NWW into ocean. Also, a few related reports before April 13, 2021, were collected. Because we analyzed Japanese domestic attitudes to discharging FDNPP NWW into ocean via Japanese media reports to the issue, research data from Japanese professional research institutes or TEPCO were not collected, such as the results change of environment monitoring of Fukushima coast. Most data and reports were from media, which had no signature.

Data were combed and analyzed qualitatively. And as needed, data were analyzed by text analysis methods in Discussion. Qualitative analysis was engaged in analyzing evaluation language in interrelated reports of three major Japanese media and local media, and restoring Japanese media's attitude and reaction to this issue objectively. And text analysis methods analyzed report highlights and contents in detail, which restoring reaction, opinion and evaluation of Japanese domestic public opinion and Japanese media's worry about environmental disruption objectively. Supporting each other with Japanese media, the analysis about correlative information published by the website of Fukushima Prefecture Government and Greenpeace Japan proofed the reason why Japanese public opinion opposed discharging NWW into ocean.

3. Results

We explored reaction and attitude of Japanese public opinion and common people of Japan to this issue through analyzing relevant Japanese media reports. Rather than the environmental harm research from discharging NWW into ocean.

3.1. Japanese media payed plenty of attentions to the issue of handling FDNPP

From April 13, 2021, when Japanese government published the decision that the NWW produced in FDNPP would be discharged into ocean, there are few papers about "FDNPP NWW discharge" from Japanese academic world. However, Japanese major media have been presenting a large number of reports all the time, and paying plenty of attentions to the issue.

Since the accident happened, Mainichi Shimbun, Asahi Shimbun and Yomiuri Shimbun reported it frequently. The average reports number per month of Mainichi Shimbun was 91.06, which was almost 3 articles per day (Table 1). And Asahi Shimbun was more than 2 articles per day (Table 1). Although Tokyo Shimbun is a local media, as the mouthpiece of capital Tokyo, it possesses higher impact than other local media. It showed the highest frequency among the six media, which was 92.92 articles per month (Table 1). For example, from December 23, 2016 to December 24, 2021, Asahi Shimbun published 6052 articles related to FDNPP, which was 100.87 articles per month. There were 4260 reports taking "FDNPP accident" as keywords. In addition, 5, 250 articles from Asahi Shimbun, 98 articles per month averagely, were published before April 13, 2021, when Japanese government announced the plan discharging NWW into ocean. Afterwards, 802 articles, 140 articles per month averagely, were published until December 24, 2021. On the basis of the dramatic rise of relevant articles, Asahi Shimbun paid more attention to this issue obviously. So did other media. Japanese major media payed plenty of attentions to the issue.

Table 1. Data statistics of related reports collected by keywords "FDNPP accident", "Fukushima treated water", "contaminated water" and "NWW".

Date	Newspaper	FDNPP accident	Fukushima treated water	Contaminated water	NWW	Average reports number per month
2011.03-2021.12.24	Mainichi Shimbun	12020	1110	1101	3	91.06
2016.12-2021.12.24	Asahi Shimbun	4260	306	666	1	70.72
2019.09-2021.12.24	Yomiuri Shimbun	413	685	61	3	41.5
2017.05-2021.12.24	Tokyo Shimbun	3580	683	922	18	92.91
2019.01-2021.12.24	Fukushima Minyou	22	123	118	0	10.52
2020.01-2021.12.24	Fukushima min Pao	360	56	15	0	33.15

3.2. Local media presented high concern to the issue

Japan is long and narrow from north to south, which is constituted by 4 islands, Hokkaido, Kyushu, Honshu and Shikoku. The biggest island is Honshu island, and Fukushima is a south prefecture located to the northeast of Honshu island, which faces the Pacific Ocean. 200 km away from capital Tokyo, Fukushima is the third largest county of Japan with the area of 13, 782.75 km^2. Moreover, there are abundant natural beauty, fish stock and famous sakes in Fukushima county. But, earthquake and FDNPP accident casted a shadow over the natural beauty. Local media, Fukushima Minyou and Fukushima min Pao, payed plenty of attentions to NWW treatment all the time. From January 2020, Fukushima min Pao published 431 reports taking "FDNPP accident" and "NWW" as keywords, which was 33.15 articles per month and 1 article per day. This indicated the concern form the local media.

However, there was an interesting phenomenon during data research. Reports about the issue from Japanese media showed distinct regionalism. Media from Tokyo and other places near FDNPP and local media of Fukushima responded intensively with all kinds of contents reported. Nevertheless, the further the media were away from the central region, the less correlative report was published. Taking southern media such as Saga Shimbun and Kumamoto Shimbun located in southern Kyushu for example, correlative reports reduced obviously, of which only about 10 reports emerged after being searched.

3.3. Japanese media strongly opposed discharging NWW into ocean

By means of text analysis methods, we explored contents of relevant articles from 6 media including Asahi Shimbun. And it's found that all media strongly opposed discharging NWW into ocean, and highly worried the succedent contamination and environmental disruption from NWW in ocean.

On April 13, 2021, when Japanese government declared the decision, Asahi Shimbun published a round-up concerning the response of all walks of life in China, whose title was "Discharge into ocean. 'Listen to fisherfolks.' Opposition of Fukushima" to query the decision directly. Thereafter, Asahi Shimbun started to follow this issue roundly. On October 16, 2020, an article about that the government discussed discharging treated NWW into ocean from Asahi Shimbun reported: "In regard to treatment methods, in February 2021, a sub-committee from Ministry of Economy, Trade and Industry composed by experts expressed that discharging NWW into ocean or atmosphere was a 'biable option', and that discharging into ocean would 'carry out reliably'. On the other hand, fisherfolks worrying damage were strongly against discharging NWW into ocean. In June, 2021, National Federation of Fisheries Cooperatives decided to "object to incomprehensible discharging stoutly". And on December 15, 2021, the federation president met the Minister of Economy, Trade and Industry of Japan, and showed a violent objection. [2]"

On August 29, 2021, Tokyo Shimbun published an article with the title "'No common people', 'despicable' government and TEPCO 'give up responsibility'. Disaster area rose to stop FDNPP NWW to be discharged into ocean", which minutely reported real reaction of common people of Japan especially Fukushima local people [3]. The article pointed out: "On April 2021, when it's the 10th anniversary of The 2011 Earthquake of the Pacific coast of Tōhoku, the government of Naoto Kan formulated the policy discharging purifying treated NWW in FDNPP into ocean 2 years later. Regardless of the local objection for triggering possible new loss of public opinion, government made a judgment, which violated past writ 'appoint' signed among government, TEPCO and fisherfolk. On the other hand, because of the impact of COVID-19, intercourse was conditioned, and via internet citizens carried out a activity of collecting objection signatures." Moreover, the article pointed out: "Government expressed that it had no deleterious effect to human body and environment even though treated NWW containing radioactive tritium was discharged into ocean. However, China and South Korea listed damages from treated NWW to environment, and criticized repeatedly. In addition, in congressional reply, Prime Minister emphasized about treated NWW discharged into ocean: 'We will keep punching to make everyone understand that safety of treated NWW is dependable.' However, it's really hard to say that it's been conveyed to the abroad adequately." At last, the article expressed pessimistically: "finishing touches of the most serious nuclear plant accident all over the world are too hard to be completed."

3.4. Japanese media reported opposing attitude of surrounding countries objectively

Greenpeace Japan believed that NWW discharged into ocean contained not only tritium but also carbon-14. Because the half-life period of carbon-14 is 5730 years, general Japanese domestic public opinion believed that discharging NWW into ocean would cause infinite detriment to ocean, and that government should find another way.

Here is the text description in regard to the processing procedure of this issue on the website of Fukushima Prefecture government. Scraping furnace means disassembling and dismantling atomic reactors. Procedure: Removing nuclear fuel→ Removing burn-off fuel inside reactors→ Disassembling. Time: 30- 40 years needed [1].

According to reports of Japanese local government, it would cost at least several decades to treat and digest contaminated water generated by FDNPP damage. While Greenpeace Japan forecasted that it would take 5730 years to digest carbon-14. This would destroy ocean and environment of surrounding countries. Hence, surrounding countries including China and South Korea expressed intense censures and rages.

Japanese media objectively reported opposing attitude of surrounding countries such as China and South Korea to discharging NWW into ocean.

4. Discussions

With respect to the contaminated water after FDNPP damage in the reports of Japanese media, it was "NWW" at the beginning. After Japanese government deciding to discharge "NWW" into ocean, "treated water" began to be engaged. "Treated water" is the wording that Japan calls the nuclear polluted water which has been treated. The water that Japanese government decided to discharge into ocean is the "treated water", which is "contaminated water" generally. Actually, Japanese media also employed the word "contaminated water". The so-called "contaminated water" means the water mixed with radioactive substances on account of unit equipment damage. Previously, there were "groundwater" flowing into ocean under FDNPP. The website of Fukushima Prefecture government introduces: "The 'groundwater' flowing into accident reactor buildings (unit 1-4) is 100-200 tons per day. There is high concentration contaminated water under reactor buildings where the 'groundwater' flows through, which results in more and more contaminated water. [1]". In order to stop the contaminated water flowing into ocean, previously Japanese and local governments took a variety of measures, such as removing pollutant sources, keeping water away from pollutant sources, preventing contaminated water leaking and so on. Simultaneously, to confirm the influence from radioactive substances discharged by FDNPP to environment, TEPCO implemented marine monitoring continually. Likewise, Japanese media pay great attention to it, because the water problem is crucial.

4.1. Japanese media were fairly anxious about discharging NWW into ocean

Up to December 24, 2021, there were 681 related reports, whose keywords were "FDNPP" and "treated water" in Yomiuri Shimbun, another large newspaper in Japan. Most reports covered the local life suffered nuclear radiation.

The earliest report from Yomiuri Shimbun was the image video about doctors and patients of Fukushima Futaba Health Care Hospital transferring on March 14, 2021. Words under the video wrote: "Patients and doctors, who were from Fukushima Futaba Health Care Hospital perhaps suffering nuclear radiation in FDNPP accident, were sent to Nihonmatsu City, Fukushima Prefecture by a self-defense force helicopter, and proceeded works including measuring radioactive concentration and so on. [4]" While the latest news was the one on December 21, 2021, whose title was "'Treated water' of FDNPP. TEPCO applied for marine discharge scheme which would begin from the spring of 2023". At the beginning, this report pointed out that: "TEPCO applied to Nuclear Regulation Authority (NRA) for an implementary scheme, which was about discharging 'treated water' continually stored inside FDNPP into ocean, and carried out through subsea tunnels from the spring of 2023." Meanwhile, the news explained the 'treated water': "The 'treated water' means the waste water generated in nuclear power plant, which is purified with few radioactive substances left such as tritium. When discharging through subsea tunnels built in future 1 km away from FDNPP, TEPCO will dilute 'treated water' with abundant seawater, in which the tritium concentration will much lower than the drinking water standard of WHO. [5]". And 3 days later on December 24, 2021, correlative news reported that rice produced in Fukushima were exported to USA again: "Rice produced in Fukushima were exported to USA for the first time after the earthquake. And measures stopping importing Japanese foodstuffs for FDNPP accident were abolished. [6]"

On October 21, 2020, Tokyo Shimbun published an editorial, whose topic was "Radioactive contaminated water. Are safety precautions perfect?" At the very start, the editorial wrote: "The radioactive contaminated water from FDNPP of TEPCO will be discharged into ocean by government for the reason that 'too much to store indeed'. This is not an eligible justification to response safety and negative evaluation." At present, thousands of storage tanks stand at the scene of FDNPP, which are full of radioactive contaminated water generated here, and as much as 12.3 million m^3. Furthermore, after analyzing the volume of contaminated water, the editorial pointed out: "In regard to that radioactive contaminated water are discharged into ocean, the law only stipulates the emission concentration rather than the quantity. And Japanese government believes the thought: 'It's OK if it diffused or was diluted.' [7]". Actually, it is well known that the radiation of treated tritium is weak but not zero. Therefore, this article took a dim view of consequences of contaminated water emission: "Treated water from fusional nuclear power plant flows into ocean chronically, whose subsequent is still unclear. [7]". Moreover, the editorial analyzed with the example of minamata disease resulted from environmental pollution in last centry: "The era is different. But wasn't the minamata disease the consequence of discharging chemical plant wastewater containing mercury into ocean unceasingly because of the inordinate confidence of seawater dilution ability?" In addition, the article persented profound understanding to local fisherfolks fearing the issue, also worried that "chain reactions of other surrounding countries" would be generated. Hence, the editorial pointed out decidedly: "In accord with nuclear waste, it is necessary to treat the contaminated water." Otherwise, "An apple of discord would be generated. [7]" Also, there were other associative reports of this newspaper such as "4000 containers with unknown content. TEPCO neglected FDNPP", "Radioactive waste in FDNPP showed high risk of leakage. Storage equipment degraded critically", "In the 'absolutely opposed' voice of fishery, policy discharging NWW into ocean decided by government will start from 2023", and so on.

Other media such as BBC, Japan Broadcasting Corporation (NHK) reported the issue repeatedly as well. For example, with the decision publish of Japanese government on April 13, 2021, BBC rapidly published an article with the title "The government decided to discharge NWW of FDNPP into ocean" [8]. On April 16, 2021, the media published an article once again to comment the issue, whose topic was: "Japanese government discharges treated water of FDNPP. Two theories of praise and disapproval for support and suspense".

4.2. Japanese media took an explicit stand against the issue

The frozen soil wall operated from 2016 is one of the most important facilities in FDNPP, which stops groundwater flowing into reactor buildings and restrains NWW generation. Surrounding the buildings, the frozen soil wall is 1.5 km long, where 1568 freezing pipes are buried 30 m underground full of cryogen at -30 ℃ for freezing ambient soil. The facility cost as high as 34.5 billion yen with maintenance charges about hundreds of millions of yen per year. Unfortunately, the frozen soil wall went wrong recently. Published on December 24, 2021, the latest article from Asahi Shimbun was the comment to the frozen soil wall accident, whose title was "The monitoring system of NWW pool in prefecture has problems, and problems of frozen soil wall emerge. [9]" The article reported that: "Stopping groundwater flowing into FDNPP and restrained NWW generation, the frozen soil wall was found to partly thaw in October possibly. However, the truth and reason are still unclear. Experts pointed out that:' The local people feel anxious and restless, and the government should monitor closely and at all times.' With the problem emerging on October 28, at a meeting TEPCO declared that the temperature of a frozen soil wall in the southwest side of unit 4 of FDNPP rose again from the end of August. On October 13, some areas were confirmed with temperature higher than 0 ℃, when it should be lower than -5 ℃. [9]" Asahi Shimbun told Japanese people the truth by news reports, drew people's attention, and mediately criticized the work of TEPCO harshly.

As early as December 12, 2016 when 3 reports were published, with the title of "Unit 3 of Fukushima Daiichi Nuclear Power Plant delays fuel emission again", the first article pointed out that: "On 22nd, the government and TEPCO declared that it would be delayed again that cleaning up residual 566 kinds of fuels in used fuel pool of unit 3 of FDNPP. The radiation dose on building roofs didn't decrease, and the preparatory work such as clearing debris was delayed. So, the aim to start a new project from January 2018 is abandoned. [10]" The new project refers to formerly cleaning away disused fuels starting from January 1, 2018, which was postponed. This article was the first report from Asahi Shimbun in regard to the NWW issue. On the same day, another report focused on the issue with respect to that outward children taking refuge were discriminated and bullied because of FDNPP leakage, whose topic was "'The reason is lack of understanding', the declaration of suffered bullying resulted from nuclear accident". At the beginning, the report mentioned that: "Being bullied examples of children evacuated for FDNPP accident emerged in succession. National Liaison Group of Plaintiffs for Victims of Nuclear Power Accident (NLGPVNPA) is going to prosecute the government and TEPCO for compensating for the loss in nuclear power accident, and issues a statement that bullying is a severe problem leading to self-death, which is resulted from the short of understanding about victims' status. [11]"

Being made a general observation of Japanese media such as Asahi Shimbun, they expressed distinct attitude, and took an explicit stand against the issue.

4.3. Local media reacted violently

Fukushima Minyou is the local media of Fukushima. Initially, in regard to the contaminated water after FDNPP damage, this media used "NWW", which was replaced by "treated water" afterwards. Searched by "NWW treatment" as the core keywords, the media published about 123 interrelated reports, among which the earliest was on March 9, 2018 with the topic as "Discarded furnace has foothold finally without solutions yet. And the biggest issue is how to treat NWW". This was the first report from the local newspaper about the issue: "To decommission FDNPP of TEPCO, investigation about reactor containments of unit 1-3 was conducted, and in the meantime, method about treating treated water containing radioactive tritium was still suspensive. The attitude of TEPCO and the government toward to how to solve the escalating water volume was queried. [12]".

The latest article of the media was published on December 24, 2021, whose topic was "It is determined to discharge into ocean 2 years later. Comprehending is on the way". The article wrote: "With respect to the issue about the treated water generated during the treating procedure of waste water in FDNPP of TEPCO, on April 13, the government decided to discharge it into ocean 2 years later. On December 21, TEPCO applied to Nuclear Regulation Authority for an investigation to the implementary scheme including discharging facilities development. The discharging scheme will be initiated before the spring of 2023. Time lapses ceaselessly, while comprehending to discharging waste water is still on the way. [13]". With regard to the attitude of local fisherfolks, the media pointed out: "Fisherfolks are fully anxious about compensation and rumors. Although TEPCO expressed: 'We will be understood increasingly', so far, the distrust about the countermeasure blocks between both sides." At the same time, the local media specifically demanded TEPCO to "take a sincere action as the party of nuclear accident" [13].

Fukushima min Pao also published a good deal of relevant articles, took an explicit stand against the issue, and expressed exceeding anxious.

4.4. Objective reports about opposite voices of surrounding countries

About the decision discharging contaminated NWW into ocean, Japanese and TEPCO explained several times, and vowed solemnly that the NWW had been treated and did no harm to environment. So, was it really as safe as Japanese government described that treated NWW of FDNPP was discharged into ocean?

The research report "Treated water discharging into ocean. Radioactive carbon causes damage to human DNA possibly—Greenpeace Japan report 'contaminated water crisis 2020 of FDNPP, TEPCO'" published by Greenpeace Japan, definitely warned: "If Japanese government proceeded as the plan, radioactive carbon (carbon-14) in contaminated water would enter the Pacific Ocean. With the half-life period as 5730 years, carbon-14 would exist in environment for thousand years. And as the basic constitutive element of creatures, carbon-14 would enter all living creatures. In the long run, this would be the principal factor of radiation dose. Hence, carbon-14 does harm to human DNA possibly. In regard to this query, Japanese government and TEPCO continued to explain that 1 million and 230 thousand tons of NWW stored in storage tanks had been 'treated completely' with only tritium left. [14]" Furthermore, this report pointed out: "It's been 10 years since the earthquake. TEPCO and Japanese government has been covering up the real crisis of FDNPP. They have already got the detailed information including radioactive substances in NWW for years, which they hided designedly. To Fukushima people and neighboring countries, Japanese government didn't explain the danger level of discharging NWW into the Pacific Ocean, as well as carbon-14. Together with other radioactive substances in NWW, carbon-14 would cause genetic damage, and be dangerous in thousand years. [14]"

Japanese media also reported reactions of surrounding countries such as China to Japanese government synchronously. On April 13, 2021, when Japanese government declared the decision, Chinese government protested it mightily. Asahi Shimbun published a round-up concerning the response of all walks of life in China, whose title was "Discharge into ocean. 'Listen to fisherfolks.' Opposition to Fukushima": "People's Daily (overseas edition), the newspaper of the Communist Party, pointed out this decision would 'threaten the common security and benefits of surrounding countries and international community'. Moreover, the newspaper criticized: 'Japanese government lacks responsible public spirit evidently.' At the same time, Chinese SNS hotspot also reported: 'The decision from Japanese government also raised concerns in China.' On the morning of 13th, the topic 'Japanese government decided to discharge waste water of FDNPP into ocean officially' took the top position of 'Weibo', the Chinese SNS. Internet users made numerous critical comments such as 'This is a selfish decision. Ocean does not only belong to Japan', and 'Marine products from Japan are no longer edible'", and so on [15].

The objective report of Japanese media about responses of all walks of life in China to Japanese action discharging NWW into ocean, transmitted the distinct opposing attitude to Japanese government.

4.5. Limitation of this study

It has been almost 11 years since March 2011, when the great earthquake occurred in Fukushima and that FDNPP got damaged caused subsequent problems. The issue that how to deal with the FDNPP damaged severely drew tremendous attention of academic world. There were 10753 papers taking "FDNPP accident" as keywords from National Diet Library of Japan. And there were 53 papers in 2021. The latest paper was "Contamination of radioactive substances in FDNPP—take Motomiya, Yomogida-dake and Koyasan in 2020 for examples" published on "Geography Education and Scientific Exercises" [16]. As well as, there were 2104 papers taking "Fukushima nuclear" as keywords. From April 13, 2021 when Japanese government published the decision that the NWW would be discharged into ocean, Japanese media reported in succession. But surprisingly, there were no paper taking "Fukushima NWW" or "Fukushima NWW discharge" as keywords from National Diet Library of Japan, which showed a striking contrast with abundant reports of Japanese media.

With respect to research in China, there were 81 papers taking "Fukushima NWW" as keywords from CNKI. While since April 13, 2021 when Japanese government published the decision that the NWW in FDNPP would be discharged into ocean, there were as many as 8 specialized papers. Among these papers, as many as 4 discussed how to solve the issue by law. The latest paper was "Legal issues study of decision of Japanese NWW discharged into ocean from the perspective of ocean dumping" published on "Future and Development" in December [17].

Making a general observation of academic world of China and Japan, there has never been study taking relevant reports of Japanese media as sample analysis. By analyzing the relevant reports from Japanese media, we figured out the Japanese domestic reactions and attitude towards to the issue, which endowed this study brand new significance. Nevertheless, there are limits of this study. Firstly, data collection needs improvement, because collection range is limited in media reports. Next, text analysis about report contents scratched the surface. In the future study, more data and report original text will be collected and analyzed in depth, in order to further objectively restore the opposing attitude of Japanese domestic public opinion to Japanese government's decision discharging NWW into ocean.

5. Conclusions

TEPCO and Japanese government took many measures to rescue and improve FDNPP damaged in earthquake as far as possible, but first of all, the disaster brought about huge damage to Fukushima local, and there were numerous difficulties blocking the way to prosper the disaster area: waste water leak in FDNPP was exposed frequently, over a hundred thousand victims couldn't go home, reconstruction lacked labor forces resulted from the external migration of the prefectural young people, it's difficult to sell local agricultural, sideline and aquatic products, and so on. This would recover and be solved after several decades or longer time possibly. In the next place, although the nuclear contaminated water was treated, residual radioactivity including not only tritium but also carbon-14 still existed. If discharged into the Pacific Ocean, it would lay profound hidden danger about environmental pollution to not only local fisherfolks, but also surrounding countries and international community. Japanese media strongly opposed discharging NWW into ocean, and highly worried the succedent contamination and environmental disruption from NWW in ocean. Judging from reactions and attitudes of Japanese media, the Japanese people took a quite opposing attitude to the issue.

Acknowledgement
This study was funded by the Highland for Innovative Development of Vocational Education built by Ministry and Province (Grant NO.: GD65), and the City-School integration development Strategy Project of Jinan City in 2021 (JNSX2021026).

Reference

[1] *Fukushima Prefecture government* https://www.pref.fukushima.lg.jp/site/portal/
[2] 2020-10-16 The government adjusts treated water in Fukushima Nuclear Power Plant discharging into ocean *Asahi Shimbun* https://www.asahi.com/articles/ASNBJ3PW9NBJULBJ003.html
[3] Ichikawa C, Nakane M 2021-8-29 'No common people', 'despicable' government and TEPCO 'give up responsibility'. Disaster area rose to stop FDNPP NWW to be discharged into ocean *Tokyo Shimbun*
[4] Otake Toyoda 2011-3-14 Video from Photo Department of Tokyo headquarters filmed by Otake Toyoda *Yomiuri Shimbun* https://www.yomiuri.co.jp/stream/article/02170/
[5] 2021-12-21 Treated water' of FDNPP. TEPCO applied for marine discharge scheme which would begin from the spring of 2023 *Yomiuri Shimbun* https://www.yomiuri.co.jp/national/20211224-OYT1T50206/
[6] 2021-12-21 Rice produced in Fukushima were exported to USA for the first time after the earthquake. And measures stopping importing Japanese foodstuffs for FDNPP accident were abolished *Yomiuri Shimbun* https://www.yomiuri.co.jp/national/20211224-OYT1T50206/
[7] 2020-10-21 Editorial： Radioactive contaminated water. Are safety precautions perfect *Tokyo Shimbun* https://www.tokyo-np.co.jp/article/631690
[8] 2021-4-13 The government decided to discharge nuclear waste water of FDNPP into ocean *BBCNEWSJAPAN* https://www.bbc.com/japanese/56728227
[9] Fukuchi K, Kasai T 2021-12-24 The monitoring system of NWW pool in prefecture has problems, and problems of frozen soil wall emerge *Asahi Shimbun*
[10] 2016-12-23 Unit 3 of Fukushima Daiichi Nuclear Power Plant delays fuel emission again *Asahi Shimbun*
[11] 2016-12-23 The reason is lack of understanding, the declaration of suffered bullying resulted from nuclear accident *Asahi Shimbun*
[12] 2018-3-9 Discarded furnace has foothold finally without solutions yet. And the biggest issue is how to treat nuclear waste water *Fukushima Minyou Shimbun* https://www.minyu-net.com/news/sinsai/serial/07/06/FM20180309-250995.php
[13] 2021-12-23 It is determined to discharge into ocean 2 years later. Comprehending is on the way *Fukushima Minyou Shimbun* https://www.minyu-net.com/news/news10/FM20211223-675143.php

IOP Conf. Series: Earth and Environmental Science 1087 (2022) 012004 doi:10.1088/1755-1315/1087/1/012004

[14] Ban H, Ohuma J 2020-10-23 Treated water discharging into ocean. Radioactive carbon causes damage to human DNA possibly—Greenpeace Japan report 'contaminated water crisis 2020 of FDNPP, TEPCO'

[15] *Asahi Shimbun* https://sitesearch.asahi.com/sitesearch/?Keywords=%E6%A0%B8%E5%BB%83%E6%B0%B4&Searchsubmit2=%E6%A4%9C%E7%B4%A2&Searchsubmit=%E6%A4%9C%E7%B4%A2&iref=pc_gnavi

[16] Chiba S 2021 Contamination of radioactive substances in FDNPP—take Motomiya, Yomogida-dake and Koyasan in 2020 for examples *Geography Education and Scientific Exercises* **87** 9

[17] Cheng G, Shun Y 2021 Legal issues study of decision of Japanese NWW discharged into ocean from the perspective of ocean dumping *Future and Development* **45** 12

Comparison and study of several methods for solving the hydraulic jump values in pumping well

Xuezhen Zhang[1,2], Aidi Huo[1] and Jucui Wang[1]

[1] School of Water and Environmental, Chang'an University, Xi'an, 710054, China.

[2] Corresponding author's e-mail: zxz-2000@126.com

Abstract. This paper was envisaged to provide deeper insights of the hydraulic jump value in pumping well in the loess aquifer. Herein, we discussed the research status and problems of the value, compared the differences between the analytical solution method and the empirical formula method. Subsequently, we proposed a fitting method with improvement and the tangent method to estimate the hydraulic jump value, and quantitatively analyzed the linear variation rule of the water level curve around the well. The results show that the value calculated by the traditional empirical formula was about one tenth of the theoretical value in constant flow and the water level around the well was found to be logarithmic in the range of N_1~N_3. When the range was extended, the water level curve changed to a parabolic nature and power function in turn. The hydraulic jump value obtained by *h-r* extension of the modified water level curve fitting equation was improved greatly, and the average error was found to be less than 1 m. The stability of the hydraulic jump value using the tangent method was poor, and the estimated horizontal distance was larger than 3 m. The results show that the reliability of the improved curve fitting method was better, while the empirical formula method and the tangent method exhibited a larger error and poor stability. When the error correction was performed, the improved curve fitting method could be used to estimate the hydraulic jump value under the same conditions, and can replace the theoretical calculation value.

1. Introduction

The existence of the hydraulic jump value plays an important role in the hydrogeological parameter estimation and groundwater resource evaluation. Studies on the hydraulic jump value were mainly concentrated in the 1960's to 1980's, and a relatively few were reported in the last 20 years [1, 2]. Some professinals explored the mechanism of the values, but it is difficult to obtain common research conclusions [3, 4, 5]. Another proup of professionals simplified the hydrogeologic conditions, often ignored the existence of the value in the study of hydrogeologic models [6, 7]. In some studies the computing software was used to fit and to estimate the hydraulic jump value through graphic method and cyclic operation; however, the mechanism of the occurrence of the hydraulic jump value was not elucidated [8, 9]. In general, due to the complexity of the influencing factors and the difficulty of research, there are more qualitative and less quantitative studies on the hydraulic jump values, which make it difficult for the researchers to reach a common correct consensus, and there are few representative case studies.

In our earlier studies, we have reported a preliminary exploration on the estimation of hydraulic jump value [10]. In the existing research experience, a few methods were put forward to calculate the characteristics of the water jump. Under certain conditions, the value exhibits a linear relationship with the water output, but this linear relationship does not always exist. Although the proposed curve fitting method has its rationality, the error was slightly larger.

In practical applications, it is difficult to observe the water level of pumping wells under in-situ conditions, while the layout of observation wells is relatively easy. Use of the water level curve fitting of observation wells to intercept the hydraulic jump value can replace the theoretical calculation and simplify the tedious calculation process. The main purpose of this study was to find a practical method to improve the estimation of the value by improving the fitting method and adjusting the direction of curve cutting, based on the existing research experience

2. Site location

The test site was located in the irrigation area about 10 km north of the Weinan city, China. The pumping well was a general agricultural pumping well. The aquifer belonged to the submersible type with a thickness of H=35 m. The hydraulic gradient of the groundwater was less than 0.5%, which was similar to the recognized level. Based on the borehole data available, the aquifer may be considered homogeneous and isotropic. The water table in the initial state was close to zero. The layout of the pumping test site is shown in the Figure 1.

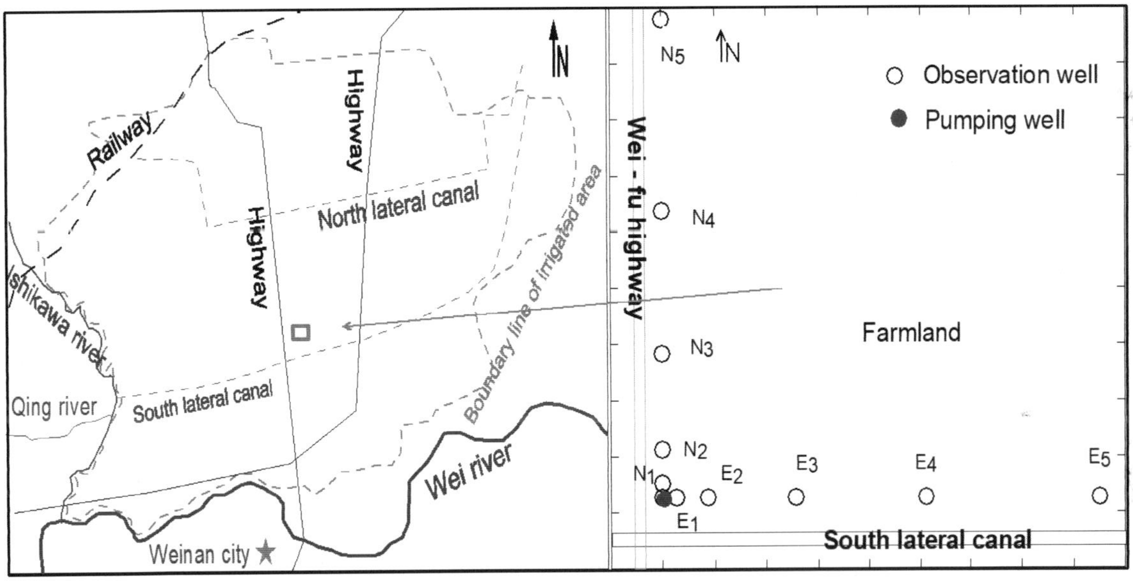

Figure 1. Schematic diagram of the pumping well.

In this study, the fixed flow pumping process was devised in 4 stages considering the pumping test design. The time interval of each stage was about 12 hours, each pumping process was relatively independent. The relevant data selected for the pumping test is shown in Table 1. From the data it was found that the effective water area of the filter pipe in the pumping well F=54.978 m^2.

3. Methods

From the past research experience, the estimate of the hydraulic jump was based on the analytical solution. The empirical formula method, the fitting method and the tangent method were selected to calculate the value, so as to select a better and convenient method of estimation of the hydraulic jump value, which is a supplement to the analytical method.

Table 1. Data sheet of the partial pumping test.

Test phaset_i	Water flow from the main well Q_i (m³/h)	Measured water level drop of main well S (m)	Permeability coefficient K (m/d)	Observation well number	Horizontal distance from observation well to main well	Depth of each observation well was lowered when steady flow pumping (m)			
t_1	16.74	4.300	2.953	N_1	3	1.490	2.227	2.08	1.975
t_2	28.98	5.740	3.916	N_2	10	0.961	1.599	1.381	1.311
t_3	33.77	8.820	3.122	N_3	30	0.635	1.030	0.900	0.845
t_4	43.92	13.860	2.810	N_4	60	0.355	0.435	0.525	0.495
				N_5	100	0.222	0.215	0.305	0.280

3.1. Analytical method for calculating hydraulic jump value

In practice, we can use the analytical solution of groundwater steady flow pumping theory that is based on the Joupuis equation, and deduce the calculationEquation (1) of a single well with one observation well. Theoretically, Equation (1) can be used to effectively calculat the value.

$$\Delta h = S_w - S = S_w - H + \sqrt{H^2 - \left(2HS_1 + {S_1}^2 - \frac{0.732Q}{k}\right)\lg\frac{r_1}{r}} \tag{1}$$

where, k is the permeability coefficient, Q is the flow rate, S is the theoretical depth of the pumping well, H is the aquifer's thickness, r is the well's radius, r_1 is the horizontal distance to the pumping well, S_1 is the depth drop value of the observation hole N_1, and $\triangle H$ is the hydraulic jump value.

However, in many cases, due to different well conditions, the pumping well equipment often limits the measurement of pumping well water level, and it is difficult to obtain the measured pumping well water level drop S_w. In order to facilitate this issue, a special method was adopted to observe the water level drop of the pumping well.

3.2. Empirical formula method for calculation of hydraulic jump value

The theoretical and experimental research on the hydraulic jump value is scanty. The empirical formula of the hydraulic jump value widely recognized in China is the Abramov equation proposed by the former Soviet Union scholars (Wang Jiansheng, 1985). Abramov equation was obtained under a borehole pumping condition, and the calculated hydraulic jump value was larger than the actual observation based on the laboratory conditions. The effect of the calculation under the *in-situ* condition is not mentioned.

Abramoff equation, as a promising approach to calculate the value, can be expressed as:

$$\Delta h = 0.01a\sqrt{\frac{QS}{KF}} \tag{2}$$

where, Δh is the hydraulic jump value (m), S is the water level drop depth in the pumping well (m), Q is the water inflow of borehole (m³/d), K is the permeability coefficient (m/d), F is the effective area (m²) of the filter pipe, and $F=2\pi r_0 h_0$, where R0 and h0 are the radius and effective height of the filter pipe respectively. a is the empirical coefficient of filter pipe structure (a=20 for mesh and gravel filters). The values of K and a are not discussed.

3.3. Improved curve fitting method

It is easy to obtain the hydraulic jump value calculated by Equation (1). In actual field conditions, the water level around pumping wells exhibits an unstable flow state, and there is an error that cannot be ignored by applying the analytical method [11, 12, 13]. In previous studies, the fitting method was adopted to screen out the water level curve, by taking advantage of the quick advantage of computing software. By extending to the well wall, the water level could be estimated, thus indirectly confirming the jump value, supplementing and modifying the analytical solution [10]. However, the available data for fitting the water level curve around the well is limited, the fitting curve equation is not representative enough, the position curve near the pumping well changes dramatically, and the extension direction is difficult to control.

In this study, the fitting method with improvement and the difference method were used to identify the *r-h* type of the water level curve around the well. Subsequently, based on the observation data close to the main well, the fitting curve was debuggled using the Grapher software and the fitting equation with high correlation coefficient was screened. In this way, the lack of fitting points can be moderately resolved.

3.3.1. Determination of the type of the curve method and difference method. The water level curve fitted with the water level data of limited observation wells around the well can be regarded as a curve arc. It can be observed that the curve of the water level around the well varies greatly, and the fitted curve arc may change with an increase in the horizontal distance around the well. From prior research experience, the curve method was adopted to distinguish the curve type [14, 15]. Specifically, two points were selected from the fitting curve, and the observation data near the observation wells N_1 and N_2 were preferentially selected to calculate the curve value. The calculation formula is as follows:

$$n = \frac{\lg r_2 - \lg r_1}{\lg h_2 - \lg h_1} \tag{3}$$

where, h is the observed well water level depth drop (m), r is the horizontal distance to the pumping well (m). If n<1, n=1, 1<n<2, n=2, n>2, the corresponding *r-h* curve types can be preliminarily determined as data error, straight line, power curve, parabola and logarithmic curve.

According to the existing research experience (Chen Qiaohong, 2019), the *r-h* curve type of pumping test was preliminarily determined by calculating the curve value of (3). In order to improve the accuracy, the difference method should be further used to determine the type of curve. Curve fitting error c calculated by difference method can be determined by the Equation (4).

$$c = \frac{\Delta_2 - \Delta_1}{(\Delta_2 + \Delta_1)/2} \tag{4}$$

where, $\triangle_1$ and $\triangle_2$ are the first order difference quotient, and are marked as the order of difference quotient. It is generally believed that the smaller the fitting error is, the more suitable is the curve type. If it is a logarithmic curve, the first-order difference quotient can be calculated as follows:

$$\Delta_1 = \frac{r_2 - r_1}{\lg h_2 - \lg h_1}, \quad \Delta_2 = \frac{r_3 - r_2}{\lg h_3 - \lg h_2} \tag{5}$$

3.3.2. Curve fitting method was used to intercept the hydraulic jump value. In previous studies, the type of water level curve around the main well was preliminarily identified. Then, the fitting curve of the identified type was debutted by using the Grapher software, and the fitting equation of the water level curve around the well h-r was screened out, which was extended to the position of the pumping shaft wall to compare the measured depth drop of the main well, so as to estimate the hydraulic jump value.

Considering that the curvature of the water level curve around the well varies greatly, this study believed that the closer the well is to the main well, the greater the curvature of the water level curve around the well. Therefore, in pursuit of the accuracy of curve fitting, the location data of observation wells near the main well need to be selected first, and the properties of data lines should be the same.

After preliminary debugging, the line nature of water level data fitting curve at N_1, N_2 and N_3 was logarithmic, while the line nature of water level data fitting curve at N_4 and N_5 was slightly distant is close to parabolic, which is not included in this fitting debugging. This result is different from the previous studies [10]. The curve fitting process is shown in Figure 2.

3.4. Tangent method

This method is an extension of Section 3.3, that is, the tangent line of the curve close to the main well is used to replace the curve arc, and the tangent line is extended to the shaft wall to estimate the hydraulic jump value. The tangent lines are shown in Figure 2.

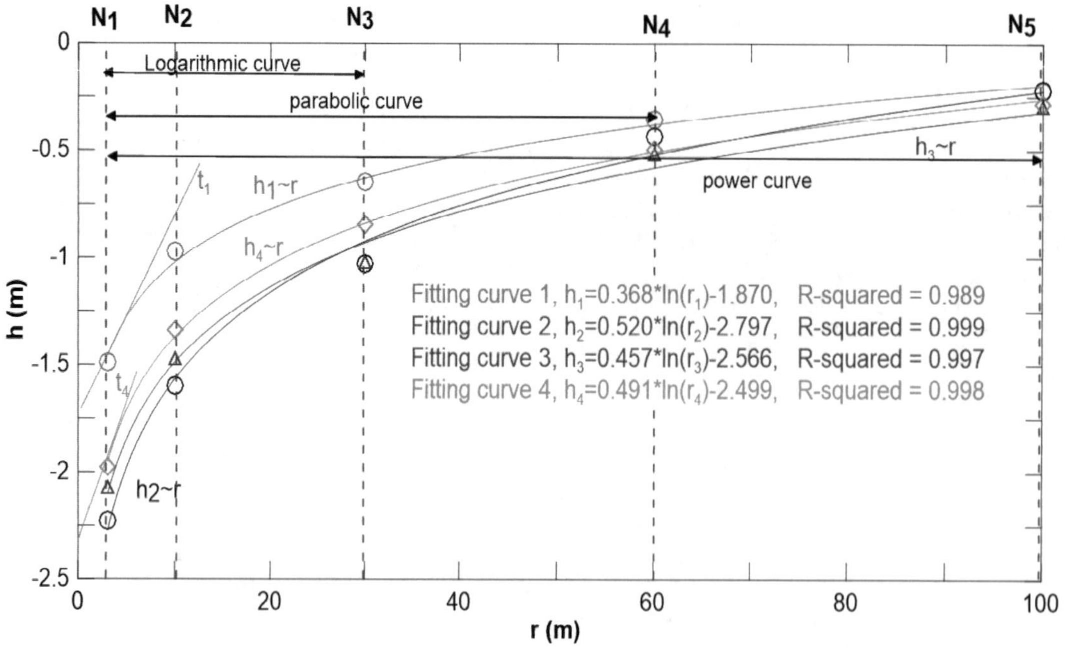

Figure 2. Schematic representation of Δh value estimated by the improved fitting method.

There exists an abnormal infinite value, when the logarithmic curve r is close to 0, which is only of mathematical significance and does not conform to the seepage mechanism. Therefore, the reliability of the improved curve fitting method to intercept the hydraulic jump value is questioned. In order to facilitate comparison, the tangent line was taken at the position of observation well N_1 on the fitted curve selected in 3.3, and then extended to the position of the main well wall to indirectly obtain the hydraulic jump value.

The hydraulic jump value obtained by Section 3.3 and 3.4 methods were verified and compared with each other, so as to find a better way to obtain the hydraulic jump value. The advantage of this method is that the water level variation around the well is considered and the interference of the linear change of the water level curve, when it is far from the main well is eliminated.

4. Results and discussions

4.1. Contrast between the analytical method and empirical formula method

According to the pumping theory, the hydraulic jump values from analytical solution were taken as the reference object. The reliability of the values estimated by traditional empirical formula method were verified. According to equation (1) and equation (2), the values calculated in four stages are listed in Table 2.

Based on the above analysis, the values estimated by the empirical formula method were found to be only one tenth of the former, which is significantly smaller than one order of magnitude. This result shows that the traditional empirical formula method is not reliable in calculating the hydraulic jump value and cannot be used as an estimation method of the hydraulic jump value.

Table 2. Statistical data of the calculation results of hydraulic jump value Δh (m).

Water flow Q_i	Analytical method and empirical formula methods used to calculate the values		Estimation by original curve fitting method	Fitting method with improvement	
	Analytical solution computes the value	Empirical formula method for valuation		Fitting method to estimate	Tangential valuation
Q_1	1.951	0.133	4.340	2.380	20.922
Q_2	2.087	0.176	6.947	3.518	7.929
Q_3	5.616	0.263	6.168	3.199	4.067
Q_4	6.806	0.397	5.812	3.479	-1.963
Average	4.115	0.242	5.817	3.144	7.739

4.2. Comparison of estimation results using the fitting method with improvement

4.2.1. Linear determination of fitting curve. The first thing to consider was, whether the curve value was stable, and the kind of linearity was determined. Two points were selected from the curve arc at the left end of the fitting curve to calculate the curvature value of each curve arc segment, as shown in Table 3.

The calculated results show that the curve arc in the range of $N_1 \sim N_3$ near the main well-2 is a logarithmic curve arc. In the range of $N_1 \sim N_4$, the average value of curve n is about 2, which can be identified as a parabolic curve arc. Similarly, in the range of $N_1 \sim N_5$, the average value of curvature n is less than 2, which can be identified as a power function curve arc.

According to the variation characteristics of curve value of water level curve, logarithmic curve needs to be preferred, when the fitting curve extends to the main shaft wall.

Table 3. Calculation results of the curvature method and difference method.

Curvature value calculation results (n)						Difference method curve fitting error (c)			
Curve arc	Q_1	Q_1	Q_3	Q_4	linear	Curve arc	Δ_1	Δ_2	c
$N_1 \sim N_2$	2.812	3.634	3.545	3.089	logarithmic	$N_1 \sim N_2$	37.641	112.575	0.988
$N_1 \sim N_3$	2.750	2.986	3.276	2.812	logarithmic	$N_2 \sim N_3$	48.654	104.706	0.731
$N_1 \sim N_4$	2.088	1.834	2.176	2.165	parabolic				
$N_1 \sim N_5$	1.842	1.500	1.827	1.795	Power curve				

In order to test the reliability of curve arcs, the difference method was further adopted to select two curve arcs, $N_1 \sim N_2$ and $N_2 \sim N_3$, at the left end of the water level fitting curve, which exhibited a same linearity. Curve fitting error c was then calculated, as shown in Table 2. There was a little difference between the two calculations. According to the curve fitting error requirement of difference method, the linear corresponding to $N_2 \sim N_3$ curve arc with smaller value was selected as the reference to fit the curve equation of water level.

4.2.2. Improved fitting method is used to estimate the value. It was observed that the water level curve type around the main well conforms to the logarithmic curve within the range of horizontal distance N_1~N_3. When the range was extended, the water level curves transits into a parabola and power function. Comprehensive analysis shows that the curvature increased with the proximity to the shaft wall, which was in line with logarithmic linear curve. Therefore, in order to improve the accuracy, the water level data of the observation wells N_1, N_2, N_3 and N_3 close to the main well were selected for this curve fitting, in order to ensure the same linear lines of selected points. The fitted curve equation is shown in Figure 2, and the intercepted hydraulic jump value after the corresponding curve arc that was extended to the well wall is shown in Table 2.

From this method, it was found that:

A: Q_1-Q_4 was the fitting curve equation *h-r* of the water level curve arc corresponding to 4 flows, and the correlation coefficient r was close to 1, indicating that the fitting effect was very good.

B: The water level curve arc was extended to the well wall, and the intercepted hydraulic jump value was generally stable, which is generally close to the theoretical calculation value, and the average error was less than 1 meter. Compared with the empirical formula method, it showed a better effect and could replace the theoretically calculated hydraulic jump value after considering appropriate modification.

4.3. Estimation of the water jump using the tangent method

This method is a continuation of Section 4.2. The basic idea is that the closer the main well is, the greater the curvature of the curve, and more prominent is the logarithmic nature of the curve. Using the water level data of N_1 observation well position, the tangent equation of each curve arc was obtained using Equation (6).

$$\begin{cases} r_1 - 3 = 0.123(h_1 - 1.49) \\ r_2 - 3 = 0.173(h_2 - 2.227) \\ r_3 - 3 = 0.152(h_3 - 2.08) \\ r_4 - 3 = 0.164(h_4 - 1.975) \end{cases} \tag{6}$$

Equation (6) is the tangent equation of the fitting curve arc at the observation well position N_1, when the pumping quantity is Q_1~Q_4; r_i is the horizontal distance and h_i is the water level.

Compared with the estimated results in Table 2, The stability of the values estimated from the tangent method was poor, and the estimated horizontal distance was more than 3 meters. Compared with the values estimated by the fitting method with improvement, the error was larger.

5. Conclusions

The present study was undertaken to find a practical method to improve the value's estimation by improving the fitting method and adjusting their direction of curve cutting. Following conclusions were obtained from the study:

(1) The analytical solution of derived from pumping theory was taken as reference, the values estimated by traditional empirical formula method were not reliable.

(2) The water level curve around the pumping well was logarithmic in the range of N_1~N_3. When the range was extended, the water level curve changed to a parabolic form and power function.

(3) The improved water level curve fitting equation exhibited a good effect. The intercepted hydraulic jump value was generally stable, and the average error was less than 1 meter. After considering the error correction, it could approximately replace the theoretical calculation value of hydraulic jump value.

(4) The stability of the values estimated from the tangent intercept method were poor, and the estimated horizontal distance was more than 3 meters, so it is not recommended for use.

In the comparison of several water jump algorithms, although there are still some problems and shortcomings, the improved curve fitting method can nearly replace the theoretical calculation of the hydraulic jump value. In common engineering practice, there are technical difficulties in the theoretical calculation for the value, and the fitting method with improvement roposed in this paper can make up for this deficiency, which is helpful to simplify the estimation of the hydraulic jump value and improve the accuracy. Our results indicate that it is indeed challenging to calculate and estimate the hydraulic jump value, and its regularity and algorithm needs a further study.

Acknowledgments
This research was strongly supported by the Project of Science and Technology of Social Development in Shaanxi [grant No. 2015TG-39]. The authors sincerely thank all helpers and the anonymous reviewers who improved the manuscript.

Disclosure statement
No potential conflict of interest was reported by all authors.

References
[1] Pozdniakov, SP, Wang, P, Lekhov, MV 2016 A semi-analytical generalized Hvorslev formula for estimating riverbed hydraulic conductivity with an open-ended standpipe permeameter *Journal of Hydrology* **540** 736-743, DOI:10.1016/j.jhydrol.2016.06.061
[2] Wang, QR, Zhan, HB, Wang, YX 2017 Single-well push-pull test in transient Forchheimer flow field *Journal of Hydrology* **549** 125-132, DOI:10.1016/j.jhydrol.2017.03.066
[3] Wen Jet-Chau, Chen Jyun-Lin, Yeh Tian-Chyi Jim, Wang Yu-Li, Huang Shao-Yang, Tian Zhong, Yu Chia-Yii 2020 Redundant and Nonredundant Information for Model Calibration or Hydraulic Tomography *Groundwater* **58(1)** 79-92 DOI: 10.1111/gwat.12879
[4] Yuanyuan Zha, Jinzhong Yang, Jicai Zeng, Chak-Hau M. Tso, Wenzhi Zeng, Liangsheng Shi 2019 Review of numerical solution of Richardson–Richards equation for variably saturated flow in soils *Wiley Interdisciplinary Reviews-Water* **6(5)** DOI:10.1002/wat2.1364
[5] Matsumoto Shinji, Machida Isao, Hebig Klaus H., Zeilfelder Sarah, Ito Narimitsu 2020 Estimation of very slow groundwater movement using a Single-Well Push-Pull test *Journal of Hydrology* 591 DOI: 10.1016/j.jhydrol.2020.125676
[6] Chen KW, Zhan HB, Zhou RJ 2016 Subsurface solute transport with one-, two-, and three-dimensional arbitrary shape sources *Journal of Contaminant Hydrology* **190** 44-57 DOI:10.1016/j.jconhyd.2016.04.004
[7] Quanrong Wang, Hongbin Zhan, Yanxin Wang 2017 Single-well push-pull test in transient Forchheimer flow field *Journal of HydrologyVolume* **549** 125-132 DOI: 10.1016/j.jhydrol.2017.03.066
[8] Wu YX, Zheng Q, Zhou AN, Shen SL 2021 Numerical evaluation of the ground response induced by dewatering in a multi-aquifer system *Geoscience Frontiers* **12(5)** DOI:10.1016/j.gsf.2021.101209
[9] Tang Youliang 2016 Drawing of hydraulic pump characteristic curves based on the least square method *Modern Manufacturing Engineering* **(1)** 147-150
[10] Zhang XZ, Huo, AD, Wang JC 2019 Experimental study on water level and absorption capacity in a radial well flow in a loess area *Water Supply* **19(5)** 1313-1321 DOI:10.2166/ws.2018.192
[11] Zhuang C, Li YB, Zhou ZF, Illman WA, Dou Z, Yang Y, Wang JG 2022 Effects of exponentially decaying aquitard hydraulic conductivity on well hydraulics and fractions of groundwater withdrawal in a leaky aquifer system *Journal of Hydrology* 607 DOI: 10.1016/j.jhydrol.2022.127439
[12] Zhang Wen, Quanrong Wang 2019 Approximate analytical and numerical solutions for radial non - Darcian flow to a well in a leaky aquifer with wellbore storage and skin effect *International Journal for Numerical and Analytical Methods in Geomechanics* **37(11)** 1453-1469 DOI: 10.1002/nag.2091
[13] Paradis CJ, McKay LD, (...), Hazen TC 2018 Perfect Edmund et al. Push-pull tests for estimating effective porosity: expanded analytical solution and in situ application

Hydrogeology Journal **26(2)** 381-393 DOI: 10.1007/s10040-017-1672-3

[14] Zhang Fengrong, Tao Jianfeng and Liu Chengliang 2018 Testing Method for Low Flow Characteristic of Variable-Speed Proportional Pump *Measurement & Control Technology* **37(5)** 12-15

[15] Tang Yue, Xiao Mei and Tang Lingdi 2017 Piecewise least squares polynomial curve fitting algorithm for tested pump performance data *Journal of Drainage and Irrigation Machinery Engineering* **35(9)** 744-748

IOP Conf. Series: Earth and Environmental Science 1087 (2022) 012006 doi:10.1088/1755-1315/1087/1/012006

Application of a modified graphite paste electrode to the detection of fluorene in fuscous latices

Yu Liu[1], Qi Luo[2], Qiping Tian[3], Wen Xu[2,5], Huili Hu[4] and Li Shi[2]

[1] College of Life Sciences, Zhejiang University, Hangzhou 310058, China;
[2] Zhejiang Light Industrial Products Inspection and Research Institute, Hangzhou 310018, China;
[3] Zhejiang Institute of Mechanical & Electrical Engineering Co., Ltd, Hangzhou 310051, China;
[4] Quality Technology Association of Hangzhou, Hangzhou 310019, China.

[5] Corresponding author's e-mail: 563275690@qq.com

Abstract. A graphite paste electrode (GPE) was obtained by immobilizing graphene, and the electrocatalytic activity of fluorene (Flu) was studied. The voltammetric behavior using graphene-modified GPE showed an irreversible redox reaction for Flu-ACN-$KClO_4$. A good linear correlation between anodic I and C_{Flu} ranged within 8–300 ng mL^{-1}. Therefore, we recommend a facile synthesis of modified GPE impregnated with Flu, which matched the concept of analytical minimalism. The electrochemical detection method combined with portable potentiostat is promising to be used for the field tests of Flu in fuscous latices as well as other polycyclic aromatic hydrocarbons.

1. Introduction

Polycyclic aromatic hydrocarbons (PAHs) are carcinogenic pollutants which pose environmental risks to ecosystems [1,2], which are generated primarily during incomplete burning of organic matters [3]. Fluorene (Flu) is frequently added to fuscous latices among latex products [4]. Present assay methods for Flu are high performance liquid chromatography (HPLC) [5], gas chromatography-tandem mass spectrometry [6] and room temperature phosphorescence [7]. However, these methods with large equipment make techniques unsuitable for field experiment and require expensive instrumentation.

The low solubility of heterocyclic compounds in water and their extraction in solvents [8] usually involve preconcentration and separation procedures for highly sensitive detection of PAHs. Based on uncomplicated pretreatment steps, graphite paste electrode (GPE) is applied for in electroanalysis due to its high sensitivity and wide potential (E) range [9-11]. Voltammetric applications of the disposable modified graphite electrodes are in combination with the selectivity and sensitivity of assay [12-14].

The aim of this paper was to evaluate the possibility of developing a sensor material for electrochemical detection of Flu combined with GPE. The electrochemical oxidization of Flu-ACN-$KClO_4$ was characterized using GPE filled with graphite paste and surface-modified by graphene. The adequate sorption of Flu on graphene-modified GPE (graphene/GPE) was conducted at open circuit potential. Various mass concentrations of Flu (C_{Flu}) were obtained by differential pulse voltammetry (DPV). The electrochemical data were exhibited by comparing with HPLC method. This portable assay for Flu was requisite to develop procedures for the DPV determination of trace concentrations. Thus, meaningful progress had been made for an electroanalytical technique of Flu in fuscous latices.

2. Experimental section

2.1. Instruments and chemicals

A Chenhua CHI1200C potentiostat (China), Thermo Fisher Vanquish Core chromatograph (USA), TESCAN MIRA electron microscope (Czech), Sineo JUPITER microwave extraction apparatus (China) and Chenhua 3-mm diameter electrode bodies were used.

High-purity graphene, graphite powder, acetonitrile (ACN), N, N-dimethylformamide (DMF), muriatic acid and potassium perchlorate were purchased from Macklin Inc. (China). Water (0.055 μS cm^{-1}) was prepared by a Ulupure-II purifying device (China).

2.2. Provision of graphene/GPE

The GPE was acquired by blending 75% (mass fraction) graphite with 25% liquid paraffin in a mortar. Then the well-mixed paste was pressed into the electrode bodies, and the surface of electrode was then polished using a flint paper. A graphene suspension was prepared by mixing 0.6% (w/w) graphene with DMF via a simple sonication. The GPE surface was modified with 15 μL modified liquid by infrared irradiation to get the graphene/GPE. The surface condition of activated electrode markedly influenced the chemical process [15]. The electrodes needed to be freshly prepared to prevent cross contamination [16].

2.3. DPV technique

To complete electrochemical oxidization of Flu, a three-electrode method was presented. The designed specifications of DPV are shown in Table 1 which are conducted under normal atmospheric conditions. The C_{Flu} at 8, 40, 80, 160, 240 and 300 ng mL^{-1} were produced using 0.1 M supporting electrolyte (soluble $KClO_4$ in 90% ACN; pH at 3.0 adjusted by muriatic acid). The graphene/GPE was submerged in aforesaid Flu standards for 10 min. The mean peak current was noted as the anodic current (I), and the electrochemical measurement was performed three times.

Table 1. Designed specifications of DPV.

Factors	Specifications
Working electrode (WE)	Graphene/GPE
RE	Ag/AgCl
CE	Pt
E range	1.1–1.7 V
Magnitude	60 mV
Pulsation	40 ms
Sweep rate	40 mV s^{-1}

2.4. Precondition and HPLC investigation

Latex (0.4 g) and ACN (20 mL) were extracted at 70 °C for 25 min using microwave extraction. Subsequently, the fluid was rotated at 2000 rpm for 5 min and passed through organic film filter. It was next added to ACN-$KClO_4$ for electrochemical sensor (ES) measure.

1.3 mL extracted solution was prepared via silica gel solid-phase extraction for HPLC analyses. Water and chromatographic grade ACN (volume fraction: 70%) were selected as the mobile phases on reversed phase C_{18} column. The sampling volume was 20 μL, and the confirming peak was determined based on retention time with column flow of 1.0 mL min^{-1}. The C_{Flu} was calculated by the integral area of LC peaks.

3. Results and discussions

3.1. Surface structure of graphene/GPE and voltammogram of Flu

The microstructures of bare GPE and graphene/GPE were examined by scanning electron microscopy. The surface of bare GPE was comparatively ordered before the deposition of graphene (Figure 1*A*), but physical alterations occurred after modification (Figure 1*B*). The graphene was decentralized in DMF, and the electron microscope images indicated a nonuniform sedimentation of graphene suspension on GPE with an irregularly rough exterior.

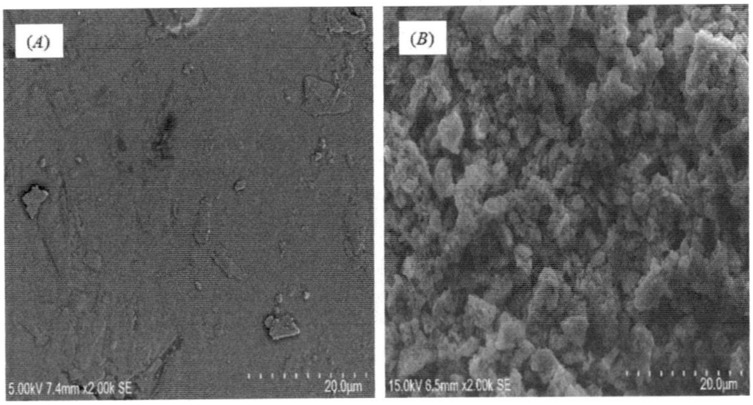

Figure 1. Scanning electron microscope images of the GPE surfaces.

The bare GPE and graphene/GPE were submerged in 40 ng mL^{-1} Flu-ACN-$KClO_4$ for 10 min, and then carried out using cyclic voltammetry. No anodic peaks appeared in 0.1 M ACN-$KClO_4$ (Figure 2*A*) and Flu-ACN-$KClO_4$ (Figure 2*B*) on GPE, while the signal at 1.34 V was detected (Figure 2*C*) owing to the oxidation of Flu on graphene/GPE. This result showed a distinct anodic peak without cathodic peaks, occurring as an irreversible redox reaction. The graphene afforded an effective sorption of analytes, causing by high specific area. The oxidation potential of Flu was a basis for identifying Flu from other PAHs.

During the immersion process, Flu was gradually accumulated on the paste electrode surface. The volume of modified liquid greatly affected the electron transport. Graphene suspensions (7–19 μL) were applied to the GPE surface. Within the volume range of 7–14 μL, the I of 40 ng mL^{-1} Flu-ACN-$KClO_4$ continuously raised (Figure 3). A maximum value of I had been acquired at 15 μL, possessing favourable response to peak current. When the volume increased from 15 to 19 μL, the I slowly reduced. This may be due to the modified layer was relatively dense, blocking the electron transport to a certain extent.

IOP Conf. Series: Earth and Environmental Science 1087 (2022) 012006 doi:10.1088/1755-1315/1087/1/012006

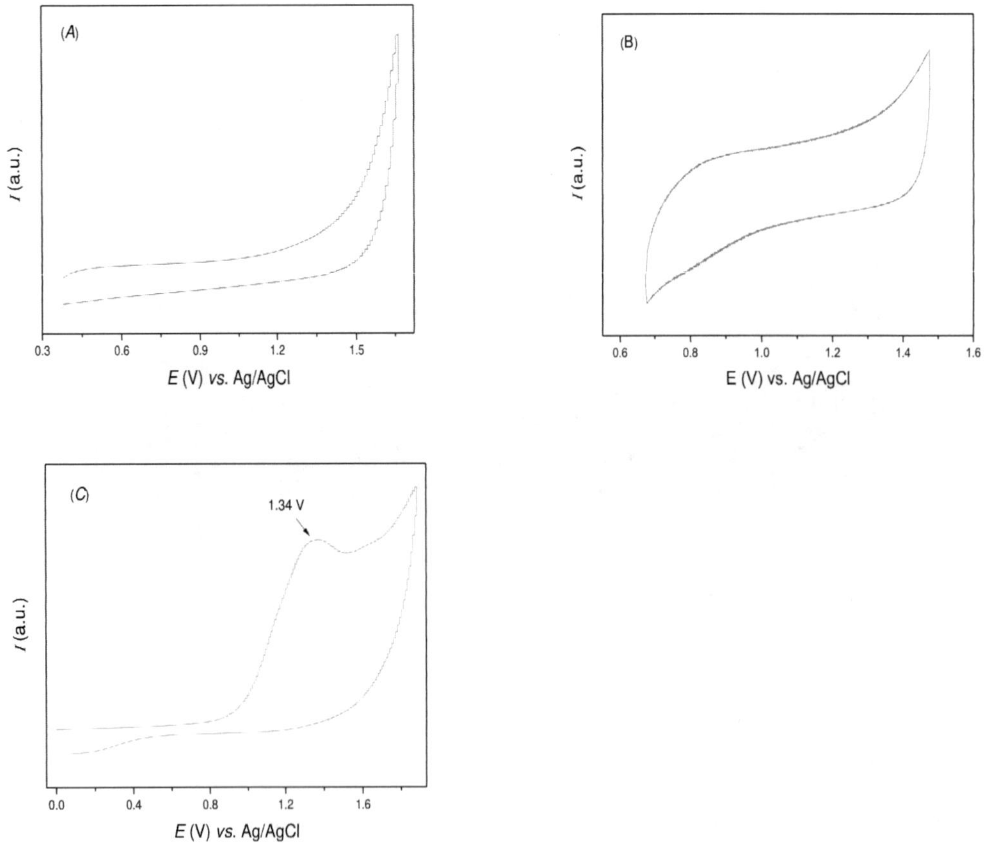

Figure 2. Cyclic voltammetry curves of supporting electrolyte (*A*) and Flu-ACN-$KClO_4$ (*B*) on bare GPE, and cyclic voltammetry of Flu-ACN-$KClO_4$ on graphene/GPE (*C*).

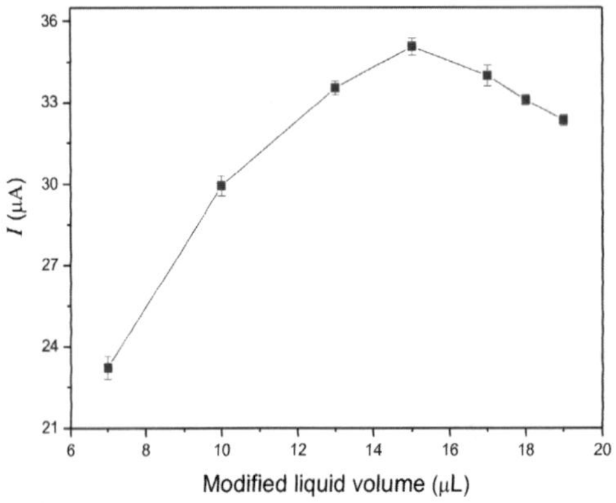

Figure 3. Influence of volume on response current of modified GPE (n = 3), and n represents repetition.

3.2. DPV detection of Flu in fuscous latices

The WE was immersed in Flu-ACN-$KClO_4$ at 8–300 ng mL^{-1}, and they were detected using DPV (Figure 4). The correlation between C_{Flu} and I was obtained: $I = 0.149C_{Flu} + 27.36$, $R = 0.9989$. The detection limit was calculated according to the following formula: detection limit = $3\sigma/k$ = 2.5 ng mL^{-1}; where σ is the standard deviation (SD) of I for residual solution (n = 10), and k is the gradient of standard plot. Therefore, the microfabricated ES can be used for rapid assay of Flu.

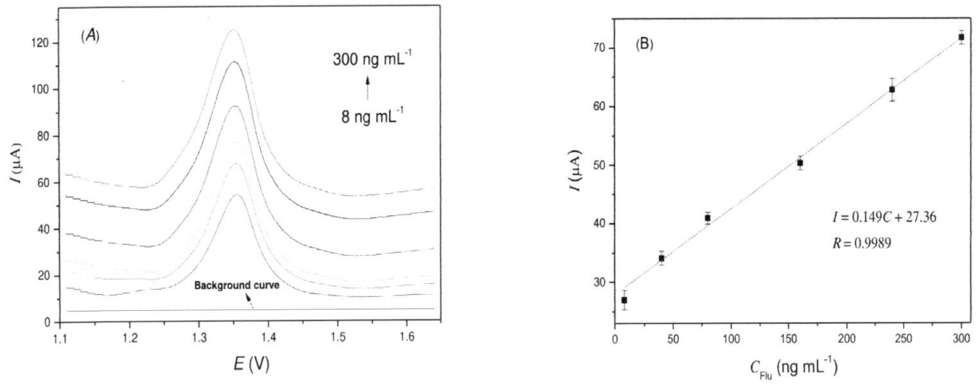

Figure 4. DPV curves of Flu on graphene/GPE involving various C_{Flu} (*A*), and standard plot for Flu standards (*B*).

The mass concentrations of spiked samples at 4, 20 and 50 ng mL^{-1} were produced. The C_{Flu} was studied by HPLC and ES separately. The mean C_{Flu} of the latices was 6.2 ng mL^{-1} via modified GPE and 6.4 ng mL^{-1} via HPLC. As shown in Table 2, the mean recoveries with ES were 92.6%–105.9%, while those of HPLC were 95.4%–109.6%. The bias was tolerable, and the ES viewed favourably for related fields of selectivity, low cost and ppb-level detection limit. The RSD of ES and HPLC was ≤4.9% and ≤4.1%, respectively. The repeatability and accuracy of the ES were fairly good.

Table 2. Comparison of two-type recoveries in terms of Flu detection.

Authentic samples	Added (ng mL^{-1})	Detected (ng mL^{-1})		Recovery (%)		RSD (%)	
		ES	HPLC	ES	HPLC	ES	HPLC
	50	56.7	53.8	100.8	95.4	4.7	4.1
Latices	20	24.3	27.2	92.6	103.0	3.6	3.7
	4	10.8	11.4	105.9	109.6	4.9	3.2

4. Conclusions

An ES methodology to identify the Flu in fuscous latices was proposed. The graphene/GPE had a higher sensitivity than bare GPE for Flu detection, which was attributed to good abilities of Flu sorption. The DPV using modified GPE showed an enhancement in current signal for Flu-ACN-$KClO_4$. The mean recoveries of Flu for latices via ES approach were 92.6%–105.9% with a low RSD. The electrochemical data conducted verified a facilely quantitative assay for Flu as well as other heterocyclic compounds. A portable detection system combined with potentiostat is advantageous to be used for the field tests of PAHs.

Acknowledgement
This work was financially supported by Zhejiang Provincial Program for Public Wel-Fare of Technology Application Research Plan (No. LGC21B050006).

References
[1] Balcioglu E B 2016 Potential effects of polycyclic aromatic hydrocarbons (PAHs) in marine foods on human health: A critical review *Toxin Reviews* **35** 98
[2] Nam T H, Jeon H J, Mo H H, Cho K, Ok Y S and Lee S E 2015 Determination of biomarkers for polycyclic aromatic hydrocarbons (PAHs) toxicity to earthworm (*Eisenia fetida*) *Environmental Geochemistry and Health* **37** 943
[3] Abdel-Shafy H I and Mansour M S M 2016 A review on polycyclic aromatic hydrocarbons: Source, environmental impact, effect on human health and remediation *Egyptian Journal of Petroleum* **25** 107
[4] Wang J, Ren T, Wang Y, He X, Liu W and Shen X 2014 Synthesis, curing behavior and thermal properties of fluorene-containing benzoxazines based on linear and branched butylamines *Reactive and Functional Polymers* **74** 22
[5] Chetiyanukornkul T, Toriba A, Kameda T, Tang N and Hayakawa K 2006 Simultaneous determination of urinary hydroxylated metabolites of naphthalene, fluorene, phenanthrene, fluoranthene and pyrene as multiple biomarkers of exposure to polycyclic aromatic hydrocarbons *Analytical and Bioanalytical Chemistry* **386** 712
[6] Gao Z W, Mu L, Xue S F, Tao Z and Zeng X 2010 Cucurbit [8]urils-induced room temperature phosphorescence of phenanthrene and fluorene *Spectroscopy and Spectral Analysis* **30** 1026
[7] Yu C, Luo L, Chen L and Cai R 2020 A screening method of polycyclic aromatic hydrocarbons in halogenated butyl rubber stoppers by GC-MS/MS *Chinese Journal of Pharmaceutical Analysis* **4** 681
[8] Maigari A U and Maigari M U 2015 Microbial metabolism of polycyclic aromatic hydrocarbons (PAHs): A review *International Journal of Scientific & Engineering Research* **6** 1449
[9] Sayahpour M, Hashemnia S and Mokhtari Z 2021 Preparation and assessment of a polysafranin/multiwall carbon nanotube-Nafion hybrid film-modified carbon paste electrode and its performance as an easy sensing probe for H_2O_2 and Cu^{2+} *Electrocatalysis* **12** 91
[10] Erdogan Z O and Kucukkolbas S 2021 Fabrication of an electrochemical biosensor based on Fe_3O_4 nanoparticles and uricase modified carbon paste electrode for uric acid determination *Monatshefte für Chemie-Chemical Monthly* **152** 309
[11] Ferrari G M, Elbardsiy H M, Silva V, Belal T S and Brownson D 2020 The influence of lateral flake size in graphene/graphite paste electrodes: An electroanalytical investigation *Analytical Methods* **12** 2133
[12] Anantha M S, Kumar S R, Anarghya D, Venkatesh K and Muralidhara H B 2021 ZnO@MnO_2 nanocomposite modified carbon paste electrode for electrochemical detection of dopamine *Sensors International* **2** 100087
[13] Biswas S, Naskar H, Pradhan S, Chen Y and Pramanik P 2020 Sm_2O_3 nanorod-modified graphite paste electrode for trace level voltammetric determination of acetaminophen and ciprofloxacin *New Journal of Chemistry* **44** 1921
[14] Nag S, Pradhan S, Naskar H, Roy R B and Bandopadhyay R 2021 A simple nano cerium oxide modified graphite electrode for electrochemical detection of formaldehyde in mushroom *IEEE Sensors Journal* **21** 12019
[15] Zhong N F, Yuan R H, He H H, Zhong W H and Wang Y 2021 Electrochemical sensor based on graphene-modified GCE for rapid quantification of benzo(a)pyrene in dark rubber materials *Journal of Physics: Conference Series* **1978** 012020
[16] Che X H, Yuan R H, Tang L C, Xiong Y D, Wu M F and Zhu M H 2021 Study of catecholate-functionalized zinc complex on carbon steel and its application to corrosion resistance *Journal of Physics: Conference Series* **1978** 012009

Sorption studies of Pb(II) onto montmorillonite clay

Yuting Chuang[1], Jiajun Chen[1,2,4], Jianzun Lu[1,2], Lingcheng Su[1], Sabrina Yanan Jiang[3], Yijia Zhao[1], Chiu Hong Lee[1], Zhihui Wu[1] and Huada Daniel Ruan[1,4]

[1] Environmental Science Program, Division of Science and Technology, Beijing Normal University-Hong Kong Baptist University United International College, Zhuhai 519087, China;
[2] Department of Chemistry, Faculty of Science, Hong Kong Baptist University, Hong Kong 999077, China;
[3] National Observation and Research Station of Coastal Ecological Environments in Macao, Macao Environmental Research Institute, Macau University of Science and Technology, Taipa 999078, Macao, China.

[4] Corresponding authors' e-mail: gzchenjiajun@163.com; hruan@uic.edu.cn

Abstract. In this research, the lead (Pb(II)) removal rate of montmorillonite (Mt) was studied in aqueous solution. The adsorption conditions (contact time, pH, and temperature) were adjusted to study the effect of those factors on the removal rate of Pb(II) adsorption by Mt. The results indicated that the effect of temperature on the Pb(II) removal rate of Mt was not significant. The Pb(II) removal rate was found to be pH-dependent, improving with increasing pH in the range of 3 to 7. The equilibrium time of adsorption was attained after around 1 hour. The adsorption kinetics were analyzed by fitting to three models (pseudo-first-order (PFO) kinetic model, pseudo-second-order (PSO) kinetic model and intraparticle diffusion model. In comparison, R^2 value (0.9952) of PSO kinetic model is better than PFO (0.9718), supposing that the adsorption process of Pb(II) onto Mt was chemisorption. The intraparticle diffusion model fitting showed that the Pb(II) adsorption process by Mt was controlled by three steps. Langmuir, Freundlich, Redlich-Peterson and Sips isotherm models were used to describe the adsorption mechanism of Pb(II) adsorbed onto Mt. The experiment results showed the adsorption isotherms are a perfect fit to the Sips model, indicating both homogeneous monolayer adsorption and heterogeneous multilayer adsorption were occurred in the Pb(II) adsorption process by Mt, and the adsorption was mainly controlled by heterogeneous multilayer adsorption.

1. Introduction

Lead (Pb(II)) pollution has caused a great deal of harm to humans and the environment, which has garnered a great deal of attention in environmental issues. Currently, Pb(II) is one of the major industrial pollutants, which acts as a high concealment (dissolving capacity), easily accumulated, hard to remove and significantly toxic [1–3]. It has been proven harmful to both human health and ecological systems; and the concentration of Pb(II) has exceeded safety levels in various places around the world, which has led to toxic matter being absorbed by rice, aquatic animals, and as a consequence, entering the human body through food chains [4,5]. Of particular note is that Pb is able to be absorbed through the skin, alimentary tract and respiratory tract, leading to accumulation in tissues and resulting in anemia, neurological disorders and kidney damage [6]. Precipitation, condensation, adsorption and ion exchange are common and traditional physical-chemical methods in removing Pb(II) in aqueous solution [7–10].

Currently, adsorption technology has been paid more intensive attention, due to its low cost and high efficiency [11–13]. Some scientists have found that montmorillonite (Mt) is a good adsorbent in removal of pollutants [14–17]. Mt is a layered clay mineral composed of extremely fine hydrous silicates with the formula $(Na,Ca)_{0.33}(Al,Mg)_2[Si_4O_{10}](OH)_2 \cdot nH_2O$ [18]. The structure of Mt is one Al octahedral sheet sandwiched by two Si tetrahedral sheets. Due to the isomorphous substitution of Si^{4+} in the tetrahedral layer and Al^{3+} in octahedral layer by metals with lower charge, Mt has strong negative charge surface, resulting in strong attraction to cation pollutants [19]. The intermediate layer charge density of Mt was in the range of 0.2-0.6 per half unit cell [18]. As an adsorbent, Mt has the following merits: (1) Mt is not expensive as there is a rich reserve in nature [20]; (2) Mt has a large surface area [20]; (3) Mt has high ion exchange capacity providing better adsorption ability [21]. Natural Mt can adsorb a variety of impurities (especially metal cations) by ion exchange and complexation (acting with terminal hydroxyl groups).

In this research, the montmorillonite adsorption towards Pb(II) removal under the different adsorption conditions (contact time, pH, and temperature) were investigated. The adsorption mechanisms of Pb(II) by Mt were achieved via experimental data fitting to three kinetic models (pseudo-first-order, pseudo-second-order, and intraparticle diffusion adsorption kinetic models) and four isotherm models (Langmuir, Freundlich, Redlich-Peterson, and Sips adsorption isotherm models). Furthermore, the goal of this study is to investigate if montmorillonite and montmorillonite-containing mineral waste materials can be applied in the fabrication of adsorbents, which can be used to remove pollutants, especially heavy metals in the aquatic systems.

2. Materials and method

2.1. Materials

The chemicals (including lead nitrate in AR grade (99% purity), hydrochloric acid, and sodium hydroxide) used in the experiment were purchased from Tianjin Damao Chemical Reagent Co, LTD. The montmorillonite (Mt) was purchased from the Jingdezhen Clay Mineral Science and Technology Ltd., China, and the origin of Mt is in Inner Mongolia, China.

2.2. Characterization of Montmorillonite

The element composition of montmorillonite was determined by IRF model 550 energy dispersive X-ray spectroscopy (EDX). The X-Ray Diffraction (XRD) patterns of samples were obtained using the PANalytical X'Pert Pro MPD Power diffractometer with CuKα radiation at 40 kV and 40 mA. The sample was scanned from 3° - 70° (2θ) at a step size of 0.0167°. The Fourier Transform Infrared (FTIR) spectra of samples were obtained using Bruker Lumos IR spectroscopy. Before analysis, the sample was mixed with potassium bromide to make a platelet. The platelet was scanned from 4000 to 600 cm^{-1} for 32 times with a spectral resolution of 4 cm^{-1}. Nitrogen adsorption-desorption isotherm of Mt sample was determined using SA3100 surface area analyzer under -196℃. Before determination, the Mt sample was degassed at 120 ℃ for 15 hours under vacuum condition.

2.3. Pb(II) removal (adsorption) under different conditions

For the study of removal (adsorption) rate, 50 mL of varying concentrations (10, 20, 30, 40, 50 mg/L) of Pb(II) solution was transferred into 50 ml centrifuge tubes at room temperature (~25 ℃), which contained a specific amount (0.02 g) of Mt adsorbent. The pH of the solution was adjusted from 3 to 7 (3, 3.5, 4, 4.5, 5, 5.5, 6, 6.5, 7) either by sodium hydroxide or chloride acid. The condition of pH to alkalinity were not adjusted for excluding the influence of lead hydroxide precipitation. The temperature of the solution was adjusted by a constant temperature oscillator. After that, the mixture was shaken using a benchtop shaker at 200 r/min at different temperatures (30, 40, 50, 60, 70℃) within different time intervals (10s, 20s, 1min, 2min, 5min, 30min, 60min, 120min, 240min and 480min). After shaking, the mixture was filtered with a 0.22 μm filter. Each sample was conducted in triplicate. The concentration of the remaining Pb(II) in solution was measured using Inductively Coupled Plasma Optical Emission Spectrometer (ICP-OES), P.E. OPTIMA 2100 DV.

The removal rate (%R) of Pb (II) was calculated according to the following equation.

$$\%R = \left(\frac{(C_0 - C_t)}{C_0}\right) * 100\%$$

where C_0 is the initial concentration of Pb(II) (mg/L); and C_t is the remained concentration of Pb (II) after adsorption at t time after adsorption (mg/L).

The adsorption isotherm study was carried out by adding 50 ml Pb (II) solution with different concentrations (10, 20, 30, 40 and 50 mg/L) to 0.02 g Mt and shaken at 200 r/min at room temperature (~25 ℃) for 1h. Then, the mixture was filtered by a 0.22 μm filter. The concentration of the Pb (II) in the solution was measured using ICP-OES.

The adsorption amount of Pb(II) onto Mt was calculated by the following equation.

$$q_t = \frac{V(C_0 - C_t)}{m}$$

where q_t is the adsorption amount of Pb(II)at t time (mg/g); V is the volume of the solution (L); and m is the mass of the adsorbent (g).

3. Results and discussions

3.1. Characterization of Mt

As shown by the EDX data (Table 1), the most abundant elements present in Mt were O, Si and Al, whereas C, Fe, Mg and Ca were found only in small quantities (less than 10%). According to XRD pattern (Figure 1), it shows that Mt displays a typical reflection of montmorillonite and contains a very small amount of quartz. The (001) reflection of Mt appears at 2θ = 5.73°, and the corresponding d-value is 1.54 nm. The FTIR spectrum of Mt is given in Figure 2 and the assignments of the vibrational bands [22–24] are listed in Table 2. According to the EDX data, XRD pattern, FTIR spectrum, and the BET surface area results, the Mt is proved to be a highly pure sample with only a very small amount of quartz. The samples were the highly pure montmorillonite with a relatively high surface area of 61.1 m^2/g, supposing that the sample could adsorb large amount of Pb(II) in the aqueous environment.

Table 1. The element composition of Mt determined by EDX.

Element	Mass of Mt (wt.%)
O	45.21
Si	28.48
Al	10.07
C	5.01
Fe	4.8
Mg	3.87
Ca	2.55
Total	100

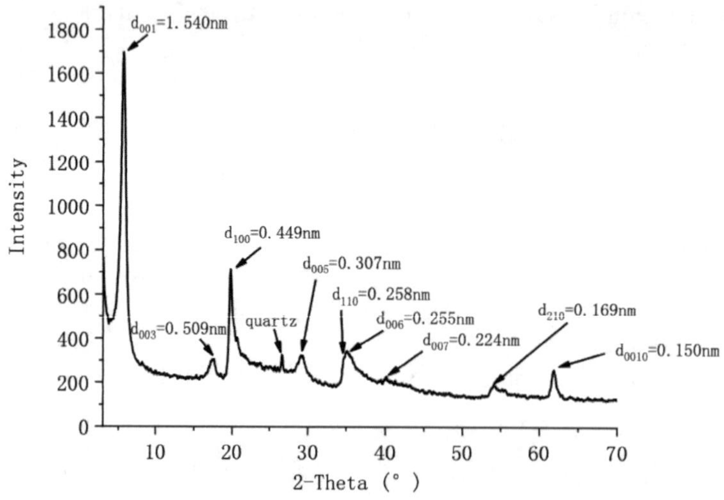

Figure 1. The XRD pattern of Mt.

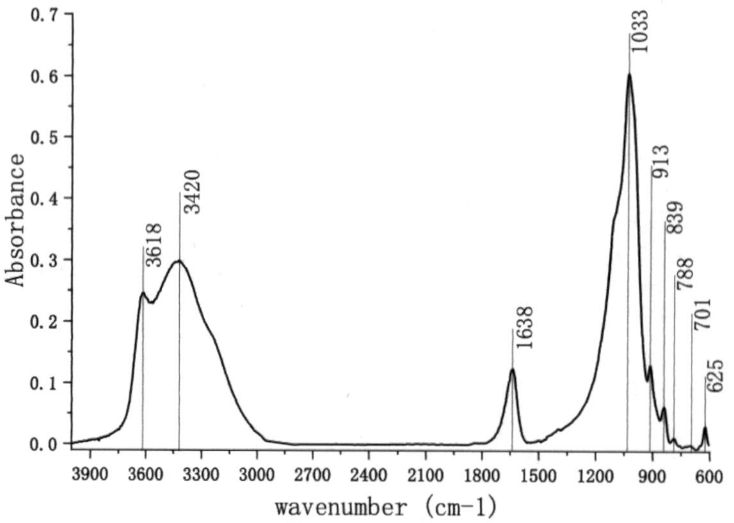

Figure 2. The FTIR Spectrum of Mt.

Table 2. The assignments of vibrational bands in the FTIR spetrum of Mt.

Position(cm^{-1})	Assignments
3618	-OH stretchingof Al-OH and Mg-OH
3420	Stretching vibrations of H-O-H of adsorbed water
1638	Bending vibrations of H-O-H of adsorbed water
1033	Si-O stretching vibrations
913	Al-Al-OH group deformation
839	Al-Mg-OH group deformation
625	Coupled Al-O and Si-O out-of-plane vibrations

3.2. Conditions affecting Pb(II) removal rate of Mt

3.2.1. Effect of contact time on Pb(II) removal rate. Figure 3 indicates that the Pb(II) removal rate increased rapidly and reached 54.62% within 5 mins to almost the equilibrium point of adsorption. After 5 mins, the removal rate increased slowly until a maximum rate of 56.84% was reached. After 60 mins, the removal rate remained stable. Thus, it took about 1h for the reaction to reach equilibrium with maximum adsorption amount of Pb(II). The previous studies also reported that an similar equilibrium time was consistent with the results in this study [25–27].

The dramatically increase of Pb(II) removal rate in the first 5 mins is due to the higher concentration of Pb(II) in solution, providing a strong driving force for Pb(II) ions to interact with the montmorillonite. In addition, there were relatively higher free adsorption vacancies on the surface of montmorillonite, where a strong adsorption affinity to Pb(II) was generated [28]. As time went on, the Pb(II) ions gradually occupied the adsorption sites from Mt until the equilibrium was finally reached between adsorption and desorption [29].

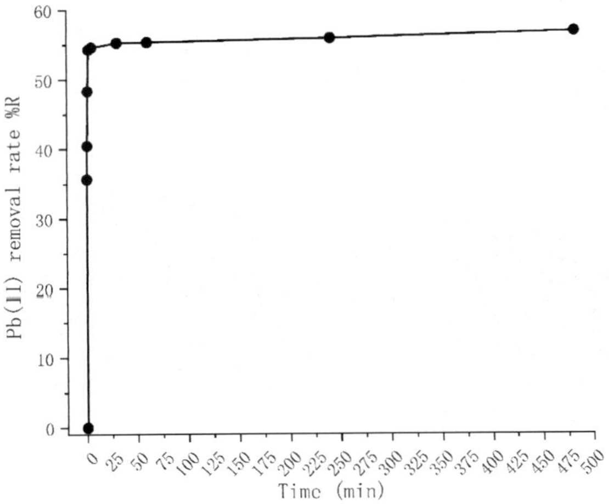

Figure 3. Relationship between time and Pb(II) removal rate for Mt.

3.2.2. Effect of pH on Pb(II) removal rate. Figure 4 shows that the removal rate of Pb(II) increased rapidly in the pH range of 3.5 to 5, and became gradual in the pH range of 5 to 7. The removal rate reached its maximum when pH was around 6.5.

The lower removal rate of Pb(II) in association with the lower pH could be attributed to the fact that hydrogen ions competed with Pb(II) for adsorption sites, and Si-O- and Al-O groups were gradually protonized, weakening the binding ability of montmorillonite to cations [7]. One of the possible reasons for increased removal rate of Pb(II) in higher pH is the deprotonation of hydroxyl groups, providing adsorption sites for Pb(II) attachment to hydroxyl groups on the surface of Mt [25]. The change of chemical environment of Pb(II) in different pH is another reason for the removal rate of Pb(II) affected by pH. The equilibrium constant (logK) for hydrolysis from Pb(II) to $Pb(OH)^+$ is 6.48, so, colloidal micelles of $Pb(OH)^+$ become the dominant specie in the solution at pH values over 6 [25,30]. The $Pb(OH)^+$ colloidal micelles carrying positive charges are easily attracted by the negatively charged Mt to form precipitate [25,31]. Therefore, when the pH is over 6, the removal rate of Pb(II) increases. The trend of relationship between pH and Pb(II) removal rate for Mt in this study are consistent with those reported in the literature [25,31,32].

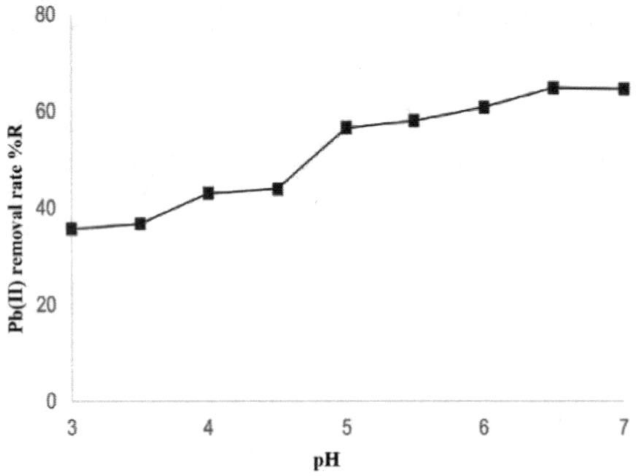

Figure 4. Relationship between pH and Pb(II) removal rate for Mt.

3.2.3. Effect of temperature on Pb(II) removal rate. Figure 5 shows that Pb(II) removal rate increased slowly and changed gently with the increase in temperature. When the temperature rose from room temperature to 60 ℃, the removal rate increased by less than 10%, indicating that the temperature has less effect on the Pb(II) removal rate of Mt. As the temperature effect is insignificant on the removal rate and using high temperature for adsorption is energy-consuming and impractical, so it is better to select room temperature (~25 ℃) as the optimal temperature.

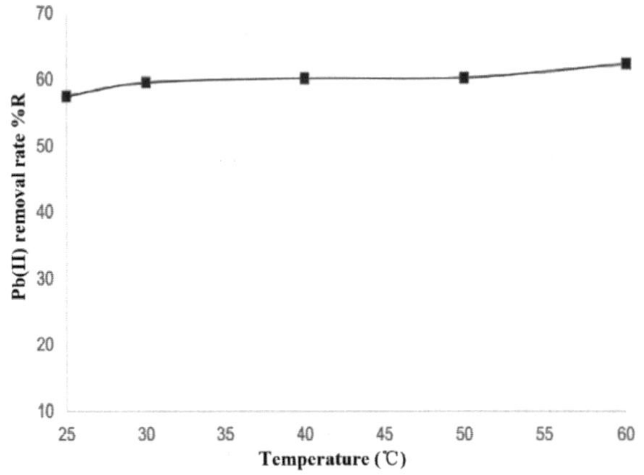

Figure 5. Relationship between temperature and Pb(II) removal rate for Mt.

3.3. Adsorption kinetics

The adsorption data were fitted to the PFO and PSO kinetic models for studying the adsorption process and adsorption mechanism. The PFO reaction occurs when one reacting substance is great excess or remained constant in concentration compared with the other, so the adsorption diffusion and physical adsorption are domain. In the PSO model, chemisorption is involved in the adsorption to attract the adsorbate onto the adsorbent surface.

The PFO kinetic equation is given as below [33]:

$$q_t = q_e(1 - e^{-k_1 t})$$

where q_e is the adsorption amount of Pb(II) onto Mt at equilibrium time, (mg/g); k_1 is the rate constant (min^{-1}).

The PSO kinetic equation is given by the equation below [33]:

$$q_t = \frac{k_2 q_e^2 t}{1 + k_2 q_e t}$$

where k_2 is the rate constant (min^{-1}).

The fittings of kinetic data to these two models in non-linear form are depicted in Figure 6 and the relevant parameters were shown in Table 3. It was apparent that the adsorption data was only slightly better fitted to PSO model ($R^2 = 0.9952$) than PFO model ($R^2 = 0.9718$), so the adsorption process of Pb(II) onto Mt was supposed to be chemisorption [34] associated with the adsorption diffusion and physical adsorption at different stages of adsorption [33]. The high adsorption rate (k_1) could be due to the adsorption diffusion and physical adsorption, while the low adsorption rate (k_2) due to chemisorption.

Table 3. Parameters derived from PFO and PSO fittings.

Models	Paramters	Value
	k_1	5.24±0.68
PFO	q_e	67.78±1.48
	R^2	0.9718
	k_2	0.13±0.01
PSO	q_e	69.67±0.67
	R^2	0.9952

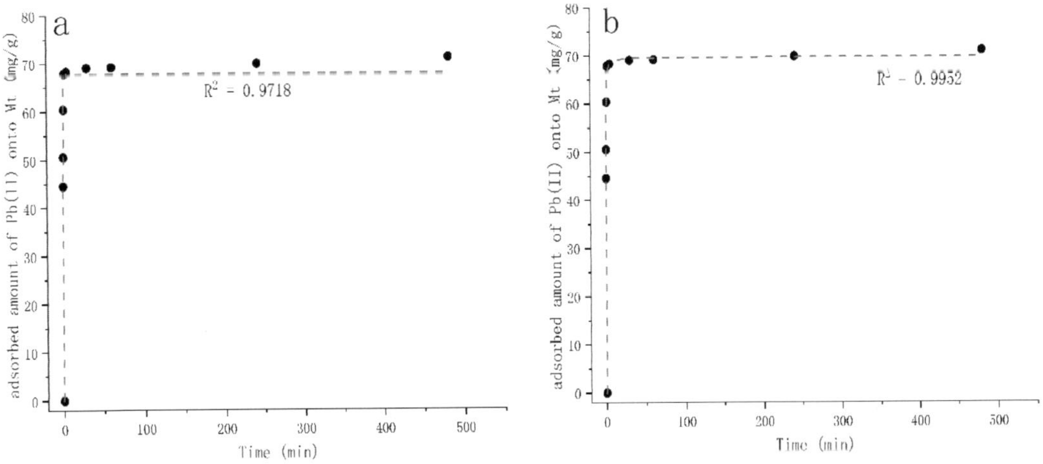

Figure 6. The fittings of kinetic data to PFO kinetic model in non-linear form (a) and PSO kinetic model in non-linear form (b).

The intraparticle diffusion model was used to further investigate the diffusion process of Pb(II) onto Mt. The intraparticle diffusion model is studied using the Wber-Morries equation, which is shown as below [35].

$$q_t = k_p t^{1/2} + C$$

where k_p is the intraparticle diffusion rate constant ($mg/(g\bullet min^{1/2})$); and C is the film thickness (mg/g).

The plot of q_t vs. $t^{1/2}$ is shown in Figure 7 and the relevant parameters are shown in Table 4. The plot showed three linear segments, indicating that the adsorption process occurred in three steps. The order of adsorption rate of these three stages was $k_{p1} > k_{p2} > k_{p3}$. The higher value of k_{p1} indicates that at the beginning of adsorption, Pb(II) was instantaneously adsorbed by the exterior surface of Mt. When the exterior surface was saturated with adsorbed Pb(II), the Pb(II) ions were diffused into the pores of Mt. The diffusion was gradually resisted during Pb(II) being diffused into the pores, resulting in decrease in the diffusion rate (k_{p2}). At the final stage, the adsorption almost reached the equilibrium, during which the diffusion rate (k_{p3}) slowed down. Overall speaking, the adsorption process of Pb(II) by Mt was controlled by more than one-step, showing a decreasing trend of adsorption rates that corresponded to physical adsorption with a high rate and a chemical adsorption with a relatively low rate.

Table 4. Parameters derived from intraparticle diffusion model.

	Intraparticle diffusion model	
	value	R^2
k_{p1}	91.53±16.65	0.9679
k_{p2}	20.68±1.56	0.9942
k_{p3}	0.13±0.01	0.9515

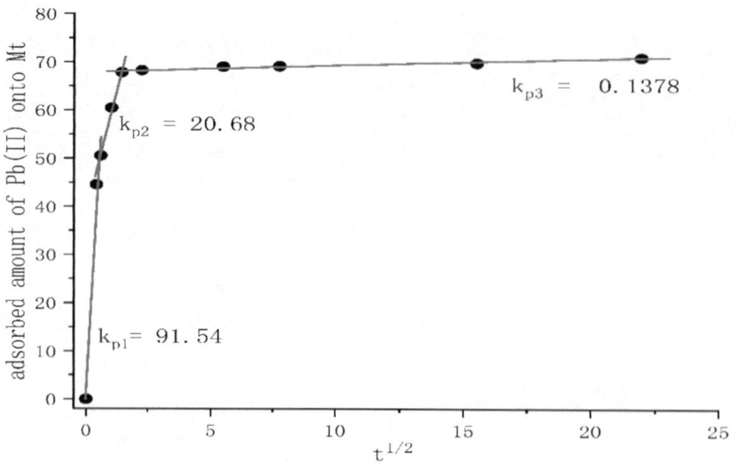

Figure 7. Intraparticle diffusion plot for Pb(II) adsorption onto Mt.

3.4. Adsorption isotherms

The overall amount of Pb(II) increased coinciding with its increased concentration in solution (Figure 8). The adsorption isotherm can be used to discuss the equilibrium relationship between adsorbent and adsorbate. The Freundlich model assumed that the adsorption enthalpy is non-uniformly distributed on the surface of the adsorbent.

The Freundlich isotherm equation is described as the follows[36]:

$$q_e = k_F C_e^{\frac{1}{n}}$$

where K_F is the Freundlich constant related to the sorption capacity; Ce is the concentration of Pb(II) in the solution at equilibrium time (mg/L); and $\frac{1}{n}$ is the sorption intensity.

The Langmuir theory indicates that it is monolayer adsorption and the adsorbate adsorbed homogenously onto the surface where the adsorption site had an equal affinity to the adsorbate [37,38].

The non-linear form of Langmuir isotherm equation [37,38] is written as:

$$q_e = \frac{Q_0 C_e K_L}{(1 + C_e K_L)}$$

where Q_0 is the monolayer adsorption capacity (mg/g); and K_L is the constant related to free adsorption energy (L/mg).

The fittings of experimental adsorption data to Langmuir and Freundlich models are shown in Figure 9 and the relevant parameters in Table 5. Since the adsorption isotherm was fitted only slightly better to the Freundlich model (R^2 = 0.9985) than the Langmuir model (R^2 = 0.9903), the adsorption on the surface was considered to be homogeneous monolayer adsorption at a low concentration [39], while the multilayer adsorption was performed at a high concentration in association with the indication of non-uniform distribution of adsorption enthalpy on the surface of the adsorbent [36].

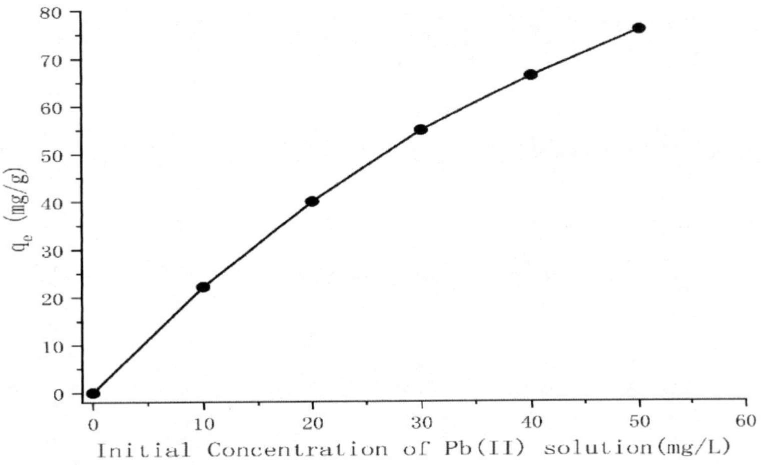

Figure 8. Relationship between different concentrations of Pb(II) in solution and the adsorption amount of Pb(II) onto Mt.

Table 5. Parameters derived from Langmuir and Freundlich fittings.

Models	Paramters	Value
Langmuir	Q_0	91.08 ± 5.83
	K_L	0.20 ± 0.04
	R^2	0.99031
Freundlich	K_F	22.44 ± 0.82
	1/n	0.41 ± 0.01
	R^2	0.99854

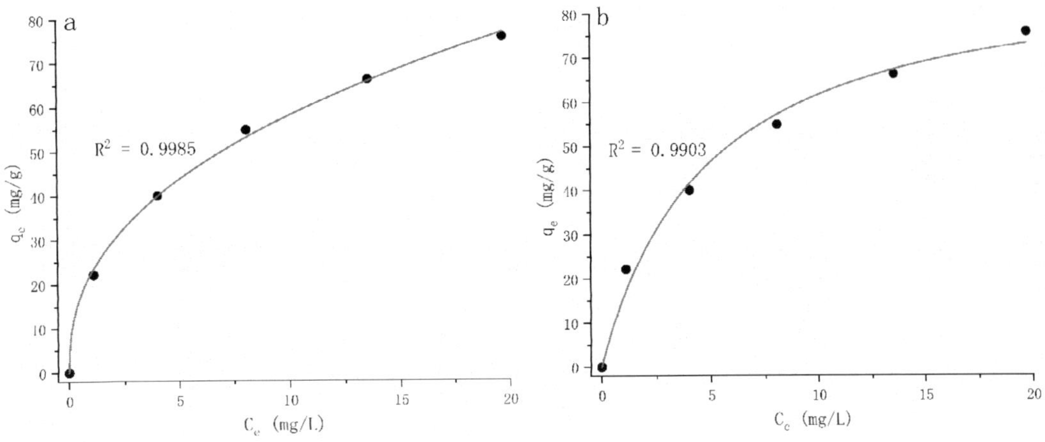

Figure 9. Adsorption isotherms of Pb(II) by Mt fitted to Freundlich equation (a); and Langmuir equation (b) in non-linear forms.

According to the fittings to Langmuir and Freundlich models, the adsorption of Pb(II) onto Mt is a combination process. Thus, Redlich-Peterson isotherm and Sips isotherm were used to further study the adsorption process.

The Redlich-Peterson isotherm is a combination of the Langmuir and the Freundlich isotherms. The benefit of this model includes the Henry region at infinite dilution [40]. As this model is combined with Langmuir and Freundlich isotherms, the adsorption mechanism is different based on the parameters after fitting and the adsorption condition.

This model is defined by the following expression [41,42].

$$q_e = \frac{K_{RP}C_e}{1 + a_{RP}C_e^{\beta_{RP}}}$$

where K_{RP} is Redlich-Peterson isotherm constant (L/g); a_{RP} is constant $(mg/L)^{-\beta}$; and β_{RP} is exponent in the range of 0 and 1.

At infinite concentration of adsorbate ($C_e \rightarrow \infty$), the equation can be reduced to the Freundlich equation:

$$q_e = \frac{K_{RP}}{a_{RP}} C_e^{1-\beta_{RP}}$$

where $\frac{K_{RP}}{a_{RP}}$ is K_F and $(1-\beta_{RP})$ is $\frac{1}{n}$.

When β_{RP} is closed to 0, the equation become Henry's law equation which is $q_e = \frac{K_{RP}}{1+a_{RP}} C_e$, where $\frac{K_{RP}}{1+a_{RP}}$ is the Henry's constant.

On the other hand, when β_{RP} is closed to 1, the equation reduces to Langmuir equation:

$$q_e = \frac{K_{RP}C_e}{1 + a_{RP}C_e}$$

where K_{RP} is $K_L Q_0$ and a_{RP} is K_L.

The Sips isotherm is another combination equation of the Langmuir and Freundlich isotherm models. The equation is given as followed [43]:

$$q_e = \frac{Q_s K_s C_e^{\beta_s}}{1 - K_s C_e^{\beta_s}}$$

where Qs is the monolayer adsorption capacity, mg/g; K_s is the energy of adsorption (L/mg); and β_s is Sips exponent. The above equation can be reduced to Langmuir isotherm when β_s is closed to 1, or Freundlich isotherm when C_e is closed to 0.

Table 6. The parameters from Redlich-Peterson and Sips fitting results.

Models	Parameters	Value
Redlich-Peterson	K_{RP}	60.44 ± 13.69
	a_{RP}	1.93 ± 0.59
	β_{RP}	0.67 ± 0.02
	R^2	0.9997
Sips	K_s	0.12 ± 0.01
	Q_s	189.40 ± 23.72
	β_s	0.56 ± 0.02
	R^2	0.9998

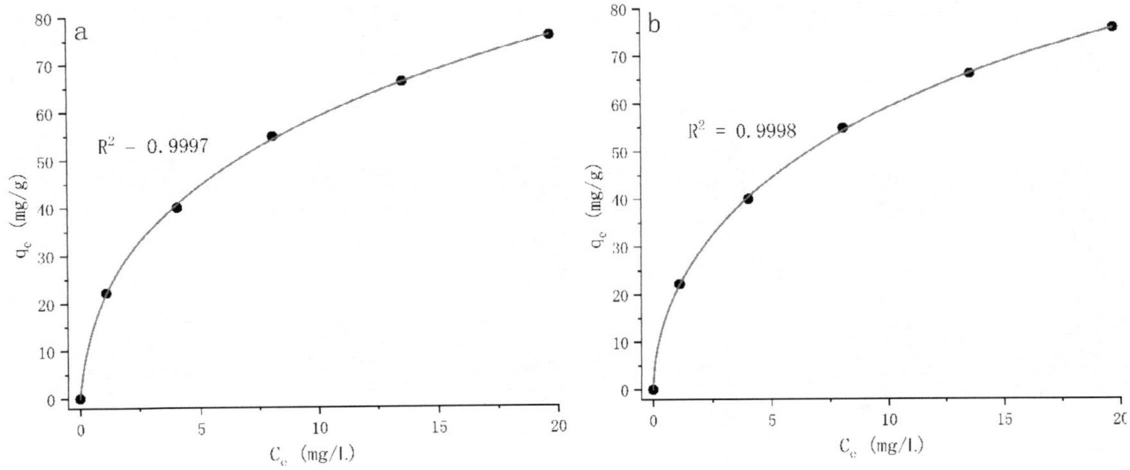

Figure 10. Adsorption isotherms of Pb(II) by Mt fitted to Redlich-Peterson equation (a); and Sips equation (b).

Table 7. The maximum Pb(II) adsorption capacities of various adsorbents.

Material	Maximum adsorption Capacity of Pb(II) (mg/g)	References
Inner Mongolia Montmorillonite	91.08	This study
Activated bentonite	6.535	[44]
bentonite	68.84	[31]
Bentonite based composite	95.88	[31]
Bentonite	19.19	[26]
Activated bentonite	1.73	[26]
Egyptian Na-activated bentonite	5.435	[32]
MX-80 bentonite	68.58	[25]
Apricot stone activated with sulphuric	21.38	[45]
nature kaolinite	2.35	[46]
nature zeolite	78.6	[47]
activated carbon modified by nitric acid under microwave heating	143.45	[48]
mesoporous activated carbon adsorbent	20.3	[49]
magnetic activated carbon incorporated with amino groups	104.2	[50]
Magnetic EDTA-modified chitosan/ SiO_2 /Fe_3O_4 adsorbent	123.5	[51]
Activated carbon from olive stones (COSAC)	84.3	[52]
activated carbon supported nanoscale zero-valent iron composite	59.35	[53]
EDTA-functionalized bamboo activated carbon	123.45	[54]
Theobroma cacao activated carbon	47.17	[55]
Theobroma-Tithonia activated carbon	46.95	[55]

The fittings to Redlich-Peterson and Sips isotherm models were shown in Figure 10 and the relevant parameters were shown in Table 6. Compared to Langmuir, Freundlich, and Redlich-Peterson, the fittings of adsorption data showed very close high values of R^2 while both Redlich-Peterson and Sips model even showed almost the same highest ones. The highest R^2 and the value of β, which is in between

0 and 1, in the fittings of both Redlich-Peterson and Sips isotherms indicate that both homogeneous monolayer adsorption and heterogeneous multilayer adsorption were occurred in the adsorption process of Pb(II) by Mt, and the adsorption was mainly controlled by heterogeneous multilayer adsorption [41,42,56].

3.5. Comparison montmorillonite on Pb(II) removal with different adsorbents

Table 7 shows the maximum adsorption capacity of adsorbents towards Pb(II). It shows that the Pb(II) adsorption capacity of MMt in this study is better than most of clay minerals and some activated carbon, though poorer than a few modified adsorbents.

4. Conclusions

The adsorption condition study revealed the following characteristics: Pb(II) adsorption is pH-dependent, the removal rate increased as the pH of the solution increased from 3 to 7; Pb(II) removal rate increased slowly with increasing temperature showing an insignificant effect on Pb(II) adsorption. The optimum conditions can therefore be concluded as: pH of the initial solution is around 6.5; the contact time for equilibrium is around 1 h; and the temperature is 25 ℃.

Compared with the PFO kinetic model, the PSO model is only slightly better in describing the adsorption of Pb(II) by Mt in aqueous solution. The Pb(II) adsorption mechanism of Mt is ion-exchange chemisorption in combination with adsorption diffusion and physical adsorption. According the fitting results of intraparticle diffusion model, the Pb(II) adsorption process by Mt was controlled by more than one-step. Although the Freundlich isotherm model is slightly better to describe the adsorption of Pb(II) by Mt in aqueous solution than the Langmuir one, R^2 values of them are similar. It indicates that the adsorption process might be a multi-process. This is proved and explained by the better fitting results from Redlich-Peterson model and Sips model than Langmuir model and Freundlich model. It indicates that the adsorption might be both homogeneous monolayer adsorption and heterogeneous multilayer adsorption, but mainly controlled by heterogeneous multiplayer adsorption.

This study provides preliminary information in applying montmorillonite and in further applying montmorillonite-containing mineral wastes to fabricate adsorbents for the removal of metal pollutants in aquatic systems.

Acknowledgement

This work is supported in part by Guangdong Department of Education (2020-2023) of the "Research Platform and Research Project" No. 2020ZDZX1022, and UIC Research Grant No. UICR5202018 at BNU-HKBU United International College, Zhuhai, PR China; Guangdong High Education Youth Creation Grant No. 2020KQNCX099, and UIC Research Grant No. UICR5202016 at BNU-HKBU United International College, Zhuhai, PR China; and UIC Research Grants Nos. R202011, R25202016, R5202018, and R201805 at BNU-HKBU United International College, Zhuhai, PR China.

References

[1] Wang H T 2021 Current status of soil heavy metal pollution and application of remediation technology *South. Agric. Mach.* 64–6

[2] Zhou Y 2018 *Pollution characteristics, sources and health risks of heavy metals in urban atmospheric PM_(2.5)* (Master's thesis, East China Normal University)

[3] Zhou Y, Xu B F, Wang P Y, Wu Y C and Zhao Q L 2019 Adsorption of trace arsenic and lead in water by microplastics *J. Zhejiang Univ. Technol.* 648–53

[4] He J Q 2018 Research on increasing water metal pollution status and treatment methods *China Met. Bull.* 242–4

[5] Elsheikh M A, Muchaonyerwa P, Johan E, Matsue N and Henmi T 2018 Mutual Adsorption of Lead and Phosphorus onto Selected Soil Clay Minerals *Adv. Chem. Eng. Sci.* **08** 67–81

[6] Purwiyanto A I S, Suteja Y, Trisno, Ningrum P S, Putri W A E, Rozirwan, Agustriani F, Fauziyah, Cordova M R and Koropitan A F 2020 Concentration and adsorption of Pb and Cu in

microplastics: Case study in aquatic environment *Mar. Pollut. Bull.* **158** 111380
[7] Ajmal M, Rao R A K, Ahmad R and Khan M A 2006 Adsorption studies on Parthenium hysterophorous weed: Removal and recovery of Cd(II) from wastewater *J. Hazard. Mater.* **135** 242–8
[8] Alghamdi A A, Al-Odayni A-B, Saeed W S, Al-Kahtani A, Alharthi F A and Aouak T 2019 Efficient Adsorption of Lead (II) from Aqueous Phase Solutions Using Polypyrrole-Based Activated Carbon *Materials (Basel).* **12** 2020
[9] Günay A, Arslankaya E and Tosun İ 2007 Lead removal from aqueous solution by natural and pretreated clinoptilolite: Adsorption equilibrium and kinetics *J. Hazard. Mater.* **146** 362–71
[10] Cooper C, Jiang J-Q and Ouki S 2002 Preliminary evaluation of polymeric Fe- and Al-modified clays as adsorbents for heavy metal removal in water treatment *J. Chem. Technol. Biotechnol.* **77** 546–51
[11] Li L, Zhu F, Ren T, Ou W, Lu Y, Li H and Zhai B 2016 Adsorption kinetics of Cd(II) ions from water by surface imprinted polymers *Environ. Chem.*
[12] De Gisi S, Lofrano G, Grassi M and Notarnicola M 2016 Characteristics and adsorption capacities of low-cost sorbents for wastewater treatment: A review *Sustain. Mater. Technol.* **9** 10–40
[13] Carolin C F, Kumar P S, Saravanan A, Joshiba G J and Naushad M 2017 Efficient techniques for the removal of toxic heavy metals from aquatic environment: A review *J. Environ. Chem. Eng.* **5** 2782–99
[14] Ju L 2012 The Research Progress of Bentonite Modification Methods *Guangdong Chem. Ind.*
[15] Pablo L De, Chávez M L and Abatal M 2011 Adsorption of heavy metals in acid to alkaline environments by montmorillonite and Ca-montmorillonite *Chem. Eng. J.* **171** 1276–86
[16] Chen C, Liu H, Chen T, Chen D and Frost R L 2015 Applied Clay Science An insight into the removal of Pb (II), Cu (II), Co (II), Cd (II), Zn (II), Ag (I), Hg (I), Cr (VI) by Na (I) -montmorillonite and Ca (II) -montmorillonite *Appl. Clay Sci.* **118** 239–47
[17] Uddin M K 2017 A review on the adsorption of heavy metals by clay minerals, with special focus on the past decade *Chem. Eng. J.* **308** 438–62
[18] Brigatti M F, Galán E and Theng B K G 2013 Structure and Mineralogy of Clay Minerals *Developments in Clay Science* vol 5, ed F Bergaya and G Lagaly (Developments in Clay Science) pp 21–81
[19] Rathnayake S I, Xi Y, Frost R L and Ayoko G A 2015 Structural and thermal properties of inorganic-organic montmorillonite: Implications for their potential environmental applications *J. Colloid Interface Sci.* **459** 17–28
[20] Gardolinski J, Lagaly G and Czank M 2004 On the destruction of kaolinite and gibbsite by phenylphosphonic, phenylphosphinic and phenylarsonic acids: evidence for the formation of new Al compounds *Clay Miner.* **39** 391–404
[21] Gardolinski J and Lagaly G 2005 Grafted organic derivatives of kaolinite: II. Intercalation of primary n-alkylamines and delamination *Clay Miner.* **40** 547–56
[22] Chinoune K, Bentaleb K, Bouberka Z, Nadim A and Maschke U 2016 Adsorption of reactive dyes from aqueous solution by dirty bentonite *Appl. Clay Sci.* **123** 64–75
[23] Qin Z, Yuan P, Zhu J, He H, Liu D and Yang S 2010 Influences of thermal pretreatment temperature and solvent on the organosilane modification of Al13-intercalated/Al-pillared montmorillonite *Appl. Clay Sci.* **50** 546–53
[24] Rathnayake S I, Xi Y, Frost R L and Ayoko G A 2016 Environmental applications of inorganic-organic clays for recalcitrant organic pollutants removal: Bisphenol A *J. Colloid Interface Sci.* **470** 183–95
[25] Xu D, Tan X L, Chen C L and Wang X K 2008 Adsorption of Pb(II) from aqueous solution to MX-80 bentonite: Effect of pH, ionic strength, foreign ions and temperature *Appl. Clay Sci.* **41** 37–46
[26] Kul A R and Koyuncu H 2010 Adsorption of Pb(II) ions from aqueous solution by native and activated bentonite: Kinetic, equilibrium and thermodynamic study *J. Hazard. Mater.* **179**

332–9
[27] Chen Y, Peng J, Xiao H, Peng H, Bu L, Pan Z, He Y, Chen F, Wang X and Li S 2017 Adsorption behavior of hydrotalcite-like modified bentonite for Pb 2+ , Cu 2+ and methyl orange removal from water *Appl. Surf. Sci.* **420** 773–81
[28] Parolo M E, Avena M J, Pettinari G R and Baschini M T 2012 Influence of Ca2+ on tetracycline adsorption on montmorillonite *J. Colloid Interface Sci.* **368** 420–6
[29] Zhu H, Sun N and University P 2019 Application Status and Research Progress of Modification Technology of Montmorillonite *Henan Chem. Ind.*
[30] Weng C-H 2004 Modeling Pb(II) adsorption onto sandy loam soil *J. Colloid Interface Sci.* **272** 262–70
[31] Randelović M, Purenović M, Zarubica A, Purenović J, Matović B and Momčilović M 2012 Synthesis of composite by application of mixed Fe, Mg (hydr)oxides coatings onto bentonite - A use for the removal of Pb(II) from water *J. Hazard. Mater.* **199–200** 367–74
[32] Taha A A, Shreadah M A, Ahmed A M and Heiba H F 2016 Multi-component adsorption of Pb(II), Cd(II), and Ni(II) onto Egyptian Na-activated bentonite; Equilibrium, kinetics, thermodynamics, and application for seawater desalination *J. Environ. Chem. Eng.* **4** 1166–80
[33] Largitte L and Pasquier R 2016 A review of the kinetics adsorption models and their application to the adsorption of lead by an activated carbon *Chem. Eng. Res. Des.* **109** 495–504
[34] Sheng Z L 1998 Experimental Researches on the Treatment of Lead Containing Wastewater with Activated Zeolite *Environ. Monit. CHINA*
[35] Weber W J and Morris J C 1963 Closure to "Kinetics of Adsorption on Carbon from Solution" *J. Sanit. Eng. Div.* **89** 53–5
[36] Freundlich H M F 1906 Over the Adsorption in Solution *J. Phys. Chem.* **57** 384–470
[37] Langmuir I 1916 THE CONSTITUTION AND FUNDAMENTAL PROPERTIES OF SOLIDS AND LIQUIDS. PART I. SOLIDS. *J. Am. Chem. Soc.* **38** 2221–95
[38] Langmuir I 1917 The constitution and fundamental properties of solids and liquids. Part II.— Liquids *J. Franklin Inst.* **184** 721
[39] Tong D S, Wu C W, Adebajo M O, Jin G C, Yu W H, Ji S F and Zhou C H 2018 Adsorption of methylene blue from aqueous solution onto porous cellulose-derived carbon/montmorillonite nanocomposites *Appl. Clay Sci.* **161** 256–64
[40] Ayawei N, Ebelegi A N and Wankasi D 2017 Modelling and Interpretation of Adsorption Isotherms *J. Chem.* **2017** 1–11
[41] Brouers F and Al-Musawi T J 2015 On the optimal use of isotherm models for the characterization of biosorption of lead onto algae *J. Mol. Liq.* **212** 46–51
[42] Tran H N, You S J, Hosseini-Bandegharaei A and Chao H P 2017 Mistakes and inconsistencies regarding adsorption of contaminants from aqueous solutions: A critical review *Water Res.* **120** 88–116
[43] Gil A, Assis F C C, Albeniz S and Korili S A 2011 Removal of dyes from wastewaters by adsorption on pillared clays *Chem. Eng. J.* **168** 1032–40
[44] Pawar R R, Lalhmunsiama, Bajaj H C and Lee S M 2016 Activated bentonite as a low-cost adsorbent for the removal of Cu(II) and Pb(II) from aqueous solutions: Batch and column studies *J. Ind. Eng. Chem.* **34** 213–23
[45] Mouni L, Merabet D, Bouzaza A and Belkhiri L 2011 Adsorption of Pb(II) from aqueous solutions using activated carbon developed from Apricot stone *Desalination* **276** 148–53
[46] Jiang M qin, Jin X ying, Lu X Q and Chen Z liang 2010 Adsorption of Pb(II), Cd(II), Ni(II) and Cu(II) onto natural kaolinite clay *Desalination* **252** 33–9
[47] Wang S, Terdkiatburana T and Tadé M O 2008 Adsorption of Cu(II), Pb(II) and humic acid on natural zeolite tuff in single and binary systems *Sep. Purif. Technol.* **62** 64–70
[48] Yao S, Zhang J, Shen D, Xiao R, Gu S, Zhao M and Liang J 2016 Removal of Pb(II) from water by the activated carbon modified by nitric acid under microwave heating *J. Colloid Interface Sci.* **463** 118–27

[49] Asuquo E, Martin A, Nzerem P, Siperstein F and Fan X 2017 Adsorption of Cd(II) and Pb(II) ions from aqueous solutions using mesoporous activated carbon adsorbent: Equilibrium, kinetics and characterisation studies *J. Environ. Chem. Eng.* **5** 679–98

[50] Fu R, Liu Y, Lou Z, Wang Z, Baig S A and Xu X 2016 Adsorptive removal of Pb(II) by magnetic activated carbon incorporated with amino groups from aqueous solutions *J. Taiwan Inst. Chem. Eng.* **62** 247–58

[51] Ren Y, Abbood H A, He F, Peng H and Huang K 2013 Magnetic EDTA-modified chitosan/SiO2/Fe3O4 adsorbent: Preparation, characterization, and application in heavy metal adsorption *Chem. Eng. J.* **226** 300–11

[52] Bohli T, Ouederni A, Fiol N and Villaescusa I 2015 Evaluation of an activated carbon from olive stones used as an adsorbent for heavy metal removal from aqueous phases *Comptes Rendus Chim.* **18** 88–99

[53] Liu X, Lai D and Wang Y 2019 Performance of Pb(II) removal by an activated carbon supported nanoscale zero-valent iron composite at ultralow iron content *J. Hazard. Mater.* **361** 37–48

[54] Lv D, Liu Y, Zhou J, Yang K, Lou Z, Baig S A and Xu X 2018 Application of EDTA-functionalized bamboo activated carbon (BAC) for Pb(II) and Cu(II) removal from aqueous solutions *Appl. Surf. Sci.* **428** 648–58

[55] Eletta O A A, Ayandele F O and Ighalo J O 2021 Adsorption of Pb(II) and Fe(II) by mesoporous composite activated carbon from Tithonia diversifolia stalk and Theobroma cacao pod *Biomass Convers. Biorefinery*

[56] Jeppu G P and Clement T P 2012 A modified Langmuir-Freundlich isotherm model for simulating pH-dependent adsorption effects *J. Contam. Hydrol.* **129–130** 46–53

Electrochemical determination of formamide for EVA splicing mats using activated CPE

Wenhan Zhong[1], Lei Lin[1], Zhiyuan Yin[2], Yan Li[3,4], Xiubing Du[2] and Yi Yang[1]

[1] Zhejiang Light Industrial Products Inspection and Research Institute, Hangzhou 310018, China;
[2] Sunon Technology Co., Ltd, Hangzhou 310020, China;
[3] Hangzhou Maysun Biotechnology Co., Ltd, Hangzhou 310030, China.

[4] Corresponding author's e-mail: liyan9664@163.com

Abstract. Electrochemical properties of formamide were investigated using cyclic voltammetry (CV) and linear sweep voltammetry (LSV) on an activated carbon paste electrode (CPE). A distinct oxidation peak was at 0.51 V in light of the electrochemical behaviour of formamide-KCl. This indicates an improved current response towards the electrooxidation of formamide due to the high charge-transfer ability of carbon paste. The LSV presented a linear dependence on the concentration of formamide-KCl in the range of 5–110 μM whose limit of detection was 0.16 μM. In addition, this dependable method can be explicitly designed to fit with formamide detection in EVA splicing mats.

1. Introduction

To increase the soft comfort of the floors, splicing mats are widely used by children in family playgrounds, kindergartens, shopping malls and fast food stores. Most of them are made of polyethylene and ethylene vinyl acetate (EVA) [1]. These mats are usually produced using azodicarbonamide, which can decompose into formamide during the foaming process and is harmful to the human body [2,3]. With the addition of a foaming agent, manufacturers can reduce the weight of mats and cut costs. Formamide is a polar solvent that is included in the list of highly concerning substances [4]; the European Union chemical administration classifies formamide as a reproductive toxicant. Formamide can be absorbed through the skin, affecting blood and nervous systems [5]. Long-term inhalation of formamide will poison organs, especially producing respiratory and liver damage [6,7]. Hence, the determination of formamide is of practical importance.

With the continuous improvement of living standards, mat products have undergone rapid developments especially in recent years. The main raw materials of indoor splicing mats are EVA copolymers. Previous quantitative approaches of formamide quantification included solid-liquid extraction combined with gas chromatography [8] and gas chromatography-mass spectrometry analyses [9,10]. However, nearly all of them touch upon pricey and large instruments, making these skills undesirable for functional microanalysis. Few studies have been performed on the formamide in splicing mats, particularly for the common problem of EVA mats. When these mats are under enclosed conditions for a long time, the formamide is released with potential hazard in daily life.

A portable method for electrochemical determination of formamide is necessary and has social benefits, which attempts to achieve the simplified, economic and miniaturised detection assays [11]. In the present work, the electrooxidation with CPE and the electrocatalysis of formamide were examined using cyclic voltammetry (CV) and linear sweep voltammetry (LSV). The formamide in EVA splicing mats was effectively identified by electrochemistry (EC) and gas chromatography/mass spectrometry (GC/MS). An EC method was proposed for the quantification of formamide with acceptable sensitivity.

2. Experimental section

2.1. Instruments and materials

A KST CS310X potentiostat (Wuhan), a Thermo Fisher Trace ISQ 7000 gas chromatograph/mass spectrometer system (USA), an Avanti J-HC high-speed centrifuge (USA) and a JC Electronic 4-mm diameter electrode shell were used.

Water (18.25 MΩ cm) was derived from Thelab-F purifier (USA). Ethylene glycol (99.5%), graphite powder (99.95%), liquid paraffin, hydrochloric acid, methanol, carrene, ethyl acetate, tetrahydrofuran, cyclohexane, ethanol, acetone and potassium chloride were purchased from Sinopharm (Beijing). Formamide (5000 μg mL^{-1} in H_2O) was derived from BePure Agentia.

2.2. Fabrication of activated CPE

The activation of CPE involved the following steps: 1.0 g graphite powder was mixed with 0.5 mL liquid paraffin in an agate mortar [12]. Subsequently, the homogenized paste was packed into the tip of a syringe equipped with Cu wiring. In the potential step, the working electrode was then submerged in 1.0 M HCl for 3 min, and it was increased step-wise from open-circuit condition to 0.4 V for 30 s to obtain the activated CPE. The surface state markedly influenced the redox behavior. Thus, the as-prepared electrodes used were freshly fabricated to avoid cross-contamination, and they should be polished before each measurement [13].

2.3. LSV studies

For electrooxidation of formamide, the LSV was studied within the range of 0.2–0.9 V at 40 mV s^{-1}. A three-electrode cell was made up of CPE, saturated calomel electrode (SCE) and Pt electrode.

Concentrations of formamide (C_f) at 5, 10, 20, 40, 60, 80, 100 and 110 μM were acquired with 0.1 M KCl electrolyte. The activated CPE was submerged in formamide-KCl for 20 min, and the oxidation current (I) of LSV measurements was fetched. The background concentration of the EC test without electrochemical activation was used to normalize the blank group [14].

2.4. Specimen pre-treatment and MS assay

Aliquots of EVA mats were cut into small pieces. Mat (0.1 g) with extractive solvent (10 mL) was incubated in an ultrasonic extractor at 50 °C for 35 min. The fluid was next revolved at 2800 rpm (1006×g centrifugal force) for 4 min with filtration using a 0.22 μm filter membrane. It was finally shifted to 0.1 M KCl for the LSV investigation.

1 mL extract in a chromatographic sampling bottle was analysed (Table 1). The GC area was related to Rt and *m*/*z* [15]. The integral areas were used to calculate the C_f of EVA mats.

Table 1. GC/MS parameters.

Instrumentation	Parameters
GC	Column: TG-WAXMS (30 m × 0.32 mm × 0.25 μm)
	Gas: 99.999% helium at 1.0 mL min^{-1}
	Temperature programme: 80 °C (3 min) at a rate of 20 °C min^{-1} to 230 °C (6 min)
	Injection volume (mode): 1 μL (splitless)
MS	Mode: 70 eV electron impact
	m/z: quantitation peak at 45.0; confirming peak at 44.0
	Ionisation temperature: 240 °C
	MS interface: 270 °C
	Postponement: 5 min

3. Results and discussions

3.1. Redox behaviour of formamide on the CPE

The CPE were immersed in formamide-KCl for 20 min as a distinctive feature of CV. No redox signals were examined in electrolyte on activated CPE (Figure 1*A*), while a well-defined peak of 5 μM formamide-KCl was observed (Figure 1*B*). The peak at 0.51 V was attributed to the oxidation of sorptive formamide, and a reduction peak at 0.31 V emerged on negative potential scan. The activated CPE has good electrocatalytic activity, and the redox potential can differentiate formamide from other amides to guarantee selectivity.

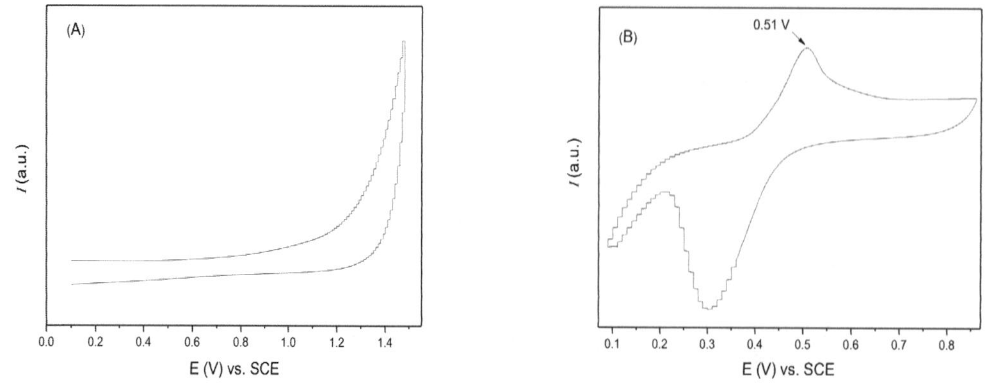

Figure 1. CV curves of 0.1 M supporting electrolyte (*A*) and 5 μM formamide-KCl (*B*) on activated CPE.

3.2. Effect of extractive solvent and extraction time

Carrene (1), methanol (2), ethyl acetate (3), tetrahydrofuran (4), cyclohexane (5), ethanol (6) and acetone (7) were tested as extractive solvents using positive EVA mats. As shown in Figure 2, under the same conditions (temperature 50 °C, ultrasound 35 min, extraction dosage 10 mL), the extraction efficiency of formamide in mats via (2) was higher than those of other organic solvents. Thus, methanol was selected as the extractive solvent.

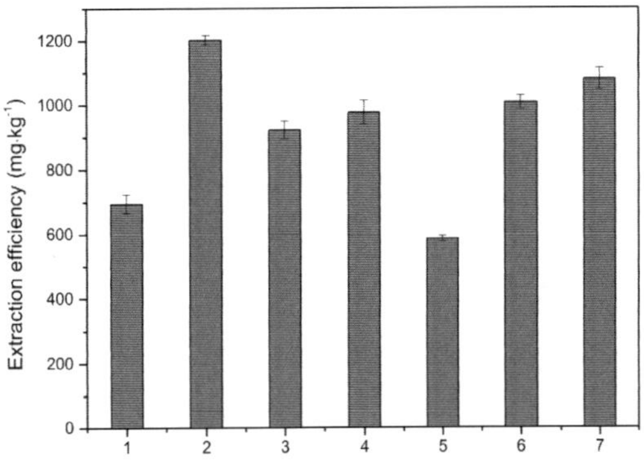

Figure 2. Extraction efficiency of formamide in splicing mats with different solvents (n = 3), and n stands for the repeat number.

The extraction time affected the content of formamide observed, and it was an essential factor for achieving the redox behaviour of formamide-KCl. After the addition of 10 mL methanol, the specimens were ultrasonically heated in the range of 5–45 min (Figure 3). When the extraction time was ≥30 min, the formamide extracted from EVA mats approximately reached a saturation state. When regarding the complex ingredients of mats, we evaluated the methanolic extracts at 35 min.

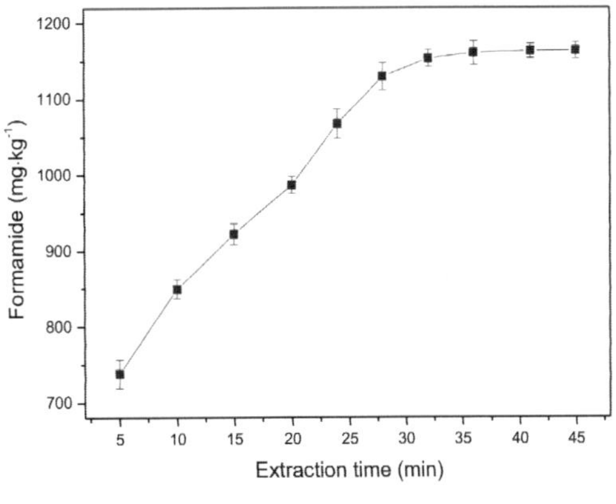

Figure 3. Influence of methanol extraction time on formamide from splicing mats (n = 3).

3.3. EC detection of formamide in EVA splicing mats

The activated CPE was constructed in formamide-KCl at a range of 5–110 μM, and the formamide standards were analysed using LSV. The relation of I (μA) and C_f (μM) was $I = 0.319C_f + 20.07$, and the correlation coefficient was 0.9973 (Figure 4). The limit of detection (LOD) was counted as 0.16 μM by this equation: LOD = $3\sigma/b$; where σ is the STDEV of I, and b is the straight slope.

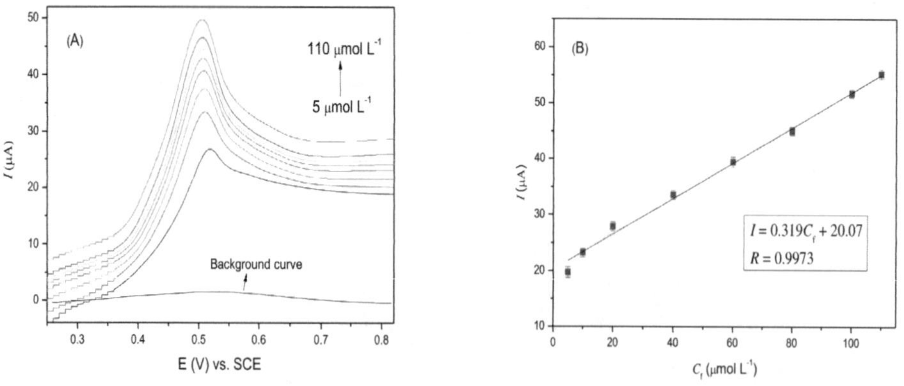

Figure 4. LSV responses of formamide-KCl on activated CPE at different C_f.

The concentrations of actual marked specimens at 5, 30 and 60 μM were correlated with the C_f of positive EVA mat. The C_f of the positive EVA mat was 10.3 and 10.8 μM using EC and GC/MS, respectively. The mean recoveries with EC were 93.0%–102.1%, while they were 94.5%–107.9% for GC/MS (Table 2). Their difference was slight, and the redox current was created as a convenient technique. The RSDs of the two analyzing methods were ≤6.1% and ≤5.5%. The precision and reproducible datum of CPE were satisfactory in light of electroanalytic. Besides the pre-treatment time, the analysis time of GC/MS was ca. 8 min, while the analysis time of EC was ca. 4 min for each datum.

Table 2. Recoveries of two analyzing methods for formamide.

Specimen	Added (μM)	Detected (μM)		Recovery (%)		RSD (%)	
		EC	GC/MS	EC	GC/MS	EC	GC/MS
Positive EVA mat	60	69.2	76.4	98.4	107.9	4.8	5.5
	30	37.5	42.7	93.0	104.6	6.1	4.3
	5	15.6	14.9	102.1	94.5	5.0	5.2

4. Conclusions

(1) In summary, a simplified detection assay was employed for the adsorption and subsequent oxidation of formamide.

(2) Surface functionalities were investigated using electrochemical behaviours of formamide-KCl on CPE. Analytical procedures for the LSV of formamide at activated CPE were developed, wherein the extractive solvent and extraction time played prominent roles. The mean recoveries of formamide in EVA splicing mats via EC method were 93.0%–102.1%, and the RSDs were ≤6.1%.

(3) The designed CPE promises to open a new perspective in the domain of amides monitoring, and CPE for formamide determination exhibits simple preparation and fast response.

Acknowledgement

This work was financially supported by NQI Project of Zhejiang Administration for Market Regulation (No. 20210121).

References

[1] Ren L, Wang M, Yan H, Dong Z, Dong W and Bai L 2014 Determination of formamide in

foam plastics by GC-MS *Plastics Science and Technology* **42** 99

[2] Rezaeian I, Jafari S H, Zahedi P, Ghaffari M and Afradian S 2010 Improvements of physical and mechanical properties of electron beam irradiation-crosslinked EVA foams *Polymers for Advanced Technologies* **20** 487

[3] Spina R 2015 Foam injection moulding of PE/EVA blends with an azodicarbonamide agent *Key Engineering Materials* **651** 863

[4] Adam Z R, Hongo Y, Cleaves H J, Yi R and Fahrenbach A C 2018 Estimating the capacity for production of formamide by radioactive minerals on the prebiotic earth *Scientific Reports* **8** 265

[5] Zheng L, Zhang W and Guo J 2020 Electrodeposition of Ni(II)–alizarin red complex on graphene-modified carbon paste electrode for formamide oxidation *Ionics* **26** 449

[6] Chaabouni H, Diana S, Nguyen T and Dulieu F 2018 Thermal desorption of formamide and methylamine from graphite and amorphous water ice surfaces *Astronomy and Astrophysics* **612** 47

[7] George J D, Price C J, Marr M C, Myers C B and Jahnke G D 2002 Evaluation of the developmental toxicity of formamide in new zealand white rabbits *Toxicological Sciences* **69** 165

[8] Fu T, Zheng G, Zheng W J, Sun Y and Li J Y 2012 Quantitative analysis of dimethyl formamide residues in acrylic fiber by GC *Advanced Materials Research* **554** 1901

[9] Sakano K, Oikawa S, Hasegawa K and Kawanishi S 2001 Hydroxyurea induces site-specific dna damage via formation of hydrogen peroxide and nitric oxide *Japanese Journal of Cancer Research* **92** 1166

[10] Wollinger W, Monteiro T M, Campos É G, Rego E C P and Cunha V S 2016 Child safety: A validated GC-MS method to determine formamide in EVA foam mats *Analytical Methods* **8** 5228

[11] Zhong N F, Yuan R H, He H H, Zhong W H and Wang Y 2021 Electrochemical sensor based on graphene-modified GCE for rapid quantification of benzo(a)pyrene in dark rubber materials *Journal of Physics: Conference Series* **1978** 012020

[12] Patil R H, Hegde R N and Nandibewoor S T 2009 Voltammetric oxidation and determination of atenolol using a carbon paste electrode *Industrial & Engineering Chemistry Research* **48** 10206

[13] Liu W H, Yuan R H, Teng Y J and Ma C A 2013 Electrochemical SERS of self-assembled monolayer of thiosalicylic acid adsorbed on activated gold electrodes *Acta Physico-Chimica Sinica* **29** 2599

[14] Che X H, Yuan R H, Tang L C, Xiong Y D, Wu M F and Zhu M H 2021 Study of catecholate-functionalized zinc complex on carbon steel and its application to corrosion resistance *Journal of Physics: Conference Series* **1978** 012009

[15] Zhu H Q, Lou C Y, Liu F R, Jiang G J, Ye X Y and Yuan R H 2021 Nanosized S-doped TiO_2 with effective visible-light degradation of BTEX from wood-based panels *Key Engineering Materials* **880** 139

Apoptosis of NG108-15 cells induced by fluorine combined with aluminum exposure and the effect of curcumin intervention

Xuemei Yang[1,4], Xingang Wang[1,4], Dandan Zhang[1], Yue Zhang[2], Dan Tao[1] and Chun Xie[1,3]

[1] School of Public Health, The Key Laboratory of Environmental Pollution Monitoring and Disease Control, Ministry of Education, Guizhou Medical University, Guiyang 550025, China;

[2] School of Basic Medical Sciences, Department of Pharmacology, Guizhou Medical University, Guiyang 550025, China.

[3] Corresponding author's e-mail: 1009207189@qq.com;
[4] These authors contributed equally to this work.

Abstract. The mechanism by which the combination of fluorine and aluminum (FA) cause learning and memory impairment is still unclear. Curcumin is a yellow polyphenol extracted from the rhizome of the plant curcuma longa linn, which has anti-apoptotic and antioxidant functions; studies have shown that the PKC-NMDAR pathway plays an important role in the process of cell apoptosis. Based on the above, we designed an in vitro study on the role of curcumin in cell apoptosis caused by coexistence of FA. In this study, the cell viability of NG108-15 cell was detected by CCK-8; cell apoptosis was detected by AO/EB dual fluorescence staining kit; mRNA expressions of PKC and NMDAR were evaluated by qRT-PCR and protein expressions of caspase3, Bax, Bcl-2, PKC and NMDAR were evaluated by WB. Our results showed that curcumin alleviated FA-induced apoptosis of NG108-15 cells, possibly by inhibiting the increase of caspase3 protein and Bax protein, and the decrease of Bcl-2 protein, thereby upregulating the expression of PKC and NMDAR to exert neuroprotective effect.

1. Introduction

Both fluorine (F) and aluminum (Al) exposure alone can cause abnormal changes in neurons, lead to excessive neuronal apoptosis and impair learning and memory [1, 2]. There are many fluorine and aluminum (FA) coexisting environments in life, such as coal-burning type endemic fluorosis where coalmixed with kaolin before burning and drinking water type endemic fluorosis where using aluminum salts for defluoridation, etc [1, 3, 4]. Aluminum fluoride easily accumulates the brain through the blood-brain barrier, causing structural changes and neuronal damage [5, 6]. Early research showed that FA can impair spatial learning and memory in offspring rats, but mechanism is not fully understood.

Learning and memory are the high-level neural activities of the brain. Learning is the biological process of acquiring new knowledge from the outside world; memory is the ability to integrate short-term memory into long-term memory. The main area of the brain that processes long-term learning and memory is the hippocampus, hippocampal neurons play a central regulatory role in it [7]. The previous research of our group showed that the FA exposure could impair the spatial learning and memory ability of the offspring rats, resulting in the abnormal structure of hippocampal neurons [8]. Most neurological diseases are related to abnormal changes in hippocampal neurons [9]. Hippocampal neurons reduced attention and cognition by regulating apoptosis-related proteins, ultimately leading to Alzheimer's disease (AD) [10].

Apoptosis is an autonomous programmed death process strictly regulated by related genes [11], which is one of the mechanisms of cell death in neurodegenerative diseases [12]. Apoptosis and learning and memory are closely related, manganese-induced hippocampal apoptosis can impair learning and memory in rats [13]. Apoptosis is mainly regulated by caspase family proteins and Bcl-2 family proteins [14]. Caspase3 can be activated to trigger cell apoptosis when cells are stimulated by apoptosis signals [15]. The mice deficient in the caspase3 gene have severe defects in the development of the nervous system [16]. Bax and Bcl-2 are the most representative apoptotic proteins, the ratio of the two is often used as a valuable indicator to measure the degree of apoptosis [17].

Protein kinase C (PKC) is involved in the regulation of multiple signal pathways in the brain and induces the regeneration of nerve synapses [18]. PKC is regarded as a family of "memory kinases", which is critically involved in memory maintenance [19]. When the body is stimulated, the intracellular Ca^{2+} concentration increases, which activates PKC and enhances the excitability of neurons [20]. NMDAR is one of the excitatory ionotropic glutamate receptors highly expressed in the hippocampus [21], which is involved in various physiological functions such as nervous system development, cell survival, excitatory synaptic transmission, etc [22]. Excessive activation of NMDAR or PKC will transfer Mg^{2+} in the NMDAR channel, thereby reducing the blocking effect of Mg^{2+} on the NMDAR channel, which allows Ca^{2+} to flow in, triggering intracellular calcium overload and cascade reactions, causing cell damage or death [23], this process is important pathogenesis of many neurological diseases. Meanwhile, NMDAR activation is a vital mechanism for the occurrence of LTP or LTD that induces synaptic plasticity [24]. The study has found that mice with NMDAR1 gene knockout have significantly reduced LTP in the hippocampus [25]. Besides, it was found that learning and memory ability of rats was impaired and accompanied by down-regulation of the NMDAR gene and protein expression in the Bisphenol A model [26]. The PKC-NMDAR pathway is essential in the formation of neuronal apoptosis and synaptic plasticity of LTP [27]. Based on the above, we hypothesized that learning and memory dysfunction may be related to the abnormal expression of PKC and NMDAR.

Curcumin is the main component of the traditional Chinese medicine turmeric, it has anti-oxidation, anti-apoptosis, anti-inflammatory and other pharmacological activities [28]. Curcumin could enter the brain through blood-brain barrier to enhance cell viability and synaptic plasticity, and increase neuron regeneration, thereby improve neurodegenerative diseases [29]. Curcumin protected the rat PC12 cells against damage induced by Aβ(25-35), which was associated with increased of the NMDAR2A expression [30]. In addition, curcumin improves the cognitive impairment of aging mice by increasing NMDAR1 expression in the hippocampi of mice [31]. In summary, we speculated that FA exposure may cause excessive apoptosis of nerve cells, and whether curcumin can protect nerve cells from damage caused by FA through the PKC-NMDAR pathway is worthy of in-depth study.

NG108-15 cell has neuronal characteristics in biology, which is widely used in the study of neurodegenerative diseases, such as AD [32, 33]. Therefore, in this study, the model of NG108-15 cells treated with FA with or without curcumin was established so as to explore the role of curcumin in FA-induced neuronal apoptosis by observing the effects of curcumin intervention or not on apoptosis and the PKC-NMDAR pathway. This study may provide a scientific basis for the prevention and treatment of FA-induced learning and memory impairment.

IOP Conf. Series: Earth and Environmental Science 1087 (2022) 012009 doi:10.1088/1755-1315/1087/1/012009

2. Materials and methods

2.1. Cell lines and agents
NG108-15 cell (ATCC Cell Bank, Shanghai, China), sodium fluoride and aluminum trichloride (Sigma-Aldrich, St. Louis, MO, USA). Curcumin (Yuanye, Shanghai, China). Primary antibodies targeting caspase3, Bax, Bcl-2, PKC, NMDAR1, NMDAR2A, NMDAR2B and β-actin (OH, USA).

2.2. Cell culture
NG108-15 cells were grown in DMEM culture medium supplemented with 10% fetal bovine serum in an incubator with 5% CO_2 at 37 °C.

2.3. Cell grouping and treatment
NG108-15 cells were treated with 20, 40 or 80 mg/L NaF and 160 mg/L $AlCl_3$ in the culture medium. The control (blank control) and experimental groups (FA groups) were generated by adding NaF and $AlCl_3$ to the medium at the following concentrations: (0, 0), (20, 160), (40, 160), and (80, 160) mg/L, noted as: control, low FA, medium FA and high FA groups. According to the results of the preliminary experiment, 16 μmol/L curcumin was selected for intervention. After 24 hours of FA treatments, 16 μmol/L curcumin was added to each of exposure group, curcumin was dissolved by DMSO (Watson. BIO, China).

2.3.1. Cell viability analysis. Cell viability was assessed by a cell counting kit-8 (CCK-8) (Dojindo Laboratories, Tokyo, Japan). NG108-15 cells were cultivated with medium in 96-well plates and treated with various concentrations of NaF and $AlCl_3$ mentioned above. After 24 h of FA treatments, 16 μmol/L curcumin was added to each of exposure group. A Multiskan GO spectrophotometer (Thermo Scientific, USA) was used to measure the absorbance at a wavelength of 450 nm.

2.3.2. Acridine Orange/Ethidium Bromide Assay. Apoptotic cells were detected by acridine orange/ethidium bromide (AO/EB) dual fluorescence staining kit (Yuanye, Shanghai, China). After FA treatment and curcumin intervention, the cells were digested with 0.25% trypsin and washed two times with PBS. AO/EB working solution (1:1 v/v) was added to the cell suspension and cellular morphology was evaluated by a fluorescent microscope (Nikon, Japan).

2.3.3. Real-time fluorescence quantitative PCR (qRT-PCR). RNA was isolated using TRIzol reagent (Invitrogen, Carlsbad, CA, USA). Preparation of the RNA and cDNA according to the previous method of the research group [8]. The levels of gene expression were calculated by the $2^{-\Delta\Delta Ct}$ method. The forward and reverse primers were AGAGGTGCCATGAGTTGGTTA and GGCTTCCGTATGTGTGGATTTT for PKC, ATGCACCTGCTGACATTCG and TATTGGCCTGGTTTACTGCCT for NMDAR1, ACGTGACAGAACGCGAACTT and TCAGTGCGGTTCATCAATAACG for NMDAR2A, GCCATGAACGAGACTGACCC and GCTTCCTGGTCCGTGTCATC for NMDAR2B, and AAGAAGGTGGTGAAGCAGG and GAAGGTGGAAGAGTGGGAGT for GAPDH.

2.3.4. Western blotting
Protein was extracted using RIPA buffer (Solarbio, Beijing, China). Preparation of protein samples according to the previous method of the research group [8]. The selected endogenous control was β-actin. The proteins were analysed using the ImageJ (Bio-Rad, CA, USA) software.

2.4. Data analysis
All the data were presented as the means ± standard deviation (SD). Statistical analysis was carried out using SPSS 20.0. The means of multiple samples were compared by single factor extraordinary analysis, the least significant difference (LSD) test was used to compare between groups with

homogeneous variance, Dunnett test was used to compare with heterogeneous variance, respectively. A *P*-value < 0.05 was considered statistically significant.

3. Results

3.1. The effects of curcumin intervention on viability of NG108-15 cells exposed to FA
In Table 1, cell viability was decreased in FA groups as compared to control and increased in each curcumin intervention group as compared to corresponding FA group.

Table 1. The cell viability of NG108-15 cells.

Groups	Cell viability (%)
control	100.01±1.07
DMSO	99.36±0.48
curcumin	100.97±1.64
low FA	81.89±0.99 [a]
medium FA	61.39±0.84 [a]
high FA	21.35±0.83 [a]
low FA + **curcumin**	87.59±0.72 [b]
medium + **curcumin**	72.81±0.77 [b]
high FA + **curcumin**	69.24±2.17 [b]

[a] Denotes significant difference (P <0.05) as compared to control group; [b] Denotes significant difference (P <0.05) as compared to corresponding FA groups (n = 6).

3.2. The effects of curcumin intervention on apoptosis in NG108-15 cells exposed to FA
In Figure 1B, the percentage of apoptotic cell death was increased in all FA and curcumin intervention groups as compared to control. The rates were decreased in low and high FA + curcumin intervention groups as compared to corresponding FA group.

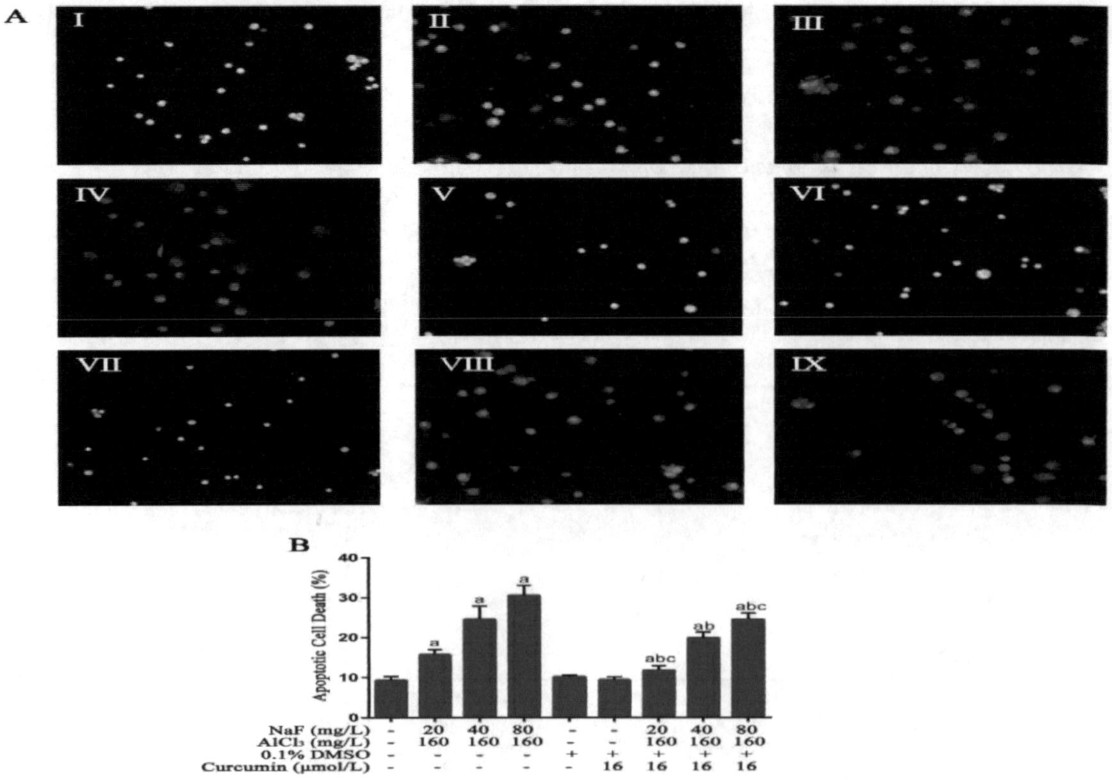

Figure 1. The effects of curcumin intervention on apoptosis in NG108-15 cells exposed to FA.
(A) Apoptosis of NG108-15 cells after FA exposure and curcumin intervention (100×). Ⅰ: control group; Ⅱ: low FA group; Ⅲ: medium FA group; Ⅳ: high FA group; Ⅴ: DMSO group; Ⅵ: curcumin group; Ⅶ: low FA + curcumin intervention group; Ⅷ: medium FA + curcumin intervention group; Ⅸ: high FA + curcumin intervention group. AO/EB was used to measure apoptosis. Yellow-green fluorescence: normal cells, orange-red fluorescence: apoptotic cells. (B) The percentage of apoptotic cell death of NG108-15 cells. [a] Denotes significant difference ($P < 0.05$) as compared to control group; [b] Denotes significant difference ($P < 0.05$) as compared to curcumin group; [c] Denotes significant difference ($P < 0.05$) as compared to corresponding FA groups (n = 6).

3.3. The effects of curcumin intervention on expression of caspase3, Bax and Bcl-2 protein in NG108-15 cells exposed to FA

In Figure 2A, all FA and curcumin intervention groups showed increased caspase3 and Bax proteins expression as compared to control, while the expression of Bcl-2 protein decreased. The caspase3 proteins in medium and high FA + curcumin intervention groups exhibited lower expression than corresponding FA groups. The Bax protein expression was decreased in the low FA + curcumin intervention group as compared to corresponding FA groups, but all curcumin intervention groups showed enhanced expression of Bcl-2 protein expression. In Figure 2B, Bax/Bcl-2 was increased in all FA and curcumin intervention groups as compared to control, and each curcumin intervention groups were decreased as compared to corresponding FA groups.

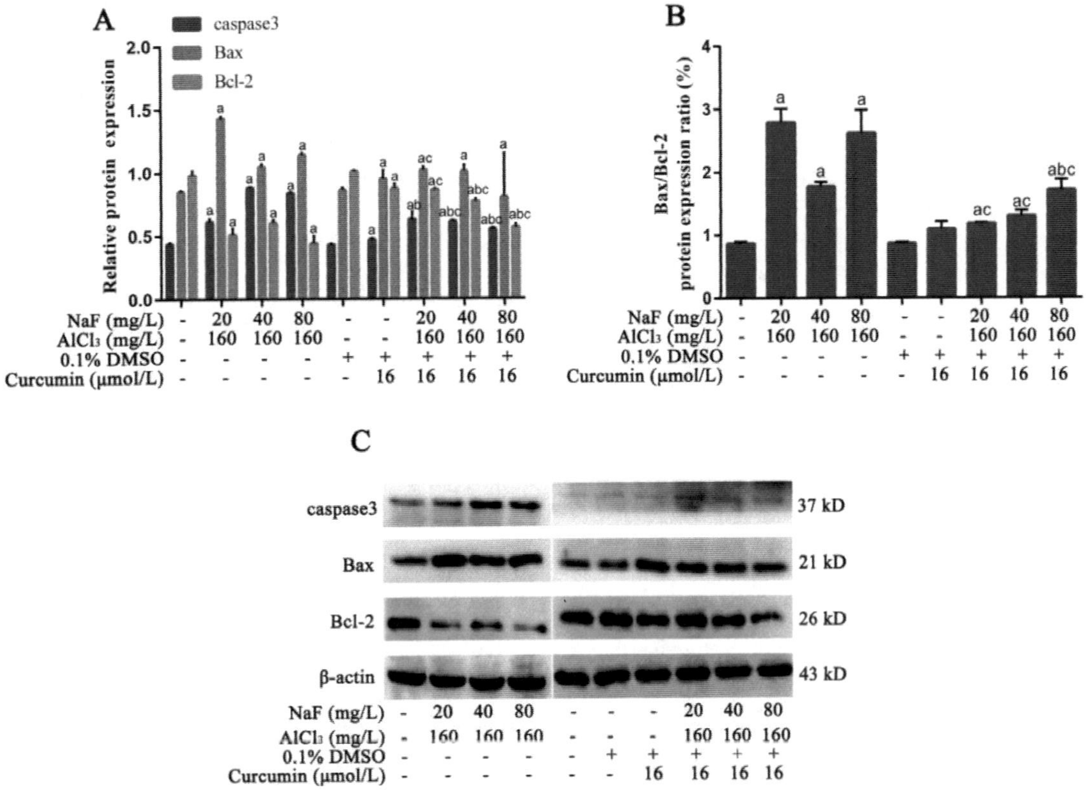

Figure 2. The effects of curcumin intervention on caspase3, Bax and Bcl-2 protein expression in NG108-15 cells exposed to FA.

(A/B/C) The expression of caspase3, Bax and Bcl-2 protein in NG108-15 cells. [a] Denotes significant difference ($P < 0.05$) as compared to control group; [b] Denotes significant difference ($P < 0.05$) as compared to curcumin group; [c] Denotes significant difference ($P < 0.05$) as compared to corresponding FA groups ($n = 6$).

3.4. The effects of curcumin intervention on expression of PKC and NMDAR mRNA in NG108-15 cells exposed to FA

In Figure 3A, the PKC mRNA expression was decreased in all FA and high FA + curcumin intervention groups as compared to control, it was increased in each curcumin intervention groups as compared to corresponding FA groups; NMDAR1 mRNA expression was significantly decreased in medium and high FA groups as compared to control, it was increased in medium and high FA + curcumin intervention groups as compared to corresponding FA groups. In Figure 3B, the NMDAR2A mRNA expression was decreased in medium and high FA groups, and increased in low FA group, low and medium FA + curcumin intervention groups as compared to control; NMDAR2B mRNA expression was decreased in all FA and high FA+ curcumin intervention groups as compared to control; NMDAR2A and NMDAR2B mRNA expressions were increased in each curcumin intervention group as compared to corresponding FA group.

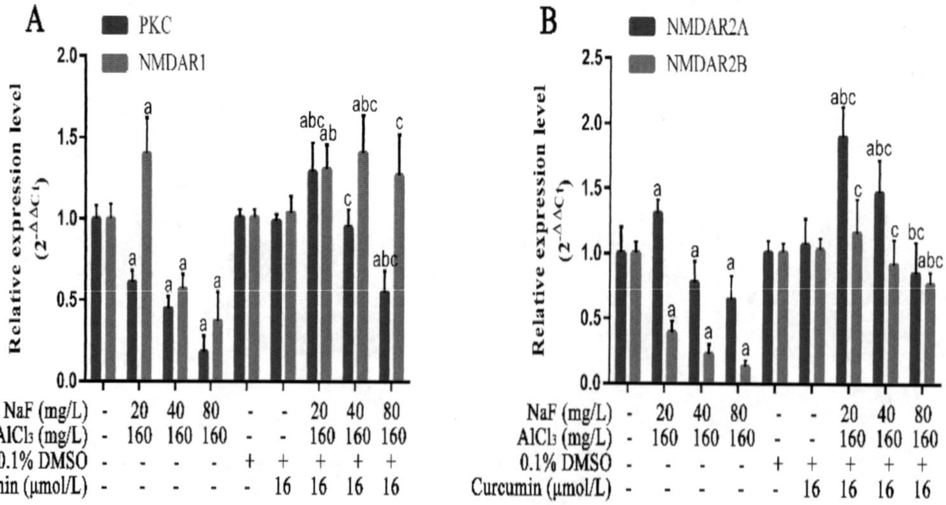

Figure 3. The effects of curcumin intervention on expression of PKC and NMDAR mRNA in NG108-15 cells exposed to FA.

(A/B) The expression of PKC and NMDAR mRNA in NG108-15 cells. [a] Denotes significant difference (P <0.05) as compared to control group; [b] Denotes significant difference (P <0.05) as compared to curcumin group; [c] Denotes significant difference (P <0.05) as compared to corresponding FA groups (n = 6).

3.5. The effects of curcumin intervention on expression of PKC and NMDAR protein in NG108-15 cells exposed to FA

In Figure 4A, the PKC protein expression was decreased in medium and high FA groups as compared to control, it was increased in each curcumin intervention group as compared to corresponding FA group; NMDAR1 protein expression was increased in all FA and curcumin intervention groups as compared to control. In Figure 4B, NMDAR2A protein expression was increased in all FA and curcumin intervention groups except medium and high FA groups as compared to control, it was increased in medium and high FA + curcumin intervention groups as compared to corresponding FA group; the NMDAR2B protein expression was decreased in FA groups and increased in curcumin intervention groups as compared to control, it was increased in each curcumin intervention group as compared to corresponding FA group.

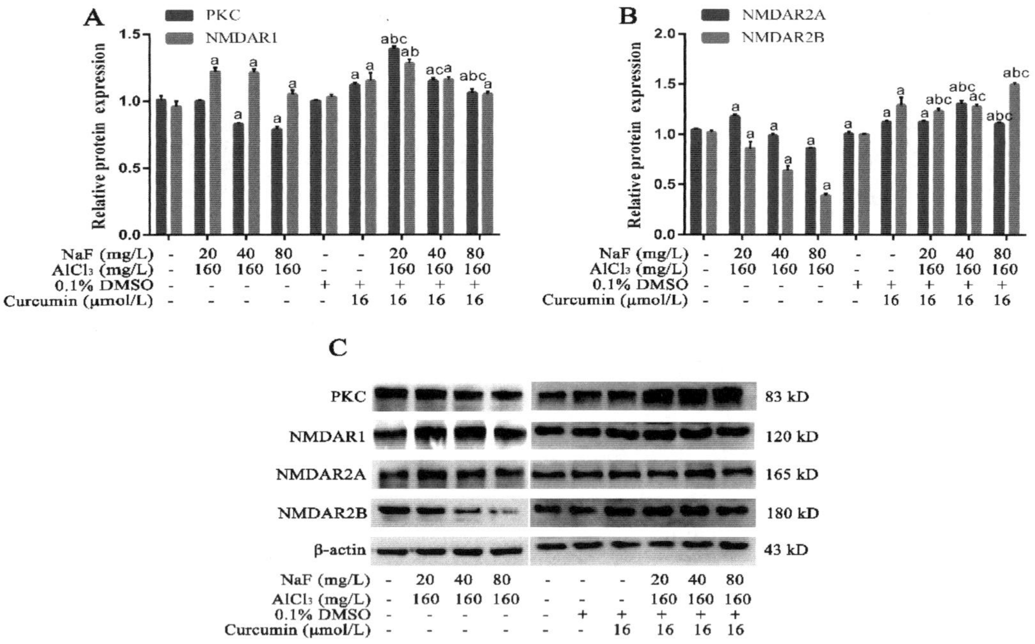

Figure 4. The effects of curcumin intervention on expression of PKC and NMDAR protein in NG108-15 cells exposed to FA.

(A/B/C) The expression of PKC and NMDAR protein in NG108-15 cells. [a] Denotes significant difference (P <0.05) as compared to control group; [b] Denotes significant difference (P <0.05) as compared to curcumin group; [c] Denotes significant difference (P <0.05) as compared to corresponding FA groups (n = 6).

4. Discussions

In this study, we found that FA reduced the viability of NG108-15 cells. However, the cell viability was increased after curcumin intervention. In the neuronal PC12 injury model, curcumin could improve cell viability [34]. It showed that curcumin could improve the cell viability of FA-suppressed NG108-15 cells. Both F and Al are neurotoxic substances, either alone or combined, which can accumulate in the hippocampus and lead to cell apoptosis and abnormal nervous system function [35, 36]. It indicated the mechanism of FA induced cell damage may be related to apoptosis.

Caspase3 and Bcl-2 family proteins are representative apoptotic proteins [37]. The results of this study showed that FA exposure increased the percentage of apoptotic cell death of NG108-15 cells. Meanwhile, the expressions of pro-apoptotic proteins caspase3 and Bax were significantly increased, the apoptosis-inhibiting protein Bcl-2 was decreased and Bax/Bcl-2 was increased in each exposure group. In the vascular dementia (VaD) rat model, at the same time as the apoptosis rate of nerve cells increased, and the expression levels of caspase3 and Bax protein were also increased [38]. In a model of cerebral ischemia-reperfusion injury, rat neuron over-apoptosis is accompanied with upregulation of caspase3 and Bax expression and downregulation of Bcl-2 expression [39]. The above suggested that FA exposure could exacerbate the degree of apoptosis of NG108-15 cells, possibly by upregulating the expression of pro-apoptotic protein Bax and downregulating the expression of apoptosis-inhibiting protein Bcl-2 to activate the expression of caspase3 protein to aggravate cell apoptosis.

Owing to its anti-oxidation, anti-apoptosis, anti-inflammatory, curcumin can be used to accelerate neural recovery and help treat nerve injury [28, 40]. Our results showed that, compared with the corresponding FA groups, caspase3 protein expression in medium and high FA + curcumin intervention groups were significantly downregulated, and Bax protein decreased in low FA + curcumin intervention group, Bcl-2 protein was upregulated in each curcumin intervention group, and the ratio of Bax/Bcl-2 was significantly reduced. Curcumin can increase the cell viability of SH-SY5Y cells in an in vitro model of cerebral ischemia reperfusion, reduce LDH activity, downregulate the expression of caspase3 and Bax protein, and upregulate the Bcl-2 protein, thereby inhibiting the occurrence of cell apoptosis [41]. Besides, curcumin can inhibit ischemia-induced mitochondrial apoptosis by inhibiting the activation of caspase3 and Bax, upregulating the anti-apoptotic protein Bcl-2 expression, thereby promoting the survival of neurons in the body [42]. It was suggested that curcumin intervention may reduce the apoptosis of NG108-15 cells to a certain extent by downregulating the expression of caspase3 protein and Bax protein and upregulating Bcl-2 protein, thereby exerting neuroprotective effects.

PKC can participate in the regulation of multiple signal pathways in the brain, which regulates synaptic plasticity and cell apoptosis; NMDAR induces Ca^{2+} to enter neuronal cells and promotes the activation of PKC, which contributes significantly to the formation of different types of memories and the expression of LTP [29, 43]. In this study, FA exposure decreased the expression of PKC, NMDAR2A and NMDAR2B protein in NG108-15 cells. In the AD rat model, the memory loss of rats was found to be accompanied with a decrease of PKC protein expression, and resveratrol could improve rat memory by upregulating the expression of PKC [44]. In the Al-exposed PC12 cells model, Al may inhibit mRNA and protein expressions of PKC and NMDAR subunits, thereby affecting learning and memory, which is similar to the results of this study [45]. It indicated that the FA exposure may affect the learning and memory function by inhibiting the PKC-NMDAR pathway.

Curcumin has a positive effect on the death of hippocampal neurons and can reverse memory deficits and cognitive impairment [46]. In this study, PKC and NMDAR expression in NG108-15 cells from each curcumin intervention group were increased as compared to corresponding FA group, which showed curcumin intervention attenuated FA-induced PKC and NMDAR reduction. Curcumin inhibits NMDA receptor and NMDAR-mediated intracellular Ca^{2+} increase, and inhibits the phosphorylation of CREB, thereby inhibiting Aβ-induced neuronal damage and apoptosis [47]. Curcumin can protect hippocampal neurons by increasing the level of NMDAR2A [31]. In a rat model of dementia, curcumin intervention improves the learning and memory function of rats and promotes the expression of NMDAR2B [48]. The result suggested that curcumin may show a positive effect on nerve cells by upregulating the expression of PKC and NMDAR.

5. Conclusions

In summary, this study found that FA exposure led to cell apoptosis increased, the expression of caspase3 and Bax upregulated, and Bcl-2 downregulated, which inhibited the PKC-NMDAR pathway. However, curcumin can alleviate FA-induced apoptosis of NG108-15 cells, possibly by inhibiting the increase of caspase3 protein and Bax protein, and the decrease of Bcl-2 protein, thereby upregulating the expression of PKC and NMDAR to exert neuroprotective effect.

Statements & Declarations

Funding: This work was supported by the Natural Science Foundation of China (Grant No. 81860563).

Competing Interests: The authors declare that they have no conflict of interest.

References

[1] Choubisa SL, Choubisa D 2016 *J. Environ Sci Pollut Res Int* **23** 7244–7254
[2] Qin X, Li L, Nie X, Niu Q 2020 *J. Biol Trace Elem Res* **197(2)** 571-579
[3] Chowdhury CR, Shahnawaz K, Kumari D, Chowdhury A, Bedi R, Lynch E, Harding S, Grootveld M 2016 *J. Perspect Public Health* **136(6)** 353-360
[4] Fan Z, Gao Y, Wang W, Gong H, Guo M, Zhao S, Liu X, Yu B, Sun D 2016 *J. J Epidemiol* **26** 57–63
[5] Kinawy AA 2019 *J. Environ Sci Pollut Res Int* **26(11)** 10951-10960
[6] Nalagoni CSR, Karnati PR 2016 *J. Interdiscip Toxicol* **9(2)** 78-82

[7] Shin JD, Tang W, Jadhav SP 2019 *J. Neuron* **104(6)** 1110–1125.e1117
[8] Ge QD, Tan Y, Luo Y, Wang WJ, Zhang H, Xie C 2018 *J. Environ Toxicol Pharmacol* **63** 60-68
[9] Noël A, Zhou L, Foveau B, Sjöström PJ, LeBlanc AC 2018 *J. Cell Death Differ* **25(7)** 1319-1335
[10] Hu X, Song C, Fang M, Li C 2018 *J. Exp Ther Med* **15(2)** 1795-1802
[11] D'Arcy MS 2019 *J. Cell Biol Int* **43(6)** 582-592
[12] Chi H, Chang HY, Sang TK 2018 *J. Int J Mol Sci* **19(10)**
[13] Nkpaa KW, Amadi BA, Wegwu MO, Farombi EO 2019 *J. Toxicology* **418** 51-61
[14] Ying J, Xu J, Shen L, Mao Z, Liang J, Lin S, Yu X, Pan R, Yan C, Li S, Bao Q, Li P 2017 *J. Biol Trace Elem Res* **179(1)** 59-69
[15] Liu HC, Zhang Y, Zhang S, Xin T, Li WH, Wu WL, Pang Q, Chen YZ 2015 *J. Eur Rev Med Pharmacol Sci* **19(18)** 3459-3467
[16] Imbriani P, Tassone A, Meringolo M, Ponterio G, Madeo G, Pisani A, Bonsi P, Martella G 2019 *J. Int J Mol Sci* **20(14)**
[17] Mahdavi S, Khodarahmi P, Roodbari NH 2018 *J. Hum Exp Toxicol* **37(3)** 321-328
[18] Lanuza MA, Just-Borràs L, Hurtado E, Cilleros-Mañé V, Tomàs M, Garcia N, Tomàs J 2019 *J. Cells* **8(12)**
[19] Sun MK, Alkon DL 2014 *J. Prog Mol Biol Transl Sci* **122** 31-59
[20] Newton AC 2018 *J. Crit Rev Biochem Mol Biol* **53(2)** 208-230
[21] He F, Ran Y, Li X, Wang D, Zhang Q, Lv J, Yu C, Qu Y, Zhang X, Xu A, Wei C, Chou CJ, Wu J 2020 *J. Bioorg Chem* **103** 104109
[22] Parsons MP, Raymond LA 2014 *J. Neuron* **82(2)** 279-293
[23] Murphy NP, Cordier J, Glowinski J, Prémont J 1994 *J. Eur J Neurosci* **6(5)** 854-60
[24] Sachana M, Rolaki A, Bal-Price A 2018 *J. Toxicol Appl Pharmacol* **354** 153-175
[25] Burnashev N, Szepetowski P 2015 *J. Curr Opin Pharmacol* **20** 73-82
[26] Yu H, Ma L, Liu D, Wang Y, Pei X, Duan Z, Ma M, Zhang Y 2020 *J. Environ Pollut* **266(Pt 1)** 115055
[27] Pradier B, Lanning K, Taljan KT, Feuille CJ, Nagy MA, Kauer JA 2018 *J. J Neurosci* **38(25)** 5750-5758
[28] Laabbar W, Elgot A, Elhiba O, Gamrani H 2019 *J. J Chem Neuroanat* **100** 101654
[29] Choi GY, Kim HB, Hwang ES, Lee S, Kim MJ, Choi JY, Lee SO, Kim SS, Park JH 2017 *J. Mediators Inflamm* **2017** 6280925
[30] Qian W, Li H, Pan N, Zhang C 2018 *J. Med Sci Monit* **24** 2693-2699
[31] Sun CY, Qi SS, Zhou P, Cui HR, Chen SX, Dai KY, Tang ML 2013 *J. Pharmacol Biochem Behav* **105** 76-82
[32] Kubickova J, Hudecova S, Csaderova L, Soltysova A, Lichvarova L, Lencesova L, Babula P, Krizanova O 2016 *J. Neuroscience* **325** 100-110
[33] Oshita J, Okazaki T, Mitsuhara T, Imura T, Nakagawa K, Otsuka T, Kurose T, Tamura T, Abiko M, Takeda M, Kawahara Y, Yuge L, Kurisu K 2020 *J. Neurol Med Chir* **60(2)** 83-93
[34] Farkhondeh T, Ashrafizadeh M, Azimi-Nezhad M, Samini F, Aschner M, Samarghandian S 2021 *J.* Curr Mol Pharmacol **14(6)**1146-1155
[35] Rifaai RA, Mokhemer SA, Saber EA, El-Aleem SAA, El-Tahawy NFG 2020 *J. J Chem Neuroanat* **107** 101795
[36] Peng Z, Yang X, Zhang H, Yin M, Luo Y, Xie C 2021 *J. Ecotoxicol Environ Saf* **224** 112658
[37] Liu Q, Si T, Xu X, Liang F, Wang L, Pan S 2015 *J. Reprod Health* **12** 65
[38] Qi YY, Feng XF, Qiu L, Yang F 2019 *J. Eur Rev Med Pharmacol Sci* **23(10)** 4474-4480
[39] Zhou T, Liang Y, Jiang L, Yu T, Zeng C, Tao E 2017 *J. Biochem Biophys Res Commun* **486(4)** 1005-1013
[40] Moharrami Kasmaie F, Jahromi Z, Gazor R, Zaminy A 2019 *J. Excli j* **18** 653-665
[41] Wu L, Jiang C, Kang Y, Dai Y, Fang W, Huang P 2020 *J. Int J Mol Med* **45(4)** 993-1004

IOP Conf. Series: Earth and Environmental Science 1087 (2022) 012009 doi:10.1088/1755-1315/1087/1/012009

[42] Xie CJ, Gu AP, Cai J, Wu Y, Chen RC 2018 *J. Brain Behav* **8(2)** e00921
[43] Rodríguez-Durán LF, Escobar ML 2014 *J. Behav Brain Res* **266** 58-62
[44] Zhang Y, Dong R, Yang Q, Zhang L, Li J, Zhao H 2019 *J. Wei Sheng Yan Jiu* **48(2)** 269-278
[45] He C, Zhao X, Li H, Wang F, Zhang J, Wang Y, Han Y, Yuan C, Niu Q 2021 *J. Neurotox Res* **39(3)** 634-644
[46] Xu Y, Lin D, Li S, Li G, Shyamala SG, Barish PA, Vernon MM, Pan J, Ogle WO 2009 *J. Neuropharmacology* **57(4)** 463-471
[47] Huang HC, Chang P, Lu SY, Zheng BW, Jiang ZF 2015 *J. J Recept Signal Transduct Res* **35(5)** 450-457
[48] Dong J, Lu DX, Pan R, Tang HM 2008 *J. Xi Bao Yu Fen Zi Mian Yi Xue Za Zhi* **24(4)** 328-331

IOP Conf. Series: Earth and Environmental Science 1087 (2022) 012010 doi:10.1088/1755-1315/1087/1/012010

Management and suggestion of ecological flow guarantee of key rivers and lakes in the Yangtze River Basin

Zhimin Deng[1,2], Dandan Shen[1] and Keke Zhang[1]

[1] Changjiang Water Resources Protection Institute, Hubei Wuhan 430051, China;

[2] Corresponding author's e-mail: whudzmgx@163.com

Abstract. In this paper, inter-provincial key rivers and lakes (KRLs) and their ecological flow (eco-flow) targets that have been issued and implemented, were taken as the research object. Based on the daily data of eco-flow at each control section, the eco-flow guarantee situation of the KRLs from 2020 to 2021 was analyzed, showing that except Dashadian section, the eco-flow at other sections was all up to the standard. And the reason of target dissatisfaction is that the discharge of two power stations located in the upstream of Dashadian section did not meet the requirements due to their maintenance. Moreover, the accomplishments and problems of eco-flow management were summarized. Furthermore, in the terms of dispatching management, monitoring and early warning, supervision and management, and further research, suggestions and measures were put forward to provide reference for the follow-up eco-flow management.

1. Introduction

Ecological flow (eco-flow) is an important index to maintain the ecological functions of rivers and lakes and to control the water resources development and utilization. It is also an important basis for the overall planning of water use for life, production and ecology and the optimal allocation of water resources. Eco-flow guarantee is related to the recovery of the eco-environment of rivers and lakes, the construction of ecological civilization and high-quality development, as well as the implementation of the relevant requirements of the Yangtze River Protection Law of the People's Republic of China. As the mother river of China, the Yangtze River spans 19 provinces, cities and autonomous regions. There are many rivers and lakes in this basin, and they are rich in natural resources [1-2]. In recent years, due to human activities, the ecological functions of lakes and wetlands in the middle and lower reaches of the basin have been degraded, and the intermittent river cutoff occurred in some reaches [3-4].

The Ministry of Water Resources attaches great importance to the target determination and guarantee management of eco-flow, and has carried out the compilation of guarantee schemes. From 2019 to 2020, 62 control sections of 32 inter-provincial KRLs in the Yangtze River Basin and their eco-flow guarantee targets were determined by the river basin management agency, and implementation plans for eco-flow guarantee was completed. Moreover, from 2019 to 2020, 11 provinces (including municipalities and autonomous regions) in this basin issued and implemented eco-flow guarantee targets for 230 major control sections of 102 KRLs within each province. From different levels of the Ministry of Water Resources, the river basin management agency, local provinces and cities, eco-flow guarantee management has been promoted, and the aquatic eco-environment has been significantly improved.

In this paper, inter-provincial KRLs and their eco-flow targets that have been issued and implemented, were taken as the research object. Based on the daily flow (or water table) data, KRLs in the Yangtze River Basin were evaluated, and their eco-flow guarantee situation and related issues were analyzed. Furthermore, suggestions of the subsequent eco-flow guarantee management were put forward.

The purpose of this study is to grasp the status of eco-flow management in the Yangtze River Basin, find out main problems and their causes, and then put forward corresponding countermeasures. The novelty and major contribution to eco-flow management is to systematically evaluate eco-flow guarantee status of KRLs from the perspective of large scale, and make some inquiries about eco-environment status in this basin.

2. Eco-flow guarantee situation

2.1. Control section

From 2019 to 2020, eco-flow guarantee targets of 32 inter-provincial KRLs in the Yangtze River Basin were issued and implemented. There are 30 rivers and 2 lakes with 62 control sections in total. Thereinto, eco-flow was taken as guarantee index for river sections, while ecological water table was taken as guarantee index for lake sections. Figure 1 shows the distribution of KRLs and their control sections.

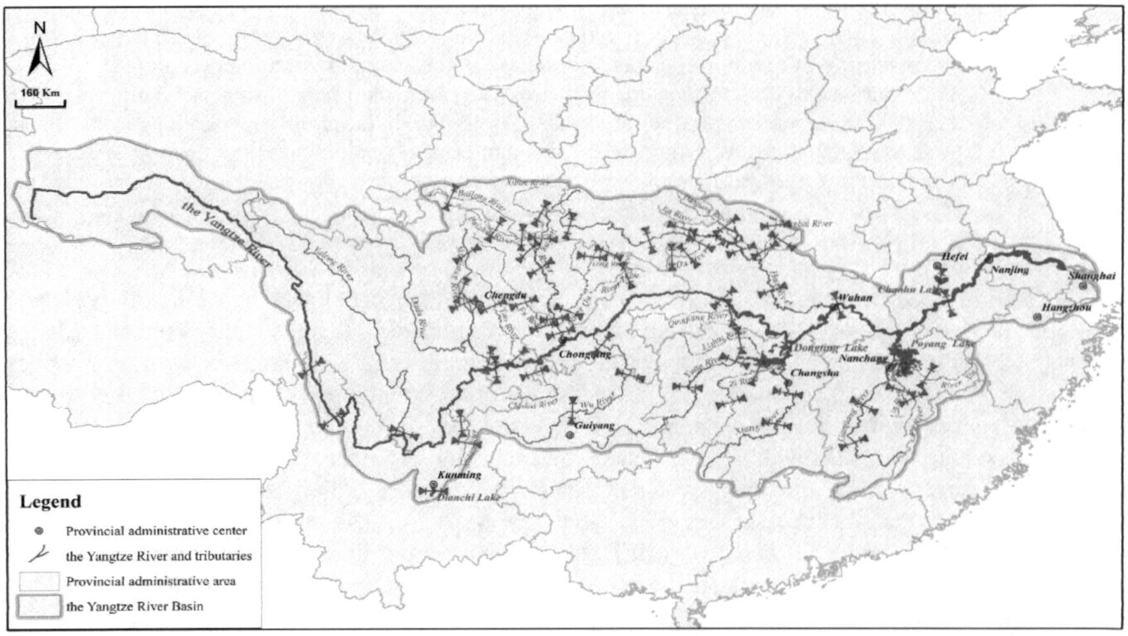

Figure 1. Distribution of control sections of KRLs in the Yangtze River Basin.

2.2. Eco-flow guarantee situation

(1) Data sources

The data of daily flow (or water table) in this paper are collected from river basin management agency and relevant provincial water administrative departments during. Table 1 shows eco-flow targets for KRLs and their daily satisfaction degrees of representative control sections on the main stream of the Yangtze River and its major tributaries during 2020~2021.

Table 1. List of eco-flow targets for KRLs and their daily satisfaction degrees of representative control sections on the main stream of the Yangtze River and its major tributaries during 2020~2021.

Number	KRLs	Section	Eco-flow (m^3/s)	satisfaction degree of eco-flow	
				2020	2021
1	Jinsha River	Benzilan	234	100%	99.73%
2		Shigu	298	100%	99.73%
3		Xiangjiaba	1200	100%	100.00%
4	Main stream of Yangtze River	Yichang	5500	100%	100.00%
5		Hankou	7026	100%	100.00%
6		Datong	10000	100%	100.00%
7	Niulan River	Huanglishu	13	100%	100.00%
8		Dashadian	19	**83.90%**	99.45%
9	Yalong River	Tongzilin	422	100%	100.00%
10	Min River	Zhenjianguan	7.4	100%	100.00%
11		Doujiayan	68.4	100%	100.00%
12		Control gate of Jinma River	15	100%	100.00%
13		Gaochang	551	100%	100.00%
14	Tuo River	Sanhuangmiao	13.3	100%	100.00%
15		Fushun	35.2	100%	98.63%
16	Jialing River	Guangyuan	25	98.90%	100.00%
17		Cangxi	124	100%	98.90%
18		Wusheng	157	100%	98.90%
19		Beipei	257	100%	100.00%
20	Wu River	Wujiangdu	72	100%	100.00%
21	Wu River	Yanhe	80	100%	99.73%
22		Wulong	224.5	100%	100.00%
23	Chishui River	Chishui	269	100%	100.00%
24	Xiang River	Hengyang	59	100%	97.53%
25		Xiangtan	155	99.70%	99.73%
26	Zishui River	Lengshuijiang	333	100%	98.63%
27		Taojiang	56	100%	98.08%
28	Yuan River	Pishizhen	107	100%	100.00%
29		Taoyuan	176	99.70%	99.73%
30	Lishui River	Shimen	300	99.20%	98.08%
31	Qing River	Gaobazhou	70	99.20%	99.18%
32	Han River	Hanzhong (second)	46	/	100.00%
33		Ankang	9.48	96.40%	99.73%
34		Huangjiagang	66	100%	100.00%
35		Huangzhuang	174	100%	100.00%
36	Fu River	Liaojiawan	4.9	100%	100.00%
37		Loujiacun	4.5	99.20%	99.18%
38	Xin River	Meigang	32	100%	100.00%
39	Gan River	Xiajiang	17.9	100%	100.00%
40		Waizhou	57	100%	100.00%
41	Lake Dian	Haigeng	221	100%	100.00%
42	Chaohu	Zhongmiao	281	100%	100.00%

(2) Evaluation method

Frequency method was used to evaluate the eco-flow guarantee situation of KRLs, and the calculating formula is as following:

$$CRi = Ai/Bi \times 100\%$$

Where CRi is the satisfaction degree of eco-flow at control section, %;

Where Ai is the sample number of daily average flow greater than or equal to the target value in the evaluation period;

Where Bi is the total number of daily average flow samples in the evaluation period; Normally, the total number of samples is 365 (366), and in other cases, the total number of samples shall be determined according to the true state; For the abnormal operation of hydrological monitoring facilities, the actual monitoring sample number is used as the total number of samples.

When the daily satisfaction degree of eco-flow is greater than 90%, it is considered that the eco-flow guarantee assessment of this section reaches the standard; otherwise, it fails to reach the standard.

According to the assessment requirements of the strictest water resources management system, 42 control sections of 20 KRLs in the basin were evaluated in 2020. And 62 control sections of 32 KRLs were evaluated in 2021, thereinto, Wansengsi, Xijie and Hekou (Zhouhe) sections were newly built, that were not included in the evaluation scope. Dongbei section was also not included in the evaluation scope due to relocation.

(3) Evaluation results and discussion

Our calculation results show that in 2020 the satisfaction degree of eco-flow only at Dashadian section was less than 90%, i.e. the eco-flow at Dashadian section was not up to the standard. And the others' satisfaction degrees were all greater than 90%, thereinto, satisfaction degrees of 34 sections were 100%, accounting for 81%. The eco-flow at Dashadian section was not up to the standard for 59 days, mainly resulting from the maintenance and discharge dissatisfaction of Xiangbiling and Xiaoyantou hydropower stations in the upstream of this section.

In 2021, the satisfaction degrees of all sections in the evaluation scope were greater than 90%, i.e. eco-flow guarantee of all sections were all up to the standard. Thereinto, satisfaction degrees of 30 sections were 100%, accounting for 66%.

3. Accomplishments of eco-flow management

According to the information we have gathered and sorted out, we know that in recent years, water administrative departments at all levels in the basin have earnestly implemented the new ideas of water resource management in the new era, fulfilled their duties, and steadily promoted the management of eco-flow, basically established a new pattern of guarantee and management of eco-flow. They have achieved certain results, that are mainly reflected in the following aspects.

(1) A target system for eco-flow guarantee has been basically established. Over the last two years, according to the arrangement of Ministry of Water Resources, the river basin management agency in conjunction with the related provinces (autonomous regions and municipalities) have identified (in batches) eco-flow guarantee targets of 62 control sections of 32 inter-provincial KRLs in the Yangtze River Basin during 2019 ~ 2020, steadily deployed the management of inter-provincial KRLs, basically covered the whole basin. Meanwhile, provinces (autonomous regions and municipalities) in the Yangtze River Basin also actively promoted eco-flow management, and have formulated three-year eco-flow promotion plans. At present, provincial water administrative departments in the Yangtze River Basin have determined eco-flow guarantee targets for 230 control sections of 102 KRLs, that have been issued or authorized by provincial people's governments.

(2) The implementation plan of eco-flow guarantee has been carried out. Adopting the approach of "one section, one plan" that for each control section, there are an eco-flow target, a dispatching plan of eco-flow, a monitoring and early warning scheme, and a responsibility subject, measures for section monitoring, early warning and regulation of eco-flow were defined. Besides, responsible entities and supervision entities were assigned. Further, plans for eco-flow guarantee of KRLs in the Yangtze Basin have been formulated and implemented.

(3) The dispatching mechanism for eco-flow guarantee has been initially established. In recent years, taking the Three Gorges Project as the core, the river basin management agency has carried out the joint dispatching of 107 hydro-projects along the main and tributaries of the Yangtze River. Ecological dispatching experiments of the Three Gorges Reservoir, reservoirs in the upper reaches of the Yangtze River and Danjiangkou Reservoir in the Han River are ongoing, and results are monitored accordingly. Meanwhile, provinces (autonomous regions and municipalities) in the basin have also actively promoted the management of water resources dispatching. For example, Sichuan province has issued and implemented the Measures for the Management of Water Resources Dispatching in Sichuan Province (trial), and Chongqing Municipality has issued the Measures for the Management of Reservoir Ecological Dispatching in Chongqing (Trial).

(4) The mechanism of daily supervision and management of eco-flow is being formed. At the end of 2019, the river basin management agency established a real-time monitoring platform for eco-flows, for carrying out real-time monitoring, early warning and disposal of eco-flows at the control sections of inter-provincial KRLs, and initially established a supervision and communication mechanism with provincial water administrative departments. At the same time, the agency strengthened the supervision of the discharge of hydro-projects and carried out monthly and annual assessment of eco-flow guarantee. Provinces (autonomous regions and municipalities) also strengthened eco-flow management, such as Hunan province issued "Notice on strengthening eco-flow management in the whole province" and developed an eco-flow monitoring and management platform; Chongqing municipality issued "Notice on carrying out eco-flow monitoring of control sections of main rivers", etc.

(5) The monitoring pilots of eco-flow were jointly carried out. In order to test the effectiveness of eco-flow guarantee, the Ministry of Water Resources organized and carried out the monitoring pilots of eco-flow in 2020. The river basin management agency jointly carried out the pilot work of aquatic eco-environment monitoring with relevant provinces (autonomous regions and municipalities), as well as water quality monitoring.

(6) Measures for eco-flow management have been introduced. The river basin management agency formulated and issued the Supervision Measures for River and Lake Ecological Flow for Changjiang Water Resources Commission * of the Ministry of Water Resources (Trial), which for the first time enacted special legislation on eco-flow management at the river basin level, established management norms for eco-flow, and provided a solid and strong institutional support for eco-flow guarantee.

In general, eco-flow management in the Yangtze River Basin is orderly being carried out. Eco-flows in the main stream, major tributaries and major lakes of the Yangtze River have been guaranteed, and eco-environment has been effectively improved. Moreover, functions of the river and lake ecosystems have been restored quickly, wetland areas have recovered gradually, and the biodiversity has increased significantly.

4. Problems of eco-flow management

(1) Basic work related to eco-flow

First, records of the eco-flow management of hydro-projects are not completed. There are numbers of water conservancy and hydropower projects, water extraction projects and gate dams in this basin. At present, their distribution and the requirements for eco-flow have been basically mastered. Whereas, related facilities of discharge and monitoring and dispatching plans are still unclean.

Second, the foundation of sensitive eco-flow determination is weak. The basin is important for ecological protection in China, where eco-flow requirement is complicated. There are different sensitive eco-flow demands for important lake wetland, rare and endemic fish habitat, spawning ground, bait field, overwintering field of migratory fish, as well as water supply security and so on.

* The Yangtze River Water Resources Commission is the Yangtze River Basin management agency dispatched by the Ministry of Water Resources, PRC.

The determination of sensitive eco-flow demand needs scientific and reasonable analysis and long-term observation and evaluation, meanwhile, the relevant research basis needs to be reinforced.

(2) Eco-flow dispatching and control measures

Some hydro-projects have not established conventional eco-flow dispatching mechanism, not built special eco-flow discharge facilities or discharged according to the requirements, so it is difficult to guarantee eco-flows of rivers. For example, at Dashadian section of Niulan River, the satisfaction degree of eco-flow in 2020 was 83.9%, and mainly affected by Xiangbiling and Xiaoyantou hydropower stations. Overall, eco-flow dispatching and control measures need to be implemented urgently.

(3) Eco-flow monitoring and early warning mechanism

First, eco-flow monitoring capacity is insufficient to meet management needs. At present, based on hydrometric station, the monitoring period, frequency and method of hydrology station can not fully meet requirements. Eco-flow monitoring capacity of more than 30 sections is insufficient, which is difficult to meet management needs of comprehensive dynamic monitoring.

Second, monitoring and early warning mechanism is not perfect. Monitoring section can not achieve full coverage, and monitoring methods and accuracy can not meet requirements. Besides, early warning mechanism has not yet been established, so it is difficult to meet management needs of dynamic monitoring.

(4) A long-term mechanism for eco-flow monitoring and management

First, the responsibility of eco-flow management needs to be implemented. River and lake eco-flow management involves provincial and local government agencies, hydro-projects, shipping and other fields. Coordination and linkage between river basins, regions and departments should be established to form efficient regulatory synergy, so as to ensure the implementation of eco-flow management.

Second, the daily supervision and assessment of eco-flow need to be strengthened. From the perspective of management procedures, the focus is still on the target determination, additionally, the supervision and assessment of eco-flow is in the exploratory stage, with immature experience and shortage of management. Particularly, the effectiveness of supervision methods, assessment standards and methods in the supervision practice is still need to be confirmed.

5. Suggestions on eco-flow guarantee

(1) It is suggested to deepen the eco-flow guarantee and determine eco-flow targets in a scientific and reasonable way. First, it needs to complete the establishment of eco-flow target, plan and implementation of full coverage of KRLs. Second, according to relevant characteristics of rivers and lakes, eco-flow targets and technical requirements of sensitive eco-flow demand should be proposed during the different period, and the applicable scope and conditions of the calculation method of eco-flow demand should be unified and standardized.

(2) It is suggested to strengthen the management of water resources dispatching and improve the ability of eco-flow guarantee. It needs to continuously deepen unified dispatching of inter-provincial rivers, strengthen the dispatching management of important hydro-projects, strict annual water resources dispatching plan, strongly promote eco-flow regulation, research and organize the implementation of ecological operation experiment, so as to ensure healthy aquatic eco-environment, and provide a solid support for promoting high-quality development.

(3) It is suggested to strengthen monitoring and early warning, and comprehensively improve our capacity for intelligent supervision. It is required to upgrade eco-flow monitoring facilities, improve precision and automation levels, to strengthen eco-flow real-time early warning management, establish sound system of eco-flow information submitted and build a complete joint response mechanism of early warning [5-7] and a coordination mechanism of data review in the eco-flow platform.

(4) It is suggested to consolidate the responsibility of eco-flow management and improve monitoring and management mechanism. It is needed to establish a sound assessment mechanism for eco-flow guarantee, carry out in-site supervision and inspection as well as problem rectification when

appropriate [8-10], improve coordination and communication mechanisms cross-departments and regions, to ensure smooth work procedures and information sharing.

(5) It is suggested to intensify research on eco-flow and improve the level of scientific and technological support. It is needed to accelerate the research on eco-flow demand and guarantee measures under multi-objective conditions, strengthen the optimization of ecological dispatching and its effect evaluation, and the optimization scheme of eco-flow guarantee based on water supply security, bloom risk prevention and control, and water environment protection.

References

[1] Jia S W 2018 Carbon storage distribution and its dynamic changes of forest vegetation in Yangtze River Basin based on continuous forest resources inventory *Journal of Ecology and Rural Environment* **34** 997

[2] Feng X H, Zhong Y X 2015 Causal analysis on economic disparity in Yangtze River Basin *World Regional Studies* **24** 100

[3] Pan B Z, Liu X Y 2021 A review of water ecology problems and restoration in the Yangtze River Basin *Journal of Yangtze River Scientific Research Institute* **38** 2

[4] Liu L S, Huang G X, Wang F 2020 Main problems, situation and countermeasures of water eco-environment security in the Yangtze River Basin *Research of Environmental Sciences* **33** 1081

[5] Wang J P, Li F P, Sun J 2020 Current status of ecological flow in key rivers and lakes in China and regulation suggestions *China Water Resources* **1** 14

[6] Wang G J, Liu Z, Lang M X 2022 Considerations on strengthening regulation of river ecological flow *China Water Resources* **1** 19

[7] Tu M, Yi R 2019 Practice and suggestions of ecological flow management in the Yangtze River Basin *China Water Resources* **17** 64

[8] Zhang J Y, Wang X H, Yang Q 2017 Study on safeguard measures of ecological water demand for major rivers and lakes in China *China Water Resources* **23** 8

[9] Zhang H B,Yin X, Li W 2019 System of ecological flow safeguard measures for rivers and lakes in China *Journal of Economics of Water Resources* **37** 13

[10] Li Y Y, Liao W G, Zhao Z N 2019 Definition of river and lake ecological flow and safeguarding mechanism in the new period *China Water Resources* **17** 13

Effects of organic matter on the distribution pattern of bacterial community in Zaohe River

Zhihan Wang[1], Yuanjie Wang[1], Ke Han[1], Yuan Liu[1], Mingbo Zuo[1], Guangwen Li[1,5], Hao Zhang[3,4], Youyi Zhao[2], Xisheng Tai[2] and Tianpeng Gao[1,2,3,5,6]

[1] College of Biology and Environmental Engineering, Xi'an University, Xi'an 710065, China;
[2] College of Geography and Environmental Engineering, Engineering Center for Pollution Control and Ecological Restoration in Mining of Gansu Province, Lanzhou City University, Lanzhou 730070, China;
[3] College of Civil Engineering and Mechanics of Lanzhou University, Lanzhou 730030, China;
[4] Key Laboratory of Mechanics on Disaster & Environment in Western China, Ministry of Education of China;
[5] Academy of Eco Xi'an, Xi'an 710065.

[6] Corresponding author's e-mail: zkgtp@163.com; ORCID: 0000-0001-6821-3836

Abstract. This study revealed the bacterial community structure and the effects of carbon and nitrogen on the distribution pattern of bacterial community in the sediments of the Chang'an section of the Zao River in Xi'an. And we expect to provide a theoretical basis for the biological monitoring and bioremediation of the Zao River by this study. Therefore, based on 16SrRNA high-throughput sequencing technology, we analyzed the characteristics of carbon, nitrogen and bacterial community structure in Chang'an section of Zaohe River. The results showed that in the Chang'an section of Zaohe River Basin, Proteus and firmicum were tolerant bacteria, and the dominant bacteria were Proteus (24%~45%). At the gate level, the difference of bacterial community composition was significantly correlated with pH, TN, NO_2^- and TOC.

1. Introduction

Bacteria are widely distributed in soil and water, and coexist with other organisms as well. Bacteria can sensitively capture subtle changes in the environment and adapt to the environment through various physiological activities [1]. River sediments usually contain many organic and inorganic substances. Some scholars have found that the organic matter in sediments will directly or indirectly affect the bacterial community structure and diversity of sediments [2]. After investigating the bacterial diversity in the Pearl River Estuary in China, Tao et al found that ammonia oxidizing microorganisms mainly exist in the sediments of the Pearl River Estuary, belonging to ammonia oxidizing archaea and ammonia oxidizing bacterial community [3]. After analyzing the community structure and diversity of microorganisms in the bottom mud of Jiulong River in Southeast China, Liu et al. found that the content of nitrogen and phosphorus in sediments is an important factor affecting microbial diversity and community composition [4]. Many scholars have found that physical and chemical indexes such as pH, organic matter, total potassium, $NH4^+$, $NO3^-$, and $NO2^-$ are important factors affecting bacterial communities in sediments [5].

IOP Conf. Series: Earth and Environmental Science 1087 (2022) 012011 doi:10.1088/1755-1315/1087/1/012011

In recent years, with the continuous development of economy, various chemical industries have developed rapidly. The treatment of industrial wastewater is not complete and does not meet the standard, and river basin pollution events such as organic matter occur from time to time. Therefore, it is necessary to study the impact of carbon and nitrogen in river basin sediments, evaluate and warn the pollution caused by carbon and nitrogen in river basin through the change of bacterial community structure, and put forward constructive suggestions for water environment protection [6].

As a tributary of Weihe River [7], Zaohe River flows through Chang'an District, high tech Zone, Beilin District and Weiyang District of Xi'an city, Shaanxi province. Water eutrophication has attracted much attention as to 64 industrial enterprises along the line but few studies on the sediments of Zaohe River in recent years. The Manuscript analyzed the relationship amongst bacterial community structure, sediment diversity, and sediment organic matter (pH value, TN, TP, NH4 +, NO3 -, NO2 - and total potassium) at four sampling points of Zaohe River by 16SrRNA Gene Sequencing Technology based on the of Zaohe River sediments, hoping to provide basis for the ecological restoration of water environment in Zaohe River Basin.

2. Methods and materials

2.1. Sampling points

In order to reveal the relationship between bacterial community structure composition and diversity upon carbon and nitrogen in sediments in Zaohe River Basin, the cross-section sediment of Chang'an section of Zaohe River was sampled in December 2020. Four sampling points were arranged successively from upstream to downstream according to the river flow direction. The location information of the sampling points is as follows in Table 1.

Table 1. Sludge samples location information.

Sampling point	longitude	latitude
S1	108°50′55″E	34°14′25″N
S2	108°50′43″E	34°19′19″N
S3	108°50′48″E	34°22′27″N
S4	108°50′52″E	34°22′39″N

Note: Use GPS positioner for positioning during sampling.

2.2. Sampling method and pretreatment

The tubular sampler is used for sampling. Before sampling, the sampler is disinfected with 75% ethanol, and the collected sediment is stored in a brown glass bottle sterilized with a high-pressure sterilizer. After sampling, it was transported back to the laboratory immediately under the conditions of avoiding light with sealing and low temperature, and stored in the refrigerator at -80 ℃.

2.3. Detection method of sediment related properties

Analysis method of carbon and nitrogen in sediment: pH potentiometric method (NY / t1377-2007) [8]. Organic matter is determined by potassium dichromate volumetric method [9]. Total nitrogen is determined by Kjeldahl method (HJ717-2014). Total phosphorus is determined by alkali fusion molybdenum antimony resistance spectrophotometry (HJj632-2011) [10]. Ammonium nitrogen was determined by Indophenol Blue Colorimetry [11]. Nitrate nitrogen was determined by UV spectrophotometry [12]. Nitrite nitrogen α- Naphthylamine Colorimetry [13].

2.4. *Sequencing experiment flow*

2.4.1. DNA extraction and PCR amplification. DNA extraction kit was used to extract DNA from the sample. General amplification primers, 338F (5'-ACTCCTACGGAGGCAGCAG-3') and 806R (5'-GGACTACHVGGTWTCTAAT-3') were used to PCR amplify the 16rna gene of the sample. Specific primers with barcode were synthesized according to the specified sequencing region. Reaction system: 10×Buffer 2μL, 2.5mmol/ L dNTPs 2uL, primer 338F/ 806R 0.8uL each, Taq enzyme 0.2μL, BSA 0.2μL, formwork DNA 10ng, supplement ddH2O to 20 μ L. The PCR instrument was Abi gene map 9700 (ABI Company). All samples were conducted according to the formal experimental conditions. Each sample was replicated 3 times. The PCR products of the same sample were mixed with 2% agarose gel electrophoresis. The PCR product was recovered by Gel Extraction Kit (AXYGEN Company, American) and Tris was recovered by AxyPrepDNA gel (Shanghai Yihui Biotechnology Co., Ltd). The PCR products were detected and quantified by quantifluor st blue fluorescence quantitative system (Promega company, American), and then mixed in corresponding proportion according to the sequencing volume requirements of each sample, and the miseq library was constructed by TruSeq DNA sample prep kit. Stay Qubit@2.0 Library quality was evaluated on thermos scientific and Agilent Bio-analyzer 2100 systems. Finally, the library was sequenced on illuminahiseq2500 platform. 16SrRNA Gene amplicons were sequenced and analyzed by Majorbio Technology Co., Ltd. (Shanghai, China).

2.4.2. Sequencing and species data analysis. The pereads obtained from illumine PE250 sequencing were first spliced according to the overlap relationship, and the quality control and filtering sequence quality were performed at the same time. After distinguishing the samples, uparse7 0.1090 was used for Operational Taxonomic Unit（OTU）cluster analysis and species taxonomic analysis, and rdpclassifier 2.11 was used for sequence classification notes. Flash database 1.2.11 was used to splice the double ended sequence, and the data analysis was processed on the Majorbio cloud platform. According to the number and abundance of the species in the communities, and diversity indices were calculated by qiime1.9.1 to evaluate the bacterial community diversity.

2.5. *Statistical analysis*

Use qiime1 9.1 generate water abundance tables for each taxonomy and calculate beta diversity. Using rdpclassifier 2.11 make sequence classification notes. Principal component analysis (PCA) was used to analyze the differences of bacterial community structure in Zaohe River. Analyze Spearman correlation and heat map by using Majorbio cloud platform. In addition, redundancy analysis (RDA) was used to study the relationship between bacterial community and environmental factors and was processed on Majorbio cloud platform and plotted. This was followed by one-way ANOVA test ($P< 0.05$), using SPSS 20.0 software to evaluate the alpha diversity among samples and the difference of environmental factors among samples.

3. Results and analysis

3.1. *Microbial alpha diversity*

Bacterial diversity in environmental samples can be estimated from two aspects: species richness and species uniformity. Chao1 index and Ace index represent the richness of bacterial community, and Shannon and Simpson index represent the diversity and uniformity of bacterial community. The bacterial alpha diversity (Chao l, Ace, Shannon, and Simpson) of four sampling points was calculated. The results of bacterial diversity at the four sites are shown in Table 2. According to Chao1 index and ACE index, the order of bacterial community richness from high to low was S4 > S2 > S3> S1. According to Shannon and Simpson indexes, the order of bacterial diversity from high to low was S2 > S4 > S1 > S3.

Table 2. Results of microbial diversity at four sites.

Sample	Shannon	Simpson	Ace	Chao1
S1	5.77 ± 0.02^{c}	0.01 ± 0.01^{b}	2554.05 ± 241.10^{c}	2540.47 ± 245.22^{c}
S2	6.17 ± 0.09^{a}	0.01 ± 0.001^{c}	3268.43 ± 107.83^{b}	3234.02 ± 157.40^{b}
S3	5.14 ± 0.04^{d}	0.02 ± 0.001^{a}	3057.14 ± 508.70^{bc}	2826.50 ± 228.67^{c}
S4	5.93 ± 0.10^{b}	0.01 ± 0.01^{b}	3906.49 ± 59.47^{a}	3868.62 ± 23.88^{a}
p	P<0.001	P<0.001	P<0.001	P<0.001

Note: the mean and standard deviation are from 3 biological replicates (i.e., 3 samples were collected and analyzed for one locus. Two different lowercase letters in the same column indicate significant differences ($P < 0.001$).

3.2. *Community structure analysis*

It can be seen from Figure 1 that at the phylum level, the bacterial community structure of the four samples was mainly composed of Proteobacteria (24% ~ 45%), Actinobacteria (15 ~ 27%), Firmicutes (5% ~ 24%), Bacteroidota (4% ~ 13%), etc. The relative abundance of community structure was different in each sample, but the dominant phyla still had high similarity. β diversity can reflect the differences of bacterial community structure in different places. Principal component analysis (PCA) on each sample at the gate level was shown in Figure 2. The different shapes in the figure represent bacterial samples at four locations, and the distance of each plot is negatively correlated with the similarity of microbial community structure. There were significant differences in bacterial community composition among the four positions, and the bacterial community composition of S1 and S2 were very similar.

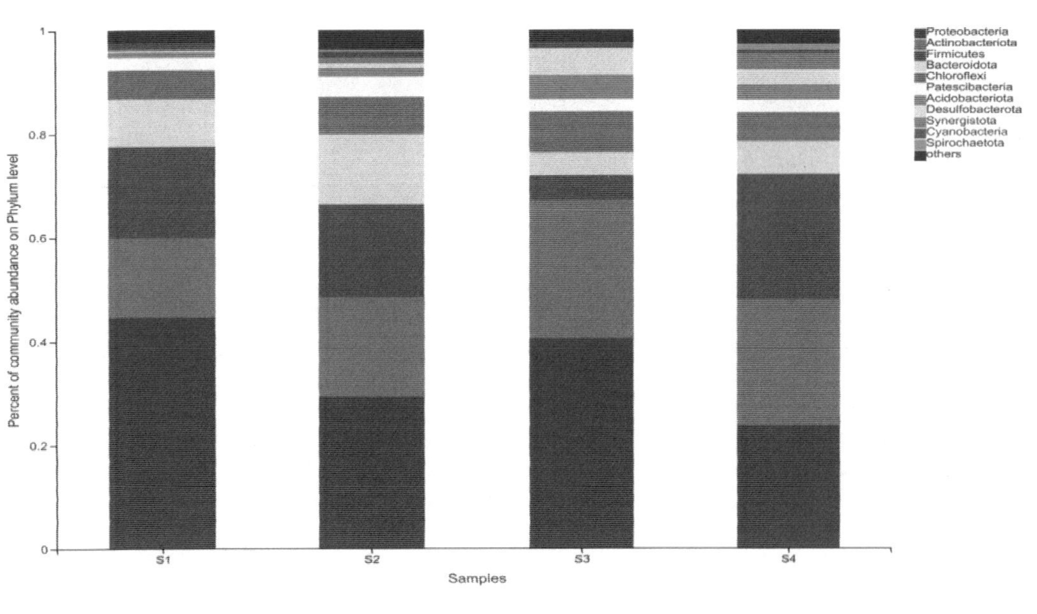

Figure 1. Stacking diagram of bacterial community structure.

In order to further analyze the difference of bacterial community composition in the four samples, non metric multidimensional scaling analysis (NMDS) was carried out on the bacterial samples based on the gate level, as shown in Figure 3. The results showed that the difference of bacterial community structure was closely related to the difference of loci. In contrast, plot S3 has stronger repeatability, closer aggregation and higher similarity of bacterial components, and the bacterial community structure is the most different from other plots. The samples at plot S1 have poor repeatability, loose aggregation and low similarity of bacterial components. The similarity order of bacterial community composition in different sites was S3 > S4 > S2 > S1.

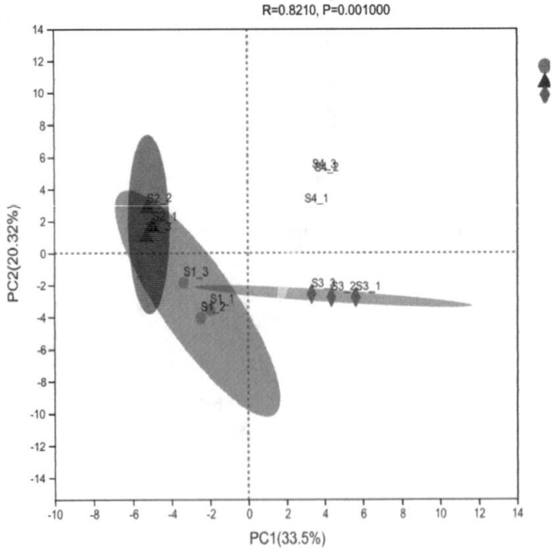

Figure 2. Principal component analysis diagram.

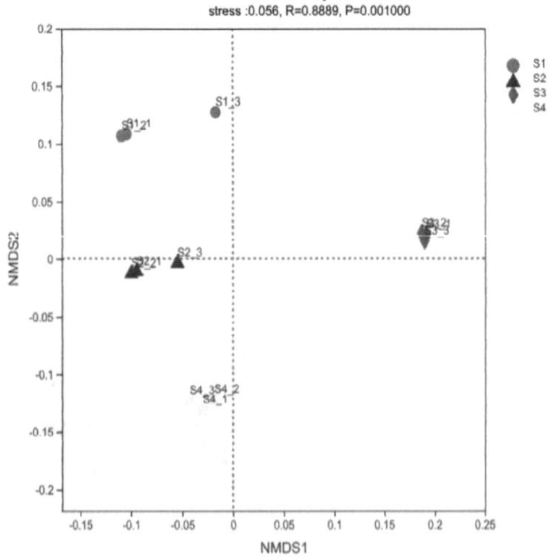

Figure 3. Non metric multidimensional scaling analysis chart.

3.3. Environmental factor analysis

Table 3. Physical and chemical indexes of different sites in Zaohe, Xi'an.

Sample	S1	S2	S3	S4	P
pH	7.57±0.06^b	7.55±0.038^b	8.42±0.11^a	7.67±0.142^b	P<0.001
TOC(mg/kg)	5.96±0.036^c	9.25±0.0153^a	2.13±0.03^d	8.71±0.201^b	P<0.001
TN(mg/kg)	0.541±0.02^b	0.666±0.014^a	0.122±0.007^d	0.44±0.075719^c	P<0.001
TP(mg/kg)	0.268±0.007^b	0.31±0.002^a	0.251±0.018^b	0.153±0.003^c	P<0.001
NH_4^+(mg/kg)	73.68±0.087^d	186.16±0.0153^a	154.6±0.306^b	77.71±0.11^c	P<0.001
NO_3^-(mg/kg)	12.69±0.035^a	7.47±0.095^c	2.63±0.0153^d	10.93±0.0473^b	P<0.001
NO_2^-(mg/kg)	1.58±0.0322^a	1.22±0.0361^b	0.43±0.0153^d	0.95±0.01^c	P<0.001

Note: the mean value comes from three biological repetitions, and for one locus, each environmental factor was analyzed three times in parallel. Two different lowercase letters in the same column indicate the difference ($P < 0.001$).

According to Table 3, the pH value of S3 is the highest, reaching 8.42. The TOC, TN and NH_4^+ content of S2 was the highest, reaching 9.25, 0.666, 186.16mg/ kg. The contents of NO_3^- and NO_2^- in S1 were the highest, reaching 12.69 and 1.58mg/ kg respectively. We speculate that the higher indicators of TOC, TN and NH_4^+ in the river may be caused by the discharge of domestic waste and artificial fertilizer.

3.4. Relationship between bacterial community structure and environmental factors

The heat map between 7 organic indicators (pH, TOC, TN, TP, NH_4^+, NO_3^-, NO_2^-) and bacteria with relative abundance greater than 1% is shown in Figure 4. As can be seen from the figure, pH value is positively correlated with verrucous flora, actinomycetes, acid flora and desulfurization flora, and NO_3^-, TN and NO_2^- are negatively correlated with verrucous flora, actinomycetes, acid flora and desulfurization flora. NH_4^+ was positively correlated with Pasteurella and chloroform. There was a significant positive correlation between TP and cyanobacteria. PH, NO_3^-, TN and NO_2^- have a great impact on the bacterial community in sediments. PH value mainly has a positive impact on bacterial community structure, while the increase of NO_3^-, TN and NO_2^- will have a negative impact on bacterial community structure.

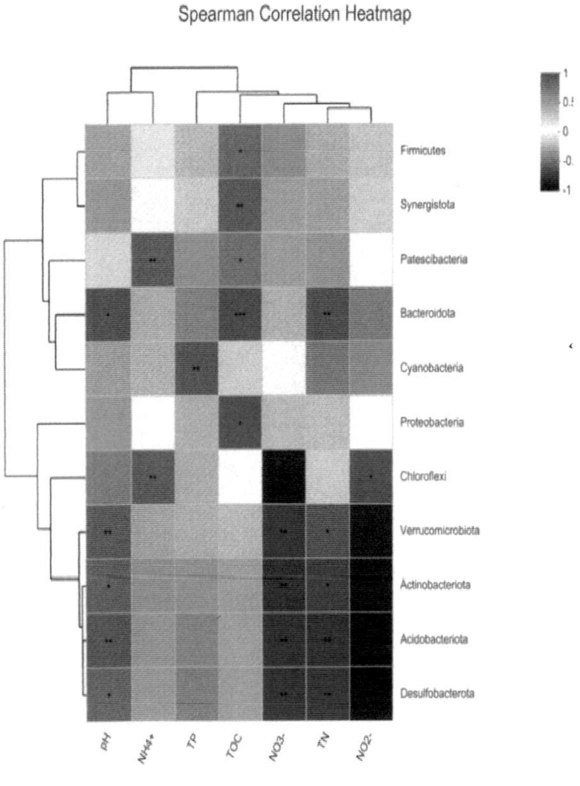

Figure 4. Correlation analysis between bacterial community and sedimentary physicochemical indexes heat map.

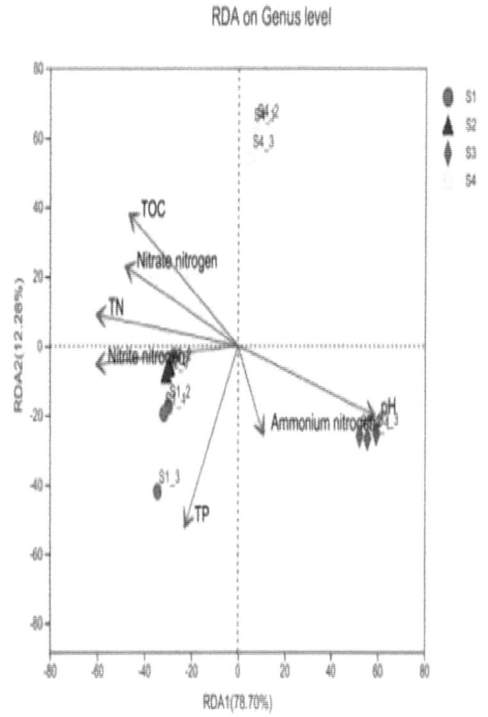

Figure 5. Redundancy analysis of bacterial flora.

The explanations of the first axis and the second axis of the redundancy analysis (RDA) were 78.73% and 12.42% respectively. The 7 organic matter indexes had a significant impact on the bacterial community ($P < 0.05$), among which the three sedimentary physicochemical indexes of pH value, TN and NO_2- had a very significant impact on the bacterial community (Figure 5, $P < 0.01$).

4. Discussions

At the phylum level, the dominant phyla of bacterial community in the four samples include Proteus bacteria 24% ~ 45%, Actinomycetes 15% ~ 27%, Firmicutes 5% ~ 24% and Bacteroides 4% ~ 13%. This is similar to the research results of microbial community structure in the sediments of Taihu Lake [14]. Some studies have shown that NH_4^+ and pH value are important factors determining the composition of bacterial community in sediments [15], and TOC and TP are the main environmental factors affecting the structure of microbial community [16]. The content of organic matter will affect the diversity of bacterial community to a great extent. Other works have shown that TP has a significant impact on the structure and distribution of microbial communities [17]. Through Spearman correlation coefficient and PCA analysis, we found that at the phylum level, the composition of bacterial community had a very significant relationship with pH value, TN, NO_2^- and TOC. Proteobacteria and Fimictes were tolerant bacteria in the Zaohe River Basin and have strong adaptability to the environment of the river basin. The results of diversity analysis showed that the bacterial community structure of the four loci in the study had certain similarity. Among them, the bacterial species at S3 and S4 sites are quite different from those at other sites. This may be due to the different contents of organic matter (TOC, TN, TK, NH_4^+, NO_3^-) in the sediments at this point, in which TP, TOC and TN have a greater impact. According to the content of physical and chemical indexes at each point, we can see the TOC, TN, TP and NH_4^+ at S2 were higher than other points. However, there is a large amount of domestic waste accumulation near point S2, and the organic components in the domestic waste can be decomposed by some microorganisms in the river sediment. Therefore, the contents of TOC, TN, TP and NH_4^+ at point S2 are directly related to the accumulation of domestic waste near this point. It can be seen from the community abundance map that Proteobacteria accounts for the highest proportion in S2. Therefore, we speculate that Proteus is easier to survive in TOC, TN, TP and NH_4^+ environments, and has a certain ability to oxidize and decompose organic matter.

5. Conclusions

Proteus and firmicum in the sediments of Chang'an section of Zaohe River in Xi'an was salt tolerant bacteria, and the dominant bacteria was Proteus (24% ~ 45%). The difference of bacterial community composition in sediments was significantly correlated with pH, TN, NO_2^- and TOC. The increase of NO_3^-, TN and NO_2^- had a negative impact on the bacterial community structure. Proteus and Candida are tolerant bacteria in Zaohe River Basin, which have strong adaptability to the environment of the river basin.

6. Expectation

Subsequently, we expect to screen some local bacteria with high nitrogen and phosphorus removal capacity from the bottom mud of Zaohe River, and inoculate them in the aquatic plants of the river, so as to improve the eutrophication of the river through the interaction between aquatic plants and microorganisms.

Conflicts of interest

The authors declare that the research was conducted in the absence of any commercial or financial relationships that could be construed as a potential conflict of interest.

Author Contributions

Wang Z H, Wang Y J, Han K, Liu Y: conceptualization, writing, and review. Zuo M B contributed to the review and editing of this perspective. Zhang H, Zhao Y Y, Li G W, and Tai X S were responsible for sampling, Gao T P: supervision.

Acknowledgments

This work was financially supported by the National Natural Science Foundation of China (NSFC 31860176), Key Research and Development Program of Gansu Province (20YF3FA037), Key Research and Development Program of Shanxi Province (2020ZDLSF06 06) and XAWLKYTD012. We are grateful to all anonymous reviewers whose comments improved the quality of the manuscript. Many thanks should be given to Xi'an Key Laboratory of plant stress physiology and ecological restoration technology, Key Laboratory for ecological restoration and high quality development of Qinling Mountains in the upper and middle reaches of the Yellow River for their help.

References

[1] Pascual-García A, Tamames J, Bastolla U 2014 *BMC microbiology* **14(1)** 1-16
[2] Zhao S Y, Ni C Y, Fu W C, Wang J F 2016 *Journal of Jiangxi Normal University (Natural Science Edition in Chinese)* **40(02)** 194-199
[3] Jin T, Zhang T, Ye L, Lee O O, Wong Y H, Qian P Y 2011 *Microbiology and Biotechnology*

90(3) 1137-1145
[4] Liu L, Yang J, Zhang Y 2011 *Hydrobiologia* **678(1)** 113-125
[5] Wen T, Zhong H, Fu G, Zhong Y 2021 *International Journal of Environmental Science and Technology* 1-12
[6] Wu J W, Tan F X, Qi M, Zhang X Q, He J Q, Wu C Z, Zhang F, Chai Y 2021 *Journal of Hydroecology (in Chinese)* **(3)** 1-9
[7] Wang D, He Y, Liang J, Liu P, Zhuang P 2013 *China Environmental monitoring and assessment (in Chinese)* **185(2)** 1041-1053
[8] Wang R K 2018 *Shanxi Chemical Industry (in Chinese)* **38(03)** 64-65
[9] Wu Y, Chen J 2016 *China High-Tech Enterprises (in Chinese)* **(26)** 11-12
[10] Chen Y S, Hou M T, Ma D, Han X H, Zhang R Y, Zhang X Z 2018 *China Standardization (in Chinese)* **(S1)** 193-196
[11] Chen X Q, Sun Y F, Luo H J, Zhang X J 2010 *Journal of Agriculture and Forestry Science and Technology (in Chinese)* **(03)** 26-27
[12] Yi X L, Li K Y, Han L F 1983 *Soil bulletin (in Chinese)* **(06)** 35-40
[13] Shi A H 2002 *Shanxi architecture (in Chinese)* **(03)** 105-106
[14] Tang X, Li L, Shao K, Wang B, Cai X, Zhang L, Gao G 2015 *China Canadian Journal of Microbiology (in Chinese)* **61(1)** 22-31
[15] Lu Z, Liu Z, Zhang C, Wei Q, Zhang S, & Li M 2021 *Microbiology and Biotechnology* **105(5)** 1979-1989
[16] Gao H Q, Liu L, Fang Z J 2011 *Journal of Hehai University (Natural Science Edition In Chinese)* **39(04)** 361-366
[17] Wang B B, Liu Y X, Lu G H, Wang Y H 2021 *Environmental Science and Technology (in Chinese)* **34(03)** 1-6

Effects of feeding methods on ecological soil wastewater infiltration systems for sewage treatment

Zefang Jiang[1,3], Yuqing Xia[1,3], Ziqi Li[1,3], Yuxi Zhang[1,3], Lin Li[1], linlin Fan[1], Shiyao Wang[1] and Jing Pan[1,2]

[1] College of Life Science, Shenyang Normal University, Shen Yang.

[2] Corresponding author's e-mail: crystalpan78@126.com
[3] These authors share first authorship due to dedicating equally to this research

Abstract. Intermittent feeding method and continuous feeding method were adopted in two experimental form ecological soil wastewater infiltration systems (ESWISs), respectively. The results showed that COD, NH_4^+ and TN removal efficiencies were enhanced from 81.8%, 70.5% and 45.8% in continuous feeding ESWIS to 90.0%, 88.5% and 72.7% in intermittent feeding ESWIS. The same TP removal efficiency was achieved in two feeding mehtods. Matrix oxidation reduction potential (ORP) was enhanced through the intermittent feeding ESWIS, especially above 60 cm depth of the matrix, which encouraged nitrification. The metabolic activity and reproduction of microbes were improved by intermittent feeding method, such as actinomyces, bacteria, nitrifying bacteria, fungus. Meanwhile, denitrifying bacteria and nitrate reductase (NR), nitrite reductase (NIR) activities were also increased. PCR-DGGE analysis confirmed that intermittent feeding method improved bacterial diversity. These results suggest that intermittent feeding method is viable to heighten the performance of ESWISs in sewage treatment.

1. Introduction

Ecological soil wastewater infiltration system (ESWIS) is a widely applied method to handle decentralized wastewater when it flows past unsaturated soil matrix of infiltration system [1]. It has many merits compared with the conventional wastewater treatment process, such as low construction, operation and maintenance cost, excellent removal performance for phosphorus and organic matter, eco-environment protection [2]. Nowadays, ESWIS technology has been a fashionable sewage treatment process in rural areas [3].

More and more researchers have dedicated to study ESWISs, such as soil matrix [3], operating load [2], pollutants removal [4], soil matrix clogging of ESWIS [5]. The removal of pollutants is unsatisfactory with continuous feeding method under high hydraulic load rate [2]. Intermittent feeding method (flooding-drying cycles) as the substitute of continuous feeding method is widespread applied in ecological soil aquifer treatment, which can well improve aerobic and anaerobic conditions [5]. There are very few attentions to focus on the application of feeding method and its effects on an ESWIS. The influence of intermittent feeding or continuous feeding on matrix environment, actinomyces, bacteria, fungus, denitrifying bacteria, nitrifying bacteria populations, nitrate reductase (NR), nitrite reductase (NIR) activities and bacterial diversity relating to pollutants removal is still blank.

The main targets of this study were: (1) to assess the effects of feeding methods on pollutants removal and matrix redox environment, (2) to investigate the effects of feeding methods on actinomyces, bacteria, fungus, denitrifying bacteria, nitrifying bacteria populations, nitrate reductase (NR), nitrite reductase (NIR) activities, (3) to explore the effects of feeding methods on bacterial diversity with PCR-DGGE technology.

2. Experimental materiasl and experimental methods

2.1. Experimental system and its operation

Two simulated ESWISs (A and B) were operated indoors of Experimental Central of Shenyang Normal University, northeast China. These ESWISs were made of clear plexiglass pillars, which were 50 cm of internal diameter and 120 cm of length. Matrix sampling wells were set in 40, 60, 80 and 100 cm starting at the top of each ESWIS to measure the number of microbes, enzyme activities and bacterial diversity relating to pollutants removal. ORP electrodes were laid in depth of 40, 60, 80 and 100 cm in the centre of pillars. Gravels (10-20 mm, diameter) of 10 cm thickness at the root brace the matrix. In 50 cm depth of matrix was the distributing pipe. At the bottom of the pillars was the delivery port, which collected treated wastewater. The same matrix consisted of 80% black earth and 20% boiler burning cinder by weight ratio filled each simulated ESWIS. Boiler burning cinder is byproduct of anthracite combustion, mainly contained of SiO_2, Fe_2O_3, CaO and MgO. ESWIS A (SA) was continuously fed with wastewater with 0.06 $m^3/(m^2\ d)$ hydraulic load. Intermittent feeding method was applied in ESWIS B (SB) with 0.12 $m^3/(m^2\ d)$ hydraulic load. Each cycle of the intermittent feeding included a feeding period (with feeding sewage) of 24 h (between 7:00 AM and 7:00 AM the next day) and a non-feeding period (without feeding sewage) of 24 h. Indoors temperature was kept at 23 ± 1℃ in the whole experimental period

2.2. Sewage characteristics

Two ESWISs were fed with domestic sewage which was pretreated by a septic tank before inflowing into each ESWIS. The qualities of domestic sewage to be treated were in Table 1.

Table 1. The qualities of domestic sewage.

Item	COD	TP	NH_4^+	TN	pH
Influent (mg/L)	185.37-262.44	4.62-8.38	30.39-44.61	33.95-46.72	6.6-7.3

2.3. Water, matrix sampling and analytic procedures

Water samples were taken from influent and effluent of each ESWIS to estimate removal performance of organics, phosphorus and nitrogen every 10 days on the basis of the standard methods [6]. Matrix samples of the number of microbes, enzyme activities and bacterial diversity were collected every 20 days. Water and matrix samples were stored in pretender box and taken to the laboratory to measure immediately. Significant differences were made at $P<0.05$ for all analyses using SPSS 19.0 (n=10).

The number of actinomyces, fungus and bacteria was analyzed by plate counts with Gause's Synthetic Medium, Czapek Agar Medium and Nutrient Broth No.5 Medium [3]. The number of nitrifying and denitrifying bacteria was analyzed by the most probable number (MPN) calculation [2].

Urease activity was determined by incubating 10 g matrix with 20 mL 10% urea solution and 40 mL citrate buffer with pH of 6.7 for 24 h at the temperature of 25℃. The formation of ammonium was measured at 578 nm by a spectrophotometer and the urease activity was shown as NH_4^+-N μg/(g d). Nitrate reductase (NR) activity was determined with 20 mL 1% KNO_3 and 100 mL potassium phosphate buffer with pH of 7.5 at the temperature of 25℃ for 24 h. The concentration of NO_2^--N produced was measured at 540 nm by a spectrophotometer and NR activity was presented as the formation of NO_2^--N μg/(g d). Nitrite reductase (NIR) activity was analyzed by incubating 10 g matrix with 50 mL potassium phosphate buffer with pH of 7.4 and 10 mL 1% $NaNO_2$ at the temperature of 25℃ for 24 h. NIR activity was measured by a spectrophotometer at 540 nm [3].

2.4. Bacterial diversity analysis

Matrix samples were collected in the depth of 40, 60, 80 and 100 cm (sample number of SA was A40, A60, A80, A100; sample number of SB was B40, B60, B80, B100) in two ESWISs after the experiments to carry out PCR-DGGE analysis. Comparison was the matrix of two systems before treating sewage, whose number was CK. DNA was drawn from 1 g matrix sample fractions by a FastDNA™ SPIN Kit for Soil (MP, US) on the basis of the instructions presented by manufacturer. The universal forward primer F357GC: 5-CGCCCGCCGCGCGCGGCGGGCGGGGCGGGGGCACGGGGGGCCTACGGGAGGCAGCAG-3 and reverse primer R518: 5-ATTACCGCGGCTGCTGG-3 was used to amplify bacterial 16S rRNA genes. The detail procedures of PCR and DGGE were drawn on the experience of Pan and Yu [7].

The structural diversity of the bacterial community is assessed by the Shannon-Weaver index of general diversity *H*. *H* is counted according to peak height from the different bacteria. The Shannon-Weaver index and the richness index (R) are counted according to the formula of (1) and (2), respectively.

$$H = -\sum[P_i \times \ln(P_i)] \tag{1}$$

where P_i represents the ratio between especial band strength and total strength of all bands in a strip sample.

$$R_i = L_i / L_T \tag{2}$$

where R_i represents the richness index of sample i; L_i represents the band number of sample i; L_T represents the number of different bands in all samples [8].

3. Results and discussions

3.1. Matrix ORP and removal performance in two ESWISs

Figure 1 shows ORP in the matrix of two ESWISs. The similar tendency was observed for all depths in SB, i.e. matrix ORP descended during the feeding period but increased slowly during the non-feeding period. For instance, matrix ORP in the depth of 60 cm descended to 40 mV in the feeding period owing to oxygen expenditure mainly caused by microbial oxidation of organic matter, but increased to 196 mV as the maximum value in the non-feeding period slowly. In the same depths, matrix ORP values of the intermittent feeding ESWIS were higher than those of the continuous feeding ESWIS. The matrix ORP of 40 cm increased to about 300 mV during the non-feeding period. The matrix ORP results indicated that the oxidative condition was improved through intermittent feeding method, especially above 60 cm matrix depth.

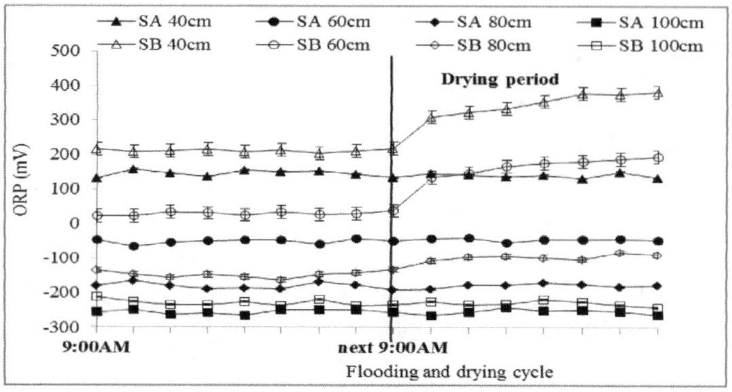

Figure 1. ORP of matrix in two ESWISs.

As seen from Table 2, average COD concentration in the effluent was 22.0 mg/ L and average COD removal efficiency was 90.0% in SB. Compared with COD removal efficiency of SB, average removal efficiency of COD in SA was 81.8% with a notable decline by 9.2%, which was in consistent with other study [3]. Intermittent feeding method has two roles in an ESWIS. One was to aerate naturally, the other was to increase the infiltration potential for the next application period. Intermittent feeding method greatly enhanced microbial aerobic and anaerobic decomposition of organic matter capacity [3]. Previous study reported that adverse matrix oxygen environment constrained organic matter degradation in infiltration wastewater treatment [9]. Therefore, intermittent feeding method in SB improved the matrix oxygen, which was propitious to microbial aerobic and anaerobic organic matter removal.

Table 2. Pollutants removal in SA and SB.

	Item	COD	TP	NH_4^+	TN	NO_3^-	NO_2^-
	Influent	219.5±23.5	6.6±0.9	38.1±3.5	41.1±2.7	1.3±0.1	0.07±0.02
SA	Effluent (mg/L)	38.9±2.9	0.3±0.02	11.2±0.6	18.3±1.3	10.7±0.3	0.06±0.02
	Removal efficiency(%)	81.8±0.9	95.8±0.4	70.5±1.9	48.8±1.1	--	--
SB	Effluent (mg/L)	22.0±4.5	0.3±0.03	4.4±0.3	11.2±0.8	5.6±0.3	0.2±0.04
	Removal efficiency (%)	90.0±1.3	95.9±0.4	88.5±1.5	72.7±3.3	--	--

More than 95% TP was removed in SA and SB. Phosphorus removal in an ESWIS is physical sedimentation and chemical adsorption of matrix, which accomplishes right away [10]. Intermittent feeding and continuous feeding methods had no influence on TP removal.

Average NH_4^+ removal efficiency increased from 70.5% in SA to 88.5% in SB. Intermittent feeding method enhanced the matrix oxidative condition, which was in favor of NH_4^+ oxidation of nitrification process in SB. Intermittent feeding method was the most effective alternative to substantially improve NH_4^+oxidation of the ESWIS [3].

NO_2^- was the intermediate product during nitrification and denitrification process, which mainly derived from NH_4^+ oxidation, denitrification of NO_3^- and influent. From Table 2, it can be seen that there were significant differences between SA and SB in NO_2^- concentrations of the effluent ($P<0.05$). Average NO_2^- concentrations in the effluents were 0.06 mg/L in SA and 0.2 mg/Lin SB. In the effluent, NO_2^- concentration was in the range of 0.1 mg/L and 0.5 mg/L, which represented nitrification process and denitrification process finished effectively in an ESWIS [2]. Average NO_2^- concentration of 0.2 mg/L was obtained in the effluent of SB, which was attributed to intermittent feeding method.

NH_4^+ flows in the ESWIS, which is adsorbed on the matrix first of all. Then, the adsorbed NH_4^+ is oxidized into NO_3^- in nitrify process. When nitrification process was finished, NH_4^+ concentration in the effluent would decrease and NO_3^- concentration in the effluent would rise [4]. The NO_3^- was eliminated from wastewater through denitrification process [9]. Insufficient organic carbon source and excess oxygen were the main limiting factors to denitrification process [7]. Average NO_3^- concentrations were 10.7 mg/L in the effluent of SA and 5.6 mg/L in the effluent of SB. Average TN removal efficiency was 48.8% in the effluent of SA and 72.7% in the effluent of SB. The low TN removal efficiency in SA is due to poor nitrification. NH_4^+ concentration in the effluent of SA (11.2 mg/L) was significantly higher than that in the effluent of SB ($P<0.05$). In SB, nitrification process and denitrification process were completed efficiently.

3.2. Microbial population characteristics and enzyme activities of two ESWISs

The number of actinomyces, fungus and bacteria presents the similar tendency, which decreased with the increase of depth in two ESWISs according to Table 3. In the same depth, the number of actinomyces, fungus and bacteria in the matrix of SB were higher than that in the matrix of SA. The results could further interpret the higher COD removal in SB. Organic pollutants are mainly oxidized in upper matrix of infiltration systems with sufficient oxygen [7]. In this study, profiting from wastewater intermittent feeding, oxygen could arrive at the upper matrix in non-feeding period, which was in favor of the metabolic activity and reproduction of actinomyces, fungus and bacteria.

With depth increasing, the number of nitrifying bacteria decreased and the number of denitrifying bacteria increased according to Table 3. In this study, nitrifying bacteria were more active above 60 cm and the denitrifying bacteria were dominant below the matrix of 80 cm, which profited from the oxygen spreading availability in the upper matrix. In the same depth, the number of nitrifying bacteria in the matrix of SB was 100 times than that in the matrix of SA from 40 to 60 cm. However, the number of denitrifying bacteria in the matrix of SB was 10 times than that in the matrix of SA from 80 to 100 cm. The results could further interpret the higher removal performance of NH_4^+ and TN in SB. Nitrification was the first limiting step for nitrogen removal process in an ESWIS [3]. The oxidative environment of upper matrix was enhanced during the non-feeding period of intermittent operation, which was propitious to the nitrification process and promoted the metabolic activity and reproduction of nitrifying bacteria in the matrix of SB. Moreover, there had been little change in the oxidative environment below 80 cm in intermittent feeding in the matrix of SB and the nitrification process was better, which was beneficial to the metabolic activity and reproduction of denitrifying bacteria.

Table 3. Microbial populations and enzyme activities in two ESWISs.

	40 cm		60 cm		80 cm		100 cm	
Item	SA	SB	SA	SB	SA	SB	SA	SB
BN	$(9.4\pm0.2)\times10^8$	$(4.9\pm1.0)\times10^9$	$(5.6\pm0.3)\times10^8$	$(3.2\pm0.1)\times10^9$	$(2.3\pm0.6)\times10^5$	$(6.6\pm0.2)\times10^6$	$(6.1\pm0.5)\times10^4$	$(7.5\pm0.3)\times10^4$
FN	$(4.8\pm0.1)\times10^6$	$(7.3\pm1.1)\times10^7$	$(3.6\pm0.8)\times10^6$	$(4.2\pm0.3)\times10^7$	$(6.7\pm1.0)\times10^4$	$(3.1\pm0.4)\times10^5$	$(5.7\pm0.6)\times10^2$	$(2.3\pm0.5)\times10^2$
AN	$(5.1\pm1.2)\times10^6$	$(2.1\pm1.0)\times10^7$	$(6.4\pm0.4)\times10^5$	$(3.4\pm1.1)\times10^6$	$(2.6\pm0.6)\times10^3$	$(4.5\pm0.1)\times10^3$	$(2.3\pm0.3)\times10^2$	$(2.2\pm0.1)\times10^2$
NBN	$(3.3\pm0.2)\times10^5$	$(5.3\pm0.4)\times10^7$	$(8.5\pm0.3)\times10^4$	$(2.9\pm1.2)\times10^6$	$(1.4\pm0.7)\times10^4$	$(7.2\pm0.1)\times10^5$	$(3.6\pm0.8)\times10^3$	$(4.4\pm0.4)\times10^3$
DBN	$(5.6\pm0.7)\times10^2$	$(3.4\pm0.5)\times10^2$	$(6.3\pm0.2)\times10^3$	$(2.8\pm0.2)\times10^3$	$(4.0\pm0.4)\times10^6$	$(4.1\pm0.1)\times10^7$	$(3.2\pm0.3)\times10^7$	$(7.2\pm0.5)\times10^8$
UE	2.15 ± 0.11	2.08 ± 0.12	2.41 ± 0.17	2.48 ± 0.15	1.52 ± 0.15	1.56 ± 0.08	1.14 ± 0.09	1.18 ± 0.11
NRE	0.62 ± 0.05	0.66 ± 0.08	0.95 ± 0.13	1.04 ± 0.17	1.45 ± 0.09	1.95 ± 0.12	1.18 ± 0.11	1.37 ± 0.12
NIRE	0.12 ± 0.12	0.14 ± 0.11	0.21 ± 0.02	0.24 ± 0.05	0.51 ± 0.11	0.65 ± 0.3	0.42 ± 0.01	0.56 ± 0.15

BN: The number of bacteria in the matrix (cfu/g), FN: The number of fungus in the matrix (cfu/g), AN: The number of actinomyces in the matrix (cfu/g), NBN: The number of nitrifying bacteria in the matrix (cfu/g), DBN: The number of denitrifying bacteria in the matrix (cfu/g), UE: urease activity in the matrix (mg/(g d)), NRE: nitrate reductase activity in the matrix (mg/(g d)), NIRE: nitrite reductase activity in the matrix (mg/(g d)).

There was no difference in urease activity in the same depth matrix of two ESWISs. Because of high organic nitrogen concentration above 60 cm in the matrix of two ESWISs, the higher urease activity was obtained here. Urease catalyzed the hydrolysis of organic nitrogen into ammonia or ammonium ion [3]. Organic nitrogen and NH_4^+ was the main state in domestic sewage. In this study, NH_4^+ was the dominant form in experimental wastewater. Therefore, feeding method did not affect urease activity.

NR and NIR activities in soil indicate anaerobic nitrate reduction which catalyzed the first step by reducing NO_3^- to NO_2^- and the second step by reducing NO_2^- to N_2 or N_2O, respectively [3]. As shown from Table 3, matrix depth affected NR and NIR activities. The sequence of matrix NR and NIR activities in the depth was: 80 cm> 100 cm > 60 cm > 40 cm in two ESWISs. In the same depth, NR and NIR activities in the matrix of SB were greater than those in the matrix of SA, which was consistent with the number of denitrifying bacteria. Li et al. [3] and Pan et al. [7] reported NIR activity was positively correlated with TN removal in an ESWIS. Therefore, higher TN removal efficiency of SB was obtained compared with SA.

3.3. PCR-DGGE for bacterial diversity in two ESWISs

Analysis of bacterial communities from matrix was carried by PCR-DGGE after experiments. The 16S rRNA gene PCR-DGGE amplicons with banding patterns were shown in Figure 2. Table 4 showed the number of bands in each lane, the Shannon-Weaver and richness indexes (calculated by using the DGGE profiles). The diverse community of bacteria was obtained in two ESWISs compared with CK and that might affect the final effluent quality. Mohan et al. reported that the microbial diversitywas affected by the ORP, porosity and organic matter content of matrix. ORP of the soil could affect the metabolic activity and reproduction of microorganism. The number of bands, Shannon-Weaver and richness indexes were higher in SB than those in SA. The reason was intermittent feeding in SB improved the ORP of the matrix. It concluded that intermittent feeding method could enhance bacterial diversity, which was advantageous to pollutants removal.

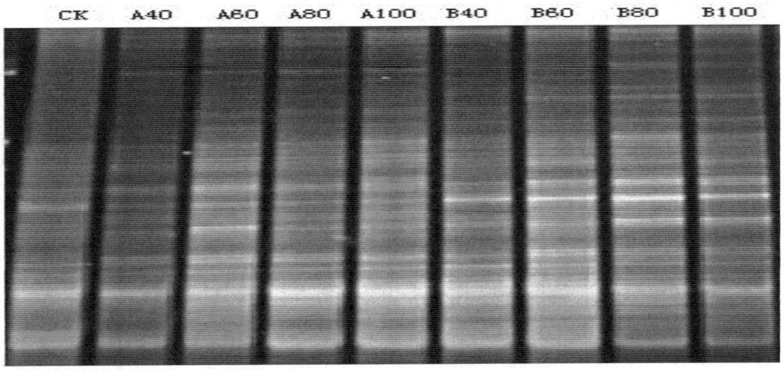

Figure 2. DGGE profiles of matrix bacterial population in two ESWISs.

Table 4. Number of bands, Shannon-Weaver and richness indexes in two ESWISs.

No.	CK	A40	A60	A80	A100	B40	B60	B80	B100
Bands	17	19	21	19	19	21	22	21	20
Richness index	0.65	0.73	0.81	0.73	0.73	0.81	0.85	0.81	0.77
Shannon-Weaver	1.225	1.272	1.311	1.288	1.271	1.311	1.333	1.316	1.291

4. Conclusions

Average COD, NH_4^+ and TN removal efficiencies were increased from 81.8%, 70.5% and 45.8% in continuous feeding ESWIS to 90.0%, 88.5% and 72.7% in intermittent feeding ESWIS. The same TP removal efficiency was achieved in two feeding methods. Matrix ORP results showed that oxidative environment was improved by intermittent feeding, especially above 60 cm depth, which encouraged nitrification. Intermittent feeding promoted the metabolic activity and reproduction of actinomyces, fungus, bacteria, nitrifying bacteria and denitrifying bacteria. NR and NIR activities in intermittent feeding ESWIS were higher than those in the same depth of continuous feeding ESWIS. PCR-DGGE results demonstrated that intermittent feeding method enhanced bacterial diversity. Intermittent feeding offers an appealing choice for wastewater soil matrix of infiltration system.

Acknowledgements

This study is funded by Shenyang Science and Technology Project (21-108-9-36), Shenyang Normal University's Major Original Program (ZD201904); The Ninth Batch of Education and Teaching Reform Project of Shenyang Normal University (JG2021-YB099); College Students Innovation and Entrepreneurship Training Program (X202110166142; X202110166145; X202110166148; 202206017; 202206013).

References

[1] Fan JL, Zhang B, Zhang J, Ngo HH 2013 *Bioresour. Technol.* **141** 117-122
[2] Liang CL, Li Y, Chai BB, Wu HM 2019 *Bioresour. Technol. Rep.* **5** 12-19
[3] Li HB, Li YH, Sun TH, Wang X 2012 *Ecol. Eng.* **38** 105-109
[4] Song WL, Dong X 2020 *Bioresour. Technol.* **314** 123715
[5] Li GH, Tang ZW, Mays L, Fox P 2000 *Water Resour. Manag.* **14** 13-33
[6] APHA 2003 Water Environment Federation. Washington DC
[7] Pan J, Yu L, Li GZ, Huang LL 2013 *Ecol. Eng.* **61** 127-132
[8] Ma L, Liu W, Tan QY, Zhou QH 2019 *Ecol. Eng.* **134** 101-111

IOP Conf. Series: Earth and Environmental Science 1087 (2022) 012012 doi:10.1088/1755-1315/1087/1/012012

[9] Pang JL, Pan J, Tong DL, Fu X, Sun B 2020 *Ecol. Eng.* **146** 105747
[10] Brooks AS, Rozenwald MN, Geohring LD, Lion LW 2000 *Ecol. Eng.* **15** 121-132

Analysis uneven selective catalytic reduction flow field of coal-fired units under the situation of double carbon

Fanliang Meng[1], Xueli Wang[1], Huahai Qiu[1] and Qianwei Feng[2,3]

[1] Huadian Zhangqiu Power Generation Co., Ltd; Jinan 250000, China;
[2] Huadian Electric Power Research Institute Co.,LTD, Hangzhou 310030, China.

[3] Corresponding author's e-mail: fqw0528@sina.com

Abstract. With the implementation of China's carbon peaking and carbon neutralization strategy and the accelerating transformation of energy structure, higher requirements are put forward for the carbon reduction and operation and maintenance capacity of coal-fired units. Based on the analysis of the uneven Selective catalytic reduction flow field of a coal-fired unit in a power plant, the field tests under different conditions are used to study the inlet flow field distribution and outlet NOx concentration distribution of the denitration device of the unit, and the targeted optimization, adjustment of ammonia injection branch pipe is carried out, which effectively solves the uneven Selective catalytic reduction denitration flow field and can provide reference for similar units.

1. Introduction

China is the world's largest energy producer and consumer. In the long-term and rapid development of social economy, China has formed an energy production and supply system with coal as the main body, electric power as the center and oil, natural gas, new energy and renewable energy developing in an all-round way [1,2]. With the implementation of China's carbon peaking and carbon neutralization strategy and the accelerating transformation of energy structure, higher requirements are put forward for the carbon reduction and operation and maintenance capacity of coal-fired units [3-7]. As the mainstream process of flue gas denitration in China's coal-fired power plants, Selective catalytic reduction（hereinafter referred to as SCR） still occupies an absolutely dominant position in this round of denitration ultra-low emission transformation technology [8, 9]. From the operation after the transformation of SCR denitration ultra-low emission, the problem of uneven flow field still exists after the transformation of denitration devices of some units, which affects the stable and up to standard emission of denitration devices and adversely affects the safety, stability and economic operation of units [10].

2. Brief introduction of SCR denitration system in a thermal power plant

A thermal power plant with an installed capacity of 330MW adopts selective catalytic reduction (SCR) process, and 0ne boiler is equipped with two SCR reactors. The three-layer honeycomb catalyst is used, and the semi telescopic rake steam soot blower and acoustic soot blower are used at the same time. Liquid ammonia is used as the reducing agent. Ammonia comes from the ammonia preparation area of the public system, and is mixed with the air provided by the dilution fan according to the volume ratio of 5%, through the ammonia air mixer, and then injected into the reactor through the ammonia spraying grid. The ammonia supply pipeline of each ammonia / air mixer is equipped with ammonia flow measuring device and flow regulating valve respectively. The injection amount of ammonia is calculated according to the inlet and outlet NOx, O_2 concentration and unit load of SCR reactor. The ammonia injection control is adjusted through the ammonia injection flow valve [11].

The main design parameters of SCR denitration system are shown in Table 1.

Table1. Main design parameters of SCR denitration system of 330MW unit in a power plant.

project	numerical value
Denitration efficiency /%	≥89
Ammonia escape concentration /10^{-6}	≤3
SO_2/SO_3 Conversion rate /%	≤1
Pressure loss of denitration system /Pa	≤1200
Mass concentration of fly ash /($g \cdot m^{-3}$)	<42

3. Existing problems and test points

During the actual operation of the unit, it is found that there is a deviation in the NOx concentration monitoring data between the outlet A and B of the denitration device reactor, and there is an inverted phenomenon with the NOx concentration at the inlet of the chimney, indicating that the flow field of the denitration device is uneven. In order to better find out the root cause of uneven flow field, the test and evaluation shall be carried out under the working conditions of high group load (100% load), medium load (75% load) and low load (65% load), with reference to the requirements of relevant performance acceptance test standards [12-15], and the test positions are the inlet and outlet sections of SCR reactor (see Figure 1). The concentration of flue gas composition is detected online by gas analyzer, and the flue gas flow rate is calculated by pitot tube, micro manometer, thermometer, etc. the distribution uniformity of NOx concentration and flue gas flow rate are determined by calculating their relative standard deviation according to the detection results of the above equipment.

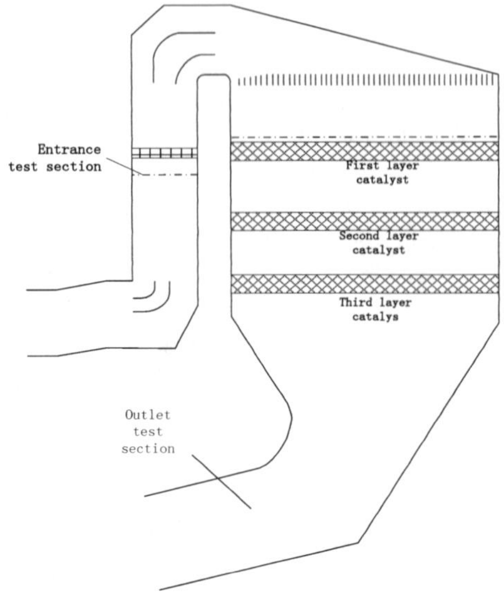

Figure 1. Location and layout of test points.

4. Analysis and discussions

4.1. Flow field distribution at the inlet of denitration unit

The test results in Figure 2 show that under the normal working condition of 100% load rate, the average flue gas flow rate on side a of SCR reactor is 12.91m/s, with a relative standard deviation (Hereinafter referred to as C_V) of 26.8%, and the average flue gas flow rate on side B is 13.90m/s, with a C_V of 31.8%. The flue gas velocity distribution shows the trend of low on both sides and high in the middle, and increases with the increase of test depth. The test results in Figure 3 show that under the condition of 75% load rate, the average flue gas flow rate on side a of SCR reactor is 11.47m/s, with a C_V of about 25.2%, and the average flue gas flow rate on side B is 13.33m/s, with a C_V of about 25.6%. The flue gas velocity distribution shows the trend of low on both sides and high in the middle, and also shows an increasing trend with the increase of depth. Figure. 4 test results show that under the condition of 65% load rate, the average flue gas velocity of the A side of the SCR reactor is 10.34m/s, the C_V is 25.7%, the average flue gas velocity of the B side is about 12.72m/s, and the C_V is about 29.4%. The flue gas velocity distribution shows the trend of low on both sides and high in the middle, and also shows an increasing trend with the increase of depth.

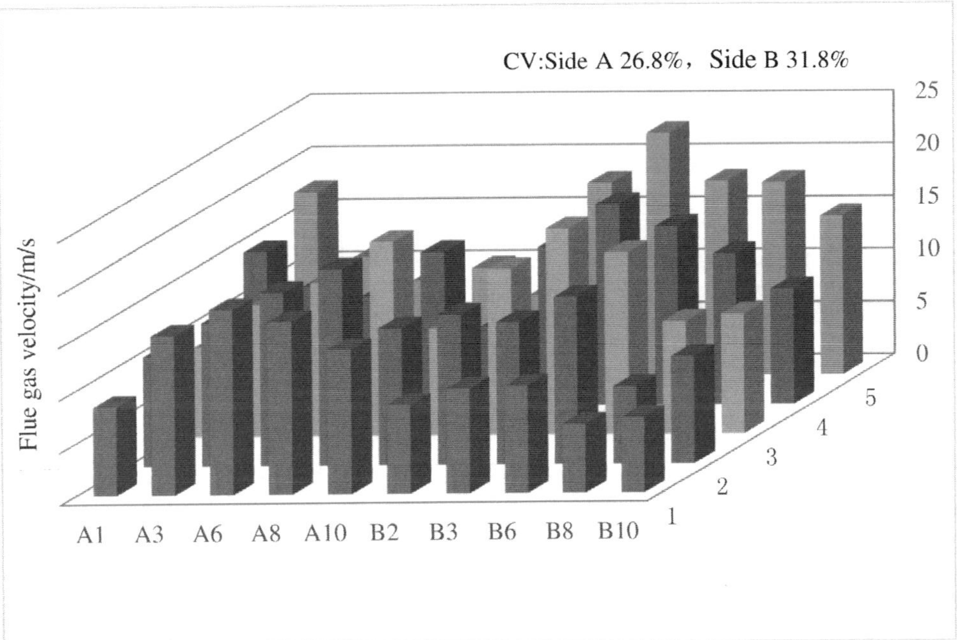

Figure 2. Flue gas velocity distribution at denitration inlet under 100% load rate.

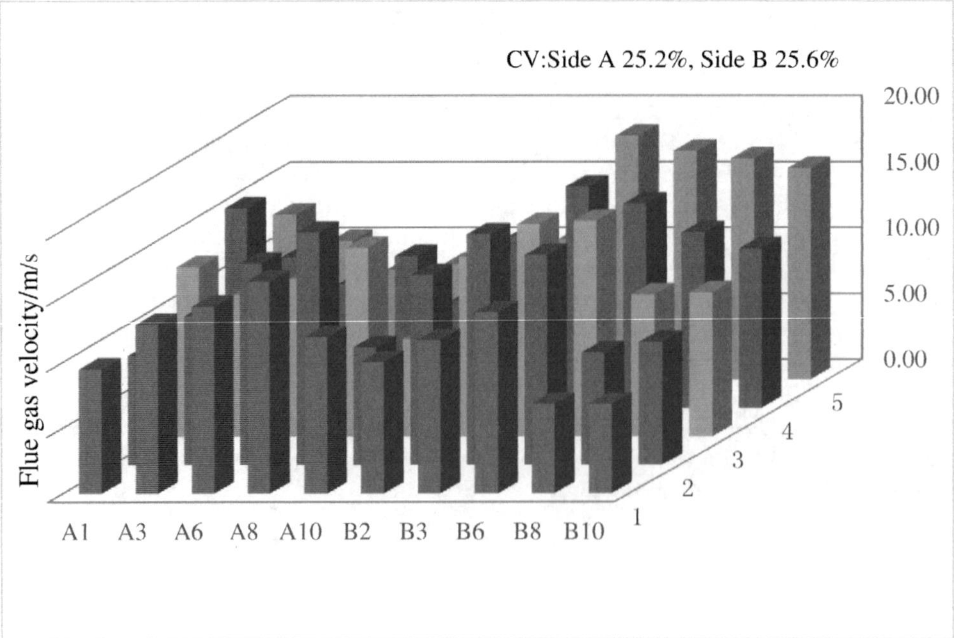

Figure 3. Flue gas velocity distribution at denitration inlet under 75% load rate.

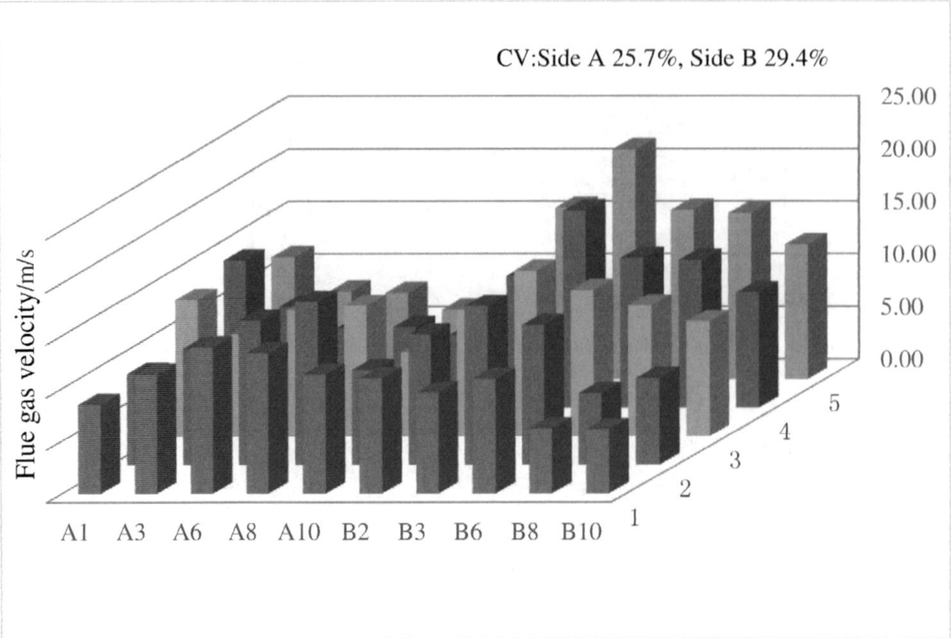

Figure 4. Flue gas velocity distribution at denitration inlet under 65% load rate.

4.2. NOx concentration distribution at the outlet of denitration unit

Figure 5 test results show that under the normal working condition of 100% load rate, the average NOx concentration at the outlet of side A of SCR reactoris 34mg / m^3, and the C_V of NOx concentration distribution after oxygen conversion is 26.0%; The average NOx concentration of side B is 41mg / m^3, the local NOx concentration peak reaches 62mg / m^3, and the C_V of NOx concentration distribution after oxygen conversion is 35.2%. Figure 6 test results show that under the condition of 75% load rate, the average NOx concentration of side A of SCR reactoris 46mg / m^3, the local NOx concentration peak reaches 56mg / m^3, and the C_V of NOx concentration distribution after oxygen conversion is 23.5%; The average NOx concentration of side B is 38mg / m^3, the local NOx concentration peak reaches 59mg / m^3, and the C_V of NOx concentration distribution after oxygen conversion is 36.6%. Figure 7 test results show that under the condition of 65% load rate, the average NOx concentration of side A of SCR reactoris 29mg / m^3, and the C_V of NOx concentration distribution after oxygen conversion is 19.0%; The average NOx concentration of side B is 45mg / m^3, the local NOx concentration peak reaches 66mg / m^3, and the C_V of NOx concentration distribution after oxygen conversion is 36.9%.

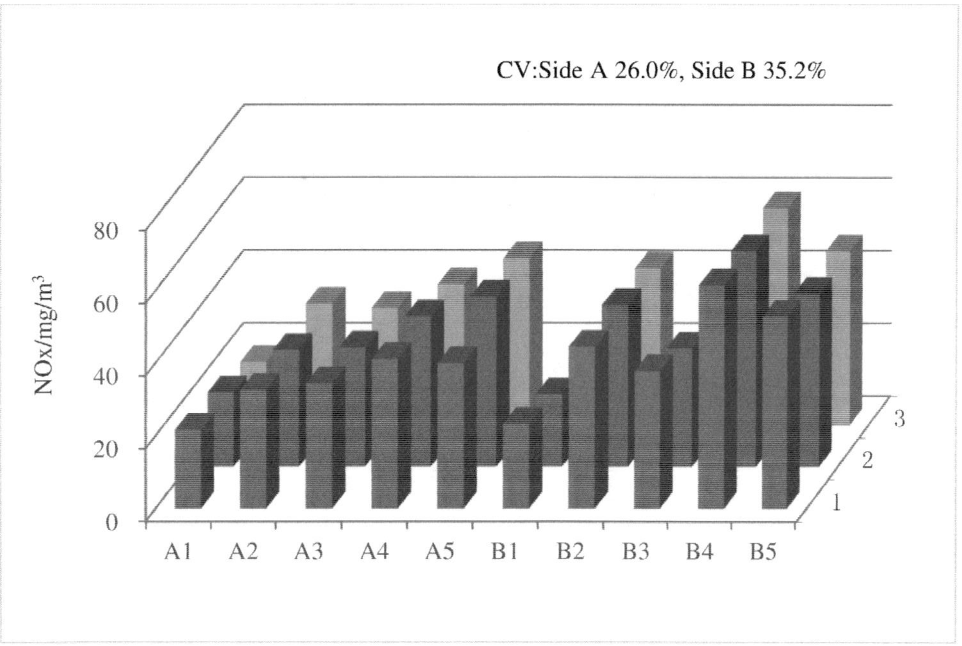

Figure 5. NOx distribution at denitration outlet a and B sides under 100% load rate.

4.3. Solutions

According to the above analysis, the ammonia injection branch pipe is optimized and adjusted [16-19]. Through theoretical calculation and actual test data analysis, the ammonia injection amount of each ammonia injection branch pipe is estimated. Through multiple rounds of optimization and adjustment of the manual regulating valve opening of the ammonia injection branch pipe, the uniformity of NOx concentration distribution at the SCR outlet is significantly improved, and the problem of NOx concentration inversion at the chimney inlet has also been effectively solved. It shows that the optimization, adjustment of ammonia injection grid valve can effectively solve the problem of NOx concentration deviation between A and B sides of reactor outlet of denitration device, and the inversion of NOx concentration at chimney inlet.

5. Conclusions

By analyzing the uneven flow field of SCR denitration in a power plant, this paper studied the flow field distribution characteristics of the denitration device of the unit. Through the targeted optimization and adjustment of ammonia injection branch pipe, the uneven NOx concentration at the reactor outlet of denitration device and the inversion of NOx concentration at the chimney inlet of the unit could be effectively solved, which could provide reference for similar units.

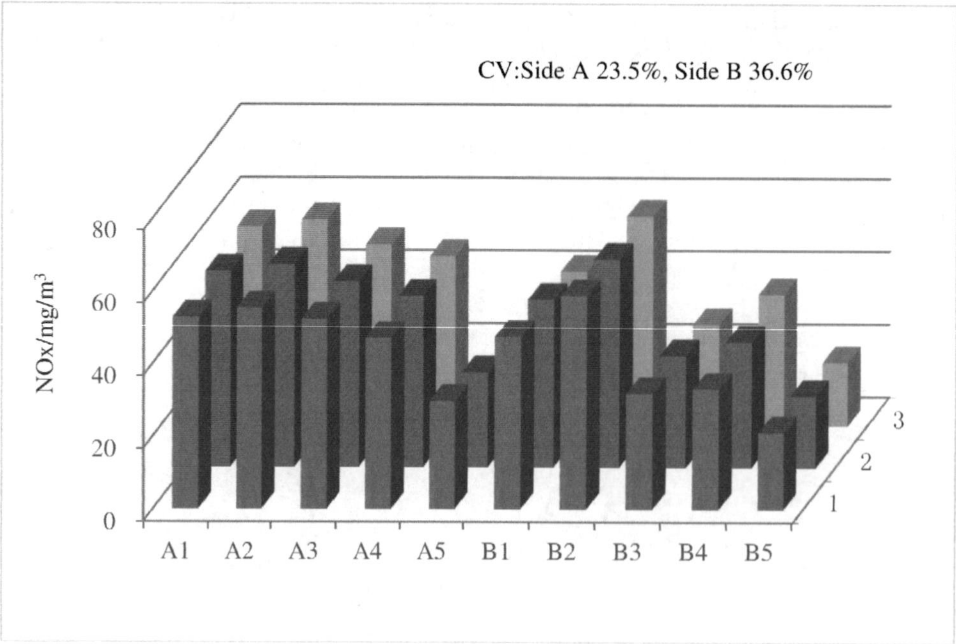

Figure 6. NOx distribution at denitration outlet a and B sides under 75% load rate.

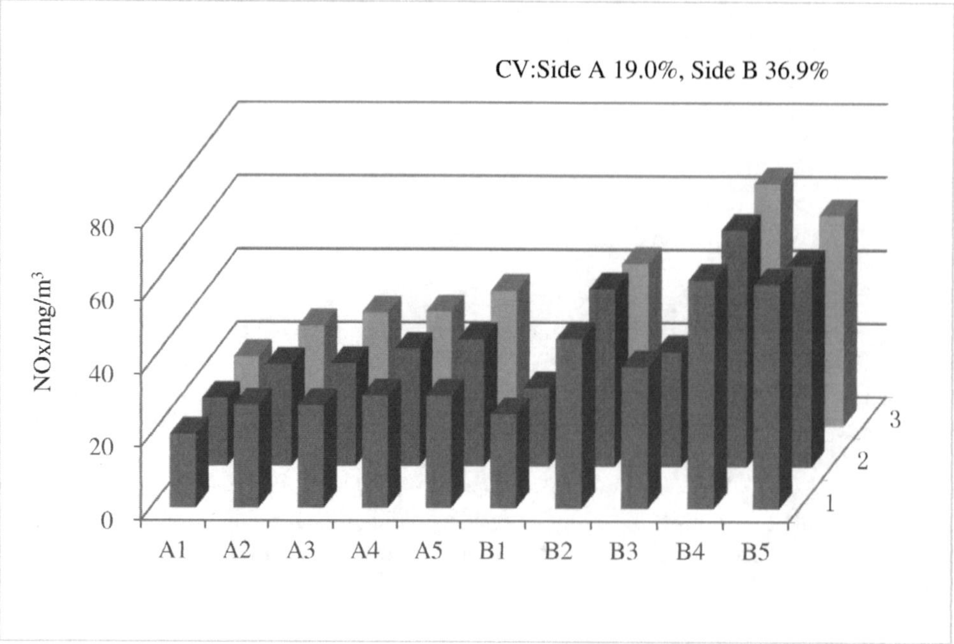

Figure 7. Distribution at denitration outlet a and B sides under 65% load rate.

References

[1] National Bureau of statistics. China Energy Statistics Yearbook[M] *Beijing: China Statistics Press* 2020
[2] National Development and Reform Commission, Ministry of environmental protection, national energy administration. Notice on printing and distributing the action plan for upgrading and

transformation of coal power energy conservation and emission reduction (2014-2020) [A/OL]

[3] Yang Yongping, Yang Zhiping, Xu Gang, et al. 2017 Energy consumption status and Prospect of thermal power generation in China[J] *Chinese Journal of electrical engineering* **33(23)** 2-4

[4] Roger Jorden, Jom Maulbtsch 2002 Zero Discharge Concert Rannle with Reality[J] *PowerEngineering* **(6)** 107-110

[5] Cui, He Rui, Chen, Lu, Qiu, Da Fang 2017 Fuzzy Comprehensive Evaluation on the Thermal Power Plants Ability of Energy Conservation and Emission Reduction[J] *Advanced Materials Research* **281** 211-214

[6] Rao Qinghua, Qiu Yu, Xu Lizhong 2011 Energy conservation and emission reduction index system and performance evaluation[J] *Environmental science research* **9(1)** 1067-1063

[7] Zong Hongsan 2019 Technical discussion on energy conservation and consumption reduction in thermal power plants[J] *Power system equipment* **(07)** 92-93

[8] ROY S, HEGDE M S, MADRAS G. 2009 Catalysis for NOx abatement[J] *Applied Energy* **86** 2283-2297

[9] Jiang Jian ping, Yang Yong long, He Yong bing, et al. 2019 Analysis on the Key Technology of Ultra-low Emission System Evaluation of Coal-fired Power Plant [J] *Electric System* **20** 71-72

[10] FENG Qianwei, ZHANG Yang, WANG Fengji, et al. 2016 Performance Comparison and Analysis before and after SCR Ultra-low Emission Transformation of Coal-fired Units[J] *Proceedings of the CSEE* **36(16)** 4308-4318

[11] Zhang Qiang 2007 SCR Flue Gas Denitration Technology and engineering application in coal-fired power plants[M] *Beijing: Chemical Industry Press*

[12] GB/T 16157-1996 Determination of Particulate Matter in Stationary Pollution Sources and Sampling Methods of Gaseous Pollutants[S]

[13] DL/T 260-2012 Specification for performance acceptance test of flue gas denitrification device in coal-fired power plant[S] *Beijing: China Electric Power Press*, 2012

[14] GB/T 21508–2008 Test method for performance of coal-fired flue gas desulfurization equipment [S] *Beijing: Standards Press of China*, 2008

[15] DL/T 998-2016 Specification for performance acceptance test of limestone-gypsum wet flue gas desulfurization unit[S] *Beijing: China Electric Power Press*, 2016

[16] Wang Yanpeng, Dong Liwei 2021 Evaluation of ammonia injection optimization effect of denitration device of a 600 MW Coal-fired Unit[J] *Cleaning the world* **(12)** 46-49

[17] LIU Xiaomeng, LIU Peiqi, ZHANG Ran,et al.2020 Industrial test of SCR denitration subarea ammonia injection based on linear control[J] *Energy Environmental Protection* **34(5)**41-46

[18] Wang Jianhao, Bai Yuan 2021 Experimental study on ammonia injection optimization and adjustment of SCR denitration system in coal-fired power plant under variable load conditions[J] *Electric power technology and environmental protection* **37(05)** 46-52

[19] Guo Lei, Xu Yongyi, Tian Yingqiang, et al. 2021 Analysis and evaluation of optimal adjustment effect of ammonia injection in SCR unit of a 660MW coal-fired unit[J] *Shandong industrial technology* **(03)** 41-45

IOP Conf. Series: Earth and Environmental Science 1087 (2022) 012014 doi:10.1088/1755-1315/1087/1/012014

Big data analytics for adaption to climate change

Haijing Wang[1,2,3] and Yaping Wang[1,2]

[1] Wenzhou University, Wenzhou, 325035, China;
[2] Wenzhounese Economy Research Institute, Wenzhou, 325035, China.

[3] Corresponding author's e-mail: wang.haijing@wzu.edu.cn

Abstract. Big data analytics (BDA) has the 5-v benefit of volume, variety, veracity, velocity, and value and has the potential to help adaptation to climate change. This paper reports the preconditions for designing the BDA system specifically for adapting to climate change, the designing of this BDA system, and the actual performance. In terms of layers, the components of the BDA system have five integrated layers: the persistent layer, the distributed data storage indexing layer, the web-based data analytics layer, the web-based layer responding to web-based requesting portals, and the web portal layers for specialist administrators to make responses. The BDA system includes the primary database, the sectoral database, and the integration process. The BDA generally met the preconditions and the goals of improving the adaptation capacity to climate change. The trial use of this BDA covered seven provinces in terms of geographic areas and covered preventing flood, tracing diseases caused by climate change, zoning and planning urban areas, and other fields in terms of content. The actual performance indicates that this BDA system functions well as to the speed of processing, the soundness of analysis, and the accuracy of data. However, there is still a need to improve its capacity to spot and detect erroneous but non-conflicting data.

1. Introduction

The traditional data processing techniques are incapable of speedy and accurate big data processing. Persons and organizations often use a rule-of-thumb analysis process on complex issues without the capacity to collect and process big data. Big data analytics (BDA) was developed to improve the collection and processing of data [1-2]. Based on BDA, computer-based analytic systems are developed for particular areas to discover and analyze problems [3]. BDA has the 5-v benefit of volume, variety, veracity, velocity, and value [4-6]. Volume refers to the amount of data, variety refers to different sources and nature of data, veracity refers to the importance of data quality, velocity refers to the speed of data processing, and value refers to the importance of BDA. There is rich literature on the design of BDA [7-8] and its benefit [9]. There is also rich literature on the design of BDA for general issues of climate change [10]. BDA has been used to study how to adapt agriculture to climate change [11-12], however, the studies on BDA specifically for the adaption to climate change are still insufficient. The current BDA systems for the adaption to climate change also have problems of not allowing inputting and visiting data with smartphones and problems of not allowing simultaneous inputting and visiting data by a large number of people at different places. These problems make the current BDA systems unfriendly to under-educated people who both suffer from climate change and can input their local knowledge on adapting to climate change. In addition, the current BDA systems lack appropriate information barriers to prevent privileged or personal information from leaking. Therefore, the current BDA systems need to be further improved.

This paper proposes a design of a BDA system specifically for the adaption to climate change, including the preconditions to be met, the components of the systems, and then reports its actual performance, and finally points out further improvements needed.

2. Designing a BDA system specifically for the adaption to climate change

2.1. Preconditions to be met

Because the adaption to climate change affects all aspects of life, the input and output of a BDA system specifically designed for the adaption to climate change need to consider certain preconditions. The preconditions to be met for our BDA system include: the database should be stored in the cloud or web-based, the input end should allow the supply of data with smartphones, iPads, and other mobile devises, the out-put end should also allow the access with smart phones, iPads, and other mobile devices in addition to desktop computers. The system should allow a large number of people to input information simultaneously at different places and should also allow a large number of people to get information from the system simultaneously at different places. The system should set appropriate information barrier to prevent personal or privileged information from leaking. The BDA should have a decision support system (DSS) allowing multiple-levels of administrators with different authorities. The system should be prepared for occasions of erroneously supplying information. It should be able to detect conflicting or missing data and report such conflicting or missing data to the administrators.

2.2. Components of the BDA system

Figure 1 shows the components of the BDA system. In terms of layers, the components of the BDA system designed have five integrated layers: (1) the persistent layer, (2) the distributed data storage indexing layer, (3) the web-based data analytics layer, (4) the web-based layer responding to web-based requesting portals, and (5) the web portal layers for specialist administrators to make responses.

The functions of these layers are as follows. The first layer stores climate data of different types in their original formats. The second layer indexes spatiotemporal data and speeds up spatiotemporal queries. The third layer supports integrated data related to climate change, the needed adaption to climate change, and the potential responses to address the adaption needs. The fourth layer uses generally available search engines. The fifth layer enables the administrators to interact with the data, computing resources, and analytical operations. The design of these layers is partly modelled on ClimateSpark [13]. These layers may interact in a certain way.

In terms of databases, similar to other BDA systems [14-15], this BDA system includes the primary database, the sectoral database, and the integration process. The primary databases include climate, geographic, and demographic information. They are accessible by all governmental agencies and partly available to other people, although the demographic information does not include individual-level information, which is only available to the police. Various governmental agencies, including the police department, the water affairs agency, the agricultural agency, establish their respective sectoral databases. The police department has the data of individual citizens, the water affairs agency has the data on lakes, rivers, and underground water, and the agricultural agency has the data on agriculture. These data are collected in the daily work by the staff of these agencies. The integration process is the key to BDA. The integration process integrates the primary database and various sectoral databases.

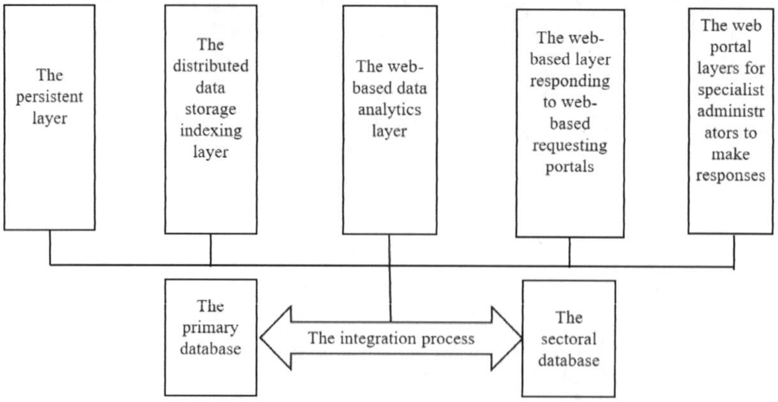

Figure 1. Design of the BDA system.

3. Actual performance

This system was put into trial use and its actual performance generally met the expectation. The trial use covered seven provinces including Hubei, Zhejiang, Fujian, Shanghai, Sichuan, Shanxi, and Henan. The trial use included preventing flood, tracing diseases caused by climate change, zoning and planning urban areas, and other fields. In Wuhan, Hubei Province, the BDA system reduced damages caused by flooding and improved water supply. In Fuzhou, Fujian Province, the BDA system helped to detect an endemic disease caused by the erosion of soil, which was caused by climate change. This paper focuses on the following three aspects of its performance.

3.1. Accuracy of data

This research finds that the data collected by big data technology are more accurate than the data collected with traditional method. Traditionally, the data were collected by requiring different agencies to fill in numerous tables. Data so collected is fragmental and sometimes conflicting. Ideally, the conflicting data should be addressed at the earliest time but they are sometimes not addressed and cured because they are scattered and hard to detect. The use of big data collection techniques provides a potential to address this problem. This system has the capacity to spot conflicting data and report to the administrator to verify the data and delete incorrect data. However, the system has more difficulty to spot and report erroneous but non-conflicting data. That is to say, if the data are erroneous but do not conflict with other data, this system has more difficulty to detect and report. As the availability and accuracy of data are the bases of big data analysis [16-17], this situation is still a big concern.

3.2. Speed of data processing

The BDA system generally improves the speed of data processing. The improvement of speed is more because of the interconnection and interaction among databases than the computing capacity. The web-based computing facilities are connected with distributed data and significantly increase data processing speed.

3.3. Soundness of analysis

The soundness of analysis mainly depends on the sufficiency of data, the soundness of analytic methods, and the detection of gaps and conflicts of data [18-19]. This research finds that the analysis based on BDA has access to more data and process the data with more advanced analytical tools. The analysis based on BDA can also detect the gaps and conflicts of data. Once the gaps and conflicts are identified, researchers can input more data to fill in the gap or to check the data to solve the conflicts. This is confirmed in our interview. Some interviewees said that some hidden risks arising from climate change were previously overlooked but now are spotted.

4. Conclusion and further improvements

The BDA generally met the preconditions and the goals of improving the adaptation capacity to climate change. The actual performance indicates that this BDA system functions well as to the speed of processing, the soundness of analysis, and the accuracy of data. However, there is still a need to improve its capacity to spot and detect erroneous but non-conflicting data. We plan to improve the BDA in two aspects to enhance the capacity to spot and detect erroneous but non-conflicting data: the first targets the web-based data analytics layer and the second targets the sectoral databases.

References

[1] Grover V, Lindberg A and Benbasat I 2020 The Perils and Promises of Big Data Research in Information Systems *Journal of the Association for Information Systems* **21** 9-293

[2] Kar A K and Dwivedi Y K 2020 Theory Building with Big Data-driven Research–Moving Away from the "What" towards the "Why" *International Journal of Information Management* **54** Article 102205

[3] Power D J and Kaparthi S 2002 Building Web-based Decision Support Systems *Studies in Informatics and Control* **11(4)** 291-302

[4] Manogaran G and Lopez D 2018 Spatial Cumulative Sum Algorithm with Big Data Analytics for Climate Change Detection *Computers and Electrical Engineering* **65** 207-221

[5] Addo-Tenkorang R and Helo P T 2016 Big Data Applications in Operations/Supply-chain Management: A Literature Review *Computers & Industrial Engineering* **101** 528-543

[6] Papadopoulos T, Gunasekaran A, Dubey R, Altay N, Childe S J and Wamba S F 2017 The role of Big Data in explaining disaster resilience in supply chains for sustainability *Journal of Cleaner Production* **142(2)** 1108-1118

[7] Hose K, Romero O, and Song I 2022, Trends in Design, Optimization, Languages, and Analytical Processing of Big Data (DOLAP 2020) *Information Systems* **104** 101929

[8] Anadiotis A, Balalau O, Conceição C, Galhardas H, Haddad MY, Manolescu I, Merabti T, and You J 2022 Graph integration of structured, semistructured and unstructured data for data journalism *Information Systems* **104** 101846

[9] Wang Y, Kung L, Terry, and Byrd A 2016 Big data analytics: Understanding its capabilities and potential benefits for healthcare organizations *Technological Forecasting and Social Change* **26** 3-13

[10] Schnase JL, Duffy DQ, Tamkin GS, Nadeau D, Thompson JH, Grieg CM, McInerney MA, Webster WP 2017 MERRA Analytic Services: Meeting the Big Data challenges of climate science through cloud-enabled Climate Analytics-as-a-Service *Computers, Environment and Urban Systems* **61** 198-211

[11] Siloko IU, Ukhurebor KE, Siloko EA, Enoyoze E, Bobadoye AO, Ishiekwene CC, Uddin OO, Nwankwo W 2021 Effects of some meteorological variables on cassava production in Edo State, Nigeria via density estimation *Scientific African* **13** e00852

[12] Ukhurebor KE 2020 The Role of Biosensor in Climate Smart Organic Agriculture toward Agricultural and Environmental Sustainability *Agrometeorology* DOI: http://dx.doi.org/10.5772/intechopen.93150

[13] Hu F, Yang C, Schnase JL, Duffy DQ, Xu M, Bowen MK, Lee T, Song W 2018 ClimateSpark: An in-memory distributed computing framework for big climate data analytics *Computer and Geosciences* **115** 154-166

[14] Nerini F, Sovacool B, Hughes N, et al. 2019 Connecting climate action with other Sustainable Goals *Nature Sustainability* **2** 674-680

[15] NDRC 2016 National Strategy on Energy Production and Consumption Revolution (2016-2030) http://www.ndrc.gov.cn/fzgggz/fzgh/ghwb/gjjgh/201705/W020170517397451808076.pdf (in Chinese)

[16] Publicity Office of the State Council 2021 China's Policies and Actions in Response to Climate Change http://www.gov.cn/zhengce/2021-10/27/content_5646697.htm (in Chinese)

[17] Liu S 2018 Research on the Urban Sustainable Development in Response to Climate Change with Big Data Governance *Journal of Southwest University of Ethnic Minorities* **3** 205-213
[18] Guo H 2020 Big Data on the Earth for Sustainable Development of China *Beijing: Science Press* 178-193
[19] Deng X and Wang J 2014 Smart City Climate Chart Based on Big Data *Times* Shanghai Urban Management **23(04)** 33-36

IOP Conf. Series: Earth and Environmental Science 1087 (2022) 012015 doi:10.1088/1755-1315/1087/1/012015

A study of the urban heat island effect in Guangzhou

Chi Fong Tang[1,2]

[1] Faculty of Humanities and Social Sciences, Macao Polytechnic University, R. de Luís Gonzaga Gomes, Macao.

[2] Corresponding author's e-mail: p1103445@mpu.edu.mo

Abstract. As the world economy continues to grow and energy consumption increases, the global climate is gradually warming. Among them, the economic activities in metropolitan areas are more intense compared to other areas, forming an urban heat island effect where the temperature in the urban area is higher than that in the surrounding areas. This article selects Guangzhou, a rapidly developing city in southern China over the past two decades, as a case study. Surface temperature data for five time periods, namely 2000, 2005, 2010, 2015 and 2020, are collected to analyse how Guangzhou is affected by the urban heat island effect and its relationship with land use status and population density. The results of the article found that the urban heat island effect in Guangzhou is very obvious, mainly in the central and southern parts of the city, where the spatial distribution and evolution of land use type and population density are significantly related to the intensity of the heat island effect. In the future, Guangzhou can alleviate the impact of climate change and urban heat island effect by rationally planning urban layout, optimizing urban ecological environment, improving the material of urban underlying surface and actively developing renewable energy.

1. Research background

After the industrial revolution, human beings have started to use fossil fuels in large quantities, emitting increasing amounts of carbon dioxide, and global temperatures have been on an upward trend, continuing to have a significant impact on the Earth's ecosystem and socioeconomic systems. In response to climate change, the international community adopted the United Nations Framework Convention on Climate Change in 1992, which is the starting point and basic framework for cooperation and negotiations among countries around the world to address climate change. In order to strengthen the implementation of the Convention, the Kyoto Protocol adopted in 1997 was the first time to quantify the greenhouse gas emission limits of industrialized countries in the form of an international legal document. However, with the intensification of differences among parties in the climate negotiations, the Paris Agreement, adopted in 2015, formally arranged the international mechanism to address climate change after 2020, establishing a bottom-up emission reduction path with autonomous contributions from all parties. This marks a new stage in the global response to climate change by achieving the long-term goal of controlling global temperature rise. It is true that climate change is a global issue, but since the sources of greenhouse gas emissions that cause climate change come from different geographical areas of each country, especially in certain metropolitan areas. Local governments need to be gradually considered as stakeholders, so that cities can play a key role in the process of addressing global climate change and implementing the Paris Agreement.

With the further acceleration of urbanization in countries around the world, cities have become increasingly prominent in social development. At the same time, cities also consume the most natural resources of the entire country, and are the most important source of environmental pollution. Currently, global cities cover only 3% of the earth's land area, and more than 50% of the population lives in urban areas, contributing about 70% of GDP, which in turn consumes more than 60% of energy and generates 75% of carbon emissions, leading to a surge in carbon dioxide and the formation of the greenhouse effect [1]. The IPCC states that greenhouse gas emissions caused by human activities are the root cause of global warming [2]. At present, the global climate is experiencing significant changes characterized by warming, including a range of issues such as sea level rise, frequent extreme catastrophic weather, and reduced biodiversity, which have had and will continue to have significant impacts on natural and human systems [3]. As the number of urban populations rises to 6.7 billion by 2050 [4], the energy consumption and carbon emissions of cities will increase dramatically, and the frequency and severity of extreme weather and natural disasters brought about by climate change will also increase, making cities highly sensitive and vulnerable areas to climate change. How to cope with climate change is a major challenge for cities. Among them, urban heat island (UHI) effect caused by the change of the urban surface structure and the emission of man-made heat is one of the most prominent climate change issues.

Therefore, in order to alleviate the impact of the heat island effect brought by climate change to the city, this study takes Guangzhou, the core city of the Guangdong-Hong Kong-Macao Greater Bay Area, one of the four major bay areas in the world, as a research case, and collects the surface temperature of five periods in the past 20 years, including 2000, 2005, 2010, 2015 and 2020, to analyze the situation of the heat island effect in Guangzhou, and to analyze its interaction with land use status and population density, with a view to providing urban planning, ecological planning and sustainable Development policy provides decision-making and basis.

2. Literature review: urban heat island effect

The study of the urban heat island effect can be traced back to the first half of the 19th century. Howard, by observing and comparing the temperature data of urban and suburban areas in London during the same period of time, found that the temperature in urban areas was significantly higher compared with that in suburban areas, which became the beginning of the study of urban heat island effect [5]. Manley gave the first definition of urban heat island [6]. Oke defined the difference between urban and suburban temperatures as the urban heat island intensity, reflecting the correlation between the intensity of heat island and the degree of urban development [7].

The formation of urban heat island effect is related to many factors. The external factors are mainly global warming, geographic location and weather conditions; the internal factors are mainly caused by the process of urbanization [8]. Among them, the change of surface structure caused by urbanization is the main reason for the formation of heat island effect. Because the natural surface of urban land is constantly replaced by artificial impermeable surfaces, it affects the heat exchange process between the soil and the atmosphere, thus causing the ground temperature rises and leads to the formation of urban heat island [9]. According to statistics, urban heat islands have been found in more than 1,000 cities of different sizes in the world, covering all latitudes in the southern and northern hemispheres [10]. Although the urban heat island effect does not directly cause major natural disasters, it often changes the local climate characteristics of the city (temperature, humidity, convection, precipitation pattern), and indirectly harms human production and life [11]. At the same time, the urban heat island effect and atmospheric pollution interact to increase the concentration of air pollutants, thus increasing the morbidity and mortality of respiratory and cardiovascular diseases [12]. Under the framework of the Paris Agreement, how to alleviate the urban heat island effect from the national level to the local government level has attracted much attention. One of the important ways is to improve the urban heat storage capacity. Vegetation, wind and water are the main factors to alleviate the urban heat island effect [13]. Reasonable planning and increasing urban parks and green spaces are effective means [14]. The reason is that parks can contribute to the improvement of urban thermal environment and microclimate through the process of transpiration of vegetation and lake water, and play the role of the city's "cold island", which is the key to mitigating the urban heat island effect.

With the development of space science and technology, scholars from different countries have explored and researched the heat island effect of cities in different regions using different methods. At present, the research methods of urban heat island effect are mainly based on meteorological observation method and remote sensing inversion method. The meteorological observation method is a traditional method for analyzing the urban heat island, and it provides the ability to directly assess the intensity of the urban heat island effect. However, as temperatures change over time and space, there can be significant differences in temperatures at different times and under different surface types. The remote sensing inversion method is one of the most common methods used to analyze urban heat island, and this method enables large scale spatial detection by using remotely sensed data that presents a faceted view. Currently, research on urban heat island effects using remote sensing methods is relatively mature. Among them, remote sensing images such as Landsat and MODIS provide an effective means to study the changing patterns of urban thermal environment and the distribution characteristics of heat island effect.

In 1972, Rao first used TIROS-1 satellite remote sensing data and inversion of the surface temperature for urban heat island effect analysis, thus opening up a new era of remote sensing methods applied to urban heat island effect research [15]. In 2001, Janos et al. used a combination of MODIS and TM data to analyze the state of the thermal field in eight large Asian cities, including Tokyo, Beijing, Shanghai, Seoul, Pyongyang, Bangkok, Manila and Ho Chi Minh City, and investigated the corresponding spatial distribution patterns of the thermal field, diurnal and seasonal variation characteristics in relation to vegetation conditions and the city's own thermal radiation [16]. Nichol in 2005 studied the diurnal variability of the urban heat island effect in Hong Kong using ASTER and Landsat data and concluded that urban parks have the effect of suppressing the urban heat island effect [17]. Rita et al. in 2006 used MODIS data to analyze the heat effect in the 10 most populous cities in Hungary and compared the characteristics of the changes in heat field distribution under different seasonal and macro-environmental parameter conditions [18]. Manju et al. in 2013 used MODIS and Landsat data to study the urban heat island characteristics of New Delhi, and found that the most severe urban heat island intensity was found in the dense urban and commercial areas [19]. In 2015, Khaled Abutale et al. used Landsat data to study the urban heat island effect in Cairo, and found that the local urban heat island effect was significant and mainly distributed in densely populated neighborhoods [20].

Thus, the urban heat island effect is an urban problem that is pervasive across the globe. In China, the rapid economic development in recent decades has been accompanied by various environmental problems, one of which is the heat island effect. At present, studies by scholars on the urban heat island effect in China are mainly focused on major cities such as Beijing and Shanghai, reflecting that the overall intensity and extent of the urban heat island in China is showing a gradual increase, bringing more difficulties and hazards to people's life and work. Therefore, it is imperative to combat the urban heat island effect.

3. A case study of the urban heat island effect: Guangzhou, China

In the past 30 years, with the rapid economic development of China, urbanization has entered an unprecedented stage of development. The urbanization rate has increased from 17.9% in 1987 to 60.60% in 2019 [21].

3.1. Introduction to Guangzhou

Guangzhou City, located in the south of mainland China, at the northern edge of the Pearl River Delta, is the third largest city in China and the political, economic, scientific, educational, and cultural center of Guangdong Province. The city has 11 districts with a total area of 7434 km^2, a built-up area of 1249.11 km^2, a resident population of 15,305,900, and an urbanization rate of 86.46% [22]. Guangzhou has a subtropical monsoonal maritime climate with an average annual temperature of 23.1°C, an average relative humidity of 78%, and a rainfall of 2,336 mm [23]. The geomorphic environment of Guangzhou is characterized by hills and plains, and the terrain basically slopes from northeast to southwest. The hills are mainly located in the northeast and are low hills below 250 m above sea level, while the plains are mainly located in the west and south, with river and sea silting plains in the south. In the past 20 years, Guangzhou has experienced rapid economic development and urban expansion, but at the same time, it has brought about a series of meteorological disasters and ecological environment problems such as local climate warming, air and water pollution, which constitute a serious threat to the sustainable development of the city.

3.2. Spatial distribution of urban heat island effect in Guangzhou

This article analyses the distribution of urban heat island in Guangzhou during summer (July to August) from 2000 to 2020 (once every five years) based on remote sensing and GIS techniques. The remote sensing data were obtained using Moderate-resolution imaging spectroradiometer (MODIS) data, a NASA optical remote sensing instrument on board the Earth Observation System (EOS) satellite. The morning star Terra (transit time 10:30 local time) and the afternoon star Aqua (transit time 13:30 local time). This article uses daytime Land Surface Temperature (LST) data from the MOD11A2 product of Terra, downloaded from the NASA website, (https://ladsweb.nascom.nasa.gov/search). To facilitate statistical analysis, the mean-standard deviation method is used to classify urban heat island levels. This method uses a combination of the mean surface temperature and a multiple of its standard deviation to classify surface temperature and to define the urban thermal environment. For heat island intensity, the mean-standard deviation method is less sensitive to the number of classifications than the equal-spaced method, and the mean-standard deviation method is also better than the equal-spaced classification method in terms of the spatial distribution of heat islands and the detail expression of temperature variation. Based on the above reasons, using u (mean value), 0.5std (standard deviation), and 1std as the dividing points, the surface thermal field in Guangzhou is divided into 5 grades, which are high LST zone, sub-high LST zone, medium LST zone, sub-low LST zone, and low LST zone.

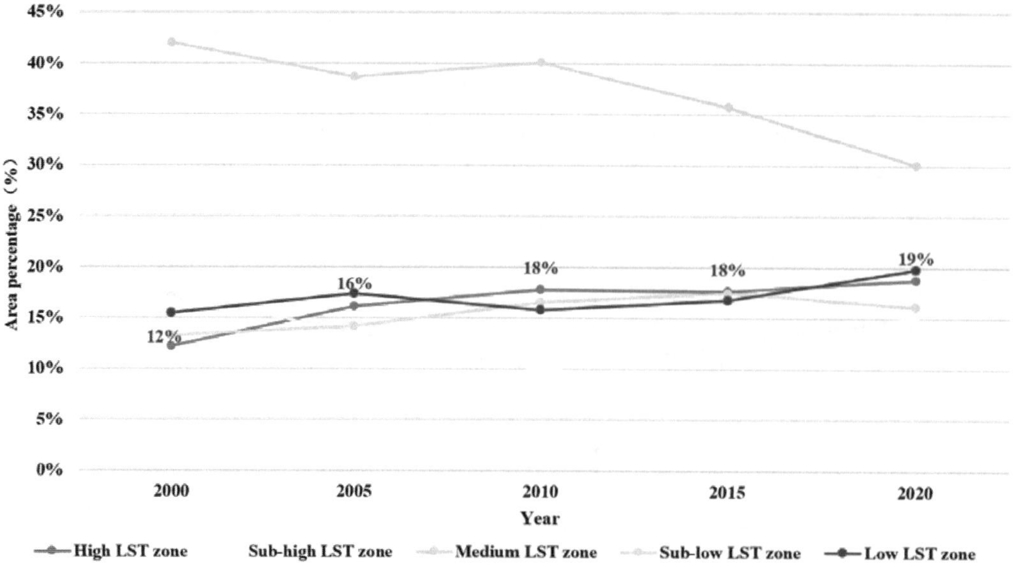

Figure 1. Percentage change curve of area affected by urban heat island effect in Guangzhou from 2000 to 2020.

On the whole, by analyzing the spatial distribution map of urban heat island in Guangzhou from 2000 to 2020, it can be seen that in the past 20 years, the urban heat island space in Guangzhou showed an outbreak and then basically stabilized situation, that is, the spatial expansion trend of urban heat island before 2010 was more obvious than that after 2010. Among them, the area of the low LST zone gradually increased, the area of the medium LST zone decreased significantly, and the area of the sub-high LST zone slowly decreased, while the area of the high LST zone and the sub-low LST zone tends to increase gradually (See Figure 1).

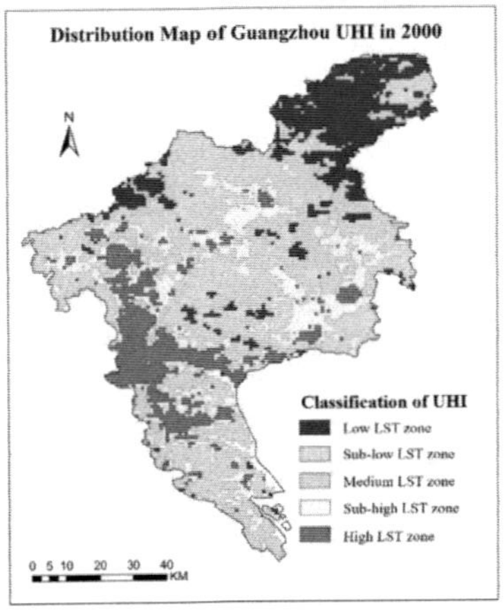

Figure 2.

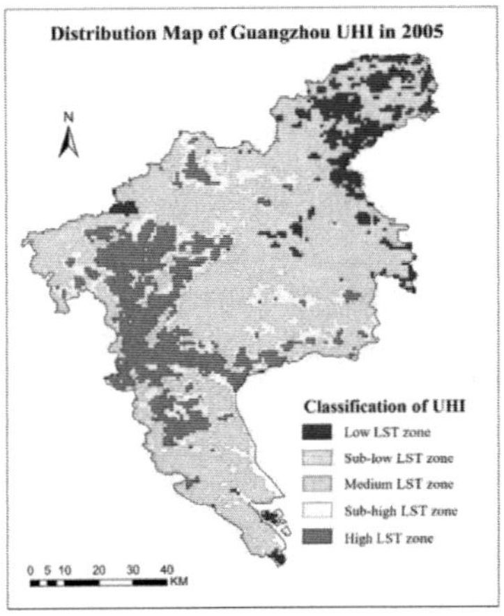

Figure 3.

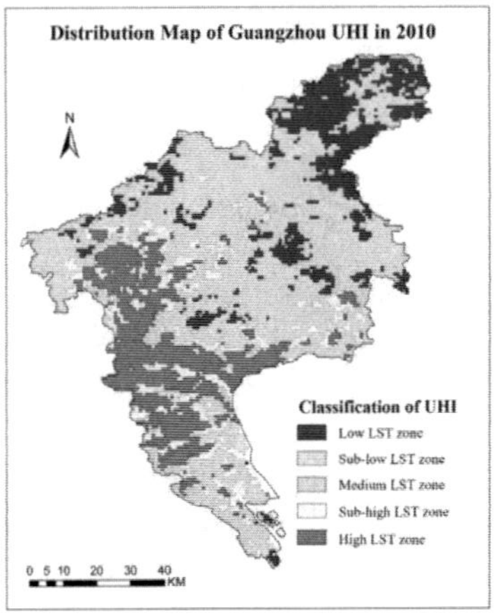

Figure 4.

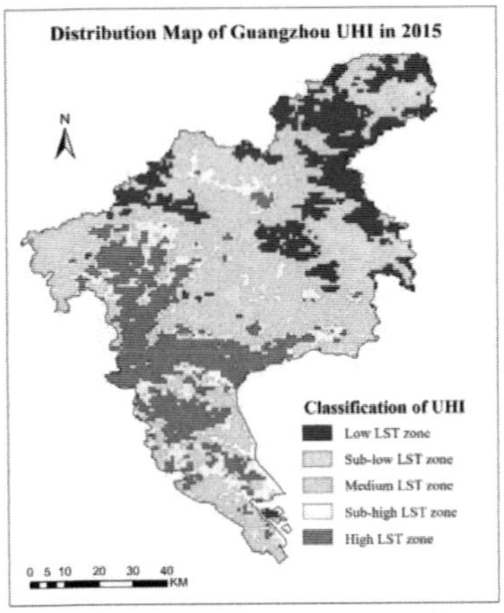

Figure 5.

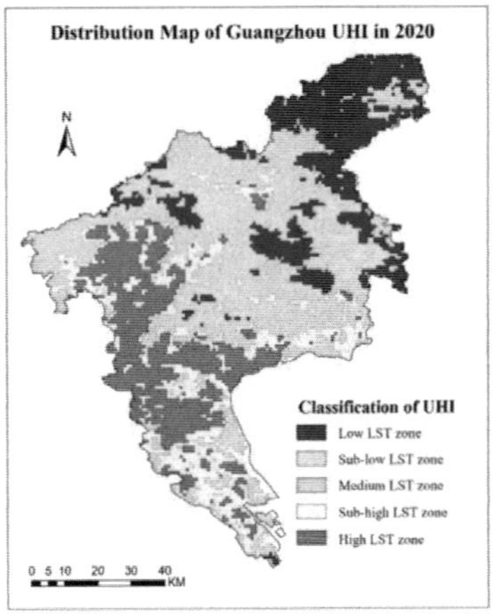

Figure 6.

Figure 2 to 6. Spatial distribution map of urban heat island in Guangzhou from 2000 to 2020 (the order of Figure 2 to 6 is 2000, 2005, 2010, 2015 and 2020).

* Data is sourced from the Moderate Resolution Imaging Spectroradiometer.
* Please see the end of this article for a larger version of these images.

The spatial distribution of urban heat island in Guangzhou is relatively concentrated, showing a gradual expansion trend from north to south and west to east, and basicall conforms to the law that the high temperature effect is obvious in the area with high building density and high population density, and the low temperature effect is obvious in the green area, water area as a whole (See Figure 2 to 6). They are mainly distributed in the central and southern plains with the highest degree of urbanization, namely Liwan District, Yuexiu District, Haizhu District, Tianhe District and Huangpu District, which are urban functions such as administrative districts, central business districts, residential areas, urban-rural transition zones and industrial zones. These areas are the most urbanized areas in Guangzhou, while the medium and low LST zones are mainly found in the northern mountainous area and the central low hill area, where the forests are concentrated and the surface temperature is relatively low due to the transpiration of vegetation.

3.3. Correlation between urban heat island effect, land use status and population density in Guangzhou

As seen from the urban land use map of Guangzhou from 2000 to 2020 (once every 5 years), the types of land use in Guangzhou are mainly construction land, cultivated land, woodland and waters, with construction land is concentrated in the central part, including Yuexiu District, Haizhu District, Liwan District District, Tianhe District, Baiyun District, Huangpu District, the cultivated land is distributed in Baiyun District, Huadu District, Conghua District, Zengcheng District, Panyu District, Nansha District, and the forest land is concentrated in the northern part, including Baiyun District, Huadu District, Conghua District and Zengcheng District, and waters are scattered in various regions. In terms of temporal trends, Guangzhou witnessed rapid urbanization from 2000 to 2005, showing a significant increase in the percentage of construction land and a significant decrease in the percentage of cultivated land (See Figure 7). The construction land extends significantly from the central city to the cultivated land in the east, south and north, forming several satellite cities, while the period from 2005 to 2020 is mainly based on the existing construction land to gradually develop to the surrounding cultivated land (See Figure 8 to 12). By comparing the spatial distribution maps of urban heat island, we can find that the distribution of urban heat island is positively correlated with the distribution of construction land. Due to the high building density and dense population in the central city of Guangzhou, the human-generated domestic energy consumption releases a large amount of heat, and it is difficult for airflow to flow within the city, resulting in a weakened heat dissipation function in the urban area. In addition, most of the urban buildings in Guangzhou are composed of masonry, asphalt, cement and concrete, which have small reflectivity to solar radiation and can absorb more solar radiation energy. Especially in the central business district, a large number of buildings adopt glass curtain wall structure, and the multiple reflection and absorption between the walls also aggravate the formation of urban heat island.

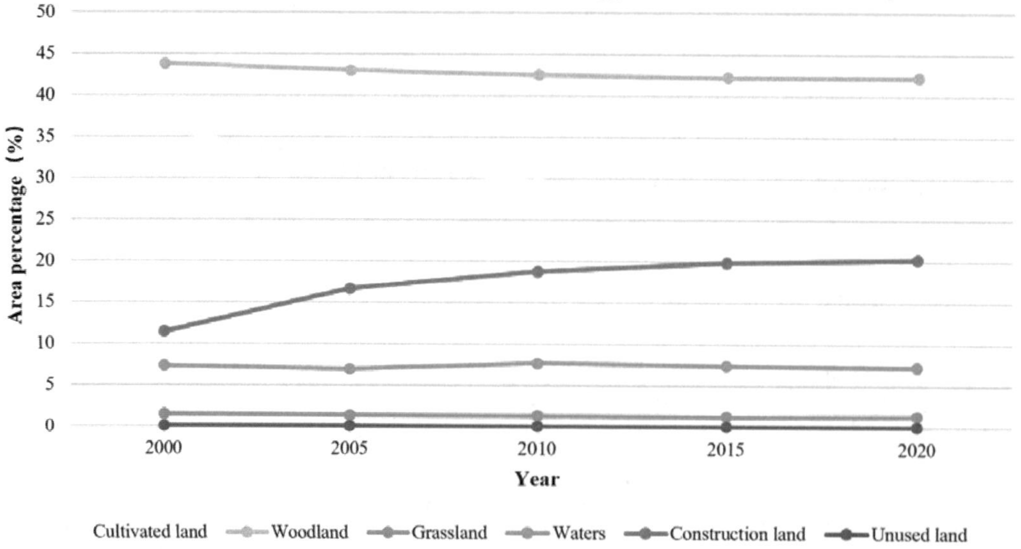

Figure 7. Percentage change curve of land use status in Guangzhou from 2000 to 2020.

From the population density distribution map of Guangzhou from 2000 to 2020 (once every 5 years), the resident population of Guangzhou is mainly concentrated in the central part, including Yuexiu District, Haizhu District, Liwan District, and Tianhe District. In terms of temporal trends, the population of Guangzhou moved substantially from the central city to the sparsely populated east, south and north from 2000 to 2005, while the 15 years after 2005, people mostly based on the densely populated areas for further concentration (See Figure 13 to 17). By comparing the spatial distribution maps of urban heat island, it can be found that the distribution of urban heat island shows a positive correlation with population density. Since it is relatively less difficult to develop land in low elevation areas, the scale and intensity of land development in these areas continue to increase, and most of these newly developed lands are built with high-rise, high-density residential buildings, commercial buildings, and hotels etc., which in turn promoted the continuous expansion of the population. The concentration of migration to these areas is accompanied by a continuous increase in energy consumption, which in turn leads to an increase in the intensity of the urban heat island effect.

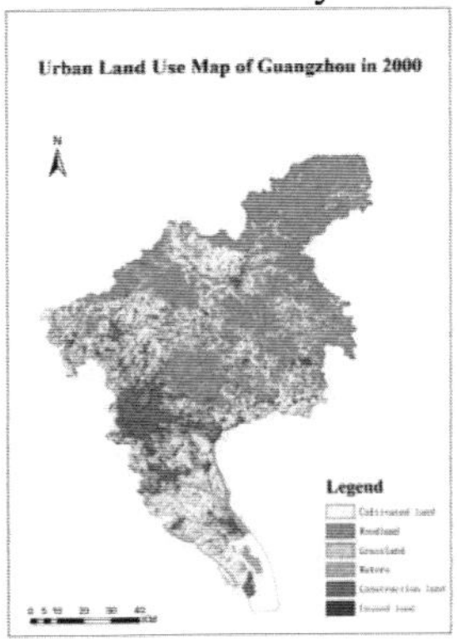

Figure 8.

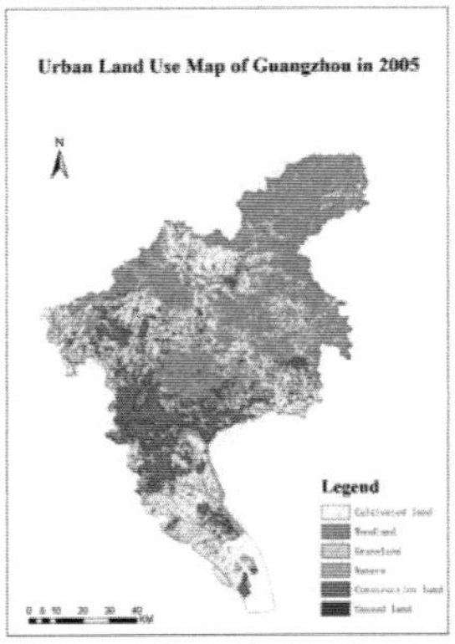

Figure 9.

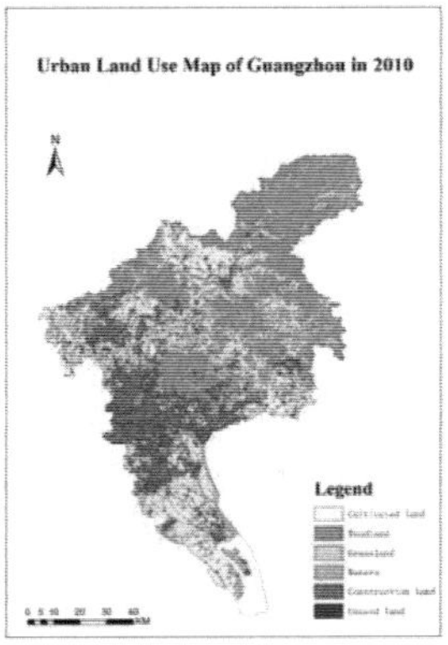

Figure 10.

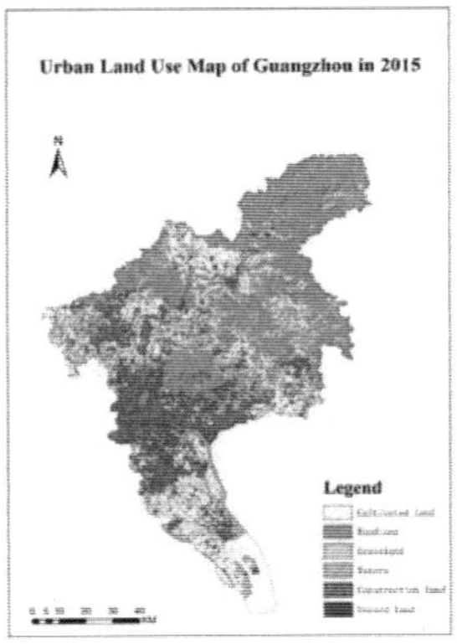

Figure 11.

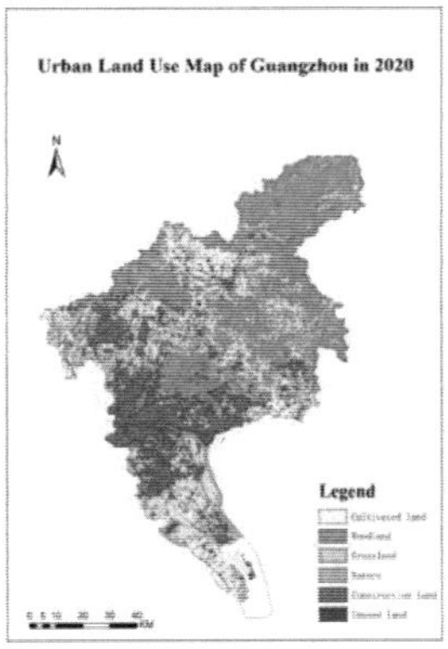

Figure 12.

Figure 8 to 12. Urban land use map of Guangzhou from 2000 to 2020 (the order of Figure 8 to 12 is 2000, 2005, 2010, 2015 and 2020).

* Data is sourced from the Geographic State Monitoring Cloud Platform.

3.4. Analysis of the causes of the urban heat island effect in Guangzhou

(1) Increasing man-made heat sources

Based on the fact that Guangzhou needs to consume a lot of energy for production, living and transportation travel, and Guangzhou is still in the stage of coal-based power generation, the energy conversion rate of this power generation method is not high, which not only causes waste of resources, but also results in a large amount of heat loss, intensifying the heat island effect and causing significant impact on air pollution. In the high-density commercial and residential areas of Guangzhou's urban centers, including Liwan District, Yuexiu District, Haizhu District, Tianhe District and Huangpu District, which are all major energy-consuming areas, especially during the summer months, there is a large increase in demand for cooling, and this energy consumption not only increases electricity expenditure, but also further increases anthropogenic heat emissions, leading to a rise in temperature and forming a vicious circle with the heat island effect.

Figure 13.

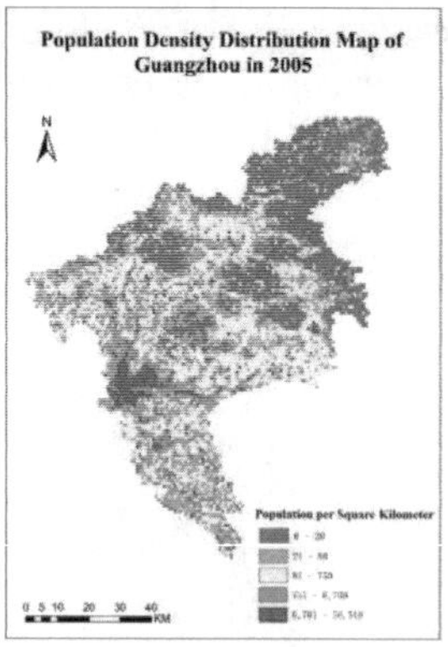

Figure 14.

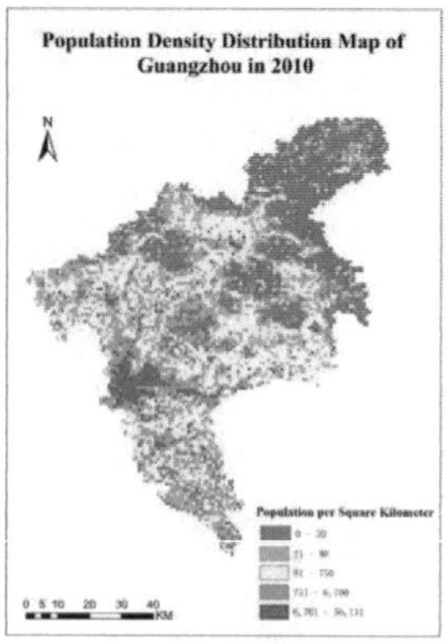

Figure 15.

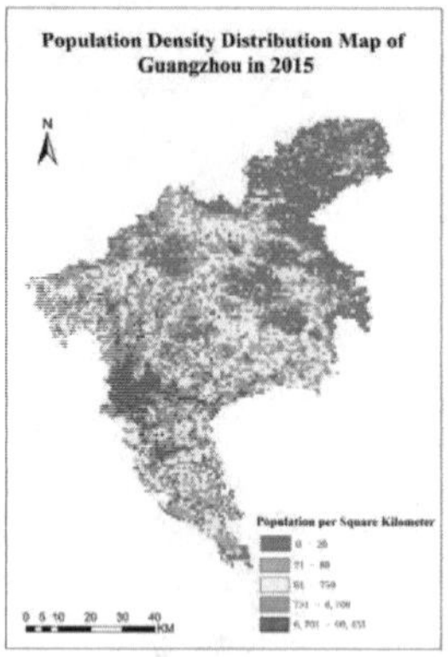

Figure 16.

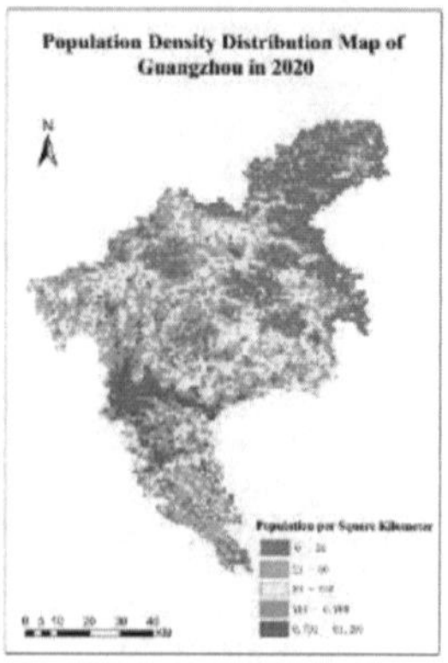

Figure 17.

Figure 13 to 17. Population density distribution map of Guangzhou from 2000 to 2020 (the order of Figure 13 to 17 is 2000, 2005, 2010, 2015 and 2020).

* Data is sourced from WorldPop.

(2) Changes in the underlying surface of the city

In the process of urbanization in the past two decades, the natural ecological environment elements such as mountains, forests, green spaces and waters in Guangzhou have been continuously reduced, resulting in a decrease in the water content of the urban surface. At the same time, a large number of concrete structures and asphalt roads have been increased, which has changed the properties of the underlying surface of the city. These artificial structures have the attributes of fast heat absorption and low heat capacity, making the urban hard surface absorb more sunlight than the natural surface under the same solar irradiation environment, absorbing more solar radiation and in turn more heat to the air, accelerating the generation of heat island effect.

(3) Continuous development of high-rise buildings

Guangzhou's central district is characterized by a high density of buildings of varying heights. The heat from solar radiation reflects between the varying buildings and between the buildings and the ground, undergoing multiple reflections and repeated absorptions, and the air pollutants are also adept in absorbing ground radiation, causing the temperature in the urban area to rise continuously, creating a high temperature and low-pressure zone over the city, resulting in the heat island effect in the central district.

(4) Increasing pollutant emissions

In Guangzhou, the lack of awareness of environmental protection in many industrial manufacturing enterprises has led to the emission of waste gases from industrial production into the atmosphere without timely cleaning and filtering. At the same time, the large number of people and motor vehicles in Guangzhou's urban centers generates a large amount of particulate matter, dust and harmful gases. These substances absorb a lot of energy from the heat radiation, which in turn generates a greenhouse effect and further warms the atmosphere in the city center.

3.5 Suggested strategies for mitigating the urban heat island effect in Guangzhou

(1) Scientific and rational planning of urban layout

The extent of the urban heat island effect is directly proportional to the density of urban buildings. The denser the urban building layout, the less efficient the air circulation and the more intense the heat island effect. Therefore, in the future urban planning and layout of Guangzhou and a series of construction processes such as the reconstruction of old urban areas, the government should consider the environment as the primary factor, vigorously develop public transport systems, reduce exhaust emissions from private cars, and organically integrate building design with topography and landscape so that buildings are both practical and ecologically compatible. In addition, the government can determine the width of the road network, the orientation of buildings and the shape of buildings according to the geographical environment of the city, including latitude, topography, wind direction, wind speed, sunshine, radiation conditions, etc., so that buildings in the city can receive reasonable sunshine and radiation, and also facilitate natural ventilation, and then form urban breezeways to attract wind into the city and increase atmospheric circulation, which is beneficial to the defense against urban heat disasters and the improvement of local microclimate.

(2) Optimizing the overall ecological environment of the city

At present, a large amount of urban land in Guangzhou is planned as commercial land and residential land, but the area of urban green space is gradually decreasing. It is true that green plants can not only provide human beings with the oxygen we need to breathe, but also their transpiration can effectively regulate and improve the microclimate of local areas. On the one hand, Guangzhou can increase the greening rate of the city by introducing the concept of three-dimensional greening, which will result in less heat absorption on the building walls and at the same time will have a significant effect on the cooling of the ground surface. On the other hand, as the heat capacity of water is greater than that of soil, it also needs to absorb a large amount of heat in the evaporation process. Therefore, in the urban layout and development of Guangzhou, the protection and management of wetlands and rivers around the city can be strengthened, and ecological buffer zones such as forests, lakes and grasslands can be constructed between different areas of the city to enhance the coverage of the natural ecological environment.

(3) Improve the material type of the underlying surface of the city

In order to alleviate the urban heat island effect of Guangzhou, one of the directions is to adjust the underlying surface of the city and change its thermal properties. As traditional paving materials such as lime stone slabs, granite and marble have a relatively low heat capacity, they are highly susceptible to exacerbate the formation of the urban heat island effect in the city, especially in hot weather. The environmentally friendly, high specific heat capacity materials such as colored rubber stone and permeable concrete can better solve this problem, while combined with the planting of landscaping and the reasonable arrangement of the vertical structure of vegetation, so that the composite vegetation structure can be reasonably matched in the city, gradually improving the radiation absorption and heat dissipation of the city's lower bedding surface, effectively regulating the local microclimate in the city center, while effectively absorbing harmful gases and improve air quality.

(4) Active development of renewable energy

At present, Guangzhou is dominated by fossil fuels including coal, natural gas and petroleum as sources of energy for living and production. However, the exploration, production and combustion of these fossil fuels emit large amounts of greenhouse gases, and the resulting global climate crisis has put countless lives at risk. The Guangzhou government should therefore accelerate its energy transformation and reduce carbon emissions at source by eliminating the economic model that relies on fossil fuels. Renewable energy sources such as solar, wind, tidal, hydro and geothermal energy have a much lower carbon footprint than fossil fuels in the process of power generation. Therefore, if renewable energy can be mainstreamed into society through energy transformation in the future, it will help to phase out fossil fuels and mitigate the impact of the heat island effect.

4. Conclusions

Currently, the Earth's climate system is undergoing a significant change with warming as the main feature. At the same time, with the accelerated urbanization, the population and industries continue to concentrate in cities, resulting in a phenomenon that the temperature in urban areas is higher than the surrounding area, thus forming the urban heat island effect. Taking Guangzhou as a case study, this article analyzes the relationship between the spatial distribution of urban heat islands and different land use types and population densities, and discusses key points for mitigating climate change. The results show that the spatial distribution of urban heat island in Guangzhou is mainly concentrated in the highly urbanized central and southern plain areas, while the medium and low temperature areas are mainly distributed in the densely forested northern mountainous and central low mountainous areas. At the same time, the type of land use and population density in Guangzhou directly change the surface temperature and influence the formation of urban heat islands. In the future, Guangzhou can alleviate the impact of the heat island effect by rationally planning the urban layout, optimizing the urban ecological environment, improving the material of the urban underlying surface, and actively developing renewable energy.

References

[1] Xinhuanet 2019 Cities have become the main battlefield for human sustainable development from http://www.xinhuanet.com/politics/2019-01/22/c_1124022617.htm
[2] IPCC 2018 Special Report on Global Warming of 1.5 °C
[3] Thomas R K, Kevin E T 2003 Modem Global Climate Change *Science* **302(5651)** 1719
[4] United Nations 2019 World Urbanization Prospects: The 2018 Revision
[5] Howard L 1833 The climate of London, deduced from meteorological observations
[6] Manley G 1958 On the frequency of snowfall in metropolitan *England Quarterly Journal of the Royal Meteorological Society* **84(359)** 70-72
[7] Oke T 1982 The energetic basis of the urban heat island *Quarterly Journal of the Royal Meteorological Society* **108(455)** 1-24
[8] Nakayama T, Hashimoto S 2011 Analysis of the ability of water resources to reduce the urban heat island in the Tokyo megalopolis *Environmental Pollution* **159(8-9)** 2164-2173
[9] Kimura F, Takahashi S 1991 The effects of land-use and anthropogenic heating on the surface temperature in the Tokyo Metropolitan area: A numerical experiment Atmospheric Environment Part B Urban Atmosphere **25(2)** 155-164
[10] Arnfield A J 2003 Two decades of urban climate research: A review of turbulence, exchanges of energy and water, and the urban heat island *International Journal of Climatology* **23(1)** 1-26
[11] Grimm N, Faeth S, Golubiewski N, Redman C, Wu J, Bai X, Briggs J 2008 Global change and the ecology of cities *Science (New York, N.Y.)* **319(5864)** 756-760
[12] Araujo R, Albertini M, Costa-Da-Silva A, Suesdek L, Franceschi N, Bastos N, AllegroV 2015 São Paulo urban heat islands have a higher incidence of dengue than other urban areas Brazilian *Journal of Infectious Diseases* **19(2)** 146-155
[13] Taha, Akbari, Rosenfeld 1991 Heat island and oasis effects of vegetative canopies: Micro-meteorological field-measurements *Theoretical and Applied Climatology* **44(2)** 123-138
[14] Jauregui E 1990 Influence of a large urban park on temperature and convective precipitation in a tropical city *Energy and Buildings* **15(3)** 457-463
[15] Rao P K 1972 Remote sensing of urban heat islands from an environmental satellite **53** 647-648
[16] Ja'nos U, Zolta'n S, Judit Z 2001 Temperature cross-section features in an urban area *Atmospheric Research* **58** 117-127
[17] Nichol J 2005 Remote sensing of urban heat islands by day and night *Photogrammetric*

Engineering and Remote Sensing **71(6)** 613-621
[18] Rita P, Judit B, Zsuzsanna D 2006 Remotely sensed thermal information applied to urban elimate analysis *Advances in Space Research* **37** 2191-219
[19] Manju M, Yukihiro K, Gurjar R 2013 Assessment of urban heat island effect for different land use-land cover from micrometeorological measurements and remote sensing data for megacity *Delhi Theoretical and Applied Climatology* **112(3)** 647-658
[20] Khaled A, Adeline N, Ahmed D 2015 Assessment of Urban Heat Island Using Remotely Sensed Imagery over Greater Cairo Advances in Remote Sensing **04(01)** 35-47
[21] National Bureau of Statistics 2020 National Data from https://data.stats.gov.cn/index.htm
[22] Guangzhou Statistics Bureau 2020 Administrative Divisions from http://tjj.gz.gov.cn/gzsq/content/post_2786710.html
[23] Guangzhou Meteorological Bureau 2019 Guangzhou Climate Bulletin

IOP Conf. Series: Earth and Environmental Science 1087 (2022) 012016 doi:10.1088/1755-1315/1087/1/012016

Nitrogen and phosphorus load estimation of inflow rivers to Yuqiao Reservoir based on GWLF model

Zeli Li[1], Kai Gao[1, 3], Xiaochao Gu[1], Pengyu Mei[1], Zhen Zhang[1] and Yuqiu Wang[2]

[1] Tianjin Eco-Environmental Monitoring Center, Tianjin 300191, China;
[2] College of Environmental Science and Engineering, Nankai University, Tianjin 300350, China.

[3] Corresponding author's e-mail: kaigao123@sina.cn

Abstract. Watershed model technology was an important tool for non-point source pollution prevention and control. In order to figure out the total nitrogen (TN) and total phosphorus (TP) load of inflow rivers to Yuqiao Reservoir in 2021, GWLF model was used to simulate the hydrologic process in its sub-basin. Firstly, the hydrological parameters of the model were determined by the measured meteorological and hydrological data. Secondly, the hydrological parameters were extended to the whole basin, and the three main river inflows to Yuqiao Reservoir were simulated in 2021. Lastly, the nutrient loads were estimated by multiplying the simulated inflow as mentioned above by the observed TN and TP concentration data at the same site. The results showed that the TN and TP load of inflow rivers was 7 106.9 t and 80.0 t, respectively, of which the flood period from July to August contributed 55.3% and 69.3%, respectively. GWLF model was effective in simulating hydrological processes in the study area. The Nash efficiency coefficients of the calibration and the validation period were 0.89 and 0.91, respectively. GWLF model was suitable for hydrological simulation and estimation of non-point source pollution load in small and medium-sized basins.

1. Introduction

The vast proliferation of cyanobacteria in lakes and reservoirs has become a growing aquatic environmental problem. Yuqiao Reservoir is an important water source for potable water of Tianjin city, which has a population of over 14 million [1]. However, the eutrophic degree intensified and cyanobacteria blooms occurred frequently in summer and autumn in the past 5 years, which had adversely affected the safety of potable water supply and the reservoir management [2].

Determined the source and load of nitrogen and phosphorus nutrients into the water body is the primary task to control eutrophication [3]. At present, the watershed model technology is widely used in watershed non-point source pollution prevention and control research internationally and domestically [4], which comprehensively considered meteorology, underlying surface conditions, hydrological cycle process and pollutant transformation mechanism. GWLF (Generalized Watershed Loading Function) model, as one of the models recommended by EPA and the Ministry of Ecology and Environment of China, is suitable for simulation of hydrology, sediment and pollutant transport processes in small and medium-sized watersheds and can provide reliable simulation results on a monthly scale [5]. Meanwhile, compared with SWAT, HSPF and MIKE and other distributed models with strong mechanical rationality, GWLF model requires less spatial data and fewer parameters, and has been successfully applied in many countries and regions such as the United States [6-9], Europe [10-12], China [13-19].

Nitrogen and phosphorus non-point source pollutants carried by rainstorm runoff into water body are one of the most important reasons for lake and reservoir eutrophication [20-21]. In 2021, there were more extreme precipitation in the Yuqiao reservoir basin. GWLF model was employed to figure out the total nitrogen (TN) and total phosphorus (TP) load of inflow rivers to Yuqiao Reservoir, so as to provide technical support for water environment management of local government.

2. Materials and methods

2.1. The study area

Yuqiao Reservoir is located 4 km east of Jizhou District, Tianjin, with a normal water storage area of 86.8 km^2 and a watershed area of 2 060 km^2. The basin belongs to the temperate semi-arid and semi-humid monsoon climate, with four distinct seasons. The average temperature for many years was 12.5℃, and the average precipitation was 670 mm, mainly concentrated in June to September. The main inflow rivers to Yuqiao Reservoir were Shahe River, Lihe River (concurrently serving as a channel for water diversion from Luanhe River) and Linhe River, with three water quality monitoring sections namely Shahe Bridge, Lihe Bridge and Linhe Bridge, respectively (Figure 1). Among them, the Shuipingkou hydrological station located in the lower reaches of Shahe River controls an area of 811 km^2, with relatively complete hydrological data. The sub-basin was dominated by natural falling aquatic flow with little human intervention. Therefore, it was selected as the target area for the hydrological parameter calibration of GWLF model.

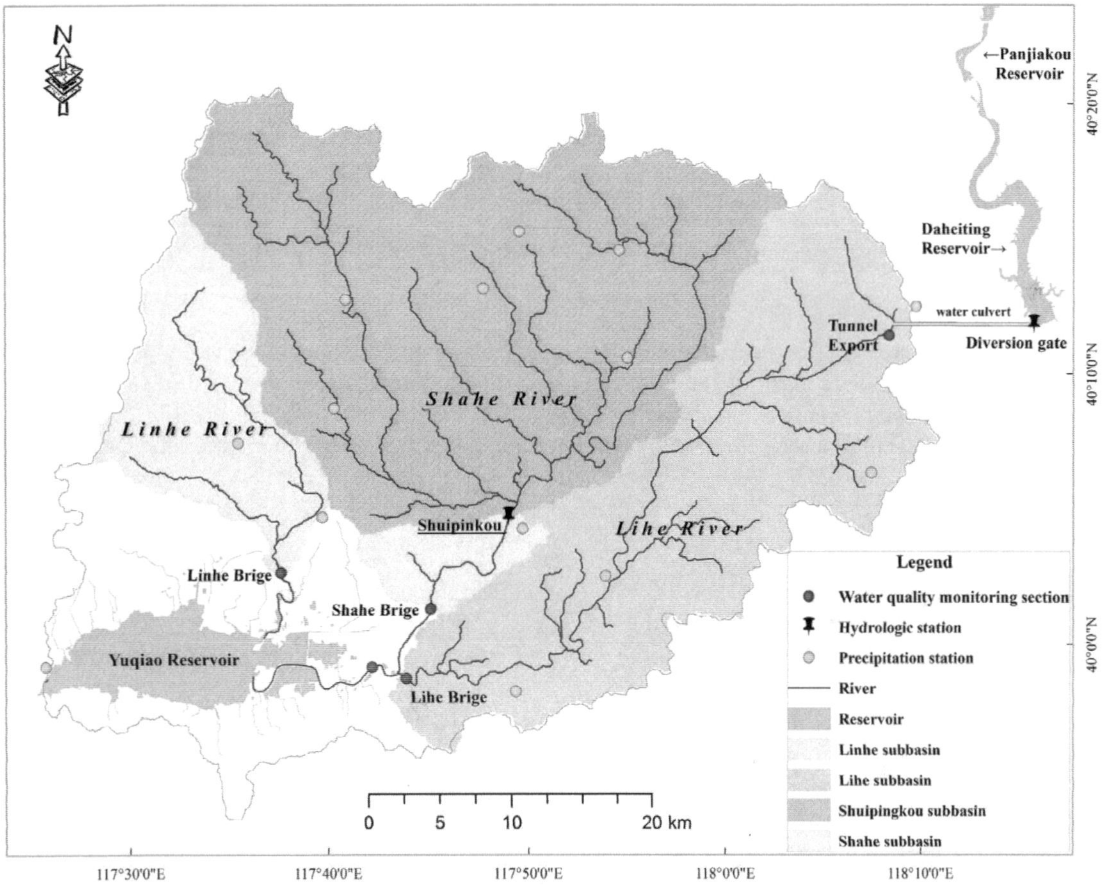

Figure 1. River network and sub-basins in Yuqiao Reservoir Basin.

2.2. Principle of model hydrological module

GWLF model, jointly developed by Professor Haith from Cornell University and Professor Shoemaker from Pennsylvania State University, is a semi-empirical lump-parameter watershed pollution load model, which can provide simulation results of discharge, sediment amount, and soluble and insoluble nitrogen and phosphorus load in the study area at monthly and annual scales [6].

GWLF model hydrological simulation needs to provide daily precipitation and daily mean temperature of the basin, and runoff curve number (SCS-CN) equation is used to simulate surface runoff. According to the daily soil hydrological balance conditions, the daily potential evaporation and surface water infiltration were calculated. The surface below is divided into three parts: shallow unsaturated layer, shallow saturated layer and deep unsaturated layer. The water content of each layer changes with time. The shallow saturated layer can produce base flow, and the surface runoff can be combined with the river runoff and reach the outlet of the basin. The hydrological cycle process of the model is shown in Figure 2.

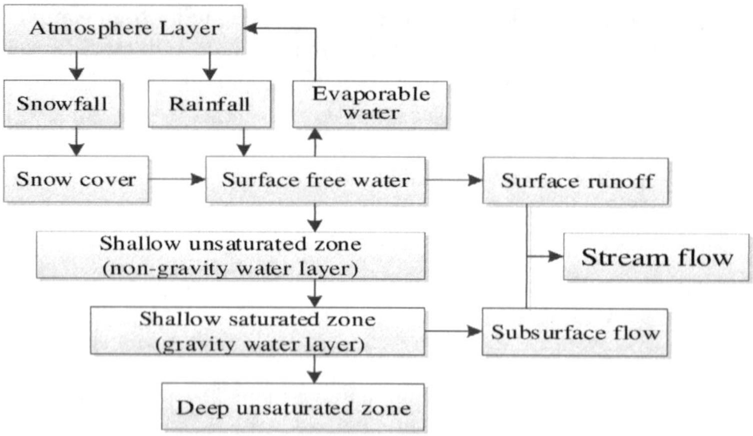

Figure 2. The hydrologic balance module of GWLF model.

2.3. Data collection and processing

The model input data for this study mainly include Digital Elevation Map (DEM), land use type map and continuous monitoring data of meteorology, hydrology and water quality within the simulation period. The original data sources are shown in Table 1.

Table 1. Original data source for GWLF model construction in Yuqiao Reservoir Basin.

Data type	Data source	Data description
DEM	Http://www.gscloud.cn	Raster data of 2009, 30 m×30 m
Land use	Tianjin Eco-Environmental Monitoring Center	Remote sensing interpretation of 2018, 1:100,000 land use map
Weather	Http://data.cma.cn/; Haihe river basin hydrology yearbook	Daily average temperature of 1 meteorological station from 2006 to 2021; Daily precipitation of 16 rainfall stations from 2006 to 2018
Hydrology	Haihe river basin hydrology yearbook; Tianjin Water Bureau	Monthly average flow of Shuipingkou hydrological station from 2006 to 2018; Monthly flow of water diversion from Luanhe River in 2021
Water quality	Tianjin Eco-Environmental Monitoring Center	Monthly TN and TP concentrations of 3 rivers in 2021

Based on the ArcGIS platform, after inputting DEM and actual river network, the Shuipingkou sub-basin was generated with the Shuipingkou hydrological station as the outlet of the basin (Figure 1). Then, the areas of various land use in sub-basins were extracted as the input data of the model. Temperature and precipitation data are input into the model with daily steps, and precipitation data is the spatial weighted average of Tyson polygon of 16 rainfall stations in the basin. In the sub-basins of Shahe Bridge, Lihe Bridge and Linhe Bridge, the same method was used to obtain model input data.

3. Model construction and simulation

The GWLF model used in this study adopts ReNuMa (Regional Nutrient Management) model version [8]. The input interface of the model is Excel software in Microsoft Office tool, which makes use of Excel's powerful plan-solving macros. One or more parameters are automatically calibrated based on monthly measured data, based on the principle of nonlinear least square method, so that the simulation results are as close to the measured results as possible (the sum of error squares is minimum).

3.1. Calibration and validation

The hydrological parameters were calibrated and validation by comparing the simulated and observed monthly time series of stream flow of Shuipingkou hydrological station. The model ran for a period of 156 months, from the January 1, 2006 up until the December 31, 2018. In order to allow model to equilibrate to approach reasonable starting values, the first year of model simulation was used as a spin-up period. Firstly, monthly aggregations data from 2006 until 2012 were used for calibration. The calibration parameters mainly include runoff curve number, field water holding capacity, evaporation coverage factor, groundwater regression coefficient and leakage coefficient. Secondly, Data from 2013 until 2018 were used for validation after all parameters were determined. The stream flow of the target area in 2021 were simulate after inputting meteorological data of 2021 and running the calibrated model again.

The initial values in Table 2, creating the rest of the important parameters, were based on the ReNuMa and GWLF user manuals and literatures surveys. The final values of these same parameters were determined by calibration processing.

Table 2. Hydrologic parameter for GWLF model calibration in Yuqiao Reservoir Basin.

Parameter	Variable	Initial value	Calibrated value
Curve number (CN)	Cultivated land	85	50
	Forest	65	40
	Grassland	60	45
	Water body	100	100
	Residential area	95	75
Recession coefficient (day^{-1})	R	0.01	0.003
Seepage coefficient (day^{-1})	S	0	0.004
Field capacity(cm)	U	10	12.9
Evaporation coverage factor	CV	0.8	0.6/0.4 [a]

[a] 0.6/0.4 was calibrated for the growing and non-growing periods, respectively.

3.2. Evaluation index of simulation results

The performance and adaptability of the model were evaluated by referring to the statistical indexes and evaluation criteria commonly used in hydrology at home and abroad [22]. Nash coefficient (NSE), whose value range is (-∞, 1). The closer it is to 1, the higher the matching degree between simulated value and measured value is. The formula is as follows:

$$NSE = 1 - \frac{\sum_{i=1}^{n} (Q_{oi} - Q_{pi})^2}{\sum_{i=1}^{n} (Q_{oi} - \overline{Q_o})^2}$$

Where Q_{oi}, Q_{pi}, $\overline{Q_o}$, n represent the observed values, simulated values, average observed values and number of samples, respectively.

The evaluation reference of the model was shown in Table 3.

Table 3. Evaluation standard of model performance levels based on monthly simulation.

Statistical indicator	Unsatisfactory	Satisfactory	Good	Excellent
NSE	(-∞, 0.5]	(0.5, 0.65]	(0.65, 0.75]	(0.75, 1]

4. Results and Discussions

4.1. Model Performance

Model predictions and observations for monthly stream flow of calibration and validation were compared in Figure 3. At the monthly time scale for the stream flow, the model captured the variation trend of runoff well and the model's performance indicated excellent predictive capability. Stream flow (presented as water depth in centimeters over the surface area of the catchment) could be predicted fairly well as reflected by the high ENS values, with 0.89 for calibration and 0.91 for validation period, respectively.

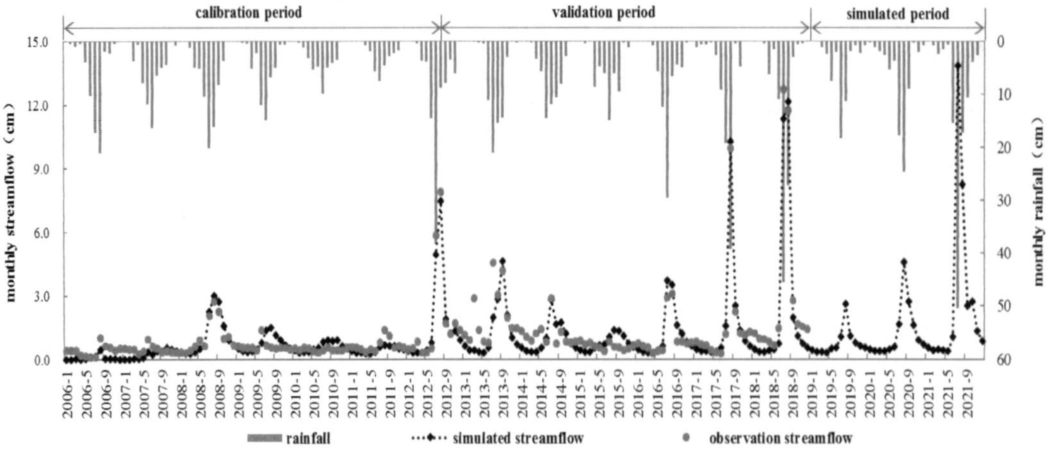

Figure 3. Precipitation process and runoff fitting effect of Shuipingkou sub-basin.

4.2. Nitrogen and Phosphorus Load Estimation

The calibrated hydrological parameters of the model were brought into the models of Shahe, Lihe and Linhe sub-basins, and monthly discharge of each sub-basin in 2021 was simulated. TN and TP concentrations were multiplied by the corresponding monthly TN and TP concentrations of Shahe Bridge, Lihe Bridge and Linhe Bridge sections respectively, and the annual TN and TP loads were summarized. The TN and TP loads of water diversion from The Luanhe River were calculated separately by multiplying the amount of water diversion by the TN and TP concentrations of the Lihe Bridge in corresponding months. The load estimation results of TN and TP were shown in Table 4.

The results showed that in 2021, the total water inflow into the reservoir of incoming rivers was 855.6 million cubic meters, and the source and contribution proportion were 33.3% from Shahe River,

26.7% from Lihe River, 25.9% from Luanhe River, and 14.2% from Linhe River, respectively. The TN input load of Yuqiao reservoir was 7 106.9 t, and the source and contribution proportion were 34.3% from Shahe River, 33.2% from Lihe River, 22.1% from Linhe River, and 10.5% from Luanhe River, respectively. The TP input load was 80.0 t, and the sources and contribution ratios were 51.7% from Lihe River, 34.8% from Shahe River, 8.7% from Luanhe River, and 4.9% from Linhe River, respectively.

Table 4. Estimation results of water yield and TN and TP load of inflow rivers in 2021.

Source	Runoff ($10^8 m^3$)	Average concentration (mg L^{-1})		Load(t)	
		TN	TP	TN	TP
Water diversion	2.218	3.33	0.03	742.8	6.93
Shahe river	2.847	9.87	0.06	2 439.7	27.8
Lihe river	2.280	9.49	0.09	2 356.9	41.4
Linhe river	1.211	13.7	0.02	1 567.5	3.89
Total	8.556	/	/	7 106.9	80.0

According to the monthly statistical results in Figure 4, the TN into the reservoir from July to August accounted for 55.3% of the whole year, while the TP accounted for 69.3%, mainly because the precipitation in July to August accounted for 64% of the whole year, and the heavy precipitation in the basin produced multiple floods, carrying a large amount of dissolved and granular nitrogen and phosphorus pollution into the reservoir.

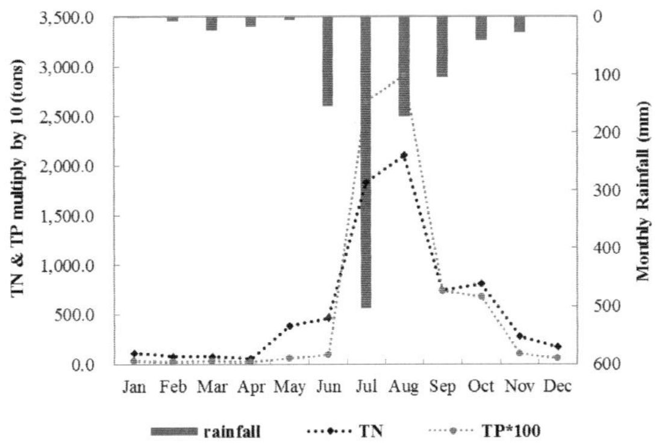

Figure 4. Comparison of monthly precipitation and pollution load estimation.

4.3. Related research results comparison

At present, the methods of river pollution load accounting mainly include direct monitoring method, discharge inventory method and mechanism model method. Direct monitoring method is mainly used to calculate the load through synchronous monitoring of water quality and water quantity, which has been applied to estimate TN and TP load around Taihu lake [23,24], Dongting lake [25] and Poyang lake [26]. The method is simple, intuitive and easy to master, but the disadvantage is that the monitoring results are mostly instantaneous and sampling is time-consuming and laborious, restricted by on-site sampling conditions.

Emission inventory method is mainly based on economic and social statistical data and emission coefficient (output coefficient) to calculate the emissions of pollution sources in the study area. Due to its simple principle and good availability of data, the results can be effectively connected with the management department, so it has been widely used. In the Yuqiao Reservoir basin, Zhang Hui [27] estimated the pollution load of domestic sewage, household garbage, human excrement, livestock and poultry excrement by using the sewage discharge coefficient method through field investigation and data collection in 2012, and obtained that TN and TP pollution load were 356.3 t/a and 125.2 t/a respectively. Wang Hanyang [28] conducted a questionnaire survey on production and domestic sewage discharge in six towns around Yuqiao Reservoir in 2017, and estimated the inflow of TN and TP pollution around the reservoir by using the pollutant discharge coefficient method. The results showed that TN and TP pollution loads were 257.2 t/a and 38.5 t/a, respectively. However, the reliability of the estimation results was questionable, since the output coefficient was an empirical parameter, most studies did not consider the response relationship between precipitation, underlying surface and pollutant transport.

As mentioned above, mechanism model takes meteorology, underlying surface conditions and hydrological cycle process into account, and is a common method used internationally to carry out pollutant load simulation in river basins. In the Yuqiao reservoir basin, Zhao Dongmei [29] estimated the annual TN and TP loading of various pollution sources to be 808.96 t and 140.07 t by combining field investigation and SCS model method in 2006. Xu Ning [30] used SWAT model to simulate TN and TP pollution load of the inflow river of Yuqiao Reservoir from 2010 to 2013, and the results showed that the average annual TN and TP loads were 2 900.6 t and 163.2 t, which were similar to in magnitude of the method adopted in this paper.

5. Conclusions

The GWLF model was applied to the runoff simulation of Yuqiao Reservoir basin in northern China, and a set of localized hydrological parameters were calibrated. The Nash coefficients of the hydrological simulation during calibration and verification periods were 0.89 and 0.91, showing good performance of the model. On this basis, the flow yield of Shahe river, Lihe River and Linhe River in 2021 was simulated to calculate TN and TP input load of the inflow rivers. The results showed that TN and TP load into Yuqiao reservoir in 2021 were 7 106.9 t and 80.0 t, and the non-point source pollution caused by rainfall should be paid more attention.

This study integrated the advantages of direct monitoring method and watershed load model, adopted GWLF model with few model parameters and medium demanding data to simulate water quantity, combined with water quality monitoring data of the same period to estimate TN and TP load, the results were relatively reliable, which could provide decision-making support for watershed water environment management departments. The model had the potential to be applied and extended in China and other similarly developing countries.

References

[1] ZHANG Chen, SONG Didi, LIAN Tiehui 2020 The Influences of the Water Diversion Structure Change on Total Phosphorus Retention and Implication for Ecological Flows in Yuqiao Reservoir Tianjin *Journal of Lake Sciences* **32** 2

[2] FANG Xu, DUAN Hongtao, CAO Zhigang 2018 Remote Monitoring of Cyanobacterial Blooms Using Multi-source Satellite Data: A Case of Yuqiao Reservoir Tianjin *Journal of Lake Sciences* **30** 4

[3] XU H, PAERL H W, QIN B Q 2010 Nitrogen and Phosphorus Inputs Control Phytoplankton Growth in Eutrophic Lake Taihu, China *Limnology and Oceanography* **55** 1

[4] LI Qian, ZHANG Peng, PENG Fuli 2019 Research Progress and Preliminary Plan of National Water Quality Forecasting and Alarming System *Environmental Monitoring in China* **35** 1

[5] LI Yajiao, SONG Jiabao, LI Jiake 2019 Research and Application Progress of Four Typical Non-point Source Pollution Model *Water Resources and Power* **37** 3

[6] HAITH D A, SHOEMAKER L L 1987 Generalized Watershed Loading Functions for Stream Flow Nutrients *Journal of the American Water Resources Association* **23** 3

[7] SCHNEIDERMAN E M, PIERSON D C, LOUNSBURY D G 2002 Modeling the Hydrochemistry of the Cannonsville Watershed with Generalized Watershed Loading Functions (GWLF) 1 *Journal of the American Water Resources Association* **38** 5

[8] HONG B, SWANEY D P 2013 Regional Nutrient Management (ReNuMa) Model, Version 2.2.1 User's Manual *New York: Cornell University*

[9] NIRAULA R, KALIN L, SRIVASTAVA P 2013 Identifying Critical Source Areas of Nonpoint Source Pollution with SWAT and GWLF *Ecological Modelling* 268
[10] LEE K Y, FISHER T R, JORDAN T E 2000 Modeling the Hydrochemistry of the Choptank River Basin Using GWLF and Arc/Info: 1. Model Calibration and Validation *Biogeochemistry* **49** 2
[11] LEE K Y, FISHER T R, ROCHELLE N E 2001 Modeling the Hydrochemistry of the Choptank River Basin Using GWLF and Arc/Info: 2. Model Validation and Application *Biogeochemistry* **56** 3
[12] JENNINGS E, ALLOTT N, PIERSON D C 2009 Impacts of Climate Change on Phosphorus Loading from a Grassland Catchment: Implications for Future Management *Water Research* **43** 17
[13] SHA J, LIU M, WANG D 2013 Application of the ReNuMa Model in the Sha He River Watershed: Tools for Watershed Environmental Management *Journal of Environmental Management* 124
[14] DU Xinzhong, LI Xuyong, ZHANG Wangshou 2014 Simulation of Total Nitrogen Load in Watershed and Analysis of Pollution Source Based on GWLF Model *Journal of Water Resources and Water Engineering* **25** 3
[15] ZHANG Ruihao, ZHU Longji, WANG Baimei 2018 Simulation of Characteristic Pollutant Load and Analysis of Pollution Source in Liuhe River Basin *South-to-North Water Science & Technology* **16** 4
[16] QI Z D, KANG G L, SHEN M 2019 The Improvement in GWLF Model Simulation Performance in Watershed Hydrology by Changing the Transport Framework *Water Resources Management* **33** 3
[17] SHA J, SWANEY D P, HONG B 2014 Estimation of Watershed Hydrologic Processes in Arid Conditions with a Modified Watershed Model *Journal of Hydrology* 519
[18] LI Z L, LIU M H, ZHAO Y 2014 Application of Regional Nutrient Management Model in Tunxi Catchment: In Support of the Transboundary Eco-compensation in Eastern China *Clean-Soil, Air, Water* **42** 12
[19] QI Z D, KANG G L, CHU C L 2017 Comparison of SWAT and GWLF Model Simulation Performance in Humid South and Semi-arid North of China *Water* **9** 8
[20] YANG Liuyan, YANG Xinyan, REN Liman 2019 Mechanism and Control Strategy of Cyanobacterial Bloom in Lake Taihu *Journal of Lake Sciences* **31** 1
[21] WANG Y P, KONG X Z, PENG Z L 2020 Retention of Nitrogen and Phosphorus in Lake Chaohu, China: Implications for Eutrophication Management *Environmental Science and Pollution Research* **27** 33
[22] MORIASI D N, ARNOLD J G, VAN LIEW M W 2007 Model Evaluation Guidelines for Systematic Quantification of Accuracy in Watershed Simulations *Transactions of the ASABE* **50** 3
[23] XU Pengzhu, QIN Boqiang 2005 Water Quantity and Pollutant Fluxes of the Surrounding Rivers of Lake Taihu During the Hydrological Year of 2001-2002 *Journal of Lake Science* **17** 3
[24] YAN Shuwen, YU Hui, ZHANG Lulu 2011 Water Quantity and Pollutant Fluxes of Inflow and Outflow Rivers of Lake Taihu, 2009 *Journal of Lake Science* **23** 6
[25] TIAN Zebin, WANG Lijing, LI Xiaobao 2014 Characteristics Analysis of Pollutant Influx and Outflux in Dongting Lake *Research of Environmental Sciences* **27** 9
[26] LUO Yong, WU Hengyun, LI Ying 2014 The Study on the Pollutant of Flux of Lake Inflow in Poyang Lake Basin *JiangXi Science* **32** 5
[27] ZHANG Hui, LIANG Qiong, GAO Rutai 2014 Analysis on Treatment Solution on External Pollution at Water Source of Yuqiao Reservoir *Anhui Agricultural Science Bulletin* **20** Z1
[28] WANG Hanyang, ZHANG Chunxue, ZHENG Xiangqun 2020 Estimation of Rural Non-point Source Pollution Load and Countermeasures in Yuqiao Reservoir Watershed *Rural Economy and Science-Technology* **31** 9
[29] ZHAO Dongmei 2007 Study on Agriculture Non-point Source Pollution and Control Measures in Yuqiao Reservoir Area *Tianjin, Tianjin University*
[30] XU Ning, GAO Jinqiang, WANG Qian 2017 Simulation of Non-point Source Pollution in Zhouhe River Basin based on SWAT Model *Haihe Water Resources* 5

The Evaluation of Carbon Peak and Carbon Neutrality Industry Based on AHP- Entropy Weight Method

Meng Peng[1], Lu Yang[2], Kun Zhang[3], Weidong Zhao[1,4] and Jiayu Xu[3,4]

[1] Institute of energy conservation and environmental protection, China Electronic Information Industry Development Research Institute;
[2] School of mathematics and statistics, Beijing Jiaotong University;
[3] School of environment, Tsinghua University.

[4] Corresponding author's e-mail: zhaoweidong@ccidthinktank.com; jiayu-xu@tsinghua.edu.cn

Abstract. In response to the carbon peak and carbon neutrality dual targets, an index evaluation system was established around green development and decarbonisation consisting of three directions: "driving force", "state", and "response". It includes 16 indicators such as the number of low-carbon technology releases, the proportion of CO2 generated from the industrial process. After collecting several current methods of constructing an index evaluation system around green development, an index evaluation system for assessing the low carbon industry process was advocated in different cities. The defined AHP analytic hierarchy process (AHP)-entropy weight method was used to determine the weights of evaluated indicators, and the fuzzy comprehensive evaluation method was adopted to complete the quantitative evaluation of data. Finally, the model was applied to 31 provinces, cities and autonomous regions in China and work out the score matching each province. According to the overall score, the development of CPCN process 31 provinces is categorized into 5 levels. The results show that a level C and a level D dominating and the level D accounting for the most significant proportion (more than 50%). Only three provinces and cities achieved the outstanding level of A. The conclusion that the carbon peak and carbon neutrality development in our country is still relatively lagging at the current stage with uneven regional distribution matches the actual situation; therefore, it validates the reliability of the method. Based on the scoring relationships between different indicators in each region, the defects and resistance of low-carbon industrial development in some provinces should be analysed to provide a foundation for further accelerating the steps towards achieving carbon peak and carbon neutrality.

1. Introduction

Currently, China is at a crucial turning point to achieve the "carbon peaking and carbon neutrality" (CPCN) goals. However, the actions of saving and reducing emissions are challenged by energy-intensive industries, as China's backward industrialization system of has prevailed for a long time at the cost of resource abuse and environmental damage, since reform and opening-up [1]. Moreover, with the rapid economic growth, the country is facing serious practical challenges such as the historical evolution of industrial sectors, the tightening of energy constraints and industrial transformation and upgrading. Under these circumstances, it is crucial to study the development of green industry, as indicated from the carbon peaking and carbon neutrality" goals initiatives: identify the issues that hinder the development of green industry in each region; and to help the industrial enterprise to break the bottleneck of "carbon peaking and carbon neutrality" development according to local conditions.

In the assessment system of the CPCN process in the industry, many domestic scholars have established quite a number of evaluation models. Nevertheless, most of them focus on evaluating the effectiveness of the existing CPCN process, lacking the assessment of the future development potential. Therefore, the evaluation model they constructed lacks the timeliness of the medium and long-term CPCN development changes. From the perspective of CPCN development in cities, we should strive to consider existing achievements and future trends comprehensively. Some scholars have constructed an integrated energy system, introducing indicators for assessing the interaction between social factors and energy systems, which allowed to conduct a comprehensive analysis of interaction capabilities between systematic carbon emission and the grid system, and to show more precisely the interaction between CPCN and the participating parties [2]. On the other hand, for the power sector, the most important source of carbon emissions under the CPCN target, some environmentalists established a multi-layer of "source, grid and load" assessment system by combining the power supply side, grid side and load side together from the scientific, systematic and operational principles for the power industry, highlighting the ability of a city's power industry to save energy and reduce emissions and develop new energy under the CPCN target [3]. Other researchers, evaluated the innovative green development of the manufacturing industry under the CPCN target, and combined four simple evaluation model and conducted assessments of compatibility and driftance to get a comprehensive evaluation system, which leads to a conclusion that in the aspects of regional manufacturing industry, the overall level of green development driven by green technology innovation is not high and varies greatly from region to region [4]. In addition, three important tools in promoting low carbon economy have been discussed, including integrating regional development characteristics, linking low carbon productions and planning a low carbon city from the low carbon consumption aspect, thus demonstrating the importance and necessity of exploring suitable carbon emission analysis methods for systems in the city-level [5]. Other researchers, established evaluation indicators in three dimensions of environment, economy, and energy efficiency and evaluated the distributed energy system using the information entropy weight method with gray correlation analysis [6].

Regarding the comprehensive evaluation method, domestic and international studies have already been relatively mature in terms of data processing, selection and calculation of evaluation indicators, and determination of weights. The current research focuses on how to establish a comprehensive, integrated, and scientific set of systems and calculation approaches based on the known issues to reflect the state of the evaluated objects. The ideal solution of TOPSIS has been applied by some researchers to conduct a comprehensive evaluation of environmental pollution caused by industrial enterprises [7]. The rationale of the TOPSIS approach is to form a relative comparison between evaluated objects by measuring and ranking the proximity of each evaluation object towards the most desirable target, which has a certain value of promotion. From a perspective of small and medium-sized banks， a study was carried out on how small and medium-sized banks, represented by city commercial banks, can play their own advantages, especially on the pre-loan risk assessment of green credit [7]. Some scholars adopted the AHP method to establish a practicable assessment model and conduct an empirical analysis to provide reliable advice for small and medium-sized banks on green credit decision-making [8]. By using risk assessment index system, they established a Delphi-AHP weighting approach that comprehensively integrated both subjective and objective factors, and introduced a new membership function based on the traditional fuzzy comprehensive evaluation model to avoid the ignorance of objective factor of other membership degree's existence, caused by the large differentiation in membership degree [8]. The new approach has been proved that it can comprehensively evaluate the safety and security strength of the source of the Yangtze River [8]. Furthermore, the importance of scale in the traditional AHP method has been improved by changing the scale method from (1 ~9) to (-1、0、1) to avoid the interactive restrict between evaluated objects [9]. In order to synthetically assess the application of the "two mountains" theory in Gansu Province, the synthetic method was employed to evaluate the level of coordination between economic development and ecological and environmental protection by introducing the weighting of coordinated development degree [10]. Therefore, it can maximize the benefits for both economy and environment and conduct a detailed and precise classification of coordinated development of economy and environment in each region [10].

Additionally, the analytic hierarchy process method and entropy weight method was firstly combined by researchers to analyze the weight of each risk factor of real estate and then conduct a synthetic risk evaluation of that project by fuzzy comprehensive evaluation model, which not only pointed out the low comprehensive risk of the project but also identified the more influential risk factors [11]. In the process of evaluating the quality of water conservancy projects, "Cuckoo search algorithm" was proffered creatively to optimize the artificial neural network (ANN) [12]. The cuckoo algorithm was used to find the optimal solution of the evaluation model by imitating the behavior of the cuckoo bird in retaining the best offspring [13]. Finally, the cuckoo algorithm presents the same result as the AHP method and fuzzy evaluation method in actual practice, while also demonstrating a faster operating rate than other models [13].

In summary, according to the relevant literature on the CPCN industry, many experts in this study area concluded that the development of regional low carbon industry needs to be combined with the existing low carbon industrial base, the government support for low carbon policies and the future low carbon industrial development potential， in order to comprehensively evaluate the progress of CPCN industry in cities. In this paper, we will establish a set of index systems for the green industry from three aspects of "driving force", "state," and "response" and assign weights to 16 indicators using the comprehensive subjective and objective weighting method of AHP-entropy weight from the perspectives of urban driving force, existing green industrial conditions, and future development force of the low-carbon industry.

We will also quantify the industrial data related to each province using the fuzzy comprehensive evaluation model to provide scientific suggestions for exploring the development paths of green industry in cities.

2. Methodology of an index system for evaluating low carbon industries

In this paper, three criterion layers of "driving force", "state" and "response" are designed.

2.1. Selection of "state" index

The level of industrial development and capacity varies significantly in our country. On the one hand, large cities and coastal cities with heavy industries have a mature industrial production system due to their early development. Therefore, during the 14th Five-Year Plan period, these cities have been more actively responding to the targets of reducing CO2 emissions and saving energy consumption in industrial production and have already achieved certain results that other cities can learn from it. On the other hand, the less industrial developed regions have a later-mover advantage, with less pressure on energy saving and emission reduction and more space for balancing high-paced development and green development. In order to reasonably measure the state of green development of cities with different industrial levels under CPCN goals, we select four three-level indicators under the " state " index: CO2 emission ratio of industrial processes, industrial energy consumption value-added, CO2 emission of industrial value-added, and electrification rate.

2.2. Selection of "driving force" index

The steady growth of an industrial economy is a significant driving force for promoting regional green development of industry. Therefore, the contribution of green industrial growth is an essential indicator for constructing a comprehensive evaluation system of the green industrial development level. This indicator can be used to measure three aspects: the development status of local industry, research and development condition and level of government support. To further refine the indicators of green industrial growth contribution, from the economic perspective, we select the proportion of green financial scale, industrial value-added, and growth rate of industrial value-added; from the innovation perspective, we count the number of low-carbon technology releases, green patent applications in prefecture-level cities, and the number of employees above the scale; from the infrastructure development perspective, we take the number of new renewable energy installations/consumption generated during the 14th Five-Year Plan.

2.3. *Selection of "response" index*

Table 1. CPCN Evaluation Index System.

Target Layer	Criterion Layer	Indicator Layer
Low-carbon Industry	Driving Force	The number of low-carbon technology releases
		New renewable energy installations/consumption generated during the 14th Five-Year Plan
		The proportion of green financial scale
		Green patent applications in prefecture-level cities
		industrial value-added
		Growth rate of industrial value-added
		The number of employees above the scale
	State	CO_2 emission ratio of industrial processes
		Industrial energy consumption value-added
		CO_2 emission of industrial value-added,
		Electrification rate.
	Response	Number of low-carbon policy releases, update frequency, update time ranking, etc.
		The number of green demonstration parks
		The number of green factories
		The number of green designed products
		The number of demonstration enterprises of green supply chain management

National policy releases play a critical role in promoting industrial development. From the policy release aspect, the number of low carbon industrial policy release can be used to measure the development of the green industry. The development of the green industry also reflects the current level of green transformation in regional industries and its potential sustainable development ability. Therefore, the green industrial development level plays an important role in measuring our progress. We select four three-level indicators to measure industrial green development: the number of green demonstration parks; the number of green factories; the number of green-designed products; and the number of demonstration enterprises of green supply chain management.

Based on the above, we obtained a hierarchical evaluation index system for the targets of CPCN and low carbon industry, as show in Table 1.

3. Determination of evaluation index weights based on analytic hierarchy process (AHP) and the entropy weight method

The weight is a value that measures the relative importance of each evaluation index of the evaluated object. Reasonable determination of weights is crucial to reflect the process of CPCN in cities accurately. To avoid the limitations in the traditional way of using single subjective or single objective evaluation method, in this paper we combined the subjective weighting approach of AHP with the objective weighting approach of entropy weight method to determine the combination weights of indicators that can comprehensively reflect the actual situation and make the results to reflect the empirical judgements from the experts on CPCN cities and to present the objective scientific nature of the evaluation. The main process is shown in Figure 1. The evaluation system of the CPCN process is multi-layered, and each layer should follow the procedure as shown in Figure 1.

The first step is to identify the AHP subjective weights and the entropy weight method of objective weights, respectively, and then assess the relative importance of objective weight and subjective weight with least-squares as the optimization objective. Then we can get the composite weight of each factor for the target layer. To avoid excessive deviation caused by one-signer expert's experience, the one-one comparison matrix should be used to calculate the subjective weights more reasonably as follows: Each expert should independently complete the questionnaire relates to the comparison of the importance of indicators, then construct a one-one comparison matrix for each questionnaire, and perform consistency tests and weight calculations for each of them. The final weights are obtained by averaging the weighting results of all experts.

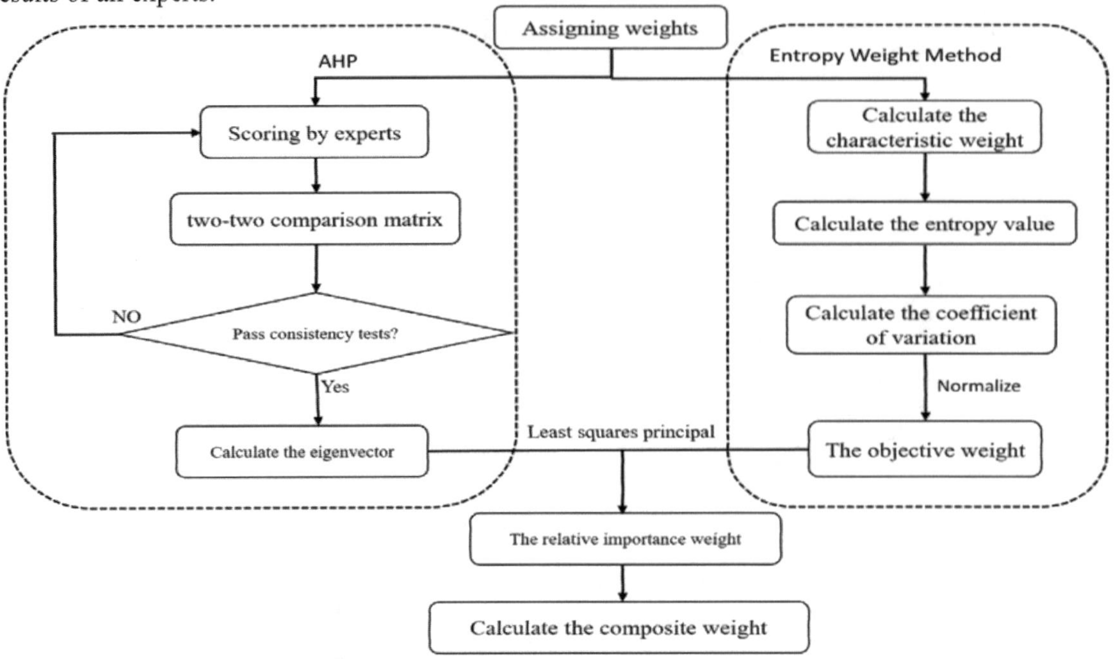

Figure 1. AHP-ENTROPY WEIGHT METHOD calculation flow chart.

3.1. Assigning weights to indicators based on the entropy weight method.

The entropy weight method is an objective empowerment method. The rationale of the entropy weight method is that if the degree of variation of an evaluated object is smaller in a particular indicator, the information reflected by the indictor will be smaller and thus lower the relative weight of the indicator [14]. The main calculation method is below:

Supposing x_{ij} is the actual value of singer indicator of alternative object j under the evaluation factor i. There are n evaluation indicators and m evaluation objects. First, indicators need to be normalized. In general, all indicators of low carbon city evaluation can be divided into the maximum index and minimum index etc. The normalisation process is as follows:

•for maximum index, transform:

$$x_{ij}^{*} = 0.999 \cdot \frac{x_{ij} - \min\limits_{1\le j\le m} x_{ij}}{\max\limits_{1\le j\le m} x_{ij} - \min\limits_{1\le j\le m} x_{ij}} + 0.001$$

•for minimum index, transform：

$$x_{ij}^{*} = 0.999 \cdot \frac{\max\limits_{1\le j\le m} x_{ij} - x_{ij}}{\max\limits_{1\le j\le m} x_{ij} - \min\limits_{1\le j\le m} x_{ij}} + 0.001$$

Calculate the characteristic weight of evaluated object i under the indicator j:

$$p_{ij} = \frac{x_{ij}^{*}}{\sum_{i=1}^{n} x_{ij}^{*}} \quad (i = 1,2,\dots,n; j = 1,2,\dots,m)$$

Calculate the entropy value ej of the jth indicator as：

$$e_j = -\frac{1}{\ln m}\sum_{i=1}^{n} p_{ij}\ln p_{ij}$$

Calculate the coefficient of variation g_j of the jth indicator as:

$$g_j = 1 - e_j$$

Normalize the coefficient of variation to obtain the weight coefficient w_{2j} of the jth indicator：

$$w_{2j} = \frac{g_j}{\sum_{k=1}^{m} g_k}$$

Based on the above calculations, we can obtain the objective weight value of the upper upper-level criterion on the factors of its lower-level indicators.

3.2. Assigning weights to indicators based on the AHP method

Analytic Hierarchy Process (AHP) is a subjective empowerment method. The basic principle of AHP is to evaluate the importance of each indicator, on the basis of dividing the decision-making factors into the target layer, criterion layer and indicator layer [15]. The main calculation process is as follows：

- The scaling method is used by introducing a scale number from 1 to 9, to evaluate the relationship of relative importance between two indicators of indicator layers under each criterion layer. A number of one-one comparison matrix $A = (a_{ij})_{n\times n}$ are formed, where a_{ij} represents the relative importance of "ith" element and "jth" element with respect to a factor in the criterion layer.
- Consistency tests are completed to ensure that all the element compared in A are of transitive nature, and whether the element involved in the two-two comparison almost satisfy:

$$a_{ij} = \frac{1}{a_{ij}}$$

First, calculate the coincident indicator CI of the matrix:

$$CI = \frac{\lambda_{max}(A) - n}{n - 1}$$

Among which, $\lambda_{max}(A)$ represents the maximum eigenvalue of matrix A. According to the random consistency index RI and the dimension number of A introduced by Saaty, the random consistency ratio CR of the Paired comparison matrix A is calculated by

$$CR = \frac{CI}{RI}$$

if CR < 0. 1, it means that paired comparison matrix A has good consistency.

- Calculate the eigenvector W which corresponds to the maximum eigenvalue of the matrix A

$$A \cdot W = \lambda_{max}(A)$$
$$W = (\omega_1, \omega_2, \dots, \omega_n)^T$$

After normalization, each element of the W vector represents the relative weights of each indicator under the corresponding criterion-layer's index. By using the same method, the relative subjective weight of each criterion-level indicator on a subordinate factor can be obtained.

In a multi-layered evaluation index system, when calculating weights and assessing consistency, it is necessary to consider the weights and consistency of its upper level in an integrated manner. Therefore, we should deal with the urban CPCN evaluation problem by using a combined relative weight approach and a combined consistency test. The calculation of the combination weighting should be processed as follows: Assume we have m*n matrix W,

$$W^{(p)} = [\omega_{(1)}^p, \omega_{(2)}^p, \dots, \omega_{(n)}^p]_{m\times n}$$

is the weights of the m th element of level p to the elements on level p-1, and the column vector $\omega_{(i)}^p$ is the level-p elements' relative weight vectors to the ith element. Thus the combined weight vector of the level-p element to the target layer is

$$\omega^{(p)} = W^{(p)}\omega^{(p-1)}$$

, among which $\omega^{(p)}$ represents level-p elements weight vector to the highest layer (target layer). The assessment of combined consistency should be processed in the following steps: Assume the consistency indicator of level p are $CI_1^{(p)}, CI_2^{(p)}, \dots, CI_m^{(p)}$, the random consistency indicator are $RI_1^{(p)}, RI_2^{(p)}, \dots, RI_m^{(p)}$, thus the combined consistency indicator of level p is

$$CI^{(p)} = (CI_1^{(p)}, CI_2^{(p)}, \dots, CI_m^{(p)})\omega^{(p-1)}$$

The combined random consistency indicator of level p is

$$RI^{(p)} = (RI_1^{(p)}, RI_2^{(p)}, \dots, RI_m^{(p)})\omega^{(p-1)}$$

The combined random consistency ratio of level p is

$$CR^{(p)} = CR^{(p-1)} + \frac{CI^{(p)}}{RI^{(p)}}$$

If the combined consistency ratio of the bottom level CR is less than 0.1, then the entire layer passes the consistency assessment.

3.3. *The evaluation of composite weight*

We assume α as the relative importance of the subjective weighting approach and the objective weighting approach. A greater value of α indicates a greater tendency towards opinion based on experts' experience and expertise, while a smaller value of α suggests a higher preference for decision making relying on scientific and data based information. Assume ω_{1j} is the weight of the jth indicator obtained by AHP approaches, ω_{2j} is the weight of the jth indicator obtained by entropy weight method, thus the combined weight ω_{1j} of jth indicator is:

$$\omega_j = \alpha \cdot \omega_{1j} + (1-\alpha) \cdot \omega_{2j}$$

In this paper, we choose the minimum square loss function as our loss function to determine the value α, thus balance the tendency of objective weight and subjective weight:

$$\alpha_0 = argmin \sum_{i=1}^{m} [(\omega_{i1} - \omega_i)^2 + (\omega_{i2} - \omega_i)^2]$$

The derivation of the above equation can obtain a minimum value while $\alpha_0 = 0.5$, which means the subjective weight and objective weight are equally important, thus the composite weight of the indicator j can be calculated by:

$$\omega_j = 0.5\omega_{1j} + 0.5\omega_{2j}$$

4. Determination of raw data matrix using the membership function

The membership function is the function to map set to unit rational interval [0,1], and it is used to verify if an element belongs to a specific set indicator function, which means changing from using indicator vector 0 or 1 to represent the relationship between elements and set to be a value within the range of [0,1] to represent the "credibility "of an element belongs to a fuzzy set. On the basis of the CPCN evaluation system, indicators are divided into maximum type, minimum type and mediate type, and adopt different levels of membership function for these three types of indicators.

For the indicators preferring smaller values, we use the following membership function:

$$\mu_A = \begin{cases} 1, & x < a \\ \left(\dfrac{b-x}{b-a}\right)^k, & a \le x \le b \quad k = 1,2 \\ 0, & x > b \end{cases}$$

For the indicators preferring larger values, we use the following membership function:

$$\mu_A = \begin{cases} 0, & x < a \\ \left(\dfrac{x-a}{b-a}\right)^k, & a \le x \le b \quad k = 1,2 \\ 1, & x > b \end{cases}$$

5. Numerical Analysis

The basic types of industry's CPCN evaluation data in our country mainly include statistics yearbook in 2021, environmental monitoring data and documentary data. Statistical yearbook data are collected from national, provincial, and municipal statistical yearbooks, Statistical Communiqué and specific bulletins; environmental monitoring data are collected from the China General Environmental Monitoring Station, and literature is collected from various relevant newspapers and periodicals. Based on the aforementioned model, we evaluate the development of low-carbon industries in 31 provinces, cities and autonomous regions of China, following the requirements of indicators.

First of all, 16 relevant indicators of 31 regions were transformed by the membership function and the weights were calculated by the AHP-entropy weight method, and the importance discriminations used in this paper were calculated to pass the consistency test. The detailed subjective and objective composite weights are presented as follows.

Table 2. Indicator weights in the evaluation system.

Target Layer	Criterion Layer	Indicator Layer
Low-carbon Industry	Driving Force (0.407)	The number of low-carbon technology releases (0.0675)
		New renewable energy installations/consumption generated during the 14th Five-Year Plan (0.0867)
		The proportion of green financial scale （0.1001）
		Green patent applications in prefecture-level cities(0.0730)
		Industrial value-added(0.0584)
		Growth rate of industrial value added(0.0831)
		The number of employees above the scale(0.0412)
	State (0.278)	CO2 emission ratio of industrial processes(0.0554)
		Industrial energy consumption value-added(0.0896)
		CO2 emission of industrial value-added(0.0293)
		electrification rate(0.0464)
	Response (0.315)	Number of low-carbon policy releases, update frequency, update time ranking, etc. (0.0128)
		The number of green demonstration parks(0.0252)
		The number of green factories(0.0420)
		The number of green designed products(0.0940)
		The number of demonstration enterprises of green supply chain management(0.0951)

The specific weights were weighted and summed with the indicators processed by the membership function to obtain the scores of CPCN process for each province. The results are shown in table 3 below:

Table 3. Stratification of CPCN scores by province.

Categories	**Region**
Outstanding	Guangdong, Jiangsu, Zhejiang
Good	Beijing, Shandong, Henan
Fair	Fujian, Shanghai, Shaanxi, Anhui, Xinjiang
Relatively Poor	Sichuan, Hubei, Jiangxi, Hunan, Tianjin, Qinghai, Shanxi Inner Mongolia, Guizhou, Hebei, Yunnan, Gansu, Tibet, Chongqing, Liaoning, Guangxi
Poor	Heilongjiang, Jilin, Ningxia, Hainan

In accordance with the results of the overall score, the development of the industrial CPCN process in 31 provinces, municipalities and autonomous regions of China has been divided into five categories, from high to low: "excellent", "good", "fair", "relatively poor" and "poor", corresponding to the five levels of A-E respectively. As shown in the table, the classification results lead to the conclusion that the overall level of China's CPCN industrial development is not high, with a level C and a level D dominating and the level D accounting for the most significant proportion (more than 50%), as presented in Figure 1. According to the result, the average score for the 31 provinces in China is 6.11, and only 11 provinces and cities exceed the average. Moreover, 16 provinces and cities are at the relatively poor level of D, accounting for 51.6%; among them, the worst performance is Hainan province, with a score of 3.31. On the other hand, only three provinces and cities achieved the outstanding level of A; among them, the best one is Guangdong province, with a score of 13.37.

Due to the space limitations, Table 4 can only show part of the index's' ranking for some regions.

As Table 4 shows, Guangdong, Jiangsu and Zhejiang provinces ranked 1st to 3rd place in terms of their overall score. These three provinces have a relatively good traditional industrial base as well we obvious location advantage and high government attention. Taking Guangdong province as an example, the province ranks top in many indicators such as low carbon technology release, industrial value-added, electrification rate, and green factory. It indicates that Guangdong Province is at the upper level of industrial development in the country, with greater investment in the design and development of green products as well as solid and efficient abilities in using science and technologies to drive industrial economic growth by producing green products.

Beijing, Shandong and Henan Province's rankings are at 4th to 6th places, indicating that these three provinces have started to yield preliminary results in promoting green industry. Beijing has many talents engaged in science and technology, so it has a more mature application of technological tools and ranked high in science and technology aspects such as green industry and green design product factor. Nevertheless, Beijing has decentralized a large volume of non-capital functions to the surrounding areas in recent years, including a significant number of general manufacturing and polluting enterprises, causing Beijing to rank low in terms of industrial value-added and growth factors. Henan and Shandong provinces ranked relatively low in the proportion of CO_2 in the industrial processes and its value-added. This ranking indicates that the industrial structure in these two provinces is aging, with high energy consumption generated from producing processes and a high number of highly polluting industrial industries, therefore cannot adapt to the trend of green development in the new era.

The ranking of Xinjiang and the other five regions are 7th to 11th place, which means that the CPCN industrial development is at the beginning stage. Due to the practical situation and development characteristics in these regions, there are different kinds of resistance affecting the promotion of CPCN development. Xinjiang autonomous region is one of the worse western regions in developing CPCN industry. As shown by the scores in Table 4, Xinjiang has a severe shortage of enterprises and talents engaged in CPCN industrial development, and the ranking of its low-carbon technology releases is at the bottom level in the country. The capacity for science and technology innovation and sustainable development in Xinjiang is also relatively weak. However, under the impact of the "The Belt and Road Initiative" and the "China's Western Expansion", Xinjiang has ranked among the top five in the country in electrification and value-added growth rates, reflecting a rapid growth momentum in CPCN industrial development. In the next step, the government in Xinjiang should further promote green development and economic efficiency based on the two-handed strategy to form a good trend of joint progress.

Table 4. The index weights in the evaluation system.

	Partial index score ranking						
Region	Overall score	Number of green industries	Industrial value-added	industrial energy consumption value-added	CO2 emission of industrial value-added (0.0293)	Electrification rate	Number of low carbon technology releases
Guangdong	1	1	19	2	2	1	1
Zhejiang	3	4	4	4	3	4	4
Beijing	4	11	22	1	1	12	3
Shandong	5	2	3	18	15	7	5
Xinjiang	11	12	24	28	25	5	27
Sichuan	12	10	8	14	14	21	9
Heilongjiang	28	22	26	22	20	24	22
Hainan	31	30	31	25	28	29	29

Regions with a relatively low level of CPCN industrial development in China are Sichuan, Hubei, Jiangxi and other 16 provinces and cities. This result means that these regions have a certain level of the industrial base, but their low-carbon industrial development is at a relatively low level, with significant deficiencies in specific factors that can affect the CPCN process. Take Sichuan as an example; the overall evaluation score of Sichuan's CPCN industrial development level is ranked 12th. On the other hand, the scores of low-carbon technology releases and the number of green patent applications of prefecture-level cities have been ranked in the top 10 nationwide. It indicates that Sichuan's overall industrial economic development is at the middle or upper level of the country, with a relatively high capacity for research and development of new technological products and scientific research and innovation. However, the indexes of industrial value-added, CO2 emission, electrification rate level, and the number of industrial low carbon policies released in Sichuan ranked 26th, 21st and 26th in the country, respectively. There is still considerable room for improvement in energy saving in industry and pollution treatment, and the ability of end-of-pipe control is a big concern.

Heilongjiang, Jilin, Ningxia and Hainan are the regions with the poor level of CPCN industrial development in China. In general, these regions have significant deficiencies in many aspects, including industrial growth, science and technology innovation level, research intensity, government support, industrial energy consumption and governance efficiency. Also, the green industrial development in those regions has a certain gap compared with other types of regions mentioned above and has great potential for improvement. Take Hainan province as an example; the overall evaluation score of Hainan Province's CPCN industrial level is the lowest in China, but its index of CO2 emission ratio generated from the industrial process and the index of the number of industrial low-carbon policy releases ranked 9th and 8th respectively in China. The ranking proves that the local government clearly recognizes the greening industry and achieved a good outcome in end-of-pipe control. However, other factors are ranked relatively low in the country, as indicators such as industrial value-added, number of employees above the scale, green factory and green designed products are all out of the 30th, which reflects a weak foundation of industrial development in Hainan and an overall low level of industrialization. While the geographical location of being across the sea from China's mainland increases the inconvenience of transportation, making Hainan province more challenging to introduce talents and advanced industrial enterprises, the process of greening industry needs to be further strengthened.

6. Conclusions

In this paper, we established an evaluation model for assessing the process of the CPCN industry in cities based on the AHP-entropy weight method, and we conducted an empirical analysis of 31 provinces in China, which draws a conclusion that China presents many limitations, including a relatively backward development level of the CPCN process, uneven CPCN development among regions. Next, China has to coordinate resource allocation of environmental conservation efficiently as well as implement stricter management on the CO2 emission, and some province should be provided foundation to accelerate the whole CPCN process. The evaluation results are basically consistent with the reality, indicating that the AHP-entropy weight method has good applicability in establishing the evaluation system of the industrial CPCN process. Furthermore, we can introduce or remove new indicators based on the current index system in order to adapt the development of CPCN.

References

[1] Xu, H., Tang, A. and Pittock, J. 2021 The dark side of ambition: side-effects of China's climate policy *Environmental Research Letters* **16(11)** 111001

[2] Qian, M., Dou, Z., Chen, H., JIa, X., Wang, S. and Li, D 2021 Research progress on the evaluation index system and methodology of comprehensive energy system under the carbon peak and carbon neutrality target *Electrotechnical Application* **40(12)** 72–79

[3] Li, X. and Niu, S 2021 A study on the multi-layered source-grid-load evaluation system under carbon peak and carbon neutrality target *Proceedings of the CSEE* **41(S1)** 178–184 doi:10.13334/j.0258-8013.pcsee.211576

[4] Yin, S., Wang, M. and Xu, J. 2022 Evaluation on Green Innovation-driven Development of Regional Manufacturing Industry under Carbon Peak and Carbon Neutrality Target *The Border Economy and Culture* **03** 30–32

[5] Li, H. and Yang, N. 2010 Exploration of Low Carbon Economy and Carbon Emission Evaluation Methods *Academic Exchanges* **(04)** 85–88

[6] Han, Z., Qi, C., Xiang, P., Liu, M. and Wang, S. 2018 Benefit analysis and comprehensive evaluation for distributed energy system *Thermal Power Generation* **47(02)** 31–36 doi:10.19666/j.rlfd.201705040

[7] Qu, X. 2014 Comprehensive Evaluation of Environmental Pollution in China's Industrial Sector - An Empirical Analysis Based on Topsis *Industrial Economic Research* **(04)** 51–59 doi:10.13269/j.cnki.ier.2014.04.006

[8] Wang, L. 2021a Research on green credit risk assessment of small and medium-sized banks based on AHP analysis method *Changchun University* doi:10.27912/d.cnki.gcdcx.2021.000059
[9] Wang, Y., Fang, G., Zhang, Y. and Huang, X. 2022b Risk Assessment of River-based Water Sources Based on Improved Fuzzy Comprehensive Evaluation *South-to-North Water Transfers and Water Science & Technology* 1–15
[10] Zhang, H., Wu, S. and Li, H. 2022 Construction and Evaluation of the Comprehensive Evaluation Index System for Building 'Lucid Waters and Lush Mountains are Invaluable Assets' in Gansu *Research on Development* 1–14
[11] Wang, L., Liang, X., Shen, S. and Shi, L. 2022a Fuzzy Comprehensive Evaluation of High Standard Farmland Construction Benefits Based on Improved AHP Method *Agricultural Science and Engineering in China* **34(02)** 74–78 doi:10.19518/j.cnki.cn11-2531/s.2022.0046
[12] Xu, C. and Li, Y. 2022 Qualitative Evaluation of Water Conservancy Projects on the Basis of Optimal Artificial Neural Network and TOPSIS *Technical Supervision in Water Resources* **(04)** 60–62
[13] Wang, S. 2021b Risk Assessment of Real Estate Project A Based on the AHP and Entropy Power Method Hebei University of Economics and Business doi:10.27106/d.cnki.ghbju.2021.000723.
[14] Zhu, Y., Tian, D. and Yan, F. 2020 Effectiveness of entropy weight method in decision-making. *Mathematical Problems in Engineering* **2020** 3565835
[15] Vaidya, O.S. and Kumar, S. 2006 Analytic hierarchy process: An overview of applications. *European Journal of operational research* **169(1)** 1-29

IOP Conf. Series: Earth and Environmental Science 1087 (2022) 012018 doi:10.1088/1755-1315/1087/1/012018

Simulation study on pollutant emission characteristics of 1000MW coal-fired boiler blended with sludge

Linfeng Chen[1,2], Fangqin Li[1], Kejun Li[1], Xin Liu[1] and Jianming Cai[1]

[1] School of Energy and Mechanical Engineering, Shanghai University of Electric Power, Shanghai 200090.

[2] Corresponding author's e-mail: 563335241@qq.com

Abstract. Fluid calculation simulation software was used to simulate the sludge blending in a 1000MW coal-fired power station boiler to study the effect of different blending ratios on the generation and emission of NOx, SO2, and CO. In this paper, the total calorific value of blending is kept constant for blending. The results show that the furnace chamber temperature decreases with increasing blending ratio. NOX concentration also decreases with increasing blending ratio. SO2 and CO concentrations gradually increase at the burner nozzle of the blended sludge, and the SO2 concentration at the furnace outlet is on the rise.

1. Introduction

China's energy structure is dominated by coal [1]. The combustion of coal-fired boilers will generate a large amount of SO_2 and NO_x, which will cause environmental pollution as the flue gas is discharged into the atmosphere. As the national environmental protection policy becomes stricter, the pollutant emission limits decrease yearly [2]. Since sludge contains many toxic germs, heavy metals, and dioxins, direct emissions will seriously damage the ecology [3]. Coal-fired units coupled with sludge combustion for power generation can recycle the calorific value of sludge and degrade harmful substances simultaneously, reducing the impact on the environment [4].

Park J M et al. tested the mixed combustion of dry sludge and pulverized coal. When the blending ratio was 20%, the TG curve of the composite sample was close to that of the coal sample, indicating that the combustion properties of the mixed coal and dry sludge combustion are comparable to those of the single pulverized coal combustion. This also shows the feasibility of the blending of sludge and pulverized coal [5]. Meng Tao et al. concluded by simulation that the outlet NOx concentration increased substantially with the increase of excess air coefficient. Under different air distribution methods, the NOx concentration at the air distribution outlet around the waist is the lowest while maintaining high furnace chamber temperature and burnout rate [6]. Zhang Cheng found that NOx emissions increased as the sludge moisture content decreased, but boiler combustion improved [7].

In this paper, for a 1000MW four-corner tangentially coal-fired boiler, the combustion characteristics and pollutant emissions of sludge under different mixing ratios (0%, 5%, 10%, 15%, 20%) were simulated and analyzed. In addition, the environmental impact of sludge blending in coal-fired boilers was explored.

2. Boiler overview

The research object is a 1000MW ultra-supercritical power station boiler, which adopts a single chamber tower structure arrangement, a medium-speed coal mill primary air positive pressure direct blowing type pulverizing system, a four-corner tangent circle combustion method, a swing type burner, a typical LNTFS burner arrangement for pulverized coal, and air classification combustion. The boiler has three groups of burners, and each group of burners includes two layers of burners. There are six layers, from bottom to top is ABCDEF. When the boiler is running at full load, the upper 5 burner layers are open. Each burner layer has 2 primary air nozzles (A1, A2), 1 fuel auxiliary air nozzle (AO), 1 end secondary air nozzle (AA), and 1 secondary air nozzle (AB). On top of the upper burner layer, there are 2 layers of closed-coupled over fired air (CCOFA) and 6 layers of separated over fired air (SOFA). The burner arrangement and boiler cross-section are shown in Figure 1.

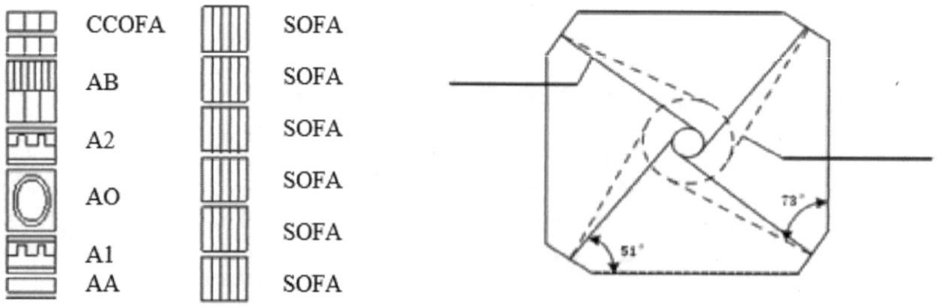

Figure 1. Schematic diagram of the burner arrangement and boiler cross-section.

The industrial analysis, elemental analysis, and calorific value of the coal-fired and sludge are shown in Table 1. Sludge has high volatile content, high ash content, low calorific value, and low fixed carbon. Sludge has higher nitrogen and sulfur content than coal.

Table 1. Industrial and elemental analyses of coal and sludge.

	Industrial Analysis /wt%				Elemental Analysis /wt%					Heat generation /MJ • kg^{-1}
	M_{ar}	A_{ar}	V_{ar}	FC_{ar}	C	H	O	N	S	
Coal	14	12	27	47	61.45	3.61	7.80	0.71	0.43	23.4
sludge	40	33.64	23.56	2.8	13.85	1.94	7.87	1.86	0.84	4.92

M: Moisture; A: Ash; V: Volatile; FC: Fixed carbon; ar: As-received

3. Mathematical models and calculation methods

3.1. Mathematical models

Realizable k-ε turbulence model was used to simulate gas-phase turbulence; Kinetic/ diffusion limit reaction rate model was used for coke combustion; P1 radiation model was used for radiation heat transfer calculations; Discrete random walk model was used to simulate the motion of pulverized coal particles and sludge particles. Two-competing-rate model was used to simulate the pyrolysis processes of coal and sludge. Component transport and combustion were simulated using the eddy dissipation model (EDM). It is assumed that sludge moisture and volatile are detached from the coal together and that volatile combustion is a two-step reaction. Their molar masses are set and the volatile combustion process is defined based on industrial and elemental analysis of coal and sludge. The reaction coefficient and standard enthalpy of production are determined by equations (1) and (2).

$$\text{volatile} + aO_2 \rightarrow bH_2O + cCO + dCO_2 + eN_2 + fSO_2 \tag{1}$$

$$CO + 0.5\,O_2 \rightarrow CO_2 \tag{2}$$

NOx generated by boiler combustion is mainly divided into thermal NOx, fuel NOx, and prompt NOx. Due to the tiny percentage of prompt NOx, it is generally not considered. Thermal NOx is generated according to the zeldovich mechanism, and fuel NOx is divided into NO from volatile fraction and NO from coke according to the De Soete mechanism.

IOP Conf. Series: Earth and Environmental Science 1087 (2022) 012018 doi:10.1088/1755-1315/1087/1/012018

3.2. Grid division

The zoned meshing method decomposes the numerical calculation area into the cold ash hopper, the burner, and the upper area of the furnace chamber. The quadrilateral structured mesh was generated for the furnace cross-section using the Map method and the hexahedral mesh was generated along the height direction using the Cooper method. The primary and secondary air inlets in the burner area were appropriately encrypted to improve the accuracy of the simulation. After verifying mesh irrelevance, the total number of meshes was determined to be 1.79×10^6. The gridding of the longitudinal section of the furnace and the burner section is shown in Figure 2.

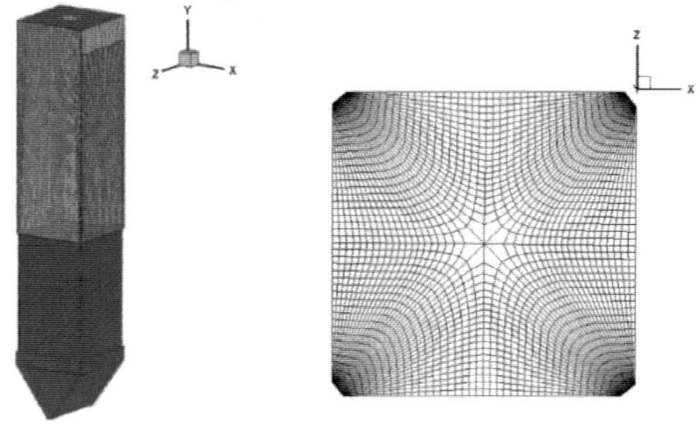

Figure 2. Schematic diagram of the gridding of the longitudinal section of the furnace and the burner section.

3.3. Boundary conditions

The discrete method uses first-order upwind format. Velocity-inlet boundary conditions were used for both primary and secondary air. The primary airspeed is 27.3m/s, and the primary wind temperature is 353 K. The secondary airspeed and wind temperature are 63m/s and 607 K. The primary and secondary air rates are 16.6% and 80.1% respectively. The outlet boundary type is outflow. Standard wall equations are used for the furnace wall without slip boundary conditions. Wall temperature is 700K, and wall emissivity is 0.7. The boiler excess air coefficient is 1.2. The total amount of fuel is 94.37 kg/s. Table 2 shows the amount of coal and sludge input to the F-layer burner, and the sludge blending ratio. The elemental analysis of coal and sludge shows that sludge has a higher sulfur content than coal. Maintaining the total calorific value for blending, as the blending ratio increases, the input coal decreases, the sludge increases and the overall sulfur content increases. The high ash and moisture content of the sludge makes for poor combustion.

Table 2. Fuel input at different blending ratios.

Working conditions	Amount of coal（kg/s）	Amount of sludge（kg/s）	Blending ratio（%）
1	18.88	0	0
2	17.936	4.492	10
3	16.992	8.983	20
4	16.048	13.475	30
5	15.104	17.966	40

4. Simulation results and analysis

In this paper, the total calorific value was kept constant for blending, and the uppermost F-layer burner was selected for sludge blending. The numerical simulation studied the five operating conditions of 0%, 5%, 10%, 15%, and 20% blending ratios.

4.1. Effect of sludge blending ratio on boiler combustion characteristics

Figure 3 shows the temperature clouds of the central section of the F-layer burner for sludge blending and the central section of the furnace chamber. As can be seen from the diagram, the central cross-section of the burner is well-formed at different blending ratios. The high-temperature region is mainly in the path of the pulverized coal jet, and the temperature is well distributed. It can be seen that the high-temperature region gradually becomes smaller as the blending ratio increases. This is because the sludge contains a large amount of water, which rapidly evaporates and absorbs heat after entering the furnace chamber, decreasing temperature levels.

The temperature clouds from the central section of the furnace chamber show that the most intense combustion and the highest temperature levels are found in the main burner area. After blending in the F-layer burner, the overall boiler temperature level decreases, but the change is not significant, indicating that 20% sludge blending in the upper burner is feasible.

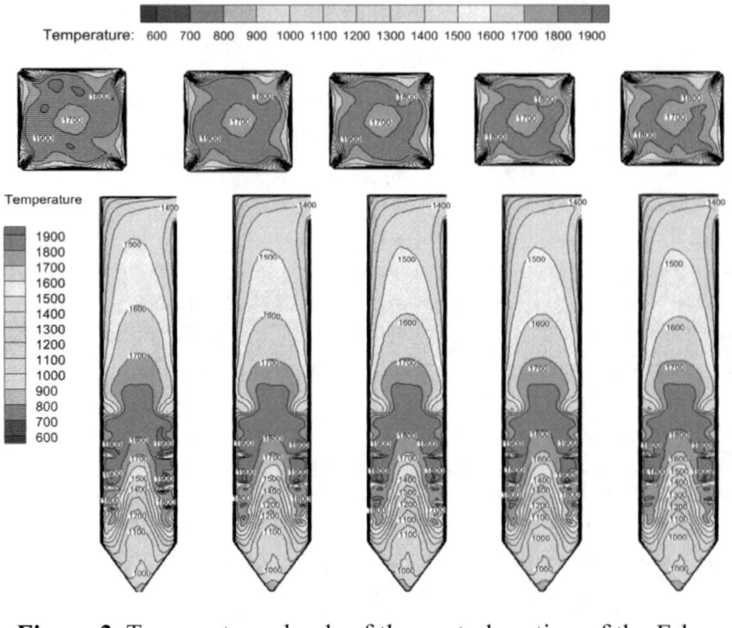

Figure 3. Temperature clouds of the central section of the F-layer burner for sludge blending and the central section of the furnace chamber.

Figure 4 shows the temperature distribution of the central section in the direction of the furnace height for different blending ratios. As shown from the figure, when the blending ratio increases, the overall combustion pattern is similar to that for single coal combustion. The temperature level decreases at the corresponding height of the burner (40-42m) for sludge blending. This is due to a large amount of water in the sludge, which absorbs heat, and the significantly higher ash content in the sludge than pulverized coal, making the combustion less effective and leading to a drop in temperature.

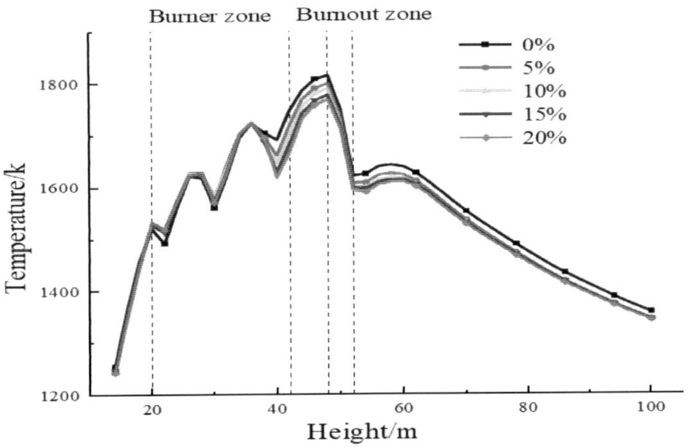

Figure 4. Temperature distribution of the central section in the direction of the furnace height for different blending ratios.

4.2. Effect of sludge blending ratio on boiler pollutant emissions

Figure 5 shows the NO_x distribution in the height direction of the furnace chamber. As can be seen from the diagram, the burner zone is the area where NO_x is mainly generated. This is due to the high temperature in the burner zone, which produces a large amount of thermal NO_x. NO_x concentrations first decrease, then increase, and finally plateau in the burnout zone. This is due to the input of the combustion air diluting the NO_x and providing a large amount of oxygen-enriched combustion and an increase in NO_x generation in the burnout zone. The distribution pattern of NO_x concentration along the height of the furnace is generally the same for the blending and simple coal combustion conditions. As the blending ratio increases, NO_x decreases above the F-layer burner area due to the decrease in temperature and the reduction in thermal NO_x generation after sludge blending. At the same time, most of the combustible part of the sludge is volatile burning rapidly. The intermediate product HCN is produced in increasing amounts, which causes the fuel NO_x to decrease as a reducing gas.

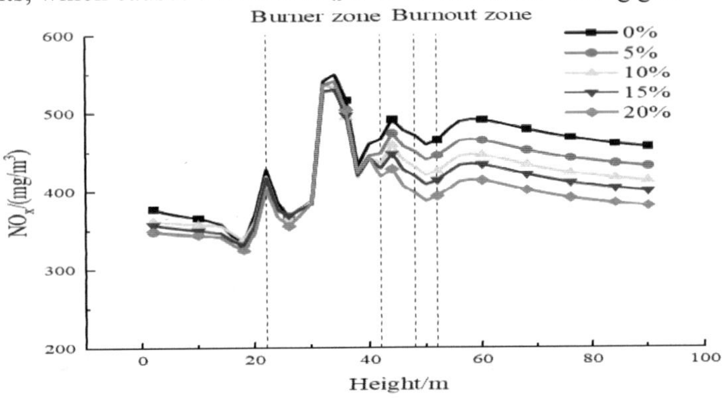

Figure 5. NOx distribution in the height direction of the furnace chamber.

Figure 6 shows a cloud graph of the SO_2 concentration in the furnace chamber at different blending ratios. The SO_2 concentrations in Figure 5 are 0%, 5%, 10%, 15%, and 20% from left to right, respectively, and are high at the burner inlet. This is because when the fuel enters the boiler, the volatile fires and burns quickly, producing a high amount of SO_2. This corresponds to the previous equation (1). The lower concentrations between the burner inlets are due to the input of secondary air, which reduces the SO_2 concentration here. At the height corresponding to the F-layer burner, the SO_2 concentration increases significantly with the increasing blending ratio. This is where the highest SO_2 concentrations are found. SO_2 concentrations are up to 0.00151 (0%), 0.00177 (5%), 0.00197 (10%), 0.00209 (15%) and 0.00223 (20%). This is because SO_2 is generated from the oxidation reaction of the sulfur-containing compounds in the fuel during the combustion of the fuel. The amount of SO_2 generated depends on the sulfur content of the fuel. According to the industrial analysis of the fuel, the elemental S content of the sludge is much higher than that of coal. So as the proportion of sludge increases, the sludge mass increases and the overall sulfur content of the F-layer burner fuel increases, leading to an increase in SO_2 concentration.

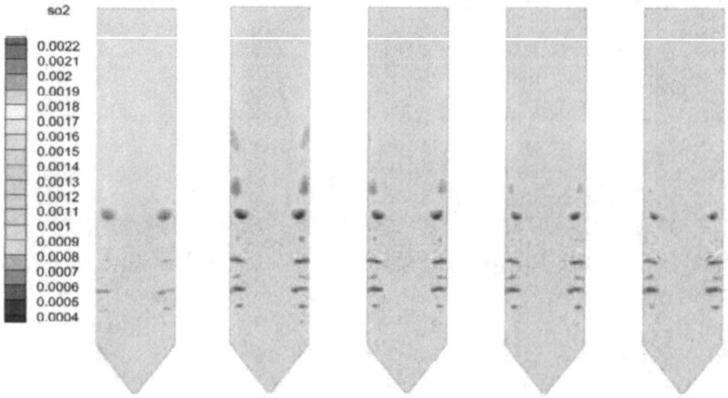

Figure 6. The cloud graph of the SO2 concentration in the furnace chamber at different blending ratios.

Figure 7 shows the cloud graph of CO concentration in the furnace chamber under different combustion ratios. The CO concentrations in Figure 7 are 0%, 5%, 10%, 15% and 20% from left to right. CO generation is mainly concentrated in the burner zone. This is because the excess air coefficient in the burner zone is less than 1, the pulverized coal is not combusted sufficiently, and more CO is generated. The burner zone is in the state of oxygen deficiency and CO cannot be fully converted to CO_2. In the burnout zone, because of the input of the combustion air, the unburned particles and CO are completely burned and the CO is oxidized to CO_2. This results in an almost zero CO concentration in the upper part of the furnace chamber. The highest CO concentrations are found in the direction of the F-layer burner height. CO concentrations are up to 0.004509 (0%), 0.005124(5%), 0.005464 (10%), 0.005702 (15%) and 0.006595 (20%). Because the input sludge mass increases, the sludge combustible material is mainly volatile, and more CO is generated. At the same time, a large amount of water and ash in the sludge makes the combustion less effective, which also leads to an increase in CO concentration.

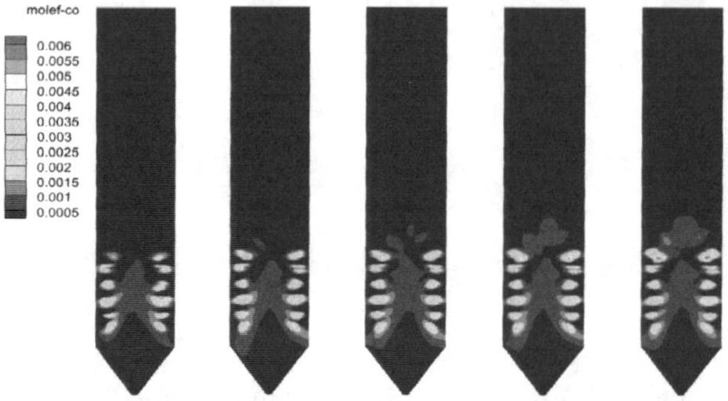

Figure 7. The cloud graph of CO concentration in the furnace chamber under different combustion ratios.

The pollutant parameters at the furnace outlet for different blending ratios are shown in Table 3. As the blending ratio increases, the overall sulfur content of the fuel increases, resulting in an increase in

SO_2 production and an increase in the concentration of SO_2 at the furnace outlet. In contrast, the concentration of NO_x decreases with increasing blending ratio under the influence of temperature levels and intermediate reducing gases.

Table 3. The pollutant parameters at the furnace outlet.

Simulation conditions	SO_2 concentration (mg/m^3,6%O_2)	NO_x concentration (mg/m^3,6%O_2)
0%	1248.132	446.7509
5%	1335.984	425.8686
10%	1423.982	405.5084
15%	1522.454	392.5143
20%	1617.944	373.514

5. Conclusions

(1) Different percentages of sludge blending were carried out in the F-layer burner. The results show that the high-temperature region of the sludge blending burner section decreases, and from there, the temperature level decreases along with the furnace height. However, the overall combustion pattern is close to simple coal combustion, indicating that a certain percentage of sludge blending is feasible.

(2) As the blending ratio increases, the temperature level decreases in the F-layer burner area, and the amount of thermal NO_x generation decreases. At the same time, the reduction of gas HCN generated from the sludge volatile fraction reduces the production of fuel NO_x, which also leads to a reduction in the NO_x concentration at the furnace outlet.

(3) As the blending ratio increases, CO production increases in the F-layer burner area. The SO_2 concentration gradually increases due to the high sulfur content of the sludge, resulting in an increase in the SO_2 concentration at the furnace outlet.

A single-layer burner for 20% sludge blending is safe and feasible. Coal-fired boilers coupled with sludge for power generation can be an effective way of treating sludge and can mitigate the environmental impact of sludge. This is also a flexible way of using the boiler. As the blending ratio increases, the SO_2 concentration at the furnace outlet increases and the NO_x concentration decreases, so a low proportion of sludge blending ratio is recommended. This study provides some basis for the actual blending of sludge in power plants.

References

[1] Yingxi Zhang, Jiechang Xia 2021 Research on potential risks and prevention strategies of a new round of energy crisis *China Energy* **43(12)** 7-13

[2] Wenjing Sun, Jianchao Qi 2017 Analysis and Discussion of Air Pollutant Emission Standard of Boiler *Renewable Resources and Circular Economy* **10(11)** 29-31

[3] Yanchao Bi 2019 Research status of application of oily sludge treatment technology *China Petroleum and Chemical Standards and Quality* **39(09)** 247-248

[4] Hui Jiang, Chuanhua Liao 2019 Current Situation and Research Progress of Sludge Chemical Conditioning Technology *China chemical equipment* **21(01)** 10-13+31

[5] Park J M, Keel S, Yun J, et al. 2017 Thermogravimetric study for the co-combustion of coal and dried sewage sludge *Korean Journal of Chemical Engineering* **34(8)**

[6] Tao Meng, Xiaolin Xing, Jie Zhang, et al. 2021 Numerical simulation study on the effect of air distribution on sludge mixing in coal-fired boilers *clean coal technology* **27(01)** 263-271

[7] Cheng Zhang, Tianyu Zhu, Libao Ying, et al. 2015 Experiment and Numerical Simulation of Sludge Blending Combustion in 100MW Coal-fired Boiler *Combustion Science and Technology* **21(02)** 114-123

IOP Conf. Series: Earth and Environmental Science 1087 (2022) 012019 doi:10.1088/1755-1315/1087/1/012019

Nitrogen and phosphorus concentration zoning control method of surface water in the Haihe River Basin

Maoqing Duan[1], Shilu Zhang[1], Qi Liu[1], Wei Gao[1], Mingxia Xu[1] and Jun Zhang[1,2]

[1] Center of Eco-environmental Monitoring and Scientific Research, Administration of Ecology and Environment of Haihe River Basin and Beihai Sea Area, Ministry of Ecology and Environment of PRC, Tianjin 300170, China.

[2] Corresponding author's e-mail: haihezhangjun@163.com

Abstract. Various uncertain influencing factors and incomprehensible mechanisms have posed daunting challenges to the management and treatment of regional surface water environment quality. The unified measures and the "one-size-fits-all" management approach limit the treatment effectiveness. Therefore, considering natural and human activities which are major factors affecting the surface water environment quality, the present study proposed a set of zoning management and control schemes for nitrogen and phosphorus concentrations in surface water based on its natural attributes. Selecting DEM, rainfall, vegetation type, soil type, and land use, and employing "grid transformation", "data extraction", "attribute superposition" of GIS software and "correlation analysis", "cluster analysis", and "principal component analysis" of SPSS software, the Haihe River Basin was divided into the prevention zone, the control zone, and the non-control zone with different natural attribute sets by the correlation coefficient R2 and the nitrogen and phosphorus pollution therein. Based on the nitrogen and phosphorus data of 276 surface water quality monitoring sites, the multiple nonlinear stepwise regression analysis was conducted to construct the relationship between a single water quality indicator and its natural attributes in the three zones. The results are of essential practical significance to surface water environment quality zoning management and survey in the Haihe River Basin. Meanwhile, it provides innovative insights into environmental zoning management in other regions.

1. Introduction

With the investment increase in ecological and environmental protection in recent years, the surface water environment quality has been significantly improved in China [1][2]. As of 2020, the surface water quality proportion of class I ~ III in China was 83.4%, up by 3% from 2016 [3]. Since the start of the 14th Five-Year Plan Period, the eco-environmental quality evaluation system has been continuously improved with the previous 21 indicators no longer monitored. Instead, the "9 + X" pattern has been applied for monitoring while the "5 + X" pattern has been utilized for evaluation. During the process of water environmental quality evaluation in special regions, the impact of environmental background values on water quality monitoring results is taken into full account [4][5][6]. A series of measures to strengthen the control over pollution have achieved remarkable results.

Nowadays, the "national-basin-administrative region" hierarchical water environment management system has been implemented in China. However, various factors that affect the water environment quality, especially the surface water, are more susceptible to changes in the external environment [7]. Therefore, surface water environment quality zoning management under the dual influences of nature and society should be taken into consideration. The spatial differences and homogeneity of natural and social attributes should be fully considered without regard to the administrative region. According to the spatial distribution characteristics of nitrogen and phosphorus concentrations in surface water and the natural influencing factors, this paper investigated the nitrogen and phosphorus concentrations zoning control method of surface water in the Haihe River Basin.

2. Materials and Methods

2.1. Study Area

There are 7 major watersheds in China, among which the Haihe River Basin borders the Bohai Sea in the east, Taihang Mountains in the west, the Yellow River in the south, and Mongolia Plateau in the north. With a total area of 320,000 square kilometers, it covers Beijing City, Tianjin City, most areas of Hebei Province, the east of Shanxi Province, the north of Shandong Province and Henan Province, Liaoning Province, and part of Inner Mongolia Autonomous Region [8]. The overall terrain of the basin is high in the northwest and low in the southeast, with plateaus and mountains accounting for 60% and plains for 40%.

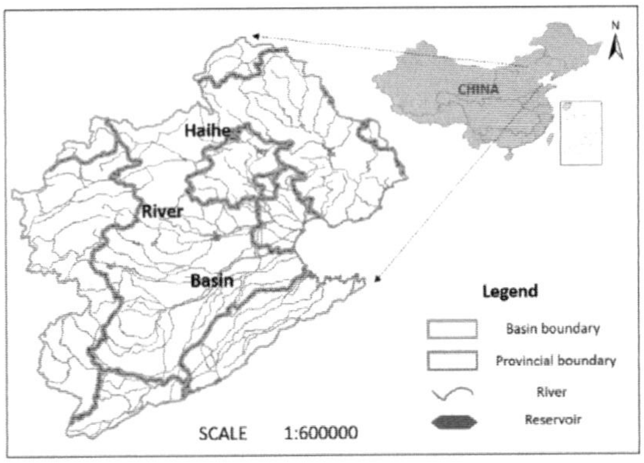

Figure 1. Geographical distribution map of the Haihe River Basin.

2.2. Data Source

The monitoring data of nitrogen and phosphorus concentrations in the surface water was obtained from China National Environmental Monitoring Centre. There were 276 monitoring sites each month. In this paper, the applied monitoring data are the average values in 2020, with the multi-year average rainfall in the basin obtained from the rainfall stations data measured between 2011 and 2019, land use data from Landsat 8 Operational Land Imager (OLI) satellite remote sensing image interpretation, elevation data (DEM) from GDEMDEM 30M resolution digital elevation data, vegetation type data from the spatial distribution data of 1 million vegetation types in China, and soil type data from the second national land survey.

2.3. Method

Zoning Method. The spatial heterogeneity of its influencing factors was fully considered regarding the control over the surface water nitrogen and phosphorus concentrations zoning of the Haihe River Basin. Under the principle of attribute superposition and spatial autocorrelation and taking the grid data as the object, the spatial superposition analysis was conducted on the thematic maps of various attributes that affect nitrogen and phosphorus concentrations via clustering technology and the spatial analysis function of GIS. Ultimately, a multi-dimensional attributes grid data layer that reflects the impact of multiple attributes was constructed [9]. The formation and output mechanism of surface water nitrogen and phosphorus concentrations depends on the comprehensive influence of attributes on different scales. Thus, weights were introduced into the cluster analysis to reflect the differences among various attributes.

First, the attributes involved in the clustering in the Haihe River Basin were transformed into 30 × 30m grid data. According to the attribute values extracted from each grid data center to the central point, a multi-dimensional attribute vector point set was established. Afterward, each participating spatial clustering index was standardized. the nitrogen and phosphorus concentrations of 276 monitoring sites were spatially interpolated, where the above operations were repeated. Hence there was a congruent relationship between the nitrogen and phosphorus concentrations and the various attributes, between which a simple correlation analysis was conducted with the average proportion of correlation analysis results taken as the weight value in cluster analysis. Cluster analysis was performed on vector points with multi-dimensional attributes after weighted multiple values, whereas the classification results were assigned to each object that participated in clustering, which could be expressed in GIS (Figure 2).

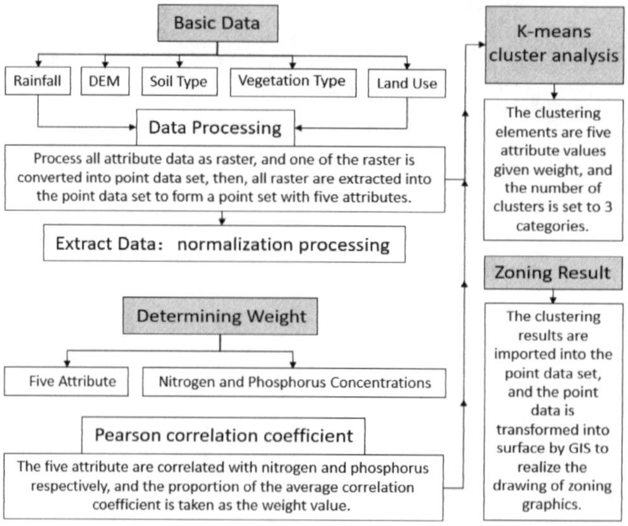

Figure 2. Zoning method technical roadmap.

Control Scheme. Taking the zoning results as the control unit and the surface water nitrogen and phosphorus concentrations as the object, water environment quality pollution conditions were qualitatively analyzed through the pollution index method. Afterward, combined with the pollution conditions, the quadratic polynomial stepwise regression analysis was performed applying the five attribute values and their nitrogen and phosphorus concentrations to define the nature of the zones.

3. Results and Discussions

3.1. Zoning Results

Based on the multi-year average rainfall grid data of the Haihe River Basin, the grid data were transformed into a point data set. Meanwhile, GIS software was applied to extract DEM, vegetation type, soil type, and land use grid data for the point data set. Considering the difference in each attribute's impact on the surface water environment, its weight was assigned. The weight was determined according to the Pearson correlation analysis results [10], whereas the correlation between the nitrogen and phosphorus concentrations and attributes was analyzed via SPSS, with the proportion of correlation coefficient in the total correlation coefficient applied as the weight value. The weight value of each attribute is presented in Table 1.

The principal component analysis method was adopted to determine the optimal classification number [11], as presented in Table 2. The total interpretation cumulative variance of principal components 1 and 2 was 95.2%, hence the optimal clustering number was classified into three categories. The attributes were normalized to eliminate the magnitude influence on the clustering results, to each of which the weight was assigned. Utilizing SPSS software, K-means cluster analysis was performed on the point data set based on five attribute values.

Table 1. Correlation coefficient and weight value.

Attributes	Correlation coefficient(R^2)		Proportion(%)		Weight value
	Nitrogen	Phosphorus	Nitrogen	Phosphorus	
Rainfall	0.177	0.034	0.207	0.117	**0.162**
DEM	0.316	0.092	0.369	0.316	**0.342**
Land use	0.054	0.005	0.063	0.017	**0.040**
Vegetation type	0.181	0.153	0.211	0.526	**0.368**
Soil type	0.129	0.007	0.151	0.024	**0.087**
Sum	0.857	0.291	/	/	

Table 2. Principal component analysis results.

Explained total variance			
Component	Initial eigenvalue(a)		
	Total	Variance%	Accumulate%
1	.029	81.372	81.372
2	.005	13.823	95.196
3	.001	2.424	97.620
4	.001	1.871	99.491
5	.000	.509	100.000

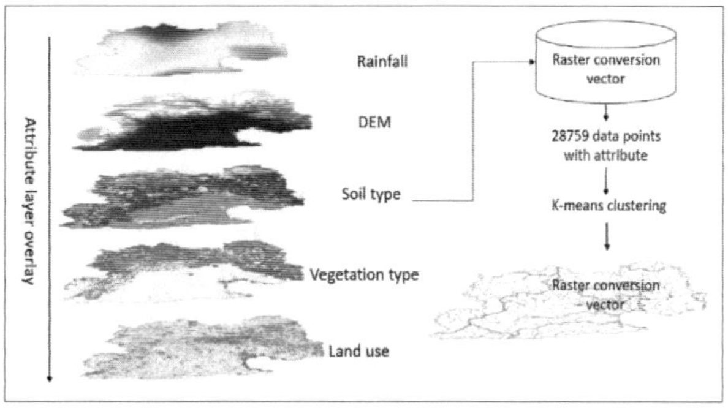

Figure 3. Data conversion, processing, extraction, and cluster analysis.

The clustering results indicate that the first zoning area (red) accounts for 19.2% of the total area, the second zoning area (green) for 32.7%, and the third zoning area (blue) for 48.1%. Among the 276 water quality monitoring data sites in the Haihe River Basin, the first zoning area contains 16 monitoring sites, the second contains 49 monitoring sites, and the third contains 209 monitoring sites.

3.2. Control Scheme

The nitrogen and phosphorus concentrations in the three zones were evaluated employing the pollution index method [12], which could better reflect the category and parameters of exceeding the standard compared with the single factor evaluation method. Due to its simplicity, it has been one of the commonly applied environmental assessment methods. The mathematical expression is as follows:

$$\mathrm{P} = \frac{c_i}{c_s} \tag{1}$$

P denotes the pollution index of the *i* indicator, c_i represents the monitoring concentration value of the *i* indicator, and c_s signifies the standard value of the *i* indicator. For reverse indicators (e.g.

nitrogen and phosphorus), $P<=0$ indicates that the water quality was unpolluted by the indicator; $P>0$ implies that it is polluted, and the larger the P value, the greater the pollution degree. The results demonstrated that the nitrogen and phosphorus concentrations and the pollution degree in surface water in the second zone were the lowerest, followed by the third zone, and were the highest concentration in the first zone.

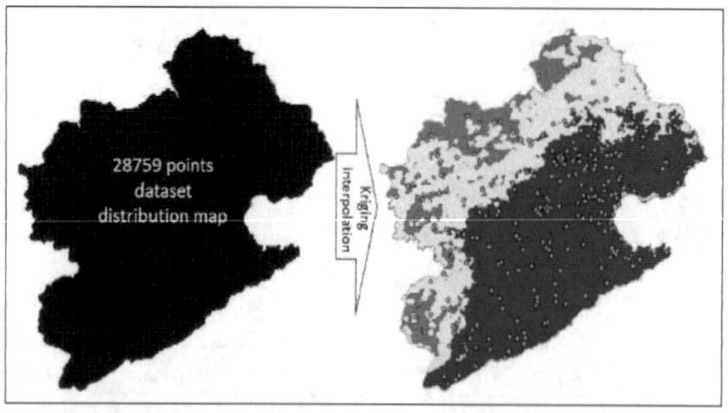

Figure 4. Partition results based on natural attributes.

Table 3. Nitrogen and phosphorus pollution degree in three zones.

Zones	Pollution index		Pollution degree
	Nitrogen	Phosphorus	
First	4.86	0.098	severe
Second	3.53	0.061	mild
Third	3.89	0.087	severe

The nitrogen and phosphorus concentrations in the first zone were at a severe pollution level, and the quadratic polynomial stepwise regression equation was established via DPS 7.05: $Y=172.7-2044.9X_1-0.27X_2+0.32X_3-1168.8X_1X_1-1.93X_4X_4+0.1X_1X_3+200.7X_1X_4-0.7X_1X_5$, the coefficient of determination R^2 being 0.98. Observed from Figure 4, X_1(DEM), X_2(Rainfall), and X_4(Soil type) were respectively negatively correlated with the phosphorus concentration while X_3(vegetation type) and X_5(land use) were not. The quadratic polynomial stepwise regression equation of nitrogen was as follows: $Y=654.8-0.48X_2-63.7X_4-2.9X_5+0.44X_1X_1+0.04X_2X_4+0.28X_4X_5$, the coefficient of determination R^2 being 0.99. The nitrogen concentration was negatively correlated with X_1(DEM) and X_2(rainfall) but was positively correlated with X_3(vegetation type), X_4(Soil type), and X_5(land use) to varying degrees. The results indicated that natural factors, e.g. DEM, rainfall, and soil type, were the main factors that affected the nitrogen and phosphorus concentrations in regional surface water.

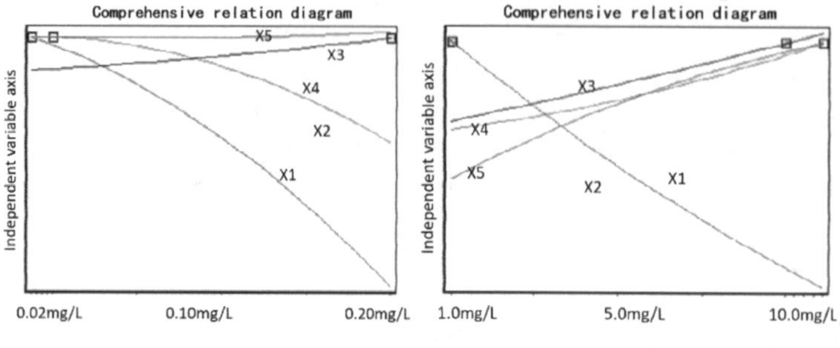

Figure 5. Comprehensive relation curve in the first zone (Left is phosphorus and attributes; right is nitrogen and attributes).

The nitrogen and phosphorus concentrations in the surface water of the second zone were at a mild pollution level. DPS 7.05 was utilized to establish the quadratic polynomial stepwise regression equations as follows: $Y(phosphorus)=15.5-34.7X_1+0.03X_2-0.11X_3+0.55*X_4+0.14*X_5+89.8*X_1*X_1-0.04X_1X_2+0.1X_1X_3+0.74X_1X_4-0.19X_1X_5$, the coefficients of determination R^2 being 0.38; $Y(nitrogen)=-25.6-1.63X_1+0.09X_3+1.84X_4-0.02X_5+0.07X_1X_1-0.03X_1X_4$, the coefficients of determination R^2 being 0.35. There was a certain correlation between the nitrogen and phosphorus concentrations and individual attributes, the R^2 being too small and the reliability of fitting being not high. The response of nitrogen and phosphorus concentrations to natural attributes in the second zone was not dramatic.

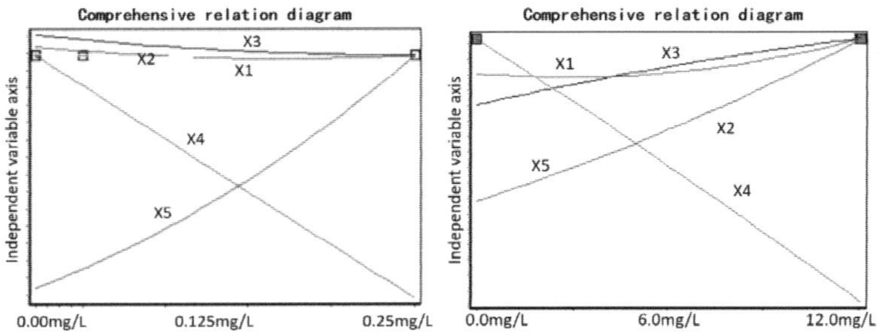

Figure 6. Comprehensive relation curve in the second zone (Left is phosphorus and attributes; right is nitrogen and attributes).

The nitrogen and phosphorus concentrations in the surface water of the third zone were at a severe pollution level. DPS 7.05 was employed to establish quadratic polynomial stepwise regression equations as follows: $Y(\text{phosphorus})=41.7-25.2X_1-0.14X_3-0.55X_4+0.14X_5-10.4X_1X_1-0.02X_1X_2+0.01X_1X_3+1.3X_1X_4-0.01X_1X_5$, the coefficients of determination R^2 being 0.22; $Y(\text{nitrogen})= 38.7+0.1X_1-0.14X_3-0.15X_4+0.1X_5+0.03X_1X_1$, the coefficients of determination R^2 being 0.26. Thus, the natural attributes were not the main factors that caused the pollution of nitrogen and phosphorus concentrations in the third zone.

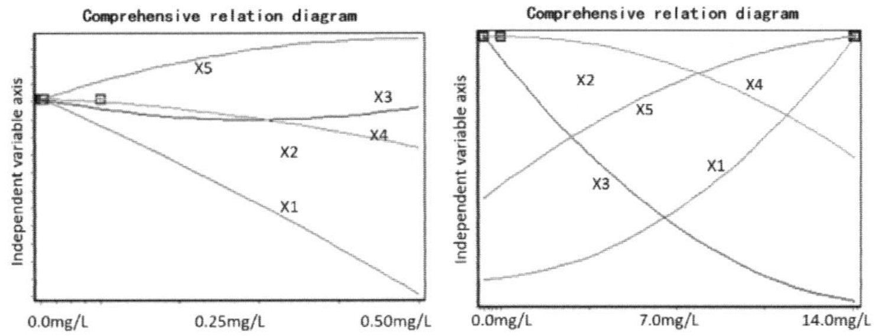

Figure 7. Comprehensive correlation curve in the third zone (Left is phosphorus and attributes; right is nitrogen and attributes).

The zoning and regression analysis results indicated that the higher nitrogen and phosphorus concentrations in surface water were significantly correlated with natural environmental factors. Therefore, the zone was defined as the "prevention zone", suggesting that more attention should be paid to natural factors therein, e.g. soil and water loss caused by rainfall rather than human social production activities. The nitrogen and phosphorus concentrations in surface water in the second zone were relatively low, which could be therefore defined as the "uncontrolled zone". It was suggested that only routine water quality monitoring should be conducted to prevent the deterioration of water quality without measuring management for nitrogen and phosphorus pollution. There was serious nitrogen and phosphorus concentrations pollution in the third zone, but natural environmental factors were not the main influencing factor. Human activities exerted remarkable impacts on surface water quality, hence the third zone was defined as the "control zone", implying that a series of measures for human activities should be conducted to reduce the nitrogen and phosphorus loads into the river.

4. Conclusions

Considering actual management needs, the differences in natural environments among regions, and the pollution characteristics of nitrogen and phosphorus concentrations in surface water, the following conclusions were drawn based on the powerful geographic information processing of GIS software and the analysis ability of SPSS software.

1) With the rainfall, DEM, soil and vegetation type, and land use in the Haihe River Basin as the input conditions, a fully covered point data set with multi-attribute values was constructed through grid conversion, layer superposition, and data extraction. The principal component and cluster analyses were employed to classify the point data set, and the classification results are presented in the form of the surface through assigned value and interpolation. Thus, the Haihe River Basin was divided into three types of regions with different natural factors.

2) The pollution index method was applied to evaluate nitrogen and phosphorus pollution conditions in surface water, which was at severe, mild, and severe levels in the first, second, and third zones respectively. Combined with the multiple nonlinear regression analysis results, they could be defined as the prevention zone with severe pollution and substantial correlation with natural environmental factors, the uncontrolled zone with mild pollution, and the controlled zone with severe pollution and no correlation with natural environmental factors. Accordingly, corresponding suggestions were proposed for the management and control of nitrogen and phosphorus concentrations in surface water.

References

[1] Peng G, Zhou X, Xie B, et al. 2020 Ecosystem stability and water quality improvement in a eutrophic shallow lake via long-term integrated biomanipulation in Southeast China – ScienceDirect *[J]. Ecological Engineering*

[2] Pan Y, Yuan Y, Sun T, et al. 2020 Are the Water Quality Improvement Measures of China's South-to-North Water Diversion Project Effective? A Case Study of Xuzhou Section in the East Route *[J]. International Journal of Environmental Research and Public Health* **17(17)**

[3] Ministry of ecological environment of the people's Republic of China. Bulletin on China's ecological environment. (http://www.cnemc.)

[4] Duan M, X Du, Peng W, et al. 2019 Necessity of Acknowledging Background Pollutants in Management and Assessment of Unique Basins *[J]. Water* **11(5)** 1103

[5] Duan M , Du X, Peng W, et al. 2019 A Revised Method of Surface Water Quality Evaluation Based on Background Values and Its Application to Samples Collected in Heilongjiang Province, China *[J]. Water* **11(5)** 1057

[6] Duan M, Du X, Peng W, et al. 2020 Quantitative assessment of background pollutants using a modified method in data-poor regions *[J]. Environmental Monitoring and Assessment* **192(3)**

[7] Mehboob M S, Kim Y. 2021Effect of climate and socioeconomic changes on future surface water availability from mountainous water sources in Pakistan's Upper Indus Basin *[J]. Science of The Total Environment* **769(3)** 144820

[8] Luo Y, Xu L, Rysz M, et al. 2011 Occurrence and Transport of Tetracycline, Sulfonamide, Quinolone, and Macrolide Antibiotics in the Haihe River Basin, China *[J]. Environmental Science & Technology* **45(5)** 1827-33

[9] Duan Maoqing, Du Xia, Peng Wenqi, et al. 2020 Research and application of characterisation using background value method for surface waters in special areas *[J]. China Environmental Science* **40(11)** 13

[10] F Zhang, Xie Z H, Cheng J T, et al. 2015 Based on the Pearson Correlation Coefficient of Vector *[J]. Fire Control & Command Control*

[11] Mahapatra S S, Sahu M, Patel R K, et al. 2012 Prediction of Water Quality Using Principal Component Analysis *[J]. Water Quality Exposure & Health* **4(2)** 93-104

[12] Wang xiaoxiao, Zhang Miaoxian, Xubingbing 2012 Combination and application of fuzzy identification index and correspondence analysis method in water quality evaluation *[J]. Acta Scientiae Circumstantiae* **32(5)** 9

Study of *Unio* shells in the Quaternary deposits, southern Iraq

Mohanad Hamid Al-Jaberi[1,2,4], Harith A. Al-Saad[1], Ghazi A. Hussain[2] and Hiba Khalid Lafta[3]

[1] University of Basrah- College of Science- Geology department;
[2] A lKarkh University of Science – College of Energy and Environmental Science;
[3] University of Basrah- College of Science- Geology department.

[4] Corresponding author's e-mail: jaberi76@yahoo.com

Abstract. Shells are useful in evolutionary biology and paleobiology. Freshwater shells compose the main types of mollusks, and can use as a key for ecological variations. Shell samples in the present study were collected on the highway of Nasyriah city- Samawa city, southern Iraq. The study area contains a large accumulation of shells at 6-7m depths, especially at 7m depth. The main types of these shells are *Unio*, *Corbicula*, and *Melanoides tuberculata. Unio* was the most species common in the study area. *Unio* species were used to detect the paleoenvironment in Quaternary deposits, in southern Iraq. Calcite and aragonite are the common minerals in the *Unio* shells. According to the distribution and percentage of calcite and aragonite in the *Unio* shells, it was found the calcite at 6m is more than 7m, whereas the aragonite at 7m is highest compared to 6m. Under X-Ray Fluorescence analyzers, found that calcium oxide, silica, magnesium oxide, and phosphor oxide are the major oxides in the *Unio* shells, while Zn, Sr, Pb, and Rb are the main trace elements. Strontium (Sr) has a high concentration in the *Unio* shells at 7m than 6m, while magnesium (Mg) is found in a high concentration at 6m than 7m. According to trace and some of the major elements concentrations, the paleotemperature, paleosalinity, and water contamination of rivers at 6m are more than 7m, and the river water at 7m depth was fresh water with a high abundance of nutrients and climate was suitable to growth mollusks shells like *Unio* shells.

1. Introduction
The bivalve shells are used as environmental indicators [1]. Trace elements are incorporated into the bivalve shells according to the concentration of these elements in the water [2]. This is also affected by water salinity and temperature [3]. The observation of heavy metals in the aquatic environment involves detecting the metal in biota, sediments, and water [4,5,6,7]. The tolerance of mollusks' bivalves has made them an ecosystem monitoring [8]. The level of trace metal contamination of aqueous organisms is a result of a balance between metal loss and metal uptake [9]. The palaeoenvironmental recorders can be detected by the use of mollusks. The three-layered shell of *Unio* species of an outer thick periostracum, a prismatic layer, and a thick nacreous layer, as well as the aragonite mineralogy [10, 11, 12]. A number of authors have studied the mineralogy of *Unio* shells, such as the study of Gregoire (1961) [13] have revealed that *Unio* shells had a few crystals collection of aragonite. Taylor et al. (1969) [14] found two arrangements of the nacreous layers in several species of *Unionidae*. A study by Dauphin et al. (2017) [15] concluded the mineralogy and composition of a nacroprismatic bivalve species: *Unio* pictorum and explained that the prismatic layer of *Unio* is aragonite, and the inner structure of the prismatic units strongly differs from those of the calcitic layers. Lyubas et al. (2019) [16] used *Unio* freshwater shells

as indicators of paleoclimate changes compare to the present in the Pleistocene and Pliocene deposits of ancient rivers in Russia. The main objective of this study was the reconstruction of some parameters of the aquatic environment in the Quaternary period, in southern Iraq.

2. Location of the study area

The study area is located in the north of Nasyriah city. The study area consists of Quaternary sediments in southern Iraq (Figure 1).

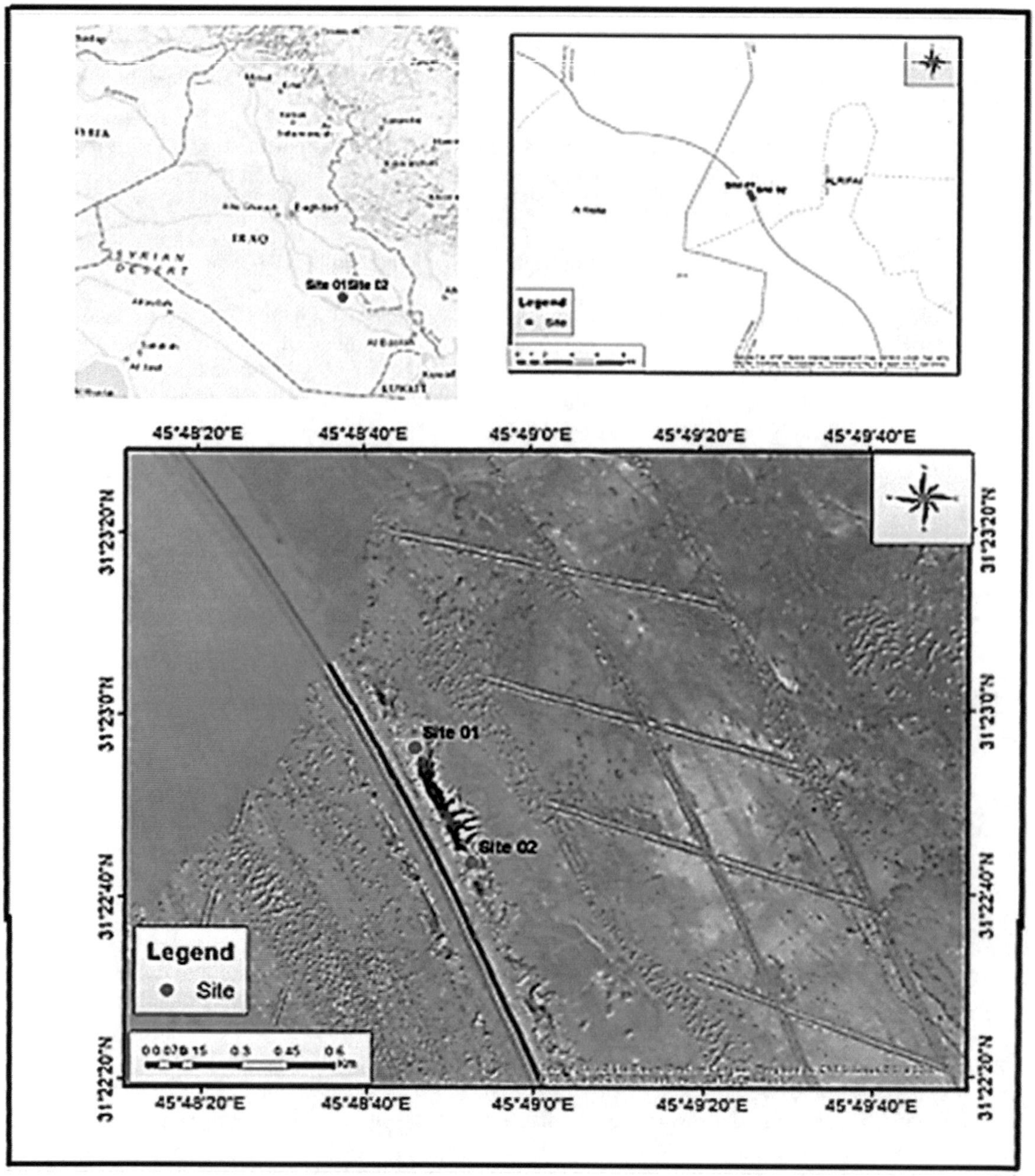

Figure 1. Map of the study area.

3. Materials and methods

Subfossil shells are the materials for this study, these shells are freshwater bivalve mollusks, belonging to the genera *Unio* [16]. The depths of 6 and 7m for two sites (S1 and S2) of outcrop Quaternary deposits (Figure 2,3,4) are chosen to study the *Unio* shells in the highway of Nasyriah - Samawa cities. The deposits in the study area are terrigenous sediments composed of sand, silt, and clay (Figure 4). The *Unio* samples were put in an acid-washed polyethylene bag. Grinding the shells by a glass mortar. X-ray fluoresces (XRF) and X-ray diffraction (XRD) are applied in this study to reveal of minerals and chemical composition of *Unio* shells.

Figure 2. *Unio* shells Shells at 6m.

Figure 3. *Unio* shells Shells at 7m.

4. Results

4.1. Mineralogy study

Calcite and aragonite are the common minerals in the *Unio* shells. According to the distribution and percentage of calcite and aragonite in the *Unio* shells, found the calcite at 6m is more than 7m, whereas the aragonite at 7m is highest compared to 6m.

4.2. Geochemical study

According to X-Ray Fluorescence (XRF) analyzers for *Unio* shells, it was estimated that calcium oxide, silica, magnesium oxide, and phosphor oxide are the major oxides, while Zn, Sr, Pb, and Rb are the main trace elements (Table 1).

Figure 4. The sequence layers from the top to the bottom in the study area.

Table 1. The XRF analysis results of *Unio* shells for the studied sites.

	S1		S2	
Elements	**6m**	**7m**	**6m**	**7m**
SiO_2 %	0.2188	0.0016	0.2672	0.0022
MgO %	0.2199	0.1394	0.2596	0.1282
P_2O_5 %	0.2119	0.2065	0.2155	0.2172
CaO %	55.49	55.95	55.6	55.86
MnO %	0.05792	0.02919	0.05438	0.02715
Zn ppm	12.131	11.087	12.86	10.581
Rb ppm	0.641	0.366	0.683	0.384
Ba ppm	0.00858	0.00831	0.00849	0.00824
Pb ppm	1.946	1.265	1.862	1.386
Sr ppm	1152	1380	1173	1376

5. Discussions

Mollusk is the second most phylum on the earth [17]. The shell shapes reflect the environment and their phylogenetic history [18, 19, 20, 21]. The chemical composition of bivalves shells can be used as monitoring contamination and detect the different environmental conditions [22,23,24,25,4,5]. The adaptability of the mollusks has made them a preferred organism as an indicator of the ecosystems [7,8,9]. XRF analysis of hard body parts of bivalves can give a piece of information for biomineralization and the interaction between the surrounding environment and organisms. XRD analysis revealed two main layers in the *Unio*, the outer layer constitutes calcite and the inner layer constitutes aragonite. The high abundance of *Unio* shells at 6 and 7m depths gave an indication of the freshwater environment. The high accumulation of shells at 7m gave an indication of suitable conditions in that period of Quaternary Formation to grow and to provide the appropriate nutrients for reproduction. [25,26] The rate of *Unio* shells in the river was present in a high accumulation in the nutrient silty clay channels. Major and trace elements in the bulk *Unio* shells are displayed and summarized in Table 1, which emphasized that calcite is higher in the *Unio* shells at 6m depth compared with 7m, which may probably be due to the increase of calcium ions in the river water, these ions uptake by *Union* shells to build their skeleton. The aragonite layer in the *Unio* shells at 7m is highest than 6m. The elements contained in the mollusk shells are suitable to concentrate in the calcite or aragonite according to their ionic radii [27]. Strontium (Sr) has a high concentration in the *Unio* shells at 7m than 6m, while magnesium (Mg) is found in a high concentration at 6m than 7m. [28,29,30] Indicated that the Sr element is more suitable to concentrate in the aragonite than the calcite layer. [31,32] Mentioned that Sr^{+2} substitutes more readily for Ca^{+2} in the aragonite than calcite. [27] The magnesium showed more preferential concentration in calcite rather than aragonite in shell growth. The Mg^{+2} content in the outer calcite layer is increased with increased environmental temperature, but not so regular as Sr concentration. The Mg content in the skeleton may be used for paleotemperature reconstruction, decrease the calcium elements and increase magnesium giving an indicator of high paleotemperature at 6m depth compare with 7m depth. An increase in the calcite layer with an increase in the Zn, Pb, Si, Mn, and Rb presented in large concentrations in *Unio* shells at 6m than 7m. [29] Mentioned the Zn, Pb, and Si which have small ionic radii and are present in a higher concentration in the calcite than aragonite. [29] Showed the concentration of Pb, Zn, and Mn, in the mollusks shells are higher when increasing the water salinity. In addition, the high accumulation of Mn, Zn, Pb, and Ba may have been attributed to sediments and water contamination at 6m than 7m, all of these showed the river at that time in 7m depth was fresh water with a high abundance of nutrients and climate was suitable to growth mollusks shells like *Unio* shells.

6. Conclusions

(1) The high abundance of *Unio* shells at 6m and 7m depths gave an indication of a freshwater environment, and the great number of shells at these depths gave a clear evidence to appropriate conditions for growth and generation.
(2) The aragonite layer in the *Unio* shells at 7m depth is more than 6m, while the calcite presence was reversed.
(3) Strontium has a high concentration in the *Unio* shells at 7m than 6m, while magnesium is found in a high concentration at 6m than 7m.
(4) Strontium was found in the aragonite layer, while magnesium is concentrated in the calcite layer of *Unio* shells.

References

[1] Al-Dabbas, M., and Al-Jaberi, M. 2015 Elemental distribution in the calcite and aragonite layers of Chione californiensis from the Iraqi Coasts—North Arabian Gulf. *Arabian Journal of Geosciences* **8(11)** 9433-9440
[2] Censi, P., Spoto, S. E., Saiano, F., Sprovieri, M., Mazzola, S., Nardone, G., Di Geronimo, S. I., Punturo, R., Ottonello, D. 2006 Heavy metals in coastal water systems: A case study from the northwestern gulf of Thailand. *Chem* **64** 1167–1176
[3] Al-Dabbas, M., Hubbard, F.H, and McManus, J. 1983 The proportion of aragonite in the shells of Mytilus edulis from the Tay estuary *JAWRR* **2(2)** 73–79
[4] Szefer, P., Frelek, K., Szefer, K., Lee, C.B., Kim, B.S., and Warzocha, J., Zdrojewska I., Ciesielski T. 2002 Distribution and relationships of trace metals in soft tissue, byssus and shells of Mytilus edulis trossulus from the southern Baltic *Environ Pollut* **120** 423–444
[5] Liu, J.H., and Kueh, C. 2005 Biomonitoring of heavy metals and trace organics using the intertidal

mussel Perna viridis in Hong Kong coastal waters. *Mar Pollut Bullet* **51** 857–875

[6] Hamed,A.H. and Emara, A.M. 2006 Marine molluscs as biomonitors for heavy metal levels in the Gulf of Suez *Red sea*. J Mar Syst **60** 220–234

[7] Darvish, A. 2007 The survey of heavy metals (Cd, Cu, Ni and Hg) in water and soft tissue of Saccostrea cucullata in intertidal zone of Hormoz islans. Persian Gulf, MS thesis *Environmental Engineering* Azad University of Ahvaz, Iran.140

[8] Conti , M., and Cecchetti , G . 2003 A biomonitoring study: trace metals in algae and molluscs from Tyrrhenian coastal areas *Environ Res* **93** 99–112

[9] Goodfriend , G., Magaritz , M., and Gat, J.R. 1989 Stable isotope composition of land snail body water and its relation to environmental waters and shell carbonate *Geochimica et Cosmochimica Acta* **53** 3215– 3221

[10] Dauphin, Y., Cuif, J.P., Cotte, M. and Salomé, M. 2012 Structure and composition of the boundary zone between aragonitic crossed lamellar and calcitic prism layers in the shell of Concholepas concholepas (Mollusca, Gastropoda) *Invertebr. Biol* **131** 165–176

[11] Dauphin, Y., Cuif, J.P., Castillo-Michel, H., Chevallard, C., Farre, B. and Meibom, A. 2013 Unusual micrometric calcite–aragonite interface in the abalone shell Haliotis (Mollusca, Gastropoda). *Microsc. Microanal* **20** 276–284

[12] Schmidt, W.J. 1924 Die Bausteine des Tierkörpers in polarisiertem. *Lichte. F. Cohen, Bonn* 528

[13] Gréegoire, C. 1961 Sur la structure submicroscopique de la conchioline ´ associée aux prismes des coquilles de mollusques *Bull. K. Belg. Inst. R. Sci. Nat. Wet* **37** 1–34

[14] Taylor, J.D., Kennedy, W.J. and Hall, A. 1969 The shell structure and mineralogy of the Bivalvia. Introduction. Nuculacae – Trigonacae. Bull. Brit. Mus. Nat. Hist. *Zool* **3** 1–125

[15] Dauphin, Y., Luquet, G., Salome, M., BELLOT‐GURLET, L., and Cuif, J. P. 2017 Structure and composition of Unio pictorum shell: arguments for the diversity of the nacroprismatic arrangement in molluscs *Journal of microscopy* **270(2)** 156-169

[16] Lyubas, A. A., Kabakov, M. B., Kriauciunas, V. V., Obada, T. F., Nicoara, I. N., and Tomilova, A. A. 2019 Freshwater mollusks from Neogene-Quaternary Dniester and Prut riverine deposits as indicator paleoenvironments: chemical composition of shells and its palaeoecological interpretation *Arctic Environmental Research* **19(1)** 35-42

[17] Bank, R. A., Bieler, R. Bouchet, P., Decock, W., Dekeyzer, S., Gofas, S., Kroh. A., Marshall, B., Neubauer, T.A., Neubert, E.G., Rosenberg, G., Sartori, A.F., Schneider, S., Trias-Verbeeck, A., Vandepitte, L.A., and Vanhoorne, B. 2014 Mollusca Base – announcing a World Register of all Molluscs. In White T. S. (ed.) Book of Abstracts EUROMAL 2014 The 7thCongress of the European Malacological Societies, 7th - 1th September 2014 *The Malacological Society of London, Cambridge*

[18] Bogan, A. E. 2008 Global diversity of freshwater mussels (Mollusca, Bivalvia) in freshwater *Hydrobiologia* **595** 139–147

[19] Sparks, B. W. 1961 The ecological interpretation of Quaternary non-marine Mollusca. *Proceedings of the Linnean Society of London* **172** 71–80

[20] Graf, D. L. and Cummings,K.S. 2007 Review of the systematics and global diversity of freshwater mussel species (Bivalvia: Unionoida) *Journal of Molluscan Studies* **73** 291–314

[21] Aldridge, D. C., Fayle,T. M. and Jackson, N. 2007 Freshwater mussel abundance predicts biodiversity in UK lowland rivers *Aquatic Conservation: Marine and Freshwater Ecosystems* **17** 554–564

[22] Lydeard, C., Cowie, R.H., Ponder, W.F., Bogan, A.E., Bouchet, P., Clark, S.A., Cummings, K.S., Frest, T.E., Gargominy, O., Herbert, D.J., Hershler, R., Perez, K.E., Roth, B., Seddon, M., Strong E.E.F.G., Thompson, F.G. 2004 *The global decline of nonmarine mollusks. Bioscience* **54** 321–330

[23] Dalbeck, P.C. 2008 Crystallography, stable isotope and trace element analysis of Mytilus edulis shells in the context of ontogeny *Unpubl. MSc thesis, University of Glasgow* 235

[24] Hubbard, F., McManus, J., and Al-Dabbas, M. 1981 Environmental influences on the shell

mineralogy of Mytilus edulis *Geo-Marine Letters* **1(3-4)** 267
[25] Boening, W. 1999 An evaluation of biomonitors of heavy metals pollution in marine waters *Environ Monit Asses* **55** 459–470
[26] Zając, K., Zając, T., and Ćmiel, A. 2018 What can we infer from the shell dimensions of the thick-shelled river mussel Unio crassus? *Hydrobiologia* **810(1)** 415-431
[27] Al-Jaberi, M. H. 2015 Study of the clastic and shells in selected area at NW of Arabian Gulf-South Iraq *Baghdad university, published thesis* 249P
[28] Al-Dabbas, M. A. 1980 An examination of shell fragment distribution and geochemical features of Mytilus Edulis in the Tay estuary *PhD thesis, University of Dundee* 173
[29] Broman, D., Lindqvist, L., and Lundbergh, I. 1991 Cadmium and zinc in Mytilus edulis L. from the Bothnian Sea and the northern Baltic proper *Environmental Pollution* 74(3) 227-244.
[30] Deer, W.A., Howie, R.A., and Zussman, J. 1992 An Introduction to the Rock-Forming Minerals, London (Longman Scientific & Technical), *Mineralogical Magazine* **56(385)** 617-619
[31] Yan, H., Chen, J., and Xiao, J. 2014 A review on bivalve shell, a tool for reconstruction of paleo-climate and paleo-environment *Chinese Journal of Geochemistry* **33(3)** 310–315 https://doi.org/10.1007/s11631-014-0692-0
[32] McMillan, E,A., Fairchild, I., Frisia, S., Borsato, A., and McDermott, F. 2005 Annual trace element cycles in calcite–aragonite speleothems: evidence of drought in the western Mediterranean 1200–1100 yr BP *J Quat Sci,* **20(5)** 423–433

Calculation and scenario prediction of methane emissions from agricultural activities in China under the background of "carbon peak"

Zhongming Li[1], Wei Fu[1,3], Mingcan Luo[1], Jiancheng Chen[2] and Long Li[1]

[1] School of Economics and Management, Southwest Forestry University, Kunming Yunnan, China;
[2] School of Economics and Management, Beijing Forestry University, Beijing, China.

[3] Corresponding author's e-mail: fuw@swfu.edu.cn(FW)

Abstract. Climate warming has become a difficult problem all over the world. As the most important component of greenhouse gas, methane plays a vital role in the earth's climate system. CH_4MOD model and emission factor method are used in calculating and analyzing the methane emission from agricultural activities in China. The scenario analysis method is used in predicting the methane emission from agricultural activities in China. The results show that: (1) The changing trend of total methane emission from agricultural activities in China from 2001 to 2020 increased firstly, then decreased, and finally tended to steady fluctuation, in which the contribution of livestock intestinal fermentation emission was the largest. (2) From 2001 to 2020, $GHGI_1$ of methane in agricultural activities in China showed a downward trend, while $GHGI_2$ fluctuated. (3) China's methane emissions from agricultural activities are predicted to increase under different scenarios from 2021 to 2030. To meet China's goal of reaching a carbon peak by 2030, the strict emission reduction scenario is more suitable for China's future methane emission development trend than other scenarios.

1. Introduction

As the world's largest emitter of greenhouse gases, China's emission reduction has attracted the attention of global climate change [1]. At present, China is the world's largest carbon emitter, accounting for 29% of global emissions [2]. Therefore, the Chinese government promised to reach the peak of carbon dioxide emissions around 2030 and try its best to reach the peak ahead of time [3]. At the same time, each class government has already taken steps to implement national policy and formulated carbon dioxide emission reduction target.

Following carbon dioxide, methane in the atmosphere is the second factor that causes global warming [4]. The total methane emissions of 2015 are 1834 ppb and are 2.5 times higher than before industrialization [5]. Compared with carbon dioxide emissions, China's non carbon dioxide greenhouse gas emissions are still not fully valued, particularly in the national policy framework and associated emission reduction targets [6]. But China's methane emissions greatly contribute to global climate change, and China is the world's largest methane emitter [7]. Considering the importance of methane in the overall greenhouse gas emission list, an increase in methane emissions can affect efforts to reduce greenhouse gas emissions in China [8]. Extensive research helps China methane emissions accounting and covers major sources such as agriculture [9]. Therefore, the study of methane emission from agricultural activities is of great significance for China to promote the "carbon peak" goals and the construction of beautiful China.

The increase in atmospheric methane concentration is mainly attributed to anthropogenic activity [10]. In 1980, the biggest contributor to methane emissions was rice cultivation, but since 2005, this contribution has been replaced by coal mining. Increase in methane emissions by coal mining and domestic animals promoted the increase in total methane emissions [11]. China is also rich in research on methane emissions from agricultural sources. Zhang et al [12]. considered that from the perspective of spatial distribution, there are obvious differences in methane emissions across regions, among which the emissions in central and southern China are the largest, followed by East China, southwest, northwest and Northeast China, and the emissions in North China are the least. Li et al [13]. thought from the perspective of LMDI factor decomposition, economic factors and population size factors had a positive effect on China's agricultural methane emission, efficiency factors and structural factors had a mitigating effect on China's agricultural methane emission. About rice planting, Guo et al [14]. concluded that the methane emission in the single rice area of Southwest China was the highest through the experiment. With regard to livestock breeding, Yao [15]. believed that the technical, economic and demographic factors of livestock methane emission had a diminishing marginal contribution.

China is a major rice-producing country, with 2017 accounts for 19% and 28 percent of the world's rice area and production [16]. In China, about 90% of paddy fields are fully irrigated and produce anaerobic conditions conducive to methane production by methanogens. The production and management practices are affecting food safety, water shortage and climate change. It can be predicted that in order to meet the needs of population growth, the rice planting area is expected to increase, which may lead to more methane emissions [17]. Methane emissions from rice cultivation are an integral part of the carbon cycle between plants and soil systems. The growth development of rice is an important regulator of methane emission [18]. Previous studies have revealed that methane emissions in rice fields are affected by agricultural type [19], nitrogen fertilizer use [20], organic input [19], rice varieties [21], planting area [22] and straw combustion [23].

Enteric methane is the energy loss of ruminants and has become a major environmental concern due to its contribution to global warming [24]. As reported by FAO, increasing the number of livestock causes 60% more methane emissions in 2030, unless the number of livestock is controlled. With the rapid growth of the Chinese economy, the number of livestock has almost tripled in the past 30 years, leading to an increase in methane emissions by intestinal fermentation and manure management. The type of domestic animals (cattle, sheep, etc.) and their nutrition also changed, and the methane emission changed [25]. Livestock greenhouse gas emissions are mainly through two processes: practice of intestinal fermentation and manure management. Methane and N_2O are two kinds of greenhouse gases discharged from livestock. Almost 90% of methane excretion is the result of intestinal fermentation. Within the whole world, intestinal fermentation resulted in 37% artificial methane emissions [26].

The calculation of methane and other greenhouse gas emissions from agricultural activities mainly focuses on Huang's model [27], CH_4MOD model [28], IPCC inventory method [29], etc. China is committed to reaching the carbon peak by 2030. The emission of non-carbon dioxide greenhouse gases such as methane is also crucial to the ecological environment. Therefore, China's methane emissions from 2001 to 2020 are estimated and analyzed based on agricultural sources. The scenario is set according to the policy to predict the methane emission from agricultural sources from 2021 to 2030, and provide a theoretical basis for the implementation of the "carbon peak" goal and the construction of beautiful China.

2. Data and method

IPCC has pointed out that the production of methane in the atmosphere mainly comes from human beings. CH_4MOD model and emission factor method are selected to calculate the methane emission from agricultural sources in China, and the methane emission intensity of two types of agricultural sources calculate. Finally, the scenario analysis method is used in predicting the methane emission from agricultural sources in China from 2021 to 2030.

2.1. Study area
China is a large country of rice production, with a large rice yield and wide distribution. According to different natural and social conditions, rice in China can divide into early rice, medium rice, one season late rice and double season late rice. In the double-cropping rice region of South China, rice has obvious comparative advantages over corn. Therefore, early rice and late rice in China mainly concentrate in East China, South China and southwest China. China's rice planting is mainly in the Yangtze River Basin and the southeast coast. It has gradually formed three major production advantage areas in the northeast, Southeast and Yangtze River Basin. At the same time, China is also a large animal husbandry country, and the breeding varieties are mainly cattle, sheep, pigs, horses, mules, camels and so on. With the promulgation of a series of policies, China's animal husbandry is developing rapidly to a large-scale and professional model.

2.2. Data sources
Based on the classification of methane emission sources from agricultural activities in the 2006 IPCC guidelines for national greenhouse gas inventories [29], the methane emission of China from 2001 to 2020 from three aspects is estimated and analyzed: rice planting, livestock intestinal fermentation and livestock manure management. The calculation method adopts the CH_4MOD model and emission factor method. The data mainly comes from official public data such as China statistical yearbook. The emission factors are the values provided in the guidelines for preparing provincial greenhouse gas inventories [30] (Table 1.).

Table 1. Sources of methane emission inventory from agricultural activities in China.

Emission source	Data sources	Calculation method	Emission factor
rice planting	China statistical yearbook (2002-2021)	CH_4MOD model	guidelines for the preparation of provincial greenhouse gas inventories
livestock intestinal fermentation	China statistical yearbook (2002-2021)	emission factor method	guidelines for the preparation of provincial greenhouse gas inventories
livestock manure management	China statistical yearbook (2002-2021)	emission factor method	guidelines for the preparation of provincial greenhouse gas inventories

The agricultural added value required to calculate methane emission intensity of agricultural activities in China comes from the China rural statistical yearbook (2002-2021), and the agricultural land area comes from the China statistical yearbook (2002-2021).

2.3. Calculation method of methane emission from agricultural activities
CH_4MOD model with high accuracy is selected to calculate the methane emission from rice. The methane emissions from livestock intestinal fermentation and livestock manure management calculate by the emission factor method.

2.3.1. Methane emission from rice planting. CH_4MOD model proposed by Zhang et al [28]. is used for methane emission from rice planting. CH_4MOD model has the advantages of few input parameters and is easy to obtain and can be used to estimate methane emission from rice under different farming systems and water management methods. The calculation formula is as follows:

$$E_{CH_{4rice}} = EF_{rice} \times A_{rice} \times 10^{-7} \quad (1)$$

Where $E_{CH_{4rice}}$ is the methane emission of rice (10^4t). EF_{rice} is the seasonal average in China of methane emission factors of different rice in different country regions provided by the guidelines for the preparation of provincial greenhouse gas inventories, which is 209.925kg/hm^2. A_{rice} is the seeded area of rice (hm^2).

2.3.2. Methane emission from livestock intestinal fermentation. In addition to the methane emission from rice planting, with the development of livestock and poultry industry, the methane emission from livestock intestinal fermentation cannot be underestimated, because the rumen of livestock will form methanogens to synthesize methane and emit it outside the human body [31]. The emission factors (Table 2.) recommended in the guidelines for the preparation of provincial greenhouse gas inventories [30] are used in calculating the methane emission from livestock intestinal fermentation:

$$E_{CH_4} = \sum EF_{CH_4,entenic,i} \times AP_i \times 10^{-7} \quad (2)$$

Where E_{CH_4} is the total emission of methane from livestock intestinal fermentation (10^4t/a). $EF_{CH_4,entenic,i}$ is the methane emission factor of livestock i (kg/(head·a)). AP_i is the number of the livestock i (head).

Table 2. Methane emission factors from livestock intestinal fermentation.

type	cattle	horse	donkey	mule	pig	sheep
$EF_{CH_4,entenic,i}$	80.13	18.00	10.00	10.00	1.00	8.23

2.3.3. Methane emission from livestock manure management. Livestock manure management is another methane emission source for the livestock and poultry industry. Livestock manure produces methane because it is produced during storage and treatment before adding to the land as fertilizer. Similarly, emission factors (Table 3.) are used in calculating methane emission from livestock manure management:

$$E_{CH_4,manure,i} = \sum EF_{CH_4,manure,i} \times AP_i \times 10^{-7} \quad (3)$$

Where $E_{CH_4,manure,i}$ is the total methane emission from livestock manure management (10^4t/a). $EF_{CH_4,manure,i}$ is the methane emission factor of the livestock manure i (kg/(head·a)). AP_i is the number of the livestock i (head).

Table 3. Methane emission factors from livestock manure management.

type	cattle	horse	donkey	mule	pig	sheep
$EF_{CH_4,manure,i}$	3.75	1.64	0.90	0.70	4.18	0.51

2.4. Calculation method of methane emission intensity from agricultural activities

Two types of emission intensity (GHGI) calculate for methane emission intensity from agricultural activities, namely, the emission intensity per unit of agricultural added value ($GHGI_1$) [32] and the emission intensity per unit of agricultural land area ($GHGI_2$) [33]. The calculation formulas are as follows:

$$GHGI_1 = \frac{E}{G} \quad (4)$$

$$GHGI_2 = \frac{E}{A} \quad (5)$$

IOP Conf. Series: Earth and Environmental Science 1087 (2022) 012021 doi:10.1088/1755-1315/1087/1/012021

Where the unit of $GHGI_1$ is t/10^4yuan, the unit of $GHGI_2$ is t/hm^2.E is the methane emission from agricultural activities (t). G is the agricultural added value (10^4yuan). A is the area of agricultural land (hm^2), namely, the area of cultivated land plus grassland.

2.5. Prediction method of methane emission from agricultural activities

At present, the prediction methods of greenhouse gas emission mainly include the STIRPAT model, scenario analysis method, system dynamics, time series method and so on. The scenario analysis method is used in predicting methane emission from agricultural activities. The scenario analysis method is an intuitive combination of qualitative and quantitative prediction method [34]. To make predictions, the scenario analysis method can formulate several feasible scenarios according to the actual situation to achieve the vision of the future and meet the needs of methane emission prediction under different scenarios.

3. Empirical analysis

According to the above calculation methods, the methane emission from rice planting, livestock intestinal fermentation, livestock manure management and $GHGI_1$, $GHGI_2$ in China from 2001 to 2020 calculate, respectively. It is used to analyze the methane emission from agricultural activities in China. It uses the scenario analysis method to predict the methane emission from agricultural activities in China from 2021 to 2030.

3.1. Analysis on methane emission from agricultural activities in China from 2001 to 2020

GH_4MOD model and emission factor method are used in estimating the methane emission and emission intensity of agricultural activities in China from 2001 to 2020. The methane emission from agricultural activities in China analyzes from the perspective of the time change.

3.1.1. Total methane emission. From 2001 to 2020, methane emissions from rice planting totalled 12436.46×10^4t, showing a trend of the first decline, then rise and finally fluctuate. In 2015, the methane emission from rice planting was the highest, reaching 646.23×10^4t. The emission in 2003 was the lowest, accounting for 89.49% of the average annual emission. The reduction of emissions was mainly due to the reduction of rice planting area: in 2001, China's rice planting area was 2881.20×$10^4$$hm^2$, while in 2020, China's rice planting area was 3007.60×$10^4$$hm^2$, an increase of 126.4×$10^4$$hm^2$. The average annual contribution rate of methane emission from rice planting to China's agricultural activities is 31.08% (Figure 1.).

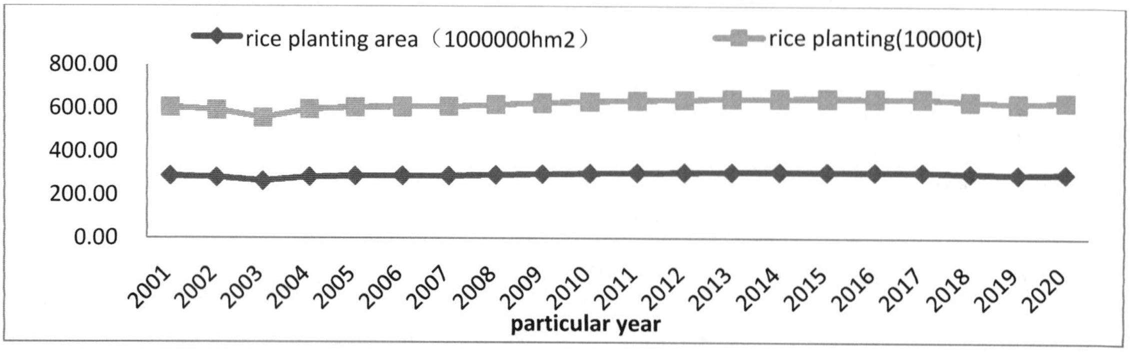

Figure 1. In this case simply justify the caption so that it is as the same width as the graphic.

From 2001 to 2020, the methane emission from livestock intestinal fermentation increased steadily, then decreased sharply, then fluctuated and finally stabilized, which is closely related to the number of livestock in that year. In 2005, the number of large livestock such as pigs, cattle, sheep and horses decreased sharply, only 85.52% of that in 2004. From 2001 to 2020, the total emission of methane from livestock intestinal fermentation in China was 22719.10×10^4t, with an average annual emission of 1135.96×10^4t, accounting for 56.79% of the average annual emission of methane from agricultural activities. Among them, the emission of methane fermented in the bovine intestine was the largest, which was the main emission source of methane in the livestock intestine.

From 2001 to 2020, 4852.79×10^4t of methane were emitted from livestock manure management, accounting for 12.13% of the annual average emission of methane from agricultural activities. Similarly, affected by the number of livestock, the methane emission from livestock manure management increased first, then decreased, and finally stabilized. Among them, the highest emission in 2004 was 274.03×10^4t, which was 112.94% of the annual average of methane emission from livestock manure management. The lowest emission in 2019 was 180.24×10^4t, accounting for 74.28% of the annual average methane emission from livestock manure management (Figure 2.).

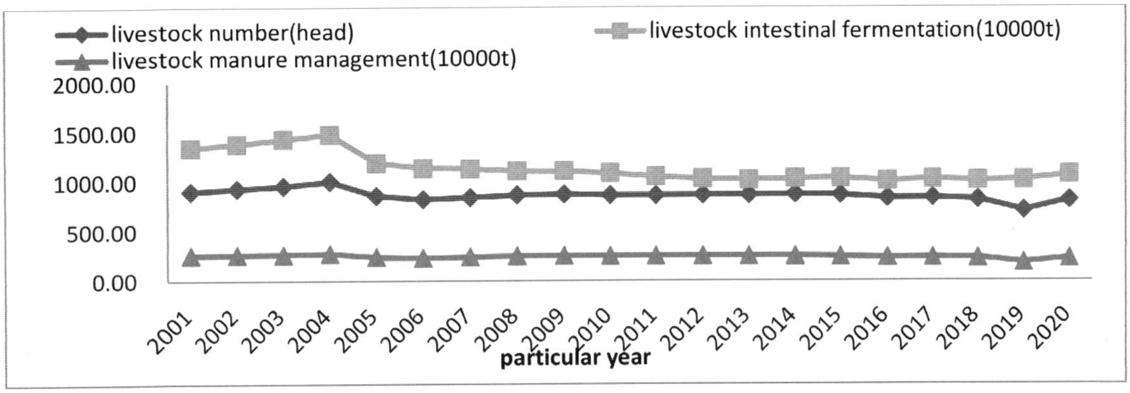

Figure 2. Change of methane emission from livestock in China from 2001 to 2020.

From 2001 to 2020, China's total methane emissions from agricultural activities have experienced three stages: gradual rise, rapid decline and stable fluctuation. From 2001 to 2004, total emissions increased slowly, from 2208.60×10^4t in 2001 to 2349.18×10^4t in 2004, with an average annual growth rate of 2.09%. In 2005, it decreased rapidly by 13.20%, reaching 2039.07×10^4t. From 2006 to 2020, China's total methane emissions from agricultural activities fluctuated in the range of $1800\text{-}2000\times10^4$t. With the gradual optimization of China's economic structure, the focus of industrial development has gradually shifted from the primary industry to the secondary and tertiary industries, resulting in a significant decline in the amount of methane emitted by agricultural activities (Figure 3.).

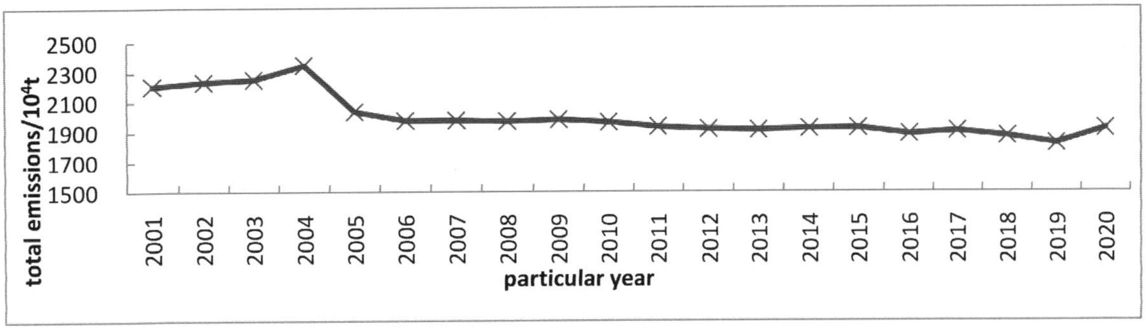

Figure 3. Change of total emission from agricultural activities in China from 2001 to 2020.

3.1.2. Methane emission intensity. China's $GHGI_1$ showed a downward trend, from 1.40 t/10^4yuan in 2001 to 0.24 t/10^4yuan in 2020. The greenhouse gas cost per unit of agricultural added value gradually decreases, as shown in Figure 4. Since the agricultural land area didn't count in some years in the China statistical yearbook, the $GHGI_2$ value is vacant in some years. China's $GHGI_2$ fluctuated. It increased steadily from 2007 to 2015, decreased firstly and then increased from 2015 to 2017, and decreased sharply from 0.054 t/hm^2 to 0.046 t/hm^2 from 2017 to 2019. It tended to be stable from 2017 to 2019, as shown in Figure 5.

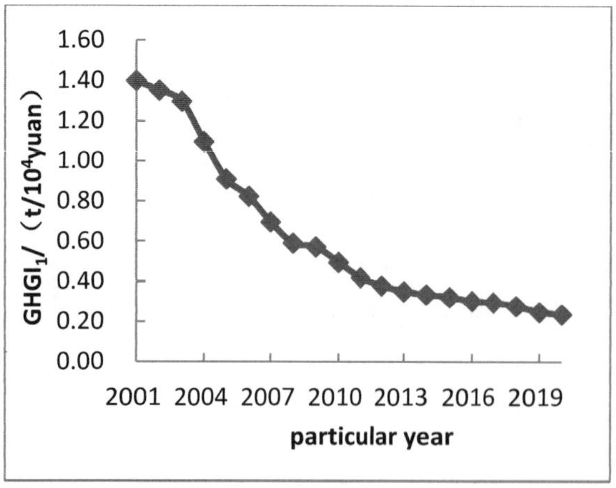

Figure 4. Interannual variation of methane emission intensity $GHGI_1$ from agricultural activities in China from 2001 to 202.

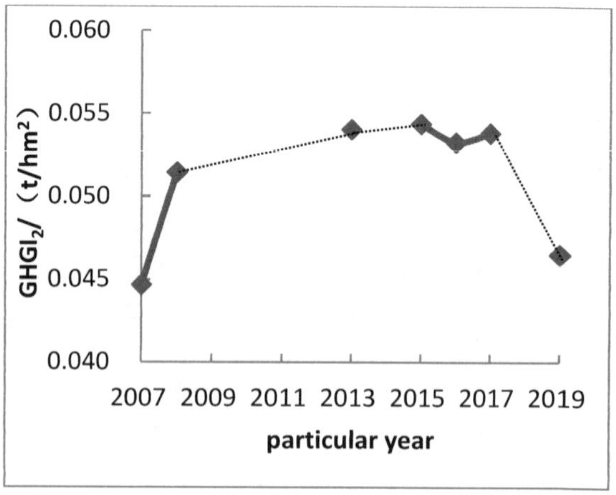

Figure 5. Interannual variation of methane emission intensity $GHGI_2$ from agricultural activities in China from 2001 to 2020.

3.2. Prediction of methane emission from agricultural activities in China from 2021 to 2030
The scenario analysis method is used in predicting the methane emission from agricultural activities in China. By predicting the changing trend of influencing parameters under different scenarios, the change of methane emission obtains.

3.2.1. Scenario parameter setting. According to the existing research of Cao et al [35], combining with the relevant policies, plans of China and the development trend of various parameters, the scenario in line with the actual situation of China is set. There are three scenarios: benchmark scenario (developing according to the existing relevant policies), optimization scenario (the emission reduction policy strategy is higher than the existing planning) and strict emission reduction scenario (the government puts forward stricter emission reduction requirements to achieve the carbon peak by 2030). Among them, the emission factors of each emission source remain unchanged, and the methane emission from agricultural activities in China from 2021 to 2030 is predicted based on the data of 2020. Specific parameter settings show in Table 4.

3.2.2. Analysis of methane emission under different scenarios. According to the parameter settings of different scenarios in Table 4, the methane emission from agricultural activities in China under the benchmark, optimization and strict emission reduction scenarios from 2021 to 2030 are calculated, respectively. The changing trend show in Figure 6.

It can see from Figure 6. that the methane emission from agricultural activities in China from 2021 to 2030 increases year by year under the benchmark scenario, and its emission increases from 1935.98×10^4t in 2021 to 2070.41×10^4t in 2030, with an average annual growth rate of 0.75%. Under the optimization scenario, the methane emission also increases with year's growth, with an average annual growth rate of 0.68%, and the methane emission will be 2055.49×10^4t by 2030, 14.92×10^4t methane emission reduction compared with the benchmark scenario. Under the strict emission reduction scenario, methane emission is the lowest. It reaches the peak in 2030, with a peak of

2047.66×10^{4}t, 22.75×10^{4}t lower than the benchmark scenario, 7.83×10^{4}t lower than the optimization scenario.

Table 4. Parameter settings of different scenarios.

Emission source	Main influencing parameters	scenario model	Rate of change (%)	Setting basis
rice planting	rice field planting area	benchmark	-0.28	China statistical yearbook
		optimization	-0.37	
		strict emission reduction	-0.80	
livestock intestinal fermentation	cattle stock	benchmark	+0.52	China statistical yearbook
		optimization	+0.34	Opinions on stabilizing pig production and promoting transformation and upgrading
		strict emission reduction	+0.25	
	large scale breeding rate	benchmark	+0.40	Opinions of the ministry of agriculture on accelerating the standardized large-scale breeding of livestock and poultry
		optimization	+0.42	
		strict emission reduction	+0.47	
livestock manure management	the utilization rate of livestock waste	benchmark	+2.53	Notice of the general office of the ministry of agriculture and rural areas and the general office of the ministry of finance on the resource utilization of livestock and poultry manure in 2020
		optimization	+2.85	
		strict emission reduction	+3.61	

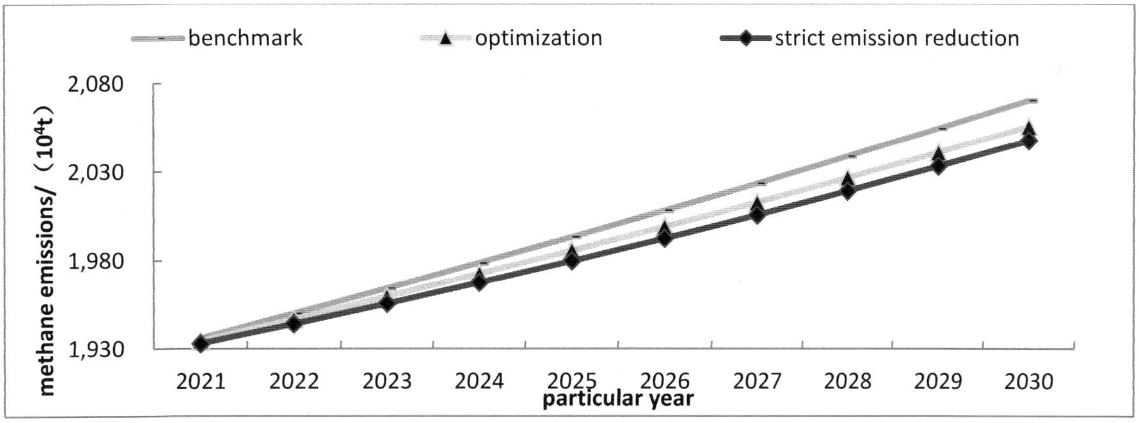

Figure 6. Prediction of methane emissions from agricultural activities in China under different scenarios from 2021 to 2030.

4. Discussions

(1) China is the world's largest developing country and one of the world's agricultural powers. At present, China's methane emissions from agricultural sources tend to be stable, but to achieve the "carbon peak" goal, China still needs to continue to reduce emissions. The main emission sources of methane emissions from agricultural activities are planting and livestock husbandry. Therefore, emission reduction measures should be put forward from the perspectives of rice planting and livestock breeding. China is a country with a large population. It is not advisable to reduce methane

emissions by reducing the planting area or quantity of rice. Selecting low emission rice varieties (low emission factor), adjusting rice irrigation methods (delayed sowing can reduce methane emission, but may have an impact on yield) and promoting large-scale rice planting (methane emission from cultivated rice seedlings is the lowest) can effectively reduce rice planting emissions. The methane emission reduction of livestock breeding can be started from three aspects: reducing the number of breeding, promoting the scale of breeding and adjusting the breeding structure.

(2) By setting different scenarios and predicting the methane emission and change trend under various scenarios, it can see that under the strict emission reduction scenario, China's methane emissions will reach a peak in 2030. If the relevant planning strictly implements, it believes that China's methane emissions can reach a peak before 2030.

(3) There are some difficulties and problems in the calculation and scenario prediction analysis of methane emissions from agricultural activities in China: CH_4MOD model does not consider the impact of chemical fertilizer application on methane emissions, so it will affect the results of methane emissions from rice planting. The selection of emission factors is based on the average value of each region and season in the guidelines for the preparation of provincial greenhouse gas inventories, which is not fully in line with the current situation and is easy to cause a certain degree of deviation. Scenario prediction method is a method that combines past scenarios to predict the future. This study sets scenario parameters through national policies and regulations and the change trend of methane emission in China in recent 20 years. The setting of scenario parameters is subjective to a certain extent, so there is uncertainty.

5. Conclusions

Taking China as the research object, the emission and emission intensity of methane from agricultural activities from 2001 to 2020 are analyzed, and using the scenario analysis method to predict the emission and change of methane from agricultural activities from 2021 to 2030 under different scenarios:

(1) The total methane emission from agricultural activities increased firstly, then decreased and finally tended to steady fluctuation. The proportion of methane emission from livestock intestinal fermentation to the total methane emission has always been the largest. In 2004, the total methane emission in China was the largest, 2349.18×10^4t. Among them, the methane emission from rice planting, livestock intestinal fermentation, and livestock manure management were 595.75×10^4t, 1479.41×10^4t and 274.03×10^4t respectively. In the past 20 years, the total methane emission in 2019 was the lowest, 1824.26×10^4t, which was related to the rice seeding area and the number of livestock breeding in that year.

(2) From 2021 to 2030, China's methane emission is the highest under the benchmark scenario, followed by the optimization scenario, and the lowest under the strict emission reduction scenario. In each scenario, animal intestinal fermentation is the largest methane emission source in China. Methane emissions under the three scenarios are increasing year by year. To meet China's goal of reaching the carbon peak by 2030, the strict emission reduction scenario is more suitable for China's future methane emission development trend than other scenarios.

Acknowledgement

This study was supported by department of education of Yunnan Province (Grant no.2022Y622), Yunnan philosophy and social science planning office (Grant no. YB2020054) and national social science foundation project (Grant no. 18BGL052).

References

[1] Liu Z, Davis S J, Feng K, et al 2016 Targeted opportunities to address the climate–trade dilemma in China *Nature Climate Change* **6** 2

[2] Quere C L, Peters G P, Andres R J, et al 2014 Global carbon budget 2013 *Earth System Science Data Discussions* **7** 2

[3] Xinhua net 2015 China eyes carbon emission peak at around 2030
[4] IPCC 2013 Climate Change 2013: the Physical Science Basis: Contribution of Working Group I to the Fifth Assessment Report of the Intergovernmental Panel on Climate Change Cambridge University Press
[5] WMO 2016 WMO Greenhouse Gas Bulletin - No. 12: the State of Greenhouse Gases in the Atmosphere Based on Global Observations through 2015
[6] Zhang B, Chen Z M, Qiao H, et al 2015 China's non-CO2 greenhouse gas emissions: Inventory and input–output analysis *Ecological Informatics* 26
[7] Zhang B, Chen G Q 2014 Methane emissions in China 2007 *Renewable and Sustainable Energy Reviews* 30
[8] Zhang B, Chen G Q 2014 China's CH4 and CO2 emissions: Bottom-up estimation and comparative analysis *Ecological indicators* 47
[9] Ju Y, Sun Y, Sa Z, et al 2016 A new approach to estimate fugitive methane emissions from coal mining in China *Science of the Total Environment* 543
[10] Riddick S N, Hancock B R, Robinson A D, et al 2018 Development of a low-maintenance measurement approach to continuously estimate methane emissions: A case study *Waste Management* 73
[11] Peng S, Piao S, Bousquet P, et al 2016 Inventory of anthropogenic methane emissions in mainland China from 1980 to 2010 *Atmospheric Chemistry and Physics* **16** 22
[12] Zhang X, Wang J, Zhang T, et al 2021 Assessment of methane emissions from China's agricultural system and low carbon measures *Environmental Science & Technology* **44** 3
[13] Li Y, Chen M 2021 Influencing factors of methane and nitrous oxide emissions from agricultural sources in China *Acta Scientiae Circumstantiae* **41** 2
[14] Guo J, Song Z, Zhu Y, et al 2017 The characteristics of yield-scaled methane emission from paddy field in recent 35-year in China: A meta-analysis *Journal of Cleaner Production* 161
[15] Yao W 2020 Macroeconomic Analysis of Methane Emissions from Livestock in China *Ecological Economy* **36** 9
[16] Nie T, Chen P, Zhang Z, et al 2020 Effects of irrigation method and rice straw incorporation on CH_4 emissions of paddy fields in Northeast China *Paddy and Water Environment* **18** 1
[17] Zhang B, Tian H, Ren W, et al 2016 Methane emissions from global rice fields: Magnitude, spatiotemporal patterns, and environmental controls *Global Biogeochemical Cycles* **30** 9
[18] Chen D, Wang C, Shen J, et al 2018 Response of CH_4 emissions to straw and biochar applications in double-rice cropping systems: Insights from observations and modeling *Environmental Pollution* 235
[19] Chen H, Zhu Q, Peng C, et al 2013 Methane emissions from rice paddies natural wetlands, lakes in China: synthesis new estimate *Global change biology* **19** 1
[20] Yan C, Han M, Liu Y, et al 2021 Household CH_4 and N_2O footprints of major economies *Earth's Future* **9** 10
[21] Zhang Y, Jiang Y, Li Z, et al 2015 Aboveground morphological traits do not predict rice variety effects on CH4 emissions *Agriculture, Ecosystems & Environment* 208
[22] Kai F M, Tyler S C, Randerson J T, et al 2011 Reduced methane growth rate explained by decreased Northern Hemisphere microbial sources *Nature* **476** 7359
[23] Jiang Y, Qian H, Huang S, et al 2019 Acclimation of methane emissions from rice paddy fields to straw addition *Science advances* **5** 1
[24] Wang R, Wang M, Zhang X M, et al 2019 Evaluation of interval between measurements and calculation method for the quantification of enteric methane emissions measured by respiration chamber *Journal of dairy science* **102** 7
[25] Chamberlain S D, Boughton E H, Sparks J P Underlying ecosystem emissions exceed cattle-emitted methane from subtropical lowland pastures *Ecosystems* **18** 6
[26] Kumari S, Dahiya R P, Naik S N, et al 2016 Projection of methane emissions from livestock through enteric fermentation: A case study from India *Environmental Development* 20

[27] Huang Y, Zhang W, Zheng XH, et al 2004 Modeling methane emission from rice paddies with various agricultural practices *Journal of Geophysical Research* **109** 8
[28] Zhang W, Huang Y, Zheng XH, et al 2004 Modeling methane emission from rice paddies: model and modification *Act a Ecologica Sinica* 11
[29] IPCC 2006 Guidelines for national greenhouse gas inventories Intergovernmental Panel on Climate Change
[30] Climate Department of national development and Reform Commission 2011 Guidelines for the preparation of provincial greenhouse gas inventories (for Trial Implementation)
[31] Iqbal M F, Cheng Y F, Zhu W Y, et al 2008 Mitigation of ruminant methane production: current strategies, constraints and future options *World Journal of Microbiology and Biotechnology* **24** 12
[32] Li B, Liu XQ, Wang K 2018 Study on Carbon Effects and Spatial-Temporal Evolution Trend based on the Changes of Agricultural Land Use in China *China Land Science* **32** 3
[33] Lin BQ, Mao DX 2014 Characteristics of China's Carbon Emission Intensity in Different development stages *Financial Research* 8
[34] McBurney P, Parsons S 2003 Chance discovery and scenario analysis *New Generation Computing* **21** 1
[35] Cao GL, Chu RN, Yu T 2021 Calculation and scenario prediction analysis of CH_4 emissions in Shaanxi Province *Journal of Safety and Environment* **22** 2

Management of inorganic nitrogen compounds in a closed aquaculture system during imitation for nursery of *Litopenaeus vannamei* under brackish condition

Haihong Huang [1,3], Chaoyun Li[2,3], Yanju Lei[1] and Pinhong Yang[1]

[1] Hunan University of Arts and Science, Changde 415000, China;
[2] Shandong Vocational Animal Science and Veterinary College, Weifang 261061, China.

[3] Corresponding author's e-mail: shinkanh@nwsuaf.edu.cn; cyzyli@163.com

Abstract. It was aimed to estimate the efficiency of management of inorganic nitrogen compounds under a closed system condition, via addition of carbon source. For this purpose, an imitation experiment for nursery of *Litopeanaeus vannamei* postlarvae under a brackish condition was conducted, in two closed systems with a salinity of about 18‰: systems with (CS) and without (control) addition of carbon source (molasses) at a carbon to nitrogen ratio (C:N) of 20:1. The levels of total ammonia nitrogen (TAN, 1.39±0.14 mg L^{-1}), nitrite (0.023±0.003 mg L^{-1}) and nitrate (0.49±0.06 mg L^{-1}) in CS system were significantly lower than those in the control system (6.92±0.60, 0.809±0.050 and 2.39±0.47 mg L^{-1}, $P < 0.05$). The results indicated that carbon source addition to the closed system imitating nursery of *L. vannamei* postlarvae under a brackish condition well managed the level of inorganic nitrogen compounds during experiment, making it possible to environment-friendly culture *L. vannamei* at inland areas under a closed condition.

1. Introduction

The most widely cultural euryhaline crustacean species, the Pacific white-leg shrimp (*Litopenaeus vannamei*) in the world, could be raised in brackish system [1]. In China, this species has been attracted to culture in freshwater earthen ponds in inland zones [2]. However, culture of marine shrimp in non-saline inland zones would result in adverse effects on local ecosystem, even under a near-freshwater or brackish condition [3, 4]. Hence, it is necessary to culture *L. vannamei* in inland areas with closed aquaculture system.

During shrimp culture, accumulation of inorganic nitrogen compounds is the main factor to deteriorate water quality and to harm shrimp health and growth, especially the accumulation of toxic ammonia and nitrite, due to bacterial degradation of organic matter [5], such as uneaten feed rich in protein and feces from continuous input of high protein content aquafeed. Traditionally, a large amount of water is exchanged to alleviate water quality problem. However, this problem would be more serious in a closed system without water exchange.

In this current study, the development of inorganic nitrogen compounds in the closed systems during imitation experiment for nursery of *L. vannamei* postlarvae (PL) under a brackish condition was investigated, and the management of those compounds in this system by addition of carbon source was also conducted.

2. Material and method

2.1. Preparation of brackish water
Water (salinity 18‰) was prepared with artificial sea salt (Jiangxi Yantong Technology Co., Ltd., Xingan, China) for experiment. The pH value was adjusted to near 8.0 with $NaHCO_3$ (China National Pharmaceutical GroupCo., Ltd., Beijing, China). Then, water was sterilized with chlorinedioxid, followed by being neutralized with ascorbic acid after 24h [6].

2.2. The experiment design
High density polyethylene tank (HDPE, 40×86×64 cm) was filled with water (100 liters) prepared above. Aeration was conducted by an HG-750-S air pump (Sunsun Group, China) with porous stones. There two closed systems were assigned, systems with (CS) and without (control) addition of carbon source, with three replicates for each.

For imitation of the nursery of *L. vannamei* PL, a 28-day experiment was designed. During this experiment, milled and sieved (100 mesh size) shrimp artificial feed (protein 40.0%, Alpha corporation, Jiangmen, China) was fed twice a day to each tank of both systems with a rate of 0.55 g d^{-1}, being determined according to an average feeding rate of 5% (from 12% at the beginning to 3% at the end), and the stocking density of 1000 PL (0.01 g) per m^3. For the CS system, molasses (carbon content 35.0%, Xinsheng food factory, Shantou, China) was used to be carbon source, with a carbon to nitrogen ratio (C:N) of 20:1, according to the method of [6] for determination of C:N. There was no water exchange during the period of experiment, except complementation of evaporating water (about 1% of the total water volume per week).

2.3. Water parameters
Temperature, DO (dissolved oxygen), pH value and salinity were measured by using (YSI-550A (Yellow Springs) and electric analyzers AZ8371 (Hengxin technological Co., Ltd., China) every day. Total ammonia nitrogen (TAN), alkalinity, hardness, total suspended solid (TSS), nitrite and nitrate were monitored once a week, according to ref [6]. Floc volume was determined every week by using an Imhoff cone. Total bacteria counts in waterbody was measured every week with plate counting method by using LB broth agar plates, and expressed as colony forming units each milliliter (CFU mL^{-1}).

2.4. Statistical analysis
Data were analyzed with the SPSS (IBM Co., USA). In brief, normality and homogeneity of data were detected with Shapiro-Wilk's test and Levene's test, respectively. Then, the student's *t*-test was executed. Or else, Mann-Whitney non-parametric test was applicated for significant difference detecting (such as pH, TAN and nitrate), respectively. Logarithm transformation for the total bacterial counts were submitted before analysis. A P value < 0.05 was considered to be significant.

3. Results
The concentrations of TAN (1.39±0.14 mg L^{-1}), nitrite (0.023±0.003 mg L^{-1}) and nitrate (0.49±0.06 mg L^{-1}) in the closed system with addition of carbon source (CS system) were significantly lower than those in the control system (6.92±0.60, 0.809±0.05 and 2.39±0.47 mg L^{-1}, Table 1). The developments of TAN, nitrite and nitrate along with time elapsing in both systems were showed in (Figure 1).

Table 1. Inorganic nitrogen compounds levels in the closed systems with (CS) and without (control) addition of carbon source during the imitating experiment for nursery of *L. vannamei* postlarvae.

Compounds	control	CS	*P*-value
TAN (mg L^{-1})	6.92 ± 0.60^b	1.39 ± 0.14^a	< 0.001
Nitrite (mg L^{-1})	0.809 ± 0.050^b	0.023 ± 0.003^a	< 0.001
Nitrate (mg L^{-1})	2.39 ± 0.47^b	0.49 ± 0.06^a	0.002

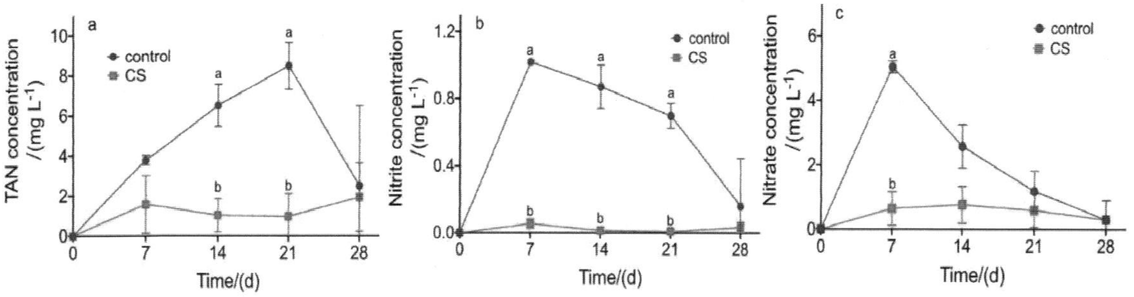

Figure 1. Development of TAN (a), nitrite (b) and nitrate (c) along with the time elapsing in closed systems with (CS) and without (control) adding carbon source during the imitating experiment for nursery of *L. vannamei* postlarvae.

The level of pH (8.01±0.01) and dissolved oxygen (9.34±0.03 mg L^{-1}) in the CS system were significantly lower than both parameters in the control system, respectively (8.13±0.03 and 9.61±0.07 mg L^{-1}, Table 2). However, the significantly higher settleable solids (24.0±4.8 mL L^{-1}) and total bacterial counts ($20.75\pm9.2\times10^4$ CFU mL^{-1}) in CS system were found, when compared to those in control system (1.1±0.3 mL L^{-1} and $0.42\pm0.19\times10^4$ CFU mL^{-1}, Table 2). Carbonate alkalinity, hardness, temperature, salinity and total suspended solids were not significantly different between both systems ($P > 0.05$, Table 2).

Table 2. The water parameters in closed systems with (CS) and without (control) adding carbon source during the imitating experiment for nursery of *L. vannamei* postlarvae.

Parameters	Control	CS	*P*-value
Carbonate alkalinity (mg $CaCO_3$ L^{-1})	113.3±12.0	119.6±4.7	0.672
Hardness (mg $CaCO_3$ L^{-1})	179.0±9.0	184.1±9.9	0.608
Dissolved oxygen (mg L^{-1})	9.61 ± 0.07^a	9.34 ± 0.03^b	0.002
pH	8.13 ± 0.03^a	8.01 ± 0.01^b	0.004
Temperature (°C)	27.47±0.43	27.6±0.14	0.757
Salinity (‰)	17.79±0.13	18.02±0.06	0.200
Settleable solids (mL L^{-1})	1.1 ± 0.3^b	24.0 ± 4.8^a	< 0.001
Total suspended solids (mg L^{-1})	217.3±44.8	329.4±19.0	0.110
Total bacterial counts ($\times10^4$ CFU mL^{-1})	0.42 ± 0.19^b	20.75 ± 9.2^a	0.015

4. Discussions

As expected, a high level of TAN (6.92±0.60 mg L^{-1}) exceeding the safety level (2.44 mg L^{-1}) for shrimp under brackish condition with a salinity of 15‰ [7] was found in the control closed system. Whereas, nitrite and nitrate were low in this system. During the process of nitrification transformation, organic matters, such as residual feed and feces, are decomposed first, to form ammonia [8]; next, nitrifiers oxidize ammonia to nitrite, and then nitrite is oxidized to nitrate [9]. Hence, these results mentioned above suggest that the nitrification was not well established in control system, which might be due to the short-term experimental duration.

Similarly, there no obvious accumulation of TAN, nitrite and nitrate was found in the closed system with addition of carbon source (CS system). Moreover, in this system, the TAN level (1.39±0.14 mg L^{-1}) was in the range of safety level, which might be not harmful for the growth and survival of shrimp in actual aquaculture practice under the low salinity condition [10]. It is considered that addition of exogenous carbon source could to adjust the C:N in the water to a high level of 15-20. And this operation would promote the growth of heterotrophic bacteria to accelerate assimilation of inorganic nitrogen compounds, consequently leading to a low TAN level even under a closed condition [3, 9]. Meanwhile, the dissolved oxygen level and pH value in the waterbody of CS system decreased, as a result of oxygen exhausting and CO_2 accumulating due to respiration of a large number of bacteria in this system. Furthermore, the blooming-growth heterotrophic bacteria excrete extracellular polymeric substances, which consolidates microorganisms, organic particles, residual feed and feces to form "bioflocs" [11], making the level of settleable solids significantly increased in the CS system of the present study, when compared with the control system.

5. Conclusions

Addition of carbon source to the closed system well managed the total ammonia nitrogen level during the imitation experiment for nursery of *L. vannamei* PL at a brackish condition, making it possible to environment-friendly shrimp culture in inland areas in a closed aquaculture system. Next, more details would be studied to further improve the feasibility of adoption of this approach.

Acknowledgements

This work was financially supported by the Microbial Technology Innovation Team of HUAS and Changde Research Center for Agricultural Biomacromolecule.

References

[1] Van Wyk P, Davishodgkins M, Laramore R *et al.* 1999 *Farming marine shrimp in recirculating freshwater systems*. Tallahassee: FDACS
[2] Bureau of Fisheries, PRC 2020 *China fishery statistical yearbook*. Beijing: China Agriculture Press
[3] Avnimelech Y 2015 *Biofloc technology: a practical hand book (3rd ed)*. Baton Rouge,: The world aquaculture society
[4] Ray A J, Drury T H, Cecil A 2017 *Aquacult. Eng.* **77** 9-14
[5] Esparza-Leal H M, Cardozo A P, Wasielesky W 2015 *Aquacult. Eng.* **68** 28-34
[6] Huang H-H, Liao H-M, Lei Y-J, *et al.* 2022 *Aquaculture* **546** 737239
[7] Lin Y, Chen J, 2001 J. Exp. Mar. Biol. Ecol. **259** 109-119
[8] Luo G, Gao Q, Wang C *et al.* 2014 *Aquaculture* **422** 1-7
[9] Ebeling J M, Timmons M B, Bisogni J J 2006 *Aquaculture* **257**(1–4) 346-358
[10] Jaffer Y D, Saraswathy R, Ishfaq M, *et al.* 2020 *Aquaculture* **515** 734561
[11] Wei G, Shan D, Li G *et al.* 2020 *Aquaculture* **529** 735632

On the Legal Mechanism of Nuclear Safety in China under the Background of Environmental Protection

Jiu Liu[1,3] and Le Shen[2]

[1] Associate Professor, College of Humanities and Social Sciences, Harbin Engineering University, Nantong Street No. 145, Nangang District, Harbin 150001, China;
[2] Master Student, College of Humanities and Social Sciences, Harbin Engineering University, Nantong Street No. 145, Nangang District, Harbin 150001, China.

[3] Corresponding author's e-mail: nine699@sina.com

Abstract. Background: Peaceful use of nuclear energy is of great significance for ensuring energy security, reducing greenhouse gas emissions, and achieving sustainable development environmental-friendly. Nuclear safety is the premise and foundation for the development of civil nuclear power industry. As a country with the largest scale of nuclear power plants under construction, preventing accidents properly is the best way to ensure nuclear safety. However, under the background of environmental protection nowadays, China has only officially implemented two laws related to nuclear safety - the Law on Prevention and Control of Radioactive Pollution and Nuclear Safety Law. It cannot fully meet the needs of nuclear safety supervision in China.
Methods: To highlight the problems in the current legal system pertaining to nuclear safety, a legislation study is used to analyze the current content of related legislation and regulations. Comparative methodology is also adopted in this paper to analyze the legislative and administrative experience of other countries with or without nuclear power industry and summarize the problems in the current legal system in China.
Result: Currently, the development of nuclear safety law system reflects the historical process of the development and application of nuclear technology in China. China's nuclear safety-related legislation closely fits the strategy of China's nuclear technology development and application under the background of environmental protection. Even though, there are still deficiencies about China's nuclear safety legal system now, for example: the atomic energy law is still absent in China, so pertaining measures must be taken to develop and improve the nuclear safety legal system in China as for not only promoting the development of nuclear industry but also avoiding radioactive pollution incidents and protecting environment. Moreover, there is a conflict between the concept of "nuclear safety" in the Nuclear Safety Law and its content.

1. The Definition and Content of Nuclear Safety

The exact meaning of the concept of "Nuclear Safety" has different definitions in different occasions, organizations, management levels, application scenarios and knowledge contexts.

First of all, "nuclear safety" is a kind of "safety". The Occupational Health and Safety Management System defines "safety" as a state in which there is no acceptable risk of harm. International Civil Aviation Organization defines "safety" as a state of the system, that is, after continuous identification and control of risks, the risk of personal injury and property loss are reduced to an acceptable level or even below. To sum up, the meaning of the word "safety" can be simply summarized as: controlling risks to an acceptable level and avoiding hazards.

On January 14, 2015, the National Nuclear Safety Culture Policy Statement held that: "Nuclear safety means taking necessary and adequate safety measures such as monitoring, protection, prevention and mitigation for nuclear facilities, nuclear activities, nuclear materials and radioactive substances to prevent accidents due to any technical, man-made or natural disasters and to minimize the radiological consequences in the event of an accident, thereby protecting workers, the public and the environment from undue radiation hazards. "[1] The IAEA Safety Glossary issued by the International Atomic Energy Agency (IAEA) in 2018 defines "nuclear safety" as "realizing the normal operation of nuclear power plants, preventing accidents or mitigating the consequences of accidents, thus protecting workers, the public and the environment from excessive radiation hazards". This is basically consistent with the definition of the Policy Statement of China.

Article 31 of China's National Security Law stipulates that: "China adheres to the peaceful use of nuclear energy and technology, strengthen international cooperation, prevent nuclear proliferation, improve the non-proliferation mechanism, strengthen the safety management, supervision and protection of nuclear facilities, nuclear materials, nuclear activities and nuclear waste disposal, strengthen the construction of nuclear accident emergency response system and emergency response capacity, prevent, control and eliminate the harm of nuclear accidents to citizens' lives and health and ecological environment, and continuously enhance the ability to effectively respond to and prevent nuclear threats and attacks. " This is different from the definition of nuclear safety in the Policy Statement and the Safety Glossary of IAEA. It is a broad sense of nuclear safety, and it is basically consistent with the ultimate goal of nuclear safety.

IAEA have put forward the "3S safety" of nuclear energy. The concept of "3S" was formally mentioned for the first time in the Handbook of Nuclear Law: Implementing Legislation published by IAEA in 2010. “3S” means safety, security and safeguards in order to ensure the peaceful use of nuclear energy, prevent the use of nuclear energy for non-peaceful purposes and establish an adequate legislative and regulatory framework for ionizing radiation. In the concept of 3S, nuclear safety of nuclear power plant means that in the process of reactor design, construction and operation, we must consider and take corresponding measures to eliminate the possibility of accidental occurrence and development of uncontrollable chain fission reaction. [2] Therefore, in the operation of nuclear power plants, we should make every effort to ensure the safety of workers and avoid being hurt by nuclear energy risks. Concurrently, the operation of nuclear power plants should take more heed to environmental protection.

2. The Significance of Applying Legal Mechanism to Guarantee Nuclear Safety

Although in the process of building and operating nuclear power plants China has always adhered to a high level of safety, and no serious nuclear incidents have occurred, the safety situation is still worthy much attention. We must make certain the safe operation of nuclear facilities and prevent the occurrence of nuclear accidents all the time, and finally gain the trust of the public so as to achieve smoother development.

Firstly, currently China is the country with the maximum scale of nuclear power plants under construction. In order to further improve the nuclear safety, it is urgent to lay down and improve the nuclear safety legal system to regulate the safe operation of nuclear-related projects and facilities. [3] Nuclear facilities and materials have potential radiation hazards. Applying related legislation and regulations is conducive to preventing hazards and providing legal guarantee for nuclear safety. A large number of nuclear facilities have brought great challenges to China's nuclear safety and financial situation. The environmental and health issues have become increasingly prominent, which urgently need to be regulated and guided by law.

Nowadays China’s nuclear energy safety supervision mainly depends on Nuclear Safety Law, Law on Prevention and Control of Radioactive Pollution, seven administrative regulations and twenty-seven departmental rules. Even though, a scientific and reasonable nuclear safety legal system has not been established, which is far from meeting the needs of China's current nuclear safety requirement. Such a current legal system in China lags behind the rapid growth of nuclear energy industry and is not conducive to strengthening the safety supervision and social supervision in the whole process of nuclear power plants construction, production and operation.

Secondly, after the Fukushima nuclear accident in Japan, the call for establishing laws related to the safety of civil nuclear industry in China is soaring. This is an inevitable response from the worldwide concern about nuclear safety. What the public really cares about is the protection of nuclear safety, environmental safety and public health. In this case, only when the state formulates and implements the nuclear safety legal system guided by the Nuclear Safety Law and shows the national position that safety is paramount can it conform to public opinion. Using laws to regulate the safe operation of nuclear facilities can not only guide the public's scientific cognition of China's nuclear safety and eliminate the public's nuclear phobia, but also effectively guarantee China's nuclear safety and fundamentally establish public confidence in the growth of civil nuclear industry.

Thirdly, applying legal mechanism to guarantee nuclear safety is just following the trend of international nuclear safety legislation, which is a necessity for a state with the whole nuclear power industry chain as China. Safety is the top priority for the development of civil nuclear power industry. Since the development of civil nuclear power industry in the mid-20th century, all countries in the world have been actively building their own nuclear safety legal system. At present, at least 36 countries and regions have enacted officially their Atomic Energy Law or Nuclear Safety Law, or Radiation Protection Law, including countries with developed civil nuclear power industry such as the United States, Russia and the developing countries without developing civil nuclear power industry such as Myanmar and Indonesia. At least nine countries have enacted their own Nuclear Safety Law; 12 countries have issued decrees establishing atomic energy committees and/or radiation committees. Australia has both the Atomic Energy Law and the Nuclear Safety Law. As a significant member of International Atomic Energy Agency and a party of relevant international nuclear conventions, China is having the mission of international cooperation in nuclear safety. Therefore, the establishment and implementation of Nuclear Safety Law and related regulations are conducive in establishing a good international image of China, strengthening confidence of the international society in China's nuclear safety, and also good to the implementation of China's "nuclear power going out" strategy. [4]

3. Current Nuclear Safety Law System in China under the Background of Environmental Protection

Up to now, China has officially implemented only two laws related to nuclear safety, namely, the Law on Prevention and Control of Radioactive Pollution and the Nuclear Safety Law. The Law on Prevention and Control of Radioactive Pollution, passed in 2003, is the first law in the field of nuclear and radiation safety in China, of which quite few articles are related to nuclear safety. The Nuclear Safety Law is the first law in China to specifically adjust the nuclear power industry and completely aim at nuclear safety.

The promulgation of the Law on Prevention and Control of Radioactive Pollution can be engaged in prevention before and supervision after the event to avoid the occurrence of radioactive pollution. In order to correctly control radioactive pollution and promote the healthy development of China's national economy, the law summarizes the experience of China's nuclear energy development and utilization in the past decades, draws lessons from the relevant international measures to deal with radioactive pollution, and takes "prevention first, combination of prevention and control, strict management and safety first" as the policy, and clarifies the goal of China's prevention and control of radioactive pollution.

In 2017, State Council replied to the Request for Instructions of the 13th Five-Year Plan for Nuclear Safety and Radioactive Pollution Prevention and Control and the Long-term Goals for 2025 issued by the Ministry of Environmental Protection (now the Ministry of Ecology and Environment). [5] Through the implementation of the plan, the Nuclear Safety Law was passed in 2017 and have come into force in 2018. The promulgation of the Nuclear Safety Law marked that China has entered

IOP Conf. Series: Earth and Environmental Science 1087 (2022) 012023 doi:10.1088/1755-1315/1087/1/012023

the development stage of systematically coping with and preventing nuclear safety risks. The Nuclear Safety Law first clearly defines the concepts involved in nuclear operation in the general part, and stipulates a number of specific contents such as nuclear facility safety and nuclear accident emergency.

In addition to the above two special laws, other laws closely related to nuclear safety supervision are also stated in Table 1 below. Moreover, there are more than ten administrative regulations regulating nuclear safety in China at present, such as Regulations on Safety Supervision and Administration of Civil Nuclear Facilities in 1986, Regulations on Administration of Nuclear Materials in 1987, Regulations on Safety Administration of Transport of Radioactive Materials in 2009, Regulations on Safety Administration of Radioactive Waste in 2011 and etc.

Table 1. China's Laws and Administrative Regulations on Nuclear Safety.

Specialized Legislation:
Nuclear Safety Law, 2017
Law on Prevention and Control of Radioactive Pollution, 2003
Related Legislation:
Environmental Protection Law
Marine Environmental Protection Law
Air Pollution Prevention Law
Water Pollution Prevention Law
Environmental Impact Assessment Law
Solid Waste Pollution Prevention Law
Land Administration Law and Sea Area Use Management Law
Electricity Law
Energy Conservation Law
Renewable Energy Law
National Security Law
National Defense Law
Safety Production Law
Military Facilities Protection Law
Special Equipment Safety Law
Administrative Regulations:
Regulations on Safety Administration of Radioactive Waste, 2011
Regulations on Safety Administration of Transport of Radioactive Materials, 2009
Regulations on Administration of Nuclear Materials, 1987
Regulations on Safety Supervision and Administration of Civil Nuclear Facilities, 1986

China has always been actively participating in international communications in this field. As one of the thirteen designated members of the International Atomic Energy Agency, China actively cooperate with the International Atomic Energy Agency in the fields of nuclear material security, radioactive source supervision, nuclear and radiation emergency response, etc., and join the International Network of Nuclear Emergency and Corresponding Rescue of the International Atomic Energy Agency, and continuously donate to its Nuclear Safety Fund to support the nuclear safety capacity building between China and other countries in Asia. So far, China has ratified, signed or acceded eight related international conventions.

4. The Characters of the above Nuclear Safety Legalization in China

With the understanding of the importance of nuclear safety, China's risk prevention and response to nuclear safety has been increasingly improved and deepened. In recent years, nuclear safety legislation has made a lot of progress. By combing the pertinent laws and regulations in the field of nuclear safety above, it can be found that China's current nuclear safety legal system presents the following characteristics:

Firstly, the development of nuclear safety law system reflects the historical process of the development and application of nuclear technology in China. The application of nuclear technology in China began in the mid-1950s. At that time, the application of nuclear technology in China was still in its infancy, and nuclear safety had not yet become the main focus of attention. In 1984, China officially became a member of the International Atomic Energy Agency. China had paid more attention to preventing safety risks in the application of nuclear technology. During that period, a series of administrative regulations and departmental rules had been formulated and enacted one after another. These regulations are guided by responding to practical needs in practice and provide guidance and norms for preventing risks in the application of nuclear technology in China. Therefore, China's nuclear safety law system has been characterized by administrative regulations and departmental rules for a period of time. This way of giving priority to formulating administrative regulations and departmental rules according to the actual needs is conducive to solving the urgent problems in practice efficiently and conveniently. Thus, the development of history also proves that this model is suitable for China. Years of safe operation have confirmed the role of these legislation and regulations. In this way, China has formed a relatively complete safety supervision system, which has accumulated a lot of valuable experience before the promulgation of the Nuclear Safety Law. In 2018, the Nuclear Safety Law was officially implemented, which filled the gap in China's legislative field and marked a new stage in the development of China's civil nuclear industry.

Secondly, China's nuclear safety-related legislation closely fits the strategy of China's nuclear technology development and application under the background of environmental protection. China's nuclear safety legislation adopts a top-down legislative model. It took less than five years to complete the formulation, giving full play to the characteristics of the top-down legislative model, which is conducive to the collaborative work of all relevant units and departments and avoids the passive situation of procrastination and inaction in nuclear legislation.

While shortening the legislative time, Nuclear safety legislation never forgets to closely follow the national security strategy. Nuclear Safety Law not only includes all the narrow nuclear safety contents, but also includes the "nuclear security" in the broad nuclear safety. And it not only pays attention to risk prevention, but also pays attention to emergency response in case of accidents, as well as responsibility implementation and damage compensation. Moreover, it also pays attention to information disclosure and accepts and welcomes the supervision of the public. This is manifested in fact that the Nuclear Safety Law absorbs the advanced experience summed up by the long-term practice of China's nuclear safety administrative regulations and departmental rules, including but not limited to nuclear safety licensing, qualification management, quality assurance, planning restricted area protection, radiation protection, nuclear accident emergency preparedness and response, etc.

5. The Deficiencies of Current Nuclear Safety Legal Mechanism in China

5.1. The Atomic Energy Law is still absent

There are two laws in the field of nuclear safety in China, which have greatly enriched the content of China's nuclear safety legal system. However, the above laws both have their own emphases, and the positioning of these in China's legal system is not the basic laws of utilization of atomic energy. Therefore, China will adopt the parallel model of establishing both Atomic Energy Law and Nuclear Safety Law, between which Atomic Energy Law is relatively more basic and its adjustment scope is wider than that of nuclear safety law. However, China's Atomic Energy Law is still absent. [6]

The Atomic Energy Law is the basic law governing the field of nuclear energy in China. The formulation of the Atomic Energy Law is for promoting the healthy and sustainable development of civil nuclear industry and ensuring the safety of nuclear energy research, development and utilization. Looking back the historical process, from 1984 till now, the establishment of China's Atomic Energy Law has lasted for nearly 37 years. On the one hand, it reflects China's high standards and requirements for the legislative quality of Atomic Energy Law, and on the other hand, it also reflects that the disputes involved in Atomic Energy Law are relatively complex and difficult to be solved.

In terms of nuclear safety, the absence of Atomic Energy Law is not conducive to coordinate the relationship among nuclear safety law, prevention and control of radioactive pollution law and the administrative regulations and departmental rules in the nuclear power industry. And without Atomic Energy Law the responsibilities and obligations of the government and operators of nuclear facilities required by IAEA cannot be clarified by law. Therefore, as the basic law in China's nuclear field, Atomic Energy Law should be the most important law for civil nuclear power industry.

5.2. There is a conflict between the concept of "nuclear safety" in the Nuclear Safety Law and its content

Numerous domestic laws, international conventions, regulatory documents, expert commentaries and guidelines have emphasized that safety is the primary requisite for the use of nuclear energy and the applications of ionizing radiation. [7] In different nuclear legislation, the concept of safety often has different meanings for different adjustment objects, and the interpretation of nuclear safety can be divided into broad sense and narrow sense. [8] In the broad sense, "nuclear safety" includes three aspects: "nuclear safety", "nuclear security" and "nuclear guarantee", while in the narrow sense, nuclear safety refers to ensuring the conduct of all safe activities for peaceful use of nuclear energy and prevention of ionizing radiation.

Article 31 of National Security Law of China uses the broad meaning of nuclear safety. As the Nuclear Safety Law formulated in line with China's overall national security strategy, its provisions do not clearly define nuclear safety. Only Article 2 of this Law implies the concept of nuclear safety. However, compared with its clear definition of nuclear facilities, nuclear materials and radioactive waste in the latter articles, the first sentence of Article 2 more like indicates the scope of application of the Nuclear Safety Law. As a result, the meaning of nuclear safety in this Law can only be seen from the specific provisions of this Law. However, the specific provisions of the Nuclear Safety Law show that the nuclear safety in China's Nuclear Safety Law goes beyond the narrow concept of nuclear safety. For example, its Articles 11 and 12 involve the content of nuclear security, which is more in line with the broad concept of nuclear safety in the National Security Law. The content in this law does not match this concept(definition), so there is conflict between such wide definition and narrow adjustment scope. The scope of the Nuclear Safety Law focuses on civil nuclear facilities and nuclear materials, which are only a part of the nuclear fuel cycle and nuclear activities, and does not apply to the application of nuclear technology and the development of radioactive minerals. Although this has the advantage of avoiding duplication with the Law on Prevention and Control of Radioactive Pollution, it limits the scope of the Nuclear Safety Law. In addition, the content of the Nuclear Safety Law is limited to the safety supervision and management of nuclear facilities, which cannot cover the whole process of nuclear fuel cycle safety supervision. Moreover, in the distribution of responsibilities, the management and supervision responsibilities of the Energy Bureau are only stipulated in principle by the Nuclear Safety Law.

5.3. There is still room for improvement in the nuclear safety supervision mechanism

In terms of supervision methods, there are four main supervision methods in China at present, which are nuclear safety license system, radioactive waste safety management system, radiation protection system and regular safety evaluation system. Among them, the nuclear safety licensing system focuses on setting certain entry barriers by examining the qualifications of these engaged in activities related to nuclear energy utilization, so as to ensure that the nuclear energy utilization practitioners as a whole have higher qualifications and levels, so as to achieve the purpose of preventing and controlling nuclear safety risks from the source. The radioactive waste safety management system focuses on the classified disposal supervision of various radioactive wastes in order to realize the reduction and harmless disposal of radioactive wastes. The radiation protection system focuses on strengthening the supervision of radiation exposure prevention and control responsibilities of nuclear facility operating units, ensuring that nuclear facility operating units implement radiation exposure control responsibilities and ensure that radiation exposure remains at a reasonable, feasible and as low as possible level. The regular safety evaluation system focuses on the long-term dynamic supervision of nuclear facilities in operation, and grasps the safety status of nuclear facilities in operation in time through regular safety evaluation of nuclear facilities, so as to contain risks in the bud.

Under this mechanism, China's nuclear safety supervision has the problem of neglecting nuclear safety cultural construction. China's Nuclear Safety Law has the articles of nuclear safety culture, and assigns the responsibility for the construction of nuclear safety culture to the nuclear safety supervision and administration department, the competent department of nuclear industry and the competent department of energy under the State Council, requiring that nuclear safety culture be integrated into all aspects of production, operation, scientific research and management. The construction of nuclear safety culture is of great significance to improve the safety awareness and quality of nuclear energy practitioners and units. Maintaining a high level of nuclear safety culture is conducive to the long-term healthy development of China's nuclear safety cause. Therefore, the cultivation and construction of nuclear safety culture should be included in the scope of nuclear safety supervision mechanism.

6. Conclusions

The generation and development of civil nuclear power industry are accompanied by danger and potential harmfulness. Therefore, nuclear safety legal system must be professional and technical, and be as perfect as possible. First of all, the promulgation of the Nuclear Safety Law does not mean that the basic laws are useless, and the Atomic Energy Law has its independent significance. Including the lack of the Atomic Energy Law, there are still problems of China's nuclear safety legal system nowadays. So with the development, in the future legal system, the Atomic Energy Law will be an authoritative and guiding basic law, which will guide and standardize the existing laws, regulations and rules. Moreover, the Nuclear Safety Law should clarify the definition of "nuclear safety", which is convenient for specific application in practice. Therefore, legislation should clearly distinguish between nuclear safety and nuclear security. Here, we can learn from the practices of other countries, such as South Korea, which only regulates nuclear safety in its nuclear safety legislation, and matters related to nuclear security are stipulated by other laws. Thirdly, China's nuclear safety supervision also should be improved. Such as the Nuclear Safety Law should further clarify the main body of supervision, clarify responsibilities and it is necessary to construct nuclear safety culture and nuclear security culture all over China.

Funding

This research was funded by Fundamental Research Funds for the Central Universities of Harbin Engineering University, whose grant number is 3072022WK1312.

References

[1] Notice on Issuing the Policy Statement on Nuclear Safety Culture, 15 January 2015, http://www.caea.gov.cn/n6759295/n6759296/c6781703/content.html
[2] Hu Bangda 2019 Research on Safety Principles in Nuclear Law *Beijing: Law Press* 223
[3] Zhang Yunchuan 2017 Explanation on the Nuclear Safety Law of the People's Republic of China (Draft) *China National People's Congress Network*
[4] Zhang Yunchuan 2017 Explanation on the Nuclear Safety Law of the People's Republic of China (Draft) *China National People's Congress Network*
[5] See http://www.gov.cn/zhengce/content/2017-03/23/content_5179622. htm
[6] Zhao Wei 2011 Research on Atomic Energy Legislation *Journal of Law* **10** 16
[7] Carlton Stoiber, Alec Baer, Norbert Pelzer, Wolfram Tonhauser 2003 Handbook on Nuclear Law *IAEA* P101
[8] Chen Gang 2018 Some Aspects of Nuclear Safety Law *Nuclear Safety* 21

IOP Conf. Series: Earth and Environmental Science 1087 (2022) 012024 doi:10.1088/1755-1315/1087/1/012024

An investigation of past and present stormwater management practices within the City of Johannesburg

Qumba B.[1], Nkhonjera G.K.[2,3] and Alowo R.[2]

[1] The University of Johannesburg, School of Civil Engineering and Built Environment, Civil Engineering Science Department, PO Box 524, Auckland Park, 2006, South Africa;
[2] University of Johannesburg, School of Civil Engineering and Built Environment, Civil Engineering Technology Department, PO Box 17011, Doornfontein, 2028. South Africa.

[3] Corresponding author's e-mail: germann@uj.ac.za

Abstract. The City of Johannesburg (CoJ) has implemented a new stormwater design manual which aims at treating stormwater as a resource and to render urban spaces usable for urban environment through design networks that convey stormwater safely to the environment. The Soil and Water Assessment Tool (SWAT), SWATCup, and Rational Method, were used to investigate and assess the past (1886 to 1900) and present (1990 to 2016) stormwater management practices within the City of Johannesburg. The current stormwater management infrastructures were investigated as green and grey, with a focus on the benefits of green infrastructure. The study showed a lower simulated quantity of runoff in the pre-development era and higher simulated quantity of runoff in the post-development era, with varying land use land cover percentages. The post-development time shows that 35% of land use land cover is used for settlements and 65% is for open fields and other uses. However, in the pre-development time, only 10% of this land was used for urban settlement while 90% was dedicated to farmland. Pre-development condition also showed an annual precipitation of 757 mm of which 28% ended up as surface runoff while 45% was lost due to evaporation and transpiration. On the other hand, the post-development time showed 772 mm of annual precipitation with 29% of this precipitation resulting in surface runoff and 50% lost due to evaporation and transpiration. This study also confirms that indeed the LULC changes over the years have affected the quantity of runoff in CoJ. This paper is essential in assisting CoJ in determining the practicality of utilising the new stormwater design manual.

1. Introduction

Stormwater management is increasingly gaining attention and any investigation on the current stormwater practices is essential in order to sustainably manage stormwater and floods. Reduced runoff and improved water quality are the goals of sustainable stormwater management. Area topography, plants, ground conditions and elements of nature that soak-up and sieve stormwater on to the ground all contribute to maintaining natural drainage patterns [1]. Sustainable Stormwater Management (SSWM) objectives and measures are becoming increasingly popular among countries and localities not only in the world but also in South Africa as well. SSWM can be used to meet a number of hydrological, ecological, community-based, and financial goals, and literature [2] adds that it can also be used to replace a significant portion of traditional drainage infrastructure especially in

Johannesburg. According to [3], in developing countries, adopting a sustainable approach to urban stormwater management is challenging due to a lack of data, unrestrained urban growth, a shortage of regulatory tools, a lack of understanding and technical data about infiltration and storage systems, and water quality issues. With increased runoff and water quality degradation, urbanization substantially affects water bodies. While quantity implications may be simple to handle, quality effects are significantly more complicated. Current water quality protection strategies are mostly unsuccessful and led by long-held beliefs, with a primary focus on effluent discharge point remedies [4].

In the City of Johannesburg (CoJ), South Africa, decisions on settlement planning and design that have been made since 1800, have little regard for the physical limitations imposed by the pre-existing natural forms and characteristics on which the city is constructed [5]. Looking at the study of [6], it is clear that CoJ is continually evolving including climate change therefore ancient planning methods may not always be suitable and power disparities in the city are noteworthy, especially where politics, leadership, and development issues are experienced. As if this is not enough, [7] adds that unrestrained flood water and discharge during heavy rainfall events determine the hazards of urban floods in the metropolis, which occur often and are expected to rise in occurrence and size due to climate change. Better stormwater management practices in accordance with worldwide-recommended practices, is therefore needed to avoid additional degradation, ameliorate existing environmental damage, and avoid societal health issues associated with poor water quality [8]. While [9] notes that there is a need to treat stormwater as a water resource, as South Africa is facing water shortages, this comes with its challenges and issues such as political, funding and society acceptance may stand in the way.

These issues can be addressed by a mixture of academic and administrative initiatives aimed at promoting institutional integration and knowledge transfer from the research sector to the general public. The importance of going beyond traditional structural interventions and the critical role that urban stormwater design can play in protecting urban water ecosystems can therefore be highlighted [3, 4]. Different Land Cover and Land Use (LULC) have major effect on the runoff properties of a catchment, which affects the land surface and groundwater availability, leading to more LULC changes [10]. An opportunity to thoroughly investigate watershed study area with regards to streamflow and stormwater runoff can be presented by numerous computer applications such as ArcGIS, SWAT, and ENVI. The approach to investigate pre-development (1886 to 1900) and post-development (1990 to 2016) runoff quantity and LULC, as related to changes in stormwater management systems, can yield different results directly affecting the City's water ecosystems. Therefore, it is against this background that the primary objective of this paper is to investigate the impacts of past and present stormwater practices on the runoff quantity of the City of Johannesburg.

2. Methods and Data

2.1. Study Area

The study area (Figure 1) was the City of Johannesburg central business district (CBD) and its southern catchment in the Vaal River basin. The study area can be located between the latitudes (26.1° and 26.5°) and longitudes (27.7° and 28.5°). The city of Johannesburg itself is situated on a continental ridge called the Hillbrow-Linksfield-Roodepoort ridge, separating the Limpopo River basin to the north and the Orange River basin to the south. This means that rainwater flowing down north ends up in the Indian Ocean while water flowing down south ends up in the Vaal River then into Orange River and eventually into the Atlantic Ocean. However, this study will only look at the part of Johannesburg that is in the Vaal River catchment. The land use within the study area and surrounding areas comprise outdoor recreation, non-essentials such as parklands, transportation (highways, railroads, airstrips), Farmland, Dwellings and Commercial companies and factories [11]. The soil type composition within the area shows 55% sandy loam soil and 45% sandy clay loam soil according to Food and Agriculture Organization (FAO). According to the South African Weather Service data, the study area receives

600 to 800 mm annual precipitation, 750 mm pre-development and 795 mm post-development with the majority of rainfall occurring between December and January of every year.

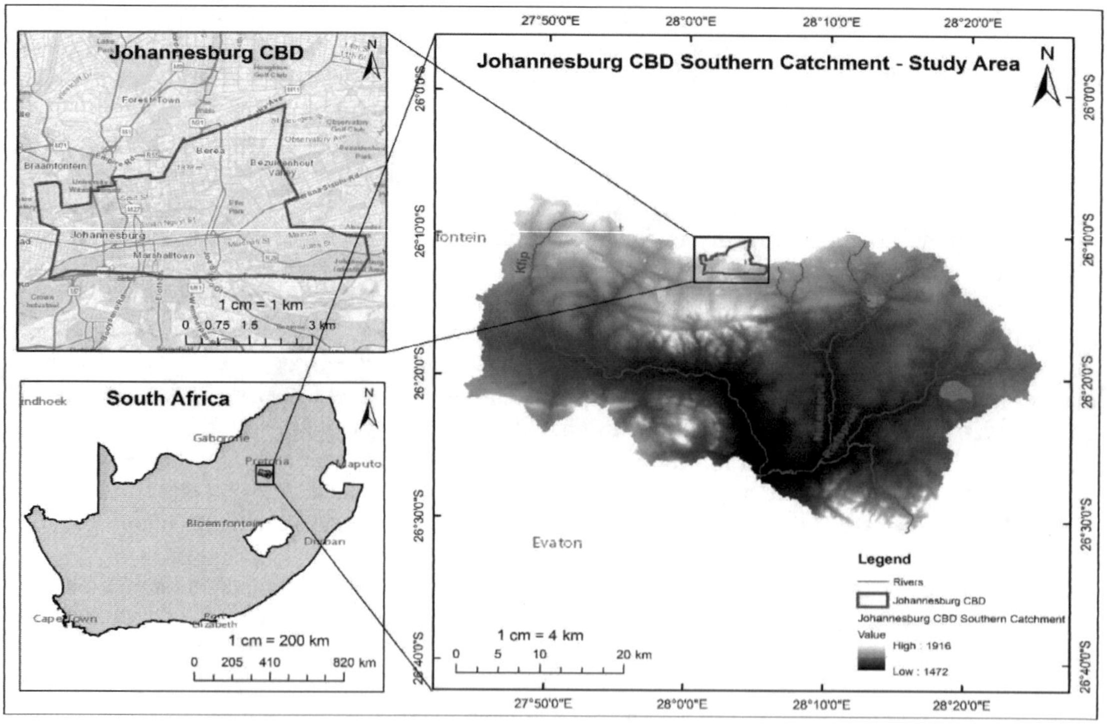

Figure 1. City of Johannesburg CBD and Southern catchment in the Vaal River basin.

2.2. Data

The data used in this study was obtained from credible service providers that are used by researchers worldwide. For example, soil data was obtained from the Food and Agriculture Organisation (FAO), LULC from Department of Environmental Affairs, historical weather data from South African Weather Services, and the digital elevation model (DEM) came from the United States Geological Survey (USGS).

The NASA Shuttle Radar Topography Mission (SRTM) Version 3.0 Global Digital Elevation Model (DEM) used in this work was downloaded from the United States Geological Survey (USGS) database using the Earth Explorer tool, with coverage of 1° x 1° tiles at 1 arc-second (approximately 30 meters) resolution. This includes the satellite images used for color bands (NIR, NDVI & natural color 4,3,2) in order to get maximum likelihood for extrapolated land use and land cover in ENVI software [12]. The ENVI software was used to process the LULC data. The ENVI software uses layerstacks and region of interests (ROI) to determine the maximum likelihood of each ROI in the area. The ROIs are manually selected through the imagery of color bands and give an indication of LULC in the area of interest.

The South African Department of Forestry, Fisheries and Environment Affairs (DEA) provided the LULC maps, namely South African Land Cover Re-projected Universal Transverse Mercator (UTM) 35 South 2013 GTI 35Classes. These were then extracted by the study area watershed shapefile. The LULC maps used were prepared for pre-development and post-development conditions of the study area. The 2013 LULC (DEA) maps were used for the post-development (1990 to 2016) study and band colours or satellite images used in ENVI (image processing and analysis software) maximum likelihood simulation were used to extrapolate and depict pre-development (1886 to 1900) LULC maps. The digitized soil map and data on relevant soil characteristics for the research region were obtained from FAO. The four weather stations located within and surrounding the study area (Table 1) provided the climate data used in this study. There were four major climate/weather variables that were used to inform the SWAT model namely; maximum and minimum daily temperature, and daily precipitation, while solar radiation was simulated by the SWAT model. The post-development weather data, from 1980 to 2020, was provided by South African Weather Services (SAWS), while the pre-development data was extrapolated using Microsoft Office Excel.

Table 1. Weather stations and their locations.

Station Number	Weather Station Name	Location		Elevation (m)
		Latitude	Longitude	
0476044 0	Turffontein	-26.2347	28.0421	1710
0476040 3	Zoological Gardens	-26.1685	28.0376	1651
0476399 0	INT WO	-26.1430	28.2346	1692
0475879 0	JHB BOT TUINE	-26.1566	27.9991	1624

2.3. SWAT Setup and Simulation

Any SWAT model's configuration starts with the watershed's delineation. The watershed is separated into many sub-basins and Hydrologic Response Units (HRU) during this procedure. The DEM, LULC map, and soil map of the research region are all prominently used in the analysis as input data. The DEM was used to construct slope maps for the watersheds. Each slope, land use, and soil map was initially divided into a number of categories. By overlaying slope, land use, and soil layers, the identified watersheds were categorized into HRUs with uniform features. HRUs with small regions were minimized using a threshold level, and such small HRUs were blended in with the surrounding HRUs. A stream network is frequently burned (an ArcSWAT feature that forces generated streams to follow existing stream reaches) in watershed delineation to force generated streams to follow existing stream reaches.

The calibration method was used to fine-tune model-sensitive input parameters identified by the sensitivity analysis. Calibration, according to [13], is the act of selecting a set of parameters to increase the consistency of observed/measured hydrological responses and simulated results. The purpose of model calibration is to determine the link between the simulated results (river water discharge) predicted by the SWAT model and the actual data (river water discharge). For SWAT, a variety of calibration strategies have been developed, including manual calibration processes as well as automated calibration procedures based on the shuffling complex evolution method and other standard methods. SWAT-CUP is a decision-making framework that includes a semi-automated technique (SUFI2) that includes both manual and automated calibration as well as sensitivity and uncertainty analysis. Only automated calibration was used in this research outcome [14]. As a result, output data such as runoff flow, catchment and sub-basins, and area, were analysed using input data acquired in the study region such as DEM, soil map, LULC, meteorological data, and slope to inform the SWAT model. The runoff flow data were then calibrated based on the time periods of pre-development (1886 to 1900) and post-development (1990 to 2016), with the first two years serving as warm up periods for both models to account for unknown initial characteristics.

2.4. Rational Method

This study also used the rational method to calculate the Quantity of runoff using the mentioned parameters while utilizing the South African National Roads Agency Limited (SANRAL) design manual guidelines. The Rational Method describes the link between rainfall intensity and catchment area as independent factors, and peak flood discharge as the dependent variable. The method has been in use for more than 150 years. It was excellent for this study because of its accuracy for small drainage areas less than 809 000 m², available information for this study and the fact that it is used in most drainage manuals in South Africa. In this manual, it is clearly stipulated that to calculate quantity of runoff using the rational method, the following equations (Eq 1 to Eq 7) and parameters should be used. For this study, 20- and 50-year return storm were used.

Peak flow,

$$Q = \frac{CIA}{3.6} \tag{1}$$

Where;

Q = peak flow (m³/s)
C = runoff co-efficient (no units)
I = average rainfall intensity over catchment (mm/hr)
A = effective area of catchment (km²)
3.6 = conversion factor

Average Slope,

$$S_{av} = \frac{H_{0.85L} - H_{0.10L}}{(1000)(0.75L)} \tag{2}$$

Where;

$H_{0.10L}$ = elevation height at 10% of the length of the watercourse (m)
$H_{0.85L}$ = elevation height at 85% of the length of the watercourse (m)
L = length of the watercourse (km)

Time of concentration,

$$Tc = \left(\frac{0.851^2}{1000S_{av}}\right)^{0.385} \tag{3}$$

Where;

Tc = time of concentration (hours)
L = length of the watercourse (km)
S_{av} = average slope (m/m)

Runoff co-efficient rural and urban,

$$C_T = \alpha C_{1T} + \beta C_2 \tag{4}$$

Where;

C_T= total catchment runoff co-efficient rural and urban
C_{1T} = total rural catchment co-efficient in relation to average slope, average soil permeability and average vegetal growth.
C_2 = runoff co-efficient for urban area
α= percentage of rural area on the catchment
β = percentage of urban area on the catchment

Rural catchment co-efficient,

$$C_{1T} = C_s + C_p + C_v \tag{5}$$

Where;

C_s = runoff co-efficient according to average catchment slope
C_p = runoff co-efficient according to average soil permeability
C_v = runoff co-efficient according to average vegetal growth

Point Intensity,

$$P_{it} = \frac{P_t}{T_c} \tag{6}$$

Where;

P_{it} = point intensity for different return periods (mm/hr)
P_t = precipitation depth for a duration of t minutes and a return period of T years in (mm)
Tc = time of concentration (hours)

Rainfall intensity

$$I = \text{Pit}\left(\frac{ARF}{100}\right) \tag{7}$$

Where;

I = rainfall intensity (mm/hr)
Pit = point intensity for different return periods (mm/hr)
ARF = area reduction factors (percentage)

The rational method was used to determine the quantity of surface runoff of pre and post-development; the results were used to conclude, by comparison, on quantity of surface runoff between the two time slices.

3. Results and Discussions

3.1. Pre-development LULC in Johannesburg

Figures 2 and 3 were developed using ArcGIS/SWAT and ENVI (Maximum Likelihood) by assessing pictures of the catchment pre-development and using NASA earth images for types of bands and determining what the land was used for predominantly between 1886 and 1900. For post-development LULC the current LULC was taken from the 2018 LULC shape file provided by the Department of Forestry, Fisheries and Environment and analyzed using ArcGIS. There is a clear contrast between pre-development and post-development land cover. Currently there is about 35% of the study area occupied by urban/settlements and the rest is waterbodies, rural, agricultural purposes, etc. From this study we see a 25% increase of urban/settlements within the area.

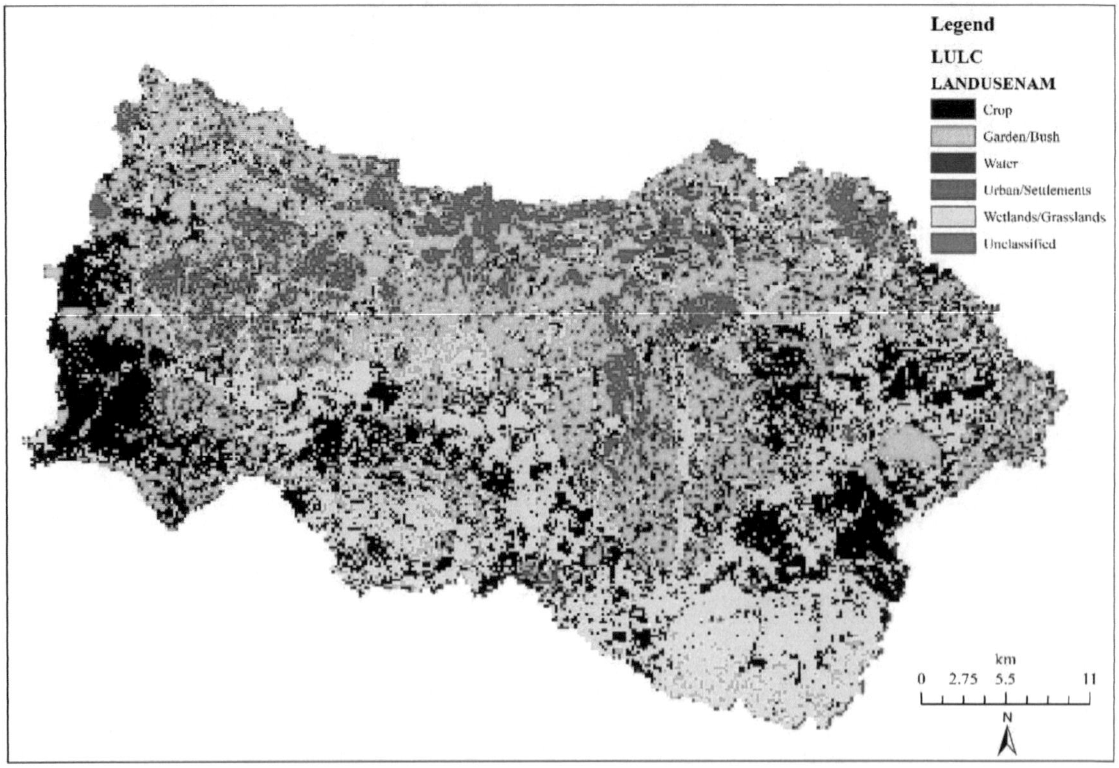

Figure 2. Pre-development Catchment LULC Map.

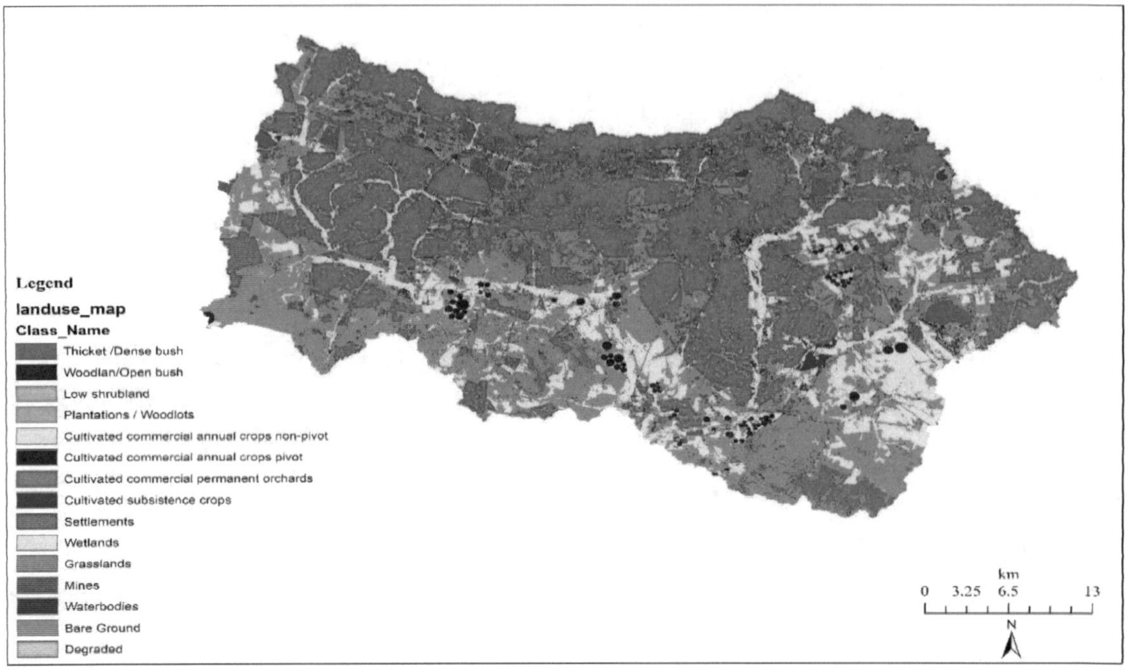

Figure 3. Post-development Catchment LULC Map.

3.2. *SWAT Simulated quantity of runoff*

Considering the Johannesburg CBD area, in the post-development time (1990 to 2016), 35% of land use land cover is used for settlements and 65% is used for open fields and other uses. However, in the pre-development time (1886 to 1900), only 10% of the land was used for urban settlement and 90% was farmland. Pre-development (Figure 4) condition also showed that there was 757 mm annual precipitation in the area, of which 28% of it was lost due to surface runoff and 45% was lost due to evaporation and transpiration. On the other hand, the post-development (Figure 5) time showed 772 mm of annual precipitation and 29% of precipitation was lost due to surface runoff and 50% was lost due to evaporation and transpiration. Figure*s 4 and 5* show that runoff was higher for *post*-development runoff compared to *pre*-development condition. Figure 4A in Appendix A and Figure 5B in Appendix B show that calibrated runoff was low in pre-development and high in post-development with the most increased runoff of flow of 31,38 m³/s pre-development and 62,63 m³/s post-development. For this study, the calibrated runoff was used as bases for realistic values. Figure 6C in Appendix C shows daily observed and simulated flows respectively.

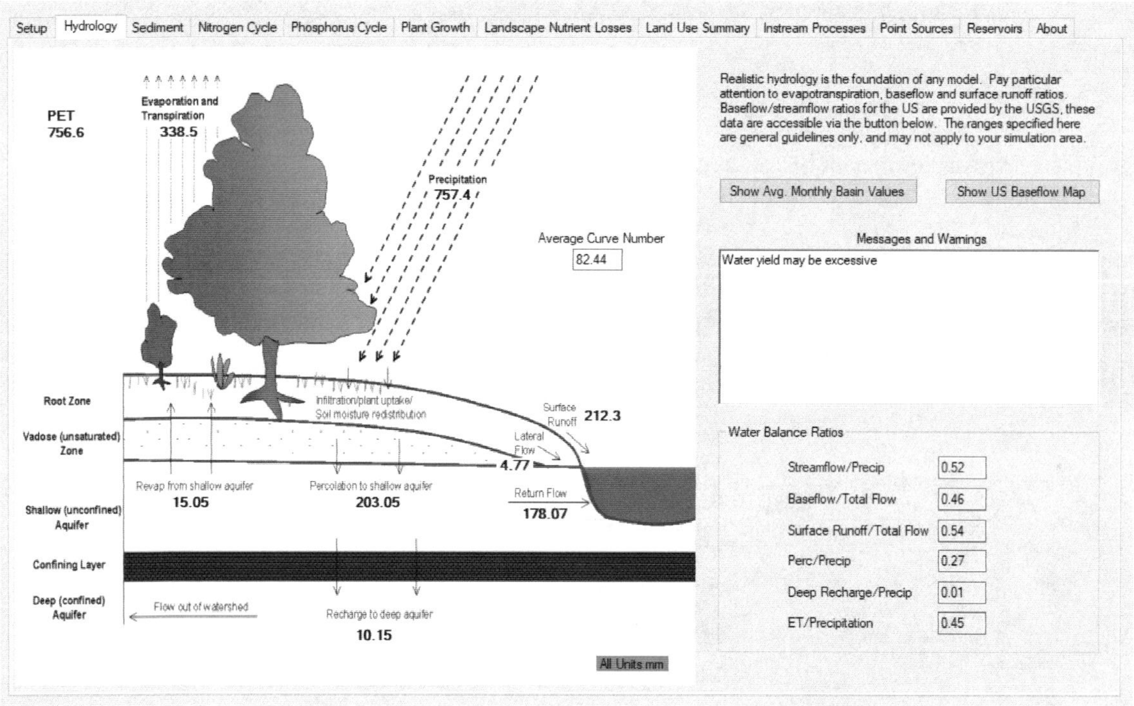

Figure 4. Pre-development hydrological parameters.

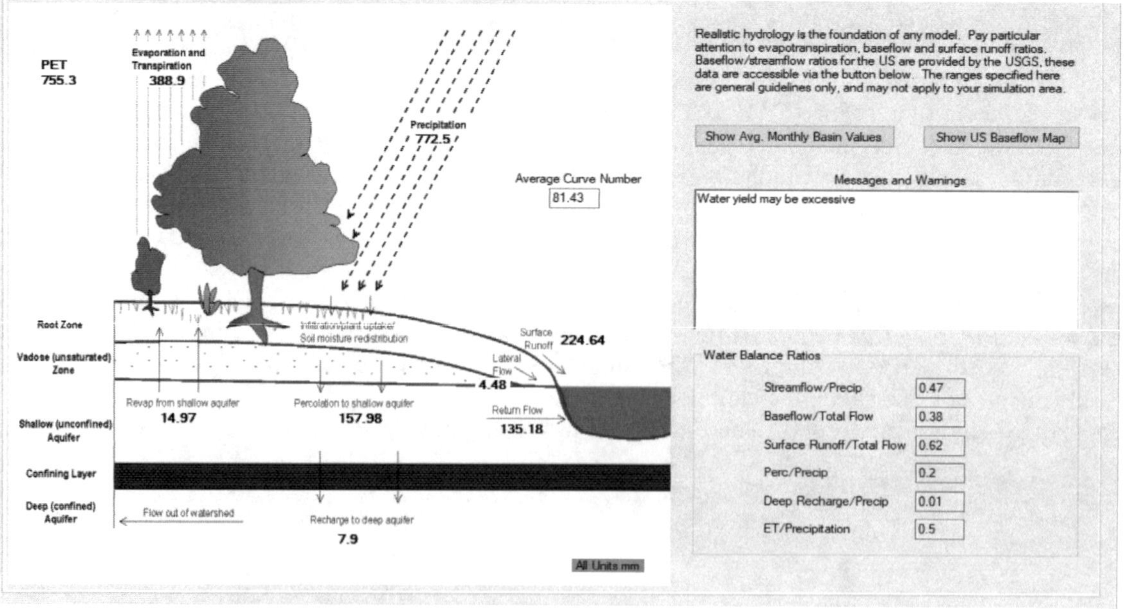

Figure 5. Post-development hydrological parameters.

3.3. Pre- and Post-development quantity of runoff

Figure 6 shows the 20-year as well as 50-year peak discharge rate for pre- and post-development time in each subbasin. The average peak discharge rate for pre-development 20-year return period is 156 m^3/s while that of a 50-year return period is 236 m^3/s. Likewise, the post-development 20-year and 50-year return period peak discharge rate are 232 m^3/s and 346 m^3/s respectively. It is evident that with time the quantity of peak discharge rate has increased relative to changes in LULC where impervious pavements have seen a drastic increase overtime.

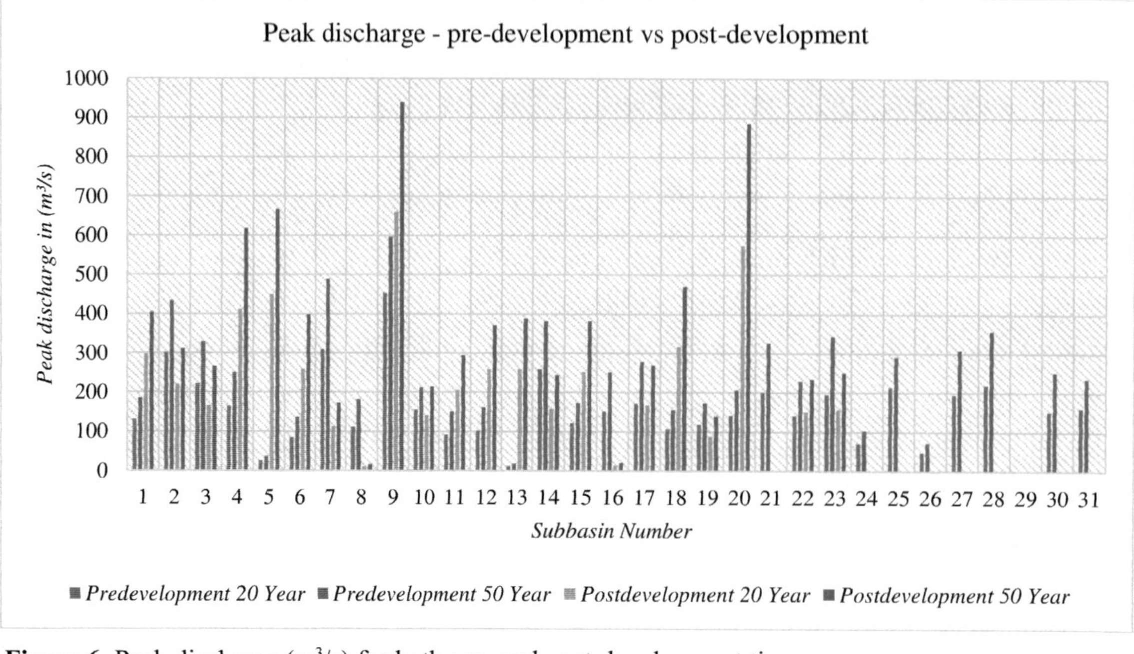

Figure 6. Peak discharge (m^3/s) for both pre- and post-development times.

4. Conclusions

The study is pivotal in highlighting the importance of sustainable stormwater management and its practicality within the City of Johannesburg as the City hopes to achieve sustainable stormwater management through their newly implemented stormwater design manual. The study took an investigative approach to stormwater management within the City of Johannesburg. Tools such as ArcGIS, SWAT (calibrated results), ENVI, and Rational method were used to derive conclusions. Weather data had limitations as data from 1886 to 1900 was not available it had to be extrapolated and LULC for pre-development was determined using earth imagery using ENVI software maximum likelihood estimation. From this study, there is a clear distinction between the pre- and post-development quantity of runoff for all the return periods considered. Currently post-development the city of Johannesburg has high quantities of runoff compared to pre-development and the current design networks do not convey stormwater safely to the environment posing an environmental and property flooding hazard. This paper has also highlighted the need for intervention by the City of Johannesburg officials and its citizens to care for infrastructure while keeping in mind the dangers associated with utilising old conventional stormwater design methods. Reasonable approach to sustainable stormwater management may address some of these challenges.

Acknowledgements

The efforts made by the University of Johannesburg by supporting the start and finishing of this study are acknowledged. We also give many thanks to the City of Johannesburg for allowing this study to be done within their jurisdiction and also to make some of the data available for the study. It is also pivotal to show gratitude to the reviewers for their time taken to study and give useful feedback for the betterment of this study.

References

[1] United States Environmental Protection Agency, "EPA," EPA, 6 August 2021. [Online]. Available: https://www.epa.gov/greeningepa/stormwater-management-practices-epa-facilities. [Accessed 15 January 2022].

[2] S. Goulden, M. E. Portman, N. Carmon and T. Alon-Mozes 2018 From conventional drainage to sustainable stormwater management: Beyond the technical challenges *Journal of Environmental Management* **219** 37-45

[3] J. Goldenfum, R. Tassi, A. Meller, D. Allasia and A. Da Silveira 2007 Challenges for the sustainable urban stormwater management in developing countries : from basic education to technical and institutional issues *NOVATECH 2007* 357-364

[4] A. Goonetilleke, E. Thomas, S. Ginn and D. Gilbert 2005 Understanding the role of land use in urban stormwater quality management *Journal of Environmental Management* **74(1)** 31-42

[5] M. Storie 2014 Changes in the natural landscape, P. Harrison, G. Gotz, A. Todes and C. Wray, *Eds., Johannesburg: Wits University Press* 137-153

[6] C. Vogel, M. Mukute, K. Coetzer and M. September 2021 Gwata Creating a climate of change in the City of Johannesburg: Co-learning to adapt to climate change *South African Journal of Science* **117** 9-10

[7] K. Hetz July 2015C ontesting adaptation synergies: political realities in reconciling climate change adaptation with urban development in Johannesburg, South Africa *Regional Environmental Chang volume* **16** 1171–1182

[8] L. Fisher-Jeffes and N. Armitage 03 July 2013 Charging for stormwater in South Africa *Water SA* **39(3)**

[9] J. Cousins 2018 Remaking stormwater as a resource: Technology, law, and citizenship *WIREs Water* **5(5)**

[10] N. Sajikumar and R. S. Remya 2015 Impact of land cover and land use change on runoff characteristics *Journal of Environmental Management* 460-468

[11] Statistics South Africa, "Stats SA," 2011. [Online]. Available:

http://www.statssa.gov.za/?page_id=4286&id=11306. [Accessed 13 January 2021]
[12] United States Geological Survey, "USGS," 2017. [Online]. Available: https://www.usgs.gov/faqs/what-are-best-landsat-spectral-bands-use-my-research?qt-news_science_products=0#qt-news_science_products. [Accessed 20 July 2021]
[13] S. Surahman, H. Zubair, A. Munir and M. Achmad 2021 Impact of land use change on groundwater flow using SWAT model, study case: Tanralili Sub Watershed *in IOP Conf. Series: Earth and Environmental Science* **807**
[14] J. Arnold, D. Moriasi, P. Gassman, K. Abbaspour, M. White, R. Srinivasan, C. Santhi, R. Harmel, A. van Griensven, M. Van Liew, N. Kannan and M. Jha 2012 SWAT: MODEL USE, CALIBRATION, AND VALIDATION *American Society of Agricultural and Biological Engineers* **55(4)** 1491-1508

IOP Conf. Series: Earth and Environmental Science 1087 (2022) 012024 doi:10.1088/1755-1315/1087/1/012024

Appendices

Appendix A: Observed flow and simulated flow of pre-development time

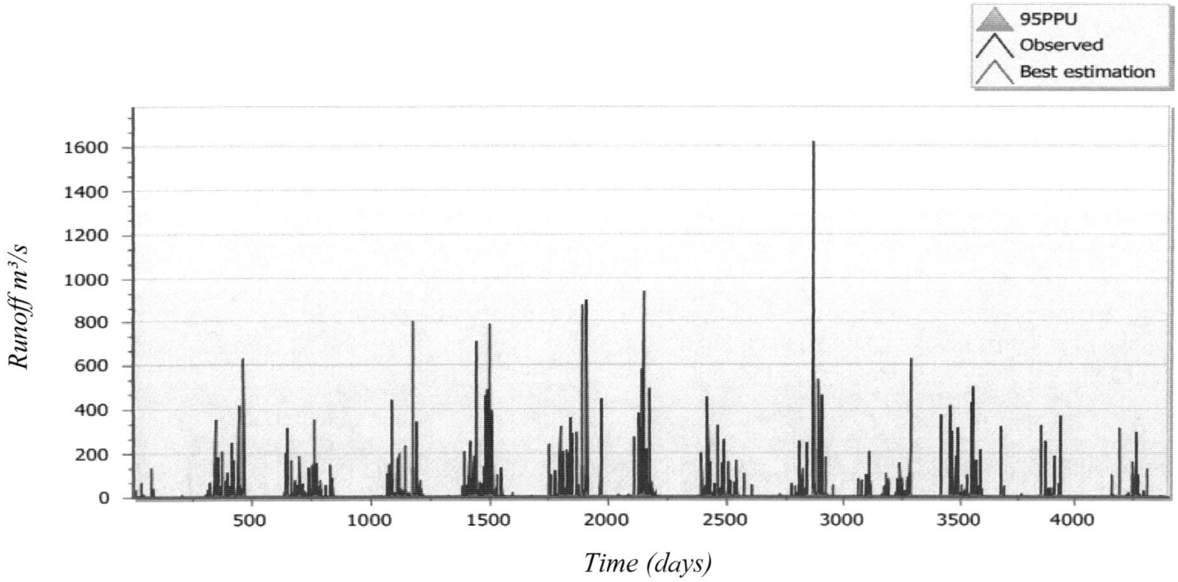

Figure 4A. Pre-development (1888 to 1900) – Observed flow vs Simulated flow.

Appendix B: Observed flow and simulated flow of post-development time

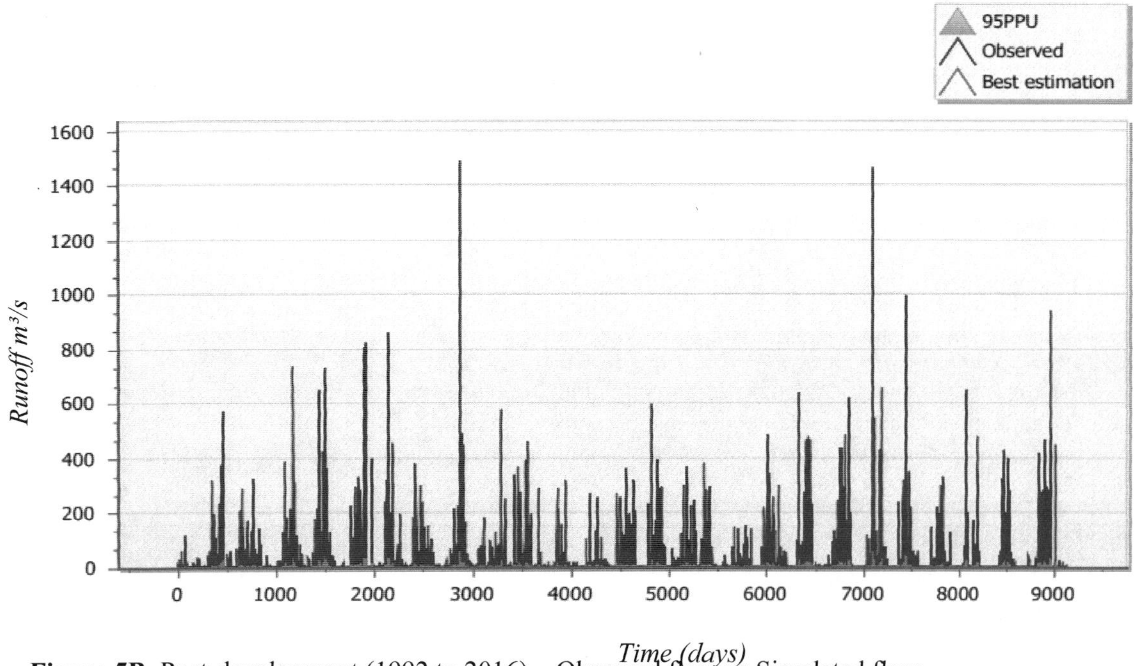

Figure 5B. Post-development (1992 to 2016) – Observed flow vs Simulated flow

Appendix C: Quantity of runoff (pre-development vs post-development times)

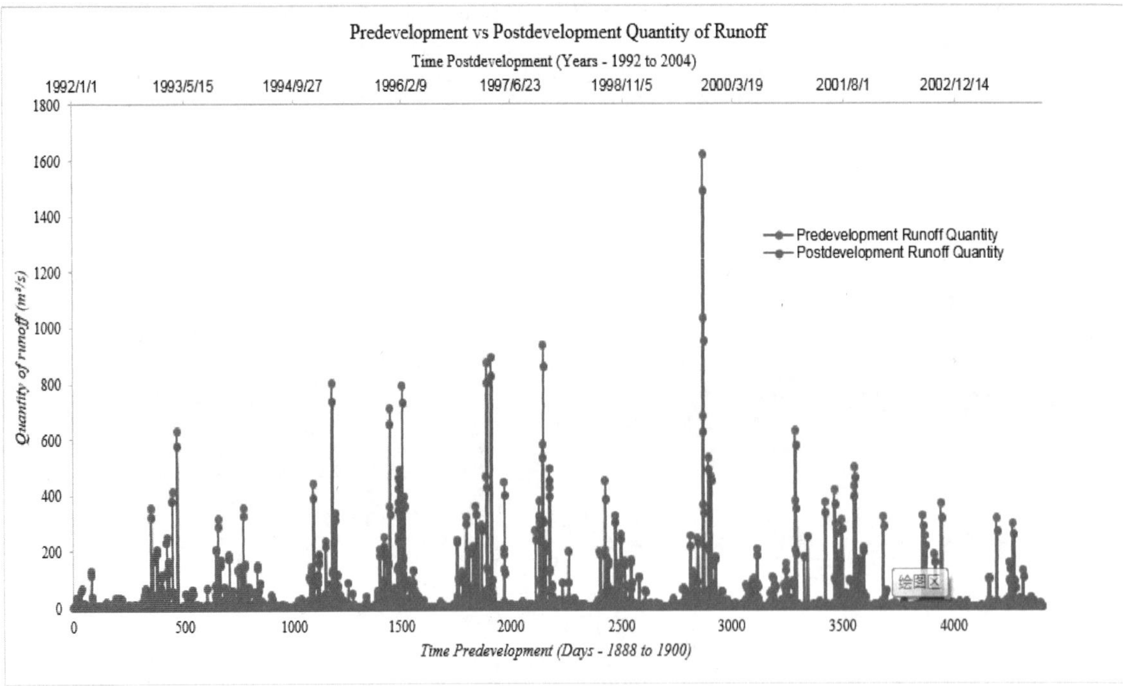

Figure 6C. Pre-development vs post-development quantity of runoff.

IOP Conf. Series: Earth and Environmental Science 1087 (2022) 012025 doi:10.1088/1755-1315/1087/1/012025

A comparative study of evaluating the ability on *Eichhornia crassipes* and *Enydra fluctuans* to remove heavy metals in water in Vietnam

Huynh The An[1,2], Chen Yi-Ching[1,4] and Tran Thi Ngoc Bich[3]

[1] Department of Environmental Engineering, Da-Yeh University, Changhua 51591, Taiwan;
[2] Department of Management Sciences, Thu Dau Mot University, Binh Duong, Vietnam;
[3] Institute of Environmental Sciences and Technology, Tra Vinh University, Tra Vinh, Vietnam.

[4] Corresponding author's e-mail: yiching@mail.dyu.edu.tw

Abstract. Aquatic plants have high effectiveness in removing organic and inorganic contaminants, and heavy metals. *Eichhornia crassipes* and *Enydra fluctuans* are both popular plants in Vietnam, with fast growth and no need to take care of because these two plants easily live in the natural environment. The focus of this study was on the utilization of *Eichhornia crassipes* and *Enydra fluctuans* to treat heavy metals in water. The research was conducted on healthy, young plants from the natural environment. Selection of qualifying plants that match the criteria, such as the absence of pests and the roots were not too old. Concentrations of 0.5 mg/L Cd, 0.5 mg/L As, 2 mg/L Pb, 5 mg/L Zn, and 5 mg/L Cu were placed in five separate pots, each containing 100 g of *Eichhornia crassipes*. Same as above with 100g of *Enydra fluctuans* in each pot. After 30 days, for all heavy metals (Cd, As, Pb, Zn, and Cu), the cleaning rates of the *Eichhornia crassipes* ranged from from 59.4–92.5%. The cleaning rate of *Enydra fluctuans* for all heavy metals ranging from 40.5 – 83.4%. Besides, Plants still grow and develop well in environments with heavy metal pollution. Based on this information, it can be inferred that *Eichhornia crassipes* and *Enydra fluctuans* are aquatic plants with a high potential to remove heavy metals in water and improve the ecological environment.

1. Introduction

The environmental concerns that have arisen as a result of globalization and fast industrialization are becoming extremely nuisance for humans. Heavy metals in wastewater and industrial effluent are a major type of environmental contamination [1]. In particular, heavy metal contamination including lead, cadmium, arsenic, copper, and zinc etc. can cause long-term risks to human health, and the ecosystem. These metals don't degrade naturally and tend to build up in living things. Acute toxic effects usually manifest very rapidly with exposure to heavy metals in certain doses. Some heavy metals become very toxic when they combine with one or more other metals under special environmental conditions. Although the appearance of some heavy metals is required for the growth and existence of living organisms, they will cause negative reactions when the concentration exceeds the allowable limit. Therefore, the application of treatment methods to reduce or remove heavy metals from the environment is necessary and increasingly concerned to protect the living environment and public health [2].

Currently, the research and finding of methods to treat heavy metals in water are suitable, easy to implement, low cost, environmentally friendly, and take advantage of locally available materials that are of interest to scientists in the world and Vietnam. Phytoremediation is an ecological treatment technology recognized as a potential solution to remove pollutants present in wastewater [3].

In the phytoremediation system, aquatic plants play a huge role in protecting water resources, Aquatic plants are species that are highly adaptable to submerged habitats and some of them can treat pollutants in water sources with great efficiency. Algae, floating plants, submerged plants, and emerging plants are the four main class types of aquatic plants. Grouping is based on the positioning of their roots and leaves [4].

The *Eichhornia crassipes* (Water hyacinth) is a fast-growing, plant native to the Amazon Bay in South America [5]. In the late nineteenth and early twentieth centuries, it was introduced into various countries outside of its natural habitat [6]. In its natural form, *Eichhornia crassipes* has the effect of absorbing organic and inorganic substances, etc so it can be used to treat environmental pollution. *Eichhornia crassipes* is used as cattle fodder, mushroom composting, and manure. Dried *Eichhornia crassipes* can make handicraft products [7,8].

Enydra fluctuans (Helencha) is a hydrophytic plant that develops from November to January in wet roadside canals and marshy wastelands. In Vietnam, Southeast Asia, Bangladesh, Malaysia, China, and Africa are all popular [9]. *Enydra fluctuans* is a plant that can grow swiftly, develop quickly, and clean contaminated water. Besides, *Enydra fluctuans* is also a potential medicinal plant used to cure diseases in some countries [9-11].

Eichhornia crassipes and *Enydra fluctuans* are both popular plants, with fast growth and no need to take care of because these two plants easily live in the natural environment. This is the ideal environment for studying how to use plants to treat environmental pollution in general as well as the use of aquatic plants to reduce heavy metal pollution in the water environment particularly.

In many scientific studies in the world and Vietnam, *Eichhornia crassipes* and *Enydra fluctuans* are mentioned as potential plants in the treatment of environmental pollution [12-14]. Most of the research on *Eichhornia crassipes* and *Enydra fluctuans* has focused on the ability of these plants to treat organic matter and nutrients in wastewater, there has been some research on these plants' ability to remove heavy metals, but only for one or two heavy metals. In fact, heavy metal polluted water often exists many types of metals at the same time [12]. Therefore, the study was carried out to compare the ability to treat heavy metals in water of two species: *Eichhornia crassipes* and *Enydra fluctuans* to evaluate the effectiveness of metal removal (Cd, As, Pb, Cu, Zn) in wastewater by aquatic plants.

2. Materials and methods

2.1. Materials

Experimental plants: *Eichhornia crassipes* and *Enydra fluctuans* have been taken from the natural environment. Plants that match certain criteria, such as no insects and roots are 3 months old, are chosen, *Eichhornia crassipes* has a 43 cm body length, and *Enydra fluctuans* has a 35 cm body length. The process of collecting wild specimens was transferred to the laboratory to ensure that the roots were kept intact. Then rinse the roots of the plant with water.

2.2. Methods

2.2.1. Experimentation setup method. The research lasted one month (30 days), and the pots included 100 g of *Eichhornia crassipes* and 100 g of *Enydra fluctuans*. Before planting, the plants were cleaned of dust and soil clinging to the plants with distilled water, stones were used to secure the plants (substrates with a low cost and low absorbency), and pots with 30 liters were used. In experiments, select experimental concentrations were assumed based on the level of content of the threshold that allows the presence of lead, cadmium, arsenic, copper, and zinc in the water environment according to Vietnamese standards QCVN 40:2011/BTNMT. In the experiment, the concentration exceeded the permitted limit by 3 to 5 times. *Eichhornia crassipes* and *Enydra fluctuans* were cultivated with irrigation water containing different amounts of Cd, As, Pb, Zn, Cu at chosen concentrations:

(1) *Eichhornia crassipes*: 0.5 mg/L of Cd (II); 0.5 mg/L of As (III); 2 mg/L of Pb (II); 5 mg/L of Zn (II); 5 mg/L of Cu (II).

(2) *Enydra fluctuans*: 0.5 mg/L of Cd (II); 0.5 mg/L of As (III); 2 mg/L of Pb (II); 5 mg/L of Zn (II); 5 mg/L of Cu (II).

(3) Control sample planted in distilled water.

2.2.2. Methods of sample analysis. After it was planted, the sample was observed and the heavy metal content in the water was measured three times, with time intervals of 10-, 20- and 30- days, to assess heavy metal content in in the treated water with *Eichhornia crassipes* and *Enydra fluctuans* over time. Inductively coupled plasma-mass spectrometry was used to detect heavy in the water.

3. Results and Discussions

3.1. Growth ability of aquatic plants after 30 days of the experiment

Eichhornia crassipes and *Enydra fluctuans* were grown for a period of 30 days to record their ability to survive and thrive in water containing heavy metals. The experimental results are shown in Figure 1. The height change of aquatic plants after 30 days of the experiment.

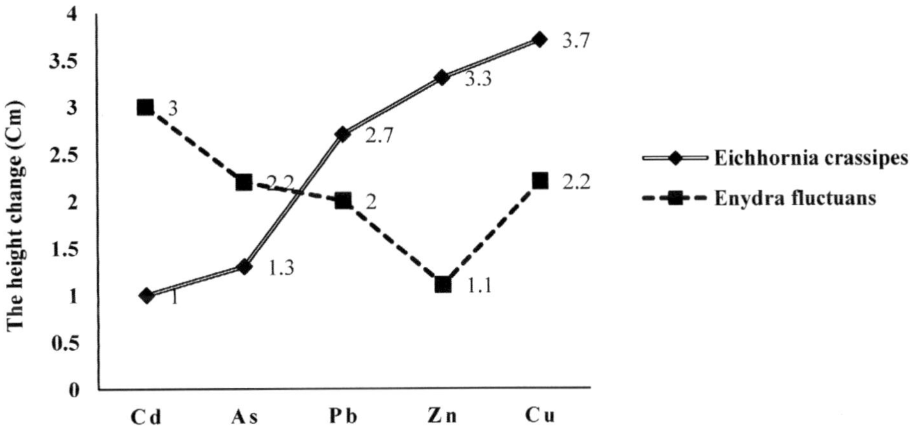

Figure 1. The height change of aquatic plants after 30 days of the experiment.

Figure 1 shows that the pace of growth in height of *Eichhornia crassipes* and *Enydra fluctuates* depending on the plants' absorption capability. The height of *Eichhornia crassipes* rises in step with the heavy metals in the water in the order Cd < As < Pb < Zn < Cu. As for *Enydra* fluctuans, in order the decreasing height corresponds to the heavy metals present in the water Cd > As, Cu > Pb > Zn.

The experiment containing these two species has a clearer color than the original watercolor, perhaps until the 15th day, it will be enough time for the suspended solids to settle to the bottom and enough for the aquatic plants to perform the process absorbs heavy metals in water. *Eichhornia crassipes* and *Enydra fluctuans* grow stably for quite a long time. However, there was a significant difference in *Enydra fluctuans* in water with zinc (Zn) content of 5 mg/L with signs of poor growth, stems, leaves had turned to be yellow. By day 20, the yellowing of leaves and stems became soft in 90% of plants. Towards the end of the experimental period, plants wilted, yellowed and no viability was observed. In which *Eichhornia crassipes* in water with zinc (Zn) content of 5 mg/L quickly grew, rising in size, but some of the leaves had turned yellow and withered in the latter days of the cycle.

Plant height is among the most main factors to consider when evaluating plant growth in a heavy metal absorption environment. Thus, the height of the *Eichhornia crassipes* and *Enydra fluctuans* in each pot experiment was considered to have grown in height. The height growth of aquatic plants, on the other hand, varied depending on the metals. Each plant reacts differently to varying levels of heavy metal exposure. Several plants are vulnerable to heavy metals, whereas others have a high tolerance. Plants have a high tolerance and may continue to grow and develop even when exposed to heavy metals [15].

3.2. The heavy metals accumulation in aquatic plants

Aquatic phytoremediation technology is a resource-efficient and money-efficient method for treating phytoremediation contaminants in a huge contaminated region. The mechanism of aquatic plants is a process by which plants absorb, accumulate, and transport pollutants and heavy metals in water [16]. The most effective and beneficial approach to removing heavy metal pollution, and toxic substances in the aquatic environment is the use of aquatic plants. In order to improve Phytoremediation technology in heavy metal treatment, it is very important to select suitable aquatic plants for heavy metal accumulation [17].

To test whether *Eichhornia crassipes* and *Enydra fluctuans* are potential aquatic plants in the phytoremediation ecological technology for heavy metal treatment [16]. As a result, after plants were planted in each pot, samples were observed, monitored and analyzed for concentrations of heavy metals in the water over 10-, 20-, 30-days intervals. From there, evaluate the ability to handle heavy metals of *Eichhornia crassipes* and *Enydra fluctuans* over time. Table 1 displays the results.

Table 1. Analysis results of heavy metals content in water over time.

Days	Cadmium (mg/L)		Arsenic (mg/L)		Lead (mg/L)		Zinc (mg/L)		Copper (mg/L)	
	EC	EF	EC	EF	EC	EF	EC	EF	EC	EF
0	0.5	0.5	0.5	0.5	2	2	5	5	5	5
10	0.4	0.3	0.4	0.4	1.5	1.6	3.9	4.1	3.9	3.3
20	0.3	0.2	0.3	0.3	0.6	1.3	2.7	3.4	2.5	2.8
30	0.2	0.1	0.2	0.2	0.2	0.8	2.0	3.0	2.0	2.1
QCVN 40:2011/BTNMT (Column B)	0.1		0.1		0.5		3		2	

*EC: *Eichhornia crassipes*; EF: *Enydra fluctuans*

The results show that the amounts of Cd, As, Pb, Zn, Cu in the water reduced with increasing time of treatment with *Eichhornia crassipes* and *Enydra fluctuans* in the experimental process. During the 30-day experiment, heavy metal concentrations (Cd, As, Pb, Zn, Cu) dropped significantly. Based on QCVN 40:2011/BTNMT, the lead content of *Eichhornia crassipes* attained the lowest concentration (0.2 mg/L), it was within the allowable levels when compared to the other elements.

Experiments on growing *Eichhornia crassipes* and *Enydra fluctuans* in water contaminated with heavy metals revealed that these plants may continue to grow and flourish in polluted environments. Thus, with this special adaptability, *Eichhornia crassipes* and *Enydra fluctuans* can not only live in heavy metal polluted environments, but they also accumulate very high Cd, As, Pb, Zn, Cu. The accumulated amount of heavy metals are many times greater than that of normal plants. From the results that we have studied, it is once again confirmed that *Eichhornia crassipes* as well as *Enydra fluctuans* is a super-accumulator of heavy metals.

Previous project-scale trials [18] using *Eichhornia crassipes* to treat contaminated river water resulted in the *Eichhornia crassipes*' ability to remove heavy metals as highly effective as removing up to 62 % Zn, 47% Cd, and 23% As in just seven hours using aquatic plants. The results showed that the heavy metal removal efficiency was very effective in the seven-hour period compared to this study, which achieved only 18.8% Cd and 21.2% As removal in 10 days. However, the very low initial metal concentrations (0.00623 mg/L Cd and 0.00211 mg/L As) were much lower than those used in this study (0.5 mg/L Cd and 0.5 mg/L As) [19]. In addition, *Enydra fluctuans* have significant removal efficiency of Pb, Cu, Zn, As, Mn, Cr, As, as reported in various recent studies [20-21]. A laboratory study at in Cachar district, India showed the ability of *Enydra fluctuans* to remove Cu and Pb from water, specifically, Cu at a rate of 98–99%, after 96 hours. Remove Pb with an average reduction of 98% [22]. The results show that the removal efficiency of copper and lead metal in 96 hours compared to this study, only 34.8% Cu and 19.2 Pb removal efficiency in 10 days. However, the initial concentrations of the two experiments are different, so the results are different.

3.3. Removal efficiency percentage of heavy metals in water by aquatic plants

Eichhornia crassipes and *Enydra fluctuans* were tested for their ability to remove heavy metals from contaminated water. At the moment, the removal efficiency percentage is the most effective and precise calculation. Calculation results are shown in Figure 2.

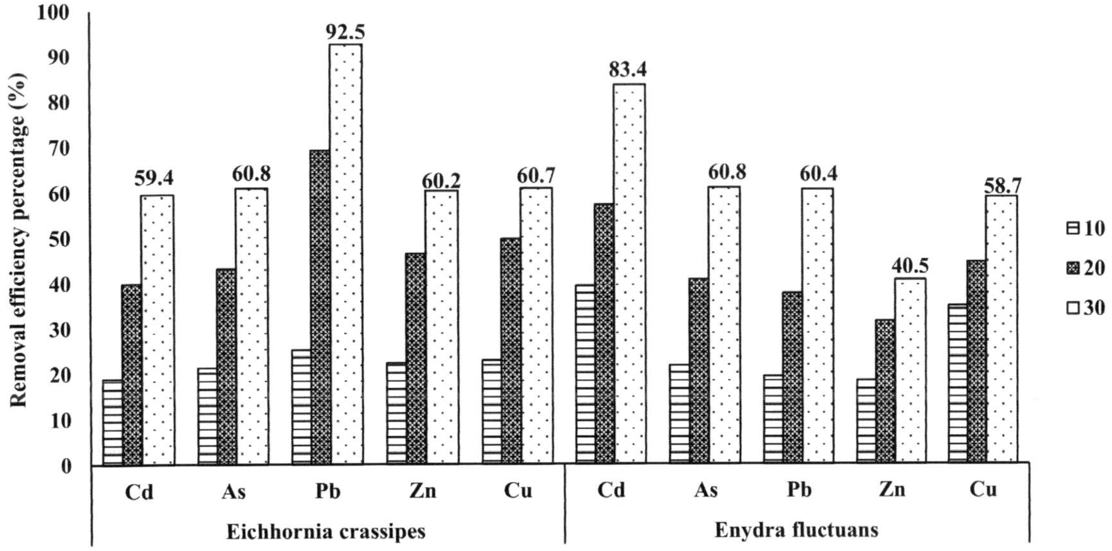

Figure 2. Removal efficiency percentage of heavy metals in water by aquatic plants.

The data show that the treatment rate of heavy metals in *Eichhornia crassipes* were largely high after a 30-days survey period, ranging from 59.4–92.5%. The cleaning rate of *Enydra fluctuans* for all heavy metals ranging from 40.5 – 83.4%. These findings suggest that *Eichhornia crassipes* and *Enydra fluctuans* can show high potential for removal of heavy metals (Cd, As, Pb, Cu, Zn) from the aqueous medium.

4. Conclusions

For the phytoremediation of a huge contaminated region, aquatic phytoremediation technology is a cost-effective and resource-effective technique. Technology to treat heavy metal pollution in natural, environmentally friendly conditions, allowing high efficiency, low cost, and stability. Besides, it also increases the value of biodiversity, improves the local landscape, the ecological environment, and brings high economic potential to the locality.

The experiment of growing *Eichhornia crassipes* and *Enydra fluctuans* for 30 days in water containing high concentrations of heavy metals, but both plants still have the ability to grow and develop well at certain pollution levels. Analysis results of heavy metals: Cd, As, Pb, Zn, Cu in water tended to decrease gradually over time of the experiment. The removal efficiency percentage for high contamination treatment for *Eichhornia crassipes* is from 59.4–92.5%, and for *Enydra fluctuans* it is from 40.5 – 83.4%. Thus, both plant species have the potential to purify heavy metal-contaminated water very well. Beside the function of phytoremediation, *Eichhornia crassipes* and *Enydra fluctuans* can also bring high socio-economic potential in Vietnam.

References

[1] Gunatilake S K 2015 *Methods* **1(1)** 14
[2] Tran L M 2012 Research on the treatment of heavy metals in water with plant-based materials Doctoral dissertation Hanoi University of Science and Technology
[3] Rezania S, Ponraj M, Talaiekhozani A, Mohamad S E, Din M F M, Taib S M, Sairan F M 2015 *Journal of environmental management* **163** 125-133
[4] Dhir B 2013 Phytoremediation: role of aquatic plants in environmental clean-up
[5] Bolenz S, Omran H, Gierschner K 1990 *Biological Wastes* **33(4)** 263-274
[6] Cock M, Day R, Herren H, Hill M P, Julien M H, Neuenschwander P, Ogwang J 2000. *Biocontrol News and Information* **21(1)** 1-8
[7] Zhu L 2007. *Zhejiang University* Zhejiang China
[8] Biswas P 1988 *Indian Veterinary Journal* **65** 496-500
[9] Rahman M 2015 Phytochemical Investigation of methanolic extract of Enhydra fluctuans (Fam-Asteraceae) Doctoral dissertation, East-West University
[10] Ali R, Billah M, Hassan M, Dewan S M R 2013 *Research Journal of Pharmacy and Technology* **6(9)** 927-929
[11] Kona S, Rahman A H M M 2016 Inventory of medicinal plants at Mahadebpur upazila of Naogaon district Bangladesh
[12] Le T T, Nguyen T M *Sci. J. Hong Duc Univ.* **5** 133–142
[13] Tangahu B V, Sheikh A S R, Basri H, Idris M, Anuar N, Mukhlisin M 2011 *International Journal of Chemical Engineering* 2011
[14] Zimmels Y, Kirzhner F, Roitman S 2004 *Water environment research* **76(3)** 220-230
[15] Sumiahadi A, Acar R 2018 *Earth and environmental science* **142** 012023
[16] Pratas J, Paulo C, Favas P J, Venkatachalam P 2014 *Ecological Engineering* **69** 170-176
[17] Ali S, Abbas Z, Rizwan M, Zaheer I E, Yavaş İ, Ünay A, Daim M N A, Jumah M B, Hasanuzzaman M, Kalderis D 2020 *Sustainability* **12(5)** 1927
[18] Jones J L, Jenkins R O, Haris P I 2018 *Sci. Rep.* **8** 1–15
[19] Huynh A T, Chen Y C, Tran B N T 2021 *Processes* **9(10)** 1802
[20] Shuvaeva O V, Belchenko L A, Romanova T E 2013 *International journal of phytoremediation* **15(10)** 979-990
[21] Mahalakshmi R, Sivapragasam C, Vanitha S 2019 *Springer*, Singapore
[22] Parven S, De A, Gupta A 2021 *Research Square* 1-17

Analysis of electric energy substitution potential and research on calculation method of carbon reduction under "carbon neutrality" background

Wei Tang[1,2], Zhilin Liu[1], Qingkun Tan[1] and Yu Zhang[1]

[1] State Grid Energy Research Institute Co., Ltd.

[2] Corresponding author's e-mail: 814197488@qq.com

Abstract. Electric energy substitution is an important means to accelerate energy low-carbon transformation, promote sustainable socio-economic development, and achieve the goal of "carbon peak, carbon neutrality". Firstly, this paper systematically and comprehensively combs the development status of electric energy substitution, and analyzes the outstanding problems existing in the actual work. Secondly, it analyzes the potential space of electric energy substitution, focusing on the impact of "carbon neutrality" on the development of electric energy substitution, and predicts the development trend of electric energy substitution potential. Thirdly, the calculation method for carbon reduction by electric energy substitution is proposed, and the important contribution of electric energy substitution in the implementation of the "carbon peak and carbon neutral" goal is further quantified. Finally, in order to ensure the sustainable development of electric energy substitution, relevant suggestions such as improving the market mechanism are suggested. The viewpoints and conclusions obtained from the analysis can provide theoretical support for the provinces to carry out the "carbon peak" action.

1. Introduction

The utilization efficiency of electric energy is obviously higher than that of fossil energy such as coal and oil [1]. For every 1% increase in electrification level, the energy intensity can be reduced by about 3% [2]. Clean energy can be efficiently utilized by converting it into electric energy, and vigorously promoting electric energy substitution can effectively promote the utilization of clean energy and improve energy efficiency. It is an important means to accelerate the low-carbon transformation of energy and promote the sustainable development of economy and society [3]. The energy security situation requires that China must vigorously promote the implementation of electric energy substitution in the terminal energy sector [3]. With the high-quality development of economy, there is still great potential for the development of electrification level and electric energy substitution in the future. There are many literatures on electric energy substitution, and the literatures before 2020 mainly focus on the calculation of electric energy substitution potential and the comparison of the technical and economic benefits of electric energy substitution, for example, reference [4-5] calculates the substitution potential by different methods, reference [6-7] analyzes the economics of electric energy substitution technology from different perspectives. In September 2020, the Party Central Committee put forward the mission goal of "carbon peak, carbon neutrality", which put forward higher requirements for clean energy and low-carbon transition. According to the goal of reaching 25% of non-fossil energy consumption in 2030, the total installed capacity of new energy sources such as wind power and solar energy will reach 1.2 billion kilowatts [8]. About 90% of China's non-fossil energy is mainly used through conversion into electric energy. Therefore, under the vision of "carbon neutrality", it is an inevitable trend to accelerate the promotion of clean energy supply and electrification of energy consumption. The implementation of Electric Energy Substitution and the promotion of the electrification of terminal energy consumption is an important way to realize the "Carbon peak, Carbon neutrality" goal. As a major influencing factor, the "Carbon peak, Carbon neutrality" target will have a major positive impact on the development of electric energy substitution in the future [8]. The aims of study were to analyse the potential space of electric energy substitution, the effect of carbon reduction and the demand of policy need. Based on the analysis of the energy consumption of subdivided industries under the background of "carbon peak", this paper calculates the potential of electric energy substitution. According to the potential calculation results, the carbon emission reduction replaced by electric energy is calculated by using the carbon emission coefficient. The paper's two methods, logic link is compact, data source is clear, carbon emission reduction quantitative conclusions can support the current work around the carbon reduction.

2. Analysis of current situation of electric energy substitution

During the 13th Five-Year Plan period, China has thoroughly implemented the new energy security strategy of "Four Revolutions and One Cooperation" proposed by General Secretary Xi Jinping, focused on building a clean, low-carbon, safe and efficient energy system, vigorously promoted the substitution of electric energy, carried out clean substitution on the supply side, and replaced coal and oil with electricity on the consumption side. A total of 290,000 electric energy substitution projects have been carried out, replacing 826.3 billion kilowatt hours of electricity. This has promoted an increase in the proportion of electric energy in terminal energy consumption by 2.8%, and has made positive contributions to promoting social energy conservation and emission reduction and improving the ecological environment [3].

From the perspective of alternative policy support, the National Development and Reform Commission, the Energy Administration and other ministries and commissions jointly issued some policies, and a series of local support policies such as environmental protection, subsidies and prices have been issued at the local level. As the electric energy substitution work has entered the deep water area, easily implemented alternative projects have been basically implemented in the early stage, facing problems such as insufficient support policy momentum and technical bottlenecks [9].

The economic problems of partial substitution technology are prominent [10]. First, the initial investment of some projects is large, the primary investment of alternative projects such as electric boilers and electric kilns is high, and the cost recovery cycle is long, resulting in relatively large economic pressure on users. Second, the operation cost of some technologies is high. Compared with coal, gas and other energy sources, the cost of electric energy is relatively high, which affects users' willingness to use. Third, the energy price ratio needs to be optimized. Environmental costs are not reflected in the development and use of energy in China, and the potential advantages of clean, safe and easy control of electric energy cannot be transformed into practical promotion advantages.

Alternative technology bottlenecks need to be broken through [9]. First, there is a lack of innovation in key technologies. The research and development of new technologies, new materials and new products lags behind, and the reserves are insufficient. There are still technological gaps in some alternative fields, and the advancement of technologies needs to be improved. Second, the technical standards and specifications in emerging fields need to be improved, and related standards in some fields are not unified, resulting in low product compatibility.

Policy support needs to be strengthened [10]. First, the current fund subsidy policies are relatively few and unsustainable. In some areas, the subsidy policies of "coal to electricity", port electricity and electric vehicles tend to "retreat", which directly affects the sustainable promotion of key areas of electric energy substitution. Second, policies need to be further optimized and adjusted. For example, some environmental protection policies have been limited effected, and clean heating subsidies need to be optimized and adjusted.

3. Spatial analysis of electric energy substitution

The space of electric energy substitution is usually quantified by substitution potential. The substitution potential can be divided into four levels. Among them, theoretical potential refers to replacing all kerosene consumption for energy consumption with electric energy according to the current situation of coal and oil consumption and the development trend of energy consumption in various industries [3]. Technical feasibility potential refers to the alternative potential obtained by considering the feasibility of the current alternative technology based on the technical level and application status of electric energy substitution in China. The feasible potential of environmental protection refers to the alternative potential obtained by considering the treatment cost of environmental pollutants and regional environmental protection requirements. Economically viable potential refers to the potential electricity quantity that is technically feasible and can be replaced under current price and policy conditions.

There are many factors affecting the substitution of electric energy, including energy policy, international trade disputes, macroeconomic operation and so on. The replacement potential of electric energy changes dynamically with the change of technology, policy environment and economic development. On the one hand, with the deepening of the replacement process, the replacement of stock shows a gradual decline, and the replacement of surplus stock becomes more difficult. On the other hand, with technological progress and other factors, the potential may also increase. In addition, there are a variety of alternative devices, and a technology may have different characteristics and advantages in different regions, so the replacement potential varies greatly.

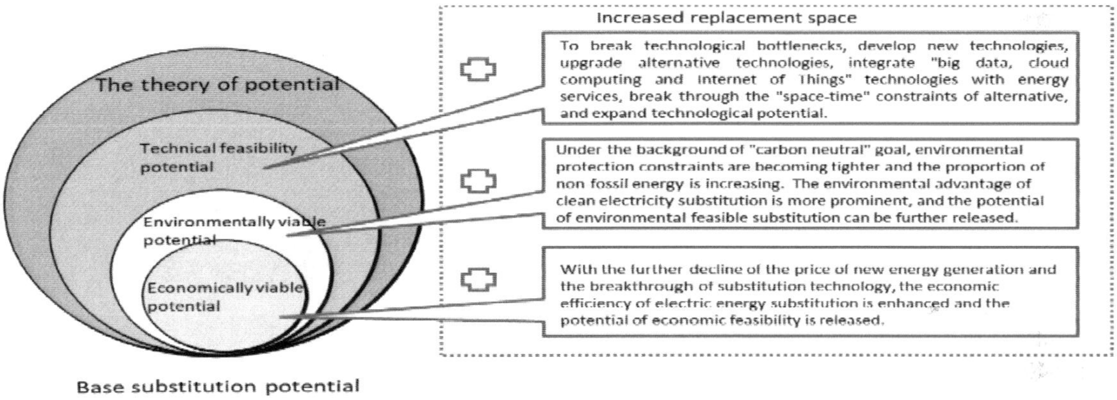

Figure 1. Sketch map of potential expansion space for electric energy substitution.

The "Carbon peak, Carbon neutrality" target has a great impact on electric energy substitution, mainly in two aspects. First, it is policy support. Environmental restriction policy is an important means to ensure the smooth progress of electric energy substitution. Under the background of the "Carbon peak, Carbon neutrality" target, ministries and commissions such as the Ministry of Ecology and Environment actively make arrangements. Governments at all levels will implement strict carbon constraints, decompose tasks layer by layer, and transfer pressure to industries and enterprises. As an effective tool for low-carbon transformation, electric energy substitution will be encouraged and supported by policies. Second, it is technological progress. Under the background of "Carbon peak, Carbon neutrality", the investment in relevant technologies will be increased, research and development strength will be strengthened, and some clean technologies will become cost-effective and permeability will continue to increase. Breakthroughs in partial electric energy substitution technologies have been accelerated, and the application of new technologies and models has increased. In the future, energy transformation will accelerate the development of clean source industry with new energy as the core, terminal electrification industry with power consumption as the core and power grid equipment manufacturing industry [10]. The field of electric energy substitution technology will also be further expanded. At the same time, technological breakthroughs and large-scale application will bring economic improvements. Under the background of "carbon neutrality", the potential of electric energy substitution is technically feasible, environmentally feasible and economically feasible.

According to the data of energy Statistical Yearbook released in October 2020 (energy data is the data of 2018), the initial value of alternative theoretical potential is determined, and the benchmark alternative potential at the end of 2020 is determined after excluding the parts that have been replaced from 2018 to 2020. All kinds of influencing factors in the planning period are analyzed to determine the incremental replacement power and the storage replacement power, and the replacement power in the planning period is obtained. With technological progress, the tighter environmental protection constraints and the enhancement of clean power economy, the stock potential of electric energy replacement are further released, and the incremental space potential is increased. The 14th Five-year-Plan framework for spatial calculation of potential electric energy substitution is prented in Figure 2. During the 14th Five-Year Plan period, high-efficiency electric to steam and electric furnace steel were applied on a large scale in the industrial field, technologies such as high-efficiency electric refrigeration / heating, clean electric heating and high-efficiency smart home appliances in the construction field were gradually popularized, and the penetration of electric vehicles, electric ships and other electric vehicles in the transportation field was accelerated, Green hydrogen will be gradually applied in transportation fields such as fuel cell electric vehicles and industrial fields such as hydrogen energy steelmaking and green hydrogen chemical industry.

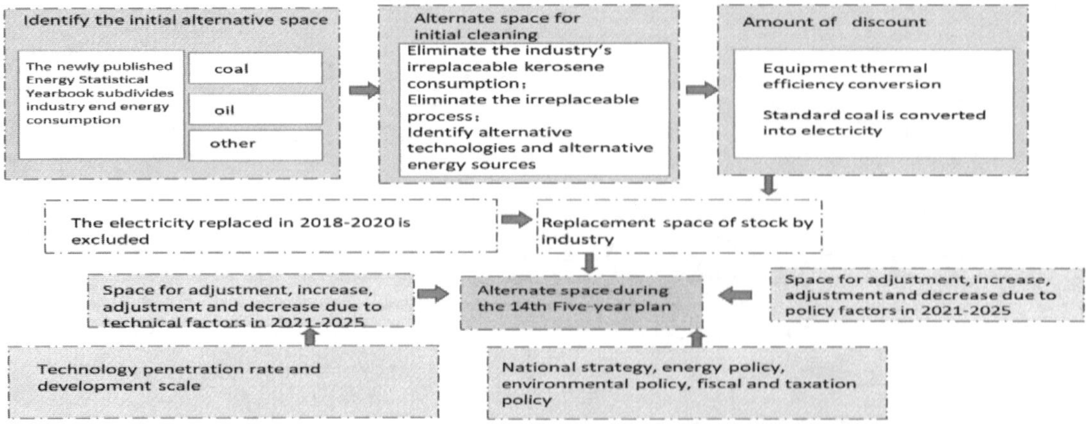

Figure 2. 14th Five-Year Plan" framework for spatial calculation of potential electric energy substitution.

Taking technology, policy and other factors into consideration, the electric energy replacement potential of key fields or industries such as industry, construction, transportation, rural agriculture is predicted.In the industrial field, the main calculation reference data in the construction field is the newly added electric heating area, and the main reference data in the rural agricultural field is the electrification technology of agricultural production (such as electric drying, electric drainage and irrigation, electrified planting and breeding technology) penetration rate, etc. It is estimated that during the 14th Five-Year Plan period, the national electric energy substitution power will exceed 700 billion KWh, of which the electric energy substitution in the business area of the State Grid Corporation of China can exceed 600 billion KWh [3].

4. Analysis of carbon reduction effect of electric energy substitution

At present, the concept that electric energy substitution can reduce carbon has been accepted by the public, but the quantitative methods for reducing carbon emission has not been unified. The data of electricity substitution emission reduction published by some websites are calculated by using the coal consumption method of power generation, that is, how much coal is consumed by the substitution electricity, and the carbon emission of these coal is taken as the carbon reduction value. This calculation assumes that all electric energy substitution is clean substitution, which currently only partly comes from clean electricity. In the context of "carbon neutrality", the whole society and industry are in the carbon reduction system. From a systematic perspective, the whole process of carbon reduction needs to place electric energy substitution as a carbon reduction measure in the overall carbon reduction system to truly represent its carbon reduction contribution.

From the power supply side, decarbonization technologies are being applied continuously in the power generation industry, while more and more electricity is coming from zero-carbon energy sources (clean energy substitutes for fossil energy) [8]. From the perspective of energy consumption terminal, the implementation of electric energy substitution project reduces the terminal's direct use of fossil energy such as coal, oil and gas, and the direct carbon emission of the terminal energy consumption sector, so that part of the carbon emission is transferred from the terminal energy consumption link to electricity. The power system not only bears the pressure of carbon emission reduction transferred by other industries, but also faces the challenge of new energy grid integration on a larger scale. However, the obvious carbon reduction effect of electric energy substitution is still an important means for energy utilization to gradually move towards zero-carbon power supply and zero-carbonization of terminal energy consumption. Through the substitution of electric energy, other industries are strongly supported to achieve a significant reduction in terminal energy carbon emissions and boost the realization of the goal of "Carbon peak, Carbon neutrality".

The main substitution objects of electric energy substitution are coal and oil. Natural gas is not considered in this method for the time being. If relevant departments explicitly include natural gas in the substitution category, the calculation method is similar to kerosene. The original carbon emission calculation of the replaced coal and oil is shown in a formula below (1).

$$p1 = C * Z_c * P_C + O * Z_o * P_o \tag{1}$$

Where, $p1$ is the total carbon emissions corresponding to coal and oil before substitution. C Is the physical consumption of coal, Z_c is the conversion coefficient of coal from physical quantity to standard coal, O is the physical consumption of oil, Z_o is the conversion coefficient of oil from physical quantity to standard coal, P_C is the carbon emission factor of coal, P_O is the carbon emission factor of oil.

In the actual substitution process, the more accurate measurement is only the measurement of electricity (from the meter measurement), while the measurement of coal and oil may be roughly estimated or missing. Therefore, in order to verify the amount of coal and oil substituted in electric energy substitution, the total value of standard coal can be calculated first by the electrothermal equivalent method. Firstly, the total value of standard coal can be calculated.

$$C_b = T * Z_t / U_t \tag{2}$$

Where: C_b is the amount of standard coal, T is the substitution of electric energy for electric energy, Z_t is the conversion coefficient of electric energy into standard coal, and U_t is the thermal efficiency of electricity consumption.

If the physical quantities of coal and oil are further estimated, electric energy replacement projects should be divided into two types: "electricity for coal" and "electricity for oil", and the thermal efficiency of coal and oil should be determined respectively.

$$C = ((T_c * Z_t / U_t) / U_c) / Z_c \tag{3}$$

Where, T_c is the electric energy replacing coal with electricity, and U_c is the efficiency of coal burning.

$$O = ((T_o * Z_t / U_t) / U_o) / Z_o \tag{4}$$

Where, T_o is the electric energy substituting electricity for oil instead of electricity, and U_o is fuel efficiency.

Since the carbon emission of clean energy power generation can be identified as zero, and the carbon emission of power generation mainly comes from the carbon emission of thermal power, the calculation formula for calculating the carbon emission substituted by electric energy is shown below (5).

$$p2 = T * A * P_h \quad (5)$$

Where: $p2$ is the carbon emission of electric energy replacing electric quantity, $p2$ is electric energy replacing electric quantity, A is the power generation proportion of fossil energy in electric energy replacing project, P_h is the carbon emission coefficient of thermal power.

The calculation formula of carbon reduction is as follows.

$$p = p1 - p2 \quad (6)$$

Under the background of "Carbon peak, Carbon neutrality" goal, the penetration rate of decarbonization technology in the power generation industry is increasing, and the proportion of clean energy power generation is increasing, and the carbon emission of electric energy replacing electricity will be smaller and smaller. In the future, in order to reduce the replacement cost of electric energy and absorb new energy, more electric energy replacement projects will participate in the direct transaction of wind and light power in the trough. The estimated proportion of clean energy generation should be higher, and the corresponding carbon emission reduction effect of electric energy replacement will be more obvious [11].

As a matter of fact, the replacement capacity of electric energy in some areas with high energy consumption and high emissions is limited. Hydrogen energy, as a kind of chemical energy carrier, is expected to fill the shortage of electric energy in the long term and help the energy consumption side to decarbonize deeply. In the future, with the decrease of the cost of renewable energy generation, the efficiency and economy of electric hydrogen production technology will gradually improve. It is expected that after 2030, electric hydrogen production will gradually enter the stage of large-scale promotion [12]. However, at present, the implementation scale of electric hydrogen production is relatively small, and the carbon reduction effect has not been fully played. After the 14th Five-Year Plan, with the large-scale application of electric hydrogen production, electric hydrogen production will be included in the category of electric energy replacement, and the calculation method can be further optimized.

In 2020, China's installed capacity of coal-fired power accounts for 49.1% of the total installed capacity [8]. According to the trend of clean energy generation on the power supply side and the absorption and integration trend of electric energy substitution and renewable energy, it is tentatively determined that fossil energy will account for 50% of electric energy substitution during the 14th Five-Year Plan period. According to relevant references, the thermal efficiency of coal and fuel oil is determined to be between 0.3 and 0.5, and the electric thermal efficiency is 0.95. According to IPCC 2006 National Greenhouse Gas Inventory Guide 2019 Revision, literature and other relevant literature and international standard, carbon emission parameters are set. The provisional carbon emission factors of coal and oil are 2.77 tons of carbon dioxide/ton of standard coal and 2.15 tons of carbon dioxide/ton of standard coal respectively. Thermal power emission factor is 823g CO_2/KWh. Reference coefficients of various energy sources are derived from the appendix of China Energy Statistical Yearbook. In the previous chapter, it is estimated that 600 billion KWh of electric power will be replaced by electricity in the operation areas of State Grid Corporation during the 14th Five-Year Plan period. In agriculture and transportation, oil will be replaced by electricity, while coal will be replaced by electricity in other areas. According to the rough estimation of kerosene thermal efficiency of 40%, the replacement of coal and oil is about 123 million tons and 61 million tons respectively. According to formula (1), it is estimated that the carbon dioxide emission corresponding to kerosene is 340 million tons and 130 million tons respectively. According to the above calculation method, it is concluded that the carbon dioxide emission of other industries can be reduced by replacing the contribution of electric energy by 470 million tons (250 million tons of which will be transferred to the power generation industry). Due to differences in parameter values or settings, there will be deviations in the calculated carbon reduction value, but the trend of replacing carbon reduction by electric energy remains unchanged.

5. Conclusions

Electric energy substitution is an important means to achieve the goal of "Carbon peak, Carbon neutrality". This paper analyzes the current situation of electric energy substitution, and analyzes the potential space of electric energy substitution and carbon reduction effect. The proposed estimation results of alternative potential space can provide support for distribution network planning of power grid companies, and the carbon reduction measurement method can provide support for local governments to determine the implementation plan of "Carbon peak, Carbon neutrality".

Acknowledgments

This work was supported by the State Grid science and technology projects under Grant 5400-202111493A-0-5-ZN. **(**Research on quantitative analysis and aggregation simulation technology of emission reduction effect of electric energy substitution in key sectors facing urban "carbon neutralization**)**.

References

[1] Wei Yiming, Liao Hua 2014 China Energy Report[M] *Beijing: Science Press*

[2] Zheng Peng, Wu Kaibin, Yue Mengmeng 2021 Applicability analysis and promotion measures of electric energy substitution technology in Fujian Province[J] *Energy and energy conservation* **(01)** 109-111

[3] Wei Tang, Chao Xu, Baoguo Shan, Yu Zhang, Peng Wu, Yuelong Jia, Chuncheng Zhang, Qingkun Tan Electric energy Substitution planning based on Energy Structure Characteristics of Subsector Industry[J] *China Electric Power* **54(07)** 36-45

[4] Shan Baoguo, Zhao Jia, Jia Dexiang, Cao Fang, Sun Yi 2018Analysis method of electric energy substitution potential based on STIRPAT ridge regression[J] *Power supply and consumption* **(01)**

[5] Sun Yi, Zhou Shuang, Shan Baoguo, Jia Dexiang, Cao Fang 2017 Analysis on the potential of electric energy substitution under multiple scenarios[J] *Power grid technology* **(01)**

[6] Yuan Xinrun, Wu Liang, Zhang Jian, Xiang Tianchun 2015 Electric energy substitution situation and economic analysis of electric heating in Tianjin[J] *DSM* **(05)**

[7] Zeng Huan 2017 Technical and economic analysis and Policy Research on electric energy substitution in rural areas of Beijing Tianjin Hebei[D] *North China Electric Power University*

[8] Wei Tang . Development trend analysis of energy and electric power in the 14th Five-year Plan under the background of carbon neutrality *Oil & Gas & New Energy* **33(02)** 13-17.

[9] Cheng Ling 2020-12-22 To push forward the electric energy substitution in many fields, it is necessary to continuously tackle key problems and new technologies to improve efficiency and benefit[n] *State Grid news* **(008)**

[10] Tang Wei 2020 Prediction of current situation and development trend of electric energy substitution [J] *China energy* **42 (02)** 35-38

[11] Shu Yinbiao, Zhang Liying, Zhang Yunzhou, Wang Yaohua, Lu gang, Yuan Bo, Xia Peng 2021-12-22 Study on carbon peak and carbon neutral path of electric power in China. *Chinese Engineering Sciences* 1-14. HTTP: Kns Ki. netkcmscdetail 11.4421.G3.2021117.1003.002. HTML

[12] Zhang Yunzhou, Lu Gang, Wang Peng, Weng Yuyan, Wu Shengyu, Liu Jun 2020 Analysis on the improvement path of energy cleanliness rate and terminal electrification rate under new energy security strategy[J] *China electric power* **53(02)** 1-8

IOP Conf. Series: Earth and Environmental Science 1087 (2022) 012027 doi:10.1088/1755-1315/1087/1/012027

Research on the measurement of contribution value of scientific and technological talents in China's mining industry——Innovative application based on "Cobb Douglas production function method"

Tianheng Xie[1] and Xueyi Zhu [1,2,3]

[1] Nantong Institute of Technology, Nantong, Jiangsu 226002;
[2] China University of Mining and Technology, Xuzhou, Jiangsu 221116.

[3] Corresponding author's e-mail: xyzhu@cumt.edu.cn

Abstract. This paper aims to determine the contribution value of scientific and technological talents from Mining to national economic development, with data collecting on the correlations between the national Mining and the national economic development from 2005 to 2019, comprising the added value, the capital formation and the total wages and growth rate of the Mining. Through the Cobb-Douglas production function method, the contribution index of Mining talents was confirmed. Combining this index with high-tech talents, the contribution value of Mining science and technology talents was calculated and determined by using big date and informatization method. The results show that in 2005, the contribution value of scientific and technological talents in China's mining industry in enterprise scientific and technological research and development was 2288.4 yuan / person year, which increased to 3821.1 yuan / person year in 2018, with an average annual increase of 4.01%, 16 years average 7011.1 yuan / person year. This value is much lower than the contribution of capital, and its output elasticity (index) is only 7.42% of capital. However, this is the result of the scientific and technological talents in the mining industry with the value of selfless dedication. It is the real motivation and purpose of the author to write this paper: appeal to the state, society and enterprises to increase the scientific and technological investment in the mining industry and constantly improve the conditions and treatment of scientific and technological talents in the mining industry.

1. Introduction

China's mining industry is composed of five industries: mining and washing of coal, extraction of petroleum and natural gas, mining and processing of ferrous metal ores, mining and processing of non-ferrous metal ores, mining and processing of non- metal ores. It is an extensive production unit in China. China's extensive production units mainly rely on capital investment and strong labor input, and their production and operation and the technical content of their products are insufficient. However, the Chinese government and the market continue to guide mining enterprises to increase investment in science and technology, especially to give full play to the role of scientific and technological talents. The writing motivation and purpose of this paper is to analyze how the role of scientific and technological talents in extensive units such as China's mining industry is played? At the same time, in the previous research process, the role of scientific and technological talents is often reflected by the index of "labor force", which can not truly reflect the contribution of scientific and technological talents.

So, this paper discusses the contribution of scientific and technological talents in China's mining industry to scientific and technological research and development from the perspective of value indicators.

2. Research on the Performance of Scientific and Technological Talents in Mining

On CNKI.cn, the research results on the contribution performance of scientific and technological talents in Mining mainly involves two aspects: research content and research methods.

The first point is to study the contribution performance of talents as well as research results of other scientific research performance. Through studying the level of scientific research, scientific innovation ability and the situation of the combination of production study and research in Mining in Henan Province, Ye Jing (2017) constructed the evaluation system of scientific and technological innovation performance, including scientific and technological talents, funds for scientific research projects, papers and monographs, cooperation and transformation of production, education and research and intellectual property rights [1]. With the applying of data of 2006-2016 Compilation of Science and Technology Statistics of China's mining industry and in reference to the 2015 scientific research efficiency level of Mining in 12 Provinces in the west of China, Fang Lijuan et al. (2018) managed to evaluate of the scientific research efficiency level of Mining in Ningxia Province. The measurements include: first, scientific research investment: Scientific researchers, research and development staffs; second, Patents for inventions and sales revenue of new products [2]. Taking Beijing Mining as an example, Yan Jian et al. (2016) designed measurements such as science and technology funds, investment on science and technology manpower, rewards for achievement, science and technology subjects, ownership and authorization of patents, licensing and transferring of patents, etc. The data of input and output of science and technology from 2004 to 2013 in 15 Mining in Beijing were analyzed through empirical analysis. The input and output constructure of Beijing has been deemed reasonable. Scientific and technology achievements is related to yearly total input, number of scientific and technological personnel and so on in Chinese [3].

The second point is to study the contribution performance of talents through the same evaluation method adopted by other research achievements. In addition to the clustering method adopted by above-mentioned Yan Jian, connecting the S&T innovation during the period of the Twelfth Five-Year, Wang Xiaomeng (2018) finished the research of her own performance (including talent contribution performance) in Chinese [4]. Through data 2011-2014 studying and the using of fixed-effect model, Meng Fanrong et al.(2017) compared the performance of scientific and technological innovation among enterprises, Mining and scientific research institutions, and it has been confirmed that the quantity of scientific and technological human resources of Mining and scientific research institutions had a major significant effect on scientific and technological innovation: the contribution from enterprises is mainly reflected on the competition with each other; The contribution from Mining is mainly embodied as the inclusiveness of other research contents in Chinese [5] . Wu Dandan et al. (2016) conducted empirical analysis on the performance of scientific and technological innovation in Mining by using nonlinear regression analysis based on the scale independent zone in Chinese. It is believed that the changes of regional GDP, input and output of scientific and technological innovation in Mining have significant non-linear effects on the performance of scientific and technological innovation in Mining in Chinese [6].

In their study, Varrichio et al. (2012) emphasise the role of synergy in future innovation, i.e. that the future trend in innovation management is to promote and improve performance through the coordination of linkages between research institutions, production sectors and government [7]. Sadegh Rast (2013) used a collaborative approach to assess the assignment of perceptual parameters in five types of Mining and research institutions in Malaysia to derive a preliminary framework for influencing performance across parameters [8].

Although the above research results are innovative in multiple aspects, there are still three deficiencies: firstly, the research only pays attention to data of researchers involved in this research, or "full-time equivalent" data, but ignores the value attribute of researchers; secondly, it did not integrate foreign research methods on this subject; thirdly, measurements were selected only from Mining in a limited number of provinces or regions, did not draw an overall picture. This paper aims to evaluate the contribution value of scientific and technological talents in Chinese Mining by studying all the relevant data of Chinese Mining through the "Cobb-Douglas production function method".

3. Confirmation of mining talent contribution and scientific research contribution value

3.1. The determination of contribution rate of mining talents

The scientific research talents in Mining are the talents in Mining. The contribution rate of mining talents is contribution from the human resource in Mining, and normally adopt the Douglas production function method.

Table 1. Application of "Cobb-Douglas Production Function Method" in China's Mining from 2004 to 2019[a].

year	Added value of Mining (100 million yuan)	Growth rate of added value of Mining	Capital formation rate in GDP under national expenditure method	Capital formation amount calculated by Mining (100 million yuan)	Total wages of Mining (100 million yuan)	Growth rate of total wages in Mining	Output elasticity of mining talents	Contribute index of mining talents
	1	2	3	4=2×3	5	6	7	8=7×6÷2
2004	7628.3		41.97%		821.2			
2005	10318.2	35.26%	40.27%	4155.5	1031.2	25.58%	0.0581	0.0421
2006	12082.9	17.10%	39.88%	4818.8	1259.6	22.14%	0.0581	0.0752
2007	13460.7	11.40%	40.42%	5441.0	1500.5	19.13%	0.0581	0.0975
2008	19629.4	45.83%	42.43%	8327.9	1847.3	23.11%	0.0581	0.0293
2009	20283.0	3.33%	45.47%	7605.2	2089.1	13.09%	0.0581	0.2284
2010	20936.6	3.22%	46.97%	9833.5	2458.8	17.70%	0.0581	0.3191
2011	27225.6	30.04%	47.03%	12804.0	3174.2	29.10%	0.0581	0.0563
2012	25093.0	-7.83%	46.19%	11589.4	3600.7	13.43%	0.0581	-0.0996
2013	25467.6	1.49%	46.14%	11749.7	3833.2	6.46%	0.0581	0.2513
2014	23417.1	-8.05%	45.61%	10681.1	3728.2	-2.74%	0.0581	0.0198
2015	19104.5	-18.42%	43.03%	8221.2	3318.2	-11.00%	0.0581	0.0347
2016	18514.8	-3.09%	42.66%	7897.5	3038.1	-8.44%	0.0581	0.1589
2017	21380.1	15.48%	43.17%	9230.2	3208.6	5.61%	0.0581	0.0211
2018	22592.3	5.67%	43.96%	9931.8	3413.4	6.38%	0.0581	0.0654
2019	23695.5	4.88%	43.07%	10205.2	3388.2	-0.74%	0.0581	-0.0088

Data source: China Statistical Yearbook from 2005 to 2021

The calculation formula is as follows:

$$GDP = A(t) L^{\alpha} K^{\beta} \mu$$

In the formula, *GDP* is the added value of Mining; *A(t)* is the comprehensive technical level, and *L* originally meant the number of labors invested (the unit is ten thousand person or person), which is often reflected with the total wages by later generations, but this paper uses the total wages in the Mining to reflect; *K* is the invested capital, including fixed capital and working capital, which is reflected with the "gross capital formation" in GDP calculated by expenditure method; α is the elasticity coefficient of labor output, β is the elasticity coefficient of capital output, and μ represents the influence of random interference. When $\mu = 1$, β is the complement of α.

When the above formula is applied, take the logarithm on both sides, and get:

$$LnGDP=LnA(t)+\alpha LnK+\beta LnL$$

See Table 1 for the application of "Cobb-Douglas Production Function Method" in China's Mining from 2004 to 2019.

3.1.1. Descriptive statistics. Using the data of "added value of mining industry", "calculated capital formation of mining industry" and "total wages of mining industry" in Table 1, operate spss26 0 statistical software for descriptive statistics, and the descriptive statistics table is shown in Table 2:

Table 2. Descriptive Statistics.

	N	minimum value	Maximum value	average value	standard deviation
Added value of Mining(100 million yuan)	15	10318.20	27225.60	20213.42	4961.28
Capital formation amount calculated by Mining(100 million yuan)	15	4155.50	12804.00	8940.63	2542.77
Total wages of Mining(100 million yuan)	15	1031.20	3833.20	2725.95	947.25
Number of valid cases (in columns)	15				

Results in Table 2 imply that the maximum value-added of China's mining industry is 272.560 billion yuan, the minimum value is 1031.82 billion yuan, and the average value is 2021.342 billion yuan. The calculated coefficient of variation is 24.5% (standard deviation 4961.28 ÷ average value 20213.42); The maximum value of capital formation calculated by China's mining industry is 1280.4 billion yuan, the minimum value is 415.55 billion yuan, and the average value is 894.063 billion yuan. The calculated coefficient of variation is 28.4% (standard deviation 2542.77 ÷ average value 894.63); The maximum value of total wages in China's mining industry is 383.320 billion yuan, the minimum value is 103.120 billion yuan, and the average value is 272.595 billion yuan. The calculated coefficient of variation is 34.7% (standard deviation 947.25 ÷ average value 2725.95). All indicators are in the normal range. The coefficients of variation of the three indicators are 24.5%, 28.4% and 34.7% respectively, which are in the range of "low variation" (10% - 40%), further indicating that the data dispersion is low and the reliability is strong.

3.1.2. Multicollinearity test. Take logarithms from the data of "added value of mining ", "capital formation amount calculated by mining" and "total wages of mining" in Table 1, and operate spss26 0 statistical software, conduct "multicollinearity" analysis, and obtain the multicollinearity test table, as shown in Table 3:

Table 3. Collinearity Diagnostics.

Dimension	Eigenvalue	Condition Index	Variable	Tolerance	VIF
1	2.932	1.000	Logarithm of added value		
2	0.056	7.260	Logarithm of capital formation	0.292	3.424
3	0.013	15.174	Logarithm of total wages	0.292	3.424

Results in Table 3 imply that the eigenvalue of each variable is greater than 0 and there is no collinearity; The condition index is less than the criterion 30, and there is no collinearity; VIF is less than the criterion 10, and there is no collinearity, so regression analysis can be carried out.

3.1.3. Regression analysis of Douglas production function model. Using the data in Table 1 to output Douglas production function by using EVIEWS least square method, see the results in Table 4:

Table 4. Output Douglas production function results table by using EVIEWS least square method [a].

Variable	Coefficient	Std. Error	t-Statistic
LnA(t)	2.3372	0.1871	12.4934***
LnK:	0.7830	0.0418	18.7362***
LnL:	0.0581	0.0327	1.7785*
R-squared		0.9972	
Adjusted R-squared		0.9934	
F-statistic		1054.8380	

[a] Note: *, * *, * * * mean significant correlation at the levels of 10%, 5% and 1% respectively (significance coefficients are 0.0000, 0.0000 and 0.1006)

The regression equation obtained from Table 2 is:

$$LnGDP = 2.3372 + 0.7830 LnK + 0.0581 LnL$$

The coefficient of *LnL* in the regression equation 0.0581 is "the output elasticity of mining talents", that is, the elasticity coefficient of mining talents (the value reflected by the total wages) to the GDP of Mining is 0.0581, far less than the elasticity coefficient of capital to the mining industry of 0.7830. It can be seen that China's mining industry is still in the stage of capital led industrial development, and the contribution elasticity of scientific and technological personnel is only 5.81%.Fill 0.0581 in the rows of "output elasticity of mining talents" column in Table 1; the formula [9] for calculating the "contribution rate of mining talents" (accurately defined as the "contribution index of mining talents") according to the relevant data in Table 1 and Table 2 is as follows (fill the calculation results in Table 1):

Contribution Index of Mining Talents = Output Elasticity of Mining Talents × Growth Rate of Total Wages in Mining ÷ Growth Rate of Added Value in Mining.

3.2. Determination of the Value of contribution of scientific researchers in Mining

The original meaning of the production function model based on Cobb Douglas is that: the left side of the equation is the output, and the right side of the equation is the amount of capital and labor input (the amount of labor force [10]). when extending this formula, the author restores the "labor force quantity" and applies the "full-time equivalent of R & D personnel in mining industry", and believes that the meaning of the Scientific and technological talents Contribution Index is (take the actual data in 2018 as an example): the GDP of the Mining increases by 1%, and the contribution value of Scientific and technological talentss increases by 6.54% （0.0581×6.38%÷5.67%）. By applying this contribution index to the mining industry scientific researchers (quantity of scientific research and technical labor force is converted into full-time equivalent), he is able to calculate the value of contribution of scientific researchers in mining industry. The formula is as follows:

Value of contribution of scientific researchers in Mining = Scientific and technological talents Contribution Index × Full time equivalent of R & D personnel in mining industry.

The full-time equivalent of scientific and technological personnel and R & D personnel in China's mining industry from 2005 to 2019 is shown in Table 5:

IOP Conf. Series: Earth and Environmental Science 1087 (2022) 012027 doi:10.1088/1755-1315/1087/1/012027

Table 5. Full time equivalent of scientific and technological personnel and R & D personnel in China's mining industry from 2005 to 2019 unit: person year.

year	mining and washing of coal	extraction of petroleum and natural gas	mining and processing of ferrous metal ores	mining and processing of non-ferrous metal ores	mining and processing of non-metal ores	Total mining
2005	33473	18500	515	858	1011	54357
2006	38249	21140	588	980	1155	62112
2007	36051	26335	528	1842	1591	66347
2008	45955	26086	2108	2348	1710	78207
2009	50001	26105	1496	2194	2065	81861
2010	44636	26473	1413	2230	1949	76701
2011	50763.3	32371.8	1942.3	3147.5	3404.8	91629.7
2012	46917	24027.3	2270.3	3686.5	2791	79692.1
2013	53713	25487.2	2725.4	3955	3131	89011.6
2014	53027.7	28445.9	2840.9	3997.7	3864.6	92176.8
2015	43819	23040	3404	3833	2946	77042
2016	40193	24483	3296	4631	3314	75917
2017	41987	21463	2236	4420	3255	73361
2018	33477	13818	2425	5206	3495	58421
2019	33947	15512	3047	5689	4498	62693

Using the data in Table 1 and table 3, Calculate value of contribution of scientific researchers in Mining and fill in Table 6.

4. Research conclusions

The mining industry is China's basic production enterprises such as energy and steel, which plays a supporting role in the development of the national economy. As the mining industry involves the exploitation of mineral resources, it not only needs to invest a lot of assets to provide mining conditions, but also needs to invest a lot of manpower to mine resource products. These people have the spirit of hard work and are determined to work in difficult situations. They are labor heroes in the mining industry, especially scientific and technological workers. They give up the superior environment to work in mining enterprises, which is a lifetime of selfless dedication.from the perspective of "value" of economic management, this paper first uses the "Cobb Douglas production function method" to determine the contribution of mining workers to national GDP, and obtains the "output elasticity of mining talents" (coefficient) of 0.0581, which is only 7.42% (0.0581 ÷ 0.7830 × 100%) of the capital output elasticity of 0.7830), and then creatively apply the elasticity coefficient of talent output to the contribution of scientific and technological talents in the mining industry to find out the "contribution value of scientific and technological talents in the mining industry". Novel coronavirus pneumonia in China has increased its contribution value to 4.01% in the 2005-2020 years. The highest year in 16 years is 2010, which is 24475.3 yuan / person year, the lowest in 2019, which is 1540.0 yuan / person year (covid-19), 16 years average age of 7011.1 yuan per person.

Table 6. Measurement of value of contribution of Mining science and technology personnel in China's Mining 2005-2019.

year	Contribute index of mining talents 1	Full time equivalent of R & D personnel in mining industry (person year) 2	Value of contribution of scientific researchers in Mining (thousands/person) 3=1×2
2005	0.0421	54357.0	2288.4
2006	0.0752	62112.0	4670.8
2007	0.0975	66347.0	6468.8
2008	0.0293	78207.0	2291.5
2009	0.2284	81861.0	18697.1
2010	0.3191	76701.0	24475.3
2011	0.0563	91629.7	5158.8
2012	-0.0996	79692.1	2288.4[a]
2013	0.2513	89011.6	22368.6
2014	0.0198	92176.8	1825.1
2015	0.0347	77042.0	2673.4
2016	0.1589	75917.0	12063.2
2017	0.0211	73361.0	1547.9
2018	0.0654	58421.0	3820.7
2019	-0.0088	62693.0	1540.0[b]

[a] 2012 is the year when supply exceeds demand, product prices are reduced and downtime occurs in China's mining industry. The contribution value of scientific and technological talents in mining industry is not calculated according to the conventional formula, but determined according to the minimum value of 2288.4 in 2005 (the data calculated according to the formula -7940.9 is replaced by 2288.4).

[b] In 2019, China was covid-19, and the contribution of mining talents to science and technology was not calculated according to the regular formula, but was approximately determined as 1545.0 according to the lowest value of 1545.2 in 2017 (the data calculated according to the formula - 550.2 is replaced by 1545.0)

From 2005 to 2018, the contribution value of scientific researchers in Chinese Mining has been increasing with the changes of the contribution index of scientific researchers and mining talents, from 2288.4 yuan / person year in 2005 to 3821.1yuan / person year in 2018, with an average annual increase of 4.01%, From 2005 to 2019, 16 years average 7011.1 yuan / person year.

5. Novelty of the paper

There are three innovations in this paper: first, the index data of model application is innovative. In the past, only the "number of R & D personnel" or "full-time equivalent of R & D personnel" was used to study the role of scientific research talents, without considering the value contribution made by R & D personnel. By using the "Cobb Douglas production function method", we first measured the contribution of mining workers to GDP, and then combined with mining scientific and technological talents to determine the contribution value of mining scientific and technological talents, which has raised a research level compared with previous research; Second, the application of "Cobb Douglas production function method" is innovated. It not only calculates the output elasticity coefficient of scientific and technological talents (0.0581), but further deepens the application of this coefficient and creates two formulas: the contribution index of mining talents (output elasticity of mining talents × growth rate of total Wages in mining ÷ growth rate of added value in mining) and value of contribution of scientific researchers in mining(scientific and technological talents contribution index × full time equivalent of R & D personnel in mining industry),the application of "Cobb Douglas production function method" has been deepened and developed; The third is the comprehensiveness of the research samples. In the past, when Chinese scholars studied the contribution of scientific and technological talents in the mining industry, the selected index samples are often a province or a region or several mining enterprises. This paper takes China's five major mining departments as the whole of China's mining industry as the research samples. The research conclusions are comprehensive and improve the credibility of the paper.

Acknowledgements
Jiangsu University Philosophy and Social Science Foundation Project "Research on financial performance of applied private undergraduate colleges" (2017SJB1279); Nantong Institute of technology bidding project "Research on financial performance of applied private undergraduate colleges" (2016002); Key discipline construction project of Jiangsu Provincial Industrial and commercial administration in the 14th five year plan (SJYH2022-2 / 285).

References

[1] Jing Ye 2017 Research on the construction of science and technology innovation capability and performance evaluation system of talents in higher education institutions - a case study in Henan Province. *Science and Technology Information* **20** 197-198
[2] Fang L J, Geng SQ 2018 Evaluation of scientific and technological innovation performance of Ningxia University. *Value Engineering* **19** 207-209
[3] Yan J, Zhang L, Wang Z 2016 An empirical study on science and technology performance of universities in the context of innovation drive - a case study of local universities in Beijing. *China University Science and Technology* **12** 38-41
[4] Wang X M 2018 Assessment and comparison of science and technology innovation performance of southwest universities during the 12th Five-Year Plan. *Science Management Research* **5** 66-69
[5] Meng F R,Wang H,Chen Z T 2017 The impact of science and technology human resources on science and technology innovation performance:a comparison based on firms and university institutions. *Science and Technology Management* **9** 173-180
[6] Wu D D,Wang Z C,Zheng S M 2016 Research on the scientific and technological innovation performance of regional universities based on scale-independence. *Management Modernization* **4** 60-63
[7] Varrichio P., et al. 2012 Collaborative Networks and sustainable business: a case study in the Brazilian System of Innovation. *Procedia - Social and Behavioral Sciences* **52** 90-99
[8] Sadegh R, Navid K, Aslan A.S 2012 Evaluation Framework for Assessing University-Industry Collaborative Research and Technological Initiative. *Procedia - Social and Behavioral Sciences* **40** 410-416
[9] Dong J, Zhu X Y 2012 Study on the economic pull effect of education factor elasticity. *Research on Higher Financial Education* **3** 1-4
[10] Hu X N 1998 Introduction to Contemporary Western Economics (2nd ed.) Beijing: *Party School of the Central Committee of C.P.C* 53

Industrial applications of the composite process "biodegradation+ activated carbon adsorption" in neutralizing the waste gas emissions

Guolin Jing[1], Xin Cheng[1,3] and Zhongqing Wang[2]

[1] College of Chemistry and Chemical Engineering, Northeast Petroleum University, Heilongjiang Daqing 163000, China;
[2] Petrochina Daqing Petrochemical Company, Heilongjiang Daqing 163714, China.

[3] Corresponding author's e-mail: cxabc1225@163.com

Abstract. In order to absorb and degrade odor gas and volatile organic compounds from sewage plant of refineries, waste gas was treated by the technology of compound "biodegradation and activated carbon adsorption". Through industrialization and production calibration, emission concentration of pollutants met the GB31571-2015 and GB14554-93, and statistics suggested the cost of utilities was 57.6 yuan per 10000m3, and fixed cost within reasonable boundaries. Therefore, this compounded technology can treat the waste gas from sewage plant of refineries.

1. Introduction

Since the implementation of the new Environmental Protection Law in 2015, all industries are facing severe environmental pressure, especially in wastewater, waste gas and waste residue treatment for refineries. In July 2015, the Ministry of Environmental Protection and the General Administration of Quality Supervision, Inspection and Quarantine of the People's Republic of China released the latest Emission Standards of Pollutants for Petroleum Chemistry Industry (GB31571-2015) with stricter control for non-methane total hydrocarbons, benzene, toluene and xylene. Meanwhile, the standard requires that volatile organic gases should be sealed in wastewater facilities and malodorous gases generated should be connected to the exhaust gas treatment facilities. After the implementation of this standard, it has become a challenge for environmental protection to meet the standard from the sewage field of refineries in an organized manner. In particular, the waste gas has a large amount of benzene and non-methane total hydrocarbons which are difficult to be biodegraded [1-2], so new environmental protection projects are needed to ensure the compliance of waste gas with the standard emission and to ensure that the enterprises carry out their production activities in accordance with the law. Traditional treatment of volatile organic compounds mainly consists f destructive method, non-destructive method, and the combination of the two. Destructive method mainly comprises direct combustion, catalytic combustion, thermal oxidation, biological oxidation and other techniques; non-destructive method mainly uses such techniques as carbon adsorption, absorption, membrane separation and etc; plasma, semiconductor catalysis and other emerging techniques have also been widely researched and developed. Traditional destructive treatment technology has the difficulty to deal with secondary pollutants, and non-destructive treatment technology faces the issue of high cost of filter replacement in industrial applications. And semiconductor catalysis and other emerging technologies including effectiveness, consumption, costs and other aspects still worth considering in industrial production and operation.

This study explores the effect of the new waste gas treatment facilities in certain refinery's wastewater field, and discusses the effect and cost of operating waste gas treatment facilities.

2. Device profile

This pilot study mainly focuses on the wastewater field in a refinery, where the exhaust gas treatment facilities primarily treat raw material tank, neutralization basin, sludge thickener, sludge pit, anoxic basin and sludge drying device, etc. After being treated by the treatment device, the exhaust gas is sent to the exhaust pipe through the fan to meet the emission standards. The ultimate emission meets the requirements of Emission Standards of Pollutants for Petroleum Chemistry Industry (GB31571-2015) and Emission Standards for Odor Pollutants (GB14554-93). The exhaust treatment facilities mainly involve biological processing and deep processing. The exhaust collection system includes exhausted enclosure, pipeline and valve. The designed volume of exhaust gas is 20000m^3/h.

The exhaust treatment facility principally includes biological treatment where pollutants such as hydrogen sulfide and ammonia are degraded through the metabolism of microorganisms, and activated carbon fiber adsorption where pollutants that are difficult to biodegrade such as benzene and non-methane total hydrocarbons can be efficiently adsorbed through the adsorption of activated carbon fiber. This waste gas treatment facility is equipped with activated carbon fiber regeneration facilities, which can regenerate the activated carbon fiber at regular intervals to ensure the activated carbon fiber's effective and long-term absorption of pollutants.

3. Process flow

Firstly, the exhaust gas enters into the biological treatment equipment after which it moves into the activated carbon fiber adsorption tank, and then it is sent to the exhaust pipe through the induced draft fan. The hydrocarbons adsorbed by the activated carbon fiber adsorption tank enter the condenser after steam desorption, and the produced condensate enters the liquid collection well and is then sent to the grease trap through the transfer pump, thus completing the circulation of the whole process. Process flow is prensented in Figure 1.

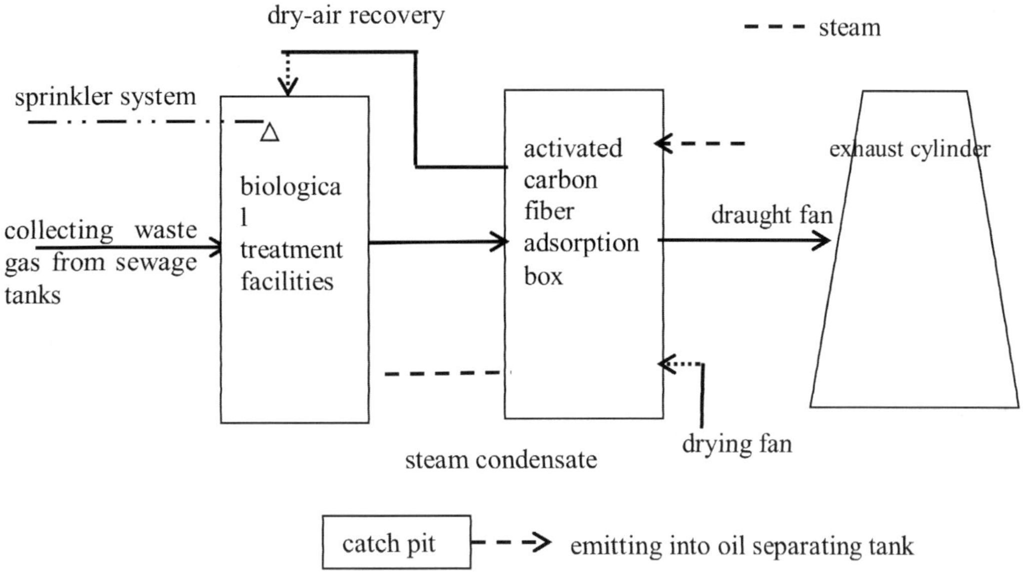

Figure 1. Process flow chart of the waste gas emission.

IOP Conf. Series: Earth and Environmental Science 1087 (2022) 012028 doi:10.1088/1755-1315/1087/1/012028

4. Technology principle of exhaust gas treatment facilities

4.1. Composition and process selection of exhaust gas

The characteristics of sewage plant in refineries are that the sewage contains higher oil content and complicated composition. Besides, the volatile gas of wastewater contains high concentration of hydrogen sulfide and ammonia, and the toxic gases such as hydrogen cyanide can be volatilized in the wastewater containing nitrile. As for the process of exhaust gas treatment facilities, the biological method and adsorption method are combined to improve the treatment effect and save the operation cost of the device. On the one hand, the biological method is characterized by low costs of operation and maintenance, while it is highly selective for pollutants and less effective for highly toxic gases [3], benzenes and non-methane total hydrocarbons. On the other hand, the adsorption method uses activated carbon fiber, characterized by high specific surface area of activated carbon fiber of 1300m2/g, high adsorption efficiency and low selectivity for pollutant adsorption, which can efficiently treat hard-to-degrade pollutants, but the raw material cost is high. The combination of biological method and adsorption method can ensure the high efficiency of treating pollutants to meet the emission standards [4], reduce the pollutant load of activated carbon fiber and save the operation cost.

Currently, the common treatment methods in processing exhaust gas include condensation, adsorption, absorption, combustion, biological method, membrane separation, etc. In view of the complex composition, high concentration and high explosion-proof level of the surrounding devices of the waste gas generated by the sewage plant in refineries, the combination of biological method and adsorption method suits better.

4.2. Biological treatment section

The biological treatment mainly processes the waste gas from the exhaust and regulating tank, accident tank and aeration tank in the grease trap preparation. The exhaust gas is pumped to the biological purification device through the fan, and is biologically purified by the bacteria strains attached to the biological filter.

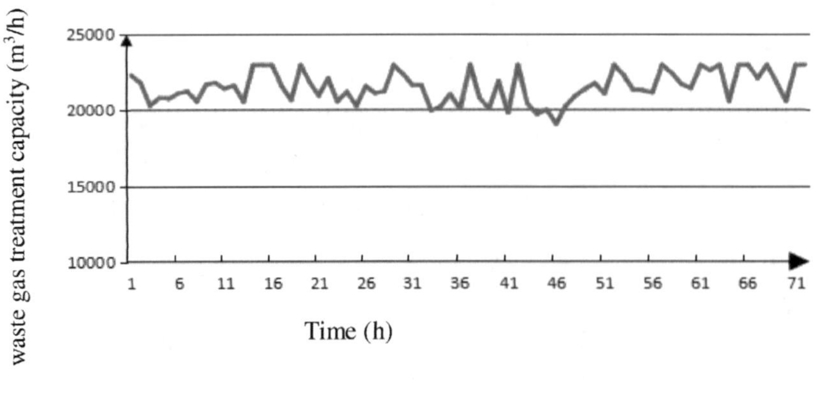

Figure 2. the amount of processing of waste gas emission.

The complete set of biological treatment adopts the screened biological filter, on which the collected microbial strains are fixed. Multiple microorganisms, in the form of populations, coexist in a single environment, where the characteristics of microorganisms are both similar and different, and different pollutants have microorganisms that degrade them in nature. Therefore, multiple pollutants can be treated and purified simultaneously in a single device [5]. According to their ability to assimilate various carbon-containing compounds, microorganisms that can degrade gaseous pollutants are classified into three categories: autotrophic, parthenogenic, and heterotrophic. Autotrophic bacteria have a complete enzyme system and can obtain the energy needed for growth in the oxidation of S,

H2S, CH3SH, NH3, Fe, etc. The carbon for their survival is provided by carbon dioxide or carbon from carbonate through the Calvin cycle [6]. Autotrophic bacteria are suitable for inorganic transformation with low bioburden. In contrast, heterotrophic bacteria obtain nutrients and energy through the oxidation of organic matter, which is suitable for organic matter transformation. Under proper nutrient conditions, temperature, pH and aerobic conditions, such microorganisms can complete the degradation of pollutants more quickly. If the hydrogen sulfide load is low, the strains are dominated by thiobacillus thiobacillus autotrophicus and naplethiobacillus [7]. Photoenergetic autotrophic sulfur bacteria belong to the genus Erythrobacter, which performs anaerobic photosynthesis, and their common species include Blastocystis, Octococcus, Thiobacillus, Thiobacillus, Entamoeba, and Exosulfobacillus, which are typically characterized by the use of hydrogen sulfide as a hydrogen donor for fixed carbon dioxide, and these microorganisms generally grow under anaerobic, anoxic, or slightly aerophilic conditions and at near neutral pH [8]. Photoenergetic autotrophic sulfur bacteria belong to the genus Erythrobacter, which performs anaerobic photosynthesis, and their common species include Blastocystis, Octococcus, Thiobacillus, Thiobacillus, Entamoeba, and Exosulfobacillus, which are typically characterized by the use of hydrogen sulfide as a hydrogen donor for fixed carbon dioxide, and these microorganisms generally grow under anaerobic, anoxic, or slightly aerophilic conditions and at pH values near neutral [8]. Chemoautotrophic microorganisms contain two groups of sulfur bacteria and sulfur bacteria. Sulfobacteria belong to the families Bezhiaceae and Leuconostocidae of cytophagales. The common species are Bezhia atrophylla, hyaline tremella, braided sulfur bacteria, bright hair bacteria and hair sulfur bacteria. Sulfur bacteria can oxidize hydrogen sulfide into sulfur particles and store them in the cells.

The biological filter is supported by corrosion-resistant plastic grids to achieve uniform air distribution. The microorganisms are attached and fixed on the inert efficient filler with large specific surface area. The organic compounds in the exhaust gas provide the nutrients such as carbon, nitrogen, phosphorus and some trace elements needed by the microorganisms. When the exhaust gas passes through the biological filter on which the screened efficient bacteria biologically oxidize the odor-causing components in the exhaust gas and thus purify the gas. Other metabolites such as sulfuric acid and nitrite produced in the oxidation decomposition of the exhaust gas enter the cycle water devices and are finally discharged together with the metabolites. [8-15]

4.3. Deep Treatment Section

The deep treatment section mainly adsorbs hydrocarbons that cannot be completely processed in the biological treatment section. The adsorption stuff used in the experiment is activated carbon fiber, with specific surface area higher than $1300m^2/g$. The cartridge filling method can increase the filtration area of activated carbon fiber effectively, which is suitable for the adsorption treatment of low concentration tail gas.

The deep treatment section is operated by normal temperature adsorption and temperature-rise desorption. The activated carbon fiber stuff is regenerated by steam temperature-rise desorption to increase temperature of the adsorbent, which can desorb the adsorbed components.

5. Production calibration of waste gas treatment facilities

In order to investigate effectiveness of the waste gas treatment facility, a production calibration of the waste gas treatment facility was conducted, concerning the capacity of the facility, pollutant concentration and utility consumption.

5.1. Background of production calibration

During the production assessment period, the production load of the three ethylene units in the chemical zone of the refinery was about 99% in terms of the instantaneous ethylene production, and the downstream supporting units were almost operating at full load. So the composition and content of VOCs in the exhaust gas indicate the treatment conditions under the full load of the units.

5.2. Analysis of waste gas volume

The exhaust gas collection system uses wind hoods to collect the exhaust gas from each unit and transport it to the exhaust gas treatment system under the negative pressure in the pipeline. This project mainly collects toxic gas from such seven sites as raw pool, neutralization basins, sludge thickeners, sludge pits, anoxic basins, mixing chambers and scum basins, with a total exhaust gas volume of 20,000 m^3/h. According to the material balance, the exhaust gas flow rate is 20,000 m^3/h. The processing volume data of the unit is obtained through 72-hour production calibration. The amount of processing of waste gas emission is prensented in Figure 2.

Through data analysis, the treatment capacity fluctuates between 19066m^3/h and 23000m^3/h, with an average value of 21549m^3/h and individual data below 20000m^3/h, which is a normal fluctuation and meets the requirements of the designed production treatment capacity.

5.3. Effectiveness analysis of exhaust gas treatment

In this study, the pollutant emission concentrations were explored, and sampling was conducted at the inlet and drain of the exhaust gas treatment facilities to analyze and study the treatment effects of the facilities. The main study objects are six major VOC and odor gas pollutants. The concentration of pollutants is prensented in Table 1.

Table 1. Inlet and emission concentration of pollutants.

pollutants	inlet gas concentration (mg/m^3)	effluent gas concentration (mg/m^3)	emission load	pollutant discharge standard (40m chimney)
hydrogen sulfide	0.078	0.06	6.91×10^{-3} kg/h	≤2.3kg/h
ammonia	1.11	0.32	6.91×10^{-3} kg/h	≤35kg/h
non-methane total hydrocarbon	46.28	7.45	-	≤120mg/m3
benzene	20.75	1.501	-	≤4mg/m3
toluene	5.452	0.002	-	≤15mg/m3
xylene	0.592	0.002	-	≤20mg/m3

5.4. Operation cost analysis of waste gas treatment facilities

The operation cost of facilities mainly includes human resource cost, utility cost and operation and maintenance cost.

5.4.1. Human resource cost. The waste gas treatment facility is all-automatic and unattended. The waste gas treatment facility is integrated with the inspection operation and maintenance of the sewage works. Four people are required to rotate shift inspections in accordance with the 24-hour continuous operation of the device to account the costs accurately. There is one post with four people. The labor cost of 135,000 yuan per person per year, and the annual labor cost is 540,000 yuan.

5.4.2. Operation and maintenance cost. This exhaust gas treatment facility unit requires periodic replacement of the biological stuff and activated carbon fibers. The replacement period of the biological filler is 15 years, with the cost of $50,000 each time and the construction cost is $30,000. The activated carbon fiber has to be replaced each 8 years, with the cost of $200,000 and a construction cost of $3,000 each time. In regular production, the activated carbon fiber is regenerated by steam heating, which can ensure the recycling of activated carbon fiber and reduce the cost. The maintenance cost of fans, pumps, etc. is 50,000 yuan per year. The equivalent cost of materials and equipment maintenance, etc. is 80,600,000 yuan.

5.4.3. Utility cost. After the exploration of utility consumption of the exhaust gas treatment facility, it primarily relates to electricity and steam. Meanwhile there is no auxiliary chemical dosing, and spraying water comes from the sewage reuse system, which is not costed. Steam consumption capacity is counted according to the meter, and electricity is counted according to equipment parameters and actual operation.

The main use of steam is for the regeneration of activated carbon fiber. According to statistics, its daily consumption is 3.8-4.2t/d, with the average value 4.01t/d, the unit price of steam 126 yuan/d, and the annual steam cost 184,400 yuan. The consumption of steam is prensented in Figure 3.

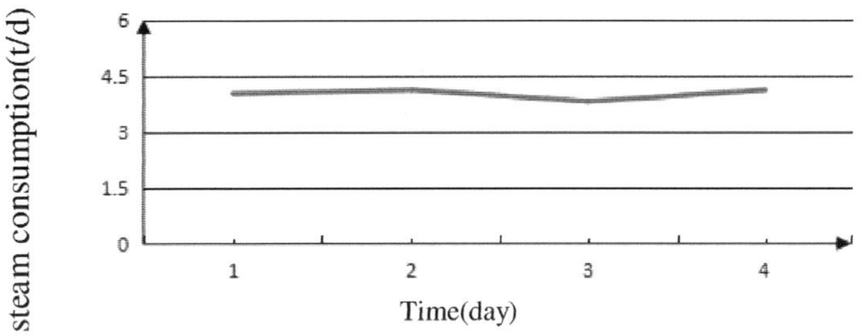

Figure 3. The consumption of steam.

The main electrical equipment of the device is blower, drying fan, condensate pump and biological spray pump. The blower is the key high-power equipment of the device, and its main function is to pump the waste gas from sewage facilities into the waste gas treatment facility, and finally discharge into the atmosphere through the chimney stack. The function of the drying fan is to drum dry air into the activated carbon fiber box which has completed the regeneration process, to discharge the steam from the activated carbon fiber box, and to cool down the box. The function of the condensate pump is to discharge the steam condensate into the nearby holding tank. The role of the biological spray pump is to spray water to wet the stuff in the bio-tank at regular intervals to maintain the growth environment of microorganisms. According to Table 2, the electricity consumption is 992.42Kwh/day, and the electricity price is 0.58 per degree, and the annual electricity cost is about 210,000 RMB.

The total cost of the facility including human resources costs, operation and maintenance costs, and utility costs, is 1.015 million yuan, and the annual treatment capacity is 1.752×108m3, which equals a cost of 57.6 yuan per 10,000 cubic meters of waste gas.

Table 2. Daily power consumption statistics of the device.

equipment	power (kW)	operation time per day(h)	power consumption kWh/d
blast apparatus	55	24	924
Drying fan	11	8	61.6
Condensate pump	2.2	4	6.16
Biological spraying pump	2.2	0.43	0.66
Total			992.42

6. Conclusions

(1) The exhaust gas treatment facility adopts the composite process of "biodegradation + activated carbon adsorption" to effectively remove pollutants from the exhaust gas and meet Emission Standards of Pollutants for Petroleum Chemistry Industry (GB31571-2015) and Emission Standards for Odor Pollutants (GB14554-93).

(2) The capacity of the exhaust gas treatment facilities meets the designed requirements, and the exhaust gas discharged from each cesspool can be effectively collected.
The operating cost of the exhaust gas treatment facilities is within a reasonable range and suits the industrial operation.
(3) The composite process technology of "bioldegradation + activated carbon adsorption" can effectively treat the exhaust gas containing pollutants of high concentration and difficult to degrade.
Steam heating method can extend the service life of activated carbon fiber and reduce the operating cost of activated carbon fiber adsorption method.

References

[1] WANG Honghua, XING Shubin, ZHOU Baohua, LEI Yongcong, WANG Yazhi 2010 Current situation of pharmaceutical industry pollution control and its countermeasures in Hebei *[J] Hebei Journal of Industrial Science and Technology* **27(05)** 355-360
[2] LIU Cuimian, DOU Hong, JIANG Jianbiao, GAO Yuan, ZHENG Wenqian 2018 Study on waste gas emission characteristics and treatment effect of sewage treatment process of a pharmaceutical enterprise in Shijiazhuang City *[J] Hebei Journal of Industrial Science and Technology* **35(05)** 363-369
[3] LIU Han, ZHANG Peiyu, YU Chunyan 2009 Air contamination and its treatment technology of wastewater treatment plant *[J] Environmental Engineering* **27(03)** 75-78
[4] GAO Kangning, CAO Huiyong, WANG Jin, et al. 2010 Example of an upgrading project of sewage treatment plants in North China *[J] Hebei Journal of Industrial Science and Technology* **27(5)** 286-290
[5] González Daniel, Colón Joan, Sánchez Antoni, Gabriel David 2019 A systematic study on the VOCs characterization and odour emissions in a full-scale sewage sludge composting plant *[J] Journal of Hazardous Materials* 373
[6] QIAN Xiao-yong, GUO Xiao-pin, LIN Li et al. 2013 Research methods for agriculturally emitted ammonia effects on formation of fine particulate matter (PM 2.5): a review *[J] Journal of Agro-Environment Science* **32(10)** 1908-1914
[7] DONG Qianqian, HE Qianmei, YANG Jiangyuan 2016 Design and operation of the printing and dyeing wastewater treatment project by combination of coagulation sedimentation-hydrolytic acidification with aeration tank-biological aerated filter-sand filters *[J] Hebei Journal of Industrial Science and Technology* **33(5)** 439-444
[8] GU Yinggang, WANG Boguang, YANG Jun et al. 2012 Ammonia source and emission factor in municipal wastewater treatment plants *[J] Environmental Chemistry* **31(5)** 708-712
[9] GU Binglin. Research on odor and foam control technology in the secondary pollution of wastewater treatment *[D] Anhui University of Technology*
[10] LIANG Yunping, LIN Anguo, WU Xiaodong 2017 Ammonia emission characteristics of a municipal sewage treatment plant in Beijing *[J] Environmental Monitoring in China* **33(6)** 63-68
[11] Aranzabal A, Pereda-Ayo B, GonzálezMarcos M P, et al. 2014 State of the art in catalytic oxidation of chlorinated volatile organic compounds *[J]. Chemical Papers* **68(9)** 1169-1186
[12] Pitkaho S, Ojala S, Maunula T, et al. 2011 Oxidation of dichloromethane and perchloroethylene as single compounds and in mixtures *[J] Applied Catalysis B:Environmental* **102(3)** 395-403
[13] MILLSB 1995 Review of methods of odour control *[J] Filtration and Separation* **2(3)** 147-152
[14] Lelicińska-Serafin Krystyna, Rolewicz-Kalińska Anna, Manczarski Piotr 2019 VOC Removal Performance of a Joint Process Coupling Biofiltration and Membrane-Filtration Treating Food Industry Waste Gas *[J] International journal of environmental research and public health* **16(17)**
[15] Aviraj Datta, Ligy Philip 2014 Performance of a rotating biological contactor treating VOC emissions from paint industry *[J] Chemical Engineering Journal* 251

IOP Conf. Series: Earth and Environmental Science 1087 (2022) 012029 doi:10.1088/1755-1315/1087/1/012029

Ecological study on native plant diversity in Kuwait desert ecosystem

Modi Ahmed[1,2], Abeer Al Saleh[1], and Noor Al Dousari[1]

[1] Environmental and Life Sciences Research Center, Kuwait Institute for Scientific Research, Kuwait P.O.Box: 24885, Safat 13109.

[2] Corresponding author's e-mail: Omanas.kw04@gmail.com

Abstract. The scarcity and irregularity of rainfall and the prevalence of strong northwesterly winds in Kuwait have a significant influence on the stability of the fragile ecosystem. The Aeolian landforms constitute most of the landscapes in Kuwait. Naturally grown native shrubs and trees can provide potential shelter to the soil surface in desert areas. To identify the characteristics of dominant perennial plant species, sediments of desert and saline (sabkhas) nabkha formed around different types of desert and halophytes plants are studied. Haloxylon salicornicum, Nitraria retusa, Cyperus conglomerates, Panicum turgidum, Rhanterium epapposum, Calligonum polygonoides, Stipagrostis plumosa Lycium shawii, Astragalus spinosus, Convolvolus oxyphyllus, Tamarix aucheriana, Halocnemum strobilaceum, and Citrullus colocynthis are landmarks characterizing the Arabian deserts and coastal salt marshes. The analyses of soil samples around the native plants showed that each native plants has its characteristics. The sediments of desert nabkhas differ from coastal nabkhas (Nitraria sp., Halocnemum sp., and Tamarix sp.). The sediments of coastal nabkha were characterized by lower pH, higher moisture content, total dissolved salts, and conductivity levels than desert nabkhas. Maximum mobile sediments were trapped by N. retusa, T. aucheriana, H. salicornicum, C. colocynthis, and L. shawii because of their ability to develop wider and longer nabkhas than other native plants. The vegetated dunes can trap maximum mobile sediments from 0.5 to 10.5 tons thus cost-saving per plant is estimated to be from 0.3 to 7.3 USD.

1. Introduction

The State of Kuwait covers 17,818 km^2 area, of which 85% is covered by the terrestrial environment. It is characterized by extremely hot and dry summer, and cool winter along with occasional rain. Due to hot temperatures, evaporation is very high ranging between 3 (winter) to 16 mm day^{-1} in summer. The rainfall is scanty and irregular; the rainy season extends between October and April. The average rainfall is about 110 mm. The prevailing wind direction is from northwest and average speed is 4 m sec^{-1}. The soils in the north are mainly Aridisols (calcids and gypsids) and in the south Entisols (sandy soils) [1].

Floristic research in Kuwait during the last few decades reveals that plant classification and distribution have received a lot of attention. Omar [2] developed a recent vegetated map with eight vegetation units: *Haloxyletum, Rhanterium, Cyperetum, Stipagrostieum, Zygophylletum, Centropodietum, Panicetum, and Halophyletum.*

The native plants shelter the soil from the wind's erosive force by stabilizing the soil with their roots [3] and were found to be effective in controlling aeolian activities, soil sealing [4], and soil organic carbon storage. The native plants' capability in trapping mobile aeolian sediments led to mobile sand accumulations around the plants named nabkha [5] [6]. Nabkha can be defined as aeolian accumulations around native plants [7] [8]. The nabkha formation surrounding native plants is economically and ecologically significant as it controls aeolian particles movement in desert and semi-desert regions [8-12].

Halophytes are plants that can survive and complete their life cycle with at least 200 mmhos/cm salt [13] [14], and come into contact with saline water through their roots. There are a lot of salt marshes and salt depressions located in the northern part of Kuwait. Plants grow on the margins and in the upper parts of sabkhas, forming sandy hills (Nabkha) around *Nitraria retusa (Forssk.) Asch. Halocnemum strobilaceum (Pall.) M. Beib. Suaeda vermiculata Forssk.ex J. F. Gmel.*

Desert plants are one of the unique components of the natural heritage of the State of Kuwait. The vegetation cover is poor because of the harsh climatic conditions. Examples of native vegetation are *Cyperus conglomeratus* Rottb., *Panicum turgidum* Forssk., *Rhanterium epapposum* Oliv., *Zygophyllum qatarense* Hadidi., *Calligonum polygonoides* L., *Stipagrostis plumose* (L.) Munro ex T. Anders., *Lycium shawii* Roem. & Schult, *Astragalus spinosus* (Forssk.) Muschl., *Convolvolus cephalop odus* Boiss. and *Citrullus colocynthis (L.)* Schard. This study aims to identify the unique properties of dominant perennial plants within various types of selected nabkha in Kuwait by combining ecological and sedimentological parameters.

2. Materials and Methods

Field investigations studied the sedimentological and morphological aspects of the different perennial plants developed around desert and coastal nabkhas to address the research objectives in the year 2019. Different areas in Kuwait's desert such as Kuwait Institute for scientific research station, Kabd, Jal Azzour, Al-Doha, Al-Liyah and Al-Dhabiaya were selected for such nabkhas. Soil samples were taken from the surface from representative nabkhas of each plant species to study physical parameters such as pH, electrical conductivity, soil moisture, and carbonate percentages.

Site information is determined through field examination, soil sampling and their analyses in the laboratory. A total of 100 soil samples were collected at three depths (0-20, 20-40, 40-60 cm). The samples were properly labelled with unique ID and stored in plastic containers and transferred to the laboratory for analyses. Field Tests were on site investigations which included soil moisture determined using gravimetric method through drying the soil at 105°C in the oven [15].

Laboratory Tests were done to determine electrical conductivity and pH of the soil samples, and a portion of soil was used to prepare saturated soil paste. The pH was then measured on the saturated soil paste using a calibrated (pH 4 & 9) pH meter. After measuring, the soil pH (pHs) the saturated paste was transferred to the Buckner funnels connected to the vacuum apparatus and soil saturation extract collected in collecting vials. The electrical conductivity of the extract collected from saturated soil paste (ECe) was determined using EC meter and the values reported as $\mu S\ cm^{-1}$. Organic matter was measured using oxidation method of Walkley and Black.

The morphological characteristics of a hundred nabkhas were measured (length, width, and height). To evaluate the economic value of native plants, the mass of the sand body was calculated by multiplying the plant's density by the volume of nabkha as follows:

The density of sand= 2500 Kg/m^3 (constant) (1)

The volume of nabkha= $2/3\ \pi$ (a.b.c) (2)

Where, a= width of nabkha, b= length of nabkha (it occupied half of ellipse which is equal to quarter)/4=b1, b-b1=b2/2, c = height of nabkha.

In 2013, the average cost of removing aeolian encroachments around desert infrastructures in Kuwait was projected to be around 1.32 USD per cubic meter [16].

The saved cost ($) of sand stabilization = volume of nabkha X cost ($) (3)

3. Results

When a plant grows in the direction of the wind and has a high humidity content, the plants operate as a sand trapper, accumulating sand around the vegetation and forming nabkha in the process. When a plant's growth accelerates, sand build-up accelerates as well. The young life stage refers to the phase of adaptation and balance between the plant, nabkha, and the surrounding ecological factors. The adaptation collapses and the plant dies when the root of the plant cannot reach the ground water level. The nabkhas around nine plant species were morphologically studied such as, *Nitraria retusa* (Forssk.) Asch., *Haloxylon salicornicum* (Moq.) Bunge ex Boiss., *Rhanterium epapposum* Oliv., *Astragalus spinosus* (Forssk.) Musc., *Panicum turgidum* Forssk., *Tamarix aucheriana* Decne., *Citrullus colocynthis (L.)* Schard., *Cyperus conglomeratus* Rottb. and *Lycium shawii* Roem. & Schult. The nabkhas vary in shape and size. They generally have a triangular shape due to the presence of a long south-easterly tail and shorter nose resulting from the dominant north-westerly wind. The nabkhas in Kuwait's northern and southern coasts, on the other hand, are elongated, with some having a mound shape. The width of the nabkhas' sand body is vital for its development. The wider the plant the longer the sand accumulation will be. Table 1 shows that larger nabkha formed under *Lycium shawii* Roem. & Schult, *Nitraria retusa* (Forssk.) Asch. and *Tamarix aucheriana* Decne., *Haloxylon salicornicum* (Moq.) Bunge ex Boiss. can form medium-sized nabkhas. Smaller nabkha formed under *Astragalus spinosus* (Forssk.) Musc., *Cyperus conglomeratus* Rottb and *Rhanterium epapposum* Oliv.

Table 1. Measurement of nabkha of the native plant (cm).

No	Plant	Plant		Nabkha		
		Length	Height	Length	Width	Height
1	*Nitraria sp*	369	258.5	398	295.5	88.5
2	*Haloxylon sp*	215	77.65	439.5	261	52.4
3	*Rhanterium sp*	153	70.5	197.5	174	25.85
4	*Astragalus sp*	112	53.45	275.5	125.75	26.9
5	*Panicum sp*	265.5	115.75	302.5	222.5	79.5
6	*Tamarix sp*	342	145.5	380.5	368.5	41.5
7	*Citrullus sp*	289	70.4	328.15	247	64.9
8	*Cyperus sp*	105.5	64.7	150	134.5	36.55
9	*Lycium sp*	279.1	188.4	444	304.75	40.75

Table 2, shows that Desert plants are Cyperus conglomeratus Rottb., Panicum turgidum Forssk., Rhanterium epapposum Oliv., Haloxylon salicornicum (Moq.) Bunge ex Boiss., Lycium shawii Roem & Schult, Astragalus spinosus (Forssk.) Musc., and Citrullus colocynthis (L.) Schard. characterized by highest pH, lower carbonates, conductivity level, and Total dissolved salts (TDS). Sabkha nabkhas (Nitraria retusa (Forssk.) Asch. and Tamarix aucheriana Decne.) were characterized by lower pH, higher conductivity level, and moisture content. Each native plant is growing under different environmental conditions. Cyperus conglomeratus Rottb. grows in sandy areas and is dispersed over mobile sand corridors. The presence of such a plant was a good sign of the aeolian process's greatest severity. Lycium shawii Roem. & Schult., Astragalus spinosus (Forssk.) Musc., and Citrullus colocynthis (L.) Schard., are dominantly found within drainage basins. On the main bedrock and the deposited Aeolian sand, Haloxylon salicornicum (Moq.) Bunge ex Boiss. and Rhanterium epapposum Oliv. are growing. Panicum turgidum Forssk. thrived in sandy areas with sandy corridors. Nitraria retusa (Forssk.) Asch. developed in saline sabkha because of a shallow lens of fresh groundwater that floats on top of the salty water body.

There is a relationship between the type of native plant and their ability as a hunter for aeolian sediment. The highest efficiency of native perennial plants in trapping aeolian sediments: *Nitraria retusa* (Forssk.) Asch., *Tamarix aucheriana* Decne., *Haloxylon salicornicum* (Moq.) Bunge ex Boiss., *Citrullus colocynthis (L.)* Schard., *Lycium shawii* Roem. & Schult, *Panicum turgidum* Forssk., *Rhanterium epapposum* Oliv., *Cyperus conglomeratus* Rottb. and *Astragalus spinosus* (Forssk.) Musc.

Such vegetated dunes can trap maximum mobile sediments from 0.5 to 10.5 tons thus cost-saving per plant is estimated to be from 0.3 to 7.3 USD. These results are inconvenient with results from Al-Dousari et al., [17]. Table 3, shows a relationship between plant and sand accumulation.

Table 2. Physical Properties of Sediment of Nabkha.

No	Plants	Carbonate%	Moisture %	EC	pH	TDS
1	*Rhanterium*	3.8	0.4	0.3	8.6	166
2	*Nitraria*	5.2	4.7	4.8	8.3	3072
3	*Astragalus*	9.6	0.6	0.4	8.2	284
4	*Panicum*	8.5	1.3	0.5	8.6	337
5	*Tamarix*	6.8	10.7	0.7	8.6	476
6	*Citrullus*	13.7	1	0.7	8.6	476
7	*Cyperus*	4.6	2.4	0.2	8.7	157
8	*Haloxylon*	7.6	0.2	0.5	8.6	291
9	*Lycium*	4.1	0.8	0.8	8.2	495

Table 3. The dominant perennial plant species with their mass, volume, and saved cost in Kuwait.

No	Plants	Mass ton	Volume (m^3)	Saved cost ($)
1	*Rhanterium*	0.872	0.460	0.608
2	*Nitraria*	10.470	5.528	7.297
3	*Astragalus*	0.457	0.241	0.319
4	*Panicum*	0.914	0.483	0.637
5	*Tamarix*	5.701	3.010	3.973
6	*Citrullus*	5.153	2.721	3.592
7	*cyperus*	0.722	0.381	0.503
8	*Haloxylon*	5.402	2.852	3.765
9	*Lycium*	2.105	1.111	1.467

4. Discussions

In Kuwait, the prevailing and strong Northwesterly wind especially during the summer season is consistently causing a significant movement of aeolian sediment that poses serious environmental and economic risks. Native plants playing a major role in controlling mobile sand and dust hazards. The mass of sediments deposited around 11 species of desert native plant was measured equivalent to a ton from the highest efficiency in trapping aeolian sediments: *Nitraria* retusa, *Tamarix aucheriana, Haloxylon salicornicum, Citrullus colocynthis, Lycium shawii, Panicum turgidum, Rbanterium epapposum, Cyperus conglomeratus, Convolvolus oxyphyllus, Calligonum polygonoides, Astragalus spinosus,* and Such vegetated dunes can trap maximum mobile sediments from 0.5 to 10.5 ton thus cost-saving per plant is estimated to be from 0.3 to 7.3 USD, which makes them the most resourceful keys for present and future applications in controlling aeolian processes as well as providing a significant contribution to the environment. The quality of life is affected by such rehabilitation. Soil, plant, animal, and aeolian process, are all examples of environmental indicators. The essential aspect in shielding the soil surface from erosive wind is vegetation cover. The current research has shed light on how the native plant can have a significant impact on the economy, people, and way of life. Native plants are more efficient in controlling aeolian sediments than non-native plants. The most capable natural plants in regulating aeolian sediments include *Nitraria retusa* (Forssk.) Asch., *Tamarix aucheriana* Decne., *Haloxylon salicornicum* (Moq.) Bunge ex Boiss., *Citrullus colocynthis (L.)*

Schard *Lycium shawii* Roem. & Schult. As a result, it is advised that such plants be given special attention due to their economic significance. Similarly, each native plant has its own set of morphological characteristics that have the potential to reduce expenses.

5. Conclusions

This study was focused on the morphological, ecological, and sedimentological characteristics of nabkhas around native dominant perennial plant species that were highly adapted to the harsh environmental conditions in Kuwait. The geographical distribution of nabkha depends on the suitable condition for each plant. The presence of *Cyperus conglomeratus* Rottb. is a reliable predictor of the severity of aeolian activity. *Lycium shawii* Roem. & Schult grows at the upstream edges of the drainage basins, while *Astragalus spinosus* (Forssk.) Musc. grows in the lower stages of the wadis downstream. The desert plant species are characterized by high pH values, lower electrical conductivity and moisture content percentage, while sabkha plants are characterized by lower pH, higher in moisture, total dissolved salt, and electrical conductivity level. Natural plants that are best at trapping aeolian sediments include *Nitraria retusa* (Forssk.) Asch., *Tamarix aucheriana* Decne., *Haloxylon salicornicum* (Moq.) Bunge ex Boiss., *Citrullus colocynthis (L.)* Schard *Lycium shawii* Roem. & Schult.

References

[1] Kuwait Institute for Scientific Research (KISR) 1999 Soil Survey for the State of Kuwait. KISR, Kuwait

[2] Omar S, Al-Mutawa Y, Zaman S 2000 Vegetation of Kuwait, A comprehensive Illustrative Guide to the Flora and Ecology of the Desert of Kuwait. Arid land Agriculture Department *Kuwait Institute for Scientific Research*

[3] Al-Dousari A M, Ramadan A, Al-Qattan A, Al-Ateeqi S, Dashti H , Ahmed M, Al-Dousari N. Al-Hashash N and Othman A 2020 Cost and effect of native vegetation change on aeolian sand, dust, microclimate and sustainable energy in Kuwait *Journal of Taibah University for Science* **14(1)** 628-639 https://doi.org/10.1080/16583655.2020.1761662

[4] Al-Dousari A M, Ahmed N M, Al-Dousari S, Al-Awadhi. 2019. Environmental and economic importance of native plants and greenbelts in controlling mobile sand and dust hazards *International Journal of Environmental Science and Technology* **16** 2415–2426 https://doi.org/10.1007/s13762-018-1879-4

[5] Al-Ghadban A N, Saeed T, Al-Dousari A M, Al-Shemmari H, and Al-Mutairi M 1999 Preliminary assessment of the impact of draining of Iraqi marshes on kuwait's northern marine environment. Part 1: Physical manipulation *Water science and technology* **40(7)** 75-87

[6] Al-Dousari A M, Pye K 2005b Mapping and monitoring of dunes in north-western Kuwait *Kuwait Journal of Science and Engineering* **32(2)** 119–134

[7] Al-Dousari A M, Misak R, Ahmed M, et al 2008 Characteristics of nabkhas in relation to dominant perennial plant species in Kuwait *Kuwait J Sci Eng* **35(1)** 1–22

[8] Ahmed M, Al-Dousari N, Al-Dousari A 2016 The role of dominant perennial native plant species in controlling the mobile sand encroachment and fallen dust problem in Kuwait *Arab J Geosci* **9(2)** 134 https://doi.org/10.1007/s12517-015-2216-6

[9] Al-Awadhi J M, Al-Dousari A, and Al-Enezi A 2000 Barchan dunes in northern Kuwait *Arab Gulf Journal of Scientific Research* **18(1)** 32-40

[10] Al-Dousari A M 2005a Causes and indicators of land degradation in the north-western part of Kuwait *Arab Gulf Journal of Scientific Research* **23(2)** 69-79

[11] Ahmed M, Al-Dousari A M, Baby S 2009 Chemical and morphological characteristics of phytogenic mounds (nabkhas) in Kuwait *Arab Gulf J Sci Res* **27(3)** 114–123

[12] Abd El-Wahab R H, Al-Rashed A R, and Al-Dousari A 2018 Influences of physiographic factors, vegetation patterns and human impacts on aeolian landforms in arid environment

Arid Ecosystems **8(2)** 97-110

[13] Flowers T J, Galal H K and Bromham L. 2010 Evolution of Halophytes: Multiple Origins of Salt Tolerance in Land Plants *Functional Plant Biology* **37** 604-612 http://dx.doi.org/10.1071/FP09269 [Citation Time(s):1]

[14] Doronzo D M, Al-Dousari A, Folch A, and Dagsson-Waldhauserova P 2016 Preface to the dust topical collection *Arabian Journal of Geosciences* **9(6)** 1-1

[15] Al-Dousari A M, Alsaleh A, Ahmed M, Misak R, Al-Dousari N, Al-Shatti F, Elrawi M, and William T 2019b Off-Road Vehicle Tracks and Grazing Points in Relation to Soil Compaction and Land Degradation *Earth Syst Environ.* **3** 471–482 https://doi.org/10.1007/s41748-019-00115-y

[16] USDA-NRCS 2004 Soil survey laboratory methods manual: soil survey investigations. Report No. 42, Version 4.0, January. US Government Printing Office: Washington, DC

[17] Al-Dousari A M, Ibrahim M I, Al-Dousari N, Ahmed M, and Al-Awadhi S 2018 Pollen in aeolian dust with relation to allergy and asthma in Kuwait *Aerobiologia* **34(3)** 325-336 https://doi.org/10.1007/s10453-018-9516-8

IOP Conf. Series: Earth and Environmental Science 1087 (2022) 012030 doi:10.1088/1755-1315/1087/1/012030

Rare Earth Elements behaviour in different types of mineral waters within Baikal Lake

G. A. Chelnokov[1,4], V.Y. Lavrushin[1], A. S. Aidarkozhina[1], I. V. Bragin[2], N.A. Kharitonova[2,3] and A.V. Ermakov[1]

[1] Geological Institute Russian Academy of Sciences. Pyzhevsky lane 7. bld. 1. 119017. Moscow. Russia;
[2] Far East Geological Institute FEB RAS. prospect 100-letya 159. 690022. Vladivostok. Russia;
[3] Geological Department. Lomonosov Moscow State University. GSP- 1. 1 Leninskiye Gory. 119991. Moscow. Russia.

[4] Corresponding author's e-mail: geowater@mail.ru

Abstract. The current study contains a review of the REE abundance and distribution in thermal and cold mineral waters which are widely spread around Baikal Lake. Studied mineral waters were divided into CH_4, N_2 and CO_2-rich waters by the associated gas composition. We found that the highest REE concentrations were connected with thermal waters that were accumulated in swamps (slow-moving waters). Another core process for accumulating of REE in waters is the rock weathering in CO_2 presence. This study shows that there are no correlations between REE and main ions, Eh, TDS and temperature. But dependence of REE on pH and positive correlations of REE with Al, P, Mn, V, Cu, Pb and As, which indicate common processes of microelements migration, were established. Within the study waters the light REE concentrations dominate over heavy REE. The presence of CO_2 gas increases heavy REE concentrations, and concentrations of several light elements (La, Ce, Nd). The observed differences in REE contents in studied waters can be provenance dependent and possibly relate to more complicated evolutionary processes as water circulation time, depth and mixing. This is the first investigation of REE distribution and abundance in mineral waters of the Baikal Lake.

1. Introduction

The thermomineral resources of the Baikal region include more than twenty thermal and cold mineral water manifestations (Figure 1). Most of them are poor studied in respect to the REE concentration. The purpose of this work is to establish the levels of REE and their fractionation in the mineral springs which demonstrate different water types, associated gases compositions and temperature. Studied waters are characterized by discharge temperature from 14 to 75°C and TDS varies from 0.3 to 3.0 g/l. The genesis and geochemistry of these waters were described by many authors [1, 2, 3], but the behavior of rare earth elements content was published for the first time. Many authors argue that anomalies of Ce and Eu could be considered as an indicators of water-rock interaction process or as hydrological tracers [4, 5].

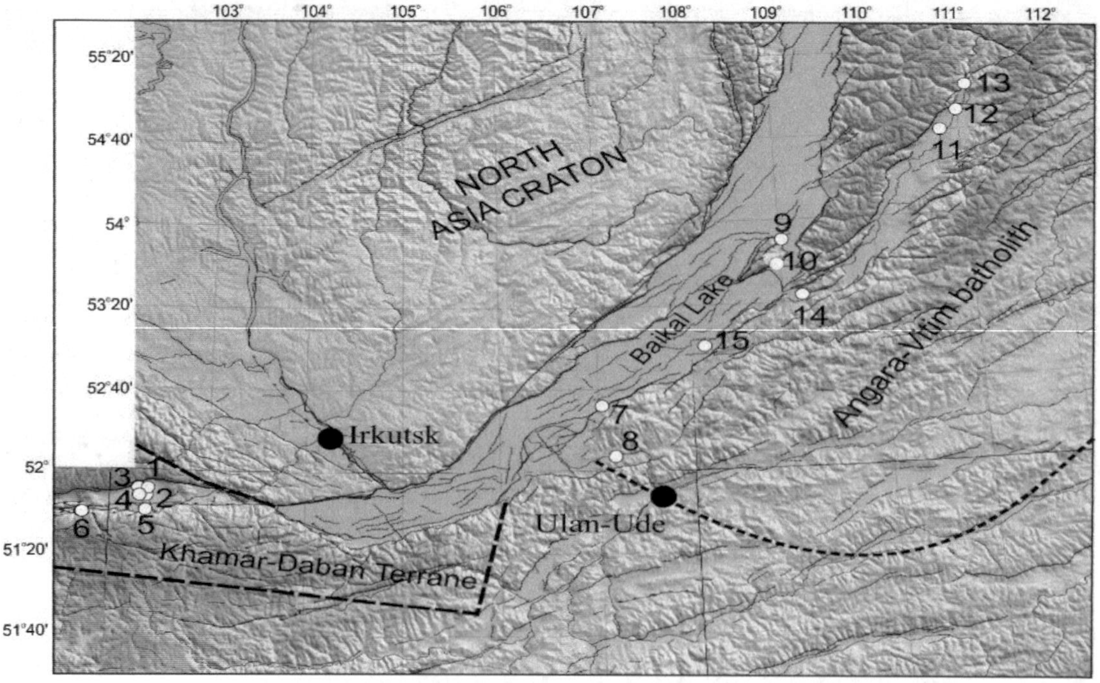

Figure 1. Location of water sampling site. Numbers of springs: 1-2– Arshan thermal; 3-4– Arshan cold; 5– Zhemchug well; 6– Nilova Pustyn well; 7 – Zagza borehole; 8 – Pitatelevsky thermal borehole; 9– Zmeevie thermal springs; 10– Kulinih Bolot thermal springs; 11– Alla thermal spring; 12 – Kucheger thermal springs; 13-Ukhmey thermal springs; 14– Gusikhinsky thermal borehole; 15 – Goriachensky thermal spring.

2. Geological conditions

Lake Baikal is located on an active continental rift in southeastern Siberia. The Baikal Rift Zone separates the Siberian craton from the Mongolian–Transbaikalian Belt (Figure 1) [5]. Angara-Vitim batholith occupies the east part of the territory. The batholith is formed in the Devonian–Early Carboniferous likely in connection to a mantle plume. Common rock types are granite with biotite and granodiorite. The south part of the study area is occupied by peneplainized Khamar-Daban Ridge which is a relatively flat plateau with the elevation of up to 1500 m and has the direction from the south-west to the north-east. Precambrian metamorphic rocks are wide spread there. The territory is characterized by increased tectonic activity and high seismicity.

The geochemistry of mineral waters located in these areas depends on lithology and circulation time. Thermal waters with T>40°C and TDS 0.35 - 3 g/l were disclosed in the depth interval from 100 to 1000 m [1]. All mineral water manifestations are associated with tectonic dislocations.

3. Sampling and analytical procedures

The cold CO_2-rich and thermal groundwaters were sampled from 15 manifestations on the south and east of Baikal Lake as a result of field works carried out in the 2020 summer season. Unstable parameters, such as water temperature, pH, Eh, alkalinity and electrical conductivity, were measured on site using the Mettler Toledo probe. Membrane filters (pore size of 0.45 μm) were used to separate dissolved fraction from suspended matter. The water samples were acidified promptly with Merk ultrapure nitric acid (to pH=1,5). Concentrations of microelements, including rare earth elements, were defined by the ICP-MS (Agilent 7500s and ELEMENT XR). The procedure took place in the Analytical Center of Far East Geological Institute, Vladivostok. During the process of ICP-MS direct determination of La, Ce, Pr, Nd, Sm, Eu, Gd, Tb, and Dy the removal of the oxide and hydroxide interferences was carried out using some mathematical technique. The method of Hathorne [6] was used to remove Ba interferences. Except for Ce and Pr an analytical precision for the REEs was < 5% RSD and close to standard deviation (RSD; 2σ). As for Ce and Pr, the precision was less 7% and less 10% RSD, respectively. The anomalous behaviors of the Ce and Eu fractionations were calculated as following: $Ce/Ce^{*}=2Ce_N/(La_N+Pr_N)$ and the $Eu/Eu^{*} = 2Eu_N/(Sm_N+Gd_N)$, where N means normalized abundance [7].

4. Results

4.1. Fluids chemistry and rare earth elements in water

The main geochemical characteristics of studied waters and their REE contents are presented in Table 1. Groundwaters were divided into three types by the gas composition. First one is CO_2-rich waters of Mg-Ca-Na-HCO_3-SO_4 and Na-Cl-HCO_3 types with TDS varying from 0.7 to 3.0 g/L and temperature from 11.8 to 54.7°C. Mineral waters of this type are locally spread in the south part of the Baikal Lake in Tunka depression. Genesis and origin of these waters were described in several works [8]. The second type is N_2-rich thermal waters of Na-SO_4 type with TDS varying from 0.3 to 1.7 g/L and temperature from 31.2 to 75.3°C. These waters are wide spread within the territory, but most part is attracted to Angara-Vitim batholith on the east (Figure 1). The third type is Na-Cl waters with TDS less than 1 g/l and methane as main associated gas. Only one borehole disclosing this type of water was sampled (Table 1).

It is well known that water-rock interaction processes drive up in the presence of CO_2 gas. It explains local enrichments of CO_2-rich waters by microelements, mg/L: Fe (1.9-4.0), B (2.4-5.2), Mn (0.05-0.2), As (0.002-0.007) and Sr (3.0-10.1). There are no correlations observed for REE and main ions, Eh, TDS and temperature (Figure2A). But ties were found for REE and pH, Al, P, Mn, V, Cu, Pb and As (Figure 2B). It could be an indicator that REE go to water during the weathering of minerals which contain these elements (for ex. biotite, plagioclase etc.), or similar processes which make microelements mobile (chemical thermodynamics). For example, the amounts of bicarbonate species available in solution are a function of the pH and the partial pressure of CO_2 and relate to the form of solid carbonates and the amount of dissolved elements present. Figure 2A demonstrates pH-dependent behaviour of REE in different waters. Figure 2B shows that REE have similar, but almost mirror image correlation with manganese. It indicates that as the pH of this system decreases (e.g., CO_2-rich waters), the solubility of carbonates (e.g., $MnCO_3$) increases rapidly. Figures reveal that during these processes REE enters in solution along with Mn^{+2}. Obviously, the behavior of REE in natural water is more complex than the simple ones represented here.

The total REE content in mineral waters varies from 0.02 to 2.5 ppb. The highest concentrations were observed in the water from the Kulinih Bolot thermal spring (#10), the Kucheger spring (#12) and Arshan CO_2-rich boreholes (#1, 2). First two thermal springs discharge in a swamp and have alkaline pH (9.5) which is not favor for REE migration in dissolved form. That allows us to suggest that a high REE concentration here is connected with colloidal particles passed through 0.45 μm filter (Table 1. Figure 2A). High REE contents in CO_2-rich waters are typically connected with more acidic condition (pH=6.6) and intensive water-rock-gas interaction [9].

The light REE (N from 57 to 64) dominate over the heavy REE (from 65 to 71) in all water types. Except thermal CO_2-rich waters (#1, 2, 5) which have maximal La, Ce and Nd concentrations, La and Pr dominate in other waters. The lowest HREE concentration is typical for water with N_2 and CH_4 associated gases. According to Kharitonova [9] the absence of CO_2 gas can be a reason for the low concentration of HREE.

The rare earth elements were normalized to Post Archean Australian Shale [10] to eliminate the Oddo Harkins effect and observed the differences between the REE distribution patterns. According to Kim [11] the REE behaviour in the aqueous environment depends on electrostatic bonding rather than covalent one, when interactions are governed by the ionic radius. This chemical features can result in fractionations between light and heavy REEs [12]. The PAAS-normalized patterns of studied mineral waters are characterized by positive Eu/Eu^{*} (1.0-6.0) anomalies, negative Ce (Ce/Ce^{*}=0.6-0.9), with weak HREE enrichment (Figure 3A, B). Ce(III) can be oxidized to Ce(IV) under typical surface conditions while most of the REEs remain in the +III oxidation state. At the same time Eu(III) can be reduced to Eu(II)under high temperature and/or extremely reducing conditions [13, 14].

Negative Ce anomaly with an average of 0.75 was observed in 7 samples. According to Och et al. [5] the fractionation of Ce is a result of an oxidation reaction which is enabled under conditions similar to the oxidation of Mn(II) to Mn(IV) [15]. That is why Ce(IV) can be eliminated from solution and included into authigenic mineral phases [16]. According to Eh, only five samples demonstrate oxidized conditions that correlate well with Ce negative anomaly. But two samples #7 and #12 show reducing conditions (Eh=-100 mV). These samples belong to water with methane and N_2-rich water from swamp. In this case we suggest that Ce depletion as well depends on microbial activity that catalyses the oxidation of Ce(III) [17].

Other eight samples show that Ce is accumulated in water. However again there is no strong correlation with redox conditions. Low Eh and high pH occur in four samples (#9, 10, 11 and 13) and allow Ce(III) to stay in solution. But the reasons for positive Ce-anomaly in thermal CO_2-rich waters (#1, 2) and N_2-rich water (#14, 15) which demonstrate oxidize conditions are not clear. Probably it could be connected with depth [18, 19] and age of the water [20].

Table 1. Chemical composition of the studied mineral water (numbers according to Figure 1).

	Unit	1-2	3-4	5	6	7	8	9	10	11	12	13	14	15
Water type		Mg-Ca-Na HCO_3-SO_4		Na-Cl-HCO_3	Na-SO_4	Na-Cl	Na-SO_4							
Gas type		CO_2			N_2	CH_4	N_2							
TDS	g/L	2.6	1.7	3	1.0	0.7	1.7	0.4	0.4	0.3	0.4	0.3	0.7	0.5
T	°C	45	14	55	40	51	65	44	43	45	40	45	75	53
pH		6.6	6.2	7.0	8.3	8.8	8.4	9.5	9.8	9.0	9.5	10	8.4	9.0
Eh	mV	155	205	228	230	-100	90	-80	-81	-54	-99	-85	299	263
La	ppb	0.076	0.046	0.006	0.047	0.038	0.014	0.017	1.439	0.062	0.331	0.026	0.017	0.009
Ce		0.104	0.006	0.009	0.006	0.005	0.002	0.001	0.094	0.006	0.035	0.002	0.002	0.001
Pr		0.018	0.029	0.002	0.020	0.018	0.005	0.006	0.320	0.020	0.117	0.008	0.006	0.003
Nd		0.092	0.006	0.008	0.006	0.005	0.002	0.002	0.051	0.003	0.018	0.002	0.002	0.002
Sm		0.025	0.006	0.01	0.002	0.003	0.004	0.001	0.009	0.001	0.004	0.001	0.002	0.001
Eu		0.012	0.011	0.008	0.005	0.006	0.003	0.002	0.050	0.004	0.017	0.002	0.002	0.001
Gd		0.048	0.002	0.013	0.001	0.001	0.001	0.001	0.005	0.000	0.002	0.001	0.001	0.001
Tb		0.008	0.016	0.002	0.003	0.008	0.002	0.001	0.026	0.003	0.009	0.001	0.001	0.001
Dy		0.066	0.004	0.01	0.001	0.002	0.001	0.001	0.004	0.001	0.002	0.001	0.001	0.001
Ho		0.018	0.013	0.002	0.002	0.006	0.001	0.001	0.013	0.001	0.005	0.001	0.001	0.001
Er		0.059	0.002	0.007	0.001	0.001	0.001	0.001	0.002	0.001	0.001	0.001	0.001	0.001
Tm		0.009	0.014	0.002	0.002	0.005	0.002	0.001	0.010	0.001	0.005	0.001	0.001	0.001
Yb		0.063	0.002	0.011	0.001	0.001	0.001	0.002	0.001	0.001	0.001	0.001	0.001	0.001
Lu		0.011	0.046	0.002	0.047	0.038	0.014	0.017	1.439	0.062	0.331	0.026	0.017	0.009
∑REE		0.816	0.658	0.092	0.123	0.121	0.049	0.042	2.57	0.138	0.751	0.057	0.042	0.024
∑LREE		0.56	0.55	0.056	0.114	0.097	0.041	0.037	2.51	0.133	0.723	0.053	0.038	0.021
∑HREE		0.23	0.26	0.036	0.009	0.023	0.008	0.004	0.062	0.007	0.025	0.004	0.004	0.003
Ce/Ce*		1.7	0.6	0.6	0.8	0.8	0.7	1.1	1.4	1	0.9	1.1	1.1	1
Eu/Eu*		0.7	3	3	2	2	6	2	0.9	2	1	1.8	4	3

Classical positive Eu anomaly corresponds to a strongly reducing condition was found only in five waters (#7, 9, 11, 12, 13). N_2-rich thermal water from the swamp (#10) reveals almost absence or slightly negative Eu anomaly. This anomaly is typical for weathering processes within host rocks or hydrogeological systems (aquifers) which are out of static equilibrium [21]. Inversely, negative Eu anomaly was found in CO_2-rich thermal waters (#1, 2) with oxidize conditions. Seven water samples (#3, 4, 5, 6, 8, 14, 15) demonstrate positive Eu anomaly (Figure 3 A, B) which is not common for fluids relating considerably more oxidized conditions. The results obtained could be connected with either local bedrock mineralogy (host rocks with Eu-rich minerals) or difference in groundwater residence time [22].

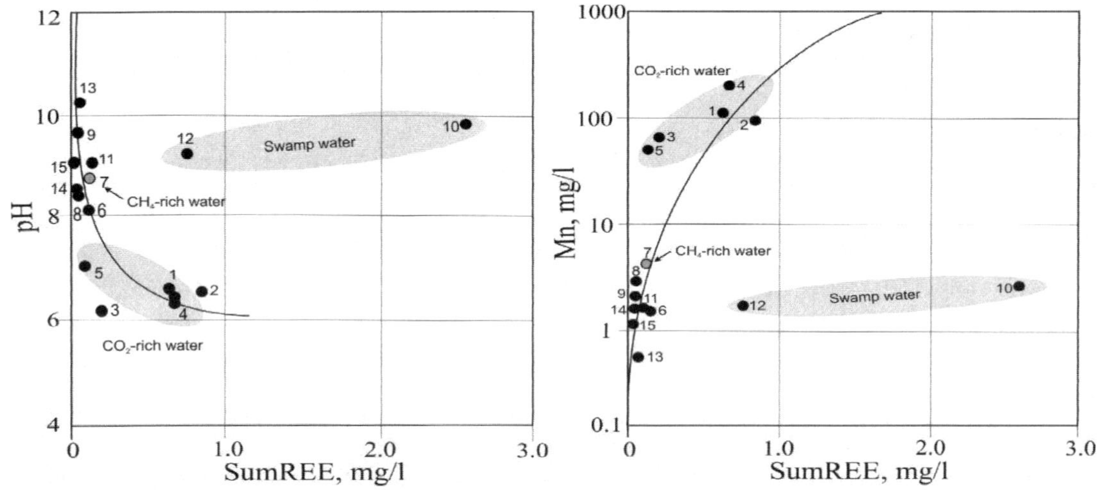

Figure 2. Concentrations of REE in thermal waters vs pH and Mn.

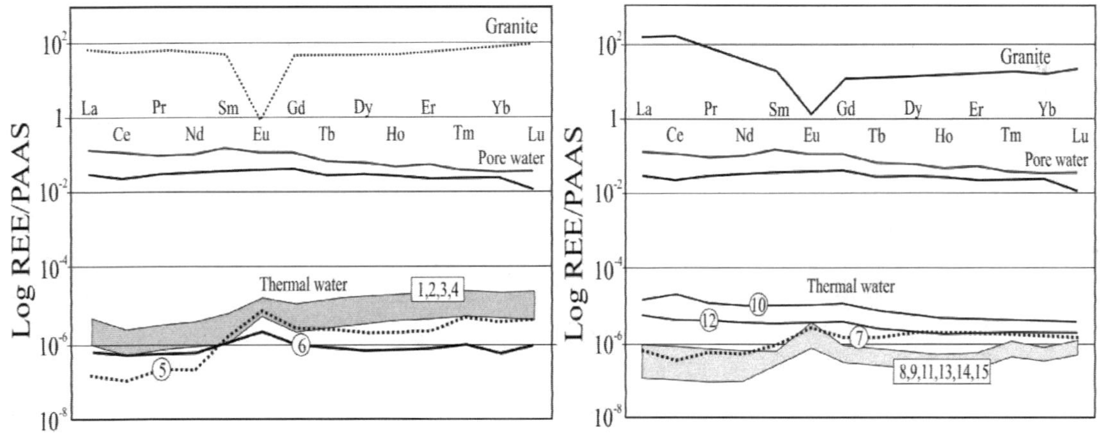

Figure 3. PAAS-normalized concentrations of REE in studied thermal and cold mineral waters. A- water from the south of Baikal Lake. B-water from the east shore of the lake. Granite- is a reference level from the south and east part of the Baikal Lake [1. 2]. Pore waters - are reference level of REE in pore water of Baikal Lake sediments.

5. Conclusions

Our investigation of mineral waters of the Baikal Lake can be summarized as follows: 1) the maximal concentrations of total REE were observed in natural thermal springs discharging in swamps and CO_2-rich thermal mineral waters; 2) there are no correlations between REE and main ions, Eh, TDS and temperature. But correlations of REE with pH, Al, P, Mn, V, Cu, Pb and As were found which indicates common processes of microelements migration; 3) the light REE concentrations dominate over heavy REE in all samples. The presence of CO_2 gas increases HREE concentration; 4) maximal concentrations of La, Ce and Nd are typical only for thermal CO_2-rich waters. La and Pr dominate in waters without CO_2 gas or in cold CO_2-rich waters. The observed differences in REE contents in studied waters can be provenance dependent and possibly related to more complicated evolutionary processes as water circulation time, depth and mixing.

Acknowledgment

The study of the behaviour of REE in N_2- and CO_2-rich manifestations of Lake Baikal was supported by a grant from the Russian Science Foundation - project No. 18-17-00245. The study was also partially supported by the Russian Foundation for Basic Research Project No. 20-05-00727.

References

[1] Tkachuk V.G., Tolstikhin N.I. (Eds.) 1961 Vol. 1 House of the USSR Acad. Sci., Moscow–Leningrad
[2] Pinneker E.V., Pisarskiy B.I., Lomonosov I.S., et al. 1968 Hydrogeology of Baikal Region. 170
[3] Borisenko I. M., Zamana L. V. 1978 Mineral'nye vody Buryatskoi ASSR 162
[4] De Grave. J.. Glorie. S.. Buslov. M. M.. Stockli. D. F.. et al. 2013 *Gondwana Research.* **23** 3
[5] Och M et al. 2014 *Chemical Geology* **376**
[6] Hathorne et al. 2012 E.C. *Geochem. Geophys. Geosyst.* **13** pp. 1-12
[7] Bau M, Dulski P. 1996 Transvaal Supergroup. South Africa Precambrian Res. **79**
[8] Pavlov S.Kh., Pinneker E.V., Pisarsky B.I. 1995 *Russian Geology and Geophysics* **36**
[9] Kharitonova N. Chelnokov G. Karabtsov A. Kiselev V 2007 *Applied Geochemistry* **22**
[10] Taylor, S.R. McLennan, S.M. The Continental Crust: Its Composition and Evolution. 1985.1-312
[11] Kim K.-H., Byrne R.H., Lee J.H 1991 *Mar. Chem.* **36**
[12] E.R. Sholkovitz 1993 *Geochim. Cosmochim. Acta.* **57**
[13] Sverjensky D.A. 1984 Earth Planet. Sci. Lett. **67**
[14] Bau M. Chem 1991 Geol. **93**
[15] Bau M. Geochim 1999 Cosmochim. Acta. 63
[16] Goldberg E.D. 1961 *Oceanography*. Am. Assoc. Adv. Sci. Publ.
[17] Moffett, J.W. 1990 *Nature* **345**
[18] Piepgras, D.J., Jacobsen, S.B. 1992 *Geochim. Cosmochim. Acta* **56(5)**
[19] German, C.R., Masuzawa, T., Greaves, M.J., etc. 1995 Geochim. Cosmochim. Acta **59(8)**
[20] German, C.R., Elderfield, H. 1990 *Paleoceanography* **5(5)**
[21] Möller P 2000 *Hydrology of crystalline rocks*.
[22] Bau M and Koschinsky A 2009 *J.Geochim.* **43**

Arboreal forests growth carbon sink and assessment of carbon storage in China's Henan Province

Bingtao Liu[1,3], Yu Shi[1], Naxin Mei[2]

[1] School of environment and municipal engineering, North China university of water resources and electric power, Zhengzhou, 450011, China;
[2] Zhengzhou preschool education college, Zhengzhou, 450099, China.

[3] Corresponding author's e-mail: liubingtao@ncwu.edu.cn

Abstract. Henan province is located in central China and the assessment of its arboreal forests' carbon sink and carbon storage is important to the sustainable use of forest resources. This study used 2018 forest inventory data to estimate forest biomass and carbon storage. The results showed that: (1) By origin, the carbon storage of natural forest is lower than that of artificial forest. (2) By type, Populus L. accounted for the highest proportion, and together with the carbon storage of Mixed broadleaf and Quercus spp. accounted for 80%. The carbon density of Populus L. is the highest, and that of Cupressus funebris is the lowest. (3) By region, the carbon storage of arboreal forest in Western and southern Henan is the highest, and that in Southeast is the lowest. (4) The carbon storage of the arboreal forest has increased to 8.891 million tons which is equivalent to 32.60 million tons of carbon dioxide equivalent. This paper's evaluation of carbon storage is a valuable reference to forest ecosystems in mitigating global climate change.

1. Introduction

The carbon cycle of forest ecosystems serves the pivotal function of regulating atmospheric CO2 concentrations. The estimation of forest ecosystem carbon storage has become an important research topic. In every endeavor to mitigate global change, forest activities have a very important and irreplaceable position and role [1]. Forest carbon sink produced by the forestry activities is a better choice to slow down the greenhouse gas increase in the atmosphere [2]. It requests to scientifically evaluate and monitor the emission reduced and sink cleared caused by these activities [3]. So Chinese government, society, and scientists all pay attention to the study of the carbon accounting of forestry activities consisting of afforestation, reforestation, deforestation, and forest management. In 2020, China would adopt more effective policies and measures to strive to reach the peak of carbon dioxide emissions by 2030 and achieve carbon neutrality by 2060 [4]. Therefore, carbon accounting for forestry is the key means to achieve this goal, which is of great significance.

Forest Inventory Data said the amount of C in China is increasing by a large amount, but the rate of increase cannot exhibit sufficient to be estimated with accuracy [5-7]. According to the ninth Henan province forest resources inventory and the framework of IPCC guidance [8], this paper evaluates and predicts the carbon sink and its potential for the main forestry activities in China's Henan province. It can provide data support for China to offset part of greenhouse gas emissions because of the quick economic development in the future, and technical support for the site choice to develop forest carbon sink projects. China has 133.7 million hectares of forested land, ranging from tropical forests in the south to forests in the north. Henan Province is sited in central of China and at the junction of south and north China, the tree growth is neither too quickly as the south area cannot be achieved due to carbon removals capacity and time limit, nor like the north as too slow so that achieve effective cycle is too long. Long-term agricultural development, forest management practices, and changing land use and forestry policies may also play important roles in the global carbon cycle. In the present research, according to the date of forest resources inventories of Henan, the forest biomass and carbon storage were estimated using a regression equation between the biomass of various tree species and volume and country-specific parameters.

2. Background of forest resources in Henan

Henan Province is located in the Yellow River, and the geographical coordinates sector is at longitude 110° 21' to 116°39' east and latitude 31°23' to 36°22' north. The length is about 580 km long from east to west, and the distance is about 530 km away from north to south. The vegetation boundary is largely on the front line of the main ridge Funiu Mountains and Huaihe River, in the north is the South temperate deciduous broad-leaved forest zone, and in the south is the North subtropical evergreen broad-leaved forest zone. The forest resources of Henan Province were mainly natural broadleaf forests, which located in the north of Taihang mountains, west of Funiu mountains, southern of Tongbai mountains, and Dabie mountains.

In 2018, the total land area of Henan province is 16.70 million hectares and the forest land is 5.21 million hectares accounting for 31.19%, and Non-forestry land is 11.49 million hectares accounting for 68.81%. The forest coverage of Henan province is 24.14%, and the forest area is 4,031,800 hectares, in which woodland area accounted for 86.38%, shrubland area accounted for 13.06%, and bamboo forest area accounted for 0.56%. The average annual net growth of forest reserves in Henan province is greater than the annual net consumption, and the total stumpage showed a rising trend. The tree species of Henan Province arboreal forests are mainly including Pinus tabuliformis, Pinus massoniana, Other pines, Cunninghamia lanceolata, Cupressus funebris, Quercus spp., Robinia pseudoacacia, Miscellaneous broadleaf hardwoods, Populus L., Paulownia Sieb., Miscellaneous broadleaf softwoods, Mixed coniferous, Mixed broadleaf, Mixed coniferous and broadleaf.

The remote sensing map of forest resources in Henan Province in 2018 is shown in Figure 1. The geographical distribution of forests in Henan Province can be divided into four regions: north Henan, east Henan, south Henan, and west Henan. North Henan includes Anyang, Puyang, Xinxiang, Hebi, Jiaozhuo. East Henan includes Shangqiu, Kaifeng, Zhoukou. Southcentral Henan includes Zhengzhou, Xuchang, Luohe, Pingdingsan, Hanyang, Xinyang, Zhumadian. West Henan includes Luoyang, Sanmenxia. The southcentral Henan region accounts for more than 44% of total biomass. It was mainly marked by needle coniferous forests such as those of Pine massoniana, deciduous oaks, and Cunninghamia lanceolata. The biomass of western Henan accounted for more than 37%, mainly including deciduous oak, Populus L., and hard broadleaf. It is related to the local geographical environment, climatic conditions, and irrigation conditions. There used to be a flood of the Yellow River with saline-alkali land and dunes everywhere. Thus, the biomass density of arboreal forests in eastern Henan is lower than the national average, as is the carbon storage density.

The total growth of Henan Province forest reserves is greater than the total consumption, and the number of forest resources is showing a rapid growth trend. The forest area and forest reserves increased rapidly, and the forest coverage has improved steadily. A net increase of 541,100 hectares of the province's forest area in five years from 2013 to 2018, and the average annual net growth rate was 3.01%. This result showed that forest area growing rapidly compared with the last five-year intervals, in which arboreal forests area increased by only 298,700 hectares. The forest coverage rate increased from 21.5% in 2013 to 24.14% in 2018, with a net increase of 2.64 percentage points in five years.

3. Data and study methods

As it is impossible to measure all trees in a forest, sampling is essential, especially sampling design. The effective sample size, shape, and dimensions need to be determined by specialized knowledge.

After the training, the sample plots were worked up and measured by Henan Provincial Forest Survey Resource Group. They use prepared field data sheets to record the data and then transfer them to a sheet or database with built-in features. These data would transfer the record into estimates of carbon and timber stock. Remote sensing and timber inventory data were also used as a measurement method.

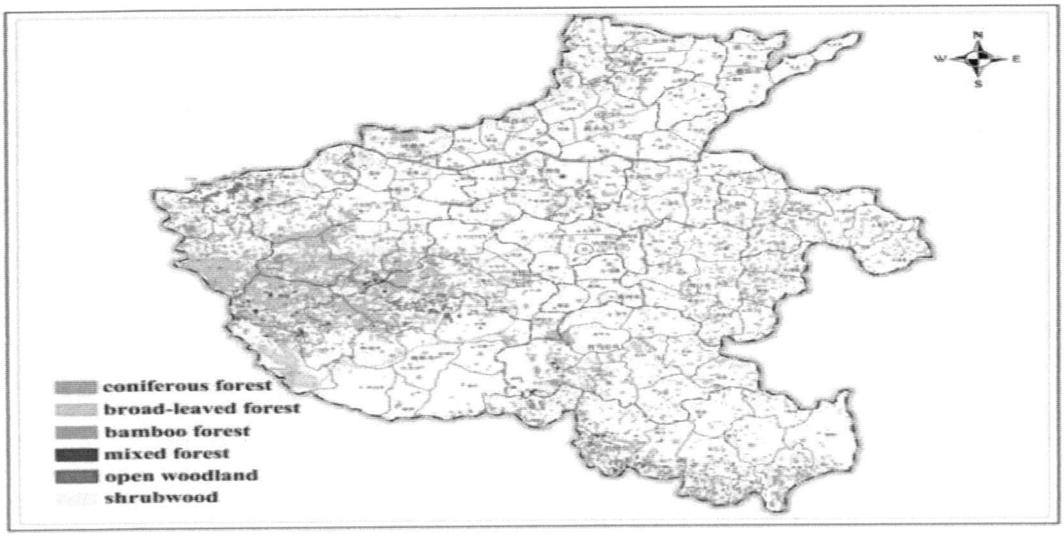

Figure 1. Remote sensing map of forest resources in Henan Province.

Table 1. Formula summary.

No.	Formula	Related parameters	Reference
(1)	$BEF = a' + \frac{b'}{X}$	X - Stem volume per unit area; a', b'-constants for a specific forest type.	Fang et al., 2001, 2003, 2006 [9-11]; Tan, K, 2007 [12]; Peng, C. H. 2009 [13] Guo et al., 2010 [14];
(2)	$B = aV + b$	B-the biomass of a certain tree species per unit area; V-the volume of a certain tree species per unit area; a, b-calculation parameters	Fang et al., 2002 [15]; Shao et al., 2017 [16];
(3)	$\Delta C_{Arbor} = A_j \times V_j \times VGR_j \times BEF_i \times WD_i \times CF_i$	A-the arboreal total area; V-volume per unit area; VGR_j-growth rate, 9.53%; BEF_i-biomass expansion factor, 1.8348; WD_i-stem volume density,0.4908t/m^3; CF_i-biomass carbon content, 0.5.	Chinese Ministry of Forestry, 2019 [17]
(4)	$C_{stock} = A_{AR,T} \times V_{AR,T} \times BEF_i \times WD_i \times CF_i$		

The research methods and formulas used in this paper are shown in Table 1. and based on the guidelines of parties to the United Nations Framework Convention on Climate Change (UNFCCC) and Henan Province forest survey resource data , we got arboreal forests total volume, the volume of the major tree species, annual growth rate (VGR), stem wood density data of Henan Province and biomass expansion factor (BEF). Stem volume density and Biomass expansion factor of 14 dominant tree species (groups) were identified in Table 2 and the arboreal forest area and volume are also shows in Table 2 for each type of tree.

Table 2. Biomass expansion formula parameters and Stem volume density and biomass expansion factor for the principal tree species in Henan's arboreal forests.

No.	Tree Species	Biomass Expansion Formula Parameters		Stem volume density (t.m^{-3})	Biomass expansion factor	Area (10^2ha)	Volume (10^2m^3)	n
		a	b					
1	Pinus tabuliformis	0.7554	5.0928	0.36	1.589	660	35269	65
2	Pinus massoniana	0.5101	1.0451	0.45	1.472	1050	70158	77
3	Other pines	0.5829	23.6859	0.528	1.687	388	18518	77
4	Cunninghamia lanceolata	0.5371	11.9858	0.295	2.357	227	15658	229
5	Cupressus funebris	0.6129	46.1451	0.492	1.732	1068	23494	38
6	Quercus spp.	1.1453	8.5473	0.676	1.355	7783	445057	73
7	Robinia pseudoacacia	0.7564	8.3103	0.515	1.586	1048	45600	9
8	Miscellaneous broadleaf hardwoods	0.7564	8.3103	0.495	1.674	803	27870	94
9	Populus L.	0.981	0.004	0.394	2.49	6236	628042	60
10	Paulownia Sieb.	0.8956	0.0048	0.256	3.499	776	35233	14
11	Miscellaneous broadleaf softwoods	0.4754	30.6034	0.443	1.586	1031	29859	21
12	Mixed coniferous	0.5168	33.2378	0.418	1.617	338	13905	14
13	Mixed broadleaf	0.8392	9.4157	0.493	1.514	10637	570166	61
14	Mixed coniferous and broadleaf	0.8136	18.466	0.442	1.609	1268	72562	103

where, n is number of samples

4. Results and discussions

4.1. Analysis by source and age group

The analysis makes it clear that Henan's arboreal forests have a total area of 3.48×10^6 ha and the entire carbon storage is 93.30×10^6 t. of that, there are 1.57×10^6 ha of natural forests and carbon density is 26.79 t/ha. These natural forests account for 45.06% of the total forest biomass and carbon storage in Henan Province. The plantation area is 1.91×10^6 ha, which is account for 55% of the arboreal forest biomass and carbon storage in Henan Province. The average carbon density of these plantations is 26.78t/ha, which is in the neighbourhood of that of Henan Province (26.79t/ha).

The age group of arbor forest is determined according to the average age of dominant tree species (groups) in the main forest layer and there are still differences between natural forest and planted forest. Figure 2 shows carbon storage in natural forests by age group. The carbon storage of young forest is the largest (54.8%), followed by middle-aged forests. The carbon content of overly-mature forests is the lowest, only 0.32%. The carbon storage of natural forests in different age groups is different. Figure 3 shows carbon storage in forest plantations by age group. Nearly-mature forest accounted for more than one-third of the total, followed by middle-aged forest, accounting for about one fourth. Overly-mature forests account for only 2.95% of the carbon storage of these plantations. This is much lower than the other four age groups. Besides, as Figure 2 shown, the existing forest carbon sink capacity of Henan Province, the young forest, and middle-aged forests the vast majority were the majority. With the continuous growth of the trees, the young forest and middle-aged forest will store more and more carbon dioxide and will play an important function as carbon sinks.

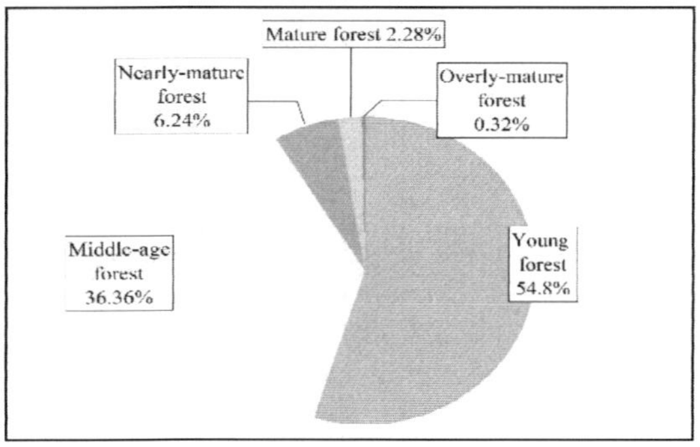

Figure 2. Natural forest carbon stores by age group.

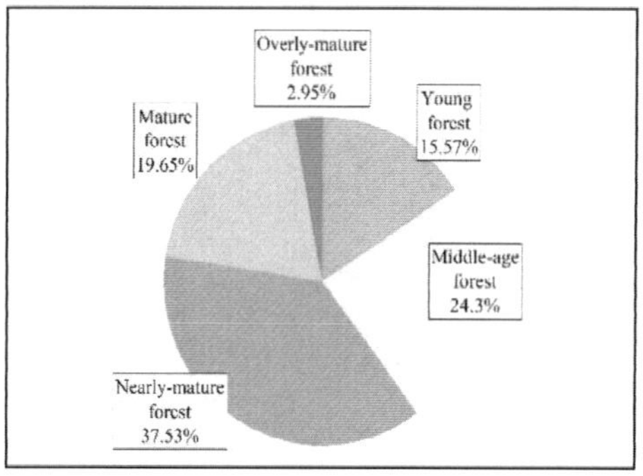

Figure 3. Planted forest carbon stores by age group.

4.2. Analysis by tree species(group)
As shown in Figure 4 and 5, 14 different types of dominant species forest were identified in ninth forest resources inventory in Henan Province. Major tree species(groups) were Pinus tabuliformis, Pinus massoniana, Other pines, Cunninghamia lanceolata, Cupressus funebris, Quercus spp., Robinia pseudoacacia, Miscellaneous broadleaf hardwoods, Populus L., Paulownia Sieb., Miscellaneous broadleaf softwoods, Mixed coniferous, Mixed broadleaf, Mixed coniferous and broadleaf, respectively. The total of these 14 types accounted for 82.63% of the forest area and 95.65% of the arboreal forest area. The total biomass and total carbon storage of these dominant species are 218.18×10^6t and 91.23×10^6t respectively, accounting for 97.79% of the total.

The largest type is mixed broadleaf, followed by Quercus spp. and Populus L. The biomass and carbon reserves of Populus L. are the largest, accounting for 30.23%. The second is Mixed broadleaf and Quercus spp., accounting for 27.45% and 21.42% of the carbon reserves of arboreal forest in Henan Province. Except for Other pines, Cunninghamia lanceolata, and Mixed coniferous, the remaining 8 types of carbon reserves accounted for 1% to 4%.

4.3. Analysis of geographical distribution patterns
From the perspective of space, Luoyang city had the highest carbon storage of 2.2355 million tons, followed by Nanyang city with 2.0825 million tons. The forest carbon storage of Hebi City is the least (Figure 6). Luoyang is one of the areas with large forest resources in Henan Province. The geographical location of Nanyang city and Xinyang City determines that they have a wide forest area, many trees in mountainous areas, and sufficient rain, which makes them grow well. Hebi, Luohe and Xinxiang have less water area and insufficient rainfall. The natural geographical location determines their less forest area because the main condition for forest growth is water. Where there is more water, trees grow well.

Because the mountainous areas of Henan province are distributed in the west and southwest which is also the mainly forest resources. It clear that the amount of carbon storage of forestry in the mountainous areas was significantly larger than that in plain areas. Although in the plain areas, the amount of carbon absorption of crops was less than that of forestry. In the future, in the development of the forestry industry in Henan province, different management measures should be taken according to different regions to make the forest a benign state of development.

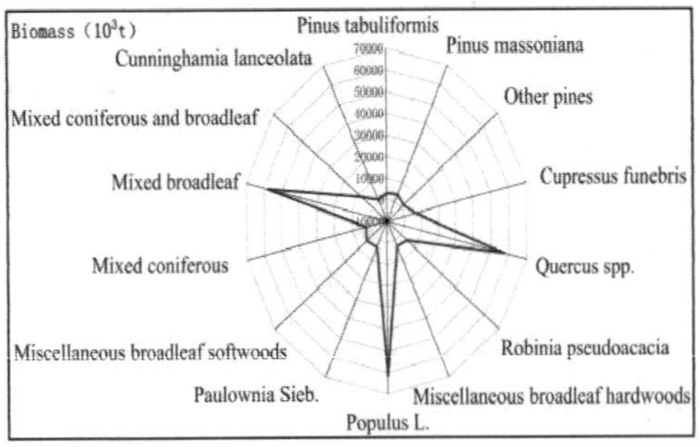

Figure 4. Subfigures of biomass of these dominant species.

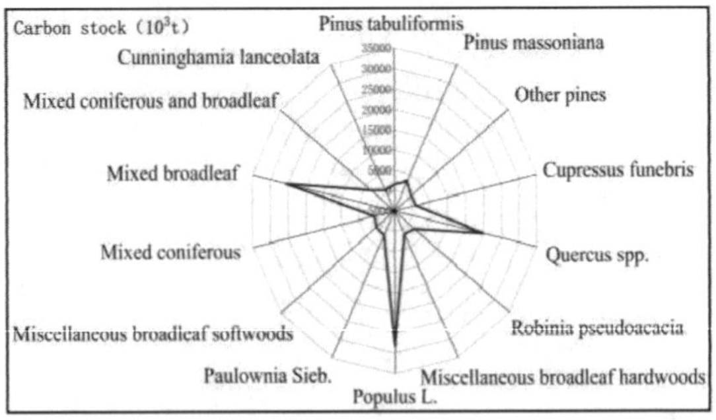

Figure 5. Subfigures of carbon stock of these dominant species.

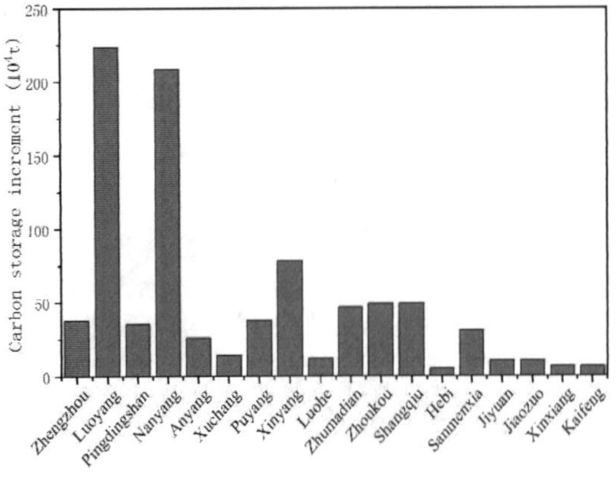

Figure 6. Forest carbon storage increment of 18 cities in Henan.

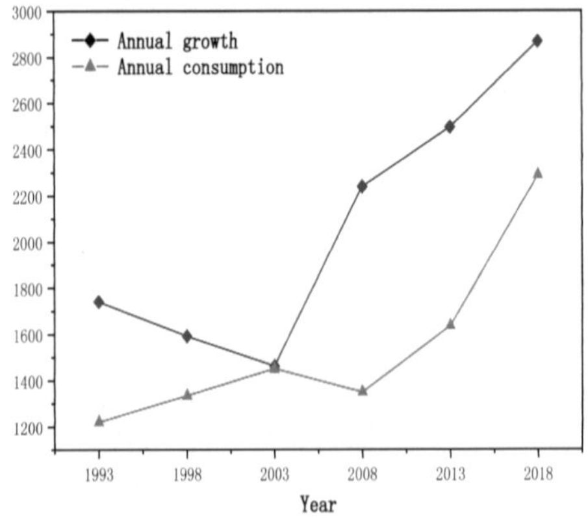

Figure 7. Dynamic change trend of annual growth and consumption ($10^4 m^3$).

4.4. Discussions

The results showed that the total stock volume of the arboreal forest was 207,191,200 cubic meters, and the area was 3,482,700 hectares. By formula (3) and (4) can be calculated that the carbon sink of arboreal forests was 8.891 million tons in 2018, and the potential carbon stock capacity was 93.29 million tons. Figure 7 shows that the annual growth of the forests is greater than the annual consumption. The annual growth of trees in 2018 was 28.6781 million cubic meters, an increase of 3.7194 million cubic meters over 2013. The annual consumption is also increasing compared with the previous years, and the growth rate from 2013 to 2018 is greater than the annual growth rate.

In summary, the carbon sink of the arboreal forests has increased to 8.891 million tons in 2018, which equivalent to 32.60 million tons of carbon dioxide equivalent. According to the average annual consumption rate of forest trees (6.82%) in 2018, we can know that consumption of carbon emissions was 6.36 million tons. Calculated in the same method, the forest growth and carbon sink were 8.005 million tons in 2013. Compared with the results of 2013 and 2018 showed that due to the increase of forest stock volume and annual growth rate, the carbon sink of arboreal forests increased 0.886 million tons in 2018 more than in 2013, and became the major contributor of forest carbon sink. It is critical to quantify carbon(C) storage and sink by forests to play the actual and potential role of making assessment of the forests in reducing atmospheric CO_2. China's forestry has the mission of responding to climate change issues. China has made deep progress in the research and evaluation of forest biomass and carbon storage. As the main body of the construction of the Central Plains Economic Zone, it is a relatively high carbon emissions district for a long time in Henan Province, and must develop forestry carbon sequestration to reduce CO_2 emissions owing to the pressure by the current severe reductions situation. Based on the previous analysis, our suggestions put forward the following two aspects. The first is that it should strengthen afforestation and pay attention to the cultivation of forest resources and accelerating the construction of forestry ecological province. The second aspect is that it should improve forest management and increase the investments in forestry technology and personnel training work, and build local public participation platform in order to provide scientific support for the rapid and healthy development of forest carbon sinks in Henan Province.

5. Conclusions

By comparing and analyzing the data of forest resources inventory, the biomass and carbon storage of arboreal forests in Henan Province are 218.18×10^6t and 91.23×10^6t respectively. At present, the average annual increase of forest carbon storage in China is more than 200 million tons. In 2018, the carbon storage increment of arboreal forests in Henan Province was 8.891 million tons, equivalent to 32.6 million tons of carbon dioxide equivalent. The carbon density of arboreal forest in Henan Province was 26.79t/ha, far lower than the national average. The proportion of biomass and carbon storage of young forests is the highest, and that of overly-mature forests is the lowest. Carbon density increased gradually and tended to be stable in the mature stage of forest development with the increase of forest age. Among the 14 dominant tree species, Populus L. had the highest biomass and carbon storage, followed by Mixed broadleaf and Quercus spp. The carbon storage of these three types account for 80% of all in Henan Province. Populus L. has the highest carbon density of all forest types. The carbon density of Cupressus funebris was the lowest. By region, Luoyang city has the largest forest carbon sink about 2.2355 million tons, Nanyang city has about 2.0825 million tons. Hebi city has a minimum of 0.0523 million tons. Therefore, the ecological projects needs to steadily expand the forest area, promote scientific forest management, improve forest quality and efficiency, and enhance forest ecological function.

References

[1] Houghton R A, Hackler J L, Lawrence K T 1999 *Science* **285** 574-578
[2] Brown S, Lugo A E 1984 *Science* **223** 1290-1293
[3] Zhang J P, Zhang L B, Hao H G, Liu C L, Wang H 2019 *Journal of Forestry Research* **30(3)** 1011-1021
[4] Tie Z, Wu S H, Zhou X R, Zhang Z H 2021 *Greening and life* **5** 4-7
[5] Fang J Y, Chen A P, Peng C H, Zhao S Q, Ci L J 2001 *Science* **292** 2320-2322
[6] Kauppi P E, Mielikainen K, Kusela K 1992 *Science* **256** 70-74
[7] Kauppi P E, Ausubel J H, Fang J Y, et al. 2006 *Proc Natl Acad Sci USA* **103** 17574-17579
[8] IPCC 2006 IPCC guidelines for national greenhouse gas inventories 2009 http://www.ipcc-nggip.iges.or.jp/public/2006gl/index.html
[9] Fang J Y, Brown S, Tang Y H, et al. 2006 *Climate Change* **74** 355-368
[10] Fang J Y, Piao S L, Field C B, et al. 2003 *Front. Ecol. Environ* **1** 293-297
[11] Fang J Y, Chen A 2001 *Acta Bot. Sin* **9** 967-973
[12] Tan K, Piao S L, Peng C H, Fang J Y 2007 *Forest Ecology and Management* **240** 114-121
[13] Peng C H, Zhou X L, Zhao S Q et al. 2009 *Global and Planetary Change* **66** 179-194
[14] Guo Z D, Fang J Y, Pan Y D, Birdsey R 2010 *Forest Ecology and Management* **259** 1225-1231
[15] Fang J Y, Chen A, Zhao S, Ci L 2002 *Acta Phytoecol. Sin* **2** 243-249
[16] Shao W, Cai J Y, Wu H B, et al. 2017 *Forests* **8(4)** 110-119
[17] People's Republic of China Ministry of Forestry 2019 National Forest Resource Statistics (2014—2018) China Forestry Publishing House, Beijing, China

Study on purification performance of nitrogen and phosphorus in landscape water by enhanced combined ecological floating bed

Minjuan Zhao[1,2,6], Yu Duan[1,2], Huan Zhang[1,2], Shanshan Liu[1,2], Yuji Li[1,2], Lin Xiao[1,2], Youyi Zhao[3], Hao Zhang[4,5] and Tianpeng Gao[1,2,3,6,7]

[1] School of Biological and Environmental Engineering, Xi'an University, Xi'an 710065, China;
[2] Xi'an Key Laboratory of Plant Stress Physiology and Ecological Restoration Technology in Qinling Mountains, Xi'an 710065, China;
[3] The Engineering Research Center of Mining Pollution Treatment and Ecological Restoration of Gansu Province, Lanzhou City University, Lanzhou, 730070, China;
[4] College of Civil Engineering and Mechanics of Lanzhou University, Lanzhou 730030, China;
[5] Key Laboratory of Mechanics on Disaster & Environment in Western China, Ministry of Education of China, Lanzhou 730030, China;
[6] Academy of Eco Xi'an, Xi'an 710065, China.

[7]Correspondence author's e-mail: zkgtp@163.com

ORCID: 0000-0001-6821-3836

Abstract. Taking Hornwort, Hydrilla verticillata, Iris, Calamus and Loosestrife as the research objects, under the static water condition of a university landscape lake in Xi'an, the combined ecological floating bed composite system of emergent plant submerged plant matrix microorganism enhancement was constructed to carry out the simulated hydroponic experiment for the investigation of the absorption to nitrogen and phosphorus in the water and the purification ability of the ecological floating bed composite system. The results showed that: (1) The best group of combined plants in the landscape lake water was calamus + Hydrilla verticillata, and their removal rate on Total nitrogen, Total phosphorus, Chemical oxygen demand, and ammonia nitrogen are 92.72%, 81.26%, 68.55%, 93.69% respectively. (2) Under the composite floating bed matrix system, the optimal substrate group and ratio are ceramsite and vermiculite (1:1), and their removal rates on Total nitrogen, Total phosphorus, Chemical oxygen demand, and ammonia nitrogen are 61.68%, 86.96%, 79.65%, and 78.12% respectively. (3) The best residence time of AM bacteria in the enhanced combination floating bed was 25 days, and the removal efficiency of total nitrogen and total phosphorus was 67.31% and 93.20% respectively.

1. Introduction

Landscape water is an important part of urban ecological environment, most of which are designed as static or slow-flow water with poor fluidity, and are prone to eutrophication, which leads to the decline of self-purification ability of water and is difficult to recover in a short term. According to the water quality survey data of major cities in China, about 80% of the landscape lakes have reached the degree of eutrophication [1]. According to the international evaluation standards of lake eutrophication [2,3]. (T-N: 0.2mg/L; T-P: 0.02mg/L), the Total nitrogen concentration of these landscape water bodies is generally 8 ~ 100 times higher than the standard, while the Total phosphorus concentration is 5 ~ 50 times higher than the standard [4]. Therefore, the in situ ecological restoration of landscape water is particularly important. Although ecological floating bed has achieved a good purification effect in the restoration of landscape water bodies, the screening of native floating bed plants, the purification mechanism of plants, the ratio of plant substrates, and the joint effect of plants and microorganisms still need further research and discussion. In the manuscript, a new ecological floating bed technology combining the traditional ecological floating bed with the microorganism of biological filler is proposed. It is hoped that it can further optimize the selection of dominant plant combinations, the optimal matrix fillers and the plant-matrix-microbial joint remediation mechanism, and provide reference for the in-situ remediation of landscape water bodies.

2. Materials and methods

2.1. Experimental material

This project intends to take the landscape lake water of a university as the experiment water, and the water quality indicators are shown in Table 1. The sampling point of test water is shown in Figure 1, B, C, D, F, E were the sampling sites.

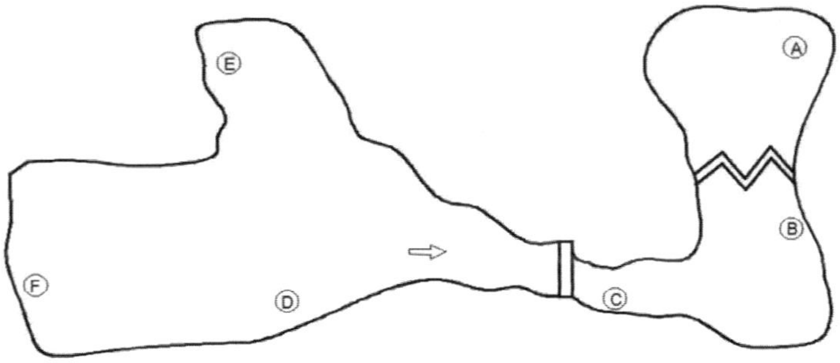

Figure 1. Schematic diagram of sampling points.

Table 1. Basic properties of experimented water samples.

water quality index	Chemical oxygen demand/(mg/L)	Total phosphorus/(mg/L)	Total nitrogen/(mg/L)	ammonia nitrogen/(mg/L)
concentration	30~70	0.5~1	8~15	5~9

Five plants with better nitrogen and phosphorus removal effects were screened out after a 40-day pre-experiment by consulting a large number of data, in-depth field investigation and comparison, including three kinds of emergent plants [5]: Iris, Calamus, and Loosestrife, two kinds of submerged plants, Hydrilla verticillata and Hornwort; ceramsite, vermiculite and quartz sand were selected as fillers; The microorganism used in the experiment was purchased from an environmental protection biological company, and AM bacteria agent was obtained after screening.

IOP Conf. Series: Earth and Environmental Science 1087 (2022) 012032 doi:10.1088/1755-1315/1087/1/012032

2.2. Experimental design
The experiment was carried out in artificial greenhouse of a university. A static hydroponic experimental water tank with a length of 64cm, a width of 44cm and a height of 41cm is selected. A floating plate of 40mm × 60mm, with six hole, each of 8 ~ 10cm diameter to fix emergent plants on the water surface in the water tank. This experiment was divided into three phases. Firstly, through indoor simulated hydroponic experiment, the growth characteristics, nitrogen and phosphorus accumulation capacity and purification effect of different aquatic plant species in landscape water were studied, and the dominant indigenous plant species suitable for local area were screened out; Secondly, determine the optimal substrate ratio through the combination of dominant plants and substrates; Finally, through the organic combination of combined ecological floating bed and biological packing, a plant-matrix-microorganism compound system was constructed to form the strengthened combined ecological floating bed. The schematic diagram of the test process is shown in Figure 2.

Figure 2. Schematic diagram of the test process

2.3. Experiment method

2.3.1. Index determination method. Monitoring was conducted every 5 days during the experiment, 100ml was sampled in each group of experiment boxes, and 100ml of blank raw water was added to the boxes at the same time, and the water evaporated by transpiration was supplemented with distilled water on time. Three repetitions are set for each combination, and blank control is set in addition. The experiment was divided into three phases. Each issue lasts 25 days. The water quality index was determined in accordance with the "Water and Wastewater Monitoring and Analysis Method" (State Environmental Protection Administration, 2002), and the concentrations of Total nitrogen, Total phosphorus, Chemical oxygen demand and ammonia nitrogen required by the experiment were determined after each sampling. Total nitrogen was determined by ultraviolet spectrophotometry with alkaline potassium persulfate digestion (HJ636-2012). Total phosphorus was determined by ammonium molybdate spectrophotometry (GB11893-89). Chemical oxygen demand was determined by potassium dichromate method (HJ828-2017). Ammonia nitrogen using Nessler's reagent spectrophotometry (HJ535-2009). Plant height, biomass, total nitrogen and total phosphorus were measured. Total phosphorus in plants was boiled with H_2SO_4-H_2O_2 and determined by vanadium molybdenum yellow colorimetric method. Total nitrogen of plants was determined by Kjeldahl method.

2.3.2. Data processing and analysis. Microsoft Excel 2007 software was used for data processing and mapping, SPSS19.0 software was used for one-way ANOVA analysis and correlation analysis, and one sided t-test ($P < 0.05$) was used to analyze the difference significance and bivariate correlation between the groups. The data of Total nitrogen, Total phosphorus, Chemical oxygen demand and ammonia nitrogen in the water were analysed by orthogonal design. Duncan's new complex extreme difference method was used for multiple comparison between different treatments [6].

3. Result and analysis

3.1. Selection of dominant plants of aquatic plants

(1) Analysis of purification effect of single plant and plant combination purification system on Total nitrogen in landscape lake water

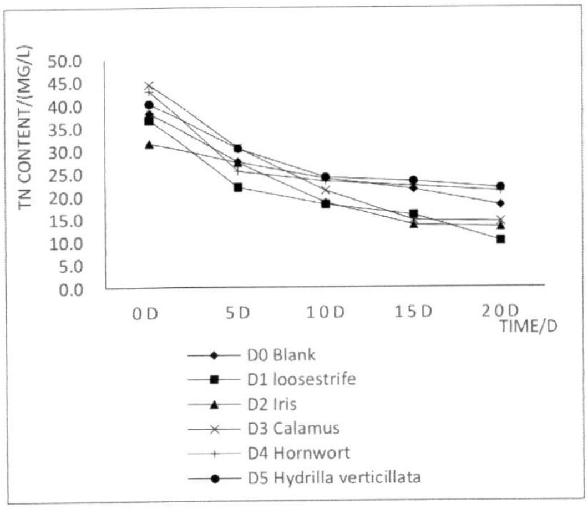

Figure 3. Changes of Total nitrogen concentration per plant with time.

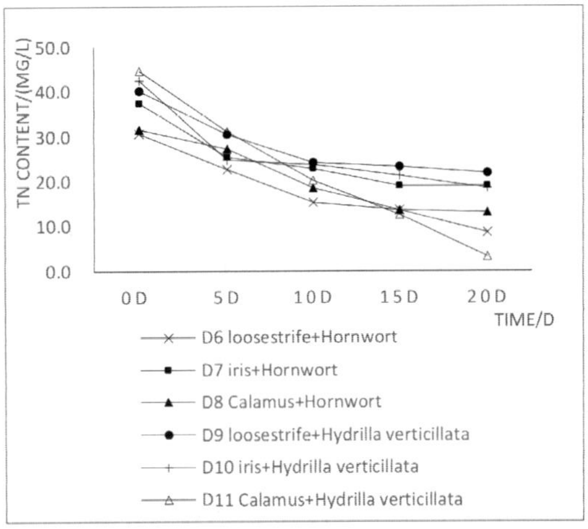

Figure 4. Changes of Total nitrogen concentration in plant combinations with time.

(2) Analysis of purification effect of single plant and plant combination purification system on Total phosphorus in landscape lake water

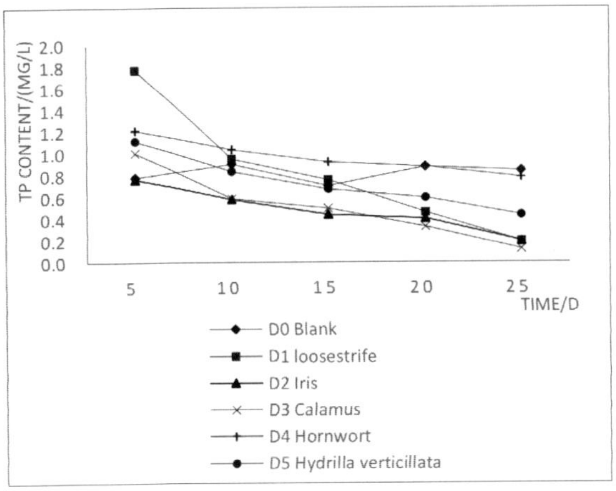

Figure 5. Variation of Total phosphorus concentration per plant with time.

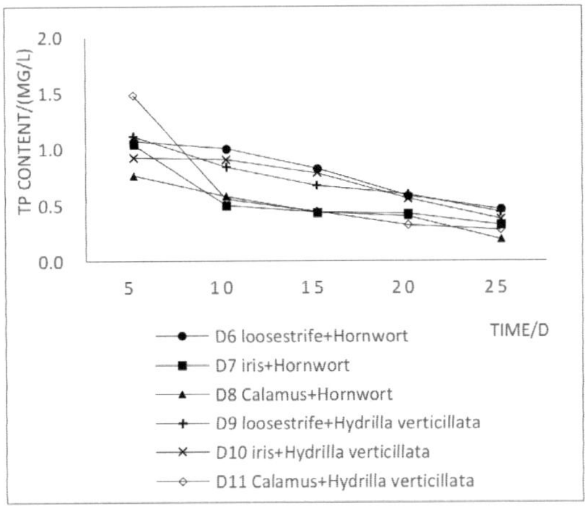

Figure 6. Changes of Total phosphorus concentration in plant combinations with time.

(3) Analysis of purification effect of single plant and plant combination purification system on Chemical oxygen demand in landscape lake water

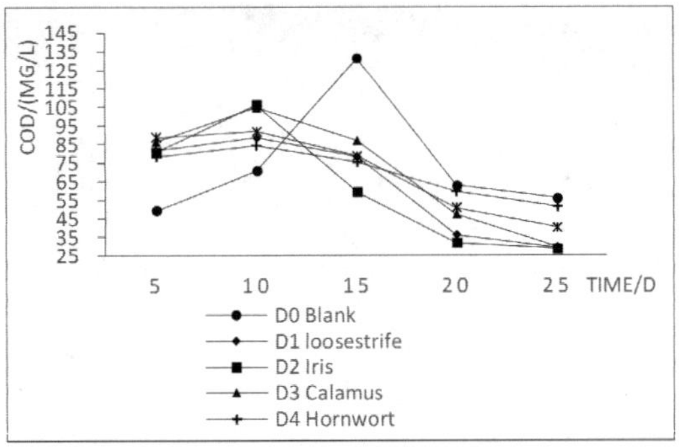

Figure 7. Variation of Chemical oxygen demand concentration per plant with time.

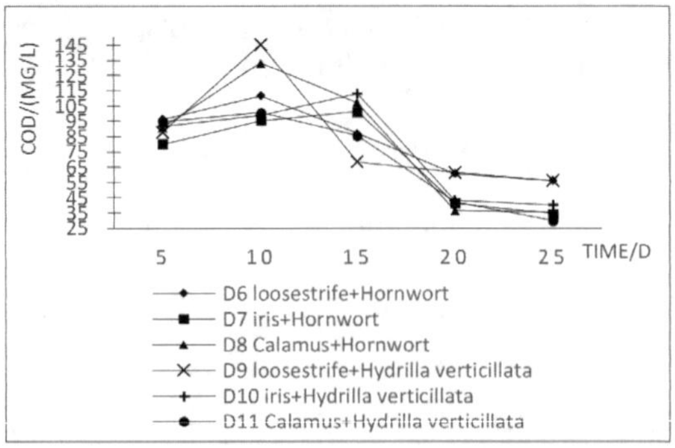

Figure 8. Variation of Chemical oxygen demand concentration of plant combination with time.

(4) Analysis of purification effect of single plant and plant combination purification system on ammonia nitrogen in landscape lake water

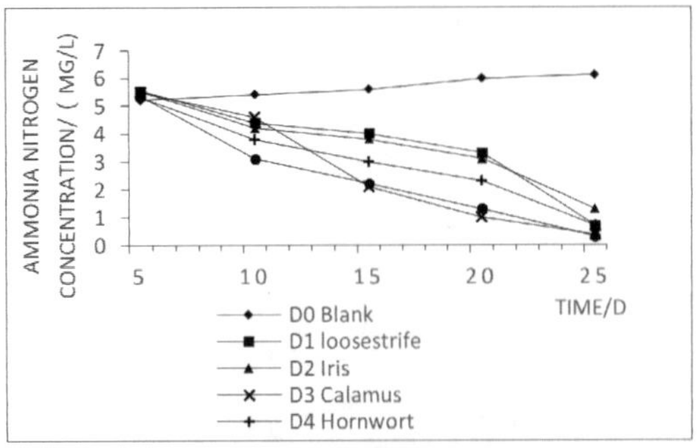

Figure 9. Variation of ammonia nitrogen concentration per plant with time.

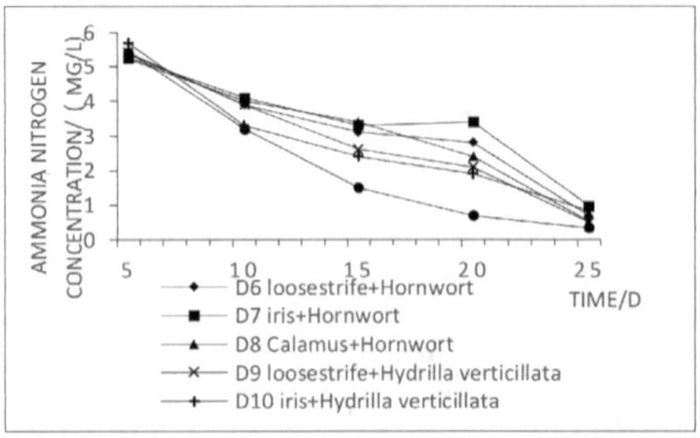

Figure 10. Variation of ammonia nitrogen concentration of plant combination with time.

(5) Plant index analysis

Table 2. Changes of plant height and biomass.

lant name	Initial state			End state			Net increment biomass dry weight /(g/m^2)
	Wet weight/ (g/m^2)	Moisture content /%	Plant height/ cm	Wet weight/ (g/m^2)	Moisture content/%	Plant height/cm	
Loosestrife	14.13	80.26	36.5	74.96	26.51	43.5	52.3
Iris	48.75	83.96	34.5	71.66	23.59	42.23	46.93
Calamus	48.47	80.04	30	63.73	19.95	33.83	41.34
Hornwort	57.4	93.74	18	65.21	17.9	31.25	49.94
Hydrilla verticillata	53.1	92.51	23	83.86	35.62	27.9	50.01

Table 3. Beginning and end of plant nitrogen and phosphorus content.

Plant name	Plant nitrogen content			Plant phosphorus content		
	Initial status /%	End status /%	Net increment nitrogen content /%	Initial status/%	End status /%	Net added phosphorus content/%
Loosestrife	1.542	2.021	0.479	0.369	0.442	0.073
Iris	1.65	2.338	0.688	0.319	0.398	0.079
Calamus	2.804	4.103	1.299	0.418	0.566	0.148
Hornwort	3.016	3.767	0.751	0.331	0.411	0.08
Hydrilla verticillata	3.7	5.314	1.614	0.668	0.911	0.243

From Figure 3 to Figure 10, it can be seen that the purification effects of five aquatic plants on Total nitrogen, Total phosphorus, Chemical oxygen demand and ammonia nitrogen in landscape water were different. The higher the species diversity of the community, the more perfect its ecological function [7]. The mixed communities composed of different emergent plants and submerged plants have different removal abilities of Total nitrogen and Total phosphorus in water, which is similar to the existing relevant research results [8, 9, 10, and 11]. The combined plants with the best removal effects of Total nitrogen, Total phosphorus, chemical oxygen demand and ammonia nitrogen in landscape lake water are Calamus + Hydrilla verticillata, and the removal rates are 92.72%, 81.26%, 68.55%, and 93.69% respectively.

It can be seen from Table 2 and Table 3 that the nitrogen and phosphorus content, plant height and biomass of the five aquatic plants have increased. Among them, the net value-added nitrogen and phosphorus content of Acorus Calamus is the highest, which is 1.299% and 0.148% respectively. The net value-added nitrogen and phosphorus content of submerged plant Hydrilla verticillata rosenbergii is the highest, which is 1.614% and 0.243% respectively. The net value-added biomass dry weight of the emergent plant is 52.30g/m^2, the net increment biomass and dry weight of submerged plant Hydrilla verticillata rosenbergii were up to 50.01g/m^2.

3.2. Selection of the best matrix combination ratio

Analysis of purification effect of dominant plants + different matrix ratio purification system on Total nitrogen, Total phosphorus, Chemical oxygen demand and ammonia nitrogen in landscape lake water

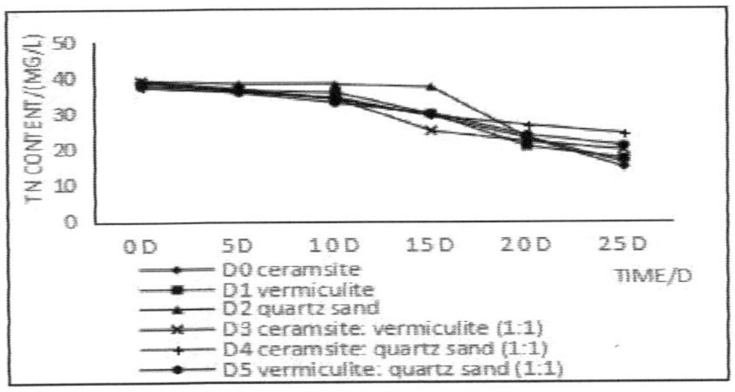

Figure 11. Variation of Total nitrogen concentration with time under different matrix ratios.

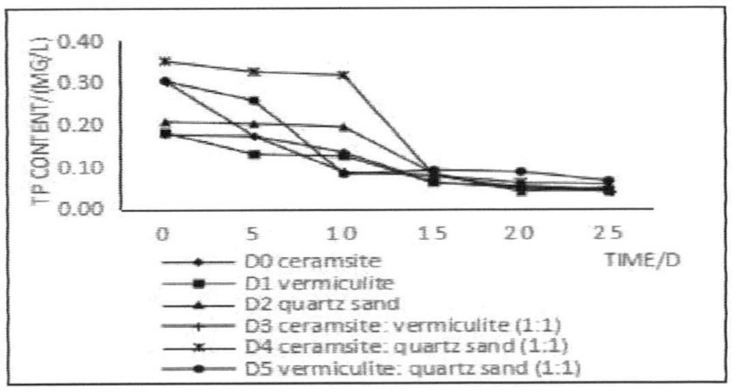

Figure 12. Variation of Total phosphorus concentration with time under different matrix ratios.

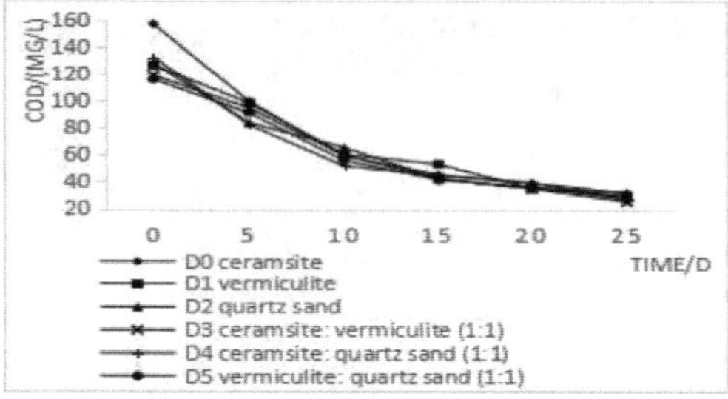

Figure 13. Variation of Chemical oxygen demand concentration with time under.

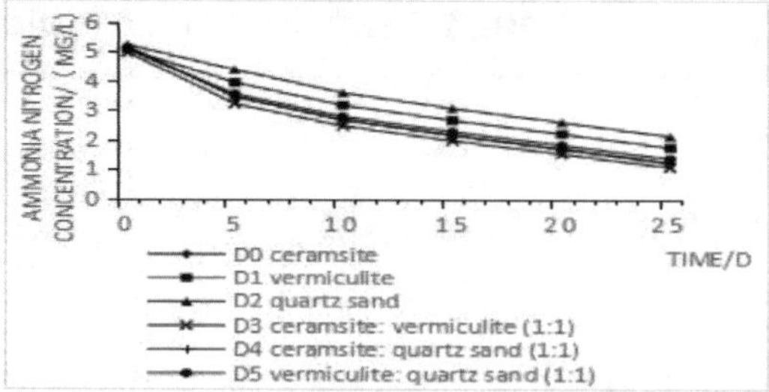

Figure 14. Variation of ammonia nitrogen concentration with time under different matrix ratios.

From Figure 11 to Figure 14, it can be seen that the contents of Total nitrogen, Total phosphorus, Chemical oxygen demand and ammonia nitrogen in the lake water showed a downward trend after 25 days of experiment. Different combinations have different effects on water purification. Specific conclusions are as follows: Calamus + Hydrilla verticillata + ceramsite had the best adsorption effect on Total nitrogen in Guanhu Lake water, and the removal rate was 61.68%; alamus + Hydrilla verticillata + ceramsite: vermiculite (1:1) had the best adsorption effect on Total phosphorus, and the removal rate was 86.96%; Calamus + Hydrilla verticillata + ceramsite had the best adsorption effect on Chemical oxygen demand and the removal rate was 79.65%; Calamus + Hydrilla verticillata + ceramsite: vermiculite (1:1) had the best adsorption effect on ammonia nitrogen and the removal rate was 78.12%.

3.3. Analysis on the effect of introducing AM bacteria agent inoculum on water quality

On the basis of the dominant plants and the best matrix ratio, AM bacteria agent inoculants were introduced.

3.3.1. Water quality effect analysis.

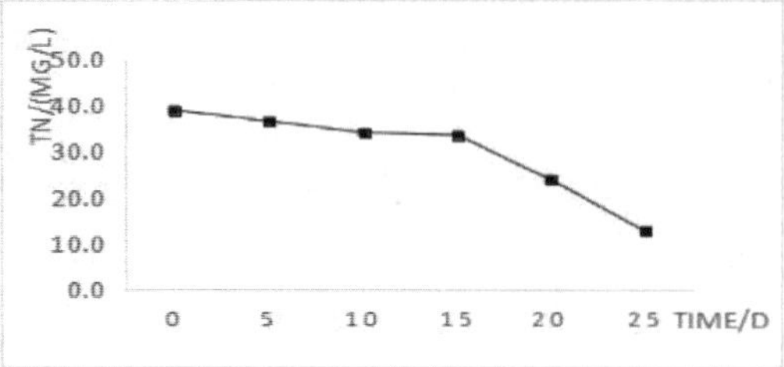

Figure 15. Total nitrogen removal efficiency of the optimal substrate after introduction of AM bacteria agent.

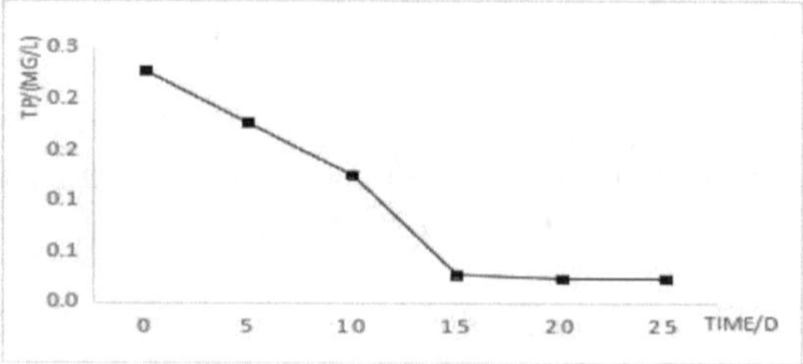

Figure 16. Total phosphorus removal efficiency of the optimal substrate after introduction of AM bacteria agent.

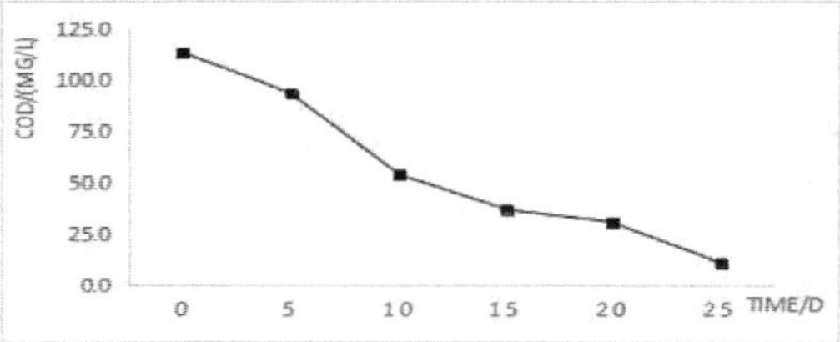

Figure 17. Chemical oxygen demand removal efficiency of the optimal substrate after introduction of AM bacteria agent.

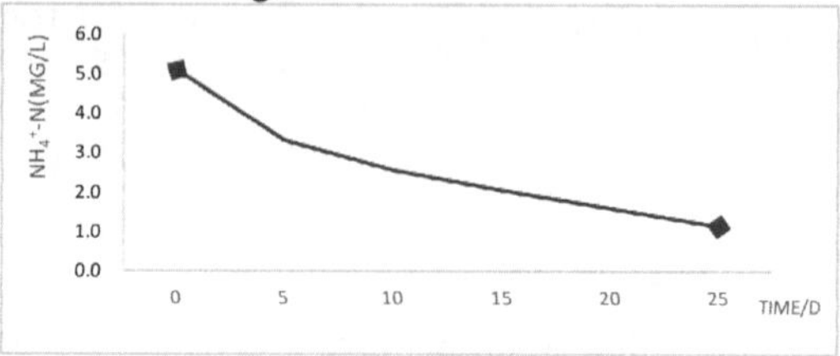

Figure 18. Removal efficiency of ammonia nitrogen after the introduction of AM bacteria agent into the optimal substrate.

From Figure 15 to Figure 18, it can be seen that the contents of Total nitrogen, Total phosphorus, Chemical oxygen demand and ammonia nitrogen in the lake water showed a downward trend after 25 days of experiment. Different combinations have different effects on water purification. Specific conclusions are as follows: Total nitrogen in landscape lake decreased from 38.877 mg/L to 12.708 mg/L, and the removal rate was 67.31%; Total phosphorus decreased from 0.227 mg/L to 0.015 mg/L, and the removal rate was 93.20%; Chemical oxygen demand decreased from 113.152mg/L to 10.800mg/L, and the removal rate was 90.48%; Ammonia nitrogen decreased from 5.050mg/L to 1.130mg/L, and the removal rate was 77.62%.

3.3.2. Stay time analysis. The degradation of organic matter in water is affected by many factors, such as water temperature, dissolved oxygen, the species and quantity of microorganisms, and it is degraded by anaerobic degradation, anoxic degradation and aerobic degradation [12]. The microorganism degrades the organic matter in the sewage through the metabolism of the microorganism so as to reduce the concentration of the organic matter; The degradation of organic matter by plants is relatively limited, mainly relying on the enrichment of microorganisms on the root system to achieve degradation, the developed root system of Calamus and Hydrilla verticillata provide conditions for the life of microorganisms, thus consuming a large amount of organic matter. In this experiment, the best retention time of AM bacteria agent was 25 days.

4. Conclusions

Through the average interval analysis of each factor, we get the main factors affecting the index and the optimal combination of each factor level. The removal effects of Total nitrogen, Total phosphorus, Chemical oxygen demand and ammonia nitrogen in different combinations were compared and analyzed. The conclusions are as follows:

(1) The results of the screening experiment of the dominant plants in the first phase show that: Calamus and Hydrilla verticillata had the best removal effect on Total nitrogen, Total phosphorus, Chemical oxygen demand and ammonia nitrogen, and the removal rates were 92.72%, 81.26%, 68.55% and 93.69% respectively;

(2) The results of the second stage of dominant plant combination + substrate experiment showed that: The best adsorption effect of floating bed on Total nitrogen and Chemical oxygen demand was ceramsite, and the removal rates were 61.68% and 79.65% respectively; Ceramsite: vermiculite (1:1) had the best adsorption effect on Total phosphorus and ammonia nitrogen, and the removal rates were 86.96% and 78.12% respectively;

(3) The results of the third phase of the experiment of introducing AM bacteria agent inoculants on the basis of the optimal plant combination and the optimal substrate ratio show that: The removal rates of nitrogen and phosphorus were 67.31% and 93.20% respectively when the retention time was 25 days.

5. Discussions and outlook

This experiment discussed the purification effect of different plant combinations, plants + different substrates and substrate ratios, and the introduction of AM bacteria on landscape water bodies, the follow-up study of this experiment still needs further carry out in situ ecological restoration pilot test, verify and revise the results of this experiment, in order to explore the optimal parameters of the combined ecological floating bed. Microorganisms that have a good effect on the removal of total nitrogen and total phosphorus can be introduced from native water bodies to improve the floating bed structure, such as DM composite bacteria, EM composite bacteria and total phosphorus degrading bacteria; the dense planting of plants will also affect the purification effect of ecological floating beds on water bodies, and the planting density of plants should be further studied. Adjust the structure of the floating bed, the three-dimensional space of the combined ecological floating bed is expanded and strengthened, and the connection among the floating bed, plants, animals and microorganisms is strengthened, which extends the food chain. Using a series of life activities such as absorption, uptake, digestion and decomposition of aquatic plants, animals, insects and microorganisms in the natural aquatic environment to achieve ecological management of eutrophicationic water bodies, constructing an upgraded version of the plant-substrate-microorganism-protozoa combined floating bed, and enhance the ecological stability of the floating bed.

Conflicts of interest

The authors declare that the research was conducted in the absence of any commercial or financial relationship that could be construed as a potential conflict of interest.

Author Contributions

Zhao M J, Duan Y, Zhang H, Liu S S: conceptualization, writing, and review. Li Y J, Xiao L, contributed to the review and editing of this perspective. Zhang H, Zhao Y Y were responsible for sampling, Gao T P: supervision.

Acknowledgments

This work was financially supported by the National Natural Science Foundation of China (NSFC 31860176), Key Research and Development Program of Gansu Province (20YF3FA037), Key Research and Development Program of Shanxi Province (2020ZDLSF06-06) and XAWLKYTD012. We are grateful to all anonymous reviewers whose comments improved the quality of the manuscript. Many thanks should be given to Xi'an Key Laboratory of plant stress physiology and ecological restoration technology, Key Laboratory for ecological restoration and high quality development of Qinling Mountains in the upper and middle reaches of the Yellow River for their help.

References

[1] Fang Xiaobo, Zhang Jianying, Chen Wei, Xu Xiangyang, Chen Yingxu 2007 *[J] Journal of Environmental Science* **(08)** 1402-1407 DOI:10.13671/j.hjkxxb.2007.08.009
[2] Zhu yiping, Zhang Haiping, Chen Ling 2007 *[J] Water supply and drainage in China* **(10)** 36-38
[3] Zhao Xiaoguang, Zhao Xue 2006 *[J] Safety and environmental engineering* **(03)** 36-39
[4] Dai Dongchao 2006 *[D] Xi'an: Xi'an University of Architecture and Technology*
[5] Wang Zhanshen, Zhao Yixi, Li Chao, Hang Yin, Geng Yongbo, Liang Wenyan 2018 **40(06)** 627-633 DOI:10.15985/j.cnki.1001-3865.2018.06.002
[6] Jin Shuquan, Zhou Jinbo, Bao Weihong, Chen Jun, Li Dandan, Li Yang 2017 *[J] Environmental Science* **38(01)** 156-161 DOI:10.13227/j.hjkx.201606135
[7] Sun Peng 2016 *[D] China University of Geosciences*, Beijing
[8] Bai Xuemei, he liansheng, Li Bicai, Meng Rui, Meng fanli, Huang Caihong, Dong Jun, Li Gang 2013 *[J] Wetland Science* **11(04)** 495-498 DOI:10.13248/j.cnki.wetlandsci.2013.04.00
[9] Yang Juan, Zheng Guoqi, Zhang Lei 2012 *[J] Northern Horticulture* **(16)** 56-59
[10] Zhang Ling, Liu Hanhu 2009 *[J] Scientific and Technological Information* **(02)** 27-30
[11] Peng wanting, Zou Lin, Duan Weibo, Li Yinxia, Pan Yuanzhi, Zeng Degang 2012 *[J] Journal of Environmental Science* **32(03)** 612-617 DOI:10.13671/j.hjkxxb.2012.03.021
[12] Li Huan, Ng Wei, Luo Fangli, Huang Wenjun, Zhang Mingxiang 2016 *[J] Wetland Science* **14(02)** 163-172 DOI:10.13248/j.cnki.wetlandsci.02.004

IOP Conf. Series: Earth and Environmental Science 1087 (2022) 012033 doi:10.1088/1755-1315/1087/1/012033

Towards mesoscale ensemble prediction for the Guangdong-Hong Kong-Macao Greater Bay Area

Kai-kwong Hon[1,2]

[1] Hong Kong Observatory, Tsim Sha Tsui, Hong Kong, China.

[2] Corresponding author's e-mail: kkhon@hko.gov.hk

Abstract. The Guangdong-Hong Kong-Macao Greater Bay Area (GBA) is amongst the most economically vibrant regions in China. Situated along the southern coast of China, the GBA region is susceptible to a variety of synoptic and mesoscale weather systems originating from the mid-latitudes or the tropical oceans, which are capable of bringing significant impact to the livelihood and economic activities of this densely populated area. This paper presents a first study on the development of mesoscale ensemble prediction capabilities aimed at providing probabilistic forecast support of high-impact weather over the GBA.

1. Introduction

The Guangdong-Hong Kong-Macao Greater Bay Area (GBA) comprises two Special Administration Regions (SAR), namely the Hong Kong SAR and the Macao SAR, as well as nine Pearl River Delta Municipalities, and covers a total area of 56,000 km^2 with a combined population of approximately 70 million (figure estimated at the end of 2017) [1]. It is one of the most economically vibrant regions in China. Situated along the southern coast of China with complex geography and orography, the GBA region is susceptible to a variety of synoptic and mesoscale weather systems originating from the mid-latitudes or the tropical oceans, which are capable of bringing significant impact to the livelihood and economic activities of the region [2]. For example, severe convective weather such as squall lines or quasi-stationary rainbands might bring damaging winds and lightning risks in addition to widespread heavy rain [3,4]. Tropical cyclones impacting or making landfall along the coastal areas of GBA may bring such hazards as sustained high winds, torrential rainfall and storm surge [5].

Weather evolution is inherently chaotic [6] (Figure 1). Short-term errors at the convective or meso-scale may also propagate upscale to influence predictability of larger scale atmospheric flows in the medium range [7]. In terms of operational forecasting, the use of global ensemble predictions systems is relatively well established around the world, despite their relatively coarse spatial resolution compared to their deterministic counterparts [8]. However, there remains considerable challenges in the implementation of mesoscale, regional ensemble prediction systems because of the associated technical demands and high computational requirements [9].

This paper presents a first study on the development of mesoscale ensemble prediction capabilities aimed at providing probabilistic forecast support of high-impact weather over the GBA. Section 2 gives an overview of the model configuration. Section 3 demonstrates probabilistic forecast products for selected high-impact weather episodes over GBA. Section 4 summarises the paper and discusses possible way forward in future development of mesoscale ensemble prediction technology.

2. Description of Mesoscale Ensemble Prediction System (MEPS)

In this section, we describe the configuration of the experimental Mesoscale Ensemble Prediction System (MEPS) including technical overview of its model settings.

2.1. Mesoscale ensemble prediction system

Figure 2 shows the forecast domain of the MEPS. It can be seen that the GBA region is located at the center of the forecast domain, which allows effective coverage of weather systems of both continental (e.g. cold fronts or northeast monsoons) and maritime (e.g. tropical cyclones or summer monsoon) origin that may potentially affect the GBA. Being a regional prediction system, the MEPS is initialised using latest available output data from the NCEP GEFS, which provides the initial and boundary conditions. To prevent statistical imbalance which may arise from the selection of a subset of parent ensemble members, all available members of GEFS are used. At the time of writing, the MEPS comprises 30 perturbed members, following the latest system upgrade of GEFS. To supplement ensemble spread, stochastically perturbed physical tendencies are applied [10]. However, the choice of physical parameterisation schemes is uniform across members, based on experience and to allow for greater numerical stability. The MEPS may be run 4 times at day (00, 06, 12, 18 UTC) up to 120 hours ahead.

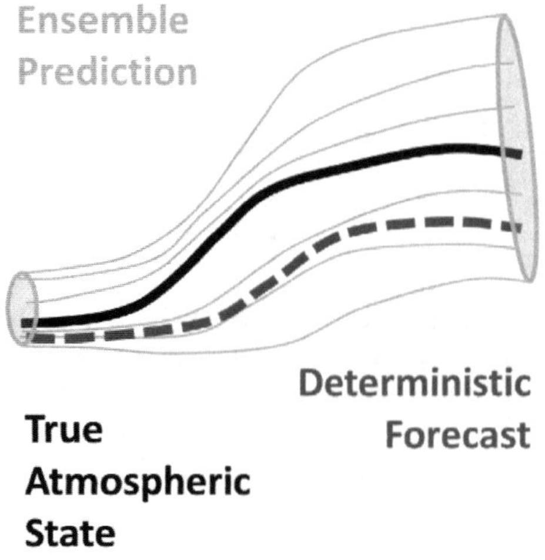

Figure 1. Schematic diagram of chaotic weather evolution and the use of ensemble prediction.

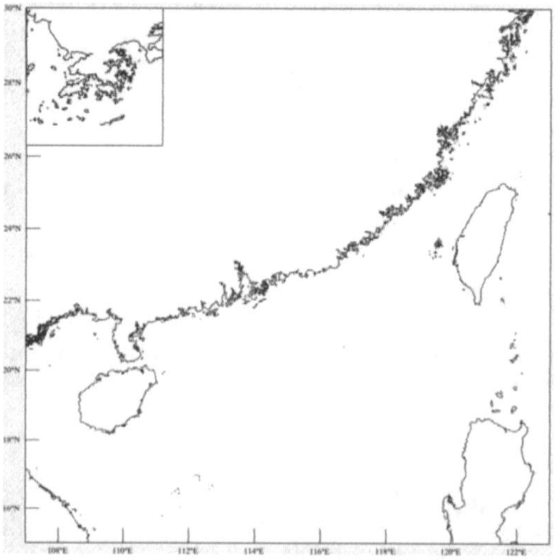

Figure 2. Schematic diagram of the forecast domain of the experimental Mesoscale Ensemble Prediction System.

2.2. Detailed model technical settings

The numerical weather prediction (NWP) model employed in the MEPS is the Weather Research and Forecast (WRF) model [11]. Version 3.9.1.1 of the Advanced Research WRF (WRF-ARW) core is used. Horizontally there are 301 x 251 grid-points at 10-km resolution. Vertically there are 36 levels with a prescribed model top of 15 hPa. To speed up computation, adaptive time-stepping is enabled and typically the integration time step is between 90 and 120 seconds, or smaller in presence of convective weather systems or high surface winds. USGS 3-second topography and MODIS land use data, both with local adaptations near Hong Kong and along the South China coastal areas to cater for urbanisation and possible coastline changes, are used.

The physical parameterisation schemes used in the MEPS are similar to those report in [12] which is a large-area 10-km resolution deterministic implementation of the WRF model spanning the Indo-Pacific region. These include the WRF Double Moment 6-Class (WDM6) scheme for microphysics

[13], the RRTMG schemes for short- and long-wave radiation [14], the Noah land surface model [15], the YSU planetary boundary layer scheme [16], and the modified Tiedtke scheme for cumulus convection [17].

3. Probabilistic forecast products of MEPS

In this section, we demonstrate some probabilistic or ensemble-based forecast products focusing on high-impact weather over the GBA.

3.1. Ensemble stamp maps

To assist the weather forecaster in making quick visual evaluation of various possible weather scenario evolution, an ensemble stamp map display of simulated satellite images has been developed (Figure 3). The simulated satellite images are computed using an in-house cloudy radiance simulation algorithm [18] based on a community radiative transfer model. These allow a quick assessment of the possible spatial extent affected by weather systems of interest including alternative scenarios, as well as facilitate direct comparison against real-time satellite observations. Stamp map display is also available for other conventional parameters such as 1-hour rainfall and simulated radar reflectivity.

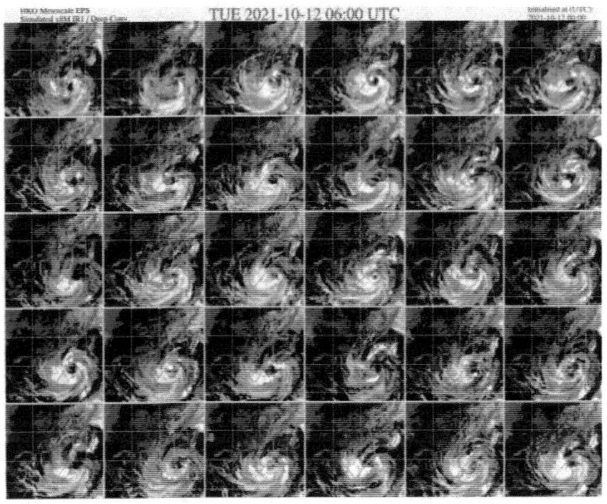

Figure 3. Sample ensemble stamp map product of simulated satellite images using the MEPS.

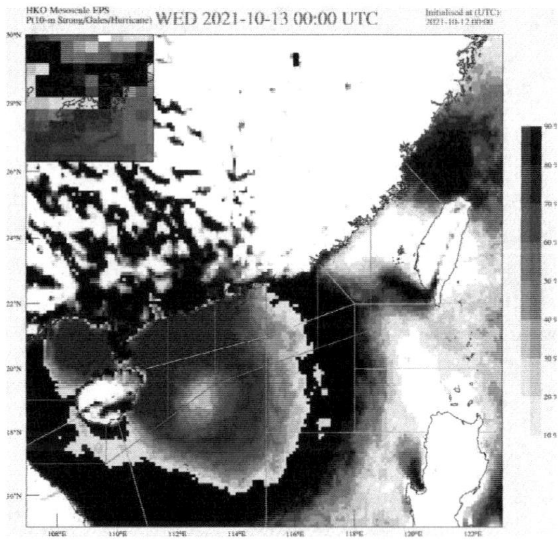

Figure 4. Sample forecast probability map of high winds using the MEPS.

3.2. Probability of high-impact weather elements

Using the output fields from individual ensemble members of the MEPS, it is possible to compute the exceedance or occurrence probability of various high-impact weather events. For example, the probability of surface winds exceeding pre-determined thresholds may be computed by counting the fraction of ensemble members reaching particular wind speeds at each grid-point (Figure 4), which is useful for assessing wind impact of tropical cyclones and monsoons over the GBA region and its adjacent coastal waters.

3.3. Interactive ensemble meteogram display

Figure 5 shows an interactive ensemble meteogram display using MEPS output at 1-hour intervals. Here the forecast time series of various weather elements of interest are generated for each MEPS ensemble member. Then they are summarised in the form of "box-and-whisker" display at grid-points of interest, for easy appreciation of location-specific hourly variations of interested parameters including the ensemble spread or uncertainty information, as well as key statistical indicators such as the minimum, maximum, and 25th/50th/75th/90th percentile values. For example, in addition to hourly rainfall, various atmospheric instability indices may be used to assess the potential for severe convective weather and thunderstorms over the GBA region [19].

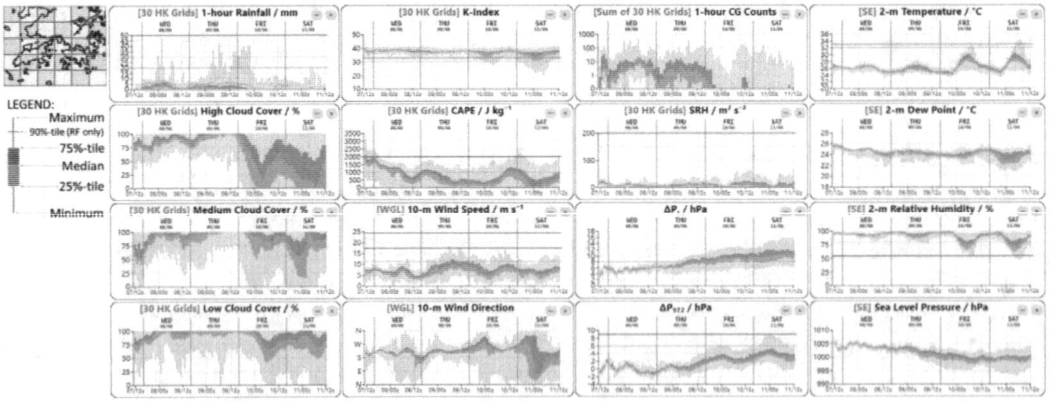

Figure 5. Sample interactive ensemble meteogram display of the MEPS for the GBA region, here focusing on grid-points around Hong Kong.

3.4. Tropical cyclone strike probability map

Probabilistic tropical cyclone forecast guidance can be generated through the identification and tracking of developing vortices in MEPS output. In particular, through compilation of forecast tracks generated from individual ensemble members, tropical cyclone strike probability map can be produced, which shows the probability of tropical cyclone occurrence at each location within the forecast time window. An example is shown in Figure 6, which shows that the MEPS prediction, despite a shorter forecast length, is on part with available major global ensemble prediction systems.

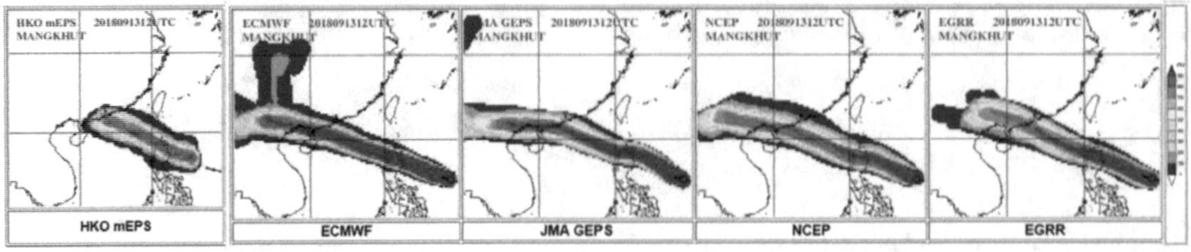

Figure 6. Sample tropical cyclone strike probability map of the MEPS (leftmost) shown alongside other major global ensemble prediction systems. Note the shorter forecast length of the MEPS.

4. Summary and discussions

Mesoscale ensemble prediction is a potentially powerful tool for addressing the forecast challenges posed by inherent uncertainties associated with mesoscale weather systems especially severe convection and tropical cyclones. The Guangdong-Hong Kong-Macao Greater Bay Area (GBA) is one of the most economically vibrant regions in China. However, given its geographical location along the southern coast of China, it is also susceptible to various types of high-impact weather which are capable of bringing significant impact to the people's livelihood and economic activities. This paper presents a first study on the development of mesoscale ensemble prediction capabilities aimed at providing probabilistic forecast support of high-impact weather over the GBA. Experimental runs and preliminary product development results indicate potentially skillful and useful application of the 10-km resolution Mesoscale Ensemble Prediction System (MEPS) output in the forecasting and assessment of high-impact weather, through probabilistic or ensemble-based approaches which are processed and visualised in compact graphical formats suitable for quick, intuitive appreciation by the operational weather forecaster.

Scientifically, mesoscale and convective-scale ensemble prediction is a rapidly developing field, which have seen significant advances in other parts of the world over the recent decade [20]. For the GBA region, apart from more detailed case studies of notable high-impact weather cases, long-term systematic validation efforts with explicit treatment of probabilistic aspects will be required to elucidate performance of the system as a whole [21]. Improvements in ensemble intialisation and perturbation techniques will allow better representation of forecast uncertainties through adequate ensemble spread or divergence [22]. With many recent advances in space-borne and ground-based remote-sensing observations, mesoscale ensemble prediction systems also provide a potential platform for greater utilisation of these spatio-temporally dense but highly non-Gaussian observations through ensemble-based, or hybrid, data assimilation techniques [23]. These areas of potential future development would no doubt contribute to the improved prediction of high-impact weather over the GBA region and beyond.

References

[1] Hong Kong SAR Government 2019 Outline Development Plan for the Guangdong-Hong Kong-Macao Greater Bay Area https://www.bayarea.gov.hk/filemanager/en/share/pdf/Outline_Development_Plan.pdf

[2] Choy C W, Wu M C, Lee T C 2020 Assessment of the damages and direct economic loss in Hong Kong due to Super Typhoon Mangkhut in 2018 *Tropical Cyclone Research and Review* **9(4)** 193-205

[3] Hon K K, Tong Y F 2011 Mesoscale analysis of an intense rainband associated with Typhoon Chanthu 25th Guangdong-Hong Kong-Macao Seminar on Meteorological Science and Technology, Hong Kong, 26-28 Jan 2011 https://www.hko.gov.hk/tc/publica/reprint/files/r942.pdf (Chinese only)

[4] Yeung W L, Chan N H, Hon K K 2014 First studies on storm motion vector computation over the South China coast 28th Guangdong-Hong Kong-Macao Seminar on Meteorological Science and Technology, Hong Kong, 13-15 Jan 2014 https://www.hko.gov.hk/tc/publica/reprint/files/r1103.pdf (Chinese only)

[5] Or M K, Hon K K, Wong W K 2013 Numerical simulation studies on Severe Typhoon Vicente 27th Guangdong-Hong Kong-Macao Seminar on Meteorological Science and Technology, Shaoguan, Guangdong, 9-10 January 2013 https://www.hko.gov.hk/en/publica/reprint/files/r1060.pdf

[6] Lorenz E N 1963 *Deterministic Nonperiodic Flow. J. Atmos. Sci.* **20** 130-141

[7] Selz T, Craig G C 2015 Upscale Error Growth in a High-Resolution Simulation of a Summertime Weather Event over Europe *Monthly Weather Review* **143(3)** 813-827

[8] Bauer P, Thorpe A, Brunet G The quiet revolution of numerical weather prediction *Nature* **525** 47-55

[9] Montani A, Cesari D, Marsigli C, Paccagnella T 2011 Seven years of activity in the field of mesoscale ensemble forecasting by the COSMO-LEPS system: main achievements and open challenges *Tellus A: Dynamic Meteorology and Oceanography* **63(3)** 605-624

[10] Ollinaho, Pirkka, et al. 2016 Towards process-level representation of model uncertainties: stochastically perturbed parametrizations in the ECMWF ensemble *Q.J. Royal Met. Soc.* **143(702)** 408-422

[11] Skamarock W C, Klemp J B 2007 A time-split nonhydrostatic atmospheric model for weather research and forecasting applications *J. Comp. Phys.* **227** 3465-3485

[12] Hon K K 2020 Tropical cyclone track prediction using a large-area WRF model at the Hong Kong Observatory *Tropical Cyclone Research and Review* **9** 67-74, https://doi.org/10.1016/j.tcrr.2020.03.002

[13] Lim K S S, Hong S Y 2010 Development of an effective double–moment cloud microphysics scheme with prognostic cloud condensation nuclei (CCN) for weather and climate models *Mon. Wea. Rev.* **138** 1587-1612
[14] Iacono M J, Delamere J S, Mlawer E J, Shephard M W, Clough S A, Collins W D 2008 Radiative forcing by long-lived greenhouse gases: Calculations with the AER radiative transfer models *J. Geophys. Res.* **113** D13103
[15] Niu G Y, Yang Z L, Mitchell K E, Chen F, Ek M B, Barlage M, Kumar A, Manning K, Niyogi D, Rosero E, Tewari M, Xia Y 2011 The community Noah land surface model with multiparameterization options (Noah–MP): 1. Model description and evaluation with local–scale measurements *J. Geophys. Res.* **116** D12109
[16] Shin H H, Hong S, Noh Y, Dudhia J 2013 Derivation of Turbulent Kinetic Energy from a First-Order Nonlocal Planetary Boundary Layer Parameterization *J. Atmos. Sci.* **70** 1795-1805
[17] Zhang C, Wang Y, Hamilton K 2011 Improved Representation of Boundary Layer Clouds over the Southeast Pacific in ARW-WRF Using a Modified Tiedtke Cumulus Parameterization Scheme *Mon. Wea. Rev.* **139** 3489-3513
[18] Hon K K 2018 Simulated satellite imagery at sub-kilometre resolution by the Hong Kong Observatory *Weather, May* **73(5)** 139-144, https://doi.org/10.1002/wea.3100
[19] Chan P W, Hon K K 2011 Application of ground-based, multi-channel microwave radiometer in the nowcasting of intense convective weather through instability indices of the atmosphere *Met. Zeitschrift* (20) 431-440, https://dx.doi.org/10.1127/0941-2948/2011/0276
[20] Stensrud D J, Xue M, Wicker L J, Kelleher K E, Foster M P, Schaefer J T, Schneider R S, Benjamin S G, Weygandt S S, Ferree J T, Tuell J P 2009 Convective-Scale Warn-on-Forecast System *Bulletin of the American Meteorological Society* **90(10)**
[21] Kunii M, Saito K, Seko H, Hara M, Hara J, Yamaguchi M, Gong J D, Charron M, Du M, Wang Y, Chen D H 2011 Verification and intercomparison of mesoscale ensemble prediction systems in the Beijing 2008 Olympics Research and Development Project *Tellus A: Dynamic Meteorology and Oceanography* **63(3)** 531-549
[22] Wang J, Chen J, Zhang H, Tian H, Shi Y 2021 Initial Perturbations Based on Ensemble Transform Kalman Filter with Rescaling Method for Ensemble Forecasting *Weather and Forecasting* **36(3)** 823-842
[23] Minamide M, Zhang F, Clothiaux E E 2020 Nonlinear Forecast Error Growth of Rapidly Intensifying Hurricane Harvey (2017) Examined through Convection-Permitting Ensemble Assimilation of GOES-16 All-Sky Radiances *Journal of the Atmospheric Sciences* **77(12)** 4277-4296

IOP Conf. Series: Earth and Environmental Science 1087 (2022) 012034 doi:10.1088/1755-1315/1087/1/012034

Research of chlorophyll-a concentration inversion at different depths in Hong Kong offshore waters based on gaussian process regression

Jie Zhang[1], Hongchang He[1,2], Donglin Fan[1], Bolin Fu[1], Shiwen Wang[1] and Shundan Dong[1]

[1] College of Geomatics and Geoinformation, Guilin University of Technology, Guilin 541000, China.

[2] Corresponding author's e-mail: hhe@glut.edu.cn

Abstract. By inverting the changes in chlorophyll-a (chl-a) concentration values, seawater's pollution level and phytoplankton's growth can be effectively assessed. In this paper, based on Sentinel-3 OLCI remote sensing image data and actual measured chl-a concentration data, the Gaussian Process Regression (GPR) model was used to invert the surface chl-a concentration in Hong Kong waters in 2018, and the average R2 of GPR model inversion was 0.68. The results show a clear vertical decreasing variation feature based on the GPR model to invert the chl-a concentrations in the surface, middle, and bottom layers of Hong Kong waters month by month in 2018.

1. Introduction

Chl-a is ubiquitous in phytoplankton, and its concentration is a reliable indicator of the biological content of aquatic phytoplankton; it is also an important indicator for monitoring seawater quality and eutrophication. In recent years, with the continuous economic development of coastal cities, due to man-made and natural factors, water pollution in coastal waters has increased, water eutrophication, and red tide outbreaks. Hong Kong waters are typical nearshore Class II waters, with special geographical locations and intensive human activities, which are crucial to the environmental monitoring of Hong Kong waters.

The inversion of chl-a concentration in various water quality environments such as open sea, offshore and inland waters has achieved good results [1-4]. For chlorophyll-a concentration inversion at home and abroad now, there has been some theoretical basis accumulated. The representative ones are (1) analysis methods, including the ratio method of the blue-green band [5], mainly used in the outer sea water bodies; near-infrared and red band ratio algorithm [6], primarily used in the study of complex water bodies of class II. (2) The fluorescence baseline method, including the reflection peak fluorescence baseline method, was first proposed by Neville et al. [7], and they all estimate the chlorophyll-a concentration by the sum of the distance between the peak and the baseline. (3) The more typical OCx algorithms, three-band algorithms [8], four-band algorithms [9], and UMOC algorithms are typical of semi-analytical methods, which have strong regional applicability. (4) Gaussian process regression (GPR) in machine algorithms has often been used to study the inversion of environmental elements in coastal and inland waters [10]; Blix K et al. used the GPR model to improve the estimation of chlorophyll content in Lake Balaton with good results [11]; Rivera Caicedo et al. used GPR with hyperspectral reflectance data to calculate chlorophyll-a concentration and evaluate the regression model's portability to other images; GPR developed a well-validated regression model with an R^2 of 0.96 [12]; At this stage, further research is needed for the inversion of chl-a concentrations for different gradients.

Therefore, to make up for the shortcomings of the above research content, this paper takes the offshore waters of Hong Kong as the research area and selects Sentinel-3 OLCI multispectral images with measured chl-a concentration data on a month-by-month basis in 2018 as the primary data source. And constructs a remote sensing inversion model with different depth gradients of chl-a concentration based on machine learning algorithms to explore the spatial and temporal dynamics of chl-a concentration in Hong Kong's offshore waters.

2. Data source

Hong Kong waters are located in the south of China, adjacent to Guangdong Province in the north, connected to the South China Sea waters in the south, and on the west coast back to the mouth of the Pearl River. It is the hub of transportation between the Pearl River inland and the South China Sea.

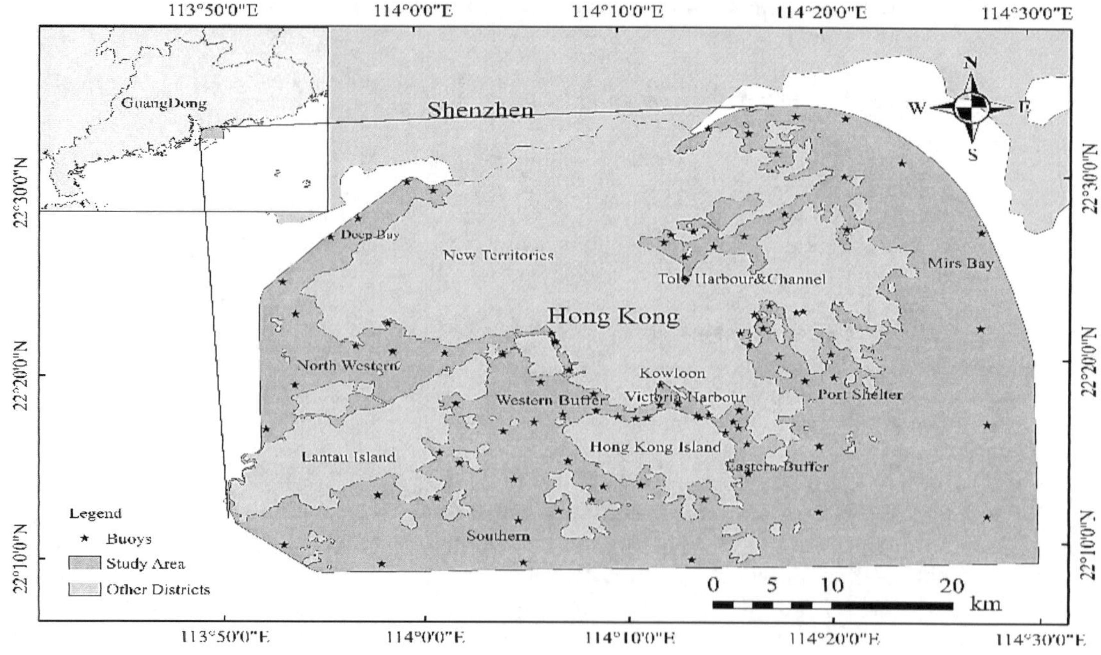

Figure 1. Location and Site Distribution of Hong Kong.

The Hong Kong Environmental Protection Department (https://cd.epic.epd.gov.hk/) provided the measured data for this investigation. The concentration of chl-a is the site data. There are 10 water quality management zones and 76 water quality monitoring sampling points in total. The specific locations are shown in Figure 1. The data measures depth: surface layer (1 m below the sea surface), mid-layer (half the depth of sea surface), and bottom layer (1 m above the seabed).

Sentinel-3 OLCI data is obtained from the official website of ESA Sentinel (https://esa-oceancolour-cci.org/). The data type is Sentinel-3 OLCI EFR (TOA radiation at panoramic resolution) data for 2018. The global coverage is completed within 2 days, with a height of 814km, a sun-synchronous orbit with an inclination of 98.6°, a width is 1300 km, the field of view is 68.5°, the resolution of the coastal zone and the overland is 0.3 km, and the cloud cover above the study area is less than 10%. A total of 12 views of quasi-synchronous or near-synchronous images were selected for each month in 2018 corresponding to the date of the measured data, where the sensor was OLCI, with a resolution of 300m. In this study, the 12 views of Sentinel-3 OLCI data images were preprocessed using ENVI5.4 software, and the preprocessing process included geometric correction, radiometric calibration, atmospheric correction, orthorectification, image alignment, and cropping.

3. Research Methodology

3.1. Spectral feature expansion

Chlorophyll has unique spectral properties. The study found that chlorophyll's remote sensing spectral rate curve has prominent absorption peaks in the visible light frequencies in the blue and red bands. There is a strong absorption peak near 443nm, a weaker absorption peak near 670nm, and chl-a concentration is more sensitive in the infrared band. The remote reflectance of different chlorophylls has peaks around 575nm and 700nm, and the height between 700 and 710nm has long been replaced by phytoplankton biomass with chl-a concentration, and the near-infrared part of the spectrum is also essential for detecting surface scum. Therefore, the 709nm and near-infrared bands are particularly important for Hong Kong waters.

According to Sentinel-3 OLCI band settings, this paper selects the reflectivity of four bands, Oa8 (665nm), Oa9 (673.75nm), Oa10 (681.25nm), Oa11 (708.75nm) as the experimental bands, and combines the bands, as shown in Table 1 below. The six groups of data with a high correlation of experimental data and the corresponding measured data were used as the characteristic variables to input into the GPR model to invert the chl-a concentration. Using the random point selection tool in ArcGIS software, the measured data were randomly divided into two parts, of which 70% were used as training data for constructing the model, and 30% were used for accuracy verification of chl-a concentration.

Table 1. Sentinel-3 OLCI Band ratios.

Band Ratio	OLCI
R(708.75)+R(681.25)	Oa11+ Oa10
R(673.75)+R(681.25)+R(708.75)	Oa9+ Oa10+ Oa11
R(673.75)*R(681.25)*R(708.75)	Oa9* Oa10* Oa11
(R(673.75)*R(681.25)*R(708.75))/R(665)	(Oa9* Oa10* Oa11)/Oa8
0.7*R(708.75)+0.3*R(681.25)	0.7*Oa11+0.3*Oa10
0.5*R(708.75)+0.3*R(681.25)+0.2*R(673.75)	0.5*Oa11+0.3*Oa10+0.2*Oa9

3.2. GPR (Gaussian Process Regression) Model

Machine learning algorithms for Gaussian Process Regression (GPR) have been shown to perform well in retrieving water quality parameters from remote sensing data. Gaussian process regression is a non-parametric probabilistic regression model that learns prior knowledge based on a Gaussian process. It is a flexible nonlinear kernel method that learns input (such as the ratio of reflectance) and output (such as chl-a concentration) through a Bayesian framework. The basic process is shown in Figure 2. The model's hyperparameters (or weights) are adjusted to minimize the error in the independent validation dataset to find the best generalization ability of the model. The kernel function in GPR actually refers to the covariance function. The covariance kernel function used in this study is a typical square exponential function (RBF kernel), See formula (1) for the kernel function.

$$k(x,x') = \sigma_f^2\, exp(-\frac{|x-x'|^2}{2l^2}) \quad (1)$$

The hyperparameters σ_f^2 and l are called the process (signal) variance and length scale, respectively.

IOP Conf. Series: Earth and Environmental Science 1087 (2022) 012034 doi:10.1088/1755-1315/1087/1/012034

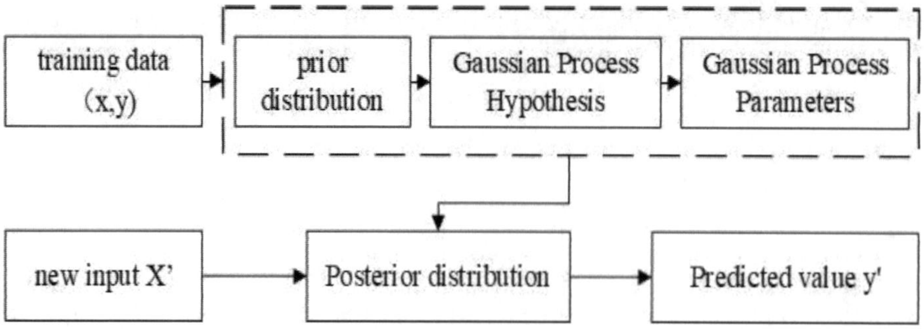

Figure 2. Block diagram of the basic process of GPR.

The GPR model is a predictive model derived by learning from historical data and calculating the optimal parameters. In this paper, the algorithm was used for the inversion of chl-a concentration in Hong Kong offshore waters in 2018. The input data are Sentinel-3 OLCI Rrs band combination data with a high correlation with the measured data, and 70% of the best band combination data values and the measured chl-a concentration data in Hong Kong waters on the corresponding dates are input to the GPR model as training samples X. The remaining 30% of remote sensing reflectance is input to the model as prediction samples for inversion, and the accurate analysis of the obtained chl-a concentration inversion values with the measured values is performed.

3.3. Inversion model evaluation index

By evaluating the pros and cons of model accuracy, analyzing the effect of model inversion of chl-a concentration to increase the accuracy of the inversion value. $P_{x,y}$ as in Equation (2) determines the common trend between the inversion value x and the measured value y. The coefficient of determination (R^2) between the inversion value of the model and the measured value as in equation (3) reflects the linear relationship between the inversion value and the measured value; root means square error (RMSE) as in Equation (4) reflects the uncertainty between the inversion value and the measured value, which can better reflect the actual situation of the predicted value error. The mean absolute error (MAE) represents the average value of the distance between the predicted value X and the measured value y, as shown in Equation (5).

$$P_{x,y} = \frac{cov(x,y)}{\sigma_x \sigma_y} = \frac{E(xy)-E(x)E(y)}{\sqrt{E(x^2)-E^2(x)}\sqrt{E(y^2)-E^2(y)}} \tag{2}$$

$$R^2 = \frac{\sum_{i=1}^{n}(\hat{x}_i-\bar{x}_i)^2}{\sum_{i=1}^{n}(y_i-\bar{x}_i)^2} \tag{3}$$

$$RMSE = \sqrt{\frac{\sum_{i=1}^{N}(y_i-x_i)^2}{N}} \tag{4}$$

$$MAE = \frac{1}{n}\sum_{i=1}^{n} |x_i - y_i| \tag{5}$$

In the formula: x is the inversion value, and y is the measured value, $\hat{x}_i$ is the inverse value of month i, $\bar{x}_i$ is the average of the inverse values, y_i is the measured value of month i, and N is January to December.

4. Experiment and result evaluation

4.1. Inversion results and accuracy evaluation based on the GPR model

The inversion of chl-a concentration by the GPR model for 12 months in 2018 and the correlation results between the GPR model and measured data are shown in Table 2. Overall the average R^2=0.68, RMSE=2.2778 μg/L, MAE=0.477 in 2018.

Table 2. Accuracy evaluation of Gaussian process regression (GPR) in 2018.

Date	GPR	Measured	R^2	RMSE(μg/L)	MAE
201801	1.930	2.073	**0.766**	1.286	0.541
201802	1.859	2.034	0.709	1.020	0.577
201803	2.401	3.153	0.688	1.489	0.618
201804	1.578	1.875	0.673	2.952	0.603
201805	3.311	2.176	0.623	3.540	0.641
201806	2.941	2.651	0.653	4.781	0.570
201807	3.267	1.784	0.747	2.662	0.617
201808	4.151	3.259	0.694	**5.183**	0.444
201809	2.507	2.792	0.672	1.598	0.332
201810	1.929	1.352	0.750	**0.771**	0.347
201811	2.328	2.592	**0.592**	1.176	0.233
201812	1.868	1.667	0.659	0.876	0.201

The correlation between the inversion of surface chl-a content in Hong Kong waters based on the GPR model and experimental data in 2018 is shown in Figure 3. "y" represents the measured value of chl-a on the vertical axis, and "x" represents the GPR model inversion value of chl-a on the horizontal axis. Overall, the correlation was relatively stable in 2018, with R^2 floating between 0.592 and 0.766. Some points in individual months were more scattered, probably due to the large measured values and the poor inversion effect of the model resulting in a lower correlation with the measured values, but the overall prediction effect of the GPR model on the surface chl-a concentration in Hong Kong waters throughout 2018 was relatively stable. In the following paper, the GPR model is used to invert the chl-a concentrations in the surface, middle, and bottom layers of Hong Kong waters, and to analyze and discuss them in time and space, respectively.

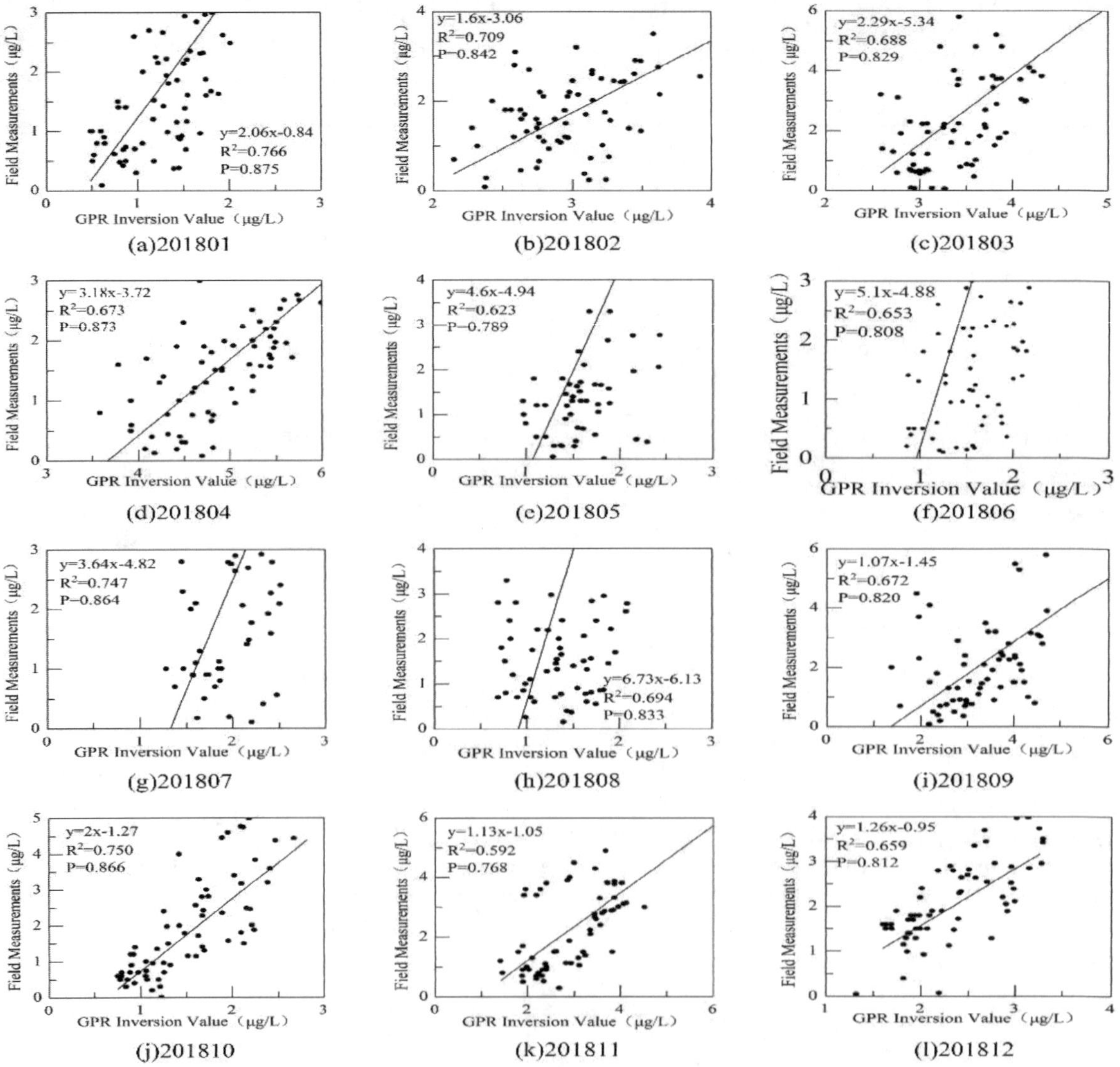

Figure 3. Correlation between GPR and measured values in 2018.

4.2. Application of different depth mapping in Hong Kong waters

The above precision analysis shows that the GPR model is suitable for inversion of chl-a concentration in the offshore waters of Hong Kong. Based on the Sentinel-3 satellite data, the GPR model was used to invert the chl-a concentrations in the surface (1m below the sea surface), middle (middle water depth), and bottom (1m seabed) of the Hong Kong offshore waters in 2018. The following is to invert the chl-a content month by month and analyze the spatial distribution characteristics from the dynamic changes in the horizontal and longitudinal directions.

The characteristics of the spatial distribution of chl-a content in Hong Kong waters from the month-by-month changes are shown in Table 3, which shows the average chl-a concentration at different depths in Hong Kong waters month by month in 2018. Figure 4 shows the distribution of chl-a concentration at different depths in the Hong Kong waters in 2018. It can be seen that the Hong Kong waters as a whole have prominent characteristics of east-west distribution differences.

In 2018, the chl-a content in the waters of Hong Kong showed apparent differences in the east-west distribution. In 2018, the concentration of chl-a in the surface layer of Hong Kong waters ranged from 0.0 to 35.0 μg/L, the concentration of chl-a in the middle layer ranged from 0.0 to 22.0 μg/L, and the concentration of chl-a in the bottom layer ranged from 0.0 to 22.0 μg/L. In 2018, the highest concentration of chl-a in the Hong Kong waters was 7.4 μg/L in the surface layer in August, and the lowest value was 0.8 μg/L in the middle layer of the Hong Kong waters in October. The highest chl-a concentration in Hong Kong waters throughout the year is concentrated in the Shelter Sea, Tolo Harbour and Chimen Strait, and the waters of the Southern District. The main reason for the high chl-a concentration in Hong Kong waters is that these three seas are close to land and human activities are more frequent. The lowest values were generally concentrated in the northwestern sea area, Houhai Bay sea area, and Dapeng Bay sea area, which is far away from the land, and the water bodies are renewed faster, resulting in lower chl-a concentration.

Table 3. Longitudinal average chl-a concentration (μg/L) in the offshore waters of Hong Kong in 2018.

Month	1	2	3	4	5	6	7	8	9	10	11	12
Surface(μg/L)	3.4	2.4	3.5	3.9	2.9	4.5	5.9	**7.4**	3.9	1.9	3.1	1.9
Middle(μg/L)	3.3	2.9	3.5	**4.3**	3.2	3.1	3.1	3.1	2.9	0.8	2.7	1.8
Bottom(μg/L)	2.3	2.2	**3.1**	2.8	2.7	2.9	2.7	2.7	2.7	1.4	2.5	1.4

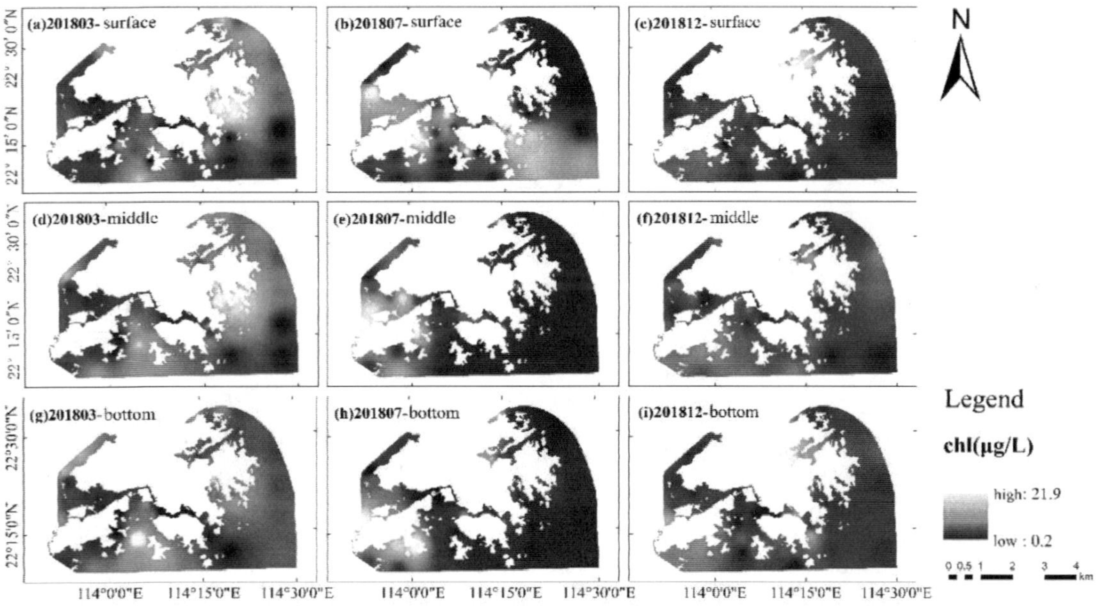

Figure 4. Chl-a concentration at different depths in the offshore waters of Hong Kong in 2018.

5. Conclusions

In this paper, we invert the surface chl-a concentration based on the Gaussian process regression (GPR) model using the 2018 data selected from the measured chl-a concentration data released by the Hong Kong Environmental Protection Department. The results showed that: The measured data in 2018 correlated well with the surface chl-a concentration in the GPR model inversion, and the model accuracy was high. The annual average determination coefficient R^2 of the GPR model is 0.68, and the monthly average R^2 varies from 0.592 to 0.766. The GPR model was used to invert monthly chl-a concentrations in the Hong Kong offshore waters' surface, middle, and bottom layers. The inversion shows that the monthly average chl-a concentration in 2018 is 1.9 to 7.4 μg/L in the surface layer, 0.8 to 4.3 μg/L in the middle layer, and 1.4 to 3.1 μg/L in the bottom layer. The difference in the average concentration among months in 2018 is not too noticeable. Spatially Hong Kong waters showed an overall trend of chl-a concentrations fluctuating from west to east. the higher chl-a concentration was concentrated in the northwestern sea area, the southern sea area, and the sea areas of Tolo Harbour and Chimen Strait close to the land, and the seasonal changes were large, and the concentration of chl-a is relatively high. The chl-a concentration values are generally low in waters like Deep Bay and Dapeng Bay, mainly because their waters are far from land and close to the Pearl River Basin and the South China Sea Basin, where the water bodies are renewed more quickly and almost like seawater, and the chl-a concentration is naturally level. In this paper, based on the Gauss regression model, the inversion of chl-a concentration under different gradients in the Hong Kong sea area is realized, which provides a theoretical basis for future related research, and also provides data support and scientific basis for the protection and restoration of marine ecosystems.

References

[1] Smith, R. C., & Baker, K. S. 1982 Oceanic chlorophyll concentrations as determined by satellite (Nimbus-7 Coastal Zone Color Scanner) *Marine Biology* **66(3)** 269–279 https://doi.org/10.1007/bf00397032

[2] Yoder, J. A., McClain, C. R., Feldman, G. C., & Esaias, W. E. 1993 Annual cycles of phytoplankton chlorophyll concentrations in the global ocean: A satellite view *Global Biogeochemical Cycles* **7(1)** 181–193 https://doi.org/10.1029/93gb02358

[3] Gordon, H. R., Clark, D. K., Brown, J. W., Brown, O. B., Evans, R. H., & Broenkow, W. W. 1983 Phytoplankton pigment concentrations in the Middle Atlantic Bight: Comparison of ship determinations and CZCS estimates *Applied Optics* **22(1)** 20 https://doi.org/10.1364/ao.22.000020

[4] Bukata, R. P., Jerome, J. H., Bruton, J. E., Jain, S. C., & Zwick, H. H. 1981 Optical water quality model of Lake Ontario 1: Determination of the optical cross sections of organic and inorganic particulates in Lake Ontario *Applied Optics*, **20(9)** 1696 https://doi.org/10.1364/ao.20.001696

[5] Robinson, I. S. 2007 Remote Assessment of Ocean Color for Interpretation of Satellite Visible Imagery: A review: By H. R. Gordon and A. Y. Morel. Springer-Verlag, Berlin, 1983. Pp. 114. DM 58 (approx. US$22.5) *Quarterly Journal of the Royal Meteorological Society* **111(469)** 872–872 https://doi.org/10.1002/qj.49711146913

[6] Carder, K. L., Chen, F. R., Lee, Z. P., Hawes, S. K., & Kamykowski, D. 1999 Semianalytic Moderate-Resolution Imaging Spectrometer algorithms for chlorophyllaand absorption with bio-optical domains based on nitrate-depletion temperatures *Journal of Geophysical Research: Oceans* **104(C3)** 5403–5421 https://doi.org/10.1029/1998jc900082

[7] Neville, R. A., & Gower, J. F. R. 1977 Passive remote sensing of phytoplankton via chlorophyll α fluorescence *Journal of Geophysical Research* **82(24)** 3487–3493 https://doi.org/10.1029/jc082i024p03487

[8] Gitelson, A. A., Dall'Olmo, G., Moses, W., Rundquist, D. C., Barrow, T., Fisher, T. R., Gurlin, D., & Holz, J. 2008 A simple semi-analytical model for remote estimation of chlorophyll-a in turbid waters: Validation *Remote Sensing of Environment* **112(9)** 3582–3593 https://doi.org/10.1016/j.rse.2008.04.015

[9] Le, C., Li, Y., Zha, Y., Sun, D., Huang, C., & Lu, H. 2009 A four-band semi-analytical model for estimating chlorophyll a in highly turbid lakes: The case of Taihu Lake, China. *Remote Sensing of Environment* **113(6)** 1175–1182 https://doi.org/10.1016/j.rse.2009.02.005

[10] Odermatt, D., Danne, O., Philipson, P., & Brockmann, C. 2018 Diversity II water quality parameters from ENVISAT (2002–2012): A new global information source for lakes. *Earth System Science Data* **10(3)** 1527–1549 https://doi.org/10.5194/essd-10-1527-2018

[11] Blix, K., Pálffy, K., Tóth, V., & Eltoft, T. 2018 Remote sensing of water quality parameters

over Lake Balaton by using Sentinel-3 OLCI. *Water* **10(10)** 1428 https://doi.org/10.3390/w10101428

[12] Jay S, Guillaume M, & Talon J. 2008 IEEE Journal of Selected Topics in applied earth observation and remote sensing *IEEE Geoscience and Remote Sensing Letters* **5(4)** 840–840 https://doi.org/10.1109/lgrs.2008.2008610

IOP Conf. Series: Earth and Environmental Science 1087 (2022) 012035 doi:10.1088/1755-1315/1087/1/012035

Changes in land use land cover within the Jukskei River basin and its implications on the water availability

Moeketsi P.[1], Nkhonjera G.K.[1,2] and Alowo R.[1]

[1] University of Johannesburg, School of Civil Engineering and Built Environment, Civil Engineering Technology Department, PO Box 17011, Doornfontein, 2028. South Africa.

[2] Corresponding author's e-mail: germann@uj.ac.za

Abstract. South Africa is just slowly emerging from its past turbulent history with some of its population still not able to access some of the basic services such as water. This study was aimed at evaluating changes in land use land cover within the Jukskei River basin and its implications on the water availability. The study used Archydro tools on ArcMap of the ArcGIS program to delineate the Jukskei River catchment, then, used the delineated catchment to clip off the land use land cover (LULC) maps for the periods 1990, 2014 and 2020. The total area of the catchment was determined to be 800 km^2, and land use practices were assessed for 1990, 2014 and 2020 LULC maps. The major land use classifications found within the catchment were agriculture, bare-land, forests, grassland, settlements (built-up area), mines, waterbodies, and wetlands. The study showed that settlements accounted for the greatest portion of the Jukskei River catchment, compared to other land use categories. The runoff was determined for the 2, 5, 10 and 20 years return periods. The study showed that runoff increased by 8.33% due to 2020 land uses (70% imperviousness), from those values calculated for 1990 and 2014 land uses which had 60.87 and 61.58% imperviousness respectively. Therefore, the results revealed that a large quantity of water (runoff) is generated in the catchment and then surrendered into the Limpopo River and eventually wasted into the Indian Ocean. The study also showed that an increase in imperviousness due to land use practices resulted in an increase in surface runoff, thus having negative implications on the catchment's water availability.

1. Introduction

Globally, water is regarded as the most important resource available at the disposal of people. The world's water resources are essential in sustaining livelihoods, and therefore all living components cannot do without water [1]. Water plays a major role in the ecological and socio-economic development of a country [2]. Furthermore, in South Africa for example, water is used for variety of purposes such as domestic use, agriculture, mining, recreation and as a transportation medium for good and wastes. Despite the many advantages, water resources are unevenly distributed and are facing depletion which has resulted in water scarcity problems in many regions across the world. The scarcity of water has been evident in South Africa and has presented itself in the major cities of Johannesburg and Cape-Town. According to [3], the scarcity of water resources in South Africa, is linked with low rainfall patterns, and the growing demand for water due to the country's increasing population, extensive urbanization, and economic growth. The study by [1] noted that urbanization is the major contributor exacerbating water scarcity in South Africa, and it is also considered the biggest environmental issue that the country is faced with. Urbanization, as noted by [4], is also regarded as a major contributor of land use and land

IOP Conf. Series: Earth and Environmental Science 1087 (2022) 012035 doi:10.1088/1755-1315/1087/1/012035

cover (LULC) changes. However, it is important that the concepts of LULC are first understood before the associated impacts on the environment are determined.

Land use, as described by [5], is the use of land and its available resources by humans for purposes such as the socio-economic growth and infrastructure development. Land use change is described as a complete modification of the natural land into another type, mainly through anthropogenic activities, [6] and [2]. These anthropogenic activities are human related actions such as agricultural and mining practices, industrialization, and human settlements which take place on land overtime due to urbanization. However, urbanization as a major driver of LULC changes, [7] reported that it causes changes that are ecological, socio-economic, and hydrological to the environment. LULC changes have its major impacts on the quality and quantity of water, through alteration of the hydrological response of catchments and associated processes such as infiltration, evapotranspiration, and recharge [2, 6]. These impacts have been evident in the Jukskei River catchment located in the heart of the highly urbanized and industrialized Johannesburg. Land use practices in Johannesburg, as noted by [8], have a negative impact of altering the hydrological response of the Jukskei River, particularly on the quantity of stormwater runoff. However, [9], found that very little is known regarding land use impacts on large river basins like the Jukskei, thus presenting literature gaps. Therefore, there is a need to investigate these impacts to redress this knowledge gap. Hence, the objective of this study was to use Geographic Information Systems (GIS) tools to delineate the catchment of the Jukskei River and then establish the potential impacts of the City of Johannesburg's (CoJ) land use practices on the runoff generated from this catchment.

2. Methods and Data

2.1. Study Area

The Jukskei River catchment is the study area and is one of the largest catchment areas in Johannesburg, Gauteng Province. The area mainly comprises the northern suburbs of Johannesburg, a preferred living area of the then South African Apartheid elite. The catchment covers an approximated total area of 800 km^2 and includes the highly industrialized and urbanized areas of Johannesburg. The areas covered within the catchment are the Johannesburg Metropolitan, Ekurhuleni, and Tshwane Municipalities [8]. The catchment has moderate climatic conditions, with erratic rainfall patterns that varies seasonally [10], and mean annual rainfall (MAP) of 713 mm [11]. Temperatures within the catchment varies seasonally, with the minimum and maximum summer temperatures ranging from 14°C to 25°C, whereas during the cold winter months the minimum temperatures run from 0°C while maximum temperatures peak to 17 °C [8].

The Jukskei River is one of the longest urban rivers within the area. The river has an approximated total length of 68 km, which runs from the source east of Johannesburg to the outlet located north of Johannesburg at the confluence with the Crocodile River [8]. Upon draining all the runoff in the northern part of Johannesburg and adjacent areas, the Jukskei joins and discharges the largest amount of water into the Crocodile River (Limpopo) Basin which eventually discharges into the Hartbeespoort Dam and the water eventually ends up in the Indian Ocean [12]. The study area is shown in Figure 1.

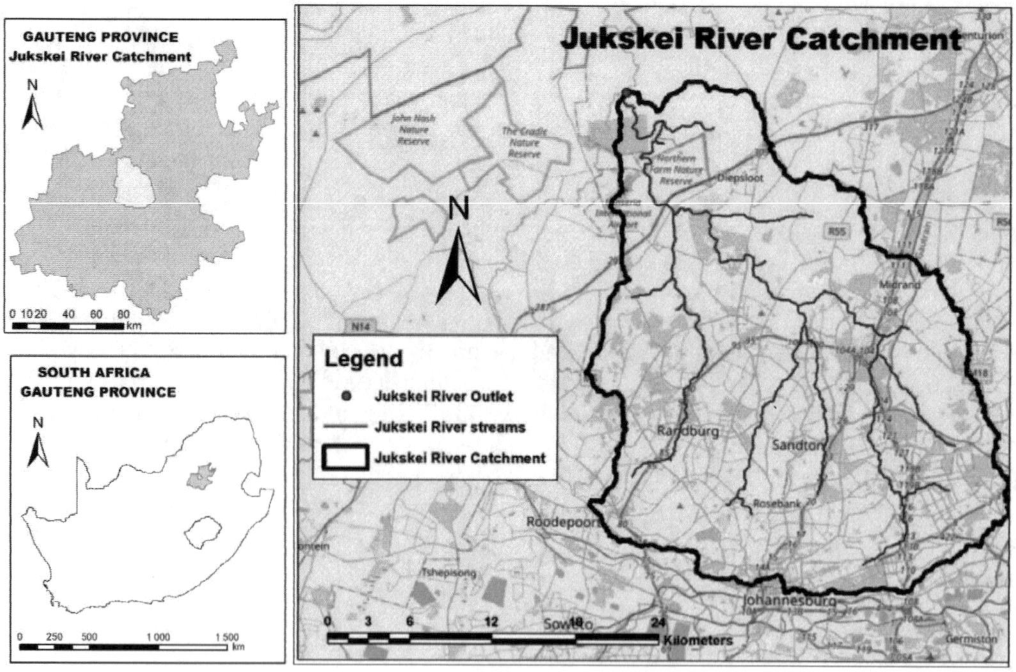

Figure 1. Location of the Jukskei River catchment.

2.2. *Watershed delineation*

A watershed which is also known as a catchment or drainage basin, is an area of land that collects water from upstream places and channel it to a common outlet point located in downstream lower reaches [13]. Hydrological modelling tools within the ArcGIS are used to delineate watersheds, and also to perform several activities such as mapping, data collection and management. Catchment delineation is done through the use of Digital Elevation Model (DEM) data. A DEM is defined as a representation of x, y, and z co-ordinates of the bare ground on the earth surface and does not signify surface features such as vegetation and man-made objects [14]. In catchment delineation, DEM are used to determine the hydrological parameters such as catchments slope, flow accumulation and direction and stream networks [15]. For this study, the catchment of the Jukskei River was delineated using the main component called ArcMap, an extension of the ArcGIS mapping tool. The DEM used were of 30- or 10-meter resolutions as recommended by [16], and these were obtained from the USGS website (https://earthexplorer.usgs.gov/). The Jukskei River catchment was delineated by adopting the procedure by [16] as shown in Figure 2.

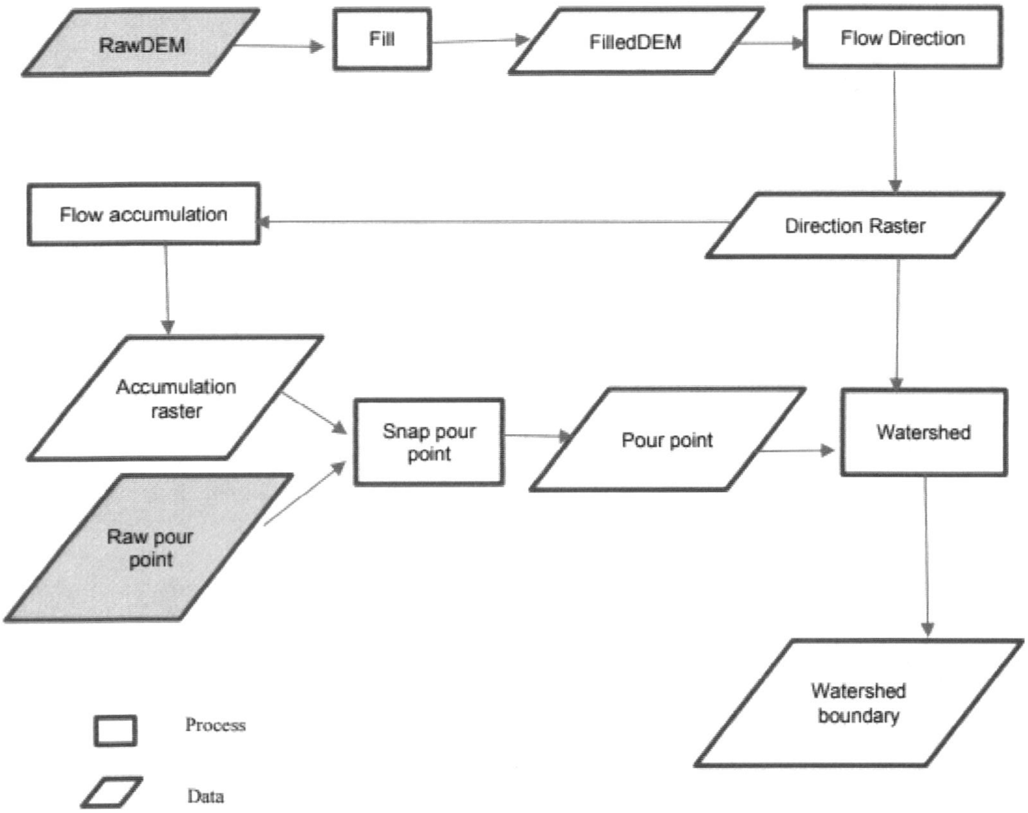

Figure 2. Watershed delineation procedure in ArcGIS [16].

2.3. Analysis of land use land cover changes

Land use and land cover (LULC) maps were evaluated to determine the extent of land cover changes across the Jukskei River catchment, and the effects of such changes to runoff generated within the catchment. These LULC maps, by their very nature, show the physical features of land such as vegetation cover, natural waterbodies, and built-up areas. The LULC maps are evaluated to show how these physical features have changed throughout the years, due to landscape modifications which are associated to urbanization related anthropogenic activities and the growing human population. For this study these maps were obtained from the South African Department of Forestry, Fisheries and Environment (DFFE) (https://egis.environment.gov.za/). The maps evaluated were for the years 1990, 2014, 2020 respectively. It was first necessary to define the maps to the same coordinate projections system used in South Africa, that is the Universal Transverse Mercator Zone 35 Southern Hemisphere (UTM Zone 35 S). The projection transformations are done to enable data to be opened simultaneously, to allow coordinates to be easily read, and to remove negative coordinates that would cause errors in distance and area calculations [17].

Various aspects were evaluated such as changes in vegetation cover, bare-land, natural watercourses, built-up areas (settlements, roads, industries). The analysis of the land use classifications assisted in the calculation of percentage imperviousness and the runoff coefficient (C) due to the catchments land-uses. This information was used to determine the runoff due to the catchments land use practices using the rational method ($Q = 0.278\,CIA$), where parameters, I is rainfall intensity in mm/hr, A is catchment area in square kilometres, Q in cubic metres per second. The term C is runoff coefficient and is dimensionless.

2.4. Water availability analysis

The Jukskei River catchment is one of the most densely populated catchment areas within the Gauteng Province [8, 11]. Overtime, there has been evidence of large-scale landscape modifications due to increased human population that arises as a result of extensive urbanization that is taking place. This has resulted in the need for more infrastructure that is impervious to water such as housing, roofs, roads, and parking lots. As a result of this, [18] found that large quantities of water are conveyed away from residential settlements during and after precipitation. This triggers changes to the components of the hydrological cycle such as evapotranspiration, interception, and groundwater recharge [8]. This contributes to the generation of more surface runoffs and increased streamflow's in urban rivers [6], which eventually affects the availability of water within the wide catchment. Therefore, to evaluate the effect on the catchments water availability, runoff due to the catchment's land uses was determined using the rational method.

3. Results and Discussions

3.1. Land use land cover maps

The extent of land use and land cover of the study area was analysed using the land cover maps for a 30-year period (1990, 2014, 2020). Table 1 shows the different land use classifications within the Jukskei River catchment, as well as their percentages as observed in the land cover maps. The major LULC classifications within the catchment are agriculture, bare-land, forests, grassland, settlements, mines, waterbodies, and wetlands. The analysis of LULC showed that settlements which includes all the built-up and paved areas, accounted for the greatest percentage of the Jukskei River catchment compared to the other land use classes. A closer look at Table 1 shows that agriculture, grassland, and settlements have undergone a major significant change due to the catchment's land use practices with regards to the evaluated LULC maps. The results obtained showed an increase in built-up areas (settlements) from 462.52 (60.35%) in 1990, 467.5 (61.00%) in 2014 and 529.7 km^2 (69.14%) in 2020, as compared to all other land use classifications for the 30 years period (1990 to 2020). The other major change observed was the decrease in natural forested and grass vegetation, with grassland cover having the greatest decrease from 20.28% in 1990, to 15.44% in 2014 and 12.77% in 2020. The analysis of LULC also showed that in 2020 almost 70% (69.14%) of the Jukskei River catchment was heavily settled with built-up of impervious surfaces (roads, roofs), as compared to the 61% in 2014 and 60.35% in 1990. Figure 3 shows a detailed representation of the impacts of the catchment's LULC changes, a result of 1990, 2014 and 2020 land use practices.

Table 1. Jukskei River catchment LULC summary for the past 30 years.

Classification Name	Area (km^2)					
	1990		2014		2020	
	Area	%	Area	%	Area	%
Agriculture	40.71	5.31	28.58	3.73	46.35	6.05
Bare-land	0.09	0.01	0.16	0.02	7.73	1.01
Forest	87.48	11.41	128.00	16.70	61.17	7.98
Grassland	155.46	20.28	118.35	15.44	97.88	12.77
Settlements	462.51	60.35	467.50	61.00	529.87	69.14
Mines	4.02	0.52	4.43	0.58	5.71	0.74
Waterbodies	1.20	0.16	1.08	0.14	2.49	0.33
Wetlands	14.93	1.95	18.26	2.38	15.21	1.98

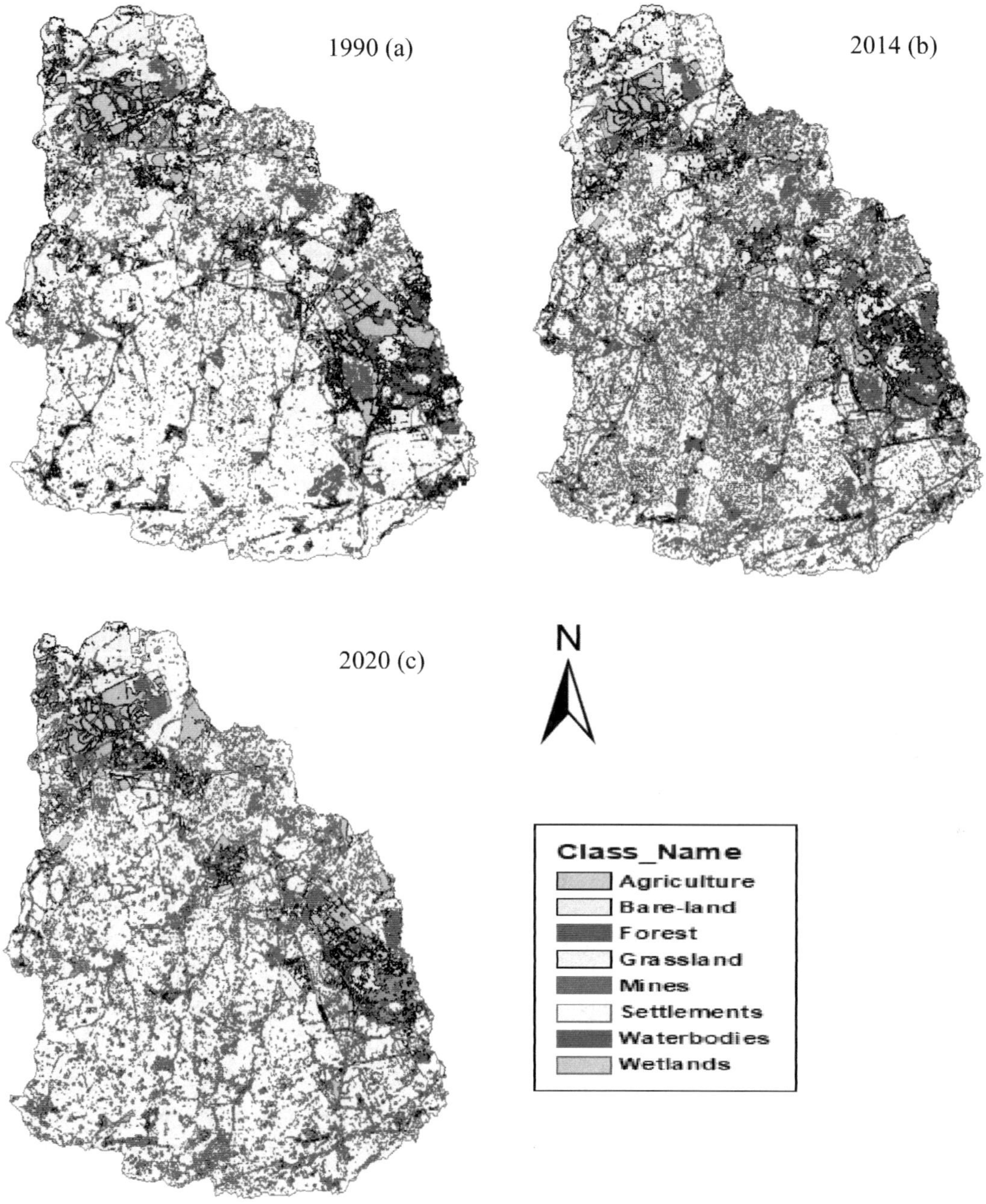

Figure 3. Land use maps (a) 1990; (b) 2014; (c) 2020.

3.2. Runoff responses to land use land cover changes

The runoff calculations using the rational method showed that there was a significant increase in the catchment's stormwater runoff due to land use practices. The major change noticed is that the catchment's runoff due to 2020 land uses (69.88% imperviousness) was much greater compared to that calculated for 1990 and 2014 catchment's land use practices, whose imperviousness was 60.87 and 61.58% respectively. Runoff was determined based on return periods (2, 5, 10, 20 years). Based on these return periods, the rainfall intensities for the rational method were calculated as 2.58, 3.61, 4.73 and 6.71 mm/hr respectively.

For the 2.58 mm/hr intensity, the calculated runoff increased from 329.55 m^3/sec in 2014 to 357.01 m^3/sec in 2020. The 3.61 mm/hr intensity resulted in runoff increase from 461.12 m^3/sec in 2014 to 499.54 m^3/sec in 2020. For intensity of 4.73 mm/hr, runoff increased from 604.18 m^3/sec in 2014 to 654.52 m^3/sec in 2020. Furthermore, for the 6.71 mm/hr intensity, the runoff increased from 857.09 m^3/sec in 2014 to 928.51 m^3/sec in 2020. These observations indicated that the catchment's runoff as a result of the four rainfall intensities had increased by 8.33% in 2020, from those values determined due to 1990 and 2014 land uses. These results are in line with those found by [19], who evaluated the impacts of land uses on the runoff generated from the urban areas of the Myslavský creek river basin in Switzerland. These authors found that the runoff from the basin was 70% higher due to 2010 land uses practices compared to that of the year 1983. The increase in the catchment's runoff is also supported by the analysis of [20], who in their study noted that impervious surfaces reduce groundwater recharge, thereby leading to increased surface runoffs as observed in the results found in the present study. Moreover, this study's results are supported by the analysis of [8], who evaluated the impacts of LULC on the surface runoff of the Jukskei River catchment using the Soil and Water Assessment Tool (SWAT model). These authors noted that the conversion of the natural land to built-up areas (imperviousness) increased the catchment's runoff depth from 70.54 mm in 1987 to 199.37 mm in 2020, thus resulting in increased catchment's surface runoff.

While these results are acknowledged, [8] also indicated that runoff methods such as rational method are mainly limited to smaller catchments. Therefore, for more accuracy in this study, perhaps the method should have been more applicable on subbasin scale where possible. Also, as argued earlier, the rational method only provides the peak runoff and unfortunately it cannot produce a hydrograph. Therefore, for more accurate analysis of runoff pattern and quantity, perhaps unit hydrograph or other methods may have been more appropriate. Likewise, the accuracy of rational method depends very much on the correct selection of runoff coefficient and delineation of catchment area. Thus, some errors inherent due to land use classification and catchment delineation, may have affected the results.

3.3. Impacts of land use land cover changes on water availability

The Jukskei River catchment is known as one of the most densely populated areas within the Johannesburg region. This is a result of the extensive urbanization within the area, which has led to a significant increase in built-up areas (settlements). The analysis of LULC changes showed that the catchment experienced an increase in impervious areas (houses, roads), from 60.87% in 1990, and 61.58% in 2014 to 69.88% due to 2020 land use practices. As such, this has a major impact on the runoff generated within the catchment, through alterations upon the infiltration, interception, and evapotranspiration processes of the hydrological cycle.

The results from rational method runoff calculations depicted that the catchment's runoff increased by 8.33% as a result of 2020 land use practices, from those values calculated for 1990 and 2014 land uses. This showed that runoff is influenced by the percentage imperviousness, which in 2020, was almost 70% of the total catchment area. This catchment was heavily settled, thus influencing the amount of runoff generated and removed away from the catchment. These observations are supported by [18] who found that large quantities of stormwater runoff are generated from Johannesburg urban areas which are covered within the Jukskei River catchment. Therefore, it is evident that land use activities had a major impact on catchment's runoff. In an area like Johannesburg which is heavily urbanised and paved, the large quantities of runoff generated, if properly managed, can have a huge influence on the water availability within the catchment.

4. Conclusions

This study evaluated the effects of land use change within the Jukskei River basin, Gauteng Province. The study involved delineating the catchment of the Jukskei River using the ArcGIS mapping tool, and then land use practices were assessed. The study also involved runoff calculations using the rational method, to determine the effects of land use and land cover changes on the catchment's water availability. The assessment of land use and land cover change was based on the 1990, 2014 and 2020 land cover maps. The analysis revealed that the major land use classifications within the Jukskei River catchment are agriculture, bare-land, forests, grassland, settlements, mines, waterbodies, and wetlands. The analysis also showed that settlements (built-up areas) accounted for the largest percentage of the Jukskei River catchment, compared to all other land use classifications. The analysis of the catchment's runoff due to land use and land cover changes depicted that the catchment's runoff increased by 8.33% as a result of 2020 land use practices (70% imperviousness) from those values calculated for 1990 and 2014 land uses which had 60.87 and 61.58% imperviousness respectively. This study's observations showed that land use practices play a very critical role in the water cycle and of a catchment. Based on the research findings and correspondence with existing literature, the objectives on this study were successfully achieved. Therefore, the results of this study contribute to existing body of knowledge and the recommendations can help counteract the negative impacts of land use on a catchment scale.

Acknowledgements
Many thanks go to the University of Johannesburg for the help rendered especially in making the ArcGIS program available for analysis. Also, the South African Department of Forestry, Fisheries and Environment (DFFE) need to be commendable for making LULC maps available for the study. Finally, we would like to thank reviewers who helped make this paper what it is.

References

[1] K. Naidoo 2005 The anthropogenic impacts of urbanization and industrialisation on the water quality, ecology and health status of the Palmiet River catchment in Durban, KwaZulu-Natal [Masters Dissertation, University Of KwaZulu-Natal], *Durban: ResearchSpace University Of KwaZulu-Natal*

[2] M. Nkosi, F. Mathivha and J. Odiyo 2021 Impact of Land Management on Water Resources, a South African Context *Sustainability* **13(2)** 701

[3] C. D. J. G. Camuto 2019 Measuring factors affecting sustainable water supply in the Province of Western Cape [Masters Thesis, University of Johannesburg] *Johannesburg: University of Johannesburg Institutional Repository (UJ IR)*

[4] C. Gerten, S. Fina and K. Rusche 2019 The sprawling planet: simplifying the measurement of global urbanization trends *Frontiers in Environmental Science* **7** 140

[5] L. Henderson, C. Mahoney, C. Mcclelland and A. Myers 2014 The effects of Land use and Land Cover on Water Quality in Urban Environments, *Natural Resources and Environmental Sciences*, Manhattan, USA: Kansas State University

[6] B. A. Mauck 2012 The impacts of future urban growth on streamflow in the Mgeni catchment [Masters Thesis] *Pietermaritzburg: ResearchSpace University of KwaZulu-Natal*

[7] E. Ndulue, C. Mbajiorgu, S. Ugwu, V. Ogwo and K. Ogbu 2015 Assessment of land use/cover impacts on runoff and sediment yield using hydrologic models: A review *Journal of ecology and the natural environment* **7(2)** 46-55

[8] T. Mawasha and W. Britz 2021 Hydrological Impacts of Land Use-Land Cover Change on Urban Flood Hazard: A Case Study of Jukskei River in Alexandra Township, Johannesburg *South African Journal of Geomatics* **10(2)**

[9] F. O. Aijbade, A. Wang, B. Adelobun, K. H. Lasisi, O. O. Fadare, T. F. Ajibade, N. A. Nwogwu, I. D. Sulaymon, A. Y. Ugya, H. C. Wang and A. Wang 2021 Environmental pollution and their socioeconomic impacts in *Microbe Mediated Remediation of Environmental Contaminants, Woodhead Publishing* 321-354

[10] A. B. Neswiswi 2014 Development of a Water Quality Index (WQI) for the Jukskei River Catchment, Johannesburg [Masters Thesis] *Wits Institutional Repository on DSPACE*

[11] W. Matowanyika 2010 Impact of Alexandra Township on the water quality of the Jukskei River [Doctoral dissertation, University of the Witwatersrand] *Wits Institutional Repository*

environment on DSpace, Johannesburg

[12] R. Malatji 2019 Towards a GIS Platform for monitoring the water quality of the Jukskei river catchment: defining a sampling network [Masters Thesis, University of Witwatersrand] *Wits Institutional Repository environment on DSpace, Johannesburg*

[13] W. Bajjali 2018 Watershed Delineation in ArcGIS for environmental and water issues. Springer Textbooks in Earth Sciences, *Geography and Environment, Springer, Cham*

[14] C. Archuleta, E. Constance, S. Arundel, A. Lowe, K. Mantey and L. Phillips 2017 The National Map seamless digital elevation model specifications *US Geological Survey, Reston, Virginia*

[15] L. G. C. Palamuleni 2009 Land cover change and hydrological regimes in the Shire River Catchment, Malawi [Doctoral Dissertation, University of Johannesburg] *Johannesburg: University of Johannesburg Institutional Repository (UJ IR)*

[16] Y. Zhang, M. McBroom, J. Grogan and I. Hung 2011 Snapping a Pour Point for Watershed Delineation in ArcGIS Hydrologic Analysis *Texas: Faculty Publications; Stephen F. Austin State University*

[17] University of Washington, ESRM 250: Introduction to Geographic Information Systems in Forest Resources *Washington: School of Forest Resources* 2010

[18] O. Adegun 2015 State-led versus community-initiated: stormwater drainage and informal settlement intervention in Johannesburg, South Africa *Environment and Urbanization* **27(2)** 407- 420

[19] M. Zeleňáková, Z. Vranayová, A. Repel and D. Kaposztasová 2018 Surface Runoff in Urban Area–Case Study *In International Conference on Urban Drainage Modelling* 152-156

[20] L. Anaba, N. Banadda, N. Kiggundu, J. Wanyama, B. Engel and D. Moriasi 2016 Application of SWAT to assess the effects of land use change in the Murchison Bay catchment in Uganda

IOP Conf. Series: Earth and Environmental Science 1087 (2022) 012036 doi:10.1088/1755-1315/1087/1/012036

Analysis of spatial-temporal atmospheric emissions from heavy-duty diesel trucks in central and southern Liaoning urban agglomeration

Yanhao Cao[1], Xiaohan Wu[1,2], Zhiwei Song[1], Baiyizhuo Chen[1], Shuai Liu[1] and Chengkang Gao[1,3]

[1] School of Metallurgy;
[2] Northeastern University, Shenyang 110819, China.

[3] Corresponding author's e-mail: gaock@mail.neu.edu.cn

Abstract. The problem of diesel vehicle pollution has aroused wide concern in central and southern Liaoning Urban Agglomeration (CSLA), as it is associated with large amounts of atmospheric emissions. This article focuses on heavy-duty diesel trucks (HDT) of Chinese National Standards for heavy-duty diesel trucks (GB) III, GB IV, and GB V. This research measures emission factors (EF) of NOx, CO, and PM by vehicle tests. Subsequently, the emission inventory has been established according to emission factors (EF), activity levels (AL), and Technic guidelines for road motor vehicle emission inventory of air pollutants (Guideline, 2015) in central and southern Liaoning Urban Agglomeration (CSLA). This study further obtained the spatial distribution map of pollutants in CSLA, 2019, using the Geographic Information System (GIS). Results exhibit NOx and CO emissions of HDT directly proportional to the amount of their ownership, and GB III was the main contributor to the PM emissions accounting for nearly 74%. It was concluded the annual emissions of NOx, CO, and PM in CSLA were 20.34, 6.46, and 0.044 million tons individually in 2019. Moreover, the emissions of NOx and CO in heating seasons were higher than in non-heating seasons, accounting for approximately 56% and 57% of the whole year respectively. Notably, Shenyang, Dalian, and Anshan were the top three cities in atmospheric emissions. In addition, the three kinds of pollutants mainly appeared in suburbs and decreased gradually from suburbs to urban areas.

1. Introduction

CSLA, located in Northeastern China is the largest heavy industry base in China, attracting people's attention to its environmental issues. Since the government has reinforced regulations on heavy industry, straw burning, and power generation plants in recent years, vehicle pollution has become the main source of urban pollution in CSLA. Among all types of motor vehicles, the pollution from HDT cannot be ignored. In 2019, CO, NOx, and PM from HDT reached 860,000, 460,700 and 36,000 tons respectively [1]. The pollution from HDT not only reduced the air quality to a large extent but also had a significant impact on public health. In addition, this area is more likely to aggravate air emissions due to its longer heating season and lower temperatures [2].

To investigate the EF of motor vehicles, current research mainly adopted bench testing methods, model calculations, real driving cycle tests, and highway tunnel test methods. Scholars in China had measured several pollutants under different driving conditions. Wang et al. [3] selected light and heavy motor vehicles (14), common in China, then measured emission factors of CO2 by using a chassis dynamometer under cold startup procedures of WLTC (World Light Vehicle Test Program) and C-WTVC (China Heavy Commercial Vehicle Transient Cycle) respectively.

Yang et al. [4] used a portable emission test system to test the emission factors of gaseous pollutants (HCRE32CO and NOx) of four different vehicle types by practical onboard experiments. Wang et al. [5] studied the emission characteristics and factors of volatile organic compounds (VOCs) of various motor vehicles via highway tunnel test methods under normal driving conditions. Gao et al. [6] obtained the emission inventory of air pollutants from motor vehicles in Henan Province through localizing coefficients of emission factors (Guideline, 2015). Moreover, scholars in other countries obtained localized emissions data of motor vehicles. Nagendra Raparthi et al. [7] studied particle and gaseous through roadway tunnel measurements in India. Barouch Giechaskiel et al. [8] summarized filter weighing and chemical group analysis instruments for motor vehicle particle emissions. Ran Tu et al. [9] utilized a portable emission measurement system (exhaust flow meter, sample conditioning system, gas analyzer system, and continuous particle number module) to measure emissions of CO, NOx, and particle number (PN) in Toronto, Canada.

At present, there is little research aiming at pollution from HDTs in CSLA. What is more, the emission factors obtained (Guideline, 2015) have a non-negligible error because of large temperature variations. Selecting CSLA as a location boundary (10 cities) and 2019 as a time boundary, this research obtained the EF of HDTs in non-heating/heating seasons based on vehicle tests and discussed the impacts of seasons on actual EF by considering this Guideline. Furthermore, based on spatial analysis of emissions using ArcGIS, this research established emission inventory combining AL of HDTs in CSLA. Suggestions for controlling emissions of HDTs in low temperature and snowy areas are also implemented at the termination.

2. Theory and methods

2.1. Scope and structure

Focusing on HDTS, this research adopted a bottom-up method to investigate the total emissions of NOx, CO, and PM in CSLA. As shown in Figure 1, traffic flow can be obtained from the field research (researchers in this study implemented a high-definition camera to collect traffic flow data on urban roads, suburban roads, and expressways in CSLA). According to (the Bulletin on National Economic and Social Development in Liaoning Province, 2019), this study obtained the annual mileage and freight volume of HDTs. Moreover, emission factors of NOx, CO, and PM were measured through vehicle experiments.

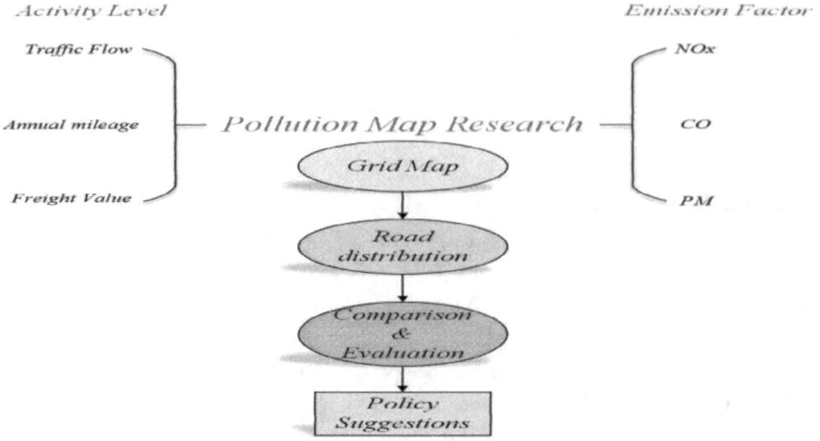

Figure 1. Flowchart of pollution map research.

Combining activity level and emission factors, total emissions can be calculated. Afterward, this study distributed emissions to HDT roads in the grided map using ArcGIS. In the final section, this research proposed policy suggestions based on the pollution distribution map of CSLA.

2.2. *Emission factor*

In the vehicle experiments, we utilized the main control computer, gas emission analyzers, portable power supply, GPS locator, and gas flowmeter as the main equipment. Measuring methods included: a non-dispersive infrared analysis method to measure CO and CO2, non-dispersive UV method to measure NO and NO2, electrochemical method to measure O2 content, and management of a flowmeter to measure tail gas flow [10]. SEMTECH-DS portable exhaust analysis system and Pegasor Mi2 particle analyzer were used to measure the instantaneous HDT emissions. Moreover, HY 100SFB particle sampler was used to collect particle matters from HDT exhaust. (SEMTECH-DS portable exhaust gas analysis system is composed of a vehicle exhaust gas analyzer and vehicle exhaust flowmeter, assisted by an air compressor and gas cylinder. The pollutant content will be marked to reach the initial value before the experiment). The size of the SEMTECH-DS analyzer is 432mm(l) 549mm(W) 355mm(H), and the weight is 35.4kg. On-board exhaust flowmeters use average pitot tube technology, satisfying most diesel and gasoline vehicle engine testing requirements. Pegasor Mi2 particle analyzer integrated the PPS-M particle sensor and is equipped with all the necessary accessories to ensure the sensor operating 24/7 to monitor particle sizes ranging from a few nanometres to 2.5 microns. The sampling frequency is 24-bit 100Hz, the sensor measurement response time is 0.2 seconds, and the particle mass and quantitative concentration can be monitored simultaneously. HY 100SFB was utilized to collect TSP/PM10/PM2.5 samples in ambient air. The flow range of HY 100SFB is set to 10-230 L/min.

Aiming at the most common HDTs of GB Ⅲ, GB Ⅳ, and GB Ⅴ (vehicle parameters, appendix), researchers in our group tested three vehicles to obtain the emission factors of three HDT types in heating/non-heating seasons (heating/non-heating seasons are six months in CSLA). The results are shown as EF_{nsh} and EF_{nwn} in the results and discussion section.

2.3. *Emissions inventory in different seasons*

Considering the significant differences of EFs between non-heating/heating seasons, this paper analyzed the temporal-distribution of CSLA emissions. The annual EF can be calculated in Eq. (1) by combining the EF in non-heating and heating seasons. Based on that, the annual total emissions can be calculated in Eq. (2).

$$EF_{ns} = \frac{(EF_{nsh}+EF_{nwn})}{2} \tag{1}$$

In Eq. (1), EF_{ns}emission factor of s pollutant annually); EF_{nsh}emission factor of s pollutant of n type of HDT in heating season (g/km)); EF_{nwn} (emissions of s pollutant of n type of HDT in non-heating season (g/km)).

$$E_{ns} = W_n \times Am_{ns} \times EF_{ns} \tag{2}$$

In Eq. (2), E_{ns} (annual emissions of s pollutant of n type of HDT (g)); W_n (holdings of n type of HDT); EF_{ns}emission factor of n type of HDT (g/km)); Am_{ns} (average annual kilometer of n type of vehicle (km)).

Eq. (3) shows total HDT emissions, and this study divided the total emissions in heating/ non-heating seasons in Eq. (4) considering the differences between emission factors.

$$E_s = \sum E_{ns} \tag{3}$$

In Eq. (3), E_s (total emissions of s pollutant annually (g)).

$$E_{nsh} = E_s \cdot EF_{nsh}/(EF_{nsh} + EF_{nwn}) \tag{4}$$

In Eq. (4), E_{nsh} (total emissions of s pollutant in heating season (g)).

2.4. Emissions inventory in different cities and roads
Also, this paper focused on the spatial distribution of CSLA emissions. Based on the total emissions (E_s), this paper allocated hourly emissions in 22 CSLA road types (neglecting sidewalks/bikeways) [11-12]. Every grid emission in the ArcGIS map were calculated as Eq. (5) and Eq. (6).

$$EF_{ts} = \frac{E_s}{L \times 365 \times 24} \tag{5}$$

In Eq. (5), EF_{ts}(emission factor of s pollutant per CSLA km/hour (g/(km· h)); L (the total length of CSLA roads (km))

$$E_{tsi} = EF_{ts} \times L_i \tag{6}$$

In Eq. (6), E_{tsi}(total emissions of s pollutant per i grid hour (g)); L_i(the total length of i grid roads (km))

3. Results and discussions

3.1. Emission factors
Emission factors of NOx, CO, and PM obtained from vehicle tests [13-14] and (Guideline, 2015) are listed in Table 1. Results showed: EF of NOx differ greatly in two methods (EF obtained from vehicle test is approximately 1.4-5.4 times as much as EF obtained from Guideline, 2015). However, the differences of EFs of CO and PM in two methods are small (EF of CO obtained from vehicle test is nearly 1.2-3.6 times as much as EF obtained from (Guideline, 2015); EF of PM obtained from vehicle test is of significant magnitude while slightly smaller than EF obtained from Guideline, 2015).

Table 1. Comparison of emission factors (g/km).

Emissions	Truck types	Guideline, 2015 Non-heating season	Guideline, 2015 Heating season	Vehicle Test Non-heating season	Vehicle Test Heating season
NOx	GBIII	5.94	5.85	8.49	10.83
	GBIV	3.37	3.32	8.32	9.68
	GBV	2.87	2.82	9.20	15.03
CO	GBIII	1.96	1.51	3.07	3.15
	GBIV	2.00	1.54	2.27	3.25
	GBV	2.00	1.54	2.39	5.49
PM	GBIII	0.13	0.29	0.04	0.03
	GBIV	0.07	0.15	0.01	0.01
	GBV	0.01	0.03	0.01	0.01

As shown in Table 1, EFs of NOx and CO of GBV are higher than those of GBIII and GB IV. In fact, GBV is the most frequently used HDT in China, especially in the transportation of iron and steel. Therefore, as freight vehicles, GBV has higher EFs of NOx and CO due to its higher deterioration degree.

3.2. Emission sharing rate of different conditions

3.2.1. Emission HDT sharing rates with different standards. This research calculated emission proportions (CO、NOx and PM) of HDTs of GBV, GBIV, and GBIII.

According to incomplete statistics, HDT holdings in CSLA were 276000 in 2019. Among them, GBIII, GBIV, and GBV account for 45.88%, 37.82%, and 16.3% respectively (neglecting GB Ⅰ and GB Ⅱ, which have been largely eliminated due to national policy impact). As shown in Figure 2, PM emissions of GBIII account for more than 70% [15]. This can be explained by the high vehicle holdings and EF(PM) of GBIII. Moreover, GBIV and GBV have further restrictions on emissions of PM. The two HDT types have taken a series of in-machine purification technologies and integrated

IOP Conf. Series: Earth and Environmental Science 1087 (2022) 012036 doi:10.1088/1755-1315/1087/1/012036

external purification (post-treatment) technologies, such as EGR (Exhaust Gas Recirculation System), DPF (Diesel Particulate Filter), and DOC (Diesel Engine Oxidation Catalyst) technologies [16-17]. Notably, GBIV and GBV can guarantee the quality of in-machine purification, as well as effectively reduce PM emissions by filtering PM through DPF. In addition, proportions of NOx and CO emissions from HDTs with different emission standards are approximately the same as holding proportions of HDTs due to small differences within EF and various HDT types. Therefore, emission levels from different HDT types are mainly dependent on their holdings.

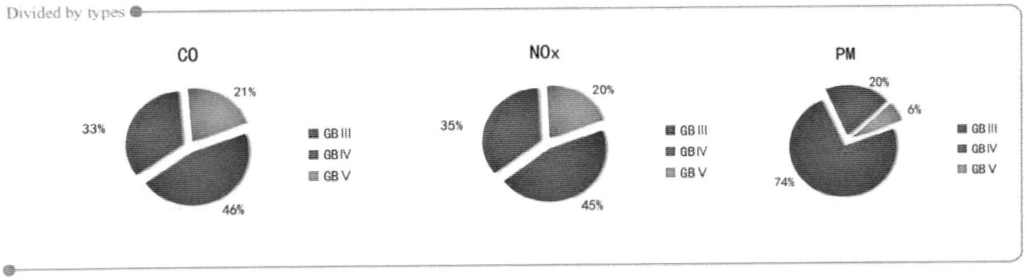

Figure 2. Emissions sharing rate of different types of HDTs.

3.2.2. Emission sharing rate of different seasons. The total emissions of CO, NOx, and PM and their proportions in heating/non-heating seasons are shown in Figure 3. NOx and CO emissions in heating seasons are higher than those in non-heating seasons, while PM emissions during heating seasons are slightly lower than in non-heating seasons [18].

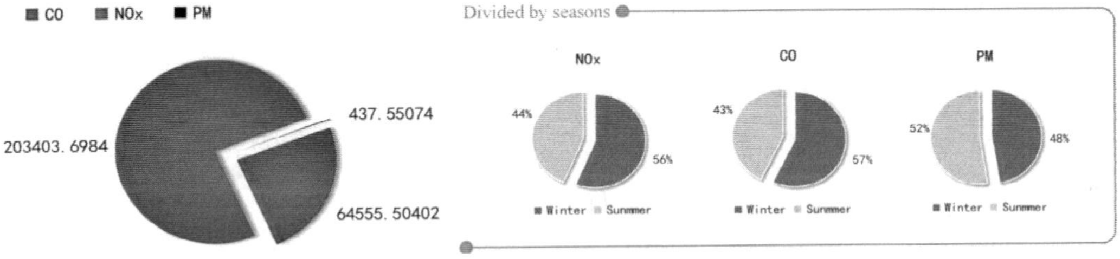

Figure 3. Emissions sharing rate of different seasons.

For NOx, low temperatures in heating seasons reduce the catalyst activity of vehicle-mounted SCR [19], resulting in low conversion percentages of NOx. However, CO emissions are often related to accelerated speed and idling situations of HDTs. Especially in heating seasons, poor road conditions will cause more idling situations, slow acceleration, and deceleration. In such cases, engines are often working at slow revolution speeds and low air intake. Therefore, the rate of residual engine gas will rise, and engineers need to supply mixed gases with higher concentrations [20]. Subsequently, the poor air-fuel ratio caused by idling situations will result in higher CO emissions in heating seasons. Different from CO and NOx, PM emissions are influenced by both environmental and driving conditions. Nevertheless, the direct influencing factor is the engine chamber temperature. High chamber temperatures are often accompanied by high concentrations of mixed fuels, which will lead to the disassociation probability of fuel molecules gradually greater than the probability of polymerization. Consequently, many free radicals will form, and groups of various chain reactions will be triggered. Subsequently, various groups will form cores of gas-exhaust particles, resulting in a rapid increase of instantaneous PM emissions [21]. Compared with heating seasons, HDTs run longer miles in non-heating seasons. Markedly, engine loads are higher, resulting in higher chamber temperatures in non-heating seasons. Therefore, HDTs will have higher PM emissions per hour in non-heating seasons with higher velocity of instantaneous emissions.

3.2.3. Emission sharing rate of different cities. As shown in Figure 4, this paper listed NOx, CO, and PM emissions among ten cities in CSLA. Results showed the pollution amount was related to economic and industrial development levels. The cities of Shenyang and Dalian are the top pollution producers. The main reasons are high economic development levels among these ten cities, resulting in large HDT holdings, large transportation volume, and lengthy road networks. Meanwhile, Anshan and Yingkou also have higher pollution for their high industrial development levels. In fact, both cities maintain large steel plants, which need larger transportation hubs for raw materials and steel. However, pollution in other cities remains a relatively lower economic development level, which leads to less HDT holdings.

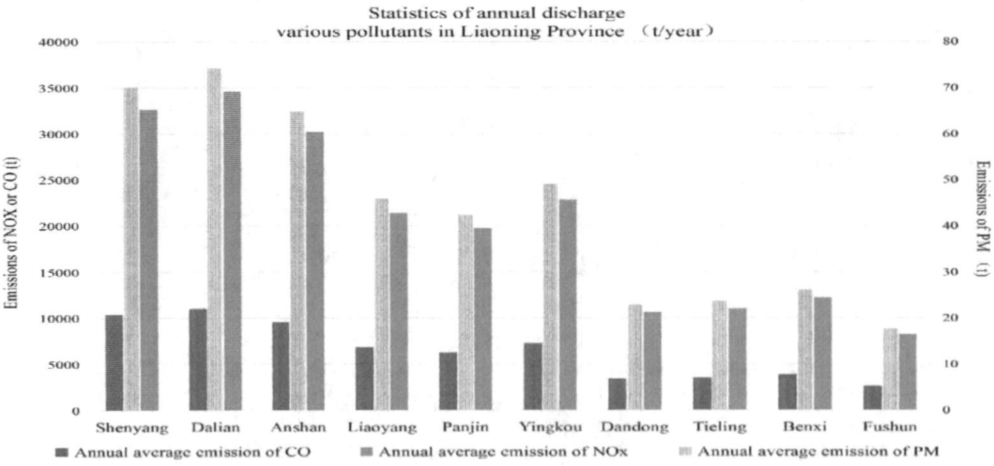

Figure 4. Emission sharing rates in different cities.

3.3. Result of spatial distribution

Based on annual NOx, CO and PM emissions, this study calculated emissions per hour, and distributed them to CSLA road networks using the ArcGIS [22], the result of spatial distribution is shown in Figure 5.

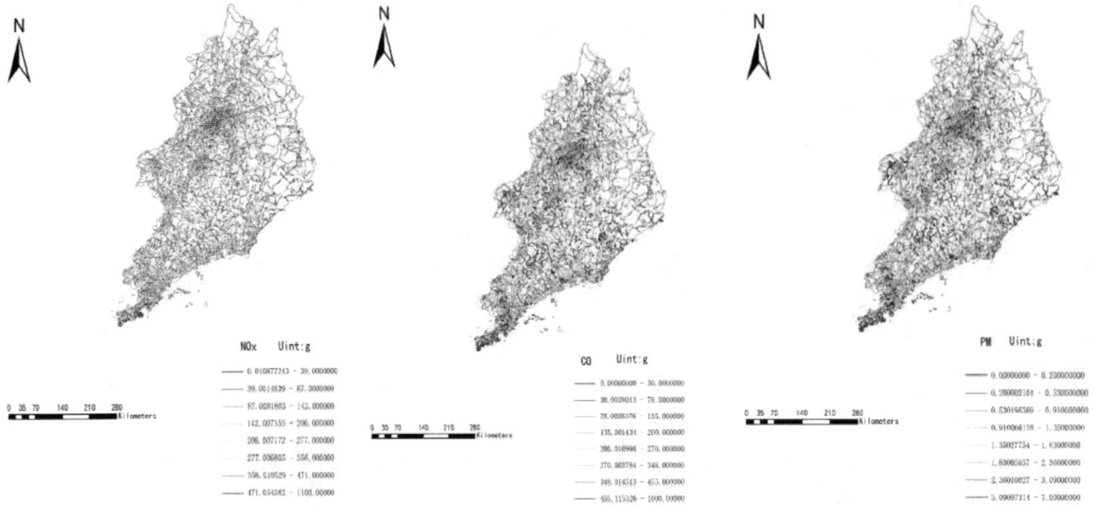

Figure 5. Spatial distribution of NOx, CO and PM.

IOP Conf. Series: Earth and Environmental Science 1087 (2022) 012036 doi:10.1088/1755-1315/1087/1/012036

As shown in Figure 5, CO, NOx, and PM emissions are higher in metropolitan and coastal areas but lower in developed economic regions. The authors have summarized these roads are concentrated around metropolises supporting large freight volume and the need for economic development; quantities of ports and freight volume are also large on coasts, which also results in a great number of HDTs [23]. However, fewer emissions remain in central areas of cities. This can be explained by the Chinese government's no-pass HDT rules (For instance, in Shenyang City, all freight vehicles are prohibited from running on some roads within the first ring road and between the first ring road and the second ring road from 6:00 to 22:00. All freight vehicles are prohibited from running on the second ring road and the roads within the second ring road between 6:00-9:00 and 16:00-19:00. HDTs are prohibited from running on the roads within the third ring road between 6:00-9:00 and 16:00-19:00. In addition, other CSLA cities also have corresponding traffic restriction rules.), which aim to reduce HDT emissions and improve environmental qualities. In Shenyang City and Dalian City, the impacts of these rules are particularly significant. What's more, in Figure 5, road colors around metropolitan and coastal areas are brighter and dense, however, road colors in underdeveloped suburbs and rural areas are duller and sparse. Although metropolitan and coastal areas maintain great numbers of HDT holdings, emissions distributed to each road are not quite high due to large road density. However, emissions allocated to each road are higher in underdeveloped suburbs and rural areas due to smaller road density.

3.4. *Result of temporal-spatial distribution*

Researchers obtained the traffic flow of HDTs in different roads (urban, suburban and expressways) within a day. Besides the combined total pollution, researchers plotted the NOx, CO and PM emissions based on traffic flow proportions within each period (6:00-9:00, 9:00-12:00, 12:00-15:00, 15:00-18:00, 18:00-21:00) per the entire day.

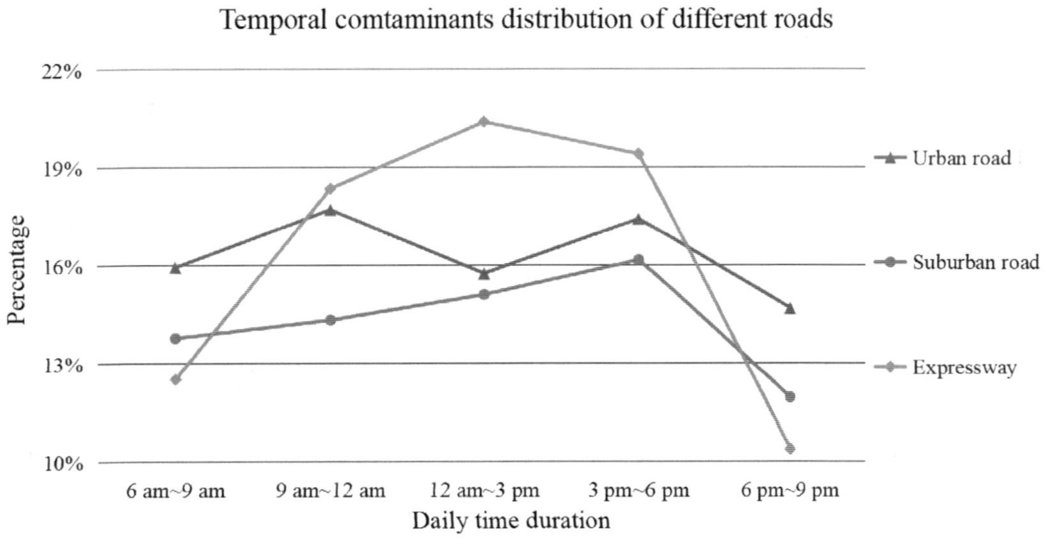

Figure 6. Temporal contaminants distribution of different roads.

As shown in Figure 6, HDT emissions changed significantly in different time periods in the expressway with peaks at noon. To be specific, the emissions are higher within the period of 12:00-15:00 and lower within 6:00-9:00 and 18:00-21:00, which is more consistent with the HDT travel times. However, two peaks of pollutant emissions emerged on urban roads from 9:00-12:00 and 15:00-18:00. This study found urban roads in most CSLA cities maintain rules of "peak limitation" resulting in higher emissions with higher HDT traffic flow after morning peaks and before evening peaks. A difference from the former, emissions slightly changed during daytimes on suburban roads with peak values in the period from 15:00-18:00. Nevertheless, emissions were lower from 18:00-21:00 on suburban roads due to poor suburban road conditions at night.

3.5. Suggestions

3.5.1. Improve diesel quality. Regulating and controlling the diesel quality is one of the primary tasks of controlling HDT pollution. Diesel quality obviously affects NOx and PM emissions without any changes to the diesel engine [24-25] but has little influence on CO emissions due to insignificant differences in diesel oxidation stability. To reduce NOx and PM emissions by regulating diesel quality, this paper offers suggestions considering the affecting diesel oil factors of distillation temperature characteristics, cetane index, aromatic content, and Sulphur content.

(1) Distillation temperature characteristics of diesel oil: engineers use T90/T95 to represent heavy components in diesel oil (large heavy components are challenging to burn off completely, resulting in the increased exhaust gas and particulate engine emissions). Literature [26] showed NOx emissions can be reduced by 3% by lowering T90/T95 by 10℃. Therefore, reducing T90/T95 can decrease NOx emissions.

(2) Diesel oil with a high cetane index, permits short combustion delay periods, lower maximum cylinder pressure, and lower pressure rising ratios. Therefore, its diffusion combustion ratio will increase while pre-mixing combustion will decrease, resulting in the decline of NOx emissions and a surge in PM emissions. However, the increased cetane index can reduce both NOx and PM emissions in an appropriate rangev (45-60) [26].

(3) Aromatic content of diesel oil significantly impacts NOx and PM emissions. Literature [26] showed a 1% increase of arene of diesel will lead to a 0.6% surge in NOx and 0.2% in PM. Nevertheless, a 1% increase in multiring hydrocarbon of diesel will lead to a 1% surge of NOx and 2% in PM. In fact, the hydrocarbon ratio is higher in arene, which can cause higher flame temperatures. Additionally, NOx emissions will ascend. What is more, arene (especially multiring hydrocarbon), as the precursor to soot formation will cause the ascension of PM emissions if its content rises. Obviously, diesel with low arene can reduce PM and NOx emissions.

(4) In addition, Sulphur in diesel oil will be transformed into sulfate in the atmosphere during the process of gas exhaust. The sulfur in the fuel is converted to sulfate at a rate of 1% to 3% as a part of PM. Therefore, low sulfur diesel should be used for all diesel engines [24].

3.5.2. Raise HDT emission standards. At present, GB III, GB IV, and GB V account for 45.88%, 37.82%, and 16.30% respectively of the total CSLA HDTs, the large proportion of GB III causes large CO, NOx, and PM emissions. Therefore, accelerating HDT elimination with old emission standards has become a top priority. Shenyang City [27] has imposed further restrictions on GB III, resulting in gradually phasing this out. Other CSLA cities should also impose restrictions on GB III. Meanwhile, since the Shenyang Municipal Government (**2018**), Pollutant emission limits and Measurement Methods of Heavy Diesel Trucks in Shenyang (The sixth Stage of China), Shenyang (in Chinese) [28] has been released, governments can accelerate HDT production and modification to meet the GB VI requirements. With that, emissions will be reduced with improved HDT standards in CSLA.

3.5.3. Add tail-gas clean up devices. In response to the State Council of the People's Republic of China (**2018**), A three-year action plan to win the battle of sky protection, Beijing (in Chinese) [29], China accelerates HDT elimination of GB III. However, due to the expansiveness of GB III, the country cannot phase this out in a short period of time. Therefore, tail-gas clean up devices can be installed on GB III to reduce emissions. In 2019, according to the announcement restricting GB III issued by Dalian City, the government will give 30% deficiency payments (no more than 10,000 CNYs) to HDT which meet emission standards after the technical transformation of pollution control devices has been installed. Moreover, these HDTs of GB III cannot be restricted through the approval of ecological and environmental authorities. With that, governments not only reduce pollution but also improve durable years and economic efficiency of GB III. However, restricting policies vary in different provinces and cities. Also, installing tail-gas clean up devices is not nationally consistent. Therefore, this paper suggests HDTs running in special-use lanes in CSLA and HDTs within Liaoning Province could be installed as tail-gas clean up devices for transition.

Tail-gas clean up technologies matured since the GB III appeared. Now, complete requirements exist for tail-gas clean up devices. Common improvements of these devices include: cooperatively using primary and standby chambers to reduce the loss of engine power caused by backpressure; designing a hollow shell (inner/outer shell) with vacuum chambers to reduce explosion possibility; adopting selective tail-gas clean up devices to improve flexibility (remove, replace, or adopt maintenance regeneration when filters are blocked by carbon accumulation); setting catalyst to thoroughly purify water and remove hazardous substances (with connections to device top and water purification pool).

4. Conclusions

This research measured emission factors of CO, NOx, and PM of three types of HDTs (GB III GB IV, and GB V) in heating/non-heating seasons in CSLA by vehicle tests. In 2019, combining with the Guideline and activity levels of HDTs in CSLA, researchers also obtained the pollution inventory of CO, NOx, and PM by reflecting proportions with different types of HDTs and their heating/non-heating seasons. Results showed: GB III is the main contributor of PM emissions in HDTs, accounting for 74%; emissions levels from different HDT types are mainly dependent on their holdings; emissions in heating/non-heating seasons are different, NOx and CO emissions are higher in heating seasons than in non-heating, but PM emissions are higher in non-heating seasons than in heating. Among ten CSLA cities, emissions in Shenyang, Dalian, Anshan, and Yingkou remain the highest due to high levels of economic and industrial development.

This paper used ArcGIS to analyze spatial characteristics of NOx, CO and PM emissions in CSLA in 2019. Results showed emissions of the three pollutant types are dominant in large cities and coastal areas, while emissions are less in areas with lower economic development levels. According to the HDT traffic flow measured on different highway types at various time periods, the researchers investigated: a noon peak of expressway emissions, which is more consistent with driving HDT times. On urban roads, two peaks dominate after the morning peak and before the evening peak, which are mainly due to restriction rules. On suburban roads, emissions fluctuate less during the daytime, and lower in the evening, mainly due to the poor suburban road and driving conditions at night.

Acknowledgments
This work was supported by National College Students Innovation and Entrepreneurship Training Program (210093), National Project of Key Research and Development Plan (2017YFC0212303-03), Second Tibetan Plateau Scientific Expedition and Research Program (STEP) (2019QZKK1003), the Based Research Projects of National Natural Science Foundation of China（41871212), Fundamental Research Funds for the Central Universities (N2025008).

References

[1] MEE (Ministry of Ecological Environment of People's Republic of China) 2020 China Mobile Source Environmental Management Annual Report, Beijing
[2] Li Y, Wang Y Q, Liu Q 2009 *Anhui Agriculture Science (China)* **37(4)** 1781-1782
[3] Wang Q, Cao F, Fu M L, Su S, Wang M W, Zhao Z Y, Lin Y Q, Zhang Y L 2021 *Journal of Nanjing University of Information Engineering (Natural Science Edition)* 1-15
[4] Yang X B, Ah Q, Wu P F, Li X 2021 *Journal of Green science and technology (China)* **23(6)** 101-105
[5] Wang L L, Ma S L, Liang J, Wang N, Wang W S 2017 *Green Technology (China)* **(22)** 4-7
[6] Gao D D, Yin S S, Gu X K, Lu X, Zhang H, Zhang R Q, Wang L L, Qi Y J 2021 *Environmental Science (China)* **42(8)** 3663-3675

[7] Raparthi N, Debbarma S, Phuleria H C 2021 *Atmospheric Environment: X* **10** 100113
[8] Giechaskiel B, Maricq M, Ntziachristos L, Dardiotis C, Wang X L, Axmann H, Bergmann A, Schindler W 2014 *Journal of Aerosol Science* **67** 48-86
[9] Tu R, Xu J S, Wang A, Zhang M Q, Zhai Z Q, Hatzopoulou M 2022 *Science of The Total Environment* **805** 150407
[10] Ge Y S, Wang A J, Wang M, Ding Y, Tan J W, Zhu Y C 2010 *Journal of Automotive Safety and Energy (China)* **1(2)** 141-145
[11] Zheng J Y, Che W W, Wang Z L 2009 *ActaSci.Circum.(China)* **29(4)** 815-821
[12] Gao C K, Gao C B, Song K H, Xing Y H, Chen W W 2020 *Process. Saf. Environ. Prot* **138** 236-245
[13] Yao Z L, Wu B B, Wu Y N, Cao X Y, Jiang X 2015 *Atmospheric Environ* **123** 1-8
[14] Li S, Liu T Y, Song W, Pei C L, Huang Z Z, Wang Y J, Chen Y N, Yan J H, Zhang R Q, Zhang Y L, Wang X M 2021 *Environ. Pollut* **280** 116972
[15] Wu B B, Shen X B, Cao X Y, Zhang W, Wu H, Yao Z L 2015 *Atmospheric Environ* **116** 216-224
[16] José M L, Felipe J, Francisco A, Nuria F 2009 *Transp Res D Transp Environ* **14(1)** 1-5
[17] Yin H D 2020 *Hebei Farm Machinery(China)* **(5)** 62-63
[18] Wang M, Tian P F, Wang L G, Yu Z R, Du T, Chen Q, Guan X, Guo Y M, Zhang M, Tang C G, Chang Y, Shi J S, Liang J N, Cao X J, Zhang L 2021 *Sci. Total Environ* **798** 149310
[19] Song Z Y, Wang M D, Tao Y F, Zhang H, Qu W D 2021 *Automotive abstracts(China)* **(9)** 49-54
[20] Zhang Y 2019 *M.S. Thesis, Chongqing University, Chongqing, China*
[21] Sun L L, Zhang N N, Huang W W, Gao M R, Miao L 2020 *Safety and environmental engineering (China)* **27(5)** 77-84
[22] Aslam S, Siddiqui G, Kazmi S J H 2020 *Reg. Stud. Mar. Sci* **33** 100956
[23] Cheng N N, Zhang C, Jing D J, Li W, Guo T X, Wang Q L, Li S J 2020 *J Environ Sci (China)* **92** 118-128
[24] Zhao L, Li K, Li B S, Li Y, Sang J, Bao X F, Yue X 2013 *Journal of environmental engineering technology (China)* **3(4)** 337-341
[25] Chen Z M, He J J, Chen H, Geng L M, Zhang P 2021 *Fuel* **304** 121360
[26] Chen W M, Wang J X, Shuai S J 2008 *Automotive engineering (China)* **(8)** 657-663
[27] SMG (Shenyang Municipal Government) 2021 Circular of Shenyang Municipal People's Government on Expanding the Low Emission Zone for Motor Vehicles and Non-Road Mobile Machinery (third-round draft for Comments), Shenyang
[28] SMG (Shenyang Municipal Government) 2018 Pollutant emission limits and Measurement Methods of Heavy Diesel Trucks in Shenyang (The sixth Stage of China), Shenyang
[29] SCPRC (State Council of the People's Republic of China) 2018 A three-year action plan to win the battle of sky protection, Beijing

Developing comprehensive resilient water systems and their driving factors for the 21st Century Society in China

Jingwei Yang[1,2], Fei Xue[1,2], Xiang Sun[3,4,6], Yao Luo[1,2] and Jian Wang[5]

[1] Shenzhen Water Planning & Design Institute Co., Ltd., Shenzhen 518001, PR China;
[2] Guangdong Provincial Engineering and Technology Research Center for Water Affairs Big Data and Water Ecology, Shenzhen, 518001, PR China;
[3] School of Environment, Harbin Institute of Technology, Harbin, 150090, PR China;
[4] School of Environmental Science and Engineering, Southern University of Science and Technology, Shenzhen, 518055, PR China;
[5] Guangdong Provincial Engineering Technology Research Center for Urban Water Cycle and Water Environment Safety, Tsinghua Shenzhen International Graduate School, Tsinghua University, Shenzhen 518055, P.R. China.

[6] Corresponding author's e-mail: 52925356@qq.com

Abstract. More frequent and severe natural disasters triggered by climate change and the increasing domestic water consumption have placed higher demands on sustainable water services. The establishment of "Resilient Cities" provides overall guidance on future directions of water affairs. Literature reviews were designed to transform the concept of resilient cities into waterworks from the dimensions and properties. Then the concept of "Resilient Water Affairs" was put forward. Evaluation indicators of resilient water affairs were summarized based on dimensions and properties to indicate the driving factors of the resilient water system. Finally, the future trends of resilient water affairs were proposed, which provides theoretical and technical support for establishing resilient water affairs.

1. Introduction

In recent years, earthquakes, tsunamis, typhoons, floods, droughts, and other disasters frequent worldwide have claimed countless lives and caused massive damage and economic loss to cities, at a time when COVID-19 is rampant. According to CRED's research (Centre for Research on the Epidemiology of Disasters), the average number of disasters worldwide from 2000-2019 was 368, while 389 in 2020, with a significant increase from previous years [1]. From UNDRR 's report (United Nations Office for Disaster Risk Reduction), the average annual damage caused by natural disasters worldwide was approximately US$314 billion in 2015. Due to increasing levels of urbanization, losses are expected to increase to US$415 billion by 2030, without taking into account social and economic shocks [2]. Water-related hazards, including flood risk, drought, storm surges, sea-level rise, and so on, account for nearly 90% of these disasters [3].

In global environmental change, the disasters are becoming more extreme and compounded while the incidence is increasing [4]. It is difficult for modern, rapidly expanding cities to cope with the threat of droughts, floods, and water pollution events through a single water supply and drainage or hydraulic project [5]. Unless countermeasures are taken, millions of people will suffer from life-threatening and poverty-stricken conditions.

2. Methodology

2.1. Research Background and State of the Art

The entire water system still suffers from multiple problems such as high energy consumption, low efficiency, fragmentation, and remedial type, including poor drainage capacity, high urban water stress, and weak disaster risk response capacity [6, 7].

To achieve flood prevention and mitigation as well as sustainable development of people and nature, the concept of "Water Sensitive Cities" has been proposed [8], emphasizing rainwater utilization, water conservation, source reduction, and ecological landscaping [9,10]. Moreover, Wang et al. [11] proposed the concept of "Sponge Watershed", combining traditional sponge cities and ecological civilization construction, taking the whole basin as the research object, and constructing urban through water system integration. However, human society, public safety, infrastructure and networks, economic resilience, and the level of public services have not yet been taken into full consideration in current water affair system.

The concept of resilient cities has provided new ideas for developing water systems. This paper draws on international experience as to how resilient cities can deliver opportunities to strengthen the anti-risk ability of the water affair system and mitigate the water affair development challenges in developing countries such as China.

2.2. Research Routes and Sources

Firstly, a review of the literature to collate articles on the topic of resilient cities concerning the water sector was undertaken, based on database of Web of Science until 2021 (Figure 1). Subsequently, we explored the definition and properties of resilient water affairs based on the core elements of resilient cities reported by WBG (World Bank Group). In addition, evaluation indicators of resilient water affairs were summarized and analyzed through different dimensions, combining engineering experience and existed evaluation methods of resilient cities to provide ideas for the future construction of resilient water affairs, especially in China.

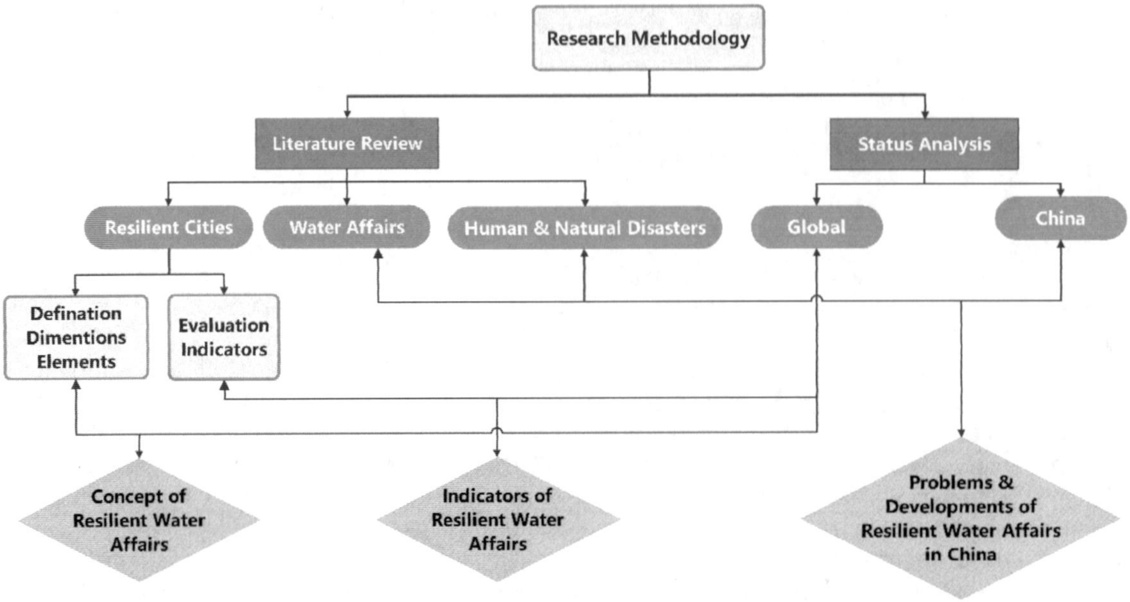

Figure 1. Methodological diagram.

3. Results

3.1. Definition, Classification and Elements

According to the ICLEI (Local Governments for Sustainability) [12], a "Resilient City" is prepared to absorb and recover from both human and natural shock or stress while maintaining its essential functions, structures, and identity as well as adapting and thriving in the face of continual change. According to Godschalk [13], resilient cities can be divided into physical systems and human communities. We made explanations and grouped it around six different topics (Table 1). The topics of water resources, water security, water environment, and water ecology involved in the physical systems of Resilient Cities are also included in the concept of Sponge Watershed and Water Sensitive Cities. Besides, resilient cities take water economy and water management into account and pay more attention to human communities. For instance, the organization functions and responsibilities, the level of public services, and economic systems, with elements such as accelerating the digitalization of water systems, early warning and forecasting capabilities, scheduling and control capabilities, rapid recovery capabilities, resident participation, and emergency training.

Table 1. Classification of a resilient city and its content related to water affairs.

Classification			Topics	Ref.
Categories	Subcategories	Detailed description		
Physical systems	Natural environmental systems	Rational development and utilization of water resources;	Water Resources	[14–16]
		Water ecological environment protection	Water Ecology Water Environment	
	Urban engineering facilities systems	Design and infrastructure of urban artificial water systems	Water Safety Water Environment	[17–19]
Human communities	Organization functions and responsibilities	Education and training; Maintain social stability	Water Management	[20, 21]
	Public services	Strengthening management and establishing networks for monitoring, early warning, community, and emergency response;	Water Management	
		Public scrutiny mechanism	Water Management	
	Economic systems	Building a sustainable water economic system;	Water Economics	[22, 23]
		Reducing economic losses due to calamities	Water Economics	

WBG (World Bank Group) summarized resilient cities in terms of five elements: robustness, redundancy, inclusion, coordination, and reflectiveness [24]. With regard to resilient water affairs, the systems should be planned and designed to ensure effective disaster resilience (robustness), adequate reserve resources (redundancy), the safety and interests of vulnerable groups (inclusion), strengthen organizational leadership and cross-functional coordination (coordination), exchange information and share best practices (reflectiveness), to continuously develop and improve urban and rural water systems. On this basis, this paper proposes the concept of "Resilient Water Affairs" regarding "Resilient Cities", which is defined as a water system that can withstand disasters and mitigate losses through planning and design, to maintain safe and stable operation of the water system, ensure public safety and social order, and continuously improve and adjust to possible disasters. Water affairs include water resources management and development, rainwater utilization, drought and flood control, urban water supply and distribution, as well as wastewater collection, discharge, treatment, and recycling.

3.2. Evaluations and Driving Factors

As "Resilient Cities" are not clearly defined in quantitative terms and may need to be inferred from indirect observations, their assessment can be challenging. Currently, several tools and frameworks for assessing the resilience of cities have been developed by a wide variety of stakeholders, including the Resilience Assessment Framework (RAF) [25], City Resilience Profiling Programme (CRPP) [26], Disaster Resilience Scorecard for Cities [27], City Resilience Framework [28], and EPA Conceptual Framework [29]. Among these, only the RAF method covers the whole process of raw water, reclaimed water, stormwater, wastewater, and drinking water. Rainwater and drinking water issues are receiving more concern by existing evaluation methods of "Resilient Cities". However, the "Resilient Water Affairs" have not been fully revealed by current indexes. This study summarized the evaluation indicators based on the existed resilience indicators and international status of "Water Affairs" (Figure 2). Those indicators were generalized from five dimensions (infrastructures, institutions, natural ecosystems, human societies, and economies), five properties (robustness, redundancy, inclusion, coordination, and reflectiveness), and five water types (raw water, stormwater, wastewater, reclaimed water, and drinking water).

4. Discussions

The current international evaluation indicators of "Resilient Cities" still lack treatment plants' energy self-sufficiency ratio, wastewater collection and treatment ratio, reclaimed water utilization rate, annual water conservation rate for residents, evaluation of water consumption structure and so on (e.g., irrigation, street cleaning, etc.) (Figure 2). In addition, some of the evaluation indicators, such as management chain reliability, financial adequacy, degree of public participation, water services evaluation, have limitations that are difficult to quantify or are not easily replicated in practice in China. It is recommended to use subjective weighting methods such as expert scoring, Delphi method, and analytic hierarchy process to describe the importance of those factors. While for the indicators that are easy to quantify, for example, water consumption, treatment rate, and rainfall, are evaluated by using the entropy evaluation method. The indicators related to infrastructure and economic systems (e.g., rainfall, treatment rate, flood control standards, etc.) are well analyzed and documented during urban water resources management. However, a few social and natural indicators are still missing from water resources bulletins and statistics yearbook, for example, days of pollution events, days of water storage capacity, etc. This, on the other hand, reflects the inadequacy of the current water management systems. Furthermore, the values of evaluation factors should take full account of climate change risks and carbon reduction targets, avoiding excessive exploitation of water resources, over construction of water facilities and waste in human resources. Also, focusing on factors such as ensuring social justice, economic friendly, and sustainable development.

In general, enhancing the resilience of water affairs systems should be according to the basis and principles of prevention-oriented, rapid response, appropriate deployment, risk mitigation at source, regulation and monitoring in the process, to reduce the chances of accidents or rapid recovery after the occurrence of disasters. At present, seven barriers are restraining the establishment of "Resilient Water Affairs". Concretely, the advanced water technologies to improve the resilience of water processes (technical barriers), sufficient equipment, funding, inspectors and other resource allocation (resource barriers), complete resilient water action plan to guide the top-level design (policy barriers), a sound interlinked resilient community as the elementary unit of the disaster prevention system (system barriers), a comprehensive safety education mechanism to strengthen public capabilities for addressing disasters (education barriers), a continuous online monitoring platform to establish a comprehensive emergency response system (platform barriers), and authoritative scientific assessment methods for "Resilient Water Affairs" (assessment barriers).

In terms of technical barriers, scientists should focus their research on improving the reliability of facilities, reducing the potential risks of the system, developing low energy consumption and resilient water technologies. With regard to resource security, waterworks engineers should scientifically draw up reserve plans to avoid wasting resources, rationally allocate funds to maximize benefits. In the case of policies, optimizing water affairs resilience strategies, increasing investment in science and education, expanding public participation channels should be implemented as soon as possible. In respect of system barriers, managers should promote resilience in governance, speed up the establishment of resilient communities, and make multi-sectoral collaborative action. As for education, more training is needed to improve operational proficiency, expand professionals' knowledge, and enhance public consciousness on disaster prevention and mitigation. Regarding platform and assessment, we should improve water affair systems' risk control and disaster warning/forecasting capabilities by accelerating establishing a digital water affair chain and an emergency information-sharing platform. In summary, water affairs systems need to be improved at three levels (facilities, systems, and people) to be resilient.

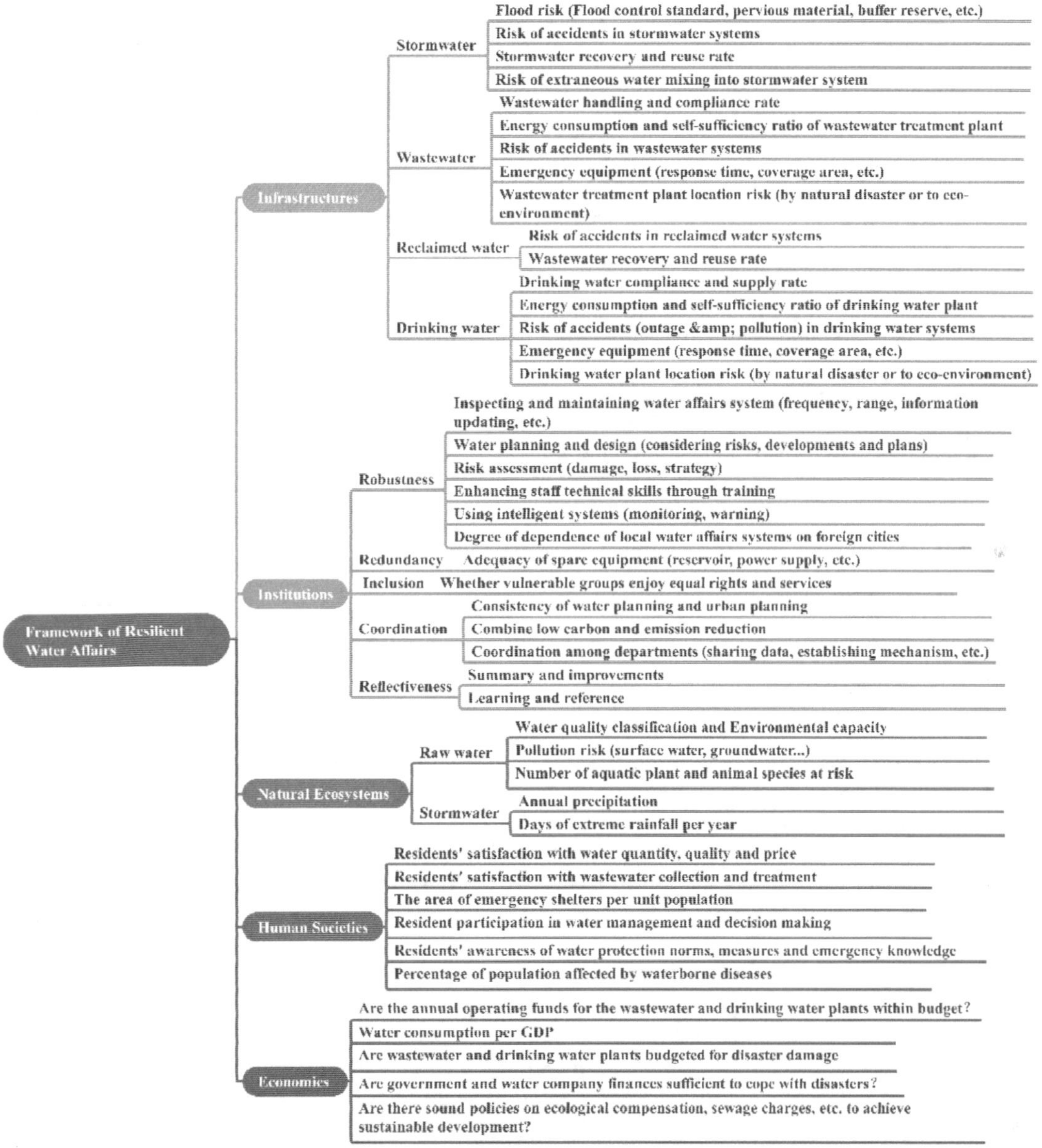

Figure 2. Indicators for evaluating the resilience of existing water affairs designs and facilities.

5. Conclusions

Increased urban resilience is an inevitable trend in the development of urban planning globally, and a realistic demand for the current transformation and development of China's cities. As an industry that produces and provides daily necessities, water affairs are crucial yet exceptionally fragile. Improving the resilience of water systems is, therefore both necessary and challenging. The development of information technology in the new era is a huge opportunity to accelerate the establishment of intelligent water systems and is strong support for resilient water affairs. In recent years, the frequent occurrence of extreme natural disaster events has attracted a great deal of attention and contributed to a positive awareness of disaster preparedness, laying the foundation for effective prevention and emergency response efforts. The COVID-19 epidemic has accelerated the establishment of joint prevention and control working mechanisms as well as group prevention and organizational control systems in urban and rural communities.

Overall, to achieve the resilient of the water affairs system, breaking seven barriers (technical, resource, policy, system, education, platform, assessment) within five dimensions (infrastructures, institutions, natural ecosystems, human societies, economics) through five properties (robustness, redundancy, inclusion, coordination, reflectiveness) are the driving factors. It is recommended that a comprehensive, unified, authoritative, and emissions-cutting goals-friendly evaluation system be established at the first opportunity. Furthermore, a technically replicable, economically implementable, socially acceptable, and environmentally friendly long-term resilient water affairs scheme and action guidelines should be established as soon as possible. These are the cornerstone for the construction and development of resilient water affairs. This paper has put forward the concept of Resilient Water Affairs, presented an overview of evaluation indicators, analyzed the opportunities and challenges take-up of such an approach in China. In general, the consideration of public safety and services, early warning and networks, economic resilience are sorely lacking in current water affair resilience systems in China. In further studies, according to the actual situation of a particular region, the detailed indicator and its rating value can be screened and increased to help policymakers to clarify the foothold of local resilient water affairs development.

References

[1] CRED & UNDRR 2021 Global trends and perspectives Executive Summary *CRED Brussels* 1-7
[2] UNISDR 2015 Global Assessment Report on Disaster Risk Reduction *New York* 1-316
[3] WWAP 2012 Managing Water under Uncertainty and Risk: The United Nations World Water Development Report *World Water Assessment Programme, Perugia* 1-909
[4] Folke C 2006 Resilience: The emergence of a perspective for social-ecological systems analyses *Glob Environ Chang* **16** 253–267
[5] WBG 2015 Investing in Urban Resilience *Washington* 1-120
[6] Miller JD, Hutchins M 2017 The impacts of urbanisation and climate change on urban flooding and urban water quality: A review of the evidence concerning the United Kingdom. *J Hydrol Reg Stud* **12** 345–62
[7] UN DESA 2019 World Population Prospects 2019 *New York* 1-46
[8] Ashley R, Lundy L, Ward S 2013 Water-sensitive urban design: Opportunities for the UK *Proc Inst Civ Eng Munic Eng* **166** 65–76
[9] Wong THF 2006 An overview of water sensitive urban design practices in Australia *Water Pract Technol* **1** 1–8
[10] Feng W and Baogang H 2017 Governance approach & practice of water sensitive city: Australia's exploration *Urban Dev Stud* **24** 86–93
[11] Wang J, Xue F, Jing R 2021 Regenerating sponge city to sponge watershed through an innovative framework for urban water resilience *Sustainability* **13** 1–36
[12] ICLEI 2019 Resilient Cities, Thriving Cities: the Evolution of Urban Resilience *Local Governments for Sustainability e.V, Bonn* 1-37

[13] Godschalk DR 2003 Urban hazard mitigation: creating resilient cities *Nat Hazards Rev* **4** 136–43
[14] Wagner I, Breil P 2013 The role of ecohydrology in creating more resilient cities *Ecohydrol Hydrobiol* **13** 113–34
[15] Agudelo-Vera CM, Leduc WRWA, Mels AR 2012 Harvesting urban resources towards more resilient cities *Resour Conserv Recycl* **64** 3–12
[16] Barry ME, Coombes PJ 2018 Planning resilient water resources and communities: the need for a bottom-up systems approach *Aust J Water Resour* **22** 113–36
[17] Jing W, Guangli Z, Xianming H 2020 Design practice of urban water system in the Netherlands based on rain and flood resilience *Sci Technol Rev* **38** 66–76
[18] Delanka-Pedige H, Munasinghe-Arachchige S, Abeysiriwardana-Arachchige I 2021 Wastewater infrastructure for sustainable cities: assessment based on UN sustainable development goals (SDGs) *Int J Sustain Dev World Ecol* **28** 203–9
[19] Ferreiro C, Villota N, de Luis A 2021 Water Reuse Study from Urban WWTPs via c-Ultrafiltration and Ozonation Technologies: Basis for Resilient Cities and Agriculture *AGRONOMY-BASEL* **11** 322
[20] Siauve S, Amorsi N 2015 Protection of the whole catchments providing drinking water - Study of practices in Europe - Part I: Cross-Analysis *IOWater, Paris* 1-35
[21] Walters D, Coglianese C 2020 Litigating EPA Rules: A Fifty-Year Retrospective of Environmental Rulemaking in the Courts *Case West Reserve Law Rev* **70** 1007–37
[22] Luthy RG, Wolfand JM, Bradshaw JL 2020 Urban Water Revolution: Sustainable Water Futures for California Cities *J Environ Eng* **146** 04020065
[23] Dieleman H 2013 Organizational learning for resilient cities, through realizing eco-cultural innovations *J Clean Prod* **50** 171–80
[24] WBG 2019 Building Urban Resilience: An Evaluation of the World Bank Group's Evolving Experience (2007-17) *Washington* 1-166
[25] Cardoso MA, Brito RS, Pereira C 2020 RAF resilience assessment framework-A tool to support cities'action planning *Sustainability* **12** 1–64
[26] UN-Habitat 2018 City Resilience Profiling Tool Guide. *United Nations Human Settlements Programme, Nairobi* 1-31
[27] UNDRR 2017 Disaster Resilience Scorecard For Cities - Preliminary Level Assessment. *United Nations Office for Disaster Risk Reduction, Geneva* 1-52
[28] ARUP 2016 City Resilience Index: Understanding and Measuring City Resilience *London* 1-51
[29] USEPA 2017 Evaluating Urban Resilience to Climate Change: A Multi-Sector Approach, *Washington* 1-675

IOP Conf. Series: Earth and Environmental Science 1087 (2022) 012038 doi:10.1088/1755-1315/1087/1/012038

Geoinformation representation of maritime knowledge flows: new frontiers of coastalization

Mikhaylova A.[1], Maksimenko D.[2], Hvaley D.[1], Maksimenko M.[2] and Mikhaylov A.[3,4]

[1] Institute of Geopolitical and Regional Studies, Immanuel Kant Baltic Federal University, Kaliningrad, 236016, Russia;
[2] Institute for Statistical Studies and Economics of Knowledge, Higher School of Economics, Moscow, 101000, Russia;
[3] Center for Geopolitical Studies, Institute of Geography RAS, Moscow, 119017, Russia.

[4] Corresponding author's e-mail: mikhailov.andrey@yahoo.com

Abstract. Marine and coastal economies are the drivers of growth for many nations around the world. Close proximity to the sea generates positive externalities being especially strong in coastal areas, whose effective governance is a challenge. The greatest challenge to date is ambiguity in delimitation of a coastal zone, as static criteria based on density of population and industry, are limited in our understanding of the functional connectivity of the sea and land. This article studies functional boundaries of a coastal zone by focusing on the geoinformation analysis of marine-related R&D. We hypothesize that maritime knowledge flows between the customers and contractors of R&D projects can outline the actual configuration of coastal zone in its knowledge production domain. The research data is sourced from ROSRID database of 2017-2019, covering 1,773 marine-related R&D projects funded or executed by entities located across 119 municipalities of 64 regions in Russia. The results of the study showed that maritime knowledge flows are not limited to coastal municipalities or even coastal regions. Some research areas, such as Fisheries, aquaculture and marine life studies, are more localized in coastal zones than others, although featuring an overall strong interconnectedness between inland and coastal territories. Functional delimitation of coastal zone management areas using dynamic data on spatial networks can enhance the effectiveness of coastal zone management.

1. Introduction

Active exploration, settlement, and intensive economic activity in the coastal zone throughout the world history has put these territories at the forefront of human civilization [1]. According to Martinez et al. [2], 41% of the world's population lives in the coastal zone within 100 km from the coastline, and for some coastal countries this figure is even higher. Scholarly research from across the globe demonstrate the positive economic [3] and social [4] effects from proximity to the sea, incl. being associated with industry clustering, the influx of human resources, the intensification of international integration of the coastal zone – a phenomenon called coastalization. A study by Lu et al. [5] on the Commonwealth of Independent States (CIS) showed that direct access to the sea is a significant competitive advantage for attracting foreign direct investments. A study on Indonesian seaports provides empirical evidence that proximity to seacoasts has a positive effect on regional development (Gross Regional Product per capita, labor productivity, poverty rate) [6]. Also, assessments of the development of maritime transport show that sea routes remain significant for the movement of people and goods. Maritime transport accounts for over 80% of world trade and in terms of value – 60-70% (2018), and for some countries, a key element of the economy, providing access to world markets [7].

The modern blue economy covers a wide range of activities: the extraction and processing of living and non-living resources of the sea (fishing, aquaculture and blue biotechnology, minerals); renewable energy production; port facilities, shipbuilding and ship repair; maritime transport, pipelines and cables, coastal and marine tourism, etc. [8, 9]. According to various estimates, the economy of coastal countries receives a considerable contribution to Gross Domestic Product from maritime sector (e.g. China – 9.4% [10]) and coastal economies (e.g. Mexico – 40% [11]). The spatial perspective on the development of maritime economy is related to certain land-based nodal elements located at different distances from the coastline, displaying the effects from the sea in multiple territorial zones [12].

In this regard, the delimitation of the coastal zone from the inland territories is becoming an increasingly important issue in the context of the potential for the manifestation of the positive effects of coastalization. Current perspectives on defining the boundaries of the coastal zone are based on an assessment of natural and environmental factors (e.g. land elevation) [13], migration processes and settlement structure [14], the location of port infrastructure [15], industrial facilities [16], etc. The range of propagation of the coastalization effect may vary depending on the country, including both local zones up to 1 km, and coastal areas up to 10, 50, 150 and even 200 km [17].

The growing importance of knowledge and innovation for sustainable socio-economic development is also relevant for coastal areas with specific spatial innovation activity forms: marine clusters [18], maritime knowledge clusters [19], smart sea ecosystems [20], smart seaports [21], etc. The accumulation of sea-related knowledge and competencies within these territorial innovation models is as important as the localization of appropriate infrastructure, the availability of human resources or the location of natural objects. Thus, accounting for maritime knowledge flows can be a new approach to the delimitation of the boundaries of the coastal zone, outlining functional coastal zone management areas (CZMAs) and the development of holistic integrated coastal zone management (ICZM). The purpose of this study is to assess the new frontiers of coastalization effects through geoinformation analysis of maritime knowledge flows for the delimitation of territorial-functional coastal zones.

2. Materials and methods

The study area is Russia – the largest maritime country with a 46 thousand km length of coastline, delineating the coasts of 3 oceans and 13 seas, and a sea shelf area of 4 million sq.km. Out of 85 regions, 23 have direct access to the sea: Atlantic – 7, Caspian – 3, Arctic – 7, and Pacific – 6. The study covers 119 municipalities located in 64 regions identified to have maritime knowledge flows.

The Unified State Information System for Accounting Research, Development and Technological Works for Civil Use (ROSRID – https://rosrid.ru) database covers all civil research and development (R&D) projects and the results of intellectual activity in Russia. The database covers 66,647 reports of 2017-2019 R&D projects, including: a) project title, keywords, OECD field of science and technology classifications, state categories on scientific and technical information; b) name and state register of legal entities code of project's customers and contractors; c) amount of R&D funding by sources.

Extraction of marine-related projects from ROSRID database had the following steps:

- marine-related fields of knowledge were defined in classifiers (e.g. Oceanography);
- 1,063 keywords from over 160 thousand keywords attributed to R&D projects were identified as marine-related according to certain stems (e.g. "marine-", "fish-");
- projects either relating to maritime field of research or containing marine keywords selected;
- irrelevant R&D projects were manually excluded.

The final database comprised of 1,772 maritime R&D projects with a total cost of 8.4 billion rubles. All the researches were divided into 3 categories according to keywords and fields of knowledge: Group A: General oceanology and ocean-related projects; Group B: Fisheries, aquaculture and marine life studies; Group C: Shipbuilding and engineering projects.

Data on customers and contractors of R&D projects was linked to SPARK-Interfax corporate reporting database via Name and State Register of Legal Entities code in order to obtain companies' addresses. The addresses of entities were geocoded using Yandex.Maps geocoder API in geopy 2.2.0 Python package. Enterprise-level spatial referenced data were aggregated on the level of municipalities.

Maritime knowledge flows are defined by projects with customers and/or contractors being located in corresponding municipalities. A wide range of organizations from the government and academic to non-profit and business sectors is identified. Spatial interconnections were visualized using QGIS (Figures 1-2). Due to the overwhelming number of projects funded by the executive authorities, federal research funds and state-owned corporations located in Moscow, the knowledge flows from all other municipalities to Moscow were mapped distinctively. The size of each pie-chart represents the total amount of projects' funding, while the color denotes the dominant category of projects defined above.

Table 1. Distribution of regions by participation in maritime knowledge flows.

Region	Customer and Contractor		Customer only		Contractor only	
	No.	% of total regions	No.	% of total regions	No.	% of total regions
Coastal	12	14.1	1	1.2	7	8.2
Inland	13	15.3	4	4.7	27	31.8
Total	25	29.4	5	5.9	34	40.0

Table 2. Distribution of municipalities by participation in maritime knowledge flows.

Region	Customer and Contractor		Customer only		Contractor only	
	total	including coastal, %	total	including coastal, %	total	including coastal, %
Coastal	40	73	1	100	7	57
Inland	39	–	5	–	27	–
Total	79	37	6	17	34	12

A number of research limitations are observed during the study. Firstly, only the national level of maritime knowledge flows is accounted for as the database used does not cover R&D with international funding. Secondly, any military-related R&D is not available in the public data, thus, a significant part of naval R&D is not considered. Thirdly, some input errors in R&D funding of numerous projects found, mistakenly multiplied by 1000 times. These observations were manually checked and validated. Finely, numerous organizations operating in different Russian regions are registered in Moscow, resulting in an even stronger divergence during spatial analysis.

3. Results

In total, in 2017-2019, municipalities of 64 regions of Russia (or 75% of all) participated in maritime knowledge flows, of which 59 regions were engaged in the marine-related knowledge production and 30 regions were involved as contractors. Among the regions involved in maritime knowledge flows, 20 are classified as coastal, which is half as much as the inland participating regions – 44. Another 3 coastal regions (Nenets and Chukotka Autonomous Okrugs, the Republic of Kalmykia) have no R&D registered. Results in Table 1 imply that the majority of coastal regions (12 out of 20) involved in marine-related R&D are both customers and contractors (both generating the demand for maritime knowledge and satisfying their needs), while the majority (27 out of 44) of inland regions are only contractors.

At the municipal level, the majority (34 out of 48) of municipalities from the coastal regions involved in marine-related R&D are also coastal, i.e. having direct access to the sea – Table 2. First of all, this is typical for those municipalities that are part of the coastal regions of the "customer and contractor" group, however, it is also true for the other two groups "customer only" or "contractor only". This demonstrates the high localization of maritime knowledge flows directly in the coastal zone of coastal regions, which must be considered when delimiting coastal areas. Also, marine-related R&D is carried out in 71 municipalities of inland regions, of which 55% are in regions that combine the functions of financing and producing knowledge. This indicates functional connectivity of coastal and inland territories for the generation of marine-related knowledge and the interregional use of competencies.

Urbanization factor has a strong influence on the geography of maritime knowledge flows. Of the 119 municipalities involved in marine-related R&D, 45.4% are the administrative-territorial centers of the regions; 20.2% belong to the administrative districts of federal cities (Moscow, St. Petersburg, Sevastopol); 19.3% are urban districts that are not the centers of their regions. Only 18 municipalities participating in maritime knowledge flows have mixed urban-rural population. These are 5 districts in the Moscow Oblast, 3 districts in the Rostov Oblast and 1 per Belgorod, Ivanovo, Leningrad, Saratov, Yaroslavl regions, the Republics of Tatarstan, Crimea, Adygea, Primorsky Krai and Krasnodar Krai. These regions are dominated by R&D contractors, however, there are also R&D customers, predominantly concentrated in coastal municipalities or municipalities located in the coastal regions.

An important assessment of the spatial distribution of maritime knowledge flows is the degree of their concentration in individual regions and their municipalities. Of the 15 regions in which the volume of completed marine-related R&D exceeded 100 million rubles, 10 are coastal regions: St. Petersburg, Primorsky Krai, Sevastopol, Rostov Oblast, Murmansk Oblast, Republic of Karelia, Krasnodar Krai, Sakhalin Oblast, Arkhangelsk Oblast, Magadan Oblast. Their total contribution was 3.9 billion rubles or 46.6% of total R&D. This is comparable to the leader of this list – Moscow with 33% of projects. In general, the distribution of the volume of marine-related R&D between coastal and inland regions is almost equal: 50.5 and 49.5%. However, as noted earlier, there are 2.2 times fewer coastal regions than inland ones; therefore, for coastal regions, the volume of R&D performed per 1 region is much higher.

While the network of contractors is distributed, the financing institutions are highly localized and have a pronounced centralization in Moscow (95.1% of customers). St. Petersburg stands second with 2.5%. Another 1.9% of R&D funding comes from coastal regions (1.5% from coastal municipalities) and only 0.5% from inland ones. This funding structure is determined by the national system of science financing, where the state is the main source, and, consequently, the customer. Thus, 49.1% of all funding comes from various ministries of Russia and the two largest state funds.

Table 3. Distribution of marine-related R&D between regions and thematic groups.

Region	Contractor						Customer					
	A[a]		B[b]		C[c]		A		B		C	
	mln rub	%	mln rub	%	mln rub	%	mln rub	%	mln rub	%	mln rub	%
coastal	2811.6	49.8	767.3	60.2	673.0	45.0	137.9	2.5	9.8	0.8	228.1	15.4
inland	2835.1	50.2	507.6	39.8	821.2	55.0	5491.2	97.5	1265.2	99.2	1251.9	84.6
total[d]	5646.6	100	1274.9	100	1494.2	100	5629.2	100	1274.9	100	1480.0	100

[a] A – General oceanology and ocean-related projects;
[b] B – Fisheries, aquaculture and marine life studies;
[c] C – Shipbuilding and engineering projects.
[d] The discrepancy between the total amount of funded and performed R&D is explained by the imperfection of the database featuring some unfilled cells.

From the standpoint of the delimitation of the coastal zone, of particular interest is the geography of customers located in coastal regions whose R&D funding done independently in accordance with their needs and not through centralized funding bodies located in the capital city of Moscow. These organizations are located (in addition to St. Petersburg) mainly in the coastal municipalities of the coastal regions: Arkhangelsk, Kaliningrad, Leningrad, Magadan, Murmansk, Rostov regions, Krasnodar Krai and Primorsky Krai, the Republic of Crimea. Among them are industrial enterprises (JSC, LLC, PJSC), universities, entities of the Russian Academy of Sciences. Also, a number of R&D

funding organizations from coastal regions are located outside the coastal zone and account for another 0.4% of total funding.

The functional connectivity of coastal and inland regions is more clearly manifested in the thematic structuring of the identified maritime knowledge flows. Table 3 gives a clear perspective on the requests for and implementation of the various studies in the coastal and inland regions.

Marine-related research in Group B turned out to be the most territorially dependent: about 60% of research held in coastal regions. However, their funding almost entirely (99.2%) comes from the inland regions. This shows that the thematic trend in this area is set mainly in the capital (Figure 1), while other regions rarely provide the need for knowledge in this area with their own funding (Figure 2).

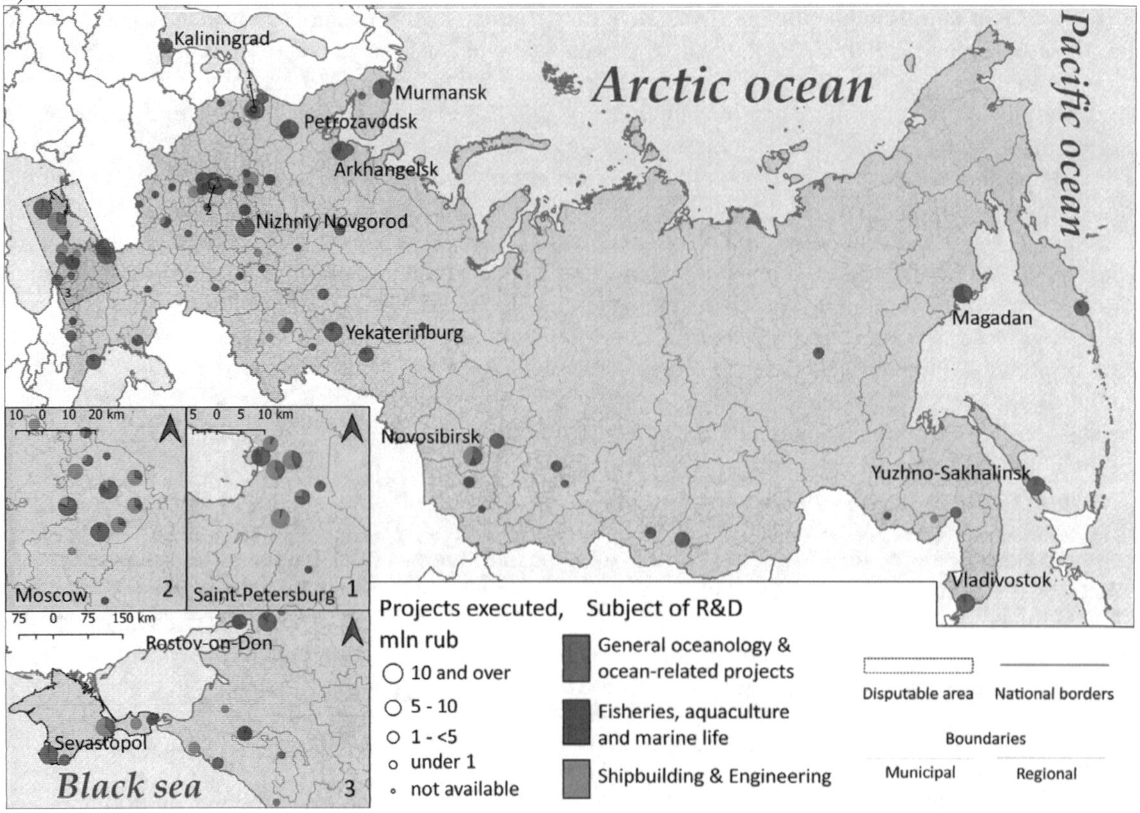

Figure 1. Maritime knowledge flows between municipalities (only with federal cities as customers), 2017-2019.

In 2017-2019, aside of capital, the interregional functional relations on Group B R&D was initiated by 9 inland regions (Voronezh, Yaroslavl, Samara, Novosibirsk, Saratov, Tyumen regions, the Republic of Bashkortostan, Tatarstan, Udmurtia) with a total amount of 9.4 million rubles, including 62% devoted for 1 project in the Voronezh Oblast to develop a technology for water purification from biological contaminants while maintaining the fish balance, as well as 8 coastal regions (Primorsky Krai, the Republic of Karelia and Crimea, Arkhangelsk, Sakhalin, Astrakhan regions, St. Petersburg, Yamalo- Nenets Autonomous Okrug) with the amount of funding of 9.8 million rubles – Figure 2.

R&D classified under the Group A occupies a leading position by project number (67.1%) and is implemented parity in inland and coastal regions – Table 3. Financing is also highly centralized (97.5% are in the inland regions, incl. 99.6% in Moscow), similar to Group B. Among the coastal regions, these are the Krasnodar Krai, the Rostov Oblast, St. Petersburg, Primorsky Krai, Leningrad Oblast and the Republic of Crimea (they set the agenda, fund research and consume new knowledge) – Figure 2.

The Group C of research projects is predominated by inland regions as R&D contractors (55%), funding is less centralized. As much as 15.4% of the funding for this group of studies comes from the coastal regions, primarily from St. Petersburg (13.1%), Krasnodar (1.2%) and Primorsky Krai (0.6%).

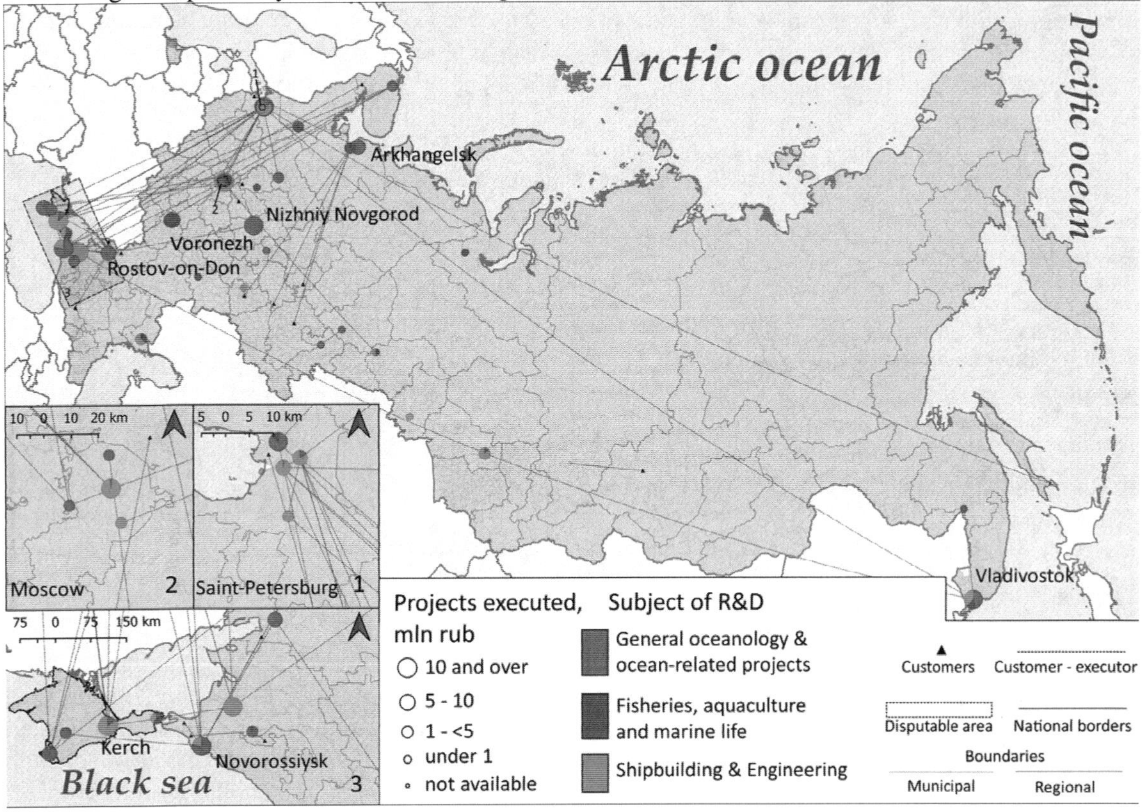

Figure 2. Maritime knowledge flows between municipalities (excluding federal cities as customers), 2017-2019.

4. Discussions

Considering the functional integrity of the coastal zone by the production, consumption, and commercial application of marine-related knowledge, we evaluated the correlation between the involvement of coastal regions in marine-related R&D and the indicators of marine economic development. Table 4 presents the calculated pairwise correlation coefficients between significant maritime indicators.

Analysis shows a strong correlation between the region's share in the national production of marine-related vehicles, machinery, and equipment (indicators II and III) and the volume of Group C R&D (both funded and executed). A significant dependence is found between all three indicators (I, II, and III) of maritime industry localization and the Group B R&Ds. Group A R&D is the least susceptible to co-location of research and business, which is reflected both in low correlation coefficients for coastal regions (Table 3) and in the high share of inland regions in R&D of this topic (Table 4).

The dependencies in the co-location of knowledge-producing organizations and the industry in coastal regions shows the interconnectedness of these processes, outlining the institutions involved in marine-related R&D as integral elements of the maritime economic ecosystem. By accounting for maritime knowledge flows in the delimitation of the coastal zone, we deepen our understanding of coastalization and enhance the methods of its registration. The traditional delimitation of the coastal zone (and coastalization) done by measuring the geographical proximity to the sea of human settlements [14,17] or the clustering of economic activity [15,16], could now be combined with an assessment of functions performed by entities. The extended understanding of the coastal economy was previously verified by Morrissey and O'Donoghue [22], who have mapped the inland communication within a maritime economic network. Our approach implies improving the territorial-functional approach to delimitation of the coastal zone in light of the knowledge economy, mapping interconnectedness of knowledge production system elements by non-spatial forms of proximity [23].

Table 4. Paired correlation coefficients between indicators of participation in marine-related R&D and marine economic activities in 21 coastal regions of Russia, 2017-2019.

Indicator	Group A		Group B		Group C	
	Contr. [a]	Cust. [b]	Contr. [a]	Cust. [b]	Contr. [a]	Cust. [b]
I. Average annual catch of fish and production of marine bioresources per employee in the field of fishing	0.236	-0.064	**0.639**	-0.064	0.477	0.511
II. Regional share in national production of machinery and equipment [c]	0.447	0.374	**0.709**	-0.063	**0.885**	**0.887**
III. Regional share in national production of vehicles and equipment [d]	0.446	0.085	**0.644**	0.037	**0.916**	**0.910**

[a] Contr. – Contractors, the volume of R&D executed in the region (based on contractors);
[b] Cust. – Customers, the volume of R&D funded by the region (based on customers);
[c] includes the production of marine engines and other components for shipbuilding;
[d] includes the construction of ships, vessels, and boats.

5. Conclusions

The study is designed to address two tasks: firstly, to assess the possibility of delimiting the coastal zone by allocating the organizations involved in the generation and consumption of marine-related knowledge resulting of focused R&D projects. Secondly, to evaluate the functional connectivity of coastal regions with inland territories within the framework of the maritime economy through the emerging interregional and interorganizational maritime knowledge flows. Based on the research results drawn from the paired customer-contractor relations for marine-related R&D in 2017-2019 across the coastal and inland regions of Russia a set of three thematic groups are outlined and studied: A – General oceanology and ocean-related projects, B – Fisheries, aquaculture and marine life studies, and C – Shipbuilding and engineering projects.

Both coastal and inland regions are involved in marine-related R&D, which created favorable conditions for the intensification of maritime knowledge flows between different territories and organizations. Inland regions predominantly play the role of R&D contractors, while coastal regions to a greater extent performed both the functions of customers and contractors. This allows, on the one hand, for the coastal regions to accumulate marine-related knowledge and competencies, and, on the other hand, to form a research agenda by driving the demand.

The network of contractors is more geographically distributed both in terms of the number of municipalities involved and the volume of R&D performed. At the same time, per 1 region, the volume of marine-related R&D performed is greater for coastal regions than for inland ones (with the exception of Moscow). Marine-related R&D customers are characterized by a higher degree of territorial and institutional concentration. Moscow is the absolute leader in terms of funding. This is a consequence of centralized policy pursued by the state to support priority topics of national interests.

This study showed a high degree of urbanization of ongoing marine-related R&D since most customers and contractors are located in cities. This is typical for both inland and coastal regions and, first of all, is associated with the system of location of research organizations. However, there are cases when maritime research funding comes from non-urban municipalities. In most cases, these are coastal municipalities or municipalities of the coastal region.

The findings support the idea that mapping the maritime knowledge flows can be used to delimit the functional boundaries of a coastal zone, which is important for their efficient governance. By considering coastal territories not as a geographically determined area, but as a functional system of networks, we step over the administrative boundaries in our understanding of coastal zone management areas (CZMAs). Thus, capturing the existing ecosystem of ties – "demand for knowledge – funding R&D – knowledge production – knowledge commercialization". Leaving maritime knowledge out of equation with the territorial and functional delimitation of the coastal space, the CZMA is incomplete, escalating the problem of asymmetry and fragmentarity of information in decision-making. This also applies to those cases when the subject of marine R&D is formed from outside, but their result should be in demand within the coastal region, by companies located in it.

The following research findings stress the importance of considering the maritime knowledge flows in defining CZMAs and elaborating on integrated coastal zone management (ICZM). Firstly, coastal regions involved in maritime knowledge flows are both customers and contractors, which allows them to accumulate relevant thematic knowledge and competencies, as well as influence the generation of new knowledge. Secondly, at the municipal level, coastal zones are most actively involved in maritime knowledge flows in coastal regions, both as R&D customers and contractors. The location of R&D contractors and customers in the coastal zone supports the efficient use of the economic and natural potential of these territories through the formation of new spatial forms of management, such as marine clusters, marine knowledge clusters, and others. Thirdly, different groups of marine-related researches differ in varying degrees of territorial conditionality, both to conduct research and apply knowledge. Of the three groups considered, R&D in the field of Fisheries, aquaculture and marine life studies (Group B) is mostly localized in the coastal zone and is associated with the industry located in them (extraction and processing of marine resources). High spatial dependence was also noted for knowledge production in the field of Shipbuilding and engineering projects and related industries (Group C). These findings fit within the framework of a single costal territorial-functional zone.

This study contributes to the development of functional delimitation models of a coastal zone. We propose to take into account the developments within the concepts of the knowledge economy and the proximity school in economic geography for a better understanding of non-spatial connectivity of coastal and inland areas.

Acknowledgments
The study was supported by the RSF grant 22-28-00022 at the Southern Federal University.

References
[1] Dalgaard C., Knudsen A. S. B., Selaya, P. 2020 The bounty of the sea and long-run development *Journal of Economic Growth* **25(3)** 259-295 https://doi.org/10.1007/s10887-020-09181-8

[2] Martínez M. L., Intralawan A., Vázquez G., Pérez-Maqueo O., Sutton P., Landgrave R. 2007 The coasts of our world: Ecological, economic and social importance *Ecological Economics* **63(2-3)** 254-272 https://doi.org/10.1016/j.ecolecon.2006.10.022

[3] Akbulaev N., Bayramli G. 2020 Maritime transport and economic growth: Interconnection and influence (an example of the countriesin the caspian sea coast; Russia, Azerbaijan, Turkmenistan, Kazakhstan and Iran) *Marine Policy* 118 https://doi.org/10.1016/j.marpol.2020.104005

[4] Kelly C. 2018 'I need the sea and the sea needs me': Symbiotic coastal policy narratives for human wellbeing and sustainability in the UK. Marine Policy **97** 223-231 https://doi.org/10.1016/j.marpol.2018.03.023

[5] Lu W., Kasimov I., Karimov I., Abdullaev Y. 2020 Foreign direct investment, natural resources, economic freedom, and sea-access: Evidence from the commonwealth of independent states *Sustainability (Switzerland)* **12(8)** https://doi.org/10.3390/su12083135

[6] Yudhistira M. H., Sofiyandi Y. 2018 Seaport status, port access, and regional economic

development in indonesia *Maritime Economics and Logistics* **20(4)** 549-568 https://doi.org/10.1057/s41278-017-0089-1

[7] Brooks M. R, Faust P. 2018 50 years of Review of Maritime Transport, 1968–2018. Reflecting on the past, exploring the future. Transport and Trade Facilitation. Series No.11. UNCTAD. 97 p. https://unctad.org/system/files/official-document/dtl2018d1_en.pdf

[8] Bocci M., Sangiuliano S. J., Sarretta A., Ansong J. O., Buchanan B., Kafas A., Vergílio M. 2019 Multi-use of the sea: A wide array of opportunities from site-specific cases across europe *PLoS ONE* **14(4)** https://doi.org/10.1371/journal.pone.0215010

[9] Roos G., Kubina N. Y., Farafonova Y. Y. 2021 Opportunities for sustainable economic development of the coastal territories of the baltic sea region in the context of digital transformation *Baltic Region* **13(2)** 7-26 https://doi.org/10.5922/2079-8555-2021-2-1

[10] Yu Z., Di Q. 2020 The coordination between maritime economies and marine carrying capacity and their spatiotemporal evolution in the cities of the bohai rim in china *Ecological Modelling* 438 https://doi.org/10.1016/j.ecolmodel.2020.109192

[11] Sevilla N. P. M., Adeath I. A., Le Bail M., Ruiz A. C. 2018 Coastal development: Construction of a public policy for the shores and seas of mexico *Coastal management: Global challenges and innovations* (pp. 21-38) https://doi.org/10.1016/B978-0-12-810473-6.00003-0

[12] Zaucha J. 2019 Can classical location theory apply to sea space? *Maritime spatial planning: Past, present, future* (pp. 97-119) https://doi.org/10.1007/978-3-319-98696-8_5

[13] McGranahan G., Balk D., Anderson, B. 2007 The rising tide: Assessing the risks of climate change and human settlements in low elevation coastal zones *Environment and Urbanization* **19(1)** pp. 17-37 https://doi.org/10.1177/0956247807076960

[14] Small C., Nicholls, R.J. 2003 A global analysis of human settlement in coastal zones *Journal of Coastal Research* **19(3)** 584-599

[15] Yan J., Xiao R., Su F., Bai J., Jia F. 2021 Impact of port construction on the spatial pattern of land use in coastal zones based on CLDI and LUT models: A case study of Qingdao and Yantai, *Remote Sensing* **13(16)** https://doi.org/10.1016/j.landusepol.2009.09.014

[16] Druzhinin A. G., Lachininskii S. S. 2021 Russia in the World Ocean: Interests and Lines of Presence *Regional Research of Russia* **11(3)** 336-348 https://doi.org/10.1134/S2079970521030035

[17] Mikhaylov A. S., Plotnikova A. P. 2021 The coasts we live in: can there be a single definition for a coastal zone? *Baltic Region* **13(4)** 36-53 https://doi.org/10.5922/2079-8555-2021-4-3

[18] Gifford E., McKelvey M., Saemundsson R. 2021 The evolution of knowledge-intensive innovation ecosystems: Co-evolving entrepreneurial activity and innovation policy in the west swedish maritime system *Industry and Innovation* **28(5)** 651-676 https://doi.org/10.1080/13662716.2020.1856047

[19] Zhou Y., Yuen K. F., Tan B., Thai V. V. 2021 Maritime knowledge clusters: A conceptual model and empirical evidence *Marine Policy* 123 https://doi.org/10.1016/j.marpol.2020.104299

[20] Meskauskiene V., Öörni A., Sell A. 2019 When the Sea meets City: Transformation towards a Smart Sea in Finland. HICSS Proceedings *IEEE Computer Society* (pp.3119-3128) https://doi.org/10.24251/HICSS.2019.377

[21] Meyer C., Gerlitz L., Henesey L. 2021 Cross-border capacity-building for port ecosystems in small and medium-sized baltic ports *TalTech Journal of European Studies* **11(1)** 113-132 https://doi.org/10.2478/bjes-2021-0008

[22] Morrissey K., O'Donoghue C. 2012 The Irish marine economy and regional development *Marine Policy* **36(2)** 358-364 https://doi.org/10.1016/j.marpol.2011.06.011

[23] Mattes J. 2012 Dimensions of proximity and knowledge bases: innovation between spatial and non-spatial factors *Regional Studies*, **46(8)** 1085-1099 https://doi.org/10.1080/00343404.2011.552493

IOP Conf. Series: Earth and Environmental Science 1087 (2022) 012039 doi:10.1088/1755-1315/1087/1/012039

Earthworm invasion in Chayu River valley of Tibet under climate change

Mingyue Jin[1], Lei Hou[1,3,5], Yihua Ren[2,3,4] and Yaozhi Zhou[2]

[1] Department of Resources & Environment, Tibet Agriculture & Animal Husbandry University, Nyingchi 860000, China;
[2] Institute of Plateau Ecology, Tibet Agriculture & Animal Husbandry University, Nyingchi 860000, China;
[3] Tibet Key Laboratory of Forest Ecology in Plateau Area, Ministry of Education, Nyingchi 860000, China;
[4] Nyingchi Ecological Monitoring Station, Tibet Ecological Environment Remote Sensing Monitoring Center, Nyingchi 860000, China.

[5] Corresponding author's e-mail: 465133704@qq.com

Abstract. Climate change affects biological invasion. To understand the invasion pattern of alien species of earthworm in Gongrigabu river watershed of Chayu river in Tibet under climate change, the participate rural appraise (PRA) was applied to investigate and analyze the peasant's perception of the appearance time and harm of alien species of earthworm in 13 villages. The results showed that the local climate was warming and drying, and climate warming was the main reason for the invasion of alien species of earthworms. Since the appearance of alien species of earthworm at an altitude of 1528 m in 2008, they had invaded the upstream area at a climbing speed of 71.43 m·a-1 and had been outbreaking in areas below 2028 m. For the climate change adaptation measures, the local peasants should change the crop layout.

1. Introduction

Earthworms belong to the Oligochaeta and are found in most ecosystems. They are the most abundant invertebrate soil animals in terrestrial ecosystems [1] and are known as "ecosystem engineers" because of their important role in material cycling and energy transfer in soil processes [2-3]. In addition, earthworms can create and sustain habitats for other soil organisms through physical and chemical modifications to the environment [2,4]. Earthworms are generally considered to have positive effects on agricultural production, improving soil structure [5], fertility [6-7], and contaminated soil [8-10]. However, once a certain limit is exceeded, earthworms can disrupt the soil ecosystem balance and cause soil loosening [11-12], which harms agricultural production. Farmers are directly engaged in labor production activities that are closely related to the natural environment and are therefore sensitive to changes in the natural environment and their impacts. During a survey on climate change perception and adaptation in the Chayu River basin in Tibet, based on farmers' own practical experience in agricultural labor production, farmers in certain villages believed that earthworm species that had not previously appeared had negative impacts on their village paddy fields, such as breaking field ridges and causing crops to fall over.

Biological invasion is one of the major environmental problems that threaten global ecological security. Along with economic globalization and climate change, China is also suffering from serious biological invasions, and more than 500 invasive species have been identified [13]. Biological invasions and climate change are two important factors affecting biodiversity and ecosystem function, and previous studies have suggested that each has an independent impact on biodiversity, but many existing studies have suggested that climate change affects the rate and extent of biological invasions and that rising temperatures, changing precipitation patterns, and the continued occurrence of extreme weather events may also affect the prevalence and distribution of invasive organisms [14].

At present, climate change has become an important issue of concern in various countries and regions, with far-reaching effects on the structure, function, and processes of various ecosystems, which are relevant to all aspects of human society [15]. Climate change has also impacted agricultural production, mainly due to the increase in temperature and changes in precipitation patterns [16-19], and the corresponding changes in biogeography [20-22]. The high average altitude of the Tibetan Plateau, with its unique geographic location and topography, has a significant impact on the climate of China, Asia, and the world [23-24]. Not only that, but the Tibetan Plateau is also an amplifier and driver of global climate change [25] and is sensitive and ahead of its time in climate change and its response [26-27]. Therefore, the Tibetan Plateau faces a more serious threat of climate change, especially the agricultural population that depends on natural resources such as agriculture, herding, and gathering, which not only increases the risk of livelihood after the impact of climate change but also may become poor as a result [28]. According to the study, under the fact of global warming [29], the Qinghai-Tibet Plateau is also accelerating [30], which also affects meteorological elements such as precipitation and humidity in different parts of the plateau [31]. The invasion of earthworm exotics to paddy fields can then be considered a threat under the influence of climate change. Climate not only determines the biological processes of earthworms themselves but also acts by changing their habitat and food supply, which in turn changes the species and distribution range of earthworms [32]. In the high-altitude region of the southeastern Tibetan Plateau, changing climatic conditions may be promoting earthworm invasion [33].

In this paper, we constructed a simple model of earthworm exotic species invasion based on the emergence time and location of the earthworm exotic species, Amynthas hupeiensis, in the Gongrigabu watershed of the Chayu River, to provide a theoretical basis for understanding the response of earthworms to climate change, preventing and controlling earthworm hazards, and enhancing farmers' adaptation to climate change.

2. Materials and methods

2.1. Study site

The study area is located in the watershed of the tributary of the Chayu River basin in Tibet, including Upper Chayu town and Lower Chayu town, which belong to Chayu County, Linzhi City, Tibet. The Chayu River is large in eastern Tibet, with double-branch rivers, the eastern branch of the Gongrigabu, and the western branch of the Sangqu, which converge into the Chayu River and flow southward out of the national border to join the Brahmaputra River (i.e., the Yarlung Tsangpo River). The Gongrigabu originates from the end of the modern glacier in Gongrigabu, and is about 170 km long, with a drop of 5140 m and an average elevation of 3980 m. The overall flow is northwest-southeast, with a watershed area of 4900 km2 [34]. Influenced by warm and humid air currents of the Indian Ocean, precipitation is relatively abundant, with annual precipitation reaching 800 mm, wet and rainy from March to September and low rainfall from October to February of the following year [35]. The average annual temperature is 12.1°C, the average wind speed is 2.1 m/s, and the frost-free period is 215 days [36]. The main ethnic groups are Tibetans, as well as Han, Naxi, Luoba, etc., living on the flatter mountainous areas on both sides of the river, with a population of about 10,600. They are engaged in agricultural, herding, and gathering production activities. Crops such as paddy and barley are grown. Rich in plant and animal resources.

a. Soil leakage b. Soil collapse

c. Plastic film wrapped soil to prevent collapse d. Plastic wrap around the soil to prevent water leakage

Figure 1. The harm and prevention of earthworm in paddy field.

The emergence and outbreak of exotic species of earthworms had damaged the soil structure of paddy fields, mainly through its activities to break through or break down the field ridge, resulting in water and fertilizer leakage or loss (Figure 1 a, b). To prevent field soil damage, soil and water fertility loss, and earthworm migration, farmers in the study area mostly adopted plastic film to wrap the field soil (Figure 1 c, d).

2.2. Research methods

The study hypothesized that earthworm exotics would invade the Gongrigabu river watershed of the Chayu River from the downstream to the upstream direction. To test this hypothesis, farmers' perceptions of the emergence time and damage of earthworm exotics were surveyed in villages along the Gongrigabu river from downstream to upstream.

2.2.1. Climate data sources. Temperature, precipitation, and relative humidity data from the Chinese meteorological data network (http://data.cma.cn/) of the Chayu station since 1969 were used to characterize the climate change conditions in the Gongrigabu watershed of the Chayu river.

2.2.2. Survey Metho;dology. To grasp the emergence time of the earthworm exotic species Amynthas hupeiensis and its harmful status to paddy fields, a special survey was conducted in May 2018 in each village of Upper and Lower Chayu towns in the Gongrigabu watershed, and 13 villages with the emergence of the earthworm exotic species were identified (Table 1, Figure 1). A participatory rural appraisal (PRA) method was adopted, including questionnaires, observation methods, and mini-symposiums. Farmers who owned paddy fields were selected for interviews. The interviewed farmers were mainly Chinese Tibetan and Mishmi people, and all of them had been engaged in agricultural activities for more than 10 years, and their knowledge about the emergence time of earthworm exotics and their hazards was representative. The farmers were generally less educated, speaking Tibetan and Chinese. To complete the questionnaire effectively, the trained Tibetan and Mishmi student investigators filled out the questionnaire. The questions included age, years of farming, the year in which earthworms were found in the village's paddy fields, and the year in which ridges in the paddy fields leaked or collapsed. For integration into the local society, the selected investigators were all born in Chayu County. The investigators did not have an ownership or interest relationship with the respondents, which can ensure the objectivity of the survey information.

Table 1. The location of the study area.

Village	Altitude (m)	Latitude and longitude	Position order
Rongyu	2028	96°42'1.33"E; 28°51'3.17"N	1
Songlin	1904	96°44'8.63"E; 28°45'39.89"N	2
Gonggu	1889	96°44'59.32"E; 28°45'48.85"N	3
Guba	1850	96°50'45.38"E; 28°40'25.82"N	4
Geyong	2077	96°51'15.98"E; 28°40'1.16"N	5
Zhaba	1783	96°54'51.66"E; 28°36'16.06"N	6
Zigeng	1732	96°56'8.92"E; 28°34'34.07"N	7
Talin	1715	96°59'23.24"E; 28°32'32.46"N	8
Xiani	1686	96°59'31.31"E; 28°31'36.01"N	9
Xin	1774	96°59'28.82"E; 28°30'57.74"N	10
Baantong	1708	96°59'59.10"E; 28°30'52.56"N	11
Gamei	1623	97°0'17.03"E; 28°30'8.57"N	12
Shaqiong	1528	97°0'45.43"E; 28°29'5.57"N	13

Note: 1-13 indicates the upstream and downstream distribution of villages along the Gongrigabu watershed of the Chayu River, with smaller numbers indicating more upstream and larger numbers indicating more downstream.

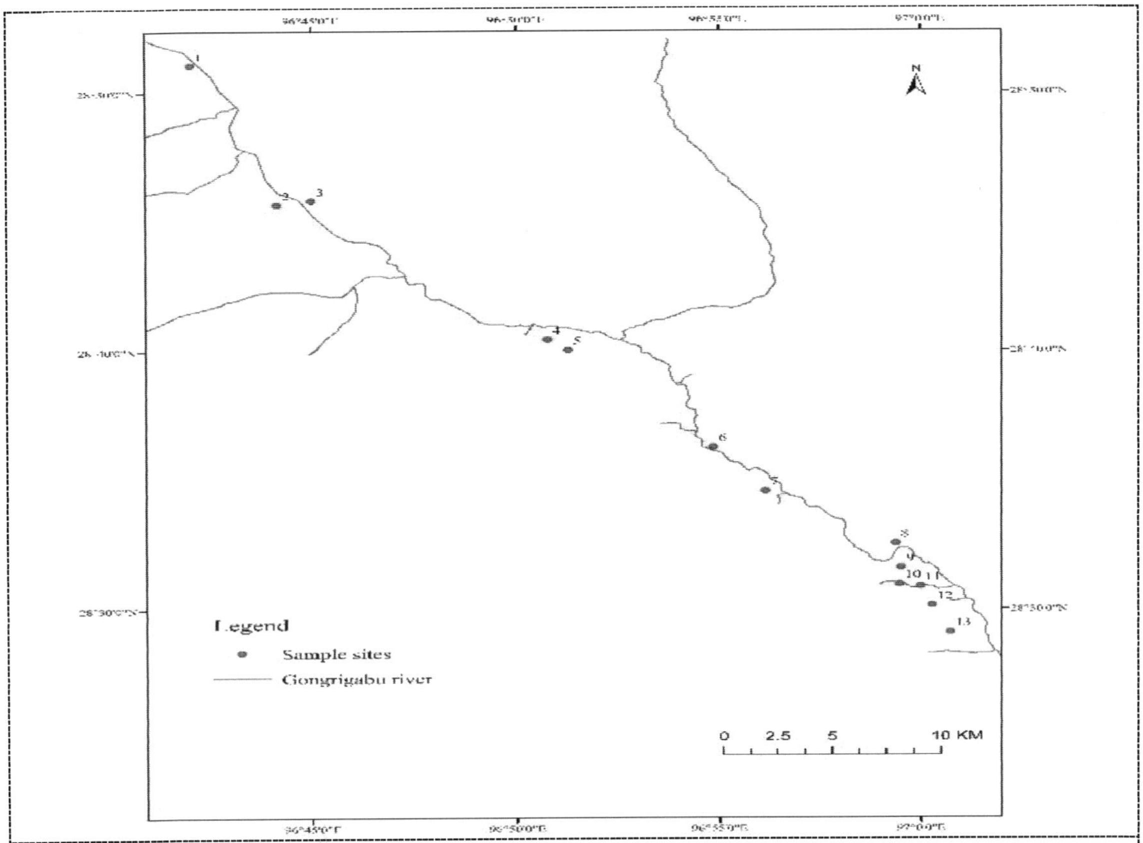

Figure 2. Location map of sample points.

Note: 1-13 indicates the upstream and downstream distribution of villages along the Gongrigabu watershed of the Chayu River, with smaller numbers indicating more upstream and larger numbers indicating more downstream.

3. Results

3.1. Climate change in the Gongrigabu watershed of the Chayu River

The trend of climate change in the Gongrigabu watershed of the Chayu River was analyzed in terms of three meteorological elements: temperature, relative humidity, and precipitation. In the past 40 years, although the temperature decreased slightly from 1969 to 1984, it had been on an upward trend after 1984. Relative humidity, in contrast to temperature, showed an increasing trend from 1969 to 1984 and had been decreasing after 1984. Precipitation had not changed much, but there was an overall decreasing trend (Figure 3). The climate as a whole tended to be warm and dry.

IOP Conf. Series: Earth and Environmental Science 1087 (2022) 012039 doi:10.1088/1755-1315/1087/1/012039

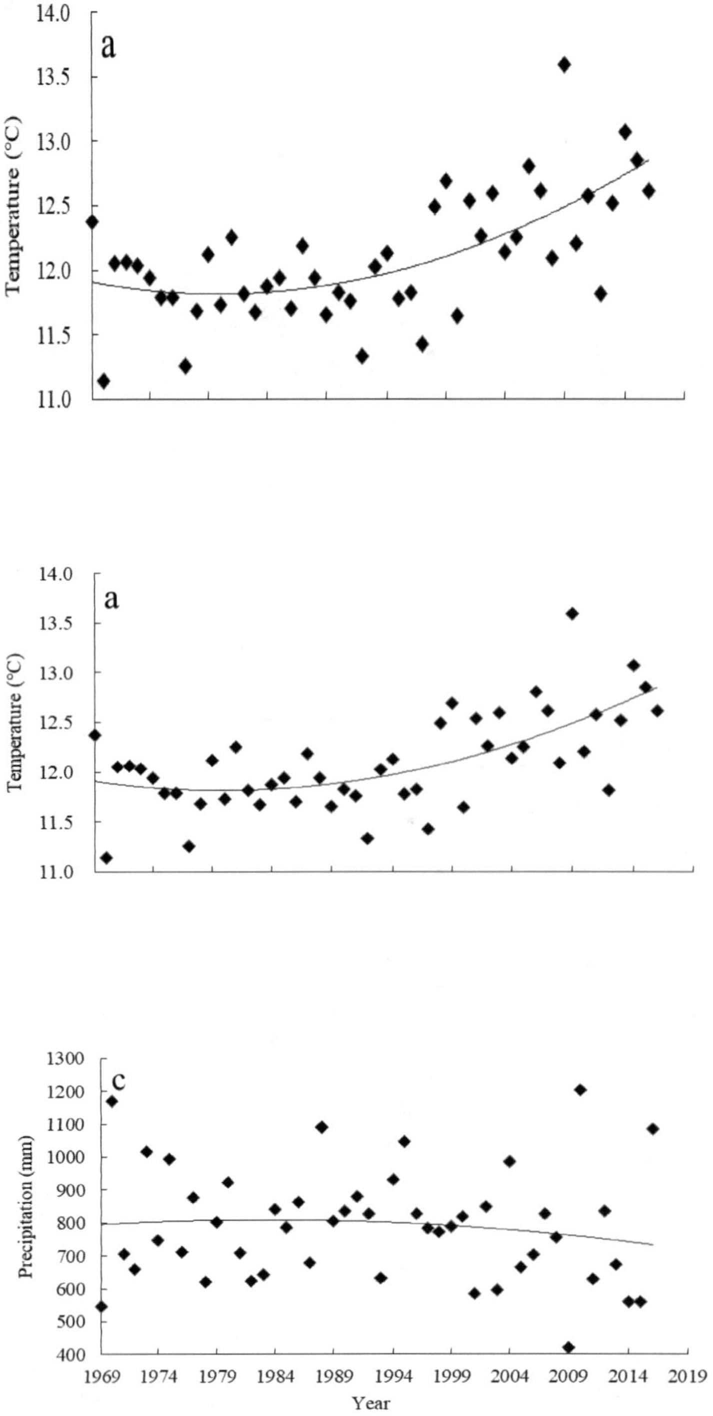

Figure 3. Trend line of average temperature, relative humidity, and precipitation in recent 40 years in Chayu river valley.

3.2. Invasion rate of earthworm exotics
The main perceived year of occurrence of the earthworm exotic species in the village was used as the perceived attitude of farmers in each village. Some villages were farther away from the river valley and higher in elevation and deviated from the trend line, but the earthworm exotic species showed an overall trend of spreading from downstream to upstream (Figure 4). Through the survey, earthworm exotics appeared from Shaqiong village to Rongyu village. Except for Rongyu village, which did not show any damage by earthworm to the paddy field, all the following villages' fields had damaged to be considered earthworm outbreaks. The villages below Shaqiong Village had not been surveyed, but because they were located at lower elevations downstream, it was inferred that earthworm exotics have already appeared and caused damage. No earthworm exotic species were found above Rongyu village. Earthworm exotics appeared in Shaqiong village as early as 2008 and as late as 2015 in Rongyu village, climbing 500 m in 7 years, with an invasion rate of 71.43 m·a-1 at altitude.

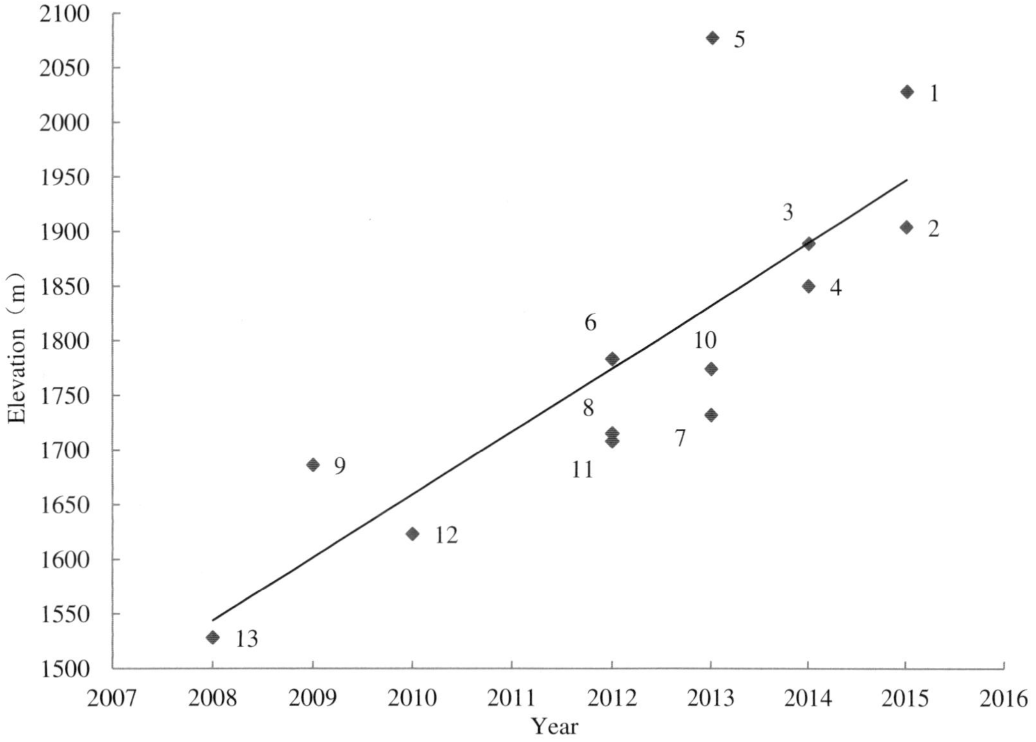

Figure 4. The invasion rate of alien species of earthworm at altitude.

4. Discussions
Global climate change is causing extensive and profound influence on human society. The Chayu River Basin in southeastern Tibet is experiencing significant climate change and its impacts, and biological invasion is one of the impacts brought about by climate change. The main processes and causes of biological invasions can be attributed to two aspects: first, the spread and proliferation of species brought about by the development of economic globalization and population movements, mainly due to human activities, intentional or unintentional[37], which can be referred to as anthropogenic introduction; and second, the ecological changes resulting from the ecological environment, i.e., natural invasion. For example, the invasion of Ageratina adenorphora has caused

serious ecological disasters and economic losses in China [38], and in Tibet, the species has also invaded Zhangmu and Jilong ports in the southern foothills of the Himalayas through economic and population activities [39], which can be considered as an anthropogenic introduction. Although this study is located in the Chayu River basin near the border of China, no border crossing has been opened, and the chance of species invasion brought about by human activities is minor, but considering that earthworm species are not as prone to carry in as plant species or other economic animals, this paper focuses on the possible effects of climate change on the spread of earthworms.

Climate change can affect the ability of invasive species to survive in new areas and alter their competitive relationships with native species. For agroecosystems, climate change weakens their defenses against invasive species, thus substantially increasing the likelihood of invasion by exotic species. Native species have been adapted to their original environment for a long time and have difficulty adapting to the new environment created by climate change in a short-term period; invasive species are generally able to adapt to the new environment more rapidly. The present study suggests that the functional richness of the earthworm community may decrease due to climate change [40] and the increase in temperature is harmful to the earthworm population [41].

Temperature is a major factor limiting the survival and reproduction of many organisms. Along with global warming, alpine forest and tree lines are extending to higher elevations [42], and as a result, some insects have shortened developmental cycles, accelerated reproduction, and are spreading to higher elevations and latitudes [43]. In terms of temperature, the Chayu River basin is experiencing a warming process, which is the same warming trend as that of the Tibetan Plateau [30,44] and southeastern Tibet [45], as evidenced by the temperature analysis of Chayu County by Li Dihua et al [36]. Continued climate warming may increase the adaptability of exotic species and expand their suitable range for survival. With regional warming, exotic species originating from warmer regions may form larger and more widely distributed dominant populations. Influenced by warm and humid air currents from the Indian Ocean, most of the Chayu River basin is a warmer region with increasing temperature suitability, which is increasingly suitable for the survival, reproduction, and spread of earthworm exotics. Some existing studies also support the possibility that warming is causing the expansion of some large pest species, including earthworms, into alpine regions [46-47]. Meanwhile, artificial warming experiments have shown that earthworms are sensitive to warming and that their species richness, density, and biomass decrease with increasing temperature [48-49]. This suggests that most earthworm species are unsuitable for warming, but this also provides opportunities for exotic species that are suitable for warming. The fluctuating resource hypothesis suggests that increased precipitation is conducive to biological invasion by increasing water availability [50]. Invasive species are likely to profit from increased water availability if they have high water requirements relative to native species. Characteristics of high water demand include short lifespan, rapid growth, and frequent aggregation, which are typical of opportunistic species [51] and are common in invasive organisms. Earthworms are a group of soil animals with high water demand [52] and are more abundant in the wet season when precipitation and humidity are ample [53-55]. The Chayu River basin is located in southeastern Tibet and is one of the areas with the highest humidity and rainfall in the entire Tibetan Plateau system. According to the meteorological data, the precipitation and relative humidity in this region show a decreasing trend, which is consistent with the study of Li Dihua [36], but contrary to the overall wetting trend of the Tibetan Plateau [53-58], and also contrary to the study of Chen Baoxiong et al [45] in southeastern Tibet. However, a survey on climate change perception and adaptation in the Chayu River basin in recent years revealed that farmers' perceptions of humidity and precipitation in 2015 and 2018 were significantly different, with those who perceived "dryness" varying from none to many, and those who perceived a decrease in precipitation increasing significantly. This indirectly confirms the trend of precipitation and humidity, indicating a drying of the climate, with a greater change after 2015. The drying of the climate in the basin, especially the larger changes in humidity, may be influenced by the increase in temperature. In the case of climate change, although the decrease in humidity and the weak trend of decrease in precipitation are unfavorable to the survival and reproduction of earthworms, the increase in temperature is large and favors the spread of earthworms. This suggests that temperature becomes the main factor affecting the invasion of earthworm exotics.

To prevent the invasion of earthworm exotics on paddy fields and reduce the loss of agricultural production, farmers in the region have taken two main measures: first, wrapping the field ridge with plastic film to prevent the spread of earthworm exotics and maintain soil fertility; second, applying pesticides for earthworm control to kill earthworms. The former measure can play a role in a short period, but the plastic film breaks down quickly in the natural environment and produces solid waste after use without treatment to pollute the soil; the latter measure can kill some earthworm exotics, but it also kills other earthworm native species and causes pollution to the soil environment. At present, it seems that the measures taken by farmers are short-term and unhelpful maladaptive strategies. According to the actual local situation, the crop layout can be changed appropriately, i.e., switching to other food crops or shifting from food crops to cash crops, such as planting fruits and herbs, to mitigate and adapt to the effects of climate change.

5. Conclusions

The climate of the Chayu River basin in southeastern Tibet is developing towards warm and dry. Rising temperatures have become the main meteorological factor for the invasion of earthworm exotics. The likelihood of human-caused earthworm invasion is extremely low. All earthworm exotics below 2028m (Rongyu) have broken out and are harmful to paddy fields. Earthworm exotics have invaded upstream areas at a climbing rate of 71.43 m·a-1 since 2008 when they were found at an altitude of 1528m (Shaqiong). To cope with the invasion of earthworm exotics, a long-term capable response might be to change the crop layout.

Acknowledgments

All authors have read and agreed to the published version of the manuscript.

Funding

This research was funded by the Key Laboratory of Forest Ecology in Tibet of the Ministry of Education, grant number XZA-JYBSYS-2020-07.

Reference

[1] Katheem K S, Ibrahim M H, Quaik S, et al 2016 Prospects of Organic Waste Management and the Significance of Earthworms Important Digestive Enzymes of Earthworm *Switzerland: Springer International Publishing* 10-122

[2] Blouin M, Hodson M E, Delgado E A, et al 2013 A review of earthworm impact on soil function and ecosystem services European *Journal of Soil Science* **64** 161-182

[3] Jang J, Xiong X, Liu C, et al 2022 Invasive earthworms alter forest soil microbiomes and nitrogen cycling *Soil Biology and Biochemistry* 108724

[4] Bernard L, Chapuis-Lardy L, Razafimbelo T, et al 2012 Endogeic earthworms shape bacterial functional communities and affect organic matter mineralization in a tropical soil *Isme Journal* **6** 213–222

[5] Huang F Z 1979 The influence of earthworm on the formation of soil structure *Acta Pedologica Sinica* 211-217

[6] Yang Y K, Lin B X, Li C X, et al 2018 Effects of vermicompost compound substrate on growth and photosynthetic characteristics of tomato seedlings Jiangsu *Agricultural Sciences* **46** 137-139

[7] Wu C T, Zhu Z J, Liu C, et al 2018 Study on the effect of earthworm manure as a rapid fertilizer improver for poor soil Modern *Agricultural Science and Technology* 191-195

[8] Cheng J, Wong M H 2002 Effects of earthworms on Zn fractionation in soils *Biology & Fertility of Soils* **36** 72-78

[9] Song F M 2013 Progress on environmental pollution control and resource application using

earthworms Guangdong Agricultural *Sciences* **40** 164-168
[10] Liu Z Z, Huang S F, Tang H 2018 Effects of earthworm activity on phytoremediation of Hg contaminated soil Environmental *pollution and prevention* **40** 866-869 + 874
[11] Liang Z J, Chen X, Ma J J, et al 2014 The occurrence and control of earthworm in facility agriculture in Korla Region Xinjiang *Agricultural Sciences* **51** 257-263
[12] Wang Z Y, Ayman A, Zhang X M, et al 2018 Earthworm population outbreak and effect of soil compaction on earthworm behavior Resource *Development & Market* **34** 1271-1275
[13] Wang F H 2012 Biological invasions: a guide to invasive alien plants in China Science Press
[14] Wu M L 2012 Global climate change and biological invasion *Bulletin of biology* **47** 4-6
[15] Li G D, Zhang J H, Chen C, et al 2013 Research progress on carbon storage and flux in the different terrestrial ecosystems in China under global climate change *Journal of ecological environment* **22** 873-878
[16] Zhang Q, Yang X W, Huang C Y 1995 Impact of climatic variation in maize productive capacity in the loess plateau area in recent 30 years *Agrometeorology in China* 19-23
[17] Ge Y N, Liu L, Xv X L, et al 2015 Temporal and spatial variations of Chinese maize production potential on the background of climate change during *Journal of natural resources* **30** 784-795
[18] Wang X Y, Yang X G, Song S, Xie W J 2015 Comparison of potential yield and resource utilization efficiency of main food crops in three provinces of Northeast China under climate change *Journal of Applied Ecology* **26** 3091-3102
[19] Zhang X F, Wang H Z, Liu L, et al 2014 Spatial-temporal characteristics of soybean production potential change under the background of climate change over the past 50 years in China Progress in Geographical *Science* **33** 1414-1423
[20] Fu Y Q, Chou J M, Dong W J 2014 Impacts of Climate Change on Crop Sown Area in China *Progress in Geographical Science* **10** 110-117
[21] Yv D 2012 Biogeographic analysis, morphological differentiation, and response to climate change of Phoxinus tench lineage *Graduate School of Chinese Academy of Sciences*
[22] Su Z H, Pan B R, Zhuo L, et al 2018 Impact of future climate change on the distribution pattern of Tamarix taklamakanensis and its conservation revelation *Arid Zone Research* **35** 150-155
[23] Ye D Z 1979 Meteorology of the Qinghai-Tibet Plateau *Science Press* 1-278
[24] Duan A M, Wu G X 2005 Role of the Tibetan Plateau thermal forcing in the summer climate patterns over subtropical *Asia Climate Dynamics* **24** 793-807
[25] Pan BT, Li J J 1996 Qinghai-Tibetan Plateau: a driver and amplifier of the global climatic change-III. the effects of the uplift of Qinghai-Tibetan Plateau on climatic changes *Journal of Lanzhou University* 108-115
[26] Wu X D, Lin Z Y 1981 Some characteristics of the climatic changes during the historical time of Qinghai-Xizang plateau *Acta Meteorologica Sinica* 90-97
[27] Yao T D, Xie Z C, Wu X L, Thompson L G 1991 Climatic change since little ice age recorded by Dunde ice cap *Science China Chemistry* **34** 760-767
[28] Zhao X Y 2014 Review on farmer's climate change perception and adaptation Acta *Meteorologica Sinica* **25** 2440-2448
[29] IPCC 2013 the physical science basis Cambridge: Cambridge Press
[30] Wu G X 2016 Characteristics of climate change over the Tibetan Plateau under the global warming during 1979-2014 *Research progress of climate change* **12** 374-381
[31] Yang C Y, Shen W S, Lin N F 2014 Climate change and its regional differences over the Tibet Plateau *Geography of arid area* **37** 290-298
[32] Phillips H R P, Guerra C A, Bartz M L C, et al 2019 Global distribution of earthworm diversity *Science* **366 (6464)** 480-485
[33] Singh J, Schädler M, Demetrio W, et al 2019 Climate change effects on earthworms-a review *Soil organisms* **91(3)** 114
[34] Sun D X, Zhong D L, Ji J Q, Tu J Y 2015 Inversion model of drainage basins' tectonic-thermal

evolution through detrital AFT ages: A case study of Chayu River in southeastern Tibet *Acta geophysica Sinica* **58** 613-627

[35] Wu P J 2006 Status and conservation of large and medium-sized mammals in the upper Zayu river basin *Journal of natural resources* 584-589

[36] Li D H 2014 Brief analysis of climate change in Chayu County in recent 32 years Tibet *Science and Technology* 44-46

[37] Guo Y, Xv L X 2004 The distributions and damages of invasive species in China *Land and natural resources* 82-83

[38] Xu H G, Ding H, Li M Y, Qiang S, Guo J Y, Han Z M, Huang Z G, Sun H Y, He S P, Wu H R, Wan F H 2006 The distribution and economic losses of alien species invasion to China *Biological Invasions* **8** 1495-1500

[39] Tu Y L, Liu L S, Liu Z Q, et al 2018 Invasion status and Countermeasures of malignant invasive plant Eupatorium adenophorum in Tibet Tibet *Science and Technology* 71-75

[40] Fourcade Y, Vercauteren M 2022 Predicted changes in the functional structure of earthworm assemblages in France driven by climate change *Diversity and Distributions* **28 (5)** 1050-1066

[41] Marcos A, Trigo D, Muñiz-González A B, et al 2021 Effects of bisphenol S on the life cycle of earthworms and its assessment in the context of climate change *Science of The Total Environment* **781** 146689

[42] Kullman L 1998 Tree-limits and montane forests in the Swedish Scandes: sensitive biomonitors of climate change and variability *Ambio* **27** 312-321

[43] Kiritani K 2011 Impacts of global warming on Nezara viridula and its native congeneric species *Journal of Asia-Pacific Entomology* **14** 221-226

[44] Li L, Chen X G, Wang Z Y, Xu W X, Tang H Y 2010 Climate change and its regional differences over the Tibetan Plateau *Research progress of climate change* **6** 181-186

[45] Chen B X, Wang J S, Ran C Q, et al 2012 Characteristics of climate change in forestry region in southeast Tibet during 1954-2009 *Research progress of climate change* **8** 43-47

[46] Bohlen P J, Scheu S, Hale C M, et al 2004 Non-Native Invasive Earthworms as Agents of Change in Northern Temperate Forests *Frontiers in Ecology and the Environment* **2** 427–435

[47] Eisenhauer N, Stefanski A, Fisichelli N A, et al 2014 Warming shifts 'worming': effects of experimental warming on invasive earthworms in northern North America *Scientific Reports* **4** 6890

[48] Zaller J G, Caldwell M M, Flint S D, et al 2009 Solar UVB and warming affect decomposition and earthworms in a fen ecosystem in Tierra del Fuego, Argentina Global Change *Biology* **15** 2493–2502

[49] Briones M J I, Ostle N J, Mcnamara N, et al 2009 Functional shifts of grassland soil communities in response to soil warming Soil *Biology and Biochemistry* **41** 315–322

[50] Davis M A, Grime J P, Thompson K 2000 Fluctuating resources in plant communities: a general theory of invisibility *Journal of Ecology* **88** 528-534

[51] Daehler C C 1998 The taxonomic distribution of invasive angiosperm plants: Ecological insights and comparison to agricultural weeds *Biological Conservation* **84** 167-180

[52] Yao B, Sun J, Jiang J B, et al 2018 The Distribution of amynthas earthworm in china and its response to hydrothermal conditions *Journal of Zoology* **53** 554-571

[53] Cao S P, Tan C, Wang H, et al 2018 Distribution of terrestrial earthworms and its influencing factors in the Nanniwan wetland Resources and environment in Arid *Areas* **32** 80-84

[54] Wang S J, Wang H, Li J H 2017 Effects of different land-use types on the spatial-temporal distribution of earthworm density and biomass *Journal of Biology* **36** 118-123

[55] Wang S J, Wang H, Li J H 2017 Spatiotemporal distribution characteristics of soil earthworm number and biomass under different forest communities *Journal of Nanjing Forestry University (Natural Science Edition)* **41** 37-42

[56] Gu Z, Chen S R, Zhou B 2018 Spatiotemporal variation of vegetation coverage over the Tibetan

Plateau and its responses to climatic factors *Journal of Ecology* **38** 3208-3218
[57] Zhao J F, Guo J P, Fang S P, et al 2011 Trends of Tibet's Dry-Wet Condition under Future Climate *Scenario Agrometeorology in China* **32** 61-66
[58] Chen D L, Xv B Q, Yao T D, et al 2015 Assessment of past, present and future environmental changes on the Tibetan Plateau *Science Bulletin* **60** 3025-3035+1-2

IOP Conf. Series: Earth and Environmental Science 1087 (2022) 012040 doi:10.1088/1755-1315/1087/1/012040

Investigating the correlation of nitrates and phosphates to the effective removal of Fats, Oils, and Grease in Two Mangaung Municipal Wastewater Treatment Plants

Mpho Gladness Sello[1,2] and Saheed A Oke[1]

[1] Civil Engineering Department, Central University of Technology, 20 President Street Bloemfontein Central Bloemfontein 9300 Bloemfontein South Africa.

[2] Corresponding author's e-mail: mphosello2018@gmail.com

Abstract. This study is about the effectiveness of wastewater treatment technologies in two treatment plants at Mangaung Municipality in South Africa. The removal of FOG, nitrogen and phosphorus is an important component of wastewater treatment. Therefore, the study aims at investigating the effectiveness of the two treatment technologies (biological trickling filter and activated sludge) in the wastewater treatment plants studied. Grab samples of the influent and effluent of the North East Wastewater Treatment Plant (NEWTP) and the Bloemspruit Wastewater Treatment Plant (BWTP) were collected per week over a period of five months. The laboratory analysis of FOG followed the description of the Environmental Protection Agency (EPA) 1664B. Hexane extraction and gravimetry-PBM method was used in the analysis. Nitrates were analysed using a SOP Chem 005 Spectroquant Nitrate test while phosphates were analysed using a Spectroquant Prove spectrophotometer in the laboratory. The results showed that the trickling filter from BWTP removed FOG by 61.36% while the activated sludge from NEWTP removed FOG by 52.81% which showed that the both of the two treatment technologies work the same and are ineffective in the removal of FOG. However, the nitrates and phosphates present in the effluent were found to be within the effluent discharge standards.

1. Introduction

Two forms of biological treatment exist which are aerobic and anaerobic. Literature has shown that microorganisms capable of degrading FOG and nutrients are filamentous organisms [1], [2]. The activated sludge and the biological trickling filter are aerobic systems. In these two methods, there are the following processes; preliminary treatment, primary treatment, secondary treatment and tertiary treatment [3].

(a) Preliminary treatment/Primary Treatment. Screening, catch basins, flotation and equalization are all methods of removing FOG in the preliminary treatment. Suspended solids including FOG can be removed using screens and filters. Other solid materials other than FOG may be paper, wood rags and hair. Additionally, wastewater passes through strainers in this stage. These strainers retain particles that are big enough to be trapped by the strainer mesh. One of the by-products of this process is FOG. Furthermore, finely suspended solids and FOG can also be removed using catch basins which utilise gravity. FOG and the fine solids will float while solids heavier than water sink. The floating FOG and scum are then removed at the top using a skimmer while a scraper is used to remove sludge from the bottom. Free floating FOG can also be removed using gravity separators in primary treatment [4].

(b) Secondary Treatment. Biological treatments use microorganisms to remove organic compounds and pathogens from the effluent. More than 90% of pollutants can be removed from wastewater using biological treatments [5]. Biological treatment occurs in the secondary treatment which includes secondary settling and the processes of aeration in an activated sludge system or active sludge systems or rotating biological contactors in the trickling filter system.

(c) Tertiary treatment. In this stage, physical and biological processes are carried out which complete the removal of both organic and bacteria before reuse of water as well as removal of nitrogen and phosphorus. The disinfection unit may also be used for bacterial removal [6].

1.1. Activated Sludge

The activated sludge technology has been the basis of wastewater treatment all over the world from the 1930 onwards [7]. However, the aeration process in the activated sludge uses high energy which may attribute for 60% of the plant running cost [8]. With the correct proportion of nutrients to mixed liquor suspended solids (MLSS), 93.3% of FOG can be eliminated using the activated sludge [9]. A study done in Shiraz Municipality Wastewater Treatment Plant in 2011 found the average removal efficiency to be 70% [10]. In another study, the accumulation of a variety of cultures was studied using FOG from a fast-food restaurant grease-trap. The activated sludge gave a more uniform removal, which was greater than 90% [11]. Despite its successful removal of FOG in other studies, a study done in 2003, by Reddy et al indicated that due to short retention times, the microflora in activated sludge systems is unable to degrade FOG effectively [12]. Further, another study done by Chipasa and Mdrzycka in 2008 indicated that the volume of the retained FOG could not be more than 15% of the original FOG [13]. The results indicated changes in the composition of single fatty acids in wastewater which had also been used by microorganisms as feedstock.

1.2. Trickling Filter

The trickling filter is a highly preferred option of biological treatment because it is less complicated in maintenance and control [14]. However, the biggest disadvantage of the trickling filter process to date, has been its poor reliability in the removal of nitrogen and phosphorus in comparison to the activated sludge [15]. Consequently, its restricted ability for nutrient elimination has led to the decline of the trickling filter although they are effective and energy-saving treatment systems [16]. In a study where an experimental plant made of primary (DAF) and secondary treatment unit (two stage trickling filters) was erected [17], around 40–50% of FOG removal was achieved from the DAF system. While the activated sludge and the trickling filter were not compared in the elimination of FOG, the FOG removal rate of the trickling filter in this study was notable.

In Mangaung, the Bloemspruit Wastewater Treatment Works (BWTW) built in 1902 which uses the biological trickling filter was hydraulically overloaded hence the construction of the North Eastern Wastewater Treatment Works (NEWTW) in 2014, which uses the activated sludge system [18]. The effectiveness and degree of the new technology, activated sludge compared to the older technology, biological trickling filter system, in the removal of FOG has no difference since both treatment plants release FOG, nitrates and phosphates in their effluents

It is observed that there is frequent occurrence of sewage blockage in Mangaung which arise from FOG accumulation in pipes. Likewise, the two municipal treatment plants release FOG into the river catchment thus negatively impacting the ecosystem within. In fact, Stams & Oude indicated that FOG not properly treated by sewage treatment works has detrimental environmental impacts [19]. In this regard, a comparison of the performance of the treatment works capability in the elimination of FOG, nitrogen and phosphorus was carried out. There are no other studies that indicate the effect of nitrates and phosphates on FOG in municipal wastewater with the activated sludge and the biological trickling filter. To address this knowledge gap, this study determined the effectiveness of the two plant treatment technologies in the removal of FOG, nitrogen and phosphorus and investigated the likely correlation between FOG, nitrates and phosphates released to the catchment after treatment. As reported by Rittman, the combination of the pollutants treated, aerobic or anaerobic treatment, the proficiency and accuracy of treatment and overhead costs are the factors that affect the wastewater treated and the efficiency of the two treatment plants [20].

2. Materials and Methods

The study began in June 2021 and ended in November 2021. Sampling was done when temperatures peaked at 17°C which was 12 pm in June, July and August, while for September and October sampling was done around 10.30 am. In both wastewater treatment plants, both influents and the effluents were evaluated for the presence of FOG, nitrates and phosphates.

2.1. The Study Area

The study area was BWTW and NEWTW situated in Bloemfontein in the Mangaung Municipality. BWTW uses the biological trickling filter technology and NEWTW uses the activated sludge technology. BWTW receives 57ML per day of wastewater of which 17ML is diverted to the NEWTW as a means to relief its system. The total influent at NEWTW is 34ML.

2.2. Data Collection

A. FOG. 20 samples of the influent and 20 samples of the effluent of BWTW and NEWTW were collected once a week and evaluated for FOG, nitrates and phosphates. For FOG, the sampling procedure of the EPA 1664B method was followed. Samples were collected in 2L glass bottles which were not pre-rinsed with sample before data collection. The sample pH was adjusted to less than 2 with hydrochloric (HCl) acid and refrigerated at 4°C if analysis was to be delayed for more than four hours. Other materials such as the stirring rod, the pH electrode, or any other materials were not dipped into the sample that would be used for FOG determination in order to avoid substances in the sample adhering to these items [21].

B. Nitrates and phosphates. Nitrate samples were collected in 1 liter plastic bottles, stored at 4°C in the refrigerator and analysed the following day [22]. Sampling for phosphates were collected in 1L plastic bottles followed to the phosphate Spectroquant test [23].

2.3. Analysis

A. FOG. The EPA 1664B uses n-hexane as a solvent. In this method, the laboratory is allowed to alter the procedure to control hindrances or decrease the expenditure on quantities, on condition that all administration regulations in this method are met and hence 500ml of wastewater was used instead of 1L [24].

B. Nitrates. The SOP Chem 005 Spectroquant Nitrate test was used to analyse nitrates. The test tubes for analysis were pre rinsed with distilled water and dried in the oven prior to analysis. 4.0ml of NO_3-^{1} was pipetted into a dry empty round cell then 0.50ml of the sample was added into the cell. 0.50ml of NO_3-2 was pipetted and the cell was closed and shook vigorously. The solution was then transferred into a corresponding rectangular cell and read [25]. The samples were analysed on the day of sampling.

C. Phosphates. The Spectroquant Prove spectrophotometer was used to analyse phosphates in the laboratory. The test tubes for analysis were pre rinsed with distilled water and dried in the oven. The pH was checked for the range 0-10. If required, sulphuric acid was added drop by drop to adjust the ph. 5.0ml of the sample was then added into the test tube. 5 drops of PO_4^{-1} were added in the test tube together with 1 level of micro-spoon PO_4^{-2}. The mixture was then shaken vigorously, transferred into a corresponding cell and read [26]. The samples were analysed on the same day of sampling.

3. Results and Discussions

3.1. FOG

Results in Table 1 imply that the removal of FOG in both wastewater treatment plants was found to be below the average removal efficiency. According to a study by Deghani et al, the FOG removal efficiency range of wastewater treatment was found to be 59% to 85% with an average of 70% using the activated sludge technology [27]. Additionally, in a study done by Porwal et al using a mixed culture of dairy sludge in the activated sludge, the removal efficiencies were over 90% [28]. On the other hand, the trickling filter is also a very effective technology in the removal of FOG from wastewater. FOG removal in a study done by Pontes and Chernicharo in a study aimed at evaluating lipids in raw sewage was substantial [29]. Furthermore, El-Masry et al found FOG removal efficiency to be 100% using the trickling filter technology in wastewater containing vegetable oil and grease.

Table 1. FOG readings in BWTW and NEWTW.

Period In weeks	BWTW Influent FOG in mg/l	BWTW Effluent FOG in mg/l	NEWTW Influent FOG in mg/l	NEWTW Effluent FOG in mg/l
1	21.12	19.76	14.62	7.90
2	21.18	19.98	14.62	9.76
3	10.66	3.26	17.42	3.02
4	8.06	6.14	3.32	1.98
5	1.72	0.70	6.62	2.66
6	7.08	6.00	16.96	5.22
7	13.6	10.86	18.88	13.12
8	12.60	10.42	6.02	3.98
9	23.52	6.40	14.40	3.98
10	23.4	5.04	10.20	4.64
11	24.06	6.34	14.24	6.02
12	21.26	5.44	8.56	5.66
13	6.22	5.38	11.48	5.80
14	7.62	6.82	8.54	6.96
15	13.28	7.84	1.32	0.42
16	6.96	1.37	3.18	2.84
17	11.54	10.82	9.52	7.28
18	14.96	13.82	7.98	6.84
19	9.04	8.54	9.02	3.34
20	13.34	11.56	6.18	5.86
Mean FOG	13.56	8.32	10.15	5.36
%removal	61.36%		52.81%	

In the present study, the trickling filter had a higher removal efficiency than the activated sludge. However, the South African wastewater effluent standards limit applicable to the discharge into a water source is ≤ 2.5mg/l, meaning neither wastewater effluent complied [30]. A similar study revealed that the removal efficiency of FOG in domestic water using the activated sludge in Wapio was 15.2% and 97.8% in Pohakapu which uses the trickling filter. Based on their findings, the present study compares agreeably with Young and Chan [31].

3.2. Nitrates and Phosphates

The Pearson correlation is a representative way to measure dependency of a variable on another [32]. The Pearson Correlation Formula and was used to evaluate the relationship between the two variables using the formula:

$$r = \frac{\Sigma(x_{i-\bar{x}})(y-\bar{y})}{\sqrt{\Sigma(x-\bar{x})^2}\,\Sigma(y-\bar{y})^2}$$

The following graphs (Figure 1-4) show the results of the study taken from the effluent of the two treatment plants.

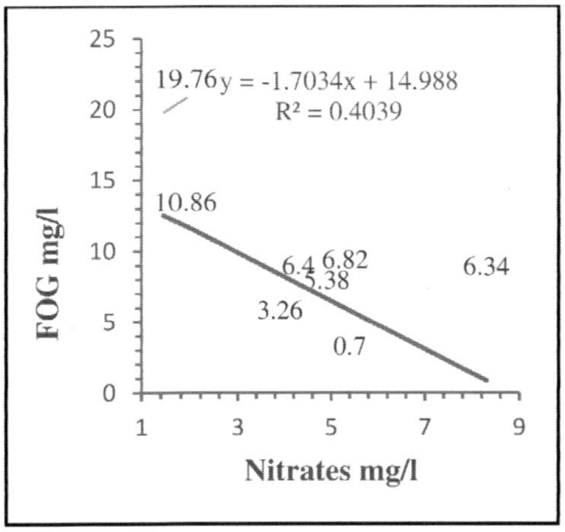

Figure 1. Correlation between FOG and nitrates in BWTW.

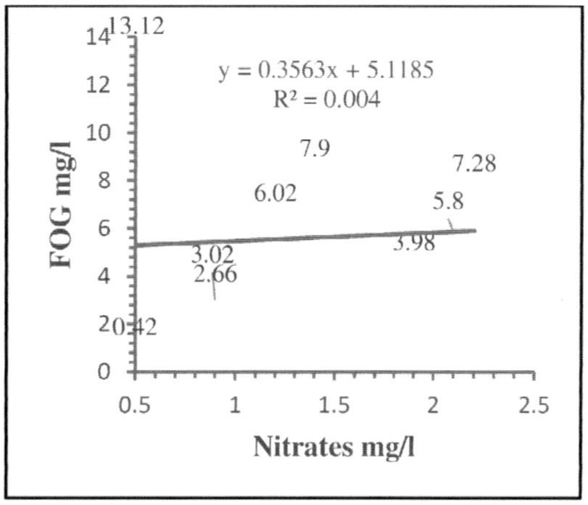

Figure 2. Correlation between FOG and nitrates in NEWTW.

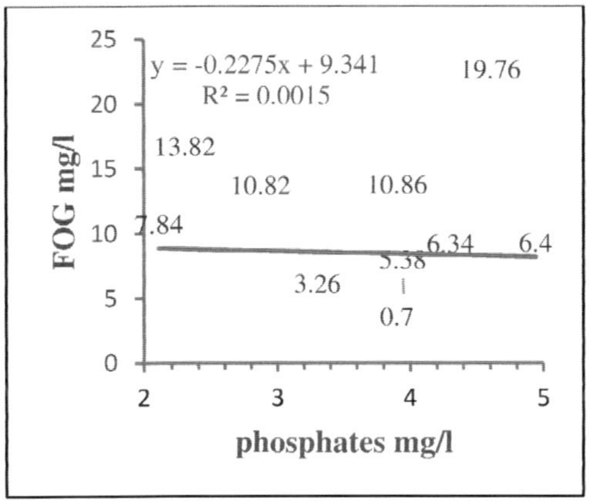

Figure 3. Correlation between FOG and phosphates in BWTW.

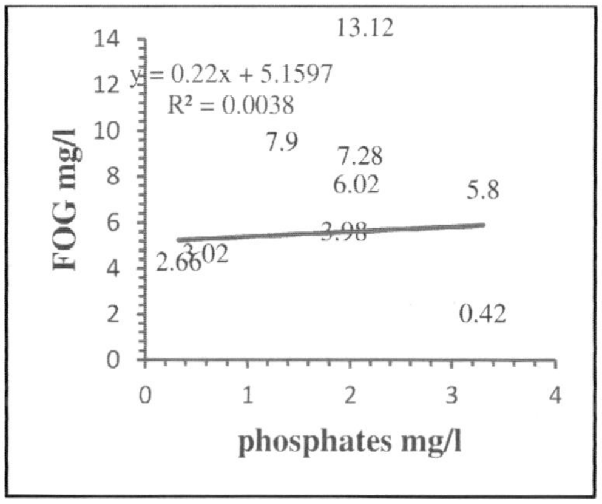

Figure 4. Correlation between FOG and phosphates in NEWTW.

The regression figure was found to be 1.000 for the BWTW which showed a strong correlation and 1.000 for NEWTW thus meaning a strong correlation between nitrates and FOG. In addition, the coefficient of determination between nitrates and FOG was 0.4039 in the BWTW showing a close to medium variance and 0.004 in the NEWTW indicating a low variance. The correlation coefficient was 1.000 in BWTW indicating a strong fit for regression and was 1.000 in NEWTW indicating a strong fit for regression as well. However, the correlation of determination for FOG and phosphates was 0.0015 for BWTW indicating a weak correlation and 0.0038 for NEWTW indicating a weak correlation as well.

From the results, the removal of nitrates and phosphates was found to be satisfactory. The average nitrates in the BWTW effluent were found to be 4.43mg/l and 1.29mg/l for NEWTW which complied with the South African discharge standards of $\leq$15mg/l. A study done by Kim showed a good performance of the trickling filter in nitrification [33]. Likewise, Almastrand et al and Perron et al, also confirmed that the trickling filter is highly efficient in the removal of nitrates [34], [35]. On the other hand, phosphate levels in the effluents of the two plants were found to be 3.6mg/l for BWTW and 1.9mg/l for NEWTW which complied with the effluent standards of $\leq$10mg/l. In a study investigating phosphorus removal in the activated sludge, the removal efficiency was found to be 97% [36]. The results of this study compare favourably to a study done by Irdemez et al which found a maximum level of phosphorus removal of 90.1% [37].

4. Conclusions

The study was conducted to compare the efficiency of the trickling filter and the activated sludge in the treatment of FOG, nitrogen and phosphorus and the possible correlation of nitrates and phosphates to FOG. The effluent discharged by the two treatment plants is above the South African wastewater discharge standards, moreover, the results showed a strong correlation between FOG and nitrates and FOG and phosphates in the two treatment plants. This concludes that the volume of FOG found in the effluents of these plants is influenced by the volume of nitrates and FOG. Moreover, the comparison of the trickling filter and the activated sludge indicated that the trickling filter is an efficient treatment method regarding treatment of FOG compared to the activated sludge. Further scientific studies for the two wastewater treatment technologies are essential to establish in depth the difference between the two technologies in the removal of FOG, nitrogen and phosphorus in municipal wastewater and their correlation. Therefore, this study has shown the discrepancy between the two wastewater treatment technologies in the removal of FOG and its correlation to nitrates and phosphates.

References

[1] Nielsen P H, Kragelund C, Seviour R J, Nielsen J L 2009 *FEMS Microbiology Reviews* **33(6)** 969-98
[2] Wakelin N G, Forster C F 1997 *Bioresource Technology* **59(1)** 37-43
[3] Templeton M R & Butler D 2011 Ventus Publishing ApS Frederickberg, Denmark
[4] Pintor A M A, Vilar V J P, Botelho C M S, Boaventura R A R 2016 *Chemical Engineering Journal* **297** 229-255
[5] Mittal S G 2006 *Bioresource Technology* **97(9)** 1119-1135
[6] Abdel-Fatah M, Sherif H O 2017 *Ain Shams Engineering Journal* **8(3)** 305-310
[7] Salgot M, Folch M 2018 *Current Opinion in Environmental Science & Health* **2** 64-74
[8] Mathew M M, Khatana K, Vats V, Dhanker R, Kumar R, Dahms, Hwang J 2022 *Frontiers in Microbiology* **12** 801051
[9] Al-Ani M Y, Al – Khalidy F R 2006 *Int J Environ Res Public Health* **3(4)** 360-8
[10] Deghani M, Sadatjo H, Maleknia H, Shamsedini N 2014 *Jentashapir Journal of Cellular and Molecular Biology* **5** 6
[11] Wakelin N G, Forster C F 1997 *Bioresource Technology* **59(1)** 37-43
[12] Reddy K, Drysdale G D, Bux F 2003 *Water SA* **29** 3
[13] Chipasa K, Medrzycka K 2008 *Journal of Environmental Sciences* **20(5)** 536-42
[14] Abou-Elela S I, Hellal M S, Aly O H, Abo-Elenin S A 2019 *Environ Technol* **40(2)** 250-260
[15] Tchobanoglous B 2003 4th edition
[16] Cramer M, Tranckner J, Kotzbauer 2021 *Environmental Technology* **17(42)** 2631-2640
[17] Li C T, Hwu N T, Whang J S 1984 Treatment of slaughterhouse wastewater by biofiltration tower
[18] Mail & Guardian newspaper Mangaung 2014 `Revitalising a central city`Mail & Gurdian Mangaung 2014 ` [Online] Available at: https://mg.co.za/article/2014-10-03-00-revitalising-a-central=city/?amp {Accessed: 18/08/2022)
[19] Stams A G Oude ESJ 1997 *Current Opinion in Biotechnology* **8** 328-334
[20] Rittman B, McCarty P L 2010 *Journal of Environmental Engineering* **136** 4
[21] United States Environmental Protection Agency 2010 Method 1664 Revision B: n-hexane Extractable Material (HEM;Oil and Grease) and Silica Gel Treated n-hexane Extractable material (SGT-HEM; Non-polar Material) by Extraction and Gravimetry pp11
[22] Nitrates Test 1.09713.0001 [Online] Available at: https://www.sigmaaldrich.com/deepweb/assets/sigmaaldrich/product/documents/925/967/109713e.pdf (Accessed: 18/08/2022)
[23] Phosphate Test 1.114842.0001-Sigma-Aldrich [Online] Available at:

https://www.sigmaaldrich.com/RU/enproduct/mm/114848 (Accessed: 18/08/2022)
[24] United States Environmental Protection Agency 2010 Method 1664 Revision B: n-hexane pp12
[25] University of Kwazulu-Natal 2017 *Pollution Research Group* **002** 1-4
[26] Supelco 2017 Analytical procedures appendices Spectroquant Prove Spectrophotometer 300
[27] Deghani M, Sadatjo H, Maleknia H, Shamsedini N 2014 *Jentashapir Journal of Health Research* **5** 6
[28] Porwal H J, Mane A V, Velhal S G 2015 **9** 1-15
[29] Pontes P P, Chernicharo C A *Environmental Technology* **32(3)** 281-287
[30] South Africa Department of Water Affairs 2013 National Water Act Revision of general authorisations pp6
[31] Young H F, Chan P L 1970 Oahu wastewater treatment plant efficiency *Journal of Water Pollution Control Federation* **42** 12
[32] Benesty J, Chen J, Huang Y & Cohen I 2009 Pearson Correlation Coefficient 37-40
[33] Kim B, Gautier M, Prost-Boucle S, Molle P, Michel P, Gourdon R 2014 *Ecological Engineering* **71** 41-47
[34] Almstrand R, Lydmark P, Sorensson F, Hermansson M 2011 *Bioresource Technology* **102(17)** 7685-7691
[35] Persson F, Wik T, Sorensson F, Hermansson M 2002 *Water Research* **36** 61439-1448
[36] Gebremariam S Y, Beutel M W, Christian D, Hess T F 2012 *Bioresource Technology* **121** 19-24
[37] Irdemez S, Demircioglu N, Yildiz Y S, Bingul Z 2006 *Sep Purif Technol* **56** 218-223

Fostering low-carbon industrial parks in Vietnam: establishment and application of an index system for Trang Bang Industrial Park

Vuong Thi Mai Thi[1], Tran Hau Vuong[2,5], Nguyen Tri Quang Hung[3], Ho Thanh Ba[3], Nguyen Kim Hue[3] and Nguyen Minh Ky[4]

[1] Institute of Environment and Resources, Vietnam National University, Ho Chi Minh City 700000, Vietnam;
[2] Ho Chi Minh City University of Natural Resources and Environment, Ho Chi Minh City 700000, Vietnam;
[3] Faculty of Environment and Natural Resources, Nong Lam University, Ho Chi Minh City 700000, Vietnam;
[4] Department of Environment and Natural Resources, Gia Lai Branch, Nong Lam University of Ho Chi Minh City, Pleiku City 600000, Vietnam.

[5] Corresponding author's e-mail: thvuong@hcmunre.edu.vn

Abstract. In recent years, Vietnam has experienced urbanization and industrialization, leading to potential air pollution and climate change risks. To reduce these negative consequences of economic development, Vietnam has promoted a low-carbon economy (LCE) by setting many goals for green growth, air pollution reduction, and increasing the absorption capacity of greenhouse gases (GHGs) in many industrial sectors. However, there is still a lack of a specified policy to promote these goals systematically. This study proposes a set of criteria that could be used to evaluate and certify low-carbon industrial parks (LCIPs). The criteria were developed using Analytical Hierarchy Process (AHP), expert opinion, data normalization, and carbon emission calculation. The set of criteria includes five groups of indicators with eighteen criteria for evaluating LCIPs, which can be applied in Vietnam's contexts. Using such an index system to Trang Bang IP in the period 2016–2020 as a case study, the results show that the Trang Bang has yet to meet the criteria of LCIPs, particularly in terms of energy consumption and GHGs emission.

1. Introduction

The direct cause of climate change (CC) is excessive emissions of greenhouse gases (GHGs), especially carbon dioxide (CO_2). GHG emissions have been investigated widely in various nations and/or regions around the globe [1, 2]. In recent years, CC has caused severe impacts leading to widespread concerns. In detail, glacial melting is one of the early indicators related to CC, indicating adverse changes [3]. Since the Industrial Revolution, the concentration of CO_2 in the atmosphere has steadily increased, owing to the excessive use of fossil fuels such as coal, oil, and gas across a wide range of industries. For detail, CO_2 concentration has grown by around 38% over natural interglacial levels [4]. To reduce the negative consequences of industrialization, it is essential to reduce the emission of CO_2, leading to the establishment of "low-carbon economy" (LCE), a concept promoted globally [5]. This concept can be understood as a set of activities that support the efficient use of fossil fuels, and ultimately reduce CO_2 emissions released into the atmosphere [6]. To foster LCE, many countries have established policies that made the components of LCE mandatory in developing IPs to support GHGs emission reduction. The "LCE" refers to the green, a friendly ecological economy based on less pollution and low energy consumption [7]. Meanwhile, the low-carbon IPs (LCIPs) significantly contribute to carbon emission mitigation and environmentally ecological protection [8]. For instance, the CO_2 emissions were directly reduced that via the 41 eco-IPs projects in South Korea analyzed by Ban, et al. [9]. This approach has led to the establishment of LCIPs, which remarked the importance of obtaining larger economic output with fewer GHGs emissions. The key feature of these LCIPs are the effective energy consumption and reduce CO_2 emissions [10], defining the core values of LCIP include the reduction of fossil resource use; altering by renewable resources; recycling wastes; promoting green energy; and increasing efficiency in natural resources consumption.

Reducing carbon emissions is an urgent need today in all countries across the globe, including Vietnam, a nation that has undergone rapid economic growth, driven by significant reforms centralized the industrial sector [6, 11]. Although having an undeniable role in the economy, the development of IPs in Vietnam is still unsustainable due to high GHGs emissions. Therefore, establishing LCIP has been the focus of recent discussions on the urgent requirements to work towards LCE for Vietnam. To achieve this objective, developing and implementing a systematic framework is necessary to evaluate and certify LCIP. However, such a toolset is unavailable as previous research has mainly focused on building criteria to assess sustainable cities [12] or on future approaches to reduce carbon emissions to cope with CC [5]. Therefore, identifying the criteria and indicators to evaluate the "low-carbon" characteristics for Vietnam's IPs will contribute to completing the LCIP model. This study aims to identify the indicators and criteria necessary to evaluate the "low-carbon" feature of Vietnam IPs. The proposed criteria were then used to evaluate the "low-carbon" characteristics in case study of Trang Bang IP, Tay Ninh province.

2. Methodology

2.1. Study area

Trang Bang IP is one of the major industrial zones in Tay Ninh province, which belongs to Vietnam's Southern Key Economic Region. This is the first IPs established in Tay Ninh province, with a total area of 190.76 ha, and the occupancy rate currently up to 98%. The total number of enterprises/projects operates over 79 registered foreign and domestic capitals. Trang Bang IP is located at the intersection of main traffic routes along with Ho Chi Minh City, Dong Nai, Binh Duong, Ba Ria-Vung Tau, Long An, Binh Phuoc, etc., has a clear advantage in transportation and circulation of essential goods.

2.2. Identifying criteria to evaluate LCIPs

Successful IPs significantly contribute to the more sustainable development of industrial operations, high-growth regions, and national economic growth [13]. To reduce overall energy and material consumption, quantitative techniques and procedures have been created to find and foster symbiotic industrial exchanges in IPs [14]. This study applied quantitative methodologies to adapt to eco-IPs networks [8, 14, 15]. The procedure to identify LCIP criteria comprises three steps as follows: Step 1–Identifying the requirement of the LCIP criteria: the list of indices to evaluate the LCIP criteria is established that based on the integrated analysis of requirements established by state agencies and from previous research on LCIP. Analytical Hierarchy Process (AHP) is applied to determine the weights of the selected indices. Step 2–Preliminary assessment of LCIP criteria: the potential criteria were selected based on domestic and international studies on LCIPs, Vietnam's policies on GHG emissions and respond to CC. Step 3–Establish a list of criteria for LCIP evaluation: from the preliminary criteria in Step 2, based on experts' opinions, conduct secondary screening to establish an index system for evaluating LCIPs.

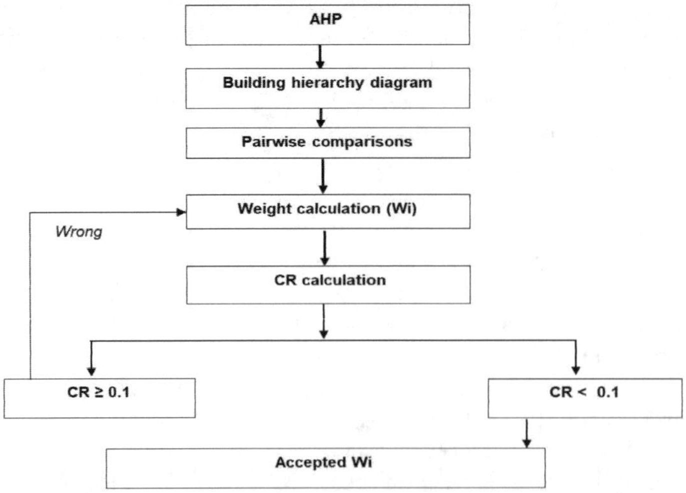

Figure 1. Procedure for determining weights by AHP.

AHP is one of the multi-criteria decision-making methods that focuses on determining each criterion's importance, thereby giving users precise results [16]. The procedure for determining weights by AHP is illustrated in Figure 1. Using the AHP method, the weight coefficient of the evaluation system was analyzed for the LCIPs [17-19]. Based on the specialist recommendation, the evaluation system's weight coefficient scores will be determined.

Step 1. Analyze the problem into small parts and build the AHP hierarchy diagram in the form of an upside-down tree where the main goal is placed on top. The second level is partial objectives that meet the main objective. Each partial objective at the second level can be decomposed into the third level objectives, and each set at each level meets the objective of the level to which they are subordinate (Figure 2).

Step 2. Build a matrix to compare the indicators. Pairwise comparisons are performed using the fundamental scale presented in Table 1. An nXn matrix is assembled, where n is the number of alternatives. Considering a given criterion, matrix A is supplemented with values a_{ij}, where i is a base alternative for comparison, corresponding to row i, and j is the alternative being compared with i. For instance, if the contribution of i to the criterion is considered of "strong importance" relative to j, a value of 5 is assigned to a_{ij}. This level of relative importance can be interpreted as a dominance of i over j regarding the criterion that is assessed. The a_{ijk} values were determined based on Table 1 of Satty's relative importance classification. The procedure was repeated for all n criteria, and the final results are Table 2 shows how to evaluate and compare pairs.

Step 3. Calculate the weight of the indicator. After determining the total matrix, the experts compare on that basis, weighting the factors according to the eigenvector method as shown in Table 3.

Step 4. Consistency check. If CR < 0.1, the result is accepted; if CR > 0.1, the result is not accepted. Regarding the value of CR is computed as CR = CI/RI. In which, CI and RI are consistency index and random index, respectively (Table 4).

$$\mathrm{CI} = \frac{\lambda_{max}-n}{n-1} \quad \text{and} \quad \lambda_{max} = \frac{1}{n}\left(\frac{\sum_{n-1}^{n} W_{1n}}{W_{11}} + \frac{\sum_{n-1}^{n} W_{2n}}{W_{12}} + \cdots + \frac{\sum_{n-1}^{n} W_{nn}}{W_{nn}}\right) \tag{1}$$

Table 1. Saaty's fundamental scale [16].

Intensity of importance	Definition
1	Equal importance
3	Moderate importance
5	Strong importance
7	Very strong importance
9	Extreme importance

Table 2. Value of pairwise comparisons of experts [16].

Comparision	The results of the assessment of the expert:						Aij
i	j	1	2	3	...	k	
TC1	TC2	a121	a122	a123	...	a12k	A12
	TC3	a131	a132	a133	...	a13k	A13
	...	...	...	...	...	...	...
	TCn	a1n1	a1n2	a1n3	...	a1nk	A1n
TC2	TC3	a231	a232	a233	...	a23k	A23
	...	...	...	...	...	...	...
	TCn	a2n1	a2n2	a2n3	...	a2nk	A2n
...	...	...	...	...	...	...	...
TCn-1	TCn	an-1n1	an-1n2	an-1n3	...	an-1nk	An-1n

Table 3. The comparison matrix of sums up the factors [16].

Criteria	TC1	TC2	TC3	...	TCn	Aij	General weight
TC1	1	A12	A13	...	A1n	X1	W'1
TC2	1/A12	1	A23	...	A2n	X2	W'2
TC3	1/A13	1/A23	1	...	An-1n	X3	W'3
...	...	...	...	1	...	...	...
TCn	1/A1n	1/A2n	1/An-1n	...	1	Xn	W'n
Total						∑X	1

Table 4. Random index for factors RI [16].

N	1	2	3	4	5	6	7	8	9	10	11	12	13	14	15
RI	0	0	0.52	0.89	1.11	1.25	1.35	1.4	1.45	1.49	1.52	1.54	1.56	1.58	1.59

This technique explores an innovative approach to prioritization from expert suggestions using the AHP [20, 21]. The questionnaire based on the list of criteria for evaluating the indicator set and pairwise comparison by the AHP method was sent to experts in the field of climate change and related managers to assess the weight of indicators. Participants (i.e., expert staff) within deep domains can utilize these prior efforts to find essential aspects. The experts will give points on the criteria for evaluating IPs (1 for passing and -1 for failing criteria). The results were then aggregated and used to identify the final criteria for LCIP evaluation. The criterion is required if the score is above seven and more than 50% of experts agree.

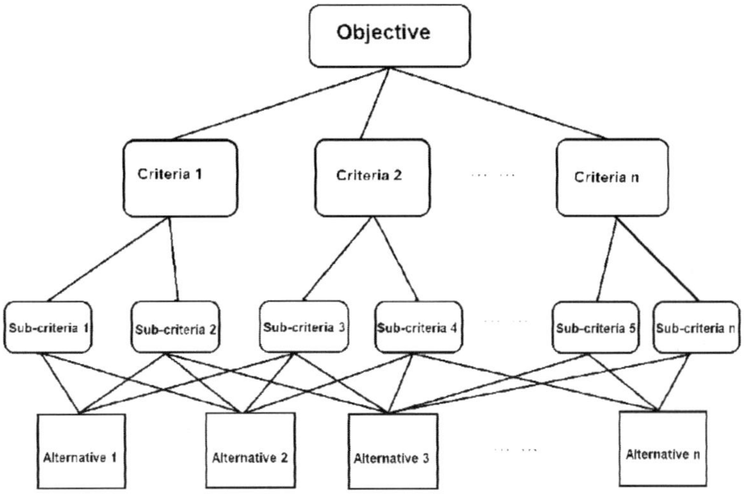

Figure 2. Hierarchy structure in the AHP method.

The normalization and data analysis for environmental footprints played an important role. Several methods are commonly applied in weighting indicators [22]. As a simple method, this is examined and based on a low-carbon sustainability index [23]. The LCIP criteria reflect many aspects of IPs development toward LCE, and each criterion has relative importance. The research on selecting a weighted sum method to evaluate LCIPs through a set of evaluation criteria with corresponding weights for each criterion.

2.3. Carbon emission estimation

Aims to find an appropriate method for investigating the IPs' carbon footprint, it needs to consider carbon emissions. Estimating CO_2 emissions has been conducted from IPs [24, 25]. Additionally, according to the guidance of IPCC (2006), there are critical methods applied to calculate GHG emissions in industrial production [26]: (i) determination by measurement, (ii) determination by material balance, (iii) determination by emissions factor. This study uses the method of calculating GHG emissions by emission factor as follows:

$$E_{CO2eKCN} = E_{CO2,fuel} + E_{CO2e\text{-}NT} + E_{CO2e\text{-}CTR} - E_{htCO2} \quad (2)$$

Where: $E_{CO2e\ KCN}$ is the total GHG emissions in the IPs (tonne CO_{2e}/year); $E_{CO2,fuel}$ is GHG emissions in the energy sector by fuel consumption; $E_{CO2e\text{-}NT}$ is GHG emissions from wastewater in IPs; E_{CT} is GHG emissions from solid waste in IPs; and E_{htCO2} is the amount of carbon absorbed by vegetation cover (plants).

3. Results and discussions

Table 5. Weight of the criteria for the evaluation of low-carbon IPs.

Matrix	Suitable for policies and laws of Vietnam	Suitable for the target of low-carbon IPs	Suitable for socio-economic conditions	Suitable for infrastructure	Effective emission reduction	Economic efficiency	Possibility	Average multiplication	Weight
Suitable for policies and laws of Vietnam	1.00	1.59	5.74	7.21	3.63	5.00	2.12	3.0635	0.3190
Suitable for target of low-carbon IPs	0.63	1.00	5.31	6.43	2.39	3.41	1.47	2.2107	0.2302
Suitable for socio-economic conditions	0.17	0.19	1.00	1.36	0.36	0.44	0.18	0.3868	0.0403
Suitable for infrastructure	0.14	0.16	0.73	1.00	0.19	0.27	0.17	0.2789	0.0290
Effective emission reduction	0.28	0.42	2.74	5.41	1.00	1.36	0.36	0.9762	0.1017
Economic efficiency	0.20	0.29	2.29	3.73	0.73	1.00	0.25	0.7093	0.0739
Possibility	0.47	0.68	5.49	6.00	2.74	4.07	1.00	1.9767	0.2059
Total								9.6022	1.0000

3.1. Evaluation criteria of LCIPs

Among the low-carbon strategies, many studies showed that applying industrial symbiosis can contribute to minimizing CO_2 emission in IPs. Also, Lehtoranta, et al. [27] has investigated environmental efficacy and carbon emissions of industrial symbioses and eco-IPs to find an environmentally efficient policy. Indicators and/or requirements of eco-IPs are critical documents to guide low-carbon/eco-IPs in their process, including 24 indicators across four primary categories, e.g., (i) economic development, (ii) resource conservation and recycling, (iii) pollution control, and (iv) environmental management [28]. To identify criteria that could be used to evaluate LCIPs, this study has developed a set of indices focusing on the following aspects: (i) the criteria should be science-based; (ii) they are suitable and effective in the context of Vietnam; and (iii) satisfying Vietnam's regulations, policies and laws on responding to CC, GHGs reduction, and economic development. To this end, this study proposed six indices to evaluate LCIP criteria: conformity with Vietnam's policies and laws, in line with the goal of LCIPs, socio-economic conditions and infrastructure, emission reduction, and economic efficiency. Based on the above set of indicators, collecting data from experts' opinions, and processing the data are demonstrated in the AHP model (Table 5).

Table 6. Criteria for evaluating LCIPs.

Indicator group	Name of indicator	Evaluation criteria	Weight	Explanation
Motivation	Use of resources and energy	The scale of IPs	0.0610	Total area for operation of IPs
		Rate of energy consumption from fuel oil	0.0565	Rate of energy consumption due to the use of oil to serve the production activities of IPs
		Rate of energy consumption from coal	0.0671	Rate of energy consumption by using coal to serve the production activities of IPs
		Rate of energy consumption from electric	0.0305	Rate of energy consumption by using electricity to serve the production activities of IPs
Pressure	GHGs rate	Rate of GHG emissions from energy consumption (due to fuel use)	0.0808	Rate of GHG emissions from energy consumption to serve production activities of IPs
		Rate of GHG emissions from non-energy consuming activities (due to waste)	0.0648	Rate of GHG emissions from non-energy activities from IPs
		Rate of CO_2 absorption	0.0580	GHG absorption rate by plants of IPs
Current status	Current status of GHG emissions	Total CO_{2e} emissions/GDP	0.0564	CO_{2e} emissions per unit of GDP
		Energy-GDP elasticity	0.0548	The ratio of growth rate due to electricity use to GDP growth rate over the same period
		Rate of electricity from solar energy in total electricity use	0.0586	Ratio of green power consumption to the total electricity consumed in IPs
		Energy intensity per unit of GDP	0.0263	The aggregate indicator shows how much energy is needed to create one unit of GDP
Impact	Impact of GHG emissions on economic development	Rate of reduction of GHG emissions in energy consumption activities (due to fuel use)	0.1147	Rate of GHG emission reduction in energy consumption activities
		Average increase in GHG emissions	0.1062	The annual increase in GHGs emissions from the operation of IPs
		Ratio of reduction in energy consumption to GDP	0.0416	The annual rate of decline of energy consumption in IPs
Mitigation	Reducing GHG emissions	Ratio of renewable energy sources in the structure of energy use	0.0500	Ratio of renewable energy sources to total energy consumed in the IPs
		The ratio of vegetation cover	0.0284	The ratio of vegetation cover area to the total area of the IPs
		Percentage of production facilities with solar equipment	0.0263	The number of establishments using solar energy equipment out of the total number of establishments operating in the IPs
		Percentage of roads in IPs using energy-saving equipment and technologies or using renewable energy for lighting	0.0183	Ratio of the length of roads to which energy-saving or renewable lighting equipment and technologies are applied to the total length of roads in the IPs

3.2. Establishing an index system for LCIPs

In recent years, the concept of industrial ecology was extended concerning the terms of eco-IPs [19]. This can be obtained by performing the circular economy, which enhances resource conservation and emission reduction [29]. This model illustrates that IPs could be profitable while following strict emission and related materials. Using the AHP application, the weight coefficient and their evaluation *system for the LCIPs* have been highlighted [17]. Also, a setup procedure and specific method were carried out to guide for index system of the LCIPs [30]. Based on a low-carbon emissions assessment, combined with legal documents issued by the state, sets of criteria for sustainable development in IPs and ecological IPs, and environmentally friendly IPs, this study proposes a list of eighteen essential criteria to evaluate LCIPs. The list of criteria for assessing LCIPs belongs to five main criteria groups: motivation, pressure, current status, impact, and mitigation [31]. In which, the dynamic group shows the criteria for the size and development of the IPs, the pressure indicator group shows the ability to emit GHGs through the use of resources and energy, and the group of criteria assessment of the current environmental status of the IPs to assess the possibility of GHGs generation from pollution. In addition, the impact group assesses the influence of factors on carbon emissions. Finally, the group of criteria meets to evaluate factors capable of reducing carbon emissions. The detailed the criteria for assessing LCIPs are shown in Table 6.

3.3. *Case study of Trang Bang IP*

Owing to accelerating urbanization and industrialization led to seriously environmental problems, e.g., water pollution, air contamination, waste management, etc. [32-35]. IPs are ususaly accompanied by some environmental disadvantages, including more carbon emissions and ecological stress/pressures [36, 37]. Recently, a case study of typical LCIPs was conducted to examine the economic and ecological benefits [38]. It could be said that the LCIPs are the new approach and model for developing sustainable IPs that respond to global challenges, e.g., CC and less GHG emissions. The LCIPs model/construction is an efficient pathway to reach coordinated development of economic, social, and environmental aspects and a promising approach to obtaining carbon reduction and energy saving. To assess the carbon emissions of Trang Bang IP (2016–2020), this study used a set of evaluation criteria, the carbon emission data, and the threshold values of the criteria in the min-max normalization. After collection, the carbon emission values were normalized, and then data was entered into the assessment criteria. The value of the carbon emission index of the IPs by the weighted sum method and the detailed calculation findings of the carbon emission index of Trang Bang IP is illustrated in Table 7. Through this investigation, most tendency datas show that Trang Bang IP was ranked in high to very high carbon emission IPs. It means Trang Bang IP has yet to meet the "low-carbon" standard. However, this index system can reflect the critical concept of the knowledge economy, ecological economy and/or circular economy [30]. LC*IPs* can enhance the essential background for the green industry initiative and Sustainable Development Goals (SDGs).

4. Conclusions

The eighteen indexes are proposed for evaluating LCIPs, which can be applied in Vietnam's conditions after the preliminary and secondary screening. A set of indices to identify the indexes has been used that focus on the following aspects: (i) suitable for policies and laws of Vietnam, (ii) consistent with the target of LCIPs, (iii) satisfying socio-economic conditions, (iv) suitable for infrastructure, (v) effective emission reduction, and (vi) economic efficiency. This index system can help promote the rapid development of LCIPs in Vietnam and provide a comprehensive reference in the future. Regarding the proposed LCIP index, this study demonstrated that Trang Bang IP had not met the "low-carbon" standard over the 2016–2020 period. However, applying the proposed LCIP index to all active IPs will provide the scientific evidence required for developing and implementing new environmental policies to foster LCE.

Table 7. Carbon emission index of Trang Bang IP.

Indicators	Evaluation criteria	2016		2017		2018		2019		2020		Tendency
		P_i	I_i	P_i	I_i	P_i	I_i	P_i	I_i	P_i	I_i	
1. Use of resources and energy	Scale of IPs	9.6	0.59	9.6	0.59	9.6	0.59	9.6	0.59	9.6	0.59	Stability
	Rate of energy consumption from fuel oil	100	5.65	100	5.65	100	5.65	100	5.65	100	5.65	Stability
	Rate of energy consumption from coal	100	6.71	100	6.71	1.0	0.07	1.0	0.07	1.0	0.07	Exceeded value, bad impact

	Rate of energy consumption from electric	1.0	0.03	1.0	0.03	1.0	0.03	1.0	0.03	1.0	0.03	Exceeded value, bad impact
	I_1		12.98		12.98		6.33		6.33		6.33	Downtrend
2. GHGs rate	Rate of GHG emissions from energy consumption (due to fuel use)	1.0	0.08	1.0	0.08	1.0	0.08	1.0	0.08	1.0	0.08	Exceeded value, bad impact
	Rate of GHG emissions from non-energy consuming activities (due to waste)	1.0	0.06	1.0	0.06	1.0	0.06	1.0	0.06	1.0	0.06	Exceeded value, bad impact
	Rate of CO_2 absorption	100	5.80	100	5.80	100	5.80	100	5.80	100	5.80	Exceeded value, good impact
	I_2		5.94		5.94		5.94		5.94		5.94	Stability
3. Impact of GHG emissions on economic development	Total CO_{2e} emissions/GDP	94.5	5.33	100	5.64	1.0	0.06	32.1	1.81	63.7	3.59	Volatility
	Energy-GDP elasticity	100	5.48	100	5.48	1.0	0.05	100	5.48	100	5.48	Stability
	Rate of electricity from solar energy in total electricity use	1.0	0.06	1.0	0.06	1.0	0.06	1.0	0.06	1.0	0.06	Not reaching the level of value, bad impact
	Energy intensity per unit of GDP	100	2.63	100	2.63	100	2.63	100	2.63	100	2.63	Not reaching the level of value, bad impact
	I_3		13.49		13.81		2.8		9.98		11.76	Volatility
4. Impact of GHG emissions on economic development	Rate of reduction of GHG emissions in energy consumption activities (fuel use)	1.0	0.11	1.0	0.11	100	11.47	10.9	1.25	1.0	0.11	Volatility
	Average increase in GHG emissions	100	10.62	100	10.62	1.0	0.11	100	10.62	100	10.62	Stability
	Ratio of reduction in energy consumption to GDP	100	4.16	100	4.16	92.1	3.83	100	4.16	100	4.16	Not reaching the level of value, bad impact
	I_4		14.89		14.89		15.40		16.02		14.89	Volatility
5. Reducing GHG emissions	Ratio of renewable energy sources in the structure of energy use	1.0	0.05	1.0	0.05	1.0	0.05	100	5.00	100	5.00	Uptrend and reaching value level
	Ratio of vegetation cover	1.0	0.03	1.0	0.03	1.0	0.03	50.5	1.44	50.5	1.44	Uptrend
	Percentage of production facilities with solar equipment	1.0	0.03	1.0	0.03	1.0	0.03	1.0	0.03	1.0	0.03	Not reaching the level of value, bad impact
	Percentage of roads in IPs using energy-saving equipment and technologies or using renewable energy for lighting	1.0	0.02	1.0	0.02	1.0	0.02	7.2	0.13	100	1.83	Uptrend and reaching value level
	I_5		0.12		0.12		0.12		6.59		8.29	Uptrend
Assessment value I_{LC}			47.42		47.74		30.6		44.86		47.20	Stability
	Rank	High carbon emission IPs		High carbon emission IPs		Very high carbon emission IPs		High carbon emission IPs		High carbon emission IPs		

Remarks: P_i: Normalized value, I_i: Ingredient value.

References

[1] Gurney, K.R., Mendoza, D.L., Zhou, Y., Fischer, M.L., Miller, C.C., Geethakumar, S., and de la Rue du Can, S. 2009 High resolution fossil fuel combustion CO2 emission fluxes for the United States *Environmental science & technology* **43** 5535-5541

[2] Shan, Y., Guan, D., Zheng, H., Ou, J., Li, Y., Meng, J., Mi, Z., Liu, Z., and Zhang, Q. 2018 China CO2 emission accounts 1997–2015 *Scientific data* **5** 1-14

[3] Thompson, L.G. 2010 Climate change: the evidence and our options *Behav Anal* **33** 153-170

[4] Forster, P., Ramaswamy, V., Artaxo, P., Berntsen, T., Betts, R., Fahey, D.W., Haywood, J., Lean, J., Lowe, D.C., and Myhre, G. 2007 Climate change 2007: the physical science basis. Contribution of Working Group I to the Fourth Assessment Report of the Intergovernmental Panel on Climate Change 212
[5] IPCC 2014 Synthesis Report. Contribution of working groups I. II and III to the fifth assessment report of the intergovernmental panel on climate change 151
[6] Snell, D., and Schmitt, D. 2012 'It's not easy being green': electricity corporations and the transition to a low-carbon economy *Competition & Change* **16** 1-19
[7] Chen, H., and Wang, L. 2017 Coproducts generated from biomass conversion processes *Technologies for biochemical conversion of biomass* 219-264.
[8] Fang, D., Chen, B., Hayat, T., and Alsaedi, A. 2017 Emergy evaluation for a low-carbon industrial park *Journal of Cleaner Production* **163** S392-S400
[9] Ban, Y.U., Jeong, J.H., and Jeong, S.K. 2016 Assessing the performance of carbon dioxide emission reduction of commercialized eco-industrial park projects in South Korea *Journal of Cleaner Production* **114** 124-131
[10] Steckel, J.C., Schwerhoff, G., and Edenhofer, O. 2017 Enabling low-carbon development in poor countries *In Sustainable Manufacturing. (Springer, Cham)* 33-44
[11] Le, V.T., and Nguyen, M.K. 2017 The study of climate change trend in Vietnam: a case study in Danang city *Hue University Journal of Science: Earth Science and Environment* **126** 19-28
[12] Komnitsas, K.A. 2011 Potential of geopolymer technology towards green buildings and sustainable cities. *Procedia Engineering* **21** 1023-1032
[13] Erkman, S., and Van Hezik, C. 2016 Global Assessment of Eco-Industrial Parks in Developing and Emerging Countries. The United Nations Industrial Development Organization, Vienna International Centre: Vienna, Austria, 2017-2002
[14] Kastner, C.A., Lau, R., and Kraft, M. 2015 Quantitative tools for cultivating symbiosis in industrial parks; a literature review *Applied Energy* 155 599-612
[15] Wang, H., Lei, Y., Wang, H., Liu, M., Yang, J., and Bi, J. 2013 Carbon reduction potentials of China's industrial parks: A case study of Suzhou Industry Park. *Energy* **55** 668-675
[16] Saaty, T.L. 2008 Decision making with the analytic hierarchy process. *International journal of services sciences* **1** 83-98
[17] Ma, X.Q., Li, M.H., Sun, Y.S., and Jin, F.Y. 2014 Application of Evaluation System for Low-Carbon Industrial Park by Using AHP Method *Trans Tech Publ* **1010** 1839-1843
[18] Yang, T., Liu, C., Côté, R.P., Ye, J., and Liu, W. 2022 Evaluating the Barriers to Industrial Symbiosis Using a Group AHP-TOPSIS Model. *Sustainability* **14** 6815
[19] Boix, M., Montastruc, L., Azzaro-Pantel, C., and Domenech, S. 2015 Optimization methods applied to the design of eco-industrial parks: a literature review *Journal of Cleaner Production* **87** 303-317
[20] Repetski, E.J., Sarkani, S., and Mazzuchi, T. 2022 Applying the Analytic Hierarchy Process (AHP) to Expert Documents *International Journal of the Analytic Hierarchy Process* 14
[21] Ishizaka, A., and Labib, A. 2009 Analytic hierarchy process and expert choice: Benefits and limitations. *Or Insight* **22** 201-220
[22] Hiremath, R.B., Balachandra, P., Kumar, B., Bansode, S.S., and Murali, J. 2013 Indicator-based urban sustainability—A review *Energy for sustainable development* **17** 555-563
[23] Zhang, M., Yang, Y., and Xia-Bauer, C. 2021 Measuring urban low-carbon sustainability in four Chinese cities. *Sustainability* **13** 12281
[24] Lvan, T., Ho, V.T.T., Thanh, H.D.T., Thong, N.T., Huynh, Q., and Nguyen, Q.L.B. 2021 Analysis of CO2 Emissions from Industrial Parks: A Case Study in Ho Chi Minh City, Vietnam. **1092** 1 Edition. (IOP Publishing) 012084
[25] Hei, W., Li, X., Gao, G., Wang, S., Zhang, R., and Wang, K. 2022 Air Pollutants and CO2 Emissions in Industrial Parks and Evaluation of Their Green Upgrade on Regional Air Quality Improvement: A Case Study of Seven Cities in Henan Province. *Atmosphere* 13

[26] Eggleston, H.S., Buendia, L., Miwa, K., Ngara, T., and Tanabe, K. 2006 2006 IPCC guidelines for national greenhouse gas inventories
[27] Lehtoranta, S., Nissinen, A., Mattila, T., and Melanen, M. 2011 Industrial symbiosis and the policy instruments of sustainable consumption and production. *Journal of Cleaner Production* **19** 1865-1875
[28] Thieriot, H., and Sawyer, D. 2015 Development of eco-efficient industrial parks in China: A review
[29] Abraham, E.J., Ramadan, F., and Al-Mohannadi, D.M. 2021 Synthesis of Sustainable Carbon Negative Eco-Industrial Parks. Frontiers in Energy Research 9
[30] Yanyan, H., and Jingjing, W. 2015 Research on Establishment of an Index System for the Low-carbon Industrial Parks in Wuhan City. In Proceedings of the 2015 International Symposium on Material *Energy and Environment Engineering (Atlantis Press)* 329-333
[31] Carr, E.R., Wingard, P.M., Yorty, S.C., Thompson, M.C., Jensen, N.K., and Roberson, J. 2007 Applying DPSIR to sustainable development. *International journal of sustainable development & world ecology* **1** 543-555
[32] Hai, V.D., Hoang, S.M.T., Hung, N.T.Q., Ky, N.M., Gwi-Nam, B., Ki-hong, P., Chang, S.W., Bach, Q.-V., Nhu-Trang, T.-T., and Nguyen, D.D. 2019 Characteristics of airborne bacteria and fungi in the atmosphere in Ho Chi Minh city, Vietnam-A case study over three years. *International biodeterioration & biodegradation* 145 104819
[33] Hung, N.T.Q., Ky, N.M., Nhung, H.T.T., Nhut, H.T., and Van Tan, L. 2021 Investigating community health effects of bioaerosol pollution in ho chi minh megacity, vietnam. *Environmental Engineering and Management Journal* 20 701-712
[34] Hoang, T.-D., Ky, N.M., Thuong, N.T.N., Nhan, H.Q., and Ngan, N.V.C. 2022 Artificial Intelligence in Pollution Control and Management: Status and Future Prospects. In Artificial Intelligence and Environmental Sustainability: Challenges and Solutions in the Era of Industry 4.0, H.L. Ong, R.-a. Doong, R. Naguib, C.P. Lim and A.K. Nagar, eds. (Singapore: Springer Nature Singapore), 23-43
[35] Nguyen, M.K., Lin, C., Hoang, H.G., Sanderson, P., Dang, B.T., Bui, X.T., Nguyen, N.S.H., Vo, D.-V.N., and Tran, H.T. 2022 Evaluate the role of biochar during the organic waste composting process: A critical review. *Chemosphere* **299** 134488
[36] Phan, C.C., Nguyen, T.Q.H., Nguyen, M.K., Park, K.H., Bae, G.N., Seung-bok, L., and Bach, Q.V. 2020 Aerosol mass and major composition characterization of ambient air in Ho Chi Minh City, Vietnam. *International journal of environmental science and technology* **17** 3189-3198
[37] Dong, H., Geng, Y., Xi, F., and Fujita, T. 2013 Carbon footprint evaluation at industrial park level: a hybrid life cycle assessment approach *Energy Policy* **57** 298-307
[38] Chen, B., He, G., Yang, J., Zhang, J., Su, M., and Qi, J. 2012 Evaluating ecological and economic benefits of a low-carbon industrial park based on millennium ecosystem assessment framework *The scientific world journal* 2012

IOP Conf. Series: Earth and Environmental Science 1087 (2022) 012042 doi:10.1088/1755-1315/1087/1/012042

Research on ice characteristics in Bohai Bay, China based on Multi-source remote sensing images

Huaiyuan Li[1], Longkun Qin[2] and Zhiyuan Han[1,3]

[1] Tianjin research institute for water transport engineering, M.O.T., Key laboratory of engineering sediment ministry of communications, PRC, Tianjin 300456, China;
[2] Information Center of Suzhou, Suzhou 215004, China.

[3] Corresponding author's e-mail: hzy0616@qq.com

Abstract. Sea ice area accounts for about 22.9% of the global ocean area, so the research of ice plays an important role in the marine field. The ice data that can be obtained by means of site investigation is very limited, and remote sensing means has unique advantages in large-scale and multi-element feature extraction. Especially in recent years, the available remote sensing image data is becoming more and more abundant, which makes it an important supplementary data for ice research. In this paper, we took the typical sea area of Bohai Bay in China as an example, analyzed the overall characteristics of the ice condition in the target sea area by collecting the winter remote sensing image data in recent 8 years. By using the methods of target recognition, edge detection and time series analysis, four main characteristics of the ice blocks were extracted: time characteristics, spatial characteristics, scale characteristics and motion characteristics. The results enriched the research data of ice condition in the target sea area.

1. Introduction

The Bohai Bay is directly affected by the cold air heading to the south from Siberia [1-2] in winter due to its geographical location, becoming the southern boundary of ocean icing in the northern hemisphere with different kind of icing every year.

The ice conditions in Bohai Bay [3-5] have significant differences in interannual and spatial characteristics. The ice conditions in Bohai Bay usually are not very serious in general years, while it becomes extremely serious in some years with a large area of sea being frozen. In general, the change of ice interannual characteristics in Bohai Bay is difficult to predict. The analysis and accurate prediction of ice situation becomes more difficult due to its temporal and spatial distribution characteristics [6-10].

Accurate ice datas can be obtained through site investigation, but it costs a lot and accompanies high risks. In the absence of measured data, identifying and retrieving sea ice from remote sensing images is an important means to obtain data. Remote sensing data is especially suitable for large-scale ice condition observation [11-15], which can realize better observation of ice period, ice volume, ice density, ice range and other factors. The main advantage of remote sensing is the wide observation range and the site risk aversion.

2. Materials

2.1. Study ares
We selected the central and western Bohai Bay as the research area, and we have rich hydrological data in this area, which is convenient for the demonstration and analysis of the experimental results. The study area is shown in Figure 1.

Figure 1. Study area in the central and western Bohai Bay, China.

2.2. Data
The satellite data we used in this study include:

1. GF-1 products: The 1-level products of GF-1 WFV satellites, which the spatial resolution is 16 meters.
2. GF-6 products: The 1-level products of GF-6 WFV satellites, which the spatial resolution is 16 meters.
3. Landsat 8 products: The 2-level products of Landsat-8 OLI satellites, which the spatial resolution is 30 meters.
4. Sentinel 2 products: The 2-level products of Sentinel-2A satellites, which the spatial resolution of used bands is 10 meters.

After excluding the data whose date and cloud amount missing requirements, we have collected 76 remote sensing images from the products listed above, covering the range from November, 2013 to March, 2021. All of the data we used in this study have red, green, blue and infrared bands.

3. Methods
Using the collected multi-source remote sensing images, we analyzed the time, spatial, scale and motion characteristics of sea ice by means of time series analysis, sea ice recognition, edge detection and other methods. The time nodes of sea ice growth, development and melting were obtained, as well as the first ice day, full ice day, melting day, final ice day and the spatial distribution of sea ice. The scale and drift characteristics of typical ice period were extracted, and the ice morphology and motion pattern were presented.

Ice remote sensing analysis mainly adopted visible, infrared and microwave bands [16-21]. The visible and infrared bands have the advantages of high spatial resolution, obvious spectral characteristics and easy availability. The relevant information such as ice distribution, ice drift, ice characteristics, ice formation and melting process can be obtained through the analysis of visible and infrared remote sensing.

3.1. Ice block Identification
We used the ice albedo index model of visible bands (red, green and blue) to retrieve the sea ice distribution, which was proposed by Hiroyuki [22] et al and defined by the following formula:

$$A_v = 0.436 \cdot b_1 + 0.227 \cdot b_2 + 0.327 \cdot b_3 \quad (1)$$

$$A_v > 0.13 \quad (2)$$

Where A_v is the albedo, b_1 is the reflectivity of blue band, b_2 is the reflectance of green band, and b_3 is the reflectance of red band.

The suspended sediment concentration in the target sea area was relatively high, and the inversion of ice image data would be disturbed. The albedo algorithm may misjudge it as ice data. In order to eliminate the influence of high suspended sediment water, threshold judgment conditions were added to the inversion model as follows:

$$\frac{b_3}{b_1} < 1.5 \tag{3}$$

Through adding the ratio of red and blue bands, the seawater with high suspended sediment concentration and the thin ice were distinguished.

3.2. Ice edge Detection

The edge structure information of the ice block was extracted through detecting the edge of the ice block data file, so as to provide the vector data for calculating the block areas. We adopted the object-oriented feature extraction method to make the edge detection of ice blocks. Based on canny edge detection algorithms [23-25], the target classification and extraction based on image texture features, spectral features and geometric features were realized.

3.3. Target recognition of Adjacent images

The motion characteristics of ice block were analyzed by extracting the front and rear position information of the same ice target between the adjacent images. In this study, it was assumed that the relatively solid ice blocks in the adjacent image periods would not change dramatically in the drift process, the texture and geometric features of these ice blocks would remain constant. Therefore, we took the ice data file after seawater separation as the data source, judged the same target in the adjacent ice data through the automatic recognition of texture features and geometric features, filtered and screened through visual recognition to ensure the accuracy of the recognition results.

4. Experimental results

4.1. Time characteristics

Through the classification and summary of 76 image, the time characteristics of different ice period in the target area were obtained, as shown in the Table 1. The results show that, it begins to freeze in the first and middle of December every year, and completely melts in the middle and late of February next year, and the ice lasts 50 to 70 days.

Table 1. Statistics of time characteristic in target ice area.

Period	First date	Last date	glacial maximum peroid	Total days
2013.11-2014.3	2013.12.13	2014.02.13	2013.12.30	63
2014.11-2015.3	2014.12.04	2015.02.09	2014.12.17	68
2015.11-2016.3	2015.12.17	2016.02.24	2016.01.24~2016.02.17	70
2016.11-2017.3	2016.12.15	2017.02.11	2017.01.25	59
2017.11-2018.3	2018.12.05	2018.02.12	2018.01.25~2018.02.05	70
2018.11-2019.3	2018.12.08	2019.02.16	2018.12.29~2019.01.19	71
2019.11-2020.3	2019.12.20	2020.02.05	-	48
2020.11-2021.3	2020.12.14	2021.02.17	2021.01.08~2021.01.23	66

4.2. Spatial characteristics

According to the remote sensing images of the most serious condition during the ice period, the periglacial line around the ice blocks was drawn as the maximum influence range in each year. The maximum outer edge line of the ice blocks in recent 8 years is shown in Figure 2, except the period of

2019.11 to 2020.3 for missing glacial maximum. The results show that the maximum offshore distance of ice blocks is 42.5 km.

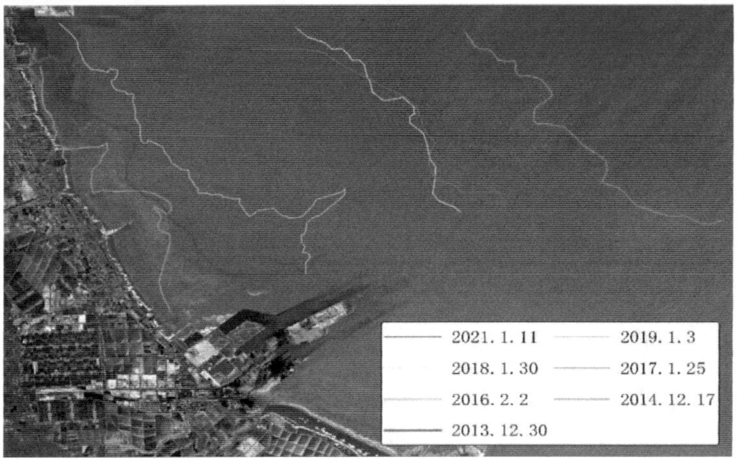

Figure 2. The maximum outer edge line of the ice blocks from 2013 to 2020 except the winter of 2019 for missing glacial maximum.

4.3. Scale characteristics

The peak date images of typical strong ice years were selected for analysis to obtain the sacle characteristics information of ice blocks. The analyzed image dates are January 3, 2019 (GF-6) and January 11, 2021 (GF-1). Figure 3 shows the ice block recognition results for the selected two images.

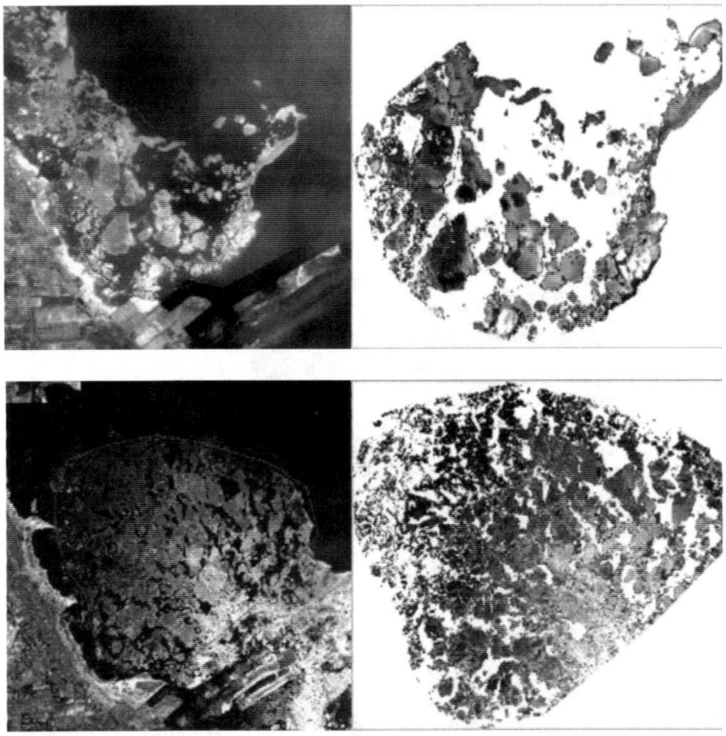

Figure 3. The ice block recognition results for January 3, 2019 (top) and January 11, 2021 (bottom).

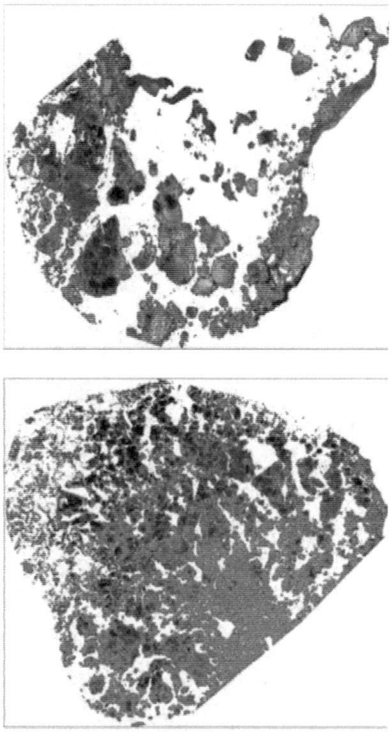

Figure 4. The edge detection results for January 3, 2019 (top) and January 11, 2021 (bottom).

Edge detection was performed on the ice blocks data files extracted in the previous step, and the results are shown in the Figure 4. The edge detection results were output as a planar vector file, which was used as a basis data to calculating block size in the next step.

Through the polygon vector file obtained in ice blocks edge detection, the scale characteristics of ice blocks were obtained by statistical analysis. Table 2 shows the size characteristics of ice blocks.

The maximum area of ice blocks on January 3, 2019 is 859136 m^2, the average area is 9259 m^2, and the number of blocks is 9946. The maximum area of ice blocks on January 11, 2021 is 1308416 m^2, the average area is 9598 m^2, and the number of blocks is 34066.

The geometry of the ice blocks was an irregular polygon. The diameter of the equivalent disc was a better way to reflect the ice block area. The diameter was calculated by the following formula:

$$d = \sqrt{\frac{4A}{\pi}} \tag{4}$$

Where d is the equivalent diameter of the ice block and A is the area of the ice block.

Table 3 shows the statistical results of occurrence frequency of ice blocks at different horizontal scales in the target area. In the two selected data, the horizontal scale of most ice blocks is in the range of 50 ~ 100m. There are few ice blocks have a scale greater than 300m and even fewer ice blocks with a scale greater than 500m. It is more likely that the maximum horizontal scale of ice blocks in this area is about 1000m.

Table 2. The size characteristics of ice blocks for two images.

date	Block number	Max area/m^2	Avg area/m^2	Total area/m^2
2019.1.3	9946	859136	9259	92091648
2021.1.11	34066	1308416	9598	326969088

Table 3. Occurrence frequency of ice blocks at different horizontal scales.

Size level	<50	50~100	100~300	300~500	500~1000	1000~1500	>=1500	Total
Percent/%	33.189	45.915	19.829	0.859	0.198	0.011	0.000	100

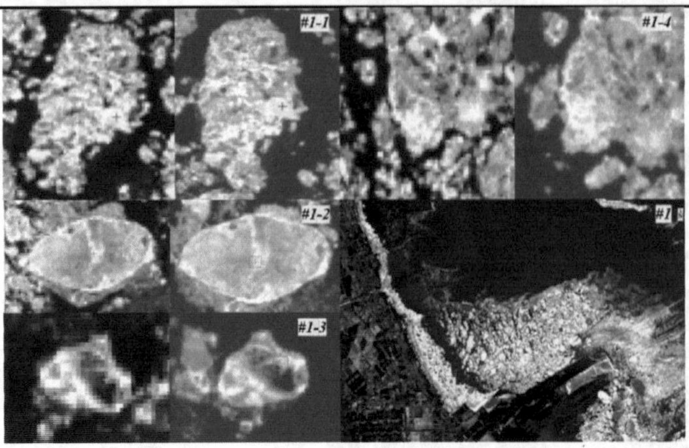

Figure 5. The recognition result of the ice blocks in group 1; #1-1, #1-2, #1-3 and #1-4 correspond to four blocks in group 1, and #1 shows the overall characteristics of group 1.

4.4. Motion characteristics

During the movement of ice blocks, some blocks would collide, disintegrate and melt, making it difficult to identify the same ice block from adjacent images. To ensure the recognition accuracy of ice block, the adjacent images were selected with condition that the time interval of two adjacent images should be less than 1.5 days. The moving characteristics of ice blocks were distinguished in these data sources. The ice blocks with obvious marks were used as the basis for estimating, and the distribution of characteristic points was relatively uniform.

Figure 5 shows the recognition result of the ice blocks in group 1. Each ice block in group 1 has an unique number.

Table 4 shows the adjacent images for estimating the moving characteristics of ice blocks.

The ice blocks in the study area drifted mainly under the influence of tidal current. Table 5 shows the statistical data of moving characteristic for each group. The maximum velocity of the ice block is 0.64m/s and the average velocity is 0.25m/s.

The statistical results show that the movement direction of the ice blocks is mainly determined by the tidal current direction, and the drift direction of the ice blocks is mainly NE-SW and ENE-WSW directions perpendicular to the coastline, NW-SE and NNW-SSE directions along the coastline.

Table 4. The adjacent images in experimental group.

Group	Two adjacent image date		Block number
#1	2016.2.7 10:47:45	2016.2.8 11:47:17	4
#2	2018.1.29 11:33:45	2018.1.30 10:56:46	3
#3	2019.1.3 11:27:04	2019.1.4 11:21:25	4
#4	2021.1.11 10:53:51	2021.1.12 11:17:08	4

Table 5. The statistical data of moving characteristic for each group.

Block number	Moving distance(m)				Moving speed(m/s)			
	1	2	3	4	1	2	3	4
#1	1465	7178	2476	3743	0.0586	0.2871	0.099	0.1497
#2	14605	12525	13340	-	0.6427	0.5357	0.5706	-
#3	2219	2484	1224	1934	0.0371	0.0416	0.0205	0.0324
#4	8000	10065	6390	5557	0.3281	0.4128	0.2621	0.2279

5. Conclusions

In this paper, we obtained the ice block characteristic data of the target area according to the remote sensing image data over last 8 years (2013-2020), and extracted four main characteristics of ice blocks: temporal characteristics, spatial characteristics, scale characteristics and motion characteristics, thus enriching the ice condition research data of the sea area.

With the increasing abundance of remote sensing data, it will be easier to obtain ice characteristics through remote sensing image inversion. Therefore, the remote sensing inversion method of ice has great application prospects. We will continue to study the combination of ice remote sensing inversion and numerical simulation, so as to accurately predict the short-term development trend of the ice.

References

[1] Liu C M, Qian Z A, Wu M C, et al. 2004 A composite study of the synoptic differences between major and minor dust storm springs over the China-Mongolia areas[J] *Tao* **15(5)** 999-1018

[2] Donaldson G C, Ermakov S P, Komarov Y M, et al. 1998 Cold related mortalities and protection against cold in Yakutsk, eastern Siberia: observation and interview study[J] *Bmj* **317(7164)** 978-982

[3] Zhang D, Xu N, Yue Q, et al. 2015 Sea ice problems in Bohai Bay oil and gas exploitation[J] *Journal of coastal research* **73 (10073)** 676-680

[4] Yun Z, Wanting M, Qiming G, et al. 2014 Detection of Bohai Bay sea ice using GPS-reflected signals[J] *IEEE Journal of Selected Topics in Applied Earth Observations and Remote Sensing* **8(1)** 39-46

[5] Yang G 2000 Bohai Sea ice conditions[J] *Journal of Cold Regions Engineering* **14(2)** 54-67
[6] Sidorenko A, Stepanov S, Petrov Y, et al. 2020 Application of a remote sensing data processing method for assessment ice cohesion in the Arctic navigation[C]//IOP Conference Series: Earth and Environmental Science *IOP Publishing* **539(1)** 012128
[7] Gilpin R R 1980 A model for the prediction of ice lensing and frost heave in soils[J] *Water Resources Research* **16(5)** 918-930
[8] Kreutz M, Alla A A, Varasteh K, et al. 2021 Convolutional neural network with dual inputs for time series ice prediction on rotor blades of wind turbines[J] *Procedia CIRP* **104** 446-451
[9] Steele M, Dickinson S, Zhang J, et al. 2015 Seasonal ice loss in the Beaufort Sea: Toward synchrony and prediction[J] *Journal of Geophysical Research: Oceans* **120(2)** 1118-1132
[10] Sutyrina E N 2020 Prediction of spring ice phenomena on lakes and reservoirs using teleconnection indices[J] *Limnology and Freshwater Biology* 946-947
[11] Kwok R 2010 Satellite remote sensing of sea-ice thickness and kinematics: a review[J] *Journal of Glaciology* **56(200)** 1129-1140
[12] Immerzeel W W, Droogers P, De Jong S M, et al. 2009 Large-scale monitoring of snow cover and runoff simulation in Himalayan river basins using remote sensing[J] *Remote sensing of Environment* **113(1)** 40-49
[13] Qi M, Liu S, Yao X, et al. 2020 Monitoring the ice phenology of Qinghai lake from 1980 to 2018 using multisource remote sensing data and Google Earth Engine[J] *Remote Sensing* **12(14)** 2217
[14] Carsey F 1989 Review and status of remote sensing of sea ice[J] *IEEE Journal of Oceanic Engineering* **14(2)** 127-138
[15] Worby A P, Comiso J C 2004 Studies of the Antarctic sea ice edge and ice extent from satellite and ship observations[J] *Remote sensing of environment* **92(1)** 98-111
[16] Kaleschke L, Lüpkes C, Vihma T, et al. 2001 SSM/I sea ice remote sensing for mesoscale ocean-atmosphere interaction analysis[J] *Canadian journal of remote sensing* **27(5)** 526-537
[17] Surdu C M, Duguay C R, Brown L C, et al. 2014 Response of ice cover on shallow lakes of the North Slope of Alaska to contemporary climate conditions (1950-2011): radar remote-sensing and numerical modeling data analysis[J] *The Cryosphere* **8(1)** 167-180
[18] Khachatrian E, Chlaily S, Eltoft T, et al. 2021 Automatic selection of relevant attributes for multi-sensor remote sensing analysis: a case study on sea ice classification[J] *IEEE Journal of Selected Topics in Applied Earth Observations and Remote Sensing* **14** 9025-9037
[19] Raup B, Kääb A, Kargel J S, et al. 2007 Remote sensing and GIS technology in the Global Land Ice Measurements from Space (GLIMS) project[J] *Computers & Geosciences* **33(1)** 104-125
[20] Wensnahan M, Maykut G A, Grenfell T C, et al. 1993 Passive microwave remote sensing of thin sea ice using principal component analysis[J] *Journal of Geophysical Research: Oceans* **98(C7)** 12453-12468
[21] Albert T H 2002 Evaluation of remote sensing techniques for ice-area classification applied to the tropical Quelccaya Ice Cap, Peru[J] *Polar Geography* **26(3)** 210-226
[22] Wakabayashi H, Mori Y, Nakamura K 2013 Sea ice detection in the sea of Okhotsk using PALSAR and MODIS data[J] *IEEE Journal of Selected Topics in Applied Earth Observations and Remote Sensing* **6(3)** 1516-1523
[23] Rong W, Li Z, Zhang W, et al. 2014 An improved CANNY edge detection algorithm[C]//2014 IEEE international conference on mechatronics and automation *IEEE* 577-582
[24] Sharifi M, Fathy M, Mahmoudi M T 2002 A classified and comparative study of edge detection algorithms[C]//Proceedings. International conference on information technology: Coding and computing *IEEE* 117-120
[25] Joshi S R, Koju R 2012 Study and comparison of edge detection algorithms[C]//2012 Third Asian Himalayas international conference on internet *IEEE* 1-5

IOP Conf. Series: Earth and Environmental Science 1087 (2022) 012043 doi:10.1088/1755-1315/1087/1/012043

Analysis of land subsidence change in Wuhan based on sentinel-1A

Yizhan Zhao[1], Lv Zhou[1,4], Jie Qin[2], Cheng Wang[1,3] and Haiquan Sheng[1]

[1] College of Geomatics and Geoinformation, Guilin University of Technology, Guilin 541004, China;
[2] Natural resources and real estate registration center of Guangxi Zhuang Autonomous Region, Nanning 530029, China;
[3] Guangxi Institute of Natural Resources Survey and Monitoring, Nanning, 530023, China.

[4] Corresponding author's e-mail: zhoulv@glut.edu.cn

Abstract. We used 52-view Sentinel-1A images to obtain the subsidence of Wuhan city by SBAS method during the period of 2020-2021. At present, Hongshan District of Wuhan is the area with the most serious subsidence, and the maximum accumulated settlement reaches 90mm. In addition, there are obvious subsidence funnels in Qiaokou, Qingshan, Jiang'an and Wuchang Districts. Spatial distribution of settlement areas in Wuhan is closely related to urban construction, and the formation of most of the subsidence funnels is related to the high-intensity urban construction activities in the region, with extensive ground subsidence along Metro Line 5 and Metro Line 11. Karst collapse and ground subsidence are more likely to occur in carbonate rock belt and soft soil covered area. Rainfall affects the total amount of groundwater and the river water level in Wuhan. Rainfall can add the groundwater volume, raise the water level of rivers and slow down the rate of ground subsidence in the major subsidence areas of Wuhan.

1. Preface

Surface subsidence is a geohazard that is common in many cities around the world [1,2]. Many countries, such as the Iran, Mexico, United States and so on, are facing the hazard of settlement [3-5]. subsidence will cause ground cracking, damage buildings, underground pipelines and other facilities, and increase urban flood risk. A United Nations study predicts that global ground subsidence will pose a threat to the survival of 1.6 billion people twenty years later [6]. China is the largest developing country in the world, and in process of large-scale and high-rate urbanization, many cities are facing the problem of ground subsidence due to over-exploitation of groundwater, industrial development and urban expansion. [7]. According to statistics, the area of ground subsidence in China is more than 120,000 square kilometers as of 2015, with 24 cities experiencing severe subsidence of more than 50 mm/year, and the area of subsidence is increasing.

Wuhan is an important central city in China, which has developed rapidly in the last decade. In the course of large-scale urbanization, large areas of ground subsidence triggered by geological formations and engineering construction have occurred in Wuhan. [8]. Records show that surface collapse in Wuhan first occurred in 1931 [9], and since then the rate and extent of surface subsidence has increased with human activity. At present, there have been many subsidence funnels in Wuhan

[10]. Wuhan mainly relies on the bench mark set in the city to monitor the surface subsidence, but leveling and GNSS are difficult to monitor subsidence on a large scale [11,12]. The time-series InSAR technique can make up for the weakness of traditional deformation monitoring methods, and can monitor surface subsidence on a large scale [13]. The first to monitor the Wuhan urban area was Bai et al in 2016. They used TerraSAR-X data to find that the ground subsidence rate in Wuhan urban area ranged from -64 to 18 mm/yr during 2009 and 2010, and that urban construction activities and karst geological conditions were the main causes of subsidence [14]. Costantini used COSMO-SkyMedSAR imagery to calculate ground subsidence between -80 and 40 mm/year for most of Hankou from 2013 to 2014 [15]. The rapid development of Wuhan after 2015 and the simultaneous start of numerous engineering projects led to severe ground settlement in the city [16,17]. A number of scholars have studied the subsidence in Wuhan during this period of time, and have found major areas of subsidence in Hankou, Wuchang, Baishazhou and Qingshan. The subsidence in Houhu is the most serious, and the maximum subsidence rate exceeds 80mm/yr.

In this paper, we used 52 Sentinel-1A images to acquire the surface deformation of Wuhan city from Jan. 2020 to Sep. 2021 by SBAS-InSAR technique. We found that the main subsidence areas after the COVID-19 were distributed in Qingshan, Jiang'an, Qiaokou, Wuchang and Qingling Districts. We found that subway and building construction are the main causes of ground subsidence by comparison, spatial distribution of subsidence is related to the distribution of Softsoils and Carbonate belts in Wuhan. In addition, we discuss the relationship between ground subsidence and changes in rainfall, groundwater and river levels.

2. Study area and datasets

2.1. Study area

The study area is shown in Figure 1a, Wuhan is located in the east of Hubei Province, at the junction of the Hanjiang plain and the mountains, with an altitude of 19 meters to 873 meters. Many rivers and lakes exist in Wuhan, and the total water area in the city accounts for 26.1% of the city's land area. The urban area of Wuhan City is distributed with soft soil layers and several carbonate belts. As seen in Figure 1b, a total of six carbonate strips are formed in Wuhan from north to south. Meanwhile, Wuhan is located at the confluence of Han River and Yangtze River, and softsoils such as silt and silt are widely distributed in the city.

2.2. Datasets

A total number of 52 Sentiel-1A images covering the Wuhan area from Jan. 2020 to Sep. 2021 were used for the study (Table 1 shows the details of Sentinel-1A images). In this study, we used the SRTM DEM with 90m resolution published by NASA as the external elevation data, which are used to remove the effect caused by topography phase. We also used POD (POD Precise Orbit Ephemerides) data provided by ESA, which are used to remove orbital errors. Further, we also acquired rainfall data, water level data, and historical Google Earth optical remote sensing images of Wuhan.

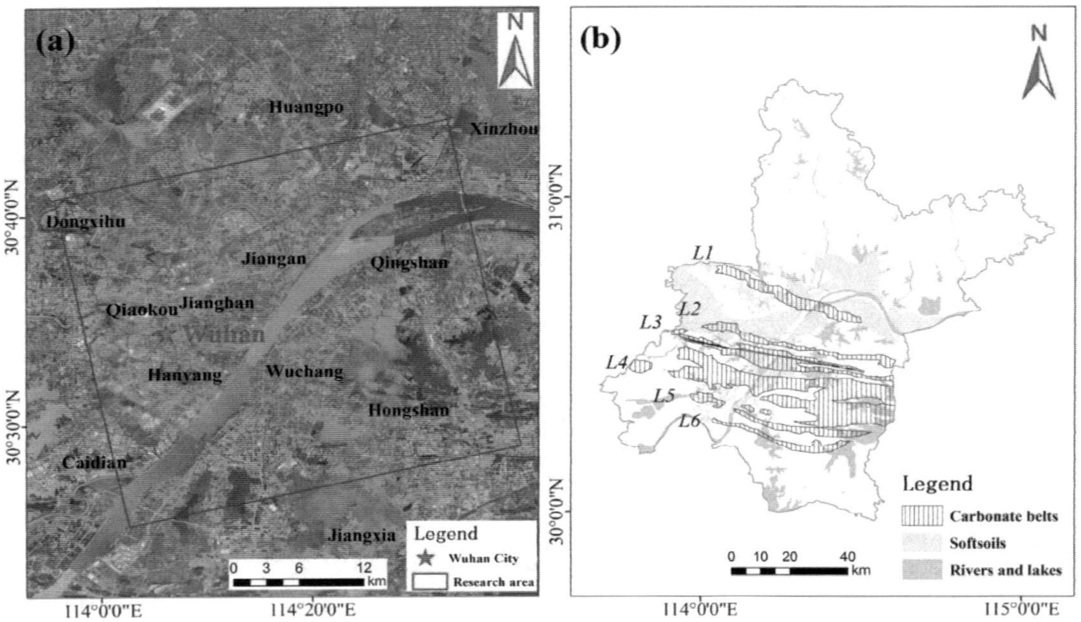

Figure 1. (a) Location of the study area; (b) Distribution of carbonate belts and softsoils in Wuhan. (The L1 to L6 are six Carbonate belts).

Table 1. SAR data parameters.

Parameter	Value	Parameter	Value
Sensor	Sentinel-1A	Wavelength	5.63 cm
Band	C	pass direction	Ascending
Incidence angle	39°	Number of scenes	52
Polarization	VH	Time range	20200103-20210930

3. Methods

The SBAS-InSAR method is based on the principle of minimum deformation rate criterion, which combines several line data into several sets to obtain small baseline interferograms. Next, multiple small baseline sets are solved jointly using singular value decomposition (SVD) or least squares methods to obtain surface deformation.

We used SARscape software to obtain the ground subsidence in Wuhan. As shown in Figure 2, the main processing flow includes steps such as Generating connection pairs, interference and phase unwrapping, refinement and reflattening, generation results. The SBAS-InSAR results are represented as a series of raster images, where each image element is a region of good coherence. In this study, the image of August 18, 2020 was selected as the primary image with a time threshold of 60 days and a spatial baseline threshold of 45%, and a total of 227 small baseline interference pairs were constructed.

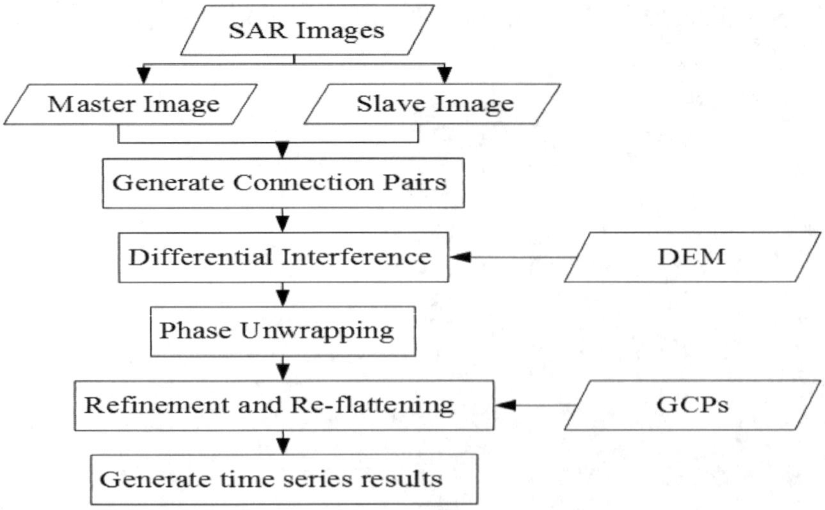

Figure 2. Flow chart of SBAS-InSAR method processing.

4. Analysis

4.1. Ground subsidence results of Wuhan

The ground subsidence rate (Line of sight direction) for Wuhan City from 2020 to 2021 obtained by SBAS-InSAR technique is shown in Figure 3. Positive values represent uplift and negative values represent decline in figure. A total of 4.57 million relevant target points were obtained for this experiment, and the maximum rate of ground subsidence in the study area was -57.08 mm/year. The ground subsidence rate at most of the relevant points in Wuhan is between -5~10mm/year, and the degree of ground subsidence is low.

In terms of spatial distribution, surface subsidence is unevenly distributed in Wuhan, and the deformation trend is northeast uplift and southwest subsidence. The main subsidence areas are located in Hongshan, Qiaokou, Wuchang, Jiang'an and Qingshan Districts. The subsidence areas in Jiang'an and Wuchang Districts are smaller, and the subsidence area in Qingshan District is the largest. The subsidence near Qingling is the most serious, and the ground subsidence rate exceeds 50 mm/yr.

4.2. Analysis of the impact of ground subsidence and urban construction

As the COVID-19 was gradually brought under control, various projects began to be built in Wuhan. Groundwater is drained during building construction operations, while buildings also compress the soil, causing the ground subsidence. Wuhan has been developing rapidly in terms of rail transportation, and the number of metro lines has been increasing. Wuhan Metro Line 5, Line 6 and Line 11 were opened one after another during the study period. Subway tunnel excavation will disturb the surrounding soil, which will easily cause ground subsidence problems.

The impact of metro construction on the ground subsidence in the Jiangan District is shown in Figure 4a, the subsidence rate in the Jiangan District is around -30mm/year. The construction of Metro Line 6 caused the formation of two subsidence funnels at the intersection of Metro Line 6 and metro line 12(Figure 4b). Figure 4c shows the subsidence of one of the subsidence funnels. The cumulative settlement reached 30mm from January 2020 to September 2021. We selected some feature points (PS1-PS4) in each subsidence location to be studied. The cumulative subsidence time change curve of the feature points shows that the cumulative subsidence near the PS2 point in Jiangan District is the largest, reaching nearly 60 mm. Overall the rate of subsidence accelerates during 2021. The subsidence changes coincide with the construction dates of Metro Line 6, it shows that the main reason of ground settlement in the Jian'an District is the construction of Metro lines

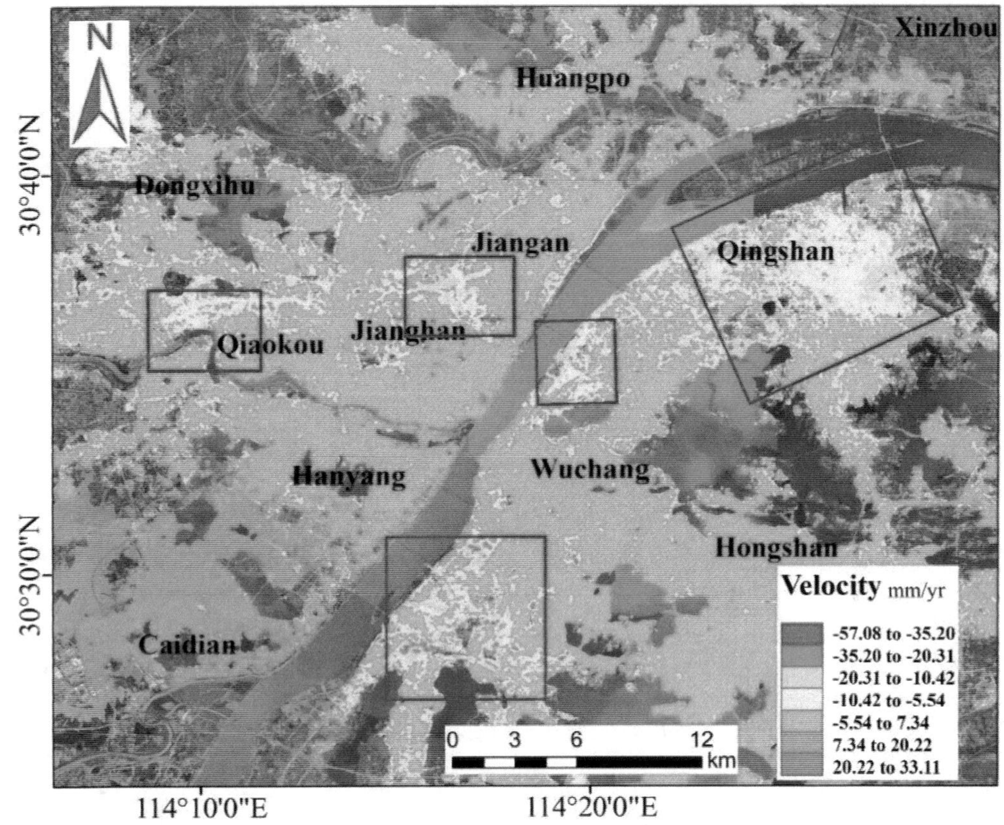

Figure 3. The annual average subsidence rate in Wuhan. (Major subsidence areas are in red boxes).

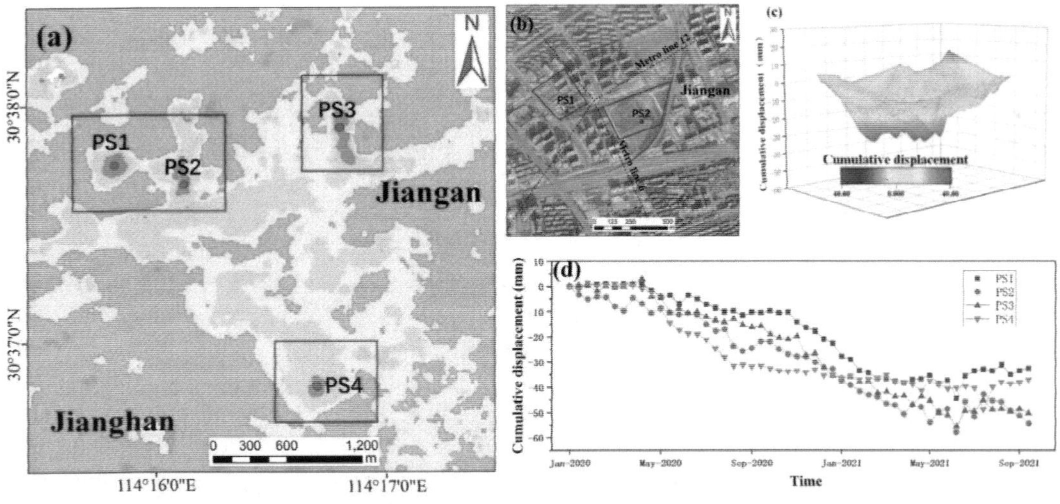

Figure 4. Land subsidence results of Jiangan District. (The red box is the subsidence area).

4.3. Analysis of the influence of ground subsidence and geological formations

Carbonate belts are widely distributed in Wuhan. Qiaokou District and Hankou District in Wuhan City are located in the Baishazhou karst belt. The Yangtze River in Wuhan area is connected to the groundwater layer, groundwater can erosion carbonate rocks, which is related to the subsidence along the Qiaokou coast and near Wuchang. Wuhan is located at the confluence of the Yangtze and Han rivers, where softsoils are widely distributed. The subsidence funnels in Qingshan and Hongshan districts are basically distributed on the soft soil layer. Mucky soil has poor bearing capacity, consolidation capacity and drainage. When the soil is not sufficiently consolidated before construction, the action of the gravity of the building will cause the soil to compress.

The ground subsidence in the Hongshan District is most serious due to the influence of the carbonate belt and soft ground (Figure 5a), and the maximum subsidence rate exceeds -50mm/year. Subsidence funnel in Hongshan District is located near Qingling, and Figure 5b and Figure 5c show the satellite images of the area. The People's Hospital of Wuhan University under construction in Figure 4c is the most serious area of subsidence, the subsidence rate reaches -52.9 mm/yr at the center of subsidence. Figure 5d shows the time variation curves of settlement for the five feature points (PS5-PS9) that we selected at the subsidence location. Tt shows that the settlement in the area of point PS6 is more serious, with the maximum cumulative settlement amount reaching nearly 100mm.

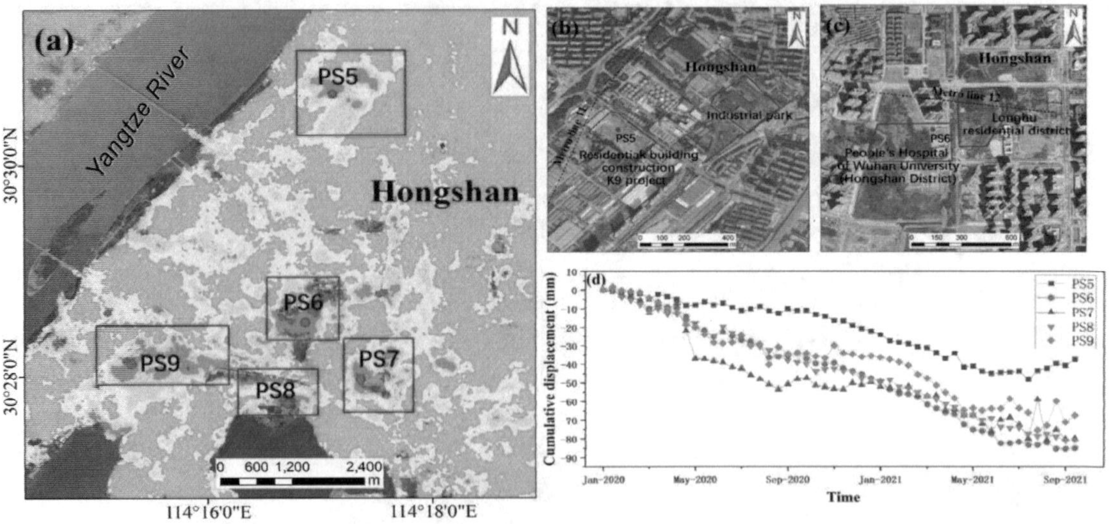

Figure 5. Land subsidence results of Hongshan District. (The red box is the subsidence area).

4.4. Analysis of ground subsidence and the effect of groundwater and rainfall

Excessive extraction of groundwater is one of the main reasons for ground subsidence. Figure 6a illustrates the water resources and rainfall in Wuhan in the last five years. After 2016, the annual rainfall in Wuhan has decreased, the total amount of water resources decreases, and the city experiences severe ground subsidence. After 2020, the rainfall in Wuhan City increases and the subsidence situation in urban areas is alleviated. In order to analyze the relationship between rainfall and ground settlement, the cumulative settlement at the feature points between 2020 and 2021 was compared with the monthly rainfall (Figure 6b), the gray lines in the figure indicate the cumulative sedimentation changes at the feature points of different sedimentation areas, and the red line is the rainfall amount. The variation trend of ground deformation in Wuhan is related to the rainfall. Wuhan has the most rainfall from May to September of the year, during which the ground subsidence rate decreases and the ground uplift slightly in some areas. After September, the rainfall decreased and the land subsidence rate began to accelerate. The results show that there is a relationship between rainfall and land subsidence. Rainfall can supplement groundwater and alleviate land subsidence.

In addition to rainfall, the change of water level will also affect the land subsidence in Wuhan. The water level changes recorded by Hankou observation station of the Yangtze River are shown in Figure 6c. Large changes in water levels during the flood season, and rising water levels in the Yangtze River will raise groundwater level along the coast and alleviation land subsidence. The seasonal variation of the Yangtze River water level is related to the rainfall in Wuhan. Overall, the change of water level has less influence on ground subsidence in Wuhan, and is more significantly influenced by the rainfall.

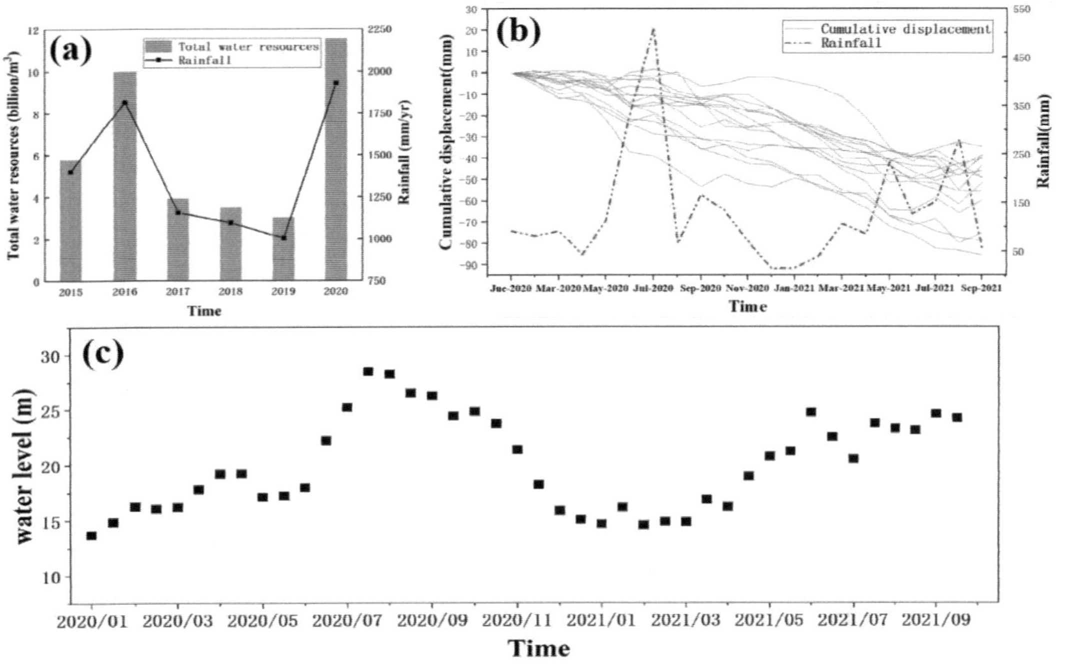

Figure 6. Relationship between rainfall, groundwater, water level and land subsidence.

5. Conclusions

We used the SBAS-InSAR method and 52-view Sentinel-1A TOPS images to acquire the surface deformation of Wuhan area from January 2020 to September 2021. The study shows that surface subsidence in Wuhan is not uniformly distributed, and there are mainly five subsidence areas with ground subsidence rates ranging from -57.08 to 33.1 mm/yr. The formation of most of the subsidence funnels is related to urban metro construction and building construction, and the construction of metro line 11 and metro line 5 has caused extensive ground subsidence along the lines. Wuhan's special geological structure is also a major factor leading to ground subsidence. Wide distribution of carbonate rock belt and large-area softsoils make karst ground collapse and uneven subsidence of foundation more likely to occur. The subsidence in the Qingling area, located on the Baishazhou carbonate belt, is the most severe, with the maximum accumulated subsidence reaching 90mm. Rainfall can change the water level of rivers. Large amounts of rainfall in Wuhan in summer can replenish groundwater resources and slow down ground subsidence. Changes in the water level will also have an impact on land subsidence in the area along rivers, but the impact is seen to be small. As more SAR satellites are launched, the variety and accuracy of radar data continues to improve, which will help us better monitor urban subsidence and study the mechanism of subsidence.

Acknowledgements
This work was supported by the Guangxi Universities Young and Middle-aged Teachers' Basic Scientific Research Ability Improvement Project (Grant No. 2020KY06032); the Guangxi Science and Technology Plan Project (Grant No. GUIKE AD19110107).

References

[1] Herrera-García G, et al. 2021 Mapping the global threat of land subsidence [J] *Science* **371** 34-36
[2] Devin L, Galloway, Thomas J 2011 Review: regional land subsidence accompanying groundwater extraction [J] *Hydrogeology Journal* **19(8)** 1459-1486
[3] Haghighi M H, Motagh M 2019 Ground surface response to continuous compaction of aquifer system in Tehran, Iran: Results from a long-term multi-sensor InSAR analysis [J] *Remote Sensing of Environment* **221** 534-550
[4] Cigna F, Tapete D 2021 Present-day land subsidence rates, surface faulting hazard and risk in Mexico City with 2014–2020 Sentinel-1 IW InSAR [J] *Remote Sensing of Environment* **253** 112161
[5] Qu F F, Lu Z, Zhang Q 2015 Mapping ground deformation over Houston–Galveston, Texas using multi-temporal InSAR [J] *Remote Sensing of Environment* **169** 290-306
[6] Gerardo H, Pablo E, Roberto T 2021 Mapping the global threat of land subsidence [J] *Science* **371(6524)** 34-36
[7] Liu Q, Yue G S, Ding X B, Yang K 2019 Temporal and spatial characteristics analysis of deformation along foshan subway using time series insar [J] *Geomatics and Information Science of Wuhan University* **44** 1099-1106
[8] Luo X J 2014 Division of "Six Belts and Five Types" of Carbonate Region and Control of Karst Geological Disaster in Wuhan [J] *Journal of Hydraulic Engineering* **45(02)** 171-179
[9] Fan S K 2006 A Discussion on Karst Collapse in Wuhan (Hubei) [J] *Resources Environment & Engineering* **2006(S1)** 608-616
[10] Bai L, Jiang L M, Wang H S 2019 Monitoring Ground Subsidence in Wuhan City with High-Resolution TerraSAR-X Images from 2013 to 2015 [J] *Journal of Geodesy and Geodynamics* **39(08)** 832-836
[11] Zhou L, Lu J J, Wen X L 2021 Spatial-temporal Characteristics Analysis of Ground Subsidence in Central Urban Areas of Wuhan Based on Small Baseline Subset Interferometric Synthetic Aperture Radar [J] *Science Technology and Engineering* **21(01)** 40-46
[12] Huang B J, Zheng F J, Bai J, Wang Y 2011 Feasibility of Land Surface Deformation Monitoring by Regional CORS [J] *Journal of Geomatics Science and Technology* **28(03)** 169-172
[13] Ferretti A, Prati C, Rocca F 2001 Permanent scatterers in SAR interferometry [J] *IEEE Transactions on Geoscience & Remote Sensing* **39(1)** 8-20
[14] Bai L, Jiang L M, Wang H S 2016 Spatiotemporal Characterization of Land Subsidence and Uplift (2009–2010) over Wuhan in Central China Revealed by TerraSAR-X InSAR Analysis [J] *Remote Sensing* **8(04)** 350-350
[15] Costantini M, Jie B, Malvarosa F 2016 Ground deformations and building stability monitoring by COSMO-SkyMed PSP SAR interferometry: Results and validation with field measurements and surveys [C] *Geoscience & Remote Sensing Symposium* **2016** 6847-6850
[16] Zhou L, Guo J M, Hu J Y 2017 Wuhan Surface Subsidence Analysis in 2015–2016 Based on Sentinel-1A Data by SBAS-InSAR [J] *Remote Sensing* **9(10)** 982-1003
[17] Xie P, Zhang S X, Zhou L 2021 Detection of the Urban Surface Deformation and New Strategy for Flood Prevention in Wuhan Central District [J] *Geomatics and Information Science of Wuhan University* **46(07)** 1015-1024

A comparative study of different deep learning models for land use and land cover mapping of flood detention basin

Nan Li[1], Jianwei Ma[1], Shifeng Huang[1,2], He Zhu[1], Yayong Sun[1] and Mengcheng Hu[1]

[1] Research Center of Flood and Drought Disaster Reduction of the Ministry of Water Resources, China Institute of Water Resources and Hydropower Research, Beijing, China.

[2] Corresponding author's e-mail: huangsf@iwhr.com

Abstract. Flood detention basin (FDB) is an important part of the flood control system in the basin. It is of great significance for scientific flood control to obtain the land use and land cover (LULC) classification map of FDB with higher accuracy quickly and accurately. In recent years, deep learning has shown great potential in LULC classification. In this study, we make LULC training dataset and explore three state-of-the-art (SOTA) DL architectures: Unet++, ResUnet++, DeepLab v3+ across Mengwa FDB. The experiments show all methods used in this study are available for LULC classification, which overall accuracy is ResUnet++(95.11%), Unet++(91.92%), DeepLab v3+ (91.04%) and kappa coefficient is ResUnet++(0.91), Unet++(0.85), DeepLab v3+ (0.83). Deep learning methods can be used for automatic extraction of LULC from high-resolution satellite images, providing data support for economic loss assessment and post disaster construction in flood storage areas.

1. Introduction

Flood detention basin (FDB) is an important part of the flood control system and an effective measure to ensure flood control safety and reduce disasters. Dynamic monitoring of LULC change in the FDB and a comprehensive understanding of the LULC situation in FDB are of great significance for coordinating the contradiction between social and economic development and flood control in the FDB, as well as the rational and scientific use and coordinated development of the FDB [1]. Remote sensing technology provides the most effective way for monitoring LULC [2]. Although many methods for LULC mapping based on remote sensing images have been developed, such as the classification and regression tree (CART) [3], support vector machine (SVM) [4], and random forest (RF) [5], the accuracy and efficiency of these approaches are insufficient to meet the increasing availability and resolution of remote sensing data.

In recent years, the potential of Deep Learning (DL) technology in mining massive remote sensing data has become increasingly prominent. With numerous training data, Deep Learning is able to extract very complex decision rules and has been successfully applied to LULC mapping[6][7]. Li et al [8] compared and studied the difference and extraction ability of CNN in land cover classification of five remote sensing images with different spatial resolutions (4 ~ 30 m) in the eastern part of Xining city. The overall accuracy was higher than 89%, and the higher the resolution, the higher the accuracy of classification results, indicating that CNN is more suitable for high-altitude resolution image classification. Karra et al [9] deployed a deep learning segmentation model on sentinel-2 data, created a global LULC map with a resolution of 10 meters, achieved excellent accuracy, and was able to automatically draw LULC maps from time series observation data.

However, there are few studies on LULC classification in FDB using in-depth learning, and there is still much room for improvement. Thus, in this study, we make LULC training dataset and explore three state-of-the-art (SOTA) DL architectures: Unet++ [10], ResUnet++ [11], DeepLab v3+ [12] across Mengwa FDB, which has shown excellent performances in several benchmarks of LULC classification [13], to evaluate the effect of DL approaches for monitoring of LULC classification in the FDB.

2. Materials and methods

2.1. Study area

In this paper, Mengwa FDB was selected as the study area. It is located by the Huaihe river in Funan County, Fuyang City, Anhui Province, with a total area of about 180.4km^2, a length of about 40km from east to west, and a width of 2~10km from north to south. It is a narrow and long zone trending southwest to northeast. Due to the warm temperate semi humid monsoon climate, there is concentrated and variable precipitation and frequent droughts and floods in the study area. The sluice was opened for flood storage 15 times in the 12 years from 1954 to 2007, which is the most frequently used flood storage area in the Huaihe River Basin. In Mengwa FDB, about more than 83% of the study area is occupied by crops. Build area comes second with 6%. Traffic embankment, accounting for about 4% ~ 5%. Forest land, canals, ponds and other areas share the remaining 1% [14]. The geographical location of Mengwa FDB is shown in Figure 1.

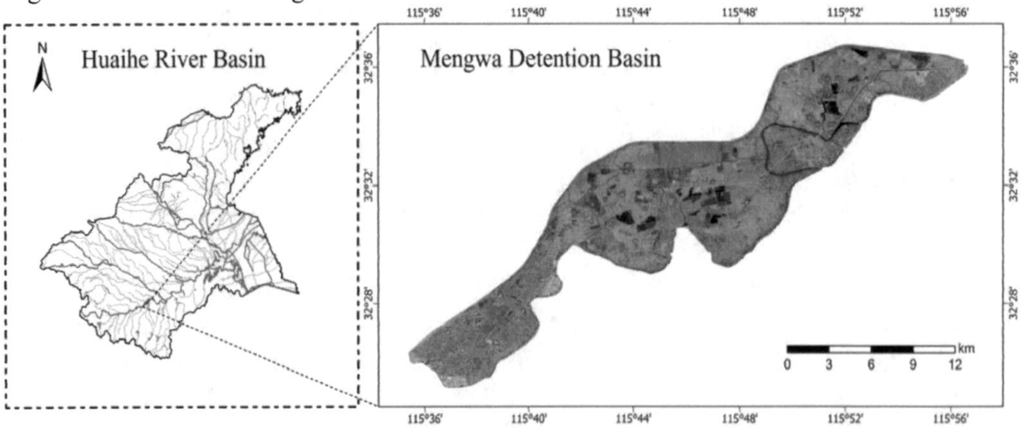

Figure 1. Location map of the Mengwa flood detention basin.

2.2. Data source and data processing

The GF-7 imagery at 0.65-m resolution was used in this study. The GF-7 constellation was launched into orbit on november3, 2019 and officially put into use in August 2020. The Gf-7 constellation is China's second-generation civilian high-resolution stereoscopic mapping satellite. It is equipped with double-line cameras (DLCs) that can obtain 0.65m backward image and 0.8m forward image, and a backward multispectral camera (MUX) that can obtain 2.6m multispectral images with four band: blue (450–520 nm), green (520–590 nm), red (630–690 nm) and near-infrared red (770–890 nm). [15].

In this study, five GF-7 images with good quality covering the study area were obtained through data screening, which are detailed in Table 1. The 2.6 m multispectral image and 0.65m panchromatic image were band fused by using gram Schmidt pan sharpening algorithm and ENVI software to obtain a 4-band 0.65m high-resolution image.

Table 1. Tile name and acquisition date of the GF-7 images covering the Mengwa FDB.

Tiles Name	Acquisition date
GF7_DLC_E115.6_N32.5_20210921_L1A0000568586	21 September 2021
GF7_DLC_E115.8_N32.5_20211223_L1A0000661807	23 December 2021
GF7_DLC_E115.8_N32.7_20211223_L1A0000661807	23 December 2021
GF7_DLC_E116.0_N32.5_20220307_L1A0000735573	7 March 2022
GF7_DLC_E116.0_N32.7_20220307_L1A0000735573	7 March 2022

2.3. *Training dataset*

Based on the preprocessed GF-7 image as the basic data, the sample database is formed through visual interpretation by ArcGIS software Feature Editing Tool. The main land use types including: built area, crop, road, water, trees and unused land. Due to the limitations of computer CPU and GPU memory, the whole image cannot be directly input into the network for training. Therefore, it is necessary to cut the preprocessed image and label into small images that can be processed by the computer. In this paper, the image and label are cropped into 512×512 pixels images by random clipping. Some images and corresponding labels are shown in Figure 2.

Figure 2. Pixels level label.

Data augmentation including horizontal flipping, vertical flipping, diagonal flipping and 90° counterclockwise rotation are performed on the dataset to improve the accuracy and robustness of the model. Each operation generates a new picture, and finally gets a sample of 500 images. The sample is randomly divided into training set, validation set and test set in the ratio of 7:2:1 as the input data of the deep learning model.

2.4. *Methods*

2.4.1. Network Architectures. Since Long et al [16] developed fully convolutional networks (FCN) for pixel-wise semantic segmentation, several network architectures for semantic segmentation have been proposed. FCN realizes end-to-end semantic segmentation by replacing the full connection layer with the convolution layer in the image classification task. It consists of three steps: multi-layer convolution for extracting feature map, deconvolution for restoring image resolution, and concatenation for fusing feature map. Researchers have improved FCN from three perspectives mentioned above. In order to evaluate and analyse the utilization of the existed semantic segmentation network in LULC classification for FDB, three methods were used: Unet++, ResUnet++, DeepLab v3+. These methods were introduced in the next paragraphs.

Unet++ is developed by Zhou et al. He improved the upsampling and skip connection methods on the basic UNet. The multi-layer UNets with dense connections are nested together to form Unet++, and the in-depth monitoring scheme is introduced to jointly train the output of each Unet. In Figure 3, $x^{i,j}$ represents the output of the image after passing through node $X^{i,j}$, i represents the down sampling layer index along the encoder path, and j represents the dense convolution layer index along the skip connection. ResUnet++ replaces the feature extractor in Unet++ with resnet34 with residual structure. The residual block was proposed by He et al [17], including an identity mapping and skip connection. This structure solves the problems of over fitting and gradient disappearance when there are too many layers of neural network.

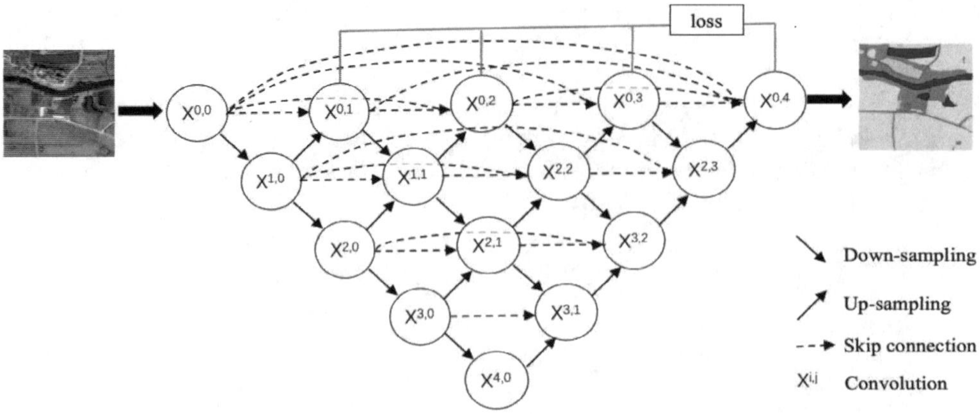

Figure 3. Unet++ architecture.

Deeplab model is proposed by Google in 2015. Till now, it has four versions: DeepLab v1, DeepLab v2, DeepLab v3 and DeepLab v3+. DeepLab v3+ is the latest network architecture in DeepLab model. It use encoder-decoder structure to extract feature and restore resolution of image. In the encoder stage, Xception is used as the feature extractor and Atrous Spatial Pyramid Pooling (ASPP) module is used to extract multi-scale features. In the decoder stage, image resolution is restored by upsampling. The architecture of DeepLab v3+ is shown in Figure 4.

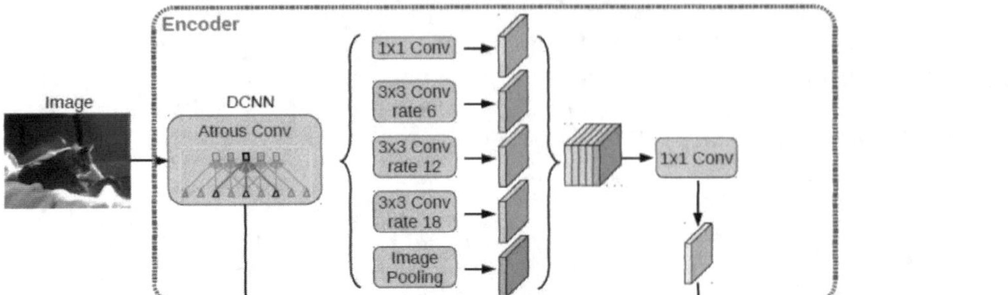

Figure 4. DeepLab v3+ architecture from the paper [12].

2.4.2. Loss function. In this study, dice loss and cross entropy loss function are used as the loss function in the process of deep learning training. Dice loss was proposed by Fausto et al [18] to alleviate the problem of data imbalance between positive samples and negative samples. Dice loss comes from dice coefficient. It is used to evaluate the similarity of two samples. The value range of dice coefficient is 0 to 1. The larger the value, the higher the similarity between the two samples. Dice coefficient is defined as follows:

$$dice = \frac{2 \cdot y \cdot y_{pred}}{y + y_{pred}}$$

In the above formula, y represents the label diagram, y_{pred} represents the model prediction diagram, and $L_{dice} = 1 - dice$.

2.4.3. Implementation details. The hardware environment of this paper is Intel (R) Xeon (R) gold 5222 processor, 128GB memory, NVIDIA GeForce RTX3090 graphics card and 24GB video memory. The deep learning software environment is pytorch1.8, cuda-11.1 and python 3.8. See the following table for the configuration of main super parameters of the network model (shown in Table 2)

Table 2. The configurations of main hyper-parameters.

Hyper-parameters	value
Epoch	100
Learning rate	1e-4
Gamma	1e-4
Optimizer	Adam
Batch size	8

3. Results

Figure 5 demonstrates the training loss tendency of the training process of Unet++, ResUnet++, DeepLab v3+ models. After 40 epochs, all three models have been sufficiently trained and acquired relatively low loss. Among them, the ResUnet++got the minimum training error (nearly 0.25) and the maximum valid accuracy (nearly 0.96).

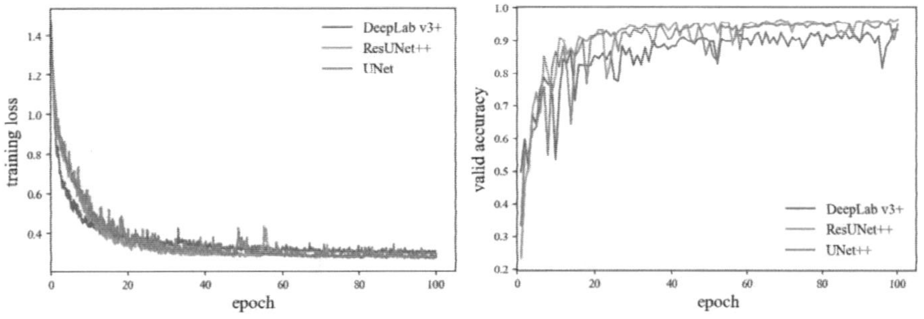

Figure 5. The training loss and valid accuracy of Unet++, ResUnet++, DeepLab v3+ for GF-7 images.

The classification results of several methods are shown in the Figure 6. It can be seen from the results that Unet++ method has a good result in water identification, but it has a poor distinction between roads and unused water. DeepLab v3+ is seriously missed in water identification. There is only a small amount of water leakage in the recognition results of the ResUnet++method. The difference between the recognition range of each feature type and the real range is small, the shape is basically consistent, and the integrity is high, which can accurately reflect the surface category and its boundary. To sum up, among the deep learning methods, the classification effect of ResUnet++ is better.

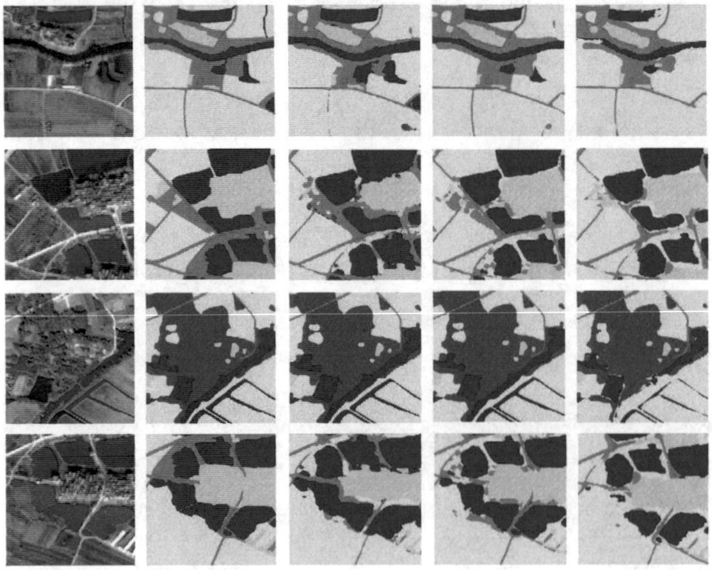

Figure 6. Classification results of different DL methods.

In order to quantify the effect of several classification methods, this paper uses the overall accuracy and kappa coefficient to measure the overall performance of the model, and uses F1 score and IOU to measure the classification accuracy of the model on different types. The four indicators can be calculated by the fusion matrix, and the results are shown in Table 3. Through comparison, the order of OA from big to small of the three methods is ResUnet++(95.11%), Unet++(91.92%), DeepLab v3+ (91.04%), and the order of kappa coefficient from big to small is ResUnet++(0.91), Unet++(0.85), DeepLab v3+ (0.83). The overall accuracy and kappa coefficient of the ResUnet++method are higher than those of the other methods. To sum up, ResUnet++has achieved high accuracy in the classification of land use types in high-resolution images, and can effectively distinguish different surface types in high-resolution images.

Table 3. Classification accuracy of different types and different methods.

Class	Unet++		ResUnet++		DeepLab v3+	
	F1	Iou	F1	Iou	F1	Iou
Built area	0.84	0.72	0.89	0.80	0.86	0.76
Crop	0.96	0.93	0.97	0.95	0.95	0.89
Road	0.75	0.60	0.86	0.76	0.80	0.67
Water	0.95	0.90	0.90	0.82	0.84	0.73
Unused	0.73	0.57	0.84	0.73	0.78	0.64
Trees	0.92	0.86	0.97	0.94	0.94	0.88
OA/%	91.92		95.11		91.04	
Kappa	0.86		0.91		0.83	

The classification result of ResUnet++ in Mengwa FDB is shown in Figure 7.

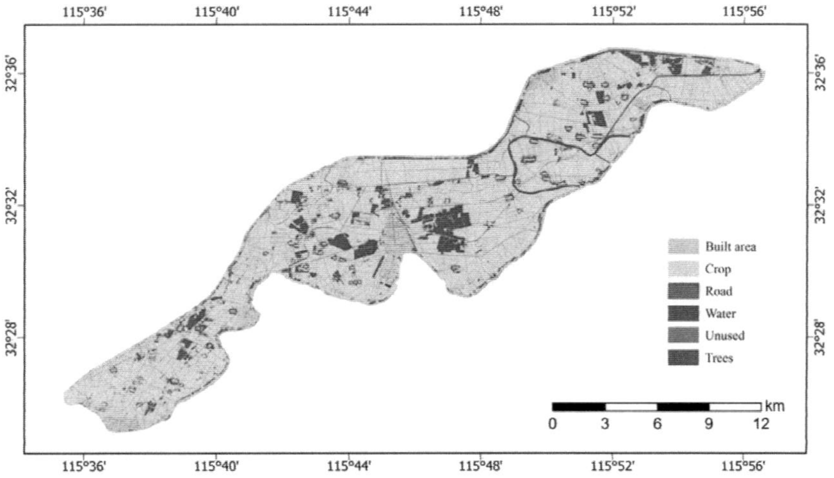

Figure 7. Land use and land cover classification using ResUnet++.

4. Conclusions

In this paper, based on GF-7 images, the land use and land cover classification of Mengwa FDB is carried out by using Unet++, DeepLab v3+ and ResUnet++ methods respectively. The experimental results show that among the depth learning methods, the classification effect of ResUnet++ is better, which overall classification accuracy is 95.11% and is 3% ~ 4% higher than Unet++ and DeepLab v3+, and the kappa coefficient is 0.91. The results indicating that the samples have high consistency with the classification results, and can meet the needs of classification accuracy.

Funding

This study is supported by the High-Resolution Earth Observation System Program of China (No. 08-Y30F02-9001-20/22); Jiangxi Province Key Research and Development Plan key project (20212BBG71008); Jiangxi Provincial Water Conservancy Science and Technology Project (202124ZDKT16).

References

[1] Cheng C, Guo Y, Luo W 2014 *Advanced Materials Research* 926-930
[2] Mora B, Tsendbazar N E, Herold M, Arino O 2014 *Land Use and Land Cover Mapping in Europe* 11–30
[3] Friedl M A, Brodley C E 1997 *Remote Sensing of Environment* **61**,399–409
[4] Melgani F, Bruzzone L 2004 *Remote Sensing* **42** 1778–1790
[5] Belgiu M, Dragut L D 2016 *Isprs Journal of Photogrammetry and Remote Sensing* **114** 24–31
[6] Zang N, Cao Y, Wang Y, Huang B, Zhang L, Mathiopoulos P T 2021 *IEEE Journal of Selected Topics in Applied Earth Observations and Remote Sensing* **14** 5372-5391
[7] Cohen N, Sharir O, Shashua A 2016 *arXiv: Neural and Evolutionary Computing* **49** 698–728
[8] Li H, Gao X, Tang M 2020 *Remote sensing technology and application* **04** 749-758
[9] Karra K, Kontgis C, Statman-Weil Z, Mazzariello J C, Mathis M M, Brumby S P 2021 *2021 IEEE International Geoscience and Remote Sensing Symposium IGARSS* 4704-4707
[10] Zhou Z, Siddiquee M, Tajbakhsh N, Liang J 2019 *IEEE Transactions on Medical Imaging* **39** *1856-1867*
[11] Jha D, Smedsrud P H, Riegler M, Johansen D, Lange T D, Halvorsen P, Johansen H D 2019 *2019 IEEE International Symposium on Multimedia (ISM)* 225-2255
[12] Chen L C, Zhu Y, Papandreou G, Schroff F, Adam H 2018 *Lecture Notes in Computer Science*

11211 833–851
[13] Rousset G, Despinoy M, Schindler K, Mangeas M 2021 *Remote Sensing* **13** 2257
[14] Zhang X, Lei T, Song Y, Yao Q, Zhou R, Li H 2019 *South-to-North Water Transfers and Water Science & Technology* **17(02)** 55-66
[15] Yang J, Wang C, Sun L, Zhu Y, Huang Y 2020 *Spacecr. Eng* **29** 61–67
[16] Long J, Shelhamer E, Darrell T 2015 IEEE Conference on Computer Vision and Pattern Recognition **7–12** 3431–3440
[17] He K, Zhang X, Ren S, Sun J 2016 *Computer Vision – ECCV 2016* 630–645
[18] F Milletari, Navab N, Ahmadi S A 2016 *Fourth International Conference on 3D Vision (3DV)*

IOP Conf. Series: Earth and Environmental Science 1087 (2022) 012045 doi:10.1088/1755-1315/1087/1/012045

Spatiotemporal characteristics analysis of ground subsidence in Wuhan based on SBAS-InSAR

Weiping Zhang[1], Lv Zhou[1], Lilong Liu[1,2] and Mingming Zheng[1]

[1] College of Geomatics and Geoinformation, Guilin University of Technology, Guilin541004, China.

[2] Corresponding author's e-mail: hn_liulilong@163.com

Abstract. Aiming at the problem of uneven ground subsidence in Wuhan, using the Sentinel-1A SAR image data of 36 scenes from July 2017 to June 2020, the SBAS-InSAR was used to extract the ground subsidence information in Wuhan. Influence of factors such as urban construction, industrial development, rainfall, and carbonate karstification on subsidence. The results show that SBAS-InSAR can effectively monitor the ground subsidence in Wuhan. There is a high correlation between ground subsidence and seasonal changes, and the subsidence time series shows a nonlinear downward trend with seasonal changes. Human activities have a serious impact on ground subsidence, and the spatial distribution of the subsidence funnel area is closely related to the geographical location of the industrial area. Among natural factors, rainfall plays a dominant role in the seasonal component of ground subsidence, and karstification has a greater impact on the area along the river. The two influencing factors jointly caused the uneven subsidence of the ground in Wuhan. The experimental results can provide a certain reference for the follow-up subsidence early warning and urban construction planning in Wuhan.

1. Introduction

Ground subsidence has become one of the geological disasters on a global scale, posing a major threat to human property [1]. Interferometric synthetic aperture radar (InSAR)technology is widely used in urban ground deformation monitoring. It improves the disadvantages of long observation time and high cost of traditional measurement methods, and can realize high-precision and large-scale earth observation. [2]. However, the bad phenomenon in the application of InSAR technology will cause the signal to close correlation and reduce the accuracy of Earth observation. To overcome limitations in signal de-correlation, scholars have developed time-series InSAR (TS-InSAR) technology [3]. Small baseline subset interferometric synthetic aperture radar (SBAS-InSAR) technology is a common method in the TS-InSAR technology, it can achieve a high-precision deformation map at the mm level. At present, TS-InSAR technology is commonly used in ground monitoring in various cities, such as ground subsidence monitoring in Xi'an [4], Beijing [5], California [6].

In 1931, ground subsidence was first discovered in Wuhan, and the subsidence directly caused the Baishazhou area to be submerged by the breach of the Yangtze River embankment. In 2000, Qingling experienced the largest subsidence in Wuhan. By 2020, the maximum accumulated subsidence in Wuhan has reached 0.88 meters. Wuhan geomatics Institute has perfected the ground subsidence monitoring network in the central urban area, but it is difficult to grasp the regional ground deformation information and the distribution of subsidence funnels due to the low sampling rate of leveling. Therefore, comprehensive and stable data and methods are required for ground subsidence monitoring.

IOP Conf. Series: Earth and Environmental Science 1087 (2022) 012045 doi:10.1088/1755-1315/1087/1/012045

Costantini et al [7] used the high-resolution X-band COSMO-SkyMed images in the Hankou area from June 2013 to June 2014 to monitor and analyze the ground subsidence. Bai Lin et al [8] analyzed the subsidence spatiotemporal characteristics of Wuhan through 27 TerraSAR images from 2013 to 2015. During the study period. In 2017, Zhou Lu et al [9] used 15 Sentinel-1A images (April 2015 to April 2016) to extract the ground deformation information of Wuhan. This study uses 36 Sentinel-1A SAR images in the period from July 2017 to June 2020 to monitor the ground deformation in Wuhan based on the SBAS-InSAR technology, combining industrial development, carbonate karstification and precipitation and other influencing factors to study the characteristics of ground subsidence.

2. Study area and data source

2.1. Study area

Wuhan is a central city in the central region, located in the transition area from the south-eastern hills to the Dabie Mountains (Figure 1). Low North high terrain South, and a residual hilly impact plain in the middle, annual precipitation of not less than 1000mm. The city is divided into three principal urban areas, Hanyang, Hankou and Wuchang, by the confluence of the Yangtze River and Han River. The area of water reaches 2,217.6 km^2, accounting for 26.1% of the total area of the city. Wuhan is crisscrossed with rivers, and the groundwater is closely related to the groundwater of the Han River and the Yangtze River, and the geological and environmental conditions are complex. As an important industrial base, science and education base and comprehensive transportation hub in China, the urbanization process is developing continuously. The construction of ground buildings, subways, high-speed railways and other large-scale infrastructure has induced problems such as ground settlement, resulting in several uneven areas of settlement in Wuhan.

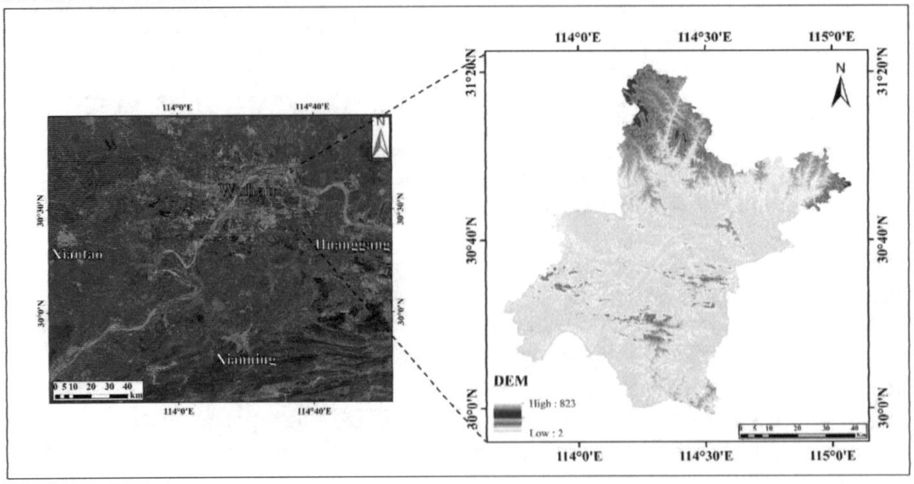

Figure 1. Location of Wuhan, China.

2.2. Data source

In this study, thirty-six SAR images (C-band) acquired by the Sentinel-1A satellite launched by the European Space Agency. The time range is from July 17, 2017, to June 7, 2020, and the monitoring period is 3 years. The Sentinel-1A SAR image data in this area has an upper orbit, a polarization mode of VV, an oblique range resolution of 2.3 meters, and an azimuth resolution of 13.9 meters. This study selects the SRTM3 elevation data released by NASA and the precise orbit data provided by the European Space Agency. Combined with the obtained precipitation data released by the Wuhan Meteorological Observatory from 2017 to 2020, a correlation analysis was carried out on the ground subsidence.

3. Methodology

SBAS-InSAR technology combines SAR image data into several subsets, uses the least squares method to solve the deformation time series of the subsets, and uses the singular value decomposition method to solve the deformation time series of the subsets. method (singular value decomposition, SVD) to jointly solve multiple small baseline sets to obtain the deformation sequence of the entire time span [10].

Considering a set of $N+1$ SAR images of the target object at acquisition time $t_0 \cdots t_N$ .the number of differential interferograms M is

$$\frac{N}{2} \le M \le \frac{N(N-1)}{2} \tag{1}$$

The j (j is an arbitrary natural number) differential interferogram generated (Figure 2) by interfering the SAR images at time t_A and time t_B, ignoring the flat phase and the interferometric phase of the terrain phase that removes DEM data errors. The interference phase at any point in the azimuth direction x and the distance direction r in the first interferometric difference image after unwrapping is:

$$\delta\phi_j(x,r)=\phi(t_B,x,r)-\phi(t_A,x,r)\approx\frac{4\pi}{\lambda}\left[d\left(t_B,x,r\right)-d\left(t_A,x,r\right)\right]+\delta\phi_j^{topo}(x,r)+\delta\phi_j^{atm}(x,r)+\delta\phi_j^{noise}(x,r) \tag{2}$$

Where $\phi(t_A,x,r)$ and $\phi(t_B,x,r)$ represent phase values of SAR images at time t_A and time t_B , respectively

$\delta\phi_j^{topo}(x,r)$ corresponds to residual phase due to inaccuracies in reference DEM. $\delta\phi_j^{atm}(x,r)$ depicts atmospheric phase error. $\delta\phi_j^{noise}(x,r)$ denotes random noise phases, $d(t_A,x,r)$ and $d(t_B,x,r)$ are respectively the deformation amount in the LOS direction that starts to accumulate at the start time t_0 at time t_A and time t_B .

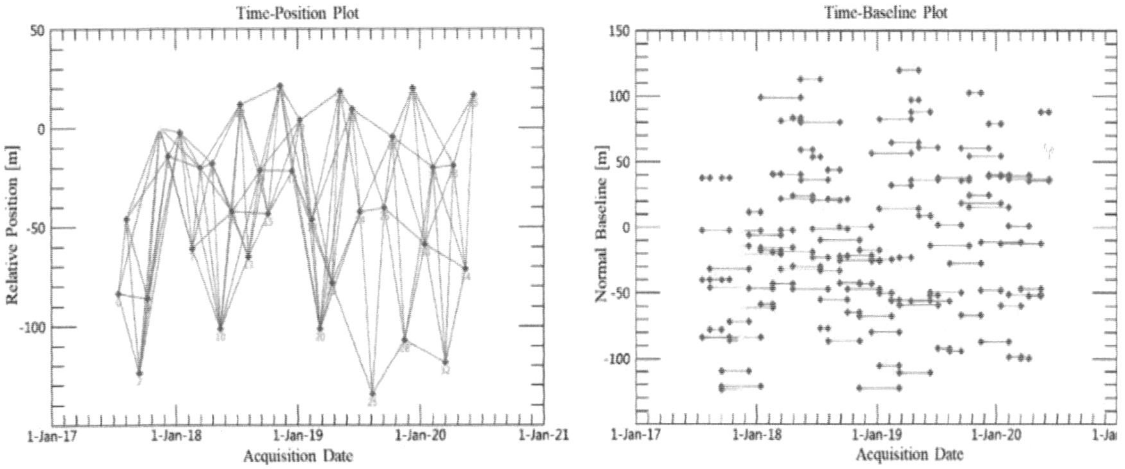

Figure 2. Time-position of Sentinel-1A image interferometric pairs and time-baseline of Sentinel-1A image interferometric pairs.

Assuming that the time period from A to B is divided into a plurality of adjacent continuous time periods that change at a constant speed, $V_{i,i+1}$ is the phase average deformation rate of the i th time interval. The cumulative deformation variables in the A 、 B time periods, are expressed as:

$$\delta\varphi_j^{disp}(x,r)=\frac{4\pi}{\lambda}\sum_{i=t_A}^{t_{B-1}}V_{i,i+1}\left(t_{i+1}-t_i\right) \tag{3}$$

Generalizing formula (3) to all interference difference maps, it can be expressed in matrix form as:

$$Av=\delta\varphi \tag{4}$$

The deformation rate is solved by the least squares method. Among them Δt is the cumulative time from the initial observation t_0 to the final observation t_B, and $\overline{v}$ is the average deformation rate during the observation period:

$$\overline{v} = \frac{\sum_{i=1}^{B} v_i \left(t_{i+1} - t_i\right)}{\Delta t} \frac{\lambda}{4\pi} \tag{5}$$

Finally, the corresponding deformation time series can be derived according to the time span between SAR acquisitions. Figure 3 shows the main steps of SBAS-InSAR method to detect ground subsidence.

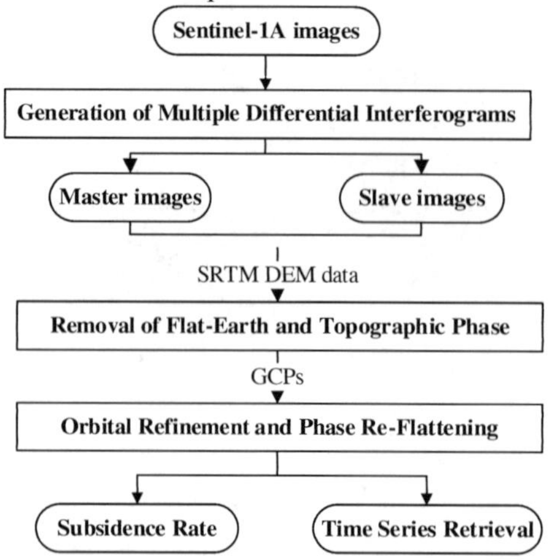

Figure 3. Flowchart of SBAS-InSAR data processing.

4. Results

4.1. Subsidence Rate Map

Figure 4 shows the 1,048,575 correlated target points were identified by the SBAS-InSAR method in the study area, and the average density of correlated target points reached 123 points per square kilometre. The annual ground sedimentation rate in the Wuhan ranged from -42.9 to 45.6 mm/ yr from 2017 to 2020. The ground sedimentation was unevenly distributed, with the sedimentation rate in most smooth areas ranging from -9.4 to 3.2 mm/ yr, and the sedimentation phenomenon was obvious in a few areas. Which the maximum settlement rate is in the Xingouqiao area of Qingshan District.

The main subsidence is distributed in four areas (Figure 5), which are located in the central city of Wuhan (S1 & S2), Qiaokou District (S3), and Qingling Township (S4). Large heavy industrial bases are concentrated in Qingshan District, such as Wuhan Iron and Steel Group, Sinopec, and Qingshan Thermal Power Plant. During the study period, Wuhan Metro Line 21 and Metro Line 5 were under construction, and excessive groundwater extraction caused soil compression and thus formed a subsidence funnel, while other subsidence areas were located along the Yangtze River.

4.2. Evolution of Ground Subsidence

The cumulative subsidence time series of the study area spanning three years was calculated using SBAS-InSAR analysis technique for SAR images starting on July 17, 2017, maximum cumulative subsidence is -118 mm (Figure 6). The subsidence of the central and eastern regions of the study area is relatively stable, with cumulative subsidence concentrated between -6 mm and 9mm. The extent and magnitude of subsidence in the study area did not change much between March and June in 2018 and 2019, and ground subsidence showed a gradual moderating trend from July to October 2018 compared with November and December of the same year. The rainy season in the Wuhan is from May to July each year, and the increased precipitation in May and June of 2019 effectively replenished the groundwater in the Wuhan with large amount of precipitation, which appropriately alleviated the groundwater exploitation in the Wuhan.

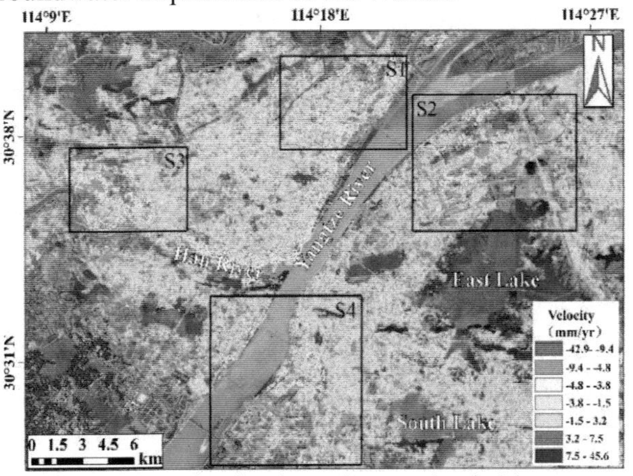

Figure 4. Distribution of ground subsidence monitoring in Wuhan.

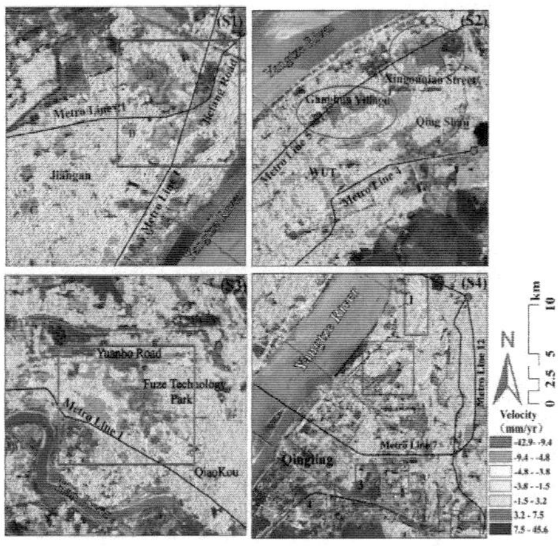

Figure 5. Settlement of four characteristic areas.

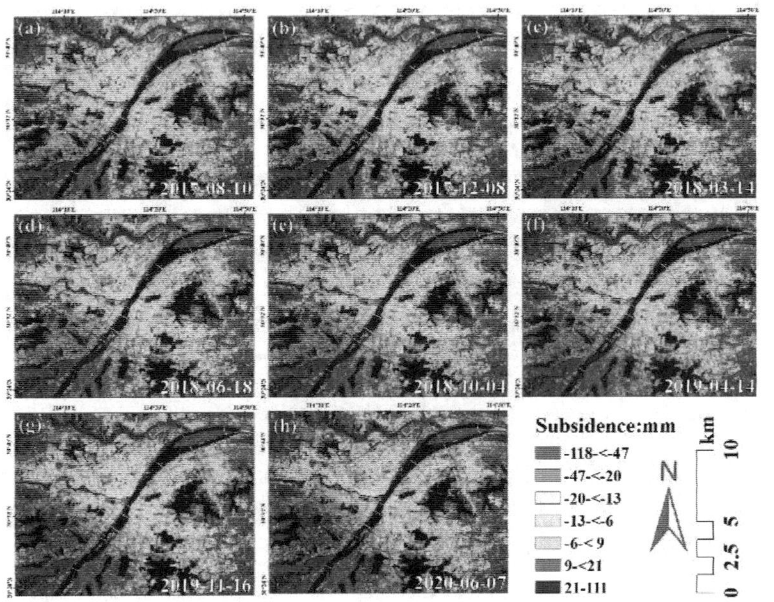

Figure 6. Time series of cumulative subsidence from 2017 to 2020 in Wuhan.

This study analyses the subsidence time-series results for four characteristic areas in the study area, and the cumulative subsidence of each area is shown in Figure 7. Which the ground subsidence trend in the S1 area is relatively gentle and stable. subsidence area. The subsidence area in the Jiangan area (S1) and Qiaokou area (S3) are mainly distributed in the Houhu area of Hankou. The subsidence funnel of Xingouqiao and Ganghuacun Street in Qingshan District (S2) are significantly larger than the other three subsidence areas in terms of surface subsidence rate and subsidence area. Hongshan District (S4) is a high-prone area for ground subsidence, and the subsidence area is distributed around the Baishazhou carbonate rock belt. The cumulative subsidence in Qingling Township reached 100mm, where the largest scale of subsidence ever recorded in Wuhan occurred.

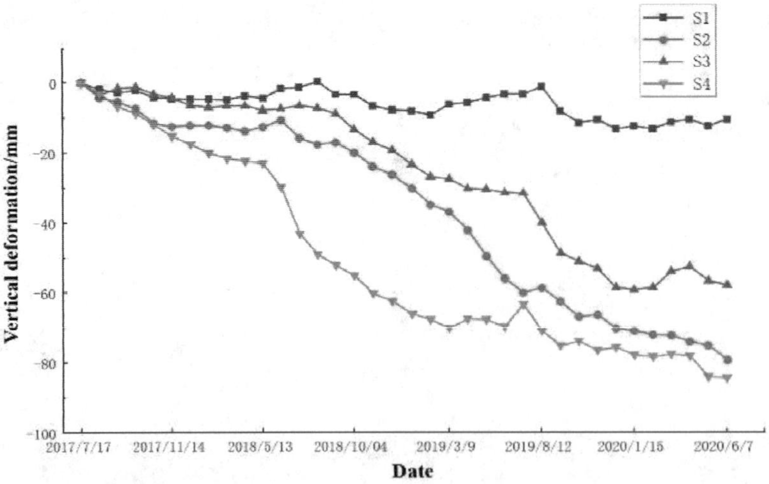

Figure 7. Cumulative settlement in each area.

4.3. InSAR Data Validation

To evaluate the inner coincidence accuracy of the subsidence rate field processed by SBAS-InSAR technology for Sentinel-1A data in the Wuhan. This section performs a statistical analysis of the standard deviation of the resulting sedimentation rates. According to the statistical results of the standard deviation of the subsidence rate of each monitoring target point in the study area from 2017 to 2020, a total of 1,048,575 relevant target points were detected, and the standard deviation of the study was 1.4mm/ yr, of which StD of the maximum subsidence average rate reached 2.3mm/ yr (Figure 8). Which the standard deviation of the average subsidence rate of 83.3% of the relevant target points is in the range of 0 ~ 1.5mm/ yr. In summary, the SBAS-InSAR technology has high reliability and stabilityin the monitoring of ground subsidence in the Wuhan.

5. Discussions

5.1. Ground Subsidence Associated with Industrial Development

Several large industrial zones in Wuhan are located in Qingshan District, including Wuhan Iron and Steel Group, Sinopec, and Qingshan Thermal Power Plant. Which WISCO Industrial Park is located in the most serious area of settlement, with a settlement rate of -40 mm/ yr. The settlement funnel of WISCO Industrial Park is basically consistent with its geographical distribution, and there is a phenomenon of severe local settlement. In this study, two representative settlement profiles are selected for settlement analysis, and the results are presented in Figure 9, shows that the Wuhan Chemical Industry Zone has gradually formed into an obvious subsidence funnel, with the maximum subsidence of -78 mm in the subsidence profile. Figure 10 shows that there are two larger subsidence funnels distributed in WISCO industrial park, and the maximum subsidence of the subsidence profile is more than -61mm. In general analysis, many large industrial parks are concentrated in Qingshan District, and the production activities in the industrial area require a large amount of groundwater extraction which will lead to ground subsidence while the reduction of pore water pressure in the soil aquifer will make the subsidence uneven. In addition, there is a high correlation between the spatial distribution of ground sedimentation funnel and the geographical location of industrial zones.

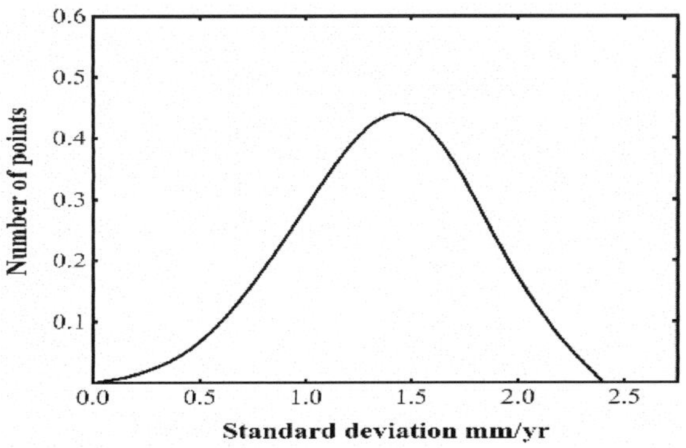

Figure 8. Distributions of standard deviations of subsidence rates.

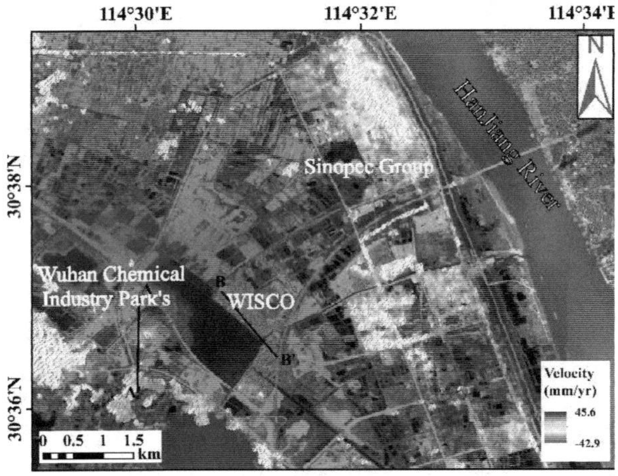

Figure 9. The vertical deformation rate in Wuhan industrial area.

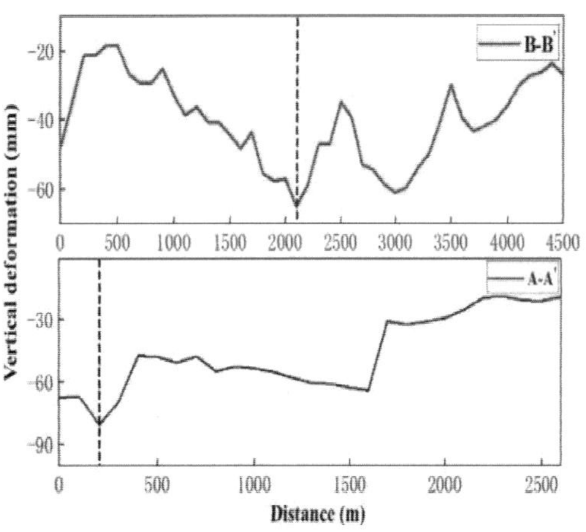

Figure 10. The vertical cumulative settlement in settlement profiler.

5.2. Ground Subsidence Associated with Precipitation

To analyse the relationship between precipitation and ground sedimentation in Wuhan, four PS points distributed in different regions were selected in this study, and the sedimentation time series of the four PS points during the study period were compared and analysed with the monthly average precipitation. The results are presented in Figure 11(a) The vertical shape variables of the four PS points accumulate with time and show a non-linear decreasing trend with the seasonal changes as a whole. Owing to the different geographical locations of the four PS points, the accumulated vertical form variables are not exactly the same. Point A (Figure 5) near the back lake area has the largest cumulative settlement among all the PS points, with a maximum settlement of -50 mm, which is related to the poor loading capacity of the silty soil layer in the back lake area.

By comparing the vertical deformation variables and precipitation during the study period, it can be found that the ground settlement in the study area is related to seasonal changes, and the vertical deformation variables have the same trend as the precipitation within a certain time threshold. The ground subsidence at the four points remained stable from April to July 2019, and the precipitation in the Wuhan during the corresponding time period was at its maximum throughout the year, and a large amount of precipitation effectively replenished the groundwater resources, slowed down the rate of ground subsidence, and the accumulated subsidence also had a tendency to lift. After August, when the precipitation drops and the temperature continues to rise, the domestic water use in the study area increases significantly and requires continuous groundwater extraction as a supplement, thus leading to accelerated ground subsidence in Wuhan. To confirm the relationship between precipitation and ground subsidence, a linear regression analysis was performed on the cumulative subsidence and the average monthly precipitation in the study area shown in Figure 11(b) There is a high correlation between the cumulative ground deposition and the average monthly precipitation, with the correlation coefficient reaching 0.7. From the above analysis, it is concluded that ground subsidence in the study area is accompanied by a seasonal trend, and precipitation plays a dominant role in the seasonal component of ground subsidence.

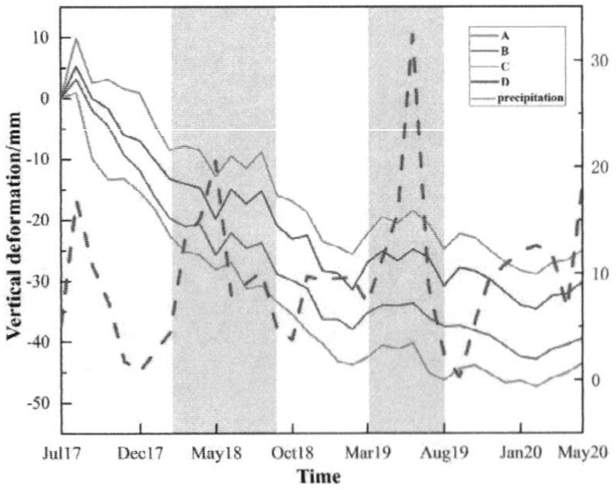

Figure 11(a). Subsidence time series versus average monthly precipitation.

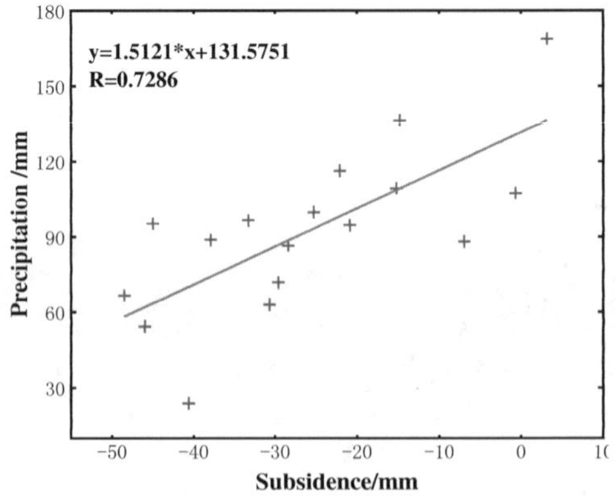

Figure 11(b). Regression analysis of monthly mean precipitation.

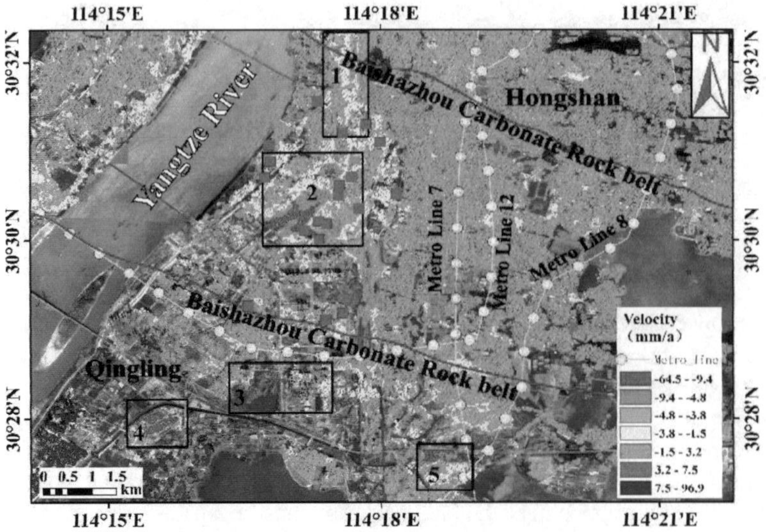

Figure 12. The vertical deformation rate in S4 area.

5.3. Ground Subsidence Associated with Carbonate Karstification

Carbonate karsts are widely distributed in Wuhan, and the first-order terrace-covered karsts are distributed along both banks of the Yangtze River, mostly concentrated in the Baishazhou and Hannan belts. Since 1931, 29 ground subsidence have occurred, and 22 of them have occurred in the S4 area [11] (indicated by the red square in Figure 12). The carbonate rock belt of Baishazhou passes through the S4 area from north to south, and the fragments of the fluid-plastic clay interlayer are directly distributed on the carbonate rock. Due to their own characteristics, crushed stone and flow-plastic clay can easily enter the underlying karst channel under natural subsidence and seepage, thereby causing the karst ground to collapse. It can be seen from Figure 12 that many subsidence areas and ground subsidences are located in the Baishazhou carbonate rock belt.

In the S4 area, 1-5 are the obvious subsidence areas, and the most serious subsidence is area 3, the subsidence rate reaches -34mm/ year, and the rest subsidence rates are distributed between -10 ~ -25mm/ year. During the study period, Metro Line 11, Metro Line 12 and river-crossing tunnels were all constructed in this area. To ensure the smooth implementation of the project, a large amount of groundwater needs to be pumped. Active water level changes will accelerate groundwater circulation and promote the development of underground karst caves, which can easily lead to ground subsidence and collapse. From the above analysis, it can be concluded that the ground subsidence in the S4 area is mainly affected by carbonate karst, and the construction of subways and river-crossing tunnels also affects the ground subsidence in the S4 area.

6. Conclusions

The analysis of the spatiotemporal characteristics of ground subsidence in Wuhan based on SBAS-InSAR through ground deformation data in Wuhan is summarized as follows:

1) Through the accuracy test results, it is concluded that the standard deviation of the average velocity in the Wuhan obtained based on the SBAS-InSAR is 1.4 mm/ yr, and 83.3% of the relevant target points had a standard deviation of the sedimentation rate in the range of 0 to 1.5 mm/ yr. From the above results, it can be seen that SBAS-InSAR has high reliability and stability in the monitoring of ground subsidence in Wuhan.

2) The subsidence distribution in Wuhan is uneven, the annual ground subsidence rate is -42.9 ~ 45.6mm/ yr, which the ground subsidence trend in the Jiangan area is relatively gentle and stable. The subsidence funnel of Xingouqiao and Ganghuacun Street in Qingshan District are significantly larger than the other three subsidence areas in terms of surface subsidence rate and subsidence area. The subsidence area in the Jiangan area and the subsidence area in the Qiaokou area are mainly distributed in the Houhu area of Hankou. Hongshan District is a high-prone area for ground subsidence, and the subsidence area is distributed around the carbonate strip of Baishazhou.

3) Ground subsidence in the study area is seriously affected by urban construction and industrial development, and the subsidence funnel of large industrial parks is basically consistent with its geographical distribution. There is a high correlation between ground subsidence and seasonal changes. Precipitation plays a leading role in the seasonal component of ground subsidence in Wuhan. Carbonate karstification also has an important impact on ground subsidence.

Natural conditions provide the basis for subsidence making subsidence possible, and human activity is the driving factor, allowing subsidence to occur. Our results provide insights into the causes of ground subsidence in Wuhan. Continuous monitoring of surface deformation is of great significance for the management of future subsidence in Wuhan and for mitigating potential hazards in this dense urban area.

References

[1] Peng G, Zheng Y C, Ai J J 2020 *IOP Conf. Ser.: Earth Environ* **569** 012045
[2] Han Y, Zou J, Lu Z 2020 *J. Remote Sensing* **12** 3788
[3] Xu X B, Ma C, Shan X J 2020 *J. Journal of Geo-information Science* **22** 2425
[4] Zhang Y M, Wang P, Luo X 2017 *J. Bulletin of surveying and mapping* **4** 93
[5] Liu K, Gong H L, Chen B B 2018 *J. Journal of Geo-information Science* **20** 128
[6] Costantini M, Jie B, Malvarosa F 2016 *Geoscience & Remote Sensing Symposium*
[7] Bai L, Jiang L M, Wang H S 2019 *J. Journal of Geodesy and Geodynamics* **39** 832
[8] Zhou L, Guo J M, Hu J 2017 *J. Remote Sensing* **9** 982
[9] Hu B, Wang H S, Sun Y L 2014 *J. Remote Sensing* **6** 3648
[10] Zhou L, Guo J M, Li X 2016 *J. Journal of Geodesy and Geodynamics* **36** 793
[11] Luo X J 2014 *J. Journal of Hydraulic Engineering* **45** 171

IOP Conf. Series: Earth and Environmental Science 1087 (2022) 012046 doi:10.1088/1755-1315/1087/1/012046

Kalman filtering for SAR satellite yaw steering

Tao Chen[1,3], Yu Li[1], Zheng Fang[1], Qiming Qiu[1], Xiaochen Yin[1], Ting Yang[1] and Hengyang Zhang[2]

[1] Key Laboratory on Avionics Integration, Aviation Industry of China (AVIC), Shanghai, China;
[2] Institute of Navigation and Information, Airforce Engineering University, Xi'an, China.

[3] Corresponding author's e-mail: tchen2avic@yeah.net

Abstract. The algorithm based on Kalman filtering was proposed to calculate yaw steering angles of spaceborne Synthetic Aperture Radar (SAR) systems. We establish the process equation and the observation equation by proper design. Performance analysis and computer simulations demonstrate that Kalman filtering may take advantages in both numerical precision and computational efficiency. It therefore can be applied to SAR satellites in a wide range of microwave bands (C, X, Ku) along noncircular orbits. In addition, an efficient algorithm based on simple interpolation may be employed in practice. Since the Kalman filter is linear in nature, the fast computation makes possible the implementation of accurate onboard processing.

1. Introduction

As well known, Doppler centroid is an important parameter of SAR systems. To a C-band radar, it may vary significantly along the satellite orbit. This property makes SAR data processing more challenging. To reduce the complexity, yaw steering was proposed in [1] and then implemented by the Earth Remote Sensing Satellite (ERS-1) [2]. The basic idea of doing so is to adjust satellite attitudes to stabilize Doppler centroid of radar echoes. Consequently, yaw steering calculation should be the challenge we have to face. To circular orbits, Raney [3] gave out the classic formula, which is accurate and in closed form. However, to SAR satellites along noncircular orbits [4], for example, SEASAT and ERS-1, this method may generate large bias. Many pioneer researchers endeavor to attack this issue. Among them, Guohua Wang [5] et. al. from Beijing University of Aeronautics and Astronautics (BUAA), derived an accurate computation to yaw steering. Their works employed appropriate approximations, this leads to a precision of 1e-2 Hz for X-band radars. Moreover, the algorithm by BUAA is mathematic tedious and hard to implement. To pursue better performance, nonlinear Kalman filtering [6] was proposed to yaw steering calculation. There are two reasons why Kalman filters may take advantages. First, Kalman filtering based algorithm may have better numerical precision compared with that of BUAA. The second reason is more important. Specifically, in the work of BUAA, they do not consider the errors by the satellite orbit as well as the errors by spacecraft attitudes. Zhongquan Wei et. al. [7] indicated that it is possible to compensate deterministic errors by proper system design. However, to random errors, it is very difficult to compensate. As a consequence, Kalman filters may take advantages by smoothing out those random errors. Extended Kalman Filters (EKF) may have good performance by linearization. Unscented Kalman Filters (UKF) uses unscented transform to estimate mean and variance of nonlinear multivariate transformation. Compared with BUAA's work, nonlinear Kalman filtering may achieve a yaw steering precision at 1e-12 Hz for the SEASAT parameters. Though EKF and UKF work well enough, the drawback of them is the increased computational complexity, thus making the onboard implementation very tough. The challenge here is to design stable and efficient algorithms based on linear Kalman filtering which is easy to engineering implementation and consequently suitable to onboard processing.

In the development of the algorithms, we try to avoid nonlinear models when possible. This can be achieved by proper parameter transform. As a consequence, we may establish both the process (state) equation and the observation equation in the linear forms. Classic Kalman filter procedures can be implemented with only trivial modifications. Since linear Kalman filtering is much more computationally efficient than nonlinear filters, it is particularly suitable to onboard processing. As can be shown, the linear Kalman filter may calculate very fast with superior precision. In addition, we also design a simple algorithm based on interpolation. This algorithm makes very much sense to engineering practice for its ease to implement. Guaranteed precision can be achieved with spline functions.

The organization of the work is as follows. In section 2, we describe the system model. In section 3, we propose the linear Kalman filter algorithm to yaw steering calculation. In section 4, we propose an efficient algorithm based on simple interpolation. We present the numerical results in section 5 and draw the conclusion in section 6.

2. System Model

Consider a spaceborne SAR system. The satellite is flying along the orbit by design. For mathematic convenience, we use the vector $\vec{R}_s$ to represent the position of the satellite, the vector $\vec{R}_t$ to represent the position of the target. The range between the radar and the target has the form

$$R = |\vec{R}_s - \vec{R}_t|$$

At the radar receiver, the Doppler frequency measured is given by [8]

$$f_d = -2\dot{R}\frac{f_0}{c}$$

in which, c represents the speed of radio wave propagation, f_0 represents the carrier frequency of the radar transmitter. The relative velocity between the radar and the target is further given by

$$\dot{R} = \dot{R}_s\cos(\gamma) - R_s\omega_s\sin(\gamma)\cos(\psi) + \omega_e R_s\sin(\gamma)[\sin(\psi)\cos(\alpha)\sin(\alpha_i) + \cos(\psi)\cos(\alpha_i)]$$

in which, $\dot{R}_s$ represents the satellite velocity along the orbit radius, γ represents the elevation angle (look angle) of the radar antenna, R_s represents the orbit radius of the satellite, ω_s represents the angular velocity of the satellite along the orbit, ω_e represents the angular velocity of earth rotation, α represents the angle that characterizes the satellite position along the orbit, α_i represents the inclination angle of the orbit plane. Finally, ψ represents the direction angle of the radar antenna, which depends on the yaw angle of satellite attitude

$$\theta_{yaw} = \frac{\pi}{2} - \psi$$

The goal of antenna yaw steering is to stabilize the Doppler centroid by proper adjustments of satellite attitudes along the orbit. To some special case, Raney [3] gave out the closed form result

$$\theta_{yaw\_steering} = \cot^{-1}\left\{\frac{1}{\cos(\alpha)}\left[\frac{\omega_s}{\omega_e\sin(\alpha_i)} - \cot(\alpha_i)\right]\right\}$$

This formula is exactly accurate to circular orbits, for example, RADARSAT-1. However, to the general case (noncircular orbit), this formula may generate bias which depends on the frequency band. Our contribution of this work is to develop a numerical calculation to yaw steering issues.

3. The Kalman Filter

In the previous works, the Least Mean Square (LMS) based algorithm [6] may provide perfect numerical precision to yaw steering. This type algorithm needs many iterations to converge to optimum. On the other hand, the complexities of EKF and UKF makes their implementation expensive. The objectives of the design include to lower the complexity and to achieve fast convergence with less iterations, while providing perfect numerical precision. To do so, we need first transform the parameters and then apply the linear Kalman filter.

Let us define the parameters as

$$\theta \triangleq \begin{bmatrix} \sin(\psi) \\ \cos(\psi) \end{bmatrix}$$

Further, it is reasonable to define the state vector as the random process, and to model its mean as

$$E[X] \triangleq [\theta_1 \quad \dot{\theta}_1 \quad \ddot{\theta}_1 \quad \theta_2 \quad \dot{\theta}_2 \quad \ddot{\theta}_2]^T$$

We may establish the process (state) equation in a linear form

$$X_k = F_{k,k-1}X_{k-1} + W_{k-1}$$

in which, we have

$$F = \begin{bmatrix} q & 0 \\ 0 & q \end{bmatrix}$$

and

$$q = \begin{bmatrix} 1 & T & \frac{T^2}{2} \\ 0 & 1 & T \\ 0 & 0 & 1 \end{bmatrix}$$

where T represents the sampling interval to yaw steering.
We also may establish the observation equation in a linear form

$$Z_k = H_kX_k + V_k$$

in which, we have

$$H = \begin{bmatrix} \omega_e R_s \sin(\gamma)\cos(\alpha)\sin(\alpha_i) \\ 0 \\ 0 \\ \omega_e R_s \sin(\gamma)\cos(\alpha_i) - R_s\omega_s \sin(\gamma) \\ 0 \\ 0 \end{bmatrix}^T$$

and

$$E[Z] = -\dot{R}_s \cos(\gamma)$$

In summary, the Kalman filter procedures are as follows,

$$\hat{X}_{k|k-1} = F_{k,k-1}\hat{X}_{k-1}$$

$$P_{k|k-1} = F_{k,k-1}P_{k-1}F_{k,k-1}^T + Q_{k-1}$$

$$K_k = P_{k|k-1}H_k^T\left(H_kP_{k|k-1}H_k^T + R_k\right)^{-1}$$

$$\hat{X}_k = \pi\left[\hat{X}_{k|k-1} + K_k\left(E[Z_k] - H_k\hat{X}_{k|k-1}\right)\right]$$

$$P_k = (I - K_kH_k)P_{k|k-1}$$

Notice $\pi[\cdot]$ represents a projection operator, such that $\sin^2(\psi) + \cos^2(\psi) = 1$, Q represents the covariance of the process noise w_k, and R represents the covariance of the observation noise v_k. It is easy to prove that we may build an accurately unbiased estimator in the MMSE sense. That is

$$E[X|Z] = E[X] + C_xH^T\left(HC_xH^T + R\right)^{-1}(Z - HE[X])$$

By taking expectation to Z, the conditional expectation reduced to the unconditional expectation, which is the defined yaw steering angle to be calculated. This proof means that we may implement theoretical calculation without any bias. In addition, the Kalman filter assumed that a digital computer has infinite word length, so the covariance propagation by recursion is accurate enough. To engineering practice, this assumption may not be feasible. Finite word length effects make the covariance propagation inaccurate, which leads to divergence phenomenon [9]. To overcome the numerical difficulties, the square root based Kalman filter is highly suggested to have a stable algorithm.

IOP Conf. Series: Earth and Environmental Science 1087 (2022) 012046 doi:10.1088/1755-1315/1087/1/012046

4. Spline Interpolation

As mentioned before, the Doppler frequency measured at the receiver depends on satellite attitude, namely, the yaw angle. Without loss of generality, we may express the dependence in the form

$$\theta_{yaw} \xrightarrow{\text{yields}} f_d$$

It is easy to see that

$$\theta_{yaw\_steering} = \lambda(f_d = \text{const.})$$

or conversely,

$$f_d = h(\theta_{yaw})$$

By the linear filtering theory, we know

$$\hat{\theta}_{yaw\_steering} = \sum_{n=-N}^{N} w(n)\lambda\left\{f_d^{(n)}\right\}$$

in which w represents the interpolation kernel. The crucial is how to design this kernel. If we define

$$J(w) \triangleq \left[h(\theta_{yaw}) - f_d\right]^2$$

By setting its derivative to zero, an optimum interpolator is obtained

$$\frac{\partial J}{\partial w} = 0 \xrightarrow{\text{yields}} w_{opt}$$

Direct calculate should be avoided when possible. This leads to the famous LMS type algorithm, i.e.,

$$w_{next} = w_{current} - \mu \cdot \text{grad}[J(w_{current})]$$

where μ denotes the step size and the gradient is given by

$$\text{grad}[J(w)] = 2 \cdot \left[h(\theta_{yaw}) - f_d\right] \cdot \dot{h}(\theta_{yaw}) \cdot \Phi$$

In the above formula, Φ represents the input vector of the interpolator. The algorithm is optimum to the Least Square (LS) criterion. It may be too complex to build in practice. To engineering implementation, a simple interpolator based on spline functions may be chosen as the alternative.

5. Simulation Results

To demonstrate the effectiveness of the proposed algorithm, we employ the system parameters of European Remote-Sensing Satellite 1A (ERS-1) and Canadian RADARSAT-1. This is because ERS-1 is the first mission equipped with the yaw steering mode. Experiment results showed that absolute values of Doppler centroid may be kept within 500 Hz in most situations. As a consequence, yaw steering makes possible precision data processing with reduced complexity. We describe the system parameters of ERS-1 in the Table 1 and those of RADARSAT-1 in the Table 2. By computer simulations, we first calculate the Doppler parameters along the satellite orbits. In both the Figure 1 (ERS-1) and the Figure 2 (RADARSAT-1), the Doppler centroid may vary significantly within one sidereal period. The goal of yaw steering is to adjust satellite attitudes to stabilize Doppler centroid. In the Figure 3, we apply the algorithm proposed by Raney to RADARSAT-1. Yaw steering angles are obtained by fast theoretical calculation. In the Figure 4, we compare the Doppler characteristics between ERS-1 and RADARSAT-1. Because of the circular orbit assumption of RADARSAT-1, Raney's formula is exactly accurate. The situation is different to ERS-1 due to the noncircular orbit assumption, its bias may reach to 300 Hz for the C-band system. In this work, we try to tackle this problem by using the Kalman filter being proposed. In the Figure 5, the Kalman filter may achieve the numerical precision at the level of 1e-12 Hz. Since its calculation is very fast, we may implement the procedures onboard without too much difficulties. In addition, the algorithm based on a simple interpolator can be used to engineering practice. An eleven points interpolator may satisfy the numerical requirements. In principle, the optimum Wiener interpolator may have perfect performance which is equivalent to the Kalman filter. However, to ease the implementation, a spline interpolator is chosen. We may find form MATLAB toolboxes the interp1 function and then apply it conveniently. In the Figure 6, the numerical precision of the interpolator based on cubic spline may reach to 1e-5 Hz for ERS-1.

Table 1. The System Parameters of ERS-1.

Parameters	Symbols	Values	Metrics
The semi-major axis of the orbit ellipse	a	7153	Kilo-meters
The eccentricity of the orbit ellipse	e	1.105e-3	NA
The inclination of the orbit plane	a_i	98.52	degrees
The longitude of the ascending node	Ω	0	degrees
The argument of the perigee	ω	90	degrees
The radar carrier frequency	C	5.3	GHz
The antenna look angle (off nadir)	γ	23	degrees
The average orbit height	H	785	Kilo-meters

Table 2. The System Parameters of RADARSAT-1.

Parameters	Symbols	Values	Metrics
The semi-major axis of the orbit ellipse	a	7167	Kilo-meters
The eccentricity of the orbit ellipse	e	0	NA
The inclination of the orbit plane	a_i	98.6	degrees
The longitude of the ascending node	Ω	0	degrees
The argument of the perigee	ω	NA	degrees
The radar carrier frequency	C	5.3	GHz
The antenna look angle (off nadir)	γ	30	degrees
The average orbit height	H	800	Kilo-meters

To more real situations, orbit perturbation should be investigated. Since the longitude of the ascending node has marginal effects to Doppler centroid, we need pay more attentions to the parameters such as the argument of the perigee and the inclination of the orbit plane. In the Figure 7 and the Figure 8, it is easy to see that errors in true anomaly and inclination do have effects to yaw steering performance. However, these biases may be kept within 100Hz, which are acceptable to the ERS-1 system.

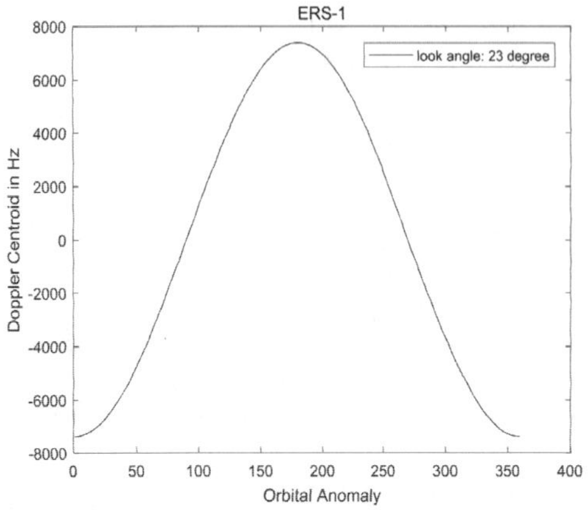

Figure 1. Doppler Characteristics of ERS-1.

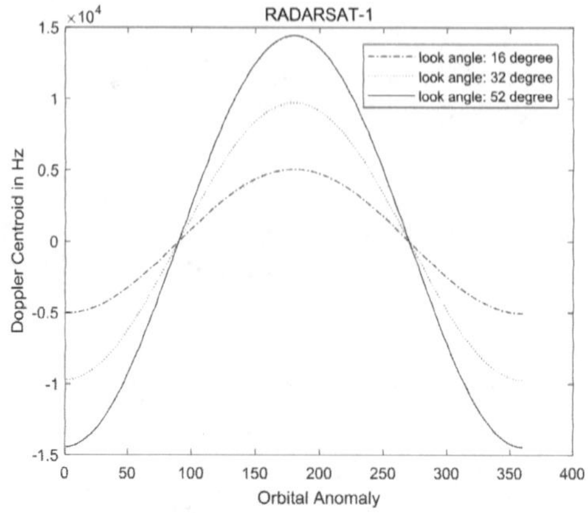

Figure 2. Doppler Characteristics of RADARSAT-1.

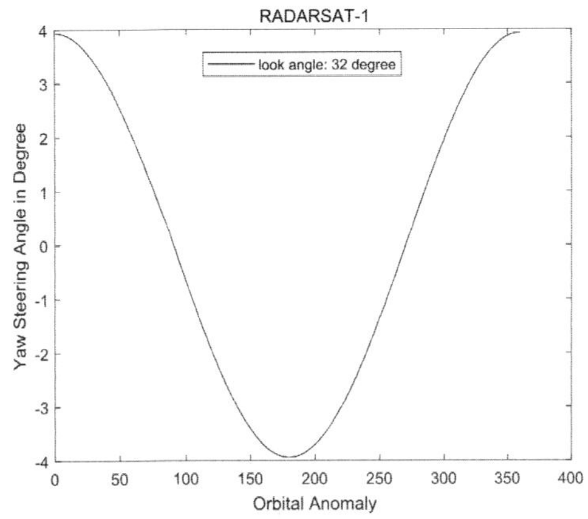

Figure 3. RADARSAT-1 Yaw Steering Calculation by Raney's Formula.

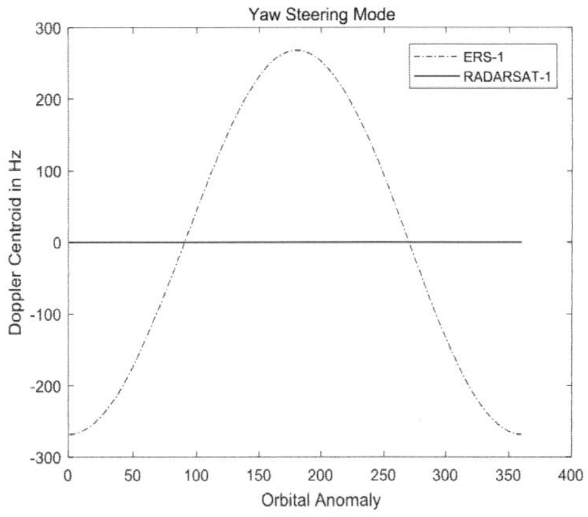

Figure 4. Doppler Characteristics in the Yaw Steering Mode (by Raney's Formula).

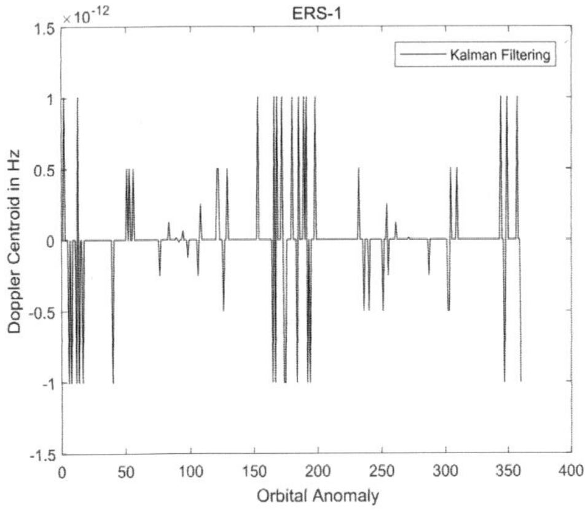

Figure 5. Doppler Characteristics in the Yaw Steering Mode (by Kalman Filtering).

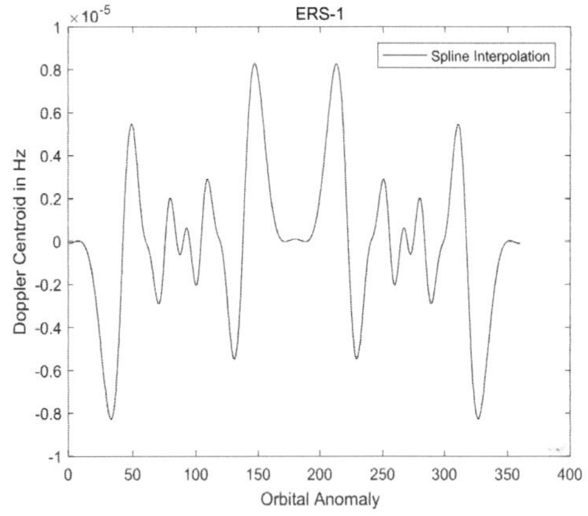

Figure 6. Doppler Characteristics in the Yaw Steering Mode (by Spline Interpolation).

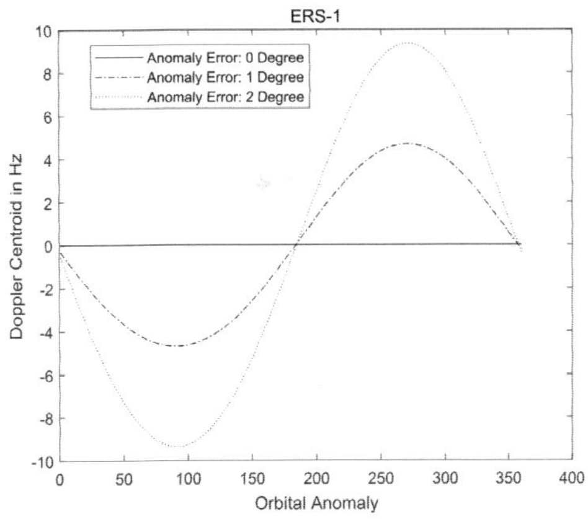

Figure 7. Yaw Steering in the Presence of Anomaly Errors.

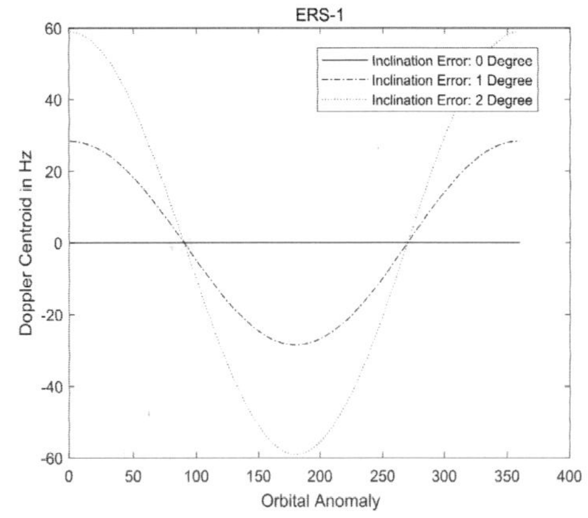

Figure 8. Yaw Steering in the Presence of Inclination Errors.

6. Conclusions

Yaw steering calculation to SAR satellites is considered in this work. Raney's classic formula is exactly accurate for circular orbits, for example, RADARSAT-1. It may generate large bias to noncircular orbits such as ERS-1. To tackle this issue, we propose the Kalman filter to achieve high precision calculation. Both the process equation and the observation equation are of linear forms by proper parameter transform. Computer simulations demonstrate that the algorithm based on Kalman filtering is very fast and accurate. This makes possible the onboard implementation. In addition, we also consider the effects by orbit perturbation and attitude errors. It is reasonable to model those errors by the process equation as well as the observation equation. To a practical system design with high precision requirements, Kalman filtering is highly suggested. To the system with reduced complexity, a cubic spline interpolator is proposed to be an alternative. The interpolator is easy to engineering practice, its precision to yaw steering may reach to 1e-5 Hz. As a consequence, both Kalman filtering and simple interpolation are suitable to future SAR missions.

Acknowledgements

This research was supported by Aeronautical Science Foundation of China (ASF) under grants 2009-55-96016 and 2013-ZC-15008. Many thanks go to Key Laboratory on Avionics Integration, Aviation Industry of China (AVIC). We also thank the colleagues from Institute of Navigation and Information, Airforce Engineering University, for the technical discussion.

References

[1] Hartmut Runge June 3-6 1991 Benefits of Antenna Yaw Steering for SAR, in the proceedings of the *International Geoscience and Remote Sensing Symposium (IGARSS'91)* **1** 257-261

[2] Dieter Just and Birgit Schattler May 25-29 1992 Doppler-Characteristics of the ERS-1 Yaw Steering Mode, in the proceedings of the *International Geoscience and Remote Sensing Symposium (IGARSS'92)* **2** 1349-1352

[3] R. K. Raney 1986 Doppler Properties of Radars in Circular Orbits *International Journal on Remote Sensing* **7(9)** 1153-1162

[4] Rui Wang and Ruliang Yang January 2004 Radar Doppler Properties Evaluation in Elliptic Orbits, in the Chinese language *Journal of Electronics & Information Technology* **26(1)** 107-111

[5] Guohua Wang, Jinping Sun, Yunneng Yuan and Shiyi Mao September 2006 Precise Computation of Yaw-Steering in Spaceborne Synthetic Aperture Radar System, in the Chinese language *Journal of Electronics & Information Technology* **28(9)** 1569-1572

[6] Tao Chen, Yongfei Ding, Ruifan Pang, Cheng Gong, Dinghai Xu and Hengyang Zhang September 26 - October 2, 2020 Yaw Steering Using Adaptive Filtering for Spaceborne SAR Systems, in the proceedings of the *International Geoscience and Remote Sensing Symposium (IGARSS'20)* 5945-5948

[7] Zhongquan Wei 2001 Synthetic Aperture Radar Satellites, in the Chinese language Beijing, *Science Press*

[8] B. C. Barber 1985 Theory of Digital Imaging from Orbital Synthetic-Aperture Radar *International Journal on Remote Sensing* **6(7)** 1009-1057

[9] Simon Haykin 1996 Adaptive Filter Theory, the third edition, Englewood Cliffs, NJ: Prentice Hall

IOP Conf. Series: Earth and Environmental Science 1087 (2022) 012047 doi:10.1088/1755-1315/1087/1/012047

3D model construction and accuracy analysis based on UAV tilt photogrammetry

Xuanyu Chen[1], Shaowei Pan[2,5] and Guanyu Chen[3,4]

[1] Natural Resources Information Center of Guangxi Zhuang Autonomous Region, Nanning 530028, China;
[2] College of Geomatics and Geoinformation, Guilin University of Technology, Guilin, 541004, China;
[3] LTD of Zhejiang Province Land Survey and Planning, Hangzhou Branch, Hangzhou 310030, China;
[4] Zhejiang Academy of Surveying and Mapping, Hangzhou 310030, China.

[5] Corresponding author's e-mail: pan.shaowei@glut.edu.cn

Abstract. UAV equipped with tilt photogrammetry platform can obtain multiple angles images and establish 3D model of tilt photogrammetry, which has unique advantages in 3D intelligent construction. The purpose of this paper is to take the playground and library of the Department of Informatics of Wuhan University as the research area, obtain regional UAV images, and evaluate the residual in the spatial direction of the model by using the measured control points and model points collected in the field, so as to evaluate and analyze the accuracy of the 3D model. By analyzing the data, the root mean square error and maximum error of plane coordinates are 0.31 cm and 0.34 cm respectively, the root mean square error and maximum error of elevation are 0.07 cm and 0.11 cm respectively, and the field measured control point coordinates. Through the accuracy evaluation and analysis of the constructed three-dimensional model, it is concluded that the maximum error in X direction is 1.26 cm, the maximum error in Y direction is 1.39 cm, and the maximum error in Z direction is 1.60 cm. The results show that the accuracy of the three-dimensional model in the study area reaches centimeter-level and meets the requirements of relevant production specifications. It can provide reliable data support for construction and urban planning industries.

1. Introduction

Nowadays, the demand for 3D smart city construction is growing day by day, traditional map data can hardly support the actual demand of today's 3D smart construction. The geographic feature information of 3D surface model can reflect the location of target features more intuitively and accurately. At present, 3D model data is widely used in infrastructure construction, tourism, cadastre, map and many other fields of Smart new cities in China [1].

Aiming at the problems of low automation, high cost and slow speed of traditional 3D modeling and production technology, the real scene 3D modeling technology based on UAV tilt photogrammetry came into being [2]. UAV tilt photogrammetry is a fast and low-cost technology to realize the establishment of large-scale urban real 3D model[3]. Conventional orthophoto stereo images of tilt photogrammetry can only be taken from two vertical angles, lacking details of the stereo side elevation images of the measured feature target. The UAV tilt photogrammetry technology can continuously take tilt images of the same feature target with a certain degree of overlap from multiple different flight angles, obtain images of different moving directions of the ground object, and provide more detailed 3D texture image information for the establishment of the whole real scene 3D model. After being processed by 3D modeling software, the required 3D real scene model can be directly generated [4]. The reconstructed 3D model data can truly reflect the ground object and surrounding environment, including a variety of attribute information such as spatial position, structure and color, so as to meet the needs of modern human stereo vision. This effectively makes up for the technical limitations of traditional tilt photogrammetry [5]. In 2016, K. Mo et al used ultracam Osprey prime multi lens camera system to collect the image of IMST area, and processed the multi view image. The errors in X direction, Y direction and Z direction of aerial triangulation were 0.035 m, 0.027 m and 0.058 m respectively [6]. In 2018, Zheng pan conducted four groups of three-dimensional modeling tests on a survey area and evaluated their accuracy respectively, which further developed the accuracy evaluation of three-dimensional model [7]. In 2018, erenoglu, RC et al proposed a 3D urban modeling method based on unmanned aerial vehicle (UAV), and compared the results with the results of RTK / GNSS, CORS and other land surveys to verify the effectiveness of modeling in accuracy and time domain [8]. Yu Lili et al collected images and conducted three-dimensional modeling of a 60 km^2 urban area by using a five lens UAV in 2021. Through checkpoint verification, the maximum error of plane position is 0.179 m and the root mean square error is ± 0.121 m; The maximum error of elevation is 0.389 m and the root mean square error is ± 0.256 m [9].

In this paper, we take playground and Library of the Department of Informatics of Wuhan University as the research area, and use the Da jiang series UAV Phantom 4 PRO to obtain the tilt image data of the research area as the original data source of this paper, and construct the real scene 3D photogrammetry model of the target ground object in the research area. In addition, we use the measured control point data to compare, analyze and evaluate the overall accuracy of aerial triangulation and 3D model, and make a technical analysis on the construction technology of real scene 3D model of UAV tilt photogrammetry, so as to verify the feasibility and accuracy reliability of this technology in constructing highly visual real scene 3D model.

2. UAV tilt photogrammetry

UAV tilt photogrammetry technology mainly refers to a popular aerial survey technology that uses tilt photogrammetry system to obtain the information of ground object target, spatial geographical location and object geometric structure. In the process of collecting the original data of ground objects, the UAV fuselage is equipped with multiple image sensors with different angles. While automatically taking multiple photos in the air, it will automatically record each spatial setting and geometric angle of the images taken by the camera through the high-elevation GPS control system and high-elevation inertial navigation control system. When the degree of longitudinal and lateral overlap in stereoscopic images reaches 66% or more, aerial triangulation can be used to automatically analyze and calculate multi-perspective stereoscopic images and analyze and calculate three-dimensional spatial stereoscopic image models. At the same time, this imaging technology mainly carries out dense point matching through the coordinate positions of multiple homologous points of multi angle images. After processing by this technology, it can directly generate dense point clouds, to achieve the comprehensive construction of irregular triangular network. In this process, the most critical technical process is aerial triangulation technology [10]. Similar to traditional photogrammetry, UAV tilt photogrammetry mainly locates images through collinearity equation and resection principle [11], as shown in the following equation.

In the ground photogrammetric coordinate system, the coordinates of the photographing center S and the ground point A are X_S , Y_S , Z_S and X , Y , Z respectively. In the image space auxiliary coordinate system, the coordinates of the ground point A are $X_A - X_S$, $Y_A - Y_S$, $Z_A - Z_S$. in the image space auxiliary coordinate system, the coordinates of the image point a are U , V , W . Since the three points S , a , A are collinear, they are transformed into matrix form according to the similar triangles theorem [12]:

$$\begin{bmatrix} U \\ V \\ W \end{bmatrix} = \frac{1}{\lambda} \begin{bmatrix} X_A - X_S \\ Y_A - Y_S \\ Z_A - Z_S \end{bmatrix} \tag{1}$$

In equation (1), is the scale factor. The relationship between the image point in the image space coordinate system and the image space auxiliary coordinate system is as follows.

$$\begin{bmatrix} x \\ y \\ -f \end{bmatrix} = \lambda \begin{bmatrix} a_1 & a_2 & a_3 \\ b_1 & b_2 & b_3 \\ c_1 & c_2 & c_3 \end{bmatrix} \begin{bmatrix} U \\ V \\ W \end{bmatrix} \begin{matrix} ① \\ ② \\ ③ \end{matrix} \tag{2}$$

Replace equation (1) with equation (2) and divide ② and ① with equation ③ to obtain.

$$X = -f \frac{a_1(X - X_S) + b_1(Y - Y_S) + c_1(Z - Z_S)}{a_3(X - X_S) + b_3(Y - Y_S) + c_3(Z - Z_S)}$$
$$Y = -f \frac{a_2(X - X_S) + b_2(Y - Y_S) + c_2(Z - Z_S)}{a_3(X - X_S) + b_3(Y - Y_S) + c_3(Z - Z_S)} \tag{3}$$

In equations (2) and (3), a_i, b_i and c_i (i = 1,2,3) are the elements of 3×3 orthogonal rotation matrix R generated by three external azimuth elements φ, ω and κ. In equation (3), x and y are the plane coordinates of ground points, and equation (2) is the transformation relationship between image plane points and ground points. The collinear equation includes 12 data. image point coordinates x and y with the image principal point as the origin; Corresponding ground point coordinates X, Y, Z; The principal distance f of the photograph and the external orientation elements $X_S, Y_S, Z_S, \varphi, \omega, \kappa$.

3. Materials and methods

3.1. Overview of research area and original data source

The playground of the Information Science Department of Wuhan University is a standard playground. Its construction area is very small. It is mostly surrounded by squares, houses and other activity areas without rivers and water systems.

In this study, 299 images of the playground and Library of the Department of Informatics of Wuhan University were obtained by using the DJI fc300s camera carried by phantom 4 Pro UAV. The camera sensor size is 6.16mm, the camera focal length is 3.54 mm, the 35 mm equivalent focal length is 20.74mm, the aperture value is F / 2.8, the exposure time is 1 /419 seconds, the ISO is 100, and the photo pixel size is 4000 × 3000 pixels. And the coordinates of 8 control points in the study area are collected by GPS-RTK to correct the aerial triangulation results. The images of the study area are equipped with GPS positioning information, and the position, height and accuracy of the control points will have a direct impact on the subsequent automatic aerial triangulation calculation.

3.2. Data processing

In this paper, ContextCapture 4.4.10 software is selected as the data processing platform, and the real three-dimensional model is constructed by using the coordinate data of 8 control points and 299 images with high-precision position information and attitude angle information. The main data processing process is shown in Figure 1.

(1) Aerial triangulation

Aerial triangulation encryption is the key technology of 3D model reconstruction of tilt photogrammetry. The purpose is to use the accurate POS data of the image and each image to calculate the tilt image matching, determine the corresponding image point between the images, obtain the multi view tilt image matching results, then carry out the bundle block adjustment to calculate the accurate elements of exterior orientation. Finally, the multi view image dense matching is processed.

In this experiment, 8 ground control points in the survey area were obtained by GPS-RTK. The distribution of control points is shown in Figure 2.

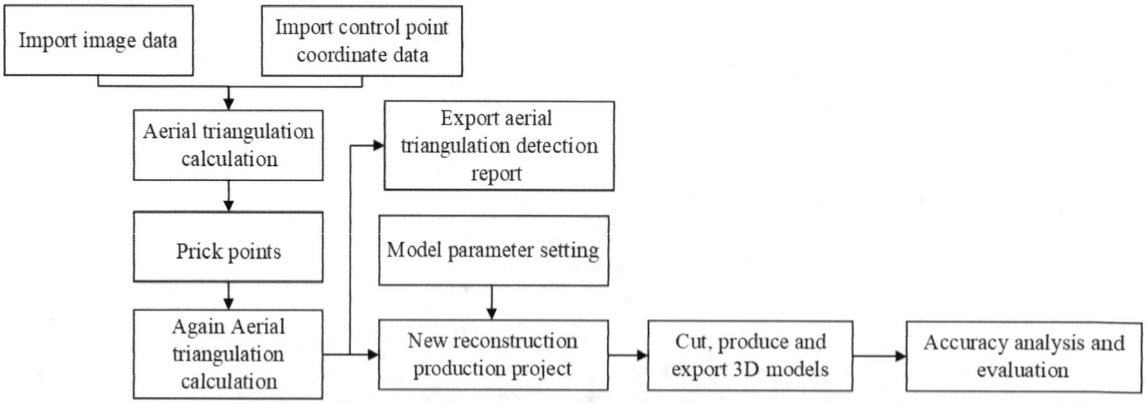

Figure 1. Data processing process.

Figure 2. Distribution control points in playgrounds.

(2) Marking points

Image marking points are divided into two steps: first, rough marking points, select several control points easy to find at the edge for rough marking points, and carry out the first aerial triangulation calculation; second, fine marking points, with the help of aerial triangulation calculation results, provide reference basis for relevant positions of fine marking points. Run the aerial triangulation calculation again until the aerial triangulation result control encryption is successful, that is, use a small number of control points to solve the interior and exterior orientation elements of the image and get the coordinates of corresponding image points on the adjacent images.

(3) Model production

According to the image dense matching processing algorithm provided by the software, a high-density 3D point cloud is generated, and then countless TIN (triangulated irregular network) are constructed from the point cloud to obtain a 3D model (white model), as shown in Figure 3. Then, through the registration of the TIN model's various triangles and corresponding texture images, the model texture mapping is automated, and the relevant 3D model is obtained. Open the S3C index file under the same geographical reference frame, and you can see the whole 3D model, as shown in Figure 4.

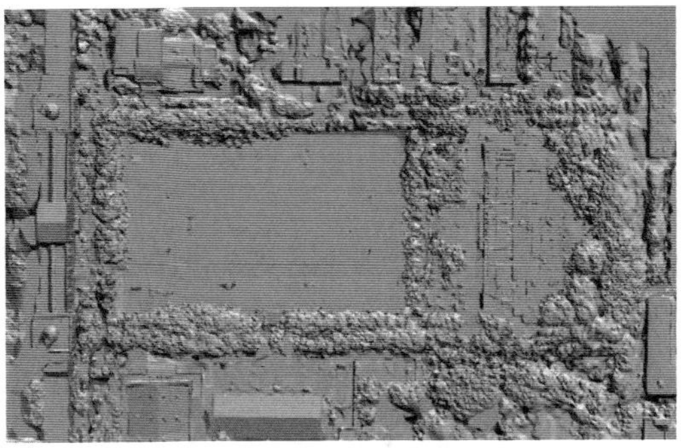

Figure 3. White model.

Figure 4. Three-dimensional model.

4. Results and discussions

After processing the image data, a complete 3D model of the library and playground of the Department of Information Science of Wuhan University has been obtained (as shown in Figure 5). The following is mainly evaluated and analyzed from three aspects: model quality and effect, aerial triangulation accuracy and model accuracy.

4.1. Model quality and effect evaluation

There are a large number of trees and grasslands around the playground and library in the study area, and there are many irregular ground objects. Light conditions are not uniform during photography, and there are blind areas for photography, leading to incomplete images of a large number of trees in the reconstruction model. By comparing the aerial images with the model, the images of the playground and library of the Department of Information Science of Wuhan University are basically consistent with the appearance of the 3D model in terms of spatial shape, and the images of regular features in the model are relatively clear. The library model of the Department of Information Science of Wuhan University presents a regular rectangular shape, indicating that the three-dimensional model constructed this time is consistent with the actual real scene in space shape. In terms of integrity, the irregular features around the playground of the Department of information science of the reconstructed model are seriously blurred, and there is a large difference between the real object and the model, but the actual features can be fully displayed, which still meets the requirements of this experiment.

The above analysis shows that the constructed 3D model can vividly and intuitively express the architectural characteristics of the playground and library of the Information Department of Wuhan University. In addition, the position of the control points have a good perspective of space, the light conditions are uniform during photography, and the 3D model constructed by the experiment is complete, with a clear level of detail and consistent tone, which lays a foundation for the subsequent comparative analysis of accuracy.

4.2. Accuracy evaluation of aerial triangulation

The accuracy calculation of aerial triangulation is the key point of later data processing of UAV tilt photogrammetry [13], and its relative solution accuracy directly affects the analysis and comparison of subsequent data processing results. Root mean square error is one of the important indicators to measure the accuracy of the three-dimensional model, which is widely used in various fields. Its equation is as follows:

$$m = \sqrt{\frac{\sum(\Delta)^2}{n}} \tag{4}$$

In equation (4), m represents the root mean square error; Δ is the difference between the observed value and the true value of the control point; n represents the number of control points.

In this experiment, the root mean square error equation is used to evaluate the accuracy of aerial triangulation results. The aerial triangulation accuracy of the model is evaluated in many aspects by calculating the model plane error, elevation error and 3D error. The accuracy of aerial triangulation is shown in Table 1.

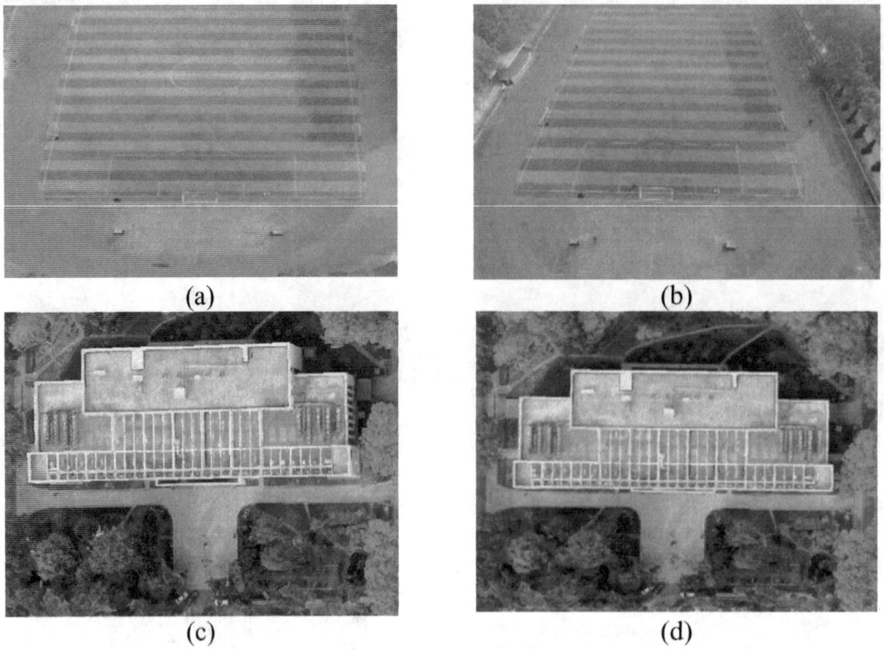

Figure 5. Comparison of 3D model and aerial photos of information science department ((a) and (c) are 3D models, (b) and (d) are aerial photos).

Table 1. Aerial triangulation accuracy of the playground of Information Department.

Control point number	Plane error (cm)	Elevation error (cm)	3D error (cm)
A	0.32	0.09	0.26
B	0.30	0.03	0.20
C	0.30	0.01	0.04
D	0.32	0.09	0.28
E	0.28	0.08	0.07
F	0.32	0.01	0.21
G	0.30	0.11	0.05
H	0.34	0.02	0.09
RMSE	0.31	0.07	0.18

It can be seen from Table 1 that the root mean square error, maximum error and minimum error are 0.31 cm, 0.34 cm and 0.28 cm respectively; The root mean square error, maximum error and minimum error of elevation are 0.07 cm, 0.11 cm and 0.01 cm respectively; The root mean square error, maximum error and minimum error of 3D error are 0.18 cm, 0.28 cm and 0.04 cm respectively. On the whole, the aerial triangulation results of the playground and Library of the Department of information science have high accuracy.

4.3. Model accuracy evaluation
In this paper, the coordinate data of ground control points measured by high-precision GPS-RTK instrument and the coordinate data of corresponding control points extracted from the model are compared and analyzed. Then judge whether the plane accuracy and elevation accuracy of the three-dimensional model meet the production requirements according to the " three-dimensional geographic information model data product specification", and evaluate the quality of the three-dimensional reconstruction model as a whole. Specifically, calculate the difference between the measured coordinates by the model and the coordinates of the control points measured in the field, count the residuals in X, Y and Z directions, and then calculate the error of the plane. So as to evaluate the accuracy of the model. The results are shown in Table 2.

Table 2. Information Department playground Control point coordinate residual statistics table.

Point number	X esidual (cm)	Y esidual (cm)	Z esidual (cm)	Plane (cm)
A	-0.22	-0.72	-0.87	0.75
B	0.32	-1.16	0.55	1.20
C	-1.26	-0.25	1.54	1.28
D	0.40	0.71	-0.74	0.82
E	0.86	-1.39	-0.47	1.64
F	-0.44	0.76	0.95	0.88
G	0.22	-1.11	-1.31	1.13
H	0.29	0.16	1.60	0.33

According to Table 2, the maximum error of the three-dimensional model in X direction is 1.26 cm, the maximum error in Y direction is 1.39 cm, the maximum error in Z direction is 1.60 cm, the maximum plane error is 1.64 cm and the minimum plane error is 0.33 cm. The root mean square errors of plane and elevation are 1.07cm and 1.08 respectively. It can be seen that the playground and library model of the Department of information science has high accuracy, the residuals in X, Y and Z directions are concentrated, and the calculated error is small. The reconstructed model obtained in the experiment has high consistency with the real model, high model integrity, and the accuracy of the produced three-dimensional model can reach centimeter rank. So far, the spatial geometric accuracy of the three-dimensional model is high, and the produced 3D real scene model meets the requirements, which can provide technical support and image basis for subsequent applications.

5. Conclusions
This paper mainly introduces the basic principle and processing flow of acquiring tilt image data and reconstructing 3D model based on the camera sensor of UAV. With the help of GPS-RTK to obtain control point data, verify and analyze the accuracy of the established 3D model, and draw the following conclusions:

(1) Compared with the actual scene, although the irregular features have a certain degree of distortion, the regular playground and library have high consistency and integrity. When analyzing the measured control points and model points, evaluating the residual value of the spatial direction of the constructed model, and verifying the errors and root mean square errors of Plane and elevation, the accuracy reaches centimeter rank and meets the production specifications of relevant factors.

(2) For the accuracy of three-dimensional model, the root mean square error, maximum error and minimum error of the playground plane of the Department of information science are 1.07 cm, 1.64 cm and 0.75 cm respectively; The root mean square error, maximum error and minimum error of elevation are 1.08 cm, 1.60 cm and -0.47 cm respectively; It can be seen that the established models have high reliability.

To sum up, tilt photogrammetry based on UAV solves the disadvantages of traditional photogrammetry technology, significantly improves work efficiency and model accuracy, and reduces the probability of human error. It provides an economic and convenient technical support for land resource investigation, disaster monitoring, industrial measurement, forestry and agriculture, and has great application value. In the future, comprehensive analysis will be carried out from the aspects of the actual survey area, topographic conditions of the survey area, surveying and mapping scale, determination of appropriate photographic scale, navigation height, meteorological conditions and other influencing factors, so as to further improve the model and improve the accuracy of the model.

Acknowledgements
This work was supported by Guangxi Science and Technology Plan Project (Grant No. GUIKE AD19110107); the Natural Science Foundation of Guangxi (Grant No. 2018GXNSFBA050006).

References
[1] Sun J, Xie W H, Bai R J 2019 UAV oblique photogrammetric system and its application *[J] Science of Surveying and Mapping* **44(6)** 145-150
[2] Ren C, Gao L M, Feng Y L, Zhang W 2019 An attempt of building 3D modeling based on UAV tilt photography *[J] Bulletin of Surveying and Mapping* **(2)** 161-164
[3] Qiu C X, Dong Q K, Liu R H, Lei L 2017 Using Tilt Images to Build 3D Model and Optimization Model *[J] Bulletin of Surveying and Mapping* **(11)** 133-138+143
[4] Yang F 2018 Experiment and Analysis of Unmanned Aerial Vehicle Oblique Photography in Constructing 3D Model of Forest Scenic Spots *[J] Forest Resources Management* **(4)** 120-124
[5] Zhang H Y, Dong C L, Wang J G, sun S J, Wang H Y, Wang H H 2019 Practice on 3D-modeling based on tilt photogrammetry and Context Capture software *[J] Bulletin of Surveying and Mapping* **(S1)** 266-269
[6] Moe K, Toschi I, Poli D, Lago F, Schreiner C, Legat, K, Remondino, F 2016 Changing the Production Pipeline - Use of Oblique Aerial Cameras for Mapping Purposes *[J] International Archives of the Photogrammetry, Remote Sensing & Spatial Information Sciences* **41(B4)** 631-637
[7] Zheng P 2018 3D Modeling and Accuracy Evaluation of UAV Inclination photogrammetry Based on Smart3D Software *[D] Chengdu University of Technology*
[8] Erenoglu, R C, Erenoglu, O, Arslan, N 2018 Accuracy assessment of low cost uav based city modelling for urban planning *[J] Tehniki Vjesnik* **25(6)** 1708–1714
[9] Yu L L 2021 Research on Urban 3D Real Scene Modeling Based on UAV Tilt Photogrammetric Technology *[J] Geomatics & Spatial Information Technology* **44 (05)** 86-88
[10] Liu B, Tang Y L, Ma C Y, Bai S Y 2019 Application of 3D modeling of UAV tilt photography in urban rain flood risk assessment *[J] Bulletin of Surveying and Mapping* **(10)** 46-50+66
[11] Chen Z H, Zhang J X, Zhang K M, Shi H L 2019 The improved method and accuracy of the establishment of three-dimensional model of UAV tilt photogrammetry scene in Yunnan expressway *[J] Bulletin of Surveying and Mapping* **(S1)** 275-279
[12] Zhang Z X, Zhang J Q 2012 Digital photogrammetry *[M] Wuhan University Press* 117-129
[13] Wang B T, Wang J 2015 Production and Quality Analysis of 3D City Modeling Using Oblique Photogrametric Technology *[J] Urban Geotechnical Investigation & Surveying* **(5)** 80-82 85

IOP Conf. Series: Earth and Environmental Science 1087 (2022) 012048 doi:10.1088/1755-1315/1087/1/012048

Global spatial distribution of vegetation and 20-years dynamic changes

Sihui Huo[1,2,5], Qing Sun[1], Yanmin Shuai[1,2,3,4], Congying Shao[1] and Jiapeng Huang[1]

[1] College of Surveying and Mapping and Geographical Sciences, Liaoning Technical University, Fuxin 123000, Liaoning, China;
[2] Xinjiang Institute of Ecology and Geography, Chinese Academy of Sciences, Urumqi 830011, Xinjiang, China;
[3] Research Center for Ecology and Environment of Central Asia, Chinese Academy of Sciences, Urumqi 830011, Xinjiang, China;
[4] University of Chinese Academy of Sciences, Beijing, 100049, China.

[5] Corresponding author's e-mail: hshyaogan@163.com

Abstract. Global spatial distribution of vegetation is determined jointly by climate, monsoon, ocean current and human activities. In recent decades, surface vegetation changes caused by global warming, extreme climate events and intensified human activities need a further understood. In this paper, we adopted available MODIS land cover dataset (MCD12Q1) to investigate the spatial distribution feature of global vegetation sand dynamics during the past 20 years. Results show that: (1) the spatial distribution pattern of global vegetation formed by the long-term ecosystem evolution exhibits stable with less variation in both latitude and vertical dimensions. Generally, the percent of forest, shrubland, grass and cropland respectively are 15.27%, 9.39%, 39.46% and 8.91% of surface covers, with a distinct land cover transition trend from rainforest at equator region to moss and lichen at the north and south pole. (2) In the past 20 years, the global forest land decreased by 3.0×10^5 km^2, with an average annual decrease of 1.255%, open shrubs experienced slight fluctuations in 2001-2017 and increased sharply by 1.6×10^5 km^2 in 2017-2018. And the grassland, croplands and cropland/natural vegetation mosaics had different degrees of growth rate, with an average annual dynamic increase of 0.021%, 0.019% and 0.502%, respectively. (3) There are about 4.2×10^6 km^2 of forest changed into grassland in West Siberia and the Amazon, and 2.5×10^6 km^2 of grassland into shrubs, as well as 1.9×10^6 km^2 of grassland vegetation into farmland, and 9.4×10^4 km^2 of forest into shrubs.

1. Introduction

Vegetation as the core of the earth's ecology, is one of the most important link between the human and nature. The life rhythm of vegetation includes seeding, germination, grow-up, maturity, senescence, and death, and the annual growth cycle of leaf-on and leaf-off with seasons' switch, which are accompanied by exchange of carbon, nitrogen, water and gas, and are modulating various geophysical and biochemical processes in the Earth system [1][2]. Various vegetation can perform different adjustment of land surface 3-D structure, alter surface texture and roughness, and meliorate the physicochemical properties of the soil [3]. In addition, vegetation frames the chain of demand-supply among different biological species on Earth.

Global vegetation has experienced a significant variation due to disturbance of extreme climate events, natural hazards, or human activities. Extreme ice storms caused severe damage to forests in eastern North America in January 1998 [4], while the tropical hurricane Katrina with surge up to 19ft as one of the most devastating extreme phenomena has caused levee collapses and massive flooding in August of 2005 [5]. Further, the impact of natural disasters on vegetation cannot be neglectable. The large bushfire occurred during the summer of 2019-2020 in Australia, swept almost 8 million hectares including 54% of the World Heritage-listed Gondwana Rainforests growing tens of millions of years [6]. Therefore, the spatial patterns and related evolution of global vegetation over recent decades need a further understood especially under the scenario of global climate change, with enhanced frequency of extreme climate events, natural disasters, and human activities.

Satellite remote sensing is the only means that can conduct vegetation research at global scale with acceptable objectiveness and effectiveness. The long-term continuous multi-source satellite observation provides the chance to timely extract global vegetation information. In the past 20 years, different Research and Development(R&D) groups around the world have developed various sporadic or operational local or global scale land cover data products, such as International Geosphere-Biosphere Programme (IGBP DISCover), Global Land Cover Dataset of the University of Maryland, Global Land Cover Data MODIS from Boston University, USA and Global Land Cover 2000 (GLC2000) of the European Union Joint Research Center, GlobCover2009 released by the European Space Agency in 2010, etc. Therefore, the development of satellite remote sensing and its standard time-series data products provides the basic data for global vegetation research.

2. Data and Methods

2.1. Data

MCD12Q1 is a 500m resolution standard data product developed by Boston University under NASA funding for a prototype DT (Decision tree) supervised classification algorithm, with an overall global classification accuracy of 75%, and jointly operated and published by NASA and GEO. The data are partitioned into $10° \times 10°$ latitude and longitude grids (tiles) using the Inter-grid Sinusoidal Grid projection and are stored in HDF format, with six classification layers (IGBP, University of Maryland (UMD), Leaf Area Index (LAI), BGC (BIOME-Biogeochemical Cycles), Plant Functional Types (PFT) and FAO-Land Cover Classification System land cover (LCCS)) and related confidence layer in Sinusoidal projection [7]. The study uses the IGBP classification mechanism recommended by the International Geosphere-Biosphere Programme. Also, a total of 5985 data MCD12Q1 files are downloaded from the NASA Land Processes DAAC website for the years 2001-2019 (https://ladsweb.modaps.eosdis.nasa.gov/), which is about 315 tiles per year.

GMTED2010, downloaded from the United States government (https://earthexplorer.usgs.gov/), is a global elevation model which is developed by USGS and NGA with three separate resolutions (horizontal post spacings) of 30 arc-seconds, 15 arc-seconds, and 7.5 arc-seconds. Absolute accuracy evaluation of GMTED2010 shows that the minimum error, maximum error, mean error, standard deviation, and root mean square error of the 15 arc-second mean elevation are -2874.18m, 570.27m, 1.89m, 29.56m, and 29.62m [8]. Its coverage, scale, quality, and vertical accuracy all own substantial improvements compared to previous data. For the areas with missing 15 arcsec, the study uses 30 arcsec data resampled to 15 arcsec, achieving the same scale of 15 arcsecond resolution size as the MCD12Q1 data. The study finally selects the highest accuracy the mean elevation data to mosaic the global DEM in ARCGIS.

2.2. Methods

Figure 1 shows the technique flow chart of the process chain with five modules (A - E). Module A (raw data) are mainly divided into global boundary vector data, MODIS global surface cover product (MCD12Q1) from 2001 to 2019, and GMTED2010 global elevation data. Module B (Data preprocessing) contains the mosaic, cropping and reprojection of the raw data which are batched processing in ARCGIS and MODIS Reprojection Tool. Module C (Data analysis) is all based on formulas (1) in Matlab:

$$P = \frac{C_{ij}}{\sum_{i=1}^{17} C_{ij}} \tag{1}$$

Where, P denotes the percentage of vegetation i in year j ; C_{ij} denotes number of pixels of vegetation, Where i represents the vegetation category and j represents the year (2001-2019). Formulas (1) can analyze the overall spatial distribution of vegetation based on the 5° latitudinal bands, vegetation occupancy ratio and vegetation area. The study uses ARCGIS to extract the global distribution of areas larger than 1000 m mountain ranges, calculate the number of vegetation categories and analyze the difference of vegetation in different regions by combining the slope and slope direction. Module D uses the land transfer matrix and land use dynamic attitude to survey the changes of vegetation in the last 20 years. Module E summarizes the spatial distribution pattern and dynamic transfer change of vegetation and grasps the changing characteristics of global vegetation, which can help the material cycle of global vegetation and subsequent development and utilization.

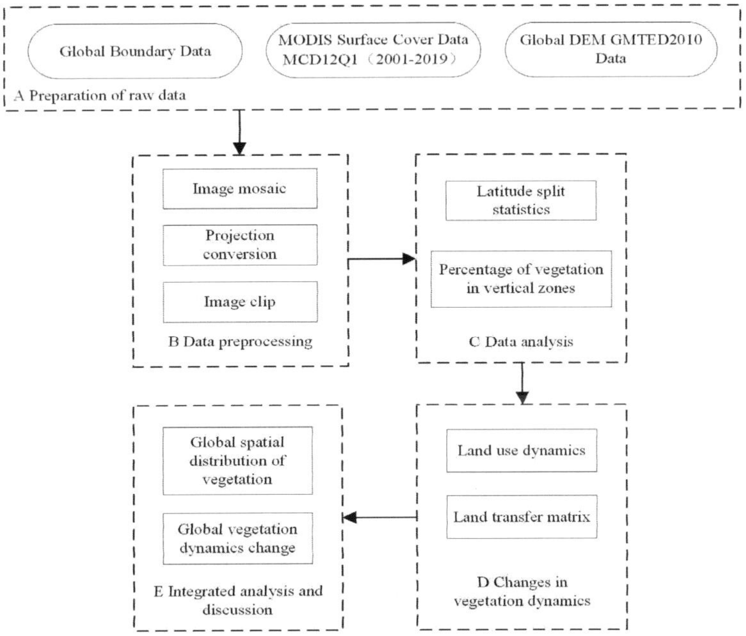

Figure 1. Flowchart of this research.

2.2.1. Land Use Dynamic Degree. Land use dynamic degree indicates the rate of change in the quantity of land resources and is mainly divided into two types: single land use type dynamic degree and comprehensive land use dynamic degree [9]. Among them, the single land use type dynamic degree expresses the quantity change of a certain land use type in a certain time range. The quantity change characteristics of vegetation change are monitored using the single land use dynamic degree, and the calculation formula is then:

$$D_i = \frac{LS_i - IS_i}{IS_i \times T} \times 100\% \tag{2}$$

Where, D_i denotes the dynamic attitude of the quantity change of a land use type in a certain time period; IS_i、LS_i denotes the quantity of a land use type at the beginning and the end of the study period, respectively; T denotes the length of the study, the unit of it is year.

2.2.2. Land Transfer Matrix. The transfer matrix can portray the structural characteristics of regional land use change and the direction of various types of change [10]. While analyzing the change process of vegetation cover, the dynamic transfer results of vegetation can be understood. The study uses the land transfer matrix to express the quantitative transfer and transfer out change details of land use types, which can reflect the structural changes of land types in detail and intuitively, and provide dynamic and rich information on the transformation of land use functions[11].

$$S_{ij} = \begin{bmatrix} S_{11} & S_{12} & \cdots & S_{1n} \\ S_{21} & S_{22} & \cdots & S_{2n} \\ \cdots & \cdots & \cdots & \vdots \\ S_{n1} & S_{n2} & \cdots & S_{nn} \end{bmatrix} \quad (3)$$

Where, S_{ij} represents the area converted from category i to category j . n represents the number of land use types discussed.

3. Results

3.1. Global spatial distribution

3.1.1. Basic Pattern of Vegetation Cover. Figure 2 shows the basic pattern of global vegetation distribution upon the 2019 data. 1.5×10^{10} km^2 land has been covered to analyse the data, and statistics on the proportion of vegetation in the overall land cover in the past 20 years can be found that the proportion of, Evergreen Needleleaf Forests (ENF), Evergreen Broadleaf Forests (EBF), Deciduous Needleleaf Forests (DNF), Deciduous Broadleaf Forests (DBF) and Mixed Forest (MF) have fluctuated at 1.68%, 8.30%, 0.32%, 2.02%, 3.13%, Closed shrublands (CS) and Open Shrublands (OS) have floated at 0.41% and 8.90%, and Woody Savannas (WS), Savannas and Grasslands have fluctuated at 7.62%, 11.03% and 20.78%, respectively. Cropland and Cropland/Natural Vegetation Mosaics (C/NVM) account for about 7.92% and 0.89%.

ENF and Mixed Forests traverse the plains of Western Siberia, with 1.6% and 3.2% of the global land cover, respectively. EBF are found near the equator where in the Amazonian Plain, the Congo Basin, and the Southeast Asian Islands. DNF are mainly found in the highlands along the Lena River and the eastern Siberian Plain. In the Atlantic Coastal Plain, the La Plata Plain, the Congo Basin, and the Northeast Asian Plain are almost covered by DBF. CS and OS are found across the Siberian Plateau, northern Europe, and the northern coastal regions of North America, as well as in the Cordillera mountains of South America, the Kalahari Desert of Africa, and the Central Desert of Oceania. WS, Savannas, and Grasslands are found in all regions of the world, mainly due to the broad definition of grasslands here, which classifies tundra in polar regions and meadows in highlands as grasslands. Croplands are mainly distributed in tropical and temperate plains, and the world's major croplands are distributed in the coastal areas of East Asia, South Asia and south eastern Europe.

3.1.2. Zonal distribution of global vegetation. Figure 3 shows the class-based are of global vegetation within each 5° latitudinal zone during 2001 to 2018. The general pattern of discussed vegetation types fallen in each 5° latitudinal zone exhibits relative stable from 2001 to 2018, in which the EBF is distributed at 45°-45°S and is mainly concentrated 15°N and 10°S; DBF is distributed in 65°-50°S, mainly concentrated in 55°-45°N; MF concentrated distribution 55°-65°N; OS is centrally distributed at 70°-80°N and 10°-30°S. Figure 3 shows that the overall distribution ratio of vegetation in each region from 2001 to 2018 is basically similar and at the same time is basically similar to the area share of vegetation band in 2019 in Figure 4, thus concludes that the overall spatial distribution of global vegetation is more stable in the last 20 years.

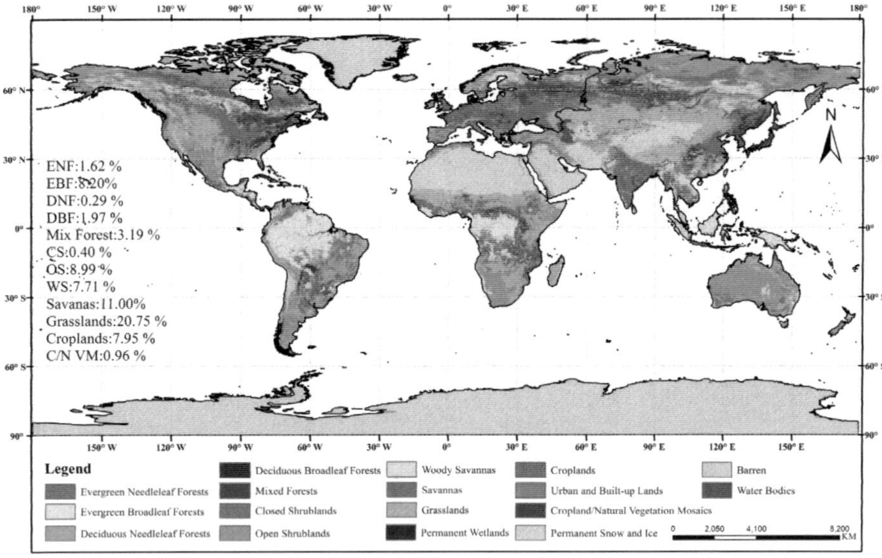

Figure 2. Global land cover distribution map in 2019.

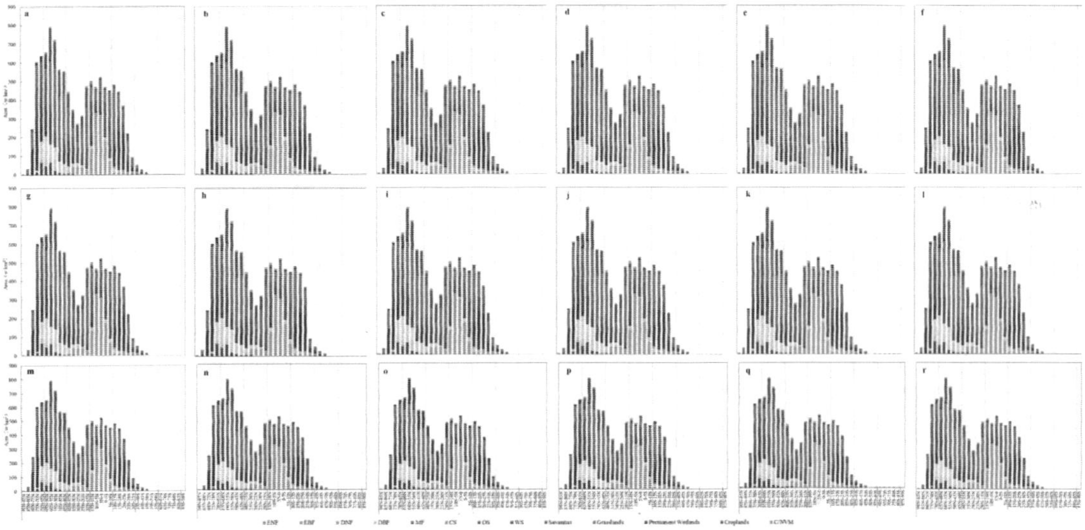

Figure 3. Zonal distribution of global vegetation types (accumulated bars) within each 5°latitudinal belt during 2001-2018 respectively in sub-figures (a-r).

EBF distributed in northern of South America and Southeast Asia within10°N-10°S occupies the largest contribution with 40.04% of land vegetation covers and an area of 5.6 × 10^6 km^2.Grasslands can be found in most of global terrestrial region with an area of 3.4 × 10^6 km^2 from 50°-55°N. According to Figure 4b, the grasslands all have latitudinal distribution area peaks at 80°-65°N, 55°-30°N, 20°-15°N, and 10°-20°S. ENF and DNF locates mainly in the northern boreal forest zone or the adjacent region extended poleward with a total area of more than2.4 × 10^6 km^2. OS as steppe and forest transition zone grows over the western Siberian plains, extending to the Arctic Circle. Global shrubs are mostly found in 80°-50°S, a total of 1.1×10^7 km^2, thereinto the southern hemisphere transition zone occupies 2.2×10^6 km^2. Among them, OS are mainly in 75°-80°N and 15°-20°S, and CS are mainly in 60°-70°N. 91.5% global cropland is mostly distributed in Asia, America and South-eastern South America, Central Africa within 65°N-40°S, totaling $(9.6+10.5) \times 10^6$ km^2.

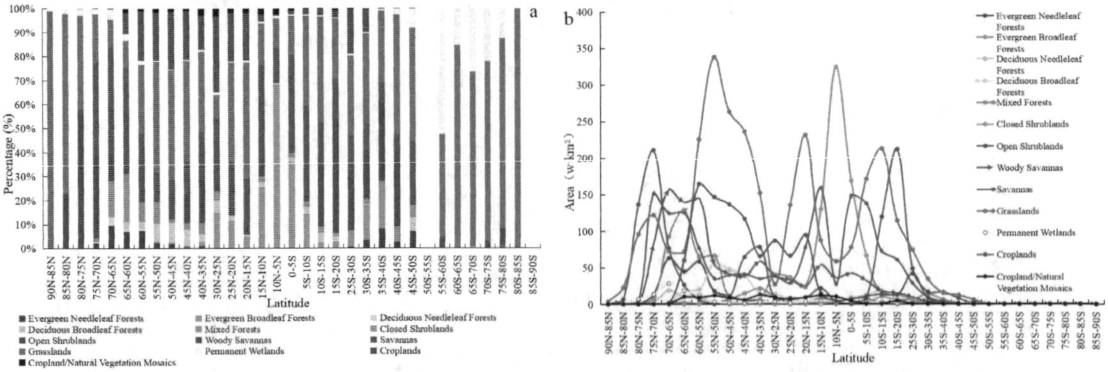

Figure 4. Percentage (a) and area (b) of vegetation cover in the 5° zonal belt in 2019.

3.1.3. Vertical zone characteristics of global vegetation. According to six typical plateaus selected globally, we examined the vertical characteristics of global vegetation over middle-, high-, and mountains higher than 1000+ meters, and achieved the following results.

1) North American Rocky Mountains (A). The nival region (above 5000 meters) is dominated by Permanent Snow and Ice (PSI) (93.473%), Barren (4.310%). The 3500-5000m area is divided into large amounts of Grasslands (62.356%), and also some Barren (13.713%). 1000-3500 m area is located in the northern and eastern Rocky Mountains, and the main vegetation consists of Grasslands (40.473%) and OS (13.021%). The north to the south through the region has been a major fire, the formation of shrub areas more shrubs wormwood.

2) The Andes of South America (B). There are EBF (21.025%) of vegetation formations at 1000-3500m elevation, Grassland (24.433%), and EBF, respectively. The Central Andes have a large number of large plateaus above 3500m eastern humid so growing OS (1.441%), Grasslands (7.007%). The sudden change in elevation of the mountain range in the middle leads to large areas of Barren (87.104%) in the region above 5,000m and eventually to the highest ecotone of PSI. Part of the plateau above 3500m due to the presence of populated plateaus with small amounts of frost-resistant crops [12].

3) The Great Dividing Range of Australia (C). Australia's overall topography is not high, with locations greater than 1,000m above sea level mainly in the Great Dividing Range in the east. The main vegetation types from low to high latitude are ENF (25.667%), Grassland (36.973%), and Grassland transition. The southern mountains of New Zealand are higher, vegetation type is not rich, in the central high-altitude areas are mainly Barren (98.190%), Grassland (1.63%) and a small area of PSI, the northern part of the mountain range has a small amount of ENF.

4) East African Plateau (D). The climate of East African Plateau is mostly Savanna, and the vegetation type from north to south transitions from Barren to WS and Savanna to OS. The vegetation in the mid-mountain region at altitudes of 1000-3500m is mostly Grassland (36.210%), Savanna (18.74%) and OS (16.03%); the vegetation in the region at 3500-5000m is basically similar to that in the middle region. Above 5000m are mainly Barren (81.188%) and PSI (15.842%), which is related to its unique landscape.

5) Asian Himalayas (E). The basal zone of the vertical zone spectrum of alpine semi-humid scrub meadows above 5000m in Asia is alpine scrub meadows, and most of the vegetation in the alpine dry zone is cold dry alpine grassland (34.593%). Most of the vegetation within the alpine arid desert steppe zone is Barren (56.896%), while the rest is mostly PSI and Water Bodies. 3500-5000m is

mainly distributed in central and western Asia, occupied mainly by Grassland (56.795%), Barren (35.188%), and PSI (1.491%). 1000-3500m is mainly Grassland (39.330%), Barren (26.832%).

6) European Alps (F). The Alps transitioning from shrub layer to steppe from south to north, mostly PSI (75.00%) on peaks over 5000m above sea level, followed by WS (16.071%). Also, due to the increase in average temperature in recent years [13], the formations richness gradually increases, but the number of communities decreases, and the vegetation transitions from west to east from WS (13.022%), MF (9.886%) and then Savanna (10.545%).

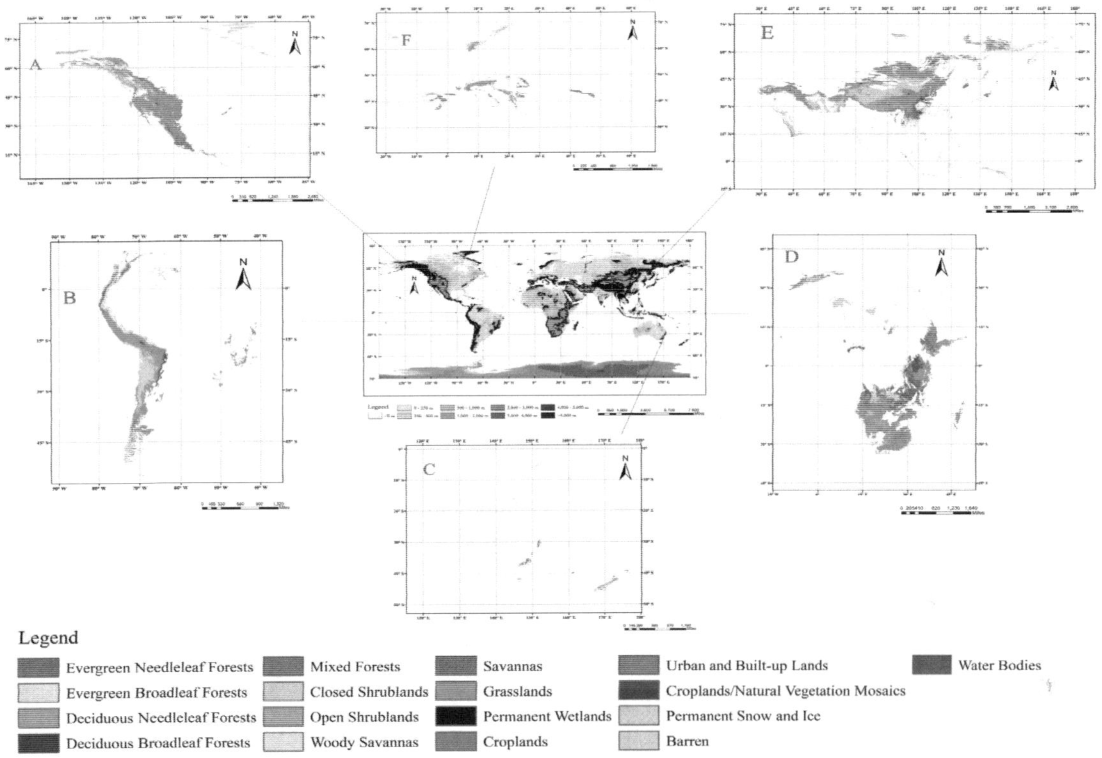

Figure 5. In this case simply justify the caption so that it is as the same width as the graphic.

3.2. *Dynamics of global vegetation*

The purpose of exploring the overall spatial distribution of global vegetation is to pave the way for further understanding of the dynamic changes of global vegetation, to quantitatively analyze the transfer of vegetation in and out through land use dynamics and land transfer matrix, and then to analyze the dynamic trends of vegetation and explore the driving factors of changes.

Firstly, the global land use dynamic index[14] was calculated by using the monitoring method of land use dynamics to clarify the rate of vegetation change in quantity over the last 20 years (Table 1). The study yielded an average annual decrease of -0.878% in DNF, -0.263% in ENF, -0.209% in DBF, -0.033% in EBF, -0.023% in Grassland, -0.010% in Savannas, while remaining vegetation types are increasing. With an average annual increase of C/NVM increased by 0.502%, MF increased by 0.128%, OS increased by 0.057%, WS increased by 0.054%, 0.024% in CS and Cropland increased by 0.019%. Among them, the obvious increase was MF, C/NVM, and forest all declined.

The results of the global vegetation transformation land cover transfer matrix (Table 2) show the transition from forest to grassland is obvious. OS is transformed into Grassland ($0.917 \times 10^6 km^2$), WS, Savannas; WS is mainly transformed into Savannas ($1.317 \times 10^6 km^2$), notably transformed into EBF, Grassland; Savannas mainly transformed into Grassland ($1.949 \times 10^6 km^2$), WS, and DBF; Grassland mainly transformed into Savannas (2.092 × $10^6 km^2$), OS and Croplands; Croplands mainly transformed into Grassland (1.116 × $10^6 km^2$), Savannas, OS.

Table 1. Global land use dynamic degree for 2001-2019.

	ENF	EBF	DNF	DBF	MF	CS
Land Use Dynamic Index	-0.263%	-0.033%	-0.878%	-0.209%	0.128%	0.024%
	OS	WS	Savannas	Grassland	Cropland	C/NVM
	0.057%	0.054%	-0.010%	-0.023%	0.019%	0.502%

Table 2. Global land cover type transfer matrix for 2001-2019.

Area ×10^6 (km^2)		2001										
		ENF	EBF	DNF	DBF	MF	CS	OS	WS	Savannas	Grasslands	Croplands
	ENF	1.967	0.033	0.002	0.003	0.074	0	0.003	0.344	0.027	0.034	0.007
	EBF	0.018	6.598	0	0.052	0.067	0.004	0.049	0.688	0.947	0.805	0.244
	DNF	0.006	0	0.286	0.005	0.024	0	0	0.152	0.042	0.001	0
	DBF	0.004	0.062	0.004	1.768	0.141	0.003	0.03	0.218	0.266	0.158	0.112
	MF	0.115	0.084	0.024	0.073	3.579	0	0.003	0.311	0.165	0.067	0.092
2019	CS	0	0.002	0	0.003	0	0.135	0.146	0.018	0.039	0.132	0.01
	OS	0.001	0.007	0.001	0.01	0.002	0.188	9.662	0.078	0.406	0.917	0.1
	WS	0.193	0.669	0.088	0.279	0.39	0.012	0.084	6.069	1.317	0.514	0.226
	Savannas	0.034	1.248	0.025	0.3	0.219	0.032	0.391	1.502	8.106	1.949	0.471
	Grasslands	0.022	0.507	0.001	0.132	0.059	0.154	1.824	0.42	2.092	21.278	1.245
	Croplands	0.005	0.157	0	0.064	0.034	0.004	0.211	0.151	0.423	1.116	8.247

Figure 6 (A-H) exhibits several representative regions with large changes. Part of the PSI in the Victoria Islands region of northern North America (A) converted to Barren and Grassland. The Amazon plains of Brazil (B) EBF and DBF transformed into WS and Grasslands, Nigeria (C) along the Gulf of Guinea was subject to capital city expansion [15], Grasslands and Savannas converted to Urban and Built-up lands. Large areas of Barren in the arid regions of northwest India and northwest China (D, G) converted to shrublands and Grasslands [16]. The Victoria Desert (E) in central Australia is expanding outwards, the surrounding Barren is increasing, the central lakes have dried up and become Barren [17], the cities and construction land in Pearl River Delta of China (F) increased on the basis of the decrease in Savanna and Barren [18]. And the Western Siberian region (H) was the region with the highest average temperature growth in the northern hemisphere in the past few decades [19].

4. Discussions

The basic pattern of global vegetation spatial distribution is determined by a combination of climate, circulation, altitude, monsoon, etc. The ecology of natural plants is most significantly influenced by climate, and regions with different climatic types and geographical location have different characteristic plants [20].

North and South America is surrounded by sea, influenced by temperate continental climate, winter and summer temperature difference is large, precipitation is low, the Cordillera mountains block the Warm North Pacific Current in western North America, so that inland dry, unsuitable for the growth of EBF. In the mountain range of low and slow areas, the summer Atlantic wet airflow can flow through the central plains, but the winter high pressure blowing dry and cold wind is still unsuitable for the growth of broad-leaved forests and then clustered shrubs.

Large areas of shrubland exist in Australia, Africa and southern of North America, where shrubs can survive in both arid semi-arid tropical desert climates and temperate continental climates. While Australia, Africa and southern North America have tropical desert climates, which are controlled by subtropical high pressure and the trade winds due to high temperatures and low precipitation throughout the year, resulting in the inability to grow broadleaf and coniferous forests with clustered shrublands. But central Australia, the Great Victorian Desert, is expanding outwards, with an increase in Barren around it and the drying up of central lakes transformed into Barren. The reduced 0.061×10^6 km^2 forested land over low-altitude Brazilian Amazonian plains is basically transformed into Grassland and Cropland. And Nigeria along the Gulf of Guinea is subject to expansion changes in the capital, and Grasslands and Savannas are transformed into urban construction land [21].

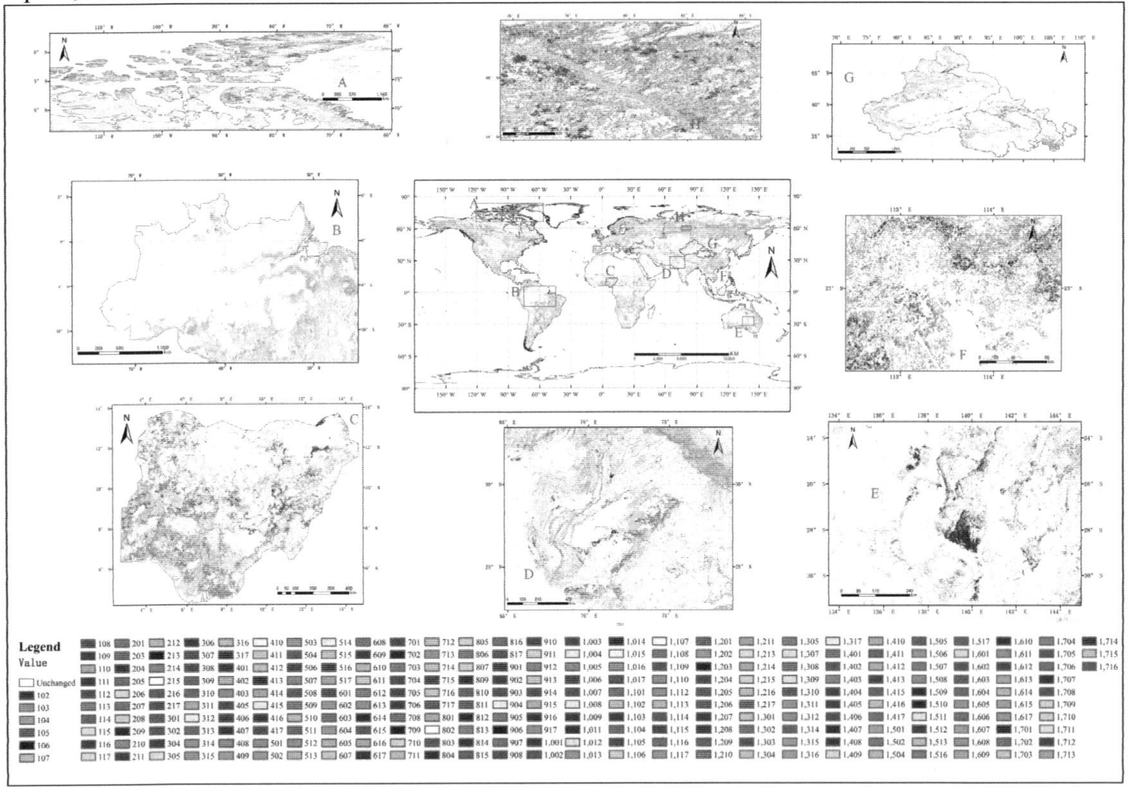

Figure 6. Spatial distribution of global land transfers from 2001 to 2019.

Note: The empty values in the legend are unchanged, 102 is the change from category 1 to 2, 110 is the change from category 1 to 10 and so on. 1-17 category values are: ENF; EBF; DNF; DBF; Mixed Forests; Closed Shrublands; Open Shrublands; Woody Savannas; Savannas; Grasslands; Permanent Wetlands; Croplands; Cropland/Natural Vegetation Mosaics; Urban and Built-up Lands; Permanent Snow and Ice; Barren; Water Bodies.

Asian, European continents and northern North America are temperate continental climate, which are located in the high latitude zone, controlled by the northeast polar wind and the subpolar low-pressure zone. Little rain and low temperature throughout the year, hence it is easy to promote the growth of large shrubs. The growth of crops needs to meet the basic water and heat requirements, and the temperature and rainfall requirements are large, therefore the Cropland is mainly located in the tropical and temperate plains. It is worth noting that large areas of Barren in arid regions of India and northwest China have been transformed into shrublands and Grasslands, with an increase of 0.099×10^6 km^2 and 0.005×10^6 km^2 of woodland and grassland in China and India, respectively, in the last 20 years [22]. The eastern side of Asia and the Rocky Mountains of North America both span three temperature zones in the northern hemisphere, but the Rocky Mountains as a whole are higher in elevation than eastern Asia, and at low latitudes the eastern part of Asia is warm in Japan, while the southern part of North America is cold in California, so the vegetation appears to be more woodland in southern Asia and less grassland in North America. In the middle and high latitudes, the cold current of Thousand Islands in the east of Asia and the warm current of Alaska in the west of North America are affected by the latitude and altitude factors, so there are grasslands, shrubs and snow in the two places respectively. Meanwhile, Western Siberia is the region with the highest average temperature increase in the Northern Hemisphere in the past decades, and spatial changes show that wetlands and mixed forests in this region have been transformed into WS and Savannas in the last 20 years, and 0.151×10^6 km^2 of forests in Russia have disappeared [23].

The causes of the above changes are directly related to climate, i.e., moisture conditions determine the main factors of vertical zone structure changes of mountain vegetation, and heat as the dominant factor in latitudinal changes becomes a sharp contrast. However, it must be pointed out that the vertical zone distribution height rises from east to west, which is still mainly the result of heat change. At the same time, dry air inevitably leads to warming, not to mention the role of heat sources and the compensation of temperature by strong radiation played by the plateau surface have an impact on the warming of the west and the growth of plants. However, the low OA of the IGBP classification (overall - MCD12Q1 product), being around 75%, is a weak point of this contribution. And in future studies, we will compare other products, elongate the timeline, and analyze the global vegetation changes this year in greater depth, so as to better protect vegetation.

5. Conclusions

Based on the valid MCD12Q1 remote sensing image data provided by NASA from 2001 to 2019, this paper analyzes the overall change of global vegetation from the basic pattern of global vegetation spatial distribution and the dynamic changes in the past 20 years, we draw the following conclusions:

(1) The vegetation formations of the global latitudinal zones have remained basically unchanged in the past 20 years, and the spatial distribution pattern of vegetation formed by the long-term evolution of ecosystems is generally stable, with forests, irrigation, grasses and farmlands accounting for about 15.3%, 9.4%, 39.5% and 8.9% of all surface cover types, respectively.

(2) The vertical distribution of vegetation in different altitude regions is obvious, with Barren, Grassland and PSI accounting for 57.703%, 33.837% and 7.968% of the global elevation above 5000m. Grassland, Barren and PSI accounting for 51.390%, 33.761% and 8.793% of 4000-5000m, respectively, while the most land cover types in the 1000-4000m area are shrubs and Grassland.

(3) In the past 20 years, the global forest land has decreased at a rate of 0.251% per year, the Grassland, shrublands, Cropland and C/NVM tended to increase, with an average annual dynamic increase of 0.021%, 0.081%, 0.019% and 0.502%. 55.1%, 14.6% of the forest are converted to WS and Savanna, 45.5% of shrublands are converted to Grassland, which occurs mainly near the high latitudes of the Arctic Circle, and a total of 48.9% of the Barren are converted to Grassland, which occurs in India and China mostly.

Acknowledgments

This work was supported by the National Key Research and Development Program of China (No. 2020YFA0608501), the National Natural Science Foundation of China (No. 42071351), Talend recruited program of the Chinese Academy of Science (No. Y938091), Project supported discipline innovation team of Liaoning Technical University (No. LNTU20TD-23), Project of Innovation and Entrepreneurship Training Program for University Students (202110147026), The Natural Science Foundation of Liaoning Province (No.2020-BS-259).

References

[1] Zu J.X; Yang J 2016 Temporal variation of vegetation phenology in northeastern China *Acta Ecologica Sinica*. **36** 2015-2023
[2] Ma X.F.; Chen S.Y.; Deng J.; Feng Q.S.; Huang X.D. 2016 Vegetation phenology dynamics

and its response to climate change on the Tibetan Plateau *Acta Prataculturae Sinica.* **25** 13-21

[3] Chang X.L; Lv S.H; Feng Z.Y. et al. 2015 Impact of topography on the spatial distribution pattern of net primary productivity in a meadow *Acta Ecologica Sinica.* **35** 3339-3348

[4] King D. J.; Olthof I.; Pellikka P. K. E. et al. 2005 Modelling and mapping damage to forests from an ice storm using remote sensing and environmental data *Nat. Hazards.* **35** 321-342

[5] Khatami R.; Mountrakis G. 2012 Implications of Classification of Methodological Decisions in Flooding Analysis from Hurricane Katrina *Remote Sens.* **4** 3877-3891

[6] Australian Government. https://www.awe.gov.au/

[7] Menashe S. D.; Friedl M.A. User guide to collection 6 MODIS land cover (MCD12Q1 and MCD12C1) product. USGS: Reston, VA, USA, 2018: 1-18

[8] Danielson, J.J.; and Gesch D.B. 2011 Global multi-resolution terrain elevation data 2010 (GMTED2010): U.S. *Geological Survey Open-File Report*

[9] Wang X.L.; Bao H.Y. 1999 Study on the Methods of Land Use Dynamic Change Research *Progress in Geography.* **1** 83-89

[10] Ran J.B.; Chen X.W. 2015 Displacement and Substitution of the Land-use Spatial Pattern *Journal of Geo-information Science.* **17** 661-667

[11] Tian P.; Gong Y.W.; Zhu Z.L.; et al. 2020 Evaluation of Land Use Change in the Ecological Barrier Zone of the Dam Area of the Three Gorges Reservoir in Recent 25 Years *J. Soil Water Conserv.* **34** 78-85

[12] Molina E G, Little A V. Geoecology of the Andes: the natural science basis for research planning *Mountain research and development* 1981: 115-144 DOI:10.2307/3673119

[13] Lamprecht A.; Semenchuk P.R.; Steinbauer K.; et al. 2018 Climate change leads to accelerated transformation of high - elevation vegetation in the central Alps *New Phytol.* **220** 447-459

[14] Feng G.Y. 2018 Study on Dynamic Changes and Driving Factors of Vegetation in the Southwest China during 2001-2016. Master Thesis. Chongqing: Southwest University

[15] Ujoh, F.; Kwabe, I.D.; Ifatimehin, O.O. 2010 Understanding Urban Sprawl in the Federal Capital City, Abuja: Towards Sustainable Urbanization in Nigeria *Journal of Geography and Regional Planning* **3** 106-113

[16] Fang J.Y.; Piao S.L.; He J.S.; et al. 2003 Vegetation activity in China has increased in recent 20 years *Science in China (Series C: Bioscience).* **33** 554-565+578-579

[17] Feng S; Wang S.R; Shen H.B. 2014 Protection and management of water environment on lakes in Australia *World Environment* **4** 67-69

[18] Li J.X.; Wang J.; Di Y.H.; Cai A.L. 2019 Change Characteristics of Coastal Wetlands in the Pearl River Delta under Rapid Urbani-zation *Wetland Science* **17** 267-276

[19] Filandysheva L.B.; Romashova T.V.; Yurkova X.D. et al. 2018 Analysis of the Dynamics of Time Characteristics of Seasons of the Year in the Subtaiga Subzone (Hemiboreal Forests) of the West Siberian Plain Over the Period From 1936 to 2015 *IOP Conference Series: Earth and Environmental Science.* 204 012010

[20] Zhou S.Z. 1980 On the Classification of Word Climates *Journal of Shanghai Normal University (Natural Sciences)* **3** 1-14

[21] Kibret. Kefyalew Sahle. Carsten Marohn. et al. Assessment of land use and land cover change in South Central Ethiopia during four decades based on integrated analysis of multi-temporal images and geospatial vector data *Remote Sensing Applications: Society and Environment* **3 (2016)** 1-19

[22] Chen C.; Park T.; Wang X.H. et al. 2019 China and India lead in greening of the world through land-use management *Nat sustain.* **2** 122-129

[23] P. R, B. P. T. V, J. V. M, V. N, S. k. C and J. T. Harish 2022 Mechanism for Predicting Forest Land Cover from Remote Sensing *International Conference on Smart Systems and Inventive Technology (ICSSIT)* 1492-1497

IOP Conf. Series: Earth and Environmental Science 1087 (2022) 012049 doi:10.1088/1755-1315/1087/1/012049

Spatial-Temporal distribution of cloud covers over China during 2001-2020

Yanmin Shuai[1,2,3,4], Yingxin Cao[1,6], Congying Shao[1], Mengjin Wu[1] and Yu Ma[5]

[1] Liaoning Technical University, Fuxin 123000, Liaoning, China;
[2] Xinjiang Institute of Ecology and Geography, Chinese Academy of Sciences, Urumqi 830011, Xinjiang, China;
[3] Research Center for Ecology and Environment of Central Asia, Chinese Academy of Sciences, Urumqi 830011, Xinjiang, China;
[4] University of Chinese Academy of Sciences, Beijing, 100049, China;
[5] RIOH Automobile Testing & Certification Technology Co., Ltd., Beijing 101103, China.

[6] Corresponding author's e-mail: 472020731@stu.lntu.edu.cn

Abstract. Information of cloud covers is the essential parameter to help better understand weather, climate, water cycle, and solar energy variation of the land surface, as well as the organization of effective satellite-based observations for numerous surface monitoring applications. Compared with traditional cloud information collected by limited ground stations, satellite remote sensing provides the opportunity to investigate the spatiotemporal features of cloud covers from above with actual measurements over global or regional scale. Few previous studies have explored the various cloud types globally or across China. In this work, the 20-year spatial and temporal distribution features and trends of various cloud types (clouds, cloud shadows and cirrus clouds) in China from 2001 to 2020 were investigated using MODIS daily surface reflectance product. Our results showed that: In terms of overall spatial distribution, the southeast was cloudy year-round with up to 75% cloud cover, while the northwest arid zone was mostly clear skies. Cloud shadows were mainly concentrated on complex undulating surfaces composed of high mountains, and cirrus clouds were mostly distributed in the northern regions of high latitudes. In recent 20 years, clouds showed a downward trend in North China, with an annual decreasing rate of more than 0.5% in Inner Mongolia, and a significant increasing trend of up to 0.6% in the middle and lower reaches of the Yangtze River. Cloud shadows had little change in trend in most parts of China. Cirrus clouds had a decreasing trend in most areas, especially in the Greater Khingan Mountains, where the decrease can reach 0.6%.

1. Introduction

A cloud is a mass of water drops or ice crystals suspended in the atmosphere, which usually is formed when water vapor evaporating from the ground or transferring from other areas condenses within a cool enough circumstance in the atmosphere [1]. Clouds driven by wind can transport water vapor from one place to another, with the consumption of energy in the movement or transition of water phases. Further higher cloud covers can reflect more solar shortwave radiation directly back into space at daytime and prevent longwave radiation escaping from land surface at daytime. Cloud types are determined by the condensation nuclei, temperature, wind, and other conditions where a cloud forms.

Cloud shadows are due to the sensor and the ground in the radiation transmission, blocked by the cloud layer, thus forming a dark shadow area at the surface. The existence of cloud shadows make it difficult to obtain accurate information on the ground [2]. High clouds usually form several kilometers above the ground, while low clouds just locate within 1-2 kilometers of Earth's surface. Compared with cumulus clouds and stratus clouds, cirrus clouds are often curved by the wind with thin and wispy features. Cirrus clouds are one of the most common cloud types made up of ice crystals and sit above the troposphere [3], their formation is mainly due to the lifting of the atmospheric system or deep convection, which determines the earth-atmosphere radiation balance. Thus, clouds are an indispensable part of the land-atmosphere hydrological cycle, and a key factor affecting the interception of solar radiation by the surface and the acquisition of clear satellite remote sensing observations of the earth.

At present, cloud monitoring is mainly achieved through conventional ground-based observations and satellite data [4]. Traditional cloud investigation usually works on ground observations collected by limited climate stations. A further interpolation is required when spatial distribution information is needed [5]. Compared with traditional cloud information collected by limited ground stations, satellite remote sensing provides the opportunity to investigate the spatiotemporal features of cloud covers from above with actual measurements over global or regional scale. Satellite data products such as ISCCP have been used to analyze global cloud variability [6], but fewer studies have used MODIS satellite remote sensing information to analyze cloud variability, which has the advantages of multispectral and higher resolution compared with ISCCP and can better achieve a quantitative analysis of large scale clouds. Most published studies mainly focused on the various characteristics of total cloud cover in local areas [7-9], but less on the characteristics of various cloud types, especially over the whole region of China. In this work, we used MOD09GA daily data to investigate the status, multi-year frequency distribution, and potential trend of clouds, cloud shadows and cirrus clouds in China from 2001 to 2020.

2. Study Area and Data

China located in east Asia possesses vast and diverse landscapes dominated by dry seasons and wet monsoons, which lies between latitudes 20° and 54° N, and longitudes 73° and 135° E [10]. The terrain is high in the west and low in the east, presenting a step-like distribution, with complex and changeable landforms, mainly mountains, plateaus, and hills. Due to China's vast territory, the monsoon climate is significantly influenced by its latitudinal position and the distribution of land and sea. There are mainly five climate types in China: the southernmost tropical monsoon climate, the Qinling-Huai River as the boundary between the north and south divided into subtropical monsoon climate and temperate monsoon climate, the northwest inland area is temperate continental climate, the Tibet Plateau area is a plateau mountain climate with perennial low temperature. According to the differences in climate and natural environment, China can be divided into seven geographical regions as shown in Figure 1 [11].

We organized MODIS daily product to serve for the investigation of spatiotemporal distribution and variability of clouds. MOD09GA, which provides bands 1-7 in a daily gridded level-2 product. The 1km observation and geolocation statistics provide a band for Quality Assurance (QA), storing 11 state flags at the pixel scale including clouds, cloud shadows, land/water, aerosol quantity, cirrus, internal cloud algorithm, internal fire algorithm, snow/ice, pixel is adjacent to cloud, salt pan, internal snow mask [12]. Totally, 7260 images of the daily MODIS surface reflectance dataset MOD09GA from 2001 to 2020. In this study, all image retrieval, mosaicking, and clipping are completed through JavaScript interface using Google Earth Engine (GEE).

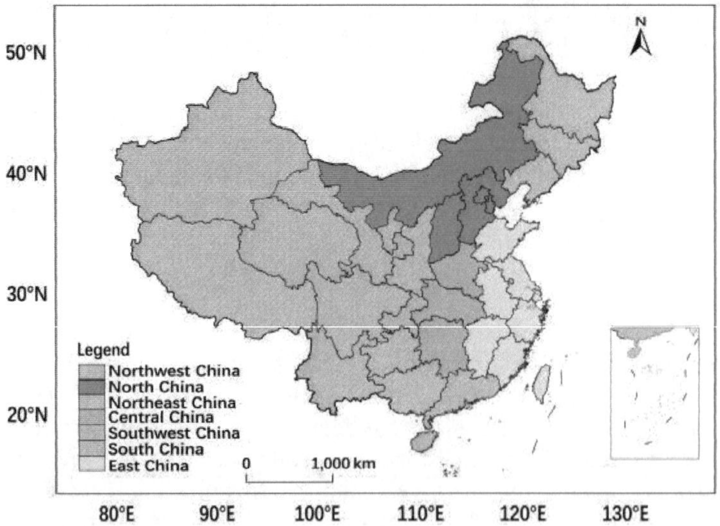

Figure 1. Distribution of the seven geographical divisions of China.

Note: All figures were produced based on the standard map GS (2016)1569 from the Ministry of Natural Resources Standard Map Service website, with no changes to the base map. boundaries. (http://bzdt.ch.mnr.gov.cn/)

3. Methods

3.1. Frequency statistics

The QA band is used to identify pixel-based cloud from clear days. Pixel-level QA is generated from product code to assess product quality at the finest granularity [13]. This information is useful for users who need to decide on the quality to retrieve at any particular pixel. Since their flags are stored as a binary sequence of 0s and 1s, the different mass states that each image element is labelled with can be identified by a bitwise operation, taking the two rightmost bits. We extracted three quality flags in the QA band recognized different cloud types, denoted clouds with bits 0-1, cloud shadows with bits 2, and cirrus clouds with bits 8-9. The occurrences of clouds, cloud shadows, and cirrus clouds for each pixel during the period 2001-2020 were obtained by accumulating, and the cloud occurrence frequency was calculated for 20 years, analyzing the spatial distribution characteristics of cloud types in different geographical regions.

3.2. Regression analysis

By analyzing the quantitative relationship between the independent and dependent variables, the historical trend of cloud covers in the study area can be simulated by using univariate linear regression. This regression analysis method first plots a set of data points on the X and Y axes, with dependent variables on the vertical Y axis and independent variables on the horizontal X axis. The analyst using the least square method will generate a best-fit line that explains the underlying relationship between the independent and dependent variables. We used the least square method to perform a pixel-by-pixel linear fitting for the interannual time series from 2001 to 2020, with the year as the independent variable and the frequency of cloud occurrence as the dependent variable, to establish the one-dimensional linear regression equation and find out the slope to obtain the interannual variation trend of cloud amount. By building a one-dimensional linear regression model to analyze the trend of each pixel, we can quantitatively study the degree of variation in different cloud types over time. The calculation formula is as followed:

$$\beta = \frac{\sum_{i=1}^{n}(X_i - \overline{X})(Y_i - \overline{Y})}{\sum_{i=1}^{n}(X_i - \overline{X})^2} \tag{1}$$

The formula has two variables X and Y, n is the number of sequences, Xi is the year, Yi is the frequency of clouds corresponding to the year Xi, β is the trend, β greater than 0 indicates an upward trend of clouds over time, β less than 0 indicates a downward trend of clouds over time.

4. Results & Analysis

4.1. Spatial distribution

Figure 2 shows the annual frequency distribution of clouds, cloud shadows, and cirrus clouds from 2001 to 2020 over China. In general, cloud covers in China exhibited in Figure 2. (a) shows a decreasing geographical distribution from the southeast coast region to the northwest inland over the past two decades. China's cloudy areas are mainly concentrated in the south of the Qinling Mountains-Huaihe River Line, the cloud amount is above 65%, especially in Chongqing, Guizhou, and Guangxi, where high values exist all year round, reaching more than 75%. The low value areas of cloud covers are mainly distributed in western Inner Mongolia, western Tibet and Tarim basin, where cloud amount is below 40%. In northwest China, the specific geography makes the cloud distribution more on both sides and less in the middle. In the north of Xinjiang, the wind field movements cause the surrounding water vapor to gather here to form a convergent transport area, while in the central, the Tarim Basin and the Qaidam Basin have large airflow dispersion areas, where water vapor spreads in all directions [14], making the cloud covers less, below 35%.

The frequency distribution of cloud shadows (Figure 2. b) shows that cloud shadows are mostly distributed at complex undulating surfaces consisting of high mountains, in the Altai-Tianshan-Kunlun-Qilian Mountains area in northwest China, and in the Daxing'an-Xiaoxing'an-Changbai Mountains in northeast China, where high value areas of cloud shadows exist, up to 35%. The frequency of cirrus clouds in China (Figure 2. c) is greater in the north than in the south, with high values occurring in northern and northwestern China, reaching over 20%, and a maximum of 30% in the Tarim Basin. Low values of cirrus clouds are mainly distributed in the low latitudes, with cirrus being particularly rare in Guangxi and Guangdong, reaching a minimum of 5%. The spatial distribution of cirrus is closely related to the location of latitude, showing a decreasing distribution characteristic from high latitudes to middle and low latitudes. Since the low latitudes are mainly located in the tropics and subtropics, typical frontal cyclone activities are rare [15], so it is not conducive to the formation of cirrus clouds.

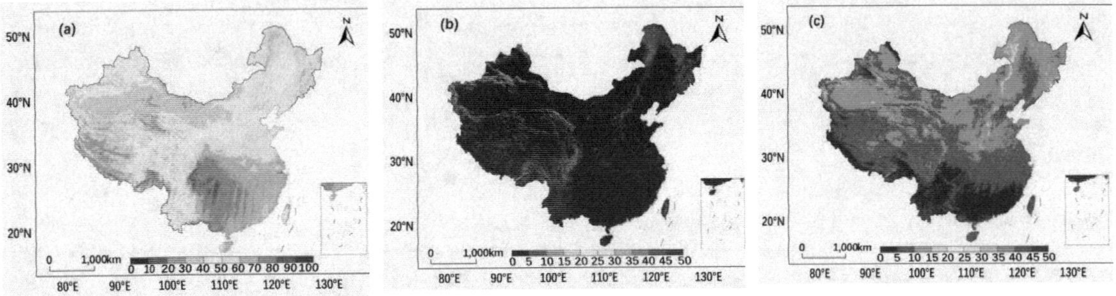

Figure 2. Annual frequency distribution of clouds (a), cloud shadows (b), and cirrus clouds (c) from 2001 to 2020 over China (unit:%).

IOP Conf. Series: Earth and Environmental Science 1087 (2022) 012049 doi:10.1088/1755-1315/1087/1/012049

4.2. Changing trends

Figure 3 gives the annual cumulative trends of clouds, cloud shadows, and cirrus clouds from 2001 to 2020 over China, the slope is used to represent the trend, with a positive value indicating an increasing trend over time and a negative value showing a decreasing trend over time. In the last two decades, cloud covers (Figure 3. a) in Inner Mongolia, Yunnan and most of Tibet has shown a decreasing trend, with the most significant decrease in central Inner Mongolia, reaching 0.5% per year. In northern Xinjiang, central northwest China, and most of southeastern China, cloud covers show an increasing trend, with the most pronounced increase in eastern China reaching over 0.4% per year and even up to 0.6% in the coastal areas. Due to the small base of cloud shadows, the overall change trend (Figure 3. b) is not obvious, with a slight decrease of up to 0.2% per year in the Himalayas and along the Hengduan Mountains. As can be seen in Figure 3. (c), cirrus clouds show a decreasing trend in most of China, especially in the Greater Khingan Mountains of Inner Mongolia, where the decrease can reach 0.6% per year. There is no obvious variation trend of cirrus clouds over central and north China. Cirrus clouds in northern Xinjiang, Tibet and coastal areas along the middle and lower reaches of the Yangtze River show a slight increase trend, with an annual growth rate of 0.2%.

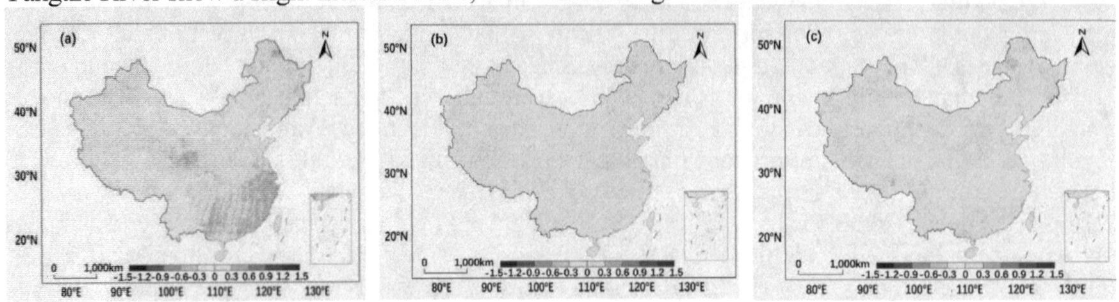

Figure 3. Annual cumulative trends of clouds (a), cloud shadows (b), and cirrus clouds (c) from 2001 to 2020 over China (unit:%).

5. Conclusions

In this work, we extracted clouds, cloud shadows and cirrus for the last 20 years and analyzed their perennial spatial distribution and trends, using the GEE platform. The results show the spatial distribution of cloud covers in China is mainly related to the local substratum environment and water vapor conditions, with a general distribution characteristic of more south and less north. Cloud shadows are mainly distributed in mountainous areas with complex topography and large fluctuations. Influenced by the southeast Pacific monsoon, a large amount of water vapor will condense into clouds as the air lifts, and it is difficult for the huge number of clouds to cast shadows on the surface. Cirrus clouds are more distributed in the north than in the south, mainly concentrated in high altitude and high latitude areas. In most areas of east-central China, clouds tend to increase, while in Inner Mongolia tend to decrease. Cirrus tend to decrease in most areas, with a more pronounced decrease in the Greater Khingan Mountains and a small increase in northern Xinjiang. In the following study, other cloud data such as ISCCP are needed to further analyze the characteristics and trends of cloud distribution.

Acknowledgments

This work was supported by the National Key Research and Development Program of China (No. 2020YFA0608501), the National Natural Science Foundation of China (No. 42071351), Project supported discipline innovation team of Liaoning Technical University (No. LNTU20TD-23).

References

[1] A. Slingo, Guangling Mao 1999 The influence of clouds on climate *J. Meteorological Science*

and Technology **04** 36
[2] Mingzhu Chen, Xianjun Gao 2020 Automatic detection of cloud shadowing using cloud positioning in high-definition remote sensing imagery *J. Geospatial Information* **18** 70
[3] David L. Mitchell, Anne Garnier, Jacques Pelon, Ehsan Erfani 2018 CALIPSO (IIR–CALIOP) retrievals of cirrus cloud ice-particle concentrations *J. Atmospheric Chemistry and Physics* **18** 23
[4] Yahui T, Yuquan Z, Miao C, Qianrong M 2020 The global cloud distribution characteristics based on CloudSat and CALIPSO joint observations *J. Atmospheric Sciences* **43** 917
[5] Weijian Q, Hui H 2009 Analysis of climate change characteristics of cloudiness in Guangxi over the past 45 years *J. Tropical Geography* **29** 525
[6] Yue X, Yanfeng J, Chen W 2019 Analysis of global cloud trend changes using REOF *J.Inner Mongolia Meteorology* **02** 12-16
[7] Jinli Bao 2021 Cloud distribution characteristics of the Tibetan Plateau based on CloudSat data *J. Agricultural Disaster Research* **08** 100-102
[8] Yanan Liu, Hebo Zou, Qiong Wu 2020 Least squares modeling and analysis of total cloudiness in Jiangxi Province based on FY-2G products *J. Journal of Meteorology and Environment* **36** 105-111
[9] Qinghu Ba, et al 2019 Characteristics of low cloud variability and analysis of influencing factors in Bozhou region in the past 54 years *J. Anhui Agricultural Science* **47** 202-206
[10] Can Wang, Houyuan Lu, Jianping Zhang, Zhaoyan Gu, Keyang He 2014 Prehistoric demographic fluctuations in China inferred from radiocarbon data and their linkage with climate change over the past 50,000 years *J. Quaternary Science Reviews* **98** 48-59
[11] Sanbao Wu, Ji Zhao 1995 Physical Geography of China (3rd Edition) *M. Science Press*
[12] Vermote E F, Kotchenova S Y, Ray J P 2011 MODIS surface reflectance user's guide *J. MODIS Land Surface Reflectance Science Computing Facility version*
[13] Masuoka E, Roy D, Wolfe R, et al 2010 MODIS Land Data Products: Generation, Quality Assurance and Validation *J. Land Remote Sensing and Global Environmental Change: NASA's Earth Observing System and the Science of ASTER and MODIS, Remote Sensing and Digital Image Processing*
[14] Keli WANG, Hao JIANG, Hongyan Zhao 2006 Advective and convergent transport of water vapor in northwest China *J. Advances in Water Science* **02** 164-169
[15] Rongshuo C, Xingrong C 2020 Changes in the ocean and its role in China's climate *J. China Population-Resources and Environment* **30** 9-20

Soil heavy metal Pb concentration quantitative inversion method based on hyperspectral remote sensing

Maoyu Chen[1,2], Dan Ke[1,3], Wenkai Wang[1], Wanming Zhang[1] and Xianglan Li[1]

[1] School of Environment and Resources, Xichang University, Xichang 615000, Sichuan, China;
[2] School of Geophysics and Measurement and Control Technology, East China University of Technology, Nanchang 330000, Jiangxi, China.

[3] Corresponding author's e-mail: kedan1125@126.com

Abstract. Taking a mining area in the Liangshan section of the Anning River Basin as the research area, soil samples were collected, and the collected soil samples were respectively measured for Pb concentration and soil spectral reflectance, and the correlation between soil Pb concentration and soil spectral reflectance was analyzed, using the partial least squares regression method to construct the soil Pb concentration estimation model and carry out the accuracy test. The results show that: (1) The correlation between the original soil spectral curve and the measured value of soil Pb concentration is greatly improved after the first-order differential(FD), second-order differential(SD), reciprocal first-order differential(RTFD) and inverse second-order differential(RTSD) transformation, and there are many characteristic bands. (2) The best model for estimating soil Pb concentration is the model established by second-order differential, whose coefficient of determination (R2) is 0.56 and root mean square error (RMSE) is 1.28. It provides a reference for the retrieval of remote sensing hyperspectral soil Pb concentration in this area.

1. Introduction

Soil is the cornerstone of the survival and development of all life on earth, not only an important part of the terrestrial ecosystem, but also the material basis for human survival and development [1-2]. Due to the continuous increase of human activities and the continuous acceleration of industrialization, soil pollution has become increasingly serious, especially heavy metal pollution [3]. Pb is a very harmful and toxic heavy metal. Excessive Pb concentration in soil will not only inhibit the growth of plants, but also enter the human body through the food chain and cause certain harm to human health [4-6]. Therefore, it is very necessary to carry out the monitoring of soil Pb concentration.

Although traditional soil heavy metal detection methods have high detection accuracy, they have long periods, high costs, and it is difficult to achieve large-area, continuous and dynamic monitoring [7-8]. Hyperspectral remote sensing technology has the advantages of high spectral resolution and acquisition of abundant spectral information, which can realize rapid and large-scale determination of soil heavy metal concentration [9-10]. At present, domestic and foreign scholars have made a series of researches on the remote sensing hyperspectral inversion of soil heavy metal concentration, and the inversion methods are very rich. Such as Partial Least Squares [11], Geographically Weighted Regression [12], Multiple Stepwise Regression [13], BP Neural Network [14], Support Vector Machine [15], Genetic Algorithm [16], Random Forest [17], Wavelet Analysis [18] etc. Among them, the partial least squares regression method is currently the most commonly used and the most mature inversion method [19].

In this study, a mining area in the Liangshan section of the Anning River Basin was selected as the research target area, and the soil samples collected in the study area were measured for Pb concentration and soil spectral reflectance, and the correlation between the soil spectral reflectance value and the measured value of soil Pb concentration was analyzed. The characteristic bands were extracted, and the partial least squares regression method was used to establish an inversion model, which provided a basis for the inversion of area-based soil Pb concentration based on remote sensing hyperspectral images.

2. Data collection and processing

2.1. Overview of the study area

The Anning River is about 337km long and is the mother river of Liangshan Yi Autonomous Prefecture. Originating from the northern mountainous area of Mianning County, Liangshan Prefecture, it is the largest tributary of the left bank tributaries in the lower reaches of the Yalong River. It mainly flows through three counties and cities of Mianning, Xichang and Dechang in Liangshan Prefecture, and finally merges into Yalong in Miyi County, Panzhihua City. The Anning River Basin covers an area of about 11,150 km^2. It is the most important industrial and agricultural production area in Liangshan Yi Autonomous Prefecture and the second largest granary in Sichuan Province [20]. In recent decades, heavy metal enterprises in the Liangshan section of the Anning River Basin have discharged a lot of wastewater and residues into the Anning River, among which the pollutants are mainly Pb, Zn, Fe, Mn, which have a serious impact on the animals, plants and human production and life in the entire basin (Figure 1) [21].

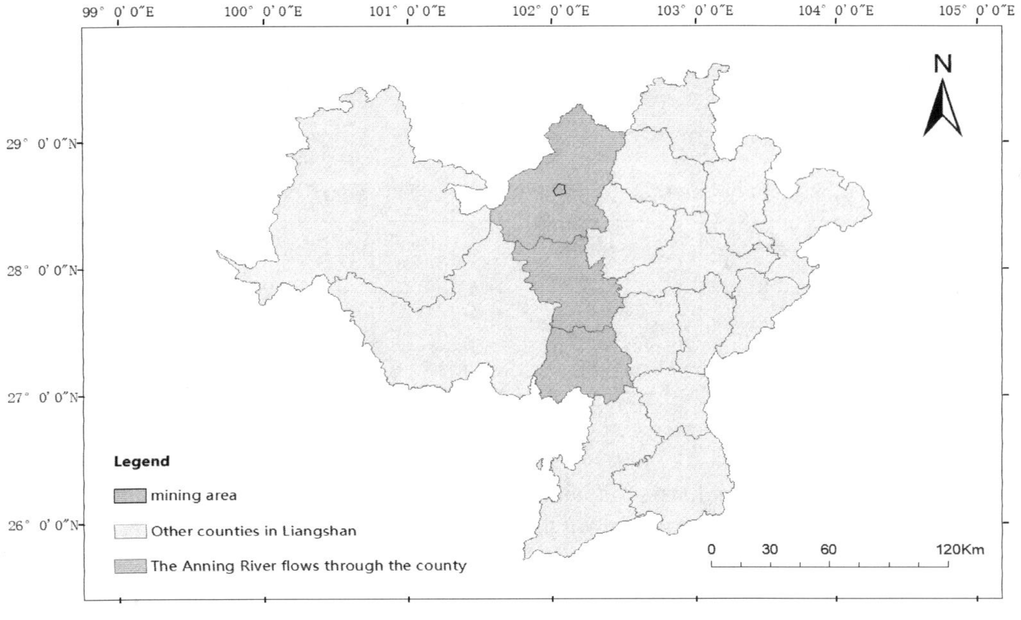

Figure 1. Overview map of the study area.

2.2. Collection and determination of soil samples

A total of 15 soil samples were collected in the study area at a depth of 0-20 cm. All soil samples were air-dried, ground and sieved through 40-mesh nylon under natural conditions, and then the soil spectral reflectance and heavy metal concentration were measured. The spectral reflectance of the soil samples was measured by the FieldSpec Pro ground object spectrometer produced by ASD Company

IOP Conf. Series: Earth and Environmental Science 1087 (2022) 012050 doi:10.1088/1755-1315/1087/1/012050

in the United States [22]. The spectrum collection range is 341.3nm-2521.3nm, and the sampling intervals are 1.4nm (341.3nm-989.3nm), 3.6nm (992.2nm-1899.8nm) and 2.4nm (1902.3nm-2521.3nm). In order to effectively extract the characteristics of soil spectral reflectance and ensure the reliability of later modeling, the adjacent average smoothing process was performed on the original soil spectral curve.

2.3. *Soil spectral curve transformation and extraction of characteristic bands*

In Origin2022 and Matlab software, the first-order differential, second-order differential, reciprocal first-order differential and reciprocal second-order differential transformation were performed on the smoothed soil spectral curve respectively, so as to enhance the characteristics of the soil spectral curve (Figure 2) [23]. From the bands that passed the significance test ($P < 0.05$), the bands with large correlation coefficients were selected as characteristic bands.

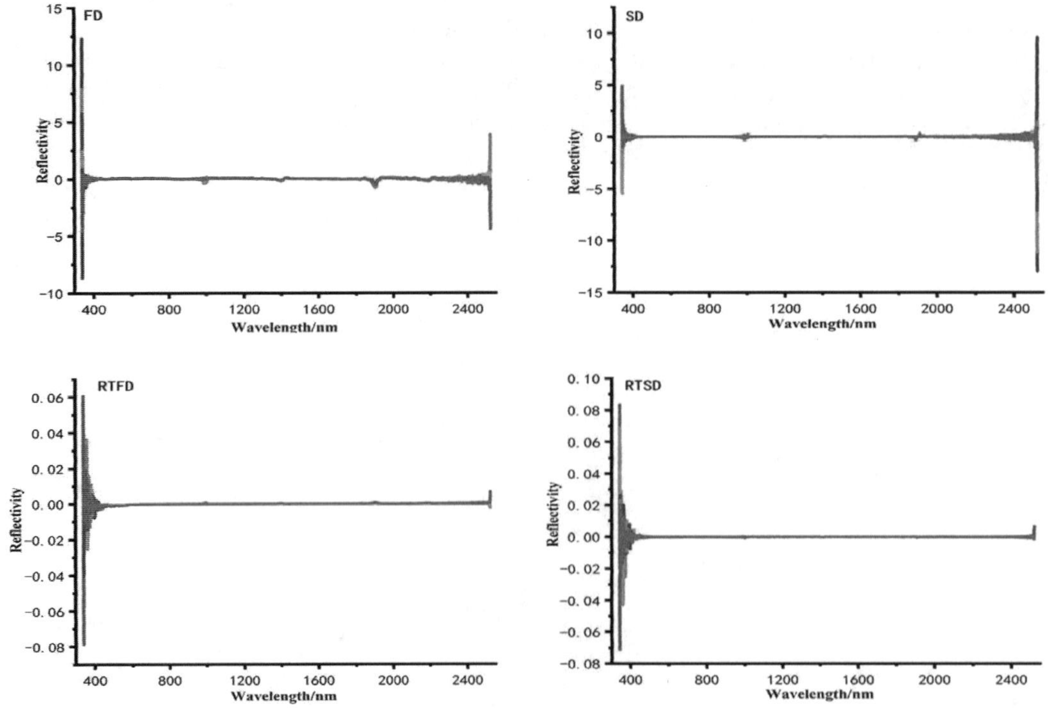

Figure 2. Soil spectrum curve after spectral transformation.

3. Model establishment and verification

The partial least squares regression method can effectively overcome the multicollinearity problem, retain the most important spectral information, and remove the invalid noise interference [24-25]. The equation for partial least squares regression modeling is:

$$y = a + b_1x_1 + b_2x_2 + b_3x_3 + \cdots\cdots + b_nx_n \quad (1)$$

Where a is a constant, b_n is a coefficient, x_n is an independent variable, and y is a dependent variable.

Set the soil spectral reflectance value as the independent variable x and the measured Pb concentration as the dependent variable y, establish a partial least squares estimation model for soil Pb concentration, and use the coefficient of determination and root mean square error to evaluate the partial least squares regression model [26].

The evaluation index R^2 reflects the fitting degree of the model, and RMSE reflects the accuracy of the model estimation results. In the overall evaluation of the model, the model with a large R^2 value and a small RMSE value is selected as the optimal estimation model. 10 samples were randomly selected from 15 soil samples for modeling, and the remaining 5 samples were used for validation.

4. Results and Analysis

4.1. Soil spectral curve analysis

It can be seen from the smoothed soil spectral curve (Figure 3) that the shapes and characteristics of the spectral curves of the 15 soil samples are basically the same. Due to the different contents of the components of each soil sample, the spectral curve of each soil sample showed a different trend. In the visible light region, the spectral curve increases with the increase of the wavelength. In the infrared band, the spectral curve tends to be flat as a whole, and the spectral noise is mainly concentrated between 2200nm-2250nm.

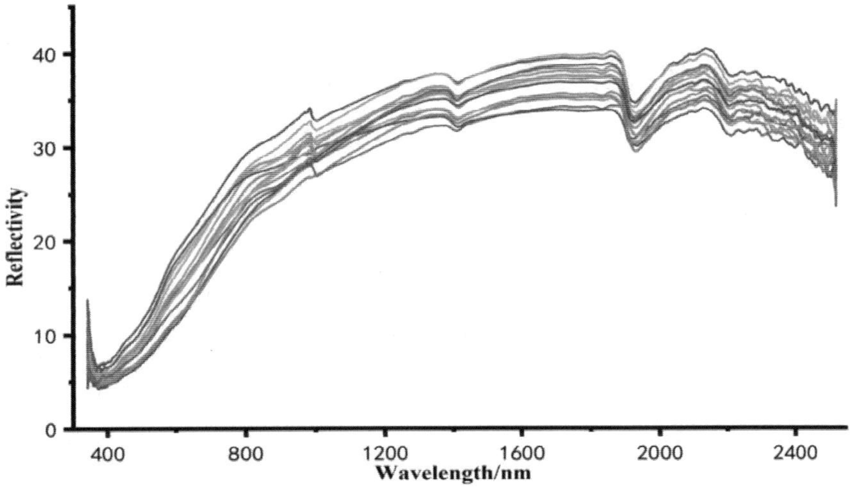

Figure 3. Smoothed soil spectrum curve.

4.2. Feature band extraction

Pearson correlation analysis was carried out between soil Pb concentration and spectral reflectance, and a significant test was carried out (*$P < 0.05$*). Table 1 shows the extracted characteristic bands and the corresponding correlation coefficients. It can be seen from the figure that the absolute values of the correlation coefficients of the selected characteristic bands are all greater than 0.5, indicating that the correlation between the selected characteristic bands and soil Pb elements is relatively large, and the range of the selected characteristic bands is similar to that of Zhou Mo [6], Gulizati Amat [9], and Ainur Maitinuri [12], etc. have roughly the same range of characteristic bands for Pb elements.

Table 1. Characteristic bands and corresponding correlation coefficients.

First-order differential		Second-order differential		Reciprocal first-order differential		Reciprocal second-order differential	
Feature band (nm)	Correlation coefficient	Feature band (nm)	Correlation coefficient	Feature band (nm)	Correlation coefficient	Feature band (nm)	Correlation coefficient
459.6	0.766	458.2	0.768	448.2	-0.637	794.4	-0.684
684.5	-0.634	720.4	0.628	923.0	0.615	1184.0	-0.731
1077.2	-0.669	1050.3	-0.519	1601.6	0.605	1724.3	0.792
1481.4	-0.620	1637.8	0.612	2201.6	0.695	2139.2	0.682
2161.8	0.741	1937.7	-0.594				

4.3. Model analysis

Taking the selected characteristic band as the independent variable, and the measured Pb concentration as the dependent variable, the partial least squares estimation models of soil Pb concentration under four spectral transformations, the first-order differential, the second-order differential, the reciprocal first-order differential and the reciprocal second-order differential, were established respectively.

It can be seen from Table 2 that the estimation model established by the four spectral transformations, the model established by the second-order differential has the best effect, the coefficient of determination is the largest, which is 0.56, and the root mean square error is the smallest, which is 1.28. The specific model is:

$$Y=24.086-29.544X_{458.2}+672.939X_{720.4}-37.897X_{1050.3}+518.586X_{1637.8}+16.866X_{1937.7}$$

Table 2. Model evaluation of soil Pb concentration.

Reflectivity	Number of bands	Modeling samples		Validation samples	
		R^2	RMSE	R^2	RMSE
FD	5	0.93	0.52	0.50	2.24
SD	5	0.84	0.80	0.56	1.28
RTFD	4	0.61	1.25	0.19	2.02
RTSD	4	0.85	0.77	0.36	2.16

5. Conclusion

After the first-order differential, second-order differential, reciprocal first-order differential and reciprocal second-order differential spectral transformation of the original soil spectral curve, the correlation between the soil spectral reflectance value and the measured value of soil Pb concentration is significant There are many bands that pass the significance test ($P<0.05$), which can be used as the basis for selecting characteristic bands.

Among the four estimation models, the estimation model established by the second-order differential has the best effect. Its coefficient of determination is the largest, which is 0.56, and the root mean square error is the smallest, which is 1.28. However, the coefficient of determination of the established inversion model is lower than that of Zhou Mo [6], Li Qiongqiong [7], Liu Xiaoqing [10], etc. The reasons may be: 1. The spectral transformation method is relatively simple, and in the follow-up research, you can try to use spectral transformation methods such as continuum removal [27], envelope removal [28], etc. based on which an inversion model is established for comparison. 2. The coefficient of determination of this study is small, which may be caused by the small number of samples. A larger sample size should be used in future studies to increase the reliability of the findings. The heavy metal estimation model is affected by different soil regions and pollution levels. Whether the established best estimation model is suitable for the estimation of other heavy metal contents in this region needs further study.

Acknowledgments

Project was supported by Xichang University Doctoral Start Foundation Program（YBZ202108）, Key Program for Academic Leader of Bureau of human resources and social security in Liangshan Yi Autonomous Prefecture（ZRS202001）and Program funded by Department of Natural Resources of Sichuan Province（Kj-2022012）. We are grateful to the anonymous reviewers for their thoughtful suggestions.

References

[1] Fu B J, Niu R Q, Wang C S 2021 Soil Nickel Metal Content Estimation Based on Hyper-spectrum in Danjiangkou Reservoir Area *Remote Sensing Information* **36(3)** 44-49
[2] Liu X, Li C, Li S Q, et al. 2019 Application Progress of Hyperspectral Remote Sensing

Technology in Soil Research *Anhui Agricultural Sciences* **47(8)** 18-21+34
[3] Guo X F, Cao Y, Jiao R C et al. 2020 Overview of Hyperspectral Remote Sensing Monitoring Method of Soil Heavy Metals *Urban Geology* **15(3)** 320-326
[4] Li X M 2019 *Nanfang Agriculture* **13(11)** 156-157
[5] He J L, Li Z Y, Li R J, et al. 2019 Study on the estimation model of soil heavy metal Pb based on spectral characteristics of soil organic matter *Environmental Pollution and Prevention* **41(7)** 753-757
[6] Zhou M, Zou B, Tu Y L et al. 2020 Hyperspectral Modeling of Pb Content in Mining Area Based on Spectral Feature Band Extracted from Near Standard Soil Samples *Spectroscopy and Spectral Analysis* **40(7)** 2182-2187
[7] Li Q Q, Liu Y L 2019 Soil Heavy Metals Estimation based on Hyperspectral in Urban Residential *Remote Sensing Technology and Application* **34(3)** 540-546
[8] Liu Y P, Luo Q, Cheng H F 2020 Application and development of hyperspectral remote sensing technology to determine the heavy metal content in soil *Journal of Agro-Environment Science* **39(12)** 2699-2709
[9] Amat G, Abliz A, Sawut R, et al. 2018 Estimating Soil Lead Content in the Eastern Junggar Coalfield Using Hyperspectral Data *Chinese Journal of Soil Science* **49(5)** 1233-1239
[10] Liu X, Liu Y L 2019 Soil heavy metals estimation based on hyperspectral in urban traffic greenbelt *Environmental Science & Technology* **42(5)** 230-236
[11] Ma L, Yan A 2019 Study on Soil Lead Content Inversion Based on Hyperspectral and Landsat8 Remote Sensing Images *Shandong Agricultural Science* **51(12)** 120-126
[12] Matnuri A, Eziz M, Turgun M, et al. 2021 Hyperspectral Estimation of Heavy Metal Pb Concentration in Vineyard Soil in Turpan Basin *Remote Sensing Technology and Application* **36(2)** 362-371
[13] Zhao Y L, Yang N N, Zhang H X, et al. 2020 *Journal of Ecological Environment* **29(4)** 819-826
[14] Wang X M, Maiming Y, Mao D L, et al. 2021 *Journal of Ecological Environment* **30(10)** 2076-2084
[15] Stazi S R, Antonucci F, Pallottino F, et al. 2014 Hyperspectral Visible-Near Infrared Determination of Arsenic Concentration in Soil *Communications in Soil Science and Plant Analysis* **45** 2911-2920
[16] Wang J J, Cui L J, Gao W X, et al. 2014 Prediction of Low Heavy Metal Concentrations in Agricultural Soils Using Visible and Near-infrared Reflectance Spectroscopy *Geoderma* **216** 1-9
[17] Meng X T, Bao Y L, Liu J G, et al. 2019 Regional soil organic carbon prediction model based on a discrete wavelet analysis of hyperspectral satellite data *International Journal of Applied Earth Observations and Geoinformation* **89**(C)
[18] Liu M L, Liu X N, Wu L, et al. 2011 Wavelet-based Detection of Crop Zinc Stress Assessment Using Hyperspectral Reflectance *Computers and Geosciences* **37** 1254-1263
[19] Shi T Z, Chen Y Y, Liu Y L, et al. 2014 Visible and Near-Infrared Reflectance Spectroscopy-An Alternative for Monitoring Soil Contamination by Heavy Metals *Journal of Hazardous Materials* **265** 166-176
[20] Shao Q F 2016 Evaluation of Eco-environment Fragility in Anning River Basin *Chengdu University of Technology*
[21] Tang J, Fu X D, Li J, et al. 2018 Temporal and spatial distribution and main factors of heavy metals in Liangshan section of Anning River *Environment and Development* **30(5)** 138-140
[22] He J L, Han C S, Wei R, et al. 2019 Research on indirect hyperspectral estimating model of heavy metal Cd based on partial least squares regression *Remote Sensing for Land and Resources* **31(4)** 96-103
[23] LI Z Y, Deng F, He J L, et al. 2021 Hyperspectral Estimation Model of Heavey Metal Arsenic in Soil *Spectroscopy and Spectral Analysis* **41(9)** 2872-2878
[24] Wen J T, Zhang X, Zhang B, et al. 2010 A Study of Band Selection Method for Retrieving Soil

Lead Content with Hyperspectral Remote Sensing Data *Advances in Earth Science* **25(6)** 625-629

[25] Lan M, Yang B, Song Q, et al. 2021 Hyperspectral Inversion of Heavy Metal Cd in Mining Area and Its Response to Concentration Gradient *Journal of Anhui Engineering University* **36(5)** 47-55

[26] Zhang X, Ding S T, Cen Y, et al. 2021 Soil Heavy Metal Pb Content Inversion Method by Combining Field with Laboratory Spectra *Geomatics and Information Science of Wuhan University*

[27] Shen Q, Zhang S W, Ge C 2019 Hyperspectral Inversion of Heavy Metal Contention Soils Reconstituted by Mining wasteland *Spectroscopy and Spectral Analysis* **39(4)** 1214-1220

[28] Peng X T, Gao W X, Wang J J 2014 Inversion of Soil Parameters from Hyperspectra Based on Contiuum Removal and Partial Least Squares Regression *Geomatics and Information Science of Wuhan University* **39(7)** 862-866

IOP Conf. Series: Earth and Environmental Science 1087 (2022) 012051 doi:10.1088/1755-1315/1087/1/012051

Permaculture inclusion in the planning of “Sansheng Space” --taking Pu Ran Farm as an example

Weinan Hong[1], Dan Ke[1,3], Chuan Zhong[2], Xianglan Li[1] and Guoqing Li[1]

[1] School of Environment and Resources, Xichang University, Xichang, China;
[2] Liangshan Yi Autonomous Prefecture Bureau of Agriculture and Rural Affairs, Xichang, China.

[3] Corresponding author’s e-mail: kedan1125@126.com

Abstract. Achieving sustainable development is not only the need for our country’s current development, but also a common challenge for all countries in the world. Therefore, Permaculture based on the concept of sustainability is of great significance to my country's land planning, and "Sansheng Space" is also the guidepost of the spatial development of China's national land. In this paper, we study the concept of Permaculture and the "Sansheng Space", and use the Pu Ran Farm as a real case to combine the concept of Permaculture with the "Sansheng Space", and apply the design principles of Permaculture to the "Sansheng Space" planning. The strategy of Permaculture suitable for the planning of "Sansheng Space" of leisure farms is analyzed and summarized, which can provide some reference for rural land planning.

1. Introduction

Permaculture is a combination of the words permanent, agriculture, and culture [1-2]. It was proposed by Australians Bill Mollison and David Holmgren in the 1970s [3], and its aim is not only to achieve the overall planning of the site, but also optimizes the interrelationship between landscape elements [4]. The essence of Permaculture is to create an ecologically sound and economically viable ecosystem by emulating nature's mode of operation [5]. The "Sansheng Space" is the basic carrier of human economic and social activities [6], a medium for human self-development of nature, which reflects the extent of human transformation and self-development of nature. The "Sansheng Space" includes living space, production space and ecological space [7-9]. The" Sansheng Space" is conducive to the optimization and intensive use of land space in China. The rational planning of the "Sansheng Space" is conducive to the optimization and intensive use of our national land space.

With the accelerated pace of urbanization and industrialization, environmental pollution and food safety problems have occurred frequently, stimulating people's desire to return to nature and pursue a harmonious coexistence between man and nature. Leisure farms are the vehicle for modern city dwellers to get in touch with nature. The author has studied the concept of Permaculture and the "Sansheng Space" that are regarded as guidelines for the planning of leisure farms, which can provide a variety of reference paths for rural ecological revitalization and development [10].

2. Overview of Pu Ran Farm

Pu Ran Farm is located in Zhangmuqin Town, Xichang City, Sichuan Province, covering an area of about 27,000m^2. The town is named "Zhangmuqin" because of the dense camphor trees and green

IOP Conf. Series: Earth and Environmental Science 1087 (2022) 012051 doi:10.1088/1755-1315/1087/1/012051

mountains, and it produces cherries, grapes, thatched rice and other crops. The farm was established under the Permaculture with the purpose of "living, production, and ecology", which includes residence, orchard, composting area, protective forest and so on, as shown in Figure 1.

The Pu Ran Farm was created in 2021 through the joint efforts of the farm staff, and was transformed from the integration of land with unreasonable farming patterns and incomplete use of resources. It is not yet officially opened to the public. When it opens, the farm can be used by appointment to hold a series of activities such as class reunions and birthday parties.

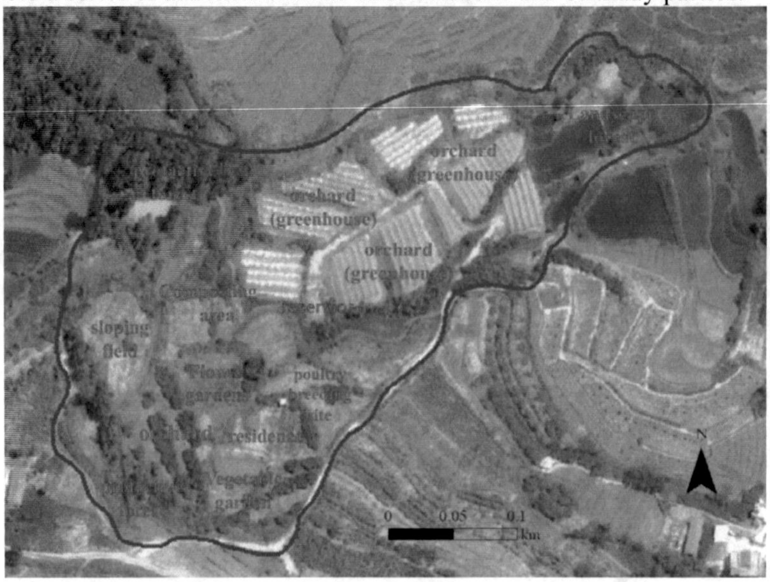

Figure 1. Farm plan.

3. Permaculture design Zoning and "Sansheng Space"

3.1. Permaculture Zoning

A good design of Permaculture can make the production system achieve low energy consumption and high output without damaging the natural environment and creating pollution [11]. Based on the basic use of the land and combined with human life, Permaculture divided the land into 5 areas and radiated outward in concentric circles. The farm is small so there is no wild area - area V - existing in its natural state for the time being, therefore, the area is divided into area 0 to area IV, such as which involves: residence, infrastructure, vegetable garden, orchards, and so on, as shown in Table 1.

Table 1. Zonal differentiation of the Permaculture Ecological Zone.

Area code	Activity intensity	Zoning Principles	Components
Area 0	Very frequent	Residence, Activity Center	Residence, infrastructure, etc.
Area I	Frequent	Around residence, areas requiring intensive management	Vegetable garden, etc.
Area II	Often	Setting up the plants and animals to be managed	Flower gardens, poultry breeding sites, etc.
Area III	Marginal	Cash crops	Orchards, etc.
Area IV	Lesser	Semi-natural area, less need for management	Protective forest, landscape forest, composting, etc.

3.2. The combination of permaculture and "Sansheng Space"
The premise of sustainable development is to harmonize the relationship between life, production and ecology [12]. Therefore, the relationship between Permaculture and "Sansheng Space" is complementary. The combination of Permaculture and "Sansheng Space" is the application of Permaculture's philosophy, methods and principles. Both Permaculture and "Sansheng Space" emphasize sustainable development, while Permaculture focuses on the harmony and unity of man and nature, and "Sansheng Space" focuses on the coordinated development of life, production and ecology. The combination of the two is of great significance to the full realization of rural revitalization in China.

3.2.1. Area 0, Area I and living area. Area 0 is the center of activity, the most controllable space with high frequency use [13]. Its periphery is Area I, which is an area that requires high intensity management, such as vegetable garden. The living area is the area that provides people with living, leisure and recreation, and is the most frequently used and most comprehensive area. The location of the living area therefore needs to be at the heart of the entire area, meet the need for people to have easy access to resources. Hence, Area 0 and Area I in the Permaculture Ecological Zone correspond to the living area in the "Sansheng Space".

3.2.2. Area II, Area III and production area. Area II and III are set up as farm areas with production functions for plants and animals that need to be managed, such as poultry and livestock, orchards, and flower garden. The production area is the area that provides people with agricultural products, service products, etc. Therefore, Area II and Area III correspond to the production areas of the "Sansheng Space".

3.2.3. Area IV and ecological area. Since the farm is small and there is no wild area - Area V - that exists in its natural state for the time being, the ecological zone here is Area IV in Permaculture, and the production zone is the area that provides ecological products and ecological services. Therefore, Area IV of the Permaculture Ecological Zone corresponds to the ecological zone of the "Sansheng Space". Ecological area serves people and things - providing ecological services for humans and a living environment for animals, with the aim of pursuing clear mountains and respect for nature support [14].

4. Design path
Permaculture design is gaining attention as a new ecological design method [15]. In the actual design of a farm, it is important to follow the 12 principles of the Permaculture design: collect and store energy, use and value renewable resources, design from scale to detail, integrate rather than separate, produce no waste, efficient energy planning, get benefits, explore and make good use of diversity, appropriate configuration, energy cycle, observation and interaction, and attention to boundary effects [16].

4.1. Living area
The living area, which is the frequently used area, including the residence, infrastructure, vegetable garden, etc., is the core area of the leisure farm and aims to meet the needs of human life and leisure.

4.1.1. Collecting and storing energy. The purpose of collecting and storing energy is to collect and store as much as possible when resources are abundant, so that resource utilization remains uninterrupted. The farm uses the roof to divert and collect rainwater, purify it and store it nearby; solar panels are set up in the park to make full use of the sun, a natural resource that is unique to the area, and the electricity generated by the solar panels can be applied to lighting, TVs, charging equipment and so on, which can reduce operating costs.

4.1.2. No waste generation. Permaculture believes that everything has its use, and as long as we are good at exploring and utilizing the value of resources, there will be no waste. Sewage from living area includes toilet sewage and washing sewage, toilet sewage is decomposed into fertilizer that can be used to fertilize the soil through waterless composting toilets; washing sewage is purified through biological methods such as natural biological treatment method and activated sludge method, using microbial metabolic functions to decompose and transform pollutants in the sewage, so that the recycled water can be used for agricultural irrigation and toilet water after reaching the standard.

4.2. Production area

The production area includes vegetable garden, fruit tree planting area and livestock breeding area. The production area is the main area of the leisure farm, which contains elements such as reservoirs, ditches, orchards, and livestock breeding.

4.2.1. Efficient energy planning. According to the topography and terrain of the study area, we can provide abundant agricultural water for the area by building ponds, constructing artificial water collection channels to extend the time that water sources stay in the area and maintain the water storage capacity of the soil; raising poultry in the orchards can ensure the demand of organic fertilizer for crops and avoid the environmental pollution caused by poultry manure, for example, geese can not only be raised as poultry, but they can also be used for biological weeding to reduce. The use of herbicides, at the same time, can solve part of the feed sources and reduce costs.

4.2.2. Getting benefits. The orchard includes grapes and cherries, which are mainly grown in large greenhouses. Not only does greenhouse farming have a high return on investment and good economic efficiency, but it also serves as an area for people to enjoy the edible landscape. With the benefits, we can promote the sustainability of the system.

4.2.3. Exploring and making good use of diversity. Confronted with different conditions, the same thing may have different effects. Companion planting is used in production areas to combine herb plants with vegetables and food plants, not only to control pests and reduce the use of insecticides, but also as ornamental plants, such as marigolds planted near western bluebells to help repel cabbage moths and dried goldenrod to repel aphids.

4.3. Ecological area

The vegetation in the ecological area can purify the air and keep the water and soil well. The ecological area not only improves the environment and increase the diversity of species in the region, but also protects against wind and fire and serves to protect production and living areas.

4.3.1. Proper configuration. The protective forest is used to block the spread of wind and sand and fire, reduce the interference of the outside world to the production and living areas, eliminate the influence of noise and other impacts on the park, and form a protective forest for ecological connotation. Ornamental shrubs and health care trees can be chosen for the landscape forest to increase the ornamental value and rest value, while also purifying the air and benefiting people's health.

4.3.2. Energy cycle. By simulating the operation of a natural ecosystem and controlling the flow of energy, a complete and sustainable ecological operation system is formed [17]. Setting up rainwater collection system in the park and reservoirs at high place, using contour lines and building artificial water collection channels to provide abundant agricultural water for the region, which can reduce groundwater exploitation, water waste and realize water recycling. At the same time, composting area is set up in the ecological zone, using materials such as fallen leaves, old limbs of trees and straws, which are fermented and decomposed for organic composting, which not only can form regional organic mulch, regulate soil temperature and reduce water evaporation, but also the long-term use of compost also serves to fertilize and transform the soil, and waterless composting toilets in living areas break down human waste into fertilizer, which is used to enrich the land, thus realizing the circulation of energy, as shown in Figure 2.

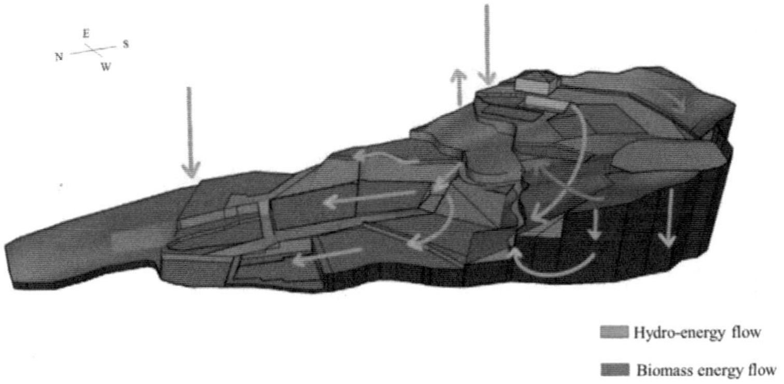

Figure 2. Energy cycle diagram.

4.3.3. Observation and interaction. David Holmgren, one of the founders of Permaculture, believes that observation and interaction are the most basic but important things that are easily overlooked, and that we need to spend more time in nature. The farm targets sloping field with no vegetation by planting green manure plants such as purple acacia and grass rhinoceros to improve soil structure and increase organic matter deposition, making the slope field a place for plants to grow again.

5. Conclusions

Through this case, we summarize the Permaculture strategy that is suitable for the planning of "Sansheng Space" of leisure farms: ① Energy strategy, the core of Permaculture design is the efficient use of energy and sustainability, so we should make full use of the site conditions, through terrain, precipitation and so on to achieve the purpose. ②Ecological strategy, when designing and planning, the characteristics and habits of various plants and animals should be given full play. For example, plants such as lavender and rosemary can not only be used as ornamental plants, but at the same time the fragrance they emit can also drive away various pests including mosquitoes. ③Site strategy, the layout of the site should be able to make full use of natural resources such as sunlight and water to increase the source of energy and reduce external demand and artificial input. The combination of Permaculture and the "Sansheng Space" is applied to the planning and design of rural land, which is conducive to solve the problems of idle and unreasonable land use and realize sustainable and high-quality development of rural land resources. At the same time, it will stimulate the development of the rural economy and attract urban workers to return to their hometowns to build beautiful villages, so as to fully realize rural revitalization.

Acknowledgements

Project was supported by the National College Students' innovation and entrepreneurship training program(202110628030), Xichang University Doctoral Start Foundation Program(YBZ202108), Key Program for Academic Leader of Bureau of human resources and social security in Liangshan Yi Autonomous Prefecture(ZRS202001). We are grateful to the anonymous reviewers for their thoughtful suggestions.

References

[1] Hu W, Zhu R, Zhou J, Li Y 2019 "Near" Nature and "Respect" Nature: The Practice and Discussion of Permaculture Development Concept in the Village of Damoyu, Yunnan *figJ]. Journal of Yunnan Agricultural University (Social Sciences)* **13(03)** 88-93

[2] Liu X, Zhou M 2016 An analysis of the ecological landscape design strategy under the concept of Permaculture--Taking the landscape design of Huali idle land as an example *[J]. Design* **(13)** 19-21

[3] Yue X, Li Y, Chang M 2016 An analysis of sustainability of urban ecovillages in the United States: the case of Los Angeles Ecovillage *[J]. Planners* **32(09)** 148-153

[4] Henfrey T, Penha-Lopes G 2015 Permaculture and Climate Change Adaptation: Inspiring Ecological, Social, Economic and Cultural Responses for Resilience and Transformation *Hampshire: Permanent Publications* 4-10

[5] Li P 2015 An eco-ethical alternative doctrine: analysis of ecological ideas in sustainable agriculture and its design *[J] Journal of Nanjing Forestry University (Humanities and Social Sciences Edition)* **15(3)** 71-78

[6] Liu C, Wang Y, He R, Wang C 2019 A spatial identification and optimization analysis framework of Sansheng Space based on residents' behavior *[J]. Journal of Natural Resources* **34(10)** 2113-2122

[7] Liu J, Liu Y, Li Y 2017 Classification evaluation and spatio-temporal pattern analysis of "Sansheng Space" in China *[J]. Acta Geographica Sinica* **72(07)** 1290-1304

[8] Huang G 2021 Reconfiguration and empowerment of "production-life-ecology" space: the case of Xiaoshan Evolution Town, Hangzhou *[J]. Creative City Journa* **(01)** 91-101

[9] Chen Z, Feng X, Hong Z, Ma B, Li Y 2021 Spatial Conflict Risk Assessment and Zoning Optimization of Land Use in Nanchang City--Based on the Perspective of "Three Life Spaces" *[J]. World Regional Studies* **30(03)** 533-545

[10] Li X, Zhong Q, Li G, Cai M, Ke D 2022 The application and development trend of the sustainable concept of Permaculture under the background of rural revitalization S*cience Discovery* **10(03)** 181-185

[11] Huang Z, Ma X, Ji L 2020 A study of edible landscapes in urban high-density communities based on Permaculture *[J]. Light Industry Science and Technology* **36(03)** 74-76+78

[12] Chen K, Chen S, Chen Z 2019 Insights into the practice of agricultural landscape tourism development based on the sustainable design of Permaculture - an example of the Pat Yan colony in Taiwan *[J]. Journal of Anyang Normal University* **(02)** 80-83

[13] Liu H, Zhao X, Shi J 2017 Study on the development strategies and approaches of sustainable agriculture in urban eco-settlements by Permaculture *[J]. Chinese Journal Agricultural Resources and Regional Planning* **38(7)** 188-194

[14] Huang J, Lin H, Qi X 2017 Advances in the study of Sansheng Space for the optimization of national land space *[J]. Progress in Geography* **36(03)** 378-391

[15] Wang Y, Zhang T 2013 The layout form of community farming garden under the concept of Permacultural design[J]. *Planners* **29(07)** 114-118

[16] Liu X 2008 Principles of Permacultural farming eco-community design and its case study *[D]. Tianjin University*

[17] Yuan F 2021 The application of Permaculture sustainable design in agricultural tourism park in the context of rural revitalization: an example of Xixiaozhuang Ecological Picking Park in Yangqu County, Taiyuan *[J]. Urbanism and Architecture* **18(15)** 178-180

IOP Conf. Series: Earth and Environmental Science 1087 (2022) 012052 doi:10.1088/1755-1315/1087/1/012052

Analysis of statistical post-processing methods for multi-model ensemble runoff forecasts in flood seasons

Bo Qu[1,2,4], Yuanhao Fang[3] and Lingqi Li[1]

[1] Yellow River Institute of Hydraulic Research, YRCC, No.45 Shunhe Road, Zhengzhou 450003, China;
[2] Key Laboratory of Lower Yellow River Channel and Estuary Regulation, MWR, No.45 Shunhe Road, Zhengzhou, 450003, China;
[3] College of Hydrology and Water Resources, Hohai University, No. 1 Xikang Road, Nanjing 210098, China.

[4] Corresponding author's e-mail: qubo_edu_hohai@163.com

Abstract. Statistical post-processing of ensemble forecasts could effectively improve their accuracy and reliability. In this study, three typical post-processing methods including equal weight (EW), model output statistics (MOS) and Bayesian model averaging (BMA) were applied to the raw multi-model runoff forecasts during the flood period (from 1 June to 30 September) of 2010-2013, and the processed results were compared and analyzed. It is shown that BMA is a promising post-processing method with highest accuracy, but this becomes completely different at lead times of 126-240 h in 2013. The main problems for the BMA performance instability were found to be outliers and low-correlation, which affect the linear regression model fitting in the bias correction procedure. Following this, a combination model of LR and EW was proposed to improve the bias correction procedure for lead times of 126-240 h. And the test results demonstrate that the combination model is effective and efficient, and it is able to lead to both accurate and reliable multi-model ensemble runoff forecasts for longer lead times.

1. Introduction

Multi-model ensemble runoff forecasts generated by coupling hydrologic models and "The International Grand Global Ensemble" (TIGGE) ensemble weather prediction models [1], have been proven to be potential for short and medium flood warning[2-4]. However, raw forecasts are generally biased and under-dispersed, as both meteorological and hydrologic models are imperfect. Thus, statistical post-processing methods are always taken to improve the forecast performance [5, 6].

The most frequently used and simple post-processing method is equal weight (EW), which indiscriminately assigns each forecast member an equal weight. Clemen [7] pointed out that the simplest EW can usually lead to more reliable results, on basis of a comprehensive review of post-processing methods. But the opponents argued that it is difficult to make considerable progress, as EW is extremely susceptible to the poor ensemble members[8]. Hence, it makes a lot of sense to combine forecast members with different weights. There are two contemporary approaches to realize variable weights: regression-based methods and performance-based methods. The regression-based methods employ the coefficients of multiple regression model as weights, one of the most widely used in the domain of meteorology and hydrology is model output statistics (MOS) [9]. And artificial neural network (ANN) with nonlinear mapping capability have also been successful applied [10]. Relatively, the performance-based methods give weights in the light of performance evaluation of different forecast members. The commonly used are based on the reciprocal (or the proportion) of errors, continuous ranked probability score, brier score and so on [11]. As an advanced method, Bayesian model averaging (BMA) is able to generate full predictive probability density function (PDF) rather than only probabilities for certain specific events, and provides more potentially valuable information for flood control decision [12]. Recently, the three types of post-processing methods above have been applied extensively in multi-model runoff forecasts [13]. However, most of relevant studies are based on too large data sample with many continuous years (the better performance in non-flood period always affect the entire results), or too small data sample with several flood events, leading to unreliable conclusion for multi-model ensemble runoff forecasts in flood season.

The aim of this study is to explore the effectiveness of the three typical statistical post-processing methods for multi-model ensemble runoff forecasts in flood season, and further proposal an effective and practical post-processing scheme. In this study, EW, MOS and BMA were applied to post-process the multi-model ensemble runoff forecasts during the flood period of 2010-2013, and the processed results were compared and evaluated. Then, the characteristics and reasons of the unreliable performance in different years were analyzed, by a deep investigation of the two key treatment procedures in the BMA model. Finally, some corresponding improvement measures were attempted, and a combination model for improving the bias correction procedure of BMA was proposed.

2. Data and methods

2.1. data

In this study, a humid catchment of 28900 km^2 in the upper reaches of Yangtze River was selected as the test case, and the summer flood seasons (from 1 June to 30 September) of 2010-2013 were selected as the test period. The observed precipitation and runoff have been collected and pre-processed into 6-hourly data series, in accordance with TIGGE data. As shown in Figure 1 that the discharge hydrographs are quite different during four years. It is clearly that flood occurs more frequently and sharply in 2013, the maximum peak discharge is about 13300m^3/s in July.

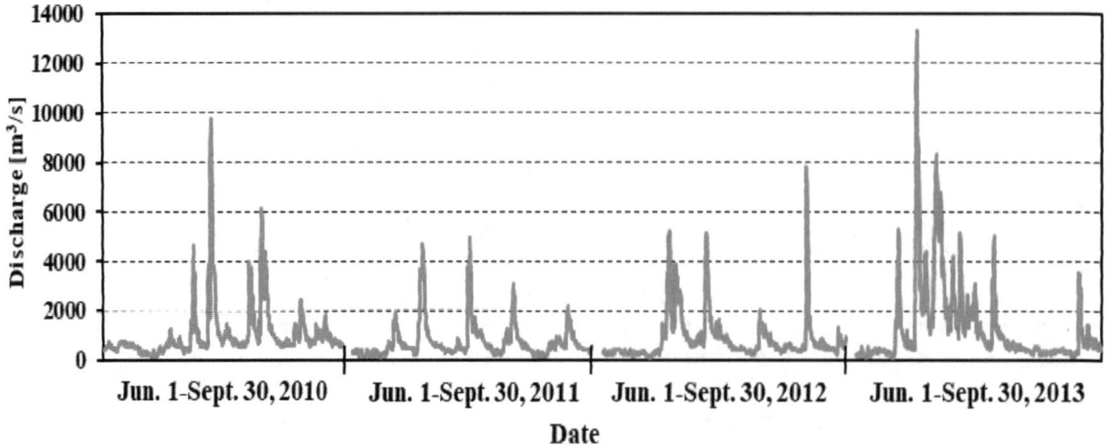

Figure 1. Observed 6-hourly discharge over the test period.

The raw multi-model ensemble runoff forecasts for the post-processing methods have been yield by running the Xinanjiang hydrologic model. The ensemble precipitation forecasts from six different TIGGE centers (CMA, CMC, ECMWF, KMA, NCEP and UKMO) were used as the main inputs of the hydrologic model, as shown in Table 1. For any TIGGE model, the 6-240 h lead times (40 steps) were considered.

IOP Conf. Series: Earth and Environmental Science 1087 (2022) 012052 doi:10.1088/1755-1315/1087/1/012052

Table 1. TIGGE ensemble precipitation forecasts used in this study.

Centre	Acronym	Ensemble members	Spatial resolution
China Meteorological Administration	CMA	14	0.56° × 0.56°
Canadian Meteorological Centre	CMC	20	0.60° × 0.60°
European Centre for Medium-Range Weather Forecasts	ECMWF	50	0.50° × 0.50°
Korea Meteorological Administration	KMA	16/23[a]	0.56° × 0.56°
National Centres for Environmental Prediction	NCEP	20	0.94° × 0.94°
Met Office	UKMO	23	0.83° × 0.56°

[a] The ensemble size of KMA was increased from 16 to 23 in early 2011.

2.2. Methods

2.2.1. Hydrologic model. The Xinanjiang hydrologic model is a conceptual rainfall-runoff model that has been widely used in the humid and semi-humid areas worldwide [14]. There are four main components in the model: evapotranspiration calculation (a generalize three-layer evapotranspiration model), runoff calculation (a saturated runoff model), three water source partition (surface runoff, interflow and underground runoff) and flow concentration calculation (unit hydrograph and Muskingum method). Its basic calculation units are different sub-catchments, which are divided from the whole basin based on natural boundary conditions and hydrologic station distribution.

In order to match the Xinanjiang model, the spatial resolution of TIGGE ensemble forecasts has been transformed into sub-catchment scale through grid mergence and area conversion. During the test period, the Xinanjiang model was run daily, generating raw multi-model ensemble runoff forecasts (143/150 members) with lead times ranging from 6 h to 240 h.

2.2.2. Post-processing methods. Three statistical post-processing methods including EW, MOS and BMA were considered. EW can be expressed as

$$y = \frac{1}{N}\sum_{k=1}^{N} f_k \quad (1)$$

where y is the predictive variable, f_k is the k_{th} forecast member, N is the number of forecast members. Sloughter et al. [15] pointed out that the median forecasts have more advantage than the mean forecasts. Thus, y was defined as the median forecasts here.

MOS is a multiple linear regression method, it can be expressed as

$$y = b_0 + b_1 f_1 + b_2 f_2 \ldots + b_k f_k \quad (2)$$

where b_k is the regression coefficient, which can be estimated using least square algorithm. In order to avoid multiple synteny problems, the stepwise regression algorithm was employed to establish the regression equation.

BMA is a way of producing probabilistic forecasts from ensemble in the form of predictive PDFs [16], it can be expressed as

$$p\left[y \mid (f_1, \ldots, f_k)\right] = \sum_{k=1}^{K} w_k g_k(y \mid f_k) \quad (3)$$

where w_k is the weight which reflects the performance of k_{th} member, and $g_k(y|f_k)$ is the conditional PDF. Note that the raw forecasts still need to be bias corrected by a simple linear regression model (LR), according to Raftery et al. [16]. BMA is primarily developed for the quantities subject to normal distribution. In this case, $g_k(y|f_k)$ be expressed as a normal PDF with mean $a_k + b_k f_k$ and standard deviation σ_k.

$$y|f_k \sim \psi\left(a_k + b_k f_k, \sigma_k^2\right) \tag{4}$$

Fraley et al. [17] noted that the ensemble members derived from the same source, should be regarded as undistinguishable and thus given equal parameters. Accordingly, the forecast members source from a same TIGGE model are exchangeable. If denote $f_{i,j}$ as the j_{th} members in the i_{th} groups, then Equation (3) can be rewritten as

$$p\left[y\middle|\{f_{i,j}\}_{i=1,\dots,I,\,j=1,\dots,m_i}\right] = \sum_{i=1}^{I}\sum_{j=1}^{m_i} w_i \psi\left(y\middle|a_i + b_i f_{i,j}, \sigma_i^2\right) \tag{5}$$

where I is the number of TIGGE models, m_i is the ensemble size of i_{th} TIGGE model. In the above equation, the parameters a_k, b_k have been determined by least square algorithm, and w_k, σ_k have been estimated by expectation maximum algorithm, based on a moving window of about 45 days of training data [18].

Additionally, runoff data are undoubtedly non-normal. In order to be able to resort to BMA relying on normal distribution, both runoff observations and forecasts must be transformed to be approximately normal. Kang et al. [19] recommend the Box-Cox transformation, which is given by

$$h_k = \begin{cases} \dfrac{f_k^{\lambda} - 1}{\lambda}, \lambda \neq 0 \\ \log(f_k), \lambda = 0 \end{cases} \tag{6}$$

where h_k is the transformed data, λ is the Box-Cox coefficient. In this study, λ has been estimated by maximizing likelihood algorithm ($\lambda = -0.33$). Figure 2 displays the normal probability plots of the original and transformed ensemble runoff forecasts at a lead time of 48 h. It is clearly that all of the transformed data is much closer to normal distribution.

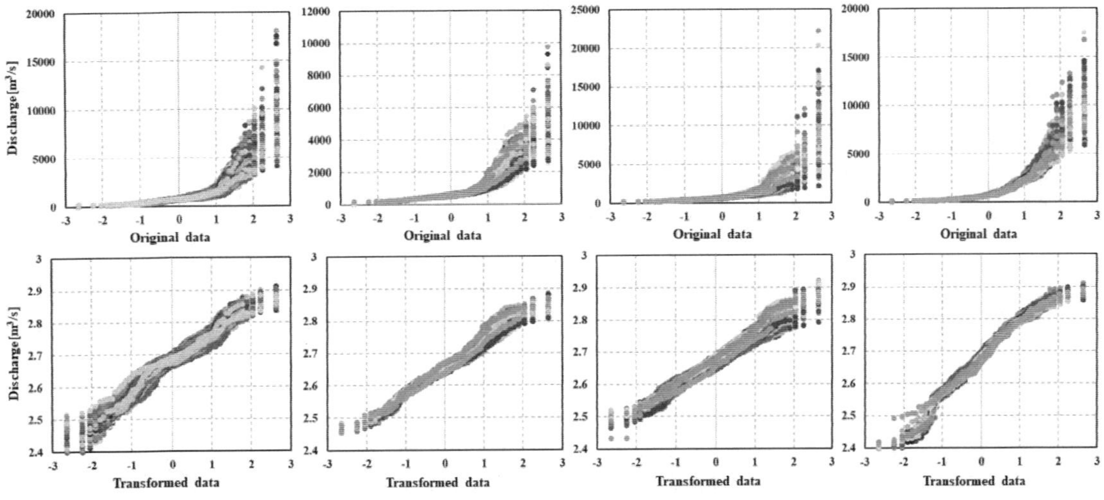

Figure 2. Normal probability plots of the original and transformed ensemble runoff forecasts at a lead time of 48 h: (a) year 2010; (b) year 2011; (c) year 2012; (d) year 2013. It can be considered to be approximately normal when all the data points appear to be a straight line.

2.2.3 Evaluation indicator. The goal of statistical post-processing are to get well calibrated and sharp forecasts. Calibration is a measure of the difference between forecasts and observations, sharpness is a measure of the width of forecast interval [20]. In this study, continuous ranked probability score (CRPS), which can address both calibration and sharpness, was applied to evaluate and examine the multi-model ensemble runoff forecasts [21]. CRPS is given by

$$CRPS = \int_{-\infty}^{+\infty} \left(F(x) - O(x) \right)^2 dx \tag{7}$$

where $F(x)$ and $O(x)$ is the cumulative distribution function of forecasts and observations, respectively. For an ensemble of m forecasts members, CRPS can be calculated by [22]

$$CRPS = \frac{1}{m}\sum_{j=1}^{m}\left|f_j - o\right| - \frac{1}{2m^2}\sum_{i=1}^{m}\sum_{j=1}^{m}\left|f_i - f_j\right| \tag{8}$$

CRPS is a negatively oriented indicator, that is, it takes smaller values for better performance. To meet the demands of BMA parameters estimation, the period from 25 July to 30 September of 2010-2013 was set as the verification period. For any verification day x, a training period of 45 days from $x-55$ to $x-11$ has been used for BMA model fitting.

3. Results and discussions

3.1. Comparison of EW, MOS and BMA

Figure 3 displays a comparison of the multi-model ensemble runoff forecasts based on three post-processing methods (EW ME, MOS ME and BMA ME). Six raw single-model forecasts (CMA, CMC, ECMWF, KMA, NCEP and UKMO) are also brought into the comparison. It is shown that MOS ME is the worst-performing for almost any year and any lead time, and even performs poorer than most of the raw single-model forecasts. Moreover, the dramatic and irregular fluctuation of MOS ME curve for different lead times seem to suggest that it is also short of reliability.

Relatively, EW ME and BMA ME are significantly improved. Due to the smoothing effect of average algorithm, EW ME performs more reliable. However, as every coin has two sides, it is also difficult to provide satisfied high-accuracy. As shown in Figure 3, EW ME is mainly at medium level in the comparison, and most of the time it has no superiority over CMC. Instead, BMA is capable of weakening and even eliminating the effects of poor members by assigning them with smaller weights. Hence, BMA ME performs best in 2010-2012. But in 2013, BMA ME becomes poorer than EW ME and part of raw single-model forecasts (CMC, ECMWF and NCEP), with a rapid performance decrease beyond lead time of about 120 h.

In summary, the comparison above demonstrates that none of the three methods can always lead to satisfied performance. (1) MOS ME is the worst of the three, and it seems to be inapplicable for the multi-model ensemble runoff forecasts; (2) EW ME is much better in both accuracy and reliability than the former, but it still belongs to a medium level; (3) BMA ME is the best performing, but this becomes completely different for lead times of 126-240 h in 2013.

3.2. Analysis of BMA performance instability

3.2.1. Investigate where the instability problems lie. Although BMA is a promising method to post-process the multi-model ensemble runoff forecasts in flood season, the instability problems make it is difficult to be adopted in practical. Thus, it is necessary to investigate why BMA has such different performance. Tian et al. [23] pointed out that the performance improvement mainly depends on two treatment procedures of BMA. One is the bias correction procedure, which corrects raw forecast members using LR. The other can be regarded as a weighted combination procedure, which gives forecast members different weights based on their performance during the training period. In order to figure out where the problem is, the intermediate outcomes before and after the two procedures were extracted and analyzed.

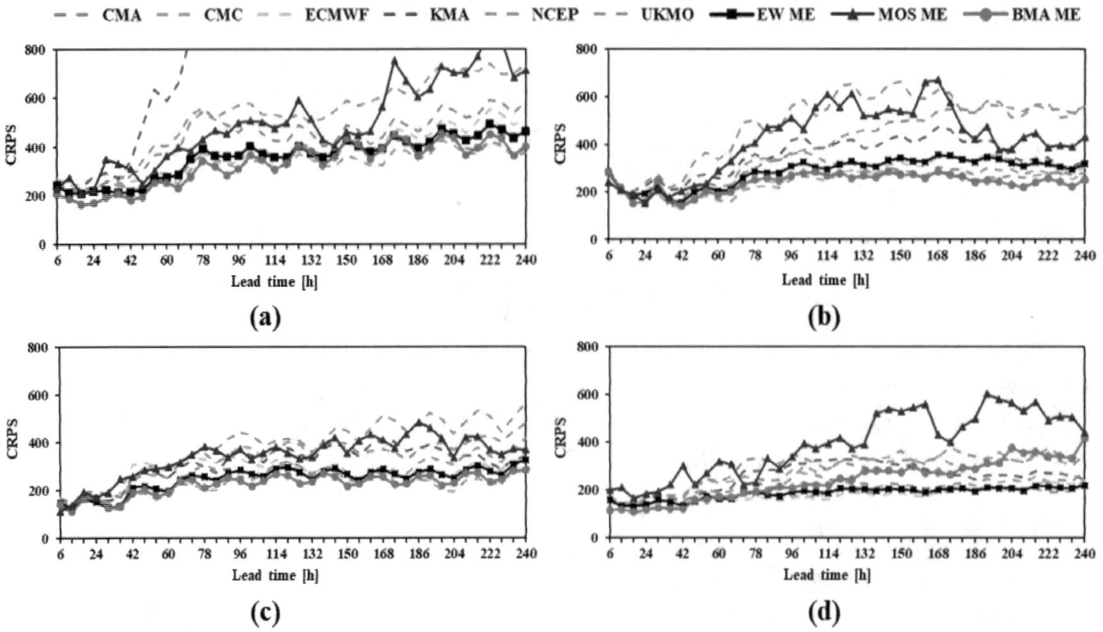

Figure 3. Comparison of the multi-model ensemble runoff forecasts based on EW, MOS and BMA at lead times ranging from 6 h to 240 h: (a) year 2010; (b) year 2011; (c) year 2012; (d) year 2013. The CRPS refers to the mean of CRPS over the verification period.

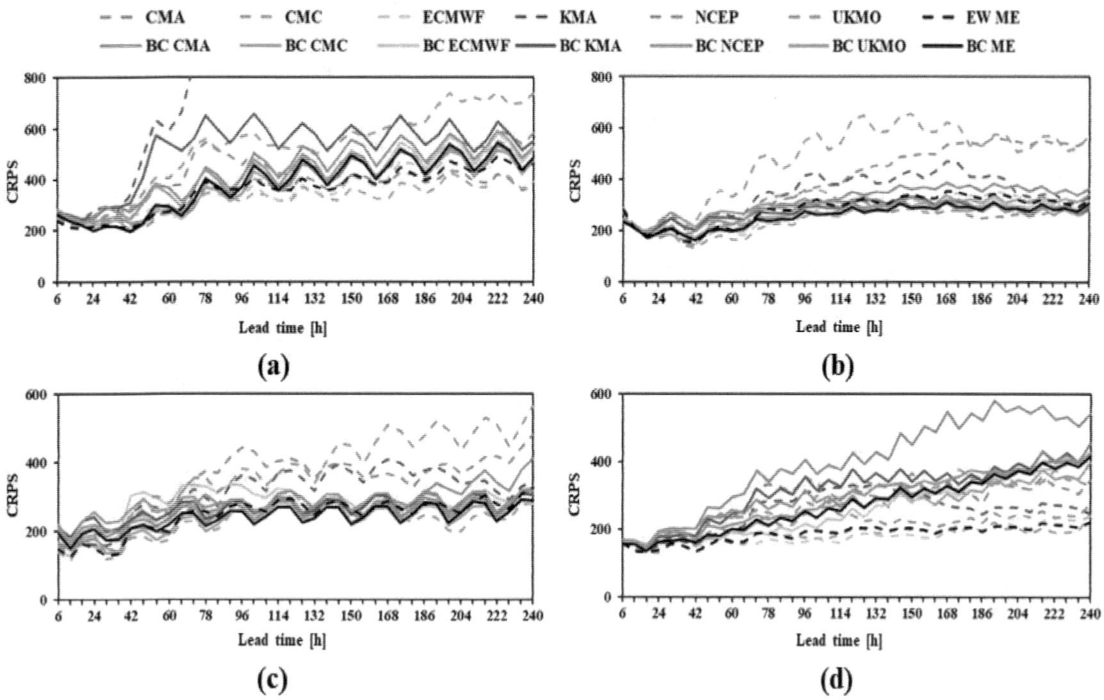

Figure 4. Comparison of the ensemble runoff forecasts before and after the bias correction procedure: (a) year 2010; (b) year 2011; (c) year 2012; (d) year 2013.

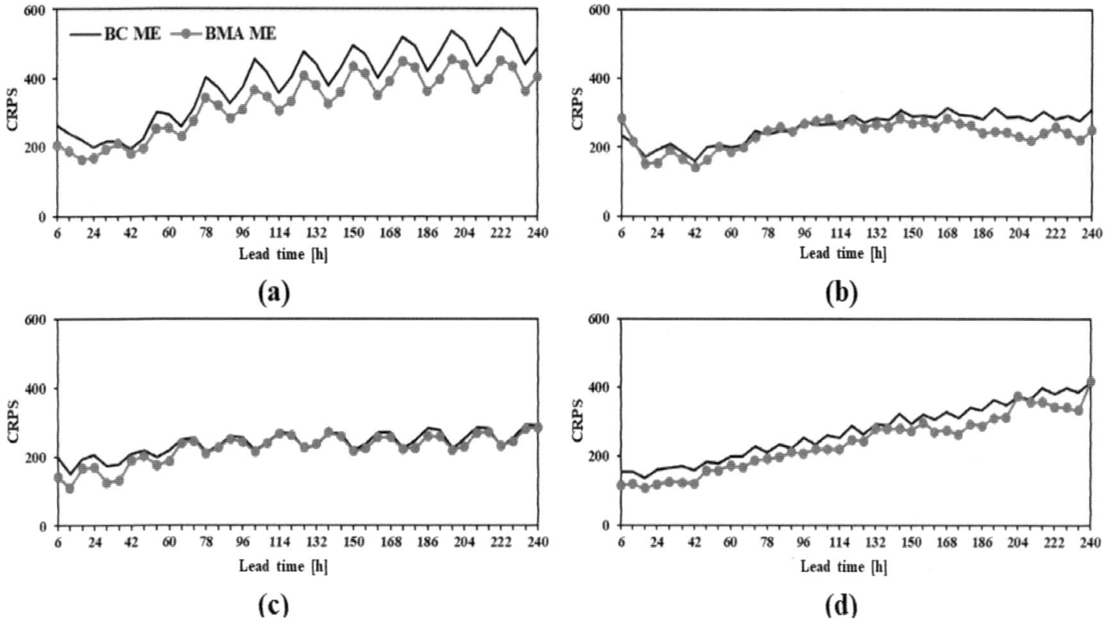

Figure 5. Comparison of the ensemble runoff forecasts before and after the weighted combination procedure: (a) year 2010; (b) year 2011; (c) year 2012; (d) year 2013.

Firstly, the bias correction procedure was investigated. Figure 4 displays a comparison between the raw forecasts and the processed forecasts (BC CMA, BC CMC, BC ECMWF, BC KMA, BC NCEP and BC UKMO). The processed forecasts have been derived from the bias correction procedure with the calibrated BMA regression parameters. It is shown that through the bias correction procedure, nearly all of the raw ensemble runoff forecasts are improved to some extent in 2011 and 2012, but they become poorer in 2013. In 2010, it takes both aforementioned, the poorer forecasts including CMA, KMA and UKMO get better distinctly, and the others get worse. Then, the weighted combination procedure was investigated. Taking into account that the inputs and outputs are not equivalent and comparable, BC ME was set as the pending data for the weighted combination procedure. BC ME is an equal-weighted combination of the six processed single-model ensemble forecasts. It is clear in Figure 5 that BMA ME performs better than or equal to BC ME for almost any year and any lead time.

In summary, the weighted combination procedure can always provide certain performance improvements. But the bias correction procedure can only offer improvements in 2011 and 2012, and it even brings about negative effects in 2013. Note that the performance of the bias correction procedure is close to the whole BMA model in the Section 3.1. Consequently, we conducted that the instability problems are in the bias correction procedure, which means that the estimates of regression parameters a_i, b_i are inappropriate.

3.2.2. Investigate what the instability problems are. To diagnose and solve the instability problems, the bias correction procedure was deeply analyzed. Previous studies have concluded that it is hard to fit a significantly changing data series using LR. This is partly due to the outliers, e.g. the flood peak, which can easily lead to regression line offset and further affect the estimation of regression parameters. It is found in Figure 1 that the discharge hydrograph fluctuates more frequently and violently in 2013 than in the other two years. The maximum peak discharge in 2013 is about 2.7 times in 2011 (4950 m^3/s), and 1.7 times in 2011 (7800 m^3/s). Therefore, the outliers might have a great negative influence on the regression parameters estimation. Additionally, the correlations between observations and forecasts tend to be weak for long lead times, and thus are difficult to meet the demands of LR. Figure 6 displays scatter plot of observation and one forecast member at lead time of 240 h. It is shown that the data points have wide and dispersed distribution for all the years. This means that it is unreasonable to use LR for bias correction at longer lead times. Unfortunately, BMA fails to take account of the outlier and low-correlation problems, resulting in unreliable performance.

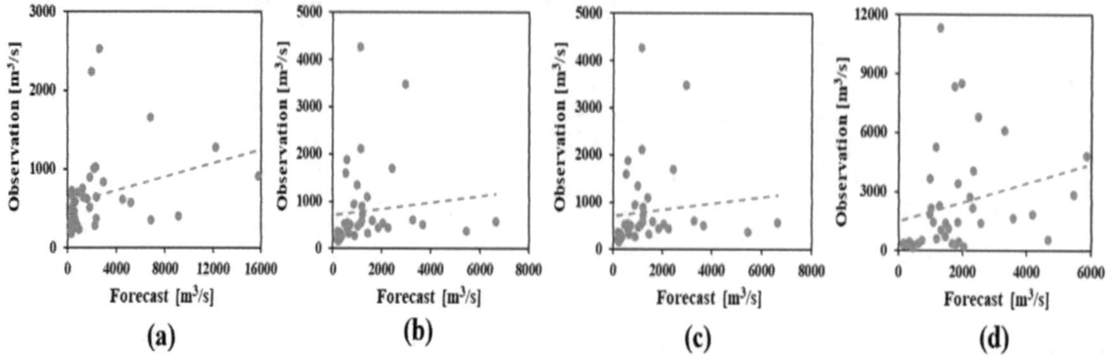

Figure 6. Scatter plot of observation and one forecast member at lead time of 240 h: (a) year 2010; (b) year 2011; (c) year 2012; (d) year 2013.

3.3. *Improvement of the BMA method*

3.3.1. Some common improvement measures. Generally, the outlier problems can be addressed using robust regression method. However, the test results show that this measure is nearly ineffective for the lead times of 126-240 h. This appear to indicate that the influence of outliers on forecasts are smaller than low-correlation. In view of this, a replacement method without high-correlation requirements are needed. Schmeits et al. [24] recommend a simple additive model, which is a special form of LR with parameter b_i=1. Following this, the bias correction procedure was improved and the multi-model ensemble runoff forecasts were regenerated (IBMA ME). Figure 7 shows a comparison between BMA ME and IBMA ME. It is obvious that IBMA ME outperform BMA ME for lead times of 126-240 h in 2013. But a key point is that IBMA ME are still poorer than ME and three single-model forecasts, that is, the improvements offered by additive model is still insufficient. To further examine the effectiveness of additive model, a comparison of the ensemble runoff forecasts based on LR and additive model was made. It is shown in Figure 8 that most of points distribute closely around the reference line for different years, which means there are no significant performance improvements by using additive model instead of LR, and such a similar conclusion also appears in Erivin et al.[25].

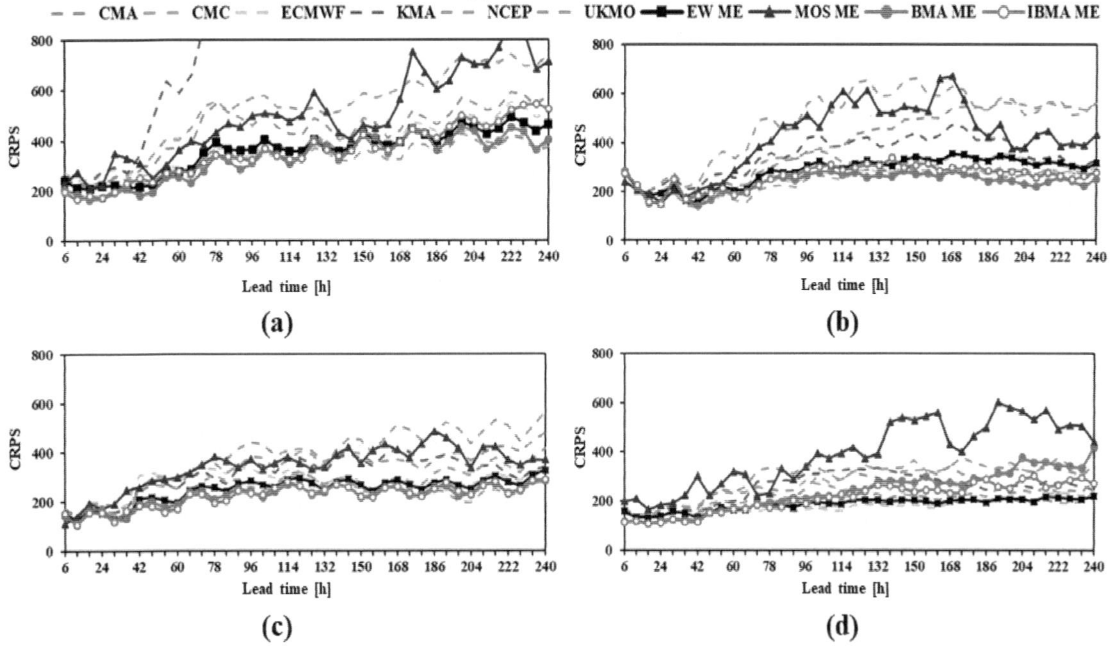

Figure 7. Comparison of the multi-model ensemble runoff forecasts based on BMA and the improved BMA: (a) year 2010; (b) year 2011; (c) year 2012; (d) year 2013.

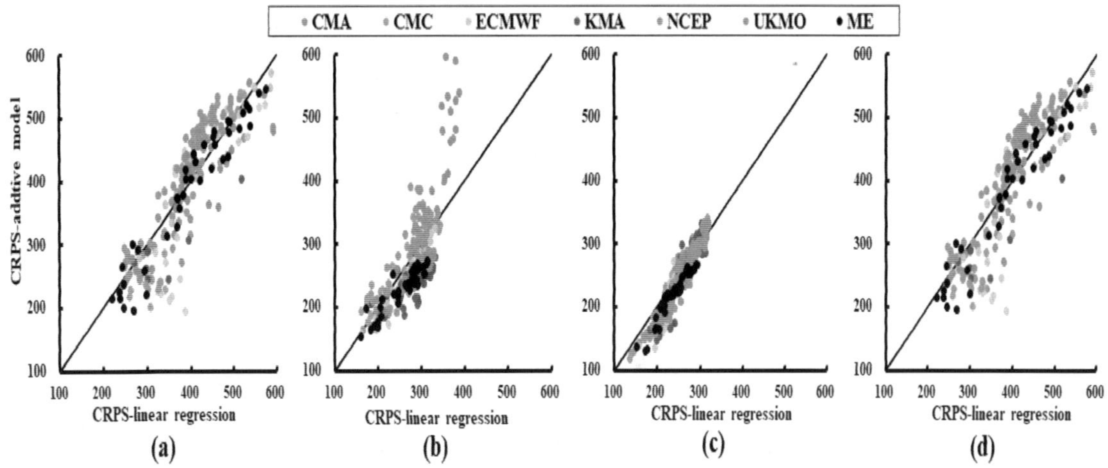

Figure 8. Comparison of the ensemble runoff forecasts based on LR and additive model: (a) year 2010; (b) year 2011; (c) year 2012; (d) year 2013. The diagonal is a reference line, when all the points stay around the line, the performance of both the methods would be pretty close.

3.3.2. A combination model of LR and EW. It is worth noting that EW ME has almost highest accuracy at lead times of 126-240 h in 2013 (Figure 7). This means that EW model may have a great potential to be used for bias correction. Hence, a combination model of LR and EW was taken into account for lead times of 126-240 h. The basic idea is to adopt the better one (LR or EW) based on the performance comparison over the training period, it can be expressed as

$$f_{i,j}^{'} = \begin{cases} a_i + b_i f_{i,j}, CRPS_{lr} \le CRPS_{ew} \\ median(f_{i,j}, i = 1,...,m; j = 1,...,m), CRPS_{lr} > CRPS_{ew} \end{cases} \quad (9)$$

where $f_{i,j}^{'}$ is the processed forecasts after the bias correction procedure, $CRPS_{lr}$ and $CRPS_{ew}$ are the mean CRPS of forecasts based on LR and EW during the training period, respectively. Figure 9 shows a comparison of the multi-model ensemble forecasts based on BMA and the combination model (CM ME). It is clearly that CM ME is significantly superior to BMA ME for all the years. And compared with EW ME and single-model forecasts, CM ME still remain distinct advantages in the vast majority of cases. This indicates that the combination model is effective and efficient for improving the bias correction procedure for lead times of 126-240 h.

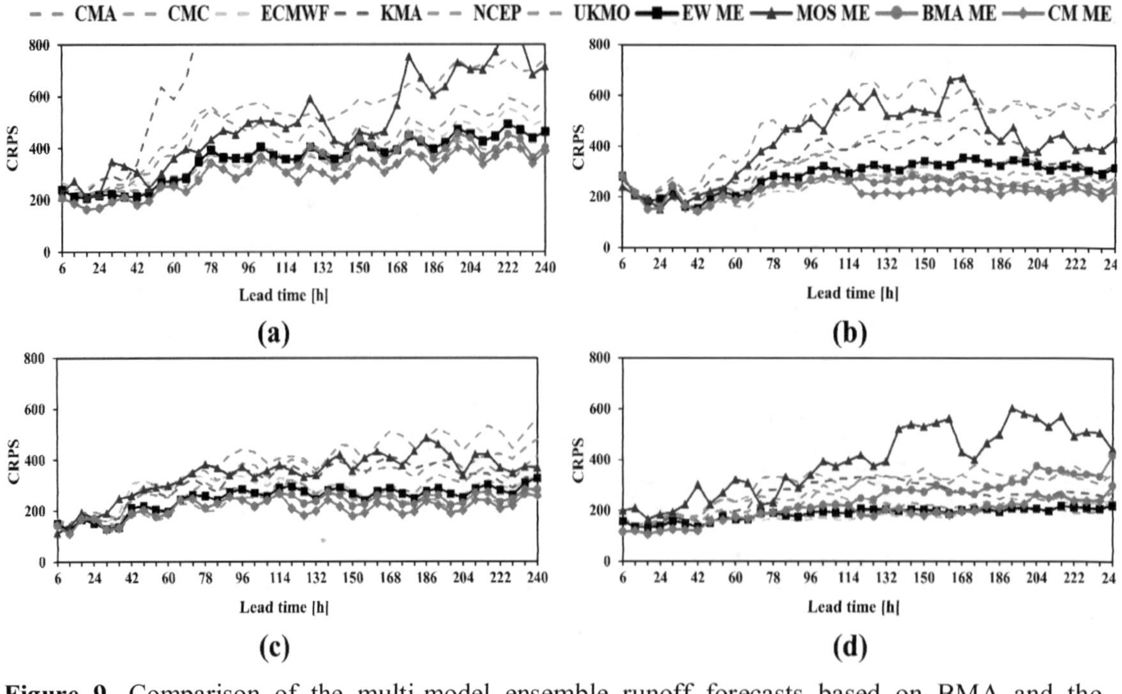

Figure 9. Comparison of the multi-model ensemble runoff forecasts based on BMA and the combination model: (a) year 2010; (b) year 2011; (c) year 2012; (d) year 2013.

4. Conclusions

In this paper, three typical statistical methods including EW, MOS and BMA were employed to post-process raw multi-model ensemble runoff forecasts in the summer flood seasons (from 1 June to 30 September) of 2010-2013. The raw runoff forecasts have been yielded by driving the Xinanjiang hydrologic model with TIGGE ensemble precipitation forecasts (143/150 members). The comparison results show that BMA has superiority over the rest two methods, but this becomes completely different for the lead times of 126-240 h in 2013. The reason for performance instability becomes clear through investigating the two treatment procedures in the BMA model. It is found that the problems are in the bias correction procedure, which has similar negative effects on ensemble runoff forecast in 2013. The further analysis of the bias correction procedure displays that outliers (e.g. the flood peak) and low-correlation are the main problems, because they can easily cause regression line offset and affect regression parameters estimation. Based on this, a combination model of LR and EW was proposed to improve the bias correction procedure for lead times of 126-240 h. It adopts the better one (LR or EW) based on the performance comparison over the training period. The test results show that the combination model is effective and efficient, and it is able to generate both accurate and reliable multi-model ensemble runoff forecasts for longer lead times.

However, this study was based on a certain amount of hydrologic data collected from a humid catchment in China, therefore the conclusion is possible to be unilateral and limited to some extent. So more relevant research on application and examination of the improved BMA (with the combination model) are urgently needed. Besides, the combination model is still imperfect, as it lacks consideration of the compatibility of EW and LR. Future work should concentrate on perfecting the combination model, or developing a new and advanced bias correction method for BMA.

Acknowledgements
This research was funded by National Key Research and Development Program (2018YFC1508400), Henan Province Science Foundation for Youths (212300410202), and Young Elite Scientists Sponsorship Program by Henan Association for Science and Technology (2022HYTP023).

References

[1] Swinbank, R., M. Kyouda, P. Buchanan, L. Froude, T.M. Hamill, T.D. Hewson, J.H. Keller, M. Matsueda, J. Methven, F. Pappenberger, M. Scheuerer, H.A. Titley, L. Wilson, and M. Yamaguchi 2016 The TIGGE project and its achievements *Bulletin of the American Meteorological Society* **97(1)** 49-67

[2] Shu, Z., J. Zhang, J. Jin, L. Wang, G. Wang, J. Wang, Z. Sun, J. Liu, Y. Liu, and R. He 2021 Evaluation and Application of Quantitative Precipitation Forecast Products for Mainland China Based on TIGGE Multimodel Data *Journal of Hydrometeorology* **22(5)** 1199-1219

[3] Liu, J., X. Yuan, J. Zeng, Y. Jiao, Y. Li, L. Zhong, and L. Yao 2022 Ensemble streamflow forecasting over a cascade reservoir catchment with integrated hydrometeorological modeling and machine learning *Hydrology and Earth System Sciences* **26(2)** 265-278

[4] Peng, T., H. Qi, and J. Wang 2020 Case Study on Extreme Flood Forecasting Based on Ensemble Precipitation Forecast in Qingjiang Basin of the Yangtze River *Journal of Coastal Research* **104(SI)** 178-187

[5] Muhammad, A., T.A. Stadnyk, F. Unduche, and P. Coulibaly 2018 Multi-model approaches for improving seasonal ensemble streamflow prediction scheme with various statistical post-processing techniques in the Canadian Prairie region *Water* **10(11)** 1604

[6] Wang, S., B. Ancell, G.H. Huang, and B. Baetz 2018 Improving robustness of hydrologic ensemble predictions through probabilistic pre - and post - processing in sequential data assimilation *Water Resources Research* **54(3)** 2129-2151

[7] Clemen, R.T. 1989 Combining forecasts: A review and annotated bibliography *International Journal of Forecasting* **5(4)** 559-583

[8] Hagedorn, R., R. Buizza, T.M. Hamill, M. Leutbecher, and T.N. Palmer 2012 Comparing TIGGE multimodel forecasts with reforecast-calibrated ECMWF ensemble forecasts *Quarterly Journal of the Royal Meteorological Society* **138(668)** 1814-1827

[9] Schick, S., O. Rössler, and R. Weingartner 2018 Monthly streamflow forecasting at varying spatial scales in the Rhine basin *Hydrology and earth system sciences* **22(2)** 929-942

[10] Rasp, S. and S. Lerch 2018 Neural networks for postprocessing ensemble weather forecasts *Monthly Weather Review* **146(11)** 3885-3900

[11] Y, D.J., Z.L. N, Z.G. P, X.F. W, and W. Z 2013 Study on precipitation multimodel super ensemble forecast technique *Meteorological Monthly* **39(6)** 691-698. (in Chinese)

[12] Xu, J., F. Anctil, and M.-A. Boucher 2019 Hydrological post-processing of streamflow forecasts

issued from multimodel ensemble prediction systems *Journal of Hydrology* **578** 124002
[13] Li, W., Q. Duan, C. Miao, A. Ye, W. Gong, and Z. Di 2017 A review on statistical postprocessing methods for hydrometeorological ensemble forecasting *Wiley Interdisciplinary Reviews Water* **4(6)** 24
[14] Ahirwar, A., M. Jain, and M. Perumal 2018 *Performance of the Xinanjiang model*, in *Hydrologic Modeling* Springer. p. 715-731
[15] Sloughter, J.M., T. Gneiting, and A.E. Raftery 2010 Probabilistic wind speed forecasting using ensembles and Bayesian model averaging *Journal of the American Statistical Association* **105(489)** 25-35
[16] Raftery, A.E., T. Gneiting, F. Balabdaoui, and M. Polakowski 2005 Using Bayesian model averaging to calibrate forecast ensembles *Monthly Weather Review* **133(5)** 1155-1174
[17] Fraley, C., A. Raftery, and T. Gneiting 2010 Calibrating multimodel forecast ensembles with exchangeable and missing members using Bayesian model averaging *Monthly Weather Review* **138(1)** 190 - 202
[18] Qu, B., X. Zhang, F. Pappenberger, T. Zhang, and Y. Fang 2017 Multi-model grand ensemble hydrologic forecasting in the Fu River basin using Bayesian model averaging *Water* **9(2)** 74
[19] Kang, Y., F. Zhang, X. Cheng, P. Chen, and S. Song 2022 A hybrid model integrating Elman neural network with variational mode decomposition and Box-Cox transformation for monthly runoff time series prediction
[20] Hemri, S., D. Lisniak, and B. Klein 2015 Multivariate postprocessing techniques for probabilistic hydrological forecasting *Water Resources Research* **51(9)** 7436-7451
[21] Han, S., P. Coulibaly, and D. Biondi 2019 Assessing hydrologic uncertainty processor performance for flood forecasting in a semiurban watershed *Journal of Hydrologic Engineering* **24(9)** 05019025
[22] Grimit, E.P., T. Gneiting, V.J. Berrocal, and N.A. Johnson 2010 The continuous ranked probability score for circular variables and its application to mesoscale forecast ensemble verification *Quarterly Journal of the Royal Meteorological Society* **132(621C)** 2925-2942
[23] Tian, X., Z. Xie, A. Wang, and X. Yang 2012 A new approach for Bayesian model averaging *Science China Earth Sciences* **55(8)** 1336-1344
[24] Schmeits, M.J. and K.J. Kok 2010 A comparison between raw ensemble output,(modified) Bayesian model averaging, and extended logistic regression using ECMWF ensemble precipitation reforecasts *Monthly Weather Review* **138(11)** 4199-4211
[25] Zsoter, E., F. Pappenberger, P. Smith, R.E. Emerton, E. Dutra, F. Wetterhall, D. Richardson, K. Bogner, and G. Balsamo 2016 Building a multi-model flood prediction system with the TIGGE archive *Journal of Hydrometeorology* **17(11)** 2923-2940

IOP Conf. Series: Earth and Environmental Science 1087 (2022) 012053 doi:10.1088/1755-1315/1087/1/012053

Mesozoic hydrocarbon accumulation model in the Northern South China sea

Guangjian Zhong[1,2], Pibo Su[1,4], Changmao Feng[1], Shenghong Chen[1], Ming Sun[1,2], Hai Yi[1], Yanlin Wang[2,3], Junhui Yu[2,3], Jing Zhao[1] and Zhongquan Zhao[1]

[1] Guangzhou Marine Geological Survey, Guangzhou, China;
[2] Southern Marine Science and Engineering Guangdong Laboratory (Guangzhou) , China;
[3] Key Laboratory of Ocean and Marginal Sea Geology, South China Sea Institute of Oceanology, Innovation Academy of South China Sea Ecology and Environmental Engineering, Chinese Academy of Sciences, Guangzhou 510301, China.

[4] Corresponding author's e-mail: spb_525@sina.com

Abstract. The ChaoShan Depression is the largest Mesozoic depression covering an area of 3.7 x 10^4 km^2 where the relict Mesozoic strata are up to 5000 m thick. It has experienced disruption since the late Mesozoic. In early survey, the oil-gas migration condition and reservoiring mechanism are poorly expounded from the Meso-Cenozoic tectonic superposition due to poor seismic imaging to the deep Mesozoic layers. New seismic surveys using long streamers and quasi three-dimensional layouts has improved obviously the deep images of the Mesozoic formations, enabling analysis of tectono-stratigraphic features and petroleum geology. Correlating to the regional Mesozoic stratigraphic and facies characteristics from the well LF35-1-1 and onshore outcrops, two sets of source layer tested with high organic carbon content are interpreted within the upper Triassic to mid-Jurassic semi-closed gulf sequences which become thicker toward the east side of Dongsha Island. Three sets of potential reservoir beds are also interpreted, which are a basin-floor fan sandstone layer of the Triassic-Jurassic boundary, a limestone layer atop the mid Jurassic, and a sandy layer in the Upper Jurassic. Over the low bulge of the ChaoShan Depression, a big anticline is found bounded by two sets of faults which played as migration passage for hydrocarbon from the deep Mesozoic source to the reservoir layers. Likely, one set of the fault act reverse barrier to block the petroleum escape, thus form the fault-bounded trap, favorable for future exploration.

1. Introduction

The northern part of the South China Sea is a key area of China's offshore oil and gas exploration. The Mesozoic Basins in the northern part of the South China Sea has become the focus of many scientists since the presence of them were confirmed in 1980s. Dongsha basin is a Mesozoic basin in the northern South China Sea (SCS), and ChaoShan Depression (CSD) is the biggest Mesozoic residual depression and has a complete Mesozoic strata, with the largest Mesozoic thickness of about 5000m (Figure 1). The study of Mesozoic geology started in the early 1990s, and LF35-1-1 proved the marine Mesozoic in this depression. Most scholars believe that it has good petroleum exploration potential [1-7]. No oil or gas has been discovered yet, conditions for oil-gas migration and accumulation are unknown due to structural superposition and deep imaging quality constraints. Now new survey data acquired by Guangzhou Marine Geological Survey (GMGS) in this area has improved obviously the imaging quality of the seismic profile. It can reveal the Mesozoic middle-deep layer structure and clarify the structural characteristics. These data include long array two-dimensional seismic data and quasi three-dimensional seismic data, the data provide rich basic data for in-depth analysis of geological evolution and petroleum conditions in the northern SCS. This study carries out systematic seismic data interpretation, identifies main seismic reflection interfaces, analyzes structural characteristics and petroleum conditions, and establishes Mesozoic oil-gas accumulation model.

2. Geological setting

CSD is the largest Mesozoic depression covering an area of 3.7 x 10^4 km^2 where the relict Mesozoic strata are up to 5000 m thick (Figure 1), it can be divided into six secondary tectonic units (Figure 1). The Mesozoic strata in CSD comprises of T_3-J_1, J_2, J_3 and K [8-17], ranging from bathyal -neritic (T_3-J_1), bathyal (J_2), bathyal-neritic (J_3), fluvial to alluvial (K) sediments.

Marine source rock has been confirmed in CSD by drill well (LF35-1-1). The type of organic matter is mainly type II_2 and III, and the total organic carbon (TOC) index is high. Geophysical investigation shows that the source rocks are widely distributed, and thick. CSD has a good oil-gas resources potential. Because the drilling structure is close to the igneous rock mass, affected by the baking of the rock mass, the physical property conditions of the reservoir are poor, and it is difficult to form reservoirs, its fields is not reservoirs. The main focus of the next step is to find new oil-gas accumulation model.

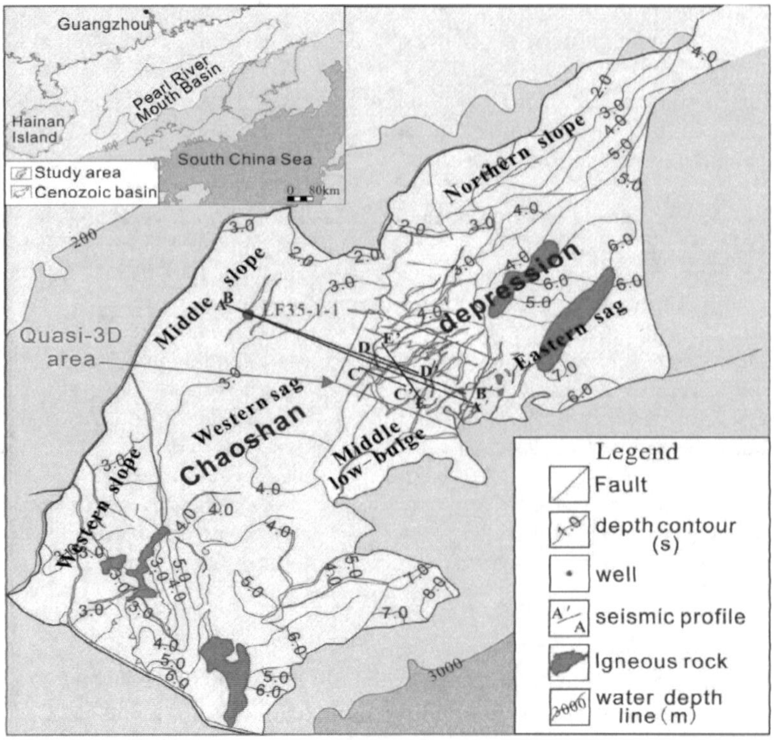

Figure 1. Location map of study area and seismic survey lines.

3. Data and methods

GMGS has acquired long array two-dimensional (2D) seismic data and quasi three-dimensional seismic data. The 2D acquisition parameters are seismic streamer with 480 channels in 12.5 m spacing, the shooting spacing is 37.5 m using source capacity of 5080 cubic inches. 2D seismic data have been processed with pre-stacked time migration.

Data processing yielded a high-quality 3D migration volume. This data processing effectively suppressed the turbulence interference, various random interference, linear interference and different types of multiples, After processing, the effective information of the shallow, medium and deep layers is highlighted, the wave train characteristics and amplitude characteristics are more clear; The quasi-3D data is evenly distributed after regularization with reasonable parameters and the lateral resolution is also improved; Complex geological structures were well imaged and reliable data were provided for the following data interpretation work.

We interpreted seismic data based on the GeoFrame software Platform, analyzed seismic stratigraphic based on correlation with LF35-1-1 well (Figure 1), and traced the reflection characteristics of mudstone or sandstone and limestone on the seismic profile.

4. Results

4.1. Structural feature

Faults in CSD can be catalogued into two generations, one occurred during Mesozoic which offset and deformed only the Mesozoic layers, and a young one during the late Cenozoic which offset and deformed the Mesozoic layers as well as the Cenozoic deposits. The old one appears with small to hardly discernable throw, as compressive in cases when in association with a few NE-trending compressive anticlines in some areas, e.g., the northwest of CSD and the south slope and southeast of the Dongsha uplift [14]. The young one, occurring ubiquitously over CSD and trending also in NE-direction, is extensional.

The faults in Cenozoic are not well developed and usually show the characteristics of tension and torsion faults. The scale of the faults is generally small and the fault displacement is short. They mainly cut the strata above the Tg unconformity surface and mostly terminate in the middle and upper part of Cenozoic. However, the faults controlling the depression or concave boundary depression are generally large in scale and fault displacement, cutting the Meso-Cenozoic strata and controlling the Cenozoic sedimentation. There are volcanic rocks near the faults. This kind of extensional fault is located in the northwest and southeast of CSD. It is mainly formed in the Cenozoic.

Large compressional anticline structural traces are developed in Mesozoic of Chaoshan depression. These wide and gentle large fold structures generally strike, forming a structural pattern of alternating arrangement of uplift zones and depression zones. Reverse faults are developed where the stress in the axis of the anticline is concentrated. These anticlines are often complicated by late tensional or tensional torsional faults with different tendencies [18-20].

4.2. Petroleum geology conditions

4.2.1. Source rock conditions. Dark flaky mud rock and muddy siltstone interbred siliceous rock were promulgated on the top part of $\mathbf{J_2}$-$\mathbf{J_3}$ by LF35-1-1, and dark flaky mud rock and muddy siltstone interbred by mud rock and carbonate rock were revealed on the bottom part of $\mathbf{J_2}$-$\mathbf{J_3}$ by LF35-1-1. Mud rocks rich organic debris, the TOC is between 1.00% and 1.48%, with an average of 1.32% [15], and this means that two sets of effective source rocks developed (Figure 2).

The corresponding relationship between mudstone and seismic facies is established by well-seismic calibration, and then the mudstone distribution in CSD can be identified by seismic data. The seismic reflection characteristics of mudstone are mainly medium-lowe amplitude, feeble continuity, low frequency or blank reflection (Figure 2). Through the identification of mudstone seismic facies, T_3-J_1 and J_3 mudstones can be traced continuously, but no more continuous mudstone intervals can be traced in the Middle Jurassic. It can be considered that T_3-J_1 and J_3 mudstones are the main source rocks in CSD. Among them, T_3-J_1 source rocks are dominated by semi-abyssal facies, and J_3 source rocks are dominated by neritic shelf facies.

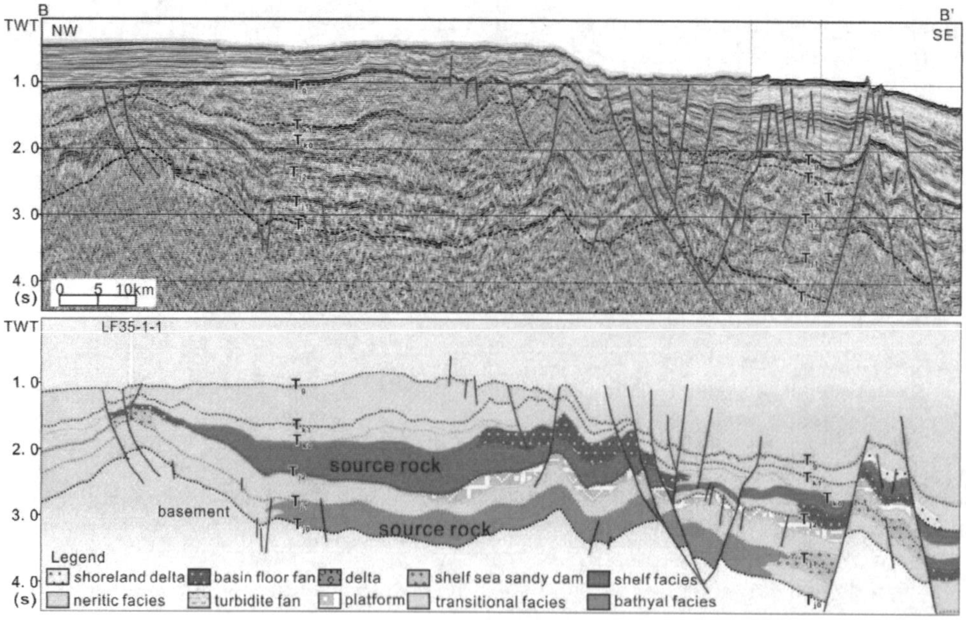

Figure 2. Two sets of source rocks in the study area (Location see Figure 1).

4.2.2. Reservoir conditions. The Well LF35-1-1 reveals that the Middle Jurassic, Middle-Upper Jurassic is 714 m thick, of which the total thickness of sandstone is 120m, and the maximum thickness of a single layer is 40 m. Favorable reservoir facies belts such as coastal deltas, reef shoals and submarine fans are developed, and most of these reservoir sand bodies are in direct contact with source rock layers.

According to the sequence of strata approved by LF35-1-1 and seismic profiles, three reservoir beds are developed CSD. One is coastal-shallow sand, slope fan and Submarine fan sand of $\mathbf{J_3}$, and the other is sand and limestone reservoirs in marine progressive overlap of $\mathbf{J_2}$. The seismic profile indicates that a carbonate rock layer with stable thickness is distributed on the top of $\mathbf{J_2}$ (Figure 3). Submarine fan developed on the top of $\mathbf{J_3}$ (Figure 4). Levee deposits can be located on southeast of the profile, and channel deposits located on northwest of the profile, with the best reservoirs properties.

The 3D seismic profile of favorable structures shows that stably distributed limestone layers develop below the T_{j2} interface, the top surface of the limestone layer (T_{j2}) is a strong-amplitude, high-continuous reflected wave group, and the top-bottom interface characteristics of limestone layers are stable and easy to trace. The thickness of the limestone layer is stable in the three-dimensional area. Between 200m and 500m, its internal reflection is blank reflection or sub-parallel weak amplitude reflection (Figure 3). The sedimentary characteristics correspond to seismic reflection characteristics of limestone. The limestone underlying strata are generally characterized by blank reflections, and are analyzed as semi-abyssal-neritic mudstone (Figure 3). The lithological assemblage of the semi-deep-neritic mudstone deposits in the lower part of $\mathbf{J_2}$ and the shallow limestone deposits in the upper part reflects the process of the depositional environment gradually changing from deep water to shallow water, which is consistent with the drilling results of Well LF35-1-1, which A thin layer of oolitic limestone is developed at the top of $\mathbf{J_2}$, which is a neritic deposit. The overlying strata of the limestone are generally characterized by blank reflection, and it is analyzed that it is semi-abyssal-abyssal mudstone (Figure 3). The limestone is overlaid with semi-abyssal mudstone, which reflects the rapid transgression in the early of $\mathbf{J_3}$. The black laminar mudstone is interposed with thin siliceous rocks (containing radiolarian and foraminiferal fossils), and it was transformed into a semi-abyssal-deep-sea sedimentary environment in the early Late Jurassic.

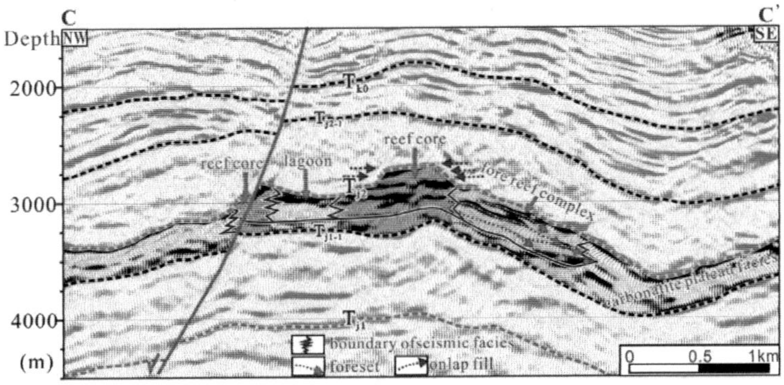

Figure 3. Architectural feature of reef limestone (Location see Figure 1).

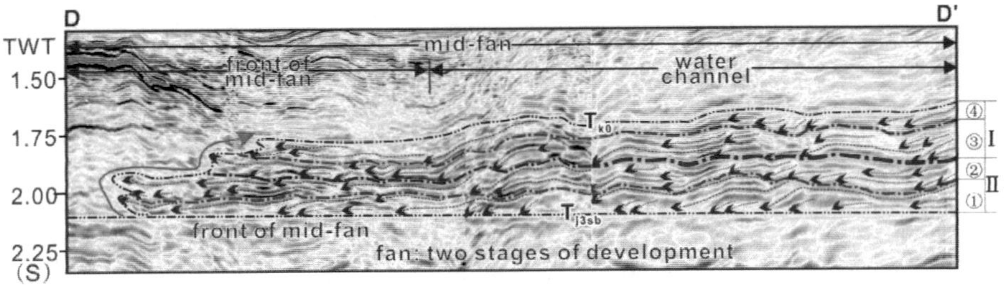

Figure 4. Submarine fan of the Upper Jurassic (Location see Figure 1).

The 3D seismic profiles of favorable structures show that the Upper Jurassic subsea fan can be tracked continuously in the 3D data volume. The bottom of the submarine fan is in abrupt contact with the underlying thick semi-deep sea mud rock, and the top is in gradual contact with the overlying thick sea-land transitional mudstone, showing a set of stable and strong amplitude reflections as a whole. Among them, along the source direction (Figure 4), the seismic section has a wedge-shaped shape, and the downward superposition feature of the forward product from SE to NW can be clearly seen. The subsea fan developed four phases, of which the second and third phases are the heydays, and the rate of sand content in the second and third phases is the highest.

The 3D seismic profiles of favorable structures show that the Upper of Lower Jurassic and bottom of Mid-Jurassic subsea fan can be tracked continuously in the 3D data volume, showing a set of stable and strong amplitude reflections as a whole. This subsea fan sand is the third set of reservoir bed.

4.2.3. Seal conditions. Basing on LF35-1-1 and seismic profiles analyses, multi-bed mudstones are thick, with 50 m thickness of individual, which have good sealing potency and act as important regional cap rocks in CSD.

5. Discussions

The oil-gas reservoirs in CSD are mainly developed in Mesozoic. The forming time of trap structure is earlier than oil-gas migration. Controlled by tectonism, the trap structures have undergone multi-stage adjustment. Although the tectonic uplift and denudation in late Jurassic and late Cretaceous-Paleogene had a great impact on Mesozoic oil-gas reservoirs, these two periods of tectonic movements had a great impact on early formed oil-gas reservoirs. The Eocene Oligocene period is very key to the oil-gas accumulation of the Mesozoic. The existence of Eocene Oligocene strata makes the Mesozoic seem to be covered. Although the internal oil and gas is damaged and readjusted, it will not be lost in large quantities. In addition, Dongsha movement also played a key action in the adjustment of

Mesozoic oil- gas reservoirs (Figure 5). This period is mainly reflected in the activity of rock mass, which is mainly manifested in the penetration and intrusion of rock mass, which may destroy and readjust the early oil - gas reservoirs in some areas.

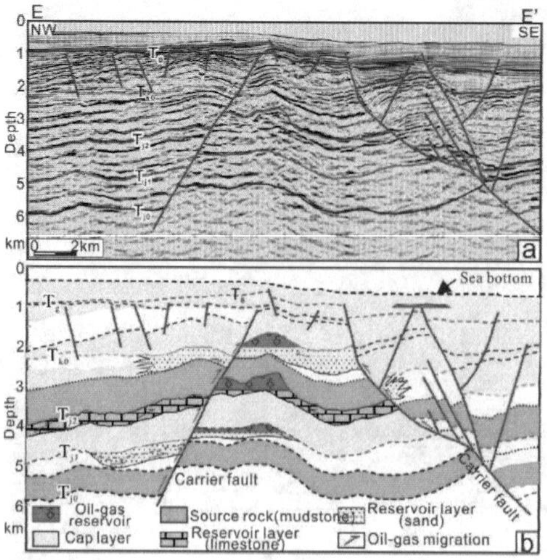

Figure 5. Oil-gas accumulation model (Location see Figure 1).

The oil-gas reservoirs in CSD are developed enough in Mesozoic, with good space-time allocation relationship of source-reservoir-cap and good preservation conditions.

The faults are the important elements for oil-gas accumulation: the upper wall of the fault mainly relies on the self-circle anticline at the structural high position Accumulation: Fractures in the footwall of the fault zone are not developed, which has strong lateral sealing ability to oil and gas, and is easy to block oil-gas accumulation (Figure 5).

6. Conclusions

1. The semi-enclosed gulf source rocks of Jurassic have relatively high organic abundance and thick, T_3-J_1 and J_3 mudstones are the main source rocks in CSD, it has strong hydrocarbon generation ability. Three sets of reservoir bed can be identified, submarine fan sandstone in J_3, limestone in top of J_2, and subsea fan in Upper of Lower Jurassic and bottom of Mid-Jurassic.

2. The fault controls the oil-gas accumulation, the source rock in the lower part of the middle low bulge is contact with sand bed in direct, and the oil and gas generated can directly migrate to sand bed through the fault, oil-gas reservoir can be formed when encountering the sand bed of the structural high point. Reverse barrier model is the effective model for oil-gas reservoir and will be the main focus of next exploration.

Acknowledgments

This research was funded by DD20221705, U1901217, 91855101, GML2019ZD0104, GML2019ZD0201 and 2019BT02H594.

References

[1] Hao H J, Lin H M, Yang M X, et al 2001 The Mesozoic in Chaoshan depression: A new domain of petroleum exploration *China Offshore Oil and Gas (in Chinese)* **15(3)** 157-163

[2] Wang P, Xia K Y, Huang C L 2000 The Mesozoic Marine sediment distribution and Geology-Geophysics characteristic at the North-Eastern of South China Continent *[J]. Journal of Tropical Oceanography* **19(4)** 28-35

[3] Liu G D 2001 The second exploitation of the oil and gas resources in China *[J]. Progress in Geophysics* **16(4)** 1-3

[4] Hao H J, Wang R L, Zhang X T, et al 2004 Mesozoic marine sediment identification and distribution in the eastern Pearl River Mouth Basin *China Offshore Oil and Gas* **16(2)** 84-88
[5] Wang L L, Cheng R H, LI F, et al 2009 The Mesozoic Sedimentary Sequences, Correlation and Geological Significance for Petroleum of the North Margin of South China Sea *[J]. Journal of Jilin University (Earth Science Edition)* **39(2)** 175-182
[6] Yao B C, Zhang L, Wei Z Q, et al 2011 The Mesozoic Tectonic Characteristics and Sediment Basins in the Eastern Margin of South China Continen *[J]. Marine Geology & Quaternary Geology* **31(3)** 47-60
[7] Ji Z Y, Zhao H Q, et al 2014 The Mesozoic Petroleum System of Chaoshan depression *Petroleum Geology and Enginerering* **28(3)** 9-15
[8] Chen D F, Su Z, Feng D, et al 2005 Formation and its controlling factors of gas hydrate reservoir in marine gas vent system *Journal of Tropical Oceanography* **24(3)** 38–46
[9] Duan J C, Mi H F 2012 Seismic Facies and Sedimentary Facies study of Mesozoic in Chaoshan Sag *[J] Resources & Industries* **14(1)** 100-105
[10] Shao L, You H Q, Hao H J, et al 2007 Petrology and depositional environments of Mesozoic strata in the Northeastern South China Sea *[J] Geological Review* **53(2)** 164-169
[11] Wu G X, Wang R J, Hao H J, et al 2007 Microfossil evidence for development of marine Mesozoic in the north of South China Sea *[J] Marine Geology & Quaternary Geology* **27(1)** 79-85
[12] Hao H J, Shi H S, Zhang X T, et al 2009 Mesozoic sediments and their petroleum geology conditions in Chaoshan sag: A discussion based on drilling results from the exploratory well LF35-1-1 *[J] China Offshore Oil and Gas* **21(3)** 151-156
[13] Zhang L, Geng A s, Wang LL, et al 2012 Assessment of Mesozoic source rocks at The Margin of South China Continent *[J] Marine Geology & Quaternary Geology* **32(1)** 99-108
[14] Zhang J Y, Sun Z, Zhang S F 2014 Analysis of Mesozoic tectonic deformation in the Chaoshan Depression of Pearl River Mouth Basin *[J] Journal of Tropical Oceanography* **33(5)** 41-49
[15] Zhang Q L, Zhang H F, Zhang X T, et al 2018 The upper Cretaceous prototype basin of the Chaoshan depression in the northern South China Sea and its tectonic setting *[J] Chinese Journal of Geophysics* **61(10)** 4308-4321
[16] Zhong G J, Wu S M, Feng C M 2011 Sedimentary model of Mesozoic in the northern South China Sea *Journal of Tropical Oceanography* **30(1)** 43-48
[17] Feng C M, Zhong G J, Sun M, et al 2022 Analysis of thermogenic gas sources of natural gas hydrates in the Dongsha waters *[J] Frontiers in Earth Science* **1** 10
[18] Zhou D 2002 Mesozoic strata and sedimentary environment in SW Taiwn basin of NE South China Sea and Peikang high of western Taiwan *[J] Journal of Tropical Oceanography* **21(2)** 50-57
[19] Zhou D, Yan J X, Qiu Y X, et al 2003 Rout for the eastern extension of Meso-Tethys in the western environs of the South China Sea *[J] Earth Science Frontiers* **10(4)** 469-476
[20] Zhou D, Chen H Z, Sun Z, et al 2005 Three Mesozoic sea basins in the eastern and southern South China Sea and their relation to Tethys and Paleo-Pacific domains *Journal of Tropical Oceanography* **24(2)** 16-25

Case study on the effectiveness of Agrometeorological Proverbs

Shuo Chen[1,5], Yize Liu[1], Yanmin Shuai[1,2,3,4], Congying Shao[1] and Yanjun Tian[1]

[1] Liaoning Technical University, Fuxin 123000, Liaoning, China;
[2] Xinjiang Institute of Ecology and Geography, Chinese Academy of Sciences, Urumqi 830011, Xinjiang, China;
[3] Research Center for Ecology and Environment of Central Asia, Chinese Academy of Sciences, Urumqi 830011, Xinjiang, China;
[4] University of Chinese Academy of Sciences, Beijing, 100049, China.

[5] Corresponding author's e-mail: 472020732@stu.lntu.edu.cn

Abstract. Many agrometeorological proverbs unique created by Chinese ancestors imply the relationship between universe, season, climate and agriculture events, which have been past considered as the valuable weather calendar to guide the agriculture events in China for thousands of years. And subject to the variation among various regions with distinguished climate or agricultural conditions, there is a requirement to examine the scientificity and applicability of agrometeorological proverbs before guiding the agriculture management. Based on the daily meteorological data of 206 meteorological stations in the Yangtze River Basin from 1966 to 2019, this paper explored the scientificity of meteorological knowledge embedded in meteorological proverbs "The Beginning of Spring at both ends resulting in hot in the dog days" and their applicability in the context of climate change. The study firstly investigated the warming rate based on the least-square method to fit the temperature data curves of the Beginning of Spring- Minor Heat in the Beginning of Spring at both Ends Years (BSEY) and Normal Years (NY), and to classify the "Fast Warming " (FW) temporal period according to the criteria that the days with a slope of warming of BSEY is greater than the slope of warming of NY in a certain temporal period, which accounts for more than 65% of the total number of days in the temporal period; Furthermore, the differences between BSEY and NY in the " Hot Dog Days "(HDD) phase were compared and analyzed according to the inter-decadal precipitation and the different " Heat levels " criteria defined by the difference of the Temperature-Humidity Index (THI) between the two years. Results shows that: (1) all regions in the Yangtze River basin, except for the plateau areas, are in line with the "FW" characteristic, accounting for 69% of the Yangtze River basin area, and the temporal period in which the "FW" characteristic appears (46 days after the Beginning of Spring) is before the the Spring Equinox. (2) There are areas in the upper, middle and lower reaches of the Yangtze River basin that meet the " HDD" characteristics, with the strongest compliance in the lower reaches, accounting for 90% of the lower reaches, and differences in the non-compliant areas, except for the Qinghai-Tibetan alpine region, mainly due to the influence of mountain ranges and other topography. In summary, the proverb is suitable for the main circulation regions in terms of applicability, and meteorological knowledge behind the proverb is also scientificity appropriate in the region where it is used. Meanwhile, this study can provide a reference for future policy development, agriculture-related research, and proper arrangement of agricultural production.

1. Introduction

As a large agricultural country, China has a long history of agricultural production activities, and agrometeorological proverbs summarized by the ancients according to the weather, phenology changes, crop growth, etc., play an undervalued role in the practice of agricultural production in China. Chinese agrometeorological proverbs have been recorded as early as the oracle bone inscriptions of the Yin and Shang dynasties [1], where each proverb was created from thousands of years of observation and summary experience of local or regional agricultural activities, so that different proverbs may be applicable to different areas or regions. The twenty-four solar terms, which are currently closely related to agrometeorological proverbs, were inscribed on the UNESCO Representative List of the Intangible Cultural Heritage of Humanity on November 30, 2016, and the world has begun to recognize traditional Chinese culture. Today's increasing global warming directly affects the climate system, mainly manifestations include frequent extreme high temperature events, heavy rainfall events, and agricultural and ecological droughts in some areas [2]. According to the latest IPCC 6th report[3], the current global average surface temperature is 1℃ higher than the pre-industrial level, and the projected average temperature change in the next 20 years is expected to reach or exceed 1.5℃. Climate change in the future will inevitably have an effect on agricultural production and livelihoods at the same time [2]. Therefore, the study of the existence of scientificity and regional aspects of agrometeorological proverbs is of great importance for the future development of agriculture in China.

Currently, an increasing number of scholars are concerned with the knowledge of proverbs related to agrometeorology, but the known studies on them basically stay at the level of the linguistic category, folk customs, agricultural cultivation, and statistical comparisons, such as Chen H [4] et al. who showed that the differences between Chinese and Western cultures affect the linguistic translation of proverbs, and SHEN Shi-pei [5]et al. who showed that the development of rice cultivation in the Southern Song Dynasty gave rise to related proverbs. Matczak P [6]et al. also found in a statistical analysis of Polish temperature-related proverbs using meteorological data that the accuracy of the proverbs was reversely proportional to chronological date (decreased with time). The exploration related to the targeted exploration of proverbs and the analysis of the scientific truth behind the proverbs through their underlying meteorological factors is still relatively lacking. Whether proverbs are still applicable in today's general environment, whether they are compatible with scientificity, and whether they are desirable in future agricultural production and life still needs further study. This paper takes the " The Beginning of Spring at both ends resulting in hot in the dog days " proverb [7], which is widely accepted and passed down in the middle and lower reaches of the Yangtze River, where farming civilization has flourished since ancient times. Based on the "FW" and "HDD" characteristics of the BSEY in the proverb, the data of 206 meteorological stations in the Yangtze River basin from 1966 to 2019 are used to study the scientificity and geographical applicability of the proverb, which provides scientific and accurate reference basis for a better grasp of climate change in agricultural production.

2. Study Area and Data

2.1. Study area

The Yangtze River basin is a vast area through which the main and tributaries of the Yangtze River flow. As the third largest river basin in the world, the total basin area is 1.8 million km^2, mainly including 11 provinces (municipalities and autonomous regions) in Shanghai, Jiangsu, Anhui, Jiangxi, Hunan, Hubei, Chongqing, Yunnan, Sichuan, Tibet, and Qinghai. The terrain of the Yangtze River basin is distributed in a stepped pattern, flowing through plateaus, basins (tributaries), hills and plains. Except for the plateau area in the west, the rest of the basin belongs to the subtropical monsoon climate, with high temperature and rain, rain and heat in the same period, and the light, heat, water and soil conditions for agricultural production are superior. The basin has more than 24.6 million hectares of arable land, accounting for 1/4 of the total arable land in China, while the value of agricultural production accounts for 40% of the country's total agricultural output, and is an important agricultural production base in China. In this paper, for the purpose of statistical analysis, as well as to consider the accuracy of the study unit, the prefecture-level cities are used as the basic analysis unit of the study (Figure 1). Areas without meteorological stations in the paper are not used as statistical analysis units.

2.1.1. Meteorological data. The meteorological data in this paper use the daily value dataset of Chinese surface meteorological data (V3.0) from 1966-2019, and the data are downloaded at http://data.cma.cn. This dataset contains daily data of many elements from 699 meteorological stations in China from 1951 to the present, in the production of this dataset, after the three-level quality control of "station-provincial-national", the quality and completeness of the data of each element are good compared with the similar data on the ground released previously, the actual rate of data of each element is generally above 99%, and the correct rate of data is close to 100%. In this paper, elemental data from 206 meteorological stations in the Yangtze River basin are selected as needed. Pre-processing using mean or linear interpolation for missing cases in the original data. Figure 1. shows the distribution of stations.

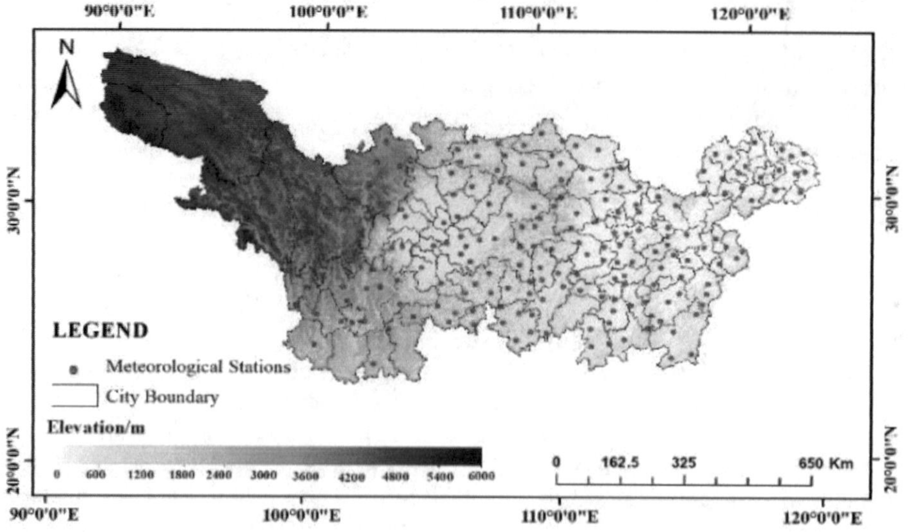

Figure 1. Overview of the study area and distribution of meteorological stations.

2.1.2. Calendar data. In agrometeorological proverbs, major solar terms such as the Beginning of Spring and Rain Water and minor ones such as the dog days are often mentioned, and the calendar data for the statistics are downloaded from https://www.rili.com.cn. To facilitate data processing, the dates of each solar term need to be counted. The twenty-four solar terms are divided according to the position of the sun on the ecliptic, there will be 2-3d changes before and after different years. To facilitate the processing of data between different years, a representative date is taken for each solar term[8] (Table. 1). After the cumulative change of anniversary movement dates, some years will have two the Beginning of Spring, commonly known as BSEY, while other years are known as NY (Table. 2). At the same time, there is a difference between leap and non-leap years in the studied years, and the treatment of leap year data follows that of Qian et al. [9], which both excludes the effects arising from the difference between flat and leap years and makes the time series of the same length for each year, which is conducive to data processing and analysis.

The timing of the dog days is determined by the Summer Solstice solar term date and the sexagenary cycle. As the Summer Solstice date varies from year to year, and the sexagenary cycle dates of the Summer Solstice are also different in the system of the heavenly stems and earthly branches, therefore the timing of the dog days varies from year to year, yet they all occur in the July-August period [10]. The start date of the dog days varies greatly from year to year and for different lengths of time, so no representative date is taken.

Table 1. The adoption date of some of the twenty-four solar terms.

Solar term	Adoption date (Month/day)	Solar term	Adoption date (Month/day)
Beginning of Spring	2/4	Rain Water	2/19
Awakening of Insects	3/6	Spring Equinox	3/21
Summer Solstice	6/22	Minor Heat	7/8

Table 2. Type of years.

Type	Year
BSEY	1966,1968,1971,1974,1976,1979,1982,1984,1987,1990,1993,1995,1998, 2001,2004,2006,2009,2012,2014,2017
NY	1967,1969,1970,1972,1973,1975,1977,1978,1980,1981,1983,1985,1986, 1988,1989,1991,1992,1994,1996,1997,1999,2000,2002,2003,2005,2007, 2008,2010,2011,2013,2015,2016,2018,2019

3. Methods

3.1. Least-square polynomial fitting

The weather changes irregularly in daily life, and the results of the temperature data analysis fluctuate and cannot be compared directly and clearly between data. To address this situation, the "FW" characteristic adopts the least-square polynomial fitting method [11] to show the warming trend. In this paper, after using mathematical statistics to obtain the multi-year average daily temperature averages for two sets of time series of 155 days for the Beginning of Spring-Minor Heat (before the dog days) temporal period of BSEY and NY. The fitting results were analyzed by comparing them several times, and it was found that the fitting results at the degree of a polynomial of 5 best met the needs of studying the warming trend, and its correlation index (R^2) was about 0.90, which passed the significance test.

Considering the variability of elements such as topography and climate in the Yangtze River basin, this paper addresses the following criteria for delineating the compliance situation of faster-than-normal warming in the years when the two ends hit spring (Equation 1).

$$\left(d_{[(k_s/k_n)>1]}/d\right) \times 100\% > 65\% \tag{1}$$

Among them, $d_{(k_s/k_n)>1}$ is the number of days when the warming slope of BSEY is greater than the warming slope of NY in a certain period, d is the number of days in a certain period, k_s is the warming slope of BSEY, k_n is the warming slope of NY.

3.2. Temperature-Humidity Index

Given that there were no modern precision measuring instruments in ancient times, the "hot" in " The Beginning of Spring at both ends resulting in hot in the dog days " indicates the temperature of human body sensation. And studies have shown that changes in temperature and humidity together affect human body temperature [12]. Therefore, this paper's proverb " HDD" characteristic cites the Temperature-Humidity Index [13] (THI) to define the dog days by daily maximum temperature, daily minimum temperature, and relative humidity. Used the Temperature-Humidity Index (THI) to evaluate the "Heat level" of the "HDD" characteristic in two different years by calculating the difference between the two spring years and the normal years at each station (Formula 2, 3). Table 3 shows the indicators of the degree of compliance with the " HDD" characteristic.

$$THI = K^{-1}\sum_{i=1}^{K} t_i - 0.55\left(1 - K^{-1}\sum_{i=1}^{K} x_i\right)(K^{-1}\sum_{i=1}^{K} t_i - 14.5) \quad (2)$$

$$D_j = THI_j^s - THI_j^n \quad (3)$$

Among them, *THI* is Temperature-Humidity Index, t_i is the temperature on the day *i* (unit: °C), x_i is the relative humidity on the day *i* (unit: 1%), *i*=1,2…, *K*, *K* is the study period. D_j is the amount of THI difference between the two years. THI_j^s is the daily maximum temperature or daily minimum temperature Temperature-Humidity Index of BSEY; THI_j^n is the daily maximum temperature or daily minimum temperature Temperature-Humidity Index of NY; *j*=*h*、*l*, where *h*: daily maximum temperature, *l*: daily minimum temperature.

Table 3. Degree of "HHD" characteristic compliance.

Difference	Degree of conformity
$D_h+D_l>0$	"HDD" strong
$D_h+D_l=0$	"HDD" weak
$D_h+D_l<0$	Doesn't fit the proverb

4. Results and Analysis

4.1. "FW" characteristic

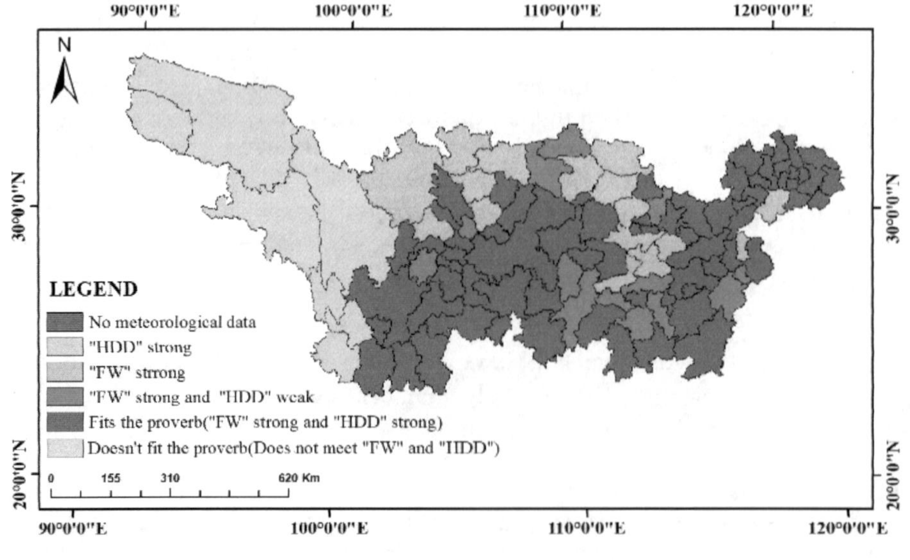

Figure 2. proverb geographical conformity type.

Figure 2 shows that all regions in the Yangtze River basin except the plateau region basically meet the " FW" characteristic, accounting for 69% of the area of the Yangtze River basin, and the time periods that meet the "FW" in the region include the Beginning of Spring - Awakening of Insects (Figure 3a) and the Beginning of Spring - Spring Equinox (Figure 3b). Regional BSEY compared to NY, the warm air carried by the warm and humid airflow from the sea in spring warms up the weather quickly (Figure 3a: e-m points, 1-14 days). By the Rain Water solar term (15 days after the Beginning of Spring, Figure 3a: m), warm and humid airflow from the ocean become active and confront the cold air masses on the mainland in intense and frequent battles, resulting in increased precipitation and temperature decrease [14] (Figure 3a: m-n points, 15-22 days). And by the Awakening of Insects solar term, precipitation is less incremental compared to the Rain Water solar term, and temperatures usher in a second warming boost (Figure 3a: n-f, 22-31 days, and Figure 3b: n-f, 22-45 days). Since the summer monsoon is mainly the southeast monsoon from the Atlantic Ocean and the southwest monsoon from the Indian Ocean, the monsoon is the first to arrive in the region in early spring, and the monsoon flow comes earlier and stays longer than other regions, so BSEY will have a faster warming rate and longer warming temporal period than other regions.

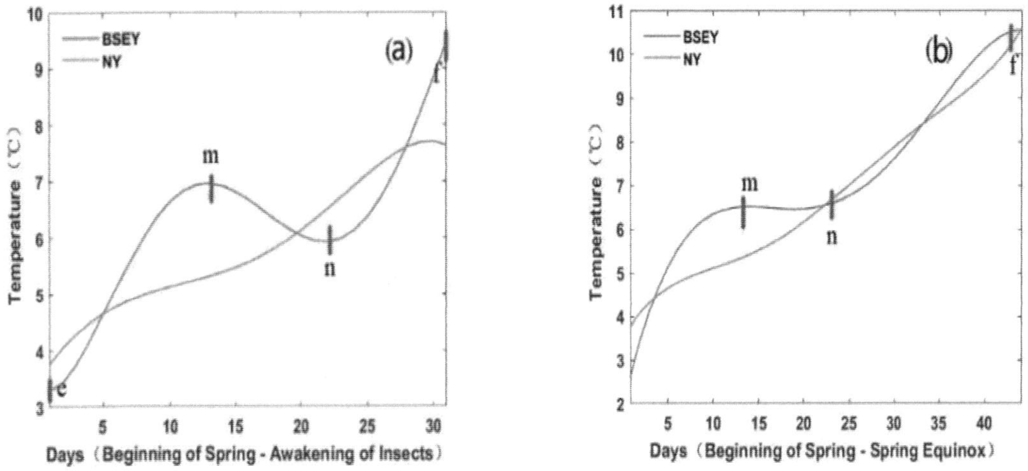

Figure 3. Warming trends in areas of strong compliance (Taking Nanjing, Zhejiang Province, as an example)

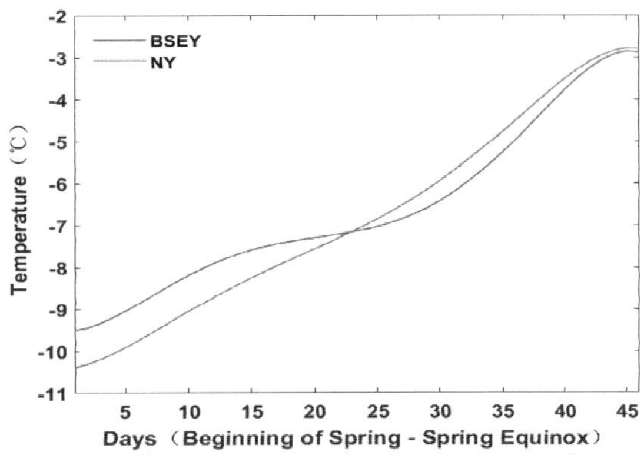

Figure 4. Warming trends in non-compliant areas (Taking Maqin, Qinghai Province as an example).

In Figure 2, there are special areas that don't meet 65% of the criteria for the "FW" characteristic due to the topography and climate of the region (Figure 4), which mainly include the Qinghai-Tibet Plateau and southern Yunnan, accounting for 31% of the Yangtze River basin area. Among them, the Tibetan region and most of Qinghai, located on the Qinghai-Tibet Plateau [15], belong to the alpine regions, because of the high terrain, the warm and humid airflow from the ocean is difficult to climb, so compared with other regions in the Pre-temperature increase temporal period will not have a significant warming trend, that is, little difference between the two different years in terms of warming after the Beginning of Spring. In contrast, the southern part of Yunnan is a low-latitude tropical valley region [16], which has the characteristics of no winter for a long time and less pronounced temperature change, the difference between years is small, so the "FW" characteristic of BSEY is not significant compared with NY.

In summary, within the Yangtze River basin, except for the Qinghai-Tibetan alpine region and the low-latitude tropical valley in southern Yunnan, which do not meet "FW" characteristic of the proverb due to climatic conditions, other regions conform to "FW" characteristic, and the region has significant warming of BSEY compared with NY in the two time periods of the Beginning of Spring - Awakening of Insects and the Beginning of Spring - Spring Equinox.

4.2. "HDD" characteristic

This paper classifies the difference between BSEY and NY during the period of the dog days (30 or 40 days) from 1966-2019 into three cases, according to the classification criteria of the degree of compliance with the characteristic of "HDD" (Figure 4). The areas in the figure where the "HDD" characteristic meets the strongest degree are those where BSEY is "HDD" than NY, mainly in the subtropical monsoon climate zone [17] of China, which are distributed in different degrees in the upper, middle and lower reaches of the Yangtze River basin, accounting for 43% of the regional area of the Yangtze River basin, with the strongest compliance in the lower reaches, where the compliance area ratio is 90%. These areas are affected by tropical marine warm and humid air currents, high temperatures, and rain in summer, the average temperature in July is up to about 28 °C, in addition, July-August because of the control of subtropical high pressure, more sunny days, long sunshine hours, high temperatures appear more frequently. Related studies have shown that the dog days region and the quasi-dog days region are located south of 40 °N in China mainly influenced by the summer wind [10]. Therefore, the subtropical region located in southern China relative meets the climatic conditions required for the second part of the proverb.

To investigate in details the reasons for the BSEY than NY "HDD" in the Yangtze River basin conforming area, and considering that precipitation affects temperature and relative humidity to some extent, this paper analyses and compares precipitation during the dog days period (30 or 40 days) 1966-2019 for two years in the region. The results showed that the overall precipitation of NY was higher than that of BSEY in the dog days period, while the detailed comparison of the precipitation of the dog days period among the eras indicated that the precipitation of BSEY was higher than that of NY only in the 1980s (Figure 5). More precipitation in the short-term will led to relatively low temperatures, and the difference between the multi-year average relative humidity of the two years in the Yangtze River basin during the 1966-2019 dog days period is small (78.5% for NY and 78.4% for BSEY), Therefore, according to the Temperature-Humidity Index, precipitation may be an important factor that causes BSEY to be " HDD" compared to NY in compliance with regional.

In Figure 2, the areas with weak (10% of the Yangtze River basin area) and non-conforming (47% of the Yangtze River basin area) "HDD" characteristic belong to the areas that do not meet the "HDD" characteristic. In contrast to the non-conforming regions that cannot meet the " HDD" characteristic due to climatic conditions, the distribution regions with weak and strong compliance are located in the subtropical monsoon zone, but also cannot meet the " HDD" characteristic due to the influence of topography and terrain. Among the regions that meet the weak influence mainly due to mountain ranges, including Hunan and Hubei areas due to the influence of the Xuefeng and Wushan mountain ranges; and Sichuan and Shaanxi areas due to the influence of the Qinling mountain range. These non-conforming areas may be influenced by the orientation of mostly northeast-southwest mountain ranges, blocking warm and humid air currents from the ocean, and precipitation under the influence of air currents varies very significantly under the influence of topographic terrain, generally more mountainous than flat, and the more windward than leeward slopes [18].

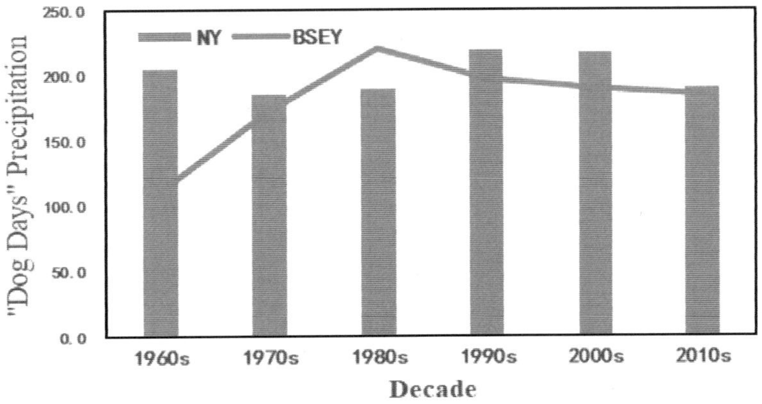

Figure 5. Comparison of inter-decadal precipitation during the dog days period.

In conclusion, within the Yangtze River basin, the conforming areas for the "HDD" characteristic are all in the subtropical monsoon region, which satisfies the climatic conditions required for the "HDD" characteristic, and according to the heat and humidity index, precipitation may be an important factor that causes the BSEY in the conforming area to "HDD" than the NY. In contrast to the conforming regions, except for the Qinghai-Tibet alpine region where the climatic conditions are not met, other non-conforming regions do not conform to "HDD" characteristic mainly due to the influence of topography and terrain.

5. Conclusions

This paper investigates the scientific significance of the meteorological proverb " The Beginning of Spring at both ends resulting in hot in the dog days " based on the elements of temperature, relative humidity, and precipitation of meteorological data from 1966 to 2019 in the Yangtze River basin, and explains the accuracy of the proverb and the main applicable regions from a scientific perspective through comparative analysis. The conclusions are as follows:

The proverbs in this paper coincide with the circulating regions, with various degrees of compliance in the upper, middle and lower reaches of the Yangtze River basin, and the strongest compliance in the lower reaches, accounting for 90% of the area of the lower reaches. The study of "FW" characteristic shows that after the Beginning of Spring, the areas where BSEY has a more obvious trend of "FW" than NY account for 69% of the Yangtze River basin area, and the warming time period are before the Spring Equinox. In addition, the analysis of " HDD" characteristic shows that the characteristics of the lower reaches of Yangtze River are the most obvious, accounting for 90% of the lower reaches, and the conforming regions of the Yangtze River basin meets the climatic conditions required for "HDD" compared to the Qinghai-Tibetan alpine region.

From the analysis of the scientificity and applicability of the proverbs, the empirical rules behind the proverbs are in line with certain scientificity and the geographical area of circulation (the middle and lower reaches of the Yangtze River basin), but they have certain limitations in terms of applicable geographical area. Proverbs are mostly local knowledge and may only apply within a certain geographical area, only in circulation as well as in areas with the same type of climate. In the context of global climate and environmental changes, proverbs should serve practice while conforming to science. While following traditional meteorological proverbs, scientific methods should be used to adjust and perfect them, so as to provide references for agricultural proverbs to better promote and arrange human agricultural production and life, and better adapt to and use climate resources.

Acknowledgments
This work was supported by the National Key Research and Development Program of China (No. 2020YFA0608501), Project supported discipline innovation team of Liaoning Technical University (No. LNTU20TD-23), Project of Innovation and Entrepreneurship Training Program for University Students (202110147036).

References
[1] Fei J 2018 Meteorological history and historical climate of China[M]//*Oxford Research Encyclopedia of Climate Science*
[2] Xiao G J ,Zhang Q ,Wang J 2007 Impact of global climate change on agro-ecosystem: a review[J] *The Journal of Applied Ecology* **18(8)** 1877
[3] IPCC: Climate Change 2021: The Physical Science Basis. Contribution of Working Group I to the Sixth Assessment Report of the Intergovernmental Panel on Climate Change *Cambridge University Press* 2021
[4] Chen H, Zhang Y 2021 The Influence of Cultural Differences between Chinese and English on Translation[J] *Cross Current Int J Peer Reviewed J Human Soc Sci* **7(4)** 67-72
[5] SHEN Shi-pei, Bao Qin 2021 On Proverbs like "A Good Harvest in Suzhou and Huzhou, a Life of Abundance for All the Country" and Low-field Rice Cultivation in Western Zhejiang during the Southern Song Dynasty[J] *Agricultural History of China* **40(02)** 15-26
[6] Matczak P, Graczyk D, Choryński A, et al. 2020 Temperature forecast accuracies of Polish proverbs[J] *Weather, Climate, and Society* **12(3)** 405-419
[7] Zhenfu Tao 2018 Agrometeorological Proverbs and Explanations [M] *Beijing: China Meteorological Press* 1-4
[8] JI C, WANG S, SHANG K 2015 Climate Changes of 24 Solar Terms in the Middle and Lower Reaches of the Yellow River during 1961-2010[J] *Climatic and Environmental Research*
[9] Qian C, Yan Z, Wu Z, et al. 2011 Trends in temperature extremes in association with weather-intraseasonal fluctuations in eastern China[J] *Advances in atmospheric sciences* **28(2)** 297-309
[10] Xia J, Yan Z 2011 Climatology of "Three Fu" in China[J] *Climatic and Environmental Research*
[11] Newville M, Stensitzki T, Allen D B, et al. 2016 LMFIT: Non-linear least-square minimization and curve-fitting for Python[J] *Astrophysics Source Code Library* ascl: 1606.014
[12] Steadman R G 1979 The assessment of sultriness. Part I: A temperature-humidity index based on human physiology and clothing science[J]. *Journal of Applied Meteorology and Climatology* **18(7)** 861-873
[13] Mather J R 1974 Climatology: fundamentals and applications[M]
[14] Xuefeng Li 2019 Twenty-four Solar Terms and Seventy-two Pentads [M]. *Beijing: China Photographic Publishing & Media Co., Ltd*. 27-29
[15] MA Z, ZHANG M, WANG S, et al. 2019 Characteristics and Differences of Temperature Rise between the Qinghai-Tibetan Plateau Region and Northwest Arid Region of China during 1960-2015[J] *Plateau Meteorology* **38(1)** 42-54
[16] Zhang W, Zheng J, Tao M A, et al. 2015 Temporal and spatial distribution and variation of extreme temperatures in Yunnan Province from 1961 to 2012[J] *Resources Science*
[17] Zhu Kezhen Subtropical zone in China[J]. *Chinese Science Bulletin* **1958(17)** 524-528
[18] Li G 2016 Progress and Prospects in Research of Mountain Meteorology in China During the Past 25 Years[J] *Advances in Meteorological Science and Technology*

Research on goaf treatment and geothermal comprehensive utilization technology in coal mining subsidence area

Yu Yang[1] **and Bo Li**[2,3,4]

[1] College of Resource and Environmental Engineering, Guizhou University, Guiyang, Guizhou 550025, China;
[2] Key Laboratory of Karst Georesources and Environment, Ministry of Education, Guizhou University, Guiyang, Guizhou 550025, China;
[3] Key Laboratory of Karst Environment and Geohazard, Ministry of Land and Resources, Guizhou University, Guiyang, Guizhou 550025, China.

[4] Corresponding author's e-mail: libo1512@163.com

Abstract. Goaf treatment and comprehensive utilization of geothermal energy are great significance to the restoration of land resources and sustainable development of energy in coal mining subsidence areas. In this paper, goaf treatment technology and geothermal energy development technology are reviewed in detail. Among them, the goaf treatment technology mainly includes: caving treatment method, filling treatment method, support method, closed isolation method and joint treatment method. Geothermal development technology mainly includes: shallow ground source heat pump technology, coaxial casing development technology, pumping and reinjection technology, directional drilling technology, U-type precision docking drilling technology. The systematic idea of filling treatment and geothermal comprehensive utilization in coal mining subsidence area is proposed, and the technology is described in detail through three cases. The research results can provide innovative inspiration for subsidence control and comprehensive utilization of geothermal energy in coal mining subsidence area, and has broad application prospects.

1. Introduction

Since the "industrial revolution", once-use fossil energy such as coal, as the main energy to support the development of the world economy, has been consumed in large quantities, making the global industry, economy and other aspects change dramatically [1]. Energy structure in China is relatively single, mainly presented as more coal, less oil and gas, so coal is heavily dependent in the development of the national economy [2]. On the one hand, a large number of coal mining has greatly guaranteed energy supply and promoted the rapid development of the national economy. On the other hand, it has caused damage to the ecological environment and caused many mine geological disasters [3]. According to relevant statistics, for underground coal mining, every 10,000 tons of raw coal mining will cause 0.2~0.33 hm^2 surface subsidence. The existing coal mining subsidence area in China has reached more than 2 million hm^2, and it is expected to reach 2.8 million hm^2 in 2030 [4]. A large area of coal mining subsidence has caused many problems, such as: landslides, surface cracking and collapse, farmland and housing damage, which seriously affect the construction of ecological civilization and living environment in China [5]. Therefore, the Ministry of Land and Resources and other five ministries issued "Guidance on Strengthening the Recovery and Comprehensive

IOP Conf. Series: Earth and Environmental Science 1087 (2022) 012055 doi:10.1088/1755-1315/1087/1/012055

Management of Mine Geological Environment" during the "13th Five-Year Plan" period, which brought mine geological environment treatment into the important measures of ecological civilization construction in China [6]. Therefore, comprehensive management of coal mining subsidence area is urgent.

Since the 21st century, China has focused on developing the economy, but ignored the ecological environment protection, resulting in frequent extreme weather, haze, temperature rise and many other environmental disasters. Therefore, since the "13th Five-Year Plan", General Secretary Xi Jinping has repeatedly put forward the assertion that "green mountains and rivers are golden and silver mountains" in the construction of ecological civilization. In order to show the determination of carbon reduction and environmental protection in China to the world, the great goals of "carbon peak" and "carbon neutralization" are put forward [7]. In the future, the proportion of fossil energy such as coal in primary energy consumption in China will be further reduced, replacing it with green and low-carbon clean energy. Among them, geothermal energy [8], as one of the emerging energy sources with clean, efficient, low-carbon, environmental protection and sustainable utilization, has gradually attracted wide attention from social circles.

Although the development and utilization of geothermal resources has many advantages, its application is greatly limited due to the high drilling cost in the early stage. However, for coal mining subsidence areas, it is necessary to fill the goaf to restore the damaged land resources for urban construction, agricultural and animal husbandry production, tourism and so on [9]. These construction operations still require a large amount of energy supply, and at present the geothermal energy can be used as a preferred energy. This is because although there are many methods for goaf treatment, the borehole grouting filling method is widely used because of its better effect, fast efficiency and dense filling [10]. When the goaf filling treatment is completed, the grouting borehole is transformed into a geothermal well with less cost, and the geothermal resources in the goaf can be developed and utilized. This not only restores a large number of land resources, but also saves the initial cost of geothermal development, and reduces the carbon emission index of construction energy, which is helpful to the early realization of "dual carbon target" in China.

In this paper, the goaf treatment technology and geothermal energy development technology are detailedly summarized. On this basis, the system idea of filling treatment and geothermal comprehensive utilization in coal mining subsidence area is proposed, and three cases are listed to elaborate the technology in detail. The relevant research can provide forward-looking path and strategic inspiration for the recovery of land resources and long-term sustainable development of energy in coal mining subsidence areas in China.

2. Goaf treatment technology

For underground coal mining, after mining, there will be huge goaf space, and the roof of coal seam will collapse to form gob. Under the long-term load of the upper stratum and building mass, the broken rock in the goaf is gradually compacted, resulting in the bending and subsidence of the upper stratum, affecting the surface and forming a coal mining subsidence area [11]. At present, the methods of goaf treatment in the world can be mainly divided into five categories: caving treatment method, filling treatment method, supporting method, closed isolation method and joint treatment method [12].

2.1. Caving treatment method

The caving treatment method uses engineering blasting and other technical means to make the surrounding rock or roof of the orebody fall naturally or forcefully, so as to fill the goaf, form a buffer cushion, and prevent the sudden large-scale collapse of the roof [13,14], as shown in Figure 1.

The caving method can be further divided into natural caving method and forced caving method. The natural caving method is mainly applicable to the mining area where the surrounding rock is weak, the strength is low, and the rock is easy to fall naturally. The technology is simple and the cost is low in this method. The disadvantage is that the natural caving law of rock is difficult to grasp. The forced caving method is mainly applicable to the mining area where the roof of surrounding rock is hard and difficult to collapse naturally. Usually, directional blasting and water injection can be used to soften the roof. This method can force the collapse of surrounding rock and roof, but it is easy to cause surface cracking and caving [15,16].

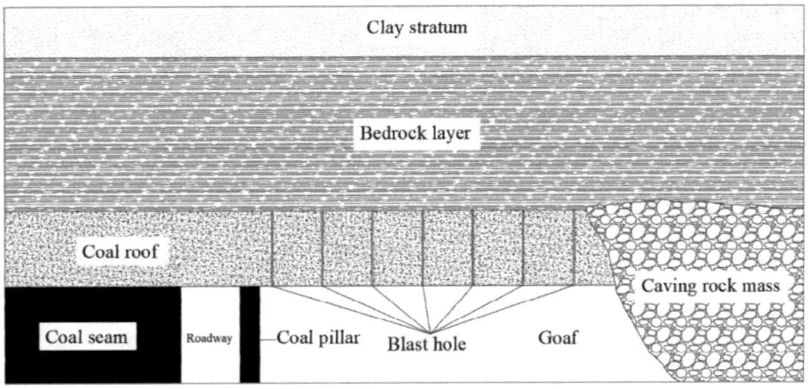

Figure 1. Diagram of the caving treatment method.

2.2. Filling treatment method

The filling treatment method is mainly to fill solid waste or wet filling material into the goaf, so as to support the surrounding rock and roof and prevent the deformation and subsidence of roof, as shown in Figure 2.

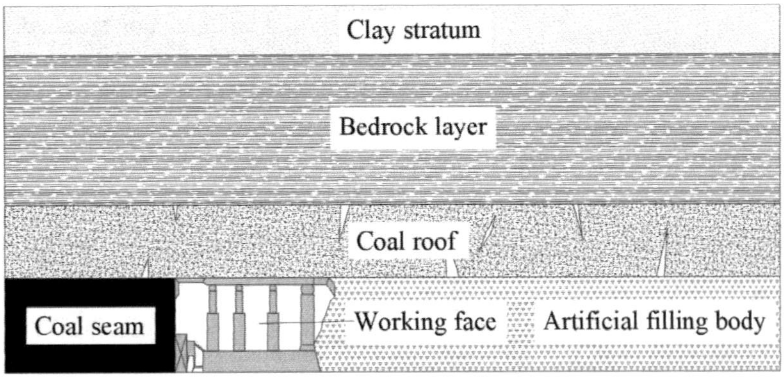

Figure 2. Diagram of filling treatment method.

Filling method can be divided into dry filling and wet filling. Dry filling mainly refers to filling construction waste and coal gangue in goaf. This method has low efficiency, high labor cost and unsatisfactory filling effect. Wet filling mainly refers to the method of grouting to the goaf after the roof collapse, which can be divided into ground grouting filling and underground grouting filling. This method has the advantages of dense filling, good treatment effect and quick efficiency, so it is widely used [17-20]. HUANG [21] studied the mechanism of rock fracture development in goaf under Hongyanhe Reservoir, and simulated it by filling treatment method. The results showed that the effect of coal mining subsidence treatment was remarkable. CHEN et al. [22] used the grouting technology of "pressureless self-flow + pressure diffusion" to strengthen and control a coal mining subsidence area, and then constructed 22-story comprehensive buildings and other large building clusters above the area, and has been put into use, and the operation effect is good. SUN et al. [23] made the mine solid waste into paste to fill the goaf. After 28 days, the uniaxial compressive strength of the rock in the gob can reach 1.5~3.0 MPa, and the filling effect is good.

2.3. Supporting method

The supporting method refers to the method of reserving permanent coal pillar or artificial pillar in the mining process so that the supporting roof does not collapse, as shown in Figure 3. This method is suitable for mining areas where the roof is stable and not easy to fall, and there is no building on the ground. This method can change the continuous large mined-out area into closed small goafs, so as to play a certain role in controlling the roof stability. However, if a coal pillar is unstable, it is easy to cause a large area of collapse, and the coal recovery rate is low, furthermore the reserved permanent coal pillar will waste a lot of mineral resources [24-26].

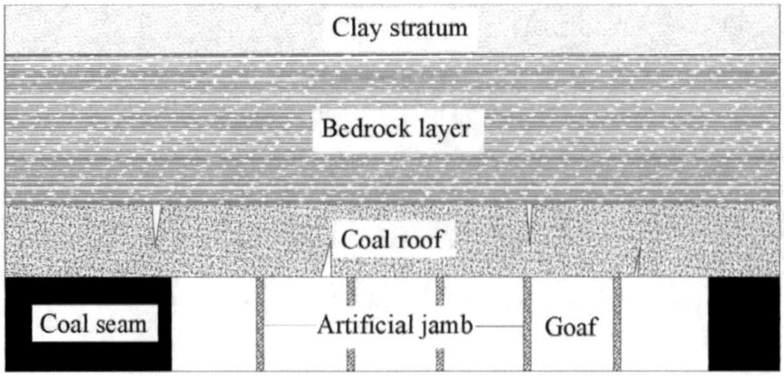

Figure 3. Diagram of supporting method.

2.4. Enclosed isolation method

The enclosed isolation method is the method of enclosing or isolating the goaf by using the construction of artificial isolation wall and other technologies, as shown in Figure 4. It is mainly including: toothed wave-resistant wall, buffer wave-resistant wall, rock wave-resistant wall and concrete wave-resistant wall [27]. This method is mainly applicable to the goaf of isolated small orebody, the goaf after end mining, and the upper goaf of large orebody that needs stratified mining. This method can effectively reduce the impact caused by the sudden fall of the upper roof, avoid the danger of construction workers entering the goaf, and is conducive to mine ventilation. However, it is difficult to ensure the effectiveness for the goaf of large and thick orebodies [28].

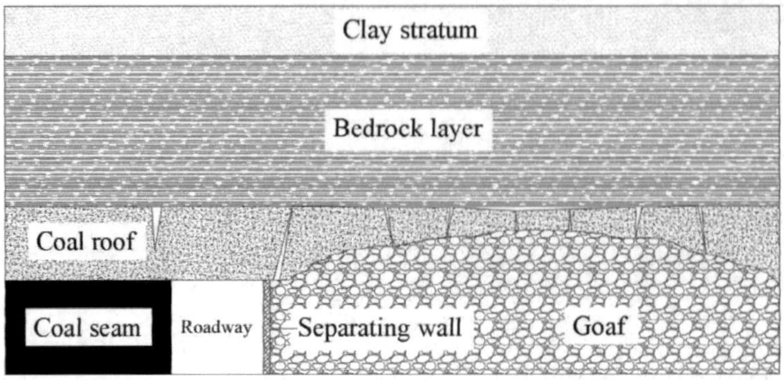

Figure 4. Diagram of enclosed isolation method.

2.5. Joint treatment method

Since the above four methods have their own application scope, advantages and disadvantages, for a large mine, the use of a single method is often not economical and effective to deal with the goaf. Therefore, it is necessary to combine two or more methods and use their advantages to achieve the most economical, effective and reasonable purpose. This method of dealing with the goaf is joint treatment method [29]. At present, the relatively mature joint treatment methods mainly include: supporting filling method, supporting collapse method, caving filling method, caving separation method, blasting caving method [30]. In short, the joint treatment method can effectively play the advantages of each methods and overcome their limitations, which is also the most reasonable goaf treatment method [31].

3. Geothermal development technology

Geothermal energy is a kind of thermal energy resource stored in the crust circle caused by magma melting and radioactive element decay [32]. At present, geothermal energy by reservoir depth can be roughly divided into: shallow geothermal energy (< 3,000 m), medium-deep geothermal energy (300~3,000 m), deep geothermal energy (> 3,000 m) [33]. The methods of development and utilization can be divided into: shallow ground source heat pump technology [34], coaxial casing development technology (also known as non-interference development technology) [35], pumping and reinjection technology [36] according to the type of underground heat exchanger. And that can also be divided into: vertical drilling technology, directional drilling technology, U-type precision docking drilling technology according to the type of geothermal well drilling [37].

3.1. Shallow ground source heat pump technology

Shallow ground source heat pump technology mainly refers to a technology that exchanges heat with the formation through a single U-type or double U-type heat exchanger within a shallow depth of 300 m [38], as shown in Figure 5. It can be seen from the diagram that the circulating medium (water or aqueous solution) flows from one end of the heat exchanger and exchanges heat with the formation and then flows from the other end. This method is mainly used in cold winter and hot summer areas. Due to the shallow depth, the underground temperature is higher in winter and lower in summer. So, this technology can be used for heating in winter and cooling in summer. It can completely replace the indoor air conditioning, and has lower operating costs than the air conditioning system [39]. However, this method has a large number of boreholes and needs to occupy large land resources in the early stage. Moreover, it requires high cooling and heating load balance throughout the year and has limited heat intake.

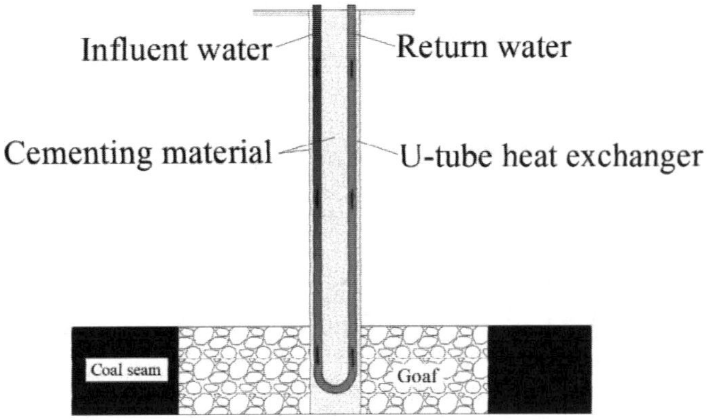

Figure 5. Technical schematic of shallow ground source heat pump.

3.2. Coaxial casing development technology

Coaxial casing development technology is mainly used in medium-deep and deep geothermal wells. The heat exchanger structure is composed of two layers of casing [40], as shown in Figure 6. The specific construction steps are: drilling, lowering outer casing, cementing with high thermal conductivity material [41], sweeping hole, lowering heat insulation and anticorrosion inner casing [42].

When the system works, the circulating medium enters from the outer casing, and flows out of the inner casing after absorbing the heat of the formation. Due to the deep depth and large heat intake, it is mostly used for heating in winter, providing domestic hot water and even geothermal power generation in cold regions [43,44]. This method only has heat exchange with the formation, does not destroy the environment, and occupies a small area with strong adaptability. However, it cannot be used for refrigeration in summer, and the cost of drilling in the early stage is high, which has become an important factor restricting the wide application of this technology [45].

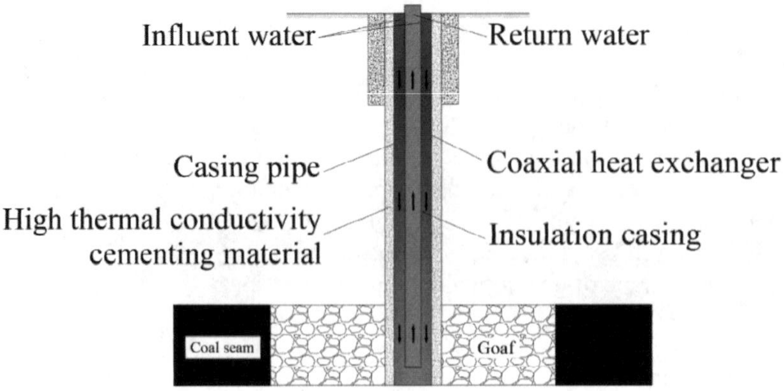

Figure 6. Diagram of coaxial casing development technology.

3.3. Pumping and reinjection technology

The pumping and reinjection technology is mainly used in the water-bearing geothermal reservoir with developed fractures, which can be divided into the types of one pumping and one reinjection, one pumping and multiple reinjection, multiple pumping and one reinjection, multiple pumping and multiple reinjection [46]. Figure 7 shows the geothermal well of one pumping and one reinjection.

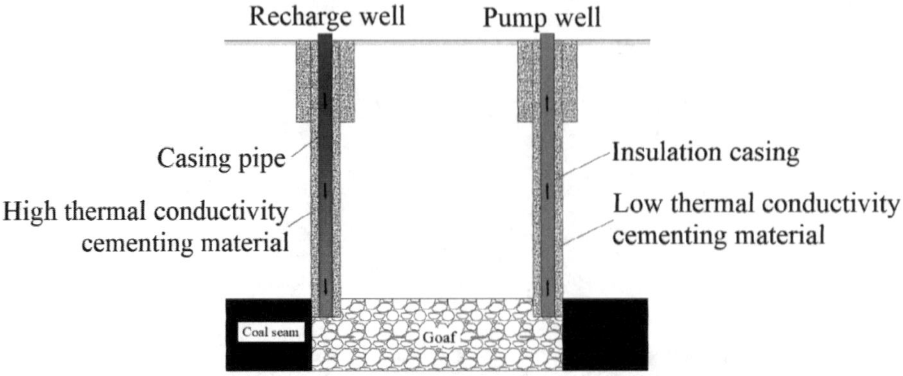

Figure 7. Diagram of pumping and reinjection technology.

This method mainly uses hot water directly from geothermal reservoirs, and then flows cold water into geothermal reservoirs through constant pressure reinjection technology, and then forms a set of circulation system through stratum heating and fracture coalescence. Due to the direct extraction of underground hot water, the effect of heat extraction is best, but the requirement for reinjection is strict. It must ensure the balance of pumping and reinjection, otherwise it will destroy the balance of groundwater system and affect the local ecological environment. In addition, this method also requires that the geothermal reservoir is an aquifer with sufficient water, and the fractures of rock are sufficiently developed. Of course, the formation can be artificially fractured by means of hydraulic fracturing [47], but the cost is high and the technical difficulty is large, so the scope of application is limited, and the prospect of popularization and application is not very optimistic.

3.4. Directional drilling technology
In recent years, the development of drilling technology has promoted the progress of geothermal energy development technology, including directional drilling technology and U-type precision docking drilling technology. Directional drilling technology is the technology of drilling holes along a specific trajectory by directional drilling tools. When this technology is used in geothermal exploitation, it becomes a directional geothermal well [48], as shown in Figure 8.

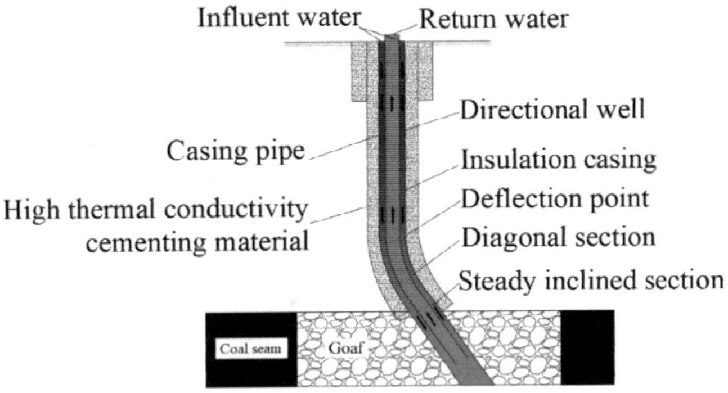

Figure 8. Diagram of the caving treatment method.

The advantage of this method is that it can avoid special formations such as structures and fracture zones. Which increases the horizontal section and the heat transfer area in disguise. At the same vertical depth, the heating effect is better than that of vertical drilling [49], but it also increases the drilling cost and the drilling efficiency is low.

3.5. U-type precision docking drilling technology
U-type precision docking drilling technology is mainly divided into two parts, as shown in Figure 9. First of all, the vertical well needs to be drilled, and the depth is slightly larger than the horizontal well. After the completion of the drilling operation, the insulating casing is lowered and the cementing material with low thermal conductivity is used for cementing. It should be noted that the casing at the docking position should be fiberglass casing. Then, horizontal well drilling is carried out. In this process, Azimuth Gamma-Ray Directional While Drilling (DWD) instrument and High-Precision Magnetic Steering Docking instrument are used to achieve precise docking of horizontal well and vertical well. After completion of docking, casing is dropped and cementing is performed with high thermal conductivity cementing material. Then, holes are swept and connectivity of U-type well is tested with water [50]. When working, the circulating medium enters from the horizontal well, after reaching the horizontal section, it absorbs a lot of heat in the geothermal reservoir and flows out through the vertical well. Due to the long heat transfer distance and high temperature in the horizontal section, the heat loss is limited by the heat insulation casing and low thermal conductivity cementing material of the return well, and the heat recovery effect of this method is extremely excellent. But the extremely difficult drilling docking technology increases the probability of drilling failure and drilling cost, so that the large and small enterprises are looking down on it.

4. Goaf treatment and geothermal comprehensive utilization technology
Based on the above detailed summary of goaf treatment technology and geothermal development technology, it can be found that the most reasonable and efficient method for the treatment of goaf that has formed coal mining subsidence area is the ground borehole grouting filling method. This method requires a lot of drilling holes, which can provide a lot of drilling resources for geothermal

development. Therefore, if the goaf treatment and geothermal development are integrated, which can realize the reuse of collapsed land resources, and develop a large number of low-carbon economic geothermal resources. The most important thing is to save a lot of repeated drilling costs, so as to maximize benefits. Then the goaf treatment and geothermal comprehensive utilization technology are described in more detail below.

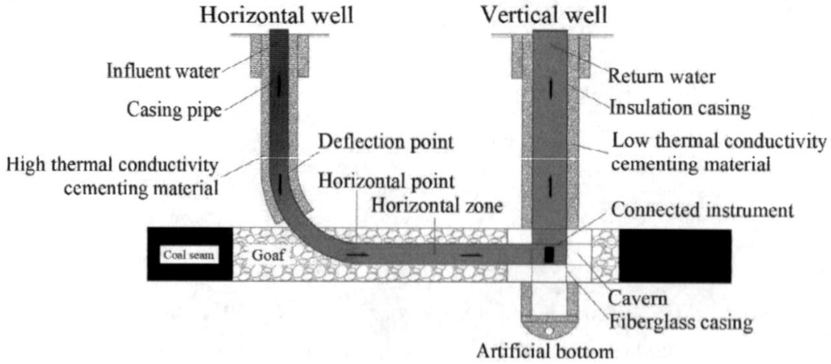

Figure 9. U-type precision docking drilling technology schematic diagram.

4.1. Based on vertical drilling technology
Vertical drilling technology is the most commonly used drilling technology. The goaf treatment and geothermal comprehensive utilization method based on vertical drilling technology need to arrange a large number of vertical drilling groups on the surface, as shown in Figure 10. The layout can be divided into square, row, plum blossom, etc., which can be determined according to the specific situation. The spacing should not be too large, and the appropriate spacing is about 20 m. The depth is 1 m from drilling to the floor of goaf. After the drilling construction is completed, if or not need casing can be selected according to the type of heat exchanger in the later geothermal development.

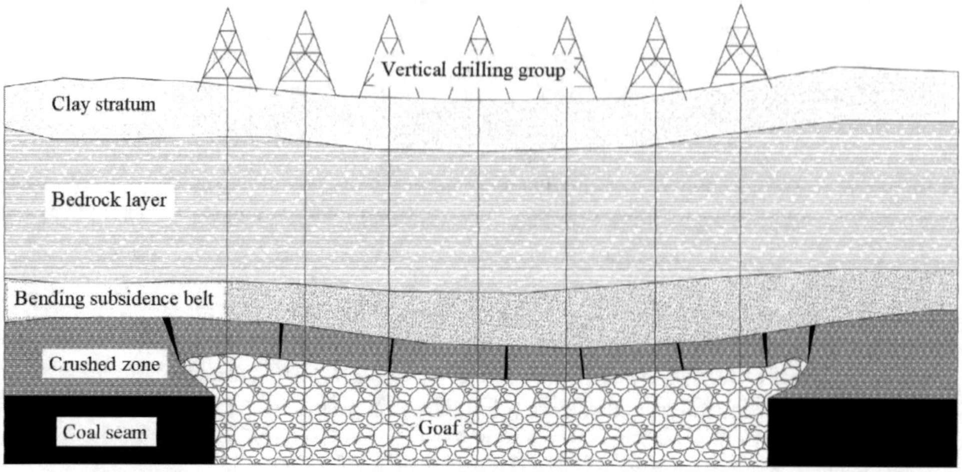

Figure 10. Goaf treatment and geothermal utilization method based on vertical drilling technology.

If the coaxial casing heat exchanger is selected, the casing is first inserted into the hole, and the flower tube is needed to pay attention to the lower part of the casing. Then the goaf is grouted and filled. After the treatment is completed, the hole cleaning operation is carried out along the inner casing, and then the inner insulation casing is inserted. At this time, the drilling hole is transformed into the coaxial casing geothermal well. The specific wellbore structure can be seen in Figure 6. If the single U-type or double U-type heat exchanger is selected, it is only necessary to enter the casing in the surface soil layer, and then the goaf is grouted and filled through bare hole. After the slurry solidification meets the technical requirements of re-drilling, the hole cleaning operation is carried out along the original borehole, and then the single U-type or double U-type heat exchanger is entered. The mud added fine sand or cement slurry is used to backfill the borehole. And then, the geothermal well is reconstructed. The specific wellbore structure can be seen in Figure 5. It should be noted that this work needs to determine the number of boreholes in combination with the load of the proposed building in the region. If the number of boreholes is insufficient, geothermal boreholes can be supplemented at an appropriate position, but it should be noted that the distance between the boreholes cannot be less than 5 m, otherwise it will cause cold and hot accumulation effect [51-52].

4.2. Based on directional drilling technology

The goaf treatment and geothermal comprehensive utilization method based on directional drilling technology are mainly applicable to the existing buildings in the region, and the building load accelerates the situation that the surface subsidence needs to be treated for the goaf. In order to avoid the destruction of buildings by drilling construction, a large number of vertical drilling groups cannot be arranged above the goaf, and the directional drilling technology can well solve this problem, as shown in Figure 11.

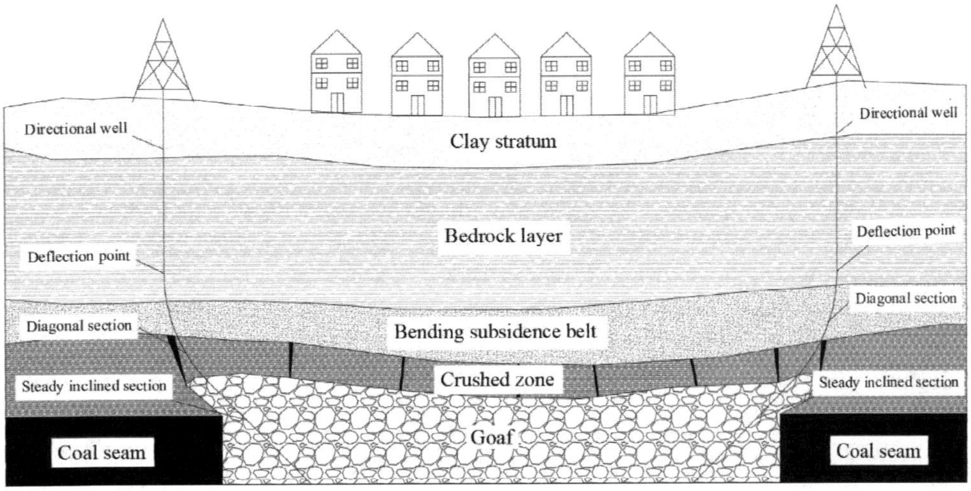

Figure 11. Goaf treatment and geothermal utilization method based on directional drilling technology.

The method is mainly to arrange boreholes around the building cluster, which is often arranged in rows. The spacing is 20~50 m. The slope shall be subject to drilling to the goaf, and the vertical depth is often deeper (> 500 m). Due to the deep depth, coaxial casing heat exchanger should be preferred, then the heating effect will be better. Therefore, after the completion of drilling construction, the casing is first inserted into the hole, and then the filling grouting is carried out to the goaf. When the slurry returns to the wellhead, the grouting is completed, so as to achieve the cementing effect. After the slurry is solidified, the hole cleaning operation is carried out along the casing, and then the inner heat insulation casing is poured into the directional geothermal well. The specific wellbore structure can be seen in Figure 8. If the existing number of geothermal wells cannot meet the construction operation load, vertical boreholes, medium-deep geothermal wells and solar energy can also be supplemented at appropriate locations for joint energy supply.

IOP Conf. Series: Earth and Environmental Science 1087 (2022) 012055 doi:10.1088/1755-1315/1087/1/012055

4.3. Based on U-type docking drilling technology
The goaf treatment and geothermal comprehensive utilization method based on U-type precision docking drilling technology can be used in areas where a large number of building clusters have been built or are proposed, as shown in Figure 12.

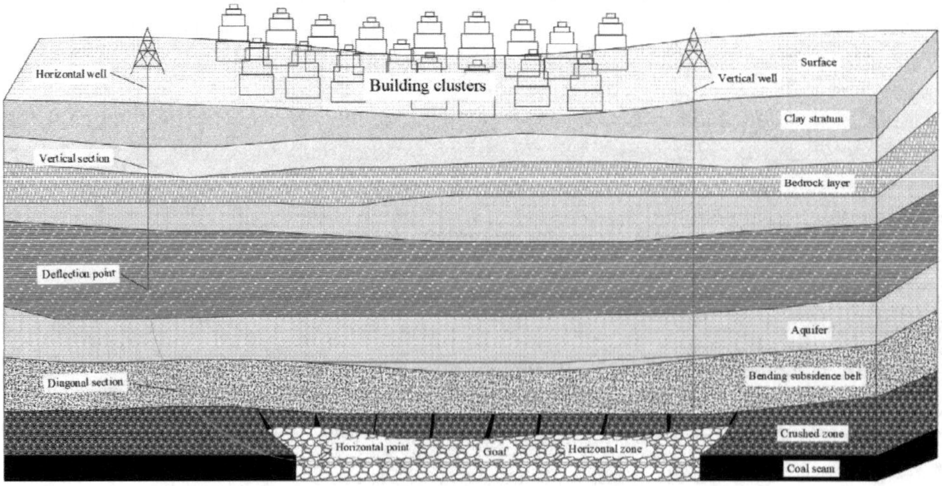

Figure 12. Goaf treatment and geothermal utilization method based on U-type docking drilling technology.

The drilling arrangement is mainly arranged in rows, the row spacing is greater than 50 m, the depth is greater than 1000 m, and the horizontal distance is greater than 300 m. This method firstly drills the vertical borehole, puts down the casing and cementing. Then the horizontal well will be drilled, the horizontal section is set in the goaf, when in the goaf it cannot be drilled, it is necessary to repeatedly grouting, sweeping and redrilling, so the construction technology is more complex. After the end of the last grouting, the grouting filling operation of the hole sequence is also completed, and then the last sweeping drilling is needed, and the vertical drilling is accurately docked. This process has the highest technical difficulty, which determines the success or failure of the whole drilling. After the successful docking is completed, the horizontal well casing is laid down and cementing is completed. Finally, the whole well section is swept and the water test is carried out. The U-type docking geothermal well is reconstructed. The specific wellbore structure can be seen in Figure 9. Due to the long horizontal heat transfer distance in geothermal reservoirs, this method has the best heat transfer effect. However, it is necessary to pay attention to the heat preservation and insulation of return wells, and it is recommended to select heat insulation casing and low thermal conductivity cementing materials, so as to minimize the heat loss.

5. Prospect
(1) Goaf treatment and geothermal development is a systematic project, which requires government overall consideration, policy support, and developers need system design to comprehensively select the most economical and efficient technical solutions according to the actual situation.
(2) The goaf treatment and geothermal comprehensive utilization method can solve the surface subsidence above the goaf and restore a large number of land resources by drilling and grouting into the goaf. Through the upgrading of grouting boreholes, the geothermal resources in the region can be developed and utilized, and the initial investment cost of separate development of geothermal can be minimized, which has a good application prospect.
(3) The rock mass is broken and there are a lot of cracks in the goaf, which is a natural water storage layer and is also conducive to the storage of heat. Therefore, it is also a natural geothermal reservoir, which can provide good heat source supply for geothermal development.

6. Conclusions

This paper mainly introduces five technical means of goaf treatment, which are: caving treatment method, filling treatment method, support method, closed isolation method and joint treatment method, respectively. And then introduces five technologies of geothermal energy development too, which are: shallow ground source heat pump technology, coaxial casing development technology, pumping and reinjection technology, directional drilling technology, and U-type precision docking drilling technology, respectively. The systematic idea of goaf treatment and geothermal comprehensive utilization is proposed, and the technology is described in detail through three examples. This method can provide innovative inspiration for subsidence treatment and geothermal energy development and utilization in coal mining subsidence areas, and has broad application prospects.

CRediT authorship contribution statement

Yu Yang: Conceptualization, Formal analysis, Methodology, Writing-Original Draft, Writing-Review & Editing, Visualization, Investigation. **Bo Li:** Formal analysis, Supervision, Resources, Investigation, Project administration.

Declaration of competing interest

The authors declare that they have no known competing financial interests or personal relationships that could have appeared to influence the work reported in this paper.

Fundings

This work was supported by Natural Science Foundation (42162022; 41702270), Guizhou Science and Technology Department Project (Qian Ke He Ji Chu [2019]1413; Qian Ke He Zhi Cheng [2020]4Y048; Qian Ke He Zhi Cheng [2020]4Y007), Guizhou Provincial Department of Education Foundation ([2018]113), Shanxi Province Coal Mine Water Hazard Prevention and Control Technology Open Fund (2020SKMS01).

References

[1] R. Shortall, A. Kharrazi 2017 Cultural factors of sustainable energy development: A case study of geothermal energy in Iceland and Japan *Renewable and Sustainable Energy Reviews* **79** 101-109

[2] H. Xie, L. Wu, D. Zheng 2019 Prediction on the energy consumption and coal demand of China in 2025 *Journal of China Coal Society* **44(7)** 1949-1960

[3] H. Xie, H. Liu, G. Wu 2012 A quantitative analysis of dependent index and contribution rate between China's economic and coal development *Journal of China University of Mining and Technology (Social Sciences)* **14(3)** 1-6

[4] J. Li, J. Yu, W. Zhang 2019 Spatial distribution and governance of coal-mine subsidence in China *Journal of Natural Resources* **34(04)** 867-880

[5] H. Liu, X. Zhu, Y. Cheng, et al. 2021 Key technology of human environment and ecological reconstruction in high submersible level coal mining subsidence area: A case study from Lüjin Lake, Huaibei *Journal of China Coal Society* **46(12)** 4021-4032

[6] Ministry of Land and Resources of China, et al. 2016 Guidance on strengthening the recovery and comprehensive management of mine geological environment *Management and Strategy of Qinghai Land and Resources* **04** 27-29

[7] G. Wang, S. Ren, Y. Pang, et al. 2021 Development achievements of China's coal industry during the 13th Five-Year Plan period and implementation path of "dual carbon" target *Coal Science and Technology* **49(04)** 1-8

[8] K. Wang, B. Yuan, G. Ji, et al. 2018 A comprehensive review of geothermal energy extraction an dutilization in oilfields *Journal of Petroleum Science and Engineering* **168** 465-477

[9] X. Zhu 2018 Study on comprehensive treatment model and subsidence land reuse of coal mining subsidence area in Huainan City *Anhui Agricultural University*

[10] K. Chen, Q. Chen, Q. Sun, et al. 2018 Application of grouting treatment technology in goaf for photovoltaic demonstration base in datong coal mining subsidence area *Safety in Coal Mines* **49(08)** 169-172
[11] J. Zhang, H. Wang, B. Zhan, et al. 2013 Exploration of mining goaf and comprehensive control technology of safety hidden dangers *Coal Science and Technology* **41(10)** 76-80
[12] Q. Chen 2009 Study on synergism of hidden danger resources mining and goafs treatment *Central South University*
[13] P. Zhu 2019 Research on caving mechanism and evolution law of roof rock of inclined thick-large ore body by caving method *University of Science and Technology Beijing*
[14] W. Li, J. Li 2014 Goafs hazard analysis and control technology of room mining in shallow coal seam *Safety in Coal Mines* **45(01)** 64-66
[15] W. Xiao, Z. Li, L. Chen, et al. 2021 Study on the formation technique of the cutting shaft by natural caving method of Pulang copper mine *Metal Mine* **11** 64-68
[16] S. Wu, Z. Zhang, Y. Wang, et al. 2021 Application of induced caving technology in formation of overburden layer for caving mining method in Longshou mine *Mining Research and Development* **41(10)** 14-21
[17] Y. Li, C. Liu, X. Li, et al. 2010 Roof control effect for gangue backfilling of goaf in thin steeply inclined seam under water body *Journal of China Coal Society* **35(09)** 1419-1424
[18] B. Zhou, Y. Pang 2018 Goaf filling and repeated coal mining technique for fully-mechanized caving faces through small coal mine *Coal Science and Technology* **46(S1)** 89-92
[19] W. Guo, X. Zhang, J. Shi, et al. 2010 Present situation of research on backfilling mining technology in mines and its application prospect *Journal of Shandong University of Science and Technology (Natural Science)* **29(04)** 24-29
[20] K. Deng, Z. Tan, H. Zhang 2008 Design method of strip grouting for old long wall mining goaf *Journal of China Coal Society* **02** 153-156
[21] J. Huang 2020 Research on control technology for underlying coal mining in Hongyanhe reservoir *China Coal Research Institute*
[22] S. Chen, L. Zhang, N. Jiang, et al. 2022 A case study of large buildings construction above old coal goaf in Shandong *Journal of China Coal Society* 1-16
[23] W. Sun, H. Wang, K. Hou 2018 Control of waste rock-tailings paste backfill for active mining subsidence areas *Journal of Cleaner Production* **171** 567-579
[24] S. Jiang, G. Fan, Q. Li, et al. 2021 Effect of mining parameters on surface deformation and coal pillar stability under customized shortwall mining of deep extra-thick coal seams *Energy Reports* **7** 2138-2154
[25] B. Xu, Y. LI, B. Lu, et al. 2022 Analysis of roof bearing characteristics and coal pillar stability of cemented backfill field *Journal of Mining Science and Technology* 1-10
[26] J. Jiang, H. Yang, D. Wang, et al. 2021 Experimental study on instability evolution mechanism of rib pillars during highwall mining *China Safety Science Journal* **31(10)** 89-96
[27] J. Zhang, S. Zeng, J. Liu, et al. 2017 Idea of combined treatment of goaf filling method and closed isolation method in a copper mine *Modern Mining* **33(05)** 215-216+219
[28] Y. Wang, Z. Guo 2015 The detriment and solution of roof caving in the mined-out area of a large mine *Conservation and Utilization of Mineral Resources* **03** 10-15
[29] J. Dai, M. Xia, L. Pan, et al. 2020 Analysis of comprehensive treatment of typical abandoned mines *Coal Technology* **39(06)** 97-99
[30] J. Huang, Z. Wu, Y. Wang, et al. 2020 Exploration and comprehensive treatment technology of overburden fracture in underlying goaf of reservoir *Safety in Coal Mines* **51(02)** 90-96
[31] X. Lan 2019 Study on the evolution law of ground pressure and the goaf treatment of gently-inclined thick ore body *Mining Research and Development* **39(12)** 68-72
[32] D.L. Gallup 2009 Production engineering in geothermal technology:a review *Geothermics* **38(3)** 26-334
[33] A. Al-habaibeh, A.P. Athresh, K. Parker 2018 Performance analysis of using mine water from

an abandoned coal mine for heating of buildings using an open loop based single shaft GSHP system *Applied Energy* **211** 393-402

[34] A.M. Omer 2008 Ground-source heat pump systems and applications *Renewable and Sustainable Energy Reviews* **12(2)** 344-371

[35] I.W. Johnston, G.A. Narsilio, S. Colls 2011 Emerging geothermal energy technologies *Ksce Journal of Civil Engineering* **15(4)** 643-653

[36] D.L. Gallup 2009 Production engineering in geothermal technology: A review *Geothermics* **38(3)** 326-334

[37] E. Barbier 2002 Geothermal energy technology and current status: an overview *Renewable and Sustainable Energy Reviews* **6(1-2)** 3-65

[38] I.A. Gondal 2021 Prospects of shallow geothermal systems in HVAC for NZEB *Energy and Built Environment* **2(4)** 425-435

[39] W. Alina, X. Li, J. Chambers, et al. 2022 Shallow geothermal energy potential for heating and cooling of buildings with regeneration under climate change scenarios *Energy* 123086

[40] L. Lamarche 2021 Analytic models and effective resistances for coaxial ground heat exchangers *Geothermics* **97** 102224

[41] Y. Yang, K. Wang, H. Zhang, et al. 2022 Investigation on the preparation, properties, and microstructure of high thermal conductive cementing material in 3500m-deep geothermal well *Geothermics* **100** 102322

[42] Y. He, M. Jia, X. Li, et al. 2021 Performance analysis of coaxial heat exchanger and heat-carrier fluid in medium-deep geothermal energy development *Renewable Energy* **168** 938-959

[43] L. Nacer, A. Abdelmalek, A. Müslüm, et al. 2021 Geothermal energy use in Algeria: A review on the current status compared to the worldwide, utilization opportunities and countermeasures *Journal of Cleaner Production* **302** 126950

[44] Y. Wang, C. Li, J. Zhao, et al. 2021 The above-ground strategies to approach the goal of geothermal power generation in China: State of art and future researches *Renewable and Sustainable Energy Reviews* **138** 110557

[45] J.C. Kurnia, Z.A. Putra, O. Muraza, et al. 2021 Numerical evaluation, process design and techno-economic analysis of geothermal energy extraction from abandoned oil wells in Malaysia *Renewable Energy* **175** 868-879

[46] N.H. Sandra, L.P. Daniele, L.J. Aída, et al. 2020 Phase segregation, boiling, and reinjection at the Los Azufres Geothermal Field, Mexico, monitored by water stable isotopes, chloride, and enthalpy *Journal of Volcanology and Geothermal Research* **390** 106751

[47] S. Li, S. Wang, H. Tang 2022 Stimulation mechanism and design of enhanced geothermal systems: A comprehensive review *Renewable and Sustainable Energy Reviews* **155** 111914

[48] B.T.H. Marbun, R.H. Ridwan, H.S. Nugraha, et al. 2021 Review of directional drilling design and operation of geothermal wells in Indonesia *Renewable Energy* **176** 135-152

[49] B.E. Harris, M.F. Lightstone, S. Reitsma 2021 A numerical investigation into the use of directionally drilled wells for the extraction of geothermal energy from abandoned oil and gas wells *Geothermics* **90** 101994

[50] G.S. Jia, Z.D. Ma, Z.H. Xia, et al. 2021 Investigation of the horizontally-butted borehole heat exchanger based on a semi-analytical method considering groundwater seepage and geothermal gradient *Renewable Energy* **171** 447-461

[51] S. Zhou, J. Li, Y. Zhang, et al. 2021 Prediction of the ground temperature variations caused by the operation of GSHP system with ANN *Geothermics* **95** 102140

[52] G. Chiriboga, S. Capelo, P. Bunces, et al. 2021 Harnessing of geothermal energy for a greenhouse in Ecuador employing a heat pump: design, construction, and feasibility assessment *Heliyon* **7(12)** e08608

IOP Conf. Series: Earth and Environmental Science 1087 (2022) 012056 doi:10.1088/1755-1315/1087/1/012056

Stability analysis of surrounding rock in a lead-zinc mine stope based on fine numerical modeling technology

Huiqiong Wang[1,2], Qibing Duan[1] and Wencheng Xing[1]

[1] Yunnan Energy Vocational and Technical College, Qujing, Yunnan 655001, China.

[2] Corresponding author's e-mail: 297382126@qq.com

Abstract. In order to achieve accurate evaluation of the stability of the surrounding rocks in complex underground mining clusters, 3DMine-Midas-FLAC3D was used to form an accurate numerical modelling analysis technique for complex mining clusters, and the technique was applied to the analysis of the stability of the surrounding rocks in complex mining clusters in a lead-zinc mine to achieve an accurate evaluation of the stability of the surrounding rocks in complex mining clusters in this mine, and to reveal the ground pressure manifestation pattern of complex mining clusters. The results of the study show that the accurate numerical modelling analysis method has better consistency with the empirical method (Laubscher method and Mathews stability diagram method) than the empirical method, and at the same time, it provides a more reliable and accurate evaluation method for the stability analysis of large complex mining area clusters from multiple perspectives and indicators, effectively compensating for the limitations of the empirical method for the stability evaluation of complex form mining sites.

1. Introduction

With the continuous development of China's mining industry, the demand for mineral resources is increasing sharply, and with it the level of mining mechanization is gradually rising, so the production capacity of many mining enterprises has risen substantially. Therefore, in the process of increasing mine capacity, the stability of the mine quarry envelope also faces greater challenges and tests. The stability of the quarry envelope is not only related to the size of the mine's capacity, but also affects the safety of construction workers, mine designers and mine equipment [1-2]. Based on this, it is crucial to carry out stability analysis of the quarry envelope.

In order to eliminate this safety hazard, a great deal of research has been done by scholars on the analysis of the stability of the quarry envelope, and have achieved a lot of fruitful research results [3-4]. The main methods of these studies are numerical simulations [5-9], nonlinear mathematical methods [10-13], Theoretical computational analysis [14-17], experimental studies [18-20] and empirical formulae approaches [21-22], As a result of the above research methods, it can be seen that the application of numerical simulation technology is more intuitive, its computing efficiency and accuracy is also higher, and also has the advantage of obvious visualization effect, therefore, many domestic and foreign mining enterprise researchers choose numerical simulation technology to study the stability of the quarry surrounding rock.

In the paper, a set of refined modelling methods for complex geological structural bodies is formed by combining the advantages of multiple parties based on 3DMine, and the method is applied to the stability evaluation of a complex underground mining hollow group in a lead-zinc mine, with a view

IOP Conf. Series: Earth and Environmental Science 1087 (2022) 012056 doi:10.1088/1755-1315/1087/1/012056

to realizing the improvement of the subjectivity and limitations brought about by the empirical evaluation method, and then accurately grasping the stability of the quarry envelope of the complex mining hollow group, achieving an overall understanding and evaluation of the stability of the quarry, and providing an accurate method for The aim is to provide an accurate stability evaluation method for the safe production of mining.

2. Fine-grained numerical modeling methods

3DMine 3D mining software is capable of constructing geometric models of 3D geological units of any form by using the advanced triangulation network modelling tools in 3DMine software, combined with the joint method of control lines and zoning lines, including powerful and comprehensive interactive generation, editing, display and calculation of entities and digital terrains, by means of flat/profile in space. The 3D geological modelling tools include a powerful and comprehensive interactive tool for generating, editing, displaying and calculating solids and digital terrain. The leading convex hull technology is used to realize the modelling of spatial point clouds and basic statistics such as projected area and thickness of arbitrary entities, which solves the problem of crossover of multi-layer models and makes 3DMine 3D mining software unique in its 3D refined geological modelling technology.

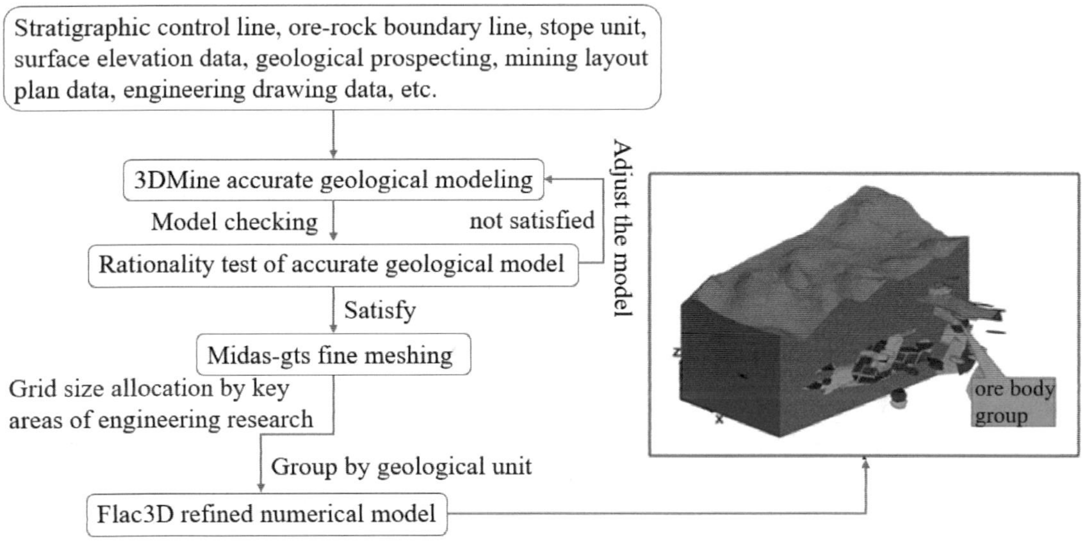

Figure 1. Fine-grained numerical calculation model construction process.

The construction of 3D accurate geological model firstly follows the 3DMine data input standard to collect and organize geological data, and then use 3DMine 3D geological modeling technology to obtain 3D geological model.

The first step: data collection and collation: focus on the data containing geological survey and mining layout plan. At this stage, the data results are mainly in the form of engineering geological plan (including topographic contours), engineering geological section and pan area layout plan in AutoCAD format. Import these results into 3DMine, and sort out and categorize the elements according to specific geological units or engineering properties (surface, stratum, ore body, structural surface structure, pan area), such as extracting topographic lines (contour lines in geological plan, surface lines in geological profile) to form individual 3DMine line objects.

Step 2: Based on the above data combing results, 3DMine 3D geological modeling technology is used to construct the surface, geological interface and ore body for each geological or engineering unit to realize the geometric representation or description of it.

Step 3: Reasonability check of the geological model. After the construction of each address unit and analysis object, the quality of the model should be checked to see if the stratigraphic structure, faults and the spatial location and size of the analysis unit meet the analysis requirements, and if not, local adjustments should be made to the model to meet the refinement modelling requirements.

Step 4: Import the established geological model (dxf/dwg file) into Midas-GTS for grid grouping dissection, and after the dissection is completed, the result file will be converted to coordinates and imported into $Flac^{3D}$ numerical analysis software for computational analysis. The refined numerical modeling process is shown in Figure 1. The refined numerical model can be used to accurately evaluate the stability of the quarry surrounding rocks under multiple indicators. The following discussion will use a lead-zinc mine engineering case to explore the application of refined numerical modeling technology in the stability of the surrounding rocks in complex mining area groups.

3. Engineering case study

3.1. Project overview

A lead-zinc mine with a design production capacity of 500,000 tons/a has mined a cumulative 120,000m^2 of lead-zinc ore in the south-east flank of the mine and the east-central flank of the mine, with a cumulative total of 99,300 tons of lead-zinc ore mined until the end of December 2018. The lead-zinc mine is tectonically complex, with no obvious folds, and the stratigraphic production is highly variable due to tectonic influence, mainly tilting north-west, with a tendency of 280°-350°and a dip angle of 12°-47°in general, with individual lots reaching 47°. The mining method used in the mine is the room-and-pillar method, and the void area has not yet been treated by filling, and the exposed area of the mining site is large.

The ore body of the mine area is divided into eight layers from bottom to top, such as Zn- 1, Zn-2, Zn-3, etc. The surrounding rocks are dolomite, and there are 1-2 layers of intercalated rocks in the sub-ore body of Zn- 1 and Zn-2. The intercalated rocks are lithological consistent with the surrounding rocks, and the lithology is mainly dolomite and silicified dolomite, and the spatial form of the intercalated rocks is mostly lenticular. The overall hydrogeological conditions of the mine are as follows: the mine is located in the subtropical to warm temperate humid monsoon climate zone, with no severe cold in winter, no extreme heat in summer, mild climate, abundant precipitation, rain and heat in the same period, with obvious dry and rainy seasons.

The ore structure is mostly blocky, with a small amount of fractured structure, with high mechanical strength and good stability, but the ore body is prone to engineering geological problems such as falling blocks during mining in the local rock fracture development section. The underlying stratum of the ore- bearing strata is the Jindingshan Group of the Lower Cambrian System. Most of the rocks are thin to medium-thick laminated, with fewer structural surfaces than laminated rocks, and the laminae are characterized by discontinuous staggered laminae or wavy laminae, which are soft rock groups with poor stability and weak water content. The quarry distribution and panel layout are shown in Figure 2.

In summary, it is clear that the basic geological conditions of the mine area are simple, all dolomite, with a single structural surface production. Therefore, the ore bodies are characterized by a wide range of ore bodies, lenticular and thin laminae development, widespread sharp extinction, poor continuity of the shallowly buried ore bodies, control of the ore bodies by the location of the drill holes, and complex distribution of geological units.

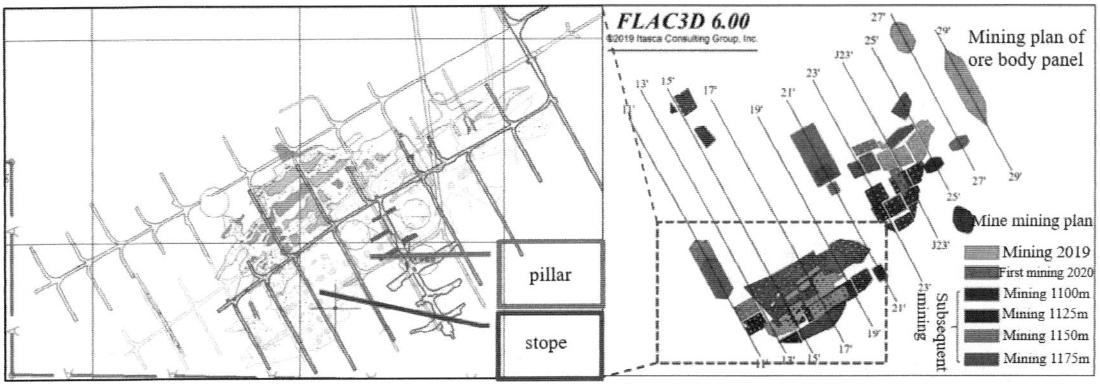

Figure 2. Distribution of quarries and layout of the pan area in the model.

3.2. Mechanical parameters of rock mass

The results of the determination of various physical and mechanical properties of the ore rocks play an important role in the analysis of quarry stability and have become an important part of the mining research. However, in mining engineering, the main difference between the engineering rock we are in contact with, mechanically speaking, and various other materials is its discontinuity and inhomogeneity, which is subjected to thermal action during rock formation and many times of violent and complex tectonic movements during the long geological years, In order to make the calculation results of engineering rock stability or safety more in line with engineering reality, the macro-mechanical parameters that can represent a large range of rock masses are needed for the calculation. Therefore, the study of macro-mechanical parameters of rock masses is one of the most important and difficult research topics in rock mechanics.

Due to the complexity of directly obtaining the macroscopic physical and mechanical parameters of the rock mass, this paper first uses the mechanical test data of the indoor rock samples, and then converts the mechanical parameters of the rock mass by discounting and converting them according to the Hoek-Brown strength criterion, and finally obtains the macroscopic physical and mechanical parameters of the rock mass, as shown in Table 1. It provides a scientific basis for the subsequent stability analysis of the quarry based on FLAC^{3D}.

Table 1. Selected values of physical and mechanical parameters of each geotechnical body.

Name of rock body	Natural Densityρ (g/cm3)	Compressive Strength (MPa)	Tensile Strength (MPa)	Modulus of elasticity (GPa)	Poisson's ratio(μ)	Internal Cohesion (MPa)	Angle of internal friction (φ/°)
Dolomite	2.878	3.07	1.110	12.415	0.23	0.925	27.92
Ore body	2.890	3.12	1.191	11.65	0.24	0.96	26.61

3.3. Numerical calculation model

Based on the aforementioned refinement numerical model construction method, a FLAC^{3D} 3D refinement numerical model for the stability analysis of the surrounding rock of the mine was constructed, as shown in Figure 3, with the bottom boundary of the model at 900m level, the model at 1150m along the X-direction and 1090m along the Y-direction, formed by a mixture of 1.2 million tetrahedral and hexahedral units. The model completely contains the key geological units and mining construction design scheme of the mine area, and the model is constructed based on the 3Dmine 3D geological model directly, which makes the structural parameters of the quarry and the size of the mining pillar almost identical to the actual mine, and its geological rationality can be guaranteed. For damage discrimination of quarry envelopes, the Hoek criterion [23] originated in the 1950s from Hoek et al. based on a summary of practice in South African mining projects, and in 1999, C.D. Martin further modified the risk rating criteria of the E. Hoek criterion based on deep burial projects in Canada and the samples of high stress damage phenomena they revealed, to establish the envelope The relationship between the conditions and the phenomenon of high stress damage and its extent is shown in Figure 4 (black marked points are samples of damage in the field).

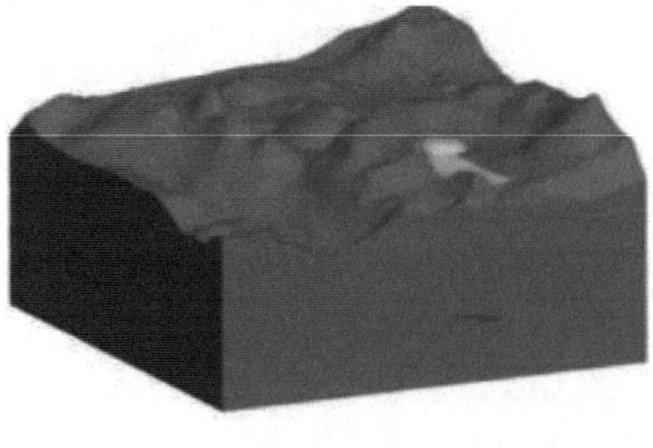

Figure 3. 3D numerical analysis model of the mine.

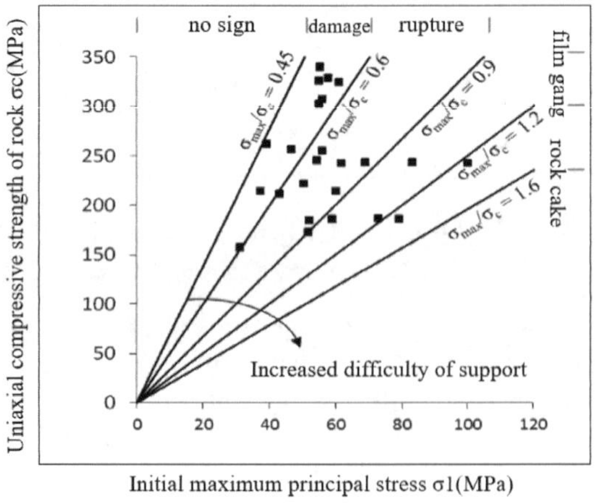

Figure 4. Hoek judgment criterion for the risk of high stress damage to the surrounding rock.

4. Evaluation of quarry enclosure stability based on numerical analysis method

4.1. Initial ground stress generation

According to the geological survey report of a Pb-Zn mine, the lithology of the whole mine area is single, all of which are dolomite, and the deposit is a dolomite sedimentary deposit. For this reason, the initial horizontal ground stress is set according to the regression equation of the ratio of the average horizontal ground stress to the vertical ground stress of sedimentary rocks in China for the model calculation in this paper [24], the initial ground stress calculation results are shown in Figure 5. In order to truly reflect the influence of the mining process on the stability of the surrounding rock, the excavation was carried out according to the real mining sequence in the simulation.

4.2. Quarry enclosure stability analysis

Underground quarry mining disturbs the rock mass in a relatively monotonous manner, with the overall performance of mining leading to unloading of the rock mass along the vertical direction.

Therefore, the response process of the rock mass in different engineering parts should also obey the monotonic nature, which can typically be manifested as the degree of deformation of the top plate in the pan area or the level of stress concentration in the ore column between the pan areas continues to increase as the excavation advances until a steady state is reached or damage is formed. It can thus be seen that the rock state corresponding to the completion of excavation is a superposition of all mining activities during the status quo phase that can form a disturbance, representing the stage of the most unfavorable condition of the surrounding rock forces, and can therefore help to analyze the basic stability characteristics of the quarry surrounding rock during the mining construction process. In addition to revealing the overall deformation and stress characteristics of the quarry rock under mining conditions and the nature of the potential problems and key engineering areas determined by this, the basic stability characteristics also directly reflect the final stability conditions of the rock after the completion of the corresponding phase of mining.

Figure 5. Calculated results of the initial ground stress field in the mine area.

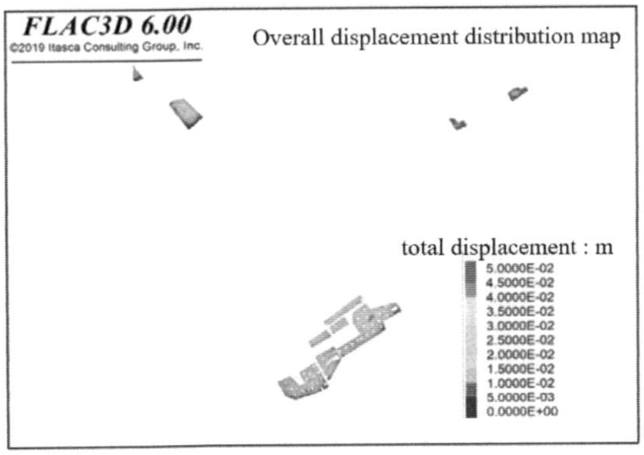

(a) Top plate displacement distribution.

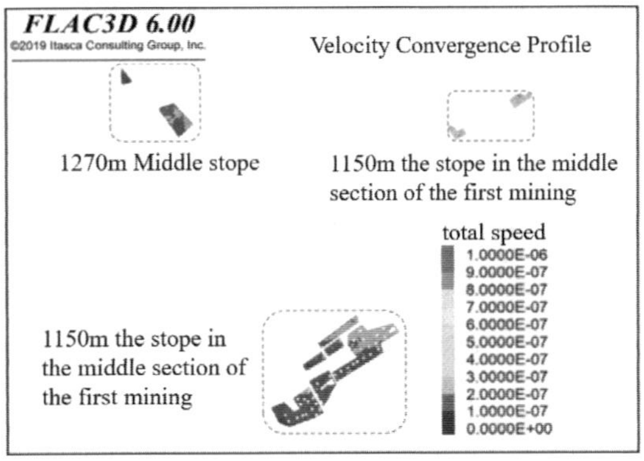

(b) Top plate velocity distribution.

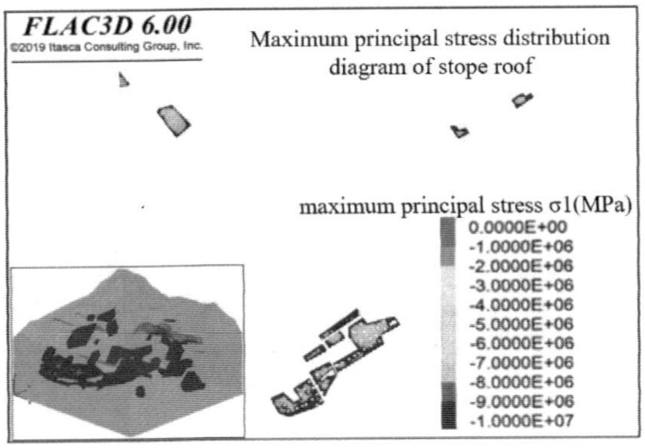

(c) Maximum principal stress distribution in the roof slab.

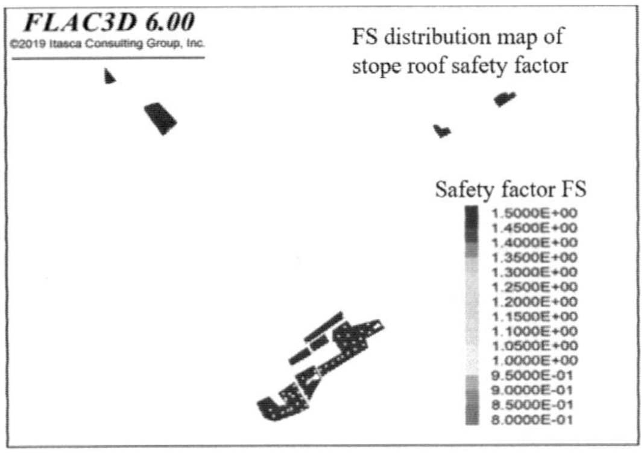

(d) Coefficient of safety distribution in the roof slab.

Figure 6. Calculated results of the mechanical indexes of the quarry surrounding rock after the completion of current mining.

Based on the refined numerical analysis model for post-mining analysis of the indicators, Figure 6 shows the excavation response characteristics of the total top slab displacement, velocity, maximum principal stress and plastic zone of the underground quarry envelope after the completion of the status quo mining.

From the calculation results, it can be seen that the overall velocity response of the roof of the quarry tends to be close to zero, which means that the risk of damage to the surrounding rock in these areas is low, i.e. there are conditions to maintain a stable state; according to the tolerance limit displacement criterion, the total displacement of the roof is between 0~50mm, and the surrounding rock of the roof can maintain the basic stability of the quarry, this conclusion is consistent with the fact that no support is taken at the site and there is no destabilization phenomenon such as falling (Figure 8). Based on the stress-based damage criterion, the current quarry, except for the smaller localized damage in the stress concentration at the quarry boundary, has an overall safety factor FS>1.0, and the quarry envelope is in a safe state in general, and FS>1.5 in a wide range, indicating that its overall safety reserve is sufficient under the current conditions; unlike this, although the mining disturbance has been completed, the top plate envelope of the quarry area still maintains a certain level of velocity, indicating that these In contrast, although the mining disturbance has been completed, the rocks in the roof slab of the extraction area still maintain a certain level of velocity, indicating that the rock in these parts has been displaced by movement, i.e., there is a relaxation disturbance, and there is a certain safety risk, and the subsequent continuous excavation disturbance will increase this risk. It can be seen that the stress response mode of roof excavation within the pan area of underground quarry is mainly manifested as stress relaxation, and rock damage is generally manifested as large deformation in reality, and block deformation can be the main damage mode under the condition of structural surface development. In addition, the Hoek high stress criterion indicates that the stress level corresponding to the appearance of damage must be $0.45\sigma c$ or 23.985MPa or more. This shows that the risk of high stress damage to the pillar is low, but still does not exclude the possibility of forming damage and rupture at localized stress concentrations (as shown in Figure 7 where the pillar has a localized risk of high stress concentration). The combined result is that the quarry envelope is in a stable state.

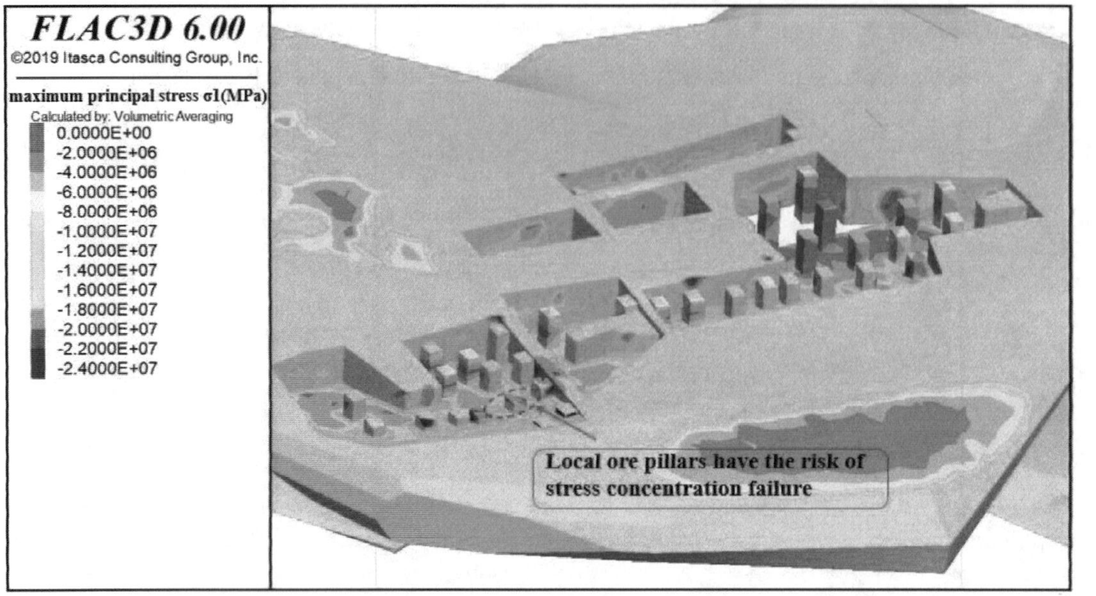

Figure 7. Characteristics of the maximum principal stress distribution in the ore column at the end of current recovery.

According to the calculation results of displacement, plastic zone, velocity and maximum principal stress after mining under current conditions, the three evaluation indexes (displacement, velocity and maximum principal stress) all reveal that the surrounding rock and roof plate are in a stable state and no large deformation instability will occur under current conditions. Combined with the actual investigation at the site, the mine roadway and quarry have not taken support measures, and the roof slab and surrounding rock of the quarry have not occurred destructive instability, nor do they store the phenomenon of falling block and bubble fall, the results of numerical analysis method and the site environment almost get the same conclusion, and the analysis results can reasonably assess the status quo stability situation, combined with the results of the site investigation and the consultation of the site staff, the mine roadway and quarry have almost not taken Support, there is no collapse and bubble roof sheeting situation, the overall stability of the surrounding rock is good.

Figure 8. Overview of site quarry and roadway envelope rock stability.

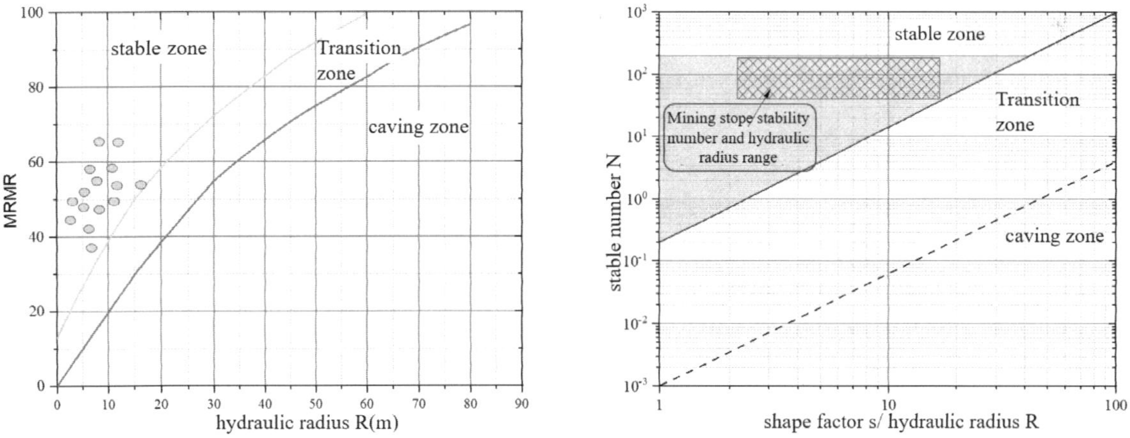

(a) Laubscher empirical diagram. (b) Mathew's stability diagram method.

Figure 9. Stability evaluation results of empirical graph method.

4.3. Evaluation method comparison results

In the evaluation of the stability of the surrounding rock, the empirical method is often used as an important reference method. In order to demonstrate the rationality and the correctness of the results of the refined numerical analysis method in the paper, the stability state and hydraulic radius of the existing quarry are plotted in the Laubscher empirical diagram method and the Mathews stability diagram method, as shown in Figure 9, the stability of the surrounding rock of the quarry proposed by Laubscher and Mathews The empirical evaluation diagram contains 2 boundary curves, and the two curves divide the whole diagram into three zones, namely the overall damage zone, the shallow emergence zone and the stability zone. The difference is that the Laubscher empirical map method judges its stability based on a given MRMR value and the morphology of the designed mining area. The Mathews stability diagram method is then evaluated according to the modified Q' value of the Q-system classification method. According to the calculated results, all existing quarries in the mine are in a stable state, which is consistent with the results of the refined numerical analysis in the paper, which laterally justifies the evaluation results of the numerical method. At the same time, the refined numerical analysis method has the significant advantage of judging the spatial location of hazardous areas and the common evaluation of multiple indicators, effectively compensating for the subjectivity and limitations of the empirical method.

5. Conclusions

(1) A refined numerical modeling method is proposed, which can accurately respond to the spatial location and size of the unit structure body, and realize the accurate modeling and grid dissection of the complex underground mining area cluster through the conversion between models.
(2) Based on the refined numerical modeling analysis, the stability of the surrounding rock of complex mining area group can be accurately evaluated from multiple indicators, providing a numerical analysis method for accurate evaluation of the stability analysis of large complex mining area group.
(3) By comparing with the evaluation results of Laubscher empirical graph method and Mathew's stability graph method, the refined numerical analysis method of quarry surrounding rock stability established in the paper can have the stability of quarry surrounding rock evaluated from multiple indicators, effectively compensating for the limitations and subjectivity brought by the single indicator of empirical method.

Fund project

Scientific Research Fund project of Yunnan Education Department (2019J0374).

References

[1] Zhang C, Song W D, Fu J X, et al. 2020 Stability analysis of rock mass under disturbance of submarine mine [J] *Journal of China University of Mining and Technology* **49(06)** 1035-1045
[2] Chen Z, Zuo Y J 2020 Stability Analysis of Surrounding Rock of Chamber and Pillar Mining in Getang Gold Mine [J] *Gold Science and Technology* **28(02)** 271-277
[3] Qi F F, Li X F, Song Z H, et al. 2021 Analysis on Mining Technology and Stope Stability of Dadong Gold Mine [J] *Mining Research and Development* **41(07)** 19-24
[4] Zheng P S, Li M, Cui B, et al. 2018 Research on the stability of a deep gypsum orebody stope [J] *Industrial Minerals & Processing* **47(10)** 40-42
[5] Ding W J, Zhou Z H, Hou T K, et al. 2021 Analysis and Researchon the Stability of Different Sublevel Height of Sublevel Room Method in An Iron Mine [J] *Nonferrous Metal Engineering* **11(08)** 113-121+133
[6] Idris M A, Nordlund E 2019 Probabilistic-Based Stope Design Methodology for Complex Ore Body with Rock Mass Property Variability [J] *Journal of Mining Science* **55(5)** 743-750
[7] Li Q H, Li S S, Geng J B, et al. 2021 FLAC~(3D) numerical simulation of open pit to underground slope and quarry stability study [J] *Nonferrous Metals (Mining Section)* **73(02)** 5-10

[8] Ma C D, Xu J Q, Liu H X, et al. 2021 Optimization of Structure Parameters of Deep Stope Based on ANSYS-FLAC3D [J] *Mining and Metallurgical Engineering* **41(01)** 20-23
[9] Zhu Q L, Shang Z H 2018 Study on Stope Stability of an Iron Mine based on Finite Difference Program[J] *Mining Research and Development* **38(11)** 35-39
[10] Hu X Y, Chen J H, Wang G M 2015 Stope stability evaluation system based on improved grey clustering analysis and dominance-based rough set [J] *Journal of Central South University (Science and Technology)* **46(01)** 223-230
[11] Zhang W G, Li H R, Wu C Z, et al. 2021 Stability assessment of underground quarry excavation based on RF and KNN [J] *Journal of Hunan University (Science and Technology)* **48(03)** 164-172
[12] Zhou K P, Wang X X, Gao F 2013 Stope structural parameters optimization based on strength reduction and ANN-GA mod [J] *Journal of Central South University (Science and Technology)* **44(07)** 2848-2854
[13] Kang Q, Wang Y M, Zhang S W, et al. 2021 Prediction of Stope Stability Using Variable Weight and Unascertained Measurement Technique [J] *Geofluids* 8821168
[14] Song W C, Liang Z C, Liu W T, et al. 2019 Theoretical analysis and experimental investigation on failure characteristics and stability of stope floors [J] *Chinese Journal of Rock Mechanics and Engineering* **38(11)** 2208-2218
[15] Zhang Z G, Tan B H, Zhong M, et al. 2021 Determination Method and Application of Stope Stability in Downdirection Slicing Drift Cemented Filling Mining Method [J] *Metal Mine* **(04)** 27-34
[16] Sun W M 2018 Stability analysis of room-and-pillar quarries in barite ore bodies [J] *Coal Mining* Technology **23(03)** 83-86
[17] Huang D, Wu P, Dong K C, et al. 2014 Study on stope structure parameters and stability for slightly inclined multi-layer ore bodies with room and pillar mining method [J]. *China Mining magazing* **23(S2)** 205-208
[18] Han B, Wu F, Meng K, et al. 2021 Influence of deep orebody filling mining on stability of hanging wall ramp [J] *The Chinese Journal of Nonferrous Metals* **31(03)** 806-814
[19] Ren S L, Tao Z G, He M C 2020 Stability analysis of open-pit gold mine slopes and optimization of mining scheme in Inner Mongolia, China [J] *Journal of Mountain Science* **17(12)** 2997- 3011
[20] Deng H W, Hu P, Zhou K P, et al. 2013 Test of orthogonal numerical simulation on sensitivity of stope structure parameters [J] *Journal of Central South University (Science and Technology)* **44(06)** 2463-2469
[21] Tao G Q, Liu Z D, Ren F Y, et al. 2010 Optimization research of stope structural parameters in sublevel caving with non-pillar [J] *Journal of China Coal Society* **35(08)** 1269-1272
[22] Xu H L, Liu Z S, Shi L, et al. 2015 Optimizing the Stope Structural Parameters Using Mathews Empirical Analysis Method and Numerical Analysis Method [J] *Metal Mine* **(12)** 19-23
[23] Hoek 1999 Putting numbers to geology - an engineer's viewpoint [J] *Quarterly Journal of Engineering Geology and Hydrogeology* **32** 1-19
[24] Zhao D A, Chen C M, Cai S L, et al. 2007 ANALYSIS OF DISTRIBUTION RULE OF GEOSTRESS IN CHINA [J] *Chinese Journal of Rock Mechanics and Engineering* **(06)** 1265-1271

IOP Conf. Series: Earth and Environmental Science 1087 (2022) 012057 doi:10.1088/1755-1315/1087/1/012057

Effects of $CaCl_2$ on concentration and speciation of soil Cd around a Pb-Zn mine

Lina Zhang[1,2], Zhicheng Dong[1,2,3], Yu Zhang[1,2], Lin Wang[1] and Chenhao Xu[1]

[1] School of Geographic Sciences and Tourism, Jiaying University, Meizhou 514015, Guangdong, China;

[2] Guangdong Provincial Key Laboratory of Conservation and Precision Utilization of Characteristic Agricultural Resources in Mountainous Areas, Jiaying University, Meizhou 514015, Guangdong, China.

[3] Corresponding author's e-mail: dongzckeyan_8@126.com

Abstract. Remediation of Cd contaminated soil around mines is still the current academic focus. In order to discuss the activation and extraction effects of $CaCl_2$ on concentration and speciation of soil Cd around a Pb-Zn mine, field investigation and soil sample collection were conducted. Batch experiment was used to study the influence of concentration and time on Cd activation and extraction. Cd contents were analysed. And Cd species were extracted by sequential extraction procedures. Results show that there are obvious concentration effect and time effect for Cd extraction by $CaCl_2$. The best choice of $CaCl_2$ concentration is 0.1 mol/L. And the best extraction time is 4h. Results also show that the percentage and content of exchangeable Cd is the lowest with the value of 9.1% and 0.16 mg/kg when the $CaCl_2$ content is 0.1 mol/L. And the percentage and content of exchangeable Cd is the lowest with the value of 10.5% and 0.18 mg/kg when the extraction time is 4h. It is found that transformation of Cd species such as exchangeable, carbonates bound and Fe-Mn oxides bound should be the main reasons of concentration effect and time effect. And the extraction of soil Cd around a Pb-Zn mine by $CaCl_2$ is a complex coupling process of dissolution, ion exchange, desorption, acidolysis, oxidation-reduction and complexation, etc. All of these above will aid to provide a basis for the research, development and application of remediation technology for Cd contaminated soil around this and other Pb-Zn mines.

1. Introduction

Remediation of Cd contaminated soil still remains research focus [1-2]. Activation and removal, together with passivation and fixation are still major directions of current research [3-4]. Chemical leaching and physical-chemical enhanced phytoremediation are effective remediation technologies for activating and removing soil Cd. compared with passivation and fixation, both of them could remove Cd from the soil and realize real remediation [5-6]. And Cd activation is the premise and foundation of leaching and phytoremediation [1, 5]. Since, it is to use activator to transform and transfer Cd from solid to liquid phase of soil though ion exchange, redox, complexation, chelation, etc. Hence, numerous researches on activators have been carried out at home and abroad [6-8]. Rich knowledge and results were gained [6, 9-10].

At present, inorganic activators (such as HCl, $CaCl_2$, $FeCl_3$, etc.), surfactants (such as sodium dodecyl benzene sulfonate, rhamnolipid, etc.) and chelating or complexing agents (such as citric acid, tartaric acid, EDTA, NTA, etc.) are three kinds of common activators [11-12]. Among them, inorganic activators such as inorganic salts or acids have the advantage of being relatively cheaper, easier to obtain and more efficient [6, 11]. For them, the main mechanism of Cd activation is to desorb soil Cd through ion exchange, acidolysis, oxidation and reduction, etc. Moreover, compared with inorganic acids, inorganic salts have less effect on soil physical and chemical properties. Li et al. studied the activation effects of 100 mmol/L KCl, NaCl, $MgCl_2$, $CaCl_2$ and $FeCl_3$ [13]. Results showed that $CaCl_2$ (about 17%) and $FeCl_3$ (38% approximately) had higher activation efficiency than other chlorides [13]. Makino et al. had similar findings [14]. Hayashi et al. found that after leaching paddy soil with 0.5 mol/L $CaCl_2$ (pH=4), Cd content of potted rice decreased from more than 0.5 mg/kg to less than 0.2 mg/kg [15]. Maejima et al. found that after leaching soil with $CaCl_2$, Cd concentration of potted soybeans decreased by 25% [16]. It is found by Makino et al. that exchangeable Cd reduced by 45% after activation with $CaCl_2$ [17]. In addition, both Xiong et al. and Christine et al. proposed that $CaCl_2$ should be the ideal extractant most suitable for activating and extracting Cd from soils with different properties [18-19]. Makino et al. proposed the reason that $CaCl_2$ has high efficiency in extracting Cd [14]. On the one hand, Ca^{2+} will reduce the pH of extracting solution by exchanging with H^+ and Al^{3+} on the soil surface [14]. And much more Cd could be extracted when the pH is lower. On the other hand, Cd-Cl complex could also promote the dissolution of Cd [14].

Xiong et al. proposed that 0.1 mol/L $CaCl_2$ could be the most adaptable extractant for acid, neutral or calcareous soil, together with soils of different properties [18]. It is also proposed by Xiong et al. that 0.1 mol/L $CaCl_2$ method could be recommended as the extraction method of paddy soil Cd in southern China [20]. However, other result shows that there is a better correlation between the extracted Cd by 0.01 mol/L $CaCl_2$ and Cd in plants [21-23]. Hence, the suitable concentration of $CaCl_2$ is still worth studying for Cd contaminated soils of different types and different properties.

It is found by Ma et al. that Cd extraction amount by $CaCl_2$ increased gradually with the increasing of extraction time [24]. Yan et al. proposed that the higher the concentration of $CaCl_2$, the longer the time required for reaction equilibrium [25]. They proposed that the best time for 0.1 mol/L $CaCl_2$ to extract Cd from acid paddy soil and moderately alkaline calcareous soil is 12h and 3h, respectively [25]. However, Xiao et al. found that 1h was the best time for 0.1 mol/L $CaCl_2$ to extract Cd from acid soil [26]. And Ji et al. advised that 1.5h is the best time for 0.1 mol/L $CaCl_2$ to extract soil Cd in Guizhou City [27]. Hence, the suitable extraction time of $CaCl_2$ is also still worth studying for Cd contaminated soils of different types and different properties.

Besides, it is found that Cd contents in soils around the Pb-Zn mine of this study were nearly two times (our unpublished result) of the risk control value of the risk control standard for soil contamination of agricultural land (GB15618-2018). Though causing no environmental harms so far, it remains to be a geochemical bomb here. And remediation strategies still remain to be studied.

Therefore, this study aims to discuss the activation of soil Cd around a Pb-Zn mine by $CaCl_2$. It also aims to analyze the effects of $CaCl_2$ on concentration and speciation of soil Cd around the Pb-Zn mine. Moreover, it aims to explain the possible mechanism and process of soil Cd extraction around the Pb-Zn mine by $CaCl_2$. It is expected to provide a basis for the research, development and application of remediation technology for Cd contaminated soil around this and other Pb-Zn mines.

2. Materials and methods

Field investigations and sample collection were conducted around a Pb-Zn mine in November 2020. This Pb-Zn mine is located near the Tropic of cancer. It is influenced by a subtropical monsoon climate, with the annual average temperature of about 21.2 °C and the annual average precipitation of about 1453.4 mm. Soil samples from 0-20 cm were taken from fields in the downstream valley of the Pb-Zn mine. The sampling and pretreatment were conducted according to the Technical Specification for soil Environmental monitoring (HJ/T166-2004). All chemical reagents were guaranteed reagent. And all the experiment water was produced by Milli-Q Reference System.

Extraction experiment were conducted at ten concentration gradient of $CaCl_2$ solution (0.02, 0.04, 0.06, 0.08, 0.10, 0.20, 0.40, 0.60, 0.80, 1.00 mol/L). The ratio of soil to $CaCl_2$ solution is 1:10.

Extraction time is 0.5h, 1.0h, 1.5h, 2.0h, 2.5h, 3.0h, 3.5h, 4.0h, 4.5h, 5.0h, 5.5h, 6.0h, 12h, 24h, 36h, and 72h. Each treatment is repeated three times.

Cd concentration was detected by ICP-MS (Agilen 7900). Cd species was separated into exchangeable, carbonates bound, Fe-Mn oxides bound, organics bound and residual fractions by the sequential extraction procedure (HJ/T166-2004). Concentration of Cd species was also analyzed by ICP-MS (Agilen 7900). Certified material was used to guarantee the accuracy of analytical quality.

All of the data processing, statistical analysis and charting were conducted by using STATISTICA 6.0 and Excel 2010.

3. Results

3.1. Concentration effect of $CaCl_2$ on Soil Cd

Figure 1 illustrates the effect of $CaCl_2$ concentrations on extracted Cd from soils around a Pb-Zn mine. As demonstrated in Figure 1, the extraction amount of soil Cd rises quickly and then declines slightly with the increase of $CaCl_2$ concentrations. More specifically, the extraction amount of soil Cd rises quickly with the increasing of $CaCl_2$ concentrations when they are less than 0.1 mol/L. In this case, the extraction amount of soil Cd varies from 0.05-0.87 mg/kg, with the average of 0.45 mg/kg. However, the extraction amount of soil Cd declines slightly with the increase of $CaCl_2$ concentrations when they are more than 0.1 mol/L. And the extraction amount of soil Cd varies from 0.77-0.87 mg/kg, with the average of 0.81 mg/kg. And especially, the extraction amount of soil Cd is the highest with the value of 0.85 mg/kg when the $CaCl_2$ content is 0.1 mol/L. Hence, there is an obvious concentration effect of $CaCl_2$ to extraction amount of soil Cd.

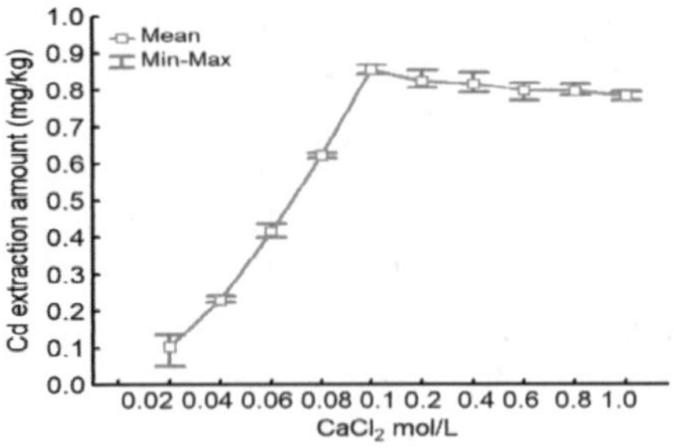

Figure 1. Effect of $CaCl_2$ concentrations on extracted Cd from soils around a Pb-Zn Mine.

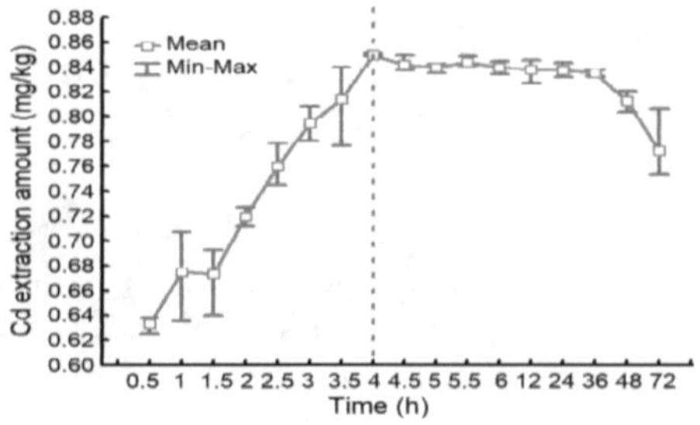

Figure 2. Effect of extraction time on extracted Cd from soils around a Pb-Zn Mine.

3.2. Time effect of $CaCl_2$ on Soil Cd

Figure 2 shows the effect of time on extracted Cd from soils around a Pb-Zn mine. As illustrated in Figure 2, the extraction amount of soil Cd increases quickly, and then declines slightly and maintains the fluctuation, and finally declines quickly with the increase of extraction time. More specifically, the extraction amount of soil Cd rises quickly with the increase of extraction time when it is less than 4h. In this case, the extraction amount of soil Cd varies from 0.63 -0.85 mg/kg, with the average of 0.74 mg/kg. However, the extraction amount of soil Cd declines slightly and fluctuates with the increase of the extraction time when it is more than 4h and less than 36h. And the extraction amount of soil Cd varies from 0.83-0.85 mg/kg, with the average of 0.84 mg/kg. And especially, the extraction amount of soil Cd declines quickly with the increase of the extraction time when it is more than 36h. Moreover, the extraction amount of soil Cd is the highest with the value of 0.85 mg/kg when the extraction time is 4h. Hence, there is an obvious time effect of $CaCl_2$ to extraction amount of soil Cd.

3.3. Concentration effect of $CaCl_2$ on Cd species

Figure 3 demonstrates the effect of $CaCl_2$ concentrations on variation of percentage and content of Cd species from soils around a Pb-Zn mine. As shown in Figure 3, both the percentage and the content of

IOP Conf. Series: Earth and Environmental Science 1087 (2022) 012057 doi:10.1088/1755-1315/1087/1/012057

exchangeable Cd decline quickly and then rise slightly with the increase of $CaCl_2$ concentrations. Especially, both the percentage and the content of exchangeable Cd are the lowest with the values of 9.1% and 0.16 mg/kg when the $CaCl_2$ content is 0.1 mol/L. Meanwhile, the percentages of Cd bound to both the carbonates and Fe-Mn oxides rise quickly and then decline slightly with the increase of $CaCl_2$ concentrations. However, the contents of these two species decline slightly with the increase of $CaCl_2$ concentrations. And the contents vary from 0.02 to 0.04 mg/kg and from 0.02 to 0.06 mg/kg, respectively. More especially, for the residual Cd and those bound to organics, both the percentage and the content show no significant changes with the increase of $CaCl_2$ concentrations.

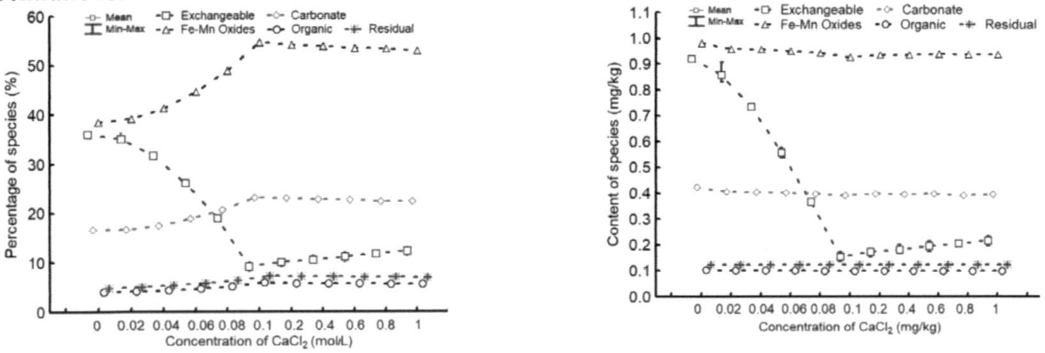

Figure 3. Effect of $CaCl_2$ concentrations on variation of percentage and content of soil Cd species.

3.4. *Time effect of $CaCl_2$ on Cd species*

Figure 4 illustrates the effect of extraction time on variation of percentage and content of Cd species from soils around a Pb-Zn mine. As demonstrated in Figure 4, both the percentage and the content of exchangeable Cd decline quickly, then decline slightly, and finally rise slightly with the increase of extraction time. Especially, both the percentage and the content of exchangeable Cd are the lowest with the values of 10.5% and 0.18 mg/kg when the extraction time is 4h. Meanwhile, percentages of Cd bound to both the carbonates and Fe-Mn oxides rise quickly and then decline slightly with the increase of extraction time. However, contents of these two species decline slightly with the increase of extraction time. And the contents vary from 0.02 to 0.04 mg/kg and from 0.02 to 0.06 mg/kg, respectively. More especially, for residual Cd and those bound to organics, both the percentage and the content show no significant changes with the increase of extraction time.

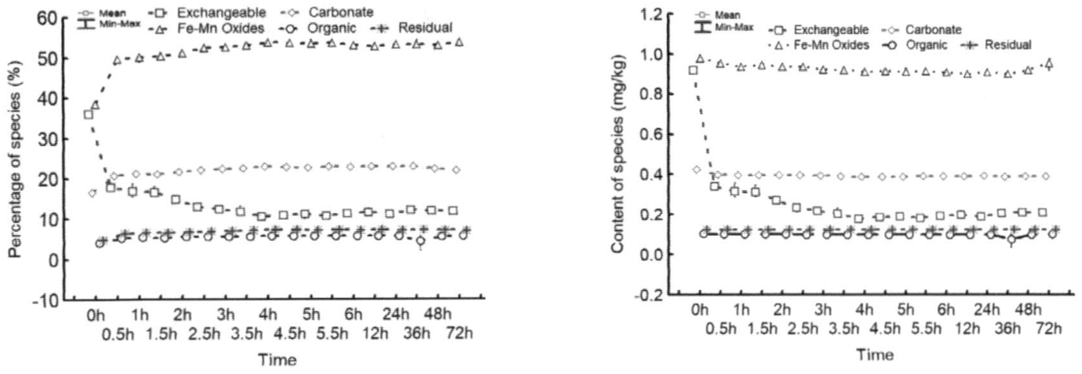

Figure 4. Effect of extraction time on variation of percentage and content of soil Cd species.

4. Discussions

4.1. Relationship between extraction amount and speciation changes of soil Cd around a Pb-Zn mine under the concentration effect

Figure 5 shows correlation between the extraction amounts of total Cd and Cd species under the concentration effect of $CaCl_2$. As illustrated in Figure 5, there is a positive correlation between the extraction amount of total Cd and Cd species, such as those bound to carbonates and Fe-Mn oxides. The correlation coefficient of the extraction amount of total Cd to exchangeable Cd is 0.96, with the multiple correlation coefficient of 0.92. Meanwhile, the correlation coefficient of the extraction amount of total Cd to Cd bound to carbonates is 0.88, with the multiple correlation coefficient of 0.78. In addition, the correlation coefficient of the extraction amount of total Cd to Cd bound to Fe-Mn oxides is 0.87, with the multiple correlation coefficient of 0.75. Hence, it could be deduced that concentration effect of $CaCl_2$ on the extraction amount of soil Cd should be caused by changing Cd species those are exchangeable or bound to carbonates and Fe-Mn oxides, etc.

4.2. Relationship between extraction amount and speciation changes of soil Cd around a Pb-Zn mine under the time effects

Figure 6 demonstrates Correlation between the extraction amounts of total Cd and Cd species under time effect of $CaCl_2$. As shown in Figure 5, there is a positive correlation between the extraction amounts of total Cd and Cd species, such as those bound to carbonates and Fe-Mn oxides. The correlation coefficient of the extraction amount of total Cd to exchangeable Cd is 0.96, with the multiple correlation coefficient of 0.92. Meanwhile, the correlation coefficient of the extraction amount of total Cd to Cd bound to carbonates is 0.74, with the multiple correlation coefficient of 0.54. In addition, the correlation coefficient of the extraction amount of total Cd to Cd bound to Fe-Mn oxides is 0.80, with the multiple correlation coefficient of 0.64. Hence, it could be deduced that time effect of $CaCl_2$ on the extraction amount of soil Cd should be caused by changing Cd species those are exchangeable or bound to carbonates and Fe-Mn oxides, etc.

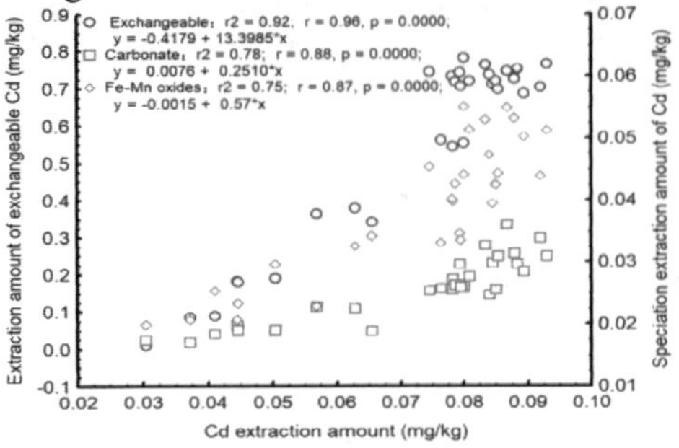

Figure 5. Correlation between the extraction amounts of total Cd and Cd species under the concentration effect of $CaCl_2$.

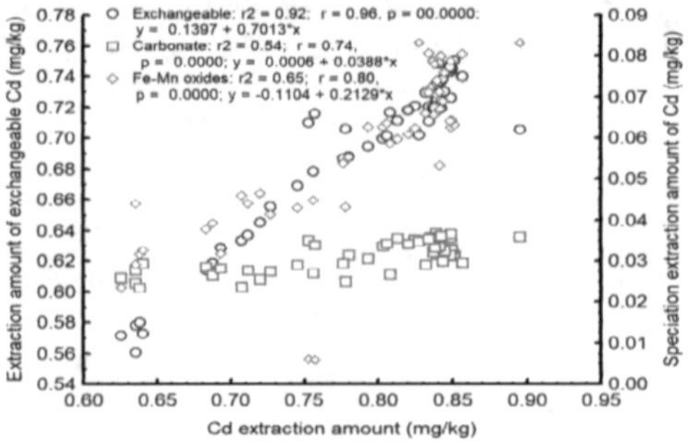

Figure 6. Correlation between the extraction amounts of total Cd and Cd species under the time effect of $CaCl_2$.

4.3. Possible mechanisms of $CaCl_2$ affecting soil Cd around a Pb-Zn mine

As mentioned previously, $CaCl_2$ promotes the transformation of Cd species those are exchangeable or bound to carbonates and Fe-Mn oxides, etc. And this causes the activation and release of soil Cd around the Pb-Zn mine. Based on geochemistry, it should be a complex coupling process of dissolution, ion exchange, desorption, acidolysis, oxidation-reduction and complexation, etc. The possible mechanisms should include but not limited to the following rules (Figure 7). 1) Dissolution reaction. Water soluble Cd will be dissolved into the extraction solution when Cd contaminated soils meet with $CaCl_2$ solution. 2) Ion-exchangeable reaction. Ca^{2+} in $CaCl_2$ solution and those cations (such as H^+ etc.) exchanged from soil colloids by Ca^{2+} could desorb exchangeable Cd ions adsorbed on soil colloids by ion-exchangeable reaction. 3) Acidolysis and redox reaction. Those desorbed H^+

exchanged from soil colloids combines with NO_3^- to form nitric acid. The nitric acid could interact with those Cd bound to carbonates and Fe-Mn oxides by acidolysis and redox reactions. It will cause transformation of these two kinds of Cd species and lead to their entering the extraction solution. 4) Complexation reaction. Soil Cd could react with chloride ion in $CaCl_2$ solution and other anions exchanged by chloride ion to form soluble complexes and enter the extraction solution. All these above should be main reasons for promoting the activation, release and entry of soil Cd into the extraction solution by $CaCl_2$. Frankly, the activation and release of soil Cd is closely related to the concentration of extractants, extraction time, ionic strength, exchange and/or complexation ability, together with physical-chemical properties and mineral composition of soil, Cd species, etc.

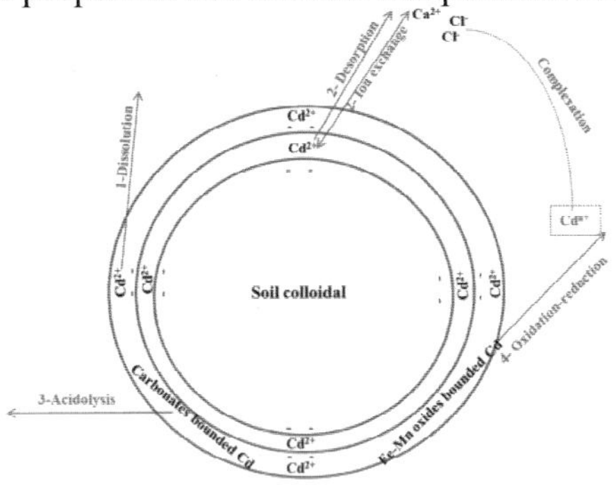

Figure 7. Possible mechanisms of $CaCl_2$ affecting soil Cd

Concentration is an important factor affecting the extraction of soil Cd by $CaCl_2$. It has been confirmed by previous studies that the extraction amount of soil Cd rise with the increase of concentrations when it is lower [25, 28]. As illustrated in Figure 1, the extraction amount of soil Cd rises quickly with the increase of $CaCl_2$ concentrations when it is lower than 0.1 mol/L. It could be due to rapider dissolution of those soluble Cd. Besides, it also could be due to the activation and release of those weakly bound Cd (such as exchangeable, carbonates bound, etc.) and even strongly bound Cd (Fe-Mn oxides bound etc.) under the effect of $CaCl_2$ of higher concentration. In this study, extraction rate of soil Cd is highest with the value of 33.5% when the $CaCl_2$ concentration is 0.1 mol/L. It is higher than the extraction rate of 26.5% found by Liu et al. [29], but lower than that of 69.8% found by other research [18]. The reasons for the difference of Cd extraction rate should be related to the concentration and species of soil Cd, soil types, together with soil physical and chemical properties such as organic matter content, pH, conductivity and redox potential [30-31], etc.

Time is another important factor affecting the extraction of soil Cd by $CaCl_2$. It has been confirmed by previous studies that the extraction amount of soil Cd rise with the increase of extraction time by $CaCl_2$ before the reaction balance [24]. And it is not obvious after balance time [25, 32]. As demonstrated in Figure 2, the extraction amount of soil Cd rises quickly with the increase of extraction time by $CaCl_2$ when it is less than 4h. It could be contributing to rapid dissolution of those soluble Cd. Besides, it also could be contributing to the activation and release of those weakly bound Cd (such as exchangeable, carbonates bound, etc.) and even strongly bound Cd (Fe-Mn oxides bound etc.) during the first 4h after the Cd contaminated soil meet with $CaCl_2$ solution. And the reaction balance time is about 4h. It is longer than 1h proposed by Ma and Yin et al. [24, 32], but within 12h found by other research [25]. The reasons for the difference of Cd extraction balance time should be also related to the concentration and species of soil Cd, soil types, soil physical and chemical properties such as organic matter content, pH, conductivity and redox potential, etc. [30-31, 33], together with concentration gradient of $CaCl_2$ solution.

5. Conclusions

$CaCl_2$ was used for activation and extraction of soil Cd around a Pb-Zn mine. It is found that there are obvious concentration effect and time effect. The best choice of $CaCl_2$ concentration is 0.1 mol/L. And the best extraction time is 4h. Meanwhile, it is found that transformation of Cd species such as exchangeable, carbonates bound and Fe-Mn oxides bound should be the main reasons of concentration effect and time effect. In addition, it is also found that the extraction of soil Cd by $CaCl_2$ is a complex coupling process of dissolution, ion exchange, desorption, acidolysis, oxidation-reduction and complexation, etc. Especially, Ca^{2+} in $CaCl_2$ solution and those cations (such as H^+ etc.) exchanged from soil colloids by Ca^{2+} could desorb exchangeable Cd ions adsorbed on soil colloids by ion-exchangeable reaction. However, it should be noted that other factors such as solid liquid ratio, extraction times, etc. could also affect the activation and extraction of soil Cd by $CaCl_2$. These all remain to be further studied.

Acknowledgments
This study was supported by Natural Science Foundation of Guangdong Province (2020A1515011077, 2021A1515012597), Scientific Research Projects of Jiaying University (2020KJY05, 2019KJY05).

References
[1] Guangyu S, Jiayuan H, Fanqi D, Shiqian L, Weilin S, Yan Chen 2022 Exogenous Pseudomonas aeruginosa application improved the phytoremediation efficiency of Lolium multiflorum Lam on Cu-Cd co-contaminated soil *Environmental Technology & Innovation* **27** 102489
[2] Niaz A, Ali R S, Subhan D, Shah F, Muhammad A A, Tayebeh Z, Valerie V, Rahul D 2021 Immobilization of Cd in soil by biochar and new emerging chemically produced carbon *Journal of King Saud University – Science* **33** 101472
[3] Zhanming L, Yi L, Hangwei H, Sabry M S, Huan Z, Filip M G T, Mengjie W, Yu-Feng L, Yuxi G, Jörg R, Jiating Z 2021 Speciation, transportation, and pathways of cadmium in soil-rice systems: A review on the environmental implications and remediation approaches for food safety *Environment International* **56** 106749
[4] Babar H, Muhammad N A, Shafeeq-ur-Rahman, A A, Jumei L, Muhammad F 2021 Cadmium stress in paddy fields: Effects of soil conditions and remediation strategies *Science of The Total Environment* **754** 142188
[5] Xing S, Min D, Jiawei Y, Lin S, Xiao T, Changsheng P, Imran A, Iffat N 2021 A critical review on the phytoremediation of heavy metals from environment: Performance and challenges *Chemosphere* **291** Part 3 132979
[6] Yao S, Zezhou Z, Yi Z, Huijie H, Jing C, Linling W, Xiaohui W, John C C 2022 Synergistic effect of floatable hydroxyapatite-modified biochar adsorption and low-level $CaCl_2$ leaching on Cd removal from paddy soil *Science of The Total Environment* **807** Part 2 150872
[7] Tasuma S, Kisyo N, Satoshi K, Junko Ta Miyu O, Masakazu N 2019 The role of anatase impurity in the leaching of Cd(II) from contaminated kaolinite *Journal of Environmental Chemical Engineering* **7** 102851
[8] McDowell R W 2022 Assessing the leaching of cadmium in an irrigated and grazed pasture soil *Environmental Pollution* **292** Part B 118430
[9] Xin K, Fei J Z, Yan Z, Hai J Z, Guan L G, Yu T 2020 Removal of Cd, Pb, Zn, Cu in smelter soil by citric acid leaching *Chemosphere* **255** 126690
[10] Iram G, Maria M, Imran H, Muhammad F B, Jean K, Muhammad A 2019 Plant uptake and leaching potential upon application of amendments in soils spiked with heavy metals (Cd and Pb) *Journal of Environmental Management* **249** 109408
[11] Jianlong H, Jixu Z, Xi Z, Simeng L, Qi L, Cunzhen L 2022 Removal of heavy metals from sewage sludge by chemical leaching with biodegradable chelator methyl glycine diacetic acid *Chemosphere* **300** 134496
[12] Peipei S, Dan X, Jingyuan Y, Yuanchen M, Shujun D, Jing F 2022 Recent advances in soil remediation technology for heavy metal contaminated sites: A critical review *Science of The Total Environment* **838** Part 3 156417
[13] Li Y, Hu P, Zhao J, Dong C 2015 Remediation of cadmium and leadcontaminated agricultural soil by composite washing with chlorides and citric acid *Environmental Science and Pollution*

Research **22** 5563

[14] Makino T, Sugahara K, Sakurai Y, Takano H, Kamiya T, Sasaki K, Itou T, Sekiya N 2006 Remediation of cadmium contamination in paddy soils by washing with chemicals: Selection of washing chemicals *Environmental Pollution* **144** 2

[15] Hayashi T, Kida T, Nanzyo M, Takahashi T, Honna M, Aikawa Y, Yoshihara K 2007 Persistence of $CaCl_2$ washing effect for amelioration of a Cd-contaminated paddy field soil *Soil Science and Plant Nutrition* **53** 721

[16] Maejima Y, Makino T, Takano H, Kamiya T, Sekiya N, Itou T 2007 Remediation of cadmium contaminated paddy soil by washing with chemicals effect of soil washing on cadmium uptake by soybean *Chemosphere* **67** 749

[17] Makino T, Kamiya T, Takano H, Itou T, Sekiya N, Sasaki K, Maejima Y, Sugahara K 2007 Remediation of cadmium contaminated paddy soils by washing with calcium chloride: Verification of on-site washing *Environmental Pollution* **147** 114

[18] Xiong L, Lu R 1992 Extracting mechanism of chemical extractants for available cadmium in soils *Environmental Chemistry* **11** 41

[19] Christine G, Sylvaine T, Michel A 2002 Fractionation studies of trace elements in contaminated soils and sediments: A review of sequential extraction procedures *Trends in Analytical Chemistry* **21** 451

[20] Xiong J, Zhu Q H, Huang D Y, Zhu H H, Xu C, Wang S, Wang H 2018 Comparison of single extraction methods for assessing Cd availability in paddy soils in South China *Research of Agricultural Modernization* **39** 170

[21] Pueyo M, López-Sánchez J F, Rauret G 2004 Assessment of $CaCl_2$, $NaNO_3$ and NH_4NO_3 extraction procedures for the study of Cd, Cu, Pb and Zn extractability in contaminated soils *Analytica Chimica Acta* **504** 217

[22] Neal W, Michael J D, Peter M K 2007 Evaluation of extractants for estimation of the phytoavailable trace metals in soils *Environmental Pollution* **145** 126

[23] He J, Lin Y, Liu P, Zhang X, Feng Y, Chen W, Zhang R 2009 Study of extraction method for available copper and cadmium in acidity soil with different extractants *The Administration and Technique of Environmental Monitoring* **5** 27

[24] Ma J 1998 The condition of extracting available Cd from soil *Journal of Hebei Normal University of Science & Technology* **12** 13

[25] Yan S,Wu C, Hu Y, Li Z, Lu Q, Chen S, Yang Q 2013 Optimization of $CaCl_2$ extraction of available cadmium in typical soils *Chinese Agricultural Science Bulletin* **29** 101

[26] [26] Xiao L, Wang G, Huang R, Zhang L, Zhang J. 2008 Extraction method for available cadmium in acid soils *Journal of agro-environment science* **27** 795-800

[27] Ji Y, Xie F, He J, Lu Y, Deng Q 2007 Study on the extracting condition for available Cd and content in cultivated soil of Guiyang city *Journal of Anhui Agricultural science* **35** 11923

[28] Sun T, Mao X, Lu K, Wang H 2015 Removal of heavy metals from co-contaminated soil by washing with citric acid *Acta scientiae circumstantiae* **35** 2573

[29] Liu Y N, Li Q S, Du Y F, Cui Z H, Li S 2011 Salinity change and its impact on heavy metals during beach soil leaching and desalination *Environmental Science* **32** 2089

[30] Singh B R, Kristen M 1998 Cadmium uptake by barley as affected by Cd sources and pH levels *Geoderma* **84** 185

[31] Cao L, Yan J 1991 Study on ecological cadmium pollution soil-crop system *Journal of Shanxi University (Nat. Sci. Ed.)* **14** 419

[32] Yin J, Liu W, Xie J, Xiao C 2000 The study on extraction conditions and extractants of soil available Cd, Hg *Journal of Agricultural University of Hebei* **23** 25

[33] Hooda P S, Alloway B J 1998 Cadmium and lead sorption behavior of selected English and Indian soils *Geoderma* **84** 121

IOP Conf. Series: Earth and Environmental Science 1087 (2022) 012058 doi:10.1088/1755-1315/1087/1/012058

Shallow geological structures building by joint waveform inversion based on curvelet decomposition

Chao Zhang[1,3] and Qingxiang Meng[2]

[1] College of Oceanography, Hohai University, Nanjing 210098, China;
[2] Research Institute of Geotechnical Engineering, Hohai University, Nanjing 210098, China.

[3] Corresponding author's e-mail: czhang18@hhu.edu.cn

Abstract. Shallow geological imaging structures often play an essential role in the field of environmental and engineering geophysics. Early arrival waveform inversion (EWI) is high-resolution method to imaging near-surface structures. We proposed a joint waveform and envelope inversion method based on curvelet transform (CJWEI) to build shallow geological structures. By taking advantage of their complementary sensitivities to near-surface structure, we can constrain both low- and high-wavenumber components of structures, as well as mitigating the cycle-skipping problem. Curvelet transform was used to decompose seismic data into different scales and provide a multi-scale inversion strategy to further reduce non-uniqueness of the inversion. With synthetic test, we demonstrate that our method can constrain the anomalies and hidden layers in the shallow structure more efficiently as well as robust in terms of noise.

1. Introduction

The near-surface imaging structures often play an essential role in carrying out high-precision shallow resources, avoiding potential natural disasters, and constructing urban underground space in engineering geophysics, which is closely related to human production and life. Early arrival waveform inversion (EWI) is an effective tool to yield accurate velocity models with complex shallow geological structures [1, 2, 3], which can obtain better-resolved earth models compared with conventional ray-based tomography.

It's well-known that some problems still exist in EWI, e.g., lacking of low-wavenumber components in observed data, suffering from severe cycle-skipping issue. Previous study proposed various methods to retrieve the low-frequency component just from available regular data with high-frequency. An important effort is to recover low-wavenumber components based on envelope due to its rich low-frequencies, which can help mitigate cycle-skipping problem. Thus, envelope inversion (EI) has been always constrained low-wavenumber structure, providing an initial model for EWI [4]. However, envelope has relative low-resolution and brings enormous challenges to invert high-wavenumber structure. Thus, joining the waveform and envelope data sets will take advantage of their different sensitivities to shallow geological structure and help mitigate cycle-skipping issue. The multi-scale inversion is considered another way to reduce non-uniqueness during inversion. It gradually inverts from low-frequency seismic data (reflecting the large-scale structure) to high-frequency data (reflecting the small-scale structure). At present, the main popular methods include multiscale waveform inversion based on frequency decomposition, wavelet decomposition and curvelet decomposition. The curvelet transform contains components of scale, angle, and position to have multi-directional characteristics [5], which extensively used in seismic migration imaging, seismic data interpolation, mixed data separation, and waveform inversion.

Motivated by the above usage of waveform envelope and curvelet transform, a curvelet-based joint waveform and envelope inversion (CJWEI) method was robust developed in this paper. We can constrain both low- and high-wavenumber components of structures by taking advantage of two types of data. Curvelet transform was used to decompose seismic data into different scales and provide a multi-scale inversion strategy.

2. Theory and Methodology

2.1. The theory of curvelet transform

The curvelet transform has been developed from wavelet transform and have been developed a multi-scale geometric transform, which can effectively characterize the edges and curves of different angles. In two-dimensional space, basis function of the curvelet transform $\varphi_j(\mathrm{x})$ is defined and its Fourier transform is $\hat{\varphi}_j(x) = W_j(\omega)$. The frequency window $W(\omega)$ of the curvelet transform defined in the Fourier domain can be written as followed:

$$W_j(\omega) = U_j(\omega)V_j(\omega) \tag{1}$$

Where $U(\omega)$ and $V(\omega)$ represent the window function of radial and angle, j is the certain scale. By rotating and translating $W_j(\omega)$, a series of curvelet basis functions $W_{j,l,k}(\omega)$ can be obtained, where θ represents the direction and k represents the position. As followed by [5], the mathematical formulation of curvelet coefficients $C(j,\theta,k)$ in the Cartesian coordinate can be seen as follows:

$$C(j,\theta,k) = \frac{1}{(2\pi)^2}\int g(\omega)W_j(S_{\epsilon_\theta}\omega)e^{i\langle(k_1\cdot 2^{-j},k_2\cdot 2^{-j/2}),w\rangle}d\omega \tag{2}$$

Where $C(j,\theta,k)$ is the curvelet coefficient, j, θ, k represent the scale, direction and displacement respectively. As shown in Figure 1, the tiling corresponds to different scales, represented by the symbol of j; The wedges correspond to different directions under the same scale, represented by the symbol of θ; and the position of points in a wedge are represented by k. These three parameters show high sparsity.

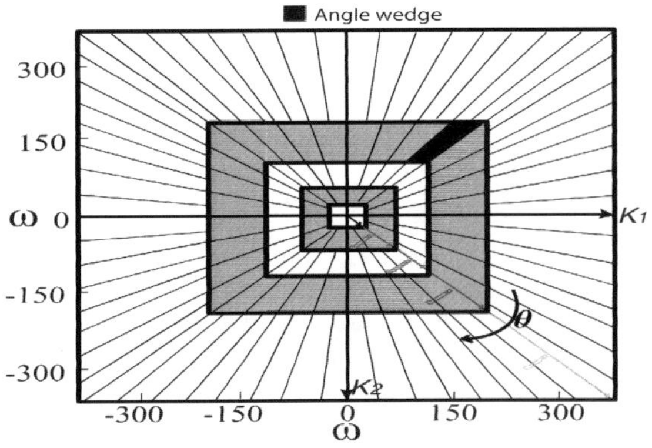

Figure 1. The schematic diagram of curvelet transform coefficient in the Cartesian coordinate.

Let the seismic data be $s(\mathrm{t, m})$, which are transformed by curvelet transform can be expressed as:

$$C(j,\theta,k) = \langle s, \varphi_{j,\theta,k}\rangle,\ where\ j > 0, 0 \le \theta \le 2^{j/2} - 1, k \in Z \tag{3}$$

Here, $C(j,\theta,k)$ represent the curvelet coefficients with scale j, direction θ and position $k = (k1, k2)$ after the curvelet transform. $\varphi_{j,\theta,k}$ is the curvelet basis function. The seismic data can be obtained from the inverse transform of the curvelet coefficients in the curvelet domain:

$$s = \sum_{j,\theta,k} \langle s, \varphi_{j,\theta,k}\rangle \varphi_{j,\theta,k} = \sum_{j,\theta,k} C(j,\theta,k)\varphi_{j,\theta,k} \tag{4}$$

2.2. *Curvelet-based multi-scale joint inversion of waveform and envelope*

For a certain scale j, the objective function of waveform inversion χ_j^W can be written as follows:

$$\chi_j^W(\mathbf{m}) = \frac{1}{2}\sum_{s,r} \int_{T_{start}}^{T_{end}} \|\mathbf{s}_j(\mathbf{x}_r, \mathbf{x}_s, t; \mathbf{m}) - \mathbf{d}_j(\mathbf{x}_r, \mathbf{x}_s, t)\|^2 dt \tag{5}$$

Where m is the model vector, χ_j^W is the misfit function for waveform inversion on the j-th scale decomposed by curvelet transform. Note that curvelet transform should be performed on both the synthetic and observed data. Similarity, the misfit function of curvelet-based multi-scale envelope inversion is defined as follows:

$$\chi_j^E(\mathbf{m}) = \frac{1}{2}\sum_{s,r} \int_{T_{start}}^{T_{end}} \left\|E_{s_j}(\mathbf{x}_r, \mathbf{x}_s, t; \mathbf{m}) - E_{d_j}(\mathbf{x}_r, \mathbf{x}_s, t)\right\|^2 dt \tag{6}$$

Where m is the model vector, χ_j^E is the misfit function for envelope inversion on the j-th scale. E_{s_j} and E_{d_j} are the envelope of $\mathbf{s}_j$ and $\mathbf{d}_j$ obtained by Hilbert transform. Thus, the objective function of curvelet-based joint inversion of waveform and envelope based on the adjoint-state method χ_j is defined as follows:

$$\chi_j(\boldsymbol{m}) = (1-\lambda)\,\chi_j^W(\boldsymbol{m}) + \lambda\,\chi_j^E(\boldsymbol{m}) \tag{7}$$

Here, λ are the weights used to balance the two data types to prevent the results from being dominated by either one. The curvelet-based multi-scale joint inversion scheme is described in Figure 2.

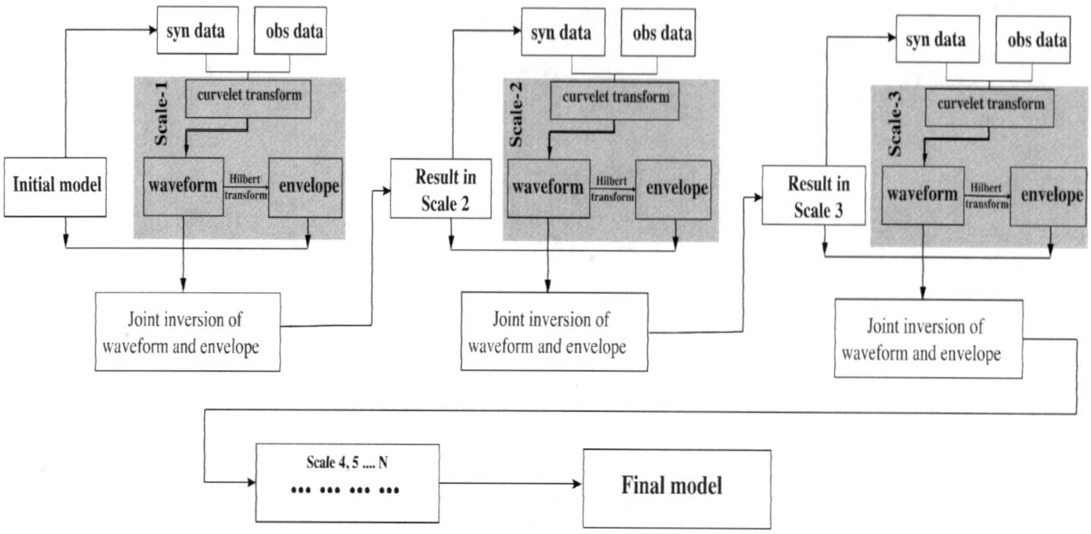

Figure 2. Workflow of curvelet-based joint waveform and envelope inversion (CJWEI).

The gray shaded box represents the process of curvelet transform which is used to generate seismic data of different scales.

3. Nonlinearity analysis

To illustrate the nonlinearity of misfit function for various types of inversion strategy, i.e., waveform inversion alone, envelope inversion alone and joint inversion, we design a constant velocity model to calculate the normalized objective functions analytically. The target model is set as homogeneous, with a constant P-wave velocity, S-wave velocity, and density (Vp = 3500 m/s, Vs = 2020 m/s, and ρ = 2800 g/m3). The numerical parameters of modeling are shown in Table 1. The model measures 9000 m in the horizontal direction, with 300 uniform mesh nodes (quadrangles in 2D), and 600 m in the vertical direction, with 20 uniform nodes. The time interval of the simulation was set to 0.0002 s to satisfy the numerical stability, and the time step was set to 10,000. The survey geometry includes a shot on the surface at location of (4500 m, 0 m) and 301 receivers with receiver spacing of 30 m. The receivers also locate at the surface. A Ricker-type source-time function with a central frequency of 10 Hz is set. Specfem2d was used to generate the waveform data, then the decomposed multi-scale seismic data was obtained based on the curvelet transform in three different scales: scale-1 (coarse-scale), scale-2 (detail-scale) and scale-3 (fine-scale). The objective functions for velocities varying from 500 to 6500 m/s, with an interval of 50 m/s for different scales is calculated, respectively.

Table 1. Parameters of numerical modeling in nonlinearity analysis.

Model size(Nx*Nz)	x-direction: 9000 m; z-direction: 600 m (300*20 elements)
Propagation time	10,000 time steps; Time discretization: 0.002s
Source Setting	Source time function: Ricker wavelet Dominant frequency: 10 Hz Source number: 101 Location: (0 m, 0m) – (9000m, 0m) spacing: 90m
Receiver Setting	Receiver number: 301 Location: (0 m, 0m) – (9000m, 0m) spacing: 30m

Take the result of scale-1 as an example. The waveform objective function clearly includes more local minima than others (Figure 3(a)). The envelope objective function has an improvement in terms of the curve for waveform fitting, but shows the least sensitivity to a solution corresponding to a low resolution. The joint inversion with a proper choosing of weighting factors gives an objective function toward to a single global minimum. It also keeps a refined sensitivity. Therefore, we combine waveform and envelope inversion, taking advantage of envelope to recover low-wavenumber model components in the near-surface area and using waveforms for the high-wavenumber model components of the near-surface structure. Figure 3(b) also depicts the misfit function of joint inversion waveform with different weight factors. The misfit curve of joint inversion with small waveform weight factor (λ=0.2) is similar to envelope inversion alone, and the objective function presents fewer local minima, which suggests that more envelope data should help the inversion recover more low-wave- number structure information with less local minima. A large waveform weighting factor (λ=0.8) should allow CJWEI to refine more details of the high-wavenumber model component but with a high risk of the local minima.

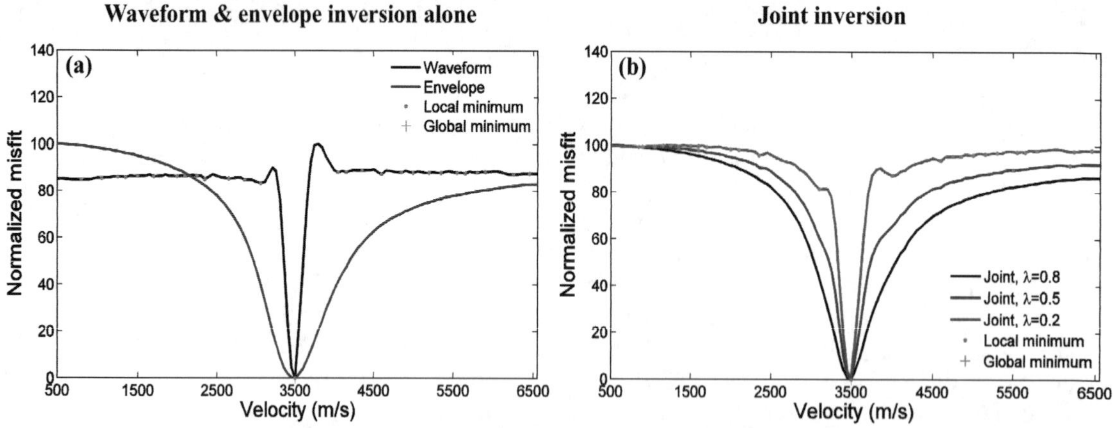

Figure 3. The normalized misfit functions for individual inversion alone at scale-1 (black: waveform inversion, blue: envelope inversion, red dot: local minimum, cross symbol: global minimum); (b) the normalized misfit functions for joint inversion with different weighting factors λ=0.8, 0.5, 0.2.

4. Synthetic Experiments

It's well-known that hidden layers (low-velocity inclusions) cannot be solved using none-linear travel time tomography. To demonstrate our method has the capacity to solve such problem, we apply CJWEI to a target model with hidden layers, as shown in Figure 4(a). It contains one high velocity layer (HVL, Vp = 3500 m/s) and two low velocity layers (LVL, Vp = 2200 m/s) distributed on both sides of HVL. Excluding the LVL and HVL, the background velocity model was provided the initial model for CJWEI (Figure 4(b)). The source time function used the Ricker wavelet and the geometry is set the same with Section 3 (detail seen in Table 1). Firstly, the synthetic data and observed data are generated by modeling. Then, we used the curvelet transform decompose both synthetic data and observed data to three scales in this experiment, i.e., scale-1 (coarse-scale with 33% curvelet coefficient), scale-2 (detail-scale with 66% curvelet coefficient) and scale-3 (detail-scale with 100% curvelet coefficient). As the increasing of curvelet coefficient, more detailed waveform information was recovered in the reconstructed shot gathers. Using these decomposed data, we obtained the inverted models (Figure 5) for different scale followed the workflow in Figure 2. Figure 5(d) is the result of traditional early arrival waveform inversion (TEWI), and Figure 5(a), 5(b), 5(c) is the results of CJWEI method in each scale.

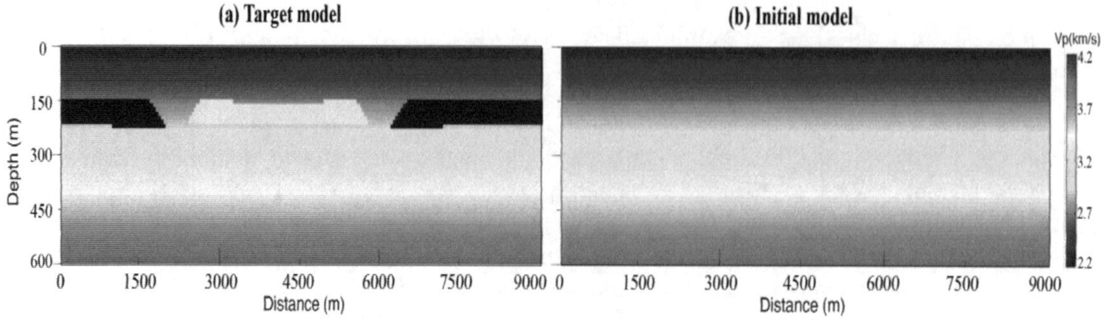

Figure 4. (a) the target model; (b) the initial model.

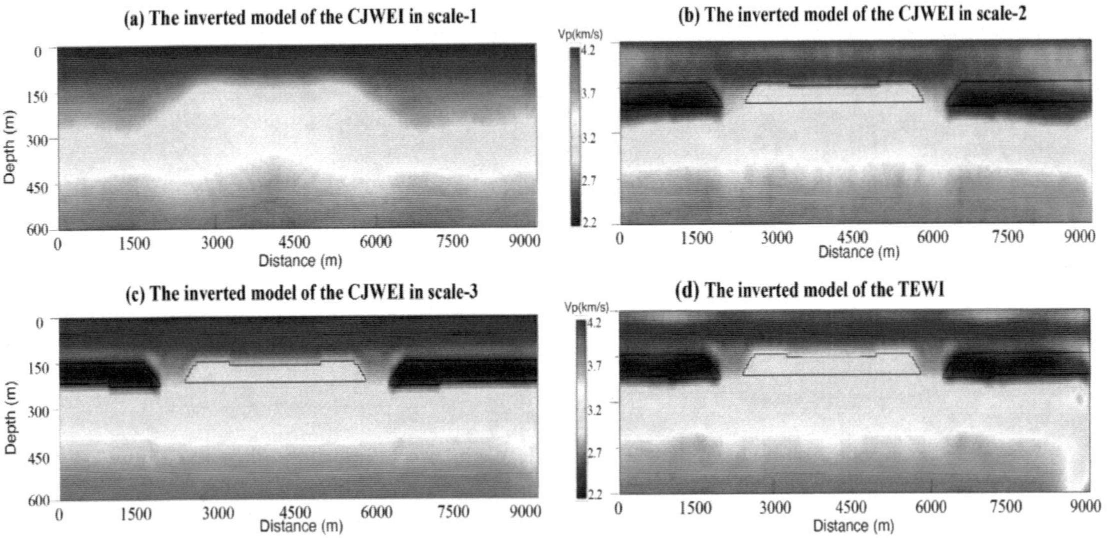

Figure 5. The inverted model of CJWEI in (a) scale-1 (with 33% curvelet coefficients); (b) scale-2 (with 66% curvelet coefficients); (c) scale-3 (with 100% curvelet coefficients); (d) The inverted model of TEWI.

The result of TEWI is acceptable, in which constrained the layers with clear contours but the position both of the LVL and HVL are lower than the real position. In addition, there are some high-speed artificial illusions in the shallow regions above the LVL and HVL. TEI can recover the LVL and HVL better in aspect of the location and constrain the layers to the accurate position. However, TEI cannot solve the velocity in deeper regions below the HVL with a low resolution. In comparison, CJWEI has a better result compared with the other inversion alone. Both the outline of LVL and HVL are clearly recovered and the layers are constrained in the right position. Also, the resolution has been improved obviously.

5. Discussions

In this study, we developed a curvelet-based joint inversion of waveform and envelope (CJWEI), for imaging near-surface geological structure with high-resolution. We build a combined misfit function including both envelope and waveform misfit, which minimizes the waveform difference, meanwhile maintains the low-frequency information from envelope and high-resolution imaging from waveform. Meanwhile, curvelet transform extract the seismic data with different scales and provide a multi-scale inversion strategy to efficiently mitigate the cycle-skipping issue. The synthetic tests revealed that CJWEI can map near-surface velocity structure more accurately, as well as capable of handling noise. The proposed multi-scale joint-inversion may provide a very powerful framework to understand the behavior of near-surface geological structure in engineering geophysics or oil exploration.

Acknowledgments

This work is supported by Natural Science Foundation of Jiangsu Province, China (BK20190499), High-level Innovation and Entrepreneurship Talents Introduction Program of Jiangsu Province of China, and National Natural Science Foundation of China (42004037).

References

[1] Sheng, J.; Leeds, A.; Buddensiek, M. et al. 2006 Early arrival waveform tomography on near-surface refraction data *Geophysics* **71(4)** 47– 57
[2] Tarantola, A. 1984 Inversion of seismic reflection data in the acoustic approximation *Geophysics*

49(8) 1259–1266
[3] Pratt, R.G.; Shin, C.; Hick, G.J. 1998 Gauss–Newton and full Newton methods in frequency–space seismic waveform inversion *Geophys. J. Int.* **133(2)** 341–362
[4] Wu, R. S., Luo, J., and Wu, B. 2014 Seismic envelope inversion and modulation signal model *Geophysics* **79(3)** 13–24
[5] Cand`es, E., Demanet, L., Donoho, D., Ying, L. 2006 Fast discrete curvelet transforms *Multiscale Model. Simul.* **5 (3)** 1–44

Effects of lime nitrogen on the fungi community of tobacco soil and its correlation analysis

Susu Zhang[1,4], Yulan Chen[2], Yangyang Tuo[3], Hongli Li[1,4] and Yan Wang[1,4]

[1] College of Ecology and Environment and School of Chemical Engineering, Zhengzhou University, Zhengzhou, China;
[2] Liangshan Prefecture Company, Tobacco Company of Sichuan, Xichang, China;
[3] Dechang Branch, Liangshan Prefecture Company, Tobacco Company of Sichuan, Dechang, Liangshan, Sichuan, China.

[4] Corresponding author´s e-mail: lihonglihn@zzu.edu.cn

Abstract. Soil microorganisms are closely related to soil environmental factors. To explore the effects of lime nitrogen, lime nitrogen combined with biocontrol agents on soil environmental factors and fungi and their correlation in tobacco fields, four treatments were set: 1_1: control group; 1_2: biocontrol agents group; 1_3: lime nitrogen group; 1_4: lime nitrogen with biocontrol agents group. Soil samples were taken during four growth periods of tobacco plants, and fungi were analyzed by high-throughput sequencing. The results showed that both 1_3 and 1_4 group increased the soil pH, OM, available nutrients, effective copper, available zinc, fungi richness and diversity of vigorous and maturity stage, urease, acid phosphatase, invertase activity, flue-cured tobacco economic characteristics, and the relative abundance of beneficial fungi such as *Chaetomium*, *Solicoccozyma* of maturity stage. At maturity stage, available zinc, available nitrogen, urease activity was significantly positively correlated with *Chaetomium*; Invertase activity showed significantly positive correlation with *Solicoccozyma*. In short, both 1_2 and 1_3 group improved physicochemical properties, more enzyme activity of the soil, which were significantly positively correlation with some beneficial fungi. Therefore, the relative abundance of certain beneficial fungi increased so that the disease resistance of tobacco plant was improved and flue-cured tobacco economic characters were better.

1. Introduction

As an important economic crop, tobacco is often cultivated continuously, which may cause continuous cropping obstacles [1-2]. For example, crop soil-borne diseases, diseases and insect pests are aggravated, soil physicochemical properties deteriorate, and soil biological environment gets worse [3-5]. Soil-borne diseases are caused by soil pathogens, which are the most common and problematic crop diseases in the world, including root rot, bacterial wilt, black shank, root-knot nematode and so on [6]. The current treatment methods for tobacco continuous cropping obstacles are: disinfecting the soil; adjusting the planting mode [7]; selecting a reasonable fertilization method; using microbial agents to repair [8].

Previous results have showed that lime nitrogen is a kind of fertilizer for controlling soil pathogens. The N element in lime nitrogen can be transformed into nutrients that can be absorbed by plants. The content of CaO in lime nitrogen is 35%, which can meet the requirement of plant for Ca. At the same time, because lime nitrogen does not contain acid root, it can prevent soil acidification. Compounds such as cyanamide and dicyandiamide produced by hydrolysis of lime nitrogen in soil can kill pathogenic microorganisms, nematodes [9], underground pests, weed seeds [10] in soil, and have the functions of preventing soil-borne diseases, promoting organic ageing, improving soil community structure [11], alleviating continuous cropping obstacles [12]. However, lime nitrogen was often used in facility agriculture such as cucumber planting to manage continuous cropping obstacles [13], but there are few studies on its application in field crops, especially in tobacco [12].

Bacillus subtilis [14], *Aspergillus Niger* [15], *Bacillus licheniformis* [16], and *Streptomyces* [17] are beneficial biological agents, which can converse soil nutrients effectively, accelerate the rate of organic fertilizer ageing, and inhabit pathogens to relieve continuous cropping obstacles. Studies have shown that the combination of a variety of biocontrol bacteria can form complementary advantages between different strains, thereby improving the control effect and stability [18]. Soil environmental factors and microorganisms are important indicators that reflect soil conditions.

There are few reports on the effects of lime nitrogen and lime nitrogen with mix microbial biocontrol agents (*Aspergillus Niger*: *Bacillus subtilis*: *Bacillus licheniformis*: *Streptomyces flavus* =3:3:3:1, 5.0×10^8 CFU/g) on the environmental factors, soil fungal communities and correlation between fungi community and the environmental factors.

So, applying lime nitrogen to tobacco fields were studied to explore the effects of lime nitrogen and lime nitrogen with mix microbial biocontrol agents on the environmental factors, soil fungal communities and their correlation, thus providing theoretical basis for lime nitrogen and microbial biocontrol agents to improve tobacco soil microorganism and environmental factors to control soil-borne diseases.

2. Materials and methods

2.1. Experiment materials

The lime nitrogen (Chemical formula: $CaCN_2$, nitrogen content:20% and CaO content:35%) was called calcium cyanamide, which is a new type of artificial agricultural nitrogen fertilizer made by Italian scientists in 1906 by using calcium carbide and nitrogen to react under high temperature conditions [12]. Mix microbial biocontrol agents (*Aspergillus Niger*: *Bacillus subtilis*: *Bacillus licheniformis*: *Streptomyces flavus* =3:3:3:1, 5.0×10^8 CFU/g) They were sourced from Hebi Runtian Biotechnology Co., Ltd., Henan Province, China.

2.2. The experiment site and treatment

On March 19, 2019, the test field was chosen from a farmer's tobacco field with hillside terraces which is located in Ninan County, Liangshan Prefecture, Sichuan Province, China, at an altitude of 1780 m (102°11′1″ N, 27°34′51″ E). The experimental site has been used for growing tobacco for ten years, with a history of tobacco root knot nematode disease and the test site is sandy loam soil. On March 22, 2019, treatment 1_3 and 1_4 were applied lime nitrogen 450 kg/ha, added water and then covered with plastic mulch. Tobacco seedings of Yunyan 87 were transplanted on April 19, then 45 kg/ha mix microbial biocontrol agents were poured in treatment 1_2 and 1_4. 45 kg/ha mix microbial biocontrol agens were poured again in treatment 1_2 and 1_4 on June 5. There are three parallel groups in each treatment, and each parallel group has an area of $66.7m^2$. Specific experimental scheme was shown in Table 1.

Table 1. Experiment design of treatments.

numeration	treatment
1_1	CK: Conventional special tobacco fertilizer, 675 kg/ha+ Biological organic fertilizer,375 kg/ha
1_2	CK+ mix microbial biocontrol agents, 90 kg/ha
1_3	CK+ lime nitrogen, 450 kg/ha
1_4	CK+ lime nitrogen, 450 kg/ha + mix biocontrol agents, 90kg/ha

Note: Conventional special tobacco fertilizer: Compound fertilizer(m(N):m(P):m(K)=1:1:3.5); Biological organic fertilizer: cake fertilizer

2.3. *Soil sampling collection and economic characteristics of flue-cured tobacco measuring*

The soil sample were collected at pretransplanting stage (on March 19, 2019) and was recorded as YN. Physicochemical properties and enzyme activity of pretransplanting stage were shown in Table 2, which were reflected the initial state of the soil. Root soil was sampled at the rosette stage (on June 3), vigorous stage (on July 20), and maturity stage (on August 27) of tobacco growth, respectively, and the samples were marked as TN, WN, CN before each numeration, respectively. Each treatment contains three parallel samples. For each sample, five-point sampling method was adopted. And each sample of 1kg soil was collected from the depth of 5-20 cm and divided into two parts. One part was immediately refrigerated at −80 °C and sent to Shanghai Majorbio Bio-pharm Technology Co., Ltd for gene sequencing and the other part was used for the determination of physicochemical indicators and enzyme activity. And the soil specific pre-treatment method was the same as that of Analysis methods Ding [19]. At maturity stage, 20 tobacco plants were randomly selected from each field, and the economic benefits of flue-cured in each plot were calculated.

Table 2. The initial physicochemical properties of the test site.

Treatment	pH	OM/ (g/kg)	AN/ (mg/kg)	AP/ (mg/kg)	AK/ (mg/kg)	Urease [mg/ (g·24h)]	Acid phosphatase [μg/ (g·h)]	Catalase (ml/g)	Protease [μg/ (g·24h)]	Invertase [mg/ (g·24h)]
YN	5.56±0.05	26.97±7.10	207.40±15.91	150.13±11.35	556.67±11.93	31.53±0.42	147.61±1.73	10.14±0.11	326.28±11.06	76.94±7.15

3. Analysis methods

3.1. *Determination of soil physicochemical properties and enzyme activity*

Soil pH, organic matter (OM), available phosphorus (AP), available potassium (AK), Alkaline Nitrogen (AN), available Fe (AFe), available Cu (ACu), available Mn (AMn), availableZn (AZn) were measured by the same way Lu used [20]. Soil pH was measured in a 1:2.5 (w/v) aqueous suspension using a pH meter (Mettler Toledo Instrument (Shanghai) Co., Ltd). Soil OM was determined by dichromate oxidation and titration with ferrous ammonium sulphate. AP) of soil was determined according to the extraction-antimony molybdenum Colorimetric Method. The AK, AFe, ACu, AMn, AZn was determined according to the flame-atomic absorption method. AN was determined according to alkaline diffusion method. Urease activity was assayed by the Phenol-sodium Hypochlorite Colorimetric Method [21]. Urease activity was expressed in μg $NH4^{+}$-Ng^{-1}. Catalase activity was determined by $KMnO_4$ titration method [22]. Catalase activity was expressed in [mL $KMnO_4$/(g dry soil)]. Sucrase activity was determined by 3,5-dinitrosalicylic acid colorimetry [23]. Protease activity was determined by ninhydrin colorimetry [24]. Acid phosphatase activity was measured following the method of disodium p-Nitrophenyl phosphate tetrahydrate Colorimetric Method [25]. Acid phosphatase activity was expressed in μg PNP g^{-1}.

3.2. Determination of soil fungi community

The diversity of rhizosphere soil fungi community was sequenced by Shanghai Major Bio-pharm Technology Co., Ltd. Total DNA was extracted according to the instructions of the E.Z.N.A. SOIL kit (Omega Bio-Tek, Norcross, GA, U. S.). The concentration and purification of DNA were detected by NanoDrop2000, and DNA extraction quality was detected by 1% agarose gel electrophoresis. The V5~V7 variable region of fungal (18S) was amplified by PCR using SSU081F (5-T TAGCATGGAATAATRAATAGAGGA-3) and 1196R (5-TCTGGACCTGGTGAGTTTCC-3) primers.

PCR products were recovered by using 2% agarose gel. The purification was performed by using the AxyPrep DNA Gel Extraction Kit (Axygen Biosciences, Union City, CA, USA), eluted by Tris-HCl, and tested by 2% agarose electrophoresis. Quantification was performed by the Quantifluor TM-ST (Promega, USA). According to the standard operating procedures of Illumina Miseq (Illumina, San Diego, USA), the purified amplified fragments were used to construct a PE2*250 fungal library. Sequencing was performed by using MiSeq PE300 platform of Illumina (Shanghai Meiji Biomedical Technology Co., Ltd.) and the original data were uploaded to the NCBI database.

Operational taxonomic units (OTUs) were used at 97% sequence similarity. The resulting sequences after quality control were analyzed through QIIME.

3.3. Statistical Analysis

Statistical analysis of physicochemical properties, enzymatic activity of this thesis was carried out by using the SPSS 20.0 statistical software package. Univariate analysis of variance (ANOVA) and student´s t test were used to analyze the community difference. The difference was considered statistically significant between treatments showed $P < 0.05$. Spearman correlation analysis was used to analyze the correlation between physicochemical properties, enzymatic activities and fungal communities [26].

4. Result and analysis

4.1. Effects on soil physicochemical properties

In order to reveal the influence of lime nitrogen and lime nitrogen with mix microbial biocontrol agents on soil effective nutrient status, the physicochemical factors of soils under different treatments were measured, and were showed in Table 3.

For 1_3 group and 1_4 group, the content of AN, AP, AK, ACu, AZn were higher than the control at vigorous and maturity stage. That is, both groups improved soil available nitrogen, phosphorus and potassium, some trace elements content in the middle and late stages of tobacco plant growth. The pH and OM content in the 1_3 group and 1_4 group were higher than the control at rosette, vigorous and maturity stage. The content of AFe, AMn of 1_3 group and 1_4 group was significantly lower than the control at rosette stage, slightly higher than the control at vigorous stage, and greatly lower than the control at maturity stage.

Compared with 1_3 group, for1_4 group, the content of AFe, AK, AP, OM decreased significantly while, pH increased significantly at rosette stage; At vigorous stage, the content of AN, AK increased significantly. All physicochemical factors in 1_3 group and 1_4 group had no significant difference at maturity stage.

Table 3. The physicochemical properties of different treatments in tobacco soils.

Treatment	pH	OM/ (g/kg)	AN/ (mg/kg)	AP/ (mg/kg)	AK/ (mg/kg)	AFe/ (mg/kg)	ACu/ (mg/kg)	AMn/ (mg/kg)	AZn/ (mg/kg)
TN1_1	5.58±0.03dA	29.03±1.58abA	211.40±18.94aA	153.23±8.79aA	567±9.54aA	63.117±1.50abB	3.736±0.56aA	25.060±6.00aB	1.180±0.21aA
TN1_2	5.97±0.03cA	22.61±3.66bB	190.4±44.8aB	145.17±5.39abB	570±16.42aB	63.598±3.15aA	4.38±0.74aA	30.283±5.52aB	1.454 ±0.44aA
TN1_3	6.44±0.09bA	37.03±3.99aA	197.87±22.8aB	137.90±8.57abB	537±28.66abB	48.797±4.64bcB	3.736±0.28aB	11.620±3.56bB	1.223±0.16aB
TN1_4	6.75±0.09aA	34.03±2.64abA	248.27±46.21aB	127.20±4.8bB	507±20.01bB	43.502±9.49cB	3.575±0.48aA	6.597±4.17bC	0.984±0.01aC
WN1_1	5.20±0.12bB	32.76±0.62aA	256.67±10.87bA	164.70±6.60aA	578±4.73bA	51.925±2.08aC	4.058±0.84aA	70.203±12.56aA	1.180±0.50aA
WN1_2	5.24±0.14bB	33.71±3.89a	298.67±32abA	167.20±7.40aA	634±30.05aA	52.166±5.91aB	5.024±0.84aA	80.316±3.64aA	1.197±0.17aA
WN1_3	5.53±0.04aB	37.33±2.75aA	335.53±44.65abA	184.00±8.52aA	631±30.2abA	52.286±0.91aB	4.863±0.28aA	75.126±2.42aA	1.274±0.04aB
WN1_4	5.61±0.12aB	36.46±1.68aA	336.47±23.48aA	181.90±9.14aA	658±4.10aA	54.934±0.55aAB	5.185±1.22aA	80.915±3.75aA	1.505±0.07aB
CN1_1	5.20±0.13bB	30.29±2.48aA	200.85±36.11bA	130.20±9.30bB	544.8±31.19bA	69.976±2.9aA	4.058±0.48aA	79.385±6.74abA	1.727±0.21aA
CN1_2	5.21±0.15bB	31.22±2.06aA	254.05±32.45abAB	143.30±9.71bB	560.6±20.16abB	72.262±3.79aA	4.057±0.48aA	95.486±10.48aA	1.785±0.31aA
CN1_3	5.57±0.03aB	34.35±4.15aA	316.59±18.64aA	177.40±9.16aA	623.7±28.18aA	63.117±3.99aA	4.863±0.56aA	65.712±11.37bA	1.912±0.25aA
CN1_4	5.69±0.1aB	32.54±2.50aA	330.59±31.22aAB	181.40±9.89aA	624.2±34.12aA	65.523±4.02aA	5.024±0.97aA	62.818±11.06bB	2.260±0.34aA

Different lowercase letters represent significant differences between different treatments at the same time (P <0.05); different capital letters represent significant differences in different stages of the same treatment (P <0.05) (the same below).

4.2. *Effects on the soil enzyme activity*

The change of soil environment and stress often lead to the change of soil enzyme activity, which is often used as the index of soil quality. The initial soil enzyme activity at pretransplanting stage were showed in Table 2.

At maturity stage, the growth of tobacco plant was slow and the enzyme activities were relatively stable. The enzyme activities of different treatments at maturity stage were showed in Figure 1.

As were shown in Figure 1, at maturity stage, compared with the control, soil urease activity in lime nitrogen group and lime nitrogen with mix microbial biocontrol agents group increased by 39.77% ($p<0.05$) and 79.60% ($p<0.05$), respectively; acid phosphatase activity increased by 28.91% and 15.82%, respectively; catalase activity decreased by 11.63% and 15.28% respectively; protease activity increased by 27.75% and 21.59% respectively; invertase activity increased by 46.94%, 51.78%, respectively. That is, lime nitrogen group and lime nitrogen with mix microbial biocontrol agent group both increased the activity of a large number of enzymes in the soil to varying degrees.

Compared with lime nitrogen group, lime nitrogen with mix biocontrol agent group significantly increased the urease activity of the soil.

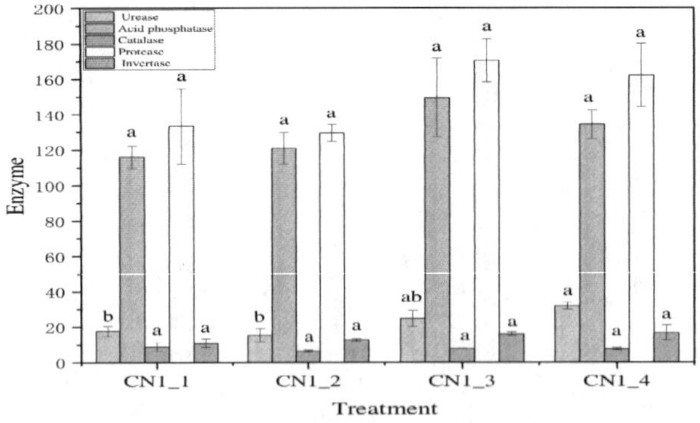

Figure 1. Soil enzyme activities at maturity stage.

4.3. *Analysis of fungal α-diversity*

α-diversity index was used to reflect the richness and diversity of species in soil. Sobs index can reflect community richness; Shannon index reflects community diversity; Shannon even index reflects community evenness, and coverage index indicates community coverage. All these indexes of each treatment were showed in Table 4. Coverage indexes are more than 99%, indicating that this sequencing can represent the real situation of soil fungi.

Table 4. Soil fungal diversity index of each treatment at OTU level.

Sample	sobs	shannon	shannoneven	coverage
YN1_1	257±29A	3.70±0.11A	0.67±0.03A	0.998±0.001
TN1_1	265±30aA	3.46±0.36aA	0.62±0.05aA	0.998±0.000
TN1_2	245±18aA	3.18±0.23aA	0.58±0.03aA	0.998±0.000
TN1_3	271±12aAB	3.13±0.17aB	0.56±0.03aB	0.998±0.000
TN1_4	248±28aA	3.02±0.49aA	0.55±0.08aA	0.998±0.000
WN1_1	245±10bA	3.34±0.18aA	0.61±0.03aA	0.998±0.000
WN1_2	255±10abA	3.21±0.20aA	0.58±0.03aA	0.997±0.000
WN1_3	288±18aA	3.58±0.24aA	0.63±0.04aA	0.997±0.000
WN1_4	269±17abA	3.57±0.20aA	0.64±0.03aA	0.997±0.000
CN1_1	226±22abA	3.31±0.18aA	0.61±0.02aA	0.998±0.000
CN1_2	206±12abB	3.16±0.12aA	0.59±0.02aA	0.998±0.000
CN1_3	250±15aB	3.37±0.05aB	0.61±0.01aAB	0.998±0.000
CN1_4	238±15aA	3.28±0.05aA	0.60±0.01aA	0.998±0.000

As were showed in Table 4, at three stages (rosette, vigorous and maturity stage), the fungal richness of lime nitrogen group was higher than that of the control and showed significant difference at vigorous and maturity stage. The fungal richness of lime nitrogen with mix microbial biocontrol agent group was lower than that of the control at rosette stage, but it began to get higher than that of the control at vigorous stage, and continued to increase significantly at the maturity stage. The fungal diversity of lime nitrogen group and lime nitrogen with mix microbial biocontrol agents was lower than that of the control at rosette stage, but was higher than that of the control at vigorous stage, and kept that trend at the maturity stage. Compared with the control group, the fungal evenness of lime nitrogen group and lime nitrogen with microbial biocontrol agent group was lower at rosette stage,

IOP Conf. Series: Earth and Environmental Science 1087 (2022) 012059 doi:10.1088/1755-1315/1087/1/012059

higher at vigorous stage, and then lower at maturity stage. The richness and diversity in lime nitrogen group were higher than that in lime nitrogen with microbial biocontrol agent group of the three stages.

Except that the richness of fungi in the lime nitrogen group at vigorous stage was significantly higher than that in lime nitrogen with mix microbial biocontrol agent group, the fungi richness and diversity between lime nitrogen group and lime nitrogen with mix microbial biocontrol agent group had almost no significant difference.

4.4. Effects on microbial community composition of fungi

The top 15 fungal genera with the mean relative abundance were selected, and the results were shown in Figure 2. This article focuses on the differences of microbial community among lime nitrogen group, lime nitrogen with mix microbial biocontrol agent group and the control, so the treatment 1_2 were not shown in the figure.

At rosette stage, the relative abundance of *Chaetomium*, *Aspergillus*, *Solicoccozyma*, *Fusarium*, *unclassified_f_Chaetomiaceae*, *unclassified_f_Hypocreaceae*, *unclassified_c_Sordariomycetes*, *unclassified_p_Ascomycota* of lime nitrogen group and lime nitrogen with mix microbial biocontrol agents group was higher than that of the control group and *unclassified_f_Chaetomiaceae* showed significant difference ($p<0.05$), while the relative abundance of *Boeremia*, *Marine_Choanoflagellates_1*, *norank_o_pleosporales* was lower than that of the control group.

At vigorous stage, compared with the control, the relative abundance of *Chaetomium*, *Solicoccozyma*, *unclassified_f_Chaetomiaceae*, *Boeremia*, *unclassified_k_Fungi*, *Naganishia*, *norank_p_Mucoromycota* of lime nitrogen group and lime nitrogen with mix microbial biocontrol agents group increased. Among them, compared with the control, the relative abundance of *Chaetomium* in lime nitrogen group increased significantly; the relative abundance of *Solicoccozyma* in lime nitrogen with mix microbial biocontrol agents group increased significantly. In lime nitrogen group and lime nitrogen with mix microbial biocontrol agent group, the relative abundance of *Aspergillus*, *unclassified_f_Aspergillaceae*, *Thysanophora*, *Fusarium*, *unclassified_o_Agaricales*, *unclassified_p_Ascomycota*, *norank_o_pleosporales* were lower than that in control group.

At maturity stage, the relative abundance of *Chaetomium*, *unclassified_c_Agaricomycetes*, *Boeremia*, *Solicoccozyma*, *norank_p_Mucoromycota* in lime nitrogen group were higher than that of the control group *and Solicoccozyma* reached significant difference ($p<0.05$), while the relative abundance of *Aspergillus*, *unclassified_o_Agaricales*, *Tolypocladium*, *norank_o_pleosporales*, *Thysanophora*, *unclassified_c_Sordariomycetes*, *unclassified_ f_Hypocreaceae* were lower than that of the control group. The relative abundance of *Chaetomium*, *Boeremia*, *unclassified_f_Hypocreaceae*, *norank_o_pleosporales*, *unclassified_p_Ascomycota*, *Solicoccozyma*, *Fusarium*,in lime nitrogen with mix microbial biocontrol agents group were higher than that of the control group *and Chaetomium* reached significant difference ($p<0.05$), while the relative abundance of *Aspergillus*, *unclassified_f_Aspergillaceae*, *unclassified_o_Agaricales*, *Tolypocladium*, *unclassified_c_Agaricomycetes*, *Metarhizium*, *unclassified_c_Sordariomycetes* were lower than that of the control group.

The group containing lime nitrogen increased the relative abundance of *Chaetomium* and *Solicoccozyma* in soil of the three periods.

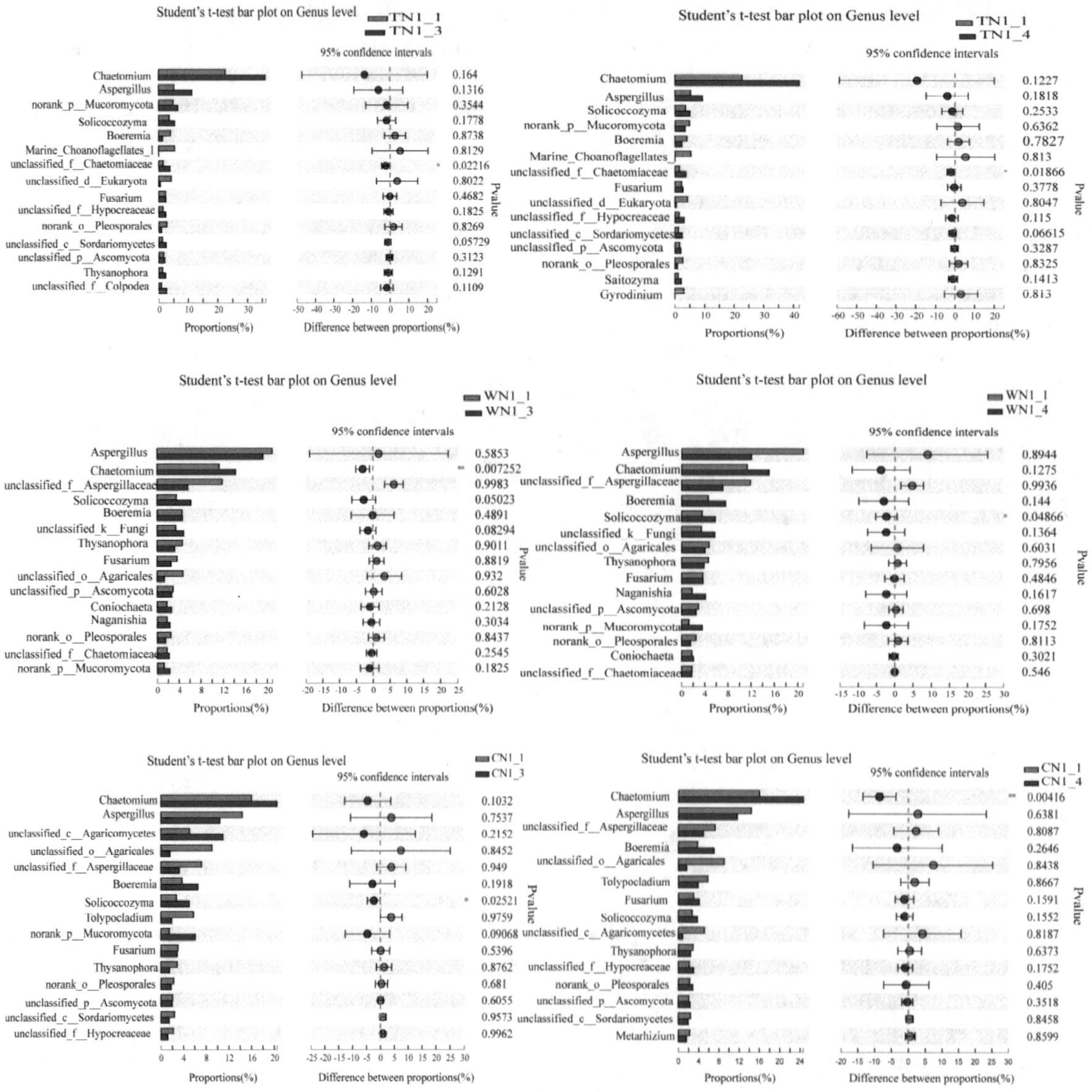

Figure 2. Species difference analysis base on student's t-test bar plot on fungi genus level of different treatments. TN: rosette stage; WN: vigorous stage; CN: Maturity stage; * $0.01 < P \leq 0.05$, ** $0.001 < P \leq 0.01$, *** $P \leq 0.001$.

4.5. Correlation analysis between environmental factors and microbial communities

Cluster analysis was conducted on the soil physicochemical properties with the 15 most abundant fungi taxa. The correlation between environmental factors and microbial communities was irregular and complicated. The results were shown in Figure 3.

Fe and Mn have a consistent correlation to the top 15 fungal genera. AFe had a significantly negative correlation with *Boeremia*, while AK and protease activity had a significantly positive correlation with *Boeremia*. AFe and AMn had a significantly positive correlation with *Tolypocladium* and *Metarhizium*, while OM, AP and enzyme activity had a negative correlation with the two generas. AZn, AN and urease activity had a significantly positive correlation with *Chaetomium*. AZn, pH, AK and urease activity had significantly positive correlation with *unclassified_p_Ascomycota*. Invertase activity had significant positive correlation with *Solicoccozyma*.

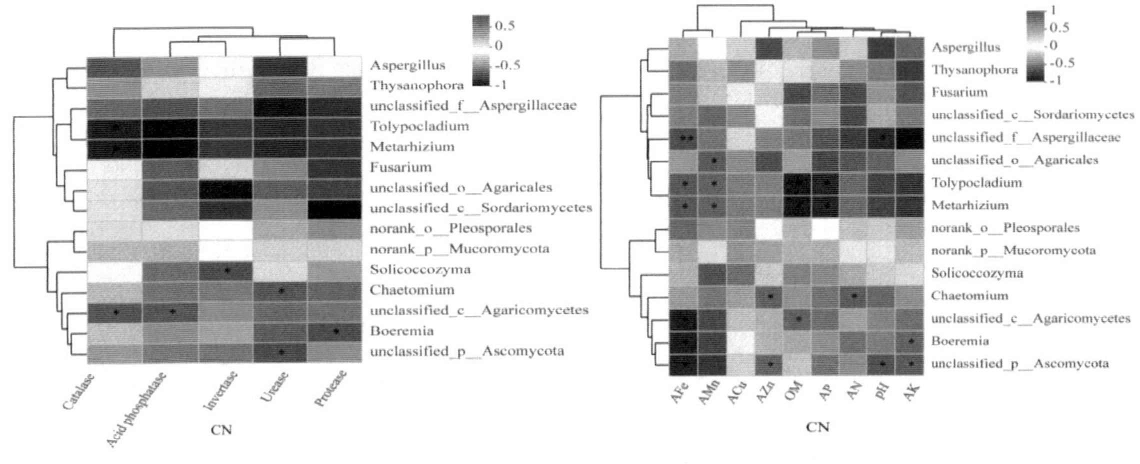

Figure 3. Spearman Correlation Heatmap between fungi and environmental factors of maturity stages on the level of genus. * $0.01 < P \leq 0.05$, **$0.001 < P \leq 0.01$, *** $P \leq 0.001$. CN: Maturity stage.

4.6. Influence of lime nitrogen and lime nitrogen with mix microbial biocontrol agents on the economic characteristics of tobacco plants

The average values of Tobacco leaf output, high-class-leaf proportion, economic output were shown in Table 5. It can be seen from Table 5 that compared with the control, the total output, yield, output value, and proportion of high-class-leaf for lime nitrogen group flue-cured tobacco increased by 30.95%, 30.71%, 36.88%, 14.55% respectively; compared with the control, these indexes in lime nitrogen with mix microbial biocontrol agents group increased by 109.52%, 57.14%, 64.77%, 5.29% respectively. For lime nitrogen with mix microbial biocontrol agent group, the values of total output, yield, output increased by 60%, 20.22%, 20.38% than that of lime nitrogen group, while, the proportion of high-class-leaf decreased by 8.09% than that of lime nitrogen group. That is, lime nitrogen group and lime nitrogen with microbial biocontrol agents group made the economical properties of flue-cured tobacco better and the combination of mix microbial biocontrol agents influenced the tobacco leaf quality.

Table 5. Economic Characteristics of flue-cured tobacco in different treatments at maturity stage.

Treatment	Total output (kg)	Yield per $667m^2$ ($kg/667m^2$)	Output value per $667m^2$ ($yuan/667m^2$)	high-class-leaf proportion（%）
CN1_1	42	140	4238	80.95
CN1_2	49	163	4915	87.76
CN1_3	55	183	5801	92.73
CN1_4	88	220	6983	85.23

5. Discussions

Soil microorganisms play an important role in soil fertility, maintaining the stability of soil ecosystems and the health of soil systems, contributing to the biochemical processes in the soil such as the formation of humus and organic matter, as well as nutrient transformation and cycling [27-29]. Soil environmental factors, including physicochemical properties and enzyme activities, play a key role in soil, which are closely correlated with soil fungi and plant growth [30-31]. As a multifunctional soil conditioner, lime nitrogen has been proved to alleviate plant continuous cropping obstacles and inhibit soil borne diseases. Pretransplanting stage, rosette stage, vigorous stage, and maturity stage are four major stages on the growth period of the tobacco plants [26]. Therefore, it is of great significance to study the effect of lime nitrogen on soil fungal community and the correlation between environmental factors and fungi at these stages.

This research showed the lime nitrogen group and the lime nitrogen with mix microbial biocontrol agents group promoted the release of AN, AP, AK, ACu, AZn during vigorous and maturity stage, this is probably due to the 20% nitrogen and 38% calcium content of the lime nitrogen used, which prolonged the effective supply period of nutrients, this shows that lime nitrogen can be used as a slow-release fertilizer, which is consistent with previous studies by scholars [32].The lime nitrogen group and lime nitrogen with mix microbial biocontrol agents group greatly increased the OM and pH of the tobacco soil at three stages, showing that the application of lime nitrogen promoted the decomposition of organic matter, which was consistent with the study by Zhang et al [33]. Organic matter is converted into carbon dioxide, water, ammonia and mineral nutrients (simple compounds or ions such as phosphorus, sulfur, potassium, calcium, magnesium) through mineralization under the action of enzymes, etc., and at the same time releases energy to provide nutrients for plants and soil microorganisms. That the formation of the intermediate products of lime-nitrogen hydrolysis $Ca(OH)_2$ and NH_3 is the main reason for the increase in soil pH, which may be the reason for the increase in soil pH in this study [12]. The content of AFe, AMn of the lime nitrogen group and lime nitrogen with mix microbial biocontrol agent group was significantly lower than the control at rosette stage, slightly higher than the control at vigorous stage, and greatly lower than the control at maturity stage. This may be because the Fe and Mn elements in the soil form hydroxides and they participate in the conversion of the intermediate product of lime nitrogen into urea [12].

The previous research showed, lime nitrogen had a significant effect in improving soil enzyme activity [34]. This research also showed the urease, acid phosphatase, protease, and invertase activity of the soil in the group containing lime nitrogen were all higher than that of the control group at maturity stage, and urease activity had reached a significant difference, which was mainly because the nitrogen element contained in lime nitrogen provided a large amount of substrate for the enzymatic reaction of soil urease, which stimulated soil urease activity. At the same time, nitrogen fertilizer improved the nitrogen nutrition of soil microorganisms to promote more soil microorganism to produce more urease [35]. Soil enzyme activity, as a powerful indicator for evaluating soil quality and microbial metabolic activity [36], mediates microbial acquisition of nutrients from organic matter [37], and is an important participant in ecosystem material cycle and energy flow. Urease and phosphatase activity were respectively involved in the conversion and absorption of nitrogen and phosphorus in the soil. The increase of urease and acid phosphatase activity might be the reason for the increase in the content of AN and AP during this period. There is a positive correlation between OM and protease

activity [38]. It might be the increase of OM leading to the increase in protease activity. Catalase activity could catalyze the decomposition of accumulated peroxides in the soil, which was closely related to the content of OM. So, the decrease of catalyze activity might be due to the increase in the content of OM during this period in the lime nitrogen group. This study showed that lime nitrogen group and the lime nitrogen with mix microbial biocontrol agent group can inhibit catalase activity. It may be that cyanamide produced by the decomposition of lime nitrogen inhibits the activity of catalase [39]. The study showed that the decrease of catalase activity in the soil will lead to an increase in hydrogen peroxide content will break crop dormancy [40]. Research showed that soil with high urease and phosphatase activity was conducive to the growth of tobacco, and invertase activity had a better effect on promoting tobacco leaf opening [41]. Therefore, the increase in urease and acid phosphatase activity might be one of the reasons for the increase in flue-cured economic characteristics of the groups containing lime nitrogen. This showed that the groups containing lime-nitrogen all increased the soil's ability to transform and absorb nitrogen and phosphorus, increased the level of soil fertility, and thereby improved the economic characteristics of flue-cured tobacco.

Compared with lime nitrogen group, lime nitrogen with mix microbial biocontrol agent group significantly increased the urease activity of the soil, so the content of AN in the soil increased, which indicated that the combined application of mix microbial biocontrol agents was more helpful to the conversion of soil nitrogen. But for lime nitrogen with mix microbial biocontrol agents, the high-class-leaf proportion of flue-cured tobacco was lower than that of lime nitrogen group. It was speculated that the accumulation of nitrogen affected the quality of tobacco leaves, which was consistent with the conclusion of Chen [41]. This study showed that lime nitrogen group and lime nitrogen with mix microbial biocontrol agents group increased the richness and diversity of fungi at vigorous and maturity stages, and the richness reached a significant difference at maturity stage. This showed that both lime nitrogen and lime nitrogen with mix microbial biocontrol agents group improved the community structure of soil fungi in the middle and late stages of tobacco plant growth, and made the fungal community structure more stable. It may be because certain fungi can use the N element provided by lime nitrogen as a nitrogen source for better growth and reproduction.

This study showed that the relative abundance of *Aspergillus* in lime nitrogen group and lime nitrogen with mix microbial biocontrol agent group were higher than that of the control group at rosette stage, but lower than that of the control group at vigorous and maturity stage. The previous study showed that *Aspergillus* are benefical to fungi, which could participate in the process of soil phosphorus dissolution and potassium dissolution [15]. and *Aspergillus* can produce secondary metabolites with biological activity, which can be used as broad-spectrum biocontrol fungi to control a variety of plant pathogenic [42]. However, the content AP in the two treatments was higher than that of the control group at vigorous and maturity stage. It is speculated that the application of lime nitrogen and lime nitrogen mix microbial biocontrol agents group made *Aspergillus* participate in the transformation of phosphorus, so the relative abundance of *Aspergillus* decreased and the content of AP increased. It had been found that some *Chaetomium* could prevent fusarium wilt, damping-off of grain seedlings and sugarcane, and could reduce the incidence of tomato and fusarium wilt caused by *Fusarium oxysporum*, apple spot disease caused by Venturia, populus root rot caused by fusarium [43-44]. At three stages, the relative abundance of *Chaetomium* of lime nitrogen group and lime nitrogen with mix microbial biocontrol agent group were all higher than that of the control group. At maturity stage, AZn, AN and urease activity had a significantly positive correlation with *Chaetomium*, indicating that this genus might participate in the process of soil Zn dissolution and nitrogen cycle. The increase in AZn, AN and urease activity in the lime nitrogen group and the lime nitrogen with mix microbial biocontrol agent group may be the reason for the increase in the relative abundance of the beneficial fungi *Chaetomium* at this stage. Research by Marcela Sarabia showed that in combination with mineral p fertilization, the application of *Solicoccozyma aeria* increased the maize shoot dry weight by 27% compared to the similar treatment without *solicoccozyma aeria* inoculation [45]. At three stages, the relative abundance of *Solicoccozyma* of lime nitrogen group and lime nitrogen with mix microbial biocontrol agent group were higher than that of the control group. Therefore, the increase in the relative abundance of *Solicoccozyma* may increase the dissolve phosphorus ability of soil. This may be the reason why the content of soil AP in vigorous and mature stage were higher than that in the control group. Phosphorus was an important life element for flue-cured tobacco and a composition of organic compounds in tobacco leave. So, the nutrient status of phosphorus directly affected the yield and quality of flue-cured tobacco. Therefore, the increase of AP content at vigorous and maturity stages might be one of the reasons why the economic properties of flue-cured tobacco in the treatment group containing lime nitrogen were higher than that in the control group. At maturity stage, invertase activity had significant positive correlation with *Solicoccozyma*. The increase in invertase activity at maturity stage may be the reason for the increase in the relative abundance of the beneficial fungi *Solicoccozyma* during this period. The relative abundance of *norank_o_pleosporales* in lime nitrogen with mix microbial biocontrol agents were higher than that in the control group. The research showed that *Pleosporales* is a biocontrol fungus of cereal cyst nematode [46]. In addition, it is worth noting that the relative abundance of *Boeremia* in the lime nitrogen group and the lime nitrogen with microbial biocontrol agent group was higher than that of the control group at maturity period. Research showed that *Boeremia exigua* was the pathogen of fruit spot of okra [47] and *Boeremia strasseri* could induce black stem and rhizomes rot of peppermint [48]. During this period, AFe had a significantly negative correlation with *Boeremia*, while AK and protease activity had a significantly positive correlation with *Boeremia*. The decrease in AFe content, the increase in AK content and protease activity may lead to the increase in the relative abundance of the pathogen *Boeremia*. Therefore, if the AFe content can be appropriately increased, AK and protease activity can be appropriately reduced by some methods, which has certain significance for controlling the diseases caused by *Boeremia*. The study showed that *unclassified_c_Sordariomycetes* were potential pathogens in tobacco soil [49]. The treatment groups containing lime nitrogen increased the relative abundance of *unclassified_c_Sordariomycetes* in the soil at rosette stage and decreased the relative abundance of *unclassified_c_Sordariomycetes* in the maturity stage. It is speculated that lime nitrogen killed some microorganisms when it was added to the soil at rosette stage, and made *unclassified_c_Sordariomycetes* enrich in the root soil. When it came to maturity stage, soil fungi's richness, diversity, physicochemical indicators, and most enzyme activities all increased, and the soil's ability to resist pathogenic microorganism increased, so the growth of pathogenic fungi *unclassified_c_Sordariomycetes* was inhibited. It showed that the treatment group containing lime nitrogen could resist the growth of some pathogenic fungi at maturity stage.

In short, the groups containing lime nitrogen increased the relative abundance of certain beneficial fungi in the soil, such as *Chaetomium*, *Solicoccozyma*, and inhibited certain pathogenic fungi such as *unclassified_c_Sordariomycetes*, which improved the soil's disease resistance and nutrient properties of the soil.

6. Conclusions

The application of lime nitrogen and lime nitrogen with microbial biocontrol agents increased the content of nutrient, trace elements, OM, pH and the richness and diversity of fungi, and the relative abundance of certain beneficial fungi such as Chaetomium, Solicoccozyma in the middle and late stages of tobacco growth, which made the soil microenvironment more stable, and nutrient more adequate. So, it was more beneficial for soil to resist the invasion of pathogenic microorganism and promote the growth of tobacco plants, and thereby improved the economic benefits of flue-cured tobacco. In addition, the combined application of mix microbial biocontrol agents could promote the release of soil available nitrogen and potassium in the middle and late stages of tobacco growth, which made the economic benefits of flue-cured tobacco higher. However, the accumulation of nitrogen reduced the proportion of high-class-leaf tobacco. It was worth noting that the application of lime nitrogen increased the relative abundance of certain pathogenic fungi such as Boeremia. Therefore, if the lime nitrogen could be applied while certain measures could be taken to regulate environmental factors related to Boeremia, it could reduce the relative abundance of pathogenic Boeremia, which might make the lime nitrogen control the tobacco continuous cropping obstacles to achieve a better effect. However, how lime nitrogen and lime nitrogen with mix microbial biocontrol agents could be better used to improve on continuous cropping tobacco soil requires further research.

Funding

This work is supported by the Science and Technology Project of Liangshan Prefecture Company of Sichuan Tobacco Company in China (Grant Nos. LSYC201803).

Availability of raw data

The raw data of Sequencing were deposited at NCBI SRA data base under the accession number PRJNA786694.

Reference

[1] Zhou X and Wu F 2012 Dynamics of the diversity of fungal and Fusarium communities during continuous cropping of cucumber in the greenhouse *FEMS MICROBIOL ECOL* **80** 469-78

[2] Chen S, Qi G, Luo T, Zhang H, Jiang Q, Wang R, & Zhao X 2018 Continuous-cropping tobacco caused variance of chemical properties and structure of bacterial network in soils *Land Degradation & Development* **29(11)** 4106-4120
[3] Ashworth, A. J., Owens, P. R., & Allen, F L 2020 Long-term cropping systems management influences soil strength and nutrient cycling *Geoderma* **361** 114062
[4] She S, Niu J, Zhang C, Xiao Y, Chen W, Dai L., ... & Yin H 2017 Significant relationship between soil bacterial community structure and incidence of bacterial wilt disease under continuous cropping system *Archives of microbiology* **199(2)** 267-275
[5] Xiong W, Zhao Q, Zhao J, Xun W, Li R, Zhang R, ... & Shen, Q 2015 Different continuous cropping spans significantly affect microbial community membership and structure in a vanilla-grown soil as revealed by deep pyrosequencing *Microbial ecology* **70(1)** 209-218
[6] Gao L, Wang R, Gao J, Li F, Huang G, Huo G, ... & Shen G 2020 Analysis of the structure of bacterial and fungal communities in disease suppressive and disease conducive tobacco-planting soils in China *Soil Research* **58(1)** 35-40
[7] Kincaid R R 1960 Crop rotation and fallowing in relation to tobacco disease control *The Botanical Review* **26(2)** 261-276
[8] Yun M A O, Feng T I A N, Zhang M, Hongli, C H E N, & Zhong J 2016 Study on Effect of Engran Series of Microbial Organic Fertilizers on Flue-cured Tobacco *Agricultural Science & Technology* **17(3)**
[9] Simujide H, Aorigele C, Wang C J, Lina M, & Manda B 2013 Microbial activities during mesophilic composting of manure and effect of calcium cyanamide addition *International Biodeterioration & Biodegradation* **83** 139-144
[10] Agamalian H 1990 Evaluation of calcium cyanamide for preplant weed control in horticultural crops *Proceedings of the Western Society of Weed Science* **43** 90-95
[11] Bletsos F A 2005 Use of grafting and calcium cyanamide as alternatives to methyl bromide soil fumigation and their effects on growth, yield, quality and fusarium wilt control in melon *Journal of Phytopathology* **153(3)** 155-161
[12] Liu S, Tang L, Shi Q, Yi Z, Zhou W, 2017 Progress on application of calcium cyanamide in agricultural production[J] *Soil and Fertilizer Sciences in China* **(05)** 1-6 (in Chinese)
[13] Bourbos V A, Skoudridakis M. T, Darakis G A, & Koulizakis M 1997 Calcium cyanamide and soil solarization for the control of Fusarium solani f. sp. cucurbitae in greenhouse cucumber *Crop Protection* **16(4)** 383-386
[14] McSpadden Gardener B B 2004 Ecology of Bacillus and Paenibacillus spp. in agricultural systems *Phytopathology* **94(11)** 1252-1258
[15] Wu Q F, Hu H B, & He L M 2020 Effect of Aspergillus niger strain XF-1 on soil nutrients and growth of Amorpha fruticose *Appl Ecol Environ Res* **18** 5079-5092
[16] Sadfi N, Cherif M, Hajlaoui M R, & Boudabbous A 2002 Biological control of the potato tubers dry rot caused by Fusarium roseum var. sambucinum under greenhouse, field and storage conditions using Bacillus spp. Isolates *Journal of Phytopathology* **150(11-12)** 640-648
[17] Gopalakrishnan S, Pande S, Sharma M., Humayun P, Kiran B K, Sandeep D, ... & Rupela O 2011 Evaluation of actinomycete isolates obtained from herbal vermicompost for the biological control of Fusarium wilt of chickpea *Crop Protection* **30(8)** 1070-1078
[18] de Boer M, Bom P, Kindt F, Keurentjes J J, van der Sluis I, Van Loon L C, & Bakker P A 2003 Control of Fusarium wilt of radish by combining Pseudomonas putida strains that have different disease-suppressive mechanisms *Phytopathology* **93(5)** 626-632
[19] Ding Y, Chen Y, Lin Z 2021 Differences in Soil Microbial Community Composition Between Suppressive and Root Rot-Conducive in Tobacco Fields[J] *Current Microbiology* **78(2)** 624-633
[20] Lu R K, 1999 Analysis Methods of Soil and Agricultural Chemistry *Chinese Agriculture and Technology Press Publisher* Beijing (In Chinese)
[21] Emami S, Pourbabaee A A, & Alikhani H A 2013 Effect of paraffin on the urease activity of

soil *Technical journal of engineering and applied sciences* **3(15)** 1526-1529
[22] KIZILKAYA R, İzzet A K Ç A, AŞKIN T, YILMAZ R, Olekhov V, SAMOFALOVA İ, & Mudrykh N 2012 Effect of soil contamination with azadirachtin on dehydrogenase and catalase activity of soil *Eurasian Journal of Soil Science* **1(2)** 98-103
[23] Narasimha G, Sridevi A, Reddy A V S, Banu M T, & Reddy B R 2012 Effect of cotton ginning mill industrial effluents on soil dehydrogenase, phosphatase, amylase and invertase enzyme activities *International journal of agricultural and food science* **2(1)** 1-6
[24] Shahriari F, Higashi T, & Tamura K 2010 Effects of clay addition on soil protease activities in Andosols in the presence of cadmium *Soil Science & Plant Nutrition* **56(4)** 560-569
[25] Dick W A, Cheng L, & Wang P 2000 Soil acid and alkaline phosphatase activity as pH adjustment indicators *Soil Biology and Biochemistry* **32(13)** 1915-1919
[26] Wang Z, Yang Y, Xia Y, Wu T, Zhu J, Yang J, & Li Z 2019 Time-course relationship between environmental factors and microbial diversity in tobacco soil *Scientific reports* **9(1)** 1-11
[27] Zhu J, Ren Z, Huang B, Cao A, Wang Q, Yan D, ... & Li Y 2020 Effects of fumigation with allyl isothiocyanate on soil microbial diversity and community structure of tomato *Journal of Agricultural and Food Chemistry* **68(5)** 1226-1236
[28] Kirk J L, Beaudette L A, Hart M, Moutoglis P, Klironomos J N, Lee H, & Trevors J T 2004 Methods of studying soil microbial diversity *Journal of microbiological methods* **58(2)** 169-188
[29] George E, Marschner H, & Jakobsen I 1995 Role of arbuscular mycorrhizal fungi in uptake of phosphorus and nitrogen from soil *Critical reviews in biotechnology* **15(3-4)** 257-270
[30] Liu S, Wang Z, Niu J, Dang K, Zhang S, Wang S, & Wang Z 2021 Changes in physicochemical properties, enzymatic activities, and the microbial community of soil significantly influence the continuous cropping of Panax quinquefolius L(American ginseng) *Plant and Soil* **463(1)** 427-446
[31] Burke D J, Weintraub M N, Hewins C R, & Kalisz S 2011 Relationship between soil enzyme activities, nutrient cycling and soil fungal communities in a northern hardwood forest *Soil Biology and Biochemistry* **43(4)** 795-803
[32] Ogihara T, Takefuji K 1954 Studies on the inorganic nitrogeneous fertilizer with the characteristics absorbed slowly by plant: massed calcium cyanamide (1) *Journal of the Science of Soil & Manure Japan*, **25** 159-162
[33] Zhang X P, Ning T Y, Yang Y, Sun T, Zhang S M, & Wang B 2015 Effects of different application rates of calcium cyanamide on soil microbial biomass and enzyme activity in cucumber continuous cropping *The Journal of Applied Ecology* **26(10)** 3073-3082
[34] Tian Y, Cao Z, Zhang X, Guo W, Mei X, & Gao L 2009 Changes of soil enzyme activities under different agricultural treatments in greenhouse and its correlation analysis *Plant Nutrition and Fertilizer Science* **15(4)** 857-864
[35] Xiao X, Zhu W, Xiao L, Deng Y, Zhao Y, & Wang J 2013 Suitable water and nitrogen treatment improves soil microbial biomass carbon and nitrogen and enzyme activities of paddy field *Transactions of the Chinese Society of Agricultural Engineering* **29(21)** 91-98
[36] Bell T H, Klironomos J N, & Henry H A 2010 Seasonal responses of extracellular enzyme activity and microbial biomass to warming and nitrogen addition *Soil Science Society of America Journal* **74(3)** 820-828
[37] Sinsabaugh R L, Lauber C L, Weintraub M N, Ahmed B, Allison S D, Crenshaw C, ... & Zeglin L H 2008 Stoichiometry of soil enzyme activity at global scale *Ecology letters* **11(11)** 1252-1264
[38] Guan S Y 1980 Soil enzymes and soil fertility *Chinese Journal of Soil Science* **(06)**41-44 (in Chinese)
[39] Or E, Vilozny I, Eyal Y, & Ogrodovitch A 2000 The transduction of the signal for grape bud dormancy breaking induced by hydrogen cyanamide may involve the SNF-like protein kinase GDBRPK *Plant Molecular Biology* **43(4)** 483-494

[40] Pérez F J, Vergara R, & Rubio S 2008 H_2O_2 is involved in the dormancy-breaking effect of hydrogen cyanamide in grapevine buds *Plant Growth Regulation* **55(2)** 149-155
[41] Chen R, S J, W J 2014 Effect of soil enzyme and edaphon on flue-cured tobacco growth and quality *Acta Tabacaria Sinica* **20(03)** 73-78 (in Chinese)
[42] Kobbe B, Cushman M, Wogan G N, & Demain A 1977 Production and antibacterial activity of malforming C, a toxic metabolite of Aspergillus niger *Applied and Environmental Microbiology* **33(4)** 996-997
[43] Soytong K 2000 Application of a new broad spectrum biological fungicide for environmental plant protection *Advanced Study on Plant Pest Biological Control (Harbin, 2000)*
[44] Walther D, & Gindrat D 1988 Biological control of damping-off of sugar-beet and cotton with Chaetomium globosum or a fluorescent Pseudomonas sp *Canadian Journal of Microbiology* **34(5)** 631-637
[45] Sarabia M, Cazares S, González-Rodríguez A, Mora F, Carreón-Abud Y, & Larsen J 2018 Plant growth promotion traits of rhizosphere yeasts and their response to soil characteristics and crop cycle in maize agroecosystems *Rhizosphere* **6** 67-73
[46] Bao L L, Occurrence and Bio-control of Cereal Cyst Nematode (Heterodera avenae) in Inner Mongolia *Inner Mongolia Agricultural University* (in Chinese)
[47] Zhao Q, Xie X W, Shi Y X, Chai A L, & Li B J 2016 Boeremia leaf and fruit spot of okra caused by Boeremia exigua in China *Canadian Journal of Plant Pathology* **38(3)** 395-399
[48] Zimowska B, Król E D, Furmańczyk A, Abramczyk B, & Okoń S 2018 Molecular characterization of Boeremia strasseri the causal agent of black stems and rhizomes rot of peppermint *Journal of Plant Pathology* **100(1)** 13-24
[49] Ding Y R 2020 Characteristics of Soil Microbial Community in Tobacco Field of Liangshan P refecture and Its Correlation with Environmental Factors *Zhengzhou University*

Interannual variation trend of hail days in hail prone areas in Xinjiang

Hongyun Fan[1], Xi Wang[2,4] and Liping Zhou[3]

[1] Xinjiang Weather Modification Office, Urumqi 830002, China;
[2] Faculty of Humanities and Social Sciences, Macao Polytechnic University, Macao 999078, China;
[3] Urumqi Meteorological Satellite Ground Station, Urumqi 830011, China.

[4] Corresponding author's e-mail: xwang@mpu.edu.mo

Abstract. Hail causes significant agricultural losses in Xinjiang every year, and the study of interannual variation of hail days and its causes can provide a climatic background for hail warning and defense. This study analyzes the trends and abrupt changes in the interannual variability of hail days in the areas with frequent hail using linear trend analysis and abrupt climate change detection methods, and explains the climatic causes of these trends using atmospheric stability. The results show that areas with frequent hails in Xinjiang are located on both sides of the Tianshan Mountains and their adjacent areas, and area with the most to the fewest hail days are in the following order: Zhaosu and Bayinbruck on the Alpine Platform, Aksu region on the south side of the Tianshan Mountains, and Bortala and Kuy-Ma Basin on the north side of the Tianshan Mountains. The interannual variation of hail days in Aksu region, Bortala and Kuy-Ma Basin from 1961 - 2016 shows a linear increasing trend, but the increase is small, only 0.2 - 0.4 d per 10 a, which is contrary to the conclusion of previous studies where the number of hail days in the Chinese region showed a decreasing trend; whereas the Alpine Platforms shows a linear decreasing trend with a larger decrease of 2.5 d per 10 a. The annual number of hail days on the Alpine Platform changed abruptly in 1993 and in the Aksu region in 1978. The linear increase (decrease) in the interannual variation of the warm season modified total index leads to a linear increase (decrease) in the number of hail days on both sides of the Tianshan Mountains and on the Alpine Platform. The trend of interannual variation in the number of hail days in the hail-prone areas in Xinjiang depends on the altitude of the elevation, with a decreasing trend at higher elevations and an increasing trend at lower elevations.

1. Introduction

Hail refers to hard, spherical, cone-shaped, or irregularly shaped solid precipitation with a diameter of more than 2 mm [1]. Hail is a serious agro-meteorological disaster that can cause serious damage to crop during the growing season, resulting in crop yield reduction or crop extinction, resulting in heavy economic losses [2-3]. Oases in Xinjiang are surrounded by mountains and water, mostly adjacent to deserts or surrounded by Gobi. This favorable topography makes the oasis more prone to hail [1]. Our statistics based on the China Meteorological Disaster Yearbook 2004 - 2015 showed that among the 29 provinces and regions, local severe convection (hail, thunderstorm, gale, tornado) caused the most serious crop damage in Hebei, Henan, Inner Mongolia, Heilongjiang. The affected area was 367.2, 327.9, 284.4, 253.7, 253.5 and 209.7 thousand hectares, respectively. Xinjiang is one of the regions with severe hail damage and the most affected area in the northwest of China.

Zhang et. al. studied the climatic characteristics of the number of hail days, the number of hailstorms and hail diameter in China using meteorological station observations [4-9]. The results indicate that hail occurs from May to September in China, concentrated in the late afternoon to evening, and there is a linear decreasing trend in the number of hail days and number of hailstorms. An analysis of the climatic characteristics of hail in Xinjiang by Xinjiang scholars using observations from meteorological stations from 1961 to 1999 indicates that hail in Xinjiang occurs mainly in the western and central mountainous areas, concentrated in the afternoon from April to October, lasting for less than 6 min [10-11]. Previous studies on the climatic characteristics of hail in Xinjiang have two shortcomings: 1) The hail data recorded at weather stations do not fully reflect the actual hailstorm conditions. There are 93 meteorological stations with long time series observation data in Xinjiang, with a relatively sparse spatial distribution. The spatial grid composed of meteorological stations is too large relative to such small and medium scale weather as hail, so they cannot yet fully capture hail weather. Thus, the analysis results using only meteorological station information differ greatly from reality. 2) Zhang et. al. showed the long-term variation of hail days in China overall, but did not show the condition of Xinjiang in detail, while the research of the Xinjiang scholars dated farther back and showed the condition of Xinjiang before 1999. Therefore, there is a great need to carry out research on long-term trends of hail and its climatic causes in Xinjiang by using hailstorm data from meteorological stations, integrated with local hail hazard data.

2. Experimental design

(1) Integration of meteorological station observations with disaster records. The 93 meteorological stations with complete information for 1961 - 2016 in Xinjiang are located on the western side of the Altai Mountains, both sides of the Tianshan Mountains, the northern side of the Kunlun Mountains, and the areas around the Junggar Basin and the Tarim Basin, respectively. Due to their relatively sparse spatial distribution, they cannot fully capture hail weather. The hail hazard records recorded by the Xinjiang Disaster System, however, were able to largely supplement the hail data observed at the meteorological stations. Based on the supplemented data, the analysis results of hail climate characteristics carried out can truly reflect the actual hail conditions. The number of hail days at each station is merged with the hail hazard records, taking out repeats, to obtain a single-station hail day series. Non-repeat merging means that if a weather station recorded hail on a certain day, and hail also occurred in the county or town to which the weather station belongs in the hail hazard record, only 1 hail day is counted at this station. In the day-by-day hail data of 93 meteorological stations in Xinjiang from 1961 - 2016, 2477 disaster records were integrated to achieve the supplementation of meteorological station observation data. The spatial distribution of hail days is statistically analyzed using the integrated data to establish areas with frequent hail.

(2) Spatial and temporal distribution characteristics of the number of hail days. The analysis of the spatial and temporal variation characteristics of the number of hail days was carried out for areas with frequent hail. Quantitatively, the magnitude of increase and decrease in the long-term variation in the number of hail days can be expressed using the slope of the univariate regression equation [12]. Detection of abrupt changes was carried out to understand the abrupt change in the long-term variation in the number of hail days. Detection was performed using the cumulative departure from the average in conjunction with Yamamoto's signal-to-noise ratio [12].

(3) Climatic causes. Atmospheric stability parameters were used to analyze the causes of long-term variability in the number of hail days. Using day-by-day 12:00 (UTC) sounding data for 1980 - 2016 from three sounding stations in areas with frequent hails, the stability parameter was calculated to explain the climatic causes of the increasing (decreasing) trend in the number of hail days based on the long-term trend in the stability parameter and the degree of correlation between the stability parameter and the number of hail days.

3. Results

The statistics of hail days were obtained from 93 meteorological stations in Xinjiang from 1961 to 2016. The hail-prone areas, where hail occurred for more than three days per year, were concentrated in Zhaosu and Bayanbulak on the Alpine Platform in the western part of the Tianshan Mountains, the Aksu region on the southern side of the Tianshan Mountains, and Bortala Mongol Autonomous Prefecture (referred to as Bortala) and the Kuytun-Manas River Basin (referred to as Kuy-Ma Basin) on the northern side of the Tian Shan Mountains. The topography and subsurface have thermal and dynamic effects on hail weather, as pointed out by Lei et al. [13] early on. In mountainous areas with high terrain and complex topography, in lowlands downwind of the plateau, in underlying areas with complex desert and vegetation that have uneven thermal properties at the intersection, and the leeward slopes of high mountains are hail-prone areas. The favorable thermal and dynamic conditions brought by the favorable geographical location make the Alpine Platform, Aksu region, Bortala and Kuy-Ma Basin as the areas where hail weather is concentrated in Xinjiang, we select these four areas as representatives of hail-prone areas in Xinjiang and analyzed the characteristics of the temporal and spatial evolution of hail. The four areas included 22 meteorological stations (Figure 1), among which four were in Bortala (Alashankou, Bole, Wenquan and Jinghe), seven were in the Kuy-Ma Basin (Wusu, Paotai, Mosowan, Shihezi, Shawan, Manas and Hutubi), two were on the Alpine Platforms (Zhaosu and Bayanbulak) and nine were in Aksu region (Wushi, Aksu, Wensu, Baicheng, Xinhe, Xayar, Kuqa, Kalpin and Awat).

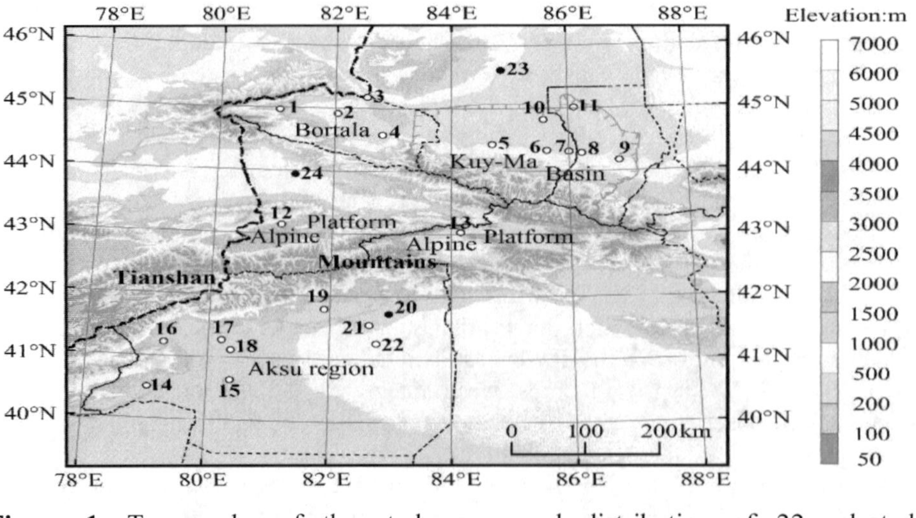

Figure 1. Topography of the study area and distribution of 22 selected meteorological stations and 2 radiosonde stations. The numbers 1 to 24 respectively represent the meteorological station: Wenquan, Bole, Alashankou, Jinghe, Wusu, Shawan, Shihezi, Manas, Hutubi, Paotai, Mosowan, Zhaosu, Bayanbulak, Kalpin, Awat, Wushi, Wensu, Aksu, Baicheng, Kuqa, Xinhe, Xayar, Karamay, Yining. ○ represents the meteorological station, and ● represents radiosonde station.

3.1. Spatial distribution and seasonal variation

The annual number of hail days at a single station on the Alpine Platform is above 9 d, with an average of 13.7 d. This is the region with the most hail in Xinjiang (Figure 2a), followed by the Aksu region, with a number of hail days between 2.1 and 3.2 d, averaging 2.6 d. Bortala has a number of hail days between 0.4 and 5.3 d, averaging 2.2 d. The Kuma basin has a number of hail days between 0.3 and 3.1 d, averaging 1.2 d. Numerous studies by scholars have concluded that the number of hail days is higher at high altitudes in eastern China [5-9] and higher in mountainous areas than in plains, indicating that in terms of spatial distribution, Xinjiang, like eastern China, also has a higher number of hail days in mountainous areas than in plains, and a higher number at high altitudes.

As seen from the multi-year average monthly hail day distribution in Figure 2b, hail in the four hail-prone regions is concentrated in April - September, and the hail days in May to August account for 80% - 92% of the year. Hail occurred the most in July in the Bortala and Kuy-Ma Basin on the northern side of the Tianshan Mountains, while it is highest in June on the Alpine Platforms and in the Aksu region on the southern side of the Tianshan Mountains. In northern China, hail occurs during April - September, with the greatest frequency in July, while south of the Yangtze River hail mostly occurs in February - April, with the greatest frequency in April [5-9]. The months of where hail is concentrated in the Tianshan Mountains and adjacent areas of Xinjiang are consistent with those in northern China. The months with the most hailfall are also consistent with those in northern China, but significantly different from southern China.

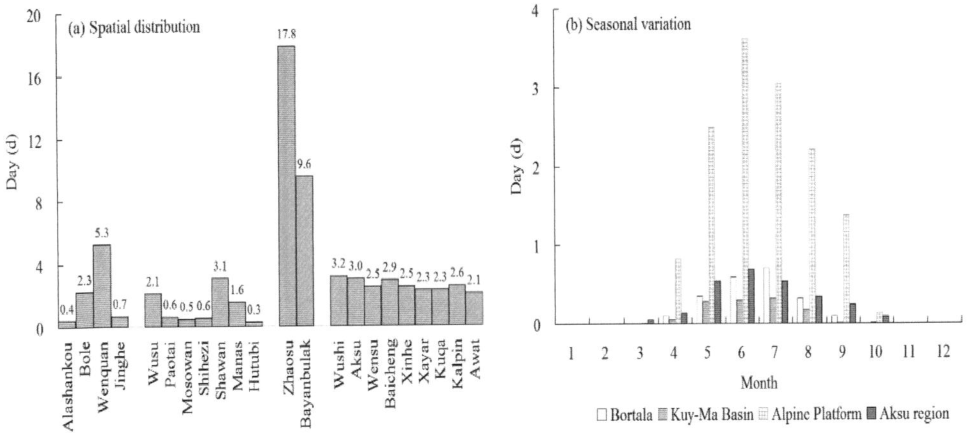

Figure 2. Spatial distribution and seasonal variation of hail in hail prone areas in Xinjiang.

3.2. Interannual and interdecadal changes

Figure 3 shows the interannual variation of hail days in four hail-prone regions of Xinjiang from 1961 - 2016. From the figure, it can be seen that the interannual variation of hail days on both sides of Tianshan and the Alpine Platform shows an opposite trend. The annual number of hail days on the Alpine Platform shows a significant linear decreasing trend, decreasing by 2.5 d per 10 a. The Aksu region, Bortala, and Kuy-Ma Basin on both sides of the Tianshan Mountains show significant linear increasing trends, increasing by 0.4 d, 0.2 d, and 0.3 d per 10 a, respectively. It can be seen that the annual number of hail days on the Alpine Platform, where hail occurs the most, is decreasing very significantly, while that in the other three hail-prone areas are experiencing tiny increase.

It is also evident from the changes in the average number of hail days in each era (Table 1) that the average number of hail days on the Alpine Platform has been decreasing since the 1960s; the Aksu region has experienced a phased change of increase → decrease → increase, Bortala has experienced a phased change of decrease → increase → decrease → increase, and the Kuy-Ma Basin has experienced a phase change of increase → decrease, with the increase trend dominating in these three regions. Previous studies have shown that there is a clear trend of decreasing hail days in most areas of China, especially after the mid-1990s [5]. This is consistent with the interannual trend in the number of hail days on the Alpine Platform in Xinjiang, but opposite to the trend on both sides of the Tianshan Mountains.

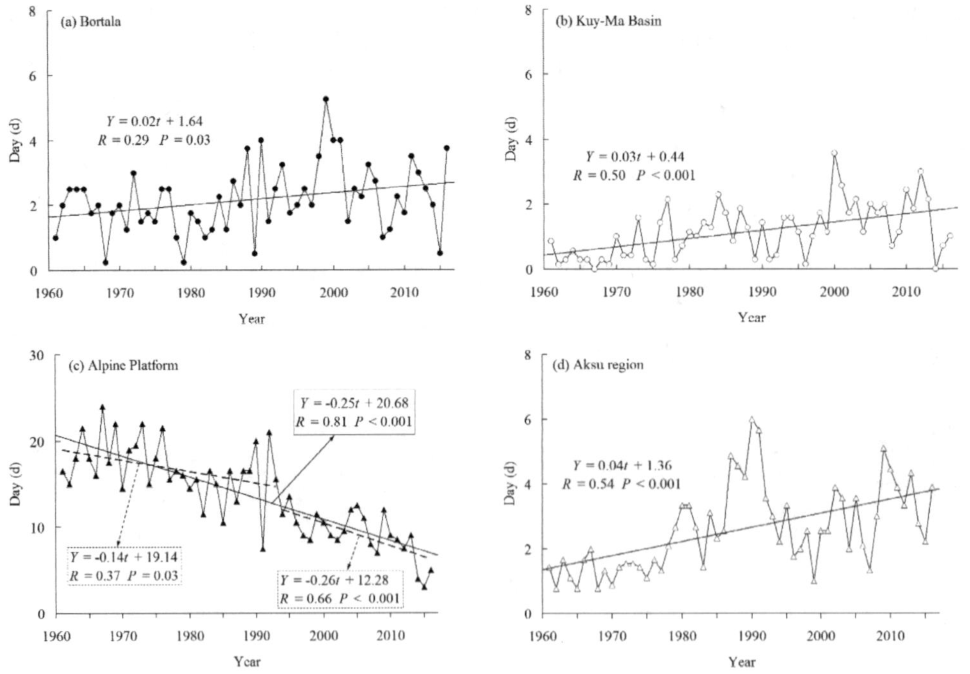

Figure 3. Interannual variation of hail in hail prone areas of Xinjiang from 1961 to 2016.

Table 1. Interdecadal variation of hail in hail prone areas of Xinjiang from 1961 to 2016 (unit: d).

	Bortala	Kuy-Ma Basin	Alpine Platform	Aksu Region
1961 - 1970	1.8	0.4	18.3	1.2
1971 - 1980	1.7	0.9	17.8	1.8
1981 - 1990	2.0	1.3	15.2	3.5
1991 - 2000	2.8	1.3	11.9	2.8
2001 - 2010	2.3	1.8	9.9	3.2
2010 - 2016	2.5	1.5	6.2	3.4
1961 - 2016	2.2	1.2	13.7	2.6

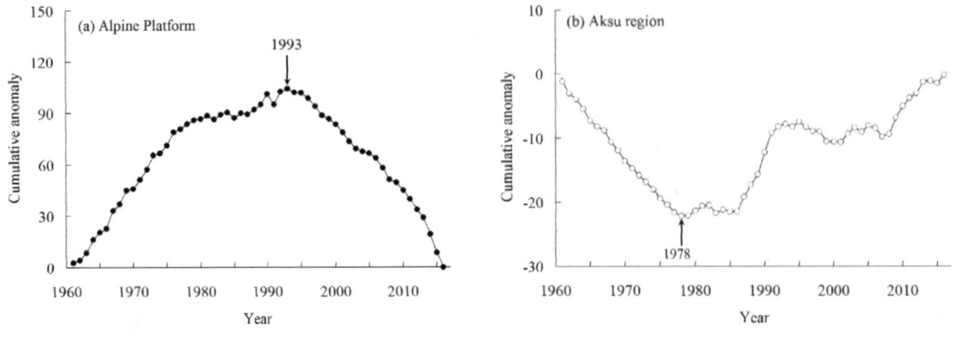

Figure 4. Cumulative anomaly curve of hail in Alpine Platform and Aksu region from 1961 to 2016.

The analysis of the cumulative departure from the average curve of the annual hail days in the four hail-prone areas shows that abrupt climate change only occurred in the Alpine Platform and the Aksu region, in 1993 (S/N 1.25) and in 1978 (S/N 1.15), respectively (Figure 4). The average number of hail days before and after the abrupt change on the Alpine Platform were 16.8 d and 9.2 d, respectively, showing a significant difference. The magnitude of decreasing hail days after the abrupt change also increased, as shown by the slope of the later section being 1.86 times that of the earlier section (Figure 3c). The average number of hail days in the Aksu region was 1.4 d before the abrupt change and 3.2 d after, showing a significant increase.

3.3. Climatic causes of long-term change trend

The variation trend of hail days can be explained by atmospheric stability parameters. The Modified Total Index (ST) reflects the stability of the atmosphere and has a unit of °C. Larger values of ST represent a more unstable atmosphere and are more conducive to hailstorms. The Karamay sounding station can be used to represent the Bortala and Kuy-Ma Basin, and the Yining and Kuqa sounding stations can be used to represent the Alpine Platform and the Aksu region, respectively.

There is a significant linear increasing trend of ST in the warm season (April - September) in the areas on both sides of Tianshan Mountain, and a significant linear decreasing trend of ST in the warm season on the Alpine Platform (Figure 5), which shows that the trends of ST on both sides of Tianshan Mountain and the Alpine Platform are consistent with the trend of the number of hail days. In addition, the correlation coefficients between ST and the number of hail days in the represented areas at all three sounding stations are positive, and the significance level of the correlation coefficients is P < 0.15. Thus, ST is a climatic influence factor of the interannual trend of hail days.

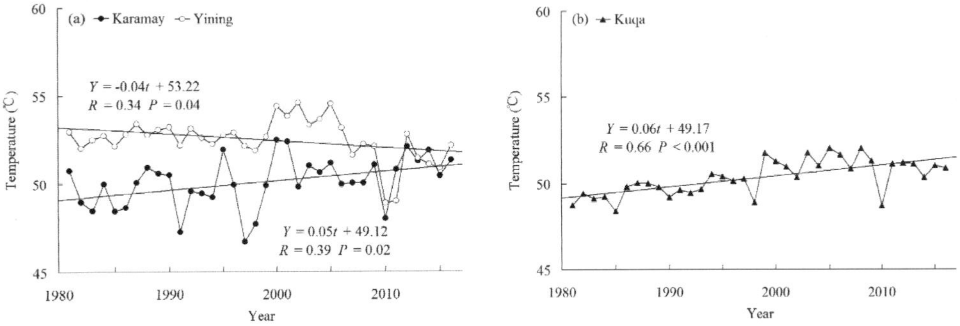

Figure 5. Interannual change of corrected total index from April to September.

4. Conclusions

In terms of spatial distribution, the hail-prone areas in Xinjiang with more than 3 d of hail per year are in Bortala and the Kuy-Ma Basin on the northern side of the Tianshan Mountains, Zhaosu and Bayinbruck on the Alpine Platform of the Tianshan Mountains, and Aksu region on the southern side of the Tianshan Mountains, with the highest number of hail days per year in Zhaosu (17.8 d), followed by Bayinbruck (9.6 d). In terms of monthly distribution, hail days in May - August in hail-prone areas account for 80% - 92% of the year, with the most hail falling in July in the Bortala and Kuy-Ma Basin, and in June on the Alpine Platform and Aksu region.

In terms of interannual variability, the annual number of hail days tends to increase on both sides of the Tianshan Mountains and decrease on the Alpine Platforms. In the Aksu region, Bortala and the Kuy-Ma Basin, although there is an increasing trend, the increase is small, only 0.2 - 0.4 d per 10 a. There is a significant decreasing trend on the Alpine Platform, with a large decrease of 2.5 d per 10 a. The number of annual hail days on the Alpine Platform changed abruptly in 1993 and in the Aksu region in 1978. The average annual number of days on the Alpine Platform before and after the abrupt change was 16.8 d and 9.2 d, respectively, decreasing by 45% after the abrupt change relative to before. At the same time, magnitude of the trend of decreasing hail days increased after the year in which abrupt change occurred. In the Aksu region, the average annual number of days before and after the abrupt change was 1.4 d and 3.2 d, respectively, and the number of hail days increased significantly after the abrupt change. Previous studies pointed out a significant decreasing trend of hail days in most of China, which is consistent with that on the Alpine Platforms but opposite to that on both sides of the Tianshan Mountains. The modified total index (ST) of the warm season can represent well the trend of hail days on both sides of Tianshan and on the Alpine Platform. Thus, ST is the main climatic factor determining the trend of hail days.

Acknowledgments
This work was supported by Guiding Planning Project of Xinjiang Meteorological Bureau (Project Name: Variation Trend and Causes of Hail Frequency and Duration in Xinjiang), and by China Special Fund for Meteorological Research in the Public Interest (Grant Number: GYHY201306047).

References

[1] Wang Y, Wang S G, Wang X, Ma Y 2019 Temporal and spatial distribution and hazard assessment of hail disasters during crop growth period in Xinjiang Transactions of the Chinese Society of Agricultural Engineering *Transactions of the CSAE* 35
[2] Raupach T H, Martius O, Allen J T, Kunz M, Zhang Q 2021 The effects of climate change on hailstorms Nature Reviews *Earth & Environment* 2
[3] Zhang Y, Zhong J I, Xue B, Wang P 2021 A novel fusion forecast model for hail weather in plateau areas based on machine learning *Acta Meteorologica Sinica* 35
[4] Xie B G, Zhang Q H, Wang Y Q 2008 Trends in hail in China during 1960-2005 *Geophysical Research Letters* 35
[5] Zhang C X, Zhang Q H, Wang Y Q 2008 Climatology of Hail in China: 1961-2005 *Journal of Applied Meteorology and Climatology* 47
[6] Xie B G, Zhang Q H, Wang Y Q 2010 Observed characteristics of hail size in four regions in China during 1980-2005 *Journal of Climate* 23
[7] Li M, Zhang Q H, Zhang F 2016 Hail day frequency trends and associated atmospheric circulation patterns over China during 1960-2012 *Journal of Climatology* 29
[8] Li M, Zhang D L, Sun J, Zhang Q H 2018 A statistical analysis of hail events and their environmental conditions in China during 2008-2015 *Journal of Applied Meteorology and Climatology* 57
[9] Li X F, Zhang Q H, Zou T, Lin J P, Kong H O, Ren Z H 2018 Climatology of hail frequency and size in China 1980-2015 *Journal of Applied Meteorology and Climatology* 57
[10] Ma Y, Wang X, Zhao B K, Feng Z W 2002 Spatial and temporal statiscal character of hail in Xinjiang *Bimonthly of Xinjiang Meteorology* 25
[11] Chen H W, Ma Y, Wang X, Yang X L 2003 Climatic characteristics of hail weather in Xinjiang *Meteorological Monthly* 29
[12] Shi N, Zhu Q G 1996 An abrupt change in the intensity of the East Asian summer monsoon index and its relationship with temperature and precipitation over East China International *Journal of Climatology* 16
[13] Lei Y S, Wu B J, Wu Z H 1978 Introduction to hail Beijing *Science Press*

IOP Conf. Series: Earth and Environmental Science 1087 (2022) 012061 doi:10.1088/1755-1315/1087/1/012061

Investigation and characterization of karst landform with GPR: a case study in Shandong, China

Qingyi Yang[1], Renjie Li[1], Hao Liu[1], Wenbo Zhang[2,3]

[1]Shandong Electric Power Engineering Consulting Institute Corporation, Jinan, Shandong, China 250013;
[2]Institute of Geophysics & Geomatics, China University of Geosciences, Wuhan, Hubei, China 430074

[3]Corresponding author's e-mail: zhwb_000@126.com

Abstract. A substation is an important infrastructure for city operation and people's life. The proposed site, 37 km northeast of Zaozhuang, Shandong province, is located on top of Cambrian limestone, most of which is directly exposed or covered by thinner Quaternary silty clay. The Karst processes on the surface contribute to the dissolution of gullies and limestone fissures in the surface topography. Directly the evidence shows that the study area is karst cave or the possibility of larger crack, the potential risk impact on subsequent substation the safety of the pile foundation installation construction. Thus led to the ground penetrating radar (GPR) investigation on the bedrock of karst development. The aim of this study was to characterize the development of karst caves in the shallow surface and delineate their locations. The 80 MHz shielding antenna was used to study and characterization of the karst features. The results of GPR method show that the application of a low-frequency antenna can satisfy a certain detection depth and well characterize the development of underground karst, which also proves that GPR method is very practical in the detection of karst landform and can be applied to identify similar geomorphic features.

1. Introduction

With the continuous expansion of the world's population, the accelerated consumption of resources, and the expansion of land, the karst landscape, as a widespread geological phenomenon, also affects the safety of land use, and decision-makers gradually realize the importance of maintaining the balance between human society and nature. The term karst landscape is derived from a carbonate outcrop region between northeastern Italy and Slovenia and refers to the unique geomorphic and hydrological features formed by a series of extensive, physico-chemical interactions on soluble rocks over millions of years on Earth [1]. The cavities and fissures formed in the surface and sub-surface of the karst process cause great safety risks to the surface buildings and human engineering activities. Therefore, to understand the development characteristics of sub-surface karst, the spatial position and geometric form of caves or fissures have attracted extensive attention from karst scientists, geomorphologists, and engineers. Geophysical methods can quickly image the underground geological structure and formation in the form of points, lines, and planes. Different methods and technologies can be used for different topographic conditions, detection purposes, and physical property differences. Among many geophysical exploration methods, Ground penetrating radar (GPR) has rapidly become one of the most widely used in the karst region over the past decade [2, 3, 4, 5]. The method solved a large number of problems related to the environment, hydrogeology, and engineering geology.

Although there are limitations to the depth of the GPR method compared to other geophysical methods, including when the bedrock is covered by loose sediment that greatly limits the depth of the GPR, the method can also provide rich near-surface information that can help determine the spatial location of underground caves, channels, and fractures. Many cases of successful detection of shallow karst caves and karst aquifers have been reported [6, 7, 8]. GPR has proven to be an excellent technique for detecting and locating near-surface cavities.

2. Geological setting of the study area

The study area is located in Zaozhuang, China (Figure 1a). The research area's terrain is typically high in the northeast and slopes southwestward. The landform is a hilly area formed by denudation, with a relatively slow slope. Cambrian limestone is directly exposed in some areas (Figure 1b&c), while other areas are covered by thin Quaternary sediments. Quaternary strata are mainly composed of Holocene diluvium, eluvium and artificial filled soil, which lithology is silty clay, gravel, and plain fill. It is found that the dielectric coefficient varies greatly (generally 5~40), which provides favorable conditions for GPR exploration.

3. Methodology

GPR method is based on the electromagnetic wave reflection, refraction, attenuation and other physical phenomenon in underground medium. Because of its rapid, non-destructive, rich information, this method has been developed and applied rapidly in recent years. It is an active source method for transmitting high-frequency short pulse electromagnetic waves (the dominant frequency is generally tens of megahertz to gigahertz) from the transmitting antenna to the underground. The electromagnetic wave returns to the ground through the subsurface layer or target body with different electromagnetic properties and is received by the receiving antenna.

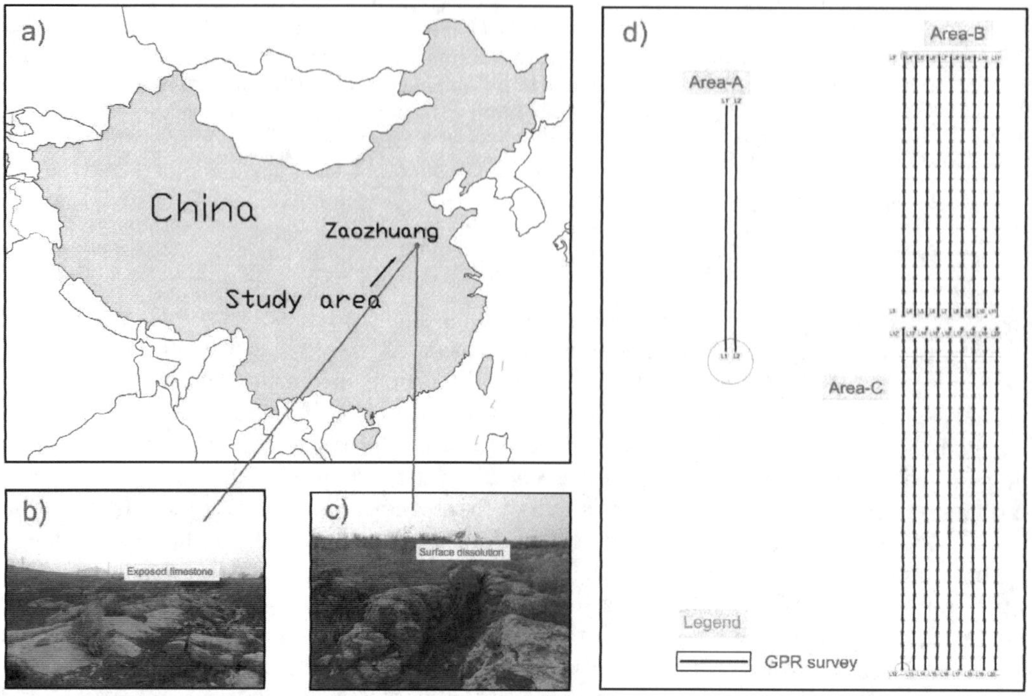

Figure 1. Arrangement of the GPR survey and the surrounding environment.

a) The location of the study area; b) The exposed limestone; c) Surface dissolution.

Figure 1d shows that GPR survey lines are mainly arranged in three areas, namely Area-A, Area-B, Area-C respectively. After fully considering the electrical, magnetic, and dielectric properties of the formation in the measurement area, we chose the 80 MHz antenna for this detection task. The GPR system is a MALA radar produced in Sweden. This survey adopted the point measurement method of self-excitation and self-collection, with the point distance of 0.1 m, the encryption of some measuring points of 0.05 m, and the time window length of 500 ns. The time-depth conversion speed is set as 0.16 ns/m. It should be noted that the complexity of the underground karst system and the difference in local materials will affect the propagation speed and attenuation of electromagnetic waves underground. The 0.16 ns/m considered in this study is an average estimate of the real situation of the underground medium.

Raw radar data was processed using Reflexw software. It is employed by many engineers and researchers as a professional software for processing GPR data. There are usually many interferences in the original data. The purpose of data processing is to enhance the effective signal and suppress the interference signal, so as to facilitate the abnormal judgment and diagram interpretation. It mainly includes the following steps: 1 Affected by unstable factors such as the acquisition environment and instrument performance, the collected data will produce systematic errors, which can be eliminated by the 'subtract-DC-shift' function. 2. Static Correction. In the field data acquisition, due to the gap between the transceiver antenna and the ground, it is necessary to eliminate the surface wave. 3. Gain. When the electromagnetic wave propagates underground, its signal strength will decay rapidly with the increase of the depth from the ground. Therefore, it is necessary to strengthen the deep signal. The enhancement coefficient can be reasonably selected according to the amplitude. This time, 0.5 is selected. 4. Subtracting average. GPR data will generally be interfered by direct waves and waves in other horizontal directions. This function can eliminate such waves by subtracting the average value of traces in a certain range. 5. band-pass filter. Interference waves outside the main frequency range need to be filtered out.

4. Results

After the raw electromagnetic signal is processed, high-resolution images of near-surface formation information can be obtained. As the survey area is located in the urban area, after filtering, there are still some interference signals on the image. Figures 2, 3 and 4 show the radar images obtained from this GPR survey. The radar survey line ran north-south and was carried out in four areas (Area-A~Area-C) with an interval of 2.8 m. There were 2 survey lines in Area-A, 9 survey lines in Area-B and Area-C.

Figure 2 shows the radar image of Area-A. L1 and L2 survey lines are parallel survey lines. Within the range of 0-20 m of the L1 survey line's mileage and the depth of 2 m, the reflection energy is relatively continuous. Very strong reflection amplitude occurs within the depth of 2-3 m, which can be inferred that the location is the interface between soil layer and bedrock. The reflection amplitude converges with the increase of depth, followed by energy attenuation. The energy accumulation area (marked by cyan lines) in the radar image is inferred to be the development of karst fissures or caves.

After 20 m of the L1 survey line, the reflection amplitude is unevenly distributed, which may be caused by the direct exposure of bedrock and widely developed fractures.
The amplitude shape of L2 survey line is very similar to L1 survey line within 0-20 m mileage and 3 m depth. L2 profile appears strong reflection aggregation at the distance of 30-40m and depth of 3-8 m, which may be caused by karst cavities and is not filled by water. If it is filled with water or other strong absorption media, electromagnetic wave energy will rapidly decay there.

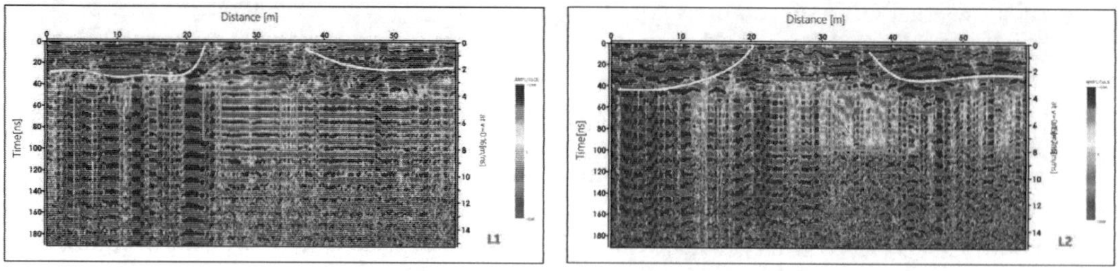

Figure 2. GPR diagram of L1 and L2.

Figure 3 shows the radar image of Area-B. L3-L11 survey lines are parallel survey lines. The reflection amplitude of the 9 profiles have certain similarities. 1 Within the depth of 2 m or 3 m, the horizontal variation of amplitude is continuous without too many strong reflections. The yellow curve roughly outlines the interface between Quaternary sediments and bedrock. 2 When the interface is relatively broken and filled with clay, the electromagnetic wave energy is absorbed to a certain extent, and the reflection amplitude is not obvious. 3 When there is strong reflection energy accumulation below the bedrock layer, it is inferred that the cracks or cavities are obviously developed, and the disordered voids and irregular shapes form multiple reflection interfaces.

When the limestone is dense and complete, there is no electromagnetic wave reflection interface in the radar image. Transverse continuous variation (areas marked by cyan rectangles and ellipses in the image) and many obvious anomalies appear in the radar profile, all of which are indicative of the complex karst environment underground.

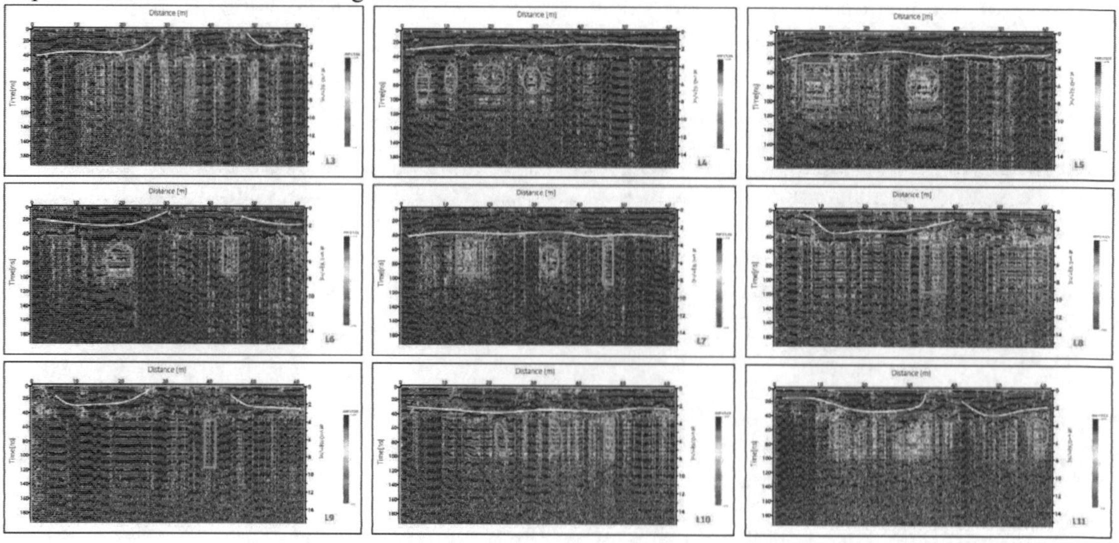

Figure 3. Radar diagram of L3-L11 survey line.

Figure 4 shows the radar line of Area-C. Area-C is located in the south of Area-B, the overall law of the radar image is similar to that of Area-B. The continuously distributed strong reflection amplitude is inferred as the interface between the overburden and the bedrock. Some survey lines have strong reflection on the surface, which basically corresponds to the exposed position of the bedrock. With the propagation of electromagnetic waves to the ground, its phase axis also appears a certain degree of distortion, which is inferred to be caused by some small cracks and karst ditches.

It is inferred that the location of karst cavity basically appears in the range of 4-8 m. With the increase of depth, the attenuation of electromagnetic wave energy and the reduction of resolution may not be enough to identify local karst. Considering that the deep bedrock is generally less weathered and less affected by corrosion, we infer that the bedrock below 8 m is relatively complete, so there is no obvious reflection amplitude.

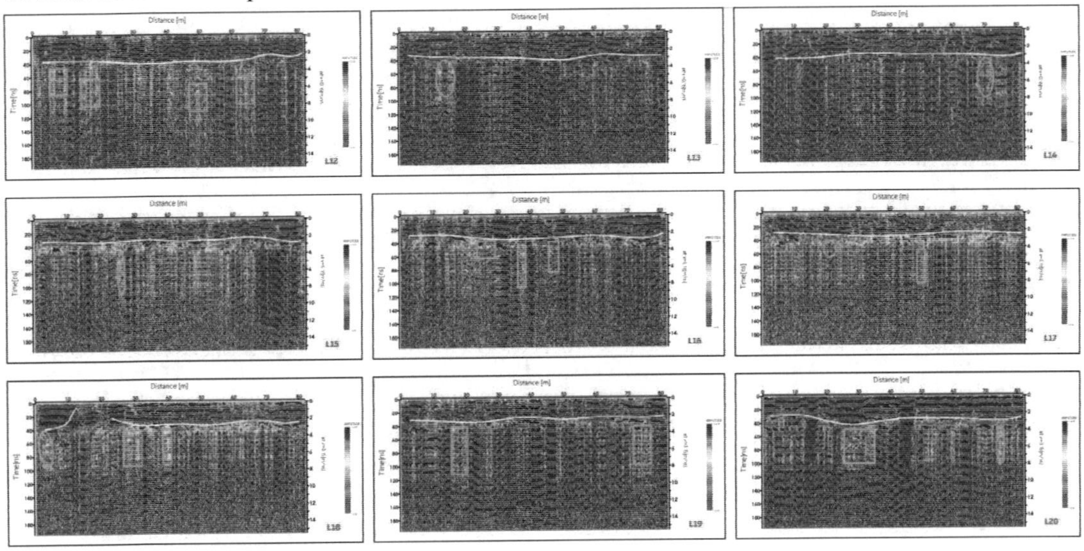

Figure 4. Radar diagram of L12-L20 survey line.

5. Discussion

In the area without any geophysical data and borehole data, GPR method can quickly, non-destructively and economically image the shallow surface karst development in the target area. Although many geophysical methods can be applied, including resistivity imaging, microgravity and so on. GPR method provides rich shallow surface information that other technologies cannot match. However, GPR method is also limited by some sites, such as thick sediments or grounded electrical facilities. The study area of this paper is located in the urban area, so it has been more disturbed. These factors should be fully considered in the application of this method.

Through the repeated interpretation and recognition of the GPR image, it can be concluded that there are obvious differences between the radar image of uniform and dense normal rock mass and that of non-uniform and dense rock mass or rock mass with underground cavities. Where there are inhomogeneous and dense parts and cavities in the rock mass, that is to say, the radar images observed above will inevitably have the phenomena of dislocation, bifurcation, arch anomaly, or dislocation of the reflection wave's in-phase axis. Because the dielectric constant of air, water, or filled clay is quite different from surrounding rock, the electromagnetic wave will produce a strong reflection signal between clay, water, or air, all these provide the basis for our information inference and interpretation. In this study, the phenomenon of reflection wave amplitude accumulation is considered to be the characteristic of karst fissures or caves. This is the principle of identifying and judging rock boundary in the geological radar image in this experiment.

6. Conclusions

A total of 20 radar profiles were obtained in this survey. Through the processing of radar images, we basically understand the strata and karst characteristics of the study area. The sediment thickness in the area is basically within the range of 2-3 m, and some areas are directly exposed by bedrock. The strong reflection continuous interface on the radar image is inferred as the interface between sediment and bedrock. It should be noted that we used a unified time-depth conversion speed, which is a comprehensive reflection of the underground based on the overall characteristics of underground materials and strata, and cannot refer to a specific or local area. Therefore, the depth location of the interface and karst cave may be different from the actual strata. Below this interface, when there is a concentration of reflection amplitude, it can be inferred as a karst cave. In the bedrock layer, it is usually difficult to determine whether the karst is filled with air, water or sediment directly from the radar image. However, based on their dielectric characteristics, we can infer that when the karst cave is not filled, the radar image may show the accumulation of strong reflection amplitude, while when it is filled with water, it is reflected in the attenuation of energy. When the resolution of electromagnetic wave decreases with the increase of depth, some small karst caves may be ignored.

The results obtained from this GPR survey clearly identified the abnormal characteristics of the stratum within 8 m of the study area, which indicates that there are certain risks in the subsequent construction in this area. It also shows that GPR can be used to identify the similar characteristics of the karst environment, which makes an important contribution to the comprehensive understanding of the karst system.

References

[1] Gutierrez F, Parise M, De Waele J, Jourde H 2014 A review on natural and human-induced geohazards and impacts in karst *Earth-Sci Rev* **138** 61-88

[2] Anchuela O P, Casas-Sainz A M, Soriano M A, Pocovi-Juan A 2009 Mapping subsurface karst features with GPR: results and limitations *Environ Geol* **58** 391-399

[3] Ceru T, Segina E, Knez M, Benac C, Gosar A 2018 Detecting and characterizing unroofed caves by ground penetrating radar *Geomorphology* **303** 524-539

[4] Liu M M, Liu Z H, Zhou D, Lan R Y, Wu H 2020 Recognition method of typical anomalies during karst tunnel construction using GPR attributes and Gaussian processes *Arabian Journal of Geosciences* 13

[5] Zielinski A, Mazurkiewicz E, Lyskowski M 2016 GPR mapping of karst formations under a historic building in Szydlow, Poland. *Geofizika* **33** 101-111

[6] Al-fares W, Bakalowicz M, Guerin R, Dukhan M 2002 Analysis of the karst aquifer structure of the Lamalou area (Herault, France) with ground penetrating radar J *Appl Geophys* **51** 97-106

[7] Mount G J, Comas X, Cunningham K J 2014 Characterization of the porosity distribution in the upper part of the karst Biscayne aquifer using common offset ground penetrating radar, Everglades National Park, Florida. *Journal of Hydrology* **515** 223-236

[8] Martel R, Castellazzi P, Gloaguen E, Trepanier L, Garfias J 2018 ERT, GPR, InSAR, and tracer tests to characterize karst aquifer systems under urban areas: The case of Quebec City *Geomorphology* **310** 45-56

IOP Conf. Series: Earth and Environmental Science 1087 (2022) 012062 doi:10.1088/1755-1315/1087/1/012062

Research on intelligent building greenhouse plant factory and "3-Positions and 1-Entity" development mode

Xinfa Wang[1,2,3], Vladislav Zubko[1], Viktor Onychko[1], Zhenwei Wu[2] and Mingfu Zhao[2,3]

[1] Sumy National Agrarian University, Sumy, Ukraine;
[2] Henan Institute of Science and Technology, School of Information and Engineering, Xinxiang, Henan province, China.

[3] Corresponding author's e-mail: wangxf@hist.edu.cn; zhaomf@hist.edu.cn

Abstract. After continuous development and evolution, the plant growth greenhouse has developed from the original heat preservation and moisture film cover to the present multi-cross intelligent solar greenhouse, which has played an important social value in agricultural production. However, in terms of the form of the greenhouse, it has not kept pace with the pace of social development, far from meeting the needs of environmental protection, energy saving, stable, intelligent, long-term use. With the wide application of big data, artificial intelligence, block chain, Internet of Things, cloud computing and other intelligent technologies in agriculture, as well as the rapid development of soilless culture and artificial light of plant lighting technology, the intelligent building greenhouse based on modern building technology can not only completely get rid of geographical location and regional climate conditions, but also durable to achieve long-term use standards and facilitate the deployment of intelligent operation equipment, which will become a more advanced and high-end form of intelligent plant greenhouse. Therefore, this paper defines the concept of "building greenhouse" for the first time, preliminarily discusses the construction idea of "intelligent building greenhouse plant factory", and puts forward the suggestion of "trinity" new development mode, in order to play a pioneer role. With the proposal of "intelligent building greenhouse plant factory", we firmly believe that the intelligent building greenhouse plant factory will gradually gain social recognition and get rapid development under the promotion of the "trinity" development mode. It can be predicted that skyscrapers style "Intelligent building greenhouse plant factory" will be everywhere in the foreseeable future, no matter in the heart of the city and other extreme weather conditions desert, ruin. In the future, people living near plant factories will be able to breathe oxygen-rich air, eat clean vegetables from nearby plant factories, and enjoy a high-quality and healthy life brought by abundant fresh plant food. What's more, the intelligent building greenhouse plant factory can better the environment of our earth to a certain extent, improve its carrying capacity, alleviate the food panic caused by outbreaks or disasters, local unrest to consolidate food security, intensive and efficient use of cultivated land resources, and rich people's pursuit of healthy plant food raw materials.

1. Introduction

Smart facility agriculture occupies a very important position in agriculture, is an important part of modern agriculture, and plays a very important role in the supply of agricultural products [1-3]. Facility plant cultivation can get rid of the limitation of natural conditions and environment of traditional agriculture by comprehensively regulating the environmental factors of crop growth, and achieve high efficiency, high yield, high quality, diversification and anti-seasonal vegetable production [4]. At present, from the early simple film mulching technology [5-6], solar greenhouse [7] and its related supporting equipment are still the mainstream of facility agriculture, and its main idea is to make full use of solar conditions for heat storage and heat preservation. After years of development, the solar greenhouse has gradually developed into a comprehensive large-scale modern intelligent greenhouse with mechanization, automation, network and intelligence [8-9]. The development of intensive facility agriculture has become the trend of modern agriculture of China [10].

In recent years, soilless cultivation technology [11-12] has developed rapidly and has been widely used in facility agriculture. Water cultivation, gravel cultivation, perlite plus peat cultivation and sawdust cultivation are commonly used in modern solar greenhouse production. Soilless cultivation is a trend of vegetable planting, which has the advantages of water-saving, fertilizer saving, labor saving, high-yield and optimal product quality [13]. It has got rid of the limitation of soil, greatly expanded the agricultural production space, and made it possible to carry out plant production on barren land, with a very excellent development prospect. Almost all plant factories adopt soilless cultivation mode [14].

With the research and practice of soilless cultivation, the development of LED energy-saving plant light source and the research of all artificial light plant factories are booming. In Japan, in recent years, the number of all artificial light source plant factories has increased rapidly, accounting for 44.2% of all Japanese plant factories. The proportion of combined solar and artificial light sources was 27.3%, and the proportion of sunlight was 28.6% [15-16]. In June 2016, the Institute of Botany of Chinese Academy of Sciences and Fujian San'an group jointly built the artificial light plant factory in Fujian Province, which is the largest single plant factory with one hundred thousand class cleanliness in the world. It starts the landing and industrialization of the theory and technology of plant cultivation by using full artificial light in China. In September 2017, the Wafangdian plant factory of China Hualu Panasonic was completed and put into operation. On May 3, 2020, the special report of "labor is the most glorious, struggle is the happiest - don't use sunlight, don't use soil, intelligent production plant factory" of CCTV news channel entered the plant factory for live broadcast. The vegetables here are cultivated in an aseptic environment, no pesticides, zero residue and can be eaten immediately after picking, which is healthy and environmentally friendly. From sowing to marketing, the whole process is visual, traceable and remote monitoring and managing. The intelligent, industrial and modern plant production mode has entered the stage of industrialization, standardization and commercialization.

With the development of urban-rural integration [17-18], the urban size is expanding, and the urban population is increasing rapidly, the agricultural population is decreasing, the land suitable for vegetable planting is decreasing due to the occupation of development and construction, the available resources, especially the water resources supply, are in short supply due to the vigorous consumption and waste, the climate environment is becoming fragile because of the green land occupied by construction and development and industrialization [19]. It is necessary to explore and develop more land-saving, labor-saving, energy-saving, environmental protection, and efficient agricultural planting mode. The emergence of all artificial light source plant factories is calling for the emergence of a more long-term, environmental protection, energy-saving intelligent building greenhouse. Although its investment in the early stage is large, its service life is very long, and the maintenance cost is relatively low in the later stage. The key point is zero pollution, zero emission and aseptic organic production. It can also break through the restrictions of geography, climate, space and other conditions. A "plant factory" of high-rise buildings can even be built in the downtown of a bustling commercial city, which provides a kind of high-quality fresh vegetables to citizens [20-22].

2. Building greenhouse, intelligent building greenhouse and plant factory of intelligent building greenhouse

2.1. Development trend of greenhouse

The original idea of planting a greenhouse is to keep the earth's surface warm. From the early simple film mulching to the plastic greenhouse [23], arched shed [24], glass greenhouse [25], ordinary solar greenhouse, sunshade momentum, single solar greenhouse [26], multi-span greenhouse [27], to large-scale intelligent solar greenhouse, although the form of greenhouse has changed a lot, its essence limited by land, sunlight and climate conditions has not changed. The stability and service life of greenhouse construction cannot meet modern building standards. With the development of soilless cultivation and full artificial plant lighting technology, the greenhouse gradually has the technical conditions to get rid of the constraints and restrictions of land and light conditions [28]. In order to achieve uninterrupted agricultural production no matter where and under what climatic conditions, the requirements for the stability, thermal insulation and durability of greenhouse building structure become more and more important. Therefore, the birth of architectural greenhouse is the call of the times, and once it appears, it will directly be the intelligent building greenhouse: the most advanced form of plant growth greenhouse. Figure 1 shows the development trend of the greenhouse form.

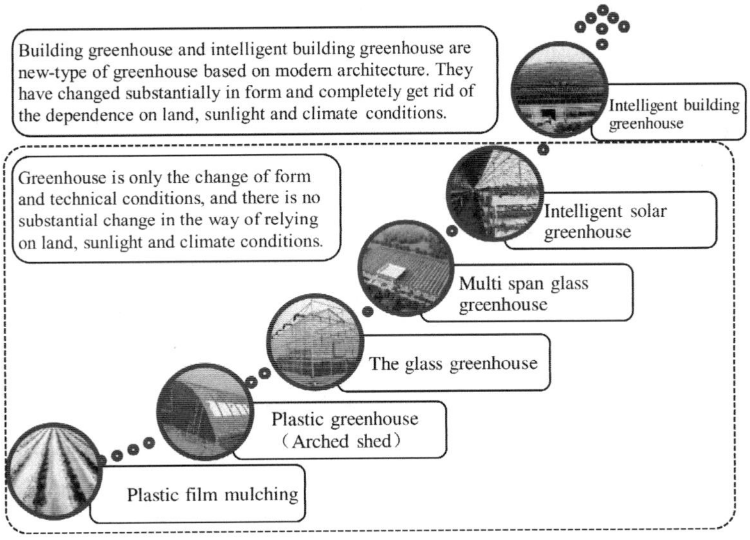

Figure 1. Development trend of greenhouse form.

2.2. *The definition of building greenhouse and intelligent building greenhouse*

General buildings refer to permanent assets formed by manual construction which are engineering buildings for people to live in, work, study, produce, manage, entertain, store goods and carry out other social activities. The landscapes and gardens, in a broad sense, are also part of architecture. More broadly, the conscious nesting of animals can also be regarded as architecture. The building greenhouse to be discussed here refers to modern buildings designed and built to provide an artificial environment for plant growth. Therefore, the following definitions are given in this paper:

Building greenhouse is built through professional architectural function design, mechanical design, aesthetic design, structural design and other comprehensive architectural design, whose foundation is specially designed and treated and is a permanent building with brick, sand and concrete or steel and concrete as the main body. In order to improve land use efficiency, greenhouses are generally designed as multi-storey or high-rise buildings.

The building greenhouse, in short, is a professional building that provides plants with the most suitable growth according to the concepts, theories and methods of modern architectural engineering design. Therefore, the building greenhouse is very different from the existing greenhouse forms such as plastic greenhouse, arch greenhouse, glass solar greenhouse and intelligent solar greenhouse and it is a new greenhouse completely different from the existing greenhouse. Its structure is more stable and its service life is longer. High quality building greenhouses can be used permanently. Moreover, constructing a building greenhouse is inseparable from the development of high technology [29]. Therefore, this paper puts forward the following concepts:

Intelligent building greenhouse refers to the comprehensive use of modern information technologies such as the Internet, Internet of things, mobile communication and mobile Internet, wireless sensor, automatic control, cloud computing, big data, blockchain and artificial intelligence in the process of building greenhouse and plant production, combined with greenhouse engineering construction technology, soilless cultivation technology and all artificial light cultivation technology. According to the plant life characteristics and growth process model, the greenhouse environment is monitored in real time and dynamically, so as to maintain the most suitable growth environment for each growth stage of the whole plant life cycle, so as to realize the high-yield and high-quality production of plants.

2.3. Plant factory urgently needs intelligent building greenhouse

Plant factories are an efficient agricultural system that can realize the continuous production of crops every year through high-precision environmental control [30-32]. The intelligent control system can automatically control the temperature, humidity, light, CO2 concentration, nutrient solution, fertilizer and other elements of crop growth environment, so that plant growth is not or rarely constrained by natural conditions. It is a high-tech intensive industrial complex. Its production mode possesses outstanding advantages that other modes of production cannot be compared with [33]. For a long time, the plant factory has been internationally recognized as the most advanced development stage of facility agriculture, and is one of the important indicators to measure the high-tech level of agriculture in a country [34]. With the high concentration of urban population and the reduction of costs of construction and operation of plant factories, plant factories will be the most effective means to solve the problems of global resources, world population and environmental degradation in the future, and it must develop rapidly.

The intelligent building greenhouse will integrate high technology, enrich the wisdom of mankind, break the traditional fetters of crop planting, and realize the high intelligence and modernization in all aspects of plant production. After special design and technical treatment, the plant production has been greatly extended in space. Good fields are greatly saved. The growth time is shortened, and the efficiency of vegetable production is improved. Intelligent building greenhouse is not limited by land, region and climate. It can be built in areas with extremely hot or cold or rapid deterioration of weather. It can also be built on barren saline alkali land, sand beach, the Gobi Desert, polar area covered with ice and snow all year round. It can also be built in the downtown of the city to be urban landmark buildings, developing tourism and entertainment or becoming a picking garden, so as to extend and expand the function of plant factories. The plant factory based on intelligent building greenhouse can start planned production or realize order production. Just press a button to start a plant production mode. The artificial intelligence system based on plant growth process model can provide the most suitable production environment for different growth stages of the plant according to the model of the target plant, so as to achieve mass production, high yield and high quality production. The advantages of intelligent building greenhouse gases have outstanding advantages, which should be the best "production workshop" of plant factory in the future.

3. The strategic significance of developing intelligent greenhouse plant factory

3.1. Consolidating food security strategy

Currently, food security, energy security and financial security, known as the three major global economic security, are the important foundations of national security [35-36]. Grain and vegetables are the most important necessities of life for people. Vegetables are essential for three meals a day. With the development of society, the growth of population, the improvement of people's quality of life and the enhancement of people's awareness of healthy diet, the demand for food and vegetables will increase rigidly. Along with urbanization, industrialization and modernization, problems such as climate deterioration, environmental pollution, poor quality of cultivated land, reduction of cultivated land, shortage of water resources, structural contradiction between food supply and demand, shortage of rural labor force and serious waste of grain have become increasingly prominent [37-40]. Food security is still affected by climate change, environmental pollution, natural disasters, turbulence, emergencies, land and water shortage, ecological damage and other factors. The novel coronavirus pneumonia has caused an emergency of food and vegetable supply, and food crisis occurs frequently. The development of intelligent greenhouse plant factories is one of the important measures of "storing grain in technology" [41-43]. It can effectively overcome the influence of many adverse factors in the process of grain and vegetable production, storage, transportation and sales, so as to strengthen and consolidate food security. The intelligent building greenhouse plant factory is based on the intelligent building greenhouse. It adopts soilless cultivation and artificial light technology to plant in a highly intelligent, modern and precise controlled and clean artificial growth environment. It does not need sunlight and soil. It can effectively avoid the impact of adverse factors such as climate change, environmental pollution, natural disasters, ecological damage and so on. Therefore, it can completely change the supply of vegetables and plants. Not affected by the seasonal changes, the supply of vegetables can be stable all year round. Moreover, this kind of plant factory does not need soil at all, which can save a lot of land resources and strengthen the protection for "storing grain in the field" in arable land. According to the population distribution and the demand for vegetables, the intelligent building greenhouse plant factory can be built in cities, deserts, densely populated areas and any other places where it is needed. It can be directly produced and sold at any time, without long-term storage and long-distance transportation. Even if there is an epidemic or other emergency, it can ensure a timely and sufficient local supply of vegetables, thus consolidating the food security.

3.2. Intensive use of land and improvement of cultivated land planting structure

Arable land is the essence of land and the foundation of grain production. It is the material basis for human survival and is a scarce resource that is not renewable. On the earth's surface, the total land area of the world is 13.081 billion hectares, of which only 11.25% is cultivated land (data from Baidu wiki). According to the situation of land use in the world, the proportion of farmland is quite low. However, the arable land is facing various severe threats and challenges, such as rampant occupation, abandonment, disaster erosion, pollution deterioration, soil erosion, returning farmland to forest, the arable land area is decreasing and the quality is deteriorating [44-46]. Especially in the development of urbanization, the rapid expansion of cities occupies a large amount of the land, most of which is high-quality arable land. In the huge pressure of the reduction and degradation of arable land, it is particularly important to make intensive use of land, improve the utilization rate of land and improve the planting structure of cultivated land. Intelligent building greenhouse plant factories can highly intensively use land, greatly raise the land use efficiency. The intelligent building greenhouse can be built into multilayer or high-rise buildings. In each layer of production greenhouse, three-dimensional soilless cultivation can be adopted. According to relevant reports, the three-dimensional planting of single-layer solar greenhouse can increase the land use rate by three times [47-48]. Moreover, under the increasing pressure of less farmland left, through reasonable layout, the intelligent building greenhouse plant factory was built on the land with poor quality or unsuitable for sowing, which not only improved the land utilization rate, but also improved and optimized the cultivated farmland planting structure, thus alleviating the pressure from the reduction of arable land resulted by social development to a certain extent.

3.3. Meet people's need for healthy diet

The ancients said: "we can have meals without meat for three days, but vegetables are needed every day." It was found that many vitamins, minerals, trace elements and related plant chemicals and enzymes are in vegetables and plants that are effective antioxidants. They are not only healthy food materials with low salt, low fat and low sugar, but also can effectively reduce the damage to the human body caused by environmental pollution and prevent many diseases [49]. It has been proved in a number of studies and dietary practices at home and abroad that dark green vegetables are good for preventing osteoporosis, obesity, diabetes, hypertension, cardiovascular disease, coronary heart disease and many kinds of cancer, and there are many benefits which have exceeded people's imagination [50]. It is demonstrated that most vegetables can be eaten raw, and it would be better to eat raw, because raw vegetables can lock the vitamins, inorganic salts, anti-cancer factors and physiological active substances inside from being damaged, which can contact human mucosal cells more effectively to the maximum extent, so as to play a better role [51]. The intelligent building greenhouse plant factory is the most suitable factory for vegetable planting. Vegetable production is carried out in a closed artificial environment. The growth environment can be strictly managed, highly sterilized. Chemical fertilizer is rarely used or no chemical fertilizer is used, and no pesticide is used. The vegetable itself is a clean vegetable that can be eaten directly. Clean vegetables will become the inherent label of intelligent building greenhouse plant factories, and will become a world famous brand to meet people's growing demand for high quality vegetables, so as to improve people's livelihood and living standards.

4. The development model of "3-Positions and 1-Entity"

4.1. The promotion of the "3-Positions and 1-Entity"

The intelligent building greenhouse plant factory has large investment, high-tech content and complex systems. In order to develop healthily and rapidly in the market environment, there must be a suitable development mode. Figure 2 shows the new development model of "3-Positions and 1-Entity" in intelligent building greenhouse plant factory. The "3-Positions and 1-Entity" mode refers to the "modern company plus intelligent building greenhouse" as the business entity and production entity of the intelligent building greenhouse plant factory, and integrates the production, operation and management of "factory production, corporate management, brand marketing" [52-53]. The modern company is the creator of the market subject, and the intelligent building greenhouse is the solid foundation of the plant production entity. The factory production, corporate management and brand marketing are the three directions of the plant factory facing the market, meeting the needs of high-grade life of consumers and leading the modern agricultural industry. The four are integrated and inseparable. Without the creativity of modern companies and the production "workshop" of plant factories with intelligent building greenhouse, there will be no foundation for the annual, large-scale, high-quality, intelligent, information-based and industrialization of vegetable production, and it will be difficult to fully realize the construction goal. Factory production, corporate management and brand marketing are useful tools for the rapid growth and development of plant factories, which, to a large extent, ensure the industrialization of production, the modernization of management, the high goal of enterprise development and the high quality of plant or vegetable products.

Figure 2. New development mode of "3-Positions and 1-Entity".

4.2. Factory production

Factories, also known as "factories" and "production enterprises", are industrial buildings used to produce goods. Modern factories generally have production lines composed of large machines or equipment. In the past, the production of agricultural products mainly depended on natural conditions such as land, sunlight and climate. These conditions had their own movement rules, and the role of human intervention was very limited, so it was difficult to popularize industrialized plant production.

The plant factory production of intelligent building greenhouse can realize plant planting, equipment mechanization, intelligent operation, process standardization, three-dimensional mode, capacity scale, production cycle and management system.

The plant production of intelligent building greenhouse plant factories will adopt the closed full artificial environment soilless cultivation. The temperature and humidity, light, water, air required for plant growth and fertilizer will be fully mechanized, fully automatic and multi factor precise control, so that the plant growth environment is always in the most suitable plant growth state in the whole plant growth process. The planting links including plant seedling, fixed planting, irrigation, fertilization, shaping, harvesting, sorting, packaging, inspection and others will be gradually realized mechanization, automation, information and intelligence. The plant factory eventually will develop into an unmanned factory, where production workers can produce at any time. With the in-depth application of artificial intelligence technology in the greenhouse plant factory of intelligent building, as long as press the start key, the expert system based on plant growth process model can realize the control from the growth environment to all planting operations, until the package is delivered to the warehouse for sale and shipment, and all operations are intelligent.

As the intelligent greenhouse building is a permanent building built by professional design to engage in efficient production of plants. In order to fully raise the utilization rate of land occupation, it is multi-storey buildings generally. Each floor has production greenhouses, which can be used for plant production. Each production greenhouse will adopt a multi-layer soilless cultivation bed or frame for stereoscopic plant production to ensure scale production and increase production capacity. Because plant production in the building greenhouse completely adopts artificial lighting which can be precisely controlled according to the needs of plant growth, thus it guarantees the plant production continuity and the periodicity, and ensure the fresh plant vegetable year-round supply.

When large-scale production is carried out in the intelligent building production greenhouse, the intelligent production control system will control the growth environment and standardize the operation according to the preliminary model of plant growth process constructed during the experimental planting. When the operation is not carried out according to the standard operation or lack of standardization, the system will send out intelligent prompting in time, so as to ensure that each operation is accurate and standard, to ensure the optimization of plant production to achieve energy saving, high yield and high quality.

4.3. *Corporate management*

The so-called corporate management is to take plant factories as a general production and operation enterprise, introduce the market mechanism and modern enterprise management mode, and construct the corporate management mode according to the form of corporatization, which includes organization optimization, system perfection, standardization and efficiency of production process, high professional quality of personnel, and large production capacity [54]. The introduction of corporate management mode in plant factories will change the situation of agricultural development supported by the government and operated by agricultural science and technology leaders, stimulate the vitality of various production factors of agricultural industry, and make plant factories move towards the development path of market-oriented, professional, high-tech, and intensive.

Agricultural modernization must be realized by agricultural industrialization [55]. Agricultural industrialization is the inevitable choice of China's agricultural modernization process [56]. It is necessary to guarantee the production of a modern agricultural economy and fine modern enterprise management system. Farmers with decentralized management are the main body of traditional planting agriculture, whose economic capacity is limited and their professional quality is generally not high, so it is difficult to become the main body of modern agricultural industrialization. In the process of agricultural modernization with commodity economy as the mainstream, the important role of the government is to guide the adjustment and development direction of agricultural structure by formulating incentive policies, so as to attract more social forces to invest in intelligent greenhouse plant factories, and it is even possible to set up factories directly with state investment to engage in modern plant production and management. However, due to the special status of the government, it can't be a part of its industrialization. Even if the government invested, the plant factory still needs to be operated in the form of corporation.

The fundamental problem of the development of modern agriculture lies in the contradiction between high investment and low investment level. The fundamental way to solve the problems and form sustainable development is to improve the input and management mechanism. We should increase agricultural input, raise investment level and improve input-output efficiency. Once the intelligent building greenhouse plant factory is born, it has the characteristics of modern agricultural industrialization, such as specialization, integration, science and technology, and intensification. Marketization, specialization, integration, intensification and standardization are the outstanding advantages of modern enterprise management system and the fixed attribute of corporatization [57]. The corporate management of intelligent building greenhouse plant factories is the specific embodiment of modern facility agriculture adapting to market demand. The corporate management mode of plant factory can attract companies with strong capital to join in, and the listed companies with outstanding performance can play a powerful role in financing, drive social capital investment, and build a larger plant factory. On the other hand, the company can efficiently combine high-quality resources in an organized and planned way, carry out product development, market expansion, industrial deepening and other development strategies, so as to rapidly promote the efficient industrialization of plant factories.

4.4. Brand marketing

Brand marketing strategy is one of the effective ways for product marketing to obtain a large market share [58]. Branding is the process of cultivating brands and improving their value. It can stabilize the consumer group and quickly occupy a larger market share. Through registered trademarks, it is convenient for consumers to identify, identify and purchase plant factory products, which is conducive to protecting the interests of consumers and promoting the consumption upgrading of plant products. Moreover, the registered trademark can protect the legitimate rights and interests, help to improve the product, standardize the production and sales behavior, and expand the product composition. The developed modern agricultural market needs the intelligent building greenhouse plant factory to establish the brand development strategy in construction planning, improve the market awareness and occupy the commanding height of the market. This requires the plant products of intelligent building greenhouse factories to meet the market demand. In the planning and construction stage, it is necessary to accurately locate position for the market and products, plan brand building from the long term, cultivate and register trademarks. and actively improve the new varieties of plants, enhance the quality of plants and perfect plant production technology and equipment, improve the service system, commit to innovation and creation, strengthen advantages, enhance corporate image, nurture corporate culture, enhance enterprise competitiveness in the activities of plant product planning, to achieve rapid prosperity and growth. The brand marketing of plant factory is market-oriented, aiming at meeting the diversified and high-quality consumption of vegetables and plants, guiding the production factors such as social capital, technical equipment and labor resources to concentrate on famous and excellent products, transferring the superior resources to superior enterprises, transforming resource advantages into quality advantages and efficiency advantages, and promoting, optimizing and upgrading the structural adjustment of modern facility agriculture industry , so as to promote the rapid development of intelligent building greenhouse plant factory.

5. Prospect

Due to its good sealing, the building greenhouse can effectively ensure high cleanliness and sterility indoors, prevent the invasion of diseases and pests, and become a "production workshop" for clean and pollution-free plant production. It also has good thermal insulation and heat preservation to cut down energy loss and reduce carbon emissions. When large-scale planting of vegetables and herbs is carried out in it, it can also clean the air to a certain extent, improve the amount of atmospheric eutrophication, and improve the surrounding microenvironment and microclimate. Moreover, the intelligent building greenhouse plant factory is built based on modern buildings, which can be built anywhere, even in the city center, and is not limited by climate conditions and geographical location. Therefore, it should be foreseen that the intelligent building greenhouse plant factory will open a new era of urban productive agriculture and become a beautiful landscape of the city.

Funding
This work is jointly funded by the Department of Science and Technology of Henan Province (Henan Science and Technology Research Project, grant number 212102110234 and 222102320080), the Department of Education of Henan Province (Key Scientific Research Project of Colleges and Universities in Henan Province, grant number 22A210013), and the Government of Xinxiang City of Henan Province (Major science and technology projects, grant number 21ZD003). The APC was funded by222102320080 and 22A210013.

References
[1] Han X, Wang R, Cui M, and Zheng F 2020 An empirical analysis on the development status and influencing factors of facility agriculture of large-scale farmers *Resources Science* **42** 4
[2] Symeonaki E G, Arvanitis K G, Piromalis D D 2019 Current trends and challenges in the deployment of IoT technologies for climate smart facility agriculture *International Journal of Sustainable Agricultural Management and Informatics* **5** 2/3 181
[3] Lia M N, Ma J, Jia W S 2016 Application of Automatics in Environment Control of Facility Agriculture *2016 ASABE International Meeting*
[4] Hu M, Jie S 2013 Analysis of Facility Agriculture Industry's Dilemma and Policy Crack Based on SCP Paradigm: Taking the Case of Yanqing County, Beijing *Advanced Materials Research* 798-799 936-940
[5] Li Y, Yang J B, Zheng S, Pan W H, Liao Y C, Li T, Qin X L 2021 Response of root traits to plastic film mulch and its effects on yield *Soil & Tillage Research* **209** 104930
[6] Zhou LF, Feng H, Zhao WZ 2021 Plastic film mulching affects the critical nitrogen dilution curve of drip-irrigated maize *Field Crops Research* **263** 108055
[7] Esmaeli H, Roshandel R 2020 Optimal design for solar greenhouses based on climate conditions *Renewable Energy* **145** 1255-1265
[8] Hemming S, Zwart F, Elings Anne, Petropoulou A, Righini I 2020 Cherry Tomato Production in Intelligent Greenhouses - Sensors and AI for Control of Climate, Irrigation, Crop Yield, and Quality *Sensors* **20** 22 6430
[9] Liu H B, Fang S Y, Guo X Q 2020 Research and Design of Intelligent Greenhouse Control System Based on AIoT Fusion Technology *IOP Conference Series Earth and Environmental Science* **474** 032036
[10] Zheng L 2021 Problems and development suggestions in facility agriculture production *Henan agriculture* **07** 58-59
[11] Zhu W M, Wang H 2013 A Brief Delopment History of Soliless Cultivation in China *Acta Horticulturae* 63-69
[12] Rathod K D, Patel A J, Chakraborty B 2021 Strawberry cultivation practices in soilless growing substrates: A review article *International Journal of Chemical Studies* **9(1)** 1253-1256
[13] Savvas D, Gruda N 2018 Application of soilless culture technologies in the modern greenhouse industry - A review *Eur.J.Hortic.Sci* **83(5)** 280-293
[14] Orsini F, Pennisi G, Gianquinto G 2020 Sustainable use of resources in plant factories with artificial lighting (PFALs) *European Journal of Horticultural Science* **85(5)** 297-309
[15] Kim Y H, Lee J H 2017 Analysis of characteristics and operation status of Japanese plant factory *KIEAE Journal* **17(6)** 125-134
[16] Wei W, Wang D X 2019 Current situation, problems and future development of Japanese plant factories *Modern agriculture* **12** 22-23
[17] Yan J M, Chen H, Xia F Z 2018 Toward improved land elements for urban - rural integration: A cell concept of an urban - rural mixed community *Habitat International* **77** 110-120

[18] Zhao P J, Wan J 2021 Land use and travel burden of residents in urban fringe and rural areas: An evaluation of urban-rural integration initiatives in Beijing *Land Use Policy* **103** 105309
[19] Shrivastava S R B L, Shrivastava P S, Ramasamy J 2017 Encouraging collaborative efforts between environment and health sector to arrest climate deterioration *Journal of Research in Medical Sciences* **22** 1 59
[20] Bon H D, Parrot L, Moustier P 2010 Sustainable urban agriculture in developing countries: A review *Agronomy for Sustainable Development* **30(1)** 21-32
[21] Orsini F, Kahane R, Nono-Womdim R, Gianquinto G 2013 Urban agriculture in the developing world: a review *Agronomy for Sustainable Development* **33(4)** 695-720
[22] Dona C G W, Mohan G, Fukushi K 2021 Promoting Urban Agriculture and Its Opportunities and Challenges - A Global Review *Sustainability* **13** 17 9609
[23] Xu W, Liu A Q, Wang Y, Ying G, Peng C H, Gu B J, Meyerson L A, Yang W, Xu B, Chang J 2011 Assessment of net ecosystem services of plastic greenhouse vegetable cultivation in China *Ecological Economics* **70(4)** 740-748
[24] Wang G D 2020 Pollution-Free and High-Yield Cultivation Techniques of Thin-Skinned Melon with Plastic Film and Small Arch Shed in Eastern Liaoning *Hans Journal of Agricultural Sciences* **10(05)** 309-314
[25] Jeong I S, Lee C G, Cho L H, Park S Y, Kim M J, Kim S J, Kim D H 2020 Development and Validation of Inner Environment Prediction Model for Glass Greenhouse using CFD *Protected horticulture and Plant Factory* **29(3)** 285-292
[26] Cao K, Xu H J, Zhang R, Xu D W, Yan L L, Sun Y C, Xia L, Zhao J T, Zou Z R, Bao E 2019 Renewable and Sustainable Strategies for Improving the Thermal Environment of Chinese Solar Greenhouses *Energy and Buildings* **202** 109414
[27] Rasheed A, Kwak C S, Wook H N, Lee J W, Kim H T, Lee H W 2020 Development of a Building Energy Simulation Model for Control of Multi-Span Greenhouse Microclimate *Agronomy* **10** 9 1236
[28] Kozai T 2018 Benefits, problems and challenges of plant factories with artificial lighting (PFALs): a short review *Acta Horticulturae* 25-30
[29] Mofidi F, Akbar H 2020 Intelligent Buildings: An Overview *Energy and Buildings* 110192
[30] Avgoustaki D D, Xydis G 2020 Plant factories in the water-food-energy Nexus era: a systematic bibliographical review *Food Security*
[31] Huebbers J W, Buyel J F 2020 On the verge of the market - Plant factories for the automated and standardized production of biopharmaceuticals *Biotechnology Advances* 107681
[32] Ares G, Birgit H, Jaeger S R 2021 Consumer attitudes to vertical farming (indoor plant factory with artificial lighting) in China, Singapore, UK, and USA: A multi-method study *Food Research International* 110811
[33] Sun J, Gao H B, Tian J, et al 2019 Development status and trend of installation horticulture in China *Journal of Nanjing Agricultural University* **04** 20-24
[34] Yang Q C 2014 Development strategy of plant factory *Science and technology guide* **32** 1020-24
[35] Hlaing K Y 2006 Globalization and Economic Security in East Asia (review) *Contemporary Southeast Asia A Journal of International and Strategic Affairs* **28(2)** 359-362
[36] Tikhomirov S N 2019 Important component of global economic security *World Economy and International Relations* **63(3)** 76-82
[37] Wang X, NIU S W, Qiang W L, Liu A M, Cheng S K, Qiu X, Li F 2021 Analysis on global food supply and demand balance and its evolution from a perspective of food trade *Journal of natural resources* **35(7)** 1659
[38] Huang J K, Wang J M, Xie W, Wang X B, Hou L L, Zhou H, Sheng Y, Liu X 2019 Modern Agricultural Transformation and Trend of Food Supply and Demand in China *Chinese Journal of Engineering Science* **21** 5 1
[39] Lee H L 2009 The impact of climate change on global food supply and demand, food prices,

and land use *Paddy and Water Environment* **7(4)** 321-331

[40] Grubor B, Kalenjuk B, Radivojša M, Ćirić M 2020 Functional food: Supply and demand in a modern society *Zbornik radova Departmana za geografiju turizam i hotelijerstvo* 195-207

[41] Carthy U M, Uysal I, Badia-Melis R, Mercier S, O'Donnell C, Ktenioudaki A 2018 Global food security - Issues, challenges and technological solutions *Trends in Food Science & Technology* **77** 11-20

[42] Jim G, James B, Byrne P F, Cassman K G, Ciampitti I, Delmer D, Habben J, Lafitte H R , Lidstrom U E, Warner D 2019 Science-based intensive agriculture: Sustainability, food security, and the role of technology *Global Food Security* **23** 236-244

[43] Lakhiar M A, Gao J M, Syed T N, Chandio F A, Tunio M H, Ahmad F, Solangi K A 2020 Overview of the aeroponic agriculture - An emerging technology for global food security *International Journal of Agricultural and Biological Engineering* **13(1)** 1-10

[44] Nath R, Luan Y B, Yang W M, Yang C, Chen W, Li Q, Cui X F 2015 Changes in Arable Land Demand for Food in India and China: A Potential Threat to Food Security *Sustainability* **75** 5371-5397

[45] Prăvălie R, Patriche C, Borrelli P, Panagos P, Roșca B, Dumitrașcu M, Nita I A, Săvulescu I, Birsan M V, Bandoc G 2021 Arable lands under the pressure of multiple land degradation processes. A global perspective *Environmental Research* 110697

[46] Wu X D, Guo J L, Han M Y, Chen G Q 2018 An overview of arable land use for the world economy: From source to sink via the global supply chain *Land Use Policy* **76** 201-214

[47] Al-Kodmany K 2018 The Vertical Farm: A Review of Developments and Implications for the Vertical City *Buildings* **8(2)** 24

[48] Despommier D 2013 Farming up the city: the rise of urban vertical farms *Trends in Biotechnology* **31(7)** 388-389

[49] Volpe S L 2019 Fruit and Vegetable Intake and Prevention of Chronic Disease *ACSM's Health & Fitness Journal* **23(3)** 30-31

[50] Griep L M O, Geleijnse JM, Kromhout D, Ocké M C, Verschuren W M M 2010 Raw and Processed Fruit and Vegetable Consumption and 10-Year Coronary Heart Disease Incidence in a Population-Based Cohort Study in the Netherlands *PLoS ONE* **5(10)** e13609

[51] Feng Q, Kim J H, Omiyale W, Bešević J, Conroy M, May M, Yang Z Y, Wong S Y, Tsoi K K, Lacey B 2022 Raw and Cooked Vegetable Consumption and Risk of Cardiovascular Disease: A Study of 400,000 Adults in UK Biobank *Frontiers in Nutrition* **9**

[52] Yuan B, Li J N, Zeng G J 2018 Corporate liquidity management and technical efficiency: Evidence from global listed hospitality enterprises *International Journal of Hospitality Management* **74** 40-44

[53] Hirsch P B 2020 A breakfast of champions: brand marketing lessons from the Great Depression *Journal of Business Strategy* **41(4)** 63-67

[54] Yekimov S, Purtov V, Buriak I, Kabachenko D, Poltorak A 2021 Improving the efficiency of corporate management of agricultural enterprises *E3S Web of Conferences* **262** 03001

[55] Kremen C, Iles A, Bacon C 2012 Diversified Farming Systems: An Agroecological, Systems-based Alternative to Modern Industrial Agriculture *Ecology and Society* **17** 4

[56] Gu X W, Zhang A P 2015 Research on the Evaluation of Coordinated Development Level of Industrialization, Informatization, Urbanization and Agricultural Modernization in China *International Journal of Smart Home* **9(10)** 1-8

[57] Arsenieva N V, Putyatina L M 2021 The marketing potential of the enterprise and the possibilities of its development in modern conditions *Vestnik Universiteta* 5-11

[58] Li S T 2019 Analysis on Brand Marketing Strategy of Perfect Diary *E-Commerce Letters* **08(03)** 107-112

IOP Conf. Series: Earth and Environmental Science 1087 (2022) 012063 doi:10.1088/1755-1315/1087/1/012063

Spatial accessibility of medical facilities based on green travel mode - Take the central urban area of Wuhan as an example

Wei Wei [1,2,3] and Jianhua Zhang[1,4]

[1] Department of Urban Planning, School of Urban Design, Wuhan University, Wuhan 430072, China;
[2] China Institute of Development Strategy and Planning, Wuhan University, Wuhan 430072, China.

[3] Corresponding author's e-mail: weiwei@whu.edu.cn;
[4] Corresponding author's e-mail: 1261056497@qq.com

Abstract. Based on the concept of "people-oriented", this paper simulates the green travel mode of Wuhan residents in the main urban area by walking and public transportation (bus and rail) based on the public transportation network and walking network of Wuhan city, combined with the POI data of residential community points and medical public service facilities. ArcGIS network analysis technology was used to quantitatively analyze the spatial accessibility distribution characteristics and qualitatively analyze the problems of medical service facilities in the main city of Wuhan. The spatial accessibility distribution characteristics of the three types of medical service facilities (General hospital, Specialist hospital and community hospital) in Wuhan city were quantitatively and qualitatively analyzed. The conclusions are as follows: under the different green travel mode (walk, by bus or take the subway), the spatial accessibility of the three types of medical facilities in the central urban area of Wuhan is good. The areas with the spatial accessibility of the three types of medical facilities within 15min account for most of the central urban area of Wuhan. However, the spatial accessibility of the three types of medical facilities shows obvious imbalance in the spatial distribution characteristics. The spatial accessibility of three medical facilities in the west of Jianghan district, the north of Jiangan district, the east of Wuchang district and the middle of Hongshan district is poor. And the spatial accessibility in these areas is more than 15min. There are distinct differentiating characteristics in areas with good and bad spatial accessibility. Based on the above analysis results, suggestions and countermeasures are put forward to improve the spatial accessibility of medical public service facilities in the central city of Wuhan.

1. Introduction

In the process of rapid urbanization, urban problems such as traffic congestion, environmental pollution and insufficient construction of public service facilities are becoming increasingly apparent. Governments at all levels are paying more and more attention to the construction of public service facilities. At the same time, as traffic congestion continues to be a problem, the government's promotion of low-carbon travel has made public transportation the primary choice for residents to commute. It is especially important to analyze the accessibility of the current public service facilities.

In the face of sudden public health emergencies such as the novel coronavirus in the post-epidemic era, the testing and consultation in community health service center and the treatment in large general hospitals are of great importance in reducing the spread of the epidemic. The accessibility of medical facilities at all levels is analyzed and the configuration and functions of public medical facilities in the city are further improved based on the analysis results.

Hansen first introduced the concept of accessibility in 1959, defining it as the extent to which the nodes in a transport network have the opportunity to interact with each other. Accessibility characterizes the potential for infrastructure in cities to provide services for consumption and is an important basis for planning and decision makers to evaluate existing services and to seek improvements in services [1]. Currently, there are many studies on the accessibility of public service facilities. In terms of object of accessibility studies, some scholars focus on the accessibility of parks and green spaces, some focus on the accessibility of medical facilities, some scholars have explored the accessibility of educational facilities in depth. In terms of accessibility study methodology, their research methods mainly include buffer zone analysis [2], improved potential model [3], network analysis [4] [5] [6] [7], two-step movement search method [8], Tyson polygon [9], etc.

Previous studies have not taken into account the disadvantaged groups. The degree of accessibility of healthcare facilities is particularly important for the quality of healthcare services received by the more disadvantaged groups. This study focuses on the disadvantaged groups of urban residents in the central city of Wuhan, including students, children, women and the elderly who are limited by their status, their physical condition and their current consumer power. Their primary choice is to use different green low-cost travel options to access healthcare services. Many previous studies on the accessibility of public services using network analysis have been based on a single mode of travel for residents. Those studies have not taken into account the reality of access to public services by different modes of transport in the context of walking. These groups do not always travel by a single mode of transport but rather by a combination of walking and other public transport modes to access specific health care facilities. This paper uses ArcGIS network analysis to quantitatively analyze the spatial accessibility of medical facilities in the main urban area based on the pedestrian road network, urban bus routes and urban metro lines. It also qualitatively analyzes the spatial distribution characteristics of the accessibility of medical facilities in the main urban area. The results of the analysis including relevant data and diagrams are used to scientifically identify areas where the accessibility of existing medical facilities is poorly distributed in space. And it is also used to identify the shortcomings of the existing medical service facilities and the unevenness of their layout. To provide reference for improving the configuration of medical service facilities and optimizing the layout of medical facilities in the main urban area in response to the analysis results. To provide recommendations for the future planning and construction of road networks at all levels in the main urban area as well as public transportation networks. To further promote the accessibility of medical facilities in the main urban area and the equalization of the configuration of medical facilities in the main urban area. To improve the convenience and quality of access to relevant medical services for the residents of the main urban area, especially the more disadvantaged groups. This will enable the disadvantaged groups in the city to receive medical treatment at a lower cost in terms of time and money and then enhance the overall sense of access and happiness of the residents.

2. The area & data of the study

2.1. The area of the study

In the *Wuhan Urban Master Plan (2010-2020)*, Jiangan, Jianghan, Qiaokou, Hanyang, Wuchang, Qingshan and Hongshan districts are identified as the main urban areas of Wuhan. The main urban areas of Wuhan are taken as the subject of this study. Taking into account the differences in the allocation standards of public service facilities between urban and suburban areas and rural areas as and the different evaluation criteria systems. The outbreak of major public health events in Wuhan in 2020 posed a great threat and challenge to the lives of the people of Wuhan. Medical public service facilities is important in responding to public health events for residents and providing medical services to the people. Improvement of medical facilities are one of the main targets of recent planning of Wuhan. As a high-density population area, the study of the spatial accessibility of medical services in the main urban area of Wuhan is conducive to improving the efficiency and quality of medical services in the city in a more targeted way. See Figure 1 below.

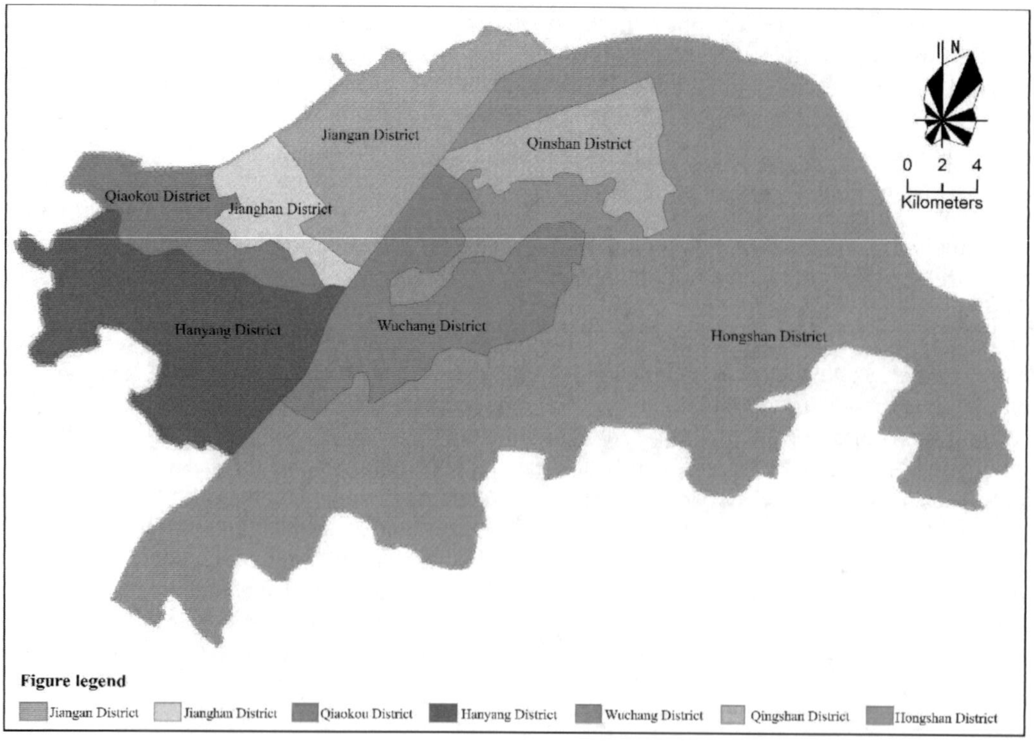

Figure 1. Diagram of the study area.

2.2. The data of the study

POI, known as Point of Interest, is a meaningful point on a map that is not geographically significant: for example, shops, pubs, hospitals, settlements and etc. It is all kinds of geospatial point data on the map. CGCS2000 is the abbreviation for the National Geodetic Coordinate System (China) 2000 which is a geocentric coordinate system established through the continuous operation of the Chinese GPS reference stations, the Space Geodetic Control Network and the joint levelling of the Astronomical Geodetic Network and the Space Geodetic Network.

In order to ensure the accuracy and timeliness of all types of auto Navi map POI data. Public transport station and line data, rail transit station and line data and pedestrian road network data, the various types of data obtained were imported into ArcGIS software. And those data were overlapped with the latest Wuhan image maps and street maps respectively to correct inaccurate data manually by deleting redundant data and adding lacking data. The point data and line data were confirmed to have a high degree of overlap with their location on the real-time HD satellite map to ensure the quality of the acquired data set.

2.2.1. POI data for residential areas, commercial buildings, educational and medical facilities. The central points of the residential areas of permanent residents in the main city of Wuhan, the workplaces of residents such as commercial and office buildings and the points of educational facilities such as various schools were used as the starting points for residents. There are 5218 residential points, 40 workplaces such as commercial and office buildings and 1560 educational facility points. The coordinate system of the data was also converted to the CGCS200 coordinate system. The final result is 6818 departure points in the main city of Wuhan. See Figure 2 below; The service points of medical facilities in the main city of Wuhan are used to represent the destinations of residents on their daily trips. The Spatial Layout Plan for Medical and Health Facilities in Wuhan (2011-2020) divides medical public service facilities into general hospitals, Chinese medicine hospitals and hospitals combining traditional Chinese and Western medicine, specialist hospitals, community health service centres and service stations, township health centres and village health offices. In this study, there are only three or four Chinese medicine hospitals and hospitals combining traditional Chinese and Western medicine in the main urban area of Wuhan which are not very meaningful to be studied as separate objects. Therefore, in this study, medical service facilities are divided into three categories: general hospitals, specialist hospitals and community health service centres which are the subjects of this study. Also in the medical facilities, the Concordia Hospital, Concordia Hospital-Outpatient Department and Concordia Hospital-Emergency Department all denote the same facility and these duplicates were removed. Leaving only one facility point characterising that hospital as the study point. This resulted in 225 general hospital facility points, 1105 specialist hospital facility points and 398 community health facility points. See Figure 2 and Figure 3 below.

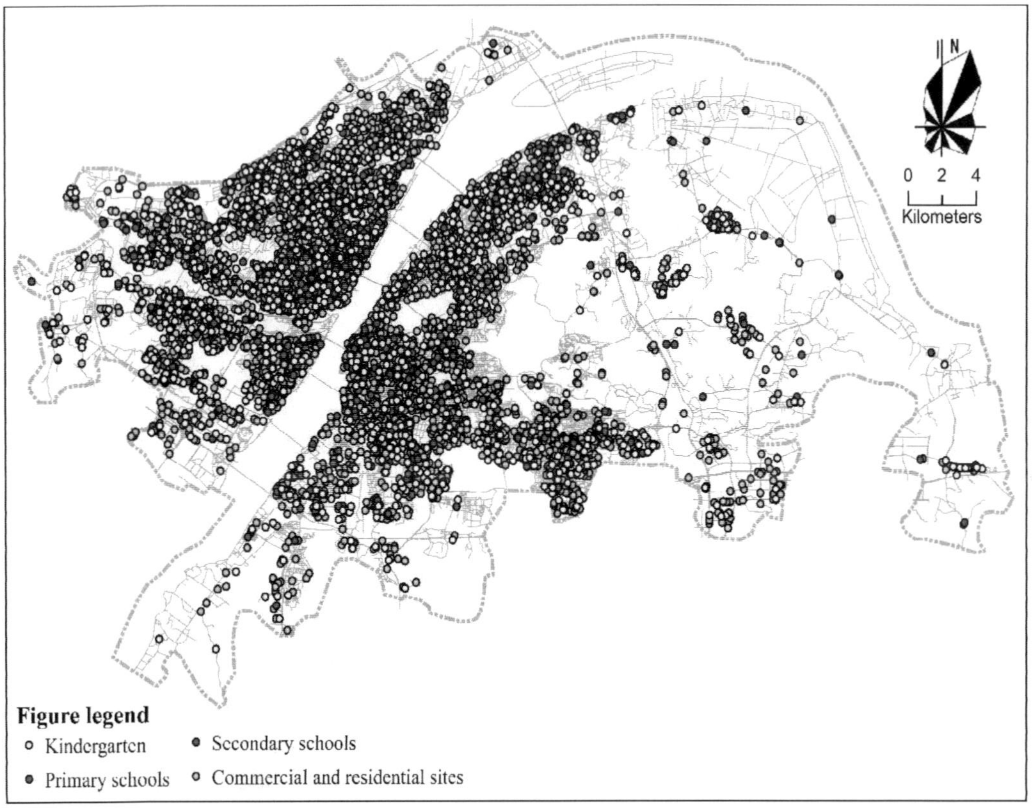

Figure 2. Diagram of the starting point.

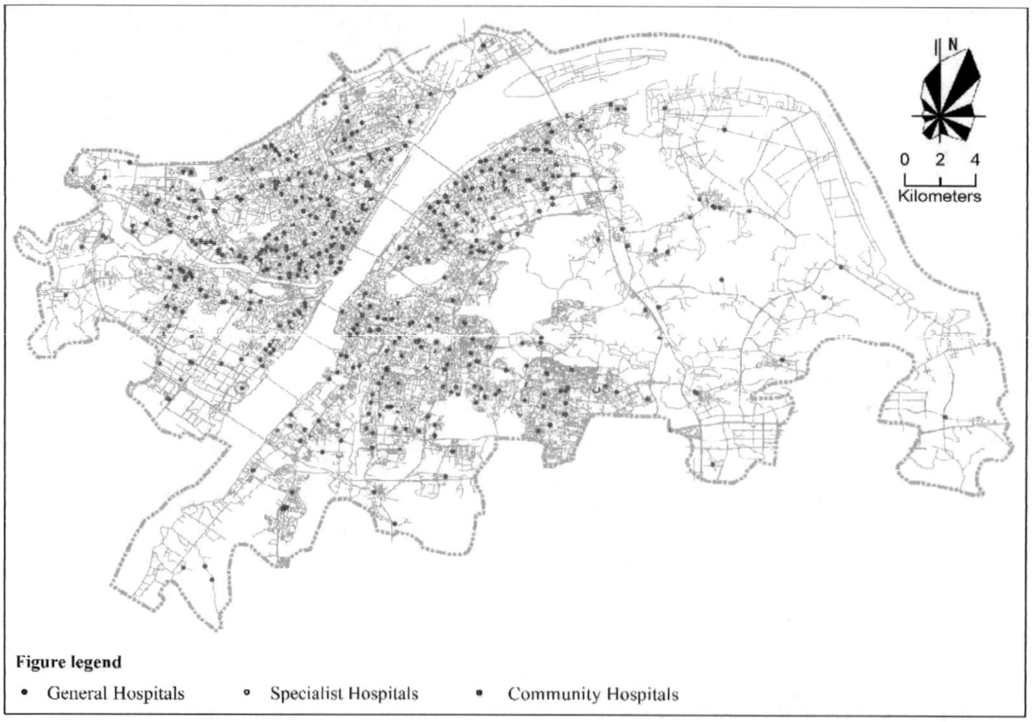

Figure 3. Diagram of the destination point.

2.2.2. Bus station & bus route data. The bus stop data is mainly from auto Navi map POI data and the bus route data is from 8486 bus network so as to obtain all the bus route data in the city. The bus stop and bus route data come from different sources, so the point element data of bus stops and the surface element data of bus routes need to be imported into ArcGIS. The CGCS200 coordinate system is used as the coordinate system for correction. At the same time, the duplicate bus stops are deleted in combination with the Gaode map and reserve the one that matches best. After obtaining and processing, a complete bus network is formed. See Figure 4 below.

2.2.3. Data of rail transit stations & rail transit lines. The source of the rail transit related data in this paper is the same as the source of the bus data. The acquired rail transit data is processed. The redundant rail transit station data is deleted against the auto Navi map. The offset rail transit network lines are adjusted and the coordinate system of the data is converted to CGCS2000 coordinate system. After processing, a complete rail network was formed. See Figure 4 below.

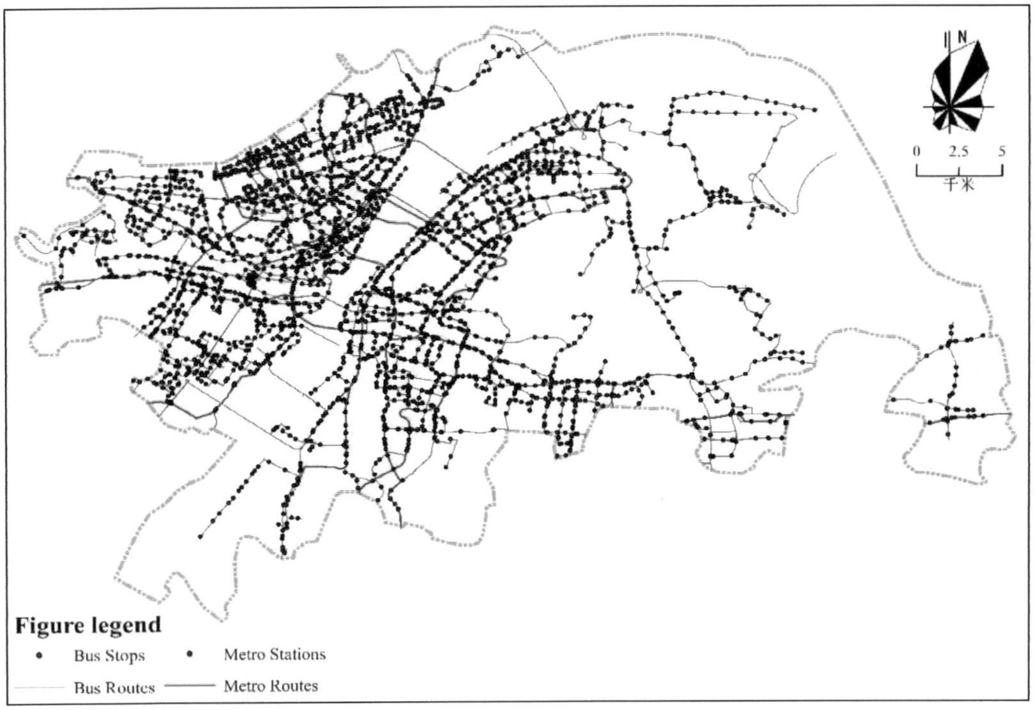

Figure 4. Bus & metro stops and routes data.

2.2.4. Pedestrian road network data. Wuhan main urban area road network data is mainly from auto Navi map. Based on the green travel mode, some road data in Wuhan main urban area does not reserve such as high-speed rail line data, highway line data. And then we delete unwanted data, adjust the offset road network data, retain two-way road network and cross street road as far as possible to simulate travel. Converting the data coordinate system into CGCS2000 coordinate system. See Figure 5 below.

2.2.5. Pedestrian and public transport connections. For the connection between walking and public transport in the walking combined with public transport mode of travel. Residents do not have access to public transport services at any location on the walking path, but can transfer to public transport at each bus stop. The ArcGIS software connects the walking path with the bus route at each bus stop so that residents can walk to the corresponding bus stop to connect with the bus route through the bus stop smoothly. Forming a combination of walking and public transport. At the same time, two different bus routes are connected at the same stop of different bus routes where residents can transfer to other bus routes. See Figure 6 below; For the connection between walking and the metro in the walking combined with urban metro travel mode. Residents can only take the metro to their destination at the metro station. At each metro station, the walking path is connected to the metro line through the metro station, forming a connection between the two modes of travel on foot and the metro. At the same time, it is possible to connect different metro lines at the interchange points so that residents can transfer to other metro lines at each interchange point. See Figure 7 below.

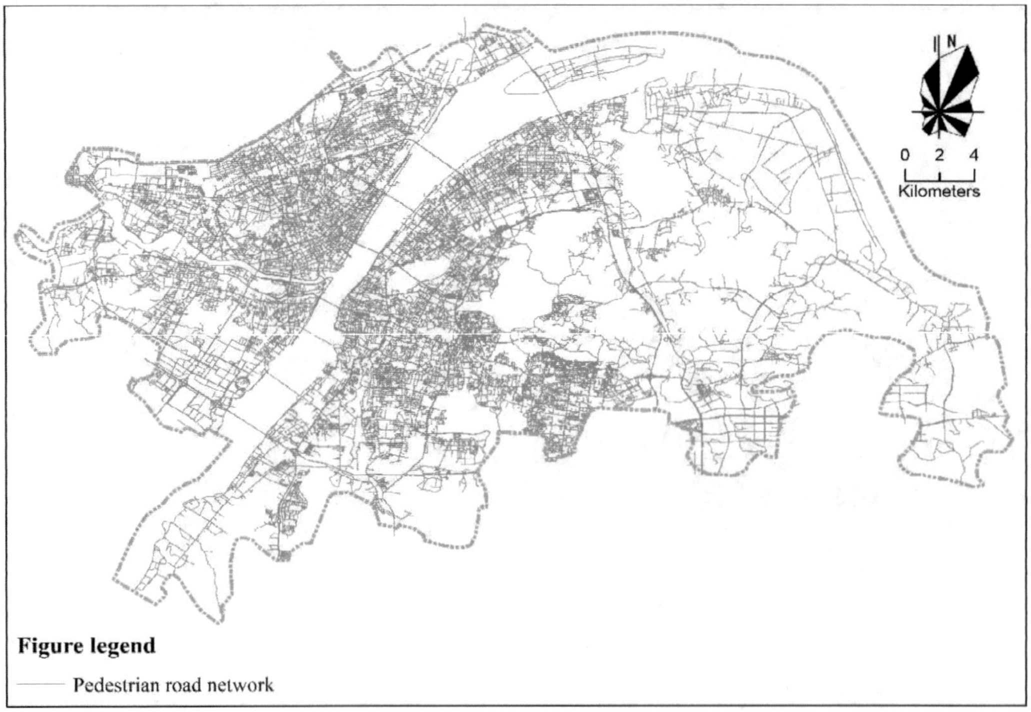

Figure 5. Central urban pedestrian road network.

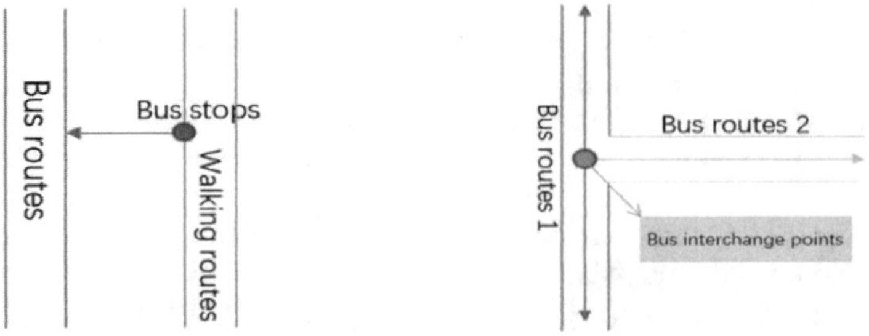

Diagram Of Pedestrian And Bus Connections Bus Route Interchange Diagram

Figure 6. Pedestrian and bus connections and interchanges diagram.

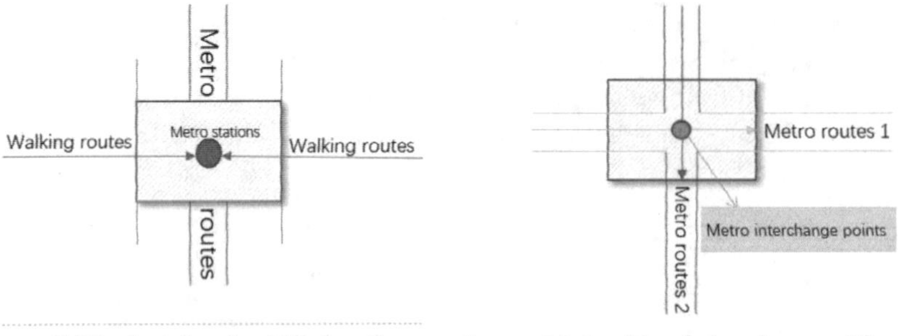

Diagram Of Pedestrian And Metro Connection Metro Line Interchange Diagram

Figure 7. Pedestrian and metro connections and interchange diagrams.

3. The research methods

3.1. ArcGIS network analysis

ArcGIS Network Analysis is a tool module for simulating and constructing geographic networks and urban infrastructure networks, mainly for the optimal allocation of resources and the improvement of network structures. The method is based on various types of public transport road networks in the main city of Wuhan and simulates the way people travel in a real state under certain resistance. Specifically, the speed attributes of the roads are set by simulating different modes of transport. In this study, the starting point is mainly the central mass point in the residential area of the city and the destination point is the three types of medical service facility points which are connected through the constructed road network.

The central city roads were used to construct a network dataset by entering them into the database through ArcGIS software. Taking the extracted community central points as the starting point, the POI of medical service facilities as the destination point. The network database is constructed with the current roads and an OD cost matrix is established between the medical-type public service facilities and residential points based on the urban green transportation network. According to *The Technical Standards for Highway Engineering of the People's Republic of China (JTGB-2003)* and *The Urban Integrated Transport System Planning Standards GB/T51328*, the speed limits for urban roads are 40 km/h to 60 km/h for main roads, 30 km/h to 40 km/h for secondary roads and 20 km/h to 30 km/h for general roads. The speed limit is set at 72m/min for people walking, 600m/min for buses and 1500m/min for railways and 30s stopping time for buses and railways at each bus stop and railways stop.

The average walking speed of residents is less affected by external influences and the speed value is basically unchanged. Metro has a specific track and its track is built under the ground. Metro has a more independent transport system compared to public transport. And metro depart according to the schedule and the stopping time of trains at stations is 30~40s, so the driving speed of the metro is basically unaffected. Bus routes within the main city of Wuhan are planned and built for buses in only a few areas and most of the bus lanes are urban motorways. During the morning and evening peak commuting hours in Wuhan, the traffic flow increases and the roads become congested. The average speed of buses is reduced from the original 36 km/h to 24~26 km/h which is about 30% lower compared to the speed during non-commuter peak hours. Based on these two conditions, the study was divided into two parts: morning and evening peak commuting hours and non-peak commuting hours to analyze the spatial accessibility characteristics of medical facilities in each case.

Analysis steps: (1) Create "OD Cost Matrix" analysis layer and load N residential points and M points of certain public service facilities as starting points and destination points respectively. (2) After running "Solve", ArcGIS automatically calculated the shortest path and distance from N starting points to M destination points, with a total of N*M records. (3) Select the connection path from N*M records to the nearest destination point and calculate the time according to the speed attribute of the road data. (4) Taking the residential area as the input point element, the shortest reachable time of each public service facility point is screened, and the shortest reachable time-space interpolation analysis diagram is obtained by inverse distance weight interpolation method.

3.2. Statistical indicators of spatial accessibility evaluation

The Shanghai Urban Master Plan (2017-2035) released in 2018 is the first in China to propose the creation of a "15-min community living circle" which means that residents will be equipped with the basic service functions and public activity spaces required for living within 15 min' walking distance. *The Technical Guide to Community Life Circle Planning* released in 2021 provides planning guidelines for urban community life circles in terms of configuration levels, service elements and spatial layout and urban community life circles can be built in two levels: "15 min and 5-10 min". With reference to the above community living circle theory, combined with the standards for the configuration and layout of public service facilities in Wuhan as set out in *The Guidelines for Public Facilities in Newly Built Areas (for Trial Implementation)*, that is, residents can achieve access to public service facilities within a 30-min walk. In this study, accessibility is divided into 7 levels: accessibility between 0-5min, accessibility between 6-10min, accessibility between 11-15min, accessibility between 16-20min, accessibility between 21-25min, accessibility between 26-30min and accessibility above 30min. The spatial accessibility of medical facilities in areas within 15 min is relatively well equipped; Those within 15-30 min are less well equipped; And those above 30 min are poorly equipped.

4. Research results & analysis

4.1. Spatial accessibility of general hospitals

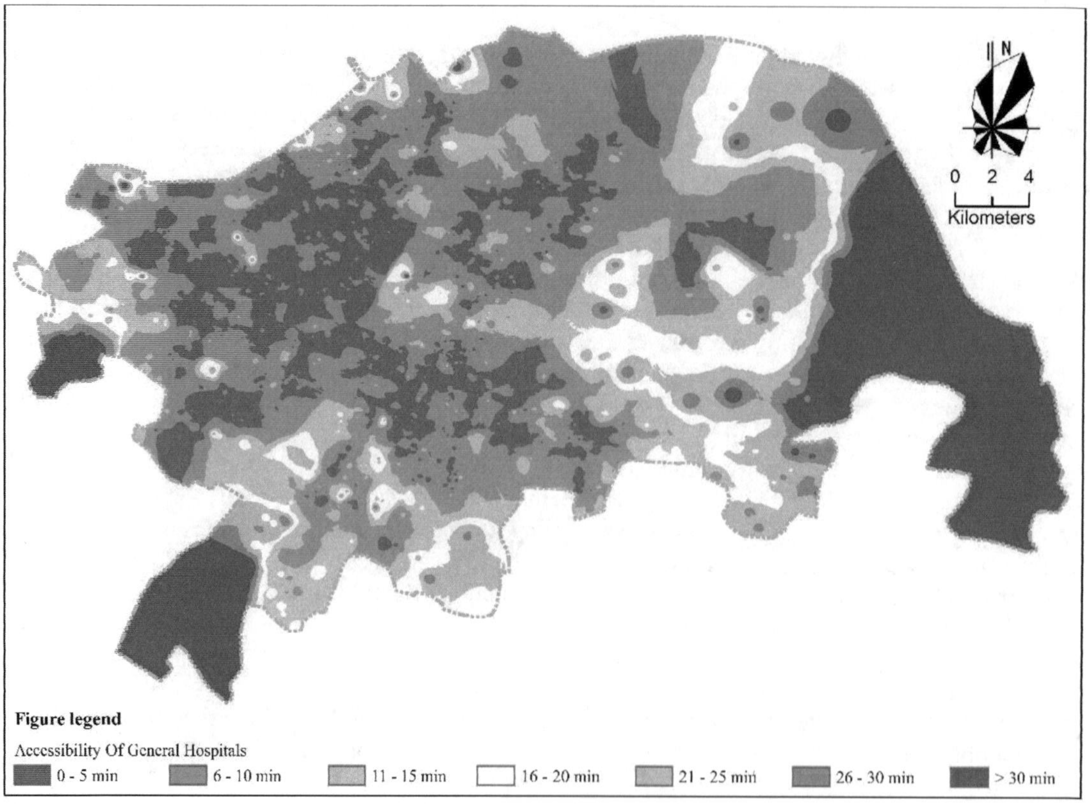

Figure 8. Accessibility of general hospitals. (During non-peak commuting hours).

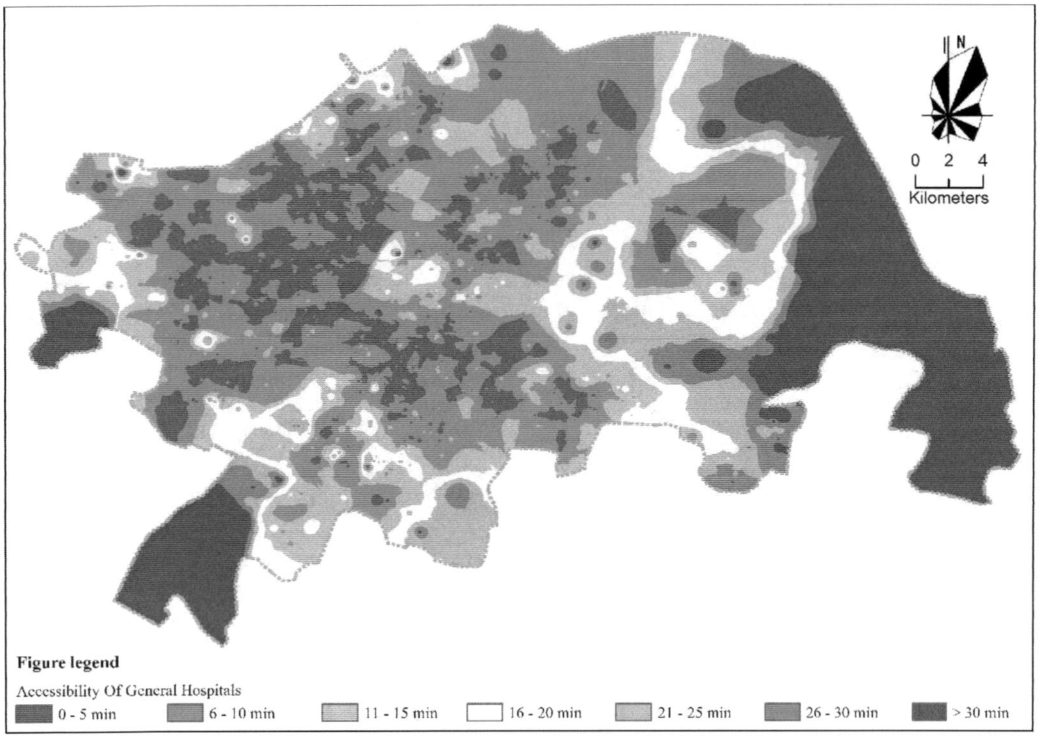

Figure 9. Accessibility of general hospitals. (During peak commuting hours).

Table 1. Statistical table of accessibility area ratio of general hospitals at all levels.

Spatial accessibility level	Impact factor (Bus speed) Non-peak(Peak) (m/min)	Accessibility ratio (Non-peak commuting hours) (%)	Accessibility ratio (Peak commuting hours) (%)
0-5 min	600(420)	16.02	12.00
6-10 min	600(420)	31.91	30.98
11-15 min	600(420)	15.95	16.01
16-20 min	600(420)	9.49	9.84
21-25 min	600(420)	6.96	7.14
26-30 min	600(420)	3.07	4.74
>30 min	600(420)	16.59	19.28

The accessibility of general hospitals is generally good during non-peak working hours. The areas with accessibility within 15 min accounted for 63.88% of the main urban area of Wuhan. With 16.02% of the study area with accessibility within 5 min, 31.91% of the study area with accessibility between 6 and 10 min and 15.95% of the study area with accessibility between 11 and 15 min. General hospitals with accessibility between 16 and 30 min accounted for 19.52% of the study area. The accessibility of general hospitals in the range of 16-30 min accounted for 16.59% of the study area. In terms of spatial distribution characteristics, Qiaokou District, Jianghan District, Jiangan District, Qingshan District and Wuchang District have good accessibility overall, with accessibility basically remaining within 20 min; The accessibility of general hospitals in the southwestern part of Yongfeng Street in Jianghan District

IOP Conf. Series: Earth and Environmental Science 1087 (2022) 012063 doi:10.1088/1755-1315/1087/1/012063

is above 30 min and the configuration of general hospitals is poor; The accessibility of general hospitals in a few areas in Chenjiaji Street, Houhu Street and the northern part of Tazihu Street is poor, ranging from 16 min and 30 min. The accessibility of general hospitals in most areas of Qingling Street in Hongshan District is poor, ranging from 16 to 30 min and the configuration of general hospitals is poor; The accessibility of general hospitals in most areas of Zhangjiawan Street is between 16 and 30 min and the accessibility is average; The accessibility of general hospitals in most areas of Hongshan Street is between 21 and 30 min and the accessibility is average; The accessibility of general hospitals in most areas of Wuhan East Lake Street is between 21 and 30 min and the accessibility is normal; The accessibility of general hospitals in Wuhan East Lake Street is poor and the configuration of general hospitals is poor. The accessibility of most areas in Hongshan Street is between 21~30min with average accessibility; The accessibility of the southern and eastern parts of Wuhan East Lake National Innovation Demonstration Zone is poor, above 30min; The accessibility of the South-eastern part of the East Lake Ecotourism Landscape is average, between 16~25min; The accessibility of most areas in the eastern part of Wuhan Chemical Industry Zone is poor, above 30min with poor configuration of general hospitals. See Figure 8 and Table 1 above.

The spatial distribution characteristics of the accessibility of general hospitals during the peak commuting hours does not change much compared to the non-peak commuting hours. But the proportion of areas with accessibility of general hospitals within 5 min decreases by about 4% compared to the former. The proportion of areas with accessibility between 6 and 10 min has decreased by approximately 1% compared to the former. Overall, accessibility has decreased. See Figure 9 and Table 1 above.

4.2. Accessibility of Specialist hospitals

The accessibility of specialist hospitals is relatively good due to their small size and relatively good configuration. During peak commuting hours, the areas within 15 min of the specialist hospital facilities accounted for 89.98% of the study area, with 43.48% of the study area within 5 min, 35.14% of the study area between 6 and 10 min and 11.36% of the study area between 11 and 15 min. In terms of spatial distribution characteristics, accessibility is generally good. The accessibility of Hanyang, Qiaokou, Jianghan, Qingshan and Wuchang districts is good and most streets in those districts within 15 min. The accessibility of specialist hospitals in the eastern part of the Wuchang East Lake Ecotourism Landscape is poor, between 16-30min; The accessibility of specialist hospitals in the north-western part of Zhangjiawan Street in Hongshan District is poor, between 16 and 20 min; The accessibility of the specialist hospital in the eastern part of Wuhan East Lake National Innovation Demonstration Zone is poor, between 16-30min. See Figure 10 and Table 2 below.

Compared to the non-peak commuting hours, the percentage of areas with accessibility within 5 min is about 10% lower in specialist hospitals. The percentage of areas with accessibility between 6 and 10 min decreases by approximately 3% and the percentage of areas with accessibility between 11 and 15 min increases by approximately 5%. There was also a small increase in the percentage of areas with accessibility between 16 and 30 min and over 30 min. Overall the accessibility of specialist hospitals has decreased. In terms of spatial distribution characteristics, accessibility is poor in the southern part of Yongfeng Street, Qingling Street, the south-eastern part of the East Lake Ecotourism Landscape, the eastern part of Wuhan East Lake National Innovation Demonstration Zone and the eastern part of Wuhan Chemical Industry Zone. See Figure 11 and Table 2 below.

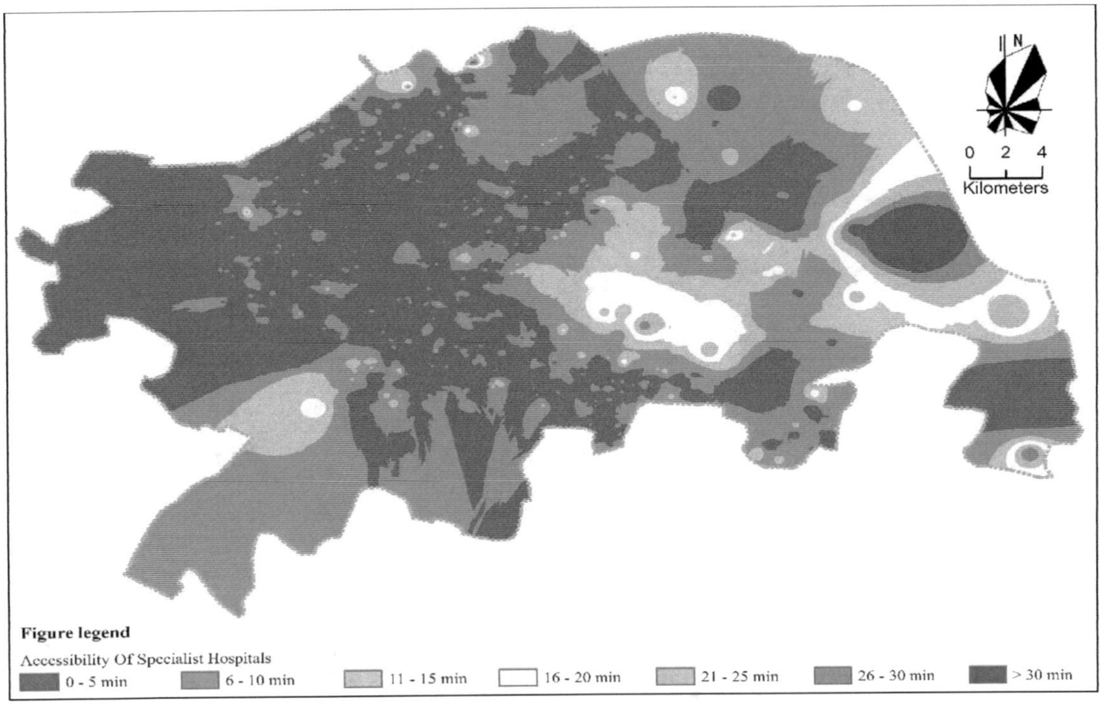

Figure 10. Accessibility of specialist hospitals. (During non-peak commuting hours)

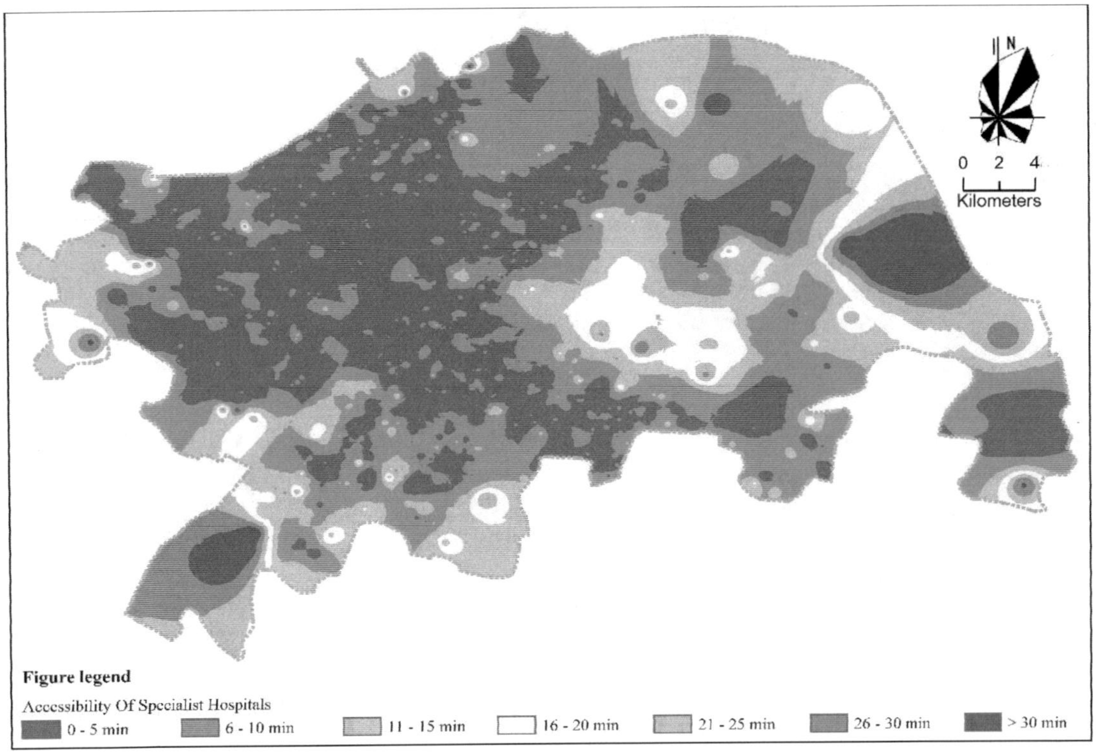

Figure 11. Accessibility of specialist hospitals. (During peak commuting hours).

Table 2. Statistical table of accessibility area ratio of specialist hospitals at all levels.

Spatial accessibility level	Impact factor (Bus speed) Non-peak(Peak) (m/min)	Accessibility ratio (Non-peak commuting hours) (%)	Accessibility ratio (Peak commuting hours) (%)
0-5 min	600(420)	43.48	32.85
6-10 min	600(420)	35.14	32.04
11-15 min	600(420)	11.36	16.82
16-20 min	600(420)	5.19	8.32
21-25 min	600(420)	2.11	3.86
26-30 min	600(420)	1.02	3.22
>30 min	600(420)	1.48	2.88

4.3. Accessibility of community medical facilities

Figure 12. Accessibility of community hospitals. (During non-peak commuting hours).

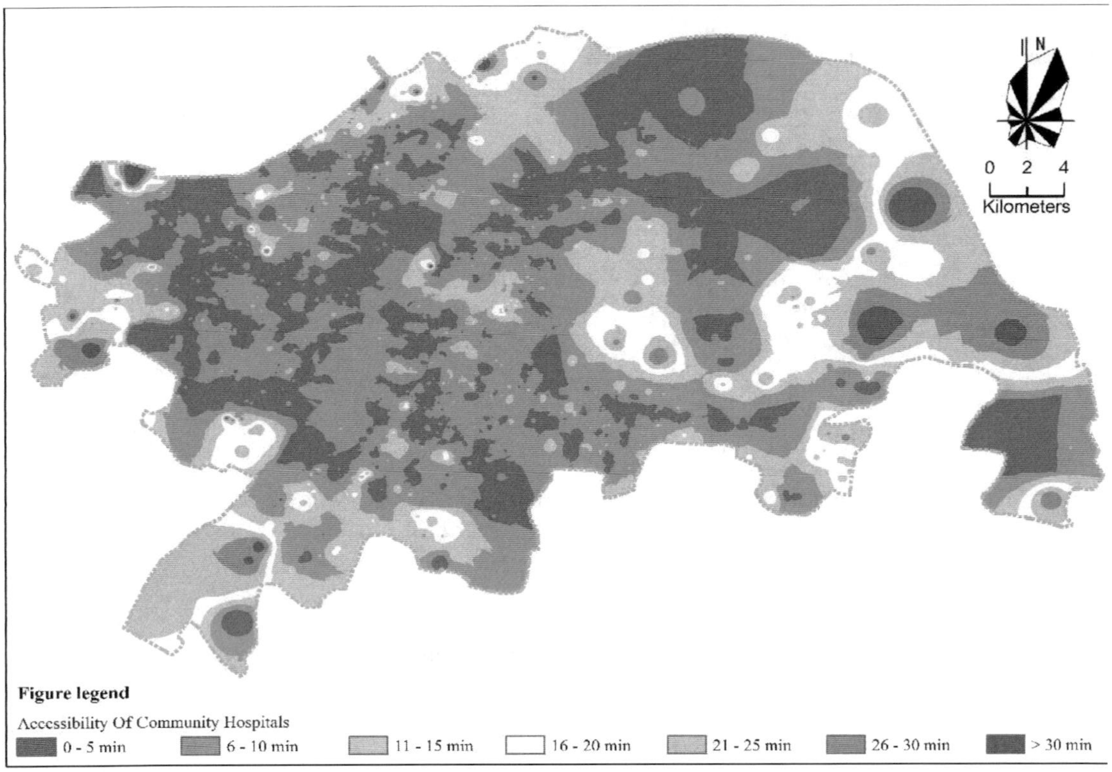

Figure 13. Accessibility of community hospitals. (During peak commuting hours).

Table 3. Statistical table of accessibility area ratio of community hospitals at all levels.

Spatial accessibility level	Impact factor (Bus speed) Non-peak(Peak) (m/min)	Accessibility ratio (Non-peak commuting hours) (%)	Accessibility ratio (Peak commuting hours) (%)
0-5 min	600(420)	29.39	24.69
6-10 min	600(420)	37.66	37.56
11-15 min	600(420)	15.47	16.24
16-20 min	600(420)	7.91	9.35
21-25 min	600(420)	6.72	7.43
26-30 min	600(420)	2.25	3.63
>30 min	600(420)	0.7	1.11

The accessibility of community healthcare facilities during non-peak commuting hours is generally good. Areas with accessibility within 15 min accounting for around 82.52% of the study area. Areas with accessibility within 5 min accounting for 29.39% of the study area. Areas with accessibility between 6-10 min accounting for 37.66% of the study area and areas with accessibility between 11-15 min accounting for 15.47% of the study area. In terms of spatial distribution characteristics, the overall spatial accessibility is good and the accessibility of community medical facilities shows an overall more equal spatial distribution. The vast majority of areas in the central city of Wuhan having good spatial accessibility to community medical facilities within 15 min. The spatial accessibility of the northern specialist hospitals in Houhu Street and Danchi Street in Jiangan District is poor, ranging from 16-30min; The spatial accessibility of community medical facilities in Jiyuqiao Street in Wuchang District is poor, ranging from 16-30min; The spatial accessibility of community medical facilities in the northeast and southeast of Wuhan East Lake National Innovation Zone in Hongshan District and the central part of Lishan Street is poor, ranging from 16-25min; The accessibility of the northern part of Yijiadun Street in Qiaokou District, the southern part of Yongfeng Street in Hanyang District and the southern part of Hanyang District is poor, ranging from 16-25min. See Figure 12 and Table 3 above.

The spatial distribution characteristics of community hospital accessibility during peak commuting hours are similar to those during non-peak commuting hours. The percentage of areas within 5min of community hospital accessibility decreased by approximately 5%, and the percentage of areas with other levels of accessibility increased by approximately 1%. Compared to the non-working peak hours, there is little change in the overall accessibility of community hospitals during the working peak hours. See Figure 13 and Table 3 above.

5. Summary & discussions

The establishment of a scientific evaluation system for urban public service facilities can better promote the healthy development of cities. And it is of great significance for the establishment of habitable cities and better access to public services for urban residents. As one of the evaluation systems, the accessibility of good public services represents the convenience of access to services for residents. This paper uses ArcGIS network analysis to study the spatial accessibility of medical public services in the main urban area of Wuhan. And the results can provide a reference for the allocation of medical services and the construction of road traffic in the main urban area of Wuhan. The findings of the study are as follows:

In terms of spatial distribution, the accessibility of various public service facilities in the central part of the main urban area of Wuhan is better than that of medical public service facilities in the western and eastern parts of the city. There are obvious imbalances and inequalities in the spatial distribution of such services. In terms of the accessibility of medical facilities, the time cost for residents to access these facilities varies considerably, with a cliff-like transition between areas with better accessibility and those with poorer accessibility. In terms of public transport systems, the planning and layout of urban bus routes is relatively well developed. The accessibility of medical services in the main urban areas is significantly improved by the combination of walking and public transport (bus and metro).

There are improvements to the study: In terms of data, the accuracy of the results of the ArcGIS network analysis depends on the accuracy of the urban road data and various POI data. In practical applications, it is important to ensure that the accuracy of the data is very high or to cooperate with relevant government departments and research institutions that can provide accurate data. In this paper, the central mass points of residential neighborhoods and various public service facilities are used as the starting and destination points of residents in each neighborhood. Those points cannot accurately represent the real starting and destination points of residents. For roads, the accuracy of the roads crawled in this paper has some missing sidewalks and the road network lacks a certain degree of completeness. In terms of analysis parameters, the waiting time of residents at bus stops and rail stations is not taken into account. And it is assumed that residents can take the bus when they arrive at the station which does not truly reflect the real travel situation of residents. In future studies, we can try to obtain route planning data to determine the best route for residents to reach the corresponding public service facilities and then take into account the waiting time for buses or trains. At the same time, we can improve the pedestrian network, bus data and subway data to further optimize the data acquisition and parameter setting. Meanwhile, this study takes into account the influence of the economic income level of urban groups on residents' travel to medical facilities. And it can be included in future studies by obtaining data related to the economic income level of urban residents to make the research results more accurate and informative.

References

[1] Meng Li 2020 Study on the Accessibility and Service Level of Urban Park Green Space Based on GIS: A Case Study of Yuelu District of Changsha City **02** 502-565
[2] Mengying Fu 2020 Study on Accessibility of Park Green Space in Binhu New District of Hefei City Based on Network Analysis **02** 615-629
[3] Shan Zhang 2020 Spatial Accessibility Analysis of Public Medical Facilities Based on GIS **02** 630-633
[4] Xue H, Zihe Shu, Li S, kaiting Ji, Xiangyu Bao 2021 Researches on Accessibility of Urban Public Park Based on GIS Network Analysis - A Case Study of Chizhou City **02** 634-636
[5] Xihai Ma 2019 Research on Spatial Layout of CommunitySport Facility Based on General Accessibility - Taking Gusu District of Suzhou as an Example **02** 634-636
[6] Bo W, Xiaopeng Jing, Meihan Wu 2021 Research on accessibility of urban sports Facilities based on network Analysis - Take Yanta District of Xi 'an as an example **02** 634-636
[7] Mengying Fu 2020 Study on Accessibility of Park Green Space in Binhu New District of Hefei City Based on Network Analysis **02** 634-636
[8] Zihan Gao 2022 Research on point accessibility of medical facilities in Yongji City based on network analysis and two-step mobile search method **02** 641-644
[9] Shaoqin Zhou, Tingting Li, Shuren Luo, Chaoping Pan, Cuilian Yang, Shaowu Zhuang, Menghua Pan, Yanning Li 2019 Research on spatial accessibility of medical service in Guangxi based on geographic information System **02** 645-647

IOP Conf. Series: Earth and Environmental Science 1087 (2022) 012064 doi:10.1088/1755-1315/1087/1/012064

Field Characteristics of Sedimentary Rocks in the Dohol Formation, Johor, Peninsular Malaysia

Alidu Rashid[1,3], Numair Ahmed Siddiqui[1], Nisar Ahmed[1], Muhammed Jamil[1], Daniel Asante Otchere[2] and Sani Ado Kasim[1]

[1] Department of Petroleum Geoscience, Universiti Teknologi PETRONAS, 32610, Seri Iskandar, Perak Darul Ridzuan, Malaysia;
[2] Department of Petroleum Engineering, Universiti Teknologi PETRONAS, 32610, Seri Iskandar, Perak Darul Ridzuan, Malaysia.

[3] Corresponding author's e-mail: alidurashid22@gmail.com

Abstract. The Dohol formation is predominantly composed of sedimentary rocks; however, the black shales attract much attention because of their potential to generate hydrocarbons. Black shales generally have a high organic matter content and can serve as good source rocks for hydrocarbon generation. Therefore, this study was conducted to characterize the sedimentary rocks, including the black shales, which are of great importance and discover new areas where these rocks can be located. Four distinctive lithofacies of shales were observed on the field; the massive black shales facies, light grey shales facies with quartzite intrusion in some areas, the red sandy shales, and the dark brown shales facies. The black shales are dark-coloured, laminated, and some parts are fissile. The second type of shale, light grey, looks similar to the dark shales, but it probably contains less organic matter. In other areas, these grey shales have intrusions of quartzite. The third type of shale facies are the red shales which are sandy, fissile, and highly weathered, and pebbly sands overlie it. Finally, the dark brown shale facies are fissile, flaky, and weathered. The sandstones encountered are massive cream-colored, and composed of mainly quartz with a few scour marks in some areas. Several new outcrop locations of these sedimentary rocks were discovered, and they are spread across Jemaluang, Kota Tinggi, and the outskirts of Mersing town in East Johor. From the outcrop trends and the sedimentary features of the shale rock sequence, the deposition environment of the shales is probably from shallow to deep marine.

1. Introduction

Sedimentary rocks cover some 80 percent of the Earth's crust. Much information obtained from stratigraphy and structural geology is based on sedimentary rock studies [1]. Common sedimentary rocks include sandstone, limestone, and shale. Sandstones are clastic sedimentary rocks formed through the deposition processes comprised of weathering, erosion, transportation, and compaction of different sand grains [2]. In recent times sandstones are of great importance because they can serve as hydrocarbon reservoirs in an unconventional manner such as tight gas. Tight gas is referred to as sandstone reservoirs with very low permeability, can only produce hydrocarbons in commercial quantities after implementing enhanced oil recovery [3]. Shale is formed from clay, silica, carbonate minerals, and organic material. Shale is usually fissile and terrigenous, with particles that are primarily of silt and clay size [4]. Even though shale formations are thought to be rich in clay minerals, the proportions of the constituents are highly variable at times [5]. Shale is based on organic content in

IOP Conf. Series: Earth and Environmental Science 1087 (2022) 012064 doi:10.1088/1755-1315/1087/1/012064

two general varieties. The Dark-coloured shales, also known as black shale formations, are rich in organic matter, while lighter-coloured formations have relatively lower organic matter. The presence of organic material in shales brings about the opportunity to generate hydrocarbon. In addition, natural gas might be produced if the organic material is preserved and adequately heated after burial [6]. Therefore, shales can serve as a good shale gas resource, and there is a need to exploit that potential in black shales as they are usually capable of generating hydrocarbons.

The Dohol formation is one of the 19 Paleozoic formations in Malaysia that contain black shale [7]. The Dohol formation comprises grey-coloured carbonaceous shale, siltstone, slate, phyllite, and schists with insignificant arenite beds, limestone, and volcanic [8]. The Dohol formation is a calcareous unit, the 'Sumalayang limestone' containing Middle Permian fusulinids deposited in an inner neritic shelf environment [9]. Another unit in South Johor called 'Linggiu formation' is continuous and like the Dohol formation [10]. However, the arenaceous section in this unit is better developed, resulting in some rhythmic bedding of shale and sandstone and the local prominence in some places of conglomeratic beds. Plant fossils found in the 'Linggiu formation' indicate a Late Permian age.

As the global significance of sandstones and shales increases due to the potential to serve as unconventional hydrocarbon resources such as tight gas, shale gas, tight oil. There is a need to fully understand the characteristics of these rocks to determine their potential. Therefore, this research aims to determine the various locations of these sedimentary rocks of the Dohol formation and provide a detailed description of the rocks.

2. Study area

The Dohol formation is east of Johor in Peninsular Malaysia (Figure 1). The Dohol formation was named from the Sungai Dohol River in east Johor. The Dohol formation falls under the eastern belt of Peninsular Malaysia. The Eastern belt comprises Carboniferous to Permian age sediments distributed from east Kelantan through Trengganu and east Pahang into east Johor in the south [11] (Figure 1a). A period of regional metamorphism, folding, and uplift likely occurred around the late Palaeozoic (Permian), followed by deposition of an older series of continental deposits such as the Murau and Redang conglomerates [12]. The pan-peninsula late Triassic orogenic event uplifted the Eastern Belt. Subsequent deposition of a younger series of gently dipping continental deposits was probably uplifted in the late Cretaceous [12]. The absence of marine sediments younger than Permian shows that the Eastern belt was uplifted towards the end of Permian, which tallies with the deposition of the Lower Triassic ignimbrite [11].

The Dohol formations are part of the Palaeozoic rocks in East Johor and are predominantly composed of metasedimentary to metamorphic, volcanic, and siliciclastics[13]. The ages are usually Carboniferous and Permian. Some of the key areas we found outcrops belonging to the Dohol formation were Kota Tinggi, Kluang, Jemaluang, and the outskirts of Mersing (Figure 1b). These rocks are grouped into seven (7) lithostratigraphic units, namely: Mersing, Marau, Dohol, Linngui, Sedili, Pengerang, and Tanjung Leman formations [13] (Figure 2).

3. Material & methods

Thorough fieldwork was undertaken to identify and describe the rocks that make up the Dohol formation and the stratigraphic units in the area. Fresh samples from the various Lithofacies were collected using the hammer. The hand lens, GPS, and camera are tools used to gather data on the rock formations successfully.

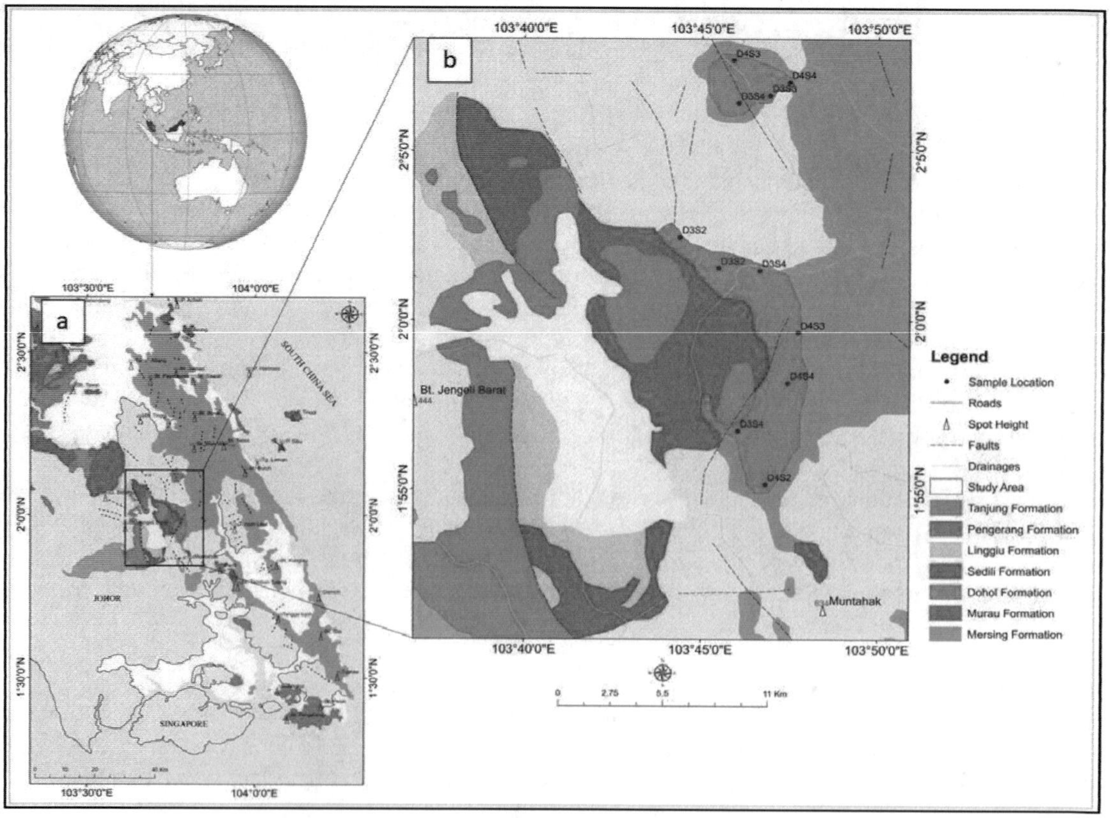

Figure 1. (a) Geological map of East Johor (b)Map of the study area.

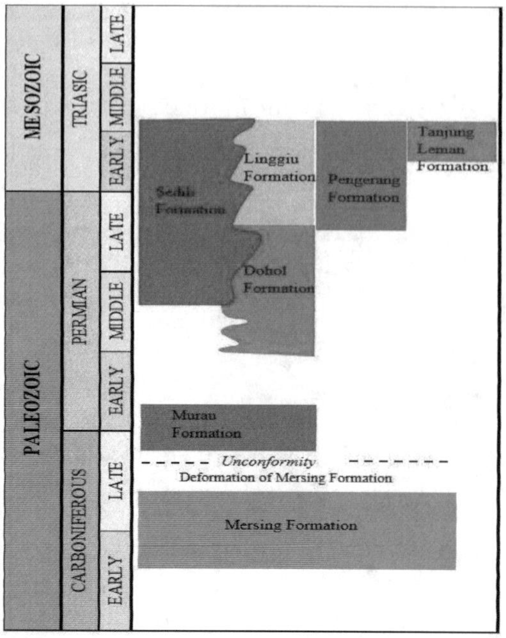

Figure 2. Stratigraphy of East Johor region with various Paleozoic formations.

4. Results and discussions

4.1. Sedimentological Observations

4.1.1. Red shale facies. These shale facies are medium to fine-grained and reddish in colour (Figure 3a). The bedding of the outcrops is made up of very thin laminae. This shale facies are highly weathered and fissile (Figure 3a). The shales have pebbly sands overlying them at some locations (Figure 3b) which indicates that the pebbly sands were deposited at the end of the flow. These shale facies are also altering to clay, and this is caused by the weathering of the shale (Figure 3c). These shale facies have been exposed by road cuttings (Figure 3d). Evidence was found which indicated that the red shale facies probably formed from the light grey shales. Outcrop of this shale facies can be found along the Jalan Felda Nitar Timur road.

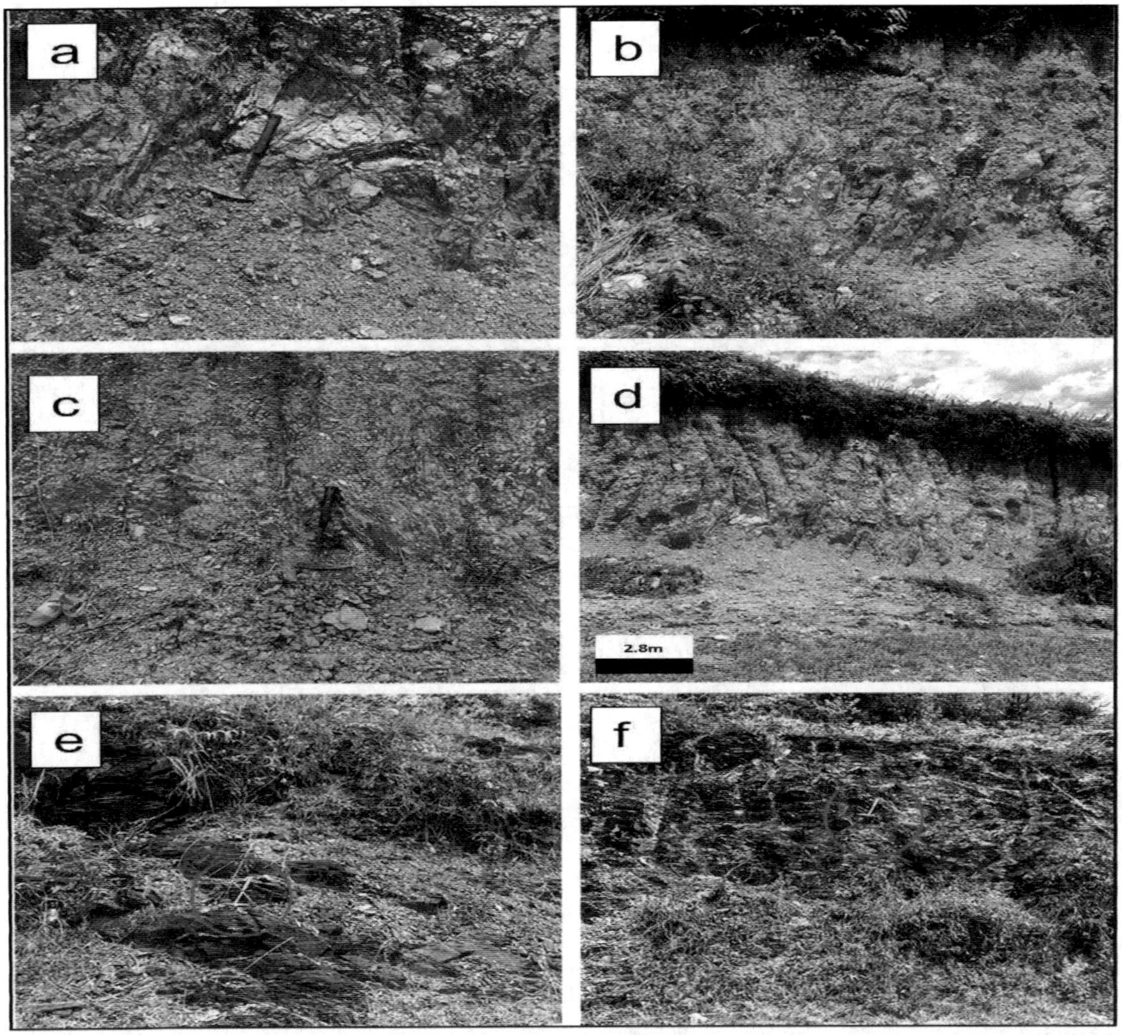

Figure 3. (a) Red weathered shale facies showing laminations (b) Massive red shale facies with sands at the top (c) Red shale facies in the process of altering to clay (d) Massive red shale facies exposed by road cutting(e) Massive black shale facies (f) black shales with silty lenses and laminations.

4.1.2. Black shale facies. These shale facies are fine-grained and dark grey in colour (Figure 3e). The bedding of the outcrops is very thin in some parts and medium bedded in other areas. These facies are thinly laminated and fissile in some parts, while others are massive (Figure 3f). The shales consisted of lots of cracks at some outcrop locations, probably because of the road cut (Figure 3e). Some of the outcrops also contained silty lenses found randomly in the samples (Figure 4a). The shales were also interbedded with pebbly sands at some locations (Figure 4b). Outcrops of this shale facies can be found in Jemaluang and along the Jalan Felda Nitar Timur road.

Figure 4. (a) black shales with silty lenses and laminations (b) Black shale facies interbedded with pebbly sands (c) Light grey shale facies showing cracks and laminations (d) Light grey shale facies boundary with sandstone (e) Light grey shale facies covered with vegetation (f) Evidence showing the light grey shales weathered to the red shale facies.

4.1.3. Light grey shale facies. These shale facies are fine-grained and light grey in colour (Figure 4c). The mineral grains are moderate to well sorted. The bedding of the outcrops comprises thin laminae in some areas and is medium bedded in other areas. These facies are thinly laminated and fissile in some parts, while others are hard. Some of the outcrops share a boundary with sandstones (Figure 4d). At some locations, the light grey shale facies were covered with vegetation, and evidence was found which indicated that the red shale facies probably formed from the light grey shales (Figure 4e & 4f). The shales consisted of lots of cracks at some outcrop locations, probably because of the road cut (Figure 4c). The light grey shales were also found to be in the process of weathering at some outcrop locations (Figure 5a). The outcrops are seen to have intrusions of quartzite (Figure 5b and 5c). Outcrops of this shale facies can be found in Kota Tinggi and along the Jalan Felda Nitar Timur road.

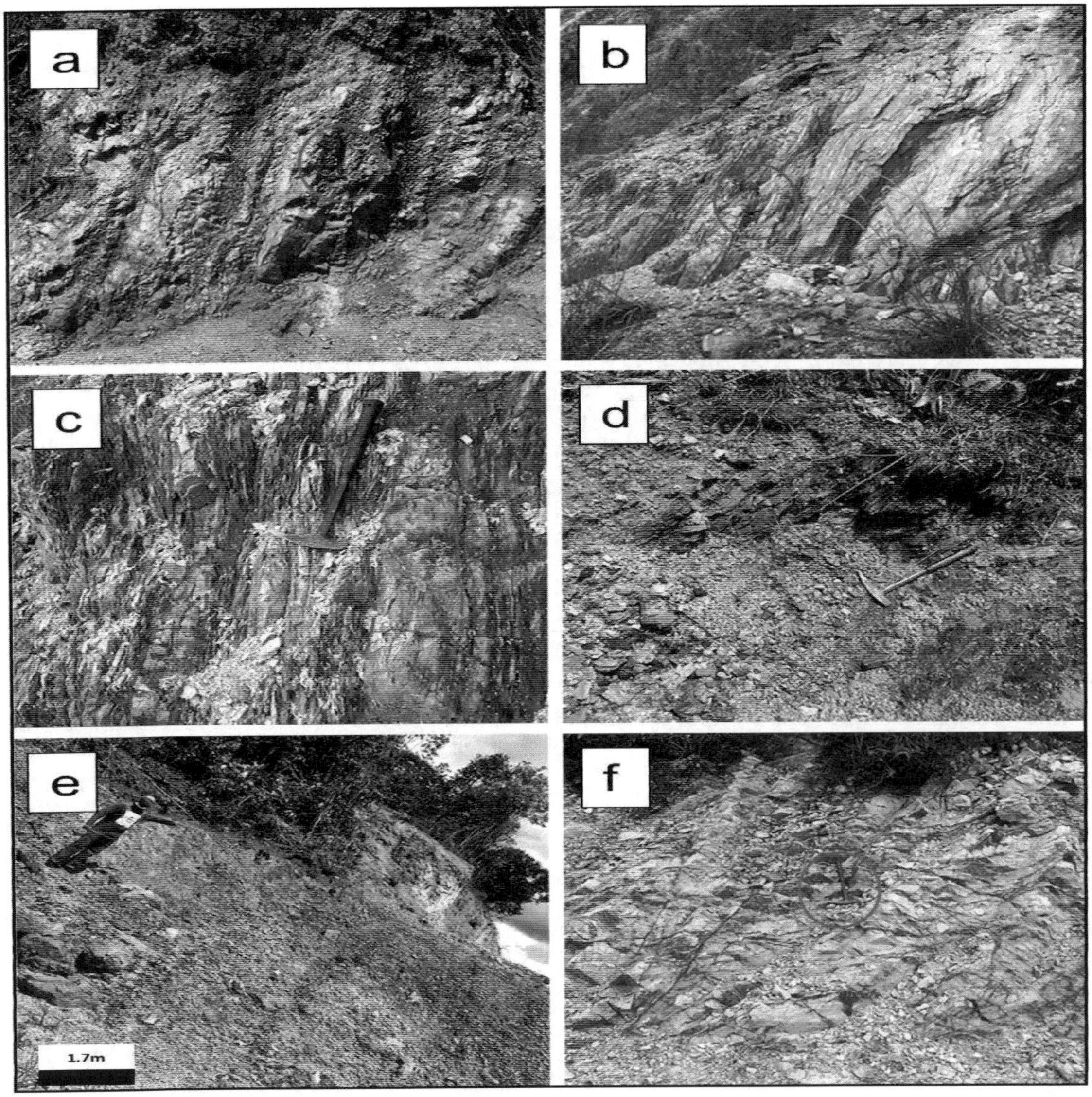

Figure 5. (a) Light grey shales in the process of weathering (b) Light grey shale facies showing thick bedding (c) Light grey shale facies with intrusions of quartzite (d) Fissile dark brown shale facies (e) Evidence showing the dark brown shales are probably weathered light grey shale facies (f) Sandstone showing abundant cracks.

4.1.4. Dark brown shale facies. These shale facies are fine-grained and dark brown in colour (Figure 5d). The mineral grains are moderate to well sorted. The bedding of the outcrops is made up of very thin laminae. These shale facies are highly weathered, fissile, and flaky (Figure 5d). The outcrop is located a few meters from light grey facies, and it is assumed that this outcrop is the weathered version of that light grey shale facies (Figure 5e). The outcrops can be found at Kota Tinggi.
4.1.5. Sandstone facies. The sandstones encountered are massive, cream coloured and composed of mainly quartz which is medium to well sorted (Figure 5e & 6a). The rocks are also highly fractured at some locations, with lots of cracks found in them probably due to the road cutting (Figure 5e). Scour marks were also observed on some of the outcrops (Figure 6b). Cross bedding was observed in some outcrops of the sandstones. The outcrops also shared boundaries with shales at some locations. Outcrops of these sandstones were in Kluang, Kota Tinggi, and along the Mersing-Jemaluang Road.

Figure 6. (a) Massive sandstone outcrop (b) Sandstone outcrop showing scour marks and cross bedding.

5. Conclusions
The preliminary study indicates the presence of four (4) different shale facies of different colours, which can be associated with weathering and probably different mineral components. The sediments of the different shale facies might have been deposited in low-energy conditions due to the uniform gradation of the sediments. The dark shale facies are assumed to have higher organic content due to the popular notion that dark shales have high organic content. This will be confirmed later through laboratory analysis. The alternating dark grey shale facies and sands could result from laminar flow deposition of the sediments. Petrographic analysis and organic geochemical analysis will be performed on the samples to classify the rocks further.

6. Acknowledgement
The authors would like to express appreciation to Fundamental Research Grant, Ministry of Higher Education (MoHE) Malaysia (Project ID 16880, Reference Code FRGS/1/12019/STG09/UTP/03/1) and YUTP Grant (015LC0-363).

References
[1] R. L. Folk 1980 *Petrology of sedimentary rocks* Hemphill publishing company
[2] S. K. Haldar and J. Tišljar 2014 "Chapter 1 - Rocks and Minerals," in *Introduction to Mineralogy and Petrology*, S. K. Haldar and J. Tišljar Eds. Oxford: Elsevier pp. 1-37
[3] C. Zou, "Chapter 5 - Shale Gas," *Unconventional Petroleum Geology*, pp. 149-190, 2013/01/01/ 2013, doi: 10.1016/B978-0-12-397162-3.00005-0
[4] H. Blatt, R. Tracy, and B. Owens 2006 *Petrology: igneous, sedimentary, and metamorphic.*

Macmillan
[5] J. Speight 2020 "Chapter 1 - Gas and oil in tight formations," in *Shale Oil and Gas Production Processes*, J. Speight Ed.: Gulf Professional Publishing pp. 3-64
[6] J. G. Speight 2016 "1 - Introduction to fuel flexible energy," in *Fuel Flexible Energy Generation*, J. Oakey Ed. Boston: Woodhead Publishing pp. 3-27
[7] H. Baioumy, Y. Ulfa, M. Nawawi, E. Padmanabhan, and M. N. A. Anuar, "Mineralogy and geochemistry of Palaeozoic black shales from Peninsular Malaysia: Implications for their origin and maturation," *International Journal of Coal Geology* **165** 90-105, 2016/08/01/ 2016, doi: https://doi.org/10.1016/j.coal.2016.08.007
[8] L. Peng, M. Leman, R. Karim, and L. Peng 2004 Stratigraphy Lexicon of Malaysia
[9] F. K. Yee 1983 The Palaeozoic sedimentary rocks of Peninsular Malaysia-stratigraphy and correlation
[10] S. S. Rajah 1986 Geology and mineral resources of the Gunong Belumut area, Johor: Dohol Formation *Geological survey district memoir* **19** 25-34
[11] C. S. Hutchison and D. N. K. Tan 2009 *Geology of peninsular Malaysia*. Published jointly by the University of Malaya [and] the Geological Society
[12] T. Khoo and B. Tan 1983 Geological evolution of peninsular Malaysia
[13] S. S. Surjono, M. S. Leman, K. R. Mohamed, and C. A. Ali 2020 A Combined Petrographic-Geochemical Provenance and Tectonic Setting Study of Palaeozoic Rocks, in East Johor Basin, Peninsular Malaysia *Indonesian Journal on Geoscience* **8(1)** 11-24, doi: 10.17014/ijog.8.1.11-24

Hydrogeological studies in the São Torcato Zone -Tabuaço with aim of optimizing the location of a special groundwater abstraction to plan a new Medical Spa in Portugal

Ferreira-Gomes L M[1,2,4], Machado Fonseca A M[3], Andrade Pais L J[1,2] and Carvalho P E M[1,2]

[1] University of Beira Interior, 6200-001 Covilhã, Portugal;
[2] GeoBioTec, Centro de Investigação;
[3] Liberal Professional, 5121-909 Tabuaço, Portugal.

[4] Corresponding author's e-mail: lmfg@ubi.pt

Abstract. The present work presents a case study, to evaluate the possibility that in an area of the territory of the interior of Portugal, there is potential for capturing special groundwater that allows its classification as natural mineral water for thermalism. In this sense, after an initial part, with the framing of the theme and the presentation of the work methodology, the geomorphological and geological elements are shown, giving more development to the hydrogeological aspects, highlighting the results of the monthly sequential hydrological balance, namely its annual surplus, crossing it with the groundwater sampling sites, and still, the existing aquifer systems and the characteristics of their waters are identified. As a main result, the conceptual geohydraulic model of the region is presented, to show that in the São Torcato zone, and particularly in the Qta do Plácido zone, there is the possibility of exploiting groundwater in an aquifer system of granitic rocks, from fissural type, which is supplied from relatively distant areas, in hydrographical basins other than São Torcato basin, provided relatively deep boreholes are to be drilled. Finally, the main conclusions are synthesized, pointing out the most favorable area to carry out groundwater abstraction with the potential to have stability and quality suitable for a medical spa. It should be noted that these elements are only previous studies, which should be considered as initial contributions, and that further work should be carried out in the future, including geophysical studies and mechanical drilling, flow tests, as well as, among others, isotopic studies in the waters of the region to assess the recharge areas and available reserves.

1. Introduction

The area under study, located in the interior of Portugal (Figure 1), is of great interest from a touristic point of view. The region where it is integrated has been classified as a UNESCO World Heritage Site, especially for its landscapes linked to the Alto-Douro wine region [1]. However, it continues to lose population and consequently, with climate change, can lead to situations of risk of desertification. Tourism is one of the possible levers, in the sense of contributing to tying up populations. Hotels are emerging in this activity segment, and to have more potential to attract more users, there is the advantage of associating Medical Spa type equipment to them. That equipment essentially constitutes a specific bathhouse with medical supervision. The specific bathhouse has swimming pools, whirlpool tanks, various types of baths, various types of showers, steam treatments, among many others. It should be noted that, for those equipments to be of the medical spa type, they must be supplied by special groundwater, called "natural mineral water" (NMW). The advantage for the society of being a medical spa is that it is mandatory to have two specific professionals accompanying them: the Technical Director (TD) and the Clinical Director (CD). The TD, a hydrogeological professional, guarantees adequate exploitation of the NMW, ensuring the quantity and quality of the same. The CD is a professional in the field of medicine who administers the type of treatment with AMN, directly or indirectly accompanying the users. Thus, the user of that hotel, if using the medical spa, he is guaranteed to have quality water and a doctor always available for advice, guidance, and other medical support.

The classification of groundwater, as NMW, follows the guidance of the definition presented in the Portuguese Law of 2015 [2], mentioning that "natural mineral waters are bacteriologically proper waters, of underground circulation, with physical-chemical particularities stable at the origin, within the range of natural fluctuations, from which possible therapeutic properties or effects favorable to health may result". It is emphasized that, from the authors' experience, the particularity of water, in physical-chemical terms, being stable at origin, i.e. at the head of the groundwater abstraction, is generally accepted when the results of the analyses (carried out on 12 samples, taken monthly, in a hydrological year), present the majority of its physical-chemical parameters, with a standard deviation in relation to the average value, less than 10%, and without showing any tendency throughout the year. Examples of classification of new NMW can be observed in the authors' works [3,4,5]. It is worth mentioning, the particularity, of the total mineralization or any other physical-chemical parameter, not have to meet a minimum value; the key is to present the stability of the various physical-chemical parameters over a hydrological year. It is also important to refer that after the classification of NMW, approved by the Portuguese State, the water can have two applications: one for bottling and another for thermalism. In the case of having potential for thermalism, the same water must be subjected to a medical-hydrological study, under the responsibility of a doctor of hydrology, in order to prove that it has beneficial effects on the health of users who undergo hydrotherapy treatments and others, in a small medical spa type bathhouse, temporary and of experimental character.

The present paper is dedicated to the study of the Qta do Plácido zone, next to the São Torcato Chapel, where a hotel unit was recently installed, near the Tabuaço urban area (Figure 1). The question arose: is it possible to extract groundwater in the hotel zone, or close to it, in order to have the quantity and quality to be classified as NMW?

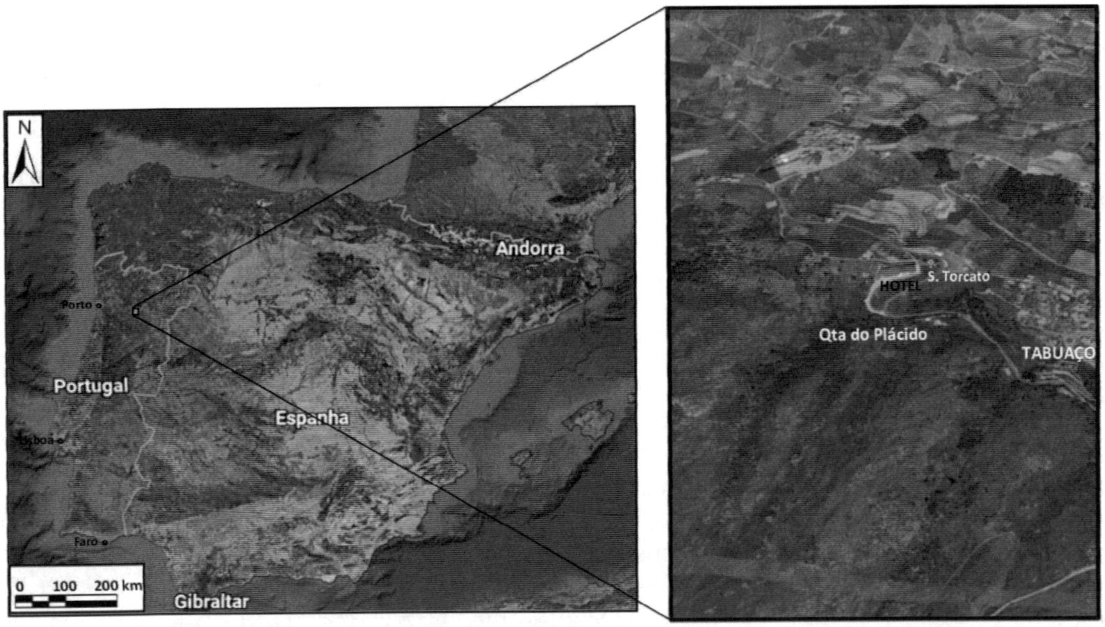

Figure 1. Localization of the study area: Qta do Plácido/São Torcato zone - Tabuaço (from [6]).

IOP Conf. Series: Earth and Environmental Science 1087 (2022) 012065 doi:10.1088/1755-1315/1087/1/012065

The present work aims to answer that question. It is worth mentioning that answering such a question is very important, because if there is no potential to explore special groundwater, then one does not proceed to the next phases, avoiding high costs, not only in geophysical studies but also in drilling with mechanical boreholes. On the other hand, if this type of study leads to positive results, it can leverage financial funds to finance studies for subsequent phases.

Regarding elements of the hydrogeological forum for the study zone, before this work, there was nothing published in the literature. The elements obtained at an early stage were organized in the unpublished internal report [7], having recently been taken up in academic scientific work in 2021 [8], also unpublished. Just some very generic elements, about the São Torcato zone, were recently presented in a work entitled Hydrogeoenvironmental studies in a region of the interior of Portugal to potentialize the birth of a new medical spa" [9].

2. Methodology

The work carried out was organized according to the following phases: i) Phase 1 – Studies on previous work; ii) Phase 2 – Initial office work; iii) Phase 3 – Field work, including geological survey, inventory of water points and their characterization by quick physical-chemical parameters; iv) Phase 4 – Advanced office work. Regarding the work of the various phases, some aspects mentioned below are worth mentioning.

About phase 1, for the Tabuaço region, it was concluded that there is no works on the theme of hydrogeology, however it is highlighted the existence of topographical maps at scale 1/25000 [10] and geological maps at scale 1 /50000 [11,12]. In the descriptive memories, of the latter there is some references to hydrogeological aspects, but to their vast region, in places far from Qta do Plácido.

On phase 2, in the office, some basic work was carried out, using the available maps, topographic cross-sections and geological sections, in order to construct a basic conceptual model to guide the following work, being worth emphasizing the decision to study in detail the small hydrographic basin of the Qta do Plácido (Figure 2).

Regarding phase 3, the fieldwork took place in May 2012, carrying out geological survey at scale 1/25000 and the survey/inventory of "water points", with in situ recording of quick physical-chemical parameters. It is considered a "water point", the place with groundwater, to the topographic surface, with conditions for measuring its physical-chemical parameters, besides, to allow sampling of its water. The "water points" of the studied zone were: 6 springs, 3 wells, 2 mines, 2 drains and 1 mini-dam. The quick physical-chemical parameters recording device was a Eutech Instruments multiparametric meter, to measure pH, redox potential (Eh), temperature (T), conductivity (C), and total dissolved solids. (TDS).

Regarding the work of the last phase, several maps were produced, namely the hydrogeological map, with the systematization of the various hydrogeological units considered, the implantation of "water points", among others. It is emphasized that, crossing all the available elements in this phase, namely topographic, geological, and hydrogeological, the conceptual geohydraulic model was produced, refined in the academic work recently carried out [8]. Finally, it should be noted that at this stage, after several scenarios have been carried out, an area is presented for drilling mechanical exploration boreholes, with guidelines on the depth to be reached, to obtain groundwater with a high possibility of being classified as NMW.

3. Results

The main results of this work are materialized: i) in the regional conceptual geohydraulic model elaborated, and ii) the detailed hydrogeological map for the Qta do Plácido zone, with the implantation of 2 sites to carry out mechanical drilling to have a high probability to reach zones with potential to obtain groundwater with adequate characteristics to be classified as NMW.

Meantime, it should be noted that the topographic and geological maps presented in this work, which present some adaptations and additions, in relation to the previous ones [10,11,12], as well as the hydrogeological elements that are totally unpublished, among others, are part of the set of the results, and for these reasons are explained in this item.

3.1. Geomorphological aspects

In geomorphological terms, the most important aspects are the fact that the Tabuaço region is morphologically framed in areas that predominantly tilt to the north, a consequence of the particularity of the Douro River, north of the study zone (Figure 2a), which percolates from east to west, for have a development at altitudes of the order of 60m. In geomorphologic terms, the various main streams in the region are predominantly percolating from south to north and converging with the Douro River on its left bank. In that context, the Távora river and the Tedo river stand out, to the east and west of Tabuaço town, respectively.

The urban area of Tabuaço is developed at altitudes of around 550m. The areas of the main basin under study in greater detail (Figure 2b), present altitudes between 550m and 925m approximately, having an average slope of about 18%. That basin, which is called São Torcato basin, consists of two small streams, which have their beginning well marked around elevation of 850m, and converge to result in a stream of NE-SW direction, crossing the area of Qta do Plácido, here with degree 2 in relation to the Strahler classification [13].

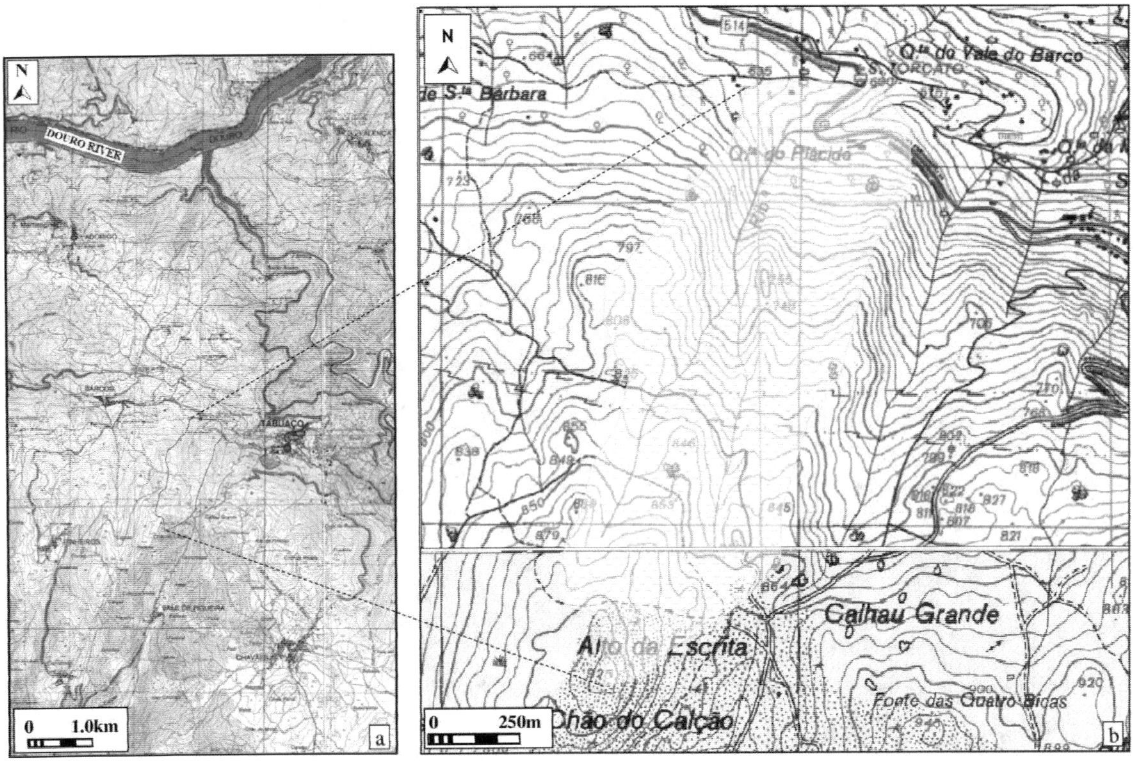

Figure 2. Global geomorphological setting of the main study zone, in relation to the Douro River (a), and details of the hydrographic basin of S. Torcato - b) (from [10]).

3.2. Geological aspects

The regional framework of the study area can be seen in Figure 3. The work zone is almost generally developed in granitic rocks, and only in the northern border occur shale rocks. In global terms, it is emphasized the fact that some extensive geological faults occur in the granitic massifs, most of them with NE-SW direction, especially the Vale Figueira Fault, which crosses an extensive region, and also the NNW-SSE faults, and in particular the Rio Tedo Fault. All these extensive faults favour groundwater recharge at great depths, also allowing percolation between several hydrographical sub-basins.

IOP Conf. Series: Earth and Environmental Science 1087 (2022) 012065 doi:10.1088/1755-1315/1087/1/012065

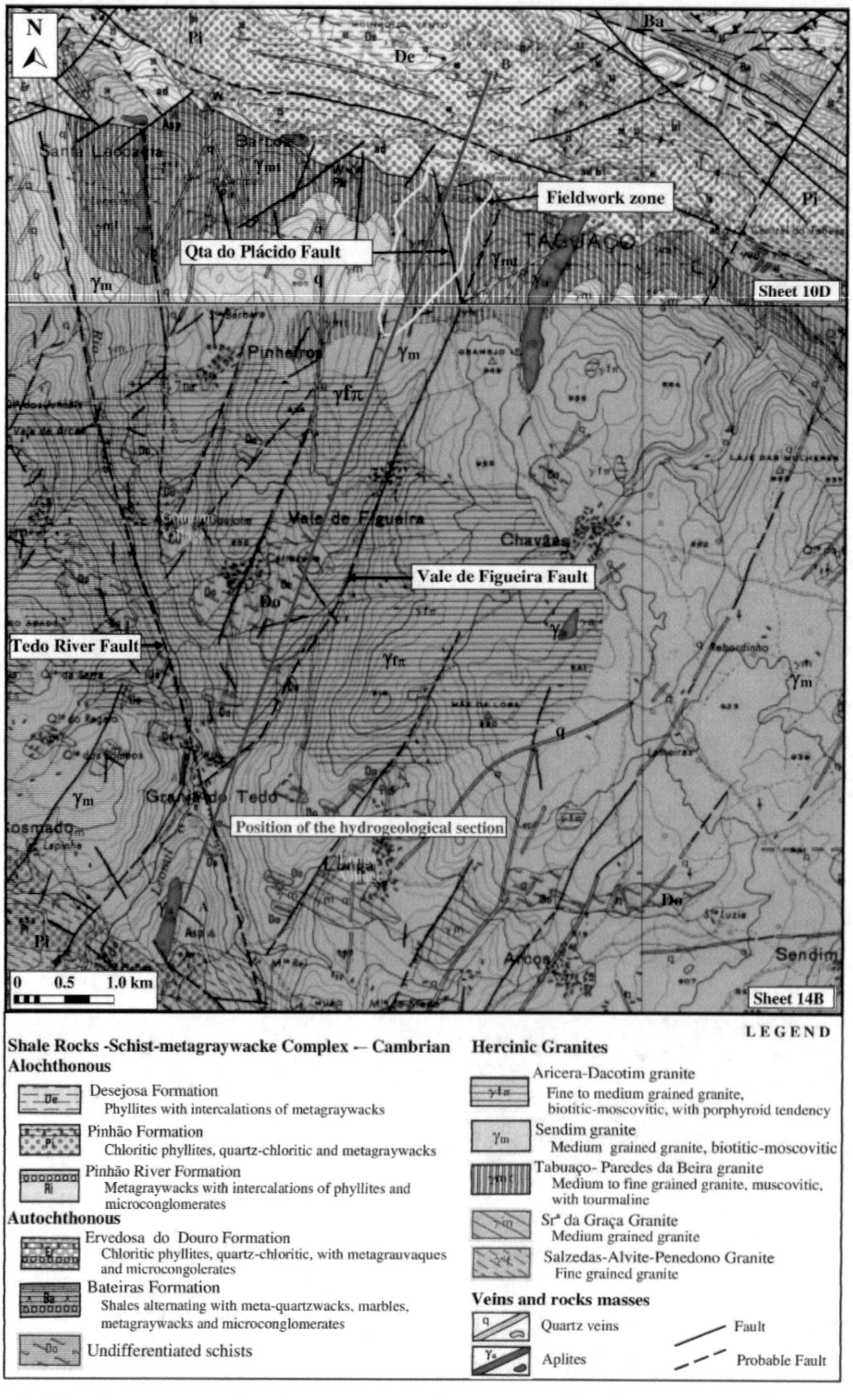

Figure 3. Geological map of the study zone and surrounding region (from the geological maps of Portugal, Sheets 10-D [11] and 14-B [12].

In local geological terms, a detailed field survey was carried out for the Qta do Plácido zone (Figure 4) the geological units, from the oldest to the most recent, according to the following:

. Pi - shale rocks, corresponding to the Pinhão Formation [11,12], consisting essentially of chlorite and quartz-chlorite phyllites and metaquartzowackes, of Cambrian age.

. γmt - medium to fine grained muscovitic granite with tourmaline, corresponding to the Tabuaço granite unit [11,12] of Hercynian age;

. γm - medium grained biotitic-moscovitic granite, corresponding to the Sendim granite unit, [11,12] of Hercynian age;

. a/Dv- alluvium and slope deposits, recent.

It should be noted that the higher altitude zones present granitic rocks, and the lower altitude zones, present shale rocks. It is mentioned that the Slope Deposits correspond to an unrepresentative and thin unit.

3.3. Hydrogeological aspects

3.3.1. Hydrological Elements. To have a notion about the groundwater potential from the recharge of rainwater with infiltrations in the basin itself, the monthly sequential hydrological balance was carried out, using the methodology proposed by Thornthwaite and Mather [14], resulting the elements presented in Table 1 and Figure 5. From all the results, it is emphasized that for the present work, the fundamental is the water surplus (WS), being in annual terms the value of 193.11 mm. The WS value is the sum of two components: surface runoff (R) and underground runoff (G); i.e. WS = R + G = 193.11 mm (in one year).

In order to have a notion of G, which is of more interest to evaluate the water recharge in depth, it is considered: i) G/WS = 35%, which is typical of situations like the present one, with granitic soils, practically in totality, with average slopes in the order of 18%, and vegetation essentially of bush and some trees, a typical situation for this type of granite massifs of the Beiras region in Portugal [4]; ii) the area of the São Torcato basin, with about 0.927 km^2. Therefore, in the scenario presented for the São Torcato basin, considering G/WS = 0.35, leads to G = 67.6 mm, which extended to the entire São Torcato basin, results in an annual volume of 62665m^3. If there were only a single outflow point for the basin, it would be possible to extract a mean virtual flow of 2.0 L/s. However, as shown in item 4.3, for the entire basin, 14 water points were inventoried, which leads to an average flow of 0.14 L/s, a situation that is initially not favorable to a mineral water uptake, as it is very low. Therefore, there is the notion, from now on, that if there are no conditions to collect groundwater with recharge in neighboring basins, it makes impossible the possibility of having in the Qta do Plácido zone, an abstraction with the potential to obtain groundwater in quantity for a medical spa.

3.3.2. Hydrogeological units. In hydrogeological terms in the study zone, the following hydrogeological units are to be considered (Figure 4):

- Unit 1: Recent deposits, consisting essentially of alluvium and slope deposits, with thicknesses generally between 0.5 and 3 m, consisting of the geological unit "alluvium/ slope deposits"; it constitutes a superficial aquifer (unconfined), of the interstitial type, generally with high permeability.
- Unit 2: Granitic rocks, including various geological units, such as γm; γmt. Two situations may occur: (i) Superficial aquifer (unconfined), fissural type, in the first approximately 100m, with more production in the lower altitude zones of the study zone; (ii) semi-confined to confined aquifer, fissural type, below about 100m depth, but only in some locations; it is admitted to find water of relatively distant origin, which may result in aquifers with some artesianism, being the places with the greatest potential, when they are associated with very extensive faults; permeability is generally low and may be high in zones of intersection of faults.
- Unit 3: Shale rocks of the Schist-metagreywacke complex, of fissural type permeability, generally very low, and can reach very high thicknesses. Generally, there is little production of groundwater in these formations.

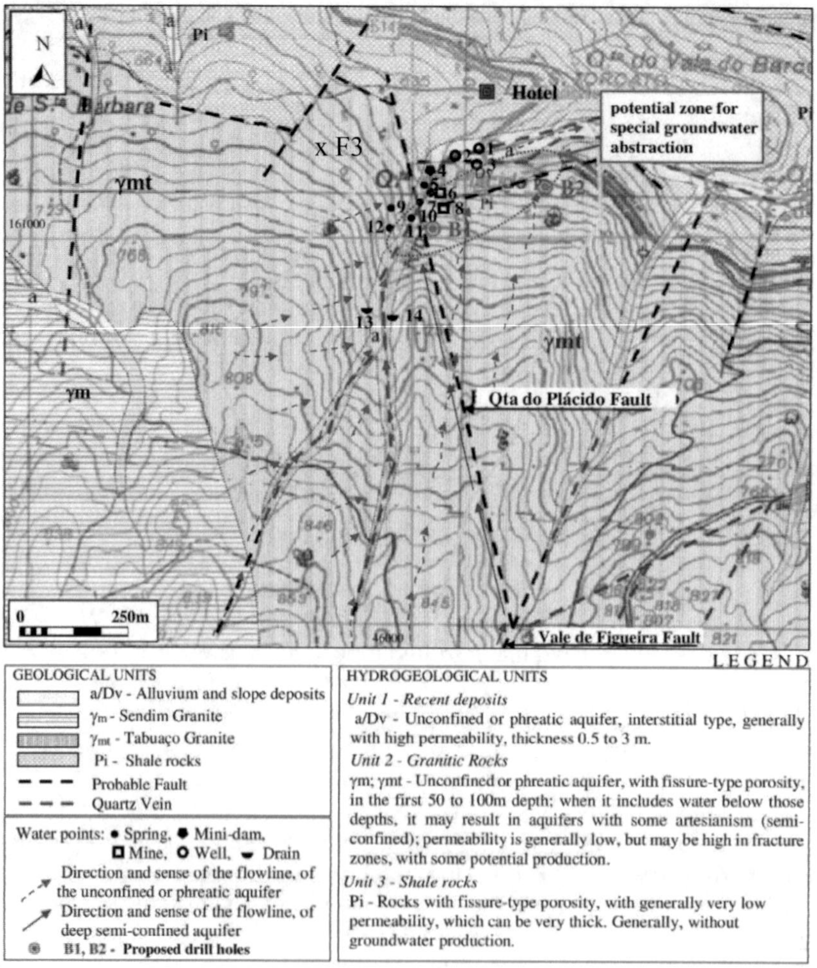

Figure 4. Hydrogeological map of the Qta do Plácido, in the São Torcato hydrographic basin.

In the hydrogeological map elaborated for the area of greatest interest in the study (Figure 4), it is highlighted that the zone of greatest potential for exploiting special groundwater is the area generically included in the ellipse represented in the proximity of the contact with the shale formations. The groundwater percolates from south to north and finds the schist, which is practically impermeable and therefore acts as a barrier, leading to a lot of groundwater in that zone.

On the relationship of the types of aquifers and water characteristics observed in the various "water points" studied, the following situations are summarized:

- the situation of superficial (unconfined) aquifers, with the majority of records in them (11 points), either associated with the recent deposits or associated with the granites in the uppermost; in these cases, low mineralized, acidic waters of low conductivity occur; the ranges of the various parameters recorded for those aquifers, are as follows: pH = 5.05 to 6.03, eH = 20.1 to 70.9 mV, T = 11.2 to 13.3°C, C = 30.5 to 42.7 μS/cm, TDS = 28.0 to 39.5 mg/L.

- the situation of confined or semi-confined aquifer systems, associated with granitic rocks, at significant depths, greater than about 100m; in the study zone, only 3 "water points" are available in this situation, corresponding to three wells (1, 2 and 3, Figure 4), with point 3, well 3, with 190m deep, which best represents this system aquifer, with the following parameters: pH= 6.41, Eh= -3.1 mV, T= 17.7°C, C= 107.3 μS/cm, and, TDS= 98.79 mg/L.

Table 1. Main results of the monthly sequential hydrological balance for the São Torcato basin[a]

Parameter	Jan	Feb	Mar	Apr	May	Jun	Jul	Aug	Sep	Oct	Nov	Dec	Annual
T (°C)	5.1	6.1	9.1	11.1	14.1	18.1	22.1	21.1	19.1	14.1	9.1	6.1	-
P (mm)	57.6	68.8	33.8	62.9	63.3	52.8	16.6	17.5	51.4	84.3	73.6	93.3	675.77
PET (mm)	12.4	15.7	33.3	46.7	72.4	101.9	134.0	118.1	90.3	55.6	26.5	15.2	-
RET (mm)	12.4	15.7	33.3	46.7	72.0	88.2	55.2	28.4	53.4	55.6	26.5	15.2	502.61
WS (mm)	45.2	53.1	0.5	16.1	-	-	-	-	-	0.0	0.0	78.1	193.11

T - monthly average temperature, P - monthly average precipitation, PET - potential evapotranspiration, RET - real evapotranspiration; WS - water surplus.
[a] In the calculation of the hydrological balance, the usable capacity by the plants was considered 100 mm, assuming full capacity in January.

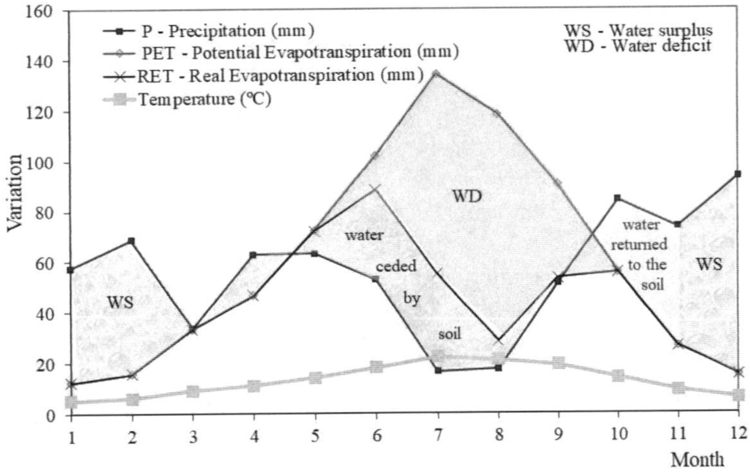

Figure 5. Results of the monthly sequential hydrological balance for the region of São Torcato.

All the physical-chemical parameters presented were recorded in May; however, according to the authors' experience with similar aquifer systems [3, 15], the order of magnitude of their values is maintained remains approximately throughout the year, with the particularity that in years with less precipitation, the springs lower their flow, drying up in summer.

Regarding the confined or semi-confined aquifer systems, although their results indicate a water origin relatively deep, from a relatively reducing medium (non-oxidized), with negative Eh, and water more mineralized than the water from superficial (unconfined) systems, it is understood that it corresponds to a situation with a mixture of water from superficial aquifers. This type of situation happens perhaps because the well does not have the upper levels properly isolated, as unfortunately is common to occur in the region in old wells. The fact that there is some artesian water in these systems is also worth noting, as evidenced by the recordings in well 2, as it spouts water freely at its head.

Groundwater with a sulfide-odor (smell of rotten eggs), with negative eH, with pH tendency above 7, or very close to it, and with total mineralization generally above 250 mg/L, of sulphureous type, are typical of deep aquifer systems in fissured granitic rocks. This situation occurs at a spring relatively close to the present study zone, in the Ponte do Fumo zone, still in the municipality of Tabuaço, about 7 km to SE of the São Torcato zone [8], having even been shown in recent work [9] that this water has similar characteristics to the waters explored in several medical spas in Portugal, such as São Pedro do Sul medical spa [16].

3.3.3. Conceptual geohydraulic model. To establish the conceptual model associated to the working area, several previous elements were crossed with the results obtained, namely the geological map of Portugal at a scale of 1/500 000 [17], in order to obtain a general notion of the main geological groups of the country and the main fault systems associated to the circulation of groundwaters with mineral potential. From this initial analysis, there does not seem to be great potential for the circulation of NMW of the sulphureous type in the study zone, in particular, because it is far from the main faults of the Portugal with associated NMW, i.e. the Vilariça Fault and the Verin-Chaves Fault. However, a careful analysis of the same map allows us to verify that there are several "water points" classified as NMW, and in places relatively distant from these large faults, as is the case of Carlão and S. Lourenço to the north of Tabuaço, Aregos to the west, and even Cavaca to the south [9,17]. To study the situation in more strictness, the geological maps of the region at the scale 1/50 000 were analysed in detail, whose extracts, which are of interest to the present study, are part of Figure 3. This map shows a set of relatively extensive faults, as presented in item 3, which may favour recharge at great depths and lead to the possibility of obtaining groundwater in the study zone, the Qta do Plácido, which could be classified as NMW and even of the sulphurous type.

Thus, the fact that the Tedo River has developed essentially along an extensive fault (Figure 3) is emphasized, enhancing the deep recharge. Then, the fact that there is a connection between the Rio Tedo Fault and the study zone is highlighted, through the Vale de Figueira Fault, which in turn, contacts to the northeast with another fault that develops until the Qta do Plácido zone. On the other hand, it is already in the zone of Qta do Plácido that occurs the contact between granites and shales. This situation is favorable to the occurrence of groundwater, at depth, in the study zone, with potential to be classified as NMW.

It should be noted that in association with the Tedo River Fault, there are 2 occurrences of groundwaters with mineral potential, where many years ago they were already ancient thermal baths [12]. One of the cases corresponds to the Baths of Dr Gomes (Águas de Nagosa). According to some authors [18] the water of those baths, called Fonte Santa de Nagosa, is sulphureous, sodium, fluoridated and hypothermal. This resurgence is located south of the village of Longa (Figure 3), along the right bank of the Tedo River, at an altitude of around 450m, already off the map.

The other case referred to in the literature [12], is located in the map area of Figure 3, further NW of the previous site, near the right bank of the Tedo River, in the village of Rib^{a} de Goujoim. This water is bicarbonate - sodic and the resurgences occur at an altitude of approximately 350m.

About the Tedo River Fault, we point out that it develops in granitic rocks to the NE and will intersect the shale zone at an altitude of approximately 150m, in the Sta Leucácia area, west of the Barcos village. This site is considered to have a high potential for exploiting groundwater with special characteristics, and its altitude is a conditioning factor for the deep aquifer system levels in the region.

Despite the last paragraph, and because there are springs associated with the deep aquifer system at altitudes above 150m, as has just been seen with the waters of the Nagosa and the Rib^{a} de Goujoim, it is admitted that in the working area, the deep water level of better quality is reached around altitude 350m, which is approximately the same altitude of the intersection between the Tedo River Fault and the Vale de Figueira Fault, which ends up placing the region in contact, in terms of groundwater.

Thus, based on all the previous analyses, results the scheme of the geohydraulic conceptual model of the region, presented in Figure 5. In the Qta do Plácido - São Torcato zone, if drill holes are started around 610m altitude, in the area and places marked in Figure 4 (B1 and B2), to reach zones with a better groundwater potential to be classified as NMW, one should ideally make abstraction with a depth of around 300m. Above this depth, the water tends to be mixed with water from the superficial aquifer. To capture only water from the deep aquifer system, it would be important to always leave open, depending on the results to be obtained in the drill holes, the isolation of the surface levels, at least up to about 200m, also enhancing the capture of groundwater with a temperature above the usual in the "water points" currently available, clearly above 20°C.

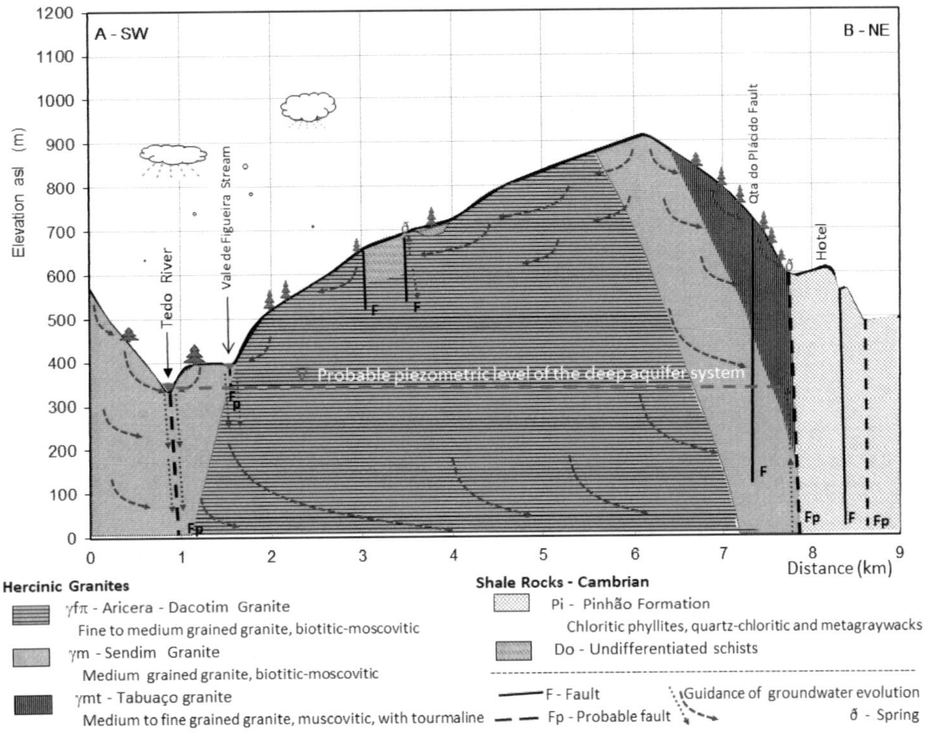

Figure 6. Sketch of the conceptual geohydraulic model of the region associated with Qta do Plácido - São Torcato zone – Tabuaço.

4. Conclusions

In the São Torcato zone (Qta do Plácido), near the town of Tabuaço, in a region that is World Heritage by UNESCO, there is a hotel unit that will have the possibility of associating a medical spa, if, in the proximity, it becomes possible to make available groundwater with special characteristics, if it is classified as natural mineral water (NMW), for thermalism.

Thus, to investigate such a possibility, the hydrogeological conditions of the region were studied, and, with more detail, the small São Torcato hydrographical basin was studied, with geological and hydrogeoenvironmental fieldwork. This basin presents only about 0.927 km^2 of area, with granitic rocks almost in its totality, contacting in its most downstream area with shale rocks, a situation that favours the groundwater storage. That small basin presents an average slope of 18%, and a density of water points (springs, mines, wells and boreholes) of 15.1/km^2 ("water points"/area). From the calculation of the annual water surplus (WS), it was evaluated for the underground flow (G) an annual global value of G= 67.6 mm; considering G/WS of 35%, it leads to an equivalent discharge per water point of 0.14L/s, during one year. These discharges correspond to waters with very low mineralization, with Total Dissolved Solids (TDS) generally lower than 40 mg/L, acid (pH generally lower than 6) and percolating in relatively oxygenated environments, with positive Eh (redox potential), generally between 20 and 71 mV, corresponding to aquifer systems essentially with granite rocks, of the unconfined type, which generally do not exceed 100m in depth. Based on several scenarios, analyzing the surrounding region, a conceptual geohydraulic model was built (Figure 6). The regional geomorphological and geological characteristics, and in particular the association of extensive geological faults intersecting several hydrographic basins favour the supply of water from the deep aquifer system. By the interpretation of the model, in the zone of Qta do Plácido, drill holes with a depth of 300m, will allow the capture of water, which in physical-chemical terms will be stable, and be possible to classify it as NMW. The flow rate admitted, considering the experience of the authors in similar situations, could be of the order of

2L/s or higher, however, to assess these values, highly developed studies are needed, namely the use of isotopic techniques, to know with strictly, the recharge areas. We point out a zone that is considered to have great potential for high productivity abstractions (Figure 4), and we propose the location of two drill holes to be carried out; the results of these drill holes, including long duration flow tests, will allow us to assess with some reliability the flows to be exploited from the deep aquifer system.

Acknowledgment(s)
This work is supported with Portuguese national funds by FCT - Foundation for Science and Technology, I.P., within the GeoBioTec - Project UIDB/04035/2020.

References

[1] UNESCO World Heritage Centre 1992-2021 *Alto Douro Wine Region*. Site visited in 2021.11.19: https://whc.unesco.org/en/list/1603
[2] Portuguese Decree 2015 Approves the bases for the legal regime of the revelation and exploitation of geological resources *Lei nº54/2015, D.R., 119, I Série* (*in* portuguese) 4296-4308
[3] Ferreira Gomes L M, Maia de Carvalho P E, Andrade Pais L J 2019 Hydrogeochemical studies of a groundwater with a view to its classification as mineral water for a new medical spa in Portugal *IOP Conf. Ser.: Earth Environ. Sci.* **221** 10, doi:10.1088/1755-1315/221/1/012030
[4] Ferreira Gomes L M 2020 Main geoenvironmental studies to launch new Medical Spa: the case of the SãoTiago Medical Spa. *Modern Environmental Science and Engineering* **6/1** 14 doi:10.15341/mese(2333- 2581)/01.06.2020/001
[5] Ferreira Gomes L M 2020 Quality of an Underground Hydro Resource in Order to Classify it as Natural Mineral Water for Thermalism: The Case of the New Resource "Termas de São Tiago". *IOP Conf. Ser.: Earth Environ. Sci.* **609** 10, doi:10.1088/1755-1315/609/1/012069.
[6] Google Earth 2021 Google Earth Pro images, with search from Tabuaço; Image Lansat/Copernicus, Data SIO, NOAA, U.S. Navy, NGA, GEBCO
[7] Anon 2012 Hydrogeological-environmental study to optimize the location of boreholes for the prospection and research/capturing of groundwater with the aim of setting up a thermal unit in Tabuaço, Internal report, *in Portuguese*, 25
[8] Machado Fonseca A M 2021 Contribution to the knowledge of the hydromineral resources of the Tabuaço region, Master's dissertation in Civil Eng. (*in* Portuguese). UBI, Covilhã, 94
[9] Machado Fonseca A M, Ferreira Gomes L M 2021 Hydrogeoenvironmental studies in an inland region of Portugal to potentiate the birth of a new medical spa, *E-book: Coleção desafios das engenharias: Engenharia ambiental*, Atena Editora, 15
[10] IGE 1993. Mapa à Escala 1/25000. *Carta Militar de Portugal*, Folhas nº 127, 138 -Tabuaço
[11] Bernardo de Sousa M, Sequeira A J D 1989 Carta Geológica de Portugal, Escala 1/50000. Notícia Explicativa da Folha 10-D, Alijó. *Serviços Geológicos de Portugal,* Lisboa
[12] Ferreira N, Bernardo de Sousa M 1994 Carta Geológica de Portugal, Escala 1/50000. Notícia Explicativa da Folha 14-B, Moimenta da Beira. *Serviços Geológicos de Portugal*, Lisboa
[13] Strahler A N 1964 Quantitative Geomorphology of Drainage Basins and Channel Networks. In: Chow, V., Ed., *Handbook of Applied Hydrology*, McGraw Hill, New York, 439-476
[14] Thornthwaite C W, Mather J. R. 1957 Instructions and tables for computing potential evapotranspiration and the water balance. Centerton, NJ: Drexel Institute of Technology - Laboratory of Climatology *Publications in Climatology* **X/3** 301
[15] Mendes E, Gomes L M F, Melo M T C 2008 Contribution to the hydrogeological characterisation of groundwater in the granitic massif of Serra da Estrela *Com.Geo* **95** 61-71
[16] Ferreira Gomes L M, Jorge A, Rodrigues L 2021 Good practices of quality control in the area surrounding of natural mineral water abstraction of São Pedro do Sul medical spa (Portugal), *IOP Conf. Ser.: Earth Environ. Sci.* **906** 10 doi:10.1088/1755-1315/906/1/012092
[17] SGP 1992 Carta Geológica de Portugal, *Serv.Geológicos de Portugal,* Escala 1/500000. Lisboa.
[18] Almeida A, Almeida J D 1975 Inventário Hidrológico de Portugal *Inst. Hidrologia* **3** 325

Estimation of forest carbon storage and carbon sink potential in Yunnan Province under the goal of "carbon peaking and carbon neutrality"

Long Li[1], Wei Fu[1,3], Mingcan Luo[1], Jiancheng Chen[2] and Zhongming Li[1]

[1] School of economics and management, Southwest Forestry University, Kunming, Yunnan 650233;
[2] School of economics and management, Beijing Forestry University, Beijing 100083.

[3] Corresponding author's e-mail: fuweiynlzd@163.com

Abstract. Through the accounting of forest carbon reserves and resource value in Yunnan Province, the future forest carbon reserves and carbon sink potential in Yunnan Province are predicted. Using the forest stock expansion method, the forest carbon reserves and their changes in Yunnan Province from 1973 to 2018 are calculated. The GM (1,1) grey prediction model predicts the forest carbon sequestration potential in Yunnan Province. The results show that: (1) the forest carbon storage in Yunnan Province increased from 485.43 million t in 1976 to 937.01 million t in 2018, an increase of 1.9 times, with an average annual growth of 10.50 million t. In 2018, the value of forest carbon reserves was 94062.92 million yuan. (2) The forest carbon sequestration in Yunnan Province varies significantly in different periods. Between 1973-1998 and 1999-2018, the average carbon sequestration was 6.22 million T / A and 16.36 million T / A respectively. (3) The predicted values of forest carbon reserves in 2023, 2028 and 2033 in Yunnan Province are 1001.41 million t, 1103.66 million t and 1216.34 million t respectively, with an average annual increase of carbon sequestration of 17.46 million T / A. According to the research results, the forest area and the carbon storage of forest resources of Yunnan Province is increasing. forest carbon sequestration in Yunnan Province has excellent development potential, and plays a more significant role in forestry carbon sequestration for the realization of the "carbon peaking and carbon neutrality" goal.

1. Introduction

The rapid development of the global economy has brought about the destruction of the natural ecological environment. The climate change caused by the greenhouse effect has increasingly become one of the core issues of general concern in the international community [1]. The increasing concentration of CO_2 in the atmosphere has brought vital harm to the human living environment. Low carbon emission reduction has become a global consensus. Countries worldwide reduce greenhouse gas emissions in the way of agreement. According to the global energy and CO_2 statistics released by the International Energy Agency (EIA), China's CO_2 emissions have increased year by year and have become the world's largest carbon emitter. Low carbon emissions reduction will become an important strategic direction for China's future development [2]. The ninth meeting of the central financial and Economic Commission in 2021 proposed that "carbon peaking and carbon neutralization should be included in the overall layout of ecological civilization construction". "carbon peaking and carbon neutrality" target means that China's CO_2 emissions will reach the peak before 2030 and strive to achieve carbon neutralization by 2060 [3]. The proposal of the "carbon peaking and carbon neutrality" goal shows China's determination to reduce low-carbon emissions. In the process of emission reduction, we should not only pay attention to the "carbon source", but also consider the "carbon sink", to balance the two aspects to continuously promote the emission reduction of greenhouse gases [4].

Among them, the carbon sink function of forest plays a vital role in the realization of "carbon neutralization" [5]. The carbon sink function of forest refers to the absorption and fixation of CO_2 in the atmosphere by forest plants [6]. The forest ecosystem is the largest in the world, and there are substantial ecological benefits in the process of realizing the "carbon peaking and carbon neutrality" goal [7]. Forest store a large number of carbon sink and have a profound impact on the global carbon cycle [8]. In 2020, the results of the global forest resources assessment released by the food and Agriculture Organization of the United Nations, the global forest carbon storage accounts for about 77% of the worldwide vegetation carbon storage. The improvement of forest carbon storage capacity plays a vital role in the realization of the "carbon peaking and carbon neutrality" goal [9]. The development of forest carbon sequestration will become one of the essential ways to deal with climate change and achieve the "carbon peaking and carbon neutrality" goal [10].

In 2017, the people's Government of Yunnan Province issued the comprehensive work plan for energy conservation and emission reduction in the 13th five-year plan of Yunnan Province, which defined the overall requirements and objectives of Yunnan Province in low-carbon emission reduction. During the 13th Five-Year Plan period, Yunnan Province has made various efforts in optimizing the industrial and energy structure, the development of circular economy, and low-carbon energy conservation and emission reduction projects and policies. The work plan for controlling greenhouse gas emissions in the 13th five-year plan of Yunnan Province points out that the role of forest carbon sink in Yunnan Province should be strengthened, the forest coverage rate should be increased to more than 60%, and the forest volume should reach 1.90 billion m^3. After the "carbon peaking and carbon neutrality" goal was put forward, Yunnan Province actively sought the development direction under the crucial advantages of green energy and forest resources, vigorously implemented key forestry projects, and continuously has improved the quality and quantity of forest [11]. Yunnan provincial Party committee and government adhere to high-quality development, build a world-class "Green three cards" strategy (Green energy brand, green food brand and healthy life destination brand), and help achieve the goal of "carbon peaking and carbon neutrality". In 2021, at the seminar on carbon neutralization at the peak of carbon and high-quality development in Yunnan, Li believed that the advantage of carbon neutralization in Yunnan Province lies in abundant forest and grass carbon sink resources, and put forward suggestions such as increasing the total amount of forest resources and establishing a carbon sink measurement and detection system. Based on the unique advantages of forest resources in Yunnan Province, forest carbon sequestration plays a vital role in the realization of the "carbon peaking and carbon neutrality" goal in Yunnan Province, and it is essential to increase forest carbon sequestration in Yunnan Province.

Under the background of low-carbon emission reduction trend and "carbon peaking and carbon neutrality" target, the current accounting of forest carbon reserves and the prediction of carbon sequestration potential in Yunnan Province can provide data support for the research on the mechanism and path of forest carbon sequestration in Yunnan Province. The accounting of forest resource value can intuitively display the economic value of forest resources in Yunnan Province, it provides a reference for improving the trading mechanism of carbon sequestration trading market and reasonably formulating carbon sequestration trading price. Firstly, taking the forest area and volume in Yunnan Province from 1973 to 2018 as the data sample, this paper uses the forest volume expansion method to calculate the forest carbon storage in Yunnan Province, and dynamically analyzes the changing trend of forest carbon storage. Secondly, convert the value of forest resources and compare the economic value of forest resources of different stand types. Finally, the first-order differential equation between variables and time variables is established according to the original data, the forest area and forest carbon storage in Yunnan Province by 2033 is predicted by GM (1,1) grey prediction model. The innovation of this paper lies in: first, the innovation of research ideas. The carbon storage of forest resources in Yunnan Province is calculated, and the dynamic change analysis is made from the age class, different origins of trees, and other forest species, to predict the potential of forest carbon sequestration in Yunnan Province. The status of Yunnan forest resources and their carbon reserves in the country is clarified, there is a clear understanding of Yunnan forest carbon reserves as a whole and locally. The second is the innovation of research methods. In the prediction part of forest volume, from the perspective of time prediction, the grey prediction model in time series prediction is used to predict the medium and short-term forest volume. The third is the innovation of research content. It enriches the carbon reserves of forest resources and forest carbon at the provincial level, the empirical research field of foreign exchange potential. Converting the actual value of forest resources carbon reserves and quantitatively displaying the economic value of the carbon sequestration is conducive to promoting the development and improvement of the carbon sequestration trading market in Yunnan Province. Finally, it provides the basis for forestry carbon sequestration and emission reduction in Yunnan Province under the "carbon peaking and carbon neutrality" goal. Also, it broadens the ideas for the research of provincial forest carbon sequestration.

2. Literature review

By combing the relevant literature at home and abroad, the overall research progress on forest carbon storage and carbon sink potential at home and abroad is as follows: it is found that the research on forest carbon storage and carbon sink potential began in the middle and late 1960s [12]. Foreign research on forest carbon storage and carbon sink potential has experienced a transition from the embryonic stage to the rapid development stage. It was the embryonic stage before 2006. Among them, Paul used the forest growth model to analyze the impact of forest intensive forest on carbon storage [13], and Nilsson et al. used the global large-scale afforestation project to explore the change of carbon cycle, the project covers an area of 345 million hm^2, and the annual carbon sequestration potential value of the overground part of the project is predicted [14]. Since 2007, it has been a period of rapid development [15]. Natural science has absorbed the theoretical framework of economics, the penetration between disciplines has become closer and closer, the research level of forest carbon storage and carbon sink potential has been extended, the global dynamic analysis model of carbon sink research has been enriched, and the economic study of international forest carbon sink has become more and more mature [16].

At present, the research methods of forest carbon storage calculation mainly include: biomass method, biomass inventory method, remote sensing estimation method, and forest stock expansion method [17]. One is the application of the biomass method. Fang et al. studied the dynamic changes of forest vegetation carbon pool in China using the biomass method and estimated the soil carbon pool in China [18]. Cheng et al. calculated the carbon sequestration and carbon density of natural forests in Gansu(Province) with the biomass method [19]. Zhang and Wang established a biomass regression model based on forest resource inventory data to calculate China's forest carbon reserves from 1949 to 2018. It is concluded that China's forest carbon reserves show a growing trend, and there are significant differences in different regions. The biomass method is widely used in the calculation of carbon reserves [20]. Its advantage is that the plan is direct and clear, and the estimation accuracy is high. However, the disadvantage is that the conversion coefficient from biomass to carbon content can not be clearly defined. Moreover, biomass calculation only considers the overground part, not the underground part. It needs to be estimated based on the sample plot, and the operation is complex. The second is the application of the biomass inventory method. Wu used the biomass inventory method to calculate the forest carbon reserves in Hunan Province quantitatively and compared the forest carbon reserves of different dominant tree species, age classes, states, and cities [21]. The application technology of the biomass inventory method is direct and clear, suitable for long-term and large-scale estimation. Still, the disadvantage is that it can not be monitored in real-time, the labor cost is high, and the comparability of estimation results is poor. The third is the application of the remote sensing estimation method. Liu applied the remote sensing estimation method to calculate the forest carbon storage in Zhejiang Province and studied its temporal and spatial evolution [22]. The remote sensing estimation method is used to calculate the forest carbon storage in the Yangtze River Basin, and its temporal and spatial changes and driving factors are analyzed [23]. The advantage of the remote sensing estimation method is that it can comprehensively, dynamically, and quickly carry out macro monitoring. The disadvantage is that it does not have universality and is difficult to obtain data. Fourth, is the expansion method of forest stock. Xi et al. novelty proposed the forest stock expansion method based on the stock method and calculated the forest carbon reserves of Heilongjiang Province in 2010 [24]. Many scholars have used this method to estimate forest carbon reserves. Huang et al. calculated the carbon sequestration potential of natural forest and artificial forest in the Daxing'an Mountains by using the forest stock expansion method and concluded that the carbon sequestration potential in Daxing'an Mountains is vast [25]. Zhang et al. used the forest stock expansion method to calculate the carbon reserves and carbon sinks of the national forest resources inventory data in 2003 and 2008, They compared the data with the national forest resources inventory data published by the state. The data are very close [26]. Yin et al. used the stock expansion method to calculate the forest carbon stock in the forest area of Inner Mongolia and calculate the forest carbon sink value of the forest area [27]. Yao et al. analyzed the forest carbon storage of 29 provinces (cities and districts) in China, they proposed a specific reverse correlation between carbon storage efficiency and regional economic development level [28]. The advantage of the forest volume expansion method is a direct and simple operation, while the disadvantage is to ignore the influence of many forest uncertain factors. The first three calculation methods belong to the category of purely natural science. While the calculation method is accurate, it is difficult to obtain cost and data. From the perspective of social science, it is not applicable. Based on the comparison of the above methods, the forest volume expansion method is simple and accurate in the calculation, so this method is selected for accounting.

In the prediction of forest carbon sequestration potential, the methods used by scholars are very different, mainly including the following methods: Chen et al. estimated the carbon storage, carbon sequestration rate and potential of forest tree layer in Qinghai province according to the inventory data of forest resources and field investigation data [29]. Xu et al. used the continuous biomass conversion factor method to predict China's forest biomass carbon sequestration potential by 2050 based on the relationship between biomass and forest age [30]. Jiang et al. indicated China's forest carbon sequestration potential under different economic backgrounds [31]. Ren et al. predicted the forest carbon stock and carbon sink potential from 2014 to 2020 based on the relationship between forest carbon density and forest age [32]. Li et al estimated the total carbon stock and carbon sink growth capacity of arbor forests in China and predicted the total carbon stock of arbor forests in China in 2050 [33]. Yuan added social economic variables and impacts to the econometric model of forest carbon sequestration, and scientifically predicted the great potential of forest carbon sequestration in China [34]. This paper novelty uses the grey prediction method model in fuzzy prediction to establish the first-order differential equation of time- variable according to the original data of Yunnan Province from 1973 to 2018 to predict the short-term carbon sink potential. Previously, the prediction of forest carbon sink potential has not been carried out from the perspective of time series. In 1986, Deng innovatively put forward the grey system theory, which is widely used in the weather forecast, flood forecast, passenger volume forecast and so on. Wu revised the GM (1,1) grey prediction model to significantly improve the model's accuracy [35] Compared with other time series prediction methods, the grey prediction model has the following advantages: it does not need a large number of samples, the prediction accuracy is high, the results of qualitative and quantitative analysis are consistent, and the short-term prediction accuracy is high [36, 37, 38, 39].

To sum up, many scholars at home and abroad have done various research on forest carbon storage and carbon sink potential and achieved rich research results. However, according to the literature review, there are still three aspects less involved. First, research methods. The calculation of forest carbon reserves and the prediction of carbon sequestration potential are mainly based on the pure natural discipline method, there is less analysis from the perspective of social science. Second is the selection of the research scope. The research area is mainly used at the national level. Although it is also involved at the provincial level, there are few specific studies at the level of Yunnan Province, and there are few studies on forest carbon storage and forest carbon sink potential prediction in Yunnan Province at different periods. Third, the value of forest resources according to different stand types is relatively small. There is little comparative analysis on the importance of forest resources of different origins, different age classes and different forest species. Based on the research of relevant scholars, this paper uses the forest stock expansion method to calculate the carbon reserves of existing forest resources in Yunnan Province, converts the actual value of forest carbon sink, makes a comparative analysis of the importance of forest resources according to the classification of different origins, different age classes and different forest species, and uses GM (1,1) grey prediction model to predict the potential of forest carbon sink in Yunnan Province. Grasp the current situation of forest carbon sequestration capacity in Yunnan Province, provide data support for the proposal of forest carbon sequestration mechanism and path in Yunnan Province, and provide policy suggestions for Yunnan Province to give full play to its advantages of forest resources and improve the contribution capacity of forest carbon sequestration under the "carbon peaking and carbon neutrality" goal.

3. Methodology

The research method of this paper is selected from three aspects: the calculation of forest carbon storage, the prediction of forest carbon sink potential and the conversion of forest resource value. Firstly, after comparing the calculation methods of forest carbon storage, the forest storage expansion method is selected as the calculation method of this paper. Secondly, for the prediction of forest carbon sink potential, the GM (1,1) grey prediction model is selected for short-term prediction. Finally, in the conversion of forest resource value, the value of the carbon sink value is calculated in combination with the most appropriate transaction price of the carbon. The forest is divided according to the origin, use and age of the forest, and the corresponding carbon storage, carbon density and carbon storage value are calculated respectively.

3.1. Forest volume expansion method

The forest volume expansion method takes the forest volume as the calculation basis, uses the volume expansion coefficient to calculate the biomass of trees (including roots and branches), then calculates the dry weight of biomass with bulk density (dry weight coefficient), and finally calculates the carbon sequestration through the carbon content rate. The carbon sequestration of forest biomass is obtained. The carbon sequestration of the whole forest is calculated according to the proportional relationship between the biomass carbon sequestration of forest land and trees and the balanced relationship between the biomass carbon sequestration of undergrowth plants and trees.

Forest carbon storage should include the carbon stock of trees, woodlands and undergrowth plants. However, because the carbon reserves of undergrowth plants and forest land change little during the year, the carbon reserves of undergrowth plants and forest land are not calculated in this study. The carbon reserves of forest trees are mainly estimated. Forest carbon sequestration includes forest trees, forest products, undergrowth plants, hummus and forest soil carbon sequestration. To avoid duplication in accounting, the logistics soil carbon sequestration part of forest products is not calculated in this paper. The overground part is mainly to estimate the carbon storage of trees. Economic forest and bamboo forest are not included in the accounting of this paper. To sum up, forest carbon reserves in this paper have forests, trees around, scattered trees and sparse forests. Forest resource carbon reserves include forest carbon reserves. Forest carbon reserves < forest carbon reserves < forest resource carbon reserves. The calculation formula of overall forest carbon sequestration is as follows:

$$CF = \sum (Sij \times Cij) + \alpha\sum (Sij \times Cij) + \beta\sum (Sij \times Cij) \quad (1)$$

$$Cij = Vij \times \sigma \times \rho \times \gamma \quad (2)$$

Where Sij represents the area of class J forest in class I area, Cij represents the forest carbon density of class J forest type in class I area, Vij represents the forest volume per unit area of class J forest type in class I area, α Is the carbon conversion coefficient of undergrowth plants, β Is the carbon conversion coefficient of forest land, σ Is the biomass expansion coefficient, ρ Is the volume factor, γ Is the carbon content. The calculation formula of forest carbon storage transformed into a specific forest is:

$$Cf = Vf \times \sigma \times \rho \times \gamma = Vf \times 1.9 \times 0.5 \times 0.5 \quad (3)$$

Where Cf is forest carbon storage, and Vf is forest carbon storage. In the process of accounting for forest carbon storage in Yunnan Province, various conversion coefficients are generally taken according to the default parameter values of the Intergovernmental Panel on climate change (IPCC). α Value 0.195, β is 1.244. σ Is the biomass expansion coefficient, normally 1.90; ρ Is the volume coefficient, generally 0.45-0.50t/m3, and 0.50t/m3 in this study; γ Is the carbon content, generally 0.50; By calculating the forest carbon storage in Yunnan Province, we can naturally calculate the forest carbon sequestration in Yunnan Province.

The forest canopy density of the statistical caliber of forest resources inventory has changed from 30% to 20% after 1994. Considering the change of statistical caliber, the carbon reserves before 1994 are converted [40]. The conversion formula is as follows:

$$TC_{0.2} = 1.122TC_{0.3} + 1.157\ (R^2 = 0.995) \quad (4)$$

Wherein, $TC_{0.2}$ is the forest carbon storage with a forest canopy density of 20%, $TC_{0.3}$ is the forest carbon storage with forest canopy density of 30%.

3.2. GM(1,1) grey prediction model

Aiming at the incomplete and less information, the grey differential prediction model is established by accumulating the original data at one time. Generally, it is more accurate for the medium-term and short-term predictions. The corresponding differential equation is as follows:

$$\frac{dx^{(1)}}{dt} + ax^{(1)} = u \quad (5)$$

Where $x^{(1)}$ is the sequence generated by one accumulation of the original data $x^{(0)}$; a Is a time variable; U is the parameter to be estimated, which is called development grey number and endogenous control grey number respectively [41]. The specific steps are as follows:

(1) Let the original sequence be $x^{(0)}: x^{(0)} = [x^{(0)}(1), x^{(0)}(2), x^{(0)}(3) \dots x^{(0)}(n)]$, do an accumulation to generate an accumulation sequence$x^{(1)}, x^{(1)}(t) = \sum_{m=1}^{t} m, t = 0,1,2 \dots n$.

(2) Solving mean series; using formula $z^{(1)}(t) = \frac{1}{2}[x^{(1)}(t) + x^{(1)}(t+1)$ Calculate the mean sequence and establish the first-order differential equation $\frac{dx^{(1)}}{dt} + ax^{(1)} = u$, The undetermined coefficients of grey GM (1,1) prediction model are estimated by using the least square method a And U. The time response function of grey GM (1,1) is obtained:

$$X^{\wedge(1)}(t+1) = (x^{(0)}(1) - \frac{u}{a})e^{-at} + \frac{u}{a} \quad (6)$$

(3) The predicted value is solved by replacing the time response function:

$$X^{\wedge(0)}(t+1) = X^{\wedge(1)}(t+1) - X^{\wedge(1)}(t) \quad (7)$$

(4) Model test, posterior the error test and relative error test. The prediction accuracy of posterior error test C is divided into four levels. $C \le 0.35$ is expected to have good accuracy, $0.35 < C \le 0.5$ is qualified, $0.5 < C \le 0.65$ is qualified, and $0.65 \le C$ is unqualified. Compared with error test, it is generally acceptable ε The smaller the degree value, the better, $\varepsilon \le 20\%$ indicates that the fitting effect is good.

3.3. Conversion of forest carbon storage value

According to different forest classification standards, the forest carbon reserves in Yunnan Province are calculated. In converting the actual value of carbon reserves, this paper adopts the statistical results of Zhang et al. [42]. The suitable price range of forest carbon sequestration in China is US $10.11-15.17/t. In this study, the ceiling price of US $15.17 is used for calculating, and converted according to the average US dollar to RMB foreign exchange rate published by China foreign exchange trading center. The data show that the average US dollar to RMB foreign exchange rate was 6.6174 in 2018. Based on this, the carbon stock and actual value of forest resources in Yunnan Province composed of stand density are calculated.

4. Empirical analysis

4.1. Overview of the study area

As a large forest Province in China, Yunnan Province has a broad forest area, large forest stock and high coverage. According to the above theoretical analysis and research methods, the study area of this paper selects Yunnan Province as the research object, and the data of nine forest resources inventory in Yunnan Province from 1973 to 2018 are chosen as the data samples for empirical analysis.

4.2. data sources

Yunnan Province, or Yunnan for short, is located on the southwest border of China. The total area reached 39.41 million hm^2, ranking eighth in the country. The forest area and forest volume of Yunnan Province are 21.06 million hm^2 and 1972.66 million m^3, ranking second in China respectively, and the forest area is 25.99 million hm^2. Green is synonymous with Yunnan Province. The existence of primitive forests makes Yunnan have a large number of rare animals and plants, as well as a large number of national forest parks. The construction of ecological civilization is at the forefront of the country. Yunnan Province is rich in forest resources. It is one of the "four forest provinces" in China. In each national forest resources inventory, it is not only in the leading position in the country in terms of forest land and forest area. The volume and coverage rate is also ahead of other provinces and cities in China. Because of the large size of forests and the support of forestry policies in Yunnan Province, the forest carbon sequestration capacity of Yunnan Province has great potential.

4.3. Data sources

The research data of this paper mainly comes from: part of the data of Yunnan Province in the Ninth National Forest Resources Inventory on the website of the State Forestry and grassland administration, including forest volume, forest area, forest land area, live tree volume, etc. The size and volume of arbor forests, different forest ages and species in Yunnan Province in China forest resources report 2014-2018; Relevant scholars' literature is used in the calculation of forest resources carbon reserves and the prediction of carbon sink potential.

The Ninth National Forest Resources Inventory takes five years as a cycle, namely 1973-1976 (1976), 1977-1981 (1981), 1984-1988 (1988), 1989-1993 (1993), 1994-1998 (1998), 1999-2003 (2003), 2004-2008 (2008), 2009-2013 (2013) and 2014-2018 (2018).

4.4. Result

According to the inventory data of forest resources in Yunnan Province, the empirical analysis is carried out through the above research methods. The empirical analysis is carried out from three aspects: forest carbon reserves, forest classified carbon reserves and value, and the prediction of forest carbon sink potential. The empirical results are as follows.

4.4.1. Forest carbon storage results in Yunnan Province. As shown in Table 1, from 1998 to 2018, the forest land area increased to 2.19 million hm2, with a growth rate of 9.2%. From 1976 to 2018, the total forest area increased by 11.50 million hm2, with an average annual increase of 0.27 million hm2. The forest coverage increased from 24.9% to 55.04%. The growth of forest area and coverage has also brought changes in the proportion of forest land, from 34.66% to 81.02%. The rise of the site has brought about the rise of standing stock, and the standing stock has increased by 2.2 times. The average annual growth is 27.23 million m3. The forest volume risen from 918.01 million m3 to 1972.66 million m3. It was increased by 2.2 times, with an average annual growth of 25.82 million m3.

The proportion of forest stock in living, the ratio of standing stock has increased. It shows that the forest protection and construction work in Yunnan Province has produced results and steadily improved.

The forest carbon storage in Yunnan Province was calculated using the forest stock expansion method. Carbon reserves before 1994 are converted by formula (1). Table 2 shows the changes in carbon reserves and carbon sinks of forest resources in Yunnan Province. The carbon reserves of forest resources in Yunnan Province increased steadily, with a total growth of 1185.44 t from 1976 to 2018, with a growth rate of 92.25%. Among them, forest carbon reserves increased by 1.9 times. Forest carbon reserves increased from 485.43 t to 937.01 t, an increase of 1.9 times. The forest carbon density in Yunnan Province has experienced a decrease in the previous decade to a steady rise in the next decade. Over the past 40 years, the average forest carbon density in Yunnan Province has been 50.75t/hm^2.

In forest carbon sequestration in Yunnan Province, there are significant differences in forest carbon sequestration in different periods. From 1984 to 1993, due to the reduction of forest area in Yunnan Province, the shift in carbon storage is also declining, and the change of carbon sequestration shows a decrease. From 1973 to 1998, the average forest carbon sequestration was 6.22 million T / A. From 1999 to 2018, the average carbon sink is 16.36 million T / A, the change in carbon sequestration shows that the forest carbon sequestration in Yunnan Province is increasing significantly, and the forest carbon neutralization capacity in Yunnan Province is improving continuously.

Table 1. Changes in forest area and volume in Yunnan Province.

Period	**Forest land area /10000 hm^2**	**forest coverage**		**forest coverage /%**	**Standing stock//10000 m^3**	**Forest volume**	
		/10000 hm^2	**Proportion /%**			**/10000 m^3**	**Proportion /%**
1976	2758	956	34.66	24.9	98860	91081	92.13
1981	2612.38	919.65	35.2	24	132131.38	109703.30	83.02
1988	2501.23	932.74	37.29	24.38	134946.76	109656.83	81.26
1993	2435.97	940.42	38.61	24.58	136640.61	110528.18	80.89
1998	2380.79	1287.32	54.07	33.64	142391.06	128364.94	90.15
2003	2424.76	1560.03	64.34	40.77	154759.4	139929.16	90.41
2008	2476.11	1817.73	73.41	47.5	171216.68	155380.09	90.75
2013	2501.04	1914.19	76.54	50.03	187514.27	169309.19	90.29
2018	2599.44	2106.16	81.02	55.04	213244.99	197265.84	92.51
1976-2018Variation	-158.56	1150.16	46.36	30.14	114384.99	106184.84	0.38

Table 2. Changes in forest carbon storage and carbon sink in Yunnan Province.

Period	Carbon storage of forest resources/10000 t	Forest carbon storage/10000 t	Forest carbon storage /10000 t	Carbon density(t/hm^2)	Carbon sequestration (10000t/a)
1976	128505.82	52688.59	48542.78	50.78	0
1981	171754.12	70420.58	58467.53	63.58	1984.95
1988	175413.73	71921.03	58442.76	62.66	-3.54
1993	177615.51	72823.77	58907.15	62.64	92.88
1998	164963.60	67635.75	60973.35	47.36	413.24
2003	179292.63	73510.72	66466.35	42.61	1098.60
2008	198358.80	81327.92	73805.54	40.60	1467.84
2013	217239.97	89069.28	80421.87	42.01	1323.26
2018	247049.65	101291.37	93701.27	44.49	2655.88
average value	184465.98	75632.11	66636.51	50.75	1003.68

4.4.2. Results of forest classified carbon storage and value in Yunnan Province. As shown in Table 3, the carbon reserves of forests and trees in Yunnan Province in 2018 were 937.01 million t and 101.29 million t, respectively. The actual value of the forest and tree carbon reserves calculated based on this was 94062.92 million yuan and 101682.31 million yuan, respectively. The carbon reserves of forests and trees are of great economic value to Yunnan Province.

According to the different origins of trees, the areas of natural forest and artificial forest are 15.99 million hm^2 and 5.08 million hm^2, respectively, and the size of natural forest is 3.15 times that of the artificial forests. The volume of natural forests and artificial forests is 1756.20 million m^3 and 216.46 million m^3. The carbon reserves of the natural forests and the artificial forests are 834.19 million t and 102.82 million t, respectively. The carbon reserves of artificial forests account for 12.33% of the carbon reserves of the natural forest. The cultivation and construction of the artificial forests in Yunnan Province have brought great results to the carbon storage of Yunnan Province. In the part of carbon reserve value, the carbon reserve value of the natural forests and the artificial forests are 83741.26 million yuan and 10321.66 million yuan. The economic value potential of the natural forests is huge, but the economic value of the artificial forests is also relatively large. It has provided sufficient confidence and support for the construction and protection of natural and artificial forests in Yunnan Province.

According to the classification of forest species, timber forest accounts for the largest area, followed by shelter forest. However, the volume of shelter forest is greater than that of timber forest. In the part of carbon storage, the carbon storage of shelter forest is also greater than that of timber forest, and the forest carbon sink effect of shelter forest is the largest among the five forest types. The value of carbon reserves is as follows: shelter forest > timber forest > special forest > fuel forest > economic forest.

According to forest age, the proportion of forest age in forest resources in Yunnan Province is 19.8% for the young forest, 45.9% for the middle-aged forest, 18.8% for the near mature forest, 11.8% for mature forest and 3.7% for the over mature forest. With the growth of forest age, the carbon storage capacity of trees increases first and then decreases gradually. According to the distribution of forest age, the storage and carbon storage of middle-aged forests in 2018 is the largest. The volume of middle-aged forest is 695.92 million m^3, and the carbon reserve is 330.56 million t. The book of the young forest is 150.39 million m^3, and the carbon reserve is 71.44 million t. The area of young and middle-aged forests accounts for a large proportion. Yunnan Province has significant potential for carbon sequestration in the future, and there is great room for development in carbon neutralization capacity. In the value of carbon reserves, the value of carbon reserves of the middle-aged forest is the largest, amounting to 33183.84 million yuan.

Table 3. Forest carbon reserves and actual value of Yunnan Province.

Stand type		**area/ 10000 hm^2**	**Storage volume/ 10000 m^3**	**carbon reserve/ 10000 t**	**carbon density (T / hm^2)**	**carbon reserve value (10000 yuan)**
Classification by origin	natural forests	1598.48	175619.57	83419.30	298.34	8374125.92
	Plantation	507.68	21646.27	10281.98	128.38	1032166.24
Classification by forest species	Protection forest	673.79	76868.72	36512.64	54.19	3665356.35
	Special forest	196.73	39011.9	18530.65	94.19	1860217.05
	Timber forest	785.04	73501.3	34913.12	44.47	3504787.00
	fuel forest	40.79	2575.2	1223.22	29.99	122794.11
	economic forest	166.52	5308.72	2521.64	15.14	253137.25
	Open woodland	51.81	1376.59	653.88		65640.40
	Scattered wood		11731.93	5572.67		559417.49
	Four-sided trees		3584.32	1702.55		170912.31
By forest age	Young forest	385.18	15039.07	7143.56	18.55	717112.94
	Middle aged forest	891.87	69592.11	33056.25	37.06	3318383.55
	Near mature forest	365.58	42752.59	20307.48	55.55	2038585.86
	Mature forest	229.36	39596.74	18808.45	82.00	1888104.42
	Over mature forest	71.08	17274.39	8205.34	115.44	823700.44
forest		2106.16	197265.84	93701.27	44.49	9406291.75
forest		2599.44	213244.99	101291.37	38.97	10168231.21

4.4.3. Prediction results of forest carbon sequestration potential in Yunnan Province. According to GM (1,1) grey prediction model, the forest carbon storage is predicted, the data processing and calculation are processed by MATLAB. In the prediction process, to ensure the accuracy of prediction, the data from 1988 to 2018 are used. This paper forecasts the forest carbon reserves in 2023, 2028 and 2033. It predicts the development trend of forest carbon reserves in Yunnan Province without considering external factors such as government policies, economic development and forest land change.

Table 4. Forest carbon reserves in Yunnan Province predicted by GM (1,1) model.

Period	Actual value of forest carbon storage /10000t	Estimate/10000t	Residual δ	Relative error ε%
1976	48542.78			
1981	58467.53			
1988	58442.76	58442.76	0	
1993	58907.15	55883.39	3023.76	5.13
1998	60973.35	61589.18	-615.83	1.01
2003	66466.35	67877.53	-1411.19	2.12
2008	73805.54	74807.94	-1002.40	1.36
2013	80421.87	82445.96	-2024.09	2.52
2018	93701.27	90863.82	2837.45	3.03
Average relative error ε				2.53
Posterior error test C				0.15

As shown in Table 4, according to the posterior error test calculated by GM (1,1) model, C = 0.15 < 0.35, the average relative error ε= 2.53% < 20%, indicating that the model has a good fitting degree, high accuracy, and meets the prediction requirements. Since the change of forest area will have a significant impact on carbon reserves, the forest area is predicted at the same time as the forest carbon reserves [43]. The prediction parameter equation of forest carbon storage and the forest area is as follows:

Time response function of forest carbon storage grey GM (1,1):

$$C^{\wedge(1)}(t+1) = 547330.95e^{0.097t} - 488888.19 \tag{8}$$

Time response function of forest area grey GM (1,1):

$$S^{\wedge(1)}(t+1) = 7677.22e^{0.1357t} - 6744.48 \tag{9}$$

Where, S ^ represents forest area; C ^ represents forest carbon storage; Values 0, 1, 2, 3, 4, Since the data in the forecast began in 1988; 0 stands for 1988, 1, 2, 3, 4 Respectively represent the value series of years after 1988.

As shown in Table 5. Under the current management and management level of forest resources in Yunnan Province, the forest area and forest carbon storage in Yunnan Province in 2023, 2028 and 2033 are predicted. In 2023, the forest area and forest carbon reserves in Yunnan Province will be 25.18 million hm^2 and 1001.41 million t. In 2028, the forest area and forest carbon reserves will be 28.83 million hm^2 and 110.37 million t. In 2033, the forest area and forest carbon reserves will be 33.02 hm^2 and 1216.34 t. Compared with 2018, the forest area increased by 11.96 million hm^2, with an average annual increase of 79712 million hm^2, and the forest carbon storage increased by 307.70 million t.

IOP Conf. Series: Earth and Environmental Science 1087 (2022) 012066 doi:10.1088/1755-1315/1087/1/012066

Table 5. Forest carbon storage and forest area predicted by GM (1,1) model.

Particular year	Actual value of forest area/10000hm^2	Actual value of forest carbon storage/10000t	Predicted value of forest area/10000hm^2	Predicted value of forest carbon storage/10000t
1976	956	91081		
1981	919.65	109703.3		
1988	932.74	109656.83	932.74	58442.76
1993	940.42	110528.18	1115.41	55883.39
1998	1287.32	128364.94	1277.47	61589.18
2003	1560.03	139929.16	1463.07	67877.53
2008	1817.73	155380.09	1675.64	74807.94
2013	1914.19	169309.19	1919.09	82445.96
2018	2106.16	197265.84	2197.92	90863.82
2023			2517.25	100141.17
2028			2882.98	110365.75
2033			3301.84	121634.27
Average relative error C			0.23	0.15
Posterior error test ε			5.4%	2.53%

5. Discussions

According to the theoretical analysis, empirical results and conclusions, it is compared with the research results of scholars. First, the decision of forest carbon storage in Yunnan Province. Comparing forests according to different classifications and trees of different origins, the carbon sink capacity of natural forests is much higher than that of artificial forests. In comparing varying forests ages, the carbon storage of middle-aged forest is the largest. Zhou J also compared the carbon sink capacity of arbor forests, natural forests and artificial forests to calculate carbon reserves of trees of different origins in Yunnan Province [44]. The results are consistent with this paper. However, calculating carbon reserves of different forest ages, the results show that the carbon reserves of young forests are the largest. The difference may be caused by the inconsistent calculation time and the growth of young forests. In the study of Deng et al., the highest carbon storage of different age classes is the middle-aged forest, which is consistent with the conclusion of this paper and confirms the reason for the error [10]. Second, in forest resource value accounting, Mao calculated the value of forest carbon reserves in Yunnan Province as 139835.88 million yuan. There are differences between forest carbon storage value accounting and this paper [5]. The differences come from two aspects: the inconsistent selection of carbon sink conversion price, and the other is the different accounting methods of carbon storage. The third is the prediction of forest carbon sequestration potential. The prediction of carbon sink potential in Yunnan Province is less, and the prediction results of Yang et al. are consistent with the results in this paper [45].

To sum up, this paper empirically analyzes the changes of forest carbon reserves in Yunnan Province, calculates the value of forest resources, predicts the potential of forest carbon sequestration in Yunnan Province, and draws a conclusion by combining theory with practice. In predicting carbon sequestration potential, the influence of external factors such as government policy, economic development and forest land change needs to be considered.

6. Conclusions

According to the results of empirical analysis, the forest area and forest resource carbon reserves in Yunnan Province have been increasing for more than 40 years. The value of carbon reserves is

IOP Conf. Series: Earth and Environmental Science 1087 (2022) 012066 doi:10.1088/1755-1315/1087/1/012066

calculated, the value of carbon reserves is converted according to different stand standards, and the potential of forest carbon sequestration in Yunnan Province is predicted. The solutions of increasing forest carbon reserves and its economic value in Yunnan Province under the "carbon peaking and carbon neutrality" are put forward. The specific conclusions are as follows:

Forest carbon storage accounting part. Forest carbon reserves in Yunnan Province have increased steadily. In 2018, forest resource carbon reserves, forests carbon reserves and forest carbon reserves in Yunnan Province were 2470.50 million t, 101.29 million t and 937.01 million t, respectively. Different stand types have other carbon storage capacity. Among them, the area, volume, carbon density and carbon storage capacities of natural forests are much higher than those of artificial forests. The carbon reserves of natural forest and artificial forest are 834.19 million t and 102.82 million t, respectively, and artificial forest accounts for 12.33% of the carbon reserves of natural forest. Among different forest types, shelter forest plays the largest role in forest carbon sink. Among different forest ages, the middle-aged forest accounts for the most significant proportion in volume and area, and the carbon reserve reaches 33183.84 million t.

Carbon storage value accounting part. In 2018, the value of forest carbon reserves in Yunnan Province was 94062.92 million yuan. Among them, the actual value of carbon reserves of natural forests is 83741.26 million yuan, accounting for 89% of the value of forest carbon reserves in Yunnan Province. Among different forest ages, the value of carbon reserves of mature forests is the highest, 18881.04 million yuan. Among different forest types, the economic value of shelter forests is the highest, 36653.56 million yuan.

Prediction of forest carbon sequestration potential. The forest area and carbon sequestration potential of Yunnan Province in 2023, 2028 and 2033 were predicted by GM (1,1) grey prediction model. In 2033, the forest carbon storage in Yunnan Province will reach 1216.34 million t. Compared with 2018, the average annual growth of carbon sequestration is 17.46 million T / A.

The improvement of forest storage level will naturally lead to increased forest carbon storage. The size of young and middle-aged forests in Yunnan Province accounts for 65.7%. With the growth of middle-aged and young forests, the area of mature forests will be larger and larger, the carbon sink capacity of forests in Yunnan Province will be further improved, and the carbon storage will be further increased. At the same time, with the development of the Yunnan forest CDM carbon sequestration project (Clean Development Mechanism), the forests in Yunnan Province will be additional managed and managed better. The area and coverage of forests will continue to increase.

The solutions of increasing forest carbon reserves and its economic value in Yunnan Province under the "carbon peaking and carbon neutrality" are as follows:

(1) Protect and increase the total amount of forest carbon sequestration

The incentive policy of Yunnan Province is to increase the proportion of carbon reserves and natural forest-related policies. Establish a stable and balanced forest ecosystem in Yunnan Province, actively use the carbon sequestration role of forests, establish a suitable mechanism and path of increasing foreign exchange, and improve the carbon sequestration capacity of forests under the "carbon peaking and carbon neutrality" goal of Yunnan Province.

(2) Actively promote the construction of carbon sequestration forest in Yunnan Province.

The prediction results of this paper show that Yunnan Province has great carbon sequestration potential, and the carbon sequestration effect of the artificial forests has also played a corresponding role. We should increase the development of forestry emission reduction and sink increase technology, select areas with crucial ecological locations, select sites suitable for carbon sequestration afforestation and carry out carbon sequestration afforestation. Strengthen the implementation of crucial forestry projects, increase the forest area, pay attention to improving forest quality and quality, and enhance the function of forest carbon sequestration and foreign exchange. Establish the forestry development strategy under the "green three cards" strategy.

(3) Promoting the balanced development of forest carbon sequestration in Yunnan Province

The optimization of forestry carbon sequestration efficiency in Yunnan Province under the "carbon peaking and carbon neutrality" goal depends on the coordination mechanism of Yunnan provincial government, such as the development of forest CDM carbon sequestration project in Yunnan Province. For different cities and autonomous prefectures in Yunnan Province, it is necessary to strengthen the balanced result of forest carbon sequestration and establish the linkage and sharing of forest carbon sequestration, learn from and share advanced forest management experience and corresponding carbon sequestration afforestation work, to promote the balanced development of forestry carbon sequestration in the whole province.

Acknowledgement
This study was supported by the science and technology innovation fund project of Southwest Forestry University (Grant no. KZ21001), Yunnan philosophy and Social Sciences Planning Project (Grant no. YB2020054) and National Social Science Foundation Project (Grant no. 18BGL052).

References
[1] Wang X, Hu S 2021 Effects of climate change cognition and environmental efficacy on Residents' low-carbon emission reduction behavior *Popular science research* **16** 3
[2] Jiang D, Zhang L 2019 Optimization path of China's low-carbon environmental policy based on multi-party cooperative game *Enterprise economy* **38** 12
[3] Zhuang G, Dou X 2021 Policy connotation and realization path of carbon emission peak under the new development pattern *Journal of Xinjiang Normal University (PHILOSOPHY AND SOCIAL SCIENCES EDITION)* 06
[4] Li Y, Chen M 2021 Provincial Agricultural Non-co sources in China_ 2 Analysis of influencing factors and peak prediction of greenhouse gas emissions *Journal of Environmental Science* 1-16
[5] Cao Q 2021 Scientific connotation of "carbon peak and carbon neutralization" and China's policies and measures *Environment and sustainable development* **46** 02
[6] Mao H 2021 Dynamic change trend and Countermeasures of forest carbon storage in Yunnan Province Forestry survey *Central South* 40
[7] Xu S 2014 Summary of forest carbon storage estimation methods *Forestry Survey and planning* **39** 06
[8] Kong R, Zhang Z, Zhang F, et al 2020 Temporal and spatial variation of forest carbon storage in the Yangtze River Basin and its driving factors *Research on Soil and water conservation* **27** 4
[9] Wang G, Sun T, Min G, et al 2021 Role and suggestions of terrestrial ecosystem carbon sink in realizing the goal of "double carbon" *China Geological Survey* **8** 4
[10] Fu W, Luo M, Chen J 2021 Research progress and Prospect of carbon footprint and its influencing factors *Forestry Economy* **43** 08
[11] Deng Y, Xiao F 2020 Study on forest carbon storage in Yunnan Province Based on forest resources planning and design survey *Forestry Survey Planning* **45** 2
[12] Li S 2005 Research on forest carbon sequestration economy *Shenyang: Northeast University* 1-7
[13] Paul Schroeder 1991 Can intensive management increase carbon storage in forests? *Environmental Management* **15** 4
[14] Nilsson S, Schopfhauser W 1995 The carbon-sequestration potential of a global afforestation program *Climatic Change* **30** 3
[15] Zhang Y, Li X 2021 Analysis of forest carbon sink potential in Beijing under the goal of carbon peak and carbon neutralization *Resources and Industry* 1-15
[16] Golub, A.; Hertel, T.; Lee, H. L.; Rose, S. and Sohngen, B 2009 The Opportunity Cost of Land Use and the Global Potential for Greenhouse Gas Mitigation in Agriculture and Forestry *Resource and Energy Economics* **31** 4
[17] Sun Q, Jia L, Liu Y, et al 2016 Research Progress on estimation of forest vegetation and soil carbon storage in China *Environmental chemistry* **35** 8

[18] Fang J, Chen A 2001 Dynamic changes of forest vegetation carbon pool in China and its significance *Acta botany Sinica* 09
[19] Cheng X, Zhang L, Li Y et al 2020 Analysis on the current situation and dynamic change of carbon sequestration of natural forests in Gansu Province Western forestry science, 49 02
[20] Zhang Y, Wang X 2021 Establishment of national forest stock biomass model and Study on carbon change *Chinese Science: Life Science* **51** 02
[21] Wu G, Zhou N 2015 Study on forest carbon storage and its economic value in Hunan Province *Journal of Central South University of forestry science and technology* **35** 8
[22] Liu T 2019 Study on temporal and spatial evolution of forest carbon reserves in Zhejiang Province Based on remote sensing and process model *Zhejiang Agriculture and Forestry University* **43** 27
[23] Kong R, Zhang Z, Zhang F, et al 2020 Temporal and spatial variation of forest carbon storage in the Yangtze River Basin and its driving factors *Research on Soil and water conservation* **27** 4
[24] Xi T, Li S 2006 Analysis of forest carbon sequestration potential in Heilongjiang Province *Forestry economic issues* 06
[25] Huang Y L, Cheng L H, Li X Q 2012 On the Basis of Forest Volume Expansion Method to Predict the Carbon Sink Potentiality in Daxing\"anling *Advanced Materials Research* 424-425
[26] Zhang Y, Zhou X, Qin Q et al 2013 Research on value accounting of forest carbon sequestration in China *Journal of Beijing Forestry University* **35** 06
[27] Yin Z, Lu C, Liu Q 2014 Research on forest carbon sequestration potential of Northeast Inner Mongolia forest region from the perspective of carbon sequestration trading *Guangdong agricultural science* **41** 19
[28] Yao R, Bian W, Fan H, Guan J 2021 Analysis on the evolution of forest carbon sink efficiency in China *Forestry economic issues* **41** 01
[29] Chen K Y, Zi H B, Ade L, et al 2018 Current stocks and potential of carbon sequestration of the forest tree layer in Qinghai Province, China Chinese *Journal of Plant Ecology* **10** 17521
[30] Xu B, Guo Z, Pu S, et al 2010 Forest biomass carbon pool in China from 2000 to 2050: prediction based on the relationship between biomass density and forest age *Chinese Science: Life Science* **40** 07
[31] Jiang X, Huang Z 2016 Analysis of China's forestry carbon sequestration potential under the new economic normal *China rural economy* 11
[32] Ren J, Xia J 2017 Prediction of forest carbon sink potential in Heilongjiang Province Based on carbon density forest age relationship *Environmental science research* **30** 04
[33] Li Q, Zhu J, Feng Y, Xiao W 2018 Carbon storage and carbon sequestration potential prediction of forest arbor forests in China *Progress in climate change research* **14** 03
[34] Yuan B 2021 Research on financial support of forestry carbon sequestration in China -- Comment on research on potential and development path of forestry carbon sequestration in China *Forestry economy* **43** 02
[35] Wu Q 1995 Modification of grey prediction model GM (1,1) *Journal of Heilongjiang business college* 03
[36] Xu Y, Wang H , Hui N L 2021 Prediction of Agricultural Water Consumption in 2 Regions of China Based on Fractional-Order Cumulative Discrete Grey Model *Journal of Mathematics* 07
[37] Ding S, Xu N, Ye J, et al 2020 Estimating Chinese energy-related CO2 emissions by employing a novel discrete grey prediction model *Journal of Cleaner Production* 259
[38] Liu J, Gao P 2021 Energy Consumption Predication in China Based on the Modified Fractional Grey Prediction Model *Journal of Mathematics* 1
[39] Gong-wei 2019 Application of GM(1,1) Grey Prediction Model in Dynamic Management of High-speed Railway Construction *Cost Value Engineering* 32

[40] Fang J, Guo Z, Pu S, et al 2007 Estimation of terrestrial vegetation carbon sink in China from 1981 to 2000 *Science of China (Series D: Earth Science)* 06
[41] Yang H, Liu J, Zheng B 2011 Improvement and application of Grey Prediction GM (1,1) model *Practice and understanding of mathematics* **41** 23
[42] Zhang Y, Pan J 2016 Study on economic accounting of forest carbon sequestration and preparation of balance sheet *Statistical research* **33** 11
[43] Zeng W, Sun X, Wang L, et al 2021 Research and construction of stock, biomass and carbon storage models of 10 main forest types in Northeast Forest Area Journal of Beijing *Forestry University* **43** 03
[44] Zhou J, Xu S 2016 Current situation and dynamic analysis of forest carbon storage in Yunnan Province *Forestry Survey Planning* **41** 1
[45] Yang Y, H, Wang X, et al 2017 Prediction of forest resources and carbon sink potential in Yunnan *Province Forestry resource management* 4

IOP Conf. Series: Earth and Environmental Science 1087 (2022) 012067 doi:10.1088/1755-1315/1087/1/012067

Analysis of the current situation and suggestion of countermeasures for international standardization for power grid industry of China

Cai Zhang[1], Yuanping Huang[1], Xiaoming Zhao[1], Lan Cao[1], Liming Zhao[2], Chaoyan Lou[2], Pengyue Zhang[2,3]

[1] State Grid Zhejiang Electric Power Co., Ltd. Research Institute;
[2] China Jiliang University.

[3] Corresponding author's e-mail:104552065@qq.com

Abstract. Facing the rapid development of China's power industry, China actively participates in the formulation and revision of international standardization and promotes the internationalization of standards in the field of power grid. But at the same time, there are also some problems in the power grid industry, such as imperfect domestic organization and insufficient international recognition of standards. Through literature investigation and statistical analysis, starting with the current situation of standardization in China's power grid industry, this paper discusses how to promote the establishment of better international technology and path of standards in China. Based on the discussion, promote the deep integration of standards and technology, actively carry out multi-party cooperation between industry, University and research, accelerate the cultivation of standardized talents, and realize the international cooperation of standards to a new level. Promote the research on the internationalization of standards in China's power grid field, and make suggestions for the standardization development of China's power grid industry, so as to promote China's power grid industry and market to better respond to opportunities and challenges, and promote the improvement of China's international influence and competitiveness in the power grid field.

1. Introduction

Electric power development is the most important basic energy industry in the development of national economy and the basic industry of national economy and people's livelihood. The 14th Five-Year Plan (2021-2025) for National Economic and Social Development and the Long-Range Objectives Through the Year 2035 states: " the modernization of economic system would be accelerated, the new development pattern would be accelerated, in which domestic and foreign markets boost each other, with the domestic market as the mainstay." The new pattern cannot be separated from basic energy services, including reliable electric power safeguard [1]. Since the founding of People's Republic of China more than 70 years ago, huge industry development and leapfrog progress in power technology, have been made. China's electricity consumption increased by 1900 times, per capita electricity consumption increased by 630 times, power generation increased by 1,600 times, installed generation capacity increased by 1,000 times, power grid lines increased by 280 times, voltage classes increased by 5 times, the technical level of the power industry has been significantly improved. In addition, international cooperation in electric power grows steadily, and the electric power system has improved drastically [2].

With the increasing progress of domestic power grid technology, standards played a pivotal role in the power industry, and the "Bringing In" and "Going Out" of power technology push the power standard to the international urgently. State Grid started to implement the internationalization strategy in 2004, and by 2007, China Electricity Standards Internationalization Project started to help the internationalization of electricity standards through the English translation of power standards and the evaluation of domestic international standards. In addition, the first New Technical Committee IEC TC115 proposed by China was established in IEC in October 2008, which greatly indicated the strength of China's power standards technology [3]. To this day, China actively participates in the international organization of power standards, leads the setting of international power standards, actively involved to international meetings, promotes the process of internationalization of China's power standards, which provides paths for the internationalization of China's power standards [4].

The internationalization of electric power standards is an important means to improve China's electric power technology in line with international standards and an important way to promote China's electricity-related trade to "go global". China has undertaken the important work of electric power standards in three major international organizations represented by IEC, ISO and ITU. In other organizations, such as CIGRE and IEEE, China also plays an important role. China is an active participant in international organizations. It has been a member of IEC since 1957 and now it has joined all four IEC international mutual recognition systems for conformity assessment: Electronic components quality assessment System (IECQ), Electrical products and Components conformity Assessment system (IECEE), explosion-proof electrical products safety Certification system (IECEx), renewable energy certification mutual Recognition system (IECRE), and fully participate in IEC conformity assessment work. State Grid's fruitful achievements in international organizations have played an important role in the internationalization of Chinese standards, the international recognition of Chinese power grid technology and the strengthening of international influence. In order to promote China's power grid standards to further go international, this paper carries out research on the status quo and analysis of problems by means of research and data analysis of relevant international documents of China's power grid industry standards [5].

2. Domestic research status and development trend

2.1. Research content

In recent years, China's electric power industry is thriving, and its technical achievements cannot be separated from the rapid development of electrical standardization. Shu Yinbiao, the 36th chairman of IEC, put forward that standardization is the catalyst to accelerate the development of scientific and technological innovation, and technological innovation based on technical standards is a means of high-level competition [5]. Figure 1 shows the number of electric power industry standards in China. According to the data of China Electricity Council, as of June 2020, there are 558 national standards, 2784 industrial standards and 245 group standards related to China's electric power industry, among which 71 are national mandatory standards [6], accounting for 12.7% of the total. There are 65 mandatory industrial standards in total. And from 2014 to 2020, a total of 300 national standards have been formulated and released, with the overall trend showing a steady rise. At the same time, China has actively participated in the formulation of international standards, and the number of international standards has steadily increased, which is conducive to promoting the process of China mastering the right to speak in international standards organizations and enhancing China's international influence. From 2010 to 2020, China has led or participated in the formulation of 84 international standards and issued 2548 enterprise standards. Figure 2 shows the State grid system revised number of international standards. Through the analysis of the years of leading or participating in the formulation of international standards. Among them, in the uhv, the advantages of the smart grid technology, accelerate the technical achievements transformation of science and technology as the international standard, develop science and technology talents with international vision and strategic thinking, established the world's first set of uhv technology and smart grid technology standard system, 1000 kv uhv ac voltage has become the international standard voltage [7].

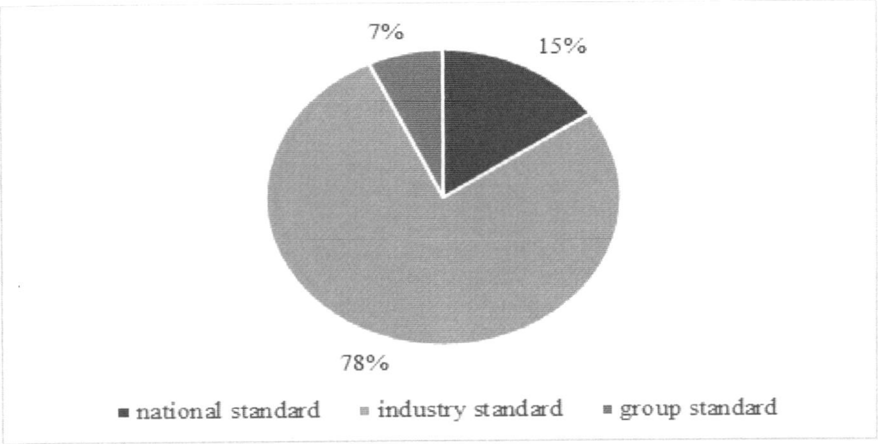

Figure 1. Electric power industry standard formulation quantity chart of China.

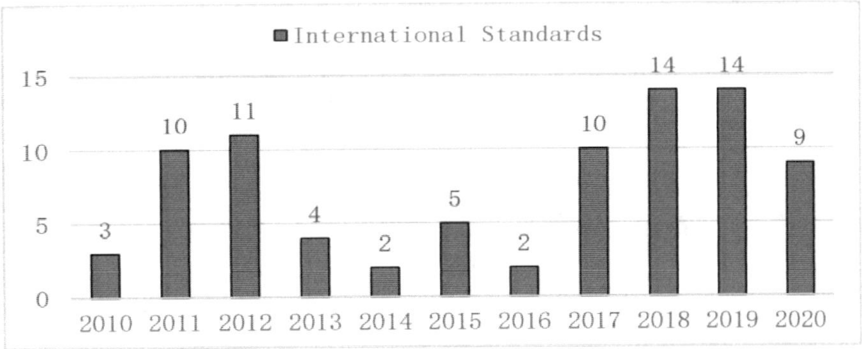

Figure 2. State Grid system revised international standard quantitative chart.

2.2. *Research methods*

Through the combination of literature review, data analysis and other research methods, as well as excel data analysis and other methods, this paper carries out research on the internationalization of China's power grid standards, so as to promote the thinking of the internationalization of China's power grid standards. Firstly, through combing, analyzing and summarizing relevant literatures of the power grid industry, the paper comprehensively presents the power grid standards and relevant international theories and researches, and puts forward countermeasures and suggestions in line with the international development of China's power grid standards based on the characteristics of the international development of China's power grid standards. Secondly, through the data analysis method, according to the research theme and requirements of this paper, according to the number of the state Grid system and revision standards, domestic issued grid standards for investigation and research. Statistical analysis of the time, quantity and content of relevant power grid standards can directly reflect the construction of international power grid standards in China, and then reflect the introduction and implementation of power grid standards and the international transformation of power grid standards. Finally, Excel statistical software is used to obtain the trend change of the analyzed table data and statistical chart to understand the situation of relevant standards, and then analyze the distribution characteristics of individual variables and the relationship between each other to understand the law of the standard release in China's power grid field.

2.3. Research results
Using literature review method, statistical analysis method, can get the research results of the status quo of the internationalization of China's power grid industry standards. Firstly, it can be seen from Figure 1 that the number of studies on international standards of State grid in China has increased significantly in the past decade, indicating that China is paying more and more attention to the internationalization of standards. Secondly, according to the number of China's standard system in recent years, it can be seen that China's power industry standards have increased in variety and quantity, and are relatively perfect. At the same time, Excel statistical analysis was conducted. As can be seen from Figure 1, electric power industry standards account for as much as 78%, while national standards and group standards only account for 15% and 7%. There are a large number of standards, most of which are released in the form of industrial standards. In addition, as shown in Figure 2, the number of international standards led or participated by State Grid of China in the past decade only accounts for 3.3% of the number of corporate system and revision, and the number of international standards participated is relatively small. The development of international standards reflects the participation of international activities. As early as 1998, China's Society of Electrical Engineering and IEES PES jointly initiated the International Conference on Power System Technology. Since 2020, the conference has been held in China and Asia-pacific countries in turn every year. The conference invites domestic and foreign scholars to conduct exchanges on new power systems and other issues, which has an important influence among international peers. In addition, in 2008, under the guidance and support of CIRED, China held an international Symposium on Power Supply and technology, and made it a regular academic conference hosted by China -- China International Power Supply Conference (CICED).

In recent years, with its advantages in power grid technology innovation and engineering practice, State Grid has actively promoted the "going global" of Chinese standards and constantly promoted the internationalization of Chinese standards. China has actively participated in the activities of international organizations, actively led the organization of relevant international conferences, strengthened international cooperation, greatly improved the international influence of China's electric power field and China's competitiveness in this field, while playing an important role in improving the quality and efficiency of China's relevant enterprises.

3. Problem analysis

3.1. Domestic related organizations are not perfect
Standardization in China's power grid sector is a long-term and systematic project, the improvement of domestic organizations and the cooperation and exchange of power enterprises is an important support for its development. The standardization technical committee was not well organized, which not only hindered the high-quality output of domestic technical standards, but also impeded improving the discourse of China's international power grid field, which was not conducive to the internationalization of China's standards and international power grid market. Some domestic enterprises in China could not understand IEC and other international organizations, failed to adopt enough international standards, and did not take the initiative to actively participate in the work of international organizations. Due to the limited technical level of the enterprise itself, lack of communication and cooperation with international experts, the unfamiliarity with the IEC standard proposal application process, they could not put forward a favorable new work proposals in organization and fail to use the technical discourse to improve international competitiveness [8]. The fact that China had only 65 technical committee chairs in the IEC/ISO committee secretariats shows that China's substantive participation in international organizations such as IEC was not deep enough, which was not conducive to enhancing China's international discourse and influence.

3.2. Low International Recognition of Standards and Lack of Domestic Expertise
In nowadays developed countries still hold the international standard discourse, the United States, the Western European countries standards are used in most countries, so Chinese standards need to be

IOP Conf. Series: Earth and Environmental Science 1087 (2022) 012067 doi:10.1088/1755-1315/1087/1/012067

recognized internationally. Explore the reasons for this, to be able to find, Because China's electricity network technical standard system and the international standard system is very different, has not yet established with international standards of advanced grid standards system, standards are not the same. In addition, because the international lingua franca is English, and most of China's relevant standards lack of foreign translation, which in all aspects seriously hinder the process of internationalization of China's standards International standards work requires personals with both a deep professional background and foreign language communication skills, who also needs to be familiar with international standards development rules and international intellectual property rights and other policies of the complex talent. At present, the proportion of high-level talents in the field of international standards in grid is small and the talent ladder has not yet been formed, thus limiting the motivation to participate in international standards work and causing a shortage of complex talents working on international standards in the grid field. Standard revision is an important process of standard maturity, China's domestic standard review cycle is generally not more than 5 years, but China's standard revision has been untimely, long cycle, the lack of international standards to track the dynamic understanding of the problem [9], resulting in China's mature standards may appear inappropriate, incompatible, unstable and other issues, is not conducive to promote China's mature standards into international standards in a timely manner. In addition to the mentioned problems, there are problems such as insufficient awareness of standards, lack of timely action and inexperience in transforming scientific and technological achievements into standards in China. By exploring these problems, it can be found that China lacks composite talents who combine standards and technology, and talents are important resources for progress, and China lacks sufficient support for the construction of a team of professional leaders in training standards, thus it has a negative impact on improving the international recognition of our country's standards.

3.3. Low Level of Leading International Standard Activities and Insufficient International Influence

As the forefront of international standard work, the depth and breadth of participation in work of ISO is a direct reflection of the strength of the influence in rule-making, and also reflects the level of a country's core competitiveness. In order to improve its international influence, China needs to proactively cooperate and exchange with other countries to promote the international competitiveness of Chinese enterprises. Taking the IEC as an example, there were 204 technical bodies as of December 2017, with China accounting for about 5.88% and 4.41% in terms of holding the presidency and secretariat. Through the degree of China's leading organizations in the standard activities can be seen, China lacks domestic and international double-loop common development, the focus is still mainly on the relevant domestic power grid field meeting activities, in the international influence of the meeting activities mainly for the participating countries, to lead the organization of international organizations meeting experience is less, is not conducive to enhancing China's leading awareness and ability [10]. In addition, because China's power grid field in the international influence is not as developed countries, in of domestic and international conferences leaded by China, such as the China International Power Supply Conference and the International Conference on Power System Technology, there may be with other countries in the innovation technology and scientific research and other aspects of the exchange and cooperation of the depth and breadth of the problem is not enough, thus hindering China's power grid field standards of high-level opening, high-quality development as well as reform and innovation. The lack of deep and wide enough exchanges and cooperation will cause a waste of human, material and financial resources, and may even reduce the enthusiasm and participation of Chinese enterprises in the international power grid sector, hindering the process of "Going Out".

4. Countermeasure and suggestions

4.1. Promoting the Standard Industry-University-Research Depth Integration, Accelerating the Organization of Institutional Reform

To solve the problems of inadequate domestic standardization organization and insufficient international competitiveness of domestic power grid enterprises, deep cooperation in production, research and development should be developed, research and standard-setting of power grid technology should be encouraged, reform of domestic power grid system mechanism should be promoted. On the one hand, the cooperation among industry, science and research in the field of power grid standardization will combine capital, technology and talents, and more professional standardization talents will have the opportunity to participate in the international standardization work in this field, which is conducive to the cultivation of international talents. On the other hand, the standardization is to give the market and technology deep integration. Marketization could not be separated from standardization, and standardization with the progress of marketization could continue to improve. Industry-academia-research cooperation, improve power grid technology to open the global market, deepen the power grid global cooperation, but also for the power grid standards to international foundation. Existing standard work mechanism need to be reformed, a benchmarking system would be established with the ISO and other international organizations, especially in aspects of standard expert team-building, standard creation norms; of three-level standard creation system, include national standards, industry standards and group standard, should be optimized and integrated, to ensure the process of standard creation universal and advanced; an efficient, authoritative international standard work coordination mechanism in the field of power grids need to be established, coordinate domestic and international resources to participate in International standards work, which would form a synergy in the national level. Therefore, the grid field standardization industry-academia-research cooperation as an important way to open the organizational gap, improve the level of technology as an important means of international competitiveness, will promote the domestic grid standardization field institutional reform as an important way to internationalization, the enterprise international participation is expected to improve [11-12].

4.2. Insisting on the Combination of "Bringing In" and "Going Out" to Carry Out the Standard Making and Revision Work

The right to speak and competitiveness is an important manifestation of China's power grid standards internationalization, enhance the right to speak and competitiveness is not only closely related to the internal organization system, but also should actively absorb the international power grid standards resources, for international standards to do reference and model. In recent years, some of the standardization activities in the field of domestic power grid through the translation of international standards, international research, the introduction of international patented technology as the main way [13]. The role of "Bringing In" is increasingly obvious, "Going Out" is not to be underestimated. Promote the translation of domestic standards, carry out professional translation of standards, actively carry out international cooperation, promote China's standards and technology with the world, and get the recognition of other countries. In recent years, China actively carry out the "one belt and one road" construction, promote cooperation with Asian and European countries, many power grid technology in the "one belt and one road" in the effective application of international cooperation is the domestic standard technology recognition of the pass, and actively seek standardization in the field of power grid Cooperation, play a "Going Out" advantage role, but also to promote the internationalization of China's power grid standards in the field of good policy [14-15].

4.3. Actively Participating in The International Cooperation of Power Grid Standards, to Enhance the Influence of China's Power Grid Standards

Cooperation is an important way of communication, is to make international awareness of China's power grid standards development and technological progress is an important way. Electricity is the industrial society and new energy, and people's lives are closely related, just participate in international activities for our country is far from enough, actively promote international activities, play our voice in international activities and leading power, in order to enhance the international influence of China's power grid standards. Figure 3 shows the Global Energy Internet Standard Internationalization Overall Thought. Based on the basis of standardization, attaching importance to the concept of standardization and combining the development status and the requirements of internationalization of standards, the general idea can be put forward. At the same time, while participating in international activities, China should show its responsibility and strength as a big country, pay attention to put forward countermeasures and suggestions for the global power grid problems, seek solutions, truly apply the experience of international activities to promote the internationalization of China's power grid standards.

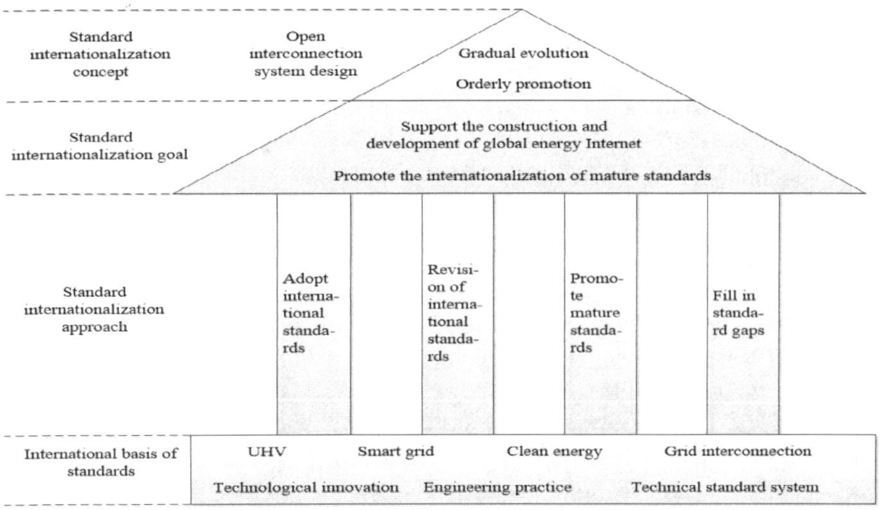

Figure 3. Global Energy Internet Standard Internationalization Overall Thought.

5. Conclusions

The power grid is the guarantee of people's livelihood and the foundation of industry. The role of the internationalization of national grid standards in promoting China's power development should not be underestimated. Facing the blueprint of socialist modernization, facing the pace of economic globalization, facing China's "One Belt, One Road" development strategy, facing the "Bringing In" and "Going Out" of the important Node, the international power industry market is playing an increasingly important role in promoting the internationalization of power grid standards is an important part. Although there are some problems in China, such as incomplete grid standardization organization, vacancy in standard formulation, low level of international recognition, and insufficient participation in international cooperation. In the future, China will contribute its "China plan" to the world and promote the internationalization of China's power grid standards by promoting the reform of the system and mechanism, enhancing the effectiveness and international influence of the standards, enhancing the voice of the international community and actively seeking international cooperation. The development prospect of the standardization internationalization in the field of power grid is bright.

Acknowledgments

Project Supported by Science and Technology Project of State Grid Zhejiang Electric Power Co., LTD (Project Name: An empirical study of IEC/ISO international standard application, Number: 5211DS20008B).

References

[1] Liu S C and Lin H C 2020 Strategic measures for the internationalization of manufacturing standards in China *[J] Macro Quality Research* **8(06)** 80-95
[2] Lu H Q, Yang M C and Zhang D D 2019 Analysis on the internationalization strategy of standards in China's electric power industry *[J] Electrical Technology* **20(09)** 7-9
[3] Lu T J and Cui Zhiqin 2021 "One Belt One Road" project integration model and China's standard internationalization strategy *[J] International Economic Cooperation* **(06)** 87-92
[4] Li W Y, Lu F and Zou C Y Statistics and analysis of world rail transit industry standards in 2020 [J/OL]. *Railway standard design* 1-11
[5] Li H, Lu C and Zhang P Y 2021 Survey on the status quo of urban rail transit standardization in countries along the "Belt and Road" *[J] Standards Science* **(S1)** 256-259
[6] Zhang J and Li B 2020 The new generation of smart grid technology standard system supports the high-quality development of power grids *[J] Power Supply* **37(03)** 2+1
[7] Zhang J, Ye J Y, Li B, Qi B and Gao Z Y 2020 Research on the internationalization strategy of smart grid standards *[J] Power Supply* **37(03)** 3-9
[8] Wang P, Liang Z and Hou J J 2019 Overview of standard knowledge system *[J] Standard Science* **(03)** 45-52
[9] Li B, Zhang J, Tian S M, Dong M Y, Qi B, Sun Y and Zhu W Y 2018 The latest progress and development trend of smart grid user domain standardization *[J] Electric Power Construction* **39(03)** 12-22
[10] Lu Z J 2019 Discussion on the evaluation system of quality service of electric power companies under the smart grid environment *[J] Times Agricultural Machinery* **46(11)** 148-149
[11] Yang Y 2021 Sharing the wisdom and achievements of digital transformation——Huawei actively participates in the formulation of smart grid standards to promote industry development *[J] China Electric Power* **(10)** 66
[12] Cai X M 2020 Promote the development of smart grid *[J] Electrical Appliance Industry* **(06)** 1
[13] He W S 2017 The development trend of the internationalization of China's power standards *[J] China High-tech* **1(05)** 24-26
[14] Shu Y B 2017 State Grid: Speeding up the internationalization of technical standards *[J] Research on Modern State-owned Enterprises* **(11)** 14-16
[15] Gao Z Y, Cao Y, Zhang J and Yu Y J 2017 China electric power **50(11)** 42-47+77

Recognition and application of single sand-body in fluvial facies

Yuhang He[1,2]

[1] Exploration and Development Research Institute of Daqing Oilfield Co. Ltd., Daqing, 163712, Heilongjiang, China.

[4] Corresponding author's e-mail: hyuhang@petrochina.com.cn

Abstract. In the ultra-high water cut stage, the remaining oil of Lasaxing oilfields in Daqing placanticline, shows the distribution characteristics of highly dispersed and locally enriched, and the remaining geological reserves are still mainly distributed in fluvial reservoirs. In order to search and excavate the remaining oil of thick reservoir, the fine description technology of fluvial reservoir under the condition of dense well pattern was studied, and the recognition technology of meandering river and braided river single sand body was formed, which deepened the understanding degree of channel sand reservoir heterogeneity. The main controlling factors of remaining oil are defined, and the distribution mode of remaining oil is established, which can effectively guide the design of horizontal wells at the top of thick reservoir and the adjustment and exploitation of in-layer remaining oil tapping.

1. Introduction

According to the data analysis of sealed core wells in Lasaxing oilfields, 26.8% of sand body thickness in river is still not washed. According to the analysis of the multi-disciplinary reservoir research results in recent years, the remaining oil is mainly distributed in the fluvial sand body, accounting for about 80% of the total remaining oil. It can be seen that the channel sand body is still the main object of the remaining oil exploration in the later period of high water cut.

The heterogeneity of reservoir is the main factor to control the remaining oil. It is the key way to understand and excavate the distribution of remaining oil and improve the development effect. At present, according to the complex internal structure of fluvial-facies reservoir, the study on the heterogeneity description of sand bodies in various channels under the condition of dense well network is carried out systematically, and some progress has been made. The technology of meandering channel hierarchical description and the braided channel combination prediction is explored, and potential exploration is made in horizontal wells, which has got good application effect.

2. Identification and application of single sand body in meandering river

Through the anatomic description of the internal structure of the reservoir in the meandering river, a method of identifying the single sand body of the point dam in the river is formed, which concluded as "determining the edge by abandoned channel, recognizing the location by thickness". The horizontal well exploitation is carried out on the basis of the point dam identification, and obtains good results.

2.1. Plane combination of abandoned channel and point dam

2.1.1. Identification and combination of abandoned channels. The point dam takes abandoned channel as boundary, so in the process of identifying abandoned channel, we should clarify the abandonment mode of the abandoned channel, speculate the abandoned channel width, and carry out the plane combination of the abandoned channel.

(1) The abandonment way of abandoned rivers. According to the abandonment mode of abandoned rivers, abandoned rivers can be divided into three types (Figure 1).

The first is chute cutting type (Figure 1(a)), which is formed by the cross flow in the point dam side accumulation due to the enhancement of flood hydrodynamic force in a certain period. Generally, it is not abandoned at the point of maximum curvature, and the form is mainly "C" shape. The extension distance of the abandoned river channel is small, the curvature is relatively small and discontinuous.

The second is neck cut type (Figure 1(b)). When the river bends more and more in the development process and the curvature reaches a certain degree, the river will be cut, straightened, and abandoned, mostly at the maximum curvature point of the river, which is generally in the form of "O" and "C". The extension distance of the abandoned river is small, the curvature is large and discontinuous.

The third is crevasse diversion type (Figure 1 (c)), which is mainly due to the crevasse of the river channel due to the enhancement of hydrodynamic force, resulting in the abandonment of the original river channel, which is generally in the shape of "s". The extension distance of the abandoned river channel is long, the curvature is general and relatively continuous.

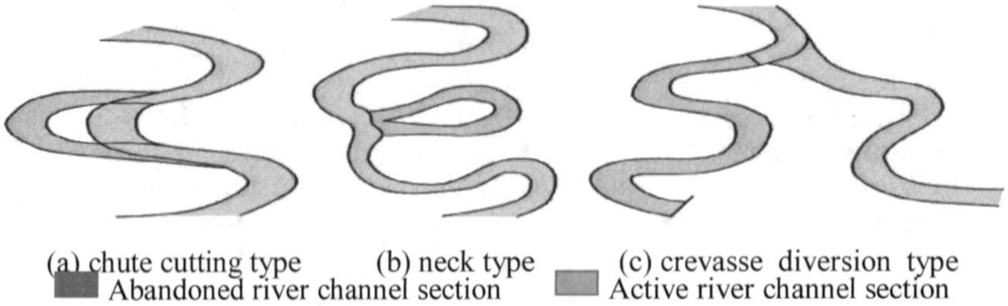

Figure 1. Schematic map of the abandoned patterns of the abandoned channel.

(2) Width prediction of abandoned river channel. Schumm empirical formula has been widely used in calculating river parameters. The main parameter chosen is the width of abandoned channel. The more reliable parameters available in the development zone (river bend span, percentage of silt content) are selected, and the channel width is calculated by using relational formula. The average silt content percentage of channel is 11.4, and all samples are around this value, with little deviation. The average thickness of channel cycle is 5.0 m, so the river depth h is (5.0)×110% 5.5m, which

$$D=255M-1.08\times h=255\times 11.4-1.08\times 5.5=255\times \text{zero point zero seven two two}\times 5.5=101. \quad (1)$$

Where, D - river width, H - river depth, M - percentage of silt content.

Therefore, the width of the abandoned river channel in the study area is 101.3 m.

According to the width measurement of abandoned channels by outcrops and modern sediments, it can be seen that there are many abandoned channels in meandering rivers with narrow width. For example, the abandoned channel in Katbank oilfield is about 15% of area. There are also a lot of yoke lakes in Songhua River， the width of which is about 100m [1-2], which is close to the calculation results.

(3) The combination mode of abandoned river. Generally, a river channel may be abandoned several times. The abandoned River in different single river channels will cross and merge with each other on the plane. It is difficult to trace and restore the original features of the abandoned river due to the complete degree of abandoned river. Therefore, we should combine the logging curve shape， the river flow direction, the sand body superposition direction, the change of sand body, etc. to make comprehensive discrimination in combination. Firstly, the abandoned channel in single wells is identified by using logging phase pattern; Secondly, the trend and range of abandoned channel are determined according to the plane distribution of abandoned river; Thirdly, according to the

development of interbank sand and the evolution trend of sand body thickness, the belonging of the abandoned river channel can be determined; Fourthly, the type of abandoned channel can be determined; Fifthly, according to the overall direction, the abandoned river should be predicted and combined in the same direction. Finally, many kinds of abandoned river can be combined.

2.1.2. Identification and combination of point dam. Point dam is the product of river lateral migration, while multiple swings of the river form composite point dam. Complex meandering zone is composed of multiple composite point dams [3-4]. There are two steps in identifying and combining point dams:

Firstly, single well identification is necessary. There are three aspects marked: the thickness of sand body, the distribution of adjacent abandoned channel and the positive rhythm of sedimentary sequence.

(1) The vertical sedimentary sequence of point dam sand body - the single well interpretation basis of point dam. The vertical section of point dam in single well shows obvious positive rhythm, which forms the main body of "dual structure" of meandering river. The point dam sand body is composed of several lateral accumulation layers. The SP and GR logging curves are mainly bell type, the amplitude difference of the microelectrode curve is large, the micro electrode curve in the developed part of the lateral deposit is obviously returned, the amplitude difference is reduced.

(2) Distribution of sand body thickness of point dam - an auxiliary sign for identifying the range of point dam. The formation process of point dam is an obvious "concave erosion and convex" process. The sand body of point dam is the largest in inner thickness of composite river, which is generally lenticular, and the distribution of this thickness can be regarded as the mark of identifying point dam.

(3) Abandoned River - auxiliary sign of the boundary of point dam. Generally, abandoned river is the last lateral migration track of point dam. Therefore, a complete point dam which is not eroded again is bounded by abandoned river. The accurate identification of abandoned river is very important to determine the distribution range of point dam and the direction of the side product of the point dam.

Secondly, the prediction of dam size is needed [5-7]. Taking 16 typical meandering reaches as the research object, regression analysis is carried out on the river width and dam span (Table 1). It is found that the two are positively correlated and have good correlation.

$$l=6.589\,4w+0.0677 \quad (2)$$

Where l is the span of point dam, m; w - full bank width of river, m.

Table 1. A table with statistics of dam parameters of different rivers.

Name of meandering river	River curvature	Full bank width of river /km	Span of point dam /km
The third Bay section of Hulunbuir, Yimin River	3.1	0.02	0.48
Jiutai, Yinmahe	2.4	0.03	0.57
Southern Branch of Nongan section of Songhuajiang River	3.5	0.04	0.21
The wuken River, a branch of Songhua River	2.7	0.07	0.55
Point dam at the intersection of Hailar River and Yimin River	1.8	0.13	0.71
Amazon River Point dam in South America	1.9	1.54	11.5

Therefore, if the width of a single channel is known under the sedimentary conditions, the size of point dam can be predicted. By using the empirical formula of Leeder (1973),

$$Log(w)=1.54Log(h)+0.83 \quad (3)$$

Where w - full bank width of river, m; H - full bank depth of river, m.

Finally, based on the single well identification and scale prediction of the point dam, the plane combination of the point dam is carried out with the abandoned River as the boundary (Figure 2).

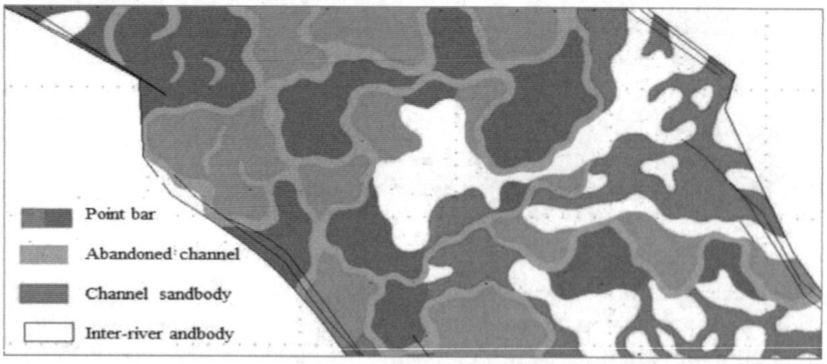

Figure 2. Distribution map of PI2 unit point dams in the western part of the central district of Saertu oilfeild.

2.2. Application example

In the ultra-high water flooding stage, the main target of horizontal well excavation of point dam is the remaining oil at the top. Horizontal well design includes 3 main technical links: selection block, fixed position and fixed track. The fixed position and track must be based on the results of internal configuration anatomy of meandering point dam. The following is an example of horizontal well X1.

Firstly, through the fine identification of abandoned river channel, the distribution profile of the sand body of the point dam is determined, and the favorable distribution area for horizontal well excavation is optimized. The abandoned channel of unit P31 in the planned horizontal well area is mainly "S" type, which can be traced in a large range. According to the distribution of abandoned channel, single point dam can be clearly identified, and four horizontal well areas are selected. The well area of well X1 is selected to design the potential exploitation of horizontal wells (Figure 3).

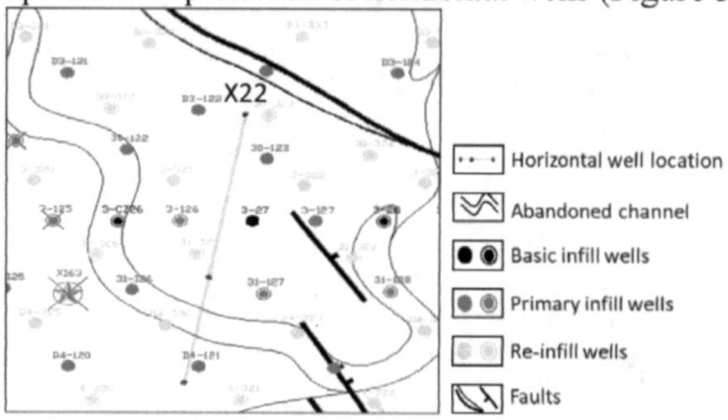

Figure 3. Horizontal well location of study block.

Secondly, after optimizing the block and determining the location of horizontal well, fine anatomy of reservoir internal configuration is carried out, and the distribution of lateral interlayer is clarified. On this basis, the horizontal well trajectory design is optimized, and the 3D geological modeling and well trajectory design of horizontal wells are carried out according to the following three principles.

(1) The track is 1 / 5 from the top of the dam body at the meandering point.

(2) The horizontal section is perpendicular to the lateral laminated direction.

(3) There are obvious compartments and contrast mark layers above the target layer in target area.

In order to excavate remaining oil at the top of point dam effectively, the horizontal well trajectory is designed below the top about 0.5-1.0m, according to the water flooded condition of adjacent well

Well X1had been drilled at the horizontal section length 472m, the drilling rate of destination layer is 100%, which is interpreted as low and submerged flooded. The well showed the daily production 127.9t, the daily oil output 20.0t, the water content 84.4%, and the flow pressure 8.49MPa at the initial stage, and the potential tapping effect of horizontal well was achieved.

3. Identification and application of braided river sand body

Shallow sandy braided river has frequent channel changes. In underground braided river reservoir, the braided channel that can be identified is only the last stage of braided river channel before its abandonment. In flood period, the sediment carried by braided river reaches the maximum. As the main stream line, braided river has great hydrodynamic force and the sediment is often eroded away. However, in the fall period, the maximum sedimentation occurs due to the decrease of water volume and flow capacity, and the braided channel accretion occurs. Shallow sandy braided river channels are all submerged in flood period, forming a single wide channel. In falling period, the core beach bar is exposed to the water, and the channel is braided. In dry period, the channel is gradually abandoned, and the last stage of the braided channel is generally very narrow. Therefore, such braided rivers generally form a "wide dam narrow channel" sedimentary pattern.

3.1. Plane combination of braided river and diara dam

In order to determine the distribution of the diara dam and braided channel in the research area, the difference of the sedimentary mechanism between the two should be clarified and the identification mark should be determined. The plane combination of the core beach dam is predicted according to the combination mode of "wide dam and narrow channel" in shallow lake basin.

3.1.1. Braided river filling style. Through the observation and description of outcrop and the core from core wells in the research area, there are three filling styles in braided river in the research area: sand filling, mud half filling and mud filling.

The braided channel filled with sand is characterized by relatively thick vertical accretion sand body because of its strong hydrodynamic force. Generally, medium fine sandstone and fine sandstone are the main ones, and the vertical upward binary structure is not obvious, but there are still some positive rhythm characteristics, which is bell shaped on the logging curve. The sedimentary top surface of this braided river is basically at the same level as the crest surface of both sides of the dam body, and the elevation difference is not obvious. When the positive rhythm of braided channel is not significant, it is difficult to distinguish from the diara dam.

When the hydrodynamic force of braided river gradually weakens, the coarse sand body is still deposited at the bottom of river. With the decrease of water energy, fine-grained suspended matter gradually deposits, which is characterized by dual structure of lower coarse and upper fine in vertical direction. The upper part of braided channel is filled with mud or silty sand, and the lower part is still medium fine sandstone and fine sandstone. The logging curve shows that the upper half of the single layer is close to the baseline or is serrated in a small range, and the lower part is bell or box type.

Because of the poor stability of braided river, frequent lateral migration of water flow, the river often changes. The braided channel that water passes through in the early stage will be abandoned quickly after the river diversion. The fine sediment of mud or silty is filled to form abandoned river. The logging curve is characterized by low amplitude toothed bell shape. When the filling mud is relatively pure, the curve is near the baseline. There is a certain elevation difference between the top surface of braided river filled by the latter two methods and the top surface of dam body, and the curve characteristics are obvious, which is easy to identify on a single well.

3.1.2. Identification and combination of diara dam. The identification marks of diara dam include three aspects: vertical sequence, relative depth of sand top and sand body thickness distribution.

(1) Vertical sequence. Diara dam is the main sedimentary unit of braided river, which is the result of flood deposition. The main feature is that argillaceous intercalation is developed inside. The vertical positive rhythm of diara dam sedimentation is not obvious, the natural potential and natural gamma curve are box shaped, and the amplitude difference of microelectrode curve is large. The braided channel is positive rhythm, and most of the electrical survey curves are bell shaped.

(2) Relative depth of sand top. The relative depth of sand top refers to the distance between the top surface of sandstone and single layer. Diara dam is formed by continuous deposition of sediment in

flooding period, which increases vertically and widens horizontally until the water surface stabilizes. Therefore, the top elevation of diara dam sandstone is higher than that of braided channel, which has been confirmed in modern sedimentation and simulation experiments. Therefore, the relative depth method can be used to identify the core beach dam. Which means, in the sedimentary part of braided channel, the relative depth of sand top is relatively large and the central beach is relatively small.

(3) Sand body thickness distribution. Considering the formation of braided river, diara dam is the result of river vertical aggradation in multiple flood periods, which is generally thick. Braided channel is mainly filled with channel, and the thickness of sand body is relatively thin. In the isopach map of sandstone, the area with relatively large thickness is the beach dam in the braided river center.

In order to divide core beach dam and braided channel in the sand body of braided flow zone, the quantitative scale of typical core beach dam must be determined. Many studies have been done on the morphological model of braided river core beach dams. Through the global satellite photos, taking 8 typical core beach dams as research objects, the regression analysis of core beach dam width and length, active channel width, braided channel width is carried out (Figure 4), which have good correlation, the quantitative scale of typical core bar and channel in modern sedimentation can provide a reliable geological basis for the plane prediction of core bar and braided channel in the study area.

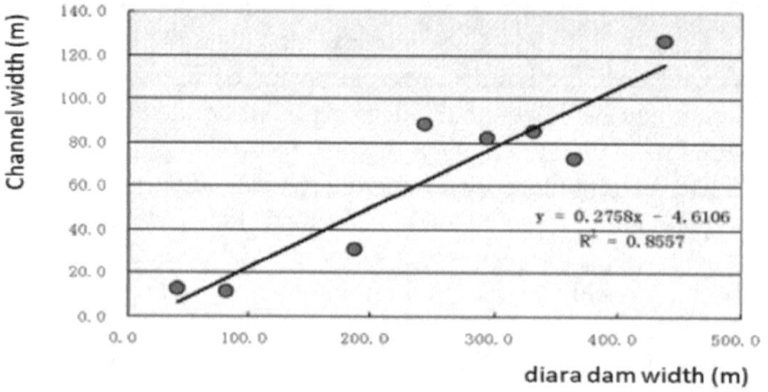

Figure 4. Relation between width of core beach and river channel width.

According to the distribution of sandstone thickness and relative depth of sand top, several possible braided channel distribution zone can be found, the relative depth of sand body is large and the thickness of sand body is thin. Based on the distribution mode of braided channel in modern sedimentation, several possible braided channels are divided according to 3D profile, and analyzes the plane combination to determine the position and shape of braided channel (Figure 5).

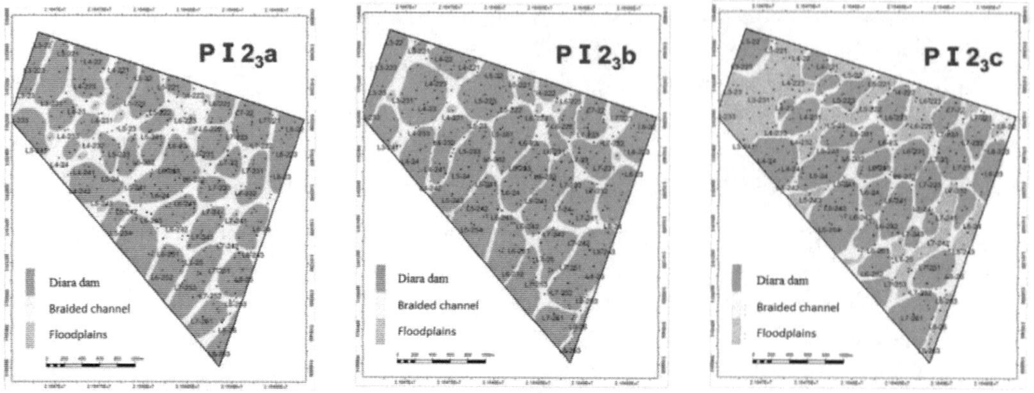

Figure 5. Distribution map of the diara dam of PI2$_3$ unit in the north and west of the in the north and west districts of Lamadian oilfeild.

4. Application example

Braided river sand body is developed in large scale, high connectivity, small in internal interlayer development scale and mainly distributed horizontally. Therefore, it is difficult to determine the remaining oil [8-9]. Through research, it is found that the remaining oil accumulation area is formed at the contact part of diara dam and braided river, which is formed at the top of core beach dam due to the mudstones blocking at the top of braided river. At the bottom, the development position of the large channel is located. The adjustment measures for controlling water potential in the layer (Table 2) are determined by using the description results of braided river. The daily liquid drop of 91t and the daily oil increase of 4t by selective perforation are determined. For example, in the vicinity of well X2, selective perforation in thick layer is carried out with the top perforated to 3.6m, the oil production 4.1t per day and water content 78.4%.

Table 2. Effect of adjustment measures in braided river layer.

Type of action	Before measures			After measures		
	Daily production /t	Daily oil production /t	Moisture content /%	Daily production /t	Daily oil production /t	Moisture content /%
Fine water plugging	152	3	98.3	61	3	95.5
Select hole filling	30	2	92.9	70	6	91.0

5. Conclusions

(1) The identification and combination of single point dam sand body can be recognized with locating boundaries by abandoned channels and determine the size by outcrop modes.
(2) Due to the waste river channel occlusion and mezzanine shading channel sand residual oil is mainly present at the top of the river channel and the boundary position of the single sand body.
(3) The fine identification of point dam and diara dam provides a reliable geological basis for horizontal well design and in-layer adjustment measures, and has achieved good results.

Reference

[1] Sui Jun 2000 Study on the river delta reservoir in Daqing Oilfield *[M].Beijing: Petroleum Industry Press* 13-20
[2] Wang Yuming 2000 30 years of technical practice in oilfield development *[M]. Beijing: Petroleum Industry Press* 64-69
[3] Miall, A.D. 1988 Architectural elements and bounding surfaces in fluvial deposits: Anatomy of the KayentaFormation(Lower Jurassic), southwest Colorado *[J]. Sediment.Geol.* **55(3)** 233-262
[4] Wang Junling, Ren Jishun, etc. 2001 Analysis of lithofacies and internal structural elements of modern river sedimentary bodies in Nenjiang River *[J]. Geoscience* **36(4)** 385-394
[5] He Yuhang, Yu Kaichun, etc. 2005 Identification and effect analysis of single channel of distributary plain facies composite sand body *[J]. Daqing Petroleum Geology and development* **2(24)** 17-19
[6] Li Yupeng, Wu Shenghe, Yue Dali, etc. 2008 The quantitative relationship between the width of modern meandering river and the length of point dam *[J]. Daqing Petroleum Geology and development* **27(6)** 19-22
[7] Lan Lifeng, Bai Zhenqiang, Yang Zhihui, etc. 2010 The method of dividing flow units of meandering river sand bodies *[J]. Daqing Petroleum Geology and development* **29(3)** 59-63
[8] Linchengyan, Sunyanbin, Dongchunmei, etc. 2013 Fine characterization of residual oils in the ultra-high aquifer period based on single sand bodies *[J]. Journal of Petroleum* **34(6)** 1131-1136
[9] Liao Chun, Qu Xinzhong, Zhao Ying, etc. 2014 Three-dimensional geological modeling technology of river delta based on reservoir configuration and flow unit- Gasikule oilfield *[J]. Journal of Petroleum Natural Gas* **36(6)** 6-10

IOP Conf. Series: Earth and Environmental Science 1087 (2022) 012069 doi:10.1088/1755-1315/1087/1/012069

Geochemical analysis of formation water as a tool for better understanding of water flooding

M.S. Shipaeva[1,2,5], V.I. Garifullina[1], R.R. Fayzetdinova[1], V.A. Sudakov[1], A.A. Shakirov[2], I.A. Nuriev[3], R.R. Khuzin[4] and D.A. Salikhov[4]

[1] Institute of Geology and Petroleum Technologies, Kazan Federal University, Kazan, Russian Federation;
[2] Geoindicator LLC, Kazan, Russian Federation;
[3] Consaltingovy Tsentr JSC, Almetyevsk, Russian Federation;
[4] Carbon-Oil LLC, Almetyevsk, Russian Federation.

[5] Corresponding author's e-mail: mariasipaeva@gmail.com

Abstract. In oil production, a reservoir flooding system is often used. This is necessary to compensate for oil production in the reservoir and maintain reservoir pressure. But the water cutting of wells is constantly increasing, that makes the exploitation less profitable, in the future this leads to the shutdown of wells. The source of water supply to the well is not always clear, and in the presence of flooded below and overlying reservoirs from which production was previously carried out, the possible number of water supply routes increases. The first aim of this work is to identify channels with high permeability by using tracer test, the second aim is to characterize geochemical state of the reservoir as a whole, and finally to choose the optimal sidetrack point using reservoir modelling.

The reservoir represents terrigenous deposits as an independent Bobrikovian horizon and the incision channel in the carbonate massif. The permeability of these zones is expected to be different. Therefore, in order to achieve the aims, the following studies were conducted:

1) Tracer test. A tracer blindly follows the fluid phase in which it is injected. Tracer tests are most often used to obtain information on the direction and velocity of the flow of brines, hydraulic conductivity, effective porosity, transport parameters (e.g., dispersivity values), and the presence of preferential flow paths. For this, we used a spectrofluorimeter.

2) Geochemical studies of the composition of all wells on the site. A unified Oil Water Contact (OWC) for carbonate and terrigenous deposits indicates the possibility of their hydrodynamic communication. To confirm this hypothesis, areal geochemical studies of the isotopic composition of reservoir waters were carried out;

3) Using the previous methods, we conducted a Field Development Analysis (FDA) and modeled the position of the horizontal wellbore in special software. One well showed better results compared to the other, as it is located in a zone with large oil reserves and the length of the horizontal wellbore was 1.3 times longer.

As a result of geochemical studies, three zones were identified based on the isotopic composition of water: the southwest - the waters of the Bobrikovian horizon, the central part - the transition zone, and the northeast - the Tournaisian waters. Zones with current reserves have been determined based on 3D modeling. For further involvement of these volumes of reservoir oil in the development, the sidetracking is proposed.

IOP Conf. Series: Earth and Environmental Science 1087 (2022) 012069 doi:10.1088/1755-1315/1087/1/012069

1. Introduction

One of the modern approaches for the effective development of small deposits is the construction and operation of wells with a complex architecture: horizontal wells (HW) and sidetracks (BS, BGS). Sidetracking is an efficient and less costly technology compared to drilling new wells, which allows to increase oil production at old fields, increase the oil recovery factor, and return to work previously watered wells [1,2]. Drilling sidetracks with horizontal ends increases the efficiency of reserves development many times over [3]. The difficulty in drilling horizontal wells is the degree of water cut during construction: a considerable percentage of such wells begin to fill with formation water, the content of which could not have been predicted in advance. The technological efficiency of sidetracking depends on the validity of the sidetrack orientation, the feasibility of drilling is determined by the amount of economically viable residual reserves [4]. It is much worthy to approach the issue of sidetracking drilling in the presence of hydrocarbon accumulations of various hydrocarbon phases (oil, oil and gas cap, gas, condensates ect.) of two groups of formations - the "normal" group and the "erosional", since they differ in the conditions of formation, the occurrence and the filtration-capacitive properties [5]. When planning the drilling trajectory, the issues of studying the hydrodynamic connectivity of such geological bodies at different hierarchical levels of the system (reservoir, layer, interlayer) are also considered using geophysical and hydrodynamic data when modeling the spatial patterns of changes in hydrodynamic parameters [6]. These parameters can be obtained using tracer test. It has been proved that at the deposits of the western slope of the South Tatar Arch of the Republic of Tatarstan, there is a hydrodynamic connection between the host carbonates of the Tournaisian stage and the terrigenous formations performing the cut- out-rocks of the the Visenian stage [7]. Consequently, the technology of sidetracking shows difficulties, due toboth the structural elements of the deposits and the technology of work. In order to reduce the risks andthe uncertainties, the work should be carried out on the basis of an integrated approach of all available data [8]. The whole complex of organizational, technical and technological solutions can be taken into account by studying the geochemical features of the produced fluid is a direct method of obtaining information about the existing hydrochemical zones of the field [9, 10]. Chemical and isotopic data for water co-extracted with hydrocarbons in oil and gas fields are commonly used to examine the source of the formation water and possible impacts on groundwater in areas of oil and gas development [11].

The geology of unconventional reservoirs is highly ubiquitous. The tightness of the formation, the heterogeneous layers distribution, mineralogy, pore throat structure, sorption, and others are responsible for its ubiquitous nature. Such a nature restricts the hydrocarbon fow in the reservoirs [12].

2. Geological settings

The vast majority of Tatarstan's oil and gas fields are concentrated within the South Tatar Arch (mainly the western slope of the STA) and in the Melekess depression (on the eastern side of the MD) (Figure 1a). Currently, some small reservoirs and deposits containing high-viscosity oil are being developed by smallindependent oil producing companies. The oil deposits of Tatarstan are characterized mainly by high sulfur content, increased high viscosity. One of such fields with high-viscosity oil is the Keremetievskoye field. In regional tectonic terms, the Keremetievskoye field is located on the southwestern side of the South Tatar Arch (Figure 1a). The geological section of the area is represented by Devoniandeposits, Carboniferous and Permian systems. Oil-bearing capacity was detected in the deposits of the Vereiskian and Bobrikovian horizons; Bashkirian and Tournaisian stages (Figure 1b). A number of small seismic uplifts are associated with the deposit, including the Keremetievskoye, Soldatskoye, Sverdlovskoe uplifts. The article discusses the Soldatskoye uplift.

The studied uplift is small, with high-viscosity oil (the viscosity is 281.5 MPa•s), trapped in the sediments of the Bobrikovian formation. The studied uplift includes the stratigraphic interval from the Tournaisian to the Bobrikovian stages (Figure 1c). At the same time, the Bobrikovian reservoirs are saturatedwith oil, and the Tournaisian that lie below are water-saturated.

IOP Conf. Series: Earth and Environmental Science 1087 (2022) 012069 doi:10.1088/1755-1315/1087/1/012069

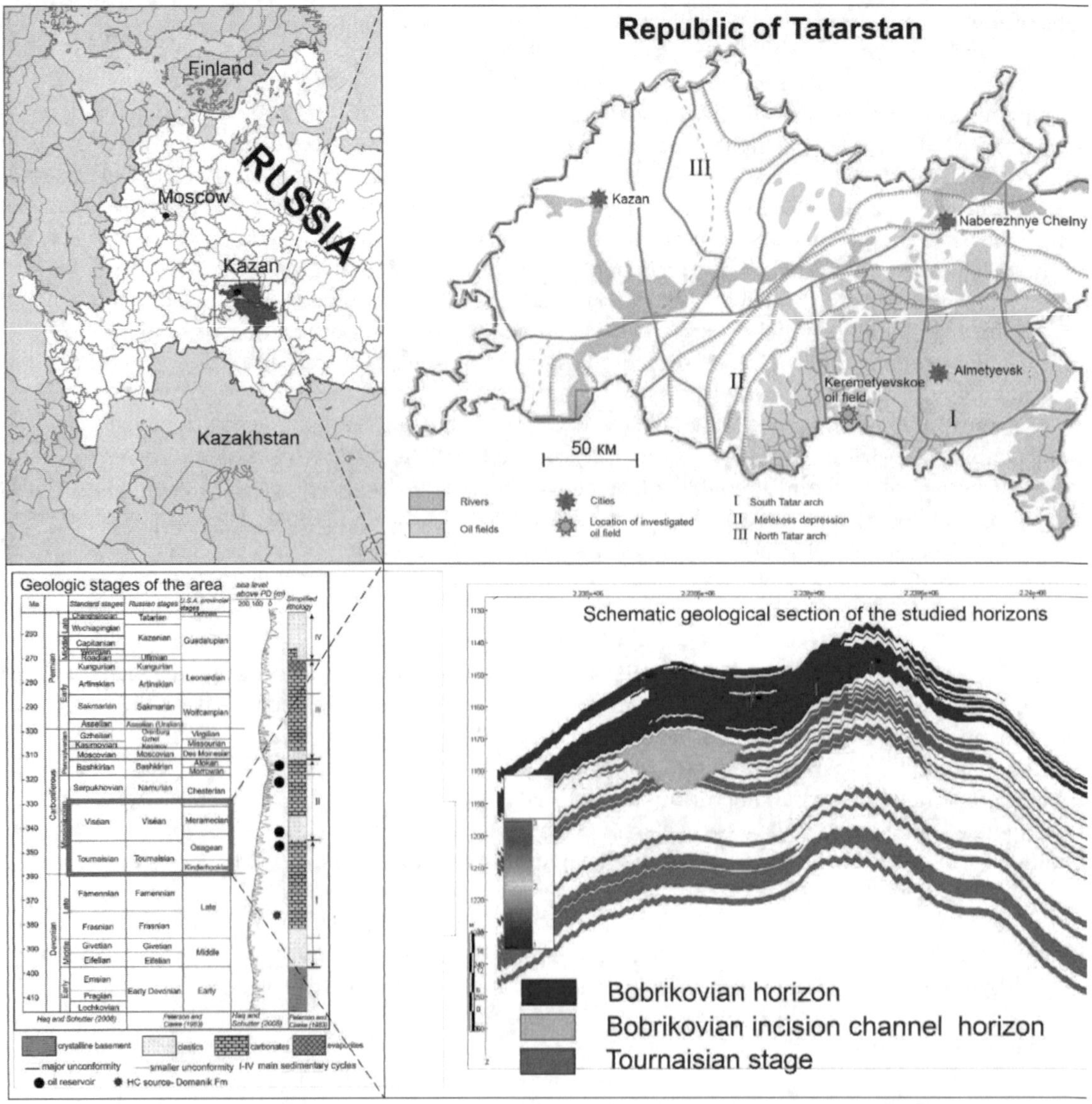

Figure 1. Simplified scheme and lithological column and geologic stages of the area of interest:

a) Location on the map of Russia;

b) Simplified stratigraphic scheme and lithological column;

c) Schematic geological section.

3. General settings about the object of research

Due to the development of the incision channel, it is assumed that the hydrodynamic connection of the carbonate Tournaisian and terrigenous Bobrikovian reservoirs and a single Oil Water Contact (OWC) of the deposit at the level of minus 1167 m (Blue color on the Figure 2). The oil deposit is confined to a complex carbonate-terrigenous reservoir (Figure 2 – Red line is incision channel in carbonate massif). The field has been developed by LLC «Blagodarov-Oil» since 2007.

The petrophysical properties of the Bobrikovian reservoir are as follows: Average porosity is 23.3%, permeability is 238 darcy, clay content is 2.9%.

The reservoir pressure maintenance system has been implemented since 2008 and injection is carried out by three wells (No. 104, 105, 1024BS) (Figure 2).

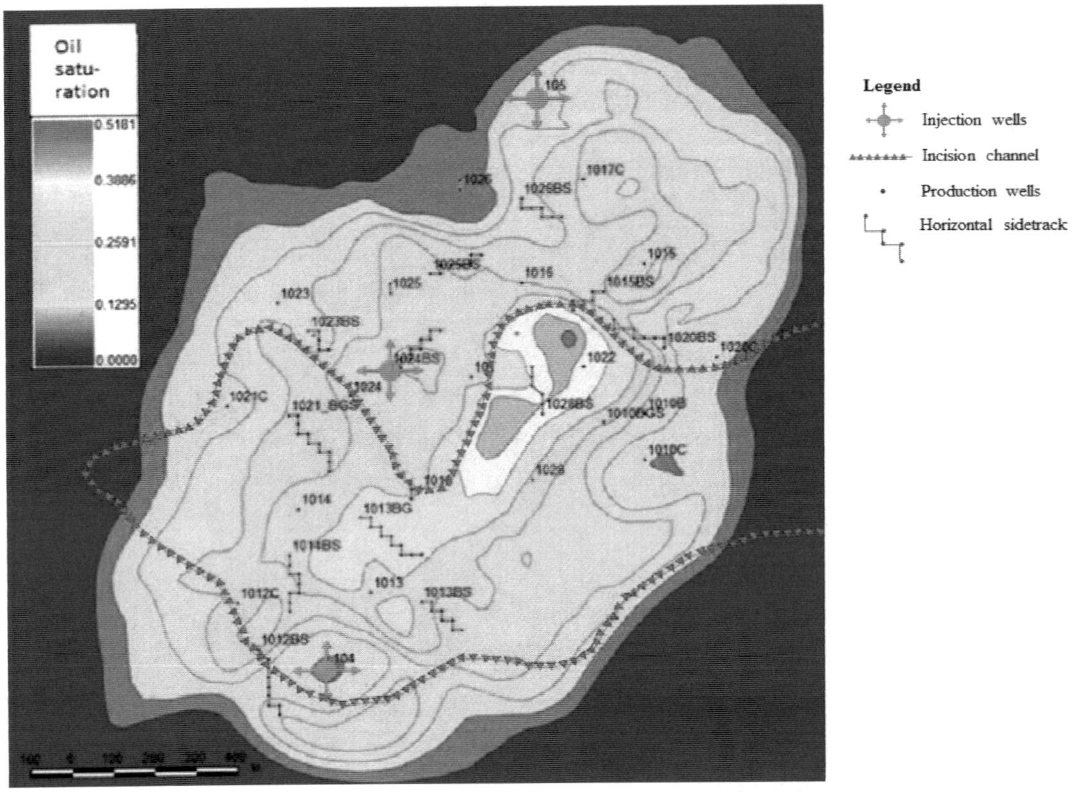

Figure 2. Oil saturation distribution map and channel location, where red color – the highest oil saturation, blue color is no oil.

Most of the wells are characterized by an intense waterflooding, which occurred in the first years of operation (Figure 3). The graph shows three groups of wells (according to colors), which show the dynamics of water cut. Dark blue (wells No. 1026, 1017C, 1020C) – there is an intense flooding in the first years of operation, light blue (wells No. 1023, 101, 1016) - there is a smooth growth, according to the development of reserves and green (wells No. 1021C, 1013, 1018) – a intense increase at the beginning of the operation, after which there is a slight decrease. In view of the premature watering of the wells, active drilling of side wellbores is taking place at the field in order to develop additional oil reserves. As of date 01.01.2020, 101 sidetracks (BS, BGS) have been drilled. It was possible to increase oil production rates by 5-10 times. For example, for wells No. 1013 before the drilling ofthe BGS, it was 8.2 thousand tons, for 1013BS, the cumulative production was 17.5 thousand tons; for well No. 1023 – 5.5 thousand tons, for 1023BGS – 11.3 thousand tons. The BGS exceeded the production of vertical wells by 1.3-8 times, because horizontal sidetracks are drilled into non-flushed zones and, due to the design features, increase the effective thickness of the formation.

The different nature of watering may indicate the penetration of different waters. It is known that the Visenian incision channel is distinguished in this area, and the reservoirs of the Tournaisian stage is saturated with water. The conducted field geophysical studies revealed the presence of behind-the-casing circulations in wells No. 104, 1010C, 1017C, 1020C, 1024, 1025, 1028. Water injection does not have a direct effect on the increase in water cut, this is also confirmed by previously conducted tracer studies (not a significant contribution to water cut production of production wells was provided only by injection well No. 1024).

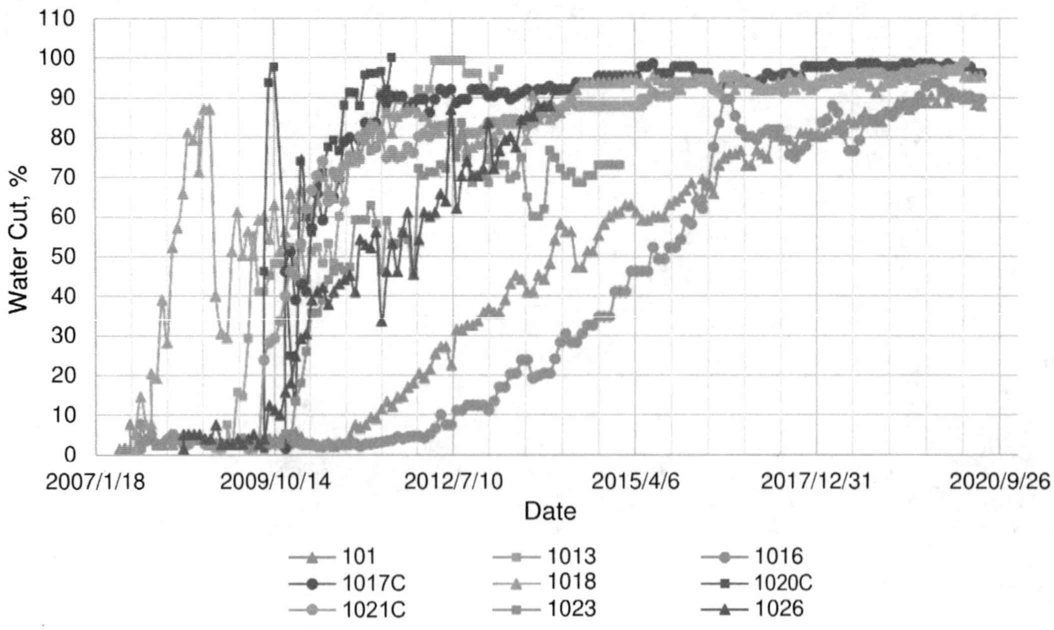

Figure 3. Dynamics of water cut: colored lines - change in water cut of wells over time.

4. Methods

3 studies were carried out: 1) tracer tests, 2) formation water geochemistry and 3) field development analysis.

4.1. Tracer test

One of the most informative and inexpensive ways to study the filtration properties of reservoirs is the indicator method. The results of the indicator method at the level of quantitative estimates reflect the characteristics of the reservoir bed in the entire interwell space [13]. The technology of the tracer test is simple and is based on the injection of a tracer (or labeled fluid) into the reservoir in an injection well. After injection, products are taken from the surrounding production wells. And the scientist is looking for where the traces came at what point in time. These samples are measured in the laboratory for the presence of an indicator. The values of the concentration and the time of the indicator output are interpreted by data, the results that are used to create a picture of the fluid movement through the reservoir and to clarify the geological structure of the site. Maps are built based on the output of the indicator and the speed of the tracer [14]. At the facility, a fluorescent tracer was pumped into injection well No.1024 G, sampling was carried out every day for 60 days from the moment of injection from all exploitation wells.

4.2. Geochemistry of formation water

Water samples from all wells of the site were explored. Mass-spectrometry methods were used in the research. Inductively coupled plasma mass-spectrometry is a type of mass-spectrometry in which inductively coupled plasma is used to ionize a sample. This method is based on the direct measurement of the mass-to-charge ratio (m/z) of elementary positively charged ions obtained from the test substance in the gas phase. Inductively coupled plasma mass-spectrometry is widely used due to its high sensitivity, that allows the determination of many elements in trace quantities, a large dynamic linear range, and the ability to perform multi-element analysis. The method also allows to determine with high accuracy the

IOP Conf. Series: Earth and Environmental Science 1087 (2022) 012069 doi:10.1088/1755-1315/1087/1/012069

ratio of isotope concentrations of various elements (Oxygen and Deuterium). Oxygen and Deuterium values are important indicators for the origin of the water [15].

4.3. Field Development Analysis

Drilling of horizontal sidetracks is planned on the site. To do this, an important point is to choose the trajectory of the horizontal sidetrack in such a way as not to get into the water-saturated part. The authors have selected the optimal trajectory of horizontal sidetrack with maximize oil rate and minimize water rates using reservoir simulation modeling - tNavigator software.

5. Results

A fluorescent tracer (or indicator) was pumped into well № 1024G, sampling was carried out every day for 60 days from the moment of injection from all exploitation wells.

The distribution of the average filtration velocities of the labeled fluid to the producing wells alongthe channels with abnormally low resistances is shown in Figure 4a. Analyzing the map, it can be seen that the average filtration rates are higher for wells 1021C and 1016.

The largest coverage of the reservoir with installed flow tubes is observed for wells No. 1022 and 1018 - the volume of channels with anomalously low resistance, through which unproductive filtrationof the injected water occurs, is 2.71 and 1.6 m^3, respectively (Figure 4b).

The inflow of the tracer to the producing wells occurs irregularly, which indicates the heterogeneity of the reservoir properties of the productive formation in the direction of the considered reacting environment.

The performed laboratory studies showed that the indicator injected into well No. 1024BGS was found in the product samples of all inspection wells (No. 101, 1014, 1016, 1018, 1021C, 1022, 1025). The tracing of filtration flows revealed from 2 to 5 systems of channels with LFR (Low Filtration Resistance).

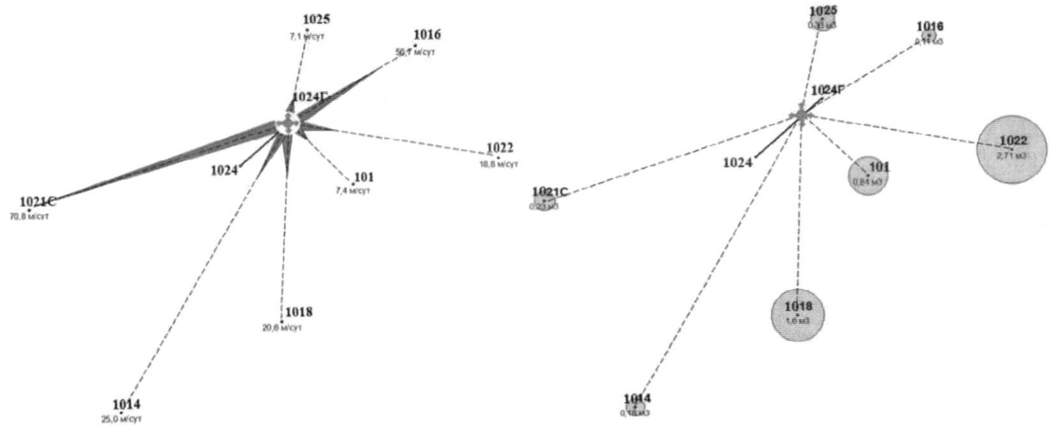

Figure 4. Maps: a) distribution of average velocities of tracer; b) distribution of volumes of identifiedsystems of channels.

The first portions of the indicator are recorded 8 hours after injection and the maximumfiltration velocity (the time for the first portion of the indicator) exceed 1800 m/day. Average filtration velocities are set at 20 meters per day (Figure 5).

IOP Conf. Series: Earth and Environmental Science 1087 (2022) 012069 doi:10.1088/1755-1315/1087/1/012069

Well No.	Forma-tion type	Portion Number	Maximum velocity of arrival of a portion, m/day	Average velocity of arrival of a portion, m/day	ΔP, MPa	L, m	Channel permeability, μm²	Channel performance, m³/day	Channels performance per well, m³/day	Channel Type	Percentage of injected water, %	Percentage of water injected into the well, %	Accepted reservoir permeability by object, μm²
1021C	C1bb channel	1	1687.7	440.3	13.8	506.3	196.30	0.025	0.032	fracture	0.023	0.030	3.311
		2	38.9	36.2			16.12	0.007		fracture	0.007		
1014	C1bb channel	1	1802.7	416.0	12.7	540.8	215.28	0.002	0.008	fracture	0.002	0.008	
		2	21.6	20.0			10.37	0.001		fracture	0.001		
		3	13.9	12.6			6.51	0.005		fracture	0.005		
1022	C1bb channel	1	1357.7	354.2	18.6	407.3	94.25	0.007	0.125	fracture	0.006	0.119	
		2	37.0	32.6			8.67	0.085		fracture	0.081		
		3	9.9	9.5			2.52	0.033		pore	0.032		
1018	C1bb channel	1	1419.0	532.1	15.2	425.7	181.11	0.005	0.078	fracture	0.004	0.074	
		2	327.5	160.6			54.68	0.006		fracture	0.005		
		3	85.1	53.2			18.11	0.012		fracture	0.011		
		4	38.7	32.7			11.15	0.013		fracture	0.013		
		5	11.5	10.4			3.53	0.043		pore	0.041		
1016	C1bb	1	972.7	162.1	19.86	291.8	28.95	0.007	0.021	fracture	0.007	0.020	
		2	36.5	14.3			4.17	0.014		fracture	0.013		
1025	C1bb	1	486.0	67.8	12.8	145.8	9.39	0.003	0.016	fracture	0.003	0.015	
		2	36.5	29.2			4.04	0.001		pore	0.001		
		3	18.2	15.3			2.12	0.001		pore	0.001		
		4	13.3	11.2			1.55	0.003		pore	0.003		
		5	3.7	3.4			0.47	0.006		pore	0.006		
101	C1bb	1	534.0	97.1	16.9	160.2	11.18	0.017	0.039	fracture	0.016	0.037	
		2	53.4	35.6			4.10	0.008		fracture	0.007		
		3	11.4	10.7			1.23	0.003		pore	0.003		
		4	3.9	3.6			0.42	0.011		pore	0.010		

Figure 5. Filtration characteristics of identified groups of channels.

6. Analysis of areas with residual reserves

In the study area, there are both flushed zones with LFR channels and zones with residual reserves (Figure 2). It is proposed to consider zones with residual reserves for their further involvement in development by drilling sidetracks. To exclude premature flooding, micro-component, macro-component and isotope composition of the produced fluid were considered, which made it possible to determine the source of its entry into the production of production wells, thereby laying the trajectory of the wells in the most favorable places. According to the isotopic composition of the water, the area distribution map was built (Figure 6). Three zones are conditionally distinguished: a) the south-west is Bobrikovian water, b) the central part is the transition zone, and the northeast is the water of the Tournaisian stage. Based on the analysis of the operation of existing wells, it is difficult to identify the most favorable zone with respect to the spread of water, however, it is noted that in the zone of no incision channel, watering does not occur so rapidly. In this area, all wells, except for No. 101, have sidetracks, therefore, when the oil production rate drops below the profitable one, the possibility of drilling the additional sidetracks in a different direction should be considered. In the area with a predominance of Tournaisian water, flooding, on the contrary, occurs rapidly, therefore, at the moment it is worth considering the area in the southwest, since it is most favorable in terms of residual reserves and in terms of water content. Isotopiccomposition of water allowed us to conditionally identify not only the zones of distribution of formationwaters, which correlate with the dynamics of water cut, but also to determine the most suitable wells forfurther geological and technical measures.

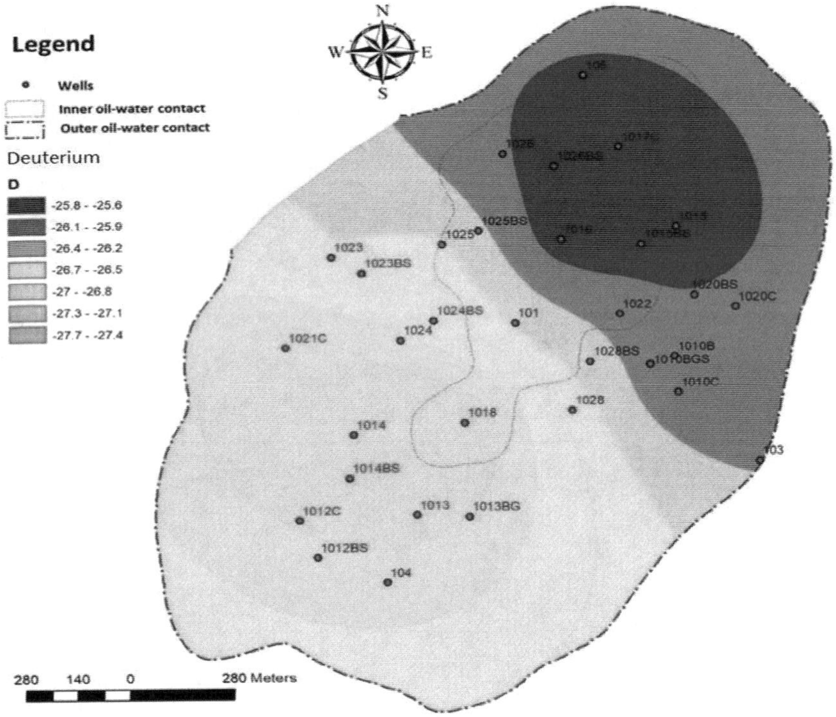

Figure 6. Isotopic distribution map (in formation water by production well sampling).

Two wells were selected for the priority introduction of the BGS - No. 1013 (change in the direction of the BGS) and 1021C (Figure 7). For well 1013BGS, it is proposed to change the trajectory of the existing sidetrack, drilling towards well No. 1021. For well No 1021, the proposed trajectory is between wells No 1014 and 1024. The length of the horizontal sidetracks is up to 150 m, the optimal length is 100 m.

Well No 1018 should be considered for drilling the sidetrack when the flow rate decreases towards residual reserves (to the north-east).

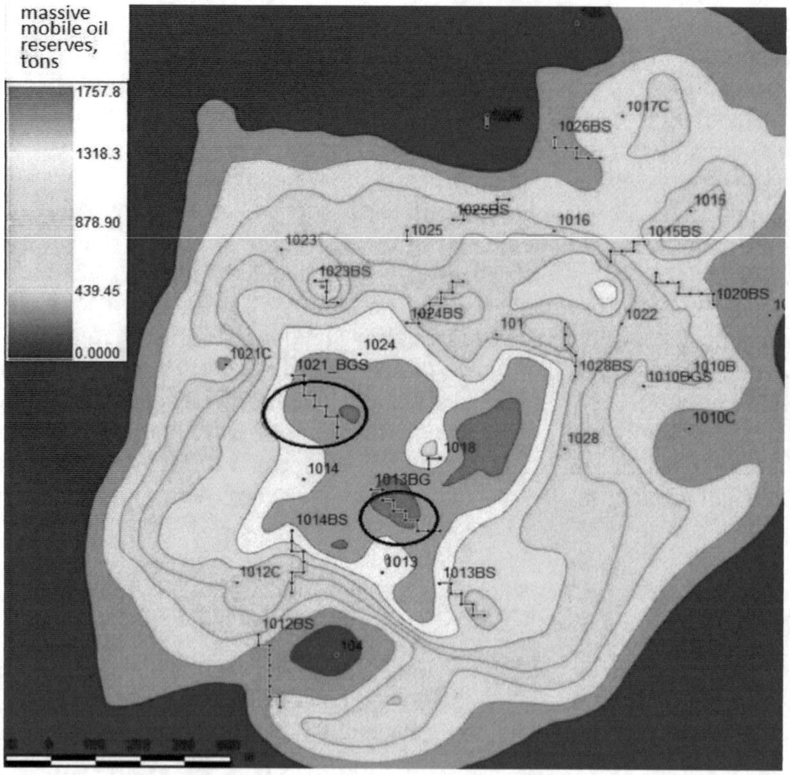

Figure 7. Distribution map of movable stocks.

When the wellbores are located in this way, the wells are expected to have the characteristics shownin Table 1 and Figures 8, 9.

Table 1. Wells operation before and after drilling the horizontal sidetrack.

Well No	Before						After				
	Bore length, m	Oil rate, t/day	Liquid rate, t/day	Water Cut, %	Cumulative oil production, thous.t.	Cumulative liquid production, thous.t	Bore length, m	Oil rate, m³/day	Liquid rate, m³/day	Cumulative oil production, thous.m³	Cumulative liquid production, thous.m³
1013	175	0,02	1,0	98,4	24,5	155,1	105	9,9	9,9	119,4	190,4
							160	12,0	12,0	144,6	
1021C	235,9	1,4	115,7	98,8	33,7	298,4	105	17,2	17,2	98,9	796,9
							160	20,3	20,3	116,1	

As of 10/01/2020, after drilling the wellbore, the following characteristics were obtained (Table 2), where we observe the success of the wellbore drilling. When comparing the data, it can be noted that the actual production rates and water cut correlate with the predicted values of the simulation model.

7. Conclusions

Consequently, comparing the data of geochemical analysis, where the zones of hydrodynamic connection of the waters of the Bobrikovian and Tournaisian deposits were identified, with a map of current reserves, horizontal wellbore drilling was recommended, and later implemented in practice. Differences in the reservoir features of the incision channel – C1bb incision channel and the reservoir zone C1bb are revealed. The reservoirs of the incision channel are more permeable. Drilling of horizontal wellbore into the incision channel allowed obtaining high production rates. The connection of the injection well with the producing ones occurs due to a sufficiently uniform propagation of the pressure front and the injected water along the pore part and channels with greater resistances than in the identified anomalous systems. The complex studies conducted on the presence of hydrodynamic connection between the layers in view of the presence of the incision channel, the filtration characteristics of these zones and the analysis of the development of the site allowed to make a conclusion about the most successful location of the horizontal sidetrack and to involve in production well-pattern dead area. We recommend to continue monitoring the study area and continue taking samples for water geochemistry once every half a year.

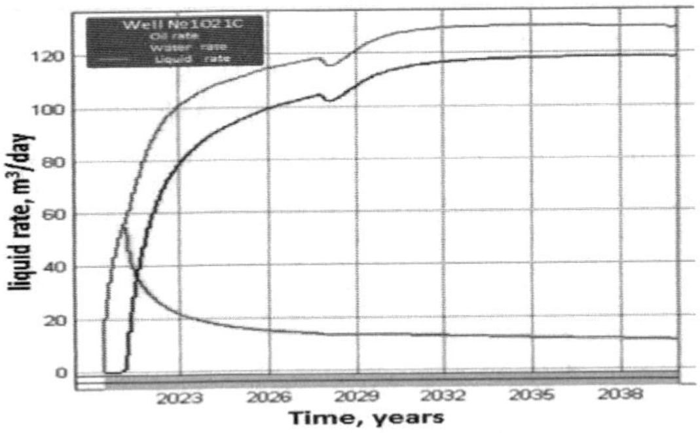

Case 1 (horizontal sidetrack length up to 160 meters)

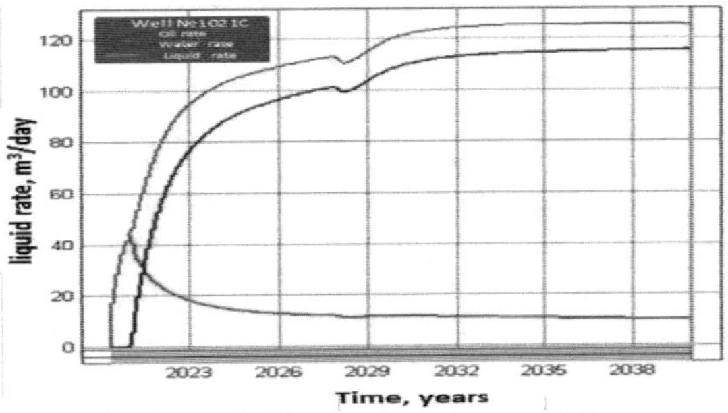

Case 2 (horizontal sidetrack length up to 105 meters)

Figure 8. Forecast of the flow rates. Well No1021C.

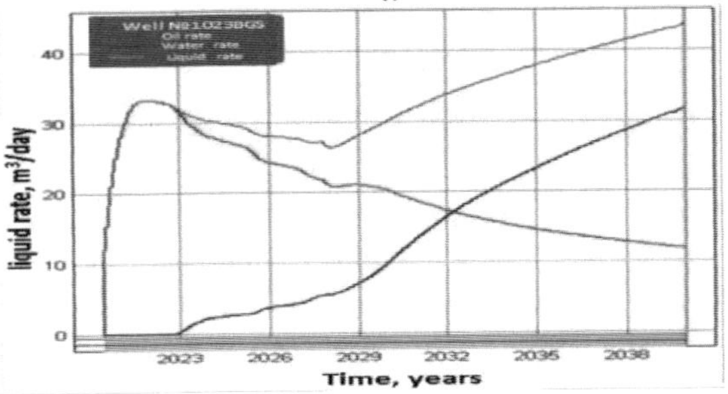

Case 1 (horizontal sidetrack length up to 160 meters)

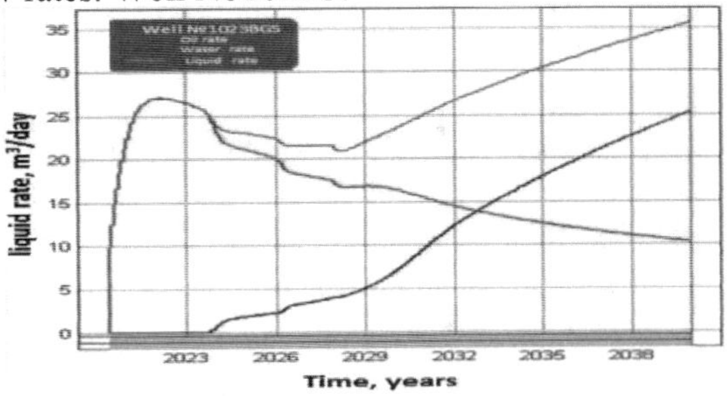

Case 2 (horizontal sidetrack length up to 105 meters)

Figure 9. Forecast of the flow rates. Well No1013.

Table 2. Actual performance of wells after drilling horizontal sidetrack.

Well number	Input oil flow rate, tons / day	Input liquid flow rate, tons / day	Water Cut, %	horizontal wellbore length, meters
1013	13,3	14,0	4,9	137
1021C	17,1	19,0	10	139,6

Acknowledgments
This work was supported by the Ministry of Science and Higher Education of the Russian Federation under agreement No. 075-15-2022-299 by 15.04.2022 within the framework of the development program for a world-class Research Center "Efficient development of the global liquid hydrocarbon reserves".

References

[1] Agzamov F.A., Akbulatov T.O., Ismakov R.A., Komleva S.F., Konesev G.V., Levinson L.M., PopovA.N., Sakaev R.M., Sannikov R .Kh., Soloviev A.Ya., Trushkin B.N., Chukturov G.K., Yangirov F.N. 2012 *Oil and gas business.* Volume 2 Drilling oil and gas wells. Study guide in 6 volumes St. Petersburg: Nedra

[2] Sabirov A.A. 2017 New developments in equipment and technology of oil production Engineering practice 1-2 82-91 (in Russian)

[3] Nizhegorodov V.A. 2019 Domestic and foreign experience in the use of sidetracking with horizontalends to improve the field development system *Scientific journal* **11 (45)** 5-8

[4] Khakimzyanov I.N. 2012 Theory and practice of oil field development with horizontal wells. Diss. dokt. tech. nauk. [Dr. engineer. sci. diss.]. Bugulma. 388 p. (In Russ.)

[5] Akhmetov A.N., Mukhametshin R.Z. 2018 Features of modeling objects complicated by erosional cuts *Modeling of geological structure and development processes - the basis for successful development of oil and oil and gas fields: materials of the International Scientific and Practical Conference.* – Kazan: Slovo 130-133 (In Russ.)

[6] Koskov B.V., Koskov V.N., Shurubor Yu.V. 2011 The problem of studying the hydrodynamic connectivity of geological bodies and methods for its solution *Bulletin of the Perm University. Geology* **4** 90–95

[7] Saifutdinov M.A., Khakimzyanov I.N., Petrov V.N., Sheshdirov R.I., Mironova L.M. 2018 Studies of the hydrodynamic connection presence between the terrigenous Bobrikovian and carbonate Tournaisian objects on the basis of the geological-technological model of the site of the field *Georesursy = Georesources* **20(1)** 2-8

[8] Samigullin V.Kh., Gilyazov R.M., Valuiskova T.N., Bikmukhametova G.I., Yumashev R.Kh. 2007Restoration of idle and marginal wells by drilling additional shafts*Oil industry* **11** 13-14

[9] Shipaeva M.S., Nuriev I.A., Evseev N.V., Miftahov T.R., Sudakov V.A., Shakirov A.A. 2020 Improving efficiency of oil recovery and finding a source of watering in multi-zone deposits by geochemical methods of research *Georesursy = Georesources* **22(4)** 93–97

[10] Kytéria Sabina L. de Figueredo, Carlos A. Martínez-Huitle, Antonio Bernardo R. Teixeira, André Luis S. de Pinho, Carla A. Vivacqua, Djalma R. da Silva 2014 Study of produced water using hydrochemistry and multivariate statistics in different production zones of mature fields in the Potiguar Basin – Brazil *Journal of Petroleum Science and Engineering* **116** 109-114

[11] Peter B. McMahon, Justin T. Kulongoski, Avner Vengosh, Isabelle M. Cozzarelli, Matthew K. Landon, Yousif K. Kharaka, Janice M. Gillespie, Tracy A. Davis 2018 Regional patterns in the geochemistry of oil-field water, southern San Joaquin Valley, California, USA Applied Geochemistry 98 127-140 https://doi.org/10.1016/j.apgeochem.2018.09.015

[12] Muther, T., Qureshi, H.A., Syed, F.I. et al. 2022 Unconventional hydrocarbon resources: geological statistics, petrophysical characterization, and field development strategies. *J Petrol Explor Prod Technol* **12** 1463–1488 https://doi.org/10.1007/s13202-021-01404-x

[13] C. Serres-Piole, H. Preud'homme, N. Moradi-Tehrani, C. Allanic, H. Jullia, R. Lobinski 2012 Water tracers in oilfield applications: Guidelines, *Journal of Petroleum Science and Engineering* **98–99** 22-39 https://doi.org/10.1016/j.petrol.2012.08.009

[14] Antonov G.P., Abramov M.A., Kubarev P.N 2015 Regulation of the process of waterflooding of oildeposits in PJS TATNEFT *Production and technical oil and gas journal "Engineering Practice"* **05**

[15] Shipaeva M.S., Nurgaliev D.K., Sudakov V.A., Shakirov A.A., Lutfullin A.A., Ganiev B.G., Minikhairov L.I. 2021 The Geochemical Survey Methods for Optimization of Oil Field Development*The Abu Dhabi International Petroleum Exhibition & Conference (ADIPEC - 2021) DOI: 10.2118/207566-MS*

Climatic changes inferred from the chroma and magnetic susceptibility of aeolian deposits in the Central Shandong Mountains over the penultimate interglacial period

Qiuyue Zhao[1,3], Xinqi Zhang[1], Yanping Sun[1], Xiuying Liu[1], Tianyan Ge[1], Ning Liu[1], Anrui Tang[1] and Rui Zhou[2]

[1] Key Laboratory of Tourism and Resources Environment in Universities of Shandong, Taishan University, 271000, Taian, China;
[2] Institute of Earthquake Forecasting, China Earthquake Administration, 100036, Beijing, China.

[3] Corresponding author's e-mail: qyzhao@cnu.edu.cn;
[4]Qiuyue Zhao and Xinqi Zhang contribute equally.

Abstract. Loess-palaeosol sediments document the climatic changes and provide a good basis for palaeoclimate reconstruction. Palaeoclimatic changes of the holocene and the last interglacial period have been widely studied, but the penultimate interglacial period has been studied scarcely. The Dongheishan (DHS) section in the western of Shandong Province (China) preserves a loess-palaeosol sequence that includes the penultimate interglacial period. The annual average temperature and precipitation are retrieved by chroma and magnetic susceptibility (MS) to reconstruct the palaeoclimatic variations since the penultimate interglacial period. The results indicate that the lightness valley values and the redness and yellowness peaks correspond to the peak MS values, which can consistently reflect climate changes; redness and MS correlated well with the annual average temperature and annual precipitation, respectively. The annual average temperature is 14.4 °C based on redness in the the penultimate interglacial palaeosol, indicating that the palaeoclimate was warmer during the penultimate interglacial period than at present. The annual average precipitation is 349 mm derived from magnetic susceptibility, showing that the magnetic susceptibility should be cautious of the inversion of precipitation in the Central Shandong Mountains (CSM) region.

Keywords: chroma, palaeosol, Central Shandong Mountains, magnetic susceptibility, climate

1. Introduction

As human society and lifestyles develop, the pollutants discharged by large-scale production are increasing, which leads to the deterioration of the global climate and the endless emergence of ecological environmental problems. The rapid deterioration of the living environment has prompted increasing attention on climate changes. The reconstruction of the palaeoclimate is important to the study of current and future climate changes [1]. The study of the interglacial palaeoclimate can not only reveal the cycles of long-term climate changes but also obtain the variations in palaeoenvironmental patterns [2]. Chinese loess is considered as a good carrier for recording quaternary environmental changes containing information on environmental evolution [3].

In recent years, MS and chroma have gradually become important palaeoclimatic proxies in loess-palaeosol researches. The MS depends on the content and type of magnetic minerals in the soil and closely related to the soil pedogenesis which is also restricted by temperature and precipitation [4]. Therefore, the interpretation of MS can be used as an indicator of precipitation research to reflect the strength of the palaeo-summer monsoon [5], and then to explore the palaeoclimate changes. Chroma can compensate for the defects of MS in humid and warm areas [6]. Previous studies have shown that variations in chroma can indicate the Asian monsoon fluctuations and global climate changes [7] and are reliable [6].

In addition to the Chinese Loess Plateau (CLP), the Shandong Province has preserved the sedimentary sequence of loess-palaeosol [3], which is far from the inland deserts, having potential in researching both global and regional climatic changes. Systematic studies on the sedimentology [8], provenance [9], chronology and palaeoclimatology [10] of the Shandong loess have achieved a relatively and unified understanding of the regional palaeoclimatology since the last glacial period. However, there is a lack of research on penultimate interglacial period palaeosols, which restricts to understand the palaeoclimatic change processes and the patterns of the period. Thus, this paper investigates the implications of MS and chroma in the DHS section in the CSM region over the penultimate interglacial period.

2. Site description and sampling

2.1. Site description

The CSM region is situated in the eastern of the North China Plain and the southern bank of the lower reaches of the Yellow River. The region is about 200 km from east to west and about 50 km from north to south, which features a series of low mountains and hills [9] (Figure 1 a). Located in the east of Chinese modern monsoon zone and the transition zone between land and sea, the region has a warm and semi-humid monsoon climate. Influenced by the warm and humid air-mass flow from the Pacific Ocean, rain and warm conditions occur in the same period with prevailing southeast winds, while dry conditions occur in cold season with prevailing northwest winds under the influence of the Siberian high-pressure system. The average mean annual temperature is 12.9 °C (range from 12.6 to 14.5 °C) [9, 11] and the mean annual precipitation is 674 mm (range from 615 to 794 mm) with 60~70% of the precipitation concentrated in the summer [9, 11].

2.2. Sampling

The DHS section (116°20′29.26″E, 36°10′21.95″N) was sampled from 478 m southwest of Sushantou village, which is located in Jinan city (Figure 1 a). The characteristics of the soil layer were described based on observations of the color of the sediment, lithologic characteristics in the field.

The DHS section mainly consists of yellow and reddish brown sediment with a thickness of 690 cm (Figure 1 b), which can be divided into five stratigraphic units comprising a horizon of cultivated soil (S_0, 0-50-cm depth), Malan loess (L_1, 50-250-cm depth), the last interglacial palaeosol (S_1, 250-440-cm depth), the penultimate glacial loess (L_2, 440-480-cm depth) and the penultimate interglacial palaeosol (S_2, 480-690-cm depth) (Figure 1 c). The ages of the samples at the 250 cm, 440 cm, 480 cm, 690 cm depth measured by the Laboratory of luminescence chronology in the Taishan University showed in the Table 1.

A total of 180 samples were collected by 2.5 cm intervals from a depth of 240 cm. At the same time, the depth, soil color, and lithologic characteristics were recorded. All samples used for MS and chroma were placed at room temperature (25 °C), protected from light and ventilation, dried naturally for 7 days and impurities were removed.

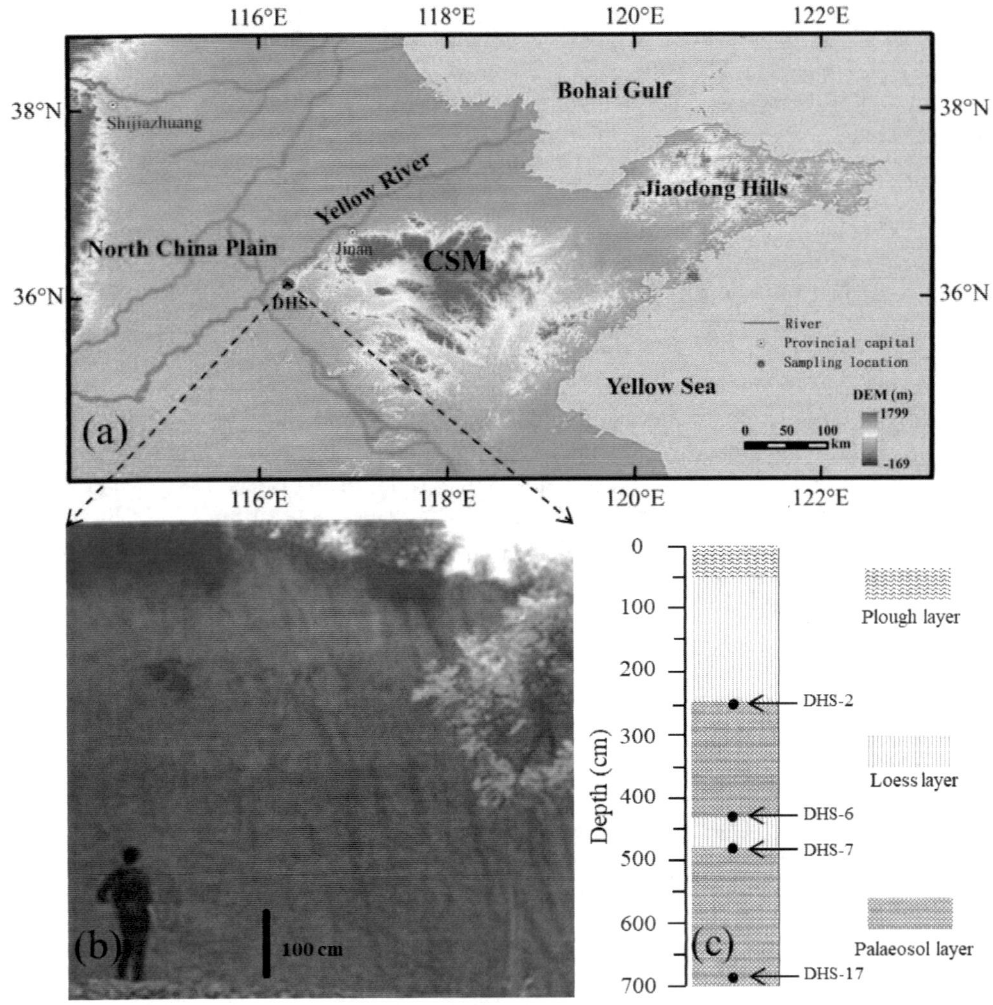

Figure 1. (a) Location of study region of the Central Shandong Mountains; (b) photograph of the DHS section and (c) the corresponding stratigraphic columns of loess section at DHS.

Table 1. Summaries of the burial depth, radionuclide concentration, water content, calculated dose rate, feldspar D_e value and luminescence age.

Sample Code	Depth (cm)	U (ppm)	Th (ppm)	K (%)	Water content (%)	Dose rate (Gy/ka)	D_e (Gy)	Age (ka)
DHS-2	250	1.94±0.50	10.90±0.30	1.99±0.03	17	3.33±0.15	314±4	94±4
DHS-6	440	2.27±0.54	12.29±0.29	2.26±0.03	8	4.02±0.18	398±6	99±5
DHS-7	480	2.76±0.59	12.43±0.31	2.08±0.03	9	4.08±0.19	730±42	179±13
DHS-17	690	2.60±0.50	12.40±0.30	2.06±0.03	11	3.77±0.17	861±50	228±17

IOP Conf. Series: Earth and Environmental Science 1087 (2022) 012070 doi:10.1088/1755-1315/1087/1/012070

3. Materials and methods

3.1. Magnetic susceptibility

MS was measured in the laboratory of Taishan University. Air-dried samples were weighed every 10 g and packed in circular non-magnetic plastic boxes. MS measurements were conducted using a Bartington MS2 meter at low-frequency (470 Hz). Then, the average of three tests was calculated as the final results.

The magnetic minerals in the sediment can react sensitively and record stably the information on environmental changes [5]. MS is considered to be related to the types, magnetic domians and contents of magnetic [4] and the strength of pedogenesis [5]. Pedogenesis was weak after the accumulation of loess dust during dry and cold climate in the glacial period, with the low contents of magnetic minerals and MS values in the loess. The leaching of carbonate minerals in the process of pedogenesis increased the content of magnetic minerals and the compaction of soil also enriched the content of magnetic minerals [4, 5] during warm and humid climate in the interglacial period, resulting the MS values increased in the palaeosol. MS value of the palaeosol layer with strong pedogenesis is much higher than that of the loess layer with weak pedogenesis in many sections of the CLP [6], indicating that the pedogenesis is an important factor for the enhancement of the MS in the palaeosol layers. In addition, MS is related to precipitation and is used as proxy of regional summer monsoon changes, especially in loess-palaeosol sequences [5].

3.2. Chroma

The CIELAB chroma system is composed of lightness (L*), redness (a*) and yellowness (b*), which results are basically consistent with the actual situation and not affected by subjective factors. The ranges of L*, a* and b* are from black to white, green to red, blue to yellow, respectively [7].

Samples were ground to below 200 mesh and stored in Ziplock polypropylene bags. The chroma was determined by a K-Minolta CR-400 colorimeter manufactured by Konica Minolta, Japan. The instrument was corrected to ensure the accuracy of the results before the samples were measured. After the 2-3 g samples were compacted inside the slides, the processed slides were placed on a standard correction white board and were measured 3 times on 3 neat and flat areas which were randomly selected. The mean values of L*, a* and b* were verified and calculated for 3 times to obtain the final results to ensure an error within 0.07 [7].

The parameter of a* is predominantly affected by the concentration of iron oxide minerals, hematite and goethite and b* is mainly controlled by the content of goethite [12]. Low a* values correspond to loess sedimentation, while high a* values correspond to the palaeosol developments, indicating enhanced pedogenesis under warm and humid climate with strong EASM [6].

3.3. Palaeotemperature and palaeoprecipitation derived from chroma and magnetic susceptibility

The study of the quantification between chroma and environment have begun since 1960s and some previous researchers explored the result of the measurement on Chinese modern soil color under different climatic conditions and some soil color parameters were used as the climatic indicators to build the functional relation between soil color and climate with different types of soils. The correlation coefficient R^2 reaches 0.73 between a* and mean annual temperature (MAT) which shows that it has good exponential relationships between a* and MAT [6]. The researchers found that the a* record can indicate the paleosol stratification in the loess section and also correspond well to the MS [6]. The formula between redness and temperature is selected to inverse the palaeotemperature situation. The relation of palaeotemperature and a* in the loess obeys the following equation:

$$\mathrm{MAT} = \begin{cases} 16.62\ln(a^*) - 25.27; \text{ red if } a^* > 0 \\ 16.62\ln(a^*) - 25.27; \text{ green if } a^* < 0 \end{cases} \quad (1)$$

MS can represent features of climate over time in specific regions and sediments. Previous researchers collected samples from different soil types in different climate zones to explore the regulation of MS with climate change. Finally, they established the regression equation of MS, annual average temperature and annual average precipitation in the CLP and its peripheral areas [13]. A positive correlation relationship exists between the MS and mean annual precipitation (MAP) and the correlation coefficient R^2 reaches 0.675. The formula between MS and precipitation is selected to inverse the palaeoprecipitation situation.The relation between palaeoprecipitation and MS in the loess obeys the following equation:

$$MAP=-22.706+11.554X-6.7166\times 10^{-2}X^2+1.8684\times 10^{-4}X^3-1.9264\times 10^{-7}X^4 \quad (2)$$

X is the magnetic susceptibility in the surrounding area of the CLP.

4. Results

4.1. Changes in the magnetic susceptibility and palaeoprecipitation

In the DHS section, the average value of low-frequency susceptibility (χlf) is 50×10^{-8} $m^3{\cdot}kg^{-1}$ within the range of $15\text{-}106\times10^{-8}$ $m^3{\cdot}kg^{-1}$. The highest MS values are observed in the last interglacial palaeosol (S_1) and the second highest MS values are observed in the penultimate interglacial palaeosol (S_2), while the lowest MS values are observed in the penultimate glacial loess (L_2). The average precipitation was 395 mm with the range of 134-646 mm derived from the MS values in the whole section (Figure 2).

The study area could be divided into three stages based on the changes in MS values and palaeoprecipitation curves, as follows:

Stage I (S_2, 480-690-cm depth, ~250-190 ka): the MS values ranged from 15 to 86×10^{-8} $m^3{\cdot}kg^{-1}$ (average 42×10^{-8} $m^3{\cdot}kg^{-1}$). The MS curves show the shape of two peaks with one valley. The mean annual precipitation was 349 mm (ranged from 134 to 582 mm) in this stage.

Stage II (L_2, 440-480-cm depth, ~190-100 ka): the MS values ranged from 18 to 83×10^{-8} $m^3{\cdot}kg^{-1}$ (average 39×10^{-8} $m^3{\cdot}kg^{-1}$). The MS is the lowest in the whole section. The mean annual precipitation (314 mm, ranged from 162 to 571 mm) was lower than that in stage I (S_2).

Stage III (S_1, 250-440-cm depth, ~100-90 ka): the MS values ranged from 37 to 106×10^{-8} $m^3{\cdot}kg^{-1}$ (average 61×10^{-8} $m^3{\cdot}kg^{-1}$). The trend was similar to the stage I. However, the mean annual precipitation (460 mm, ranged from 325 to 646 mm) was 100 mm higher than that in stage I (S_2). Therefore, our results show that the East Asian monsoon is stronger in S_1 period than in S_2 period.

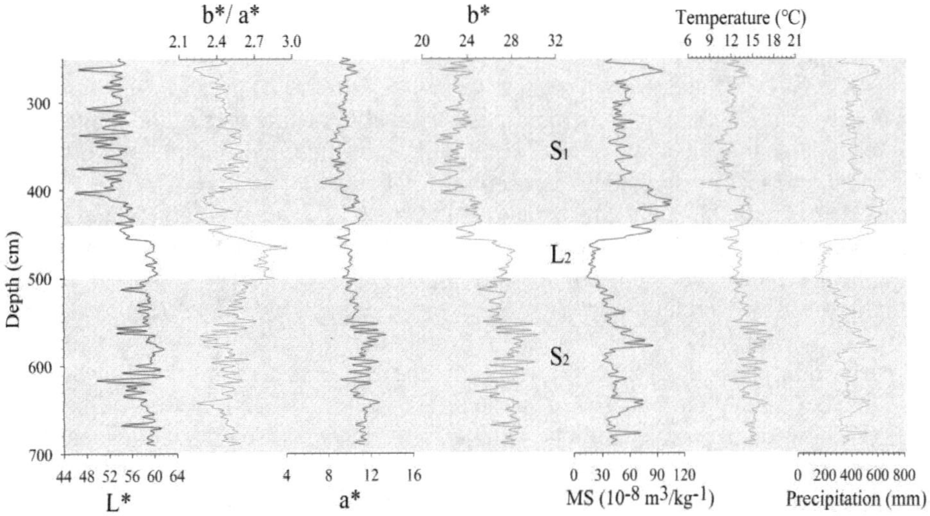

Figure 2. Chroma parameters (L*, a* and b*), b*/ a* and MS characteristics were obtained based on the depth in the DHS section. The temperature and precipitation were derived from the redness and magnetic susceptibility, respectively.

4.2. Changes in chroma and palaeotemperature

The aeolian loess layers were characterized by high L*, low a* and b* values whereas the palaeosol layers were the converse.The a* and b* values were related to the content of hematite and goethite [5], however, the increase in the values of b* was affected by the increase in the content of hematite. The parameter b* has indicative meaning for soil forming but not sensitive enough [7]. Previous researchers have found that the ratio of goethite to hematite decreased with the increase of annual temperature and precipitation, so the ratio of b* to a* (b*/ a*) was more suitable to indicate the climate changes [14]. The parameter b*/ a*, L*, a* and b* in the DHS section exhibited the ranges of 2.2-2.9, 46-62, 7-13 and 21-30 and averages of 2.5, 56, 10 and 25, respectively. The average temperature based on the parameter of a* was 13.2 °C within a range of 8.1-17.6 °C between the the penultimate interglacial and last interglacial. The study area could be divided into three stages based on the changes in values of three parameters of chroma and variation of b*/ a* curve, as follows:

Stage I (S_2, 480-690-cm depth, ~250-190 ka): the average b*/ a* value in the S_2 was 2.5 (range of 2.2-2.8). The variations of L*, a* and b*were 50-60 (average 53), 7-13 (average 10) and 24-30 (average 27), respectively. The L* values initially showed a smooth trend and the b*/ a* showed an increasing trend. The annual average temperature was 14.4 °C (range of 11.8-17.6 °C).

Stage II (L_2, 440-480-cm depth, ~190-100 ka): the b*/ a*, L*, a* and b* values revealed the averages of 2.6, 58, 10 and 26 with ranges of 2.3-2.9, 54-60, 9-10 and 23-28, respectively. The curves of L* and b*/ a* showed an obviously increasing trend with higher values. The annual average temperature was 12.5 °C (range of 10.9-13.4 °C).

Stage III (S_1, 250-440-cm depth, ~100-90 ka): the b*/ a* values varied from 2.2 to 2.8 (average 2.4). The average values of L*, a* and b* were 58, 11 and 23 with ranges of 46-56, 7-11 and 21-25. The The L* curve indicated variability similar to the b*/ a* curve with a decreasing trend. The annual average temperature was 12.1 °C (ranged of 8.1-15.0 °C).

The L* of this section exhibited higher values in loess unit (L_2) and lower values in palaeosol units (S_1 and S_2), which was similar with the b*/ a* curve and opposite with temperature curve.

5. Discussions

5.1. Magnetic susceptibility comparison and the process of palaeoclimatic changes

The average MS of the palaeosol (52×10^{-8} $m^3\cdot kg^{-1}$) is higher than that of the loess layer (39×10^{-8} $m^3\cdot kg^{-1}$), indicating that climate changes experienced from warm–wet to cold–dry to warm–wet in the studied area over the penultimate interglacial time. The fluctuation of the MS curve during the L_2 period is not obvious as the palaeosol and the values of MS is lower and more stable. The MS fluctuates obviously in the S_1 period showing a peak and valley, which is similar to the S_2 period. To further evaluate palaeolclimate changes in the DHS section of the CSM region, we compared the MS of the DHS region with that of the Xiangyang section in the coastal area of southern Anhui Province [15], Luochuan [16], Xifeng [17] and Mangshan [18] in the CLP; and Weinan [5] and Xianyang [19] in the Guanzhong Basin (Figure 3). The results suggested that the trends of the MS curves were similar among the 7 sections with some minor differences. The MS curves show a trend of one peak and one valley in the S_1 period and a small peak in the L_2 period with generally low values, but an obvious trend of two peaks and one valley in the S_2 period with a secondary peak for the MS shape.

The MS curve of the DHS section was roughly the same as that of the Xiashu loess of the Zhenjiang region [20], even though there are subtle differences. The curve of the Xiashu loess layer in Zhenjiang region showed three peaks and two valleys in the S_2 period, which is different from the two peaks and one valley in the DHS section and other places (Figure 3). The second peak in Zhenjiang area appears almost simultaneously with the secondary peak in DHS section and other places. Changes of MS in the Zhenjiang area were more obvious than that of DHS section and corresponded better to the marine oxygen isotope curve. The MS curve in the Zhenjiang region reflected more oceanic characteristics of the climate than that in the DHS section and in the CLP because it is closer to sea and more effected by the moist airflow from the ocean.

The MS values of the DHS section were much lower than those of the Luochuan section at the CLP in the same period. This phenomenon may be caused by the fact that the DHS section is closer to the ocean and is more affected by moist airflow from the ocean. The MS of the DHS section was approximately similar to that of Luochuan, Xifeng, Mangshan, Weinan, Xianyang, Xiangyang and Zhenjiang sections, indicating that the changes of MS reflected climate change in East Asia and even the global climate changes. However, the subtle differences compared with other section also indicated that the DHS section recorded the local climate changes.

5.2. Chroma and implication of palaeoclimatic variations

The values of L * are obviously lower in the S_1 than that in the L_2 period but change similarly in the L_2 and S_2 periods, so the parameter L* can not clearly distinguish the different strata. The values of a* are higher in the S_2 than that in the L_2 period with decreasing values after a brief rise in the S_1 period, which can not very well distinguish between the L_2 and S_1 strata. The values of b* are the highest in the S_2 period, the second in the L_2 period, and the lowest in the S_1 period, which can not indicate the different characteristics of the parameter b* in loess and palaeosol. The values of b*/ a* are obvious higher in the L_2 period than that in the S_2 and S_1 periods with a shape of peak and valley clearly, which can distinguish the strata of S_1, L_2 and S_2 well. We can conclude from Figure 2 that the fluctuations of the L*, a* and b* curves are not sensitively to the climate changes while the fluctuation of the b*/ a* curve shows more sensitively than the L*, a* and b* curves, which exhibits that the b*/ a* curve correlated best with the strata in the DHS section, which shows that the response of b*/ a* curve to the climate changes is the most sensitively among the four proxies. The b*/ a* values in the L_2 period are higher than those in the S_1 and S_2 periods, indicating that the climate was colder and drier in the L_2 period, and the climate was warmer and more humid in the S_1 and S_2 period. The changes of the b*/ a* values reflect the local climate changes in the DHS section and response well with the global climate changes.

The b*/ a* can correspond well to the loess layer and the palaeosol layer. In addition, our results have also been confirmed by the researches of the Bole section in Xinjiang, the Jinchuan section in Sichuan and Xuwen section in Guangdong [14].

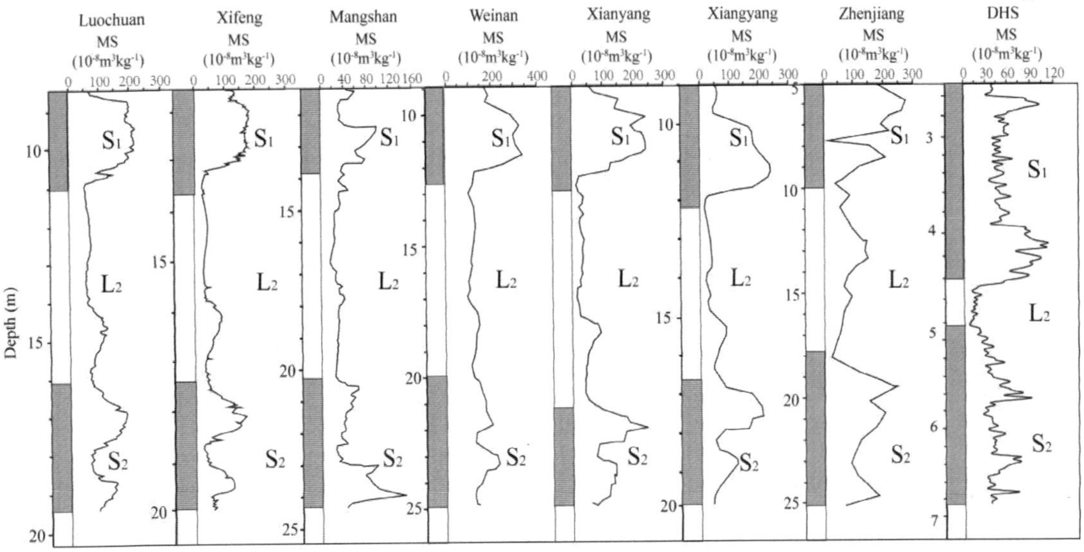

Figure 3. Comparison of the MS records of the DHS section and others on the CLP and in the Guanzhong Basin, Anhui region and Zhenjiang region: Luochuan [16]; Xifeng [17] and Mangshan [18]; Weinan [5] and Xianyang [19]; Xiangyang [15] and Zhenjiang [20], respectively.

6. Conclusions

(1) Our research presents the first luminescence ages of the CSM loss-palaeosol sediments over the penultimate interglacial periods, with large time span ranging from the penultimate interglacial to the last interglacial time, suggesting that the CSM loess was accumulated during the late quaternary.

(2) MS value changes correlate well with the studied strata. The annual average temperature (from 11.8 to 17.6 °C, with a mean of 14.4 °C) derived from the redness during the penultimate interglacial time in the DHS section is higher than the present (from 12.7 to 14.8 °C, with a mean of 12.9 °C). However, the annual average precipitation (from 134 to 582 mm, with a mean of 349 mm) from the MS inversion during the penultimate interglacial period in the DHS section is far below the present (from 550 to 950 mm, with a mean of 606 mm). Thus, researches should be careful for the precipitation derived from the MS in the CSM area, although the MS can reflect the regionally stratigraphic cycles.

(3) The variations of chroma values correlate well with the stratigraphy. The variation trend of b*/a* ratios was more sensitive than that of the L*, a* and b* in the DHS section, which can reflect evidently the climate changes. The lower values of the ratio in the S_1 and S_2 showed warmer climatic condition with strong summer monsoon and higher values in the L_2 indicated colder climatic condition with weak summer monsoon, wihch was similar to the climatic implications of L* values.

Acknowledgements
We are grateful to Junsheng Yue, Dongdong Zheng and Zhen Liu for their assistance in the field sampling. This work was supported by Shandong Provincial Natural Science Foundation (grant number ZR2021MD008), State Key Laboratory of Loess and Quaternary Geology, Institute of Earth Environment, CAS (grant number SKLLQG2134), National Nature Science Foundation of China (41602353) and the Beijing Natural Science Foundation (8224095).

References

[1] Wangyang Zhou 2021 Stalagmite record of climate change in China during different interglacial periods *[D]. Zhejiang Normal University*
[2] Boxun Cao 1990 The quarternary climate and the futhre climate and environmental change in China *[J]. Regional geology of China* **2** 97-11
[3] Tungsheng Liu 1985 Loess and the Environment *[M]. China Ocean Press, Beijing*
[4] Maher B A and Thompson R 1991 Mineral magnetic record of the Chinese loess and paleosols *[J]. Geology* **19(1)** 3-6
[5] Zhongli Ding, Zhiwei Yu 1995 Forcing mechaisms of paleomonsoons over East Asia *[J]. Sciences* **1** 63-74
[6] Shengli Yang, Xiaomin Fang, Jijun Li, Zhisheng An, Shiyue Chen, Hitoshi F 2001 Transformation functions of soil color and climate *[J]. Science in China* **44** 218-226
[7] Min Ding, Jiangli Pang, Changchun Huang, Shuzhen Peng, Jiong Yang, Dongdong Chen 2010 Chroma characteristics and its climatic significance in Holocene loess-paleosol sequence — A case study of the Holocene Liangcun profile in the western Guanzhong basin *[J]. Journal of Shaanxi Normal University (Natural Science Edition* **38(5)** 92-97
[8] Zulu Zhang, Liangjie Xin, Luguang Jiang, Ping Huang 2005 Sedimentary characteristics and genetic analysis of Zhangxia Loess in Jinan, Shandong Province *[J]. Journal of Palaeogeography* **7(1)** 98-106
[9] Shuzhen Peng, Qingzhen Hao, Luo Wang, Min Ding, Wei Zhang, Yanan Wang, Zhengtang Guo 2016 Geochemical and grain-size evidence for the provenance of loess deposits in the Central Shandong Mountains region, northern China *[J]. Quaternary Research* **85** DOI: 10.1016/j.yqres.2016.01.005
[10] Qiuyue Zhao, Shuzhen Peng, Niannian Fan, Luo Wang, Qingzhen Hao, Xiuying Liu, Rui Zhou, Min Ding, Wei Zhang, Ning Liu 2022 Luminescence chronology of loess-palaeosol deposits in the Central Shandong Mountains region: Provinces and paleoclimate implications *[J].*

Quaternary Geochronology **70** DOI: 10.1016/j.quageo.2022.101296

[11] Wei Han, Yonghong Han, Shenbin Yang 2013 Spatial-temporal change of climate resources and climatic productivity in Shandong Province during 1961-2011 *[J]. Progress in Geography* **32(3)** 425-434

[12] Schwertmann U 1971 Transformation of Hematite to Goethite in Soils *[J]. Nature* **232** 624-625

[13] Houyuan Lv, Jiamao Han, Naiqin Wu, Zhengtang Guo 1994 Analysis of Modern Soil Magnetization and Its Palaeoclimatic significance in China *[J]. Science In China (Series B)* **24(12)** 1290-1297

[14] Mengyuan Zhu, Bin Lv, Ying Guo 2022 Comparison of Haematite and Goethite Contens in Aeo lian Deposits in Different Climate Zones Based on Diffuse Reflectance Spectroscopy and Ch romaticity Methods *[J]. Spectroscopy and Spectral Analysis* **42(6)** 1684-1690

[15] Xusheng Li, Dayuan Yang, Huiyou Han 1998 Magnetic susceptibility of Aeolian-Dust-Deposition-Paleosol sequence in the south of Anhui Province *[J]. Journal of Anhui Normal University (Natural Science)* **21(1)** 64-69

[16] Junjie Zhang, Qingzhen Hao, Shenghua Li 2022 An absolutely dated record of climate change over the last three glacial - interglacial cycles from Chinese loess deposits *[J]. Geology* DOI: 10.1130/G50125.1

[17] Qingzhen Hao, Luo Wang, Oldfield F, Shuzhen Peng, Li Qin, Yang Song, Bing Xu, Yansong Qiao, Jan B, Zhengtang Guo 2012 Delayed build-up of Arctic ice sheets during 400,000-year minima in insolation variability *[J]. Nature* **490** 393-396

[18] Liye Chen, Ke Zhang, Jianli Fu, Hao Liang, Xiaoyang Li, Zhongyun Li 2018 The trace and rare earth element characteristics of Mangshan Loess since deposit of paleosol S_2 and its implications for provenance *[J]. Acta Scientiarum Naturalium University Sunyatseni* **57(3)** 14-23

[19] Leping Yue 1998 A Comparative Study between the East-Asia Paleomonsoon Climate and Glacial Climate in Chinese Loess Area *[J]. Journal of Northwest University (Natural Science Edition)* **28(2)** 157-162

[20] Xusheng Li, Dayuan Yang 2002 Magnetic Susceptibility Features and Environmental Records of the Xiashu Loess in Zhenjiang, Jiangsu Province *[J]. Journal of Desert Research* **22(1)** 27-32

IOP Conf. Series: Earth and Environmental Science 1087 (2022) 012071 doi:10.1088/1755-1315/1087/1/012071

Imaging of karst cave using ground penetrating radar and electrical resistivity tomography: a case study in Shandong, China

Xu Sun[1], **Xiangxing Lu**[1], **Kaitian Li**[2 3]

[1] Shandong Electric Power Engineering Consulting Institute Corporation, Jinan, Shandong, China 250013;
[2] Institute of Geophysics & Geomatics, China University of Geosciences, Wuhan, Hubei, China 430074.

[3] Corresponding author's e-mail: lkt802@cug.edu.cn

Abstract. A denudation-dissolution hilly area about 20km southeast of Jinan city, Shandong Province, China is characterized by karst features. The upper part of the area is covered by quaternary Holocene flood and slope deposits, and the lower bedrock is Ordovician limestone, which is a moderately weathered massive structure with developed joints and fissured fractures, and solution grooves can be seen locally. The local government to better serve the surrounding residents and the surrounding facilities, here is planned for the substation, for later security construction, electrical resistivity tomography (ERT) method and ground penetrating radar (GPR) method were used to investigate the underground karst caves. The aim of the joint use of the two methods is to better describe the characteristics of karst landform, a 160 MHz shielding antenna was employed in the GPR survey, while the frequency limits the scope of the depth of the electromagnetic waves can reach. However, it provides a lot of shallow stratigraphic information, which well represents the boundary between overburden and bedrock and the development of karst. The use of ERT method further makes up for the depth defects of the GPR method, and two-dimensional imaging of underground resistivity distribution was carried out, which also helps to further identify the characteristics of underground karst. The combination of the two geophysical methods can accurately delineate the development of underground karst, which also proves that the combination of the GPR method and ERT method has good application value in karst landform detection, and can also be applied to a similar geological survey.

1. Introduction

As the karst problem exists widely in the world, with the continuous expansion of urban land, we have to face the use of karst terrain, and the resulting engineering problems gradually emerge [1]. Karst usually occurs in limestone, dolomite, and other soluble rocks, and the existence of sub-surface and underground cracks and karst cavities are potential risks in infrastructure construction, and also cause huge treatment costs [2]. The detection of underground karst has been a wide concern by scientists and engineers. In traditional prospecting methods, boreholes are often used to directly determine the distribution of unfavorable geological bodies underground, and this is often the most intuitive way to reveal the formation. However, intensive drilling often comes with high costs, and the application of geophysical methods has greatly reduced engineering costs and enabled engineers to accurately identify subsurface features. As one of the main means of scientific research and engineering investigation in karst landforms in recent ten years, the GPR method has solved many geological problems such as engineering geology and hydrological environment [3, 4]. Its application will be limited under certain geological conditions, such as thicker quaternary coverings and high conductivity materials. The depth of GPR also has certain limitations. In recent years, many case studies of GPR combined with the resistivity method have been reported, and the combination of the two methods and technologies has been proved to be an effective means for imaging and characterization of underground karst [5, 6, 7, 8]. In this paper, GPR and ERT surveys are carried out simultaneously in the research area. The ERT method provides a more intuitive geoelectric distribution and makes up for the shortage of the GPR method in the detection depth, to delineate the occurrence of karst more accurately and expose the stratum in the proposed substation area.

2. Geological setting of the study area

Figure 1a shows the location of the study area, the composition of regional strata is relatively simple, and overlying strata are quaternary Holocene proluvial sediments, mainly for the lithology of silty clay, an average thickness of 0.5-2.8 m, the bedrock of the Ordovician moderately weathered limestone layer. Figure 1b depicts some bedrock that has been exposed directly. The joint fractures formed by weathering in the upper strata are shown in Figure 1c, and the rock mass is relatively broken, while the rock mass at other depths is intact. Figure 1d shows that the karst morphology of the proposed site is mostly developed as solution ditches or grooves. Figure 2b&c show the karst caves exposed by drilling are generally filled with silty clay and gravel.

Figure 1. Geographic location and geology of the working area.

a) Research area located in Jinan city, Shandong Province, China; b) limestone outcrop of the Study area; c) joint cracks of the limestone; d) Karren of the limestone.

The main type of groundwater in the study area is karst fissure water, which mainly occurs in karst fissures. Atmospheric precipitation is the main source of its supply, and evaporation and underground runoff are its discharge modes. According to the investigation, the perennial highest groundwater level buried depth is more than 10 m, during the survey, no groundwater was found at the proposed site.

3. Methodology

3.1. GPR method

GPR method uses a transmitting antenna to send high-frequency pulsed electromagnetic waves to the target body to detect the spatial position and distribution of the target body. The interface where electromagnetic waves are reflected usually reflects changes in the physical properties of the medium (dielectric coefficient), but changes in water content can also cause strong reflections. In this GPR survey, the GX host developed by MALA Company of Sweden was employed. After comprehensive consideration of the detection depth, resolution requirements, and physical properties of the regional interior layer, a 160 Mhz shielded antenna was used, the effective detection depth was about 8m, and the acquisition method of self-excitation and self-collection was adopted. Figure 2a shows the layout of total 8 GPR profiles.

The Reflexw software was used to process the collected raw GPR data. The following processing steps were employed: subtract-DC-shift, static correction, gain and energy decay, subtracting average, band-pass filtering. Reflexw software's batch processing feature can be utilized to process data fast because the data acquisition parameters and work area environment are similar.

3.2. ERT method

ERT method is an array resistivity method that can investigate the distribution of underground volume resistivity by observing the distribution of electric field generated by the direct current supplied to the underground. ERT method has always been one of the preferred geophysical methods to solve various engineering geological problems, hydrogeological problems, environmental protection issues and cultural relics protection [9, 10, 11], which can sensitively capture resistivity changes of underground geological bodies and help scientists and engineers better understand geological processes. The WGMD-4 current meter manufactured by Chongqing Benteng Instrument Factory was employed, which can automatically compensate for the drift of self-potential in the acquisition process. The Wenner-Schlumberger array was adopted, and the electrode distance was 2 m. Figure 2a shows the layout of the 2 ERT profiles.

RES2DINV software was used for data inversion. The underground was divided into a large number of rectangular cells in a 2-D model. The inversion algorithm used by the software is based on the smoothness-constrained least square method. Given an initial model. The damping factor can be adjusted to adapt to different types of data, so as to reduce the calculated and measured apparent resistivity [9]. The method is based on the following equation

$$\left(\mathbf{J}^T\mathbf{J}+\lambda\mathbf{F}\right)\Delta\mathbf{q}_K=\mathbf{J}^T\mathbf{g}-\lambda\mathbf{F}\mathbf{q}_K \tag{3.1}$$

where

$\mathbf{J}$ = Jacobian matrix of partial derivatives, $\mathbf{J}^T$ = transpose of $\mathbf{J}$

λ = damping factors

$\mathbf{F}=\alpha_x\mathbf{C}_x^T\mathbf{C}_x+\alpha_z\mathbf{C}_z^T\mathbf{C}_z$, α_x and α_z are respectively relative weights of the roughness filters

$\mathbf{C}_x$ = horizontal roughness filters, $\mathbf{C}_z$ = vertical roughness filters

$\mathbf{q}$ = model change vector, $\mathbf{g}$ = data misfit vector, k represents the iteration number

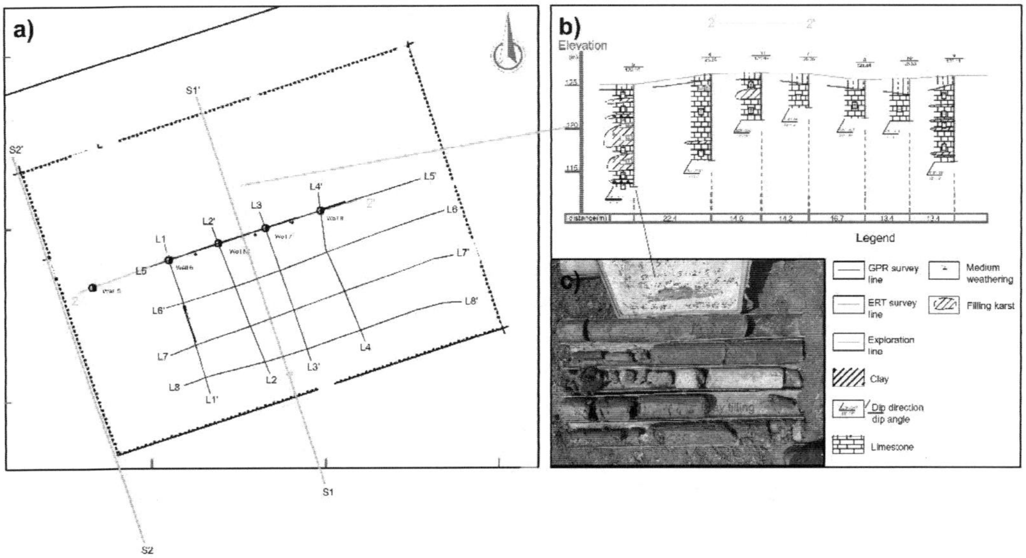

Figure 2. Map of the study area and borehole investigation results.

a) Arrangement of the ERT survey lines and GPR survey lines performed at Aida village; b) Exploration line of the study area; c) the karst cave revealed by borehole 5.

4. Results

4.1. GPR results

A total of 8 radar profiles (Profile 1-Profile 8) were obtained in this radar survey, and the radar profiles are shown in Figure 3. The maximum depth of this radar survey is about 8 m, but the effective anomaly identification is generally within 4 m. The reflection characteristics in the profile are relatively consistent, the boundary between the overlying layer and bedrock is obvious, and the in-phase axis is continuous. In the radar diagram, it is differentiated by a yellow spline curve. In the lower karst development area, the abnormal position generally reflects the reflection wave aggregation and the transverse continuity deteriorates. It is inferred that the rock fissure is developed and the solution groove is formed. The radar survey identified dozens of anomalies of varying sizes and depths.

4.2. ERT results

Figure 4 shows the inversion results of the two ERT profiles, and the inverted profiles characterize the electrical distribution within 15 m underground in the study area. ERT S1 line is 118 m long with 60 electrodes. The inversion results show that the profile can be roughly divided into three electrical layers, the first layer is about 1 m from the surface to underground, and the resistivity is generally 50-90 $\Omega \cdot m$, inferred to be the interface between quaternary strata and bedrock, mainly composed of silty clay. The second layer is located in the depth range of 1-5 m, and the resistivity is usually 170-600 $\Omega \cdot m$, the electrical layer has poor transverse continuity, and high resistance anomalies occur at 40 m, 62 m, and 100 m. Figure 4 (area A) shows their approximate position, which is inferred to be a half-filled karst cave. The third layer is approximately 5 m depth and has a resistivity of 90 to 250 $\Omega \cdot m$, the resistivity is low at 0-60 m, which indicates that the bedrock is broken and the cracks are filled with clay, leading to the increase of argillaceous content. The resistivity distribution of this layer is relatively continuous.

ERT S2 line has a length of 118 m and a total of 60 electrodes were laid. The inversion results show that the section can be roughly divided into two electrical layers, the first layer is about 5 m from the surface to the underground, and the resistivity is generally 230-600 $\Omega\cdot m$. Because of the different weathering degrees of the bedrock, the electrical layer has poor transverse continuity of resistivity. The second layer is between 5-12 m depth, with a constant horizontal distribution of resistivity that ranges from 50-125 $\Omega\cdot m$. The third layer is located in the depth range of 12-17 m, and the resistivity is generally 230-400 $\Omega\cdot m$, the distribution of ground electricity is continuous. The overall profile presents a trend of "high-low-high".

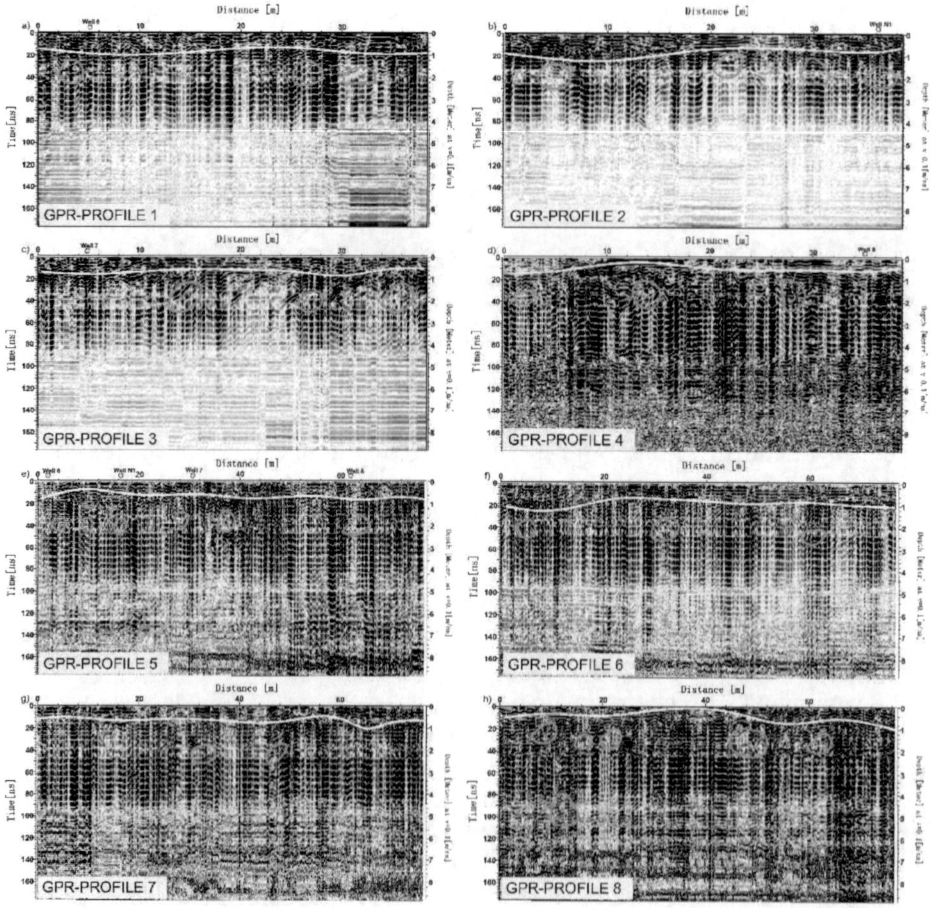

Figure 3. GPR diagram (the yellow lines are the interface between quaternary sediments and bedrock, The karst locations are marked by light blue circles).

4.3. Comprehensive interpretation

The ERT S1 Line was arranged from southeast to northwest, passing through the GPR profile 8, 7, 6 and 5 in turn. Figure 5 shows the intersection's location, which is at 42, 54, 67 and 82 meters, respectively. Meanwhile, by comparing the radar diagram and abnormal resistivity distribution, the high resistivity anomaly located at the mileage of 42 m of line S1 and the depth of about 2 m is just within the mileage of 30-32 m of GPR profile 8, the radar diagram shows strong reflection.

At the intersection of GPR profile 6, 7 and ERT S1, there is no evident resistivity anomaly and reflection amplitude change. At 82 m of S1 line, it intersects with 30 m of GPR profile 5. While the amplitude of the radar diagram changes at about 2 m, there is no apparent abnormality in the geoelectric structure. Comparing the resistivity anomaly with radar diagram, the joint interpretation provides a further understanding of karst characteristics: a) The electrical anomaly with high resistivity corresponds to the reflection with strong amplitude, and it is inferred that the karst caves are not filled or half-filled. b) The low reflection amplitude does not reflect the change of resistivity on the geoelectric profile. It can be inferred that when the cave is filled, the difference between its resistivity and surrounding rock decreases, so there is no resistivity anomaly. c) Considering the complexity of karst landform, its fracture development and filling may change within a few meters, resulting in a change of physical parameters and increasing the difficulty of interpretation.

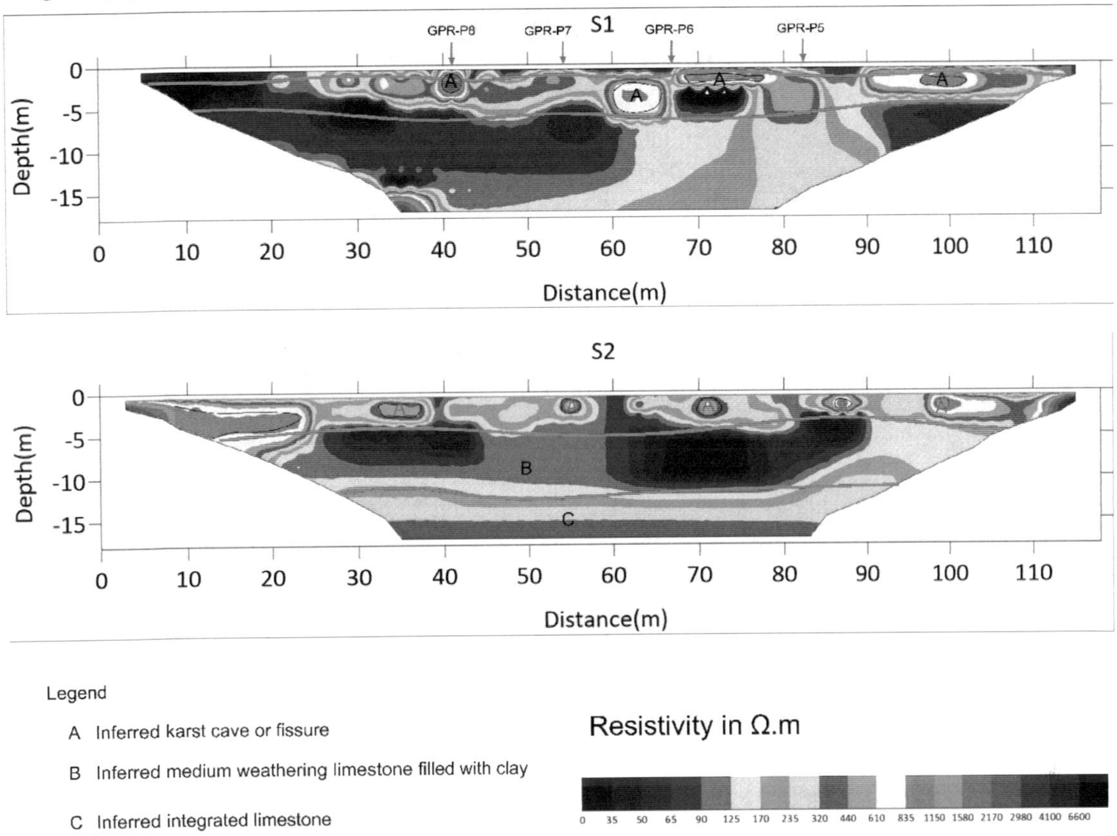

Figure 4. ERT results (Electrical layer interfaces are marked by magenta lines).

5. Discussions

Underground medium is equivalent to a complex filter, the absorption of electromagnetic wave by medium is the comprehensive embodiment of material and its inhomogeneity. The karst landform is extremely complex. Due to its special drainage system, various karst ditches are formed, and the rocks are also cut into different block shapes, thus forming many reflection interfaces. When the electromagnetic wave propagates to this point, the waveform is usually disordered, and the continuity of the phase axis becomes worse. When there is an unfilled karst cave in limestone, it is generally manifested as an arc reflected wave with strong amplitude. When there is clay filling, the electromagnetic wave energy decreases rapidly and the high-frequency part is absorbed. Same to the ERT method, resistivity imaging is directly reflected in the underground distribution of the volume resistivity. The weathering degree of the upper bedrock in the study area varies greatly, so the horizontal continuity of resistivity is poor. The overall weathering degree of the lower bedrock is weak, and the resistivity characteristics are relatively consistent. It is inferred that after the infiltration of atmospheric rainfall, a layer of stagnant water is formed in the lower weathered bedrock. When the fissures are connected, a section of relatively low resistivity is formed. Although the ERT method has the problem of insufficient resolution with the deepening of depth, the stratified characteristics of the geoelectric section also indirectly reflect the weathering degree of the stratum.

In site investigation, the detection depth of ERT method is about 15 m, and the effective detection depth of GPR method is within 8 m. The Quaternary overburden is a thin low resistivity layer in the ERT S1 line. The area with uniform and continuous amplitude on the radar diagram is also divided into the interface between Quaternary layer and bedrock. The electrical layers within 5 m are unevenly distributed laterally, and relatively high resistivity anomalies appear at different mileage. Strong amplitude reflections at the corresponding positions on the radar diagram were also found, which are inferred as unfilled or semi filled karst caves. In the area below 5 m depth of ERT profile, the distribution of electrical layers is continuous. As the current resolution decreases with the increase of depth, some small karst fractures in the deep are averaged by the volume effect, so we can't see the abnormal shape on the inversion section, which may lead to a deviation in the understanding of the actual geological situation.

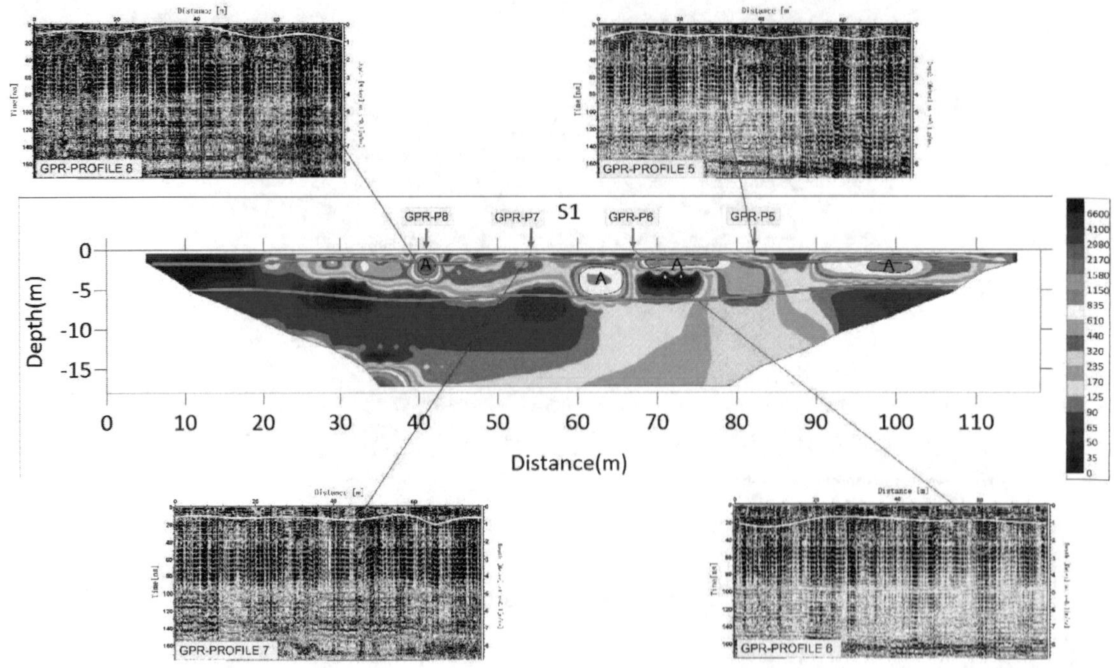

Figure 5. Joint interpretation of ERT profile and GPR diagram.

6. Conclusions

The combination of GPR and ERT has successfully delineated the location of the karst and characterized the karst landform in the study area. When the karst formed in the karst process is not filled or half filled, the relatively high resistivity anomaly in the resistivity profile directly confirms this phenomenon. Similarly, the local strong reflection in the radar image also indicates the existence of the reflection interface. However, with the increase of the detection depth, the resolution of the method is not enough to effectively identify the local karst fracture development. GPR can obtain abundant stratigraphic information on the sub-surface, and can be used to approximately map the boundaries of karst cave. but at the same time, there are some factors such as insufficient depth and easy to be restricted by the site conditions. In the application of GPR method, the thickness of clay layer, groundwater and other conditions should be fully considered.

Even though ERT method cannot accurately recognize the boundaries and morphology of the anomaly, furthermore, when there is inapparent contrast in the electrical properties between the filling materials and surrounding rock, some filled karst cave will not be well identified. However, there are reflection interfaces in these areas, and the weak reflection amplitude can also be identified on the GPR diagram. The ERT method may add a dimension of understanding to the stratum from the perspective of geoelectricity and also makes up for the lack of depth of the GPR method. The combination of the two techniques also can, and to the greatest extent, locate the location and boundary of underground karst when the detection area is completely unknown. The successful combination of the two also provides an important geophysical basis for safe construction in the later stage and provides a certain reference value for solving similar problems in the future.

References

[1] Cui F P, Li B, Xiong C, Yang Z P, Peng J Q, Li J S, Li H W 2022 Dynamic reGuizhou Province, China *Geomatics Natural Hazards & Risk* **13** 123-147

[2] Gutierrez F, Parise M, De Waele J, Jourde H 2014 A review on natural and human-induced geohazards and impacts in karst *Earth-Sci Rev* **138** 61-88

[3] Anchuela O P, Casas-Sainz A M, Pocovi-Juan A, Garbi H G, Calvin P 2014 Characterization of the Karstic Process in an Urban Environment Using GPR Surveys *Journal of Materials in Civil Engineering* 26

[4] Guo L, Lin H, Fan B H, Nyquist J, Toran L, Mount G J 2019 Preferential flow through shallow fractured bedrock and a 3D fill-and-spill model of hillslope subsurface hydrology *Journal of Hydrology* **576** 430-442

[5] Carbonel D, Rodriguez V, Gutierrez F, McCalpin J P, Linares R, Roque C, Zarroca M, Guerrero J, Sasowsky I 2014 Evaluation of trenching, ground penetrating radar (GPR) and electrical resistivity tomography (ERT) for sinkhole characterization *Earth Surface Processes and Landforms* **39** 214-227

[6] Martel R, Castellazzi P, Gloaguen E, Trepanier L, Garfias J 2018 ERT, GPR, InSAR, and tracer tests to characterize karst aquifer systems under urban areas: The case of Quebec City *Geomorphology* **310** 45-56

[7] Golebiowski T, Jarosinska E 2019 Application of GPR and ERT methods for recognizing gypsum deposits in urban areas *Acta Geophysica* **67** 2015-2030

[8] Tao M, Chen X, Cheng Q B, Binley A 2022 Evaluating the joint use of GPR and ERT on mapping shallow subsurface features of karst critical zone in southwest China *Vadose Zone Journal* 21

[9] Loke M H, Chambers J E, Rucker D F, Kuras O, Wilkinson P B 2013 Recent developments in the direct-current geoelectrical imaging method *J Appl Geophys* **95** 135-156

[10] Drahor M G 2019 Identification of gypsum karstification using an electrical resistivity tomography technique: The case-study of the Sivas gypsum karst area (Turkey) *Eng Geol* **252** 78-98

[11] Lu K, Li F, Pan J W, Li K T, Chen Y, Li Y L, Chen L, Wang H H, Ding L, Li S P, Li Z Y 2021 Using electrical resistivity tomography and surface nuclear magnetic resonance to investigate cultural relic preservation in Leitai, China *Eng Geol* 285

IOP Conf. Series: Earth and Environmental Science 1087 (2022) 012072 doi:10.1088/1755-1315/1087/1/012072

Fractal analysis on the effective heat conductivity of unsaturated soil

Shunyu Su[1,2] and Wenlanting Li[1]

[1] School of Urban Construction, Wuhan University of Science and Technology, Wuhan 430070, China.

[2]Corresponding author's e-mail: shunyusu@qq.com

Abstract. Because of the complicated structure, the traditional Euclidean geometry is difficult to accurately describe the inner distribution of pore soil. According to the fractal laws of inner structure in porous media, a popular fractal capillary model was presented to study the three phase heat transfer in unsaturated soil. Considering the soil pore size and its tortuosity, by using the fractal method about pore size distribution and tortuosity to depict the scaling relationships, the aim of this paper is to derive the effective heat conductivity of unsaturated three phase soil. The predicting values show good agreement with the experimental values of literature while the saturation values of water are among 0.2 to 0.9. Major factors that influence the effective heat conductivity of unsaturated soil were analyzed. The heat transfer mechanism of unsaturated soil was also analyzed according to the results. The conclusions of this study could help promote the improvement of the design of buried pipe heat exchanger in unsaturated soil.

1. Introduction

Soil as a porous medium has a typical fractal structure. The description and quantification of soil structure is important. Fractal methods could be applied to study the internal features of soil, such as particle size distributions in the soil. Many natural porous media were shown to meet statistically fractal scale laws, and fractal methods have been successfully applied to describe the geometrical structures such as soil fractal porous media [1]. There are many academic researches about the features of fractal porous media. Ji et al. [2] suggested that more consideration should be given to groundwater in studying statistical fractal moisture soil. Chen et al. [3] developed a novel fractal capillary model considering the hysteresis effect for multiphase flow in porous media. Qin et al. [4] presented a fractal model for porous media with different liquid saturations. It shows well effect to micro-structural porous media. Zhang et al. [5] reviewed and analyzed thirteen fractal soil models about conductive heat transfer. The advantages, disadvantages and application conditions of these models were assessed. Thus, further investigation should be conducted so as to improve prediction accuracy for geothermal applications.

Based on fractal theory, this study is focused on the analysis of the self-similarity of soil internal structure and heat conductivity of non-saturated soil. The aim of this paper is to establish a popular fractal prediction model of heat conductivity for porous medium, and analyze the influencing factors on heat conductivity of non-saturated soil. The results will be helpful to recognize the characteristics of soil heat conductivity. It can also help strengthen the heat exchange of buried pipe heat exchanger in the soil.

2. Fractal Capillary Model

Geometrical parameters, such as pore size and tortuous degree, have great effect on the multi-phase flow in non-saturated porous media. While non-saturated soil consists of solid (soil particulate matter), liquid (water), and gas (air), a popular fractal capillary model is presented [3, 5]. Figure 1 shows the representative element volume of this popular fractal capillary model. The pore routes are described by a bunch of tortuous capillaries [3]. And the major capillary is supposed to meet fractal laws.

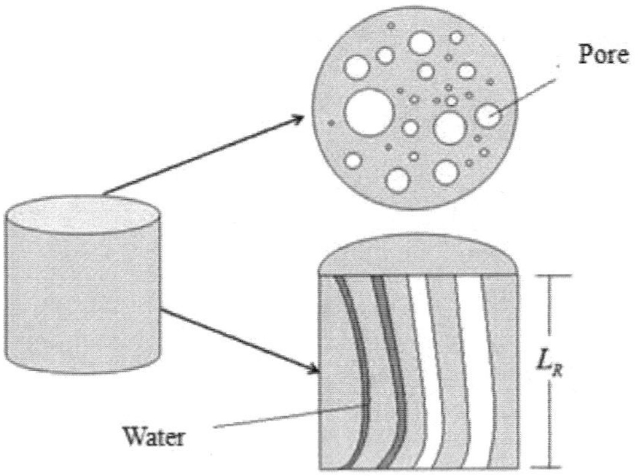

Figure 1. Diagram of popular fractal capillary model.

According to fractal theory, the soil porosity ϕ can be defined as:

$$\phi=(\lambda_{min}/\lambda_{max})^{3-D_f} \tag{1}$$

where D_f represents the pore size fractal dimensions; λ_{min} represents the minimum pore diameter of soil; λ_{max} represents the maximum pore diameter of soil.

While the pore cross sectional area is A_p, the total cross sectional area A is defined by:

$$A=\frac{A_p}{\phi}=\frac{\pi}{4}\frac{D_f}{2-D_f}\frac{1-\phi}{\phi}\lambda_{max}^2 \tag{2}$$

The representative element volume in Figure 1 is square in cross-section. The lateral area is equal to the longitudinal area. The longitudinal length of the representative element volume L_R is given by:

$$L_R=\left(\frac{D_f}{2-D_f}\frac{\phi}{1-\phi}\right)^{1/2}\lambda_{max} \tag{3}$$

The tortuous length L_t is given by:

$$L_t(\lambda)=\lambda^{1-D_T}L_R^{D_T} \tag{4}$$

where D_T represents the fractal tortuous dimension; λ represents the pore diameter of soil.

3. Calculation of Effective Heat Conductivity

The initial state in the representative element volume is full of air. The saturation of water in it rises increasingly. Supposing that there is a critical capillary diameter λ_c, it indicates whether the capillary is full of water or air [6-8]. When the diameter λ of the major capillary is less than critical capillary diameter λ_c, it indicates that the capillary is full of water.

The saturation of water S_w is written by:

$$S_w = \frac{V_w}{V_p} = \frac{\lambda_c^{3-D_T-D_f} - \lambda_{\min}^{3-D_T-D_f}}{\lambda_{\max}^{3-D_T-D_f} - \lambda_{\min}^{3-D_T-D_f}} \tag{5}$$

where V_w is the volume of water in the representative element volume; V_p is the volume of overall pore in the representative element volume.

Thus, the critical capillary diameter of pore λ_c is given by:

$$\begin{aligned}\lambda_c &= \left[S_w\left(\lambda_{\max}^{3-D_T-D_f} - \lambda_{\min}^{3-D_T-D_f}\right) + \lambda_{\min}^{3-D_T-D_f}\right]^{\frac{1}{3-D_T-D_f}} \\ &= \lambda_{\max}\left\{S_w\left[1-\left(\frac{\lambda_{\min}}{\lambda_{\max}}\right)^{3-D_T-D_f}\right] + \left(\frac{\lambda_{\min}}{\lambda_{\max}}\right)^{3-D_T-D_f}\right\}^{\frac{1}{3-D_T-D_f}}\end{aligned} \tag{6}$$

Since liquid and gas are fillers of soil pore in the representative element volume, the liquid pore and gas pore are assumed to obey fractal scaling law [5]. Therefore, heat resistance of liquid phase R_w and heat resistance of gas phase R_g are given by:

$$R_w = \frac{4\left(1+D_T-D_f\right)L_R^{D_T}}{\pi k_w D_f \lambda_{\max}^{1+D_T}\left[\left(\lambda_c/\lambda_{\max}\right)^{1+D_T-D_f} - \left(\lambda_{\min}/\lambda_{\max}\right)^{1+D_T-D_f}\right]} \tag{7}$$

$$R_g = \frac{4\left(1+D_T-D_f\right)L_R^{D_T}}{\pi k_g D_f \lambda_{\max}^{1+D_T}\left[1-\left(\lambda_c/\lambda_{\max}\right)^{1+D_T-D_f}\right]} \tag{8}$$

where k_w is the liquid conductivity; k_g is the gas conductivity.

In the representative element volume, the solid soil is non-fractal phase. The heat resistance of solid phase R_s is given by:

$$R_s = \frac{L_R}{k_s\left(1-\phi\right)A} \tag{9}$$

where k_s is the solid conductivity.

For non-saturated three phase soil, the effective heat conductivity k_e of the representative element volume could be written by:

$$k_e = \frac{L_R}{A}\left(\frac{1}{R_g} + \frac{1}{R_w} + \frac{1}{R_s}\right) \tag{10}$$

Combined with Eq. (7)-(10), k_e could be written by:

$$\begin{aligned}k_e &= k_s\left(1-\phi\right) + \frac{\pi k_g D_f \lambda_{\max}^{1+D_T}}{4\left(1+D_T-D_f\right)AL_R^{D_T-1}}\left[1-\left(\frac{\lambda_c}{\lambda_{\max}}\right)^{1+D_T-D_f}\right] \\ &+ \frac{\pi k_w D_f \lambda_{\max}^{1+D_T}}{4\left(1+D_T-D_f\right)AL_R^{D_T-1}}\left[\left(\frac{\lambda_c}{\lambda_{\max}}\right)^{1+D_T-D_f} - \left(\frac{\lambda_{\min}}{\lambda_{\max}}\right)^{1+D_T-D_f}\right]\end{aligned} \tag{11}$$

Substituting the critical diameter of pore λ_c of Eq. (6) into Eq. (11), the dimensionless effective thermal conductivity $k_e^+ = k_e / k_g$ is derived as:

$$k_e^+ = \frac{k_s}{k_g}(1-\phi) + \frac{(2-D_f)^{\frac{(D_T+1)}{2}}}{D_f^{(D_T-1)/2}(1+D_T-D_f)}\left(\frac{\phi}{1-\phi}\right)^{\frac{(D_T+1)}{2}} \times \left\{1-\left[S_w\left(1-\left(\frac{\lambda_{\min}}{\lambda_{\max}}\right)^{3-D_T-D_f}\right)+\left(\frac{\lambda_{\min}}{\lambda_{\max}}\right)^{3-D_T-D_f}\right]^{\frac{1+D_T-D_f}{3-D_T-D_f}}\right\} + \frac{k_w}{k_g}\frac{(2-D_f)^{\frac{(D_T+1)}{2}}}{D_f^{(D_T-1)/2}(1+D_T-D_f)}\left(\frac{\phi}{1-\phi}\right)^{\frac{D_T+1}{2}} \times \left\{\left[S_w\left(1-\left(\frac{\lambda_{\min}}{\lambda_{\max}}\right)^{3-D_T-D_f}\right)+\left(\frac{\lambda_{\min}}{\lambda_{\max}}\right)^{3-D_T-D_f}\right]^{\frac{1+D_T-D_f}{3-D_T-D_f}} - \left(\frac{\lambda_{\min}}{\lambda_{\max}}\right)^{1+D_T-D_f}\right\} \tag{12}$$

The validity of the presented model has been tested as Figure 2 indicates. The predicted values of k_e from Eq. (11) were compared with the measured values from reference [9]. The results of prediction agree well with the measurement while the values of water saturation S_w are among 0.2 to 0.9. In relative low saturation region, such as water saturation is lower than 0.2, there is stronger moisture adsorption from dry solid soil. Then many soil pores may not be filled with water. In this region, prediction results will deviate from the measurement of the literature.

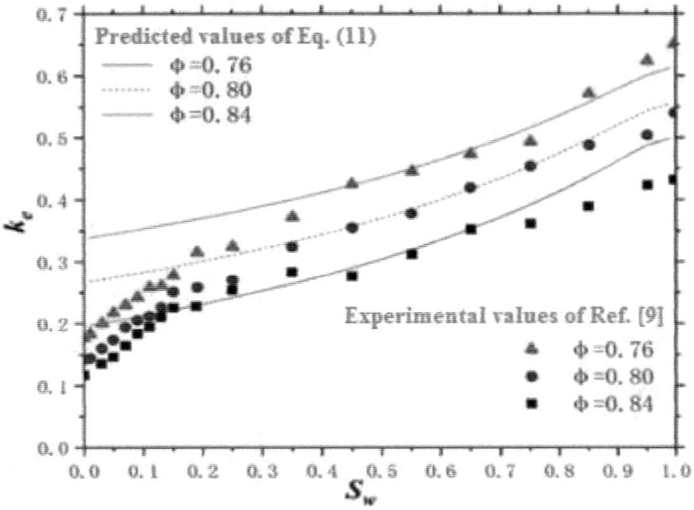

Figure 2. Comparison between current model and the measurement of literature.

4. Results and Discussions

Figure 3 to Figure 5 are the calculating results which show the changes of dimensionless effective heat conductivity with varied water saturation. They indicate the influences of non-saturated soil internal structure on soil k_e .

Figure 3 indicates the influence of soil pore size on k_e^+ while soil porosity ϕ=0.39 and fractal tortuous dimension D_T=1.1. With the ratio $\lambda_{min}/\lambda_{max}$ decreasing, the pore size range of self-similarity expands and D_f rises. It indicates that k_e^+ reduces with the soil pore size rising.

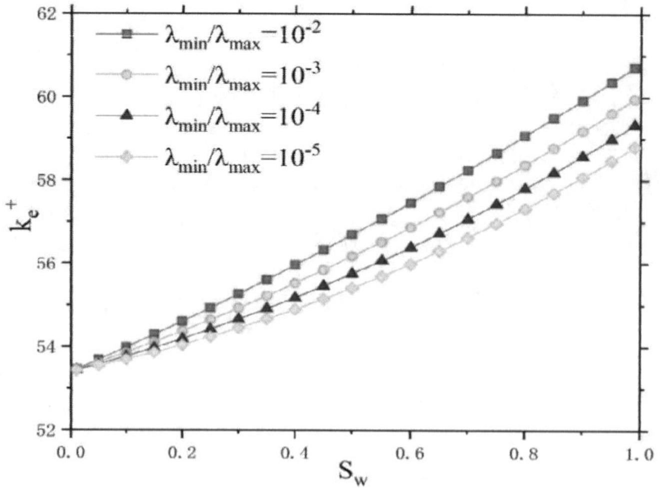

Figure 3. Influence of soil pore size on k_e^+ .

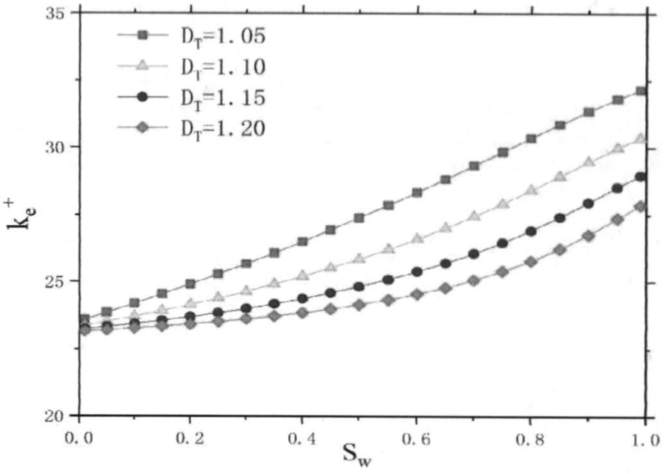

Figure 4. Influence of fractal tortuous dimension on k_e^+ .

Figure 4 shows the influence of fractal tortuous dimension on k_e^+ while soil porosity ϕ=0.5 and pore diameter $\lambda_{min}/\lambda_{max}$=0.001. With the value of D_T decreasing, the longitudinal direction of the pore becomes more curved. The length of heat transfer through gas and liquid phases will become longer and the overall heat resistance of the non-saturated soil will increase. It indicates that k_e^+ reduces with the fractal tortuous dimension rising.

Figure 5 shows the influence of soil porosity on k_e^+ while pore diameter $\lambda_{min}/\lambda_{max}$=0.001 and fractal tortuous dimension D_T=1.1. Soil porosity represents the volume ratio of pore to total porous medium. With the value of soil porosity ϕ increasing, the ratio of solid phase in the soil decreases. And heat conductivity of solid phase is bigger than heat conductivities of liquid phase and gas phase in the soil. It indicates that k_e^+ reduces with the soil porosity rising.

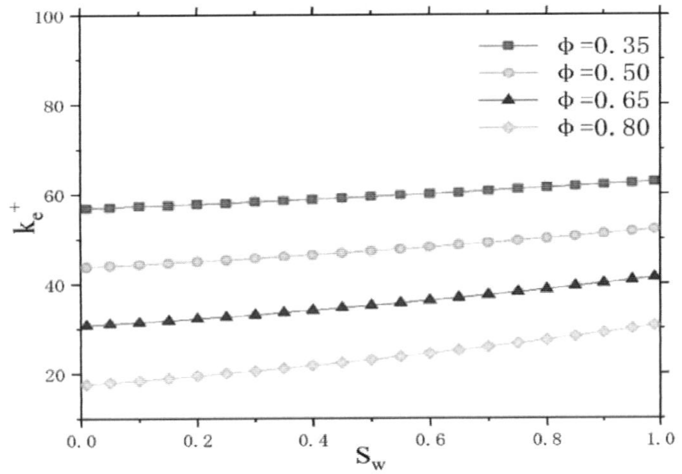

Figure 5. Influence of soil porosity on k_e^+.

5. Conclusions

Considering fractal geometry of soil internal structure, a popular fractal capillary model for non-saturated soil has been developed. According to the fractal analysis of an elementary volume in this model, the effective heat conductivity of non-saturated three phase soil was concluded. The predicting results were compared with experimental values from literature. And they show good agreement with the measurement while the values of water saturation are among 0.2 to 0.9. But further research about the characteristics of thermal conductivity in soil is still needed.

The calculating results from the proposed fractal model show that k_e^+ reduces with the soil pore size rising. And values of k_e^+ also reduce with the fractal tortuous dimension rising and the soil porosity rising. The results of this study are helpful to understand the characteristics of heat transfer in unsaturated soil. It could benefit to the design of buried pipe heat exchanger in unsaturated soil.

References

[1] P. Xu, S. X. Qiu, B. M. Yu and Z. T. Jiang 2013 Prediction of relative permeability in unsaturated porous media with a fractal approach. *International Journal of Heat and Mass Transfer* **64** 829-837
[2] X. Y. Ji, C. P. Shen and W. J. Riley 2015 Temporal evolution of soil moisture statistical fractal and controls by soil texture and regional groundwater flow. *Advancesin Water Resources* **86** 155-169
[3] C. Chen, K. Chen, M. H. Yang and P. Xu 2020 A fractal capillary model for multiphase flow in porous media with hysteresis effect. *International Journal of Multiphase Flow* **125** 103208
[4] X. Qin, J. C. Cai, P. Xu, S. Dai, Q. Gan 2019 A fractal model of effective thermal conductivity for porous media with various liquid saturation. *InternationalJournal of Heat and Mass*

Transfer **128** 1149-1156
[5] N. Zhang, Z. Y. wang 2017 Review of soil thermal conductivity and predictive models. *International Journal of Thermal Sciences* **117** 172-183
[6] H. J. Gao, B. M. Yu.Y. G. Duanet al. 2014 Fractal analysis of dimensionless capillary pressure function. *International Journal of Heat and Mass Transfer* **69** 26-33
[7] Z. L. Deng, X. D. Liu.Y. P. Huanget al. 2017 Heat Conduction in Porous Media Characterized by Fractal Geometry. *Energies* **10(8)** 1230-1238
[8] Y. Q. Shen, P. Xu.S. X. Qiu et al. 2020 A generalized thermal conductivity model for unsaturated porous media with fractal geometry. *International Journal of Heat and Mass Transfer* **152** 119540
[9] H.Q. Jin, X. L. Yao, L.W. Fan, X. Xu, Z. T. Yu 2016 Experimental determination and fractal modeling of the effective thermal conductivity of autoclaved aerated concrete: Effects of moisture content. *InternationalJournal of Heat and Mass Transfer* **92** 589-602

IOP Conf. Series: Earth and Environmental Science 1087 (2022) 012073 doi:10.1088/1755-1315/1087/1/012073

Characteristics of sedimentary system evolution and sequence stratigraphy boundary of Denglouku group in Xujiaweizi fault depression, Songliao Basin

Kunning Cui[1,2]

[1] Exploration and Development Research Institution of Daqing Oilfield Company Ltd., Daqing, China.

[2] Corresponding author's e-mail: cuikunning@petrochina.com.cn

Abstract. In the previous studies on this area, the sequence division is relatively rough. This study has improved the precision, and the Denglouku formation is divided into 10 fourth-order sequences, with the thickness of each small layer of about 70-80m. The characteristics of each small sequence boundary are described, and three third-order, four fourth-order sequence flooding surfaces are found. The stratigraphic framework is established. From bottom to top, the base level of all strata shows decline-rise-decline, which means regression-transformation-regression.

The Denglouku 1st mainly includes lacustrine and fan delta, the sources of sediments from northeast, northwest, southwest and Southeast directions. In Denglouku 2nd and Denglouku 3rd, braided fluvial delta and intermittent lacustrine are dominant, which were controlled by the sedimentary systems from east, south, north, northeast, southwest, and the northwest. In Denglouku 4th, there is mainly braided fluvial, rarely are lacustrine and braided fluvial delta plain, the main source was from the northeast, and a small amount of source is from the west.

1. Regional geological background

Xujiaweizi fault depression is located on Xujiaweizi-Bei'an fault depression belt, with Zhaodong-Chaoyanggou anticline belt and Yingshan fault depression belt in the East, Anda-Zhaozhou anticline belt in the West and Mingshui slope in the North (according to ZHAO Guo-lian, 2002) [1] (Figure 1). The North-South span is about 80km, and the longest-East-West span is 40km, with a total area of 5866km^2 (according to ZHAO LI Xing-wei, 2005) [2]. Controlled by XuXi fault and Xuzhong fault, the structural characteristics of the north-south direction are as follows: north-north-west strike, two depressions with one uplift, east-west zoning, north-south blocking, and a half graben basin with Eastern-overlap and western-fracture type. The main lithology is kind of interbeds weith gray-white massive sandstone, dark sandy mudstone, variegated sand, mudstone, and conglomerate. The bottom is sand, conglomerate, and tuff thin layer. Unconformable contact with the lower Yingcheng formation. The maximum thickness is more than 1700m. The general law of sedimentary evolution is from fan delta-lake system to river-river delta-lake system.

2. Sequence stratigraphic characteristics in Denglouku group, Xujiaweizi fault depression

Denglouku formation is a supersequence, the top and bottom boundary both are supersequence boundary. The historical study is divided into four sets of strata, which are divided into four third-order sequences from bottom to top according to the interface properties (according to SUN Dong-lai,

2012) [3]. This study based on 3D continuous seismic, field outcrop, core, and logging data, use the guide of the sequence stratigraphy theory [4], the sequence stratigraphic framework of Denglouku formation in the northern fault depression of Xujiaweizi is deeply studied.

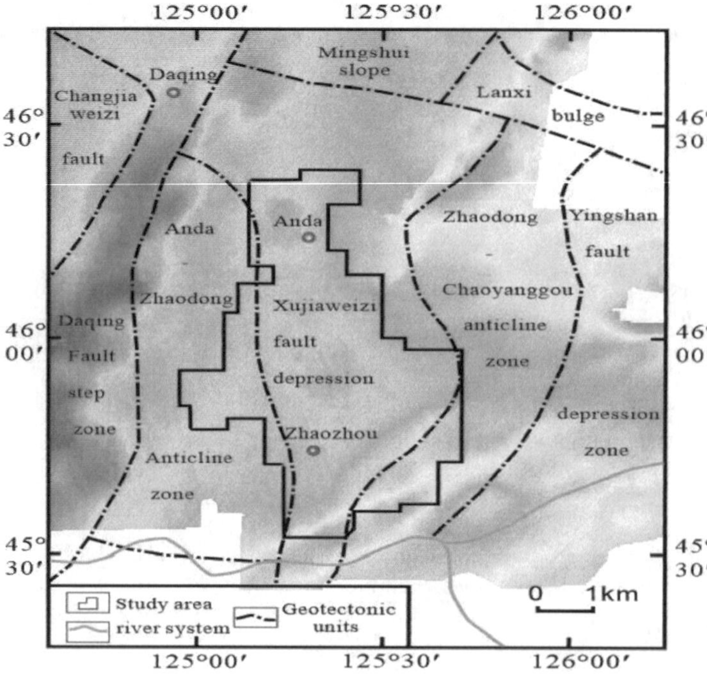

Figure 1. Tectonic units in Xujiaweizi fault depression.

2.1. Seismic reflection, lithologic and logging characteristics of sequence boundary

2.1.1. Seismic reflection, lithologic and logging characteristics of sequence boundary of the third order sequence in Denglouku formation. ***SB4 Sequence boundary:*** strong reflection, strong amplitude, good continuity, the western boundary appears onlap, while the eastern boundary appears onlap and truncation, unconformity interface. Lithology shown as the top boundary of several sets of sandstone, lake flooding surface. Below the sequence boundary, the sand (gravel)/mud ratio increases gradually, and the combination style of logging curve is progradational parasequence set. Above the boundary, the sand/mud ratio decreases gradually, and retrograde parasequence set.

SB_{32} Sequence boundary: Medium-weak reflection, strong amplitude, good continuity, onlap on the edge and integrated interface in the middle. Lithology shows a relatively stable sedimentary interface in the middle of mudstone, secondary lake flooding surface. Below the sequence boundary is a retrograde parasequence set and above the boundary is aggradation parasequence set.

SB_{31} Sequence boundary: Strong reflection, strong amplitude and good continuity. The Western onlap has went to the top of the structure. The lithology shows the bottom boundary of a large set of sandstone, lake flooding surface. Below the sequence boundary is retrograde parasequence set and above the boundary is aggradational parasequence set.

SB_3 Sequence boundary: Strong reflection, strong amplitude, good continuity, downlap in the southeast and c, which is the water regression interface. The lithology shows that the upper part is purple sand-mudstone deposition, the lower part is thicker sandstone deposition. Below the sequence boundary is retrograde parasequence set, and above it is progradational parasequence set. (Figure 2)

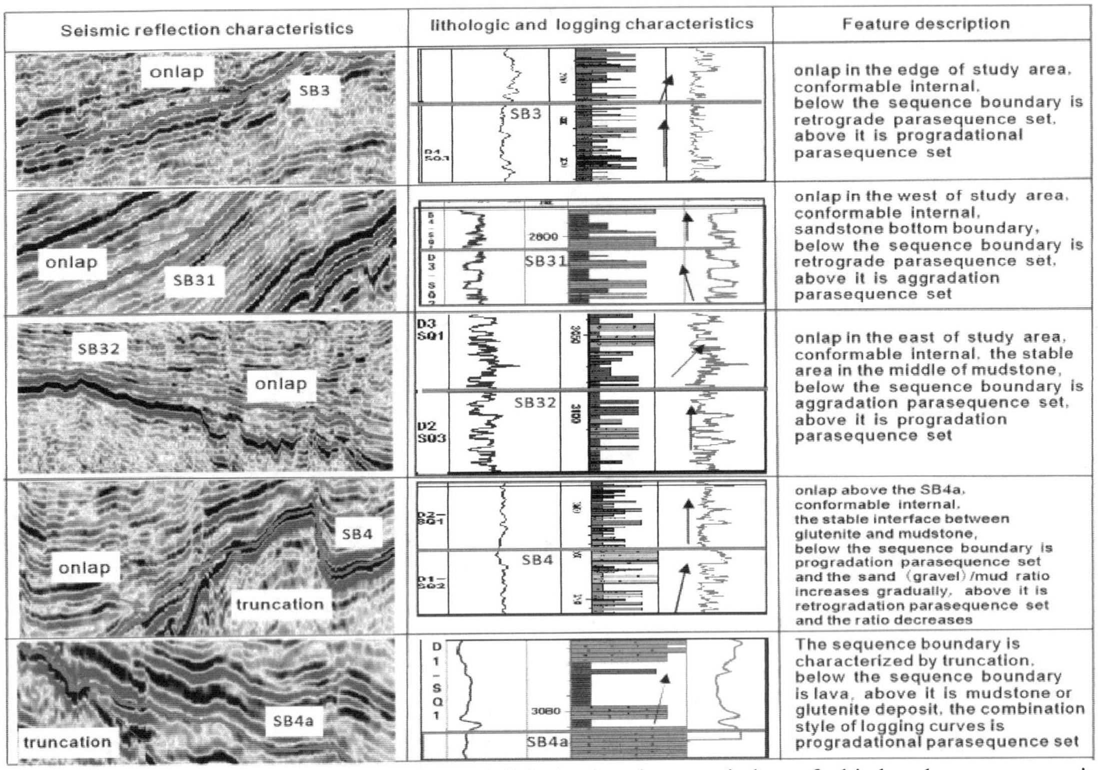

Figure2. Seismic reflection, logging and lithologic characteristics of third-order sequence in Denglouku group, Xujiaweizi fault depression.

2.1.2. Seismic reflection, lithologic and logging characteristics of sequence boundary of the fourth order sequence in Denglouku formation. **SB_4^1 Sequence boundary:** The Seismic reflection in the edge of the study area is characterized by onlap and internal integrated contact. The lithology is characterized by the top boundary of sandy conglomerate, with aggradation type on the top and retrogradation type on the bottom.

Lowstand and highstand system tract, and lake transgression are occurred below the sequence boundary, the base level above the boundary drops, and the combination relationship of logging curves is of aggradation type. A set of glutenite body at the front of fan delta is deposited (Figure 3-**SB_4^1**).

SB_{32}^2 Sequence boundary: The Seismic reflection in the edge of the study area is characterized by toplap and internal integrated contact. The lithology is characterized by thin sand mudstone, with progradation type at the top and aggradation type at the bottom. The base level drops above the sequence boundary, the combination relationship of logging curves shows progradation type, depositing a set of thick fine sandstone bodies in the distributary channel of braided river delta plain (Figure 3-**SB_{32}^2**). Two system tracts, lake transgression and highstand system tract, are occurred both above and below the sequence boundary.

SB_{32}^1 Sequence boundary: The Seismic reflection in the edge of the study area is characterized by toplap and internal integrated contact. The lithology is characterized by the middle of thin sand mudstone, with retrogradation type at the top and progradation type at the bottom. Below the sequence boundary is progradational type, dominated by braided river delta plain distributary channel deposition; The base level above the sequence boundary drops and retrograde type, depositing a set of thick fine sandstone bodies in braided river delta plain (Figure 3-**SB_{32}^1**). lake transgression and highstand system tract, are occurred both above and below the interface.

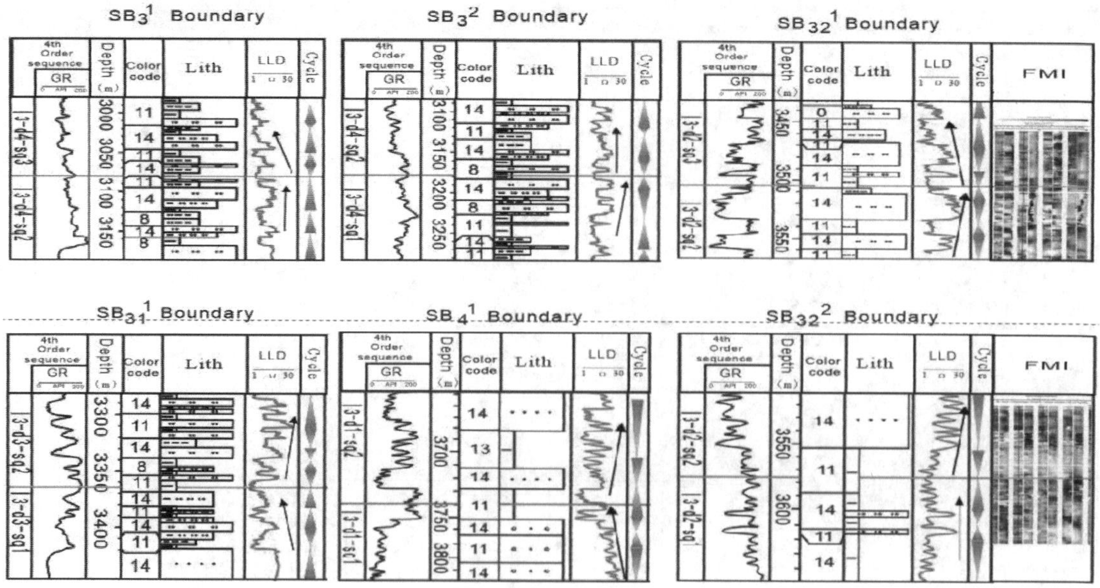

Figure 3. Logging, lithologic and cycle characteristics of fourth-order sequence in Denglouku group, Xujiaweizi fault depression.

$SB_{31}{}^{1}$ Sequence boundary: The Seismic reflection in the edge of the study area is characterized by toplap and internal integrated contact. The lithology is characterized by a relatively stable sedimentary interface in the middle of mudstone, with progradation type at the top and aggradation type at the bottom. Below the sequence boundary is retrograde type, dominated by braided river delta plain distributary channel deposition; The base level above the interface drops and progradation type, depositing a set of thick fine sand and siltstone in braided river delta plain (Figure 3-**$SB_{31}{}^{1}$**). lake transgression and highstand system tract, are occurred both above and below the interface.

$SB_3{}^2$ Sequence boundary: The Seismic reflection in the northern part of the study area is characterized by local truncation and internal integrated contact. The lithology is characterized by the bottom boundary of thin sandstone, with aggradation type on the top and progradation type on the bottom. Under the sequence boundary is aggradation type, dominated by braided river center beach deposition; The base level above the sequence boundary drops and aggradation type, depositing a set of braided river crevasse fan thin-layer silt and mudstone deposits (Figure 3-**$SB_3{}^2$**). lake transgression and highstand system tract, are occurred both above and below the interface.

$SB_3{}^1$ Sequence boundary: The Seismic reflection in the edge of the study area is characterized by toplap or local truncation, and internal integrated contact. The lithology is characterized by the middle of thin sand shale interbed, with retrogradation type on the top and aggradation type on the bottom.Under the sequence boundary is aggradation type, dominated by braided river center beach thick siltstone deposition; The base level above the sequence boundary drops and retrograde type, depositing a set of meandering river thick silty sand body (Figure 3-**$SB_3{}^1$**). lake transgression and highstand system tract, are occurred under the sequence boundary, and one lake transgression and highstand system tract are occurred on the interface.

2.2. establishment of sequence stratigraphic framework

Based on the identification of the third-order and fourth-order sequence boundaries, the fourth-order fine isochronous sequence stratigraphic framework of Xujiaweizi fault depression is established through the sequence division and correlation of sections. The whole area is further refined from the traditional 4-parts third-order sequence (According to ZHANG Shun, 2015) [5, 6] to 10-parts fourth-order sequence. The division scale of sequence stratigraphy also ranges from 150-180m of the third-order sequence to 70-80m of the fourth-order sequence. The third-order and fourth-order sequences are distributed from bottom to top as follows: **third-order sequence II_1-d_1:** The top interface is SB_4, corresponding to seismic reflection axis T_4, and the bottom interface is SB_{4a}, corresponding to seismic reflection axis T_{4a}. From top to bottom, it is divided into two fourth-order sequences, II_1-d_1-sq_1 and II_1-d_1-sq_2; **third-order sequence II_1-d_2:** The top interface is SB_{32}, corresponding to seismic reflection axis T_{32}, and the bottom interface is SB_4, corresponding to seismic reflection axis T_4. From top to bottom, it is divided into three fourth-order sequences, II_1-d_2-sq_1, II_1-d_2-sq_2, II_1-d_2-sq_3; **third-order sequence II_1-d_3:** The top interface is SB_{31}, corresponding to seismic reflection axis T_{31}, and the bottom interface is SB_{32}, corresponding to seismic reflection axis T_{32}. From top to bottom, it is divided into two fourth-order sequences, II_1-d_3-sq_1 and II_1-d_3-sq_2; **third-order sequence II_1-d_4:** The top interface is SB_3, corresponding to seismic reflection axis T_3, and the bottom interface is SB_{31}, corresponding to seismic reflection axis T_{31}. From top to bottom, it is divided into three fourth-order sequences, II_1-d_4-sq_1, II_1-d_4-sq_2, II_1-d_4-sq_3. (Figure 4).

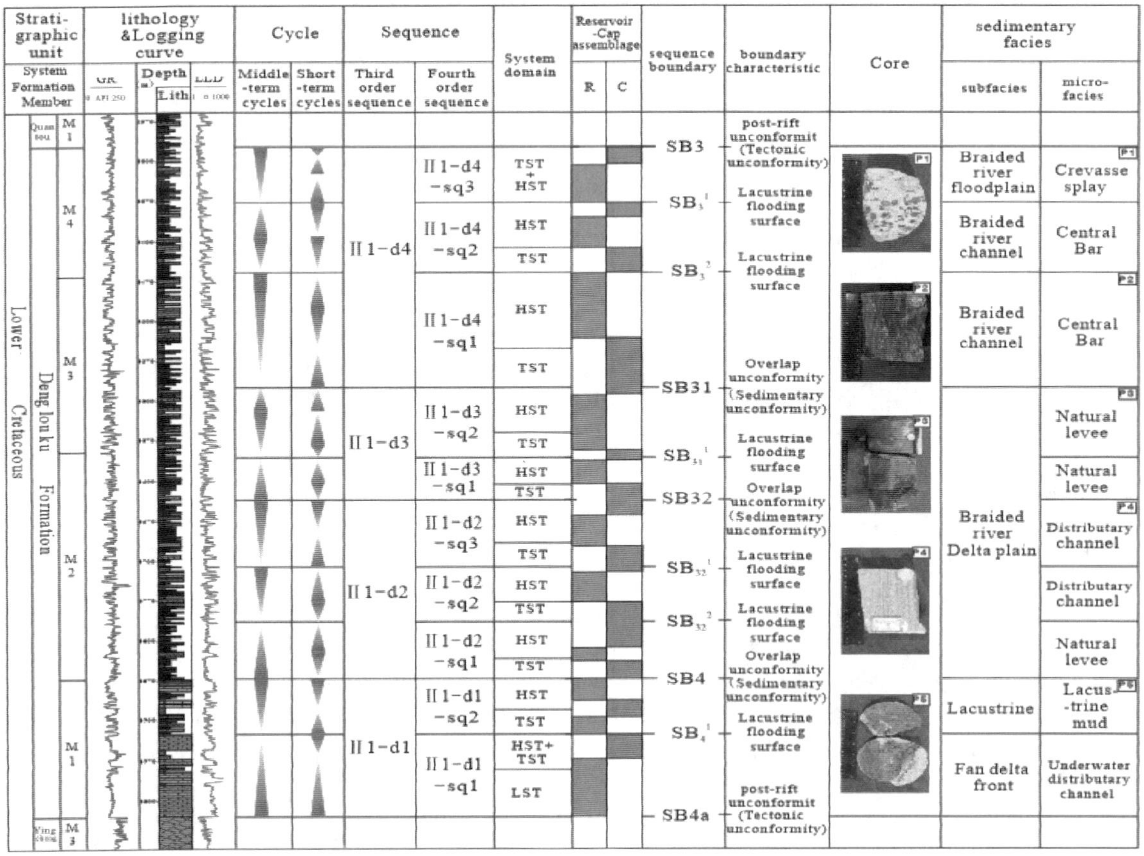

Figure 4. Sequence stratigraphic column for Denglouku group in Xujiaweizi fault depression.

3. Sedimentary facies and sedimentary system characteristics of subdivided layers of Denglouku formation in Xujiaweizi fault depression

3.1. Identification and division of logging facies, core facies and seismic facies

Effective combination of seismic, core and logging data can be used to analyze and reconstruct the sedimentary environment. Although the boundaries reflected by different information sources are different in the process, good results can also be obtained if the information can be effectively

IOP Conf. Series: Earth and Environmental Science 1087 (2022) 012073 doi:10.1088/1755-1315/1087/1/012073

combined (Figure 5). For this study, firstly, the lithofacies and sedimentary microfacies are established through core observation, the relationship between lithology and logging is studied, then the logging facies is established, and extrapolated to the non-coring well section. Finally, use the seismic data there built the "point-line-plane" research mode. Combined with the seismic attribute, fossils, grain size and other indicators, the sedimentary facies, subfacies and microfacies are divided, and established the seismic geological identification template which is suitable for study area. (according to Lemons D R, 1999, WU Yin-ye, 2015, ZOU Bolun, 2019) [7, 8, 9] (Figure 6).

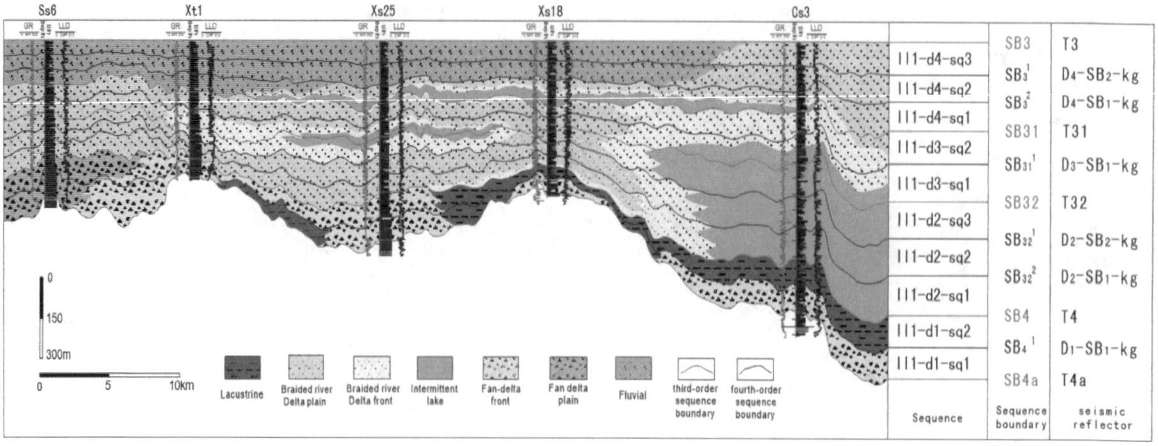

Figure 5. Sequence boundary & sedimentary facies section for Denglouku group in Xujiaweizi fault depression.

Flood plain: the core shows dark color mixed with purple red mudstone mixed with siltstone. The sandstone has fine particle size, mainly composed of siltstone. The bedding types are diverse, and the ripple marks are relatively developed. There are also cross bedding, oblique wavy bedding and horizontal bedding. The logging shows a large set of purplish red mudstone, argillaceous siltstone, domestic gray, light gray siltstone or argillaceous siltstone. The logging curve is mainly bell shaped and toothed-bell shaped in the sandstone section, and toothed in the mudstone section. The seismic character is medium amplitude and relatively continuous reflection. The seismic attribute shows a low bright zone. From D_2sq_1 to D_3sq_2.

Fan delta: the core shows grayish green gravelly sandstone, with medium gravel diameter, coarse particle size, low maturity, medium-poor sorting, roundness of circle-sub ridge, sub circle and sub ridge, and high content of filler and cement. The logging facies is represented by a large section of gray sandy conglomerate, and the logging curve is flat-toothed shape. The seismic character is intermittent reflection with medium-weak amplitude, and the seismic attribute is stripped distribution. This facies are extensively distributed in D_1sq_2 in the west of study area. (The fan delta is different from the delta. The slope of the fan delta is steep. It directly enters the lake at the tail of the alluvial fan. It has small scale, poor lithology sorting and low particle maturity. The delta is generally in gentle areas, with slow slope, low speed and large scale, high sediment sorting and maturity.)

Fluvial facies: Thick purple-gray siltstone, purple silty mudstone and mudstone. The well logging curve at the channel sediment is box type or dented box type, and the rest part are low amplitude micro dentate. There is an obvious "mud wrapped sand" feature on the profile.

Braided delta: the core shows there are mainly gray sandstone mixed with small black mud gravel, which is characterized by sand wrapped mud. The sediment particle size is medium, the sorting is good, oblique bedding is occured, and trough cross bedding is occasionally seen. The logging shows thick conglomerate and pebbly sandstone. The logging curve is bell shaped or toothed-bell shaped and finger-toothed. It is progradational reflection with medium amplitude and intermittent reflection in seismic, and the seismic attribute is stripped distribution. These facies are widely distributed in D_2sq_1 to D_3sq_2 of the study area.

Sedimentary subfacies	Seismic attribute	Seismic characteristics	Core	Sedimentary characteristics
Floodplain			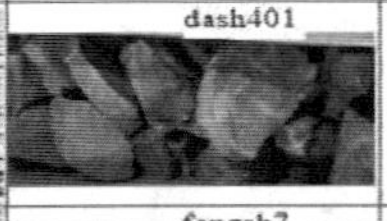	Purple mudstone is mainly deposited, blank-weak amplitude, weak reflection characteristics, discontinuous, low-energy sedimentary environment
Fan delta				Characterized by Glutenite sedimentation, progradational, medium amplitude reflection, representing a stable and medium-high energy environment
Fluvial facies		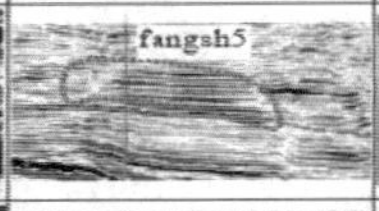		Fine sand-silt deposition, positive cycle, low frequency reflection, medium-weak amplitude, low continuous and parallel reflection characteristics
Braid delta		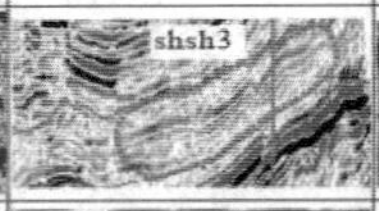		Mainly coarse-medium fine sandstone , retrogradation, low continuity reflection
Intermittent lake			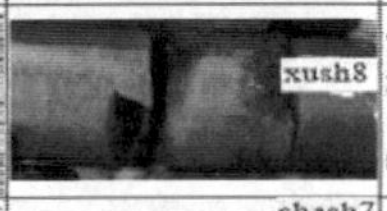	The lithology is analogue to fluvial facies. Sheet shape, and medium-strong amplitude continuous seismic facies.low-energy and stable environmental sedimentation
Lacustrine facies		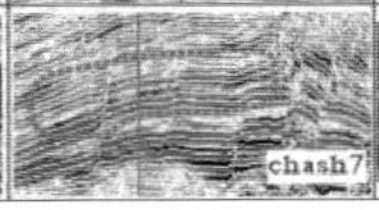	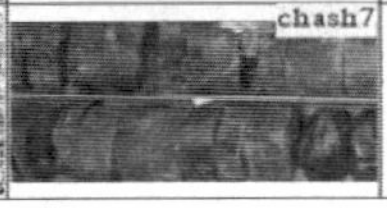	Mainly mudstone deposition, sheet shape, medium strong amplitude, continuous and parallel, low-energy sedimentary environment

Figure 6. Characteristics of lithology and sedimentary facies in Denglouku group of Xujiaweizi.

Intermittent lacustrine: according to the analysis of sedimentary characteristics of the study area, it is proposed for the first time that intermittent lacustrine (we named it) sediments are occured in this area. The sedimentary characteristics are obviously different from the common lacustrine deposits with dark mudstone, purple mudstone is widely distributed in intermittent lacustrine deposits in this area, which proves that it exists in the oxidizing environment. It is speculated that the climate and environment in the geological history period of Denglouku 2nd and Denglouku 3rd are dry, after a large amount of unloading in the flood period, the water body evaporated in a short time, so there are mudstone with oxidation characteristics widely. Other characteristics in bedding, ripple, grain size, logging and seismic reflection are not very different from normal lakes. The facies are widely distributed in the study area from D_2sq_1 to D_3sq_2.

Lacustrine facies: the core shows that it is mainly dark mudstone, locally mixed with thin sand siltstone. The thickness of single sand layer is mostly less than 1m, mainly flake and lenticular sand bodies. Horizontal bedding and corrugated bedding are particularly developed. The logging facies in the sandstone section is toothed inverted funnel shaped, and the mudstone section is flat toothed linear shaped. The seismic profile is characterized by continuous seismic reflection with medium amplitude. The seismic attribute shows a patchy low bright zone. The lacustrine facies in this area are the dominant facies in D_1sq_1 to D_1sq_2. It is also locally distributed in D_2sq_1, D_3sq_1 to D_3sq_2.

3.2. Distribution characteristics of sedimentary system

D_1sq_1 in the lower part of Denglouku 1st, is the stratum with the smallest area in the sedimentary period, the distribution range of D_1sq_2 in the upper part ofDenglouku 1st has been expanded, which is controlled by four sedimentary source systems from northeast, northwest, southwest and southeast. Lacustrine facies and fan delta facies are dominant. The stratum area in this period extends slightly to the West and North, and extends from the middle to the south to fill the whole area, and the base level is declined.

The sedimentary range in the Denglouku 2nd is further expanded, basically covering the whole area, and the deposition is the thickest in the south. Steep slope breaks are dominant and gentle slope zones are less. The lake basin was gradually reduced from bottom to top, the deposition of braided river delta expanded, and the plain channels of braided river delta were widely distributed. The source in D_2sq_1 and D_2sq_2 periods mainly came from the East and south, and was controlled by the sedimentary systems in the north, northeast, northwest and southwest in D_2sq_3 period. The sedimentary boundary moved to northward, and the base level generally shows a downward trend.

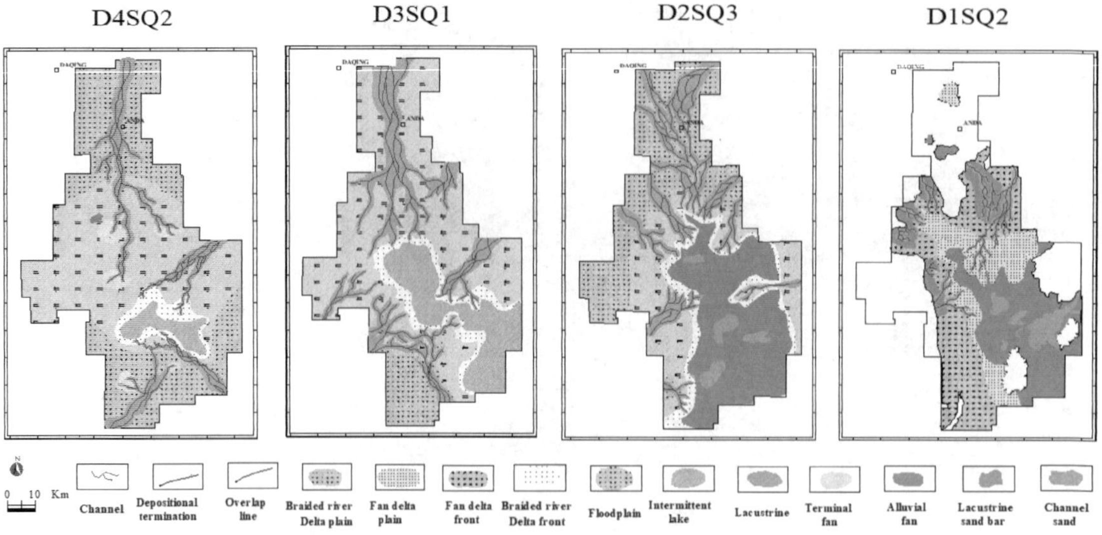

Figure 7. Evolution of sedimentary system in Denglouku group, Xujiaweizi fault depression.

D_3sq_1 and D_3sq_2 in Denglouku 3rd are distributed throughout Xujiaweizi area, and the thickness changes slowly. The strata in the southern and western regions are the thickest, and the strata in other parts gradually become thinner to the edge. The sedimentary environment of this period was controlled by the source from the north, northeast, and east, southwest and northwest. The range of gentle slope zone is gradually expanded. Due to the strong volcanic eruption and the increase of CO_2 content in the atmosphere during the period of underlying Yingcheng Formation, the sedimentary period of the first member of Denglouku formation - the second member of Denglouku formation has become the first high-temperature period in the history of basin evolution. Since the deposition of the third member of Denglouku formation, the climate has been gradually dry (according to Huang Qinghua, 1997). The Denglouku sedimentary period in the north of Songliao basin has hot and arid climate (according to Zhang Liping, 1994). It made the water body evaporates rapidly in a short time, and intermittent lacustrine sediments with distinctive characteristics of oxidizing purple mudstone are widely distributed, reaching the maximum range of the development period of the whole Denglouku formation [10]. Braided delta is also widely distributed in this period. The base level also rose to the highest level in the whole period.

D_4sq_1, D_4sq_2, and D_3sq_2 sequences of the Denglouku 4th are distributed throughout Xujiaweizi area, and the thickness changes slowly. The thickest part of the stratum is located in the middle of Xujiaweizi, and the stratum thickness in other areas gradually decreases to the edge. The inner gentle slope zone is further expanded, the sedimentary facies in the whole area is mainly braided river facies, the whole region is dominated by braided river facies, and the braided river channel is distributed in the northeast, which proves that the main source comes from the northeast. The Western source brought small braided river delta plain deposits. The base level drops to the lowest, and small shallow lakes and braided river delta plain deposits are found locally. The lake water body is very shallow, and there is basically no long-term water body, which is characterized by short-term, scattered and shallow water (Figure 7).

4. Conclusions

In this study, the Denglouku formation in Xujiaweizi fault depression is refined from the traditional 4-parts third-order sequence to 10-parts fourth-order sequence. The division scale of sequence stratigraphy also ranges from 150-180m of the third-order sequence to 70-80m of the fourth-order sequence. The third-order sequence flooding surfaces are SB_{33}, SB_{32} and SB_{31}, and the fourth-order sequence flooding surfaces are D_1-SB_1, D_2-SB_1, and D_4-SB_1.

The characteristics of sedimentary facies under the fourth-order sequence are clarified: Denglouku 1st are mainly lacustrine facies and fan delta facies, Denglouku 2nd and Denglouku 3rd are mainly braided river delta facies and intermittent lacustrine facies, Denglouku 4th are mainly braided river sedimentation, and shallow lakes and braided river delta plain sedimentation are found locally.

The sedimentary system under the fourth-order sequence framework is restored: Denglouku 1st had a short-term source, mainly discrete and coarse-grained sediments, and gradually forms River, lake and delta facies sequence system upward. Lacustrine facies were well distributed in Denglouku 2nd, the source was mainly from the East and North. The source of Denglouku 3rd and Denglouku 4th is in a circular belt, dominated by river, shallow Delta and scattered intermittent lake sediments.

And the study also restores the process of base level falling-rising-falling that is the process of Water regressive - water transgressive - water regressive.

Reference

[1] ZHAO Guo-lian, HE Shun-li 2002 Sequence development characteristics of Xujiaweizi Rift Basin in Songliao Basin *[J] Northwestern Geolog* **35(1)** 28-40

[2] LI Xing-wei, SHEN Zhan-feng, LU Shuangfang, et al. 2005 Application of comprehensive assessment method to fracture prediction of Yingcheng formation stratum of Xujia weizi fault depression *[J] Journal of Daqing Petroleum Institute* **29(3)** 10-12

[3] SUN Dong-lia, HU Yu-shuang 2012 Sequence stratigraphic framework of the Denglouku Formation in the north of the Xujiaweizi Depression *[D] Daqing, Northeast Petroleum University* 1-98

[4] WANG Jun, BAO Zhi-dong, WANG Yun-long, et al. 2017 High resolution sequence stratigraphy and sedimentary characteristics of the Member 3 of Qingshankou Formation of Upper Cretaceous in Qian'an area, south Songliao Basin *[J] Journal of Palaeogeography* **19(2)** 328-340

[5] ZHANG Shun 2015 Research Progresses of the stratigraphy and sedimentology for north Songliao Basin *[J] Petroleum Geology and Oilfield Development in Daqing* **34(3)** 1-8

[6] WANG Shi-bo, REN Yan-guang, LIN Tie-feng, et al. 2008 High-resolution sequence stratigraphic framework in Quan 4th Member and Quan 3th Member of Songliao Basin *[J] Petroleum Geology and Oilfield Development in Daqing* **27(5)** 1-4

[7] ZOU Bolun, LI Tonglin, WU Yangang, et al. 2019 Applications of gravity data in identification of faults and tectonic boundaries of a working area in Inner Mongolia *[J] Global Geology* **22(2)** 133-140

[8] Lemons D R 1999 Facies architecture and sequence stratigraphy of fine-grainedlacustrine deltas along the eastern margin of late Pleistocene lake Bonneville, northern Utah and southern Idaho *[J] American Association of Petroleum Geologists Bulletin* **83(4)** 635-665

[9] WU Yin-ye, GU Jia-yu, GUO Bin-cheng, et al. 2015 Petroleum Sequence Stratigraphy: High Quality Reservoir Analysis and Prediction Method *[M] Beijing: Petroleum Industry Press* 20-155

[10] Cross T A 1994 High-resolution stratigraphic correlation from the perspective of base-level cycles and sediment accommodation Proceedings of Northwestern European Sequence Stratigraphy *[J] Congress* 105-123

IOP Conf. Series: Earth and Environmental Science 1087 (2022) 012074 doi:10.1088/1755-1315/1087/1/012074

Research on international project of China's electric power industry standards - taking smart grid as an example

Xiaoming Zhao[1], Lan Cao[1], Cai Zhang[1], Yuanping Huang[1], Chaoyan Lou[2], Liming Zhao[2] and Pengyue Zhang[2,3]

[1] State Grid Zhejiang Electric Power Co., Ltd. Research Institute;
[2] China Jiliang University.

[3] Corresponding author's e-mail:104552065@qq.com

Abstract. International standardization has become more important because of the trend of economic globalization. Whoever has mastered the standards will have the right to speak. In developed countries, standard competition is the apex of all fields, and promoting the internationalization of domestic technology and standards is an effective way to promote the country's market leader and international status. With the continuous improvement of China's market competitiveness, the internationalization of standards related to the power industry has become more essential, therefore, it is urgent to further strengthen the development process of internationalization of standards in China's power industry. Starting from the international standard project in the field of smart grid, this paper studies the difficulties of current international standard project in the field of smart grid, in addition, systematically analyzes the key problems of the standard project in the field of power grid under the internationalization of China's standards. At last, the practical measures are given.

1. Introduction

In the field of standards, there are different international standards organizations in the international arena. Whether it is an organization, a standards body or an enterprise, the process of obtaining the recognition of international standards organizations through its own technical level and applied internationally is called standards internationalization. As far as State Grid Corporation of China (SGCC) is concerned, in its new development plan, the aim of standard internationalization is clearly pointed out, and the existing standards should be continuously improved to enhance the status of the industry and technical superiority [1]-[4]. Through continuous efforts, the SGCC had made progress in the fields of ultra-high voltage and large grid control, reaching the same level of technology as the first-class developed countries, which had laid a good foundation for China's standard nationalization work and promoted the combination of standards and industry. Based on the current situation of the domestic and international development of smart grid, this paper analyzes and summarizes the development strategy of China's standard internationalization, puts forward targeted opinions, and gives appropriate solutions and development suggestions.

2. Background analysis of standards internationalization of smart grid

With the continuous improvement of the technical level of my country's power equipment and the in-depth development of international trade, the difference between my country's power industry standards and international general technical standards is becoming a constraint on the export of my country's power equipment and affecting the participation of Chinese power companies in the construction of foreign power projects. Key factor. Promoting the internationalization of technical standards and enhancing the consistency between my country's power industry technical standards and international general standards have become an urgent matter to promote the internationalization of China's power equipment industry. Research on the development trend of smart grid standardization at home and abroad and standards that are helpful to our country Project research.

2.1. Domestic research status and development trend

Since the rapid development of China's electric power industry, electric power enterprises have actively participated in international standard organizations and carried out the internationalization of Chinese standards, among which SGCC is the leader. SGCC attaches great importance to the work of technical standards. Under the guidance of relevant state ministries and China Electricity Council (CEC), the company's technical standard work mechanism had been improved, the standard system had matured, the process of standard making and revision had been more effective, the standardization research started reaping, and standards internationalization had made innovative breakthroughs. While supporting the development of the company and the power grid, SCGG's technical standards work also helped China to improve its influence and discourse in the field of international standardization. The outstanding experts selected by SGCC had served on International Electrotechnical Commission (IEC) Advisory Committee on Electricity Transmission and Distribution (ACTAD), Ultra-High Voltage (UHV) Strategy Group, Smart Grid Strategy Group and Electric Vehicle Strategy Group. Since 2010, SGCC had led the development of a total of 84 international standards. Figure 1 shows the IEEE Smart Grid Standard System.

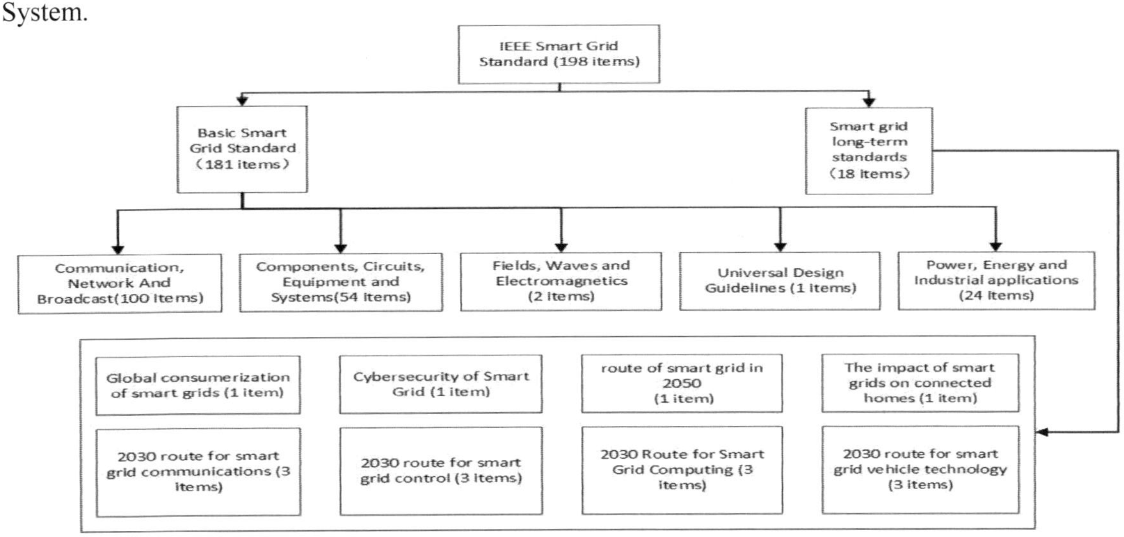

Figure 1. IEEE standards system and its industry distribution.

In October 2008, IEC TC115 International Electrotechnical Commission on High Voltage DC Transmission was established, which is the first new technical committee proposed by China independently. After several years of efforts, TC115 had made the world widely recognized China's UHVDC transmission technology. In September 2011, IEC PC118 Smart Grid User Interface Project Committee, the second one proposed by China, was established. In accordance with the prescribed procedures, PC118 organized and coordinated experts from member countries, and with joint efforts, promoted the effective development of smart grid user interface standards, which were eventually formed into IEC international standards. The establishment of IEC TC115 and IEC PC118 had enabled China to take the initiative to lead in setting international standards. In the field of UHV and smart grid technologies, remarkable results had been achieved through continuous standardization work, and China became the first country to establish a complete standard system in the world. UHV AC voltage had been internationally recognized and became the international standard voltage, which has strongly promoted the transformation of the development pattern of SGCC.

2.2. Foreign research status and development trend

As the most influential standardization organizations in the world, IEC, International Organization for Standardization (ISO) and International Telecommunication Union (ITU) had set up special standard strategic research departments (working groups) and carried out a lot of strategic research work in the fields of standard system and roadmap of electric power standards, etc. In electrical, electronic and related fields, IEC was responsible for the setting of international standards and the establishment of international mutual recognition system for conformity assessment. Its work areas also involved global hotspots with huge market potential and strategic emerging industries. In the standardization work of smart grid, in order to implement the strategic objectives, the strategies adopted by IEC mainly include: establishing the Smart Grid Strategic Working Group SG3, developing the standard system and technical route of smart grid, organizing and coordinating the internal technical committees of IEC and institutions outside IEC to jointly develop the technical standards of smart grid; since then the strategic working group had also upgraded models in research by absorbing European research results to achieve the full coverage of the field. ISO and ITU had also set up special strategic working groups to plan and carry out the development strategy of related standards [5].

With a large number of distributed energy sources, electric vehicles, energy storage, demand response devices connected to the grid, and complex power electronics, control and communication technologies applied in the grid, the U.S. grid was gradually built into a highly interconnected information-physical convergence system. With the aim of informatizing and intelligentizing the power system, the U.S. devoted more attention to upgrading its power network infrastructure and making significant use of emerging information technologies. Similar to other areas, the U.S. was still adopting a strategy of exporting "American technology concepts" in this field. In the U.S. domestic and international standardization activities, the U.S. actively developed and exported technologies to improve its competitiveness in the smart grid field through a combination of internal and external activities, and to firmly grasp the core of these technologies. Unlike the U.S. "export-oriented" concept, Japan adopted the strategy of "following in some areas and leading in the dominant areas" [6]-[8].In the field of energy storage technology and home energy network, Japan had established a secretariat or served as conveners to play a leading role in exporting technologies and methods to IEC and maintaining its influence in the international arena; while in the field of smart grid, domestic research and development and application were promoted mainly through participation in international standardization activities, and expert strategic advisory committees had been set up to conduct thematic studies. In addition, key technology areas had been selected to focus on, taking into account its own domestic conditions and the results of international competition of enterprises.

The predecessors of the IEEE were the AIEE (American Institute of Electrical Engineers), founded in 1884, and the IRE (Institute of Radio Engineers), founded in 1912. In 1963, the AIEE and IRE announced their merger and the IEEE was formed. Today, with over 400,000 members in more than 160 countries, IEEE is at the forefront of high-technology fields such as aerospace, biotechnology, computer engineering, new energy sources, geographic information systems, neural networks and wireless communications.

The IEEE Standards Association, IEEE-SA, is the world's leading standards body and has established strategic partnerships with a number of international standards organisations, including IEC, ISO (International Organisation for Standardisation) and ITU (International Telecommunication Union). The IEEE-SA has produced and revised a total of 6,523 standards, of which 1,838 are currently in force, covering aerospace, bioengineering, energy and power, transportation and other fields, Figure 1 shows the IEEE standards system and its industry distribution.

3. Strategic analysis of international project of smart grid standards

Standard internationalization is an inevitable requirement and important measure for State Grid Corporation of China to innovate its international development model and build a world-class enterprise. Looking at the world's first-class companies, more than 90% of multinational companies regard technological innovation as the main content of their corporate strategy, and master the right to speak in the market by transforming leading technologies into international standards.

3.1. *Strategic analysis*

For the SGCC, standards internationalization is the new development mode and the inevitable way to build a world-class gird. By observing cases of first-class enterprises, it is not difficult to find that the internationalization of new technology standards is the top priority of enterprise strategy, which is to transform the technology into standards so as to enhance superiority. In terms of international standards, SGCC had been planning its work with strategy. Aiming to become a key player and leader in the international electrical field, SGCC fully implemented its strategic work and improved international recognition in related fields, while maintaining spiral but steady progress [9]-[11]. In *the 13th Five-Year Plan*, it was clearly pointed out that the international standard setting work should be carried out to improve the standard setting level, enhance the international status and grasp the dominant power. Through the standard "Going Out" and then guide the development of exporting products, technologies and services.

In addition, it was clearly mentioned that the cooperation and exchange of standards with the regions along the route is the key construction content in the strategy of One Belt One Road Initiative (OBOR). And SGCC had made a positive response in participating in the construction of OBOR through investment to projects under construction overseas, in order to guarantee basic power supply relying on its own standard advantage [12]. This has greatly accelerated the process of standards internationalization.

3.2. *Tactics Research*

Tactics are especially important in the process of standards internationalization, because good tactics could reduce detours. Under the overall standard internationalization strategy of SGCC, guided by strategic experience from different fields and directions, the high point of international standards should always be set as the goal. The most of strengths should be made, to focus on the changes in international standards; reasonable introduction of domestic and foreign innovative technology, from which the relevant technologies conducive to the establishment of international standards in power grids are extracted, so as to learn and improve; active participation should be involved in standards internationalization-related interactions. Important positions in ISO/IEC and other major important international standards organizations should be strived for, and more TC\SC secretariats and serve as chairs and vice chairs should be host actively [13]-[14]. In addition, the development of SGCC and its own needs should be considered, to determine the key standard project areas and make substantial strategies for each area, so that gaps in the field of international standards for power grids could be filled. And the internationalization of mature domestic standards could be carried out in appropriate time [15]-[16].

3.3. *Key tasks for standards internationalization in different fields*

3.3.1. Intelligent power generation. For intelligent power generation, there were already 7 standards under IEC SC 8A, involving power prediction, grid compliance evaluation, voltage and frequency control, etc. The standards internationalization in these field should be actively promoted and more attention should be paid to international standards in the field of power generation; for smart hydropower plants, the IEC "Guide for Smart Hydropower Field" led by SGCC was under preparation, but the gap between international standards and domestic standards should be fully understood, it still needed actively study and continuously improvement, so that the draft to the committee could be submit at an early date.

3.3.2. Ultra-high voltage and power transmission field. For UHV AC, emphasis should be placed on the engineering examples of UHV AC transmission system, stability guidelines, substation reactive power compensation device selection and application, system maintenance, special tests and other aspects to submit new international standard proposals. For UHV DC, there are 5 developing standards, including "High voltage direct current (HVDC) substation audible noise", etc. The standard should be actively promoted to exchange and learn together with more foreign experts. For flexible AC transmission, the implementation of the technical report Performance of unified power flow controller (UPFC) in electric systems and IEEE UPFC series of standards was led by China. For flexible DC transmission, the IEC technical standard Performance of voltage source converter based high-voltage direct current transmission led by SGCC could be voted on and released this year; CIGRE SC B4.72 DC grid benchmark models08 DISTRIBUTION & UTILIZATION1 special planning for electricity supply 2020.03for system studies led by SGCC had been basically completed and was expected to be released in 2020.

3.3.3. Smart electricity distribution field. For virtual power plants, on the basis of the international standard projects of IEC virtual power plants, international standards would be proposed in due course, such as virtual power plant architecture design, data model and information interaction, operation control, etc. For DC power distribution, it is planned to organize the development of specific standards with coordinated contents. In order to accelerate the work of the standard system, the research results of international working groups (CIGRE SC6.31 International Working Group on DC Power Distribution Research and IEC/SEG4 Low Voltage DC Power Distribution Working Group) can be relied on to establish an open and common system standard system to help related industries in the international layout.

3.3.4. Smart grid field. For smart grid user interface, developing standards included IEC62939-3 Smart grid user interface Part 2: Energy interoperation services, etc. For electric vehicle charging and discharging, developing standards included IEC 61851 Electric vehicle conduction charging system series and IEC 63110 Electric vehicle charging and discharging facility management protocol, and the subsequent part of IEC 63119 Electric vehicle charging roaming information interaction would be submitted to a new international standard proposal.

3.3.5. Scheduling control and transaction field. Series of domestic standards based on D5000 smart grid dispatching technology support system were already mature, including a series of dispatching automation standard system, such as "Smart Grid Dispatching Technology Support System" (Q/GDW 680) and "Regional Smart Grid Dispatching Technology Support System Application Function Specification" (Q/GDW Z 461-2010). Some of these standards had the potential to be upgraded to international standards, mostly in the direction of power system, including dynamic message coding of power system model data, dynamic message coding of power system, simple service interface of power system, common service protocol of power system, common real-time communication service protocol of power system and other standard specifications.

4. Summary & Outlook

By summarizing the development status of smart grids at home and abroad and analyzing the development situation of smart grids in my country, combined with the development strategy of smart grids in China, nine key tasks for promoting the rapid development of smart grids in my country in the future are proposed as follows:

(1) Promote large-scale grid integration and consumption of renewable and clean energy.

Improve the policy system to promote the development of the renewable energy industry, strictly control the development and construction of the market, and coordinate the development of the market to achieve the coordinated development of various renewable energy; accelerate the development of renewable energy active support intelligent control and cluster optimization operation control technology and equipment research and development , Realize the reliable grid connection of large-scale renewable energy, and promote the substantial increase in the capacity of grid-connected renewable energy; accelerate the development of the core technology and equipment for flexible interconnection of large-scale grids, and build an energy optimization configuration platform that meets large-scale and long-distance transmission. Solve the problem of large-scale optimization and safe consumption of renewable energy.

(2) Promote the optimization and complementarity of multiple energy sources.

IOP Conf. Series: Earth and Environmental Science 1087 (2022) 012074 doi:10.1088/1755-1315/1087/1/012074

Give full play to the regulating capacity of renewable energy such as hydropower and solar thermal, promote the multi-energy complementation and application of hydropower, wind power, photovoltaic, solar thermal and other renewable energy, encourage the development of multi-energy complementary integration optimization technology and equipment research, and accelerate the enrichment of renewable energy Demonstration and construction of large-scale renewable energy multi-energy complementary projects in the region to reduce wind and wind power consumption in grid-connected areas; speed up source-grid-load sensing and coordinated control, energy and information infrastructure integration equipment, distributed energy management and other key technologies Research and develop, demonstrate and apply large-scale source-network-load friendly interaction and other advanced technologies to form a friendly and interactive source-network-load operation control mode; improve the interaction and balance of multiple energy supply and demand, explore multi-energy complementary micro-energy grids to achieve new renewable energy supply Mode, promote the construction of new energy microgrid application demonstration projects with renewable energy as the mainstay and diversified and complementary distributed power sources, and improve the comprehensive utilization efficiency of terminal energy.

(3) Improve the safe and stable operation of large power grids.

Accelerate the construction of a new generation of intelligent dispatch control system demonstration projects, explore the construction of a wide-area power generation joint control system, build a panoramic security defense system, accelerate the construction of online and real-time analysis technologies and coordinated control protection systems, and encourage the establishment of modern energy under the conditions of big data The complex large-scale power grid simulation center supports the application of high-efficiency and accurate power system simulation technology that meets the application of large-scale intermittent energy, distributed energy, intelligent interaction and large-scale power electronic equipment applications, and realizes the improvement of large-scale power grid security and stability.

(4) Promote the application of new high-efficiency energy storage devices.

Cooperate with the National Energy Strategic Action Plan to promote the demonstration application of energy storage technology, promote the demonstration application of energy storage technology in the field of renewable energy, and achieve breakthroughs in the market scale, application fields and core technologies of the energy storage industry. Enhance the flexibility and economy of energy storage peak shaving, improve energy utilization efficiency, encourage the development of distributed and microgrid energy storage application research and research, accelerate the development of large-scale energy storage module design and manufacturing and other technologies and demonstrations, and explore and master large-capacity Strategic forward-looking technologies such as new energy storage and energy conversion promote the efficient conversion of multiple types of energy, and promote the large-scale conversion, networked storage and multi-form consumption of renewable energy power.

(5)Vigorously implement the energy substitution strategy of "substituting electricity for coal and electricity for oil", promote low-voltage frequency conversion, green lighting, enterprise power distribution network management and other electric energy substitution technologies, promote "home electrification" in residential areas, and in urban heating and industrial and commercial fields Implement large-scale "replacement of coal (gas) with electricity" projects to increase the proportion of electric energy in terminal energy consumption, give full play to the advantages of convenient, safe, clean, and high-efficiency electric energy, and promote the development of clean energy.

With regard to the development of electric vehicles, we will accelerate the construction of fast charging networks for electric vehicles and the Internet of Vehicles service platform to realize the interconnection of charging facilities between cities and between cities.

With the implementation of the national "One Belt One Road" strategy, under the strong promotion of the State Grid Corporation, the internationalization of the power industry has been accelerated, enabling my country's power equipment to play an increasingly important role in the international community. The future international development of the power industry will not only be the export of products, but also the transition from "Made in China" to "Created in China". It will lead the formulation of technical standards with technological advantages and enhance my country's right to speak in the process of international standard formulation. Relying on advanced technology, export China's leading power technology standards to international standards, and enhance my country's technical competitiveness in the field of international electrical engineering. Technical standardization has become an important part of the national governance system, and technical standardization work has become a national strategy. The internationalization of my country's technical standards has a long way to go.

Acknowledgments
Project Supported by Science and Technology Project of State GridZhejiang Electric Power Co., LTD (Project Name: An empirical study of IEC/ISO international standard application, Number: 5211DS20008B).

References

[1] Zhang J, Ye J Y, Li B, Qi B and Gao Z Y 2020 Research on the internationalization strategy of smart grid standards [J] Power Supply **37(03)** 3-9
[2] Lu H Q, Yang M C and Zhang D D 2019 Analysis on the internationalization strategy of standards in China's electric power industry *[J] Electrical Technology* **20(09)** 7-9
[3] Wang X, and Wang Y 2021 State Grid Xuzhou: Standardized management to create a "four-in-one" smart grid *[J] New Finance (Government Finance)* **(12)** 37-38
[4] Qi L 2021 Recommendations for the development of communication technology standardization *[J] Brands and Standardization* **(06)** 10-12
[5] Wang P, Liang Z, Hou J J 2019 Overview of Standard Knowledge System *[J] Standard Science* **(03)** 45-52
[6] Cai X M 2020 Promote the development of smart grid [J] Electrical Appliance Industry **(06)** 1
[7] Yang Y 2021 Sharing the wisdom and achievements of digital transformation——Huawei actively participates in the formulation of smart grid standards to promote industry development *[J] China Electric Power* **(10)** 66
[8] Chen G P, Meng J B, Zhang Y, Ma L and Zhao X D 2020 Research on Smart Grid Dispatching Automation Standard System*[J] Science and Technology Wind* **(24)** 82+86
[9] Ni Y K, Zhao D W and Zhou Y 2020 The role of basic information standardization and standardization in the construction of smart grid *[J] Hubei Agricultural Mechanization* **(14)** 11-12
[10] Zhang J, Dai P, Wu T J, Yu H Y, Li B and Qi B 2020 The design and demand analysis of the new generation of smart grid technology standard system architecture *[J] Automation of Electric Power Systems* **44(09)** 12-20
[11] Zhang J and Li B 2020 The new generation of smart grid technology standard system supports the high-quality development of power grids *[J] Power Supply* **37(03)** 2+1
[12] Chen W, Yao L, Zheng B M and Chen X 2019 Research and implementation of automatic testing technology for relay protection information standardization *[J] Electromechanical Information* **(27)** 6-7
[13] Jin G Q, Chen Z H 2019 International comparative study on the basis of quality technology in the field of smart grid *[J] Quality and Certification* **(09)**
[14] Zhang G S 2019 Recommendations for the standardization of smart grid communication technology *[J] Digital Communication World* **(01)** 69.3
[15] He W S 2017 The development trend of the internationalization of China's power standards [J] China High-tech **1(05)** 24-26
[16] Gao Z Y, Cao Y, Zhang J and Yu Y J 2017 China electric power **50(11)** 42-47+77

Delineating surface sediments physical properties in Ghudai Area, Kuwait

Abeer Al-saleh[1,2], Modi ahmed[1], Noor Al-Dousari[1] and Teena William[1]

[1] Environment & Life Sciences Research Center, Kuwait Institute for Scientific Research, Kuwait.

[2] Corresponding author's e-mail: ahsaleh@kisr.edu.kw

Abstract. There are three main challenges to establishing human settlements in Ghudai area at northern Kuwait Bay, namely; sand encroachment, water floods, and coastal erosion. Therefore, the particle size distribution percentage and statistical parameters in Ghudai surface sediments at the Kuwait Bay coast were determined and delineated in an attempt to display the depositional environment and its impact on the seedling growth of native plants. Particle sizes were assessed using a centrifugal particle analyzer and 373 samples representing the top 20 cm of surficial sediments were taken from the main geomorphological zones around Nitraria retusa and Lycium shawii nabkhas as dominant species in the study area, particle size percentages, and statistical size parameters were determined and delineated. The structural parameter distribution showed the dominance of medium sand sediments, as well as their environmental consequences revealing the unidirectional signature of fluvial flow from the Jal Al-Zur escarpment and wind transportation toward Kuwait Bay. Different discrimination maps of particle size parameters and other factors were plotted. The research shows that particle size parameters are highly vulnerable to slight changes in the pattern of sediment composition where sediments are deposited and reformed in disturbed environments.

1. Introduction

Kuwait has one of the highest frequencies of dust deposition in the world, due to dust storms, rising dust, or suspended dust, Kuwait experienced an average of 255.4 dusty days each year (70% of the year). [1] were the first to investigate the sedimentology and morphology of the study region and nearby areas. Followed by other researchers [2, 3, 4, 5, 6] that employ particle-size properties of deposits to deduce the processes of transportation, and sedimentation environments. As a result, particle size analysis can reveal vital information about sediment provenance, transport history, and depositional circumstances. Direct measuring, dry and wet sieving, sedimentation, and measuring by centrifugal particle analyzer, X-ray sedigraph, and Coulter counter are some of the techniques used in particle size [7, 8]. All of the procedures entail dividing the sediment sample into a number of size fractions, which allows a particle size distribution to be created using the weight or volume percentage of sediment per size fraction [9, 10]. The large variability of statistical factors can be described through various environmental conditions and sources of materials [6]. Each sedimentary ecosystem has distinct particle-size characteristics, or "fingerprints," that differentiate them from sediments from other deposition areas [11]. Natural and anthropogenic factors lately, such as shoreline erosion, deterioration of coastal areas, sea-level rise, excessive land use, and overgrazing, have posed serious threats to Kuwait Bay's coastal zone, particularly in the Ghudai area; These actions have a substantial impact on environmental conservation and sustainable development. Kuwait Institute for Scientific Research has built a new facility that is

IOP Conf. Series: Earth and Environmental Science 1087 (2022) 012075 doi:10.1088/1755-1315/1087/1/012075

regarded as a fairly protected area in Ghudai. However, by analyzing soil data from experimental locations in Ghudai, the overall goal of this study was to evaluate correlations between plant growth and soil particle size distribution. Furthermore, to gain a better understanding of the sedimentological features of mobile sand in the research area in order to measure the heterogeneity between various aeolian deposits.

2. Methodology

Ghudai is a region on Kuwait Bay's northern side, It's at 29°26'51.62"N and 47°46'46.64"E. The area is approximately 416,695 m2. Alluvial fan sediments and fluvial deposits collecting from the mainline valleys descending from the Jal Al-Zur escarpment toward Kuwait Bay make up the geomorphology of this area. An active mobile sand corridor runs through the area Figure 1.

Figure 1. The Ghudai study area.

The goal of this research is to exhibit the distribution of particle size parameters, as well as the depositional environment and its effects on the environment and plant growth, to characterize the inherent textural and compositional variation of Ghudai region surface sediments. A total of 373 samples were collected and the sediments' textural features were determined in an endeavor to characterize them. Centrifugal particle analyzer was used to characterize the texture of bulk sediment samples, particle-size analysis was performed for a representative of the dried samples, using a standard set of sieves mounted on an electric shaker, mechanical sieving was done to all samples, this approach in which a sand sample is shaken through a nest of sequentially finer mesh sieves describes a wide range of features of size. The weight of sand retained on each sieve is customarily converted to a percentage of the overall sample weight, results of the analysis were used to classify sedimentary environments and create maps to determine the best area for plant growth, these maps are quite helpful in identifying the deposition mode of the soil. Statistical size parameters were determined using a distinct approach employing a computer application GRADISTAT for the quick analysis of particle size statistics [12] using the typical measuring procedures Folk and Ward's graphical approach (1957) [13].

3. Results and Discussions

3.1. Particle Size Distribution

The very coarse sand (V.C.S.) Figure 2(a) ranges from 0.004 %to 0.434 % with an average of 7.7% Table 1. the highest percentages of V.C.S. size fraction are recorded in the east and mid of the area. On the other hand, the west of the area is showing the lowest percentages of V.C.S. size fraction. The coarse sand (C.S.) has different distribution to very coarse size fraction, it ranges from0.099% to 0.55% with an average of 25 % Table 1 where the highest percentages of C.S. size fraction are recorded at the east of the area, while the west of the area showing the lowest percentages Figure 2(b). The medium sand (M.S.) has the highest average of 25.8% Table 1, Figure 2(c). while in Figure 2(d) the highest percentages of fine sand (F.S) size fraction is recorded in the west and mid of the area, and the lowest percentages at the east with an average of 22.2 % Table 1. The very fine sand (V.F.S) with an average of 17.2% Table 1, the highest percentage concentrated in the east and mid of the area Figure 2(e). Mud that includes two main size fractions these are (silt and clay) nearly has a similar trend with the lowest average of 2.1% Table 1, Figure 2(f).

Table 1. The average of Statistical parameters and particle size percentages.

Location	Statistical parameters (Folk and Ward 1957)				Particle size %					
	Mean	Sorting	Skewness	Kurtosis	V. C.S	C.S	M.S	F.S	V. F.S	Mud
Ghudai	1.727	1.197	0.156	0.803	7.7	25.0	25.8	22.2	17.2	2.1

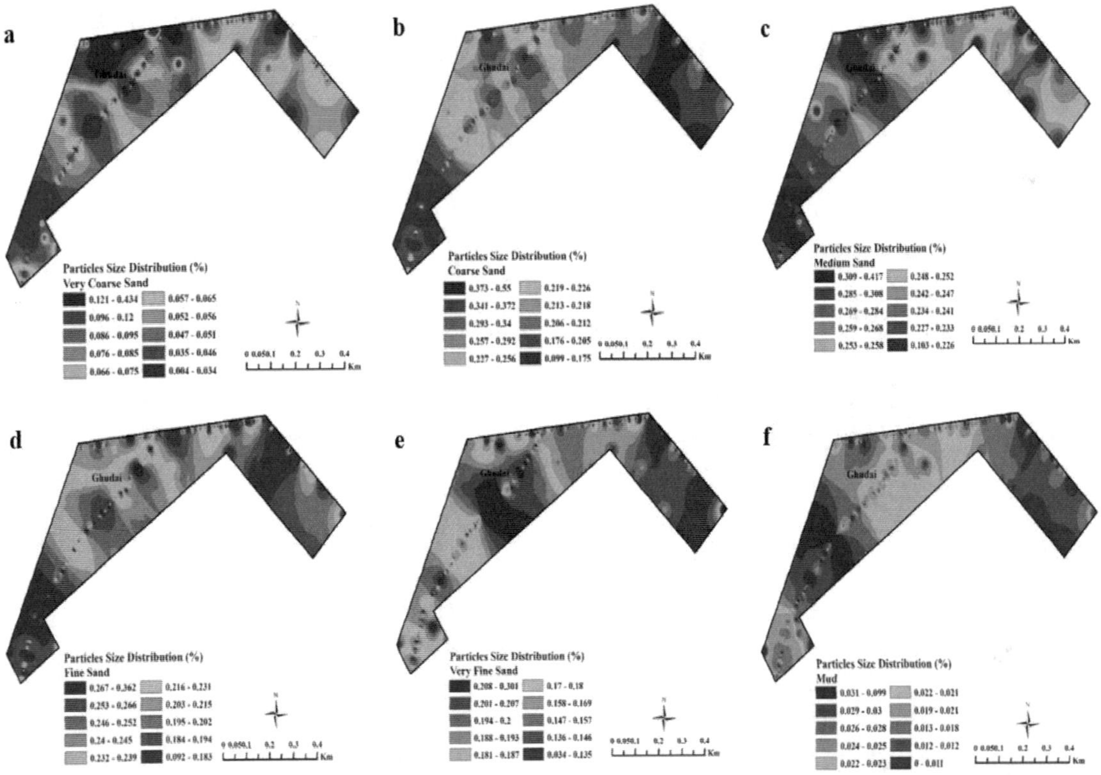

Figure 2. Percentage of Particles Size Distribution of (a) very coarse sand(b) coarse sand, (c) medium sand (d) fine sand, (e) very fine sand, (f) mud.

IOP Conf. Series: Earth and Environmental Science 1087 (2022) 012075 doi:10.1088/1755-1315/1087/1/012075

3.2. Statistical Parameters of the Particle Size Distribution

The coarsest value of mean size is recorded on the west side of the area, where The mean values range from coarse sand to fine sand with an average of 1.7Φ (medium sand). the medium sand is predominant in the area's middle and east, which will benefit plant growth. The sorting (σ_I) values depict that the east and west area is characterized by moderately sorted to poorly sorted character, with an average of 1.197Φ (poorly sorted). The skewness (SK_I) values reveal fine skewed nature for the whole area with an average of 0.156. The kurtosis (K_G) value shows that the west and east area is leptokurtic to mesokurtic, while it is platykurtic to very platykurtic in the mid area with an average of 0.803 Table 1.

3.3. Organic matter, Organic Carbon, Inorganic carbon, Moisture, Electrical conductivity, pH.

Many physical, chemical, and biological aspects of soils are influenced by organic matter, Soil structure and soil compressibility, are some of the features influenced by organic matter. It also has an impact on water storage, nutrient contributions, biological activity, and water and air infiltration average, the organic matter values are limited between 0.239 and 18.2 percent, where high levels in soil samples were noticed in the mid and east regions of the area. Organic carbon has a distribution comparable to organic matter, Carbon and soil structure are inextricably linked, Improved fertilizer and water use efficiency as well as reduced erosion, are all benefits of healthy soils with high carbon cycling, the levels of inorganic carbon, in the form of calcium carbonate, in soil samples, are higher in the mid and west of the area and lower in the east. The water capacity of the soil is known as soil moisture, has an impact on soil formation, structure, stability, and erosion, but it is most important for plant growth, the amount of moisture in the soil varies dramatically based on the location, season, soil type, and depth, the moisture content varies between about 0% and 13% in the study area Figure 3 (a, b, c, d) respectively.

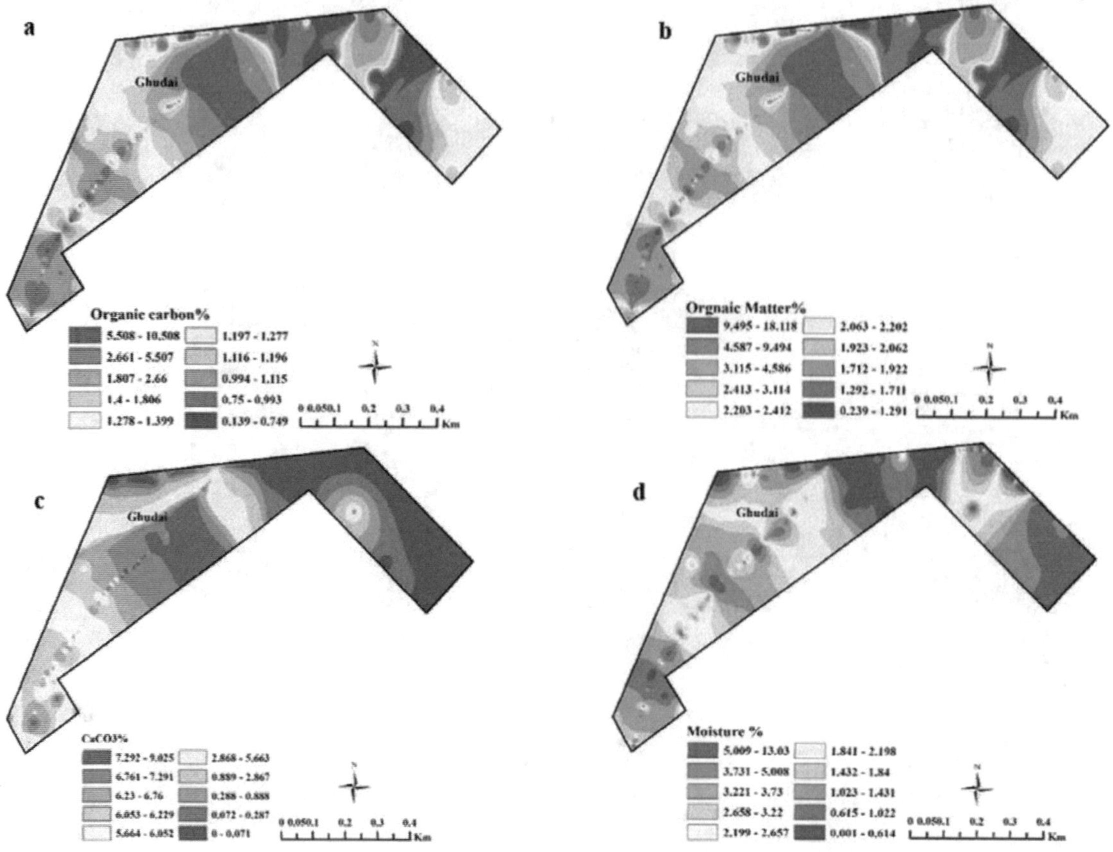

Figure 3. Percentage of Organic matter, Organic Carbon, Inorganic carbon, Moisture

A soil's acidity or basicity (alkalinity) is measured by its pH, soil pH is a distinctive feature that can be used for a qualitative and quantitative analysis of soil characteristics, the highest values of pH may be found in the east area, while the lowest can be found in the mid and west areas. Electrical conductivity (EC) is a type of electrolysis that occurs mostly through water-filled pores, soil (EC) has been used as an indirect measure of the quantity of nutrients available for plant absorption and salt levels, (EC) in the area's mid and east regions values in soil samples are high, some regions in the east and west have lower values Figure 4 (a, b) respectively.

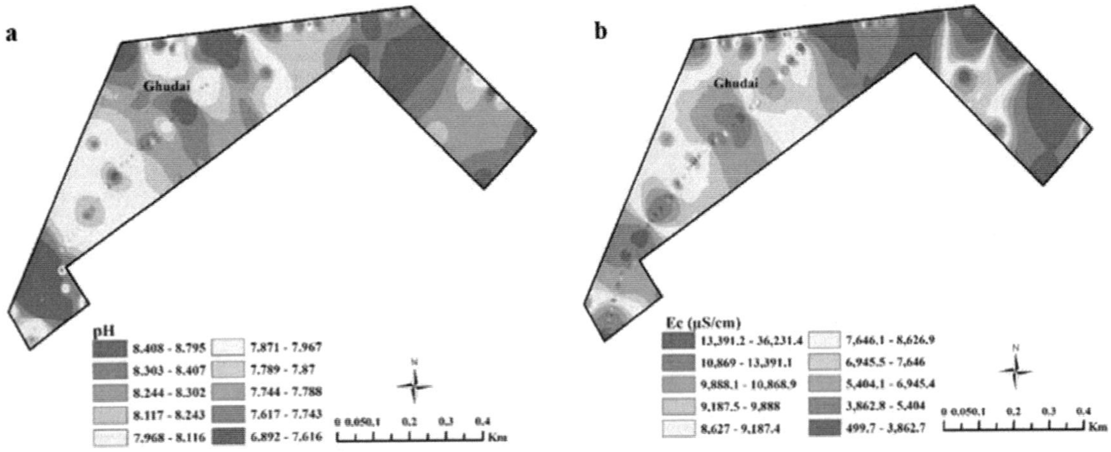

Figure 4. (a) pH and (b) Electrical conductivity.

4. Conclusions

The study demonstrates the high prevalence of moderate sorted medium sand in the research area samples, with near-symmetrical skewness and leptokurtic nature of kurtosis distribution, which corresponds to unidirectional feeding from the Jal Al-Zur escarpment fluvial flow and wind transport toward Kuwait bay, in addition to the domestic influence of the depositional environment's associated with the active environment. The high vulnerability of Particle size parameters to modest changes in the pattern of the composition of the sediments where sediments are deposited and remodeled in disturbed environments. The general results of particle size and statistical parameters indicate that the area is much suitable for the plantation of *Nitraria retusa* and *Lycium shawii.* This conclusion was also supported by many recent studies [5, 9, 14, 10].

5. Acknowledgments

This work was supported by the Kuwait Foundation for the Advancement of Science(KFAS) No. P216-44SE-02, and Kuwait Institute for Scientific Research (KISR) No. EC117C.

References

[1] Kelio, A., and M. Al-Sheikh. 1986. Geomorphological characteristics of nabkhas at Kuwait, Kuwait University press. Kuwait (in Arabic)

[2] Al-Dousari, A. M. 2005 Causes and indicators of land degradation in the north-western part of Kuwait *Arab Gulf Journal of Scientific Research* (1989) **23(2)** 69-79

[3] Ahmed M, Al-Dousari AM, Baby S. 2009. Chemical and morphological characteristics of phytogenic mounds (nabkhas) in Kuwait *Arab Gulf J Sci Res.* **27(3)** 114-123

[4] Ahmed M, Al-Dousari N, Al-Dousari A 2016 The role of dominant perennial native plant species in controlling the mobile sand encroachment and fallen dust problem in Kuwait *Arab J Geosci*

9(2) 134 http ://doi.org/10.1007/s1251 7-015-2216-6

[5] Abd El-Wahab, R. H., Al-Rashed, A. R., & Al-Dousari, A. 2018 Influences of physiographic factors, vegetation patterns and human impacts on aeolian landforms in arid environment *Arid Ecosystems* **8** 97-110 https://doi.org/10.1134/S2079096118020026

[6] Al-Dousari, A. M., Ibrahim, M. I., Al-Dousari, N., Ahmed, M., & Al-Awadhi, S. 2018 Pollen in aeolian dust with relation to allergy and asthma in Kuwait *Aerobiologia* **34(3)** 325-336 https://doi.org/10.1007/s10453-018-9516-8

[7] Al-Ghadban, A. N., Saeed, T., Al-Dousari, A. M., Al-Shemmari, H., & Al-Mutairi, M. 1999 Preliminary assessment of the impact of draining of iraqi marshes on kuwait's northern marine environment. parti. physical manipulation *Water science and technology* **40(7)** 75-87 https://doi.org/10.1016/S0273-1223(99)00586-7

[8] Doronzo, D. M., Al-Dousari, A., Folch, A., & Dagsson-Waldhauserova, P. 2016 Preface to the dust topical collection *Arabian Journal of Geosciences* **9(6)** 1-1 https://doi.org/10.1007/s12517-016-2504-9

[9] Al-Dousari A.M., Ahmed, N. M. Al-Dousari, S. Al-Awadhi 2019 Environmental and economic importance of native plants and greenbelts in controlling mobile sand and dust hazards *International Journal of Environmental Science and Technology* **16** 2415–2426 https://doi.org/10.1007/s13762-018-1879-4

[10] Al-Dousari AM, Ashraf Ramadan, Ayman Al-Qattan, Sara Al-Ateeqi , Hassan Dashti , Modi Ahmed , Noor Al-Dousari , Noof Al-Hashash & Ahmed Othman (2020) Cost and effect of native vegetation change on aeolian sand, dust, microclimate and sustainable energy in Kuwait *Journal of Taibah University for Science* **14(1)** 628-639 DOI: 10.1080/16583655.2020.1761662

[11] Al-Dousari, A.M., Alsaleh, A., Ahmed, M., Misak, R., Al-Dousari, N., Al-Shatti, F., Elrawi, M., and William, T. 2019 Off-Road Vehicle Tracks and Grazing Points in Relation to Soil Compaction and Land Degradation *Earth Syst Environ.* **3** 471-482 https://doi.org/10.1007/s41748-019-00115-y

[12] Blott, S. J., Al-Dousari, A. M., Pye, K., & Saye, S. E. 2004 Three-dimensional characterization of sand grain shape and surface texture using a nitrogen gas adsorption technique *Journal of Sedimentary Research* **74(1)** 156-159 https://doi.org/10.1306/052403740156

[13] Folk, R.l. and W.T.Ward. 1957 Brazon River bar: A study in the significance of grain-size parameters. Journal of Sedimentary Petrology **27** 3-26

[14] Al-Awadhi, J. M., Al-Dousari, A., & Al-Enezi, A. 2000 Barchan dunes in northern Kuwait *Arab Gulf Journal of Scientific Research* **18(1)** 32-40

IOP Conf. Series: Earth and Environmental Science 1087 (2022) 012076 doi:10.1088/1755-1315/1087/1/012076

Urgent need to strengthen unified management of scientific data of natural resources

Chengcheng Zhang[1,3] and Dezhu Gui[2]

[1] Chinese Academy of Surveying & Mapping, Beijing, China
[2] Development Research Center for Surveying, Ministry of Natural Resources, Beijing, China

[3] Corresponding author's e-mail: gischengcheng0914@163.com

Abstract. At present, the institutional restructuring and business restructuring of the Ministry of natural resources have been basically completed, and business integration has become the focus of the work. Earth data science is combination of earth science and data science, and is a frontier cross field of earth science research and a paradigm shift of data-driven scientific discovery. This paper reviews the progress of big earth data research in 2020 from the aspects of earth data science program, scientific data governance, and earth science data sharing. It focuses on the Deep-Time Digital Earth Science Program, Big Earth Data Science Engineering Project and achievements in global data aggregation, authentication, repository, policies, standards, and identification, as well as practices of the earth science data centers, databases, data models, data integration, and data publishing. The paper points out countermeasures and suggestions for promoting scientific management of natural resources.

Scientific data is an important strategic information resource of the country, and the scientific data center is a crucial new infrastructure of the country. In recent years, with the increasing investment in science and technology, we have witnessed a growing capacity for scientific innovation and a "blowout" growth in scientific data. To strengthen and standardize the management of scientific data, ensure the security of scientific data and improve the level of open sharing, the CPC Central Committee and the State Council issued a management measure at the national level for the first time in 2018 - Circular of the General Office of the State Council on Printing and Issuing the Management Measures for Scientific Data [1] (hereinafter referred to as the "Measures of the General Office of the State Council"), supporting the establishment of 21 national scientific data centers. Relying on them, scientific data resources in China and even in the world can continue to be gathered and integrated, and open and shared with the society.

Relatively speaking, the scientific data resources of natural resources are scattered, and the scientific data lack standardization and continuous accumulation; The construction of data center is still weak and its influence is obviously insufficient, which leads to data loss and security risks; In the post-COVID-19 era, the harmonious coexistence of man and nature has become a hot spot in global scientific research, and problems such as insufficient support for natural resource scientific data have become more prominent. It is necessary and urgent to further strengthen the management and sharing of scientific data of natural resources.

1. Connotation and significance of scientific data of natural resources

Scientific data refers to the data generated through basic research, applied research, experimental development, etc. in the fields of natural science, engineering science, etc., as well as the original data and its derivative data obtained and used in scientific research activities by means of observation and monitoring, investigation and inspection [2]. Strengthening the scientific data management of natural resources not only adapts to the new trend of scientific and technological development with big data thinking system research but also becomes an important guarantee for the development of the big data innovation service industry and the prosperity of the application market.

1.1. Scientific data of natural resources is the hot spot global scientific research after the COVID-19 epidemic

In the post-COVID-19 era, scientific research on the harmonious coexistence between man and nature, climate change and environmental evolution has become a global hotspot [3], and the large-scale and extensive restrictions on human activities brought about by epidemic control have also provided unprecedented conditions for studying the impact of human activities on nature. In the research of foreign scholars, keen attention has been paid to the changes in the natural geographical environment before and after the COVID-19 epidemic. It is expected that more scholars will conduct research on changes in various fields such as economy, society, resources, environment, and climate before and after the epidemic. To this end, there is an urgent need to strengthen the collection and collation of scientific data on natural resources in global and key regions, so as to provide the necessary support for scientific research and decision-making after the epidemic.

1.2. Natural Resources Science Data Center is an important part of new infrastructure in the field of natural resources

Every round of technological revolution is accompanied by the birth of new infrastructure construction. In today's era, the scientific and technological revolution represented by the new generation of information technology is in a state of accelerated evolution, which is followed by the accelerated development of new digital, networked and intelligent infrastructure. A scientific data center is an important component of innovative infrastructure among the three major construction contents of new infrastructure [4]. Appropriately arranging the natural resources scientific data center in advance can fully mobilize the role of the big data innovation service industry in developing and prospering the application market. For one thing, the construction of natural resources science data center, which consists of natural resources investigation and monitoring model database, biodiversity sample database, ecological protection and restoration experimental material database, time-space benchmark control point database, etc., will provide perfect experimental verification technical support conditions for the frontier technology development of natural resources, laboratory technology maturity, prototype mass production, etc. For another, under the background that the current economic development needs to improve the ability of natural resources support, increasing investment in industry data centers will help strengthen the steady growth and development potential of natural resources.

2. Trends and requirements of scientific data management

International scientific data sharing began in the 1950s, and after more than half a century of development, it shows the trend that global scientific data management is constantly advancing and China lags behind but quickly follows up, mainly including two aspects：

2.1. Strengthen the construction of rules and regulations

It is the core of scientific data management to regulate the behavior of data in management and social mobility from the legislative level. Internationally, the starting point of scientific data policy formulation in European and American countries is to set up clauses from the perspectives of administration, economy, rights and interests based on safeguarding national security and government affairs and personal privacy, and to promote the sharing and use of data by means of financial support, membership fees, market profit and voluntary crowd-funding.

Domestically, in order to improve China's scientific data management and seize the opportunity of big data development, the "Measures of the General Office of the State Council" is issued, and it is explicitly required to "establish and improve the policies and rules of scientific data management in this department (region)" to make overall planning and construction of the scientific data center of this department (region) and promote the open sharing of scientific data". Before the issuance and implementation of the "Measures of the General Office of the State Council", the departments of the earthquake, public health, agriculture, water conservancy, environmental protection, forestry, etc. have issued the scientific data management measures in this professional field; After the issuance of the "Measures of the General Office of the State Council", the Chinese Academy of Sciences and Heilongjiang, Hubei, Jiangsu, Guangxi, Anhui, Gansu, Kunming and other provinces and cities have successively issued scientific data management measures or detailed rules for their respective regions.

2.2. Strengthening the overall construction of scientific data centers

Relying on the nationally recognized scientific data center, the full life cycle management of scientific data will be carried out, the scientific data resources will be continuously integrated and aggregated, and standardized scientific data will be gradually formed. Internationally, the United States has global representation in the construction of national scientific data centers, especially the data centers in the field of earth science built by the National Aeronautics and Space Administration (NASA) and the National Oceanic and Atmospheric Administration (NOAA). There are 11 scientific data sub-centers in the center [5]. Domestically, the Ministry of Science and Technology and the Ministry of Finance jointly issued the "Administrative Measures for National Science and Technology Resource Sharing Service Platform" in 2018, which established the construction of science and technology resource platform (including scientific data, biological germplasm and experimental materials) relying on various departments of the State Council and local government science and technology management departments, and regularly evaluated the construction effect of the evaluation platform for dynamic adjustment. In 2019, 21 national scientific data centers including ocean and polar regions (one in Forestry and Grass Bureau) were identified with the Notice on Issuing the Optimization and Adjustment List of National Science and Technology Resource Sharing Service Platform [6].

2.2.1. National Marine Science Data Center. The center adopts the co construction mode of "main center + sub center + data node". The main center is the National Marine Information Center, with three sub centers and eight data nodes. Since its establishment, it has continuously improved the organization and operation management structure, strengthened the construction of systems and standards, actively absorbed high-quality data resources, carried out data exchange of science and technology projects, and tackled key technologies of data services in six aspects: data collection and preservation, collection and delivery of science and technology projects, resource mining and application, sharing and opening up and services, research and application of common technologies, and dynamic monitoring, exchange and cooperation at home and abroad, We have innovated the data sharing service mode and basically formed a marine scientific data sharing service system for the whole society. As of April 2022, a total of 12811 real name registered users, more than 1.43 million visits, and nearly 30million data services have been provided [7]. There are 29 industry fields in which users are located, and the trend is increasing year by year. Figure 1 shows the visits of marine scientific data in various industries and fields.

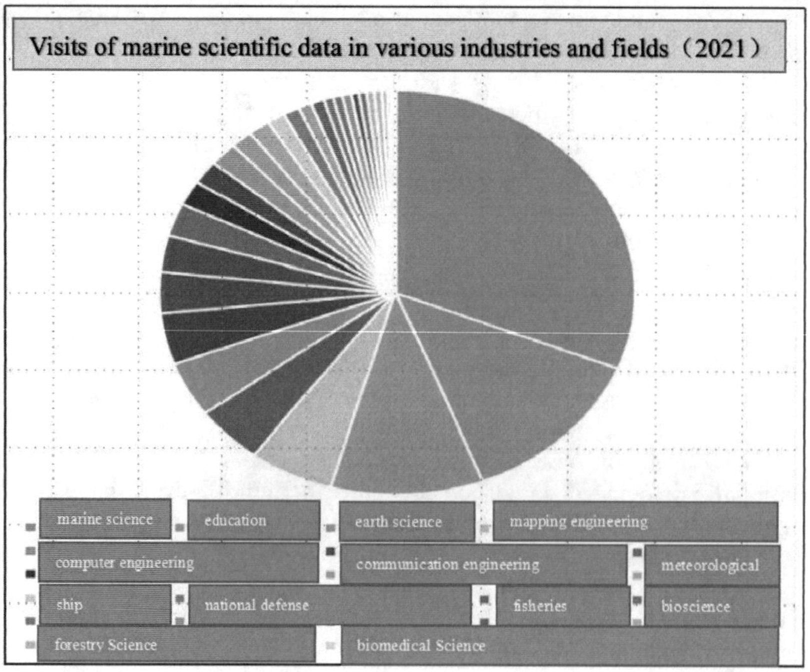

Figure 1. Marine scientific data service volume and user industry map.

2.2.2. National polar scientific data center. Relying on the construction of China Polar Research Center, the center mainly carried out the rescue of polar historical data, and thoroughly sorted out the data files collected and submitted in the history of polar investigation. A total of 80.1tb of historical data were rescued, and the timely rate of data collection and submission of polar scientific achievements was increased from 60% to 80%.

3. Status quo and shortcomings of scientific data management of natural resources

3.1. Research on scientific data of natural resources

In March, 2022, scientific data research in the field of natural resources was carried out by correspondence. A total of 2263 scientific data catalogues were generated and used by 62 units of the Ministry of natural resources. After analysis, they have the characteristics of many professional fields, many characteristic data, many data source methods, large volume, wide service range, high requirements for data security and confidentiality of nearly 1/3, high requirements for achievement management, and many shortcomings such as many data management methods, scattered data storage, and different technical standards.

First, numerous business fields. The professional fields mainly include land, geology and mineral resources, ocean polar regions, wet forest and grass, surveying and mapping geographic information, natural resources policy research and other professional fields.

Second, there are many characteristic data. In addition to digital achievements such as databases, remote sensing images, technical reports, monographs, patents, software and hardware systems, natural resource scientific data also include characteristic physical data such as geological core slices, karst samples, microbial species reference materials. These achievements are not easy to replicate and should be properly managed [8].

Third, multiple data sources. The data comes from national scientific research projects, basic scientific research business expenses of scientific research institutes, research investigation projects in special business projects, natural resources field scientific observation and research stations (there are 5 national field stations, 15 ministerial field stations and some other field scientific research bases in the Ministry of Natural Resources), as well as the original data and its derivative data generated by scientists when writing academic papers.

Fourth, diverse management modes. Scientific data of natural resources are managed by diverse modes such as national-level scientific data centers, operational data management institutions, internal offices of business units and self-management by scientists. In the marine and polar fields, a certain degree of scientific data management and sharing has been carried out in accordance with the requirements of the management mechanism of the Ministry of Science and Technology. Relevant units in the fields of land, geology and mineral resources, water resources, surveying and mapping geographic information, etc. undertake the data collection and management in this field. Since there are no relevant scientific data management responsibilities and rules and regulations to follow, no collection management has been carried out on scientific data in this field.

Fifth, large data volume and a wide range of services. Take the marine area where scientific data sharing has been carried out as an example, the total amount of data collected and managed is 1,115 TB (301TB added in 2021). 7.75TB of data from 11 co-construction units, 183TB of data from the "13th Five-Year Plan" national science and technology projects, 2.69TB of operational observation data, 4.47TB of social resources data, and nearly 20 million station times of international data were collected, forming nearly 47.828 million station times of various standard data sets, and providing nearly 30 million times of data services every year (its users are involved in 29 industries).

Sixth, few technical standards. According to the survey statistics, there is a lack of scientific data standards for natural resources, and only marine scientific data sharing standards. To this end, technical standards such as classification and grading of scientific data, security classification definition and sharing services need to be established urgently [9].

Seventh, scattered data storage. Scientific data of natural resources have not been collected and managed in a unified way. Scientific data generated by subordinate units undertaking national science and technology projects, as well as those generated by units outside the Ministry using our Ministry's financial support, are often remitted to the national scientific data center built by relevant units of foreign ministries and commissions according to the requirements of the Ministry of Science and Technology but are not remitted to relevant data management units of subordinate units. In view of the large amount of scientific data remaining outside the supervision, there are security risks and data islands have formed, restricting the application of scientific data.

3.2 There is a double gap in the system and organization of scientific data management of natural resources

First, imperfect rules and regulations. According to the survey statistics, there are 69 sets of regulations related to natural resources data management, but no rules and regulations specifically for scientific data management of natural resources have been established. Under this circumstance, the subordinate unit calls for the introduction of relevant management rules and regulations as soon as possible (including the scope of work, division of responsibilities, workflow, exchange management, sharing and confidentiality, assessment and evaluation of scientific data management), in which specific provisions in the natural resources field and the contents to be detailed and implemented in the "Measures of the General Office of the State Council" should be clearly stipulated, so as to ensure that the work can be carried out in an orderly and orderly manner. Second, the construction of a scientific data center should be coordinated. In the field of natural resources, no national or departmental scientific data center construction echelon has been formed. The existing two state-level scientific data centers are still weak on the whole, and there is still a big gap compared with the national assessment index requirements for scientific data centers and other scientific data centers. Special scientific data such as land, deep land, karst, and surveying and mapping geographic information are also well qualified to declare the National Scientific Data Sharing Platform, but their capabilities and influence need to be further improved [10].

4. Countermeasures and suggestions for promoting scientific management of natural resources

Through the first survey of scientific data of natural resources, its characteristics and shortcomings are found. It is urgent to innovate the working mechanism, strengthen the exchange management, and promote openness and sharing.

4.1. Formulate and implement rules and regulations for the management of scientific data of natural resources

First, specific rules for scientific big data management in the industry should be formulated. Priority should be given to standardizing the definition, scope, unit responsibilities, operational procedures and requirements for the collection and sharing of natural resource scientific data. Second, the main responsibilities of the legal entity that undertakes activities related to natural resource scientific data, such as data quality control, confidentiality and security, as well as the collection management, processing and sorting of scientific data centers, and the promotion of data openness and sharing, should be clearly defined. Finally, a working mechanism should be established for each department to supervise the management, submission, opening and sharing of scientific data on an annual basis.

4.2. Promote the construction of national and ministerial scientific data centers in a coordinated manner

First, a “1+N+X” operation mechanism of scientific data management of natural resources should be constructed. According to the layout principle of main center, domain center and characteristic center, “1” indicates the main center of scientific data, and N represents the domain scientific data center including land, geology and mineral resources, ocean, surveying and mapping geographic information, forest and grass wet and natural resources policy research, and “X” is the characteristic scientific data center of the polar region, deep land and karst. The above-mentioned units are entrusted with the collection and management of scientific data in professional fields and conduct a trial operation.

Second, the existing two national scientific data centers of the Ministry of Natural Resources should be strengthened. We should give full play to the role of the two centers in aggregating and integrating scientific data in professional fields, further enhance their service capacity and influence, and meet the assessment requirements of the Ministry of Science and Technology and the Ministry of Finance every two years to obtain continuous support.

Finally, a scientific and technological resource platform of subordinate units with a good foundation and distinctive features should be selected, which should be recommended to enter the sequence of national scientific data-sharing platforms according to the requirements of the Ministry of Science and Technology on platform construction standards and relevant reporting procedures.

Acknowledgment

Thanks to the research on space-time information strategy of natural resources ,two supports and one promotion" of scientific and technological innovation service for surveying and mapping geographic information (121106000000180039-2204) and the basic scientific research business fee of central public welfare scientific research institutes (AR2120) for the support of this project.

References

[1] “Circular of the General Office of the State Council on Printing and Issuing the Management Measures for Scientific Data” (Guo Ban Fa [2018] No. 17)

[2] Wang Juanle. Global scientific data governance concerns in the post-epidemic era [J] *Social Sciences Abroad* 2021, **(5)** 58-64

[3] “14th Five-Year Plan for New Infrastructure Construction”.

[4] Shuzh.GU, Taip. WU On China’s governance system for natural resources in the new era [J]. *Journal of Natural Resources* vol. **8** Beijing, China, 2020, pp. 1802-1816

[5] National Science and Technology Basic Conditions Platform Center. National Science Data Resources Development Report [M]. Beijing: *Science and Technology Literature Publishing*

House 2018

[6] Xiaoh. LIU, Xiaoj.LIU,Shub. CHENG,Construction of a national natural resources comprehensive observation system and key technologies [J] *Resources Science* vol. **10** Beijing,China, 2020, pp.1849-1859

[7] J. CHEN, H.WU,Wan z. LIU,Technical connotation and research agenda of natural resources spatio-temporal information *Acta Geodaetica et Cartographica Sinica*, in press

[8] X. GONG, C.W EN, Jiangx. GUANG, Natural Resource Service Data Governance and Reconstruction Path with Land Business Chain as the Core [J] *PLANNERS*, 2021, **37(20)** 55-59

[9] Wenb. ZHAO, S.LI, Fu. LIU, Research on Development Strategy of New Generation System Efficient Earth Observation System [J] *Strategic Study of CAE*, 2021, **23(6)** 128-137

[10] Der. LI, Xin.SHEN, Fu.LIU, Research on Development Strategy of Space-based Information Real-time Intelligent Service System in China [J] *Strategic Study of CAE*, 2020, **22(2)** 146-151

Enhancing soil fertility and lowbush blueberry (*Vaccinium angustifolium*) growth using bio-organic fertilizer

Ngo Phuong Ngoc[1], Le Van Dang[2], Le Ngoc Quynh[1] and Ngo Ngoc Hung[2,3]

[1] Department of Plant Physiology-Biochemistry, Faculty of Agriculture, Can Tho University, Campus II, Can Tho City, Vietnam;
[2] Department of Soil Science, Faculty of Agriculture, Can Tho University, Campus II, Can Tho City, Vietnam.

[3] Corresponding author's e-mail: ngochung@ctu.edu.vn

Abstract. Acid soils contain high concentrations of aluminum and ferrous ions, which limit crop growth and the available plant nutrients. The current work was to determine the effects of bio-organic fertilizer (BOF) application on soil quality and the growth of blueberry cultivated in acid soil. A completely randomized block design was arranged, with four doses of BOF (0, 2, 4, and 6 t ha^{-1} yr^{-1}). Each treatment was carried out with four replicates. In the present work, soil quality properties such as pH, soil organic matter, available phosphorus, total nitrogen, aluminum (Al^{3+}), and iron (Fe^{2+}) were measured. Additionally, the influence of organic fertilizer on the growth of blueberry (plant height, number of shoots, and leaves) was also investigated. The results indicated that both the rates of BOF (4 and 6 t ha^{-1} yr^{-1}) significantly improved soil organic matter and available phosphorus. Also, BOF application greatly decreased the soil concentrations of Al^{3+} and Fe^{2+} compared to the control treatment. Similarly, 12 months after planting, the number of shoots and leaves was improved in the pots with BOF compared to that without BOF addition. From the results of this study, we recommend using BOF as the best measure for enhancing soil health and blueberry growth.

1. Introduction

Acid sulfate soils (ASSs) are found in about one-third of the agricultural land in the Mekong Delta, which is approximately 1.60 million hectares [1]. ASS has high toxicity of Al^{3+} and Fe^{2+}, which decrease the soil pH, resulting in reduced nutrient availability [2]. According to Hung et al., the macronutrient (N, P, and K) levels in ASS are sufficient for crop growth [3]. Furthermore, the high concentrations of Al^{3+} and Fe^{2+} negatively affected plant growth and yield [4,5]. Al^{3+} and Fe^{2+} prevent cell division and elongation, which reduce the growth of crop roots, decreasing the nutrient uptake and declining plant development [6,7]. In addition, the low pH in ASS also inhibits microorganism activity, leading to decreased soil organic carbon matter decomposition [8].

There are several ways to reduce the exchangeable Al^{3+} and Fe^{2+} in ASS, such as the addition of biofertilizers, biochar, and liming, among others [2,5,9]. Biofertilizers are biological products that contain beneficial microorganisms and most of the required plant nutrients [10]. After being supplied to the soil, crop growth will be promoted by improving root biomass as well as enhancing nutrient absorption capacity [11]. Biofertilizers are considered an eco-friendly product, and they can reduce the use of chemical fertilizers [10,11]. According to Patra et al. (2021) [12], who demonstrated that biofertilizer supplementation remarkably reduced soil acidity. Moreover, treatment with biofertilizers releases organic acids that precipitate exchangeable Al^{3+} and Fe^{2+} through a chelation process, thus, being non-phytotoxic [9].

Lowbush blueberry (*Vaccinium angustifolium*) grows suitably in acid soils, wherein the soil pH has arange of 4.0–5.5 [13]. Blueberry growth and yield may be affected when presented with high concentrations of Al^{3+} and Fe^{2+} [14]. Although there are many studies globally that are related to the role of biofertilizers on soil physicochemical properties and the growth of plants that are cultivated on ASS. However, studies on the impact of biofertilizers on improving soil health in the ASSs of the VMD are limited. Therefore, this study aimed to evaluate the effects of BOF on soil quality and blueberry growth (height, number of leaves, and shoots).

2. Material and Methods

2.1. *Study site and soil used*

The experiment was carried out in a greenhouse at the Soil Science Department, College of Agriculture, Can Tho University (Can Tho city, Vietnam; lat. 10° 01' 44.8" N, long. 105° 46' 00.5" E). The study was conducted from January 2021 to February 2022.

Soil utilized in the current study was acid sulfate soil, which was collected from the Thanh Dong commune, Binh Tan district, Vinh Long province (10° 07' 04.3" N, 105° 45' 41.1" E). Soil was collected at the depth of 0–20 cm, plant residues were eliminated, and soil was naturally dried out. Twenty kilograms of soil were put into each pot. The size of a pot was 40 cm diameter × 35 cm depth, and the total area used for the trial was about 50 m^2.

2.2. *Experimental layout*

The study was conducted through a randomized complete block design, including four treatments, with four replicates, and a pot was considered as a replicate. The treatments were as follow: (i) control (not applied BOF); (ii) 2 tons of BOF (ha^{-1} yr^{-1}); (iii) 4 tons of BOF (ha^{-1} yr^{-1}); and (iv) 6 tons of BOF (ha^{-1} yr^{-1}).

Lowbush blueberry variety (*Vaccinium angustifolium*) was used in the greenhouse experiment. Because the variety of lowbush blueberry was more suitable for growing on ASSs compared with rabbiteye and highbush blueberries varieties [14]. The age of the blueberries when the study started was three months. The plant height and the number of shoots and leaves were equivalent between treatments and did not have a significant difference.

Urea (46% N), superphosphate (16% P_2O_5), and potassium oxide (60% K_2O) were used in this study. Chemical fertilizers were applied at a dose of 80 kg N, 60 kg P_2O_5, and 70 kg K_2O ha^{-1} yr^{-1}.

The BOF used in this research was Dau Trau MK N.8, which contains 20% SOM, 1 × 10^6 CFU/g phosphate-solubilizing microorganisms, 1 × 10^6 CFU/g cellulose-degrading bacteria, and 1 × 10^6 CFU/g nitrogen-fixing bacteria. BOF was applicated three times per year, including 3, 6, and 9 months after planting (MAP). Tap water was used for blueberry irrigation. The mean values of pH and EC of tap water was 7.05 and 0.35 mS cm^{-1}, respectively.

2.3. *Soil and plant collection*

To measure the soil quality properties, a soil push probe was used to collect soil at three random sites in a pot at three stages, namely, 4, 8, and 12 MAP. The soil was then dried at room temperature. The soil was crushed through 0.5 mm for soil pH measure and 2.0 mm for soil fertility (SOM, TN, and AP) and soil toxicity (Al and Fe) analysis. The chemical properties of soil were determined according to the standard procedures described by [15] and briefly described herein. The soil pH in the 1:2.5 solution (soil/water) was measured using a digital pH meter. The SOM was obtained through the Walkley–Black method [16]. The TN was obtained through the semi-micro Kjeldahl method [17]. The AP was obtained through the Bray II method [18]. The Al^{3+} was obtained through the titrimetric method [19]. The Fe^{2+} was obtained through the ferrozine method [20].

Agronomic parameters were determined in the experiment at 4, 8, and 12 MAP. Plant height was measured from the ground to the peak of the blueberry crop. The number of leaves per plant was counted as the green leaves on the plant. The number of shoots per plant was determined by counting all shoots that had above three maturity leaves.

2.4. Data analysis

The SPSS software (version 21.0) was used for data analysis in the present work. The variance analysis was carried out by analysis of variance (ANOVA), Duncan's test was used to compare the differences in the treatment means with a significant level at $P < 0.05$.

3. Results and Discussions

3.1. Initial soil characteristics before conducting experiment

The initial soil physicochemical properties (Table 1) that were obtained were as follow: pH, 4.12; EC, 0.88 mS cm^{-1}; available phosphorus (AP), 12.0 mg kg^{-1}; total nitrogen (TN), 1.20 g kg^{-1}; soil organic matter (SOM), 2.82%; Al^{3+}, 3.49 $cmol_c$ kg^{-1}; Fe^{2+}, 124 mg kg^{-1}; and soil texture of clay (the proportions the sand, silt, and clay were 4.4–33.5–62.1%, respectively).

Table 1. Initial soil quality properties in the topsoil layer (0–20 cm).

Parameter	Unit	Value
pH_{H2O}	–	4.12
Electrical conductivity	mS cm^{-1}	0.88
Available phosphorus	mgP kg^{-1}	12.0
Total nitrogen	g kg^{-1}	1.20
Soil organic matter	%	2.82
Exchangeable Al	$cmol_cAl^{3+}$ kg^{-1}	3.49
Iron	$mgFe^{2+}$ kg^{-1}	124
Soil particle	%	
Sand		4.40
Silt		33.5
Clay		62.1

3.2. Effects of BOF on soil quality properties

Using biofertilizers not only improve SOM but also enhance soil nutrients [21, 22]. In this study, BOF application significantly increased the SOM, TN, and AP compared with that of the control treatment in the three investigated stages (Table 2). By contrast, soil pH was not affected by the addition of biofertilizer. The result was in agreement with the report of [23], who showed that using biofertilizers were not changed the pH soil.

The use of BOF at the doses of 4 and 6 t ha^{-1} yr^{-1} increased the average value of SOM by 0.5% after a year of the experiment. Likewise, the concentration of TN elevated at the 4, 8, and 12 MAP stages after applying biofertilizer. The TN content in the soil when added with BOF increased about 0.4 g kg^{-1} compared to the TN in the initial soil. Similar to the SOM and TN, the AP in the soil was increased greatly after BOF application. In two treatments (4 and 6 t ha^{-1} yr^{-1}), the concentrations of AP ranged from 17.7 to 22.6 mg kg^{-1}.

Biofertilizers contain beneficial microbes that have versatile enzyme systems; thus, it is easy to conduct various nutrient transformations, resulting in increased soil nutrients [24]. In the present study, BOF accommodated a lot of microbes such as phosphate-solubilizing microorganisms, cellulose-degrading bacteria, and nitrogen-fixing bacteria. These biotas participated in processing of nutrient transformation in soil leading to increased SOM, TN, and AP levels [25, 26].

Table 2. Influence of BOF on some chemical characteristics of the acid sulfate soil.

Stages	**BOF (t ha^{-1} yr^{-1})**	**pH$_{H2O}$ (1:2.5)**	**SOM (%)**	**TN (g kg^{-1})**	**AP (mg kg^{-1})**
4 MAP	0	4.19	2.64^c	1.18^c	12.5^c
	2	4.38	2.88^b	1.36^b	15.8^b
	4	4.47	3.10^a	1.54^a	17.9^a
	6	4.41	3.09^a	1.53^a	17.7^a
	P-value	0.400	0.001	0.004	0.002
	Significance	*ns*	**	**	**
	CV (%)	*5.17*	*3.58*	*5.53*	*5.29*
8 MAP	0	4.29	2.65^c	1.16^c	12.1^c
	2	4.40	3.00^b	1.43^b	17.6^b
	4	4.50	3.19^a	1.58^a	18.8^a
	6	4.60	3.16^a	1.61^a	18.9^a
	P-value	0.369	0.001	0.003	0.002
	Significance	ns	**	**	**
	CV (%)	*5.51*	*2.38*	*4.89*	*3.75*
12 MAP	0	4.21	2.59^b	1.17^b	13.0^c
	2	4.36	3.03^a	1.39^{ab}	18.3^b
	4	4.44	3.15^a	1.54^a	21.2^a
	6	4.66	3.14^a	1.63^a	22.6^a
	P-value	0.258	0.003	0.001	0.005
	Significance	ns	**	**	**
	CV (%)	*6.71*	*3.19*	*6.62*	*6.86*

MAP, months after planting; SOM, soil organic matter; TN, total nitrogen; AP, available phosphorus; CV, coefficient of variation. Different letters indicate a significant difference among treatments at $P < 0.01$ (**); ns, not significant.

The concentration of Al^{3+} was decreased significantly at 4, 8, and 12 MAP by the addition of BOF (Figure 1). The use of BOF at 2 t ha^{-1} yr^{-1} declined the content of Al^{3+} compared with that of the control treatment. Meanwhile, the application of BOF at two doses of 4 and 6 t ha^{-1} yr^{-1} also caused a decrease in the Al^{3+} content. However, the concentration of Al^{3+} in the treatment of 2 t ha^{-1} yr^{-1} was higher than that in the treatments of 4 and 6 t ha^{-1} yr^{-1}. Hence, the supply of BOF at the rates of 4 and 6 t ha^{-1} yr^{-1} is the best measure for the mitigation of Al^{3+} in ASSs.

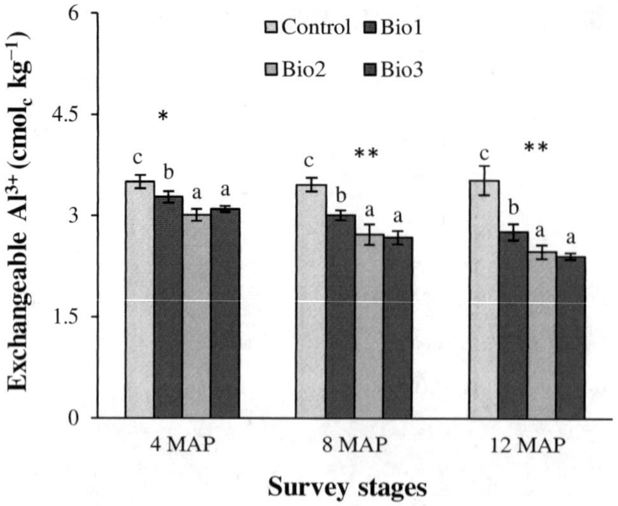

Figure 1. Concentration of Al^{3+} influenced by BOF. Control, without BOF application; Bio1, 2 tons of BOF (ha^{-1} yr^{-1}); Bio2, 4 tons of BOF ha^{-1} yr^{-1}; and Bio3, 6 tons of BOF (ha^{-1} yr^{-1}). MAP, months after planting. Different letters indicate a significant difference among treatments at * $P < 0.05$, ** $P < 0.01$.

BOF addition reduced the content of Fe^{2+} at the three stages of survey (Figure 2). In ASSs, high concentrations of Al^{3+} and Fe^{2+} lead to a decline in soil health and microbial community [21]. According to Panhwar et al. (2016) [9], using biofertilizers increased the levels of strongly-bound Al^{3+} and Fe^{2+}, resulting in fixation by the organic acids, hence, reducing the concentration of Al^{3+} and Fe^{2+} in soil.

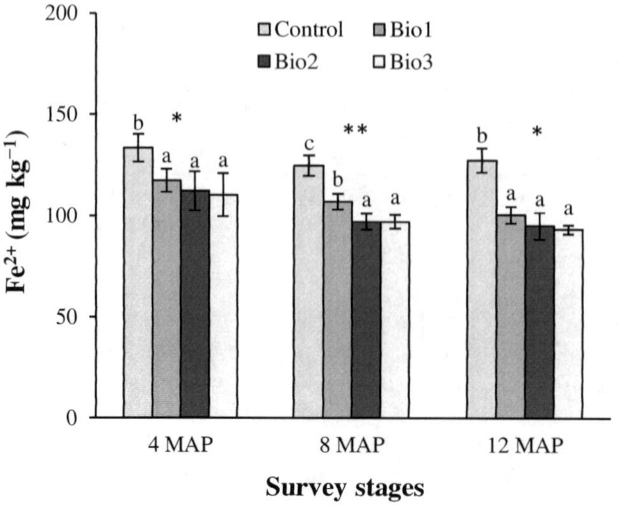

Figure 2. Addition of BOF reduced the concentration of Fe^{2+}. Control, without BOF application; Bio1, 2 tons of BOF (ha^{-1} yr^{-1}); Bio2, 4 tons of BOF (ha^{-1} yr^{-1}); and Bio3, 6 tons of BOF (ha^{-1} yr^{-1}). MAP, months after planting. Different letters indicate a significant difference among treatments at * $P < 0.05$, ** $P < 0.01$.

3.3. Blueberry growth

Table 3 presents role of BOF on lowbush blueberry growth. The plant height before and after BOF supplementation did not have a significant difference. The plant height ranged from 46.3 to 48.8 cm, 60.0 to 62.5 cm, and 73.5 to 75.0 cm in 4, 8, and 12 MAP, respectively. The treatments with BOF did not improve the number of shoots and leaves at the stages of 4 and 8 MAP. However, biofertilizer application enhanced the shoot and leaf number in 12 MAP compared with that of the control.

The results of this study were line in with that of [27] who demonstrated that biofertilizers improved crop growth. Another study [28] also revealed that blueberry growth was positively affected by biofertilization. According to Carletti et al. (1996) [29], using biofertilizer caused an enhancing the length of root, which may be increased nutrient uptake, resulting in improved plant growth. In the present work, plant height was not improved by BOF (Table 3). The results were not in agreement with the report of [30], who reported that the use of beneficial bacteria enhanced blueberry plant growth.

Table 3. Effect of BOF on the growth of lowbush blueberry.

Stages	BOF (t ha^{-1} yr^{-1})	Plant height (cm)	Number of shoots per plant	Number of leaves plant per plant
4 MAP	0	46.3	27.7	194
	2	47.5	27.8	200
	4	48.5	27.9	208
	6	48.8	28.0	217
	P-value	*0.815*	*1.00*	*0.646*
	Significance	*ns*	*ns*	*ns*
	CV (%)	*8.48*	*10.8*	*12.6*
8 MAP	0	61.0	42.0	252
	2	61.8	45.5	241
	4	62.5	48.0	257
	6	60.0	45.8	257
	P-value	*0.974*	*0.788*	*0.923*
	Significance	*ns*	*ns*	*ns*
	CV (%)	*13.1*	*10.4*	*11.8*
12 MAP	0	73.8	46.5^{c}	281^{b}
	2	74.0	55.0^{b}	324^{b}
	4	75.0	61.8^{a}	400^{a}
	6	73.5	63.3^{a}	395^{a}
	P-value	*0.989*	*0.001*	*0.03*
	Significance	*ns*	**	*
	CV (%)	*9.11*	*7.45*	*11.2*

MAP, months after planting; CV, coefficient of variation. Different letters indicate a significant difference among treatments at * $P < 0.05$, ** $P < 0.01$; ns, not significant.

4. Conclusions

Soil fertility (SOM, TN, and AP) was significantly improved by BOF addition. In fact, the concentration of Al^{3+} and Fe^{2+} in the ASSs decreased after the biofertilizer application. We observed that lowbush blueberry plant height was not affected by the addition of biofertilizer. However, the shoot and leaf number of the blueberry crop at the 12 MAP stage increased greatly compared with that of the control treatment. The results showed that BOF application mitigated toxicity content in soil, thus, increasing soil health. They are considered as beneficial option for a sustainable agriculture in ASSs.

Acknowledgement

This project was supported by the Ministry of Education and Training, Vietnam, with grant number: B2021-TCT-10. We thank Ms. Huynh Ngoc Truyen, an undergraduate student, for supporting the conduct of the measurements of the agronomic parameters.

References

[1] Hung N N 2009 Nature and processes that change soil fertility in the Mekong Delta *Agricultural Publishing, Vietnam* 471 p
[2] Shamshuddin J, Azura A E, Shazana M A R S, Fauziah C I, Panhwar Q A, Naher U A 2014 Properties and management of acid sulfate soils in Southeast Asia for sustainable cultivation of rice, oil palm, and cocoa. In Donald L. Sparks, editor: *Advances in Agronomy* **124** 91–142
[3] Hung T V, Toan L P, Dung T V, Hung N N 2017 Morphological and physicochemical properties of acid sulfate soils in Dong Thap Muoi *Can Tho University Journal of Science* **2** 1–10
[4] Choudhury B U, Malang A, Webster R, Mohapatra K P, Verma B C, Kumar M, Das A, Islam N, Hazarika S 2017 Acid drainage from coal mining: Effect on paddy soil and productivity of rice *Sci. Total Environ.* **583** 344–351
[5] Panhwar Q A, Naher U A, Shamshuddin J, Ismail M R 2020 Effects of biochar and ground magnesium limestone application, with or without bio-fertilizer addition, on biochemical properties of an acid sulfate soil and rice yield *Agronomy* **10** 1100
[6] Becker M, Folkard A 2005 Iron toxicity in rice—conditions and management concepts *J. Plant Nutr. Soil Sci.* **168** 558–573
[7] Horst W J, Wang Y, Eticha D 2010 The role of the root apoplast in aluminium-induced inhibition of root elongation and in aluminium resistance of plants: a review *Annals of Botany* **106** 185–197
[8] Feketeová Z, Hrabovský A, Šimkovic I 2021 Microbial features indicating the recovery of soil ecosystem strongly affected by mining and ore processing *International Journal of Environmental Research and Public Health* **18** 3240
[9] Panhwar Q A, Naher U A, Shamshuddin J, Othman R, Ismail M R 2016 Applying limestone or basalt in combination with bio-fertilizer to sustain rice production on an acid sulfate soil in Malaysia *Sustainability* **8** 700
[10] Dasgupta D, Kumar K, Miglani R, Mishra R, Panda A K, Bisht S S 2021 Microbial biofertilizers: recent trends and future outlook. Editor(s): Surajit De Mandal, Ajit Kumar Passari *Recent Advancement in Microbial Biotechnology* Academic Press 1-26
[11] Singh D, Thapa S, Geat N, Mehriya M L, Rajawat M V S 2021 Biofertilizers: mechanisms and application. Editor(s): Amitava Rakshit, Vijay Singh Meena, Manoj Parihar, H.B. Singh, A.K. Singh *Biofertilizers* Woodhead Publishing 151-166
[12] Patra A, Sharma V K, Nath D J, Ghosh A, Purakayastha T J, Barman M, Kumar S, Chobhe K A, Anil A S, and Rekwar R K 2021 Impact of soil acidity influenced by long-term integrated use of enriched compost, biofertilizers, and fertilizer on soil microbial activity and biomass in rice under acidic soil *J of Soil Sci. Plant Nutr.* **21** 756–67
[13] Stafne E T, Melanson R, Layton B, Silva J, Canales E 2020 Establishment and maintenance of

blueberries *Mississippi State University Extension Service* **1758** 1-21

[14] Chi N T 2021 Effects of organic fertilizers on the growth of three varieties of blueberries grown on Binh Tan's soil under net house conditions. Master thesis of Agricultural system *College of Agriculture, Can Tho University, Vietnam* 64 p

[15] Houba V J G, Vanderlee J J, Novozamsky I 1995 Soil analysis procedures (*soil and plant analysis, part 5B*)," a series of syllabi, 6th ed., Department of Soil Science and Plant Nutrition *Wageningen Agricultural University, The Netherlands*

[16] Glosolan 2019 Standard operating procedure for soil organic carbon: Walkley-Black method, *Titration and colorimetric method. FAO* 25p

[17] Sáez-Plaza P, Navas M J, Wybraniec S, Michalowski T, Asuero A G 2013 An overview of the Kjeldahl method of nitrogen determination. Part II. sample preparation, working scale, instrumental finish, and quality control *Criti. Rev. Anal. Chemistry* **43** 224–272

[18] Bray R H, Kurtz L T 1945 Determination of total, organic, and available forms of phosphorus in soils *Soil Science* **59** 39–46

[19] Abreu-Junior C H, Muraoka T, Lavorante A F 2003 Exchangeable aluminum evaluation in acid soils *Scientia Agricola* **60** 543–548

[20] Viollier E, Inglett P W, Hunter K, Roychoudhury A N, Cappellen P V 2000 The ferrozine method revisited: Fe(II)/Fe(III) determination in natural waters *Applied Geochemistry* **15** 785-790

[21] R Shivani, Sow S, Choudhury S S, Kumar S, Ghosh M 2020 Biofertilizer as a novel tool for enhancing soil fertility and crop productivity: a review *Int. J. Curr. Microbiol. App. Sci.* **11** 653-665

[22] Kumar S S, Diksha, Sindhu S S, Kumar R 2022 Biofertilizers: An ecofriendly technology for nutrient recycling and environmental sustainability *Current Research in Microbial Sciences* **3** 100094

[23] Mulyani O, Trinurani E, Sudirja R, Joy B 2017 The Effect of bio-fertilizer on soil chemical properties of sugarcane in Purwadadi Subang *ICSAFS Conference Proceedings* **2017** 164–171

[24] Ramasamy M, Geetha T, Yuvaraj M 2020 Role of biofertilizers in plant growth and soil health. In E. C. Rigobelo, & A. P. Serra (Eds.), Nitrogen Fixation *IntechOpen*

[25] Kour D, Rana K L, Yadav A N, Yadav N, Kumar M, Kumar V, Vyas P, Dhaliwal H S, Saxena A K 2019 Microbial biofertilizers: Bioresources and eco-friendly technologies for agricultural and environmental sustainability *Biocatalysis and Agricultural Biotechnology* **23** 101487

[26] Kalayu G 2019 Phosphate solubilizing microorganisms: promising approach as biofertilizers *International Journal of Agronomy* **2019** 7

[27] Kumar S, Kundu M, Rakshit R 2019 Effect of bio-fertilizer on growth, yield and quality of strawberry (*Fragaria x ananassa* Duch.) cv. Camarosa *Bull. Env. Pharmacol. Life Sci.* **8** S99-S107

[28] Jamal Q, Lee Y S, Munir S, Malik M S, Kim K Y 2018 The combined use of five bacteria including bacillus amyloliquefaciens Y1 as biofertilizer in compost improvedlow bush blueberry growth, rhizosphere bacteria and enzyme at various pH *Pak. J. Agri. Sci.* **55** 409-416

[29] Carletti S, Rodriguez C E, and Liorente B 1996 Effect of biofertilizer application on jojoba cultivation *Assoc.Adv.Industl. Crops* **1996** 53-55

[30] Yu Y, Xu J, Huang T, Zhong J, Yu H, Qiu J, Guo J 2020 Combination of beneficial bacteria improves blueberry production and soil quality *Food Science & Nutrition* **8** 5776–5784

IOP Conf. Series: Earth and Environmental Science 1087 (2022) 012078 doi:10.1088/1755-1315/1087/1/012078

Removal of Mn(II) from Acidic Wastewater on the Red Mud–Loess Mixture: Fixed–Bed Column Tests

Yisi Lu[1], Xiaofeng Liu[1], Mingxing Xie[1] and Xiaoqiang Dong[1,2]

[1] College of Civil Engineering, Taiyuan University of Technology, Taiyuan, Shanxi, China.

[2] Corresponding author's e-mail: dongxiaoqiang@tyut.edu.cn

Abstract. In this study, the purification effect of red mud–loess mixture (RM–L) as a reactive medium in a permeable reactive barrier (PRB) for groundwater contaminated by acidic wastewater (AW) containing Mn(II) was explored considering removal capacity and pH value, and the safety of Mn(II) on RM–L after purification was evaluated. The breakthrough parameters of the Mn(II) were determined and the purifying performances of RM–L toward AW were evaluated by fixed-bed column tests at various initial concentrations of Mn(II), flow rates, and packing heights. The speciation of Mn(II) loaded on the RM–L was evaluated by the Tessier method. Results show the RM–L with a removal capacity of 17.30 $mg \cdot g^{-1}$ at 100 $mg \cdot L^{-1}$ initial concentration, 1 $mL \cdot min^{-1}$ flow rate and 7 cm packing height. The reduction in initial concentration and flow rate and the increase in packing height improve the total removal percentage of Mn(II) and the removal capacity of the RM–L. Moreover, the RM–L has a long-term effective acid buffer capacity for AW containing Mn(II), ensuring the pH values of effluent could be maintained at a neutral value. Furthermore, the Mn(II) could be loaded on the RM–L after the purification in stable forms that are dominated by the iron-manganese oxidate form.

1. Introduction

Acid mine drainage (AMD) is a type of environmental pollution faced by the entire world. Acidic wastewater (AW) containing heavy metal ions generated by abandoned mines will continue to cause serious harm to the ecological environment and human health [1, 2]. Various treatment technologies have been recently developed to reduce the acidity and toxicity of AMD. Among these technologies, the permeable reactive barrier (PRB) is an extremely promising technology, which is expected to be used to remediate groundwater polluted by AMD sustainably [3].

Up to now, the active medium used in actual PRB mainly includes zero-valent iron and activated carbon. Mineral materials and industrial solid waste have gradually attracted the attention of PRB applications [4, 5]. However, most studies only focused on the removal of heavy metal ions in wastewater rather than AW, or the problem of excessively high effluent pH value was found (>12) [6, 7]. Red mud (RM) is an industrial solid waste discharged during alumina production; as an industrial by-product, RM needs a large amount of consumption. RM is used to remove heavy metal ions from wastewater and neutralize the acidity of AW due to its good heavy metal adsorption capacity and alkalinity [8, 9]. Loess (L) has a good buffer capacity for acidic solutions and has a certain capability to remove heavy metals [10, 11]. Moreover, a 7:3 mass ratio of RM and L not only ensures the efficient removal of Mn(II) but also improves the acidity of AW to the level III limit of groundwater standards (6.5-8.5) [12]; thus, RM and L are mixed to avoid the strong alkalinity of RM, which expands its wide application in practical engineering.

The feasibility of using the RM and L mixture (RM–L) under a mass ratio of 7:3 as the active material in the PRB to purify AW containing Mn(II) was explored in this study, the purpose is to analyze various operating conditions (initial Mn(II) concentration, flow rate, packing height) on the purification of AW by RM–L, obtain the column parameters, and evaluate the safety and stability of the RM–L loaded Mn(II) by Tessier method.

Nomenclature

AW acidic wastewater	AMD acid mine drainage	PRB permeable reactive barrier
RM red mud	L loess	RM–L red mud-loess mixture
t_b breakthrough time (h)	MTZ mass tranfer zone (cm)	H packing height (cm)
t_s saturation time (h)		$m_{RM\text{-}L}$ mass of RM–L (mg)
t_{total} total time of column test (h)		m_{total} total amount of Mn(II) (mg)
C_0 influent Mn(II) concentration ($mg{\cdot}L^{-1}$)		$m_{removal}$ total amount of Mn(II) uptake (mg)
C_i effuluent Mn(II) concentration ($mg{\cdot}L^{-1}$)		V_{ef} effulent volume (L)
C_b breakthrough Mn(II) concentration ($mg{\cdot}L^{-1}$)		V_b breakthrough effluent volume (L)
C_s saturation Mn(II) concentration ($mg{\cdot}L^{-1}$)		V_s saturation effluent volume (L)
R total Mn(II) removal percentage (%)		q_{exp} removal capacity ($mg{\cdot}g^{-1}$)

2. Materials and methods

2.1. Materials

The RM is a Bayer process RM, is obtained from Shanxi Liulin Aluminum Factory. The L was taken from a construction site in the Dongshan area of Taiyuan, Shanxi, with a depth of approximately 5 m underground, which belongs to the surface accumulation soil of Q4. The air-dried RM and L were dried at 105 °C for 24 h and then crushed. The selected particle size of L is 0.075 mm to eliminate impurities. The particle size of RM particles is 0.50-0.85 mm to ensure permeability. Table 1 shows the main components and contents of RM and L.

Table 1. Main chemical components of RM and L.

Component/(%)	Al_2O_3	SiO_2	CaO	Fe_2O_3	MgO	K_2O	Na_2O	TiO_2	MnO
RM	24.34	20.17	18.26	9.40	1.26	0.64	9.61	3.56	0.03
L	11.75	58.88	7.98	4.54	2.05	2.18	1.70	0.60	0.07

The analytical reagent $MnSO_4{\cdot}H_2O$ was used to produce the simulated AW containing Mn(II). And the initial pH value of AW was adjusted to 3 using dilute sulfuric acid.

2.2. Methods

The fixed-bed column used in the dynamic adsorption experiments was assembled from segmentable small polyethylene columns (inner diameter of 50 mm and height of 140 mm) according to the experimental requirements. The RM–L can obtain a permeability coefficient of 3.39×10^{-4} $cm^2{\cdot}(cmH_2O{\cdot}s)^{-1}$ under the compaction condition of a constant dry density ($1.02\ g{\cdot}m^{-3}$) [13]. A 20 mm thick permeable stone was placed on the top and bottom of the column to avoid RM–L loss and ensure the uniform distribution of the influent and effluent. The RM–L in the columns was saturated with distilled water to avoid blockage, and then the peristaltic pump was used to adjust the flow rate for constant inflow [14]. The detailed operational conditions are listed in Table 2, the initial concentration was selected on the basis of the AMD investigation in Shanxi Province, the flow rate was selected by converting the recommended permeability coefficient of PRB and the flow rate of groundwater, and the packing height was chosen in accordance with the principle that the height of RM–L is larger than the diameter of the column device to avoid preferential flow.

The effluent samples were collected at regular intervals (8:00, 12:00, 18:00, 24:00) and measured the residual objective heavy metal ion and pH value using Arcos ICP-OES (Spectro, Germany) and FE-28 pH meter (Mettler Toledo, Switzerland), respectively. According to the Tessier method, the speciation of Mn(II) loaded in the RM–L after purification of AW was analyzed [15].

Table 2. Operational conditions for fixed-bed column tests.

Test No.	Factors	Concentration /($mg \cdot L^{-1}$)	Flow rate /($mL \cdot min^{-1}$)	Packing height /(cm)	m_{RM-L} /(g)
C	Concentration	100	1	7	142.58
*M*200	Concentration	200	1	7	142.58
*M*300	Concentration	300	1	7	142.58
C	Flow rate	100	1	7	142.58
*Q*2	Flow rate	100	2	7	142.58
*Q*3	Flow rate	100	3	7	142.58
C	Packing height	100	1	7	142.58
*H*14	Packing height	100	1	14	285.16
*H*21	Packing height	100	1	21	427.73

3. Results and discussions

3.1. Calculation of column parameters

The column removal capacity of the RM–L, which is a critical indicator of column performance, can be calculated from the breakthrough curve [16]. The typical breakthrough curve is shown in Figure 1.

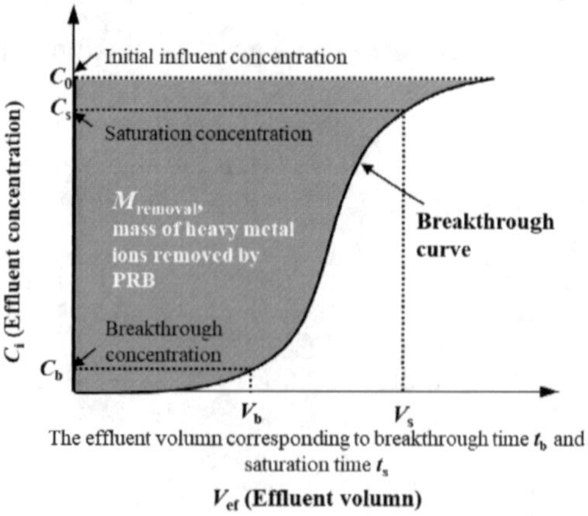

Figure 1. Representation of a typical breakthrough curve.

The length of mass transfer zone (MTZ), removal amount of Mn(II), the total amount of Mn(II), column maximum separation capacity and total Mn(II) removal percentage were calculated by the following equations (1)-(5) [16-18]:

$$\text{MTZ}=\frac{H(t_s-t_b)}{t_s} \tag{1}$$

$$m_{\text{removal}}=V_{\text{ef}}\int_{t=0}^{t=t_{\text{total}}}(C_0\text{-}C_i)\text{d}t \tag{2}$$

$$m_{\text{total}}=C_0 V_{\text{ef}} t_{\text{total}} \tag{3}$$

$$q_{\text{exp}}=\frac{m_{\text{removal}}}{m_{\text{RM-L}}} \tag{4}$$

$$R=\frac{m_{\text{removal}}}{m_{\text{total}}} \tag{5}$$

where MTZ is the length of mass transfer zone, cm; t_b and t_s are the breakthrough time corresponding to the time when the Mn(II) concentration in effluent reaches the level III limit of groundwater standard and 85% of the influent Mn(II) concentration, respectively; m_{removal} and m_{total} are the total amount of Mn(II) uptake and the total amount of Mn(II) through the column, respectively, mg; C_0 and

C_i are the influent Mn(II) concentration and effluent Mn(II) concentration, respectively, mg·L^{-1}; V_{ef} is effluent volume, L; t_{total} is the total time of column test, h; q_{exp} is removal capacity, also called equilibrium Mn(II) uptake, mg·g^{-1}; R is the total Mn(II) removal percentage, %.

Table 3. Parameters of breakthrough curves for Mn(II) removal on RM–L in fixed-bed columns.

Test No.	C	*M*200	*M*300	*Q*2	*Q*3	*H*14	*H*21
t_b/(h)	168	34	10	48	20	442	718
t_s/(h)	587	261	198	237	167	1004	1800
MTZ/(cm)	5.00	6.09	6.65	5.58	6.16	7.83	12.62
$m_{removal}$/(g)	2466	2217	2135	2092	1930	5404	9070
m_{total}/(g)	4020	4296	3996	3708	3888	7344	11700
q_{exp}/(mg·g^{-1})	17.30	15.55	14.97	14.67	13.54	18.95	21.21
R/(%)	61.34	51.61	53.42	56.42	49.64	73.58	77.53

The column parameters of breakthrough curves for Mn(II) removal on RM–L are shown in Table 3. The breakthrough time and saturation time increased with the decreasing initial Mn(II) concentration and flow rate. However, the opposite trend was observed when the packing height was increased. The removal capacity of RM–L could be reached at 17.30 mg·g^{-1} (Test No. C) when the initial concentration was 100 mg·L^{-1}, the flow rate was 1 mL·min^{-1} and the packing height was 7 cm. The maximum adsorption capacity of RM–L in the batch experiments was 7.98 mg·g^{-1} in a previous study [12], which is twice smaller than that of the fixed-bed column test. This finding may be due to the precipitation formation phenomenon of Mn(II) in an alkaline environment in the initial stage of operation. A similar result was obtained for the Mn(II) adsorption on immobilized RM beads and was explained by further adsorption of adsorbed Mn(II) [19].

3.2. Effect of initial concentration

Figure 2 shows the relationship between pH value and the Mn(II) breakthrough curve of soil column filtrate and service time at initial Mn(II) concentrations of 100, 200, and 300 mg·L^{-1}. In the initial stage, the mass of the RM–L was excessively large considering the volume of pollutants in the AW entering the column. Fortunately, with the increase in service time, the RM–L can ensure that the pH value of the filtrate is stable at a neutral value (approximately 7.00). When the initial concentration was increased from 100 mg·L^{-1} to 300 mg·L^{-1}, the breakthrough time and saturation time of Mn(II) decreased from 168 and 587 h to 10 and 198 h, respectively. With the increase in the initial concentration, the breakthrough time and saturation time of Mn(II) significantly decreased and the slope of the breakthrough curve between the breakthrough point and the saturation point in the MTZ became steep [20]. A small initial concentration realizes the closeness of the breakthrough curve to the classical breakthrough curve (S), indicating a slow removal process and a high removal capacity (17.30 mg·g^{-1}). This result is contrary to what was observed by Zhang et al for the adsorption capacity trend of MTBE adsorbed on zeolite [16], which may be explained by the hindered release of alkaline substances and the removal of Mn(II) due to serious adhesion generated from the rapid surface adsorption and micro-precipitation on the RM–L surface.

3.3. Effect of flow rate

Figure 3 shows the relationship between pH value and Mn(II) breakthrough curve of soil column filtrate and service time at initial Mn(II) flow rates of 1, 2, and 3 mL·min^{-1}.

The pH value of the filtrate first decreased and then stabilized at approximately 7.00, 7.10, and 7.30 with the increase in flow rate. This finding indicates that the RM–L still has a certain acid buffering capability despite the lost removal capability of heavy metal ions under high influent flow rates.

When the flow rate increased from 1 mL·min^{-1} to 3 mL·min^{-1}, the breakthrough time and saturation time of Mn(II) in the RM–L respectively decreased from 168 and 587 h to 20 and 167 h. The removal capacity of RM–L toward the target ion also decreased from 17.30 mg·g^{-1} to 13.54 mg·g^{-1}. This trend is due to the accelerated migration rate of Mn(II) in the column packed with RM–L with the increase in the influent flow rate, resulting in insufficient time for the pollutants to remain in the column and the reduction in contact time of RM–L with Mn(II). The flow rate reflects that a short residence time of the pollutants in the column worsens the remediation effect of the active medium on the pollutants [18]. Zhang et al [16] explained that this phenomenon is attributed to the low residency in the active medium at a high flow rate. Therefore, the permeability of the active medium and the flow rate of groundwater are crucial to the removal of pollutants. The permeability and the flow rate should not be excessively large, and the removal capacity of pollutants decreases with the increase in the flow rate.

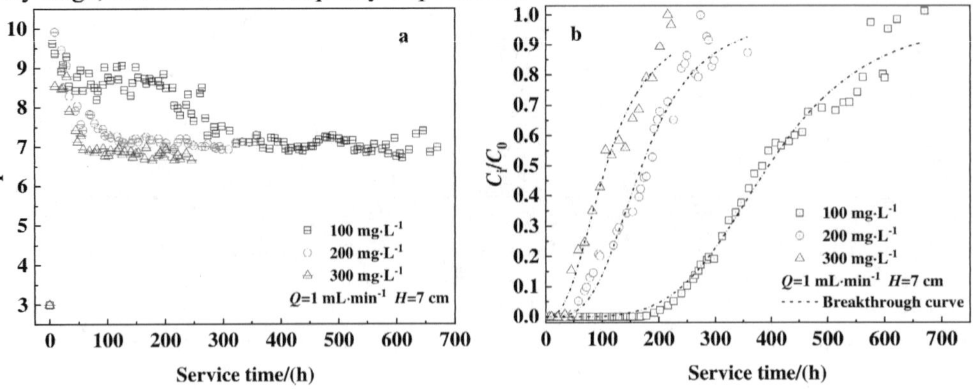

Figure 2. pH value and Mn(II) concentration of effluent from the column packed with RM–L at different initial concentrations of 100, 200, and 300 mg·L^{-1} (a. pH value, b. Mn(II) concentration).

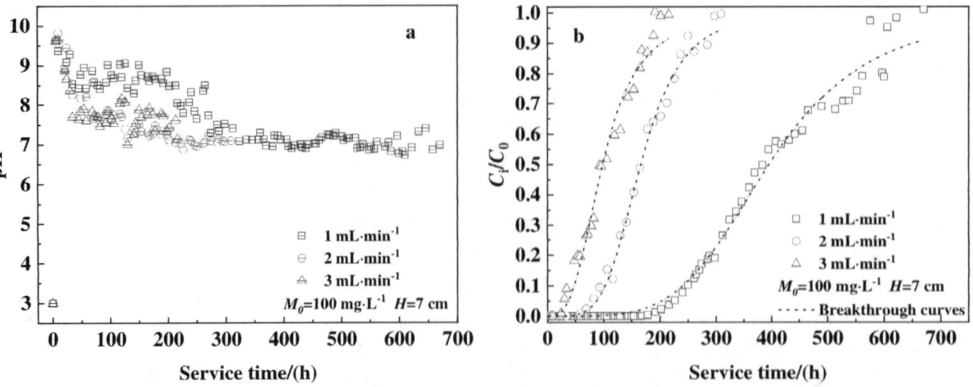

Figure 3. pH value and Mn(II) concentration of effluent from the column packed with RM–L at different flow rates of 1, 2, and 3 mL·min^{-1} (a. pH value, b. Mn(II) concentration).

3.4. Effect of packing height

Figure 4 shows the relationship between pH value and Mn(II) breakthrough curve of column filtrate and service time at packing heights of 7, 14, and 21 cm.

The pH value of the filtrate of the RM–L column decreased to approximately 7.00, 6.80, and 6.70 after a stable plateau period with the increase in packing height. In the initial stage, the pH value of the filtrate was close to 10.00, this phenomenon is due to the large mass of the RM–L in the column relative to the volume of pollutants in the AW entering the column and the additional alkaline substances in the filtrate caused by the high packing height. This phenomenon is a limitation of this study that must be improved. However, the pH value of the filtrate could eventually be maintained at a relatively stable value around neutral, indicating that the RM–L has a long-term and effective buffering capacity for AW.

The breakthrough rate and saturation rate of the RM–L slowed down with the increase in packing height, indicating that the failure time of the RM–L was delayed and the large volume of AW was treated. For the dynamic removal results of AW containing Mn(II), the breakthrough time and saturation time of Mn(II) in the RM–L column increased from 168 and 587 h to 718 and 1800 h, respectively, when the packing height increased from 7 cm to 21 cm. This trend is contributed to the increased migration distance of Mn(II) in the column, the increased contact time between Mn(II) and the RM–L, and the increased sufficient adsorption sites and effective components for the removal of Mn(II) [16].

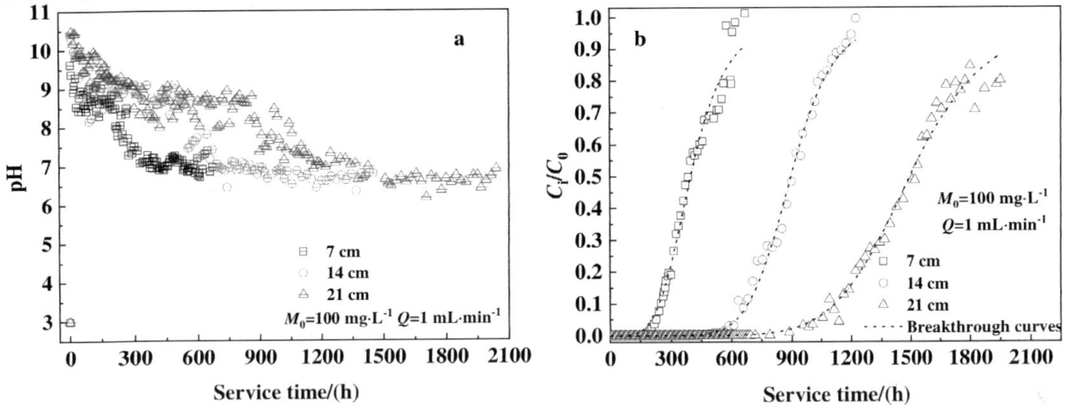

Figure 4. pH value and Mn(II) concentration of effluent from the column packed with RM–L at different packing heights of 7, 14, and 21 cm (a. pH value, b. Mn(II) concentration).

3.5. Speciation of Mn(II) in the RM–L:

Figure 5 shows the speciation percentage of Mn(II) in the RM–L after treatment. The results indicate the following: for the completely saturated RM–L column with a packing height of 7 cm by the AW containing Mn(II) with an initial concentration of 100 mg·L^{-1} and an influent flow rate of 1 mL·min^{-1}, the exchangeable form content, the carbonate form content, the iron-manganese oxide form content, the organic content, and residual form content of Mn(II) in the RM–L accounted for 9.80%, 17.17%, 50.81%. 16.04%, and 6.17%. The stability of heavy metals in the five existing forms ranges from poor to excellent [15]. The speciation of Mn(II) is dominated by the iron-manganese oxide form. The heavy metals in the bound state of iron and manganese oxides have remarkable stability due to the combination of target ions and active mediums through the action of ionic bonds. Therefore, the RM–L, which is still promising as a reactive medium of the PRB is used to purify groundwater polluted by AW. However, a part of Mn(II) will precipitate on the RM–L, which may cause blockage and reduce the permeability of PRB.

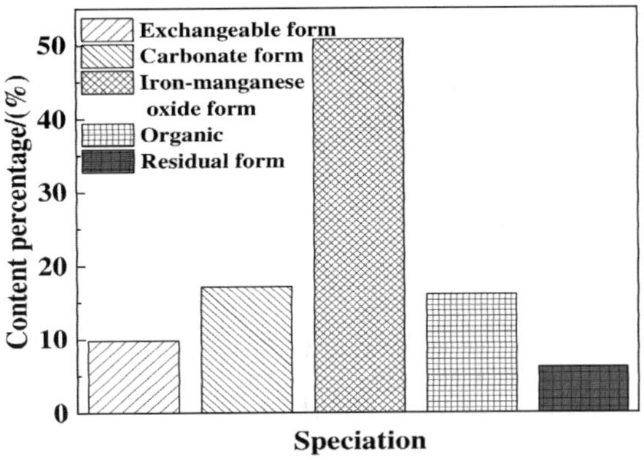

Figure 5. Speciation percentage of Mn(II) (Test No. C).

4. Conclusions

A fixed-bed column test was used to evaluate the purification performance of RM–L for Mn(II) in AW under different operational conditions, and quantitative analysis of the Mn(II) speciation was performed to clarify the safety and stability of Mn(II) on RM–L. Major conclusions can be drawn as follows:

(1) RM–L provides a cost-effective and promising active medium to purify groundwater polluted by AW containing Mn(II) as illustrated by the fixed-bed column and Tessier tests. This medium is a far-reaching move for utilizing large amounts of RM.

(2) The removal capacity increased with the decrease of initial concentration and flow rate and the increase of packing height. RM–L also has a long-term effective acid buffering capacity for AW containing Mn(II), ensuring that the pH value of the effluent is neutral.

(3) The removal capacity of the RM–L can reach 17.30 $mg \cdot g^{-1}$ at an initial concentration of 100 $mg \cdot L^{-1}$, a flow rate of 1 $mL \cdot min^{-1}$ and a packing height of 7 cm. The speciation of Mn(II) on the RM–L is dominated by the iron-manganese oxidate form, which is highly safe for the environment.

(4) This paper considered the single system solution containing only Mn(II). Thus, the removal capacity of RM–L for target ions and the buffering capacity under the coexistence of multiple ions will be explored in the future, to provide additional theoretical support for engineering applications of RM–L.

Acknowledgments

This work was financially supported by the National Natural Science Foundation of China [grant number 51978438].

References

[1] Moodley I, Sheridan C M, Kappelmeyer U, Akcil A 2018 *Miner. Eng.* **126** 207
[2] Naidu G, Ryu S C, Thiruvenkatachari R, Choi Y K, Jeong S H, Vigneswaran S 2019 *Environ. Pollut.* **247** 1110
[3] Rad P R, Fazlali A 2020 *J. Water Process Eng.* **35** 101196
[4] Song J M, Huang G X, Han D Y, Hou Q X, Gan L, Zhang M 2021 *J. Clean. Prod.* **319** 128644
[5] Phillips D H 2009 *Desalination* **248** 352
[6] Zhu F Y, Tan X F, Zhao W X, Feng L K, He S F, Wei L L, Yang L, Wang K, Zhao Q L 2022 *J. Hazard. Mater.* **424** 127605
[7] Shabalala A N, Ekolu S O, Diop S, Solomon F 2017 *J. Hazard. Mater.* **323** 641
[8] Wang M F, Liu X M 2021 *J. Hazard. Mater.* **408** 124420
[9] Li Y C, Huang H, Xu Z, Ma H Q, Guo Y F 2020 *J. Cleaner Prod.* **253** 119955
[10] Wang Y, Tang X W, Liu J J, Wang H Y, Sun Z F 2012 *Chin. J. Geotech. Eng.* **34** 2292
[11] Chen Y M, Wang Y Z, Xie H J, Jiang Y S 2014 *Chin. J. Geotech. Eng.* **36** 1185
[12] Lu Y S, Liu X F, Yu L Y, Zhang X D, Duan W, Park J B, Dong X Q 2022 *Pol. J. Environ. Stud.* **31** 6 (waiting for publication online)
[13] Al-Doury M M I 2010 *Petrol. Sci. and Technol.* **28** 1740
[14] Moradi M, Eslami A 2020 *J. Water Process. Eng.* **36** 101288
[15] Tessier A, Campbell P G C, Bisson M 1979 *Anal. Chem.* **51** 844
[16] Zhang Y H, Jin F, Shen Z T, Wang F, Lynch R, Al-Tabbaa A 2019 *Chemosphere* **220** 422
[17] Gong J L, Zhang Y L, Jiang Y, Zeng G M, Cui Z H, Liu K, Deng C H, Niu Q Y, Deng J H, Huan S Y 2015 *Appl. Surf. Sci.* **330** 148
[18] Gouran-Orimi R, Mirzayi B, Nematollahzadeh A, Tardast A 2018 *J. Environ. Chem. Eng.* **6** 2232
[19] Chen L P, Li Y C, Huang H, Xu Z, Ren B Z 2021 *Chem. Ind. Eng.* **38** 59
[20] García-Mateos F J, Ruiz-Rosas R, Marqués M D, Cotoruelo L M, Rodríguez-Mirasol J, Cordero T 2015 *Chem. Eng. J.* **279** 18

IOP Conf. Series: Earth and Environmental Science 1087 (2022) 012079 doi:10.1088/1755-1315/1087/1/012079

3D gravity inversion based on deep learning

Shuang Zhang[1], Changchun Yin[1,2], Jing Cai[1], Yunhe Liu[1], Xiuyan Ren[1], Bo Zhang[1] and Yang Su[1]

[1] College of Geo-Exploration Science and Technology, Jilin University, China, 130026

[2] Corresponding author's e-mail: yinchangchun@jlu.edu.cn;

Abstract. Gravity inversion is a typical geophysical inversion method that obtains the underground density distribution by analyzing the gravity anomaly. Normally, it can be divided into geophysics-based and deep learning based inversion. The 3D geophysics-based inversion is a time- and memory-consuming method, so 3D inversion is not routinely implemented in practical data interpretation. Here, we propose a deep learning method to transfer the 3D inversion problem to a multiple layers 2D mapping problem by decomposing the 3D target into four 2D images, including the horizontal location, vertical center, thickness and density distribution. This method is denoted as "decomposition network". By implementing synthetic experiments with regular and complex models, and comparing with the 3D U-Net inversion, the proposed network has proved can reconstruct underground targets with high accuracy and high efficiency.

1. Introduction

With the advancement of deep learning, many scientists have adapted deep learning concept into the geophysical study, including denoising [1], modeling [2], the prediction of subsurface velocity [3], resistivity [4], magnetization intensity [5], and density [6]. Unlike traditional geophysics-based inversion, the inversion based on deep learning has the advantage of no need for an initial model and can deal with large scale optimization problem. Especially for gravity inversion, without using the forward operator, the deep learning based inversion can avoid the inverted density to be concentrated near the surface. Another advantage is that the deep learning based inversion can be implemented without caring much about the theory of the geophysics once the dataset is set up.

Recently, deep learning based gravity inversion problem is becoming a hot research topic. L. Zhang [7], and R. Huang [8] proposed to use neural network to predict the 3D density structure. X. Yang built a U-Net to predict the underground CO_2 distribution by using the surface gravity data and the borehole leakage information [9]. These existed deep learning based gravity inversion methods are 2D-to-3D inversions that need much memory space to save 3D parameters and much time to train the network. To avoid this problem, we proposed a decomposition network.

2. Method

Our inversion method is aimed to simplify the 2D-to-3D inversion to a 4 layers 2D-to-2D mapping problem by dividing a 3D object into 4 components, including horizontal position, vertical center, vertical thickness, and density distribution. This method works to train a decomposition network to predict these 4 components and reconstruct the 3D models by using these 2D predicted results. The outline of this inversion method is shown in Figure 1 where the input gravity anomaly in Figure 1a and the output 2D images in Figure 1c are used to optimize this network in Figure 1b. Then, the 3D model (Figure 1d) is reconstructed from the 2D images in Figure 1c.

During the training process, the following loss function is used

$$\mathrm{L} = \frac{1}{N}\sum_{i=1}^{N}\sum_{j=1}^{4} \beta_j \left\| \boldsymbol{m}_{ij} - Net_j(\boldsymbol{g}_i;\boldsymbol{\theta}_j) \right\|_2^2, \tag{1}$$

where N represents the pairs of data used for training; For the ith pair of data, $(\boldsymbol{g}_i, \boldsymbol{m}_{ij})$ represent the gravity anomaly $\boldsymbol{g}_i$and the jth ground-truth values, including the true horizontal location $\boldsymbol{m}_{i1}$, true vertical center location $\boldsymbol{m}_{i2}$, true thickness of this anomaly $\boldsymbol{m}_{i3}$, and true lateral density distribution $\boldsymbol{m}_{i4}$. $\boldsymbol{\theta}_j$ is the learned parameter resulting from the corresponding j^{th} decoder, and β_j is the hyperparameter used to balance these 4 sub-loss terms. In this network, since the 2D images are normalized to the range of [0,1], all hyperparameters β_j are set as 1 to make the sub-losses have equal contributions to the total loss.

It is obvious that unlike traditional inversion method minimizing the difference between the observed gravity data and the gravity of the predicted model, the deep learning based inversion method minimizes the difference between the ground-truth value $\boldsymbol{m}_{ij}$ and the corresponding predicted results $Net_j(\boldsymbol{g}_i;\boldsymbol{\theta}_j)$.The Adam algorithm [10] and cosine learning rate annealing are used to train the network. It is a commonly used algorithm since it has high efficiency and fast convergence.

Once the training stage is finished, this trained network will be implemented to predict the target by using the gravity anomaly $\boldsymbol{g}_i$ as input, the predicted 2D images can be expressed as

$$\widehat{\boldsymbol{m}}_j = Net_j(\boldsymbol{g}_i;\boldsymbol{\theta}_j), j = 1,4 . \tag{2}$$

Then, the reconstructed 3D target will be

$$\boldsymbol{M}(x,y,z) = \begin{cases} \widehat{\boldsymbol{m}}_4(x,y), & if\ \widehat{\boldsymbol{m}}_1 = 1\ and\ (z \in \left[\widehat{\boldsymbol{m}}_2 - \frac{\widehat{\boldsymbol{m}}_3}{2}, \widehat{\boldsymbol{m}}_2 + \frac{\widehat{\boldsymbol{m}}_3}{2}\right]) \\ 0, & otherwise \end{cases} \tag{3}$$

Note that the predicted lateral location $\widehat{\boldsymbol{m}}_1$ is a binary image. $\widehat{m}_1 = 1$ means that there is an anomaly under this point, otherwise $\widehat{m}_1 = 0$. As can be seen from Equation (3), to do the inversion with this method, we assume that the density of the target is homogeneous in vertical direction, but there is no limitation for the distribution of the horizontal density.

By using this method, if we discrete the research area into Nx×Ny×Nz cubic grids, where Nx, Ny, and Nz represent the number of grid along x-axis, y-axis, and z-axis, the model parameters will be reduced from Nx×Ny×Nz to Nx×Ny×4. For machine learning problems with enormous training dataset, this advantage will save much memory space and dramatically improve the speed of training process. Apart from that, with the drastically reduced learned parameters, the network will be optimized easier and faster.

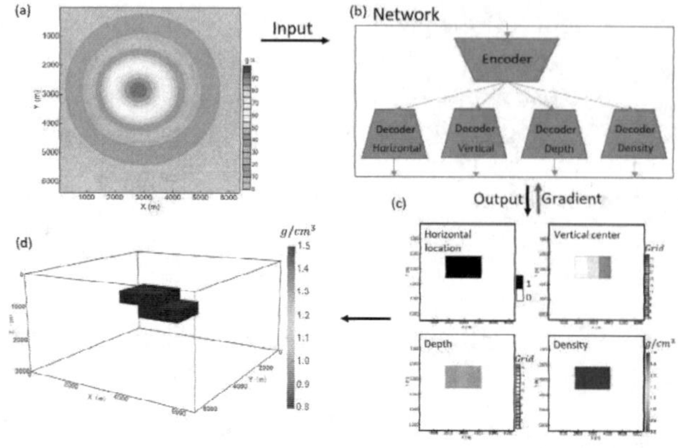

Figure 1. Inversion outline.

3. Network Structure

As the typical convolutional neural network (CNN), the U-Net is applied to build our network. The network that is shown in Figure 1b includes 1 encoder and 4 decoders that correspond to the 4 decomposed 2D images in Figure 1c. The encoder used to extract features from the input gravity anomaly is shared by these 4 decoders. Each encoder-decoder structure can be regarded as a U-Net.

The U-Net is a network modified upon the "fully convolutional network" where the pooling operators are replaced by sampling operators. It can yield more precise results with few training data and reasonable training time [11]. The structure of U-Net is shown in Figure 2. It is a symmetric architecture. The typical structure on the left side is the encoder part. The right side is the decoder part. More information about the structure of U-Net, please refer to Ronneberger et al., 2015.

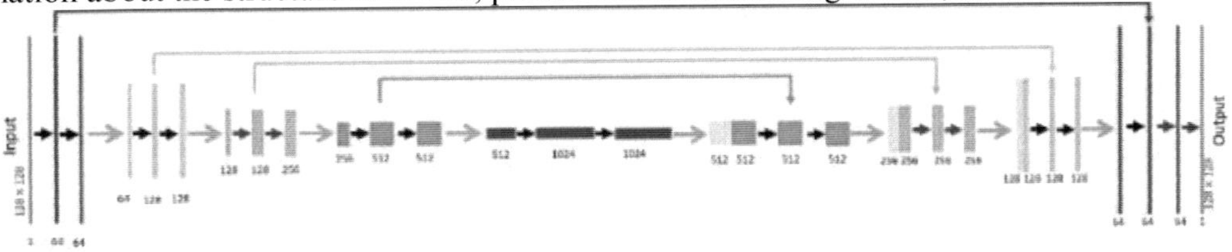

Figure 2. Simplified U-Net architecture. In this figure, the input and output size are both 128×128.

4. Experiments

4.1. Experiment 1

To demonstrate the application of our proposed network, we built a dataset with 40000 pairs of data in the 6400m × 6400m × 3000m research area. The shapes of these models used for generating this dataset include models with up to three prisms inside, stepped models, similar-rounded models, and those with inhomogeneous density in horizontal direction. To prepare this dataset, we discretize this area into 128 × 128 × 60 cubic grids and the model size are randomly selected under acceptable precision. The length along x-axis and y-axis are randomly numbers selected from [5,6,7,8,9,10] × 200m. The length along z-axis is randomly numbers selected from [5,8,11,14] × 100m. The densities of these anomalies are chosen from 0.5~1.5 g/cm^3. The observations and the 2D images are matrixes with the size of 128 × 128, which means the input and output of the network are both set as the same size. The ratio for training, validation and testing is 96:2:2. Each pair of data is composed by the gravity anomaly used as input for encoder, and four 2D images used as the outputs of these decoders, including the 2D horizontal distribution, vertical center, thickness and density.

The initial learning rate is set as 0.0001. After 50 times epochs when the loss is convergent, the training stage is stopped. This training is implemented on NVIDIA GeForce RTX 3090 (24G memory). It takes 9.77 minutes to train each epoch and it takes one second to predict the results with the trained network. The model in Figure 3a that has two stepped bodies with different densities is used for testing. The predicted 3D model is shown in Figure 3c. Figure 3b and 3d show the corresponding slices through each stepped anomalies. It can be seen that the predicted densities and locations correspond well with the true model especially the horizontal boundary. However, there is a slightly rapped boundary around the upper and bottom surfaces. Moreover, the mean square error (MSE) is also used to analyze the predicted 2D images. The MSEs for the predicted horizontal distribution, vertical center, thickness, and density are respectively 0, 0.001 m, 2.4×10^{-4} m, 4.5×10^{-7} m. This synthetic experiment demonstrate that our decomposition network can predict the underground structure with high-efficiency and high accuracy.

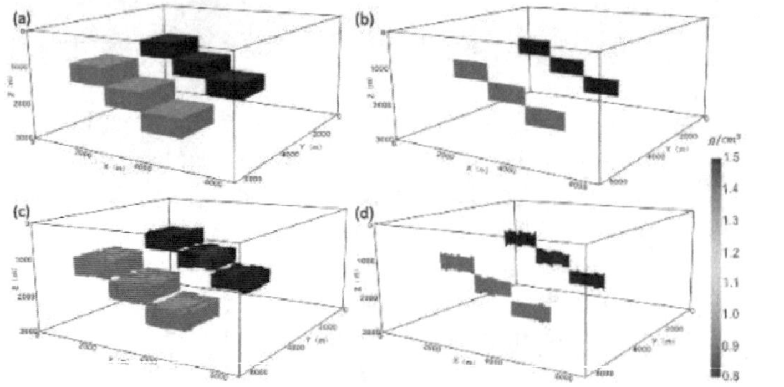

Figure 3. (a) 3D true model; (b) Slices through (a) at y= 1600m and y= 5200m (c) 3D reconstructed model; (d) Slices through (c) at y=1600m and y=5200m.

4.2. Experiment 2

To further demonstrate the application of our proposed network, we built a complex dataset with 10065 pairs of data in the research area that covers $10300\text{m} \times 8446\text{m} \times 800\text{m}$. The observations and four 2D images are used as matrix with the size of 100×82. The length along x-axis and y-axis are randomly numbers selected from $[3,4,5,6] \times 1000\text{m}$. The length along z-axis is randomly numbers selected from $[4,6,8,9,10,11,12] \times 50\text{m}$. The densities of these anomalies are chosen from [0.2,0.4,0.5,0.6,0.7] g/cm^3. This complex dataset is composed of models with irregular boundary, fluctuated depth, and inhomogeneous horizontal density distribution. The thickness and density distribution of these complex models are randomly decreased from the center to the boundary. The ratio for training, validation and testing is 96:2:2. By using these complex models for training, we get a well convergent trained network after 100 epochs. This training is implemented on NVIDIA GeForce RTX 3090. It takes 90s to train each epoch.

The complex model in Figure 4a is used for testing, and Figure 4d shows the reconstructed 3D model. The lateral slices at the depth of z=360m are also shown in Figure 4b and 4e. Figure 4c and 4f show the vertical slices of the profile existed in Figure 4b and 4e. It can be seen that the reconstructed model fits well the true model, including the densities and horizontal distribution. As for the vertical information, there are also ragged boundaries near the upper and bottom surface. Moreover, the MSEs corresponding to the horizontal distribution, vertical center, thickness and density are 0, 5.11×10^{-6} m, 2.0×10^{-5} m, 2.7×10^{-8} m. This experiment proves that our decomposition network can not only be used to predict anomalies with sharp boundary and homogeneous density distribution but also be used to detect complex anomalies with irregular inclination boundary and inhomogeneous horizontal density distribution.

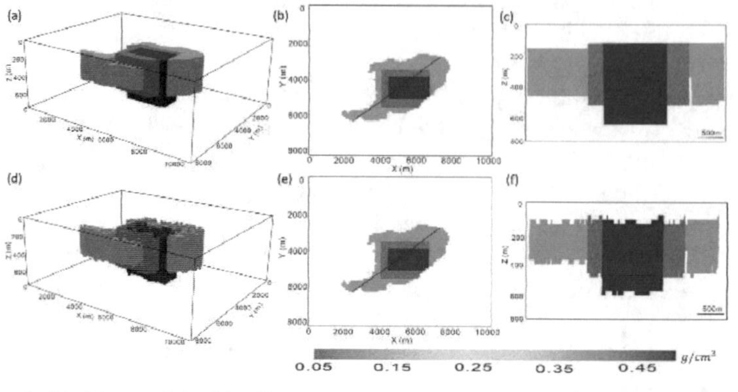

Figure 4. (a) 3D model; (b) slice at z=360m; (c) profile showed in (b); (d) 3D reconstructed model; (e) slice at z=360m; (f) profile showed in (e).

4.3. Experiment 3

To demonstrate the advantage of our method, the comparison experiment between the decomposition network and the 2D-to-3D deep learning inversion method is implemented. The 3D deep learning inversion is the network build by U-Net and the size of the input gravity anomaly is the same as that used in the decomposition network. In this experiment, the same discretized grids in Experiment 1 is used. Since the 2D-to-3D deep learning with enormous dataset needs 15 times of memory space as the decomposition network, we build 4900 pairs of data by using the model with 3 prisms embedded. Then, these two networks are optimized by the same MBGD and Adam algorithms. The initial learning rates are both set to 0.0001. Both networks are trained for 100 epochs. To train each epoch, the decomposition network takes 106s and the 2D-to-3D U-Net takes 192s. Figure 4 shows the true model and predicted results. The plane view and the vertical slice at y=3500m are also shown in the figure. It is obviously that the model predicted by decomposition network in Figure 5d fits well the true model in Figure 5a except the ragged boundary. However, for 2D-to-3D U-Net inversion, although this network can predict the existence of three prisms, there are some false low-density anomalies and the predicted density in Figure 5i is obviously higher than the true value. Generally, compared with 2D-to-3D U-Net inversion, our proposed method can predict the subsurface target with higher efficiency and higher accuracy.

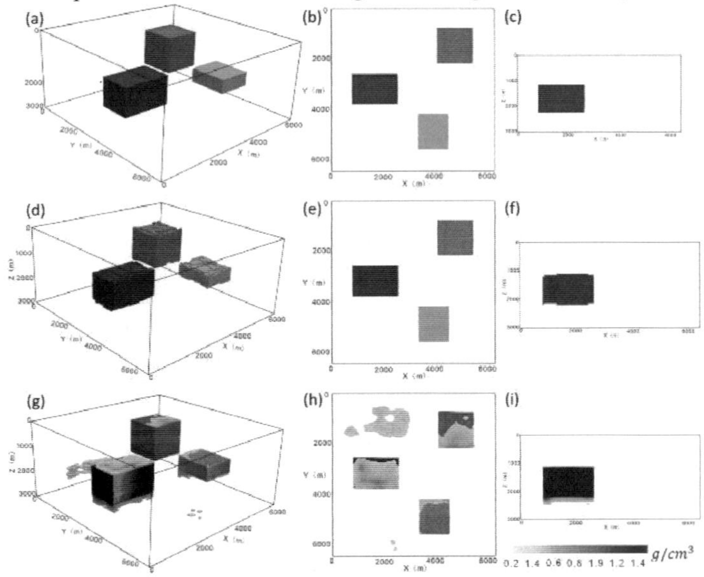

Figure 5. (a)-(c) 3D true model; (d)-(f) 3D model reconstructed by decomposition network. (g)-(i) 3D model reconstructed by 3D U-Net method; (b)(e)(h) plane view; (c)(f)(i) vertical slices at y=3500m.

5. Conclusion

To simplify the 3D gravity inversion problem, we proposed a new deep learning network to decompose the 3D gravity inversion problem to a 2D mapping problem that can save much memory space, reduce the number of predicted parameters and accelerate the inversion efficiency. By theoretical experiments and comparison with the 2D-to-3D U-Net inversion method, our proposed method has proved that it can reconstruct the underground target with high accurate and high efficiency, even the shape and density distribution of the target are complex. With sufficient training dataset, this well trained network can be used for real-time prediction. Moreover, the predicted model can be further used as a prior model for geophysics-based inversion to prevent the optimized model from being trapped in local minima and to accelerate the convergence speed of the inversion. We will further apply this network to the inversion of field data and combine the decomposition network with traditional inversion method in order to solve more complicated 3D gravity and magnetic inversion problems.

Acknowledgements
This manuscript is financially supported by the National Natural Science Foundation of China (42030806) and Key National Research Project of China (2021YFB3202104).

References

[1] B. Liu, J. Yue, Z. Zuo, X. Xu, and P. Jiang, "Unsupervised Deep Learning for Random Noise Attenuation of Seismic Data," IEEE Geoscience and Remote Sensing Letters, vol. PP, no. 99, pp. 1-5, 2021

[2] T. S. Bording, M. R. Asif, A. S. Barfod, J. J. Larsen, and E. Auken, "Machine learning based fast forward modelling of ground-based time-domain electromagnetic data," Journal of Applied Geophysics, p. 104290, 2021

[3] Z. Zhang, Y. Wu, Z. Zhou, and Y. Lin, "VelocityGAN: Subsurface velocity image estimation using conditional adversarial networks," in 2019 IEEE Winter Conference on Applications of Computer Vision (WACV), 2019: IEEE, pp. 705-714

[4] B. Liu et al., "Deep learning inversion of electrical resistivity data," IEEE Transactions on Geoscience and Remote Sensing, vol. 58, no. 8, pp. 5715-5728, 2020

[5] Z. Hu et al., "Inversion of magnetic data using deep neural networks," Physics of the Earth and Planetary Interiors, vol. 311, p. 106653, 2021

[6] Q. Yang, X. Hu, S. Liu, Q. Jie, H. Wang, and Q. Chen, "3-D Gravity Inversion Based on Deep Convolution Neural Networks," IEEE Geoscience and Remote Sensing Letters, 2021

[7] G. Z. L. Zhang, Y. Liu, Z. Fan, "Deep Learning for 3-D Inversion of Gravity Data," IEEE Transactions on Geoscience and Remote Sensing, vol. 60, pp. 1-18, 2022, doi: 10.1109/TGRS.2021.3110606

[8] R. Huang, S. Liu, R. Qi, and Y. Zhang, "Deep Learning 3D Sparse Inversion of Gravity Data," Journal of Geophysical Research: Solid Earth, vol. 126, 2021

[9] X. Yang, X. Chen, and M. M. Smith, "Deep learning inversion of gravity data for detection of CO2 plumes in overlying aquifers," Journal of Applied Geophysics, vol. 196, pp. 104507-, 2022

[10] D. P. Kingma and J. Ba, "Adam: A method for stochastic optimization," arXiv preprint arXiv:1412.6980, 2014

[11] O. Ronneberger, P. Fischer, and T. Brox, "U-net: Convolutional networks for biomedical image segmentation," in International Conference on Medical image computing and computer-assisted intervention, 2015: Springer, pp. 234-241

Assessment of geothermal resources in the west area of Taikang town in Dumeng geothermal field, NE China

Yongfa Ma[1,2,3,5], Fengtian Yang[4], Xuejun Zhou[1], Tao Zhan[1], Ling Liu[1], Nan Li[1,2], Xu Wang[1], Junling Dong[1], Yu Liu[1], Yan Liu[1], Zhenzhen Shi[1] and Chang Li[1]

[1] The Institute of Ecological Geology Survey and Research of Heilongjiang Province, 150030 Harbin, Heilongjiang, PR China;
[2] Heilongjiang 904 Environmental Engineering Survey and Design Institute Co., Ltd, 150030 Harbin, Heilongjiang, PR China;
[3] China University of Geosciences,430074 Wuhan, Hubei, PR China;
[4] College of New Energy and Environment, Jilin University, 130021 Changchun, Jilin, PR China.

[5] Corresponding author's e-mail: 80902601@qq.com

Abstract. Combined with geothermal drilling, well logging, geological and geophysical investigations, this study established a conceptual model of the Dumeng geothermal field, Songliao Basin, and evaluated the potential of the geothermal resources of western Taikang town, an area under concentrated development in the geothermal field, based on transient groundwater flow equation. The results show that the Dumeng geothermal field is part of a conductive type geothermal system. The evaluated heat reserve is 8.76×10^{17} J, the storage of thermal water is 4.94×10^{9} m^{3}, the recoverable reserve of the thermal water is 9.85×10^{5} m^{3}/a, which is equivalent to an outpower of 5.59 MW, and is comparable to a small geothermal field. This study provides a basis for scientific management of the further exploration and utilization of the geothermal resources and application of the mining rights.

1. Introduction

As the impacts of global climate change become more and more significant, renewable energies has growing fast in recent years. In fact, development of renewable energies can not only decease emission of greenhouse gas, but also mitigate harmful gases, such as particulate matter, nitrogen and sulphur oxides, which consists the main air pollutants in winter during heat supply across north China [1]. Among the various renewable energies, geothermal energy is the most practical and suitable one to be used as a base energy, for geothermal energy has a large reserve, distributes widely, thus can provide a stable supply [2-5]. To push the development and utilization of geothermal energy is of great strategic significance for the adjustment of national energy structure and the realization of "carbon peak and carbon neutral" in China and around the world. National Energy Administration of China (NEAC) has also introduced new policies to promote development of renewable energies, especially emphasized the important role of geothermal energy in building heating and mitigate smog weathers in cold regions in north China [6]. However, geothermal resources would not be "sustainable" anymore if over-exploited [7]. Assessment of the geothermal reserves and sustainable withdrawal is key for management of geothermal filed to achieve rational development [8].

IOP Conf. Series: Earth and Environmental Science 1087 (2022) 012080 doi:10.1088/1755-1315/1087/1/012080

The Dumeng geothermal field in Heilongjiang province is located in north Songliao Basin, NE China. After decades of exploration and research by geological and petroleum industries, it has been proved that the Dumeng geothermal field is abundant in low to medium temperature geothermal resources. The geothermal conditions are similar to that of the adjacent Lindian geothermal field [9-10], which is the largest one being exploited in the Songliao Basin. Previous works mainly focused on the structure of the reservoir, and survey of the resources in different blocks in the geothermal field [11-13], but the genesis of the geothermal fluid on field scale and the resource potential are still unclear [12,14].

Based on preliminary exploration and newly collected data, this paper characterizes the geological structure, strata, intrusive rocks of the Dumeng geothermal field, analyzes genesis of the geothermal anomaly based on key elements of geothermal fields, i.e., "heat, reservoir, channel and caprock"[15-16], and establishes a conceptual model of the geothermal field, evaluated the resource potential of a key block, i.e., western Tangkang Town, with significance for further exploration, exploitation and research of the geothermal resources in this area in the future.

2. Geological settings

The study area is located in the middle of Songnen Plain and the lower reaches of the Wuyuer River. Tectonically, the study area is located at the intersection of the Western Slope and the Central Depression of the north Songliao Basin, and mainly belongs to Taikang Uplift, Longhupao-Da'an Terrace and Qijia-Gulong Depression (refer to Figure 1).

Based on previous geothermal and oil drilling data [12-13], strata in the study area include the basement consists of the Carboniferous-Permian System, overlained by Jurassic System, the Denglouku, Quantou, Qingshankou, Yaojia, Nenjiang, Sifangtai and Mingshui Formations of the Cretaceous, the Yi'an Formation of the Paleogene, and the Taikang, Lindian and Daxingtun Formations of the Quaternary.

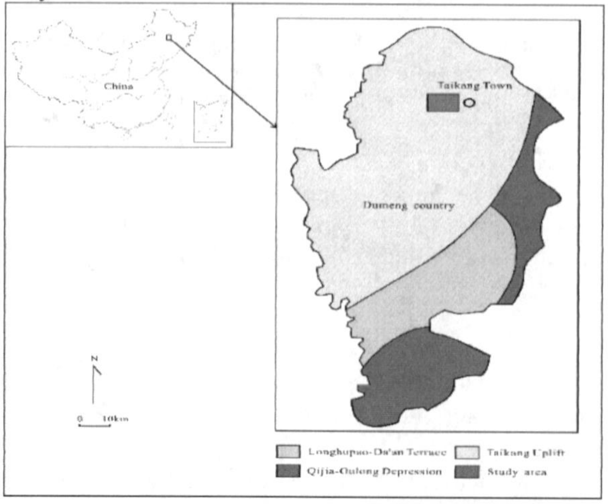

Figure 1. Location of Dumeng geothermal field.

The lithology of the Carboniferous-Permian is dominated by metamorphic marine clastic rock deposition, mainly develops hydrocarbon source rocks such as dark mudstone and phyllite. The Jurassic mainly consists of sandstone, conglomerate and sandstone, mudstone, basalt, andesite, rhyolite and tuff, and thin coal beds. The Cretaceous is well developed in this area, with a total thickness of >3000 m, with the main lithology of interbed sand conglomerate, sandstone and mudstone. The Paleogene is comprised of mudstone in the upper part and glutenite in the lower part. The Quaternary deposits are dominated by fluvial and lacustrine facies. The main lithology are gravel, fine sand, clay and loess [10,12].

IOP Conf. Series: Earth and Environmental Science 1087 (2022) 012080 doi:10.1088/1755-1315/1087/1/012080

3. Conceptual model of the geothermal field

Songliao basin is one of the basins in China with the shallowest Moho and Curie surface. The Dumeng geothermal field is located in the north Songliao basin where the Moho and Curie surface have a small burial depth (refer to Figure 2). Since there is a large area of Mesozoic granite distributed in the basement (refer to Figure 2), heat from the mantle and generated by radioactive decay flows up into the shallow crust, resulting in a high heat flow in the study area [16]. The geothermal reservoirs in the Dumeng geothermal field are mainly comprised of the sandstone layers of Quantou, Qingshankou, Nenjiang and Yaojia Formation of the Cretaceous (refer to Table 1). The burial depth of the reservoirs is ca. 1000-2000 m, and the reservoir temperature is ca. 50- 90 ℃ [12, 16]. Underlain the reservoirs are the lower and middle member of the Quantou Formation, the Denglouku Formation of the Cretaceous, the Jurassic, the Carboniferous-Permian and granite, which are of low permeable, thus serves as heat conduction layers. Lindian-halahai Fault cuts thourgh the crust, and its associated Qixi and Xiaolinke faults are good channels for deep heat conduction to the upper strata [16], whereas the Sifangtai, Mingshui Formation of the Cretaceous, the Paleogene and Quaternary containing a large proportion of muddy layers with a thickness of more than 1000 m forms a good caprock. The geothermal reservoir in Dumeng geothermal field receives recharges from atmospheric precipitation through the Paleogene, Neogene and Quaternary in the northwestern margin of the Songliao Basin [11].

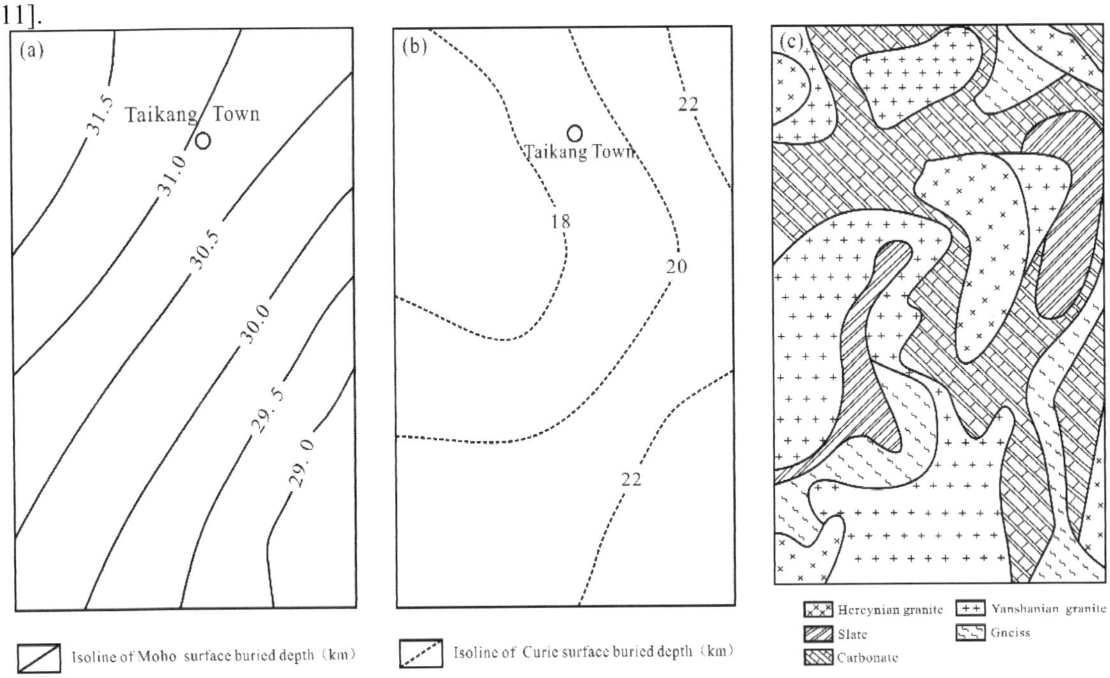

Figure 2. Contour maps for burial depth of (a) Moho, (b) Curie surface, and (c) basement lithology of the Dumeng geothermal field.

Genesis of geothermal systems can be analyzed from four aspects, i.e., "heat, reservoir, channel and caprock", which respectively refer to heat sources and geo-temperature field, characteristics of geothermal reservoir, groundwater flow channel and characteristics of caprock. According to this theory, Dumeng geothermal field is a relatively simple geothermal system consisting of the underlain heat conductive layer, the geothermal reservoir and the overlain caprock (refer to Figure 3), belongs to a low to medium temperature, conductive geothermal system, which is characterized by wide distributed sandstone reservoirs with relatively stable lithology and thickness, and simple structural conditions.

Table 1. Porosity and permeability of the main geothermal reservoirs of Dumeng geothermal field.

Stratum		φ (%)	Permeability
Formation	Member		(×10^{-3}μm^2)
Nenjiang	2-5	23.6	83.2
Yaojia	2-3	24.4	486
	1	22.5	445
Qingshankou	2-3	24.8	314
Quantou	4	21.0	154
	3	23.3	47.1

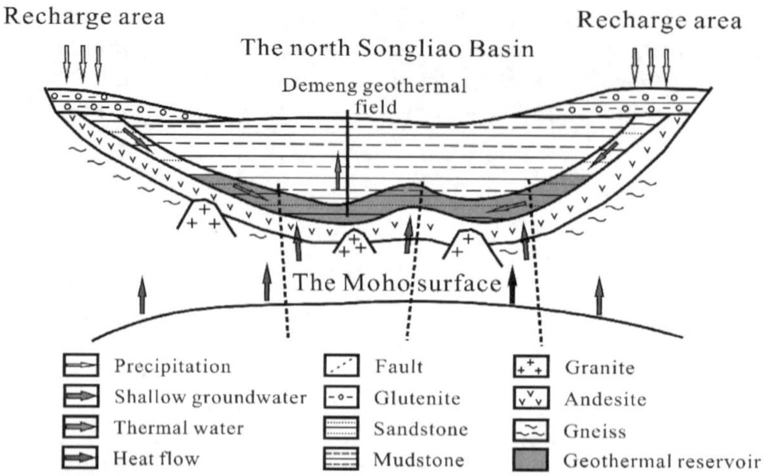

Figure 3. Conceptual model of Dumeng geothermal field.

4. Calculation of geothermal resources in western Taikang town

The chosen for potential assessment of the geothermal resources in this study is the Lianhuanhu area in the western Taikang town (refer to Figure 1), with high degree of geothermal exploration, covering an area of 159 km^2. As the geothermal field has newly been developed and lack of long time series of water level monitoring data, the resource base is evaluated mainly by the volumetric method, whereas the accessible resource base is evaluated mainly by transient groundwater flow equation, according to 'Estimation and evaluation method of geothermal resources (DZ/T 0331-2020) issued by Ministry of Natural Resources, PRC. [17]. The flow chart is shown as Figure 4.

4.1. Estimation of the resource base

4.1.1. Calculation formula. The resource base, i.e., the total amount of heat energy stored in the thermal reservoir, which is calculated by the volumetric method [17]

$$Q=Q_r+Q_w \quad (1)$$

$$Q_r=Ad\rho_r C_r(1-\varphi)(t_r-t_0) \quad (2)$$

$$Q_w=Q_L\rho_w C_w(t_r-t_0) \quad (3)$$

$$Q_L=Q_1+Q \quad (4)$$

$$Q_1=A\varphi d \quad (5)$$

$$Q_2=ASH \quad (6)$$

where Q is heat stored in a heat reservoir, J; Q_r is the heat stored in the rock, J; Q_w is heat stored in geothermal fluids, J; Q_L is the volume of thermal water stored in the reservoir, m^3; Q_1 is static storage of thermal water, m^3; Q_2 is elastic storage of thermal water, m^3; A is area of the study area, m^2; d is thickness of the thermal reservoir, m; φ is porosity of the reservoir rock, dimensionless; t_r is reservoir temperature, ℃; t_0 is average annual temperature of the study area, ℃; ρ_r, ρ_w are the density of rock and water, respectively, kg/m^3; C_r, C_w are the specific heat capacity of reservoir rock and water, respectively, J/kg/℃; S is elastic water release coefficient, dimensionless; H is pressure head calculated from roof of the reservoir, m.

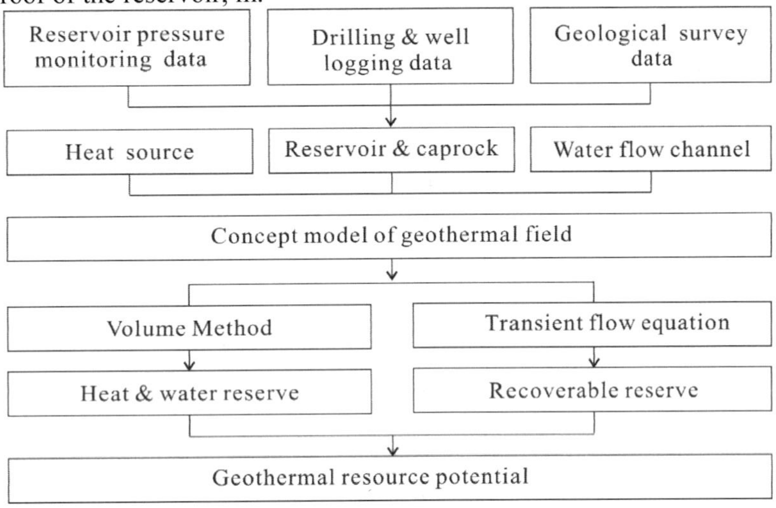

Figure 4. Flow chart of potential assessment of the geothermal resources.

4.1.2. Parameter Selection.

(1) D, φ, d, S and H

The mean values of d, φ, d, S and H of well DR1, DR6, DR8, DR9, DR10, D26, D42, D73 and D101 were selected in calculation (refer to Table 2).

Table 2. The average values of d, φ, t_r, S and H for the study area.

Well ID	d（m）	φ（%）	t_r（℃）	S	H(m)
DR1	208.4	—	52.6	—	935
DR6	—	—	48.6	8.13×10^{-4}	901
DR8	—	—	46	—	—
DR9	48.4	24.87	51.3	8.97×10^{-4}	909
DR10	—	—	44	—	—
D26	168	23.78	53.4	—	940
D42	134.4	23.72	44.1	—	900
D73	98.8	27.72	45.8	—	908
D101	83	22.59	34.7	—	870
Average	123.5	24.54	46.7	8.55×10^{-4}	909

(2) t_0
The average annual temperature t_0 in Dumeng county is 4℃.
(3) ρ_r and C_r
The empirical value of sandstone was taken, i.e., the rock density is 2600 kg/m^3 and specific heat is 878 J/kg/℃ [17].
(4) ρ_w and C_w
The density of water is estimated to be 985.7 kg/m^3, and the specific heat of water is 4.2×10^3 J/kg/℃ [17].

Table 3. Calculated heat energy and thermal water storage in the study area.

d (m)	*φ* (%)	t_r (℃)	*S*	*H* (m)
123.5	24.54	46.7	8.55×10^{-4}	909
t_0 (℃)	C_r (J/kg/℃)	*ρr* (kg/m^3)	C_w (J/kg·℃)	ρ_w (kg/m^3)
4	878	2600	4.2×10^3	985.7
Volume storage of geothermal fluids (m^3)	Elastic storage of geothermal fluids (m^3)	Total storage of geothermal fluid (m^3)	Heat stored in geothermal fluids(J)	The heat stored in the rock (J)
4.82×10^9	1.24×10^8	4.94×10^9	8.74×10^{17}	1.65×10^{15}
Total amount of heat energy in reservoir（J）			8.76×10^{17}	

4.1.3. Resource base and thermal water reserve. The total amount of heat energy stored in the geothermal reservoir in the study area is shown in Table 3.
The estimated water reserve Q_Ls shown in Table 4.

Table 4. Calculated recoverable geothermal resources in the study area.

Burial depth of static water level (m)	Maximum drawdown limit (m)	Area (km^2)	Number of deployable wells
20	330	159	9
Recoverable reserve per well (m^3/d)	Coefficient of hydraulic conductivity(m^2/d)	Elastic water release coefficient	Coefficient of pressure conductivity(m^2/d)
300	62.64	8.55×10^{-4}	7.33×10^4
Recoverable reserve	2700 m^3/d		9.85×10^5 m^3/a

4.2. Recoverable reserve of thermal water

4.2.1. Calculation model. The recoverable reserves of geothermal fluid are evaluated by uniform well distribution method of analytical model [17]

$$S_i = \sum_{i=1}^{n} \frac{Q_i}{4\pi T} W\left(\frac{r_i^2}{4at}\right) \tag{7}$$

where s_i is interference drawdown of each well to well *i*, m; r_i is distance from well *i* to the other wells, m; t is pumping duration, d; T is coefficient of hydraulic conductivity, m^2/d; a is coefficient of pressure conductivity, m^2/d; Q_i is designed water yield of well *i*, m^3/d; $S_i = W\left(\frac{r_i^2}{4at}\right)$ is well function.

4.2.2. Parameters for calculation.
(1) Calculate the exploration time and maximum drawdown limit
According to the standard [17], the exploration time for a geothermal field is 100 years. At present, the maximum lifting height of the water pump used is 350 m, thus the maximum drawdown limit is the lifting height of the water pump minus the burial depth of static water level (20 m), i.e., 330 m.
(2) Water yield per well
It was determined based on pumping test and the actual water yield of the production well, as well as the drop down of the water level during pumping. According to the study, the water yield of single well in the study area is 300 m^3/d [16].
(3) Number and spacing of wells
According to the previous observation data, continuous pumping would bring interference effects on wells with a interval of 2000 m or 3000 m. After comprehensive consideration, the well spacing in the study area is set as 4000 m. The number of deployable wells in the study area is $A/(2R)^2$, and the calculated value is nine.
(4) Coefficient of hydraulic conductivity, Coefficient of pressure conductivity
According to the pressure-lowering test data [13], the coefficients of hydraulic conductivity and pressure conductivity were calculated and determined to be 62.6 m^2/d and 7.33×10^4 m^2/d using Theis and Jacob formula.

4.2.3. The recoverable reserves. The recoverable reserves of thermal water in the study area is 2700 m^3/d, or 9.85×10^5 m^3/a (refer to Table 4).

4.2.4. Evaluation of guarantee degree of recoverable reserve. It can be seen from Table 3 and Table 4 that the recoverable reserve of thermal water in the well layout scheme is $9.85\times10^7 m^3$ in 100 years, far less than the recoverable elastic storage of $1.24\times10^8 m^3$. The maximum drawdown is 293.88 m after 100 years of exploration, which does not exceed the allowable maximum drawdown of 330 m (refer to Table 4), indicating that the well layout scheme is reasonable and the calculated recoverable reserve is reliable.

4.3. Electricity generating capacity. The electricity generating capacity can be calculated as

$$W_t=4.1868Q(t-t_0) \quad (8)$$

where W_t is equivalent of electricity generating capacity of heat energy, kw; Q is recoverable reserve of thermal water, L/s; t_r is reservoir temperature, ℃; t_0 is average annual local temperature, ℃; 4.1868 is unit conversion factor [17]. Parameters used in the calculation are shown in Table 5.

The evaluation results of the electricity generating capacity equivalent and scale of the geothermal resources in the study area is shown in Table 5. It can be judged that the electricity generating capacity equivalent of the key evaluation area is 5.59 MW, which is equivalent to small-scale geothermal field.

Table 5. Electricity generating capacity equivalent of the geothermal reservoir.

Recoverable reserve of thermal water		t_r (℃)	t_0 (℃)	Electricity generating capacity equivalent (MW)	Scale
(m^3/d)	(L/s)				
2700	31.25	46.7	4	5.59	small

5. Conclusions

The Dumeng geothermal field is abundant in low to medium temperature geothermal resources. Based on the volume method and transient groundwater flow equation, the base reserve estimated is 8.76×10^{17} J, the thermal water storage is 4.94×10^9 m^3, the recoverable thermal water reserve is 9.85×10^5 m^3/a, and the electricity generating capacity equivalent is 5.59 MW, suggesting it is a small-scale geothermal field.

It is believed that this study will provide a basis for further exploration, exploitation, management and research of the geothermal resources in this Dumeng geothermal field in the future. However, further exploitation of the geothermal resources will affect the geo-temperature field of the geothermal reservoir, so it is necessary to strengthen monitoring and adopt reinjection technologies to protect the geothermal resources.

Acknowledgements
The research work is funded by "Investigation and evaluation of geothermal resources in geothermal well concentration area of Heilongjiang Province" (2015-684) ,"Geothermal resource pre-feasibility exploration and geothermal heat transfer experiment in Nanyuan Development Zone, Qiqihar city, Heilongjiang Province" (SRKC-2019002),Scientific research fund of Geology and Mineral Resources Bureau of Heilongjiang Province, Scientific research fund of Institute of Ecological Geology Survey and Research of Heilongjiang Province (cxjj-001, cxjj-002, cxjj-003).

References
[1] Si R, Xin J, Zhang W, et al 2022 The environmental benefit of Beijing-Tianjin-Hebei coal banning area for North China *Journal of Environmental Management* **311** 114870
[2] Lin W, Liu Z, Wang W, et al 2013 Evaluation of geothermal resources and their potential in China *Geology in China* **40(1)** 312-320
[3] Zhao Z, Luo Y, Meng M, et al 2013 Preliminary study on geothermal resources in Qinghai Province and their exploration, exploitation and utilization *Qinghai Environment* **23(3)** 130-135
[4] Muther T, Syed FI, Lancaster AT, et al 2022 Geothermal 4.0: AI-enabled geothermal reservoir development- current status, potentials, limitations, and ways forward *Geothermics* **100** 102348
[5] Yang F, Wang G, Hu D 2022 Influence of water-rock interaction on permeability and heat conductivity of granite under high temperature and pressure conditions *Geothermics* **100** 102347
[6] NEAC 2021 Several opinions on promoting the development and utilization of geothermal energy. http://www.nea.gov.cn/2021-04/14/c_139880250.htm
[7] Yanto E, Firanda E, Budiyanto A, et al 2021 Exploitation Strategies to Minimize Decline Rate at Lumut Balai Area, PT. Pertamina Geothermal Energy, Indonesia. PROCEEDINGS, 46th Workshop on Geothermal Reservoir Engineering Stanford University, Stanford, California, February 15-17, 2021 SGP-TR-218
[8] Ciriaco AE, Zarrouk SJ, Zakeri G 2020 Geothermal resource and reserve assessment methodology: Overview, analysis and future directions *Renewable and Sustainable Energy Reviews* **119** 109515
[9] Zhu H 2011 Research on the sedimentary geothermal resources in North Songliao Basin Daqing: Northeast Petroleum University
[10] Su Y 2021 Genesis and rational development of typical geothermal field in the Songliao Basin: a case study of Lindian geothermal field Changchun: Jilin University
[11] Li Z, Zhang R, Shi S, et al 2004 Simulation of heat reservoir fluid migration in Dumeng area, Songliao Basin *Petroleum geology and oilfield development in Daqing* **23(6)** 28-31
[12] Shi L 2004 Study on tectono-thermal evolution and quantitative evaluation of geothermal resources in Songliao basin: A case study of Dumeng Area Guangzhou: Guangzhou Institute of Geochemistry *Chinese Academy of Sciences*
[13] Wang M, Sui X, Ni C 2010 Geothermal resources in Shirengou geothermal area, Duerbote Mongolian Autonomous County, Heilongjiang Province. Harbin, Heilongjiang: Heilongjiang Second Hydrogeology engineering Geology Investigation Institute
[14] Shi L Li Z, Shi S 2004 Formation and resource of geothermal field in Dumeng area, Songliao basin *Petroleum geology and oilfield development in Daqing* **23(3)** 26-29

[15] Rolker J, Schill E, Stober I, et al 2015 Hydrochemical characterisation of a major central European heat flux anomaly: the Bürchau geothermal spring system, Southern Black Forest, Germany *Geothermal Energy* **3** 5
[16] Zhou X, Liu L, Zhan T 2020 Investigation and evaluation of geothermal resources in concentrated area of geothermal wells in Heilongjiang Province. Harbin, Heilongjiang: Heilongjiang Second Hydrogeology engineering Geology Investigation Institute
[17] Wang G, Li M, Zhang M, et al. 2020 Estimation and evaluation method of geothermal resources (DZ/T 0331-2020) Beijing: Ministry of Natural Resources, PRC

IOP Conf. Series: Earth and Environmental Science 1087 (2022) 012081 doi:10.1088/1755-1315/1087/1/012081

Relationship between the size composition observed from the photosedimention technique and chromaticity of the fine-grained surface sediments from the modern lakes, Songnen Plain

Yangxin Su[1,2], Xinrong Zhang [3], Shun Zhang[1,2], Shi Fang[3], Xiuli Fu[1,2] and Hongjun Shao[1,2]

[1] Exploration and Development Research Institute of Daqing Oilfield Co Ltd;
[2] Heilongjiang Provincial Key Laboratory of Continental Shale Oil;
[3] Jilin University.

[4] Corresponding author's e-mail: zhangxr@jlu.edu.cn

Abstract. The relationship between clay particle size in fine-grained sediments and the enrichment of organic matter and the redox degree of their environment is conducive to the judgment and evaluation of the hydrocarbon generation potential of fine-grained sediments. The grain size characteristics of clay particles in the surface fine-grained sediments of Chaersen lake, Delong lake, Hongyuan lake, Lamasi lake, and Zhongnei lake in Songnen Plain are revealed by light sedimentation method. On the basis of the test, the relationship between the particle size of clay particles and the sediment chromaticity is analyzed, and the directivity of chromaticity to the content of organic matter and redox degree of fine-grained sediments is used for the organic content estimation. It is confirmed that the particle size of clay is negatively correlated with its organic matter content and positively correlated with its redox degree. This has important reference significance for discussing the hydrocarbon generation potential and evaluation of Cretaceous lacustrine fine-grained source rocks in the Songliao Basin.

1. Introduction

A Lacustrine deposition is the product of material and energy exchange among the atmosphere, hydrosphere and lithosphere [1,2], which contains abundant environmental information and mineral resources [3-6]. Fine-grained sediments in lakes are mainly composed of clay and silt with particle size less than 62.5 μm. The main components of sediments of lakes are quartz, feldspar, carbonate, clay and organic matter. Fine grained sediments in lakes are often related to the occurrence of oil and gas resources. With the continuous discovery of shale oil and shale gas, the deposition environment, development mechanism, and characterization methods of fine-grained sediments have attracted wide attention from academia and industry [7].What is the relationship between the clay particle size in fine-grained sediments and the enrichment of organic matter and the degree of redox in the environment directly restricts our judgment and evaluation of hydrocarbon generation potential of fine-grained sediments [8-13].

There are many methods for testing particle size components. The commonly used methods are the laser particle size analyzer method (the test range is 0.3-2000μm) and the thin-film method, but the effective test range of the two methods is relatively wide, and the expression efficiency of clay is very low (16%-20%) [14-18]. The photosedimentation grain size analysis method (measuring range 0.1~300 μm) has great advantages in measuring the grain size and expression sorting degree of clay, and is a preferred method to analyzed the grain size composition of fine sediment clay, but this method is costly and time-consuming [19]. Chroma has a strong indicator to the organic matter content and oxidation reducibility of fine grained sediments [19-22].

Cretaceous stratum lacustrine facie fine grained sedimentary strata in Songnen Plain are important hydrocarbon source rocks in Daqing Oilfield.The Songliao Basin is developed in the second subsidence zone of the Neocathaysian tectonic system. Since the late Mesozoic, it has been filling subsidence and deposited a thick fine sedimentary layer.Since the Quaternary, the region is still in a state of subsidence and filling, forming a broad group of lakes, such as The Charsen reservoir, Chaole Lake, Nierji reservoir, and other lakes in the Greater Khingan Mountains in the piedmont area of the beaded distribution;Lakes along the Nenjiang River such as Delong Lake, Hongyuan Lake and Lamasi Lake are distributed along the paleocurrent direction; sucn as Zhongnei lake distribute in the plain [22].The provenance of these lakes mainly came from the Greater Khingan Mountains, similar to the Qingshankou Formation in the Cretaceous. The latitude of Northeast China has little change Since the Cretaceous, and the climate difference caused by latitude is small. A set of modern dark argillaceous sediments have been deposited, which is an ideal carrier for an analogy between modern lakes and Cretaceous lakes [23].

Based on the systematic survey of lake bubbles in the piedmont zone of the Greater Khingan Mountains in the Songnen Plain, the coast of the Nenjiang River, and the interior of the Songnen Plain, the surface fine-grained sediment components of five lakes, including Chaersen Reservoir,Delong Lake,Hongyuan Lake,Lamasi Lake, and Zhongnei Lake, were selected for light sedimentation particle size analysis and chroma analysis;Combined with local topography and climate conditions, the coupling relationship between particle size composition and chroma of fine-grained sediment clay in the lake surface was analyzed, in order to provide basic comparative information for the study of Mesozoic and Cenozoic lacustrine source rocks in this area. This study cofirmed that the particle size of clay is negatively correlated with its organic matter content and positively correlated with its redox degree. It has important reference significance for discussing the hydrocarbon generation potential and evaluation of Cretaceous lacustrine fine-grained source rocks in the Songliao Basin.

2. Methods

2.1. Geographical background

The studied area locates around the Songnen Plain (42 ° 20 '~ 47 ° 32' N;121 ° 30 '~ 126 ° 37' E), North East China. Some of the studied lakes distribute along the northern foothills of Greater Khingan Moutains which neighboured to the Songnei Plain, and some distribute in it. Geotectonically, the Songnen Plain belongs to the north of the second subsidence zone of Neocathaysian tectonic system. The faulting activities during the Yanshan Movement resulted in some faulted lakes. Since the Cenozoic, Songnen Plain continues the sinking state, and caused many interspersal lakes in it. The Daan formation, Taikang formation and Quaternary strata are developed well in this area.

The Chaersen reservoir is located in the eastern foot of the Greater Khingan Mountains, the west side of the Songnen Plain, and the middle reaches of the Taoer River, the tributary of the Nen River. The area of the Chaersen reservoir is 46.7 km^2, and the maximum water depth is 300 m. It belongs to the cold temperate continental monsoon climate zone. The winter is cold and long, and the summer is short. The average annual temperature is 1.4-2.9°C, and the average annual precipitation is 485.7 mm [24]. Delong Lake, Hongyuan Lake, Zhongnei Lake, and Lamasi Lake are located in the alluvial plain on the east side of the Nenjiang River, belonging to the mid-temperate continental semi-humid and semi-arid climate transition zone, with less snow in spring and winter and mild summer, with an average annual temperature of 3.2°C and annual precipitation of 442mm. Delong Lake and Hongyuan Lake are shallow lakes formed by tectonic fault depressions, with a maximum depth of 4.6m and an average depth of 2.14m [25].The Zhongnei Lake is formed by the depression water, the maximum water depth is 2.5m, the average water depth is 1.67m. Lamasi Lake is an alluvial freshwater lake in the nenjiang terminal plain, with an area of 57.4km^2 and an average water depth of 1.98m [26] (Figure 1).

2.2. Experimental treatment

Lake surface sediments were collected from the beginning of March 2021, and ice breaking was obtained in the Chaersen Reservoir, Delong Lake, Hongyuan Lake, Lamasi Lake, and Zhongnei Lake, respectively (Figure 2). The holes in the east and north of the Chalsen Reservoir are analyzed respectively, and the top 10cm sediment was taken as the research object. The C2 hole in the east was gray-black silty fine sand containing organic matter, and the C3 blac-brown fine sandy clay in the north was taken as the research object. The north and east holes of Delong Lake were taken for analysis. The north D1 was gray black silt, and D2 in the East is gray-black muddy silty sand. Three boreholes in the north, middle and south of Hongyuan Lake were taken, and the top 10cm was taken as the research object. The northern H3 was gray black silty silt, the middle H5 was black green muddy silt, and the southern H 2 was black silt. A hole L1 was taken from the middle of Lamasi Lake, and the top 10cm was black green muddy silt. Four holes Z5 in the North, Z6 in the middle, Z3 in the West and Z1 in the southeast were taken from the Zhongnei Lake, and the top 10cm was taken as the research object. Z5 and Z6 were gray and dark gray silt, Z3 was gray black silt and Z1was gray black silt.

Photosedimention particle size analysis, mainly aim at < 65μm components in sediments for 2 hours effective X-ray sedimentation particle size analysis, to obtain the mass percentage of each particle fraction.In this paper, 5 g of the upper sample of the borehole was extracted, organic matter was removed with 10% H_2O_2, and carbonate was removed with 10% HCl. After full dispersion, the X-ray sedimentation granularity (SediGraph III 5125) was used for photosedimention particle size analysis,the subsidence time is 4 ~ 5 hours (effective time is 2 hours). The particle mass fraction of the 0.1-300μm part under the Stokes settlement principle is obtained, and the < 65μm part is taken for systematical analysis. At the same time, sediment cores were seperated by 1cm intervals, dried at 60°C for 24 hours. The brightness of each sample was measured by WR-18 colorrimeter at 8mm aperture, (L), red-green chromaticity (a), and yellow-blue chromaticity (b). Each sample index was repeated three times to take the average.

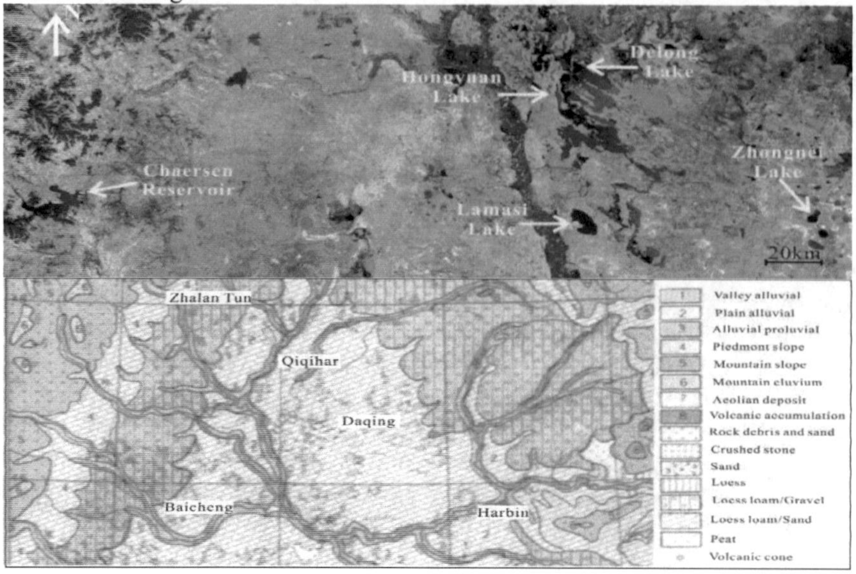

Figure 1. Quaternary Geology and Relative Position Mapof Lakes in the Study Area.

3. Results

3.1. Analysis of photosedimention particle size

The result of photosedimention size analysis of 12 surface sediments from five lakes of Chaersen Reservoir, Delong Lake, Hongyuan Lake, Lamasi Lake, and Zhongnei Lake, as Table 1. Delong Lake D1 and D2 are combined as a sample treatment due to insufficient drilling samples.

According to Table 1, it can be seen that the particle size composition of the two sampling points of the Chaersen Reservoir is quite different, and the average particle size, median, peak, and standard deviation of C3 are all smaller than those of C2. The parameters of H5 in the central position near the lake are significantly smaller than those of H2 in the southern section and H3 in the northern section. Z6 at the bay was the smallest, followed by Z1 at the outlet and Z5 at the inlet.

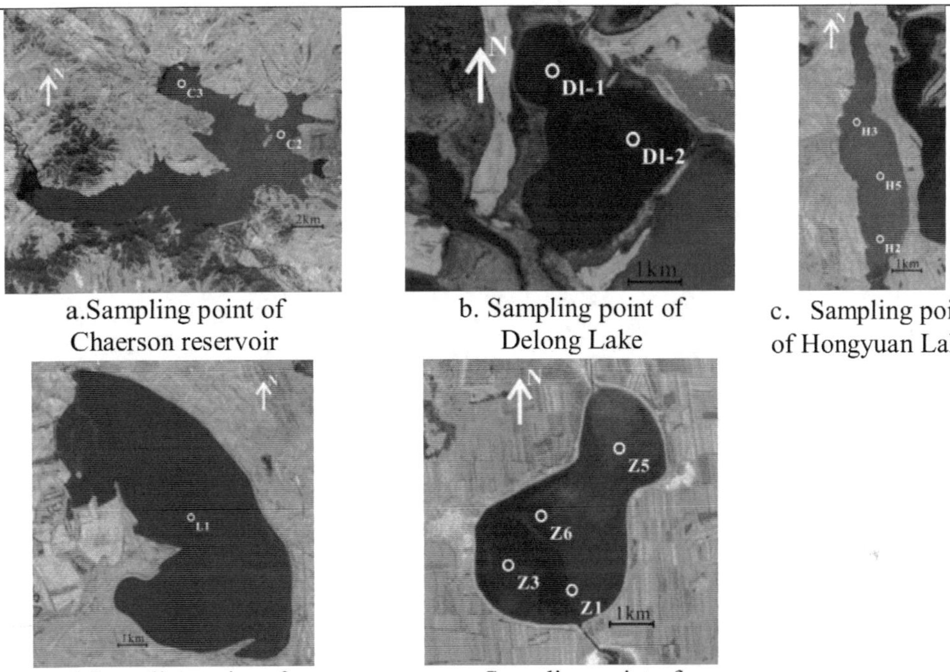

a.Sampling point of Chaerson reservoir

b. Sampling point of Delong Lake

c. Sampling point of Hongyuan Lake

d.Sampling point of Lamasi Lake

e.Sampling point of Zhongnei Lake

Figure2. Surface sediment collection locations of Chaersen Reservoir, Delong Lake, Hongyuan Lake and Zhongnei Lake.

The mean value, median value, peak value, standard deviation of grain size of Chaersen Reservoir, Delong Lake, Hongyuan Lake, Lamasi Lake, Zhongnei Lake were calculated respectively, Table 2 is obtained. Table 2 shows that the average grain size of fine sediments in the surface layer of the five lakes is Zhongnei Lake 15.41μm > Delong Lake 15.19μm > Hongyuan Lake 12.75μm > 12.61μm in Chaersen Reservoir.The characteristics of median grain size peak value of fine grain size is 34.16μm in Hongyuan Lake, which is 28.94μm in Chaersen reservoir > 28.18μm in Delong Lake > 24.76μm in Zhongnei Lake > 16.79μm in Lamasi Lake.The highest standard deviation was 15.32 μm in Hongyuan Lake, and 15.25μm in Zhongnei Lake > 14.28 in Delong Lake > 13.65μm in Chaersen reservoir > 11.65 μm in Lamasi Lake.

Table 1. Main parameters of lake surface fine sediment photodeposition grain size.

sample	mean value(μm)	Median(μm)	peak value(μm)	standard deviation
C-2	15.95	2.206	35.48	16.16
C-3	9.26	0.635	22.39	11.13
D1+2	15.19	2.889	28.18	14.28
H2	12.91	2.171	33.5	15.1
H3	14.06	1.317	35.48	16.16
H5	11.28	2.402	33.5	14.7
L1	12.02	0.954	16.79	11.65
Z1	19.9	2.192	31.62	16.68
Z3	12.04	0.286	31.62	15.17
Z5	21.91	13.49	35.48	17.95
Z6	7.792	0.692	0.316	11.18

Table 2. Average values of light deposition grain size parameters in each lake.

lakes	mean value（μm）	median（μm）	peak value（μm）	standard deviation
Chaerson reservoir	12.61	1.42	28.94	13.65
Delong Lake	15.19	2.89	28.18	14.28
Hongyuan Lake	12.75	1.96	34.16	15.32
Lamasi Lake	12.02	0.95	16.79	11.65
Zhongnei Lake	15.41	4.17	24.76	15.25

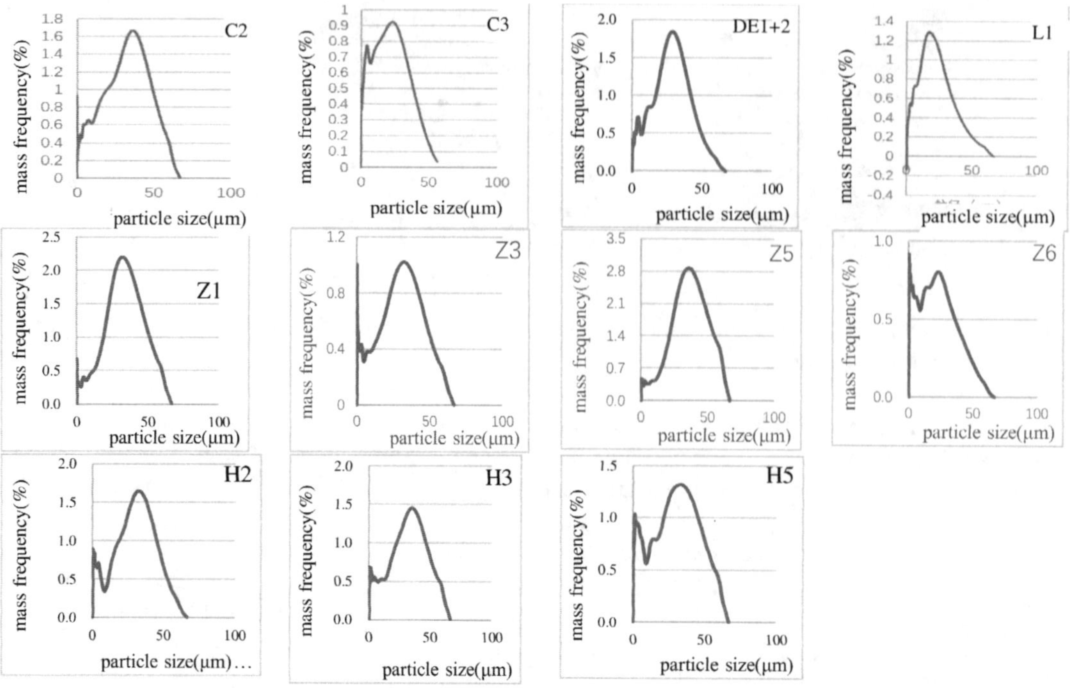

Figure 3. Mass frequency distribution of particle size components of light depositio in Chaersen reservoir, Delong Lake, Lamasi Lake, Zhongnei Lake and Hongyuan Lake.

IOP Conf. Series: Earth and Environmental Science 1087 (2022) 012081 doi:10.1088/1755-1315/1087/1/012081

The distribution of particle size mass frequency at each point is shown in Figure 3. It can be seen from the figure of the particle size of C2 and C3 sample points of Chaersen Reservoir shows obvious positive deviation, which is mainly composed of coarse-grained, and the positive deviation degree of C2 is greater than C3. Lamasi Lake slightly negative, fine component content higher than Chaersen. The degree of negative deviation of frequency curve of Delong Lake and Hongyuan Lake outlet H2 is weaker than that of Lamasi Lake.Particle Size Frequency Curve of H3 Sample in the North Section of Hongyuan Lake shows a slightly positive deviation.Positive deviation of inlet Z5, outlet Z1 and Z3 of Zhongnei Lake.

3.2. *Sediment chroma analysis results*

Chromaticity tests above 10 cm were carried out at each sampling point, and a total of 120 samples were analyzed. The brightness *L* value, redness *a* value and yellowness *b* value of each sample point showed a small fluctuation near the mean value, indicating that the surface sediments were not greatly disturbed by the external environment. The mean value of each layer of samples within 10 cm of the sample point was representative. Table 3 shows that the brightness(L) of C2 and C3 at the two sampling points of the Chaersen Reservoir are 55.97 and 57.92, respectively, which are relatively high. The brightness *L* of D2 point in Delong Lake is 47.85, which is significantly lower than that of D1 point in the estuary of Lake 56.41. The brightness *L* of H5 near the center of the lake is 41.03, which is less than 61.78 of H2 near the outlet and 57.43 of H3 near the inlet. The brightness of Z5 near the inner bubble outlet is 61.61, while the brightness *L* of Z6 near the lake bay is 49.22. Lamasi Lake L1 is in the center of the lake, the brightness value *L* is 44.09. By comparison, it is found that the maximum *L*-mean value of lake brightness is 56.95 in Chaersen Reservoir, followed by 55.47 in the Zhongnei Lake, and the Hongyuan Lake is 53.41, the Delong Lake 52.13, and the Lamasi Lake is 44.09.

The highest value of surface sediment redness *a* was 1.92 in Lamasi Lake, and the lowest value of H5 redness *a* was-0.87 in Hongyuan Lake. Among the distribution characteristics of yellowness *a*, 8.40 was the highest at the Zhongnei Lake, followed by 7.42 at the C2 sampling point of the Chaersen Reservoir, and the lowest was *b*=0.99 at the central sampling point H5 of Hongyuan Lake of Lake. The average value of redness *a* in each lake is 7.72 in Chaersen Reservoir>1.92 in Lamasi Lake>1.08 in Delong Lake>1.07 in Zhongnei Lake>0.05 in Hongyuan Lake. The mean value of yellowness *b* is 7.15 in Chaersen Reservoir>5.43 in Delong Lake>4.63 in Zhongnei Lake>4.22 in Hongyuan Lake>3.38. The redness *a* and yellowness *b* of the surface sediments of Chaersen Reservoir and Zhongnei Lake Lake are not significantly different.

Table 3. Chromaticity results of the surface sediments from the studied lakes.

Chroma parameter	Chaersen Reservoir		Delong Lake		Lamasi Lake	Zhongnei Lake				Hongyuan Lake		
	C2	C3	D1	D2	L1	Z1	Z3	Z5	Z6	H2	H3	H5
L	55.97	57.92	56.41	47.85	44.09	57.32	53.74	61.61	49.22	61.78	57.43	41.03
A	1.428	1.15	0.76	1.40	1.92	0.52	0.88	1.16	1.72	0.29	0.72	-0.85
B	7.38	6.92	6.51	4.47	3.38	3.21	2.90	8.36	4.06	5.29	6.32	1.03

4. Discussions

Few approches about the fine-grained sediment analysis has been published in the oil-shale research field. Some scholar studied the grain-sized distribution of the aeolian to investigate its provenance and depositional process [27-29] ,they used the Grain-size class-Std method and the Grain-size class-Dd methods in classifying the coarse sediment fractions. These two methods are proved equival but not good enough for the fine-grained sediment analysis [30], it might be influenced by the geomorphologic conditions [31]. For fine-grained lake sediment analysis, Dietze et al used rain-size distribution analyzed the ghe lacustrine surface sediment from the Lake Donggi Cona and obtained a clear result [32]. This enlightened us the origional distribution characteristic should be a better approach for the studied lakes in Northeast China.

4.1. Grain size distribution characteristics of lake surface sediments

Sediment particle size represents the strength of hydrodynamic and transport mode [33], which is a proxy indicator to measure the energy of sedimentary basins [34]. In general, lake sediments often show multi-component grain size characteristics [35], and the kurtosis of the frequency curve are different. The more symmetrical single-peak state represents the relatively stable deepwater environment, while the bimodal or negative and positive single-peak curves may be other processes involved in lake deposition.

The fine-grained deposition frequency distribution curves of C2 and C3 in the Chaersen Reservoir are positively skewness, reflecting that the clay particles in the surface sediments of the Chaersen Reservoir are relatively coarse (Figure 3). The single peak of C2 was obvious, and the peak particle size of 35.48 μm was significantly larger than that of 22.39 μm of C3. The peak of C3 was wide and slow, and the separation was poor, which may be the addition of fine particles. The Chaersen reservoir belongs to the riverbed reservoir, and the lake sediments are distributed in a band [36]. Combined with the topography (Figure 1), it can be seen that in the northern section of the Chaersen reservoir, three gentle valleys are imported from northwest, north, and northeast directions, which makes the source of C3 sampling point relatively complex; C2 sampling point is located in the eastern part of the lake, the lake bank is relatively steep, less outflow injection, the accumulation source is relatively simple. This may be the main reason for the different particle size frequency curves of two clay particles.

Delong Lake and Hongyuan Lake both belong to Lianhuan lake, which are medium-sized freshwater lake are formed by river, which is obviously transformed by northwest wind [26]·The particle size frequency curve of clay particles in Delong Lake fine sediment is unimodal(Figure 3), which shows that the sorting is relatively good. The single peak of the clay particle size frequency curve of Hongyuan Lake sediment is slightly positive and skewed, but compared with Delong Lake, the content of fine components is slightly higher and the sorting is slightly worse. Delong Lake water is deep, relatively closed, Hongyuan Lake water is shallow, strip shape semi-closed, which may be the main reason for the difference in clay particle size characteristics between the two lakes. Lamasi Lake is an oxbow lake on the floodplain of Nenjiang River [26]. The particle size frequency curve of clay particles in the sediment is relatively simple and slightly negative compared with Delong Lake and Hongyuan Lake (Figure 3), which indicates that clay particles have more fine-grained components and better sorting.

Zhongnei Lake belongs to wind-eroded brackish water lake [37], and the particle size frequency curve of clay particles is relatively complex. The two sampling points Z1 and Z5 are both unimodal, and they are located on the outlet and inlet channels of the lake, and the sorting is good. Due to the different transport distances of the source, the peak value of Z1 is higher than that of Z5. The Z3 sample showed similar particle size characteristics to those of Delong Lake and Hongyuan Lake clay particles, with single peak and slightly tail. Z6 has a wide and gentle peak, and the content of fine-grained components of clay particles is significantly higher, which is related to its location in the lake bay and weak water energy.

The comprehensive comparison shows that the hydrodynamic energy of the same lake is the weakest in the lake bay. In the transition from the front of the Greater Khingan Mountains to Songnen Plain, the hydrodynamic energy of Chaersen Reservoir is obviously stronger than that of other lakes. Among the lakes in the plain, the hydrodynamic strength of Delong Lake and Hongyuan Lake caused by river and lake is stronger than that of wind eroded lake. The hydrodynamic energy of Lamasi Lake caused by the floodplain is the weakest.

4.2. Chromaticity distribution characteristics of lake surface sediments

Color is an intuitive manifestation of sediment composition, generally controlled by the contents of organic matter, calcium carbonate and chromogenic minerals [38].The results of chroma analysis showed that the brightness L at the Zhongnei Lake and the Hongyuan Lake was the highest, reaching 61.61 and 61.78, respectively. The brightness *L* of the sampling point in the center of the lake is obviously smaller, such as D2 is less than D1. Z6 is smaller than other points; Hongyuan Lake H5 minimum, up to 41.03.Generally inland lake sediment brightness *L* value is positively correlated with sediment carbonate content [39], and negatively correlated with organic matter content [39-40]. The Zhongnei Lake and Hongyuan Lake in this area belong to Daqing lake group, and the brightness value of the central lake is small, indicating that the organic matter content is high. The sediment brightness *L* value of Chaersen reservoir is higher than other lakes. In addition to the negative correlation with the organic matter content, the general brightness *L* value also has a certain positive correlation with the sediment particle size [41]. It can be seen that the higher brightness *L* value of Charleson sediments may be related to the coarser sediment particle size and smaller organic matter content.

Redness *a* value and yellowness b value can reflect the redox state of fluvial and lacustrine sediments to some extent, and can provide information on lake water quantity. On the Songnen Plain, the average redness of Delong Lake, Lamasi Lake, Zhongnei Lake and Hongyuan Lake was 0.85, and the average yellowness *b* was 4.553. The redness *a* value at Z6 point of Zhongnei Lake was the highest, and the redness *a* value at the central part of Hongyuan Lake was the lowest. The redness *a* and yellowness *b* of the Chaersen Reservoir are 1.289 and 7.15, respectively, which are higher than the average value of the Songnen Plain. To some extent, they represent the relatively good oxidation conditions of the open water body and the high water quantity.

4.3. Relationship between sediment particle size and chroma

In the CIE color system, the brightness L* is measured black - white, with values ranging from 0 to 100. L* was significantly positively correlated with carbonate content, and negatively correlated with organic matter content. Redness a* is the measurement of red – green, a* is related to magnesium carbonate content and bivalent and trivalent iron minerals, where +a* is often related to trivalent iron. Yellowness b* is a measure of yellow – blue, and b* is often related to the content of iron hydroxide [42]. When redness a* is significantly positively correlated with yellowness b*, it is a good measure of the degree of environmental redox.

The average, median, peak and standard deviation of chroma parameters and particle size of sediments from Chaersen reservoir, Delong Lake, Hongyuan Lake, Lamasi Lake and Zhongnei Lake were analyzed by Bayesian correlation analysis. The results showed that the correlation coefficient between the average particle size of surface sediments and brightness *L* value was 0.795, and the correlation coefficient with yellowness was 0.721. The correlation coefficients between median particle size and brightness *L* value, redness *a* value and yellowness *b* value were 0.740, 0.541 and 0.797, respectively. The correlation coefficients between peak value and brightness *L* value and yellowness *b* value were 0.755 and 0.612, but the correlation coefficient with redness a value was 0.234 (Table 4). The correlation between redness and yellowness is obvious.

Therefore, it can be concluded that the clay particle size of surface sediments in Songnen Plain is negatively correlated with the content of organic matter and positively correlated with the degree of redox.

5. Conclusions

(1) The results of photosedimention particle size analysis show that the hydrodynamic energy of Chaersen reservoir is obviously stronger than that of other lakes during the transition from the Greater Khingan Mountains to Songnen Plain. In the plain, the hydrodynamic forces of Delong Lake and Hongyuan Lake caused by rivers and lakes are stronger than those of the bubbles in wind erosion lakes. the hydrodynamic energy of Lamasi Lake caused by floodplain is the weakest.

(2) The grain size of clay in lake surface sediments on Songnen Plain was negatively correlated with its organic matter content.

(3) The particle size of clay in surface sediments of Songnen Plain is positively correlated with its redox degree.

Table 4. Bayesian correlation coefficient of lakesurface sediment grain size and chromaticity.

	Mean value (μm)	median (μm)	peak value (μm)	standard deviation	L	a	b
mean value (μm)	1						
mean value (μm)	0.908	1					
mean value (μm)	0.865	0.719	1				
mean value (μm)	0.956	0.845	0.950	1			
L	0.795	0.740	0.755	0.798	1		
a	0.562	0.541	0.234	0.315	0.629	1	
b	0.721	0.797	0.612	0.648	0.907	0.819	1

References

[1] Ji S 2009 Progress and prospect of palaeolimnology research in China. J. Lake Science. 2009 3
[2] Sumin W, Zhenke Z 1999 New progress in lake deposition and environmental evolution in China *J. Science Bulletin* 6
[3] Jinliang Z, Huilong Y 1989 Cognition on the genesis of shallow lake sand bodies in several petroliferous basins in eastern China *Oil and Gas Geology* 1
[4] Renhou W, et al 1992 Depth and Oil and Gas Distribution of Early Tertiary Lakes in Liaohe Depression *J. acta petrolei sinica* 2
[5] Li, L, et al 2009 Lake evolution and its influence on the formation of oil shales in the Middle Jurassic Shimengou Formation in the Tuanyushan area, Qaidam Basin, NW China *Geochemistry* 1
[6] Li, Y, et al 2019 Lake Level Controls on Oil Shale Distribution in the Lucaogou Formation, Wujiawan Area, Junggar Basin, Northwest China *Energy & Fuels* 9
[7] Yinghai G,et al 2021 Research progress and prospect of fine sediment and its paleogeography *J. journal of palaeogeography* 2
[8] Hailong Y, Xiumin J 2007 Effect of particle size on combustion characteristics of oil shale *Journal of chainese society of power engineering* 4
[9] Wang, J, et al 2019 Experimental study on whether and how particle size affects the flame propagation and explosibility of oil shale dust *Process Safety Progress* 3
[10] Xinrui W, et al 2021 Research progress on characteristics and formation mechanism of fine-grained sedimentary rocks in continental lacustrine basins *Acta sedimentologica sinica* 1-38
[11] Loizeau, J.L, et al 1994 Evaluation of a wide range laser diffraction grain size analyser for use with sediments *Sedimentology* 2
[12] Peng C ,Shu G ,Xusheng L 2001 The test results of laser particle size analyzer and their comparison with sedimentation method and sieving method *Acta sedimentologica sinica* 3
[13] Aixia W , et al 2006 Errors analysis of testing sediment grain size by the methods of laser way and extinction sedimentation way *Journal of Water Resources and Water Engineering* 4
[14] Keqiang W ,Dewen L ,Youpeng W 2020 Comparison of Sediment Particle Size Characteristics in Yellow Island Beach of Shandong Peninsula Based on Image Method and Laser Method *Quaternary Research* 1
[15] Yang H, et al 2021 Research progress and discussion on fine-grained sedimentology *Unconventional Oil & Gas* 1
[16] Qingchun T,Xiaojing S, Peihong S 2020 Comparison of Sediment Particle Size Characteristics in Yellow Island Beach of Shandong Peninsula Based on Image Method and Laser Method *J.*

Quaternary Sciences 1
[17] Yingwei G 2019 The indicative significance of chromaticity in lake sediments *J. Technology Innovation and Application* 84-85
[18] Buurman P, Pape T, Muggler CC 1997 Laser grain size determination insoil genetic studies 1. Practical problems *Soil Science* **162** 211–218
[19] Yanjun L,et al 2020 Experimental study on relationship between particle size and composition of sediments and gas hydrate formation rate *Natural gas Geoscience* 2
[20] Xiaodong Z, et al 2015 Particle size endmember inversion of surface sediments in the South Yellow Sea and its implications for sedimentary dynamic environment *Chinese science: Geoscience* 10
[21] Xiqing X, YulinZhang Q, et al 2002 Particle Size Test of Mudstone in Oil and Gas Reservoir Cap and Its Application in Dayangshu Basin *Modern scientific instruments* 6
[22] Qingchun T, et al 2012 Environmental Significance and Influencing Factors of Sediment Chromaticity in Borehole BDQ0608 in Kekexili *Marine Geology & Quaternary Geology* 1
[23] Shanwen Q, et al 2014 Review on the Changes of Paleohydrological Network in Northeast China Plain and Adjacent Areas *J. Journal of Geographical Sciences* 11
[24] Hailong T, Weiping G, Gangxiu Q 2010 Problems and Countermeasures of Flood Gate Operation in Chaersen Reservoir *J. Gansu Water Resources and Hydropower Technology* 7
[25] Xiaoli W, Haifeng X, Xiayun X 2021 Environmental Changes and Human Activities Revealed by Pollen Records of Lianhuan Lake in Heilongjiang Province in about 220 Years *J. Journal of Lake Sciences*
[26] Shaopeng Y 2001 Characteristics and Classification of Lakes in Daqing *J. Journal of Harbin University* 2
[27] Guo Z, Ruddiman WF, Hao Q, Wu H, Qiao Y, Zhu R, Peng S, Wei J,Yuan B, Liu T 2002 Onset of Asian desertification by 22 Myr ago in-ferred from loess deposits in China *Nature* **416** 159–163
[28] Deng C, Shaw J, Liu Q, Pan Y, Zhu R 2006 Mineral magnetic variationof the Jingbian loess/paleosol sequence in the northern Loess Plateauof China: implications for Quaternary development of Asianaridification and cooling *Earth and Planetary Science Letters* **241** 248–259
[29] Roskin J, Katra I, Blumberg DG 2014 Particle-size fractionation ofeolian sand along the Sinai-Negev erg of Egypt and Israel *GSABulletin* **126(1–2)** 47–65
[30] YanZai Wang et al 2018 Comparison of two classification methods to identify grain sizefractions of aeolian sediment *Sciences in Cold and Arid Regions* **10**(5) 0413–0420
[31] Nottebaum V, Lehmkuhl F, Stauch G, et al 2014 Regional grain sizevariations in aeolian sediments along the transition betweenTibetan highlands and north-western Chinese desert–the influenceof geomorphological settings on aeolian transport pathways *EarthSurface processes and Landforms* **39(14)** 1960–1978
[32] Dietze E, Hartmann K, Diekmann B, IJmker J, Lehmkuhl F, Opitz S,Stauch G, Wünnemann B, Borchers A 2012 An end-member algo-rithm for deciphering modern detrital processes from lake sediments of Lake Donggi Cona, NE Tibetan Plateau, China *Sedimentary Geology* **243-244** 169-180
[33] Hongkuan H.2019 Functional Groups and Driving Factors of Phytoplankton in Saline-alkali Lakes in Western Heilongjiang Province *J. Environmental Science and Management* 11
[34] Qianli S, Jie Z, Jule X 2001 Grain size characteristics and paleoenvironmental significance of Daihai sediments *J. Marine Geology & Quaternary Geology* 1
[35] Chunhua H, et al 2005 Sediment grain size characteristics and sedimentary environment significance of Hongze Lake in the middle and lower reaches of Huaihe River *J. Scientia Geographica Sinica* 5
[36] Zhiqiang, Y, Xiaoguang, Q, Jinshui, W. U, & Bo,N 2008 Multimodal grain-size distribution charateristics and formation mechanism of lake sediments *J. Quaternary Science* 2

[37] Shouxin Y,Gangjun C, Baozhong W 2005 Sediment deposition analysis of Chaersen Reservoir *J. Water Resources & Hydropower of Northeast China* 1
[38] Jinfu L, Ronghuan X, Li Z, Xuetian L, Dongmei J 2000 Regional Difference of Lake Type Combination in Songnen Plain *J. Journal of Northeast Normal University (Natural Science Edition)* 2
[39] Kong L, Dai S, et al 2010 Climate change from 128.10 Ma to 115.30Ma recorded by sediment chromaticity in huoshizhai section of Liupanshan Group *J. Journal of Lanzhou University (Natural Sciences)* 5
[40] Balsam, W. L, B. C. Deaton and J. E. Damuth 1999 Evaluating optical lightness as a proxy for carbonate content in marine sediment cores *Marine geology* 2-4
[41] Yanhong W, Shijie L 2004 Application of Lake Sediment Chromaticity in Short Scale Paleoclimate Research *J. Advances in Earth Science* 5
[42] Dingding D, et al 2019 Sediment chroma of Selincuo Lake in central Qinghai-Tibet Plateau reflects regional paleoclimate evolution since the last ice peak period *J. Arid Land Geography* 3

IOP Conf. Series: Earth and Environmental Science 1087 (2022) 012082 doi:10.1088/1755-1315/1087/1/012082

Amplification phenomenon and mechanism analysis of ground fissure activity rate in Jiaocheng fault

Xiaofei Han[1,2,3], Xuejing Hao[1,2], Bin Dong[1,2], Shuangshuang Shi[1,2] and Xiaodong Xue[1,2]

[1] Seismological Bureau of Shanxi Province, Taiyuan 030002;
[2] National field scientific observation and research station of Taiyuan continental rift dynamics, Taiyuan, Shanxi 030025

[3] Corresponding author's e-mail: 38094048@qq.com

Abstract. Jiaocheng fault ground fissure is the longest ground fissure extending intermittently in Fen Wei fault depression zone. The occurrence and development of ground fissures are affected by rainfall, groundwater pumping, fault activity, strata and other factors. The research shows that with the rainwater flowing down and replenishing the confined water layer, there will be undercutting, scouring and hollowing, which will aggravate the development of ground fissures; In the structural area, due to the multi-stage activity of the fault, a strong tensile stress area is formed, resulting in the rapid reactivation, expansion and rebirth of the ground fracture. At the same time, there is a thicker and finer silty clay layer that is easier to lose with water, which is another incentive factor for the rapid development of the ground fracture. However, when the tectonic action causes the dislocation of the aquifer and water resisting layer and blocks the loss of formation materials, the ground fracture will lose its development conditions. Jiaocheng fault is the main controlling factor for the shaping, formation, development and continuous influence of the geological environment of Jiaocheng ground fracture, while precipitation, groundwater exploitation and natural discharge are the incentive factors, and the loss of stratum material is the necessary condition. The North basin of Tianzhuang fault cannot form ground fracture without necessary conditions.

1. Preface

The cracking of the surface rock and soil caused by tectonic action, mining, groundwater extraction, etc., we call it ground fissures [1]. Ground fissures are common geological disasters, which are distributed in the western United States and the Fenwei Basin and the North China Plain [2-3]. The Fenwei Basin is the most developed area of ground fissures in China, with A total of 520 fissures have been discovered, including 207 large and giant ground fissures with a length of more than 1 km [4].

The Xiuwen Ground Fissure in Yuci District is the earliest ground fissure that appeared in the Taiyuan Basin and is still active today [5]. The Jiaocheng Fault is the main control faultat the western boundary of the Taiyuan Fault Basin, and since the late 1970s, it has appeared along the Jiaocheng Fault, and has become the largest and longest intermittently extended ground fissure in the country, causing major threats and damage to local industrial and agricultural production and building facilities [6].

Much research has been done on the causes of ground fissures, and Zhang Jiaming (1990) believes that Xi'an fissures are caused by the bookshelf activity of stretching fractures [7]. Ding Xuewen (2000) studied the cracks in Yuci and believed that it was caused by the slight rumbling of the ground caused by the decline of the upper plate of the Bianshan Fault to form a tensile zone [8]. Su Zongzheng (2000) believed that the cracks in the Taiyuan Basin were caused by the combined effects of heavy rainfall, over-extraction of groundwater and nearby fracture activities [9]; Wu Qiang (2003) et al. used artificial neural network technology to evaluate the fractal dimension values of cracks and fractures in Yuci, and concluded that they were basically consistent [10]. Peng Jianbing found that the deep tectonic dynamics, regional tectonic stresses and local tectonic stresses of the faults in the basin formed the prototype of the ground fissures, and the fractures induced by the over-extraction of groundwater were revived and expanded, and the analysis was also applicable to the fracture formation analysis of the East African rift valley.

2. The relationship between the elastic modulus of weathered bedrock and overburden soil

By investigating the elastic modulus ratios of different lithics (combined with the existing rock modulus tables), they all obey the generalized Hooke law: the stress σ is proportional to the strain ε, that is, $\sigma = E\varepsilon$, where E is constant, called the elastic modulus or Young's modulus. After investigation, the exposed strata at the eastern edge of the West Mountain in the Taiyuan Basin where Jinci Station is located are quartz sandstone with an elastic modulus of 38-123 MPa, and the surface comprehensively considers the plastic sub-clay and light sub-clay, with an elastic modulus of 8-16 MPa, see Table 1. Due to the difference in elastic modulus, the same stress acts on different elastic media, resulting in different deformation differences, and there is an approximate relationship of 2-15 times the elastic modulus of the two media. At the same time, in practical applications, it is also found that the reaction spectrum generated by the seismic record of the station has a large frequency dispersion, the amplification coefficient βmax is between 2.1-3.3, while the response spectrum generated by the drilling hole has good consistency, and the amplification coefficient βmax advantage distribution range is around 2.5 [11].

Table 1. Elastic modulus E values of soil and some rocks [12].

Type of soil	E（MPa）	Type of soil	E (MPa)
Gravel, Pebbles	40-56	Hard-plastic clay	40-56
Coarse sand	40-48	Plastic clay	8-16
Medium sand	32-46	Conglomerate	98-1118
Dry fine sand	24-32	Quartz sandstone	38-123
Saturated fine sand	8-16	Marl	37-206
Hard molded sub-clay and light sub-clay	32-40	Limestone	206-824
Plastic sub-clay and light sub-clay	8-16	Dolomite	127-333
Hard clay	80-160		

3. Linear relationship between the activity rate of Jiaocheng fault and the amplification settlement rate of the ground fissure

3.1. The activity rate of Jiaocheng fault

The measured data of the short-level elevation (BM1-BM3) across the fault in Jinci Station show that the cumulative displacement on the east side of the Jinci fault from 1981 to 2000 (19 years) is about 21.88mm, and the corresponding annual average displacement is 1.15mm/a. According to the Forecast Center of Shanxi Provincial Earthquake Bureau, the average vertical activity rate of the Jiaocheng fault is 2.46mm/a, see Figure 1; The analysis of Liu Yuanyuan of Chang'an University show that the activity rate of Jiaocheng fault is about 5mm/a, and it has high consistency with the existing monitoring data [13]; According to Professor Jiang Wali's Jiaocheng fault filling map, the maximum Holocene settlement rate of the lower plate of the fault is 5.8mm/a. The comprehensive analysis of the above shows that the maximum vertical displacement of the Jiaocheng fault can be close to 6.0mm/a.

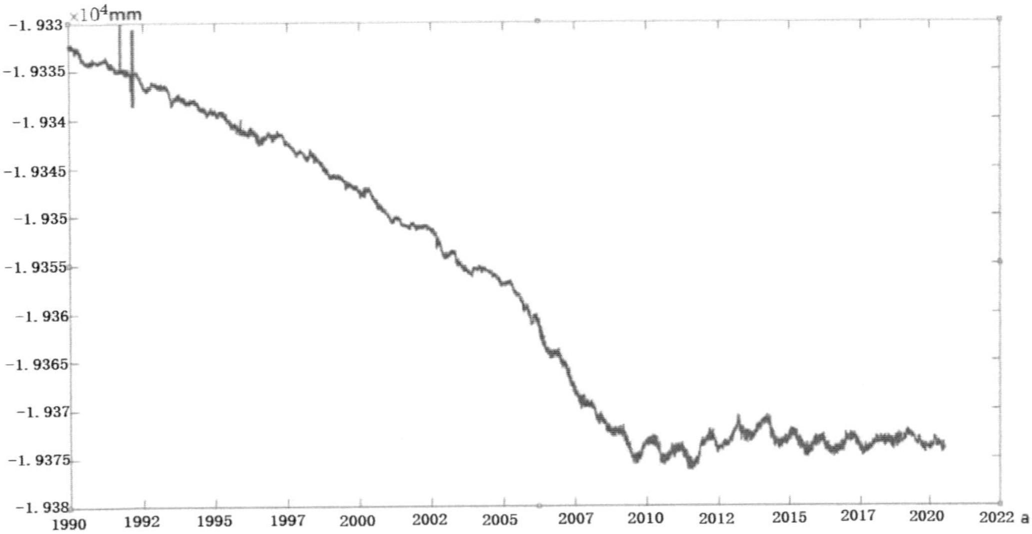

Figure 1. Cumulative deformation of Jiaocheng fault monitored by Jinci Station.

As of 2014, a total of 4 serious settlement areas were found in Taiyuan. The annual subsidence rate of Xiayuan from 2003 to 2014 was 36.1mm/a, that of Shagou Village was 31.0mm/a, that of Wujiabao was 47.0mm/a, and that of Xiaodian was 61.1mm/a. Considering that there is an approximate relationship of 2-15 times the elastic modulus of the two media of bedrock and soil layer, the average vertical activity rate of the Jiaocheng fault provided by the Forecast Center of Shanxi Earthquake Bureau is 2.46 mm/a, which is approximately 2.5 mm/a. Considering the difference in elastic modulus of the rock layer proposed in the second section and the possible magnification, it is speculated that the surface rate can be scaled up to 5-37.5mm/a, which shows that Xiayuan and Shagou Villages are mainly based on structure and soil layer amplification, and there is obvious ground subsidence caused by groundwater over-exploitation in Wujiabao and Xiaodian.

3.2. *Ground fissures activity rate of Jiaocheng fault and settlement rate of Taiyuan Basin*

From 2007 to 2011, the central and southern regions of Taiyuan, Jiaocheng County, Qingxu County, Taigu County, Qi County and Jiexiu City were key settlement areas, with an average annual settlement rate of more than 90mm/a and a maximum settlement rate of 240mm/a in Qingxu County. These rapid settlement zones correspond roughly to the over-exploited groundwater areas in the basin [13].

Using the archived data of ALOS-1 (2007-2010) and Sentinel-1A (2017-2018) to monitor ground subsidence in the Jiaocheng-Qingxu area of Shanxi Province, it is found that the maximum settlement rate on both sides of the Jiaocheng fault can reach 200mm/a. This obviously exceeds the amplification caused by the change of the elastic coefficient of the rock layer, which should be caused by groundwater over-exploitation. After effective treatment in the later stage, the settlement rate is less than 30mm/a, which is basically the settlement rate after the soil layer is amplified [14].

The longest ground fissure found in China so far is the Qingxu-Wenshui fissure in the western edge of the Taiyuan Basin, with a length of 46 km. In recent years, under the influence of over-extraction groundwater, the activity of ground fissures has become stronger, and the average activity rate of some sections has reached 44 mm/a from 2010 to 2014 [15].

Through the actual investigation and combined with "Report on the Investigation and Seismic Activity Evaluation of the Jiaocheng Fault Zone, Ground Fissures in Qingxu County, Shanxi Province" implemented by the Shanxi Provincial Earthquake Engineering Institute, and Meng Zhenjiang's 2012 master's thesis "Research on the Development Characteristics and Cause Mechanism of The Fractures in the Jiaocheng Fault Zone", the ground fissures activity rates of the Jinci-Qingxu section, the Jiaocheng section and the Wenshui section of the Jiaocheng fault zone were obtained, see Table 2,3,4.

IOP Conf. Series: Earth and Environmental Science 1087 (2022) 012082 doi:10.1088/1755-1315/1087/1/012082

Table 2. Distribution position of ground fissures along Jinci-Qingxu section of Jiaocheng fault from northeast to southwest.

Position	Horizontal activity rate	Vertical activity rate	Dextral activity rate	Basic development characteristics of ground fissures
Dongergou Village, Yaocun Town	4.4mm/a	0.6mm/a		Since 1998, the ground cracks have begun to appear in Pingquan Village and Fangshan Village, and by 2008, the length of the intermittent extension of the cracks in the Jinci-Qingxu section has reached 18.5km. The ground fissure direction is between NE30°-80°, and the dominant direction is NE 45°-70°. The ground fissure zone is composed of multiple ground cracks and has the characteristics of banding. Jinci-Qingxu structural fissure zone is composed of 1-4 ground fissures, ranging from 30m to 200m, bandwidth 30-240m, generally composed of a main ground fissure and several secondary fissures. The main ground fissure has long extension and good continuity. On the vertical main ground fissure section, the tensile tensor, vertical differential lift and horizontal torsion of the two plates of the ground fissure are the largest.
Shangguyi Village, Qingyuan Town	5.6mm/a			
Pingquan Village, Qingyuan Town	4.6mm/a	10mm/a	0.2mm/a	
Renyi Village, Mayu Township	12.5mm/a	8.3mm/a		
Dongmayu Village, Mayu Township	4mm/a	9mm/a	0.2mm/a	
Ximayu Village,Mayu Township	18.2mm/a	18.2mm/a	9.1mm/a	
Dugou Village, Mayu Township	20.0mm/a	10.0mm/a		
Dongliangquan Village, Mayu Township	0.5mm/a			
Xiliangquan Village, Mayu Township	10mm/a	10mm/a		
Dongyu Town West Highway Culvert, Dongyu Village	5.6mm/a			
Xinmin Village, 6617 factory, Dongyu Town	6mm/a	11 mm/a		
Fangshan Village, Dongyu Town	8.3mm/a	16.7mm/a	1.6mm/a	
Majiapo Village, Dongyu Town	50.0mm/a			
Kouer Village, Dongyu Zhen		30.0mm/a		
Wujiapo Village, Dongyu Town	24.0mm/a			

Note: The activity rate is calculated based on the horizontal, vertical, and dextral tension tensors from the time of ground fissures development to 2012.

Table 3. Distribution position of ground fissures along Jiaocheng section of Jiaocheng fault from northeast to southwest.

<table>
<tr><th>Position</th><th>Horizontal activity rate</th><th>Vertical activity rate</th><th>Basic development characteristics of ground fissures</th></tr>
<tr><td>Qianhuoshan Village, Lingdi Township</td><td>1.8mm/a</td><td>0.3mm/a</td><td rowspan="10">It appeared in the Bianshan belt in 1995, which is consistent with the location of the fault, and mostly exposed at 200m southeast of the boundary between the mountains and the plains, with an overall direction is NE50° and the length is 15.8 km.
Within the scope of ground fissures, school classrooms and private houses are mostly destroyed, the wall cracks are generally 1 to 5cm, or even up to 10cm, and the vertical cracks of the house are generally about 1.5 to 5cm wide.</td></tr>
<tr><td>Sanjiao Village, Guashan Cement Plant, Tianning Town</td><td>3.8mm/a</td><td>0.3mm/a</td></tr>
<tr><td>Jintaoyuan Coal Coking Group, Jiaocheng</td><td>6.3mm/a</td><td></td></tr>
<tr><td>Ciyao Village, Tianning Town</td><td>0.7mm/a</td><td></td></tr>
<tr><td>Poti Village, Tianning Town</td><td>8.8mm/a</td><td>8.8mm/a</td></tr>
<tr><td>Tianjiashan Village, Tianning Town</td><td>6.6mm/a</td><td>2.8mm/a</td></tr>
<tr><td>Wayao Village, Tianning Town</td><td>0.7mm/a</td><td></td></tr>
<tr><td>Shetang Village, Hongxiang Township</td><td>0.9mm/a</td><td></td></tr>
<tr><td>Hongxiang Village, Hongxiang Township</td><td>0.5mm/a</td><td></td></tr>
<tr><td>Guangxing Village, Hongxiang Township</td><td>3.3mm/a</td><td></td></tr>
</table>

Note: The activity rate is calculated based on the horizontal, vertical, and dextral tension tensors from the time of ground fissures development to 2012.

3.3. Linear relationship

The vertical activity rate of the Jiaocheng fault adopts the annual average vertical activity rate of the Jiaocheng fault provided by the Forecast Center of Shanxi Earthquake Bureau, which is approximately equal to 2.5mm/a. Because only the activity rate of the Jinci-Qingxu section is accurately obtained, the linear relationship is only analyzed in this section.

Table 4. Distribution position of ground fissures along Wenshui section of Jiaocheng fault from northeast to southwest.

Position	Horizontal activity rate	Vertical activity rate	Basic development characteristics of ground fissures
Beixu Village, Kaizha Town	1.8mm/a		Ground fissures in Wenshui County appeared in the 1990s. They are distributed in the northwest and southeast, and are basically parallel. The northwest side of the ground fissure is an extension of the Jiaocheng, starting from the edge of the mountain on the west side, relatively strong and active. The fissures in the northeast of Wenshui County is about NE20°, and the southernmost is north-east, parallel to the Jiaocheng fault, 100m from the geomorphological boundary, coinciding with the fault position, showing linear distribution characteristics. Both are intermittent extensions, and the ground shows an 11.7 km long, partly beaded shape. The wall is cracked horizontally, with a gap of 1 to 3cm, a band-like distribution, the widest crack is 10 to 20cm, and the southeast plate is downsloped. The rift movement in the southeastern zone farther away from the Xishan is relatively weak, and most of the cracks in the Shangxian and Madong areas tend to stabilize or gradually close after a certain period of time. The surface extension is shorter and smaller.
Zhongshe Village, Kaizha Town	0.9mm/a		
Nanxu Village, Fengcheng Town	4.0mm/a	0.3mm/a	
Goukou Village, Fengcheng Town	1.8mm/a	0.3mm/a	
Longquan Village, Fengcheng Town	7.0mm/a		
Zhangduo Village, Fengcheng Town	0.9mm/a		
Sang Village, Fengcheng Town	1.4mm/a	0.5mm/a	
Tutang Village, Fengcheng Town	0.9mm/a		
Nanyukou Village, Chengguan Town	0.5mm/a		
Wenshui County Fertilizer Factory	0.7mm/a	0.2mm/a	
Yue Village, Chengguan Town	0.7mm/a	0.2mm/a	
Xiaoziqu Village, Maxi Township	1.4mm/a		
Ma Dong, Shangxian area	1.8mm/a		

Note: The activity rate is calculated based on the horizontal, vertical, and dextral tension tensors from the time of ground fissures development to 2012.

From Table 5 and Figure 2, it can be seen that the vertical activity rate of the Jinci-Qingxu section ground fissures and Jiaocheng fault is about 4 times, and the proportion of 4 times is 60%, which is closer to the amplification coefficient recorded by the station and drilling hole reaction spectrum. This shows that there is a clear correlation between the ground fissures and the fracture activity, and the three places with obvious large multiples, Ximayu Village, Fangshan Village and Kou'er Village, are located in the vertical strong deformation area [13]. It also belongs to the area of strong groundwater extraction [15], which also proves that underground funnels formed by groundwater extraction can amplify the activity rate of ground fissures.

Table 5. Vertical rate ratio relationship between ground fissures in Jinci Qingxu section and activity rate of Jiaocheng fault.

Position	Vertical activity rate of ground fissures	Vertical activity rate of Jiaocheng fault	Rate ratio relationship
Dongergou Village, Yaocun Town	0.6 mm/a	2.5 mm/a	0.24
Pingquan Village, Qingyuan Town	10 mm/a	2.5 mm/a	4
Renyi Village, Mayu Township	8.3 mm/a	2.5 mm/a	3.32
Dongmayu Village, Mayu Township	9 mm/a	2.5 mm/a	3.6
Ximayu Village,Mayu Township	18.2 mm/a	2.5 mm/a	7.28
Dugou Village, Mayu Township	10 mm/a	2.5 mm/a	4
Xiliangquan Village, Mayu Township	10 mm/a	2.5 mm/a	4
Xinmin Village, 6617 factory, Dongyu Town	11 mm/a	2.5 mm/a	4.4
Fangshan Village, Dongyu Town	16.7 mm/a	2.5 mm/a	6.68
Kouer Village, Dongyu Zhen	30 mm/a	2.5 mm/a	12

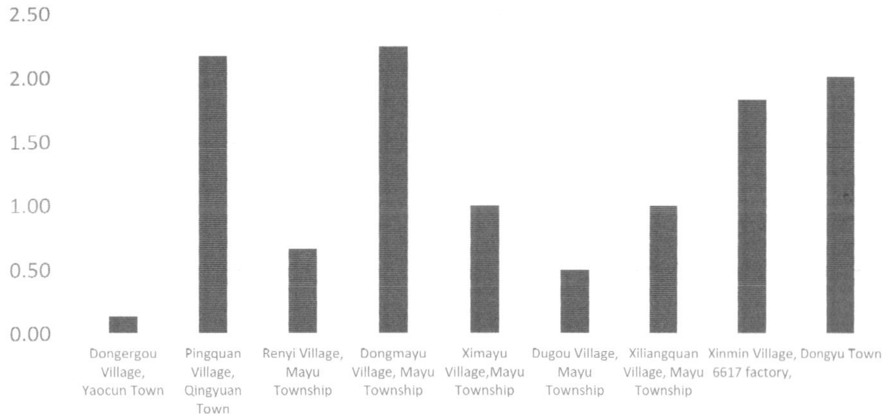

Figure 2. Relationship between relative fault activity rate and magnification of ground fissures at different locations in Jinci Qingxu section.

Zou Junjie (2020) doctoral dissertation pointed out that the dip angle of the bedrock section of Shanglan Village in Jiaocheng fault zone is 70° and the tangent value is 2.75; The Sixi Village is 60° and the tangent value is 1.73; The Longwanggou is 60° and the tangent value is 1.73. From the bedrock section, the vertical activity rate is about twice the horizontal activity rate [16].

It can be seen from Table 6 and Figure 3 that there is also an obvious linear relationship between the vertical and horizontal activity rates of ground fissures along the Jinci-Qingxu section, with an average of about 1.5 times and close to 2 times, which is consistent with the relationship between the vertical and the horizontal activity rates of the bedrock. The reason is that this section is the settlement center of the Taiyuan Basin, and the cracks in the Jinci -Qingxu section are obviously controlled by the structure, while other factors also amplify the ground fissure activity rate to 4 times according to the structural factors. However, the horizontal activity rate of ground fissures in the Jiaocheng and the Wenshui is significantly greater than the vertical activity rate. First, the structural control is weakened, and second, the influence of groundwater and surface water is increasing. Essentially, ground fissures are not only affected by the vertical and horizontal activity rates of the Jiaocheng fault, but may also be affected by the differences in the horizontal and vertical mechanical parameters of the medium and hydrodynamics.

Table 6. Linear relationship between horizontal and vertical activity rates of ground fissures along Jinci-Qingxu section.

Position	Horizontal activity rate	Vertical activity rate	Rate ratio relationship
Dongergou Village, Yaocun Town	4.4 mm/a	0.6 mm/a	0.14
Pingquan Village, Qingyuan Town	4.6 mm/a	10 mm/a	2.17
Renyi Village, Mayu Township	12.5 mm/a	8.3 mm/a	0.66
Dongmayu Village, Mayu Township	4 mm/a	9 mm/a	2.25
Ximayu Village,Mayu Township	18.2 mm/a	18.2 mm/a	1.00
Dugou Village, Mayu Township	20 mm/a	10 mm/a	0.50
Xiliangquan Village, Mayu Township	10 mm/a	10 mm/a	1.00
Xinmin Village, 6617 factory, Dongyu Town	6 mm/a	11 mm/a	1.83
Fangshan Village, Dongyu Town	8.3 mm/a	16.7 mm/a	2.01

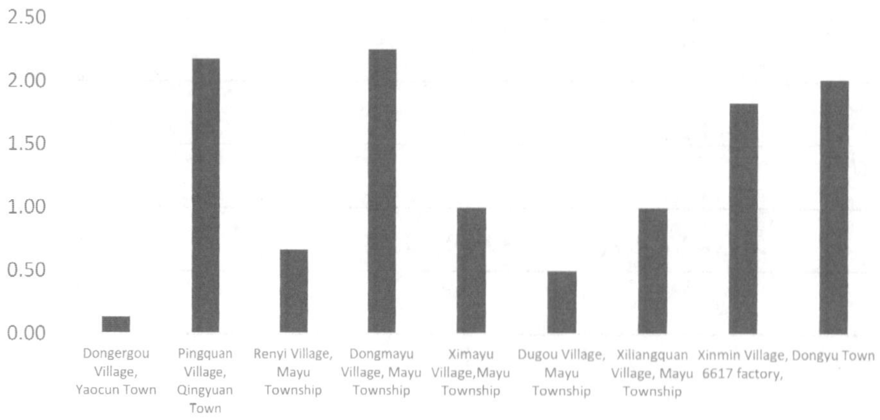

Figure 3. Ratio relationship between horizontal and vertical activity rates of ground fissures along Jinci-Qingxu section.

4. Analysis of the mechanism causing the amplification effect

Through systematic research, this paper studies the key factors controlling and affecting the development of ground fissures from the aspects of precipitation, groundwater extraction, land subsidence, Quaternary strata and basin-controlling structures.

4.1. The relationship between groundwater level change and precipitation

According to the research results of Sun Xiaohan (2016), the magnitude and growth rate of groundwater exploitation in Taiyuan Basin were very small from 1960 to 1970. From 1971 to 1985, the growth rate of groundwater exploitation increased but the magnitude was not large. From 1986 to 2004, the growth rate of groundwater exploitation was very large, and the magnitude also increased rapidly. From 2005 to 2011, the growth rate of groundwater exploitation decreased rapidly, but remained at a high level.

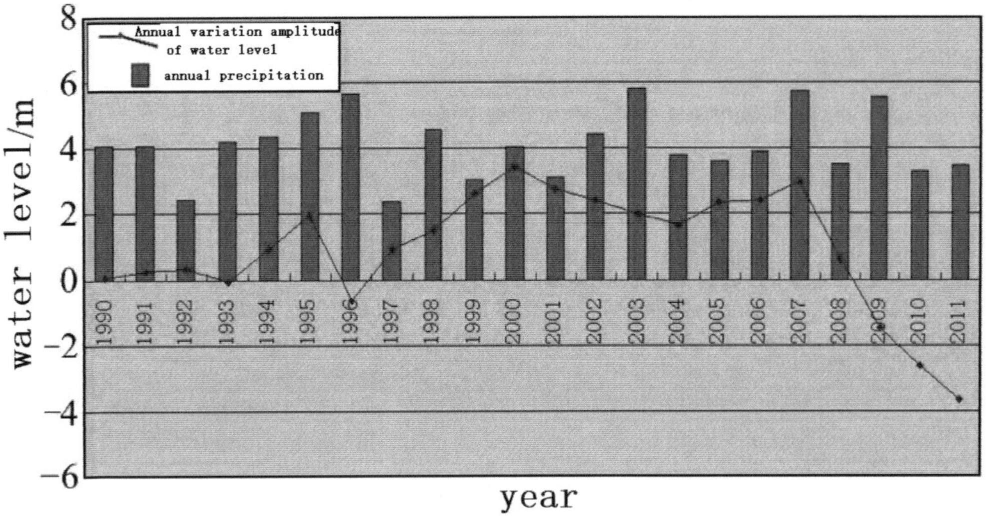

Figure 4. Comparison of annual precipitation (100 mm) and annual variation of water level (m) (Quoted from Wang Xiuwen,2018 [17]).

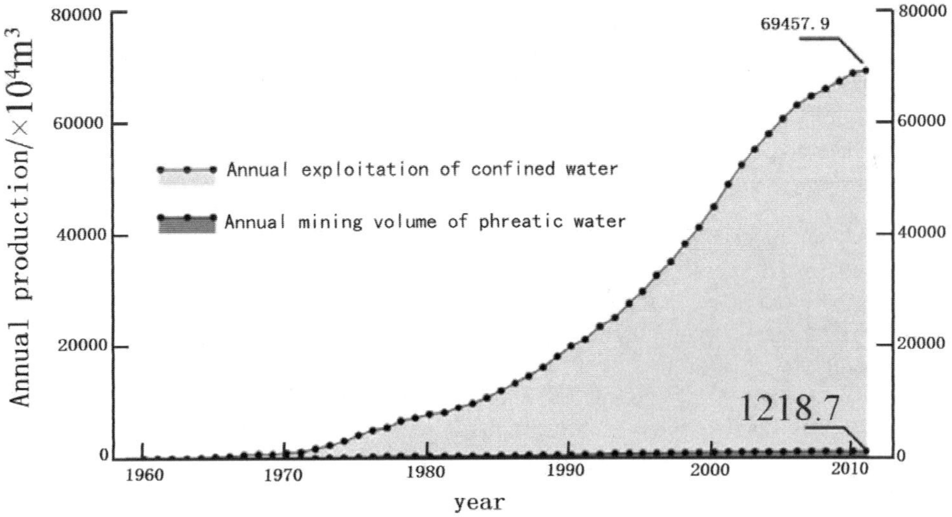

Figure 5. Groundwater exploitation over the years in Taiyuan Basin (Quoted from Sun Xiaohan,2015 [18]).

By studying the relationship between groundwater level and precipitation in Taiyuan Basin from 1990 to 2011, it is found that precipitation is not the decisive factor of groundwater level change, and sometimes there is a negative correlation, which is due to the dynamic impact of groundwater exploitation. The most convincing is that the precipitation increased year by year from 1993 to 1995. Although the exploitation of groundwater increased, the exploitation amount was less than the recharge amount, so the groundwater level rose. From 2005 to 2011, the amount of groundwater level exploitation remained high and the demand increased slightly. It can be seen from Figure 4 and Figure

5 that the precipitation recharge in 1996 has been less than the groundwater exploitation, and the precipitation recharge has increased, and the groundwater level has decreased. Since 2005, the precipitation has been generally less, but the groundwater exploitation has remained at a high level and increased at a low speed, and the groundwater level has fallen rapidly, reaching a water level difference of nearly 6 meters [17-18].

4.2. The relationship between ground fissures and groundwater exploitation time

The development of ground fissures and the development and utilization of groundwater in the Taiyuan Basin can be roughly divided into three stages [15,18]. Before 1970, it was the first stage, mainly shallow water extraction, with a total of 250 mining wells. There were no ground fissures in the Taiyuan Basin at this stage.

The second stage was from the 1970s to 2002. With the needs of economic and social development, the demand for groundwater increased sharply, and new excavation wells increased rapidly every year, mainly for the exploitation of confined water. The mining volume increased from $1082\times104m^3$ in 1970 to $7930\times104m^3$ in 1980, and then increased to $45078\times104m^3$ in 2000. The groundwater subsidence funnels in the basin are continuously added and expanded, and the confined water level shows a significant downward trend. The groundwater subsidence funnel area has expanded from Taiyuan and Jiexiu to many districts and counties in Jinzhong and Lvliang. From 1970 to 2000, there were 10, 25, and 47 ground fissures every 10 years, showing an obvious increasing trend.

After 2002 is the third stage. Through the investigation of four monitoring holes in the Taiyuan Basin from 2003 to 2013, the groundwater level showed a continuous downward trend. Among them, the Taigu hole declined slowly after 2011, and the Pingyao hole was strictly limited due to the need to protect the ancient city of Pingyao, and the groundwater level remained stable after 2008. From 2003 to 2009, Taiyuan closed a total of 413 self-provided wells in 235 units, replacing 133600 m^3/d of yellow water. The groundwater supply of the water company also gradually decreased from 393700 m^3/d to 250000 m^3/d. The groundwater level in the urban area of Taiyuan has risen, and the land subsidence rate has slowed down. As of November 2014, the water level of Jinci Spring has risen by nearly 20 meters, and it is only 9.93 meters away from the mouth of the spring. A total of 23 ground fissures were discovered at this stage. From the perspective of groundwater exploitation stage and ground fissure development stage, the two have a trend relationship.

4.3. The relationship between ground subsidence and ground fissures in Taiyuan Basin

Through the analysis of GPS and InSAR data from 2007 to 2011, the subsidence rate of Taiyuan Basin is very large, and the local rate can reach 240mm/a, while the data from 2017 to 2018 shows that the land subsidence rate is less than 30mm/a, which is close to the soil subsidence rate caused by structural factors. It shows that the overexploitation of groundwater causes the development of underground funnels, which in turn causes accelerated ground subsidence. Sun Xiaohan (2016) found that the average activity rate of some sections of the Qingxu-Wenshui ground fissures in the western of the Taiyuan Basin from 2010 to 2014 could reach 44 mm/a, which was obviously also affected by groundwater overexploitation and land subsidence.

From the perspective of the microscopic mechanism, due to the long-term over-exploitation of groundwater, the confined water level decreases year by year and is lower than the diving water level. Confined water uplifting and replenishment diving evolves into diving downwards and resupplying confined water. As the confined water level continues to drop, the current velocity and hydraulic gradient increase. The water flow continuously erodes, scours, and hollows out the new fractured zone formed by the gradual upward invasion of the fault, which can gradually form a hidden cavity or ground fissure [19-20]. If there are hidden faults or cracks formed by tectonic factors, the development of the cracks or fractures will be accelerated until they are exposed to the surface, resulting in even greater disasters.

There are two relationships between ground fissures and groundwater level changes in Taiyuan Basin, namely, temporal relationship and spatial relationship. In terms of time, the occurrence time of ground fissures in Taiyuan Basin coincides with the decline of middle and deep groundwater level; Spatially, no ground fissures have been found around the groundwater level depression funnel in the urban area of Taiyuan, which is not due to the absence of strata with ground fissures and groundwater level depression conditions, but due to the existence of East-West Sangui Horst, Qinxian Horst and Tianzhuang fault footwall uplift resistance in the Taiyuan Basin, as shown in Figure 6 [21].In other areas of the Taiyuan Basin, the edges of the groundwater decline induced funnel are all ground fissures developed to varying degrees, and most of these ground fissure development areas are related to NNE trending active faults [22].

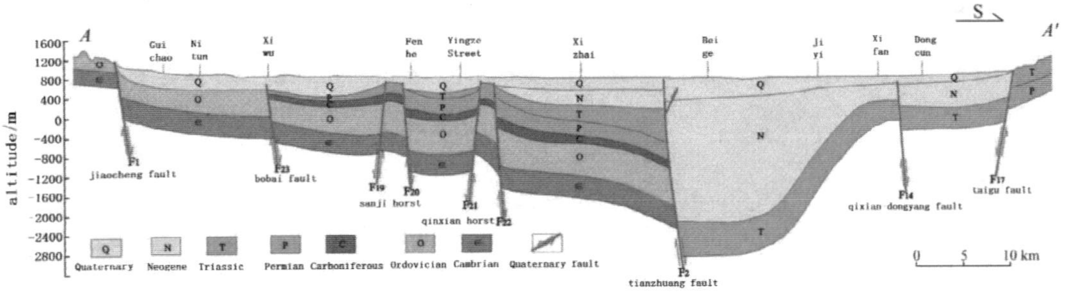

Figure 6. SN geological section of Taiyuan Basin (Quoted from Zhang Longfei,2020 [21]).

Based on the above analysis, we conclude that the drop of groundwater level in the middle and deep layers of the Taiyuan Basin is an important condition for inducing ground fissures [23], and the connection between the aquifer and aquitard should not be ignored. The aquifer and aquitard in the north of the Tianzhuang fault were dislocated, losing the ability of formation material loss. The basins south of Tianzhuang fault and west of Jiaocheng fault not only have strong historical tectonic activities, but also have strong connectivity between aquifers and aquitard, and the loss of stratum material is obvious, which provides continuous driving force for the formation and development of ground fissures. It is obvious that a local tensile stress accumulation area-rolling anticline is formed in the Jiaocheng fault zone [24], which leads to the development of ground fissures in this part, and the damage to the surface and buildings is also very serious.

4.4. *Influence of Quaternary strata on regional land subsidence*

Excessive exploitation of groundwater is the main factor that causes the groundwater level to drop and then lead to rapid land subsidence, but there are obvious spatial differences between the local groundwater level drop funnel and the center of the ground subsidence funnel, which is mainly related to the distribution of Quaternary strata, especially the distribution of clay layers.

Compared with the groundwater falling funnel, the center of the ground subsidence funnel is obviously inclined to the thicker side of the clay soil. For example, the Wanbailin subsidence funnel is shifted to the south relative to the water level funnel, the Xizhang area is obviously shifted to the east, and the Xiayuan is shifted westward, all of which are biased towards the thick clay soil side.

This is because clay soils are fine-grained and dominated by floc structures, which are more likely to be lost with groundwater flow than sandy gravel soils. In addition, the clay soil is mainly plastic deformation (non-elastic deformation), while the sand and gravel layer basically has elastic deformation as the main form, which is reversible.

5. Conclusions

Based on the analysis of the above factors, precipitation can effectively supplement the groundwater, resulting in a rise in the groundwater level, which hinders the development of ground fissures to a certain extent. However, precipitation, groundwater exploitation and natural discharge of groundwater have caused dynamic disturbances to the groundwater level, which in turn stimulated the loosening, loss and migration of soil layers near the phreatic water level and the confined water level. It plays a role in promoting the growth and expansion of the section and the formation of upward induced ground fissures. The continuous and massive extraction of groundwater causes the rapid subsidence of the ground, and at the same time induces and accelerates the development of existing structures or new fractures, until the rapid rupture to the ground forms ground fissures, which can be confirmed from the correlation between the Jiaocheng fault and the ground fissures. Thicker formations and more fine and easily lost particles have better amplification effects and are more likely to cause ground subsidence or ground fissures. Although the stress mechanism of the northern and central and southern sections of the Jiaocheng fault is similar, the area north of the Tianzhuang fault in the Taiyuan Basin did not form ground fissures under the conditions of precipitation, groundwater extraction and soil conditions, mainly due to the east-west direction. The footwall of the Sangi Horst, Qinxian Horst and Tianzhuang faults uplifted, causing the dislocation of the aquifer and aquitard, preventing the loss of stratum materials and losing the necessary conditions for the formation and development of ground fissures.

Acknowledgements
Fund Project: general nature fund project of Shanxi Provincial Department of science and Technology (No.: 201901D111462); Supported by the national field scientific observation and research station of continental rift dynamics in Taiyuan, Shanxi Province (Project No. NORSTY20-06); Scientific research project SBK-2221 of Shanxi Seismological Bureau.

Reference

[1] Pan Mao, Li Tiefeng Disaster geology [M]. Beijing: Peking University Press, 2002: 260
[2] Ye,S., Wang,Y., Wu,J., et al. Characterization of earth fissures in South Jiangsu, China [J] *Proceedings of the International Association of Hydrological Sciences*, 2015: 372, 249
[3] Bankher, K.A., Al-Harthi, A.A. Earth fissuring and land subsidence in western Saudi Arabia[J] *Natural Hazards*, 1999, **20(1)** 21
[4] China Geological Survey, Ministry of land and resources. Investigation report on land subsidence and ground fissures in China [R]. 2015
[5] Meng Ling Chao. Study on the genetic mechanism of ground fissures in Shanxi Fault Basin [D]. Xi'an: Chang'an University, 2011
[6] Meng Zhenjiang. Study on the development characteristics and genetic mechanism of ground fissures in Jiaocheng fault zone [D]. Xi'an: Chang'an University, 2012
[7] Zhangjiaming Study on ground fissures in Xi'an [m] Xi'an: Northwestern University Press, 1990
[8] Ding Xuewen, Chen Guoshun, Qi Yongsheng discussion on the causes of ground fissures in Yuci North China Earthquake Science, 2000, **18(1)** 16
[9] Su Zongzheng, Hao Helong, Hou tingai Ground fissures and their disasters in Taiyuan Basin [J] *Shanxi earthquake*, 2000, **(3) 1**
[10] Wu Qiang, Jiang Zhenquan, Li Yunlong Study on ground fissure disaster in Shanxi Fault Basin [M] Beijing: Geological Publishing House, 2003
[11] Dong Bin, Han Xiaofei, Shi Shuangshuang, statistical analysis of response spectrum amplification coefficient in Taiyuan area Shanxi architecture, 2018, **44(36)** 40
[12] Gu xiaolu et al Foundation and foundation [M] Beijing: 1993, China Construction Engineering Group, Second Edition
[13] Liu Yuanyuan. InSAR time series monitoring and mechanism analysis of comprehensive surface deformation at different scales [D] Xi'an: Chang'an University, 2018
[14] Shi Yun, Li Weixuan, Tang Yaming, et al. Application of sequential InSAR Technology in earth environment monitoring and resource management: Taking Jiaocheng as an example - Qingxu area as an example Journal of Wuhan University Information Science Edition, 2019, **44(11)** 1613
[15] Sun Xiaohan, Peng Jianbing, Cui Xiangmei, et al. Analysis of the relationship between ground fissures, groundwater exploitation and land subsidence in Taiyuan Basin, Shanxi Province Chinese Journal of geological disasters and prevention, 2016, **27(2)** 91

[16] Zoujunjie. Paleoseismic study on bedrock fault plane of Shanxi Graben System [D]. Beijing: Institute of geology, China Seismological Bureau, 2020
[17] Wang Xiuwen, Li Zihong, Shen Xiaosong, et al. Study on the influence of groundwater changes on the level deformation anomaly of Taiyuan station Geodesy and geodynamics, 2018, **28(11)** 1107
[18] Sun Xiaohan. Study on the relationship between ground fissures and groundwater exploitation in Cenozoic faulted basins - Taking Taiyuan Basin as an example [D]. Xi'an: Chang'an University, 2015
[19] Pacheco-Martínez, J., Hernandez-Marín, M., Burbey, T. J., et al. Land subsidence and ground failure associated to groundwater exploitation in the Aguascalientes Valley, México[J] *Engineering Geology*, 2013: 172
[20] Jachens, R.C., Holzer, T.L. Differential compaction mechanism for earth fissures near Casa Grande,Arizona [J] *Geological Society of America Bulletin*,1982, **93(10)** 998
[21] Zhang Longfei, Dong Bin, Shi Shuangshuang, et al. Study on the segmentation of Tianzhuang fault activity and seismic risk analysis in Taiyuan Basin Geodesy and geodynamics, 2020, **40(2)** 204
[22] Meng Ling Chao, Peng Jianbing, Lu Quanzhong, the relationship between the mechanism of ground fissure group occurrence and deep structure in Taiyuan Basin, Shanxi Province Chinese Journal of geological disasters and prevention, 2019, **30(1)** 76
[23] Guo Qinghai. Analysis of the evolution of pore groundwater system and the causes of related environmental problems in Taiyuan Basin, Shanxi Province [D] Wuhan: China University of Geosciences (Wuhan), 2005
[24] Xue Shouzhong. Study on the symbiosis mechanism of ground fissures and faults [D] Xi'an: Chang'an University, 2016

AUTHOR INDEX